11th Conference on Interdisciplinary Problems in Environmental Protection and Engineering (EKO-DOK 2019)

E3S Web of Conferences Volume 100 (2019)

Polanica-Zdroj, Poland
8 - 10 April 2019

Editors:

Bartosz Kazmierczak
Piotr Jadwiszczak
Malgorzata Kutylowska
Urszula Miller

ISBN: 978-1-5108-8832-6

Printed from e-media with permission by:

Curran Associates, Inc.
57 Morehouse Lane
Red Hook, NY 12571

Some format issues inherent in the e-media version may also appear in this print version.

No changes have been made to the content of these proceedings. There may be changes to pagination and minor adjustments for aesthetics.

Printed by Curran Associates, Inc. (2019)

For additional information, please contact EDP Sciences – Web of Conferences
at the address below.

EDP Sciences – Web of Conferences
17, Avenue du Hoggar
Parc d'Activité de Courtabœuf
BP 112
F-91944 Les Ulis Cedex A
France

Phone: +33 (0) 1 69 18 75 75
Fax: +33 (0) 1 69 28 84 91

contact-edps@webofconferences.org

Additional copies of this publication are available from:

Curran Associates, Inc.
57 Morehouse Lane
Red Hook, NY 12571 USA
Phone: 845-758-0400
Fax: 845-758-2633
Email: curran@proceedings.com
Web: www.proceedings.com

TABLE OF CONTENTS

STUDY OF VARIOUS ORGANOPHOSPHONATE ADDITIVES AS CALCIUM CARBONATE INHIBITORS FOR REVERSE OSMOSIS 1
Andrianov Alexey, Govorova Zhanna

THE ANALYSIS OF PETROLEUM HYDROCARBONS IN SOILS DERIVING FROM AREAS OF VARIOUS DEVELOPMENT 8
Badowska Ewa, Bandzierz Dawid

THE LOAD OF HEAVY METALS LEACHING FROM GRASSLAND IN CENTRAL NER RIVER VALLEY 16
Bagrowicz Tomasz, Ziulkiewicz Maciej, Mazur Janusz

PRELIMINARY RESEARCH FOR LOW-COST PARTICULATE MATTER SENSOR NETWORK 25
Báthory Csongor, Kiss Márton L., Trohák Attila, Dobó Zsolt, Palotás Árpád Bence

ANALYSIS OF OPERATING CONDITIONS OF A COOLING INSTALLATION WITH CARBON DIOXIDE 33
Bieniek Artur, Mika Lukasz, Kuchmacz Jan

INCREASING THE EFFICIENCY OF SPLIT TYPE AIR CONDITIONERS/HEAT PUMPS BY USING VENTILATING EXHAUST AIR 42
Cepinski Wojciech, Szalanski Pawel

MONTHLY MODEL OF HOT WATER CONSUMPTION 50
Chmielewska Agnieszka

STUDY ON OPTIMAL CONDITIONS OF LEMONGRASS EXTRACTION 58
Chutrtong Jaruwan, Kularbphettong Kunyanuth

GEOTHERMAL POWER BASED ULTDH FOR COOLING AND HEATING PURPOSES 65
Ciapala Bartlomiej, Janowski Miroslaw

APPLICATION OF THE COUNTER- AND CROSS-FLOW INDIRECT EVAPORATIVE COOLER FOR HEAT RECOVERY UNDER DIFFERENT CLIMATE CONDITIONS 72
Cichoń Aleksandra, Pacak Anna, Pandelidis Demis, Anisimov Sergey

DAILY VARIABILITY OF AIR POLLUTION IN LOCATIONS OF DIFFERENT POPULATION DENSITY 80
Cichowicz Robert, Stelegowski Artur

EXPERIMENTAL ANALYSIS OF WASTE TYRES AS A SUSTAINABLE SOURCE OF ENERGY 87
Czajczynska Dina, Czajka Krzysztof M., Krzyzynska Renata, Jouhara Hussam

LIFE SUPPORT SYSTEM IN THE MATURE MARTIAN COLONY FOR 1000 PEOPLE 96
Cwilichowska Natalia, Kuzma Joanna, Jurga Anna, Piszko Pawel, Nguyen Thien

INEQUALITY OF WATER CONSUMPTION FOR HYGIENIC AND SANITARY PURPOSES IN PRODUCTION HALLS 105
Dudkiewicz Edyta, Laska Marta

ENVIRONMENTAL ASPECTS OF HEAT RECOVERY SYSTEMS OF BOILER PLANTS 113
Fialko Nataliia, Navrodska Raisa, Ulewicz Malgorzata, Gnedash Georgii, Alioshko Sergii, Shevcuk Svitlana

THE USE OF SOLAR RADIATION FOR GENERATING HEAT IN A SOLAR AIR COLLECTOR IN NORTHERN POLAND 120
Fieducik Jolanta

PERFORMANCE OF SPECIAL TYPE HEAT SINK WITH AN INTEGRATED SYNTHETIC JET ACTUATOR 129
Gil Pawel

ENERGY ANALYSIS OF CHARGING INFRASTRUCTURE FOR ELECTRIC VEHICLES ON THE TEN-T ROAD NETWORK 135
Gis Maciej, Bednarski Mateusz, Orlinski Piotr

CHARACTERISTICS OF WATER SLUDGE FROM CRACOW WATER TREATMENT PLANTS – CASE STUDY 143
Gmurkowska Renata

EXPERIMENTAL INVESTIGATION OF THERMOELECTRIC COOLING SYSTEM WITH HEAT RECOVERY 151
Golebiowska Justyna, Zelazna Agnieszka

THE USE OF VARIOUS SPECIES OF FUNGI IN BIOFILTRATION OF AIR CONTAMINATED WITH ODOROUS VOLATILE ORGANIC COMPOUNDS 159
Gospodarek Milena, Rybarczyk Piotr, Brillowska-Dabrowska Anna, Gebicki Jacek

THE USE OF PULSED FLUSHING TO IMPROVE THE EFFICIENCY OF POLYMER FLOATING MEDIA FILTERS 167
Govorova Zhanna, Andrianov Alexey
REGIONAL STRATEGY OF ENERGY AND CO_2 EMISSIONS MANAGEMENT – BASE STUDY ON THE EXAMPLE OF GLIWICE REGION 173
Grelich Paulina, Moszczynska Marta, Stanczyk Grzegorz, Szczerbak Krystian
ODOUR NUISANCE ASSESSMENT OF THE FOOD INDUSTRY WASTEWATER TREATMENT PLANT 180
Grzelka Agnieszka, Romanik Elzbieta, Miller Urszula
FAILURE FREQUENCY OF WATER SUPPLY NETWORKS IN THE PODLASKIE VOIVODESHIP 187
Gwozdziej-Mazur Joanna, Swietochowski Kamil
HUMAN HEALTH RISK ASSESSMENT OF HEAVY METALS IN ROAD DUST COLLECTED IN CRACOW 195
Hajduga Gabriela, Generowicz Agnieszka, Krylów Malgorzata
THE SORPTION OF ORGANIC COMPOUNDS FROM WATER DURING TECHNOLOGICAL START-UP OF CARBON FILTERS 203
Holc Dorota, Pruss Alina, Komorowska-Kaufman Malgorzata, Madrecka Beata, Cybulski Zefiryn
CEMETERY WASTE MANAGEMENT IN POLAND. A CASE STUDY FOR WROCLAW CITY 211
Janda Anna, Marcinkowski Tadeusz
ESTIMATION OF PARTICLES EMISSIONS FROM A JET ENGINE IN REAL FLIGHT 218
Jasinski Remigiusz
ENERGY SELF-CONSUMPTION FROM PV SYSTEMS: ESTIMATIONS FOR TWO OFFICE BUILDINGS IN KRAKOW (POLAND) 227
Jurasz Magdalena, Mikulik Jerzy
EXPERIMENT DESIGN TO INVESTIGATE THE POSSIBILITY OF USING GREY WATER IN AEROPONIC CULTIVATION OF VARIOUS CROPS FOR FUTURE LONG-TERM SPACE MISSIONS 235
Jurga Anna, Kuzma Joanna, Janiak Kamil, Niederwieser Tobias
METHODOLOGY AND IMPACT CATEGORIES OF ENVIRONMENTAL LIFE CYCLE ASSESSMENT IN GEOTHERMAL ENERGY SECTOR 243
Kaczmarczyk Michal
ASSESSMENT OF THE ORTACHALA HPP LOW HEAD DAM ABUTMENT DEFORMATION AND SEEPAGE REGIME 251
Kalabegishvili Mirian, Klimiashvili Levan, Gurgenidze David, Ujma Adam
THE INFLUENCE OF PIPE DIAMETER SELECTION ON OPERATING COSTS OF HEATING INSTALLATION IN THE CONTEXT OF THE ANTICIPATED INCREASE IN ELECTRICITY PRICES 259
Knapik Maciej
CRITICAL ANALYSIS OF DECAY RATE MEASUREMENT METHODOLOGY 268
Kokurewicz Lukasz, Muszynski-Huhajlo Mateusz, Miodonski Stanislaw, Leoniak Maciej
PRELIMINARY ECOTOXICITY ASSESSMENT OF SELECTED FLAVOUR AND FRAGRANCE COMPOUNDS USING MICROTOX ASSAY 279
Kolodynski Wojciech, Piekarska Katarzyna, Strub Daniel
ACTIVE TORQUE DISTRIBUTION STRATEGY FOR ELECTRIC VEHICLE INDEPENDENT AXLE DRIVE 286
Kopczynski Artur
POWER DISTRIBUTION IN MULTI-MOTOR (AWD) POWERTRAIN OF ELECTRIC VEHICLE 294
Kopczynski Artur, Roszczyk Pawel
ECONOMIC EFFICIENCY OF BIOGAS GENERATION FROM FOOD PRODUCT WASTE 304
Koval Viktor, Mikhno Inesa, Hajduga Gabriela, Gaska Krzysztof
THE INFLUENCE OF THE DYE ADSORPTION TIME ON THE DSSC PERFORMANCE 312
Krawczak Ewelina, Zdyb Agata
COMPARISON OF DAILY FLOWS SIMULATED FOR THE YEAR 2060 ON THE KACZAWA RIVER FOR VARIOUS SCENARIOS OF CLIMATE CHANGE BY SIMPLE TIME SERIES ANALYSIS 320
Kuchar Leszek, Broszkiewicz-Suwaj Ewa, Iwanski Slawomir, Jelonek Leszek
BIOCHAR COMPARED WITH ACTIVATED GRANULAR CARBON FOR LANDFILL LEACHATE TREATMENT 328
Kwarciak-Kozlowska Anna, Wlodarczyk Renata, Wystalska Katarzyna
ENVIRONMENTAL EFFECTS OF DRIVING STYLE: IMPACT ON FUEL CONSUMPTION 337
Lasocki Jakub, Boguszewski Karol

EVALUATION OF DIFFERENT METHODS SUPPORTING SWIMMING POOL WATER DISINFECTION IN THE ASPECT OF REMOVING ORGANIC MICROPOLLUTANTS 345
Lempart Anna, Kudlek Edyta, Dudziak Mariusz
MINOR PRESSURE LOSSES FOR DIFFERENT CONNECTIONS OF PP-R AND PEX/AL/PEX INSTALLATION PIPES 353
Ligaj Kinga, Widomski Marcin K., Musz-Pomorska Anna
EFFECT ON LOUVER DAMPERS CONFIGURATION ON GAS MALDISTRIBUTION IN A HEAT EXCHANGER FOR WASTE HEAT RECOVERY 361
Ligus Grzegorz, Zajac Daniel, Panchal Jatinkumar
THE IMPACT OF OPERATING CONDITIONS ON EXHAUST EMISSIONS FROM A TWO-WHEELED URBAN VEHICLE 369
Lijewski Piotr, Szymlet Natalia, Rymaniak Lukasz, Sokolnicka Barbara, Domowicz Andrzej
THE QUALITY OF THE MICROCLIMATE IN EDUCATIONAL BUILDINGS SUBJECTED TO THERMAL MODERNIZATION 376
Lis Anna, Spodyniuk Nadiia
LIGHTING OF URBAN GREEN AREAS – THE CASE OF GRABISZYN PARK IN WROCLAW. SEARCHING FOR THE BALANCE BETWEEN LIGHT AND DARKNESS THROUGH SOCIAL AND TECHNICAL ISSUES 384
Lopuszynska Agata, Bartyna-Zielinska Malgorzata
RISK ASSESSMENT IN MUNICIPAL WASTEWATER TREATMENT PLANT 392
Łój-Pilch Magdalena, Zakrzewska Anita, Zielewicz Ewa
RADIOACTIVITY OF WASTE MATERIALS COMING FROM THE LARGEST POWER PLANTS IN POLAND 399
Lukaszek-Chmielewska Aneta, Girard Martin, Piotrowska Barbara, Wojtkowski Karol, Isajenko Krzysztof, Stawarz Olga, Godlewski Robert
MEASUREMENTS OF NATURAL RADIOACTIVITY IN SOIL SAMPLES COLLECTED IN THE KAMPINOSKI NATIONAL PARK 407
Lukaszek-Chmielewska Aneta, Girard Martin, Stawarz Olga, Piotrowska Barbara, Wojtkowski Karol, Isajenko Krzysztof
EXHAUST EMISSIONS OF AN LPG POWERED VEHICLE IN REAL OPERATING CONDITIONS 414
Merkisz Jerzy, Gallas Dawid, Siedlecki Maciej, Szymlet Natalia, Sokolnicka Barbara
MODELING CHLORINE DISTRIBUTION IN WATER SUPPLY SYSTEM UTILIZING EMPIRICALLY DETERMINED CHLORINE DECAY RATE 422
Musz-Pomorska Anna, Widomski Marcin K., Matczuk Agnieszka, Sadura Konrad
REDUCING NITROGEN OXIDE EMISSIONS IN BOILERS AT MOISTENING OF BLOWING AIR IN HEAT RECOVERY SYSTEMS 430
Navrodska Raisa, Fialko Nataliia, Presich Georgii, Gnedash Georgii, Alioshko Sergii, Shevcuk Svitlana
CROSS-BUILDING COOLING-TO-HEATING ENERGY TRANSFER 437
Niemierka Elzbieta, Jadwiszczak Piotr
POTENTIAL OF INDIVIDUAL HEAT PUMPS FOR RENEWABLE ENERGY STORAGE IN SMART GRID 446
Niemierka Elzbieta, Jadwiszczak Piotr
ISOLATION OF ANAEROBIC BACTERIA OF THE BACTEROIDES FRAGILIS GROUP FROM ENVIRONMENTAL SAMPLES 454
Niestepski Sebastian, Harnisz Monika, Korzeniewska Ewa, Osinska Adriana
THE SEPARATION OF MAXIMUM AMOUNTS OF PRECIPITATION FOR THE POLISH ATLAS OF RAINS INTENSITIES (PANDA) 462
Oktawiec Michal, Wdowikowski Marcin, Kazmierczak Bartosz, Zaleski Janusz, Licznar Pawel
FULVIC ACIDS GENERATED IN MUNICIPAL WASTE LANDFILLS IN PROMNIK 470
Orlinski Tomasz, Anielak Anna M.
THE OCCURRENCE OF ANTIBIOTIC-RESISTANT BACTERIA, INCLUDING ESCHERICHIA COLI, IN MUNICIPAL WASTEWATER AND RIVER WATER 477
Osinska Adriana, Korzeniewska Ewa, Harnisz Monika, Niestepski Sebastian, Jachimowicz Piotr
ANALYSIS OF THE MULTI-STAGE DESICCANT COOLING SYSTEM PERFORMANCE IN WROCLAW (POLAND) 486
Pacak Anna, Cichon Aleksandra, Pandelidis Demis, Anisimov Sergey
APPLICATION OF MATHEMATICAL MODELLING IN EVALUATION OF ODOUR NUISANCE FROM SELECTED WASTE MANAGEMENT PLANT 493
Pawnuk Marcin, Sówka Izabela
FORECASTING WATER CONSUMPTION IN TOWNS OF DIFFERENT SIZES 501
Piasecki Adam, Pilarska Agnieszka, Golba Radoslaw

QUASI-DYNAMIC MODEL OF THE ENERGY EFFICIENCY FOR AN AIR-TO-WATER HEAT PUMP 509
Piechurski Krzysztof, Szulgowska-Zgrzywa Malgorzata
DRUG RESISTANCE IN AIRBORNE BACTERIA ISOLATED FROM WASTE MANAGEMENT AND WASTEWATER TREATMENT PLANTS IN OLSZTYN 518
Potorski Jacek, Koniuszewska Izabela, Czatzkowska Malgorzata, Harnisz Monika
CONSTRUCTED WETLANDS AIDED WITH BIO-PREPARATION TO NEUTRALIZE WASTEWATER FROM A SMALL FRUIT AND VEGETABLE PROCESSING PLANT WITH HIGH LOAD POLLUTANTS 526
Puchlik Monika
UTILIZATION OF WASTE FOREST BIOMASS: PELLETIZATION STUDIES OF TORREFIED SAWMILL WOOD CHIPS 534
Rejdak Michal, Czardybon Agata, Ignasiak Karina, Robak Jolanta
URANS AND ACM FOR DETERMINING THE AERODYNAMIC PERFORMANCE OF VERTICAL-AXIS WIND TURBINE 543
Rogowski Krzysztof, Hansen Martin O. L.
INFLUENCE OF THE VARIABILITY OF THE ODOUR EMISSION RATE ON ITS IMPACT RANGE: A CASE STUDY OF THE SELECTED INDUSTRIAL SOURCE 551
Romanik Elzbieta, Bezyk Yaroslav, Pawnuk Marcin, Miller Urszula, Grzelka Agnieszka
ANALYSIS OF FLOW RESISTANCE IN BUNDLES OF POWER PLANT CONDENSERS 559
Rusowicz Artur, Kajurek Jakub, Baubekov Kuat
ENERGY SAVING IN THE FORMATION OF COVERED COURTYARDS 566
Rymarov Andrey
ASSESSMENT OF THE IMPACT OF THE TRAFFIC SITUATION IN URBAN AREA ON THE AMOUNT OF VOLATILE ORGANIC COMPOUNDS GETTING INTO THE CABIN OF THE CAR ON THE ROAD 574
Skretowicz Maria, Swiesciak Joanna
REDUCTION OF CHLORIDE EMISSION BY THICKENING OF METALLURGICAL WASTEWATER 582
Slowik Krzysztof, Stec Marcin, Iluk Tomasz, Czaplicki Andrzej, Kochel Mateusz
ANALYSIS OF EXPOSURE OF INHABITANTS OF POLISH CITIES TO AIR POLLUTION WITH PARTICULATE MATTERS WITH APPLICATION OF STATISTICAL AND GEOSTATISTICAL TOOLS 590
Sówka Izabela, Nych Alicja, Kobus Dominik, Bezyk Yarolsav, Zathey Maciej
ENERGY BALANCING IN GROUND HEAT EXCHANGER FOR HEAT PUMP SYSTEMS – A CASE STUDY WITH SIMULATIONS 598
Stefanowicz Ewelina, Szulgowska-Zgrzywa Malgorzata
ARTIFICIAL NEURAL NETWORK (ANN) APPROACH FOR MODELING OF SELECTED BIOGENIC COMPOUNDS IN A MIXTURE OF TREATED MUNICIPAL AND DAIRY WASTEWATER 607
Struk-Sokolowska Joanna, Ofman Piotr, Demirel Sevgi
UNIVARIATE AND MULTIVARIATE DMS CALIBRATION FOR A SINGLE ANALYTE 615
Szczurek Andrzej, Maciejewska Monika, Zajiczek Zaneta
ELECTRONIC NOSE – AN INSTRUMENT FOR ODOUR NUISANCES MONITORING 623
Szulczynski Bartosz, Gebicki Jacek
INFLUENCE OF MECHANICAL VENTILATION AND COOLING SYSTEMS ON VIBRATIONS OF HIGH PRECISION MACHINES 630
Scislo Lukasz, Szczepanik-Scislo Nina
THE MODEL OF ENERGY EFFICIENCY TRANSPORTATION OF THERMAL AND ELECTRIC ENERGY IN THE UNDERGROUND HEAT COLLECTOR 638
Titkov Dmitriy
THERMAL EFFICIENCY OF THE BUILDING ENVELOPE WITH THE AIR LAYER AND REFLECTIVE COATINGS 647
Ujma Adam, Umnyakova Nina
INVESTIGATIONS OF THE FLOW FIELD INSIDE A SQUARE CYCLONE SEPARATOR USING DPIV AND CFD 653
Wasilewski Marek, Singh Brar Lakhbir
ATTEMPTS TO ESTABLISH A REGIONAL PROBABILISTIC MODEL OF INTENSE RAINFALL FOR THE UPPER AND MIDDLE ODER RIVER BASIN 661
Wdowikowski Marcin, Kazmierczak Bartosz, Ledvinka Ondrej, Jedlicka Matej
THERMOMODERNIZATION OF BUILDINGS WITH THE UTILIZATION OF GREEN WALLS 669
Wesolowska Malgorzata, Laska Marta

ROLE OF FLOW-THROUGH LAKES IN REDUCING POLLUTANTS OUTGOING FROM RURAL AREAS 677
Wiktorska Marlena, Szymczyk Slawomir
SEASONAL CHANGES IN HYDRAULIC FLOW CONDITIONS IN OVERGROWN LOWLAND RIVER 685
Wolski Krzysztof
THE IMPACT OF CHEMICAL CONTAMINANTS ON BIOCENOSIS (ECOTOXYCOLOGICAL STUDIES) 694
Wróbel Magdalena, Stojanowska Agnieszka, Nosarzewska Martyna, Rutkowski Radoslaw, Rybak Justyna
RESEARCH AND EVALUATION OF WATER QUALITY IN OUTDOOR SWIMMING POOLS 701
Wyczarska-Kokot Joanna, Lempart Anna, Marciniak Marta
THE USE OF SLOTTED – DISPLACEMENT CEILING DIFFUSER IN ROOMS WITH STATIONARY WORKPLACES WITH COMPUTER EQUIPMENT 709
Zajac Agnieszka
CONCEPT OF REDUCING HARMFUL EMISSIONS BY ROAD TRANSPORT VEHICLES IN THE TOURIST ROUTE KARPACZ– JELENIA GÓRA 718
Zajac Pawel, Haladyn Szymon, Kwasniowski Stanislaw
RISK ANALYSIS ASSOCIATED WITH SECONDARY WATER POLLUTION IN SELECTED AGGLOMERATIONS OF THE SILESIAN VOIVODSHIP 727
Zimoch Izabela, Bartkiewicz Ewelina
24-HOUR NUTRIENT MASS BALANCE OF SMALL STORAGE RESERVOIR INCLUDED IN MUNICIPAL RAINWATER DRAINAGE SYSTEM 736
Ziulkiewicz Maciej, Górecki Michal, Fortuniak Anna, Walas Aneta, Grulke Rafal
SEASONALITY ANALYSIS OF WATER LOSSES IN A SELECTED COLLECTIVE WATER SUPPLY SYSTEM 744
Zywiec Jakub, Tchórzewska-Cieslak Barbara
Author Index

Statement of Peer review

In submitting conference proceedings to E3S Web of Conferences, I certify to the Publisher that I adhere to the **Policy on Publishing Integrity** of the journal in order to safeguard good scientific practice in publishing.

1. All articles have been subjected to peer review administered by the proceedings editors.
2. Reviews have been conducted by expert referees, who have been requested to provide unbiased and constructive comments aimed, whenever possible, at improving the work.
3. Proceedings editors have taken all reasonable steps to ensure the quality of the materials they publish and their decision to accept or reject a paper for publication has been based only on the merits of the work and the relevance to the journal.

Title, date and place of the conference

11th Conference on Interdisciplinary Problems in Environmental Protection and Engineering

8–10.04.2019
Polanica-Zdroj, Poland

Proceedings editor(s):

Bartosz Kazmierczak – Wroclaw University of Science and Technology
Piotr Jadwiszczak – Wroclaw University of Science and Technology
Malgorzata Kutylowska – Wroclaw University of Science and Technology
Urszula Miller – Wroclaw University of Science and Technology

Date and editor's signature

23.05.2019

17, avenue du Hoggar - PA de Courtabœuf – BP 112 - 91944 Les Ulis Cedex A (France)
Tél. : 33 (0)1 69 18 75 75 - Fax : 33(0)1 69 07 45 17 – www.edpsciences.org

Study of various organophosphonate additives as calcium carbonate inhibitors for reverse osmosis

Alexey Andrianov[1,*], and *Zhanna* Govorova[1]

[1]Department of Water Supply and Sewerage, Moscow State University of Civil Engineering (National Research University), Yaroslavskoe shosse 26, Moscow, 129337, Russia

Abstract. Various chemical additives are used in water treatment industry, among them – precipitation inhibitors for reverse osmosis membrane facilities. The development and synthesis of new inhibitor formulas and the study of their effectiveness is still an urgent task for many researchers working in this field. The relationship between the structure of phosphonate scale inhibitors and their effectiveness in preventing of calcium carbonate precipitation was studied. Two classes of tetraphosphonate and diphosphonate inhibitors that possess systematic structural similarities and differences have been tested on the laboratory membrane unit. All chemicals were tested at doses of 1 and 5 mg/L. The results showed that the antiscalant efficiency for both groups of inhibitors increases with elongation of methylene chain, but the longest compounds have a sharp drop in inhibition efficiency.

1 Introduction

Nowadays antiscalant dosing is assigned a leading role in the prevention of low-soluble salts sedimentation on reverse osmosis membranes [1, 2]. Every year new inhibitors are synthesized and tested in laboratory or industrial conditions. In the last 10–15 years environmentally-friendly polymers have been developed to escape environmental concerns and discharge limitations for RO concentrate [3–5]. In addition to inhibiting efficiency, it is likely that low cost, usability and availability are required for widespread industrial use. Sometimes this is a reason to the limited use of new inhibitors. Therefore sodium hexametaphosphate (HMPT) as well as antiscalants based on phosphonic acid are used widely at RO water treatment facilities. Among them the organophosphonates are dominating the global market [6–9]. They are readily soluble in water, non-toxic and effective at low treatment concentrations [10].

In addition to traditional phosphonates, more complex compounds, such as diphosphonate or tetraphosphonates, are synthesized and studied [6, 11]. Ongoing interest in the application of phosphonates is due its high efficiency in preventing the nucleation

* Corresponding author: alexei-waterlab@yandex.ru

and crystallization of many sparingly soluble inorganic salts [12, 13]. They can also affect the adhesion of inorganic crystals to critical surfaces of the water treatment system [14].

It is known that phosphonates are most effective in inhibiting the crystal growth of calcium sulfate dehydrate, but also can influence on calcium carbonate crystallization. In this work we investigate a relative effectiveness in inhibiting the crystal growth of calcium carbonate in reverse osmosis membrane element of a family of polyphosphonate additives with various length of the organic polymethylene chain.

2 Materials and methods

Two classes of phosphonate inhibitors, namely, diphosphonates (6 additives) (Table 1) and tetraphosphonates (5 additives) (Table 2), were synthesized and supplied by the Crystal Engineering, Growth & Design Laboratory, University of Crete (Greece). These two families of additives possess the same functionality, amino-*bis*(methylenephosphonic acid). Tetraphosphonates have two of these (one on each end) and diphosphonates have one (only on one end). These compounds differ in the length of the polymethylene chain. Tetraphosphonates have a $-(CH_2)_x-$ (x = 2, 4, 6, 8, 12) chain bridging the two N atoms and diphosphonates have a $-(CH_2)_xCH_3$ (x = 0, 1, 3, 5, 7, 11) non-polar side-chain [11].

Table 1. Diphosphonate scale inhibitors.

Additive abbreviation	Additive Name	Additive Schematic Structure
MBMP (C_1-D)	Methylamine-*bis* (methylenephosphonic acid)	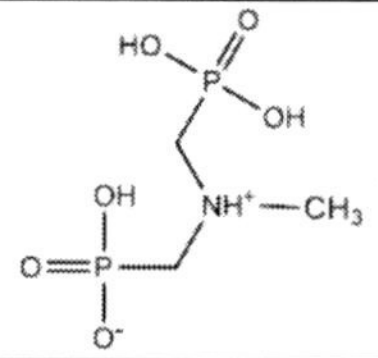
EBMP (C_2-D)	Ethyleneamine-*bis* (methylenephosphonic acid)	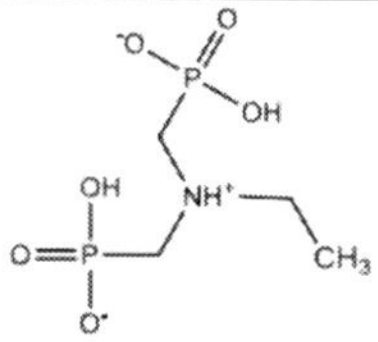
BBMP (C_4-D)	Butyleneamine-*bis* (methylenephosphonic acid)	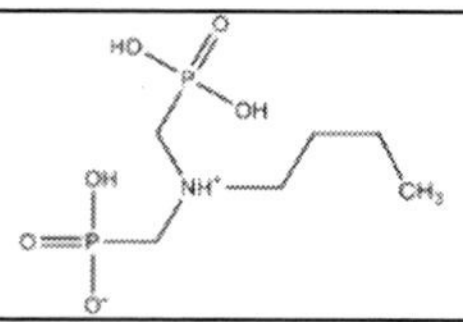
HBMP-D (C_6-D)	Hexamethyleneamine-*bis* (methylenephosphonic acid)	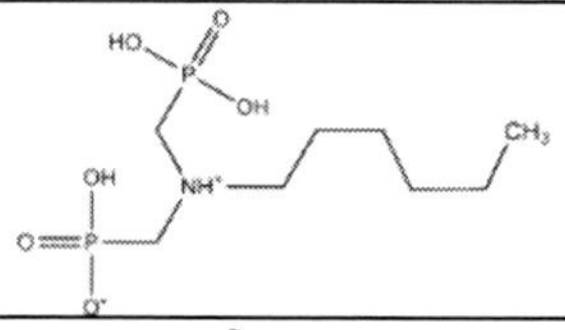
OBMP-D (C_8-D)	Octamethyleneamine-*bis* (methylenephosphonic acid)	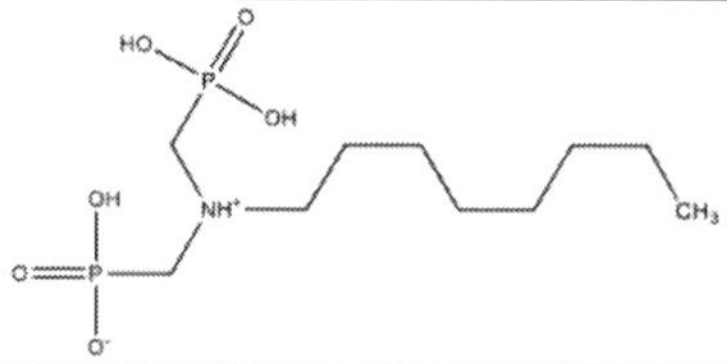

Table 1 (continuation)

DABMP (C_{12}-D)	Dodecamethyleneamine-*bis* (methylenephosphonic acid)	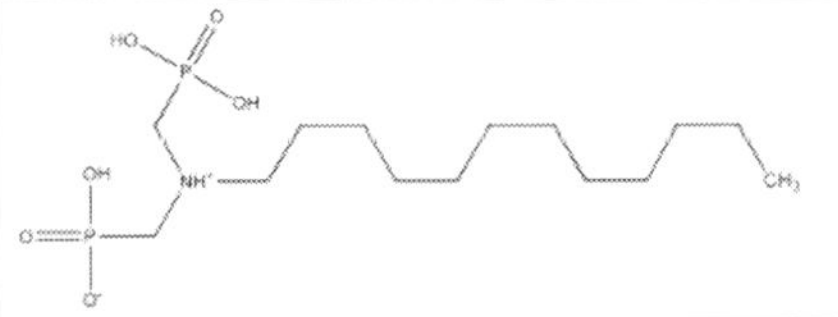

Table 2. Tetraphosphonate scale inhibitors.

Additive abbreviation	Additive Name	Additive Schematic Structure
EDTMP (C2)	Ethylenediaminetetrakis (methylenephosphonic acid)	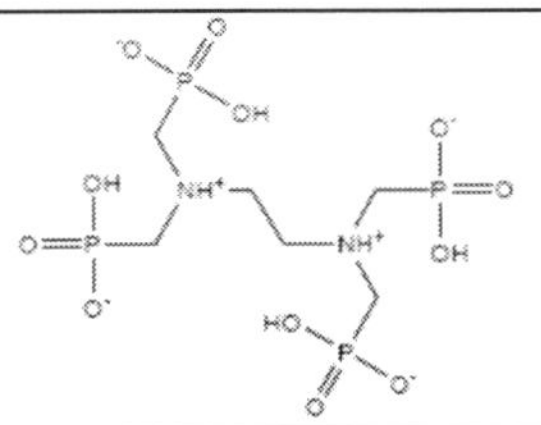
TMTMP (C4)	tetramethylenediamine-tetrakis (methylenephosphonic acid)	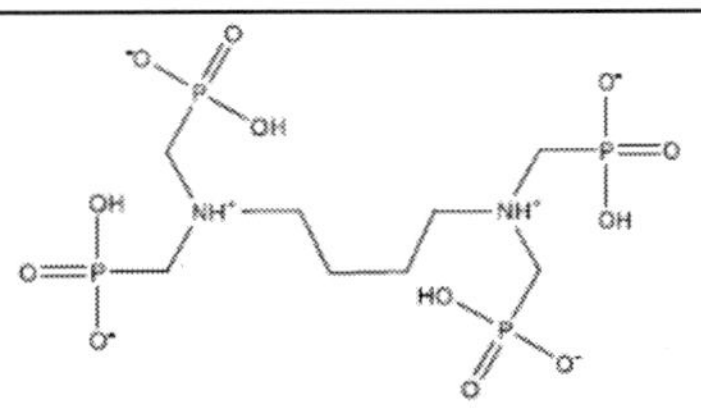
HDTMP (C6)	hexamethylenediamine-tetrakis (methylenephosphonic acid)	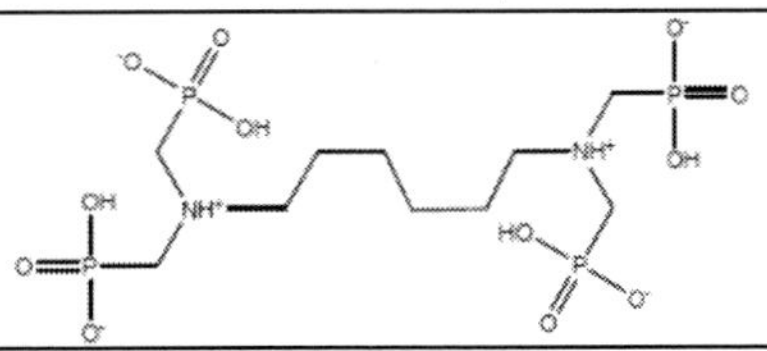
ODTMP (C8)	Octamethylenediamine-tetrakis (methylenephosphonic acid)	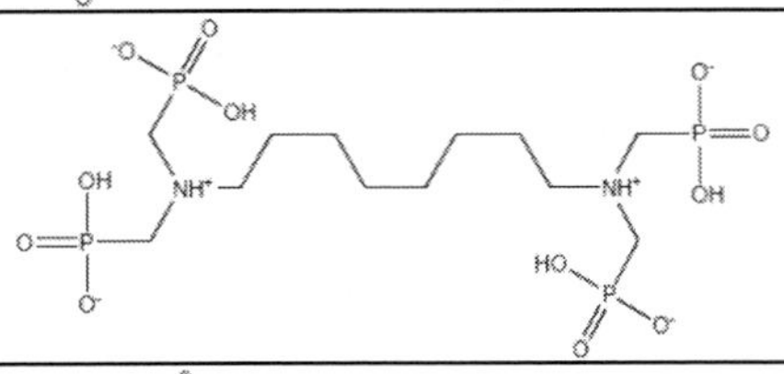
DDTMP (C12)	Dodecamethylenediamine-tetrakis (methylenephosphonic acid)	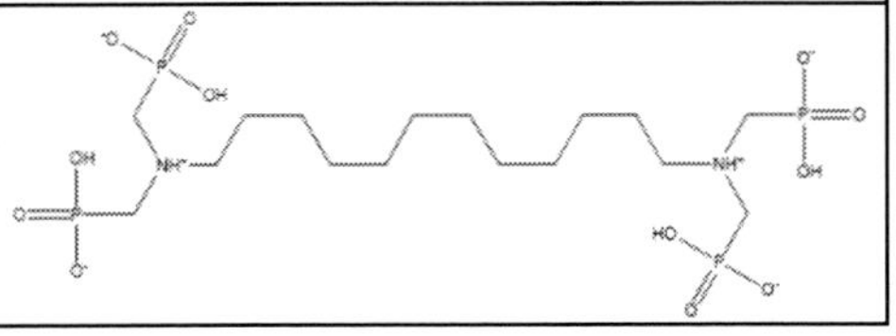

The antiscalant behavior of phosphonate compounds was investigated using lab membrane unit (Fig. 1) equipped with the spiral wound module TW 30-1812-50 (DOW Filmtec, USA) with reverse osmosis polyamide membranes with selectivity up to 98%. Scaling experiments were conducted in circulation mode whereby reject flow (concentrate) was returned to the feed water tank and permeate was collected in a separate tank. Continuous concentration of calcium and bicarbonate ions in the circulated solution provides supersaturation conditions to initiate and sustain calcium carbonate scaling throughout the experiment. Transmembrane pressure was maintained at 6 bar (the maximum pressure provided by the existing RO pump).

A model solution was prepared by careful adding and mixing of 1 M $MgSO_4$, $NaHCO_3$ and $CaCl_2$ solutions in distilled water. Final composition of feed solution had total hardness 7 mEq/L, calcium 6 mEq/L, pH = 9 and TDS was about 850 ppm. pH of feed solution was adjusted by addition of HCl aqueous solution.

Calcium carbonate scaling experiments was carried out in series for all new phosphonate scale inhibitors. Two additive concentrations were chosen: 1 ppm (low, only for tetraphosphonates) and 5 ppm (typical).

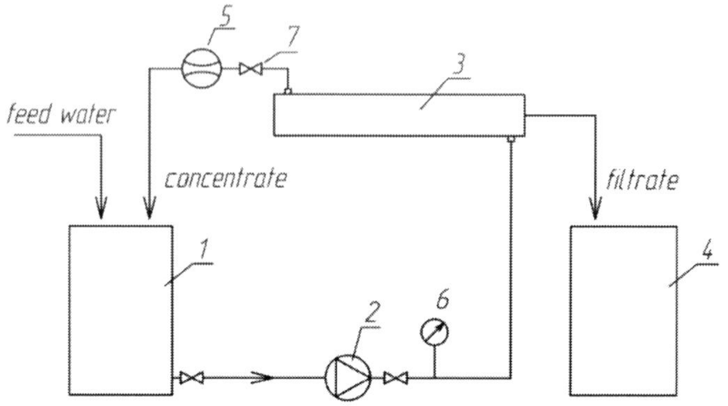

Fig. 1. Schematic diagram of laboratory membrane RO unit for membrane scaling tests:
1 – feed water tank; 2 – pump; 3 – spiral wound membrane module; 4 – permeate tank;
5 – concentrate rotameter; 6 – pressure-gauge.

The samples were taken from initial feed solution and circulated solution (for various concentration ratios) – from tank 1 (Fig. 1), and for permeate – from tank 4 (Fig. 1) (one sample characterized the average quality of product water). In all samples the following parameters are determined: temperature, TDS (conductivity), pH, total hardness, total alkalinity, and calcium. Conductivity and temperature are controlled by a laboratory conductivity meter Cond 730 (WTW inoLab®); pH value – using laboratory pH meter HI 2215 (Hanna Instruments); total alkalinity – by titration with HCl; and total hardness and calcium – by complexometric EDTA titration.

To restore the membrane element's productivity and to remove accumulated scales, chemical washing was conducted between series of experiments using a 2% solution of citric acid.

The amount of $CaCO_3$ scales accumulated in the membrane module was calculated as a difference between initial amount of calcium in the feed solution and total amount of calcium in concentrate (circulating solution) and permeate. Antiscalant performance was evaluated by comparison of scaling rate values determined throughout experiments with, and without antiscalant dosing. Antiscalant efficiency as a calcium carbonate and calcium sulphate inhibitor was calculated by using the following equation:

$$E(\%) = \frac{M_{Ca}^{blank} - M_{Ca}^{antiscalant}}{M_{Ca}^{blank}} \cdot 100 \quad (1)$$

where M_{Ca}^{blank} and $M_{Ca}^{antiscalant}$ are the mass of calcium accumulated in membrane module in the absence of antiscalant and with antiscalant dosing respectively, mg.

Scaling test for each reagent was repeated 2–3 times and the average antiscalant efficiency was calculated.

3 Results and discussion

The results show that changes in the chemical structure of an inhibitor molecule have an effect on inhibitory efficiency. Most additives improved the process and prevented scale formation in the membrane module.

Tetraphosphonates at high concentration (5 mg/L) revealed a systematic antiscaling efficiency that dependents on the length of the bridging units $–(CH_2)_x–$ (x = 2, 4, 6, 8, 12) (Fig. 2, Table 3). Specifically, the scale inhibition efficiency ranking was 2 < 4 < 6 < 8. These results reveal the tendency of decreasing of inhibition effect is proportional to the number of methylene groups on the side-chain. However the inhibitor DDTMP (C12) with x = 12 showed reduced efficiency (between EDTMP and TMTMP). Tests conducted with inhibitor concentration of 1 mg/l showed similar tendency, but not so systematic that can be partially explained by low doses of additives. In general, the experimental results with tetraphosphonates show a similar tendency obtained in [11] for calcium sulphate crystallization (with the exception of DDTMP performance).

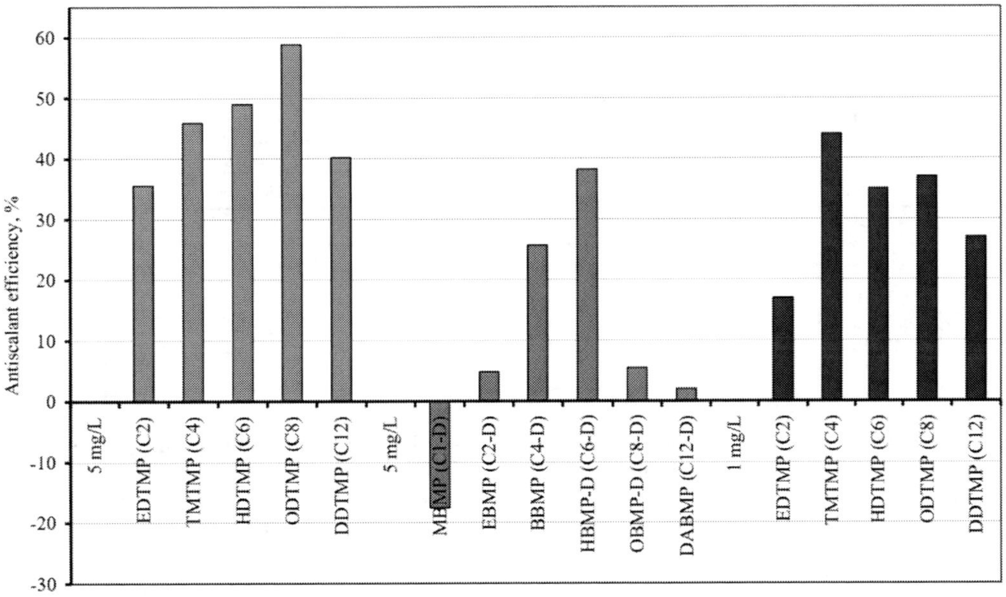

Fig. 2. Tetraphosphonate and diphosphonate scale inhibitors performance.

The similar trend was observed for diphosphonates, *i.e.* the longer the side-chain length, the higher the efficiency, but extremely long chain causes the deterioration of antiscalant efficiency (Fig. 2, Table 3). Lower antiscalant efficiency was coupled with lower RO membrane flux. Decrease in efficiency, that is an increase in the amount of calcium carbonate scale on the membrane, can be explained by a longer duration of the experiment (from 2 to 4.5 hours against 0.8–1.3 hours for other additives), which is caused by a lower flow through the membrane during these tests. Perhaps the effectiveness of OBMP and DABMP as inhibitors are high, but the reagents themselves were deposited on the membrane and reduced its productivity. This phenomenon requires verification – further scaling experiments with the SEM examination of the membrane surface. It is interesting that the first diphosphonate MBMP (C1-D) demonstrated negative efficiency – on the contrary, it increased the formation of calcium carbonate sediment on the membrane surface. At the same time, this did not cause a noticeable increase in the duration of the scaling tests as for OBMP and DABMP additives.

Since the antiscalant dosing was carried out according to the total weight of chemicals, but not the molar mass, the concentrations were recalculated to -PO(OH)$_2$, (see Table 3). Correlating corrected doses and antiscalant efficiency, it can be concluded that a decrease in the concentration for higher molecular weight compounds does not change the revealed general trends of its antiscaling behavior.

Table 3. Antiscalant efficiency in reverse osmosis tests.

Antiscalant	Dosage, mg/L	Dosage as -PO(OH)$_2$, mg/L	Average test duration, hours	Inhibitor efficiency, %
EDTMP (C2)	1.0	0.74	1.00	17
TMTMP (C4)	1.0	0.70	0.90	44
HDTMP (C6)	1.0	0.66	0.75	35
ODTMP (C8)	1.0	0.62	1.00	35
DDTMP (C12)	1.0	0.55	1.25	27
EDTMP (C2)	5.0	3.7	1.25	36
TMTMP (C4)	5.0	3.5	1.33	46
HDTMP (C6)	5.0	3.3	0.93	49
ODTMP (C8)	5.0	3.1	0.93	59
DDTMP (C12)	5.0	2.8	1.34	40
MBMP (C1-D)	5.0	3.8	1.10	-18
EBMP (C2-D)	5.0	3.6	1.07	5
BBMP (C4-D)	5.0	3.2	0.75	26
HBMP-D (C6-D)	5.0	2.9	0.84	38
OBMP-D (C8-D)	5.0	2.6	1.90	5
DABMP (C12-D)	5.0	2.1	4.50	2

Generally, diphosphonates demonstrated a much lower inhibitory efficiency in comparison with tetraphosphonates. It has been reported that the number of adsorption active centers in the molecule directly promotes inhibition process [13, 15]. Tetraphosphonates can interact by one amino-bis(methylenephosphonate) moiety with a calcium ion on the crystal surface or by two aminobis(methylenephosphonate) groups with neighboring surface calcium ions [11]. However, it is important to note that in the diphosphonate family of inhibitors, the number of active centers i.e. the phosphonate centers on the molecule backbone is the same for all diphosphonates. The polymethylene non-polar side-chain certainly does not contribute to the inhibition process, because it cannot interact with the mineral surface. Hence the systematic elongation of the side-chain undoubtedly is related to the mobility of the inhibitor molecule. Apparently, as the molecule becomes larger (due to elongation), its access to the surface of the forming nuclei becomes profoundly restricted. This is confirmed by the fact that the "longest" diphosphonates OBMP (C8-D) and DABMP (C12-D) and tetraphosphonate DDTMP (C12) are the least efficient.

4 Conclusions

Calcium carbonate scaling tests carried out on the RO unit demonstrate acceptable efficiency of most diphosphanate and tetraphosphonate inhibitors. It was found that among the family of organophosphonates the inhibitor efficiency increases with elongation of polymetylene bridging chain, but up to a certain limit. For tetraphosphonates this is a longest compound – dodecamethylenediamine-tetrakis(methylenephosphonic acid) that has a reduced efficiency. For diphosphonate family the inhibitor efficiency starting with the octamethylenediamine-tetrakis (methylenephosphonic acid) has a sharp drop. These results can be helpful for developing new classes of reagents for reverse osmosis application.

The authors would like to thank Konstantinos D. Demadis and Stefania Liakaki-Stavropoulou from Crystal Engineering, Growth & Design Laboratory of University of Crete for the kindly provided samples of phosphonate inhibitors and advices on their preparation and application.

References

1. A. Antonya, J. H. Low, S. Gray, A. E. Childress, P. Le-Clech, G. Leslie, J. of Membr. Sci. **383**, 1–16 (2011)
2. J. Mac Adam, S. A. Parsons, Rev. in Environ. Sci. and Biotechnol. **3**, 159–169 (2004)
3. D. Hasson, H. Shemer, A. Sher, Ind. Eng. Chem. Res. **50**, 7601–7607 (2011)
4. K. I. Popov, N. E. Kovaleva, G. Y. Rudakova, S. P. Kombarova, V. E. Larchenko, Therm. Eng. **63**, 122–129 (2016)
5. S. A. Ali, I. W. Kazi, F. Rahman, Desalin. **357**, 36–44 (2015)
6. K. E. Papathanasiou, K. D. Demadis, *Phosphonates in Matrices. Tailored Organic Inorganic Materials*, Ch. 3 (John Wiley & Sons, Inc., 2015)
7. M. F. B. Sousa, F. Signorelli, C. A. Bertran, J. of Petroleum Sci. and Eng. **147**, 468–473 (2016)
8. J. Tang, D. Cao, Y. Wang, C. Mohamad, L. Chen, H. Wang, Int. J. Electrochem. Sci. **12**, 11955-11971 (2017)
9. Y. Ji, Y. Chen, J. Le, M. Qian, Y. Huan, W. Yang, X. Yin, Y. Liu, X. Wang, Y. Chen, Desalin. **422**, 165–173 (2017)
10. K. S. Sorbie, N. Laing, *SPE International Symposium on Oilfield Scale, 26-27 May 2004, Aberdeen, United Kingdom*, SPE-87470-MS (2004)
11. E. Akyol, M. Oner, E. Barouda, K. D. Demadis, Cryst. Growth and Design **9**, 5145–5154 (2009)
12. E. Barouda, K. D. Demadis, S. R. Freeman, F. Jones, M. I. Ogden, Cryst. Growth and Design **7**, 321–327 (2007)
13. K. D. Demadis, E. Neofotistou, E. Mavredaki, M. Tsiknakis, E.-M. Sarigiannidou, S. D. Katarachia, Desalin. **179**, 281–295 (2005)
14. N. Abdel-Aal, K. J. Sawada, Cryst. Growth **256**, 188–200 (2003)
15. E. Mavredaki, A. Stathoulopoulou, E. Neofotistou, K. D. Demadis, Desalin. **210**, 257–265 (2007)

The analysis of petroleum hydrocarbons in soils deriving from areas of various development

Ewa Badowska[1,*], and *Dawid* Bandzierz[1]

[1]Lodz University of Technology, Institute of Environmental Engineering and Building Installations, 90-924 Lodz, Al. Politechniki 6, Poland

Abstract. The petroleum pollutants are one of the basic soil contaminants. This is due to the widespread use of petroleum products. Crude oil and its derivatives appear in the environment mainly during their storage and transport. In the research, 16 soil samples from different places were collected and analysed. The selected locations were divided into agricultural, service, recreational, residential and industrial use. The petroleum hydrocarbons were determined, in the range from C10 to C40, according to the standard PN-EN ISO 16703 [1] by means of gas chromatography. The determined results were mostly above the limited values [2] resulting from matching the soil to the group to which the specific area belongs.

1 Introduction

Petroleum substances together with heavy metals are the most common soil environment pollutants [3]. In the case of heavy metals, the occurrence of various metals in environment is analysed, depending on the possibility of their occurrence in the area. Literature data mainly concern the content of Cu (copper), Cr (chromium), Pb (lead), Zn (zinc), Cd (cadmium), Ni (nickel), but also Fe (iron), As (arsenic), Mn (manganese), or Ti (titanium) in soil. They are analysed both in industrial and urban areas [4–6]. In the case of petroleum pollutants, soil is most often analysed regarding the presence of polycyclic aromatic hydrocarbons (PAHs) [7–9] or in general considering the amount of total petroleum hydrocarbon (TPH) (in this case, depending on the used methodology, the range of compounds being tested is different).

Crude oil and its derivatives appear in the environment mainly during its storage and transport. Events such as spills oil are particularly dangerous for the environment because the contaminant fills the soil pores changing its parameters, and then it can reach the groundwater level and pollute it too [10]. In addition, petroleum pollutants limit or completely block the free flow of water and air in the soil. The situation when the content of petroleum hydrocarbons in the soil is high is particularly unfavourable for the soil. Then the bacteria and fungi present in the soil consume all available nitrogen, phosphorus and oxygen [11]. Soil contamination is associated with its impact on the ecosystem, in the case

* Corresponding author: ewa.badowska@p.lodz.pl

of plants it may be associated with their growth, in the case of animals it affects the development of diseases caused by being in a polluted environment [12].

The analyses of petroleum hydrocarbons content in soil are made mainly in places located in the vicinity of objects that affect or may have a negative impact on the soil environment. Exemplary research concerns soils from industrial areas, petrol stations, but also from farmland located near petrol activity.

The example of the analyses of soil quality in industrial areas are studies carried out in Italy in three different locations. The content of petroleum pollutants determined as THC was up to 179.33 mg/kg, however, most of the samples were below the acceptable value in Italy (50 mg/kg) [13]. On the other hand, in areas considerably contaminated, the content of TPH in the soil can be significantly higher and equal more than 80000 mg/kg [14]. The research on the total petroleum hydrocarbon content was also carried out in the Macabarie Island subantarctic area, selecting sites exposed to oil and gas (fuel farm and powerhouse). For samples from the depth of 0.5 m, the hydrocarbon content was in the range from below the level of determination to about 740 mg/kg [15]. In other studies, in soil samples from the area of fuel terminal, the hydrocarbon content in the range above C_{12} was the 32147.85 mg/kg at 0.3 m depth, while at a depth of 2.0 m was 3938.2 mg/kg [16].

In the studies of soil contamination with petroleum hydrocarbons carried out in the area of petrol stations located in various places, the maximum values of total petroleum hydrocarbon in the range from 79.9 mg/kg to 956.9 mg/kg were determined. For individual locations, the maximum values were determined from various depths [17]. In another case, soil samples originating from the areas of two fuel stations, contained 9367 mg/kg and 8075 mg/kg of petroleum hydrocarbons respectively [18].

The research of soil samples from farmland located no more than 100 m from oil wells showed that the pollution with petroleum compounds (as a sum of n-alkanes) was below 40 mg/kg, much higher values were determined for wasteland located within a radius of 20 m from the source of pollution (even 460.79 mg/kg) [19].

The analysis of the ground environment assessment is also made in case of designation changes of the area, due to a change its development. Such tests were carried out, for instance, for the area of the railway yard designated for parking, where the petroleum hydrocarbon content was 110 mg/kg [20].

The assessment of soil quality resulting from the occurrence of adverse events is equally important. According to the research [21] in Louisiana, the content of petroleum pollutants in soils, in places of earlier oil spills, was for some individual locations 42.6 mg/kg, or 85 mg/kg expressed as maximum average.

In contaminated soils, the content of petroleum hydrocarbons is often analysed depending on the depth. The amount of contaminants can diverse at individual levels. In one soil sample, depending on the layer, the values can be determined in the range from 36000 mg/kg up to 140000 mg/kg [22].

2 Methodology

The determinations of petroleum hydrocarbons in soils were made on the basis of PN-EN ISO 16703:2011 (Soil quality – Determination of hydrocarbon content in the range from C10 to C40 by gas chromatography) [1]. The range of quantification of the methodology described in the standard is from 100 mg/kg DM to 10000 mg/kg DM. Due to the lower expected values of the analysed compounds in the samples of collected soil, a calibration curve for low concentrations ranging from 200 mg/L to 8000 mg/L was made. The graph below (Fig. 1) shows the determined calibration curve.

https://doi.org/10.1051/e3sconf/201910000002

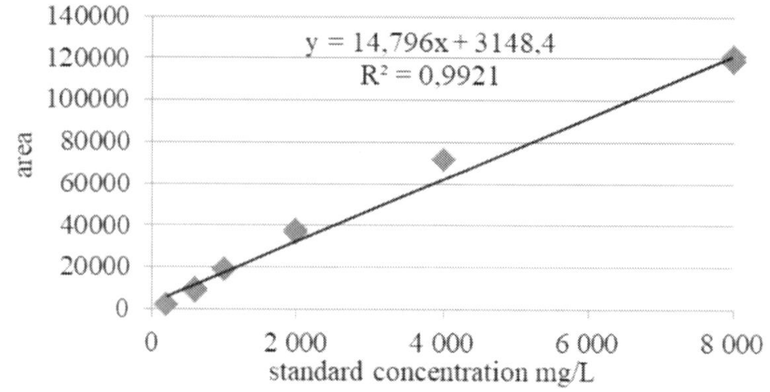

Fig. 1. The calibration curve for a mineral oil mixture standard using n-heptane as the hydrocarbon solvent.

The gas chromatograph with a flame ionization detector equipped with a Varian Capillary Column Select TM Mineral Oil LVI column 15 m long with a diameter of 0.32 mm was used for chromatographic analysis. The analysis time of a single sample was 36 minutes and proceeded according to the following temperature program: 80°C for 4 min, 10°C/min to 300°C and 300°C for 10 minutes. After receiving the results, from three replications, in the form of area under the curve, the values between which the difference was above 5% were rejected. The average was calculated from the remaining ones. The final result was calculated by means of the formula given in the standard [1] taking into account the data from calibration curve, the mass of the analysed sample and its dry matter content.

3 Research

In Polish law, the degree of soil pollution is based on The Regulation of the Minister of the Environment on the way of assessing the pollution of land surface based on land development [2]. The ground is divided there into four groups. The soil samples in the carried out research were qualified for group I (residential, service and recreational areas), groups II (agricultural areas) and groups IV (industrial and communication areas). According to this division and the depth from which the sample comes (0–0.25 m), the permissible amount of petroleum hydrocarbon is for group I 30 mg/kg DM, for group II 50 mg/kg DM and for group IV 3000 mg/kg DM.

In order to conduct a comprehensive analysis of the petroleum hydrocarbon amount in soils, depending on the land development, 16 soil sampling points were selected. All selected points were located in central Poland. The samples came from both sparsely populated and intensely urbanized areas and were divided into agricultural, service, recreational, residential and industrial use.

3.1 The soil samples from areas used for agriculture (1, 2, 3)

There were collected three soil samples deriving from agricultural areas. The first collection point was located in the vicinity of the provincial road, the second one was near wind turbines and the third one was next to the service road. The first two places are not urbanized (there are no buildings nearby). The last place can be considered as poorly urbanized because there are very few buildings in the immediate vicinity of the collection point. All analysed areas have always been used for agriculture purposes.

The analysis showed that in the first point the soil was mostly contaminated with petroleum hydrocarbons (MOI content in soil was 471 mg/kg DM).

3.2 The soil samples from service areas (4, 5, 6, 7)

The petrol stations from which the fourth, fifth and sixth samples have been taken are located by the expressway. There is a hotel in the vicinity of the first two stations. The sixth sample was taken from a place with a similar way of development to the fourth sample (at another kilometer of the same expressway). Before building the road, in each locations, there were farmlands, which are also now the basic landscape of these territories. Currently, the areas are poorly urbanized. Near the sixth place the car park for trucks is located.

The next sample point (7) connected with the service activity that could have an impact on the ground environment was a petrol station located in the medium urbanized area. The main buildings in the neighbourhood of the station are single-family houses.

The average value of petroleum hydrocarbons received from the analysed service area was 505 mg/kg DM.

3.3 The soil samples from residential areas (8, 9, 10)

In residential areas, apart from residential buildings, there is also infrastructure, the occurrence of which results from the intended use of the district. As a result, there are usually service points, car parks and public utility buildings in residential regions. The eight collected sample came from the unpaved surface , which was created after the demolition of the catering point. It is located in single-family housing. Opposite the square from which the sample was collected, there is a coal composition.

As the next locations for analysing the content of petroleum hydrocarbons in the soil, in the housing area, two extremely different points were selected. Both are located in the same housing estate, however one sample (9) came from the car park area, while the other was from the green area (sample 10). The estate is located in a highly urbanized area. In the neighbourhood of the collection points there are multi-family housing, road, public buildings and service points.

The analysis showed that minimum value of petroleum hydrocarbons from residential areas (10 mg/kg DM) was determined for green surface.

3.4 The soil samples from post-industrial and industrial areas (11, 12, 13)

Two samples from post-industrial area and one from industrial were collected. The research material from the first post-industrial area (sample 11) came from the place where the construction materials had been produced. The factory operated until the early 90's of 20th century. For many years, the area was not used in any way. In recent years, the buildings have been demolished and the area has been prepared for housing investment. The content of petroleum hydrocarbons in the soil sample from this area was 687 mg/kg DM, which is not so high value in relation to the activity which had been conducted in this area and the lack of any information regarding the reclamation of this ground.

Currently, former industrial areas are often adapted for service purposes. The area that was selected for the analysis (sample 12) was related to the textile industry. At the beginning there was a printing house for cotton fabrics, and later research centres and a textile laboratory. Presently, there are various service points on the analysed area (starting from gastronomy, through vehicle mechanics, and ending with tax services). The place is to some extent still related to a textile industry, as some of the rooms are occupied by textile companies. Not all buildings in this area have been fully renovated and being used.

The last soil sample (sample 13) came from an industrial area whose role was determined in the 60s of 20th century and has remained unchanged since then. Over the years, there have been many companies operating with various activities. A large part of the zone is occupied by storehouses. It should be noted that the sample originated from the public area, because there was no possibility to enter the area of many companies.

3.5 The soil samples from recreational areas (14, 15 and 16)

Three samples were taken from recreational areas. The common feature of all objects is the presence of a water reservoir in their area. The first object (sample 14) was created in the 70's of 20th century. The water reservoir attracted investors and currently there are numerous recreation centers on the analysed area. The area around is largely used for agriculture. There are no factories and industrial facilities in the vicinity. The second recreational facility (sample 15) from which the sample was taken is an area that is located in an urbanized area. Around the object dominate single-family houses with the needed infrastructure. The number of buildings in this area is increasing each year. The last location (sample 16) of this type of development was the area, which is surrounded by cultivated fields and a car park of a nearby supermarket. There are catering points operating in the summer.

The average value of petroleum hydrocarbons received from the analysed service areas was 464 mg/kg DM.

4 Results

The achieved findings testify the poor quality of soils in the vast majority of selected areas. Most of the obtained values exceed the permissible limits [2]. The table below presents (Tab. 1) the area development of soil sample location, the limits of petroleum hydrocarbons amount according to standard [2], the analyses results with the achieved value of MOI and the percentage of exceeding the limits.

Table 1. The area development of soil sample location, the limits of petroleum hydrocarbons amount according to standard [2] and the analyses results with the achieved value of MOI and the percentage of exceeding the limits.

sample number	1	2	3	4	5	6	7	8	9	10	11	12	13	14	15	16
area develop-pment	agriculture			service				residential			post-industrial* / industrial			recreational		
permissi-ble value mg/kg DM [2]	50			30				30			30*/3000			30		
MOI mg/kg DM	471	-	-	849	608	453	109	592	10	136	687	472	562	472	507	412
limit exceeding by %	842	-	-	2730	1927	1410	263	1873	-	353	2190	1473	-	1473	1590	1273

*Currently, due to the changed development the maximun permissible value for soil samples from post-industrial areas equals 30 mg/kg DM.

The comparison (Tab. 1) clearly shows that the standards [2] are exceeded, in most cases, by over 1000%. This situation results from the development of individual areas with facilities that had previously or have currently a negative impact on the soil environment.

The maximum value was achieved in the soil sample taken at the petrol station near the expressway (sample 4) and equaled 849 mg/kg DM.

The difference between the maximum and the minimum value in areas with a similar way of development is very large. For example, for the petrol station areas it was 740 mg/kg DM while for urban areas equaled 582 mg/kg DM. In the case of petrol stations, the difference is related to the number of users, while in the case of the urban areas it is dependent on the real (not necessarily correct) usage of the place.

Due to the fact that there are no other objects, near the first sample point collection except the road, it is very likely that the expressway has a negative effect on the adjacent farmland. This conclusion is based on the amount of petroleum hydrocarbons in the second and the third samples, in which the received amount of pollutants was below the used method of determination.

Considering the composition of petroleum hydrocarbons in soils from different locations, it is clearly visible (Fig. 2) that the fourth sample has a slightly different composition than the fifth sample (both from petrol station). However, despite various development, some of the studied locations, for example for recreation area and the area of petrol station, similar compounds appeared in analysed soil samples, which is visible in the presence of peaks in the chromatogram at the same time of analysis (Fig. 3).

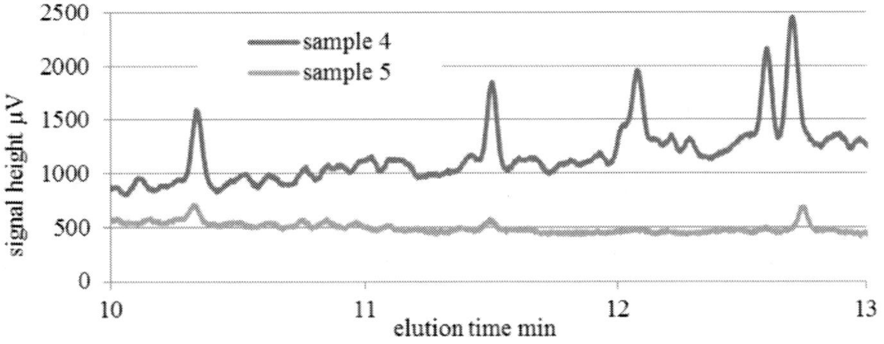

Fig. 2. The gas chromatograms of fourth and fifth soil sample in elution time between 10 and 13 minutes.

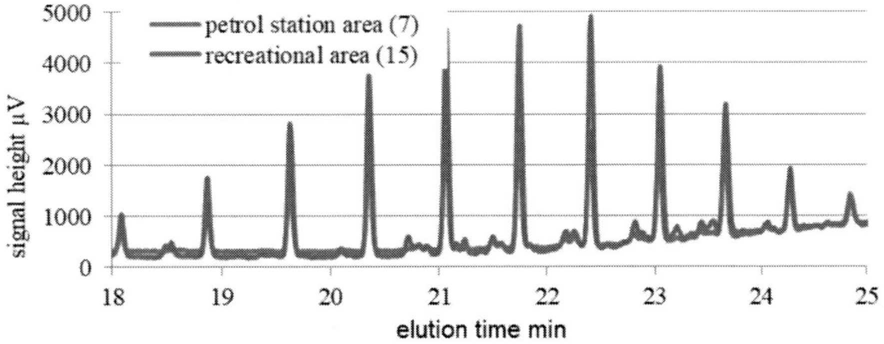

Fig. 3. The gas chromatograms for soil samples from various development area (recreational and petrol station area) in elution time between 18 and 25 minutes.

According to the legal requirements [2], some part of the analysed areas (residential area) meet the limits (green area) while the car park areas are too contaminated. All the areas have been assigned to group I according to the Regulation [2], where the maximum acceptable value of petroleum hydrocarbon is 30 mg/kg DM. It should be considered (for the 8th sample, where the value of MOI was 592 mg/kg DM) whether further use of this place as a car park will not significantly contribute to the increase of soil contamination in the future. The bumpy surface of the car park may cause the accumulation of pollutants at certain points of the square.

Due to the development change of some analysed areas (post-industrial facility) it is crucial to recultivate the soil in the area. Then the requirements [2] concerning the content of petroleum hydrocarbons will be met.

It is unfavourable for environment that in all soils samples from recreational areas the value of petroleum hydrocarbons has significantly exceeded limits (at least 382 mg/kg, by 1273%). These areas are used for the rest, relaxation and the condition of environment should be on the highest level.

According to the literature data the amount of petroleum pollution in soils from places with theoretically similar development for example for petrol stations ranges from 79.9 mg/kg to 8075 mg/kg [17]. However, in the conducted research the values range from 109 mg/kg DM to 849 mg/kg DM, which is near lower literature values.

The research indicates that the greatest risk related to the increase of pollution with petroleum hydrocarbons in the future may occur in soils near expressways (currently sample 1) that are exposed to car fluid leaks during breakdowns and damages to vehicles.

5 Conclusions

1. The current contamination level of soil environment depends on the place where the sample was collected and the depth of its collection.
2. Inadequate way of the area development (for example as unpaved parking space) and the lack of appropriate ground protection causes an uncontrolled increase in soil contamination.
3. The changes in the way of area development should make the property owners adjust the area to the effective law.
4. Monitoring of the soil environment should be constantly conducted to prevent its degradation.

References

1. PN-EN 16703:2011 (2011)
2. Regulation of the Minister of the Environment Jaw of Laws 2016, pos. 1395 (2016)
3. Report, European Environment Agency, https://www.eea.europa.eu/data-and-maps/indicators/progress-in-management-of-contaminated-sites-3/assessment (2014)
4. A. G. Caporale, P. Adamo, F. Capozzi, G. Langella, F. Terribile, S. Vingiani, Sci. Total Environ. **643** (2018)
5. P. J. Harvey, M. Rouillon, C. Dong, V. Ettler, H. K. Handley, M. P. Taylor, E. Tyson, P. Tennant, V. Telfer, R. Trinh, Sci. Total Environ. **584–585** (2017)
6. F. B. V. Silva, C. W. A. Nascimento, P. R. M. Araújo, F. L. Silva, L. H. V Lima, Int. J. Environ. Sci. Technol. **14** (2017)
7. E. J. Bielińska, B. Futa, A. Ukalska-Jaruga, J. Weber, S. Chmielewski, S. Wesołowska, A. Mocek-Płóciniak, K. Patkowski, L. Mielnik, J. Soil Sediments, **18** (2018)
8. S. N. Sushkova, T. Minkina, I. Deryabkina (Turina), S. Mandzhieva, I. Zamulina, T. Bauer, G. Vasilyeva, E. Antonenko, V. Rajput, J. Soil Sediments, **18** (2018)

9. D. Crnković, M. Ristić, A. Jovanović, D. Antonović, Environ. Monit. Assess. **125** (2007)
10. E. Mena, J. Villaseñor, M. A. Rodrigo, P. Cañizares, Chem. Eng. J. **299** (2016)
11. J. Rakowska, K. Radwan, Z Ślosorz, E. Pietraszek, M. Łudzik, P. Suchorab, https://www.cnbop.pl/wydawnictwa/ksiazki/usuwanie_substancji_ropopochodnych.pdf (2012)
12. J. Janzen dos Santos, L. T. Maranho, J. Environ. Manage. **210** (2018)
13. C. Riccardi, P. Di Filippo, D. Pomata, M. Di Basilio, S. Spicaglia, F. Buiarelli, Sci. Total Environ. **450–451** (2013)
14. E. Smith, P. Thavamani, K. Ramadass, R. Naidu, P. Srivastava, M. Megharaj, Int. Biodeterior. Biodegradation, **101** (2015)
15. I. Errington, C. K. King, S. Houlahan, S. C. George, A. Michie, G. C. Hose, Sci. Total Environ. 619–620 (2018)
16. D. Izdebska-Mucha, J. Trzciński, Geologija, **50** (2008)
17. B. H. Alharbi, M. J. Pasha, A. H. Alhudhodi, A. K. Alduwais, Soil Sediment Contam. **8** (2018)
18. J. Nowak, www.pfb.info.pl/files/kwartalnik/1_2008/nowak.pdf (2008)
19. A. Gao, J. Liang, T. Teng, M. Zhang, Appl. Soil Ecol. (to be published)
20. L. Bardel, https://bip.malopolska.pl/e,pobierz,get.html?id=1037479 (2014)
21. S. Chakraborty, D. C. Weindorf, B. Li, M. Nasim Ali, K. Majumdar, D. P. Ray, Environ. Pollut. **190** (2014)
22. G. Lassalle, A. Credoz, R. Hédacq, S. Fabre, D. Dubucq, A. Elger, Environ. Sci. Technol. **52** (2018)

The load of heavy metals leaching from grassland in central Ner River valley

Tomasz Bagrowicz[1,*], *Maciej* Ziułkiewicz[2], and *Janusz* Mazur[3]

[1]Faculty of Geographical Sciences, University of Łódź, ul. Narutowicza 88, 90-139 Łódź, Poland
[2]Laboratory of Geology, Faculty of Geographical Sciences, University of Łódź, ul. Narutowicza 88, 90-139 Łódź, Poland
[3]Laboratory of Computer and Analytical Techniques, Faculty of Biology and Environmental Protection, University of Łódź, ul. Banacha 12/16, 90-237 Łódź, Poland

Abstract. Łódź is the largest Polish city in the zone of the main watershed of the country. The watershed zone offered a dense network of small streams, mostly belonging to the Ner River catchment. The use of the Ner River as a collector of industrial sewage caused the accumulation of high amounts of heavy metals in the river sediments and arable land, intensively fertilised with the waters of the Ner River for more than 40 years. Due to the continuation of irrigations in the river valley, water with a substantially lower degree of contamination with organic substance and biogens returns to Ner from the grassland complexes. The objective of the study is the determination whether improvement of water quality subject to circulation within the grassland complexes also concerns heavy metals. The study area covers grassland complexes in the middle section of the Ner River valley, between the weir in Charbice Górne and Jeżew. Due to the variability of acidic-alkaline and oxidation-reduction conditions caused by variable moisture of the soil containing high amounts of allochtonic (post-sewage) organic matter, the observed grassland complexes show liability in terms of retention and release of the analysed metals to Ner. In summer 2018 on the analysed grassland complexes, conditions favouring the emission of all studied metals occurred. This evidences what the discontinuation of irrigation of grassland complexes, previously irrigated with sewage from Łódź, can lead to. The comparison of sites located along the course of the Ner River also provides a complex image of changes in concentrations, with evident fluctuations of metal concentrations.

1 Introduction

Łódź is the largest Polish city in the zone of the main watershed of the country. This determines a specific system of water supply and sewage management. Due to the lack of a large river that could receive sewage from the rapidly developing industrial city, sewage was directed to the local river network. The watershed zone offered a dense network of small streams, mostly belonging to the Ner River catchment. Over a short time, they became

* Corresponding author: bagrowicz.tomasz@gmail.com

degraded, and their recipient began discharging high amounts of sewage far outside the boundaries of Łódź, with a substantial contribution of contaminants originating from industrial plants. Over ten years, the share of industrial sewage was subject to considerable changes, from 51% in 1965 to 65% in the years 1985–1990, 25% in 2013, and 11.6% currently (2017) [1].

The first urban sewage treatment plant was launched in the first half of the 1930's, more than 100 years after the launch of the first mechanised factory in Łódź equipped with a steam engine. Biological treatment of sewage supplied from the city began in 2002. With the collapse of the textile industry in ŁAM, the city entered another stage of urban planning evolution [2], i.e. from the stage of an industrial city to the stage of a sanitary city. The use of the Ner River as a collector of industrial sewage caused the accumulation of high amounts of heavy metals in the river sediments [3] and arable land, intensively fertilised with the waters of the Ner River for more than 40 years. The content of heavy metals in the soils of the Ner River valley was variable, increasing with the time of irrigation. It was 2–5 times higher in soils irrigated with sewage in comparison to non-irrigated soils, and particularly depended on the distance from the point of sewage discharge to Ner, content of organic substance in the soil, and depth of the distribution ditch [4].

It is estimated that in the years 1980–1995, over an irrigated area of approximately 4300 ha in the Ner River valley, from 13% (humid year) to 34% (very dry year) of sewage discharged from Łódź was used for vegetative irrigations [4]. In the years 1994–1995, in the period of decline of the textile industry in Łódź, the daily load of metals in the outlets of the city's collectors was as follows: Hg 0.1 kg, Cd 0.23 kg, Cr(6+) 2.16 kg, Ni 3.12 kg, Pb 5.18 kg, Cu 44.48 kg [5]. The elimination of industry as the basic source of contamination of the Ner River with heavy metals caused a successive decrease in heavy metal concentrations in the river in the years 1996–2009, as well as an evident reduction of the scale of their seasonal fluctuations [6].

In 1995, granting water permits for water intake from Ner for agricultural irrigations was discontinued due to the strong contamination of the environment [7]. As a result, a process of elimination of the contemporary intensive agricultural production of diary cattle fodder began. The sense of further maintenance of the irrigation system was questioned. It was decided, however, that the system should keep functioning due to the necessity of maintenance of constant humidity conditions in the ground, in order to counteract the remission of contaminants accumulated in the soils to the Ner River, particularly biogens and toxins [4].

Due to the continuation of irrigations in the river valley, water with a substantially lower degree of contamination with organic substance and biogens returns to Ner from the grassland complexes [7]. It is important due to the significance of the river in the region (health resort aspirations of Poddębice) and in other areas (water supply in Poznań). The objective of the study is the determination whether improvement of water quality subject to circulation within the grassland complexes also concerns heavy metals.

2 Study object

The Ner River is the largest right-bank tributary of the Upper Warta River. Its total length is 129.5 km, and its catchment occupies an area of 1831 km^2. The river begins its course in the south-western suburbs of Łódź. From there, it flows westwards, and further to the north-west. The river runs down from the Łódź Hills, and then through the Łask Plateau and Koło Basin, where it flows into the Warta River. At its mouth, Ner maintains an average flow rate at a level of 13.9 $m^3 \cdot s^{-1}$. In the landscape of the southern suburbs of Łódź, Ner is a small river, with an average flow rate of 0.23 $m^3 \cdot s^{-1}$. Discharge of sewage from the Collective Sewage Treatment Plant in Łódź increases its natural flow rate nine times. An average discharge of sewage from the treatment plant in the years 2012-2017 was approximately 2.0 $m^3 \cdot s^{-1}$ (1.77–2.24 $m^3 \cdot s^{-1}$) [8]. Over the years, the amount of sewage discharged to Ner River varied from 1.4 $m^3 \cdot s^{-1}$ (1951) to 4.9 $m^3 \cdot s^{-1}$ (1992).

3 Material and methods

The study area covers grassland complexes in the middle section of the Ner River valley, between the weir in Charbice Górne and Jeżew (Fig. 1). The research concerns four neighbouring grassland complexes irrigated by means of the slope method (Table 1). Analyses of waters in the melioration system are conducted on the inlet and outlet for particular complexes, and for comparison on the Pisia River – the only natural tributary on this section of Ner. Sample collection and the accompanying measurements of water reaction, dissolved oxygen (DO) and flow rate (by means of a flow meter or floating object method) is performed once a month (starting from April 2018), on the same day in all the designated measurement-control points following the direction of water flow. Due to the currently implemented programme of meadow irrigation according to individual needs of water companies, the research terms sometimes do not correspond with the conducted cycle of interim irrigation, resulting in lack of information in certain terms. In such circumstances, measurements and sample collection are performed in the Ner channel in direct vicinity of the irrigation ditch, treating the determined hydrochemical conditions as potential and not actual.

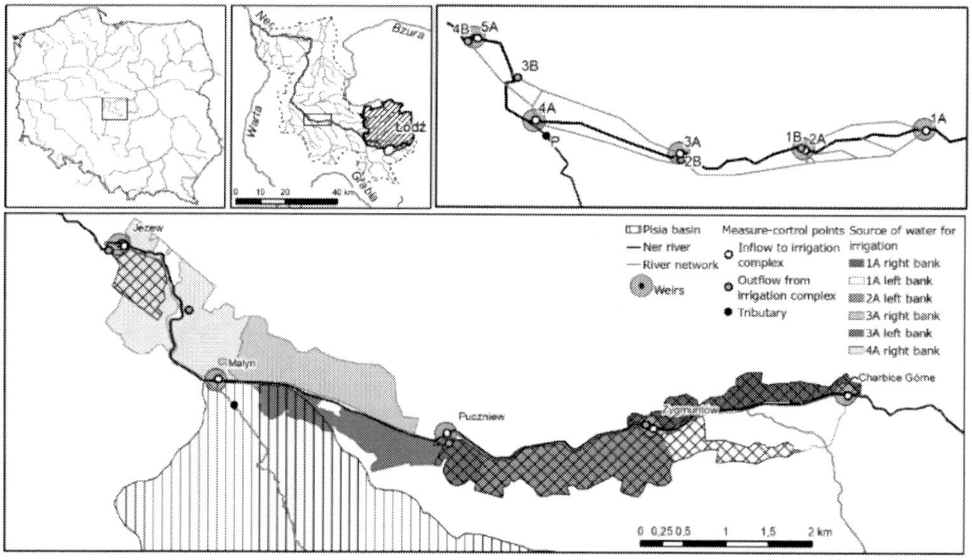

Fig. 1. Scheme of the circulation of water used for irrigation in the drainage system in the Ner River Valley and location of the balance areas.

Based on the obtained measurement of water flow rate on the inlet an outlet from the grassland complex, the load of the supplied and discharged heavy metals is determined. The scope of the chemical analysis includes the following heavy metals: Cd, Cr, Cu, Fe, Mn, Ni, Pb, Zn. The collected water samples are filtered *in situ* through a membrane filter 0.45 μm and fixed with HNO_3. Determinations are performed at the Laboratory of Computer and Analytical Techniques of the Faculty of Biology and Environmental Protection of the University of Łódź by means of the method of atomic absorption spectrometry with electromagnetic atomisation, and method of flame atomic absorption spectrometry.

Table 1. Surface areas of irrigated grassland complexes subject to the study [km^2] (A-inflow to irrigation complex, B- outflow from irrigation complex).

Complex		Source of water for irrigation:					
		Weir Charbice Górne		Weir Zygmuntów	Weir Puczniew		Weir Małyń
		Left bank	Right bank		Left bank	Right bank	
1	A		0.57				
	B		0.57				
2	A			1.16			
	B	0.35		1.16	0.22		
3	A					1.11	
	B					0.89	0.39
4	A						1.73
	B						0.34

Source: Own elaboration

4 Results

Meteorological data for station IMGW Łódź-Lublinek, referred to the multi-annual characteristics [9], present 2018 as a very warm year with a precipitation total below the multi-annual average. Mean annual temperature in 2018 exceeded the highest mean annual value from the multi-annual period by 0.3°C. The precipitation total was 510 mm, and was lower than the mean annual total from the multi-annual period by 65 mm. The highest contribution of atmospheric precipitation was observed in July and October: 21% and 18% of annual precipitation, respectively. A total of 133 days with precipitation occurred in 2018 (> 0.1 mm). The value is 34 days lower than the multiannual average. The Ner River valley in a part of which the research was conducted shows the index of climatic precipitation deficit in the period from April to October at a level from -120 mm to -80 mm [10].

Water flow on the inlet and outlet of the irrigation system of the grassland complexes is presented in Fig. 2. Only one of the studied complexes (Zygmuntów) was irrigated regularly. In Charbice, irrigations were discontinued in autumn of the same year. In Puczniew, introduction of the waters of the Ner River to fields was discontinued based on the decision of the property owner, and in Małyń water is introduced to the fields only in the scope of interim needs of farmers, because in their opinion, Ner already carries low amounts of fertiliser substances along this section.

Water reaction at sites representing river waters on the inlet to distribution ditches varied from 6.42 to 7.67. On outlets from grassland complexes reaction in a range of 6.53–7.50 was recorded. Therefore, weakly acidic, neutral, and weakly alkaline waters circulated in the system of valley irrigation. The waters of Pisia were stronger alkaline than in the Ner valley, because their reaction was within the range from 7.26 to 8.16.

Oxygen saturation of waters introduced to the grasslands varied from 12% to 81%, averaging 60%. In waters flowing out of the grasslands to the Ner River, the determined oxygen saturation showed a narrower range, namely 25%–53%, averaging 36%.

The comparison of the statistical characteristics of recorded concentrations of heavy metals at sites representing the entrance and exit from the irrigation system reveals the mosaic character of changes, expressed in the quartile range and median (Fig. 3). A decrease in concentrations of Mn and an increase in Ni was observed on all complexes. A reduction of concentrations of all analysed metals with the exception of Mn occurred on complex Zygmuntów. An increase in concentrations of Cu and Zn was recorded on the remaining complexes.

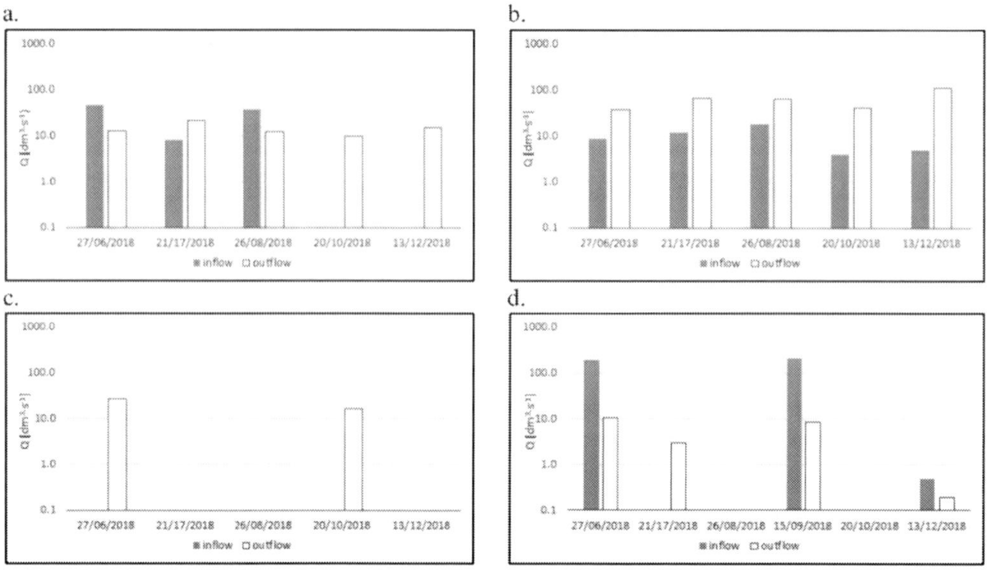

Fig. 2. Comparison of the amount of waters flowing in and out of grassland complexes during the hydrochemical research: a/ Charbice, b/ Zygmuntów, c/ Puczniew, d/ Małyń.

The comparison of sites located along the course of the Ner River also provides a complex image of changes in concentrations, with evident fluctuations of metal concentrations. It can be observed, however, that Fe and Mn gradually decreased to the mouth of Pisia, and then increased. Concentrations of Cu and Pb increased to Puczniew, and the decreased. An evident increase in the median of Cd and Cr concentrations was recorded in Małyń. The balanced unitary load of metals, calculated based on the obtained concentrations and results of water flow measurement on the inlet and outlet from the grassland complexes is presented in Table 2.

Mn is released from the analysed meadow complexes the most frequently. The remaining metals are released from practically all complexes in summer, and from Zygmuntów and Małyń also in winter. Due to the number of cases of occurrene of positive results of balancing of loads, metals are supplied to Ner from the Charbice complex the most seldom, more frequently from Małyń, and the most frequently from Zygmuntów.

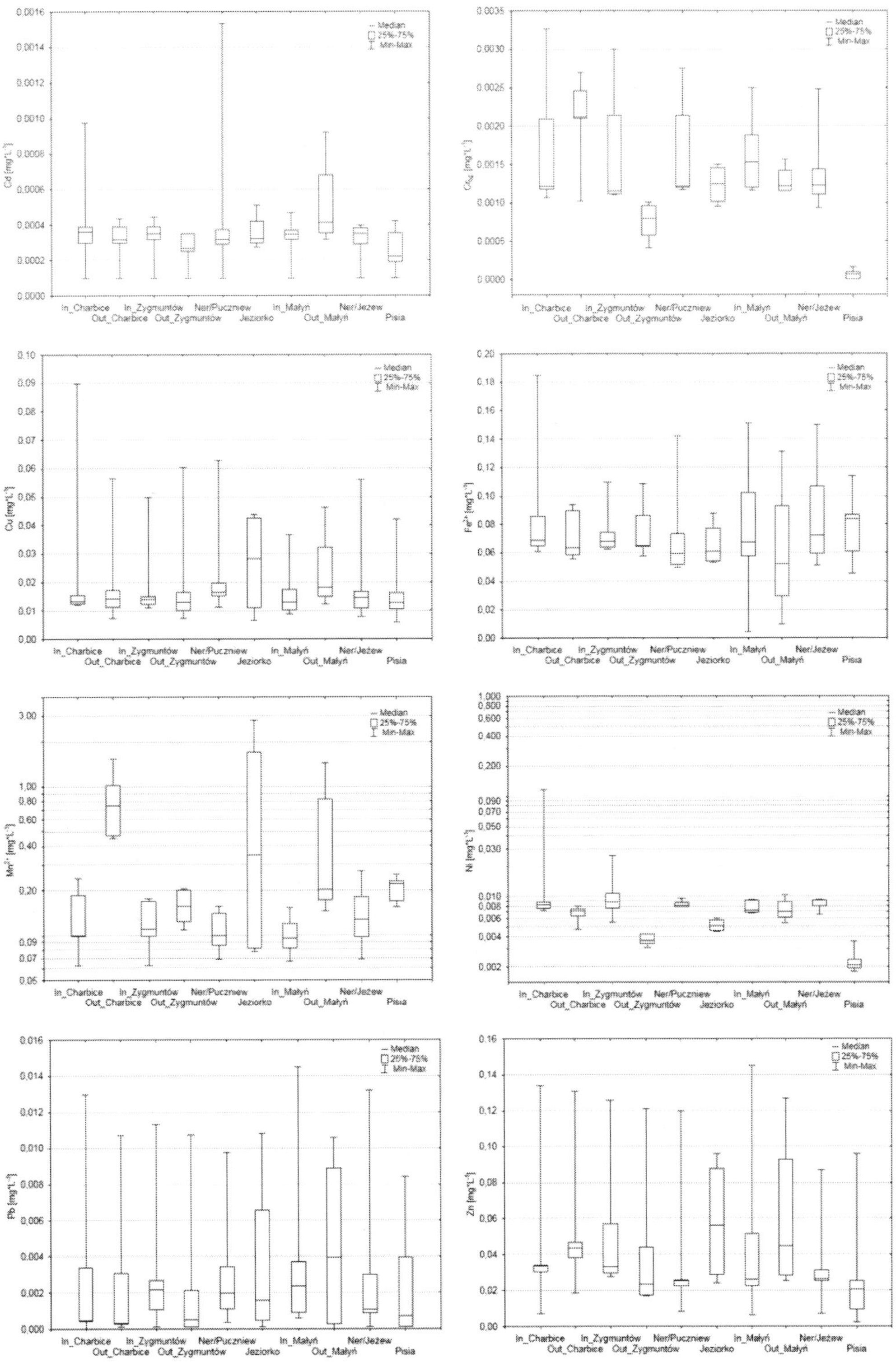

Fig. 3. Ranges of concentrations of metals at the study sites.

Table 2. Balance of unitary loads of metals in irrigated land of grassland complexes. The calculations are based on instantaneous measurements.

Complex / month	**Unitary load [g·d⁻¹·km⁻²]**						
	Cd	Cr	Cu	Fe	Ni	Mn	Pb	Zn
Charbice / VI 2018	-0.50	-6.17	-71.52	-341.42	-42.52	204.40	-2.63	-171.82
Charbice / VII 2018	0.83	5.42	30.39	122.48	14.36	4792.97	1.01	111.68
Charbice / VIII 2018	-1.16	-1.58	-47.70	-182.50	-35.73	1008.33	-2.21	-112.97
Charbice / X 2018	-0.71	-6.77	-21.66	-422.17	-16.26	1234.27	-5.75	36.21
Charbice / XII 2018	-1.99	-0.32	-146.34	-129.21	-339.06	269.96	-15.48	-114.59
Zygmuntów / VI 2018	-0.11	-2.62	-17.53	-82.07	-16.86	72.15	-1.82	-54.01
Zygmuntów / VII 2018	0.31	1.75	35.30	196.23	2.12	435.67	-0.95	35.26
Zygmuntów / VIII 2018	0.09	-1.66	-13.83	142.46	-17.47	297.25	-0.60	-54.09
Zygmuntów / X 2018	0.24	-2.51	6.74	-85.16	-2.23	125.32	0.36	67.26
Zygmuntów / XII 2018	0.52	3.28	251.02	278.38	-92.94	370.49	47.42	546.33
Małyń / VI 2018	0.67	1.06	16.65	18.26	2.47	316.08	0.44	24.69
Małyń / VII 2018	-0.12	-0.23	1.69	-16.08	-2.55	977.16	-1.53	2.47
Małyń / VIII 2018	-0.36	-1.18	-10.11	-73.19	-7.13	-66.08	-0.91	-28.37
Małyń / X 2018	-0.33	-2.46	-13.27	-148.73	-6.67	-81.13	-3.12	-6.09
Małyń / XII 2018	0.02	0.05	2.16	2.23	0.48	6.90	0.46	5.70

Notice: positive value means release to the Ner River, negative value means retention in soil

5 Discussion and conclusions

Concentrations of heavy metals in the waters of the Ner River show considerable variability – from values characteristic of natural waters to those of polluted waters. Lead concentration in non-polluted rivers is several mg·dm^{-3}. Concentrations of lead and its compounds in the Warta River in Dobrowa, above the mouth of the Ner River, are 0.0047 mg·dm^{-3}, and cadmium and its compounds 0.0005 mg·dm^{-3}, respectively. They are values within the range of class I of water cleanliness. In Koło, values of lead and cadmium point to class III and IV of waters, respectively [11]. The adopted natural content of zinc in river water was 0.01 mg·dm^{-3}. The waters of Ner reach more than 5 times higher concentrations, but the values are generally within the range of values of the waters of the Oder River catchment which are estimated as 0.03-0.165 mg·dm^{-3}, whereas higher concentrations are recorded in the waters of the Upper Silesian Agglomeration [12]. Chromium concentration in non-polluted waters is 0.0005 mg·dm^{-3}. The minimum recorded concentrations are twice as high, although they do not exceed values for class I of water quality [13].

Copper, zinc, nickel, and manganese concentrations are higher than the values recorded in the Jangcy River, where they are within the following ranges: 0.0015–0.0231 mg Cu·dm^{-3}, 0.0013–0.0457 mg Zn·dm^{-3}, 0.0004-0.0063 mg Ni·dm^{-3}, and 0.077–0.5 mg Mn·dm^{-3}. Concentrations in waters used for irrigations in China show very high fluctuations, and can exceed the values recorded in the Ner River multiple times: Cd^{-} 0.0002–3.45 mg·dm^{-3}, Cr^{-} 0.0003–4.08 mg·dm^{-3}, Mn^{-} 0.251 mg·dm^{-3}, Ni^{-} 0.007 mg·dm^{-3}, Pb^{-} 0.0005–6.86 mg·dm^{-3}, Zn^{-} 0.05–51 mg·dm^{-3} [14]. In the El Saf Region in Egypt, sewage waters used for irrigations also show higher concentrations of heavy metals: 0.004–0.08 mg·dm^{-3} cadmium, 0.02–0.28 mg·dm^{-3} lead, and 0.001–0.03 mg·dm^{-3} nickel [15].

Maximum values of Ni, Cu, and Mn concentrations exceed values recorded in the rivers of the Silesian Voivodeship: Kłodnica, Bytomka, Czerniawka, or Biała Przemsza [16]. Total load of heavy metals discharged from the Oder River to the Baltic Sea in 2009 includes: zinc – 0.48 kg·km^{-2}·year^{-1}, translating into 49.1 t·year^{-1}, copper – 0.27 kg·km^{-2}·year^{-1} i.e. 27.22 t·year^{-1}, nickel – 0.23 kg·km^{-2}·year^{-1} i.e. 23.88 t·year^{-1}, and cadmium 0.01 kg·km^{-2}·year^{-1} i.e. 0.67 t·year^{-1} [17].

The range of pH recorded in the waters circulating between the inlet and outlet of the irrigation system is neutral and weakly alkaline. Assuming the occurrence of oxygenation conditions at an average saturation of the waters at a level of 60%, such reaction should cause geochemical immobility of practically all the analysed metals. Especially that the valley environment is abundant in organic matter, loamy minerals, silica, iron and manganese hydroxides, fulfilling the role of effective sorbents [18–20]. Waters flowing out of the grasslands show a somewhat lower oxygen saturation – on a level of 36% (min. 25%). This suggests the impact of the gley environment of valley soils, developed based on alluvial and sewage sediments. In such conditions, Fe and Mn become mobile water migrants [20].
Due to the variability of acidic-alkaline and oxidation-reduction conditions caused by variable moisture of the soil containing high amounts of allochtonic (post-sewage) organic matter, the observed grassland complexes show liability in terms of retention and release of the analysed metals to Ner. The conditions occurring in the Zygmuntów complex deserve particular attention. They favour emission of Cd as well as Cu, Fe, and Zn. Cadmium and the remaining aforementioned metals are effectively sorbed by iron hydroxides [18, 19]. The disappearance of hydroxides, suggested by the release of iron in summer and winter, led to the gradual mobilisation of Cd, very hazardous for living organisms [19].

In summer 2018 on the analysed grassland complexes, conditions favouring the emission of all studied metals occurred. During hot and dry weather, a decrease in the groundwater level occurred – oxygenation of the uppermost parts of the saturation zone – oxygenation of the deposited sulphates – acidification and mobilisation of metals effectively migrating in the acidic environment. This evidences what the discontinuation of irrigation of grassland complexes, previously irrigated with sewage from Łódź, can lead to. In winter 2018, the release also occurred from complexes Zygmuntów and Małyń. This requires further investigation.

From the point of view of the quality of river waters, retention of toxic metals in irrigated land is positive. Final treatment of Ner occurs there. The river currently predominantly carries contaminants mobilised from channel bottom sediments.

The progressing accumulation of heavy metals in the grasslands excludes them from the production of safe fodders.

References

1. *Raport o stanie środowiska w województwie łódzkim. Na podstawie badań przeprowadzonych w ramach Państwowego Monitoringu Środowiska w 2017 r., 2018.* (Wojewódzki Inspektorat Ochrony Środowiska w Łodzi, Łódź)
2. S.S. Kaushal, W.H. McDowell, W.M. Wollheim, T.A. Newcomer-Johnson, P.M. Mayer, K.B. Belt, M.J. Pennino, Water **7**, 8 (2015)
3. I. Bojakowska, *Wpływ odprowadzania ścieków na akumulację metali ciężkich w osadach wybranych rzek Polski* (Państwowy Instytu Geologiczny, Warszawa, 1995)
4. J. Mosiej, *Przyrodniczo- techniczne uwarunkowania gospodarowania wodą w dolinie rzeki Ner* (SGGW, Warszawa, 1999)
5. J. Diehl, *Założenia polityki ekologicznej miasta Łodzi* (Urząd Miasta Łodzi, Łódź, 1997)
6. *Wyniki monitoringu wód powierzchniowych w ramach Państwowego Monitoringu Środowiska w latach 1996-2009* (Wojewódzki Inspektorat Ochrony Środowiska w Łodzi. https://www.wios.lodz.pl/Wyniki_badan_wod_powierzchniowych,154 updated 2018.07.02, visited 2018.11.21)
7. A. Matczak, *Ocena możliwości i skutków środowiskowych doczyszczania wód ściekowych odpływających z oczyszczalni ścieków w Łodzi poprzez nawodnienia i inne*

przedsięwzięcia ochronne [Typescript of the diploma thesis] (SGGW-KKŚiM, Warszawa, 1998)

8. http://www.gos.lodz.pl/dla-studentow/oczyszczalniadane-technologiczne/updated 2018.03.02, visited 2019.01.15.
9. A. Podstawczyńska, *Temperatura powietrza i opady atmosferyczne w regionie łódzkim w ostatnim stuleciu* (Bogucki Wydawnictwo Naukowe, Poznań, 2010)
10. A. Ciepielowski, *Podstawy gospodarowania wodą* (SGGW, Warszawa, 1999)
11. *Stan wód Warty na terenie województwa wielkopolskiego w latach 1999−2009* (Biblioteka Monitoringu Środowiska, Poznań, 2010)
12. *Biuletyn: Ocena jakości powierzchniowych wód płynących w punktach pomiarowo-kontrolnych województwa śląskiego w 2001 roku. Zlewnia Odry* (Inspekcja Ochrony Środowiska, Śląski Wojewódzki Inspektorat Ochrony Środowiska w Katowicach, Katowice, 2002)
13. *Rozporządzenie Ministra Środowiska w sprawie klasyfikacji dla prezentowania stanu wód powierzchniowych i podziemnych, sposobu prowadzenia monitoringu oraz sposobu interpretacji wyników i prezentacji stanu tych wód* (Dz.U.2004.32.284)
14. S. Cheng, Environ. Sci. Pollut. R. **10**, 3 (2003)
15. H. El Azab, A. Afify, M. Soliman, EJRS **18**, 2 (2015)
16. W. Nocoń, *Rola zawiesin w transporcie metali ciężkich w płynących wodach powierzchniowych* [Typescript of the diploma thesis] (Politechnika Śląska Wydział Inżynierii Środowiska i Energetyki, Gliwice, 2011)
17. *Stan czystości rzek na podstawie wyników badań wykonywanych w ramach państwowego monitoringu środowiska w latach 2007-2009* (Biblioteka Monitoringu Środowiska, Warszawa, 2010)
18. C. A. J Domenico, D. Postma, *Geochemistry, groundwater and pollution* (A. A.Ballkema, Rotterdam, 1996)
19. I. Bojakowska, *Wpływ czynnika antropogenicznego na procesy geochemiczne w powierzchniowych warstwach litosfery* (Państwowy Instytut Geologiczny, Warszawa, 1994)
20. F.R. Siegel, *Environmental Geochemistry of Potentially Toxic Metals* (Springer-Verlag, Berlin/New York, 2002)

Preliminary research for low-cost particulate matter sensor network

Csongor Báthory[1,*], *Márton* L. Kiss[2], *Attila* Trohák[2], *Zsolt* Dobó[1], and *Árpád Bence* Palotás[1]

[1]University of Miskolc (ME), Institute of Energy and Quality Affairs,
3515 Miskolc-Egyetemváros, Hungary
[2]University of Miskolc (ME), Institute of Automation and Infocommunication,
3515 Miskolc-Egyetemváros, Hungary

Abstract. Low-cost particulate matter (PM) sensors may be suitable for indicative air quality measurements thanks to their small dimensions and high spatial resolution. Three different sensor types were selected for investigation in this study with specific focus on a Honeywell HPMA115S0 sensor to find out its usability at outdoors, perform load and long-term tests. The load test showed that the sensor calculates PM_{10} based on measured $PM_{2.5}$ values. The analysis shows a break in calculation method at 25 μg/m^3 $PM_{2.5}$, and the calculation method for PM_{10} varies from 25 μg/m^3 by around 81 μg/m^3. Parallel test performed with different sensor types has shown that the protective cover formed by lamellar exterior does not affect the accuracy of the sensors, no accumulation or loss of sensitivity occurs. Long-term measurements have shown that the concentration values measured by the Honeywell sensor during outdoor measurements require humidity compensation, over 90% relative humidity (RH) the Pearson correlation coefficient (R) between the reference and sensor $PM_{2.5}$ concentrations decreased by 0.3.

1 Introduction

PM concentration is one of the most important indicator of ambient air quality due to its impact on human health [1], visibility [2] and climate [3, 4]. From a health point of view, the negative health effects of elevated PM concentrations include arrhythmia, heart disease, lung cancer and mortality [5]. Because of these serious consequences, the ambient concentrations of PM_{10} and $PM_{2.5}$ are regulated by The United States Environmental Protection Agency (US EPA [6]), the European Union (Air Quality Directive 2008/50/EC, AQD [7]) and the Hungarian Legislation (306/2010 (XII 23.) Government Decree on Air Protection [8]), but the World Health Organization (WHO [9]) also makes recommendations to that. Compliance with the regulations is checked by the environmental authorities in accordance with the Federal Reference Methods (FRM [10]) or the Air Quality Directive (AQD [7]). According to the standards, the appropriate size of dust is collected by pre-separating and filtering, and the daily and annual average concentrations of PM are calculated by gravimetric method,

* Corresponding author: csongor.bathory@uni-miskolc.hu

based on EN12341:2014 standard [11]. Beside FRM, there are so-called Federal-Equivalent Methods (FEM [10]) approved by EPA, which can provide hourly, quarter hourly and even minutely data for measuring PM concentration [12]. FEM-specific methods operate according to a measurement principle different from the gravimetric method of the FRM and EN12341:2014, such as optical detection [13], beta-ray absorption [14] or Tapered Element Oscillating Microbalance (TEOM) [15]), complying with strict requirements.

FEM devices cost almost the same as the FRM devices, but they are easier to handle and more practical for continuous monitoring. There are laboratory calibrated desktop and handheld devices (Calibrated Portable Devices, CPD) on the market which are suitable for detecting PM, and the price is one order of magnitude lower than the previous ones, but the results are not accepted in official procedures. Low Cost Sensors (LCS) are available with significantly lower price (<$50), some of which are calibrated, but their calibration is not well documented and unreliable. Fig. 1 shows the classification of PM measurement devices.

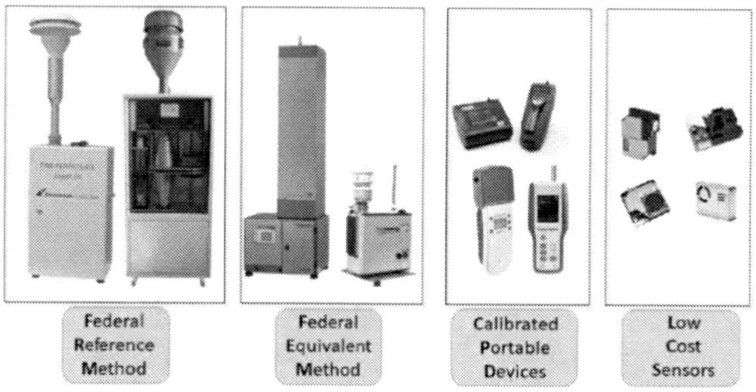

Fig. 1. Classification of PM measurement devices.

AQD gives the opportunity to use complementary techniques such as air quality models and indicative measurements. Low-cost dust sensors may be suitable for such indicative measurements. The legal and procedural background is currently being developed by EU Delegated Working Groups (CEN/TC 264/WG42 [16]). Despite many studies, there are still many unanswered questions about the accuracy, reliability, consistency and reproducibility of these sensors [17–23] and the environmental impacts are not fully clear. For example, humidity is an important parameter in the analysis of sensor values, as FEM and FRM devices provide PM concentration values at constant 50% RH [19–21].

According to the AQD, indicative measurements are measurements which fulfill less strict data quality objectives compared to fixed measurement methods. LCS belong to the group of indicative measurements, because they operate with accuracy of 10–15% under ideal conditions. In addition, the meteorological characteristics may further increase the error of measurement; however, indicative measurements can be carried out with them. Commercially available low-cost sensors provide concentration values for PM_{10}, $PM_{2.5}$ and even PM_1. According to our previous research, fan-mounted sensors were preferred during tests, because the amount of air sucked in is less influenced by the weather and they are more reliable. Similar conclusion was reported by Canu et al. [24]. The main goal of this paper is to test three different types of sensors from various aspects including reproducibility, design variations, sensors comparison and humidity influence.

2 Materials and methods

Three sensors were selected for investigation in this study, the main parameters are summarized in Table 1. Plantower can measure more than one particle size at a time, providing data on the number of particles less than 0.3, 0.5, 1, 2.5, 5, and 10 μm in a volume of 0.1 liters, plus concentrations of PM_1, $PM_{2.5}$ and PM_{10} in μg/m^3. Winsen gives PM_1, $PM_{2.5}$ and PM_{10}, while Honeywell gives $PM_{2.5}$ and PM_{10} in μg/m^3. Plantower shows extreme humidity range of 0–99% RH, but non-condensing conditions are required for all the three sensor types. The upper RH limit of Honeywell is 95% but some studies reveal that the humidity impacts the readings from 60% [25], 70–75% [26, 27] and 85–95% [28, 29].

Table 1. Main parameters of sensors selected for investigation.

Brand, Name	Plantower	Winsen	Honeywell
Type	PMS7003	ZH03	HPMA115S0
PM_1	x	x	-
$PM_{2.5}$	x	x	x
PM_{10}	x	x	x
Light source	laser	laser	laser
Flow provider	fan	fan	fan
Limit of detection [μm]	0,3	0,3	-
Measurement range [μg/m^3]	0–1000	0–1000	0–1000
Working Humidity [%RH]	0–99	0–85	0–95
Working Temperature [°C]	-10–60	-10–50	-10–50

An earlier study found that the firmware of Nova SDS011, which is another type of $PM_{2.5}$ and PM_{10} sensor, calculates the PM_{10} concentration from $PM_{2.5}$ values by an unknown interpolation [43]. So, based on the datasheets, it is not clear yet that in what range makes the sensor real measurement and how much it relies on calculations. The PM_{10} calculation method was determined in case of Honeywell sensor in the full measurement range. A test was carried out with two Honeywell type sensors in a cigarette smoky open-air environment. The typical particle size of cigarette smoke is below 1 μm [44], other PM sources were not significant. The average temperature during the test was 28 ± 0.1°C and the relative humidity was 61 ± 2.1%. The results of the two Honeywell sensor cigarette smoke tests were averaged to reduce sensor uncertainty. Inspired by meteorological stations a lamellar protective cover was made, in which the individual types of sensor can be fixed in a convertible manner. For comparability, two of the three types of PM sensors were operated by two STM32 microcontrollers. Temperature and relative humidity data were collected by BME280 sensor, which is also an LCS. Data from sensors was recorded by a Raspberry PI 3 B + microcontroller in every 6 seconds. Measurements were made on the 1st floor balcony at Miskolc, Táncsics square residential building, about 50 meters from the main road. In this area, air pollutants typically originate from the main road and from the surrounding household fires, the latter is more significant during heating season. A series of one-hour period measurements were conducted with different sensors to investigate the influence of cover as well.

Comparative measurements were carried out without a calibrated reference device, so the measurement without the protective cover serves as a reference. The Honeywell type sensor with the tested protective cover was installed at the Országos Légszennyezettségi Mérőhálózat (National air pollution measurement network of Hungary, OLM) measuring station of Miskolc, Alföldi Street. The sensor was placed at the same height as the inlet of the installed monitoring stations GRIMM EDM 180, which was the reference FEM PM device during the 13 days long measurement.

3 Results and discussion

To determine the effect of the protective cover, changes in the measured values of the same type of sensors during hourly series were compared. A total of 12 measurements were performed. Fig. 2a shows the differences of Honeywell sensors in full range, so outliers are displayed (-24.3 μg/m³ and -61.9 μg/m³).

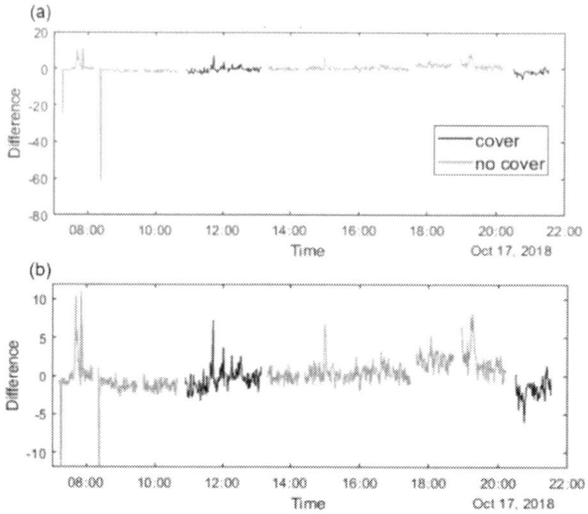

Fig. 2. Differences of $PM_{2.5}$ concentrations between Honeywell sensors with cover and without cover (a) in full range and (b) in constricted range.

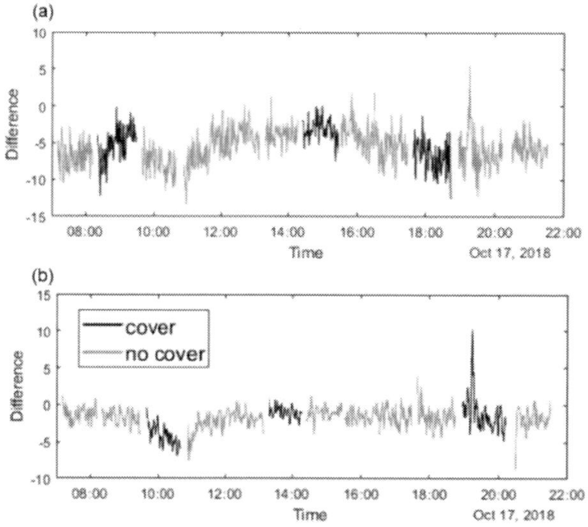

Fig. 3. Differences of $PM_{2.5}$ concentrations between (a) Plantower and (b) Winsen sensors with cover and without cover.

The outliers were recorded during the measurements as none of the Honeywell sensors were under protective cover (no cover). Fig. 2b do not shows the outliers, the differences can be seen between 11.0 μg/m³ and -6.0 μg/m³, resulting a 17.0 μg/m³ range. With the outliers this range is 72.9 μg/m³. Fig. 3 (a) shows the Plantower results, the measured difference range was 18.9 μg/m³. Fig. 3b shows the Winsen results, the measured difference range was 18.8 μg/m³. Outliers were not found during Plantower and Winsen tests and the concentration reading differences between the sensors are small. The results show that the protective cover does not affect the sensitivity of the sensors. Further analysis was carried out to determine the probability density functions of the normal distribution of the differences (Fig. 5). Compared the measurements made with cover to the measurements with no cover, significant difference can be seen only in the case of Honeywell (Fig. 5a), which is caused by the outlier values (Fig. 2a). If we ignore functions with the outliers the cover and no cover functions are near to each other expect the long term no covered function, which means that the outlier values occur in long term too. All functions of the Plantower and Winsen sensors are shifted a little under 0 μg/m³, but still inside the range of -15–15 μg/m³ (Fig. 5b and 5c). The Honeywell is less reliable and consistence than Plantower or Winsen.

The recorded $PM_{2.5}$ values were averaged to minutely and hourly concentration as simple moving average. Fig. 6 (a) illustrates the change in humidity and minutely $PM_{2.5}$ as a function of time. RH changes between 53–96% based on the measurement of OLM station. The RH was 95% or higher during 42% of the sampling period. Fig. 6 shows minutely (b) and hourly (c) $PM_{2.5}$ values of Honeywell and OLM device. It can be observed on Fig. 6 that in case of low humidity the concentration values of the sensor are close to the OLM values, while in case of high humidity it differs significantly in positive direction. The Honeywell sensor does not compensate for the change in humidity, either hardware or software, which greatly affects the size and reflective capacity of PM.

Table 3 contains the correlation coefficients between the $PM_{2.5}$ concentrations of Honeywell sensor and OLM device. The hourly values are correlating better than minutely values because of the data aggregation. Correlation coefficients of values under 90% RH are ≈ 0.3 higher than over 90% RH.

Table 3. Correlation between Honeywell sensor and OLM device in full time, under and over 90% RH.

Correlation coefficient	R	R (RH < 90%)	R (RH > 90%)
Minutely $PM_{2.5}$	0.48	0.58	0.35
Hourly $PM_{2.5}$	0.58	0.80	0.45

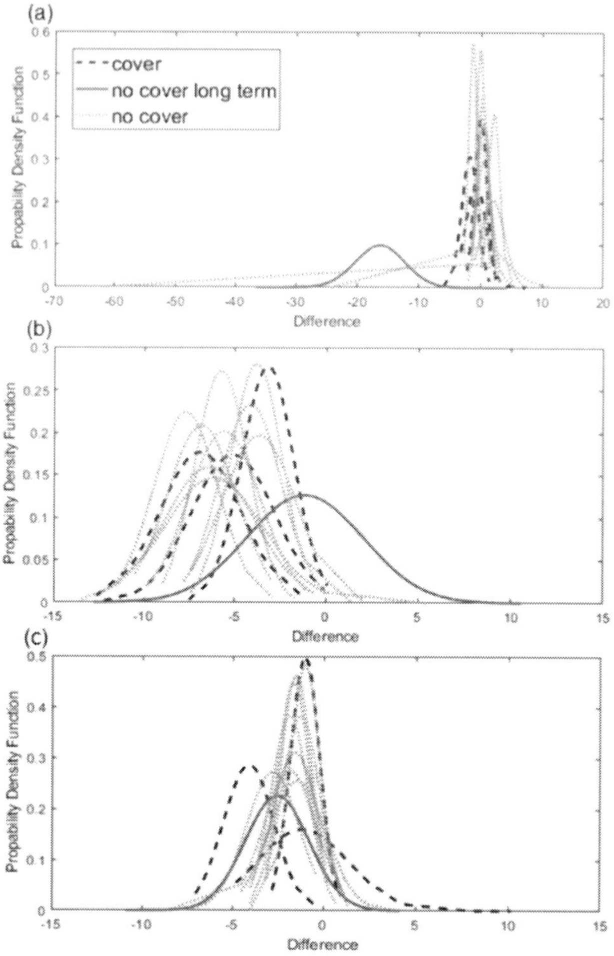

Fig. 5. Probability density functions of (a) Honeywell, (b) Plantower and (c) Winsen sensors differences with cover and without cover.

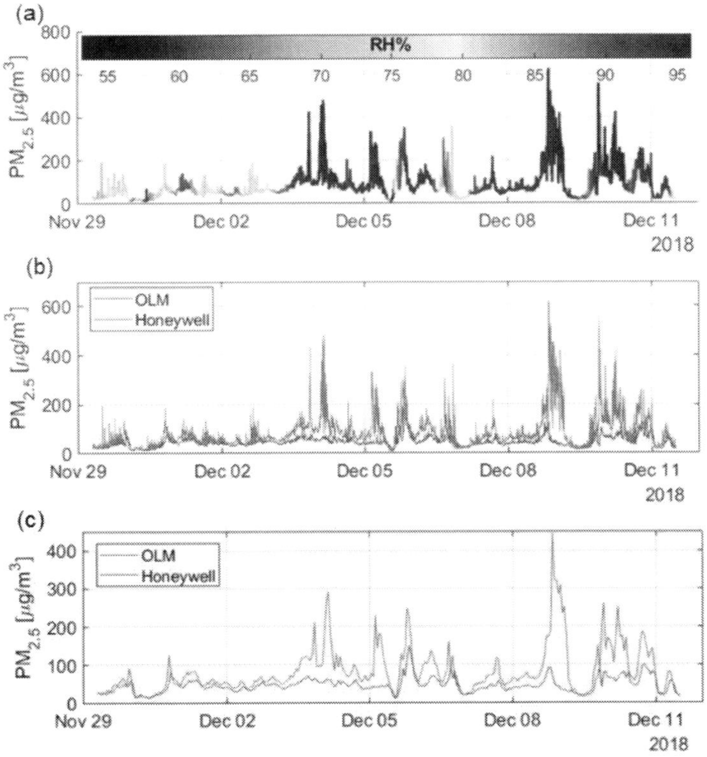

Fig. 6. Illustration of (a) Honeywell $PM_{2.5}$ concentrations and RH values, (b) comparison of minutely $PM_{2.5}$ values of Honeywell and OLM device, (c) comparison of hourly $PM_{2.5}$ values of Honeywell and OLM device.

4 Conclusion

Three different types of sensors from various aspects including reproducibility, design variations, sensors comparison and humidity influence were tested. The effect of a lamellate protective cover is negligible. The difference between the Honeywell sensors with and without the protective cover shows some outliers that were caused by the sensor failure. The Plantower and Winsen are more consistent. Long-term measurement shows that the Honeywell sensor is correlating well to the concentration values of the OLM station in case of low humidity. Future research will be focused on the effect of environmental parameters to the LCS. Our goal is to develop an artificial intelligence based on-filed calibration method that allows us to consider all environmental conditions, time of day, season, heating and non-heating periods. To have enough data for deep learning technique at least one-year long measurement needed from the examined area.

Supported by the ÚNKP-18-3-I-ME/29. New National Excellence Program of the Ministry of Human Capacities.

References

1. A. Valavanidis, K. Fiotakis, J. Environ. Sci. Health C **26**, 339–362 (2008)

2. J. Wu, W. Cheng, J. Environ. Sci. Manag. **21**, 100–109 (2018)
3. O. Jolliet, A. Antón, In. J. Life Cycle Assess. **23**, 1–19 (2018)
4. H. He, X.-Z. Liang, Atmos. Environ. **179**, 166–176 (2018)
5. J. Lepeule, F. Laden, Environ. Health Perspect. **120**, 965–70 (2012)
6. Air quality criteria for particulate matter, vol. 3: US Environmental Protection Agency, Office of Research and Development, National Center for Environmental Assessment (1996)
7. Official Journal of the European Union (2008)
8. 306/2010. (XII. 23.) Korm. rendelet a levegő védelméről, ed. Magyar Közlöny
9. WHO, *Air Quality Guidelines: Global Update 2005. Particulate Matter, Ozone, Nitrogen Dioxide and Sulfur Dioxide*: World Health Organization (2006)
10. Ambient Air Monitoring Reference and Equivalent Methods, in Federal Register 40 CFR Parts, ed: US EPA, (1997)
11. 12341:2014, Determination of the PM_{10} fraction of suspended particulate matter. Reference method and field test procedure to demonstrate reference equivalence of measurements methods, in Air Quality, ed: BSI, (2014)
12. C. A. Noble, R. W. Vanderpool, Aerosol. Sci. Technol. **34**, 457–464 (2001)
13. J. Bol, W. Heinze, *Method of measuring the concentration and/or size of suspended particles by forward scattering of light* (Google Patents, 1972)
14. E. S. Macias, R. B. Husar, Environ. Sci. Technol. **10**, 904–907 (1976)
15. H. Patashnick, E. G. Rupprecht, J. Air Waste Manag. Assoc. **41**, 1079–1083 (1991)
16. CEN–Technical Bodies–CEN/TC 264/WG 42 2018, Available: https://standards.cen.eu/dyn/www/f?p=CENWEB:7:0::::FSP_ORG_ID:2012773&cs=1FD71819F25D74834BB38751B78ACE16D
17. M. Badura, P. Batog, J. Sensors, **2018** (2018)
18. Y. Wang, J. Li, Aerosol. Sci. Technol. **49**, 1063–1077 (2015)
19. C. Borrego, A. Costa, Atmos. Environ. **147**, 246–263 (2016)
20. A. L. Clements, W. G. Griswold, Sensors **17**, 2478 (2017)
21. J.-B. Renard, F. Dulac, Atmos. Meas. Tech. **9**, 1721–1742 (2016)
22. J.-B. Renard, F. Dulac, Atmos. Meas. Tech. **8**, 1261–1299 (2015)
23. A. L. Northcross, R. J. Edwards, Environ. Sci. Process Impacts **15**, 433–439 (2013)
24. M. Canu, B. Galvis, *Understanding the Shinyei PPD24NS low-cost dust sensor* (in: IEEE International Conference on Environmental Engineering, 1–10, 2018)
25. L.-A. Tortajada-Genaro, E. Borrás, J. Environ. Monitor. **13**, 1017–1026 (2011)
26. G. Allen, C. Sioutas, J. Air Waste Manag. Assoc. **47**, 682–689 (1997)
27. R. Jayaratne, X. Liu, Atmos. Meas. Tech. **11**, 4883–4890 (2018)
28. L. R. Crilley, M. Shaw, Atmos. Meas. Tech. 709–720 (2018)
29. R. Williams, T. Conner, *Performance Evaluation of the United Nations Environment Programme Air Quality Monitoring Unit* (EPA – Office of Research and Development National Exposure Research Laboratory, Washington DC, 2017)

Analysis of operating conditions of a cooling installation with carbon dioxide

Artur Bieniek[1,*], *Łukasz* Mika[1], and *Jan* Kuchmacz[1]

[1]AGH University of Science and Technology, Department of Thermal an Fluid Flow Machines, Krakow, Poland

Abstract. In response to international regulations, natural refrigerants
such as carbon dioxide are more and more frequently used in the refrigeration industry. Due to thermodynamic properties, R-744 is used in the transcritical cycle as an individual refrigerant. In the hereby article, high pressure of CO_2 and air temperature values were analysed. The measurements were conducted on the gas cooler side and involved external air temperature values in the summer period between 1 June to
30 September 2018. The "Booster" installation was used in one of Polish supermarkets. Correlations required to determine the optimal pressure of carbon dioxide depending on ambient temperature were presented in the article. The equations presented hereby allowed to maximize the energy efficiency ratio. An optimal high pressure for one of the correlations from literature was calculated on the basis of the measurement of ambient temperature. Actual and optimal pressure values of carbon dioxide were compared in the analysed period of time.

1 Introduction

Climate changes occurring on the earth trigger the increase of interest in environmental protection, which leads, inter alia, to the reduction of greenhouse gas emissions and the elimination of harmful chemical compounds. The cooling industry is subjected to restrictions in the area of environmental care, mainly due to adverse parameters of synthetic refrigerants used in cooling installations.

The selection of refrigerants was undertaken by the European Commission. As the criterion for withdrawal of individual refrigerants, the ODP indicator was initially adopted, followed by GWP [3]. The Directive of 1 January 2017 prohibits the use of refrigerants with a GWP ratio greater than 150 in new automotive air conditioning installations [6, 11]. Due to the elimination of subsequent refrigerants, the analysis of the use of natural refrigerants such as carbon dioxide (R-744) has begun.

* Corresponding author: artbie@agh.edu.pl

As compared to synthetically obtained refrigerants, carbon dioxide is environmentally friendly, even though it is considered one of the main greenhouse gases. The GWP value for CO_2 is 1. In comparison to, for example, the refrigerant R-404A (GWP = 3300), 3.3 tons of CO_2 affects the greenhouse effect in the same way as 1 kilogram of R-404A [2].

Pursuant to the PN EN ISO 378 standard, the R-744 is qualified to the group A1, which means that it is non-flammable, non-toxic and safe during operation, unless concentration, threatening one's health or life is exceeded [9]. Despite the very good environmental properties, carbon dioxide, due to its thermodynamic parameters, creates some problems in the design and use of refrigeration systems. Its critical temperature is only 31°C, at the pressure of 73.75 bar [4]. In this work, the actual high pressure values obtained in the gas cooler were analyzed in a booster system with carbon dioxide installed in one of the Polish supermarkets. The analysis was based on the results of measurements in the summer period, from 1 June to 30 September 2018, because during this period, due to the high temperature of the air, the installation most often worked in supercritical conditions.

2 Booster systems

As an independent refrigerant, carbon dioxide can occur in both the subcritical and supercritical cycles. In Polish conditions, due to the high temperature of the upper heat source, the cooling cycle with carbon dioxide is a supercritical cycle. If the temperature of the CO_2 on the high pressure side is greater than or equal 31°C, then the installation works in the supercritical zone, at the pressure exceeding 74 bar, and the gas itself does not condense but it cools. In practice, it happens when the outside air temperature starts to exceed 20–22°C. In relation to synthetic refrigerants, in the supercritical cycle, carbon dioxide is characterized by a lower energy efficiency, mainly due to the increased work of the compressor. Unfavourable thermodynamic parameters of such a system must be taken into account by system manufacturers, increasing the size of individual system components. With the increase in the size of the carbon dioxide system components, the prices of these systems go up, well above the costs of classic cycle with synthetic refrigerant. Despite thermodynamic and economic difficulties, transcritical cycles are successfully used, among others, in supermarkets, most often in booster systems. An example of a system scheme together with the cycle on the log(p)-h diagram is shown in Figure 1.

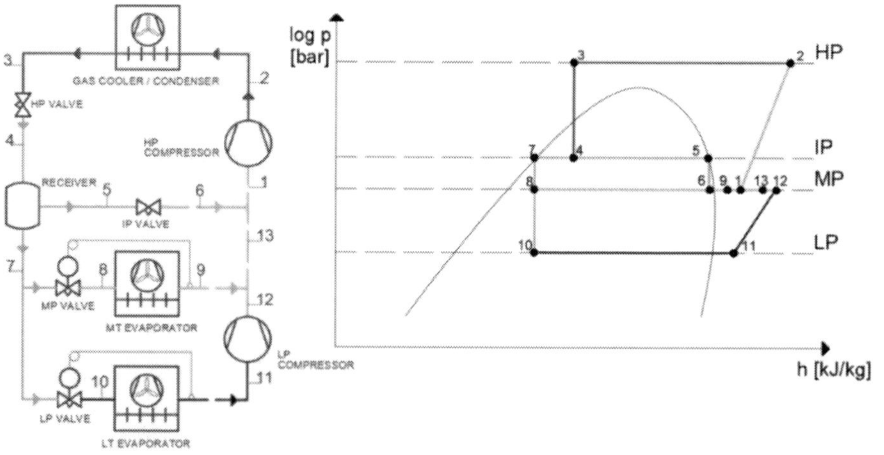

Fig. 1. Scheme of the booster installation with log(p)-h diagram.

In large supermarkets, gas coolers, through fans, are cooled by air at ambient temperature. Thus, in the summer period, the system often works in a supercritical mode. Therefore, the question arises, what should the pressure of carbon dioxide be in order to maintain the highest possible value of the Energy Efficiency Ratio (EER) or Coefficient of Performance (COP)? Do the actual values of CO_2 pressure on the second compression stage obtained under real conditions for supermarket installations coincide with the results of theoretical considerations?

3 Theoretical, high pressure of carbon dioxide

The EER or COP of the CO_2 system can be affected by factors such as the design of the gas cooler, the pressure in this exchanger and the temperature of carbon dioxide at the outlet of the gas cooler [7]. The exergy analysis shows that regardless of the operating conditions of the gas cooler, it loses about 30.7% of the total exergy of the system, which is the largest loss in the installation. In comparison, the expansion valve loses about 24.9%, and the compressor 19.5% of the total exergy [13]. When optimizing such a system, special attention should be paid to the high pressure exchanger, in particular to the high pressure of carbon dioxide, which is dependent on the temperature of the upper heat source. The literature includes many publications that describe the optimal CO_2 pressure in the gas cooler, depending on the air temperature or the outlet temperature of the carbon dioxide. Correlations are aimed at achieving the maximum energy efficiency (EER).

Y. T. Ge and S.A Tassou [8], proposed the relationship for supercritical cycle, between high pressure of carbon dioxide and the value of the energy efficiency ratio depending on the air temperature (Figure 2).

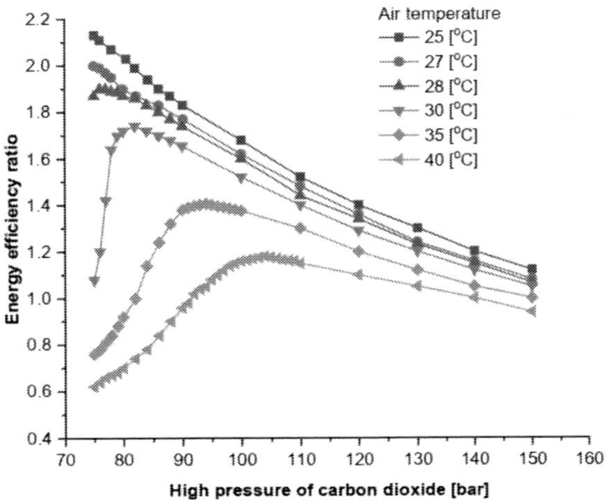

Fig. 2. Variation of energy efficiency ratio with high pressure side and air inlet temperature for transcritical booster installation [8].

In [8], it was observed that the EER decreases with increasing air temperature and for each air temperature value it reaches a different maximum at different values of carbon dioxide pressure. On this basis, an equation describing the optimal pressure (in bars) prevailing in the gas cooler was created depending on the value of the air temperature. The correlation between Ge and Tassou takes the form [8]:

$$P_{opt} = \begin{cases} 44.97, & t_{air} < 0°C \\ 1.352 \cdot t_{air} + 44.34, & 0°C \leq t_{air} \leq 20°C \\ 75, & 20°C < t_{air} \leq 27°C \\ 2.3426 \cdot t_{air} + 11.541, & t_{air} > 27°C \end{cases} \quad (1)$$

S. Sawalha published a work [12], in which he made the characteristics of the EER for supercritical cycle depending on the high pressure and outlet temperature of carbon dioxide from the high temperature exchanger. The equation describing the optimal pressure of carbon dioxide in the gas cooler was derived. The correlation which is based on theoretical analysis takes the form [12]:

$$P_{opt} = 2.7 \cdot \left(t_{air} + t_{CO_2,exit}\right) - 6.1 \quad (2)$$

Y. Chen and J. Gu in their work [5] dealt with the optimization of pressure in the booster system with the internal heat exchanger (IHX) and they proposed a correlation for calculating the optimal pressure of CO_2 [5]:

$$P_{opt} = 2.68 \cdot t_{air} + 0.975 = 2.68 \cdot t_{CO_2,exit} - 6.797 \quad (3)$$

The publication [10] describes the dependence to determine the optimal pressure of CO_2 in a high temperature exchanger. For the supercritical system, the optimal pressure of carbon dioxide depends on two parameters: the outlet temperature of the refrigerant R-744 from the gas cooler and the evaporation temperature. The dependence presented in [10] takes the form:

$$P_{opt} = (2.778 - 0.0157 \cdot t_0) \cdot t_{CO_2,exit} + (0.381 \cdot t_0 - 9.34) \quad (4)$$

4 Measurement and calculation results

The analysis of the high pressure of carbon dioxide was carried out for the refrigeration installation (booster system) of the Warsaw supermarket. Measurements of high pressure and air temperature were made during the summer, i.e. from 1 June to 30 September 2018. Data concerning air parameters for the Warsaw area came from the register made by the Institute of Meteorology and Water Management in Warsaw. The authors have assumed that an optimal pressure of CO_2 could be function of an outside air temperature (according to [8]) without any additional heating, e.g. due to solar radiation.

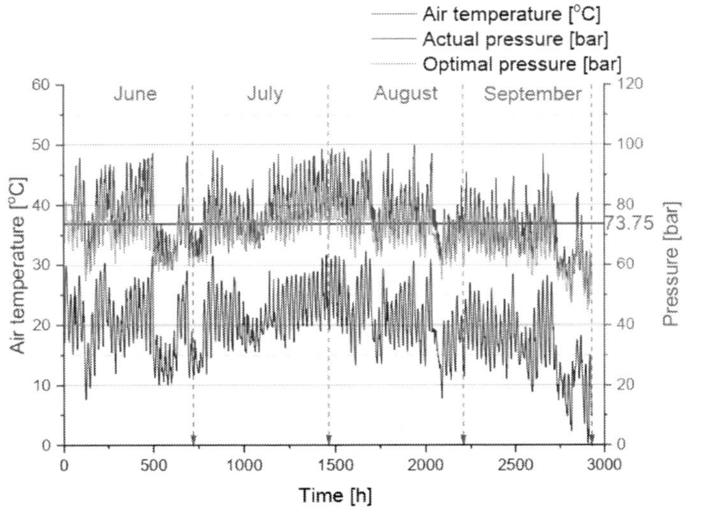

Fig. 3. Theoretical and actual high pressure values in summer period.

A diagram was drawn showing the distribution of the actual and theoretical pressure value as a function of the analyzed time. The total number of hours is shown on the horizontal axis. The analyzed period lasted 2928 hours. On the pressure axis, the pressure limit was set, above which the installation worked in supercritical cycle. The formula (1) was used for theoretical analysis because this correlation takes into account the change in CO_2 pressure as a function of the outside temperature and because it can be used for two operating states. The distribution of actual and theoretical pressure is shown in Figure 3.

The actual values of carbon dioxide pressure were significantly higher than the values recommended for maximizing the EER. According to the formula (1), the optimal pressure of carbon dioxide in the high temperature exchanger increases linearly, above the air temperature of 27°C. As it can be seen from the measurements, the optimal CO_2 pressure values are definitely lower than the empirical values. It is worth paying attention to the final phase of the diagram. The optimal and actual CO_2 pressure reaches similar values at air temperatures much lower than 20°C. In this area, the system is in a subcritical state for the majority of the analysed time.

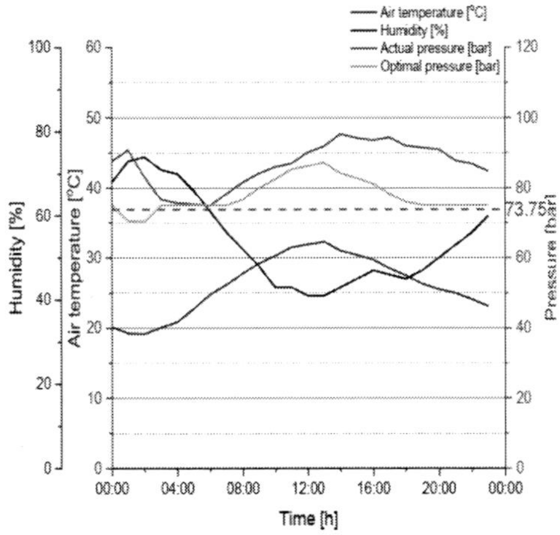

Fig. 4. Pressures values for the warmest day.

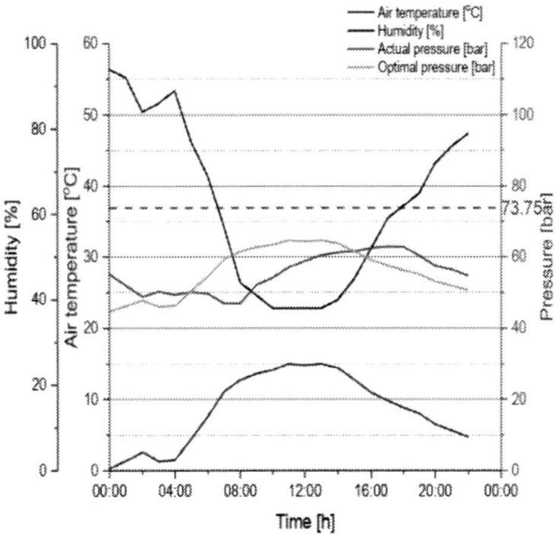

Fig. 5. Pressures values for the coldest day.

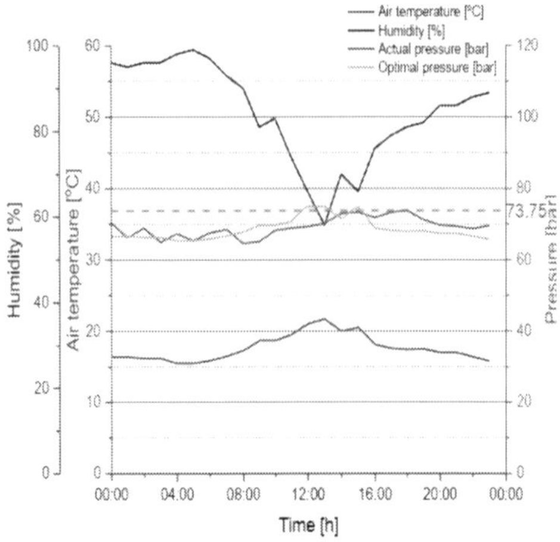

Fig. 6. Pressures values for the highest humidity day..

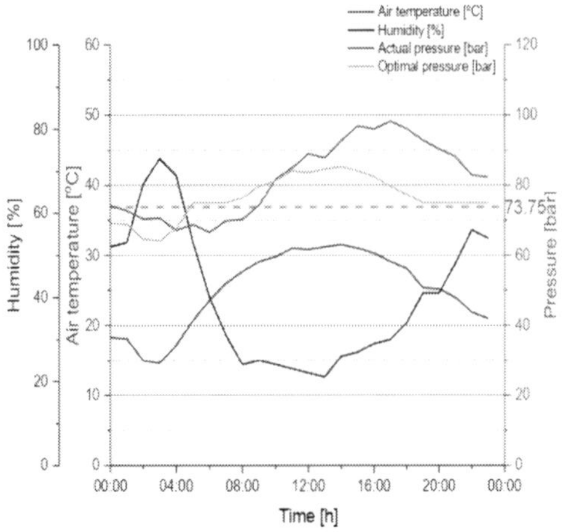

Fig. 7. Pressures values for the lowest humidity day.

The following figures show the daily distribution of carbon dioxide pressure, the pressure resulting from the maximum EER, air temperature and relative humidity. The diagrams were drawn up for the days when:

a) the highest temperature was recorded – Figure 4 (09/08/2018),
b) the lowest temperature was recorded – Figure 5 (30/09/2018),
c) the highest relative humidity was recorded – Figure 6 (09/09/2018),
d) the lowest relative humidity was recorded – Figure 7 (05/07/2018).

Analyzing the diagrams, it was noted that the discrepancy between the measured and optimal value of CO_2 pressure depends on the time of a day, and above all on the weather conditions. For high air temperatures, often above 30°C, there were significant differences between the actual and optimal carbon dioxide pressure values after 12 p.m.. In these areas of the diagrams also much lower relative humidity was recorded as compared to the rest of the day. Figure 4 shows that for the warmest day, the supercritical parameters were maintained around the clock (air temperature range from 20 to 32°C). The measured values of carbon dioxide pressure significantly differed from the correlation described by the equation (1) for most of the day (16 hours). A similar situation is shown in Figure 7. With low relative humidity and high air temperature, the installation worked for the most part of the day in the supercritical cycle and similarly to Figure 4, after 12 p.m., a growing disproportion between the optimal and the actual value of CO_2 pressure can be observed.

Figure 6 clearly shows that for the most humid day, with high humidity and low outside temperature, the actual pressure of carbon dioxide was very close to the optimal values. A similar situation took place on the day with the lowest measured temperature (Figure 5). The actual pressure of carbon dioxide oscillated towards the values resulting from the optimization relationship. Disproportion was noticed only about noon with the lowest humidity recorded on that day. In contrast to the other figures, for this day the values of actual CO_2 pressure lower than the optimal values were recorded.

Analyzing Figures 4 and 7, one should consider the actual values of outside air temperature available during the operation of the installation. The question is why the system, despite the temperature of the outside air much lower than the critical temperature for CO_2, reached such high values of carbon dioxide pressure at 12 o'clock already? The reason of this situation could be the possibility of additional warming up the outside air due to solar radiation. Gas coolers are usually mounted outside supermarkets, usually on the roof. Therefore, it could happen that air can be locally, additionally heated by various elements in the immediate vicinity of the exchanger (or roof cover), which are exposed to solar radiation. The amount of moisture in the air and the occurrence of atmospheric precipitation is also worth considering. The water vapour contained in the air can intensify the heat exchange in air heat exchangers.

Figure 8 summarizes the total number of hours for which the difference between the actual and optimal pressure of carbon dioxide was: a) below 1 bar, b) from 1 to 2 bar, c) from 2 to 5 bar, d) from 5 to 10 bar , e) from 10 to 20 bar, f) above 20 bar.

The difference between the optimal and actual pressure of R-744 was less than 1 bar for only 243 hours, which is only 8.3% of the time period studied. For 278 hours, this difference was in the range of 1–2 bar. If we assumed an acceptable increase or decrease in the high pressure of carbon dioxide at 5 bar and assumed this range as the optimal operating time of the system, the installation would work 1313 hours (45%) within the maximum EER. It is worth noticing that for about 20% of the time, the difference between the actual pressure and the theoretical pressure value was in the range of 10–20 bar, which is a significant deviation.

According to Figure 2, the deviation of the high pressure of CO_2 from the nominal point causes a reduction in the EER. Figure 9 presents the characteristics of the decrease in the EER ratio, depending on the difference between the actual and the optimal pressure. The diagram was drawn up on the basis of figure 2 and the work [8].

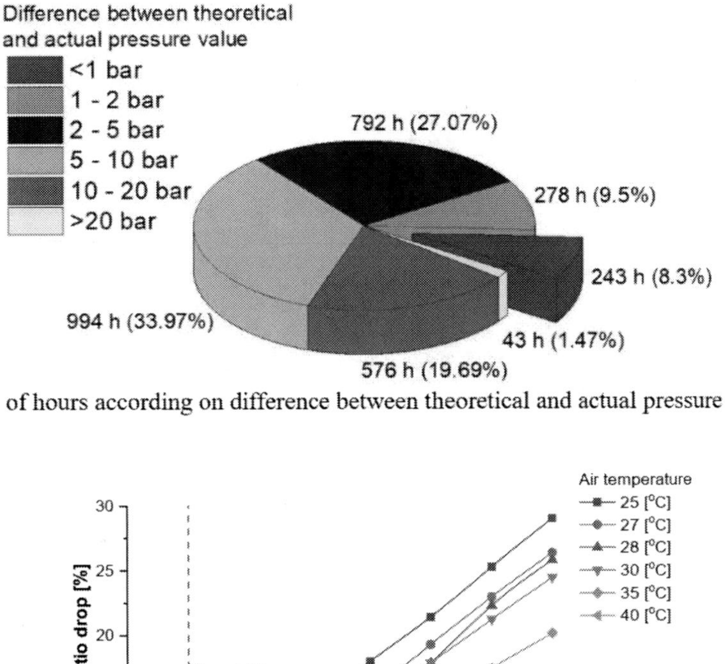

Fig. 8. Amount of hours according on difference between theoretical and actual pressure value.

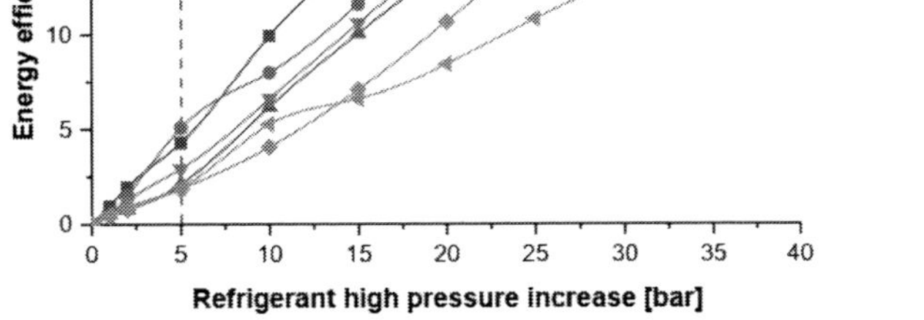

Fig. 9. Energy efficiency ratio drop.

The increase in CO_2 pressure by 1 bar decreased the EER by approximately 1%, in relation to the optimal pressure value. The increase in the high pressure of carbon dioxide by 5 bar (which was assumed to be an acceptable increase in pressure) resulted in the EER drop from 2 to 5%, depending on the air temperature. For the maximum analyzed pressure difference of 20 bar, the EER was 8 to 18% less than the theoretical value.

5 Conclusions

The work shows the impact of the measured high pressure values on an energy indicators of a "Booster" refrigeration installation. Literature sources which present correlations of an optimal pressure value of CO_2 on an ambient temperature have been shown. The analysis of the gas cooler functioning in the actual installation was made, on the basis of the meteorological data on particular days. For one of equations proposed, the optimal pressure of carbon dioxide was amounted on the basis of the outside air temperature. The convergence between actual and optimal high pressure values (observed in figures 6 and 7) reveal the assumption made in chapter 4 seem to be proper. In the result of a conducted analysis it has been proved that weather conditions have a significant influence on the high pressure value of carbon dioxide, especially temperature and humidity of the air. Gas coolers are cooled by outside air. It causes the change of parameters of an outside air that may effect parameters of a refrigerant in heat exchangers. Actual and optimal high pressure values of carbon dioxide were convergent on days with the outside air temperature below 20°C and the high relative humidity. A reverse situation was recorded for warmer days in the analysed period of time. Actual pressure values of CO_2 were much higher than the optimal pressure when the outside air temperature was above 30°C. A significant difference was noticed for the afternoon when the solar radiation is the highest. Analyzing the diagrams 8 it was noted that the system did not work efficiently for more than half the time (discrepancies between measured and optimal values of the CO_2 pressure from 5–20 bar). The EER drop reaches 2–10% at pressure discrepancies between optimal and measured values from 5–10 bar and 4–18% at a pressure discrepancy between 10–20 bar. For the largest difference between an optimal and a real pressure CO_2 the EER drop can even exceed 20%.

According to the authors' hypothesis, the main cause of the difference between actual and calculated pressures of CO_2 is the exposure of the gas cooler on a solar radiation. High-pressure heat exchangers usually are mounted outside or on the roofs of large buildings. The outside air which cools the gas cooler could be heated up by various elements of the installation or by the roof cover. In the results, warmer air gets on the inlet of the gas cooler and its temperature value is a signal for the high pressure controller. In consequence, higher temperature of air on the gas cooler inlet causes reaching higher pressure of CO_2. Authors assume that optimal pressure values depend on an outside air temperature which should be measured in the point where disruptors like a solar radiation will be eliminated. This value of the outside air temperature should be taken into consideration within the investigated heat exchanger; however in this case it cannot the point of references. According to the figure 9, the increase of the outside air temperature on the inlet to the gas cooler and indirectly a high pressure value of CO_2 causes the drop of an energy efficiency ratio. At a constant temperature of a lower heat source, which is determined by a technological demand, the increase of the temperature of an upper heat source causes the increase of a compression power and higher exploitation costs. Although the detailed analysis was conducted only for one case, this problem occurs in another cooling installations with a carbon dioxide. Therefore authors turn attention to the method of exploitation of gas coolers. One of the propositions on how to solve the problem is to choose a shaded place to install a heat exchanger. Moreover, spraying the heat exchanger with water during a high solar radiation is worth considering.

6 Nomenclature

COP – Coefficient of performance [-]; EER – Energy efficiency ratio [-]; P_{opt} – Optimal pressure [bar]; t_0 – Evaporation temperature, °C; t_{air} – air temperature, °C; $t_{CO_2,exit}$ – Outlet CO_2 temperature, °C

Reference

1. A. C Beaver, J. M Yin, C. W Bullard, P. S Hrnjak, ACRC Report CR-18 (1999)
2. A Bensafi, *CETIAT* (2007)
3. T. Bohdal, H. Charun, M. Sikora, Rocz. Och. Środ. **17**, 461 (2015)
4. J. Calm, G. Hourahan, Ref. for Susta. (2011)
5. Y. Chen, J. Gu, Int. J. of Reff. **28**, 1238 (2005)
6. Directive 2006/40 / EC. Directive 40/2006 of the European Parliament and of the Council of 17 May 2006. Documents 70/156 / EEC
7. D. K. Gupta, S. Kuma, M. S. Dasgupta, J. of Add. Res. in Mech. Eng. **1**, 147 (2010)
8. Y. T. Ge, S. A. Tassou, Int. J. of Ref. **34**, 540 (2011)
9. ISO 817:2014 Refrigerants – Designation and safety classification
10. S. M Liao, T. S Zhao, A. Jakobsen, Appl. Therm. Eng. **20**, 831 (2000)
11. Regulation 517/2014. Regulation (EC) No 517/2014 of the European Parliament and of the Council (EU) of 16 April 2014 on fluorinated greenhouse gases and repealing Regulation (EC) No 842/2006
12. S. Sawalha, Int. J. of Reff. **31**, 516 (2008)
13. Y. B. Tao, Y. L. He, W. Q. Tao, Appl. En. **87**, 3065 (2010)

Increasing the efficiency of split type air conditioners/heat pumps by using ventilating exhaust air

Wojciech Cepiński[1,*], and *Paweł* Szałański[1]

[1]Department of Air Conditioning, Heating, Gas Engineering and Air Protection, Faculty of Environmental Engineering, Wrocław University of Science and Technology, ul. C.K. Norwida 4/6, 50-373 Wrocław, Poland

Abstract. The article presents the possibility of using exhaust air from ventilation system to increase the efficiency (SCOP, COP, SEER and EER) of commonly used air conditioners with the function of a heat pump. These types of devices are very popular both in residential and in public buildings. The topic discussed in the article is very important, because the widespread increase of the energy efficiency of these devices significantly influences national electricity consumption. The possibility of increasing their efficiency by directing the exhaust air from the ventilation system to the heat exchanger of the air conditioner outdoor unit has been analysed. It has been shown that the use of the simple design solution described in the article allows for a significant increase of the efficiency of these devices (seasonal efficiency even up to 35% at 100% share of exhaust air), reducing the energy consumption and increasing their capacity and operation range. By increasing the share of exhaust air it is possible to ensure year-round operation of the device and even 100% coverage of heat demand.

1 Introduction

Nowadays it is necessary to face the dynamically growing energy demand. This is a very important task for mankind in the face of a dynamic increase of the number of humanity accompanied by the increasing amount of energy consumed by each of us, resulting from the desire to use energy-consuming comforts such as being online, using individual transport, air conditioning in flats and houses. It is our duty, not only as researchers, but also as people to constantly search for solutions that limit the consumption of the dynamically shrinking natural resources of our planet. It is necessary to increase public awareness of the negative impact of increasing energy consumption on the environment in which we live and our descendants will live. This should be associated with the widespread development of a low-energy lifestyle. However, it is a prerequisite to create technology that uses energy as efficiently as possible.

* Corresponding author: wojciech.cepinski@pwr.edu.pl, ORCID: 0000-0003-4294-0431,
Paweł Szałański ORCID: 0000-0001-9263-3486

The EU Directive [1] imposes on individual member states an obligation to reduce primary energy consumption. In the case of energy consumed by buildings, this goal can only be achieved by reducing the energy need for heating or by improving the efficiency and energy efficiency of appliances. The more common the technology is, the more important it is to increase its efficiency. The topic of estimating energy need for heating is discussed in the [2].

Due to the increase in the wealth of the societies of many countries in Europe and the reduction of the costs of various types of technologies, access to this type of goods is becoming more and more common. On the example of Poland, air-conditioning technology, which until a dozen years ago was an indicator of luxury, is now widely available and at prices similar to computers, TV sets or mobile phones. Today, room cooling is no longer a luxury good. Today, room cooling is no longer a luxury good. In Poland, 20% of the air conditioners market in 2015 was residential construction [3]. Currently, the devices of this type have the ability to reverse the cooling circuit and operate in the heat pump mode. Therefore, it is very popular to use them also in the function of room heating. That is why these devices often operate in full-year mode, which increases the impact of their efficiency on the environment.

The market for these devices has been very receptive in recent years and it can be said that Poland has been flooded with them in recent years. The Polish market of air conditioners is estimated at over PLN 1 billion annually [3]. At the price of a branded set of approx. 3–4 kW for a room of approx. 30 m^3 oscillating around 3500–5000 PLN with installation [3], it may result in 200,000–280,000 devices of this type installed annually. The forecasted popularity of air conditioners in Poland is also the growing market of new flats, as the use of these devices is already very popular not only in office buildings, but also in flats and houses [4]. Popularity of these devices in Europe is considered in [5].

Due to the popularity of these devices, it is imperative to improve their energy efficiency. In [6], an attempt to optimize the seasonal efficiency of air-to-air heat pumps is presented. In [7], the economics of using an air-to-air heat pump as an alternative to a gas-fired boiler was analyzed. The paper [8] discusses the rules for optimal heat pump sizing in a fixed site. The proper sizing of heat pumps should be based on simulations of their efficiency, the subject of which was considered in [9].

The aim of this article is to present a very simple and often cost-free solution, enabling a significant increase in the energy efficiency of cooling and heating air conditioners, using heat usually lost in ventilation systems. Heat recovery from ventilation systems requires other solutions than, for example, heat recovery from hot flue gases discussed in [10]. Popularization of the proposed solution may have a significant impact on the reduction of the national energy demand and a significant reduction in the emission of coal combustion products to the atmosphere due to the supply of electricity to these devices in Poland, mainly from coal.

1.1 Characteristics of the analysed devices

Split air conditioners consist of five basic components: compressor, expansion valve, evaporator, condenser and connecting piping. Such a system is filled with refrigerant with appropriate thermodynamic properties. When operating in cooling mode, the unit receives heat by means of a evaporator located in the cooled room. The properties and parameters of the refrigerant are so selected that when entering the evaporator in the form of liquid, it receives heat from the air around evaporator and evaporates as a result of the associated increase in its temperature. The refrigerant is then transported by a compressor to a condenser located outside the room. There, the refrigerant transfers heat to the air around condenser, as a result of which it cools down and as a result its condensation takes place.

Air conditioners with heat pump function have the ability to reverse the cooling circuit and as a result heat the room. After reversing the circuit, the unit becomes an air-to-air heat pump. The exchanger in the room becomes a condenser, and the exchanger located outside the room becomes a evaporator. As a result, the heat received from the surrounding evaporator is transported to the condenser and then transferred to the air in the room. In this situation, the air surrounding the condenser in the room is the upper source of the heat pump and the air surrounding the evaporator located outside the room is the lower source of the heat pump.

The main feature of these devices, operating in both heating and cooling mode, is the transfer of heat from a lower temperature environment to a higher temperature environment. The main feature of these devices, operating in both heating and cooling mode, is the transfer of heat from a lower temperature environment to a higher temperature environment. One of the main factors influencing the efficiency and energy efficiency of these devices is the temperature difference that the device has to overcome, that is between the temperature of the medium from which the heat is taken and the medium to which the heat is transferred. Improper heat dissipation in cooling mode and too low ambient air temperature in heating mode can significantly reduce the efficiency of the evaporator. The problem of high operating costs due to poor heat recovery is highlighted in [4].

1.2 Coefficients of performance COP, SCOP, EER, SEER

The efficiency of a heat pump under certain operating conditions (e.g. outdoor and indoor air parameters) is determined by the COP (Coefficient Of Performance). It is equal to the ratio of the heating capacity of the device to the electrical power input of the device – equation (1).

The seasonal efficiency of a heat pump under certain climatic conditions is determined by the SCOP (Seasonal Coefficient Of Performance). It is equal to the ratio of the annual heating demand supplied by the device to the seasonal electrical energy consumption of the device – equation (2).

The efficiency of a cooling unit under certain operating conditions (e.g. outdoor and indoor air parameters) is determined by the EER (Energy Efficiency Ratio). It is equal to the ratio of the cooling capacity of the device to the electrical power input of the device – equation (3).

The seasonal efficiency of a cooling unit under certain climatic conditions is determined by the SEER (Seasonal Energy Efficiency Ratio). It is equal to the ratio of the seasonal cooling demand supplied by the device to the seasonal electrical energy consumption of the device – equation (4).

$$COP = \frac{P_H}{P_{elH}} \quad (1) \qquad SCOP = \frac{Q_H}{Q_{HE}} \quad (2) \qquad EER = \frac{P_C}{P_{elC}} \quad (3) \qquad SEER = \frac{Q_C}{Q_{CE}} \quad (4)$$

where:

P_H and P_C – heating and cooling capacity output at a specific conditions, kW
P_{elH} and P_{elC} – power input at a specific conditions in a heating and cooling mode, kW
Q_H and Q_C – annual heating and cooling demand supplied by the device, kWh
Q_{HE} and Q_{CE} – annual electricity consumption for heating and cooling, kWh.

2 Description and analysis of the proposed solution

The simplest way to increase energy efficiency is to ensure the lowest possible temperature difference between these media. The temperature in rooms usually cannot be changed due to the principle of using the device, which is to provide adequate thermal comfort.

Therefore, it is only possible to ensure the most appropriate air temperature around the outdoor unit as close as possible to the room air temperature. The typical location of an outdoor unit of an air conditioner is the outdoor environment. The outdoor units are usually mounted on the external walls of buildings, on roofs or at ground level near the building. Much less common solution to increase energy efficiency is to install them in spaces with a higher air temperature, for example in indoor spaces such as unheated underground garages. There are also works showing the possibility of increasing energy efficiency by washing the exchanger of the outdoor unit with heated or cooled air from the ground exchanger [11, 12]. These solutions are effective, but can be used in specific circumstances and require the use of additional components and devices.

In contrast, the solution proposed in this article may become much more common as it does not require significant additional expenses. Its idea is presented on figure 1.

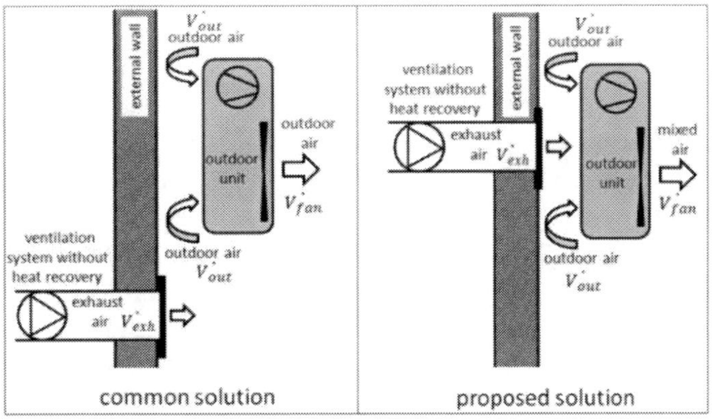

Fig. 1. Diagram with an air conditioner that operates only in outdoor air (common solution) and in a mixture of outdoor and exhaust air (proposed solution).

Typical solutions do not pay attention to the location of the air conditioning outdoor unit in relation to the air exhaust of the ventilation system. The air flow through the outdoor unit is a characteristic parameter and we have no influence on it. We can only have an influence on the temperature of this stream. Appropriate placement of the air exhaust enables directing the extracted air from rooms to the exchanger of the outdoor unit. As a result, the exchanger will be surrounded by a mixture of extract air and external air. During the heating period, the air from the room has a higher temperature than the outdoor air, so that the temperature of the mixture will be higher the higher the percentage of exhaust air will be. On the contrary, during the cooling period, the indoor air in the cooled room has a temperature lower than the outdoor air temperature. With the proposed solution, in the heating period the exchanger of the outdoor unit operates in the environment of warmer air, and in the cooling period colder. An additional advantage of the solution is the possibility of using air exhausted from rooms polluted by smell, e.g. from toilets, garages or other places where it is not possible to apply heat recovery, if this air is not destructive to the heat exchanger. The above enables the use of almost all thermodynamic advantages of the exhaust air because of its temperature and relative humidity, which increases the capacity of the device.

2.1 Comparative analysis of air conditioner operation

The assumptions and results of the analysis (figure 2–3) of the impact of the proposed solution on the parameters of an exemplary air conditioner are presented below.

For a typical and proposed solution, the following were compared:

- the dependence of the heating power of the device on the outdoor air temperature,
- the dependence of the energy efficiency coefficient of the device on the outdoor air temperature in the heating mode,
- the dependence of the energy efficiency coefficient of the device on the outdoor air temperature in the heating mode,
- the operating range of the device in heating mode,
- dependence of the cooling capacity of the unit on the outdoor air temperature,
- dependence of the coefficient of energy efficiency of the unit on the temperature of the outdoor air in cooling mode,
- the operating range of the unit in cooling mode.

The results of the analysis were presented for different values of the percentage of exhaust air in the air flow through the exchanger of the outdoor unit, defined as:

$$ARR = \frac{\dot{V_{exh}}}{\dot{V_{fan}}} = \frac{\dot{V_{exh}}}{\dot{V_{exh}} + \dot{V_{out}}} \qquad (5)$$

where:

$\dot{V_{exh}}$ – exhaust air flow from the ventilation system, m^3/h

$\dot{V_{out}}$ – the outdoor air flow through the fan of the outdoor unit of the air conditioner, m^3/h

$\dot{V_{fan}}$ – the total flow of air flowing through the heat exchanger of the outdoor unit of the air conditioner, m^3/h.

The analysis was carried out with the following assumptions:

- in a typical solution, the outdoor unit operates only in the outdoor air and COP, EER, P_H, P_C are the functions of outdoor temperature T_e (in this case $T_e = T_{ou}$),
- in the proposed solution, the outdoor unit operates on a ideally mixed outdoor and exhaust air and COP, EER, P_H, P_C are the functions of temperature of mixed outdoor and exhaust air surrounding the outdoor unit T_{ou},
- no heat recovery in the ventilation system,
- constant exhaust air temperature during the heating and cooling period of 20°C,
- for estimation of SCOP and SEER climatic data for the city of Wroclaw [13] in south-western Poland was considered using bin method (average duration of particular external temperature ranges was used),
- constant value of the total fan-flow of the outdoor unit of the air conditioner $\dot{V_{fan}}$, but different for heating and cooling mode,
- the air conditioner has been designed to cover the capacity needed for cooling during the cooling period,
- an exemplary profile of heat load of a building/room depending on heat loss by transmission and ventilation at a given outdoor temperature during the heating period, and depending on internal and solar heat gains, heat gains by transmission and ventilation at a given outdoor temperature in the cooling period,
- the heat gains from solar radiation in cooling mode depend linearly on the outdoor temperature [14],
- the device operates in ON/OFF mode with full power, adjusting its operating time to the needs,
- heating season at outdoor temperatures equal to or lower than 12°C,
- cooling season with outdoor temperatures equal to or lower than 20°C,
- the device does not operate at an outdoor temperature in the range of 12°C to 20°C,
- the defrosting process will take place below an air temperature of 5°C,

- the additional energy consumption for defrosting the evaporator in heating mode is not included in the calculations.

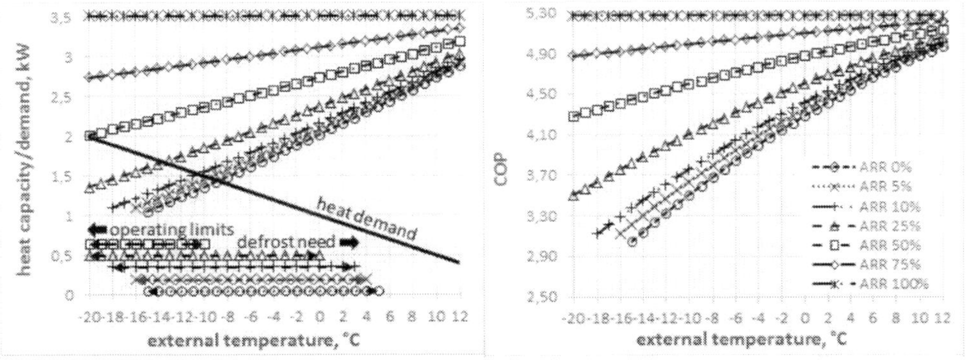

Fig. 2. Heating capacity and COP coefficient as the function of outdoor temperature.

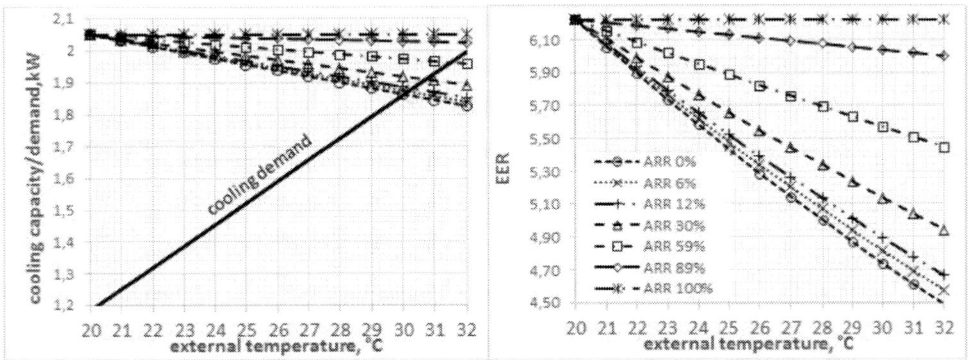

Fig. 3. Cooling capacity and EER coefficient as the function of outdoor temperature.

The following equations were used to determine the SCOP, without consideration of the auxiliary heater – equation (6) and considering auxiliary heater – equation (7). Analogical equations were used to determine SEER.

$$SCOP = \frac{\sum_{j=1}^{n} h_j \cdot pl_j(T_{ej}) \cdot COP_j(T_{ouj}) \cdot min\left(1; \frac{Q_{Hdj}(T_{ej})}{P_{Hj}(T_{ouj})}\right)}{\sum_{j=1}^{n} h_j \cdot pl_j(T_{ej}) \cdot min\left(1; \frac{Q_{Hdj}(T_{ej})}{P_{Hj}(T_{ouj})}\right)} \tag{6}$$

$$SCOP = \frac{\sum_{j=1}^{n} h_j \cdot pl_j(T_{ej}) \cdot min\left(1; \frac{Q_{Hdj}(T_{ej})}{P_{Hj}(T_{ouj})}\right) \cdot \left(1 + \left(COP_j(T_{ouj}) - 1\right) \cdot max\left(1; \frac{P_{Hj}(T_{ouj})}{Q_{Hdj}(T_{ej})}\right)\right)}{\sum_{j=1}^{n} h_j \cdot pl_j(T_{ej}) \cdot min\left(1; \frac{Q_{Hdj}(T_{ej})}{P_{Hj}(T_{ouj})}\right)} \tag{7}$$

where:

h_j – bin hours – hours per season for which an outdoor temperature interval occurs,
Q_{Hd} – heat demand of the building/room, kW
T_{ou} – temperature of air entering the outdoor unit, $T_{ou} = ARR \cdot T_i + (1 - ARR) \cdot T_e$,°C
pl – part load coefficient, $pl = \frac{T_i - T_e}{T_i - T_{ed}}$
T_i and T_e – indoor and outdoor air temperature, °C
T_{ed} – outdoor air temperature at design conditions, °C.

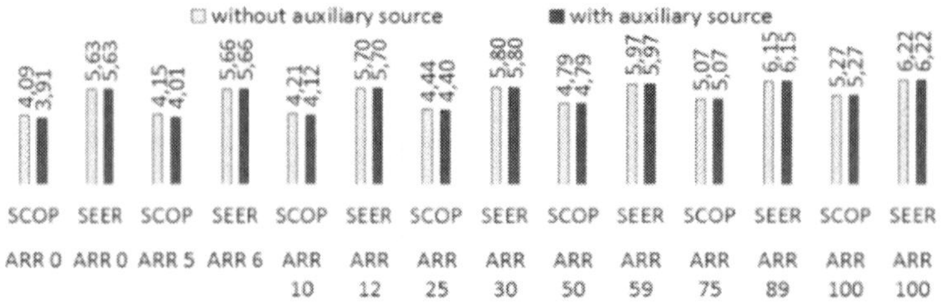

Fig. 4. SCOP and SEER in the function of ARR.

2.2 Conclusions

The results of the analysis (figure 2-4) indicate that the application of the proposed solution:

- increases the coefficients of performance in the heating mode COP (e.g. at $T_e = -20°C$ from 38% to 108% for ARR25 and ARR100 respectively) and in the cooling mode EER (e.g. at $T_e = 30°C$ from 9% to 32% for ARR30 and ARR100 respectively) at particular outdoor air temperatures,
- increases the seasonal coefficients of performance the SCOP and SEER devices (respectively even up to 35% and 11% at 100% share of exhaust air),
- may reduce investment costs for heating by enabling the use of a device with a lower nominal capacity, because as the temperature of mixed air increases (depending on the ratio of ARR), the heating capacity of the device increases - such a device will operate with a higher use of installed capacity,
- increases the temperature and power operating range of the air conditioner in heating mode (with a typical solution, the analysed exemplary device switches off below -15°C), which may eliminate the need for an additional source of heat and become a monovalent system - in the considered case, for ARR equal to or greater than 25%, the device can operate for the entire heating period, and for AAR equal to or greater than 50%, it allows to eliminate the need for an additional source of heat,
- reduces the operating time of the unit in heating mode, where defrosting of the evaporator is needed - considering the high frequency of external temperatures at which defrosting is required, therefore by postponing this necessity by several degrees Celsius an even greater increase in seasonal efficiency of the heat pump is expected than the presented analysis (defrosting can reduce heat pump efficiency in the heating period by up to 10%),
- allows to defrost the evaporator with warm air, which can also reduce the energy demand for this process, so even greater savings can be expected than presented in the analysis,
- may reduce investment costs for cooling by enabling the use of a device with a lower nominal capacity, because as the temperature of mixed air decreases (depending on the ratio of ARR), the cooling capacity of the device increases - such a device will operate with a higher use of installed capacity,
- increases the operating range of the unit in cooling mode,
- by increasing the capacity of the unit at particular outdoor temperatures, the total operating time in the season is reduced,
- reduces the operating time of the unit, which, due to the compressor's lifetime, can result in a few additional operating seasons,

In the case of using in the proposed way the air exhausted from the ventilation system with heat recovery, the effect will be much smaller, because the temperature of this air will be lower. The effect of the proposed solution will be lower the higher the efficiency of heat recovery is.

3 Summary

The presented analysis shows that the proposed solution, recommended for use in new and existing installations, may contribute to the improvement of seasonal efficiency of the device in both heating and cooling mode. Depending on the available exhaust air flow from the ventilation system, it can give significant savings and allows to reduce the primary energy consumption for the purpose of controlling the microclimate of rooms in an environmentally friendly way. This is particularly important in the climatic zone to which Poland belongs, because for these purposes almost 40% of the energy is used in this zone. The implementation of the proposed solution should have a mass scale, and the regulations should recommend its use where possible. This solution may also be considered as a form of heat recovery from the exhaust air for which, for technical reasons, heat recovery could not be applied, e.g. due to: smell or pollution. Further analyses of the proposed solution with the use of air from ventilation systems with heat recovery and for air conditioners/heat pumps with variable capacity control are planned. Laboratory tests of the proposed solution are also planned.

The work was supported by The Faculty of Environmental Energy, Wroclaw University of Science and Technology, Poland. no. 0401/0055/18.

References

1. *Directive 2010/31/EU of the European Parliament and of the Council of 19 May 2010 on the energy performance of buildings* (Official Journal of the EU, 2010)
2. P. Kowalski, P. Szałański, E3S Web of Conferences **17**, 00045 (2017)
3. *Air conditioners – market and perspectives* (in Polish), Rynek Instalacyjny **5**, (2015)
4. *Air-conditioning and ventilation market in Poland and the Silesia* (in Polish), (Park Naukowo-Technologiczny Euro-Centrum, Katowice, 2014)
5. L. Grignon-Massé, P. Rivière, J. Adnot, Energy Policy **39**, 2152–2164 (2011)
6. S. H. Lee, Y. Jeon, H. J. Chung, W. Cho, Y. Kim, Appl. Therm. Eng. **144**, 362–370 (2018)
7. G. Ala, A. Orioli, A. Di Gangi, *Energy and economic analysis of air-to-air heat pumps as an alternative to domestic gas boiler heating systems in the South of Italy*, Energy (2019) (to be published)
8. M. Dongellinia, C. Naldia, G. L. Morinia, Energy Procedia **78**, 1123 – 1128 (2015)
9. K. Piechurski, M. Szulgowska-Zgrzywa, J. Danielewicz, E3S Web of Conferences **44**, 00139 (2018)
10. E. Dudkiewicz, N. Fidorów-Kaprawy, Energy **127**, 198–208 (2017)
11. W. Cepiński, E3S Web of Conferences **17**, 00012 (2017)
12. C. Baglivo, P. M. Congedo, Energy Procedia **126**, 187–194 (2017)
13. A. Pełech, *Wentylacja i klimatyzacja. Podstawy* (Oficyna Wydawnicza Politechniki Wrocławskiej, Wrocław 2008)
14. Recknagel, Sprenger, Hőnmann, Schramek, *Kompendium wiedzy: Ogrzewnictwo, Klimatyzacja, Ciepła woda, Chłodnictwo 2008/2009* (Omni-Scala, 2008)

Monthly model of hot water consumption

Agnieszka Chmielewska[1,*]

[1]Wroclaw University of Science and Technology, Faculty of Environmental Engineering, Wybrzeże Wyspiańskiego 27, 50-370 Wroclaw, Poland

Abstract. The article presents a statistical analysis of the monthly consumption of hot water in apartment buildings. On this basis, a monthly model of hot water consumption in multi-family buildings was proposed. This model consists of two sub models: model of average daily consumption of domestic hot water and a model describing changes in consumption of hot water in individual months of the year. To construct the model of average daily consumption of hot water, data on the consumption of hot water was used out of 16 multi-family buildings from the period of two years. The data was supplemented with information on the housing structure in the building. The study of changes in hot water consumption in the following months was based on data obtained from 30 buildings. Statistical analysis showed a strong correlation between the monthly consumption of hot water, the outside temperature and holiday period.

1 Introduction

Modelling the consumption of hot water is complicated. The consumption of hot water in residential buildings depends on sociological and technical factors, which are often impossible to determine. In residential buildings the consumption of hot water depends on: the size of flats, sanitary equipment for apartments, the method of accounting for water consumption, the size and demographic structure of the family, the age of residents, the way of spending free time, lifestyle and lifestyle, the attractiveness of TV programs, etc.. Intake of domestic hot water in residential building is a random volume and is characterised by large variability, both within a period of 24 hours and on particular days of the week. Due to the noticeable need for performing detailed simulations of operation of systems preparing domestic hot water, there is a need for developing profiles of warm water consumption at various time stages. Knowing the monthly water consumption profile seems to be necessary, as it is used for various types of power simulations or selection of solar collectors. Unfortunately, a model that will allow to measure the amount of monthly collection of domestic hot water does not exist in Polish publications, thus this amount is often assumed at a fixed level during the entire year. Whereas in foreign publications, it is possible to find numerous elaborations describing the monthly distribution of domestic hot water consumption. Georg et al. (2015) [1], for instance, presented changes of seasonal consumption of domestic hot water in their elaboration, that indicate a clear drop in the use of domestic hot water in summer. Such changes are explained by the authors, who state that

* Corresponding author: agnieszka.chmielewska@pwr.edu.pl

this is due to higher outdoor temperatures in summer, which affects bathing preferences of users and temperature of water from the mains, used for bathing. Ahmed, Pylsy, Kurnitski (2015) [2] noted, based on measurements of 182 Finnish flats, that unit consumption of domestic hot water in summer amounts to 40 l per day and 46 dm^3 per day, in winter. Studies conducted by Mayer (2000) [3] based on hourly measurements from 120 flats in South Africa (including 30 traditional houses, 30 downtown houses and 30 apartments), indicated that the unit hourly consumption of domestic hot water depends on the building type, population density and month of the year. In order to conduct analyses, each group contained 10 facilities with low, average and high population density. In measurements taken by the authors, a reverse trend can be observed as compared to the previous studies (high temperatures in January and February, low temperatures in July and August), which is associated with the climate in South Africa. In such countries as Germany [4], Italy [5], Denmark [6], or Great Britain [7], measurements regarding the demand for domestic power for preparing hot water, are taken monthly. It should be noted that the intake of domestic hot water in the presented studies is closely associated with the usable area of multi-family buildings and the number of days in a given month. In Great Britain [7], the authors additionally introduced reduction indicators of hot water consumption in particular months, besides monthly intake of domestic hot water, taking into account lower intake of hot water in summer.

2 Research method

2.1 Measurement data

Measurements of hot water consumption were made in 42 multi-family buildings located in Wrocław and Zawidów. There was a total of 1376 apartments . Each of the units (the apartments) has been equipped with a residential water meter for measuring the cold and hot water consumption. The model of daily consumption of domestic hot water was made based on the measurement data from 16 multi-family buildings (773 apartments) located in Wroclaw. In addition, the analyses were supported by the data obtained from the estate manager. Comprehensive data about the apartments' surface and the annual consumption of hot water came from 626 apartments. The study covered the period from January 2012 to December 2013. Analysis of change in the deviation of daily consumption of domestic hot water in subsequent months, as compared to the average annual value was made on the basis of data from 30 multi-family buildings. The data was gathered every months. After rejecting incomplete and incorrect data, measurements from 585 apartments were used for the analysis. The study covered the period from January 2012 to December 2013.

2.2 Analysis of the measured data

At the beginning, the consumption of domestic hot water in months in each apartment was calculated. For each of the apartments the monthly consumption of the domestic hot water was calculated ($V_{M,DHWi}$), as a difference between the measured values at the beginning and at end of the month. Then, on the basis of equation 1, the deviation of the average daily consumption of hot water in the month in relation to the average daily consumption for each apartment was calculated.

$$s_{M_i} = \frac{\frac{V_{M,DHW_i}}{l_{d,m_i}}}{\frac{1}{12}\sum_{i=1}^{12}\left(\frac{V_{M,DHW_i}}{l_{d,M_i}}\right)} \quad (1)$$

The values calculated according to equation 1 were subjected to statistical analysis. It consisted in examining the significance of differences between the average daily consumption of hot water in the months for the whole year. These analyses showed the desirability of developing a model of hot water consumption in each month of the year.

The publication contains an analysis of the variability of monthly use of domestic hot water within a year. Fig. 1 presents monthly variability and changes of the deviation of domestic hot water consumption in particular buildings, as compared to the average annual value. The diagram was based on medians to eliminate the impact of the deviating values on the results of conclusions. The results presented were compared with data from publications. A model developed in a British standard, i.e. SAP 2012 [7] presents the general trend of monthly changes of actual domestic hot water consumption. However, it does not take into account the non-variable consumption of domestic hot water in the period from January to March, and reduced use of domestic hot water in August. The above discrepancies may be caused by varying school calendar arrangement or the climate itself that affect preferences of users of domestic hot water. In the case of analyses conducted with data regarding consumption of domestic hot water in a Canadian community, i.e. Solar City [1], significantly better representation of the collected data is visible but differences in the holiday season and in February are noticeable.

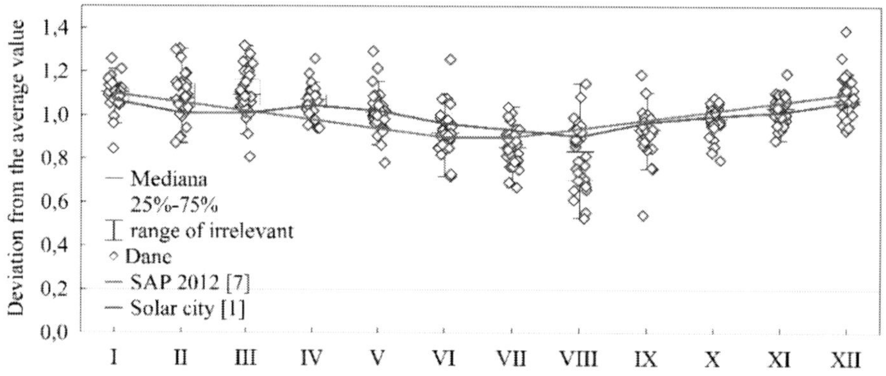

Fig. 1. Change in the deviation of daily consumption of domestic hot water in subsequent months as compared to the average annual value.

Next, the measurements were subjected to a statistical analysis, consisting of verifying whether differences in the amount of daily consumption of domestic hot water in particular months are statistically important. To assess the varied intake of water, the Kruksal-Wallis test was applied which is based on observation ranks. Because the calculated values of p test probability is 0.00, the hypothesis about equal consumption of domestic hot water in particular months can be rejected. It is also stated that there are at least two months when this value varies. Then, a test of multiple comparisons of average ranks for all samples was conducted, based on which it may be stated that there are statistical differences between January, February and March, and months from July to October, among others.

Fig. 2 and Fig. 3 present the deviation in the monthly use of domestic hot water from the average annual value, calculated on the basis of measurement of monthly consumption separately for buildings located in Wrocław and Zawidów. In both of the analysed locations, water consumption reduces in the summer period but this reduction is significantly more noticeable in Wroclaw. This may be associated with a bigger number of people going to large cities during the summer holiday, or students leaving Wroclaw for the same reason.

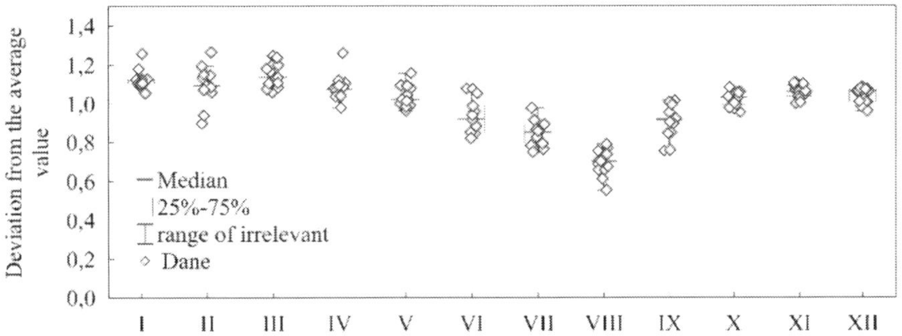

Fig. 2. A change in the deviation of daily consumption of domestic hot water in subsequent months as compared to the average annual value in Wrocław.

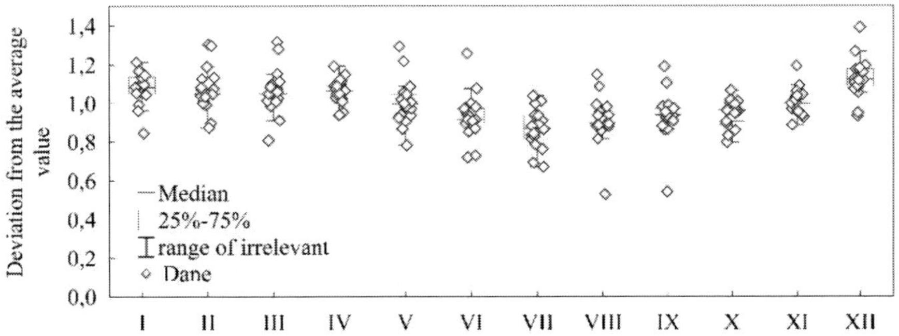

Fig. 3. A change in the deviation of daily consumption of domestic hot water in subsequent months as compared to the average annual value in Zawidów.

3 Monthly model of domestic hot water consumption

Two alternative methods were proposed for developing a model for describing the monthly consumption of domestic hot water. The difference between them consists of choosing independent variables being the basis of measurement and different approach to the method of analysing measurement data. First model ($s_{Mi}^{v.1}$) was developed using the method of multiple regression. The model enables to estimate changes in the deviation of monthly consumption of domestic hot water from the average annual value. Variables for this model were chosen using Spearman's rank correlation coefficients. In the second model ($s_{Mi}^{v.1}$) it was assumed that water consumption is related only to a given month in a year. This assumption enabled to use data regarding domestic hot water consumption from particular flats to expand the database regarding consumption of water in buildings. On this basis, a model was developed describing the changes in the deviation of monthly consumption of domestic hot water from the average annual value. The bootstrap method was used for expanding the database. Initially, the database contained data regarding consumption of domestic hot water in subsequent months for 583 flats located in 30 buildings. The data enabled to draw 1000 new configurations of buildings with various number of flats. The proposed model of monthly domestic hot water consumption was presented in equation 2. It was based on deviation of the average daily consumption of domestic hot water in particular months (s_{Mi}) from the average annual value (V_{srD}) and number of days in a given month ($l_{d,Mi}$).

$$V_{M_i} = s_{M_i} \cdot V_{srD} \cdot l_{d,M_i} \quad (2)$$

3.1 Average daily consumption of the domestic hot water in multi-family buildings

In the [8], it was evidenced that when indicating the average daily domestic hot water consumption in multi-family buildings, it is worth basing one's elaborations on the residential structure, as this enables to obtain significantly better quality of estimates. The relation described in [8] and [9] was presented in equation 3. It was based on surface of flats with a certain number of rooms.

$$V_{\acute{s}rD} = 0{,}0021 \cdot A_{M1} + 0{,}0018 \cdot A_{M2} + 0{,}0017 \cdot A_{M3} + 0{,}011 \cdot A_{M4} + 0{,}011\ [m^3/doba] \quad (3)$$

where:
A_{M1} – area one-room apartments [m^2], A_{M2} – area two-room apartments [m^2], A_{M3} – area tree-room apartments [m^2], A_{M4} – area four-room apartments [m^2].

Because information about the area of flats may be difficult to obtain, in this publication, the author proposed simplification of the model of average daily domestic hot water consumption. The proposed model based on the number of flats with a certain number of rooms was presented in equation 4.

$$V_{\acute{s}rD} = 0{,}063 \cdot n_{M1} + 0{,}084 \cdot n_{M2} + 0{,}106 \cdot (n_{M3} +\cdot n_{M4}) + 0{,}015\ [m^3/doba] \quad (4)$$

where:
n_{M1} – number of one-room apartments [m^2], n_{M2} – number of two-room apartments [m^2], n_{M3} – number of tree-room apartments [m^2], n_{M4} – number of four-room apartments [m^2].

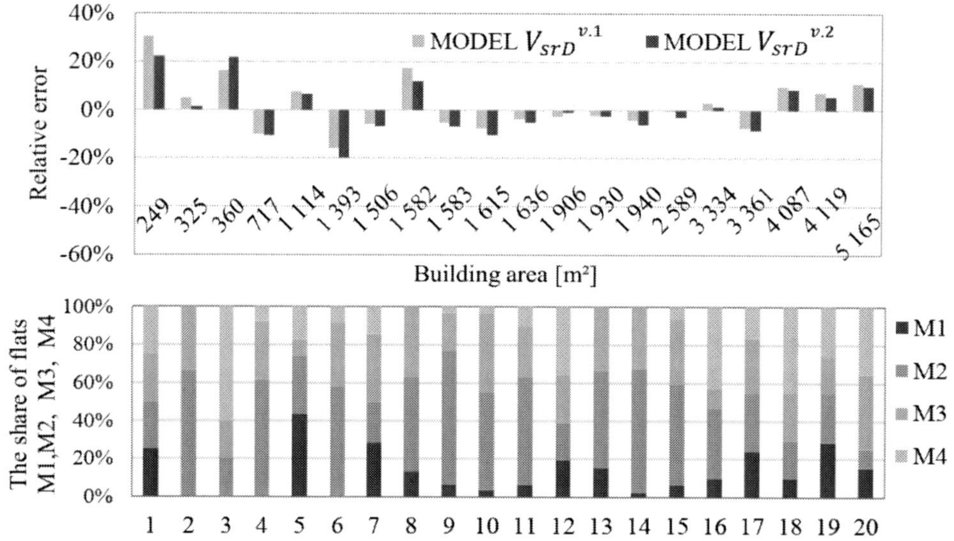

Fig. 4. Relative error of fitting the models to the actual measurement data from buildings with different housing structure.

Fig. 4. presents the relative errors of estimating daily consumption of domestic hot water indicated according to equation 5 as well as the structure of flats and area of buildings. For verifying the model, data obtained by means of the bootstrap method was used, which enabled to draw buildings with non-typical structure from measurement of 626 flats (database A). Both proposed models describe the daily consumption of domestic hot water relatively well. The biggest error can be observed in building 1 characterised by equal arrangement of flats with varying size. The smallest error in the case of both models occurs in buildings 12,

13, 15 and 16. Average unconditional error in estimating in all buildings chosen at random in model $V_{\acute{s}rD}^{v.1}$ amounts to 8.65%, while in model $V_{\acute{s}rD}^{v.2}$ 8.5%.

$$\delta_{V\acute{s}rD} = \frac{V_{\acute{s}rD} - V_{\acute{s}rD,rzecz}}{V_{\acute{s}rD,rzecz}} \tag{5}$$

The conducted analyses indicated that using both of the proposed models, it is possible to indicate daily domestic hot water consumption correctly. The following model was adopted for further analyses and as the final one: $V_{\acute{s}rD}^{v.2}$ based on the number of n-room flats. It is easier to use as it doesn't require information about the area of n-room flats, but merely the number of such flats in a building.

3.2 Model deviation of the average daily consumption of domestic hot water in particular months from the average annual value

In the construction of calculation models, it is a big problem to determine the number of independent variables (explanatory) that should be included in the model. After analysing the measurement data and literature, it was decided that the explanatory variables to be considered in the model are: share of days-off from school in a month, share of Sundays and public holidays in a month, share of holiday days in a month, share of Fridays and Sundays in a month and outside temperature. To test the link between these variables and the consumption of hot water, Sperman's rank correlation was used. The analyses showed a strong correlation between the deviation of the average monthly consumption of hot water from the annual average value and the outside temperature and the share of holiday days. These analyses also showed that the remaining variables do not affect the monthly consumption of hot water. The temperature of the outside air assumed for modelling was transformed according to the equation 6, where $T_{zewn,srM}$ is the average monthly temperature during the year, and T_{Mi} is the average monthly temperature in a particular month.

$$T_{zewn.skor.} = 1 + 0{,}2 \cdot \frac{T_{zewn,srM} - T_{M_i}}{T_{zewn,srM}} \tag{6}$$

The share of individual days in the month was calculated using the equation 7, where l_w is the number of holiday days and ld,Mi is the number of all days in the month.

$$u = \frac{l_w}{l_{d,Mi}} \tag{7}$$

The multiple regression analysis performed using Statistica package estimated the model parameters. The final model ($s_{Mi}^{v.1}$) is described by the equation 8. The coefficient of determination of the model (R^2) is 0.64.

$$s_{Mi}{}^{v.1} = 0{,}363 \cdot T_{zewn.skor.} - 0{,}184 \cdot u + 0{,}671 \tag{8}$$

In the second proposed model ($s_{Mi}^{v.2}$) to improve the quality of parameter estimation the bootstrap method was used. The bootstrap method allowed to create 1000 new buildings. The created database consists of four groups of 250 buildings, characterized by the number of apartments. The created base consists of four groups of 250 buildings with the number of apartments. In the first group there were buildings with the number of flats from 4 to 10, in the second group there were buildings with the number of flats from 10 to 30, in the third group there were buildings with the number of flats from 30 to 50, and in the fourth group there were buildings with a number apartments from 50 to 70. In each group, there was a visible relationship between the size of the building and the scatter of the data received. This means that in larger buildings we can describe the monthly distribution of hot water consumption with greater accuracy. After analysing the obtained values of medians and averages in particular groups, it was found that their values differ by a maximum of 1%. All four groups of buildings were used to build the model. On their basis, the average deviation of hot water consumption in individual months was calculated. The structure of the model is presented in table 1.

Table 1. Model ($s_{Mi}^{v.2}$) of the monthly deviation of DHW consumption from the annual average value.

Month	I	II	III	IV	V	VI	VII	VIII	IX	X	XI	XII
Model $s_{Mi}^{v.2}$	1.10	1.09	1.11	1.07	1.01	0.92	0.88	0.77	0.93	1.00	1.04	1.07
Trust -95%	1.10	1.09	1.11	1.07	1.01	0.91	0.87	0.77	0.93	1.00	1.04	1.07
Trust +95%	1.10	1.09	1.11	1.07	1.01	0.92	0.88	0.78	0.94	1.01	1.05	1.07

3.3 Verification of developed models

Next, a comparison was made between both proposed models and a profile characterized by a constant distribution of DHW consumption in all months. Fig. 5 shows the absolute error of fitting the $s_{Mi}^{v.1}$ model and the $s_{Mi}^{v.2}$ model in individual months to the actual measurement data. The measurement data were from 6 multi-family buildings. The analyses covered the period from January 2012 to December 2013.

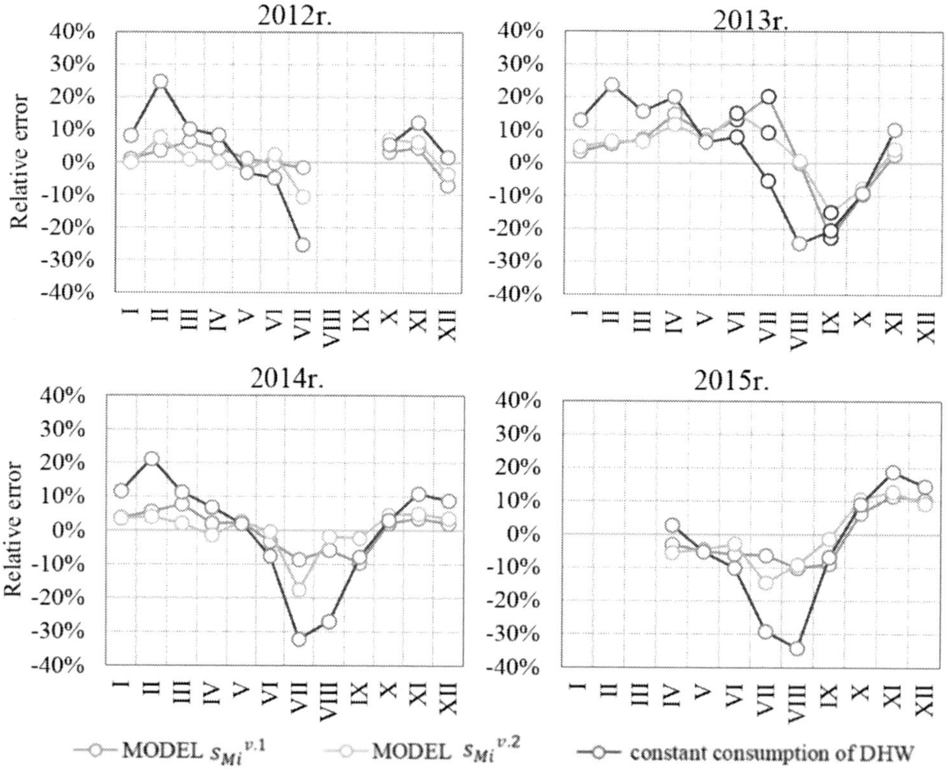

Fig. 5. Absolute error of deviation of the average daily consumption of hot water from the average annual value based on the model sMiv.2, sMiv.2 and constant consumption of hot water.

In 2013, in June, July and September, there were failures in the boiler room, which disturbed the standard distribution of DHW consumption. These months have been excluded from the analysis. The graphs in 2012, 2014 and 2015 show a trend indicating the largest mismatch of all models for measurement data in the summer period, in July and August. Relative error in this period based on the $s_{Mi}^{v.1}$ model is not greater than 10% (August 2015).

In the $s_{Mi}^{v.2}$ model the biggest error at 18% is in July 2014. The constant profile of hot water consumption has an error of 34% in August 2015. Analyses showed that the $s_{Mi}^{v.1}$

model best describes the distribution of water consumption during the holiday season. On the other hand, the $s_{Mi}^{v.2}$ model gives better results for the months like March, April and May. The adoption of a fixed profile of hot water consumption is the worst in describing the distribution of hot water consumption during the year.

$$\sigma_{s_{mi}} = s_{mi} - s_{mi,rzecz} \tag{9}$$

Table 2 presents the analysis of the mean absolute error determined based on the equation 9. The analysis of both developed models were carried out in 2012-2015. The smaller average error of estimation from the period of four years showed the $s_{Mi}^{v.1}$. This error is 5.3%. The $s_{Mi}^{v.2}$ model error is only 0.1% larger than the $s_{Mi}^{v.1}$ model and its average error is 5.4%. The model $s_{Mi}^{v.2}$ is only 0.1% less than the $s_{Mi}^{v.1}$ model and its average error is 5.4%. Unfortunately, to use the $s_{Mi}^{v.1}$ model, you need to know the average monthly outdoor temperature and the school calendar. The second model ($s_{Mi}^{v.2}$) is easier to use because the additional information is not required.

Table 2. The mean error for the models $s_{Mi}^{v.1}$ and $s_{Mi}^{v.2}$.

Error in year:	2012	2013	2014	2015	average value of the error
Model $s_{Mi}^{v.1}$	3%	7%	3%	6%	5.3%
Model $s_{Mi}^{v.2}$	4%	6%	4%	8%	5.4%

4 Conclusions

In this study, the authors demonstrated the ability to describe the consumption of hot water based on the number of flats with a given number of rooms in the flat. The study also showed that when calculating the monthly consumption of hot water, it is necessary to apply the reduction coefficients in the summer and the growth coefficients in winter. It has also been shown that the use of a simple $s_{Mi}^{v.2}$ model based on correction factors in individual months gives similar results as in the case of using the $s_{Mi}^{v.1}$ model. However, the use of the s_{Mi}^{v} model is more troublesome as it requires knowledge of the calendar of holiday days and outside temperature.

References

1. D. George, N. S. Pearre, L. G. Swan, Energy and Buildings **109**, 304–315 (2015)
2. K. Ahmed, P. Pylsy, J. Kurnitski. Energy and Buildings **97**, 77–85 (2015)
3. J. P. Meyer, Rueiued September 1999, 2000, final version August 2000
4. Verordnung uber energiesparenden warmeschutz und energiesparende anlagentechnik beigebauden (energieeinsparverordnung - enev) anlage 3 (2012)
5. Prestazioni energetiche degli edifici, parte 2 (2014)
6. Be10 beregningskerne (2012)
7. SAP 2012: The governments standard assessment procedure for energy rating of dwellings.
8. A. Chmielewska, M. Szulgowska-Zgrzywa, J. Danielewicz, E3S Web of Conferences, **17** (2017)
9. A. Chmielewska, *Modelowanie zapotrzebowania na energię użytkową do przygotowania ciepłej wody w budynkach wielorodzinnych* (Raporty Wydziału Inżynierii Środowiska Politechniki Wrocławskiej, 2017)

Study on optimal conditions of lemongrass extraction

Jaruwan Chutrtong[1,*], and *Kunyanuth* Kularbphettong[1]

[1]Faculty of Science and Technology, Suan Sunandha Rajabhat University, 10300 Bangkok, Thailand

Abstract. This research was conducted to find the optimal conditions to extract active ingredients from lemongrass for using to inhibit microorganisms. The solvents selected for testing are safe chemicals such as water, ethanol and methanol. We also studied the ratio of the lemongrass and the extract solvent and extract time which make the most extraction. Using water as an extract solvent, whether it is heating or not, the extract cannot inhibit the microbes. While using ethanol and methanol as extract solvent, no heat, the extract showed effectiveness in microorganisms inhibition at lemongrass extract concentration 50 mg/ml. Ethanol and methanol used in the extraction has been evaporated to prevent ethanol from being microorganisms deterrent. Therefore, it can be concluded that water cannot used as solvent for extraction of microbial inhibitors from lemongrass. Ethanol and Methanol are suitable solutions because of the low price, safety and the method is not complicated.

1 Introduction

Microbial contamination in the environment may affect more or less to humans health. Those germs microbes can infect human body through many way such as breathing, wound area, eating contaminated food or water. Moreover, being stuck on the surface of various appliances which exposed or infected by germs such as *Staphylococcus aureus*, *Bacillus cereus*, *Escherichia coli* and *Pseudomanas aeruginosa* can also cause infection and illness. Therefore preventing infection by cleaning on these exposed surfaces is another way to prevent infection. To control germ on the surface of the contact area or equipment, disinfect chemical substances (such as alcohol, glutaraldehyde, chlorine compounds, Iodophor, phenol) are used. These substances can cause irritation of various mucous membranes such as membranes in the mouth, pharynx, esophagus and intestinal system. They are also residues in the soil and water. The effects and dangers of chemicals using cause interest in using natural substances to replace chemicals. It was found that many herbal extracts have the ability to inhibit microorganisms. Extracts from plants are complex organic compound. They accumulate in leaves, flowers and stems. Herbal substances are easily decomposed which make them safe for users and the environment. Lemongrass (Scientific name Cymbopogon citratus (DC.) Stapf) is one of the most popular Thai herbs used for cooking. This herb originated in India, Indonesia, Myanmar, Sri Lanka and Thailand [1]. It was found that

* Corresponding author: Kunyanuth.ku@ssru.ac.th

 https://doi.org/10.1051/e3sconf/201910000008

chemicals in lemongrass essential oils can inhibit E. coli bacteria. Those active chemicals include Citral (3,7-dimethyl-2,6-octadienal), Citronellol (acyclic monoterpenoid), Geraneol (monoterpenoid and an alcohol) and Cineole. Therefore, using lemongrass extracts to inhibit bacteria is a way that can reduce chemicals using. And it is also another way to use herbs as raw materials in industrial products which will encourage farmers for increasing income and causes sustainability in their profession [2].

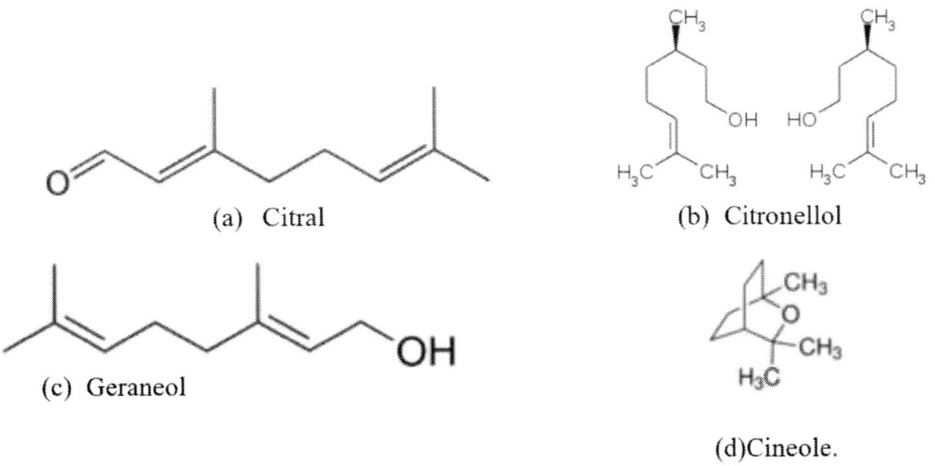

Fig. 1. Chemical structure of Citral, Citronellol, Geraneol and Cineole.

Citral, or or lemonal, is either a pair, or a mixture of terpenoids with the molecular formula C10H16O. The two compounds are double bond isomers. The E-isomer is known as geranial or citral A. The Z-isomer is known as neral or citral B.

This research objective is to find suitable conditions for lemongrass extraction. The extract from lemongrass which has the effect on microorganisms inhibiting will be developed to use as an ingredient in many product for destroying pathogenic germ without using chemicals.

2 Procedure

2.1 Crude extraction from lemongrass

2.1.1 Extraction using water as solvent

Wash the lemongrass thoroughly, dry, and cut into small pieces. Ground thoroughly. Add 20 gram of ground lemongrass to 30 ml water (ratio of lemongrass and water is 2: 3). Shake with shaking machine at 100 rpm for 3, 5 and 7 days in room temperature. Another flask of blended lemongrass with water is boiled at 100 degrees Celsius for 3 hours [3, 4]. Extracted by reflux. Filter with a thin white cloth, then filter with Whatman filter paper no.4. Collect crude extract and dry by freeze dryer. Keep the residue at 4 degrees Celsius in the refrigerator.

2.1.2 Extraction using ethanol and methanol as solvent

Wash the lemongrass thoroughly, let it dry and cut into small pieces. Ground thoroughly. Soak 500 grams of ground lemongrass in 1,000 ml ethanol and methanol (lemongrass and ethanol/methanol ratio is 1: 2). Shake with shaking machine at 100 rpm for 7 days in room temperature. Filter with Whatman Filter paper No. 4 to separate the plant parts and then concentrated all the solutions with an evaporator under vacuum condition. Get viscous substances and dry with freez dryer. Keep the residue at 4 degrees Celsius in the refrigerator.

2.2 Determination of microbial inhibiting lowest concentration by agar dilution method

Prepare the dilution of the test bacteria expected to contain 1.5 x 10^8 CFU/ml (Colony forming unit per milliliter). Add crude extract from 2.1 in bacteria dilution and mix well. Spread 0.01 ml. of bacteria dilution on nutrient agar (NA). Incubate at 35 degrees Celsius for 24 hours. Check bacterial growth on NA.

2.3 Statistical data analysis

The data from the bacterial inhibition test of the extract are analysed for variance (ANOVA) set by DMRT method.

3 Result

3.1 Extraction using water as solvent

When applying the finely ground lemongrass with water to the specified ratio, 3 set left at room temperature.

A B C

Fig. 2. Water extraction at 3 (A), 5 (B) and 7 (C) days.

Fig. 3. Water extraction by using heat

Fig. 4. Liquid separation filtrate.

Fig. 5. Evaporated crude extract.

3.2 Extraction using ethanol and methanol as solvent

When applying the finely ground lemongrass with ethanol and methanol to the specified ratio, left at room temperature. After that, filter with Whatman filter paper No. 4 and concentrate the filtrate with evaporator.

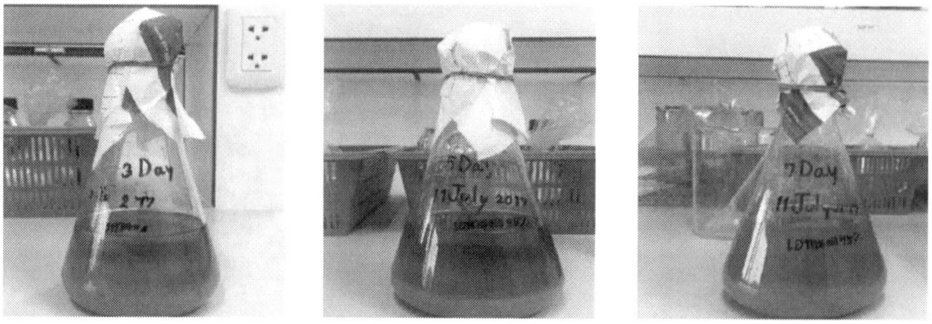

A B C

Fig. 6. Ethanol extraction at 3 (A), 5 (B) and 7 (C) days.

Fig. 7. Clear filtrate of 7 days' soak sample.

Fig. 8. Evaoporate filtrate with rotary evaporator.

3.3 Determination of microbial inhibiting lowest concentration by agar dilution method

After add different amount of prepared crude extract to the dilution of the test bacteria, set aside 5–180 minute before spread 0.01 ml. of bacteria dilution on nutrient agar (NA) and incubate at 35 degrees Celsius for 24 hours. The result of bacterial growth is show in table 1–3.

Table 1. Growth of *Escherichia coli* on NA food after adding lemongrass extract.

solvent	concentration (mg/ml)	variable time before planting on NA (min)						
		5	10	15	30	60	120	180
water	200	+	+	+	+	+	+	+
	100	+	+	+	+	+	+	+
	50	+	+	+	+	+	+	+
95% ethanol	200	+	-	-	-	-	-	-
	100	+	-	-	-	-	-	-
	50	+	-	-	-	-	-	-
methanol	200	+	-	-	-	-	-	-
	100	+	-	-	-	-	-	-
	50	+	-	-	-	-	-	-

+ = GROWTH - = NO GROWTH

Table 2. Growth of *Staphylococcus aureus* on NA food after adding lemongrass extract.

solvent	concentration (mg/ml)	variable time before planting on NA (min)						
		5	10	15	30	60	120	180
water	200	+	+	+	+	+	+	+
	100	+	+	+	+	+	+	+
	50	+	+	+	+	+	+	+
95% ethanol	200	-	-	-	-	-	-	-
	100	-	-	-	-	-	-	-
	50	-	-	-	-	-	-	-
methanol	200	-	-	-	-	-	-	-
	100	-	-	-	-	-	-	-
	50	-	-	-	-	-	-	-

+ = GROWTH - = NO GROWTH

Table 3. Growth of *Bacillus spp* on NA food after adding lemongrass extract.

solvent	concentration (mg/ml)	variable time before planting on NA (min)						
		5	10	15	30	60	120	180
water	200	+	+	+	+	+	+	+
	100	+	+	+	+	+	+	+
	50	+	+	+	+	+	+	+
95% ethanol	200	-	-	-	-	-	-	-
	100	-	-	-	-	-	-	-
	50	-	-	-	-	-	-	-
methanol	200	-	-	-	-	-	-	-
	100	-	-	-	-	-	-	-
	50	-	-	-	-	-	-	-

\+ = GROWTH - = NO GROWTH

4 Conclusion

This research study the efficacy of bacterial inhibition of lemongrass extract by different solvents. The 3 testing pathogenic bacteria are *Bacillus*, *Staph. aureus* and *E. coli*. The result showed that lemongrass extract by water could not inhibit all testing bacteria but lemongrass extract by ethanol and methanol could inhibit all 3 testing bacteria. The efficacy of extract by methanol and ethanol are similar. Both of them can inhibit *Bacillus* and *Staph. aureus* from the lowest concentration (50 mg/ml). The period of time which left substance to react with bacteria is effective for *E. coli*. The extract can inhibit *E. coli*. after 10 minute left. From this result, it may be due to different gram of testing bacteria. Gram negative bacteria have an outer membrane and periplasmic space, which are not found in gram-positive bacteria, and lipoprotein-based polysaccharides that form the outer membrane. This property are good barrier to permeability. Various substances are easier to penetrate into the cell membrane of gram-positive bacteria than gram-negative. Information from this research can use as guideline to use lemongrass extract as a component in many product such as cosmetic products or food products to reduce the amount of pathogenic bacteria or extend shelf life. In addition, lemongrass is an easily obtainable raw material in the local area of Thailand, so large amount of lemongrass extract usage will be the benefit to agricultural products. Further study is searching appropriate procedure for using lemongrass extract in products.

References

1. M. A.Suryawanshi, V. B.Mane, G. B.Kumbhar, Journal of Engineering and Technology (IRJET) **3**, 08 (2016)
2. J. Chutrtong, Procedia – Social and Behavioral Sciences **197**, 797–800 (2015)
3. F. A. Mediani, A Khatib, C. P. Tan, Molecules **18**, 10452–10464 (2013)
4. P. N. Alam, H. Husin, T. M. Asnawi, A. Disalamun, *Extraction of citral oil from lemongrass (Cymbopogon Citratus) by steam-water distillation technique* (IOP Conf. Series: Materials Science and Engineerinh, 345, 2018)

Geothermal power based ULTDH for cooling and heating purposes

Bartłomiej Ciapała[1,*], and *Mirosław* Janowski[1]

[1]AGH University of Science and Technology, Faculty of Geology, Geophysics and Environmental Protection, Department of Fossil Fuels

Abstract. Ultra-low temperature district heating systems facilitate use of waste and renewable heat sources. The article presents a possible scheme of operation and optimisation of small ultra-low temperature district heating system consisting of waste heat source, a number of heated individual dwellings and borehole thermal energy storage plant. Optimisation performed for typical meteorological year for Kraków indicate significant potential of decreasing energy amount discharged to the environment and total length of borehole heat exchangers, compared to individual heat/cold production from low-temperature geothermal resources. Meanwhile, satisfied is a set of constrains providing borehole thermal energy storage sustainability and fulfilling entire heating and cooling demands.

1 Introduction

Sufficient heating and cooling of buildings and industrial processes becomes one of the most crucial challenges in their proper design and realisation. Even more difficult is to meet requirements using environmentally friendly systems while keeping operation economically competitive. Especially, that the area and resources are usually limited. Common solution is an individual supplying in heat and cooling. It guarantees flexibility and potentially gives independency from external suppliers (which usually is illusive or no use is made of it). Simultaneously, such a solution entails entire variety of efforts including maintenance, optimal management, risk mitigation and other. Considering narrow specialisations of companies and regarding desire of comfortable and effortless maintaining internal temperature of flats and detached houses, many owners are likely to connect their buildings to the district heating. Therefore, some efforts are made to ease using district heating network not only for heating, but also for cooling [1]. Also, heat pumps in district heating gain in importance [2]. Among solutions prospective in this context, especially placed in not very densely built areas is ultra-low-temperature district heating (ULTDH) [3].

ULTDH is system, in which supply temperature is too low to prepare sanitary hot water (SHW) without an additional heat source [4]. Thus, it has potential to be both heat source and heat sink, as temperature in the pipeline may be low enough for heat discharge from many appliances [4]. Simultaneously, decreased temperature does not need to increase pipe's diameter compared to typical medium-temperature DH, as heat pumps operation may result

* Corresponding author: bciapala@agh.edu.pl

in similar temperature differences between upstream and downstream nodes. It is important to notice, that energy loss in the pipeline may be decreased in comparison with medium-temperature DH and heat boosters may be used to meet end user's requirements [5–7].

The aim of this publication is to perform test and optimisation of ULTDH operating in conditions of Kraków city in Poland and compare results with stand-alone installations covering the same heating and cooling demands.

2 Problem description

The present article is to propose ultra-low-temperature district heating system used for cooling small industrial object, space heating and sanitary hot water preparation for citizens (this solution is already described and known from real-life conditions [4]) and energy gaining/storage from borehole heat exchangers (BHE), which constitute borehole thermal energy storage (BTES) (see Fig. 1.). The energy is recovered for utilisation with heat pumps installed individually at dwellings. As mentioned buildings may be easily heated and cooled using their own systems of borehole heat exchangers, comparison with such individual solution will be shown.

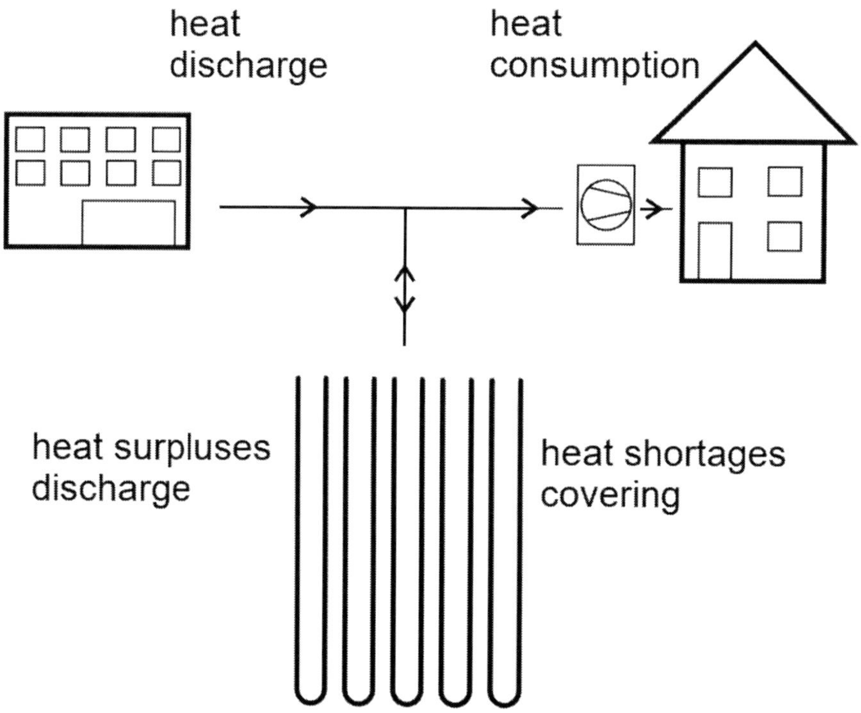

Fig. 1. Scheme of proposed system operation. Heat is discharged in a small industrial facility, while it is consumed in a dozen of individual dwellings/multi-storey building. Excess heat and its shortages are managed with borehole thermal energy storage.

Operation of such a system has been optimised in regard to the size of BTES and number of citizens whose living space heating and SHW preparation may be provided by proposed ULTDH system cooperating with small industrial facility with cooling demand (heating power discharged) varying from 64 to 210 kW. Operation schedule may be observed on the graph Fig. 2.

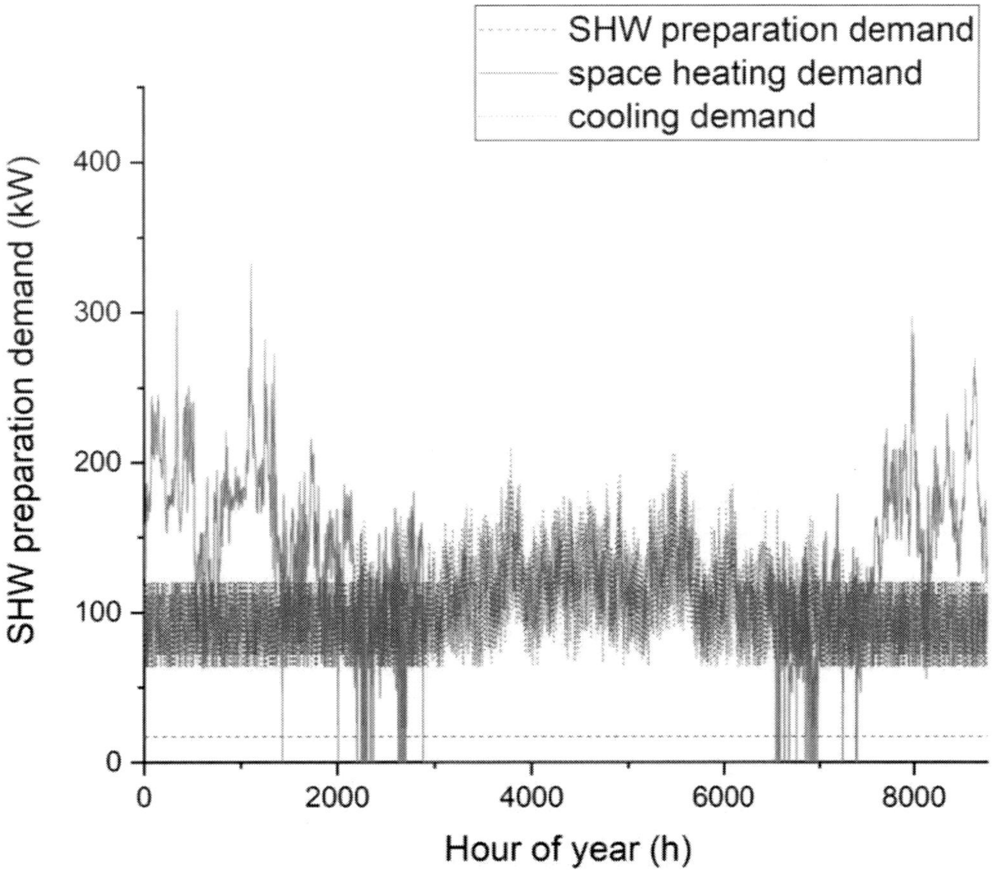

Fig. 2. Operation schedule of system receivers' demand for cold, heat and sanitary hot water preparation.

3 Mathematical description of the problem

Heat demand in hour n is described by the equation:

$$E_n^{heating} = E^{SHW} + E^{design} \cdot \frac{T_{designed}^{internal} - T_n^{external}}{T_{designed}^{internal} - T_{designed}^{external}} \quad (1)$$

Where $E^{design} \cdot \frac{T_{designed}^{internal} - T_n^{external}}{T_{designed}^{internal} - T_{designed}^{external}}$ part is used to calculate energy demand for space heating in an hour n ($E_n^{space\ heating}$).

$T_{designed}^{internal}$ has been assumed as 20°C, $T_{designed}^{external}$ as -20°C, $T_n^{external}$ is external temperature as given in typical meteorological year (TMY) for Kraków Balice meteorological station. Design heat load E^{design} equals 80W/m^2 and total heated area is optimised and amount to 27.6 m^2 of flat per capita, as mean for Poland (GUS). Space heating has been arbitrary disabled from May to September (no clear heating schedule is binding in Poland).

Heat energy required from ULTDH is calculated by the equation

$$E_n^{ULTDH\ heating} = (1 - \frac{1}{COP}) \cdot (E_n^{space\ heating} + E^{SHW}) \quad (2)$$

Optimisation has been performed with objective function stated as

$$Z = \min(H) \quad (3)$$

Where H is a total length of BHE.
With following constrains:

- Peak heating or cooling power cannot be higher than total BHE heat transfer rate potential
- BTES's energy content after an year of operation must be equal +/-10% to its initial state of charge
- In any hour, BTES state of charge cannot be lower than 0
- In any hour, ULTDH must provide entire energy demand

An examined set of acceptable combinations is presented on Fig. 3.

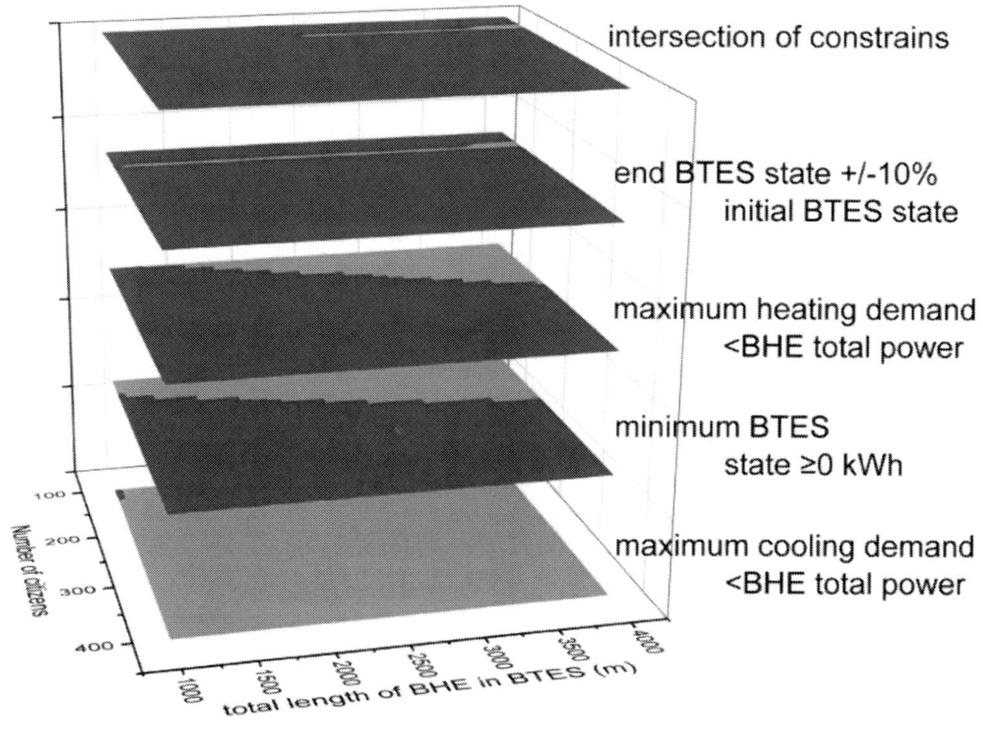

Fig. 3. Parameters' constrains. Green areas represent sets of parameters which do not violate constrains e.g. are acceptable. Otherwise – red parts.

The initial (in hour n = 1) BTES value has been set as 80 kWh per meter of the borehole heat exchanger (BHE), which is moderate value for its energy content [8]. For each and every hour, energy provided by industrial facility is compared with energy demand decreased by 25%, as heat pumps (HP) coefficient of performance (COP) equal to 4.0 has been assumed. Heat transfer rate of BHEs is assumed to be 40W/m. If any excess occur, energy is stored in BTES of 40% efficiency [9]. Otherwise, lack of energy for space heating and domestic hot water preparation (50 kg per 24 h per person, temperature shift 50°C, demand distributed steadily and equal in every hour) is covered by heat pump from BTES. For every hour n > 1, BTES energy content is described as:

$$E_n^{BTES} = E_{n-1}^{BTES} + 0.4 \cdot E_n^{cooling} - (1 - \frac{1}{COP}) \cdot (E_n^{space\ heating} - E^{SHW}) \tag{4}$$

Every hour, when

$$\frac{E_n^{cooling}}{1-\frac{1}{COP}} > E_n^{space\ heating} + E^{SHW} \tag{5}$$

BTES is charged with excessing energy, otherwise – energy is extracted from BTES with 40% efficiency and extracted heat is calculated as:

$$E_n^{BTES\ extracted} = (1 - \frac{1}{COP}) \cdot \left(E_n^{space\ heating} + E^{SHW}\right) - \frac{E_n^{cooling}}{1-\frac{1}{COP}} \tag{6}$$

ULTDH transport heat loss has been neglected, as temperature difference between medium and ground would be small, supposedly less than 15°C. Similarly, circulation pumps and heat capacity of DH medium is neglected.

4 Results

For given initial conditions, BTES consisting of about 2400 m of BHE is supposed to meet expectations. In its lowest energy content is 90 MWh, which is 47% of its initial state, while in its highest is 249 MWh and 129.5% of BTES initial energy content. In total, for typical meteorological year BTES provided 430 MWh of cooling and 180 MWh for heating purposes. Simultaneously, receivers discharged 909 MWh in cooling process and consumed 880 MWh for space heating (note: 220 MWh was produced from electricity in heat pump compressor, so total energy consumption was 880 MWh). Thus, significant amount of waste heat has been recovered and reused instead of being directly discharged into the environment, significantly reducing environmental burden. State of charge is presented in Fig. 4.

Over 8760 analysed hours, 780 BTES charging/discharging switching events occurred. The transient periods during which system is close to balance and BTES is hardly used are especially interesting, as represent system's highest efficiency. The final BTES state of charge indicates long-lasting system's durability and reliability, as its natural resource is not depleted. This leads to the conclusion that BHE might be even smaller, if its finite lifetime was assumed.

The simulation suggests that space heating and SHW preparation might be provided for 150 people. There is not a single hour in which heating and cooling demand would not be covered (what was one of the assumptions).

In comparison with individual sources of heat and cold for the same conditions, the BHE size is reduced by more than 5 times, bringing significant limitation on required space and capital costs. For typical conditions in Poland, about 60% of heat injected to BTES from cooling is dissipated in the environment. It is still not sufficient enough, yet significantly less than in case of traditional dry-coolers discharging heat into the atmosphere. Meanwhile, the heat discharged to ULTDH during space heating demand is consumed and significantly decreases conventional heat demand. Thus, heat is used efficiently, not susceptible for exergy decrease or heat storage inefficiency loss. For presented optimal conditions fulfilling all the constrains, share of heat discharged to ULTDH and directly used by heat pumps in overall heat demand is never lower than 50% (Fig. 5).

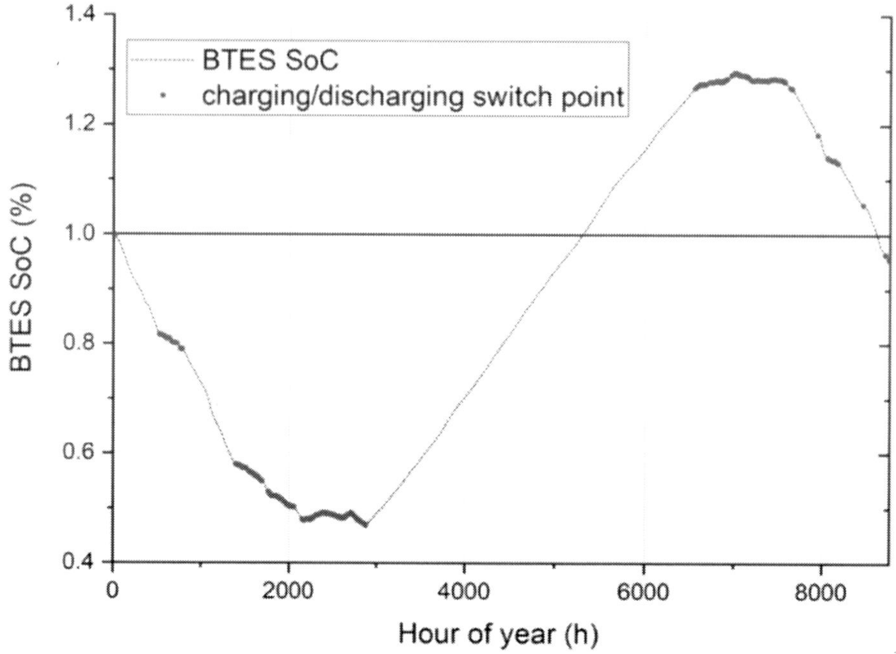

Fig. 4. State of charge of BTES. Red dots represent transient conditions, for which heat exchange in BTES occur in a very limited range, while entire ULTDH system's efficiency is highest.

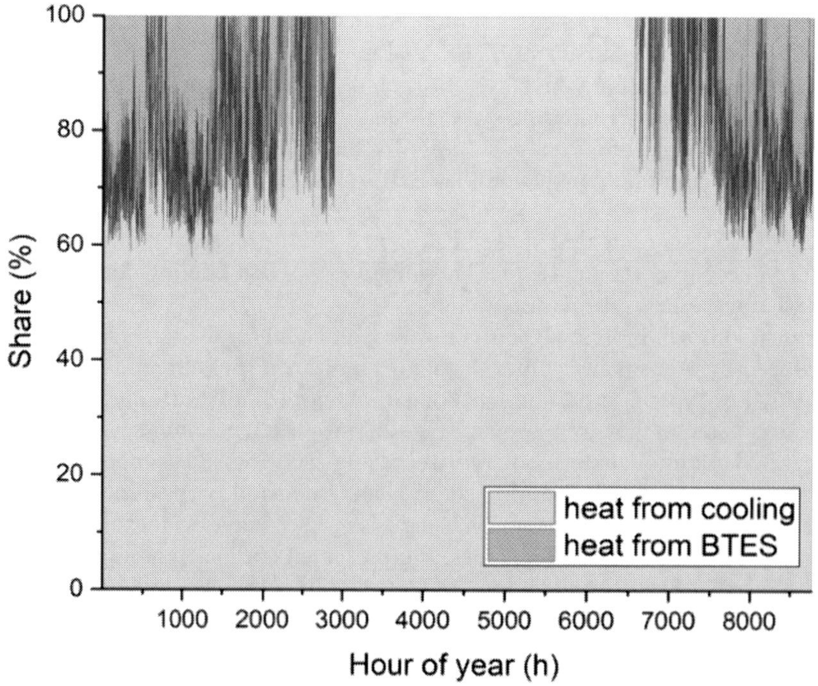

Fig. 5. Proportions of heat provided directly from cooling and from BTES.

5 Summary

The optimisation and operation mode of the geothermal power based ULTDH for cooling and heating purposes has been presented. Obtained results indicate significant benefits derived from the ULTDH system operation connecting heat sources and heat sinks, which nominally are not dependent of each other. Low temperature of used medium brings benefits in limited heat losses and increased heat sink effectiveness. Total BHE length may be reduced 5 times compared to individual heat and cold sources while satisfying entire demands during the typical meteorological year. Simultaneously, less energy is dispersed in the environment, what has a potential in reducing negative effects like urban heat islands. Use of heat pumps at dwellings has potential to facilitate use of energy-efficient low temperature heating in houses (e.g. floor heating).

Presented results are valid for assumed situation during the typical meteorological year. Yet, the methodology is applicable in real-life considerations, where more complex and diversified data would be provided. For the future work, including DH system's heat capacity and performing sensitivity analyses for longer time periods is planned.

The paper has been prepared under the AGH-UST research grant No. 11.11.140.03.

References

1. B. Rezaie, M. A. Rosen, Appl. Energy, **93** (2012)
2. M. A. Sayegh, P. Jadwiszczak, B. P. Axcell, E. Niemierka, K. Bryś, H. Jouhara, Energy Build. **166** (2018)
3. B. Ciapała, J. Jurasz, M. Janowski, 10th Conf. Interdiscip. Probl. Environ. Prot. Eng. EKO-DOK (2018)
4. O. Gudmundsson, M. Brand, J. E. Thorsen, The 14th International Symposium on District Heating and Cooling (2014)
5. X. Yang, H. Li, S. Svendsen, Energy Convers Manag. **122** (2016)
6. T. Ommen, J. E. Thorsen, W. B. Markussen, B. Elmegaard, Energy, **137** (2017)
7. B. Ciapała, M. Janowski, J. Jurasz, Tech. Poszuk. Geol. Geoterm. Zrówn. Rozw. **2** (2018)
8. A. Casasso, R. Sethi, G. Pot, Energy, **106** (2016)
9. M. Lanahan, P. C. Tabares-Velasco, Energies **10**, 6 (2017)

Application of the counter- and cross-flow indirect evaporative cooler for heat recovery under different climate conditions

Aleksandra Cichoń[1,*], *Anna* Pacak[1], *Demis* Pandelidis[1], and *Sergey* Anisimov[1]

[1]Department of Environmental Engineering, Wroclaw University of Science and Technology, Wyb. Wyspiańskiego 27, 50-370 Wroclaw, Poland

Abstract. This paper investigates the potential of applying an indirect evaporative cooler for heat recovery in air-conditioning system under various climate conditions. The counter- and cross-flow configurations of the indirect evaporative exchanger are addressed in this study in terms of their performance and applicability for different climate zones. Presented analyses are carried out with original ε-NTU model considering condensation from the product airflow and validated against experimental data. It was stated that both configurations of the indirect evaporative exchanger have a high application potential to be employed as a heat recovery device under most climatic parameters. Additionally for each climate zone considered in this paper, the air-conditioning system that consists of either the indirect evaporative exchanger and vapor-compression cooler or indirect evaporative exchanger and direct evaporative cooler is proposed.

1 Introduction

Worldwide energy demand has increased in recent years, and it is expected to further grow in coming years [1]. The significant part of global energy consumption relates to the building sector, where it is largely used for air-conditioning (AC) and cooling purposes. Nowadays, most of conventional cooling devices are driven by electric energy, which production is based on fossil fuels. One of the promising cooling technology, that is driven mainly by the process of water evaporation instead of electricity, is the indirect evaporative cooling (IEC) [2, 3].

The IEC heat and mass exchanger consists of two adjacent channels, namely the product and working air channels, arranged in repetitive series. The plates in the working air channels are covered with water that evaporates into the working airflow and as a result of water evaporation, temperature of the plate that separates the working and product air channels decreases. The product airflow flowing through the product air channels contacts with the plate of decreased temperature and so it is therefore cooled without adding any moisture.IEC exchangers can be arranged in many ways, among others in a counter-flow or a cross-flow configuration. In the counter-flow configuration, working and product airflows

*Corresponding author: aleksandra.cichon@pwr.edu.pl

move parallel in opposite directions while in the cross-flow configuration, the working airflow moves perpendicularly to the product airflow (Fig. 1).

Recent development of the IEC technology allows to apply both the counter- and cross-flow IEC exchangers as a heat recovery devices. To meet comfort requirements of the conditioned space, the IEC cooler used for heat recovery should be followed by a vapor-compression cooler (VCC) or by a direct evaporative cooler (DEC), depending on the climate conditions. The schematic design of the AC system that is addressed in this study, is presented in the Figure 1. Before the product airflow is delivered to a conditioned space, it passes through the IEC unit and then, if necessary, is additionally cooled and dehumidified in the VCC unit or cooled and humidified in the DEC unit. The exhaust airflow from the conditioned space is thereafter employed as the working airflow in the IEC exchanger. In such case, due to the low temperature of the working airflow, the process of water vapor condensation from product airflow is likely to occur and so is has to be taken into account. IEC exchanger performance was addressed in several studies that proved high potential of the IEC technology to operate under various parameters [4–7]. This paper aims to investigate the performance of the counter- and cross-flow configuration of the IEC exchanger under specific climate conditions. Based on the airflow parameters at the outlet of the IEC exchanger, it is also proposed whether the IEC unit in the AC system should be followed by the VCC or DEC unit.

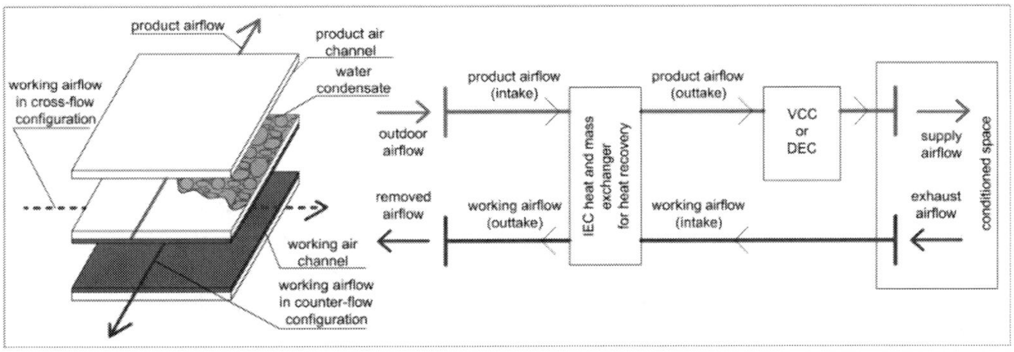

Fig. 1. Scheme of the AC system consisted of the IEC heat and mass exchanger for heat recovery.

2 Methods

In this study, ε–NTU mathematical model is used to describe heat and mass transfer in the counter- and the cross-flow IEC exchanger. Heat and mass balance is calculated under steady-state operation with ordinary differential equations for the counter-flow configuration and partial differential equations for the cross-flow configuration, solved with numerical methods. In order to determine whether the condensation has to be taken into consideration or not, the control algorithm is applied that compares the product airflow dew-point temperature and the temperature of the plate surface at each step of solving differential equations. The condensation process is taken into consideration when product channel plate surface temperature is lower than product airflow dew-point temperature.

Based on energy and mass balance equations that describe the airflow passing through the control volume of the investigated exchanger, the overall energy balance equations under non-condensation state are developed. The overall energy balance for the counter-flow configuration of the IEC exchanger is described with Eq. (1) and with Eq. (2) for the cross-flow configuration.

$$\left(\frac{dt_1}{d\overline{X}}\right)-\left(\frac{W_2}{W_1}\right)\left[\left(\frac{dt_2}{d\overline{X}}\right)+\frac{\left(i_{g2}-i_{w2}\right)}{c_{p2}}\left(\frac{dx_2}{d\overline{X}}\right)\right]=0 \tag{1}$$

$$\left(\frac{\partial t_1}{\partial\overline{X}}\right)+\left(\frac{W_2}{W_1}\right)\left[\left(\frac{\partial t_2}{\partial\overline{Y}}\right)+\frac{\left(i_{g2}-i_{w2}\right)}{c_{p2}}\left(\frac{\partial x_2}{\partial\overline{Y}}\right)\right]=0 \tag{2}$$

where: t – temperature, °C; $\overline{X}$ – relative coordinate along product airflow direction, -; $\overline{Y}$ – relative coordinate perpendicular to product airflow direction, -; W – heat capacity rate of the fluid, W/K; i – specific enthalpy, J/(kgK); x – humidity ratio of moist air, kg/kg; c_p– specific heat capacity of moist air, J/(kgK). Subscripts: 1 – related to product channel; 2 – related to working channel; g – related to water vapor; w – related to water film.

The overall energy balance under condensation state for the counter-flow configuration of the IEC exchanger is described with Eq. (3) and with Eq. (4) for the cross-flow configuration.

$$\left[\left(\frac{dt_1}{d\overline{X}}\right)+\frac{\left(i_{g1}-i_{w1}\right)}{c_{p1}}\left(\frac{dx_1}{d\overline{X}}\right)\right]-\left(\frac{W_2}{W_1}\right)\left[\left(\frac{dt_2}{d\overline{X}}\right)+\frac{\left(i_{g2}-i_{w2}\right)}{c_{p2}}\left(\frac{dx_2}{d\overline{X}}\right)\right]=0 \tag{3}$$

$$\left[\left(\frac{\partial t_1}{\partial\overline{X}}\right)+\frac{\left(i_{g1}-i_{w1}\right)}{c_{p1}}\left(\frac{\partial x_1}{\partial\overline{X}}\right)\right]+\left(\frac{W_2}{W_1}\right)\left[\left(\frac{dt_2}{d\overline{Y}}\right)+\frac{\left(i_{g2}-i_{w2}\right)}{c_{p2}}\left(\frac{\partial x_2}{\partial\overline{Y}}\right)\right]=0 \tag{4}$$

The mathematical model accuracy for the counter-flow IEC exchanger, under condensation state, was verified by authors with experimental data obtained at the dedicated test station [8]. It was concluded that the model can predict outlet product airflow temperature with average accuracy (mean absolute percentage error) equal to 2.5% and outlet product airflow relative humidity with average accuracy equal to 0.9%. The model accuracy under non-condensation state for the counter-flow configuration, was verified with the existing experimental study carried out by Riangvilaikul et al. [9]. It was found that the model can predict outlet product airflow temperature with average accuracy equal to 1.8%. The maximum discrepancy does not exceed 5%. It is assumed that accuracy of the mathematical model predictions for the cross-flow IEC exchanger, are similar for those of the counter-flow unit. The similarity results from the analogy in the mathematical model equations used for describing heat and mass transfer in both exchangers.

3 Results and discussion

In this section simulations results are presented and discussed. The geometric parameters of the counter- and cross-flow configuration of the IEC exchanger investigated in this study are given in the Figure 2 and Table 1. Numerical simulations based on the mathematical model presented in previous section, are carried out for inlet product airflow parameters characteristic for different climatic locations listed in Table 2. Inlet product airflow parameters (temperature and humidity ratio) are defined based on the climatic design conditions given by ASHRAE [10]. Each location in Table 2 represents one of the climate zone in the Köppen-Geiger climate classification system (following climate zones: Dwd – monsoon-influenced extremely cold subarctic climate, Dsd – extremely cold, subarctic climate, ET – tundra climate and EF – ice cap climate are not taken into consideration in this study due to the low population density ratio and therefore scarce data available) [11]. Inlet working airflow parameters are: temperature t_{2i}=25°C and relative humidity RH_{2i}=50%.

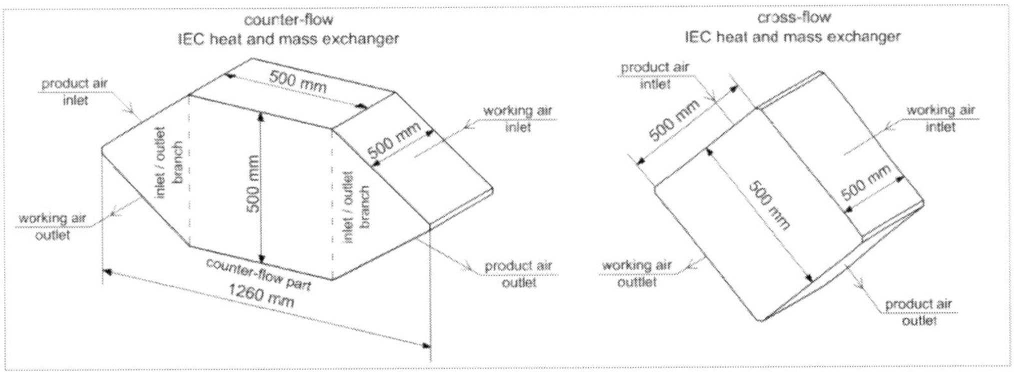

Fig. 2. General dimensions of the IEC heat and mass exchanger in different configuration.

Table 1. Characteristic parameters of the IEC heat and mass exchanger in different configuration.

Parameters	Counter-flow	Cross-flow
Wall thickness, mm	0.3	0.3
Channel length/width/height, m	0.5/0.5/0.0034	0.5/0.5/0.0034
Number of product air channels	68	68
Working to product air ratio/NTU_1†	1/6	1/6
Airflow rate, m^3/h	500	500
Total sheet surface, m^2	52	34
Heat exchange sheet surface, m^2	34	34

3.1 IEC exchanger applicability under different climate conditions

In this study, different performance indicators were used to evaluate performance of the IEC heat and mass exchanger under various climatic conditions. For the counter- and cross-flow configuration of the exchanger, average temperature drop is calculated with Eq. (5). In case of water vapor condensation from the product airflow, the average humidity ratio drop is described with Eq. (6). Average total cooling capacity, given by Eq. (7), is the sum of average sensible cooling capacity and average latent cooling capacity related with condensation process. The average values $\bar{t}_{1o}$, $\bar{x}_{1o}$, $\bar{i}_{1o}$ relate to the outlet section of the exchanger. Calculated performance indicators are listed in Table 3.

$$\Delta\bar{t}_1 = t_{1i} - \bar{t}_{1o},\ ^\circ\mathrm{C} \tag{5}$$

$$\Delta\bar{x}_1 = x_{1i} - \bar{x}_{1o},\ \mathrm{g/kg} \tag{6}$$

$$\bar{Q}_1^T = V_1 \cdot \rho_1 \cdot \left(i_{1i} - \bar{i}_{1o}\right) = \bar{Q}_1^S + \bar{Q}_1^L,\ \mathrm{kW} \tag{7}$$

where: Q – cooling capacity, kW; V – airflow rate, m^3/s; ρ – density, kg/m^3. Subscripts: i – inlet parameters; o – outlet parameters; T – related to total heat transfer; S – related to sensible heat transfer; L – related to latent heat transfer.

† NTU – number of transfer units, - (NTU = αF / (Gc_p)); α – convective heat transfer coefficient, W/(m2 K); F – surface area, m^2; G – mass flow rate, kg/s.

Table 2. Climatic parameters for cooling design conditions [11] for different climate zones [10] (*DB*– dry-bulb; M-i - Monsoon-influenced).

Symbol	Climate zone	City, country	t_{DB}, °C	x, g/kg
Af	Tropical rainforest	Singapore	33.5	18.9
Am	Tropical monsoon	Miami, Florida, USA	32.7	17.4
As	Tropical savanna, dry	Lihue, Hawaii, USA	29.1	15.9
Aw	Tropical savanna, wet	Caracas, Venezuela	33.1	21.8
BWk	Hot semi-arid	Damascus, Syria	38.1	5.0
BWh	Cold semi-arid	Cairo, Egypt	36.9	9.8
BSk	Hot deserts	Denver, Colorado, USA	32.9	4.1
BSh	Cold desert	Nicosia, Cyprus	37.7	7.0
Cfa	Humid subtropical	Buenos Aires, Argentina	30.0	11.6
Cfb	Temperate oceanic	London, United Kingdom	25.6	10.0
Cfc	Subpolar oceanic	Auckland, New Zealand	24.3	11.9
Csa	Hot-summer Mediterranean	Izmir, Turkey	35.8	8.9
Csb	Warm-summer Mediterranean	Cape Town, South Africa	29.8	9.8
Csc	Cool-summer Mediterranean	Bolmaceda, Chile	22.9	5.8
Cwa	M-i, humid subtropical	New Delhi, India	40.8	10.7
Cwb	Subtropical highland	Mexico City, Mexico	28.1	2.6
Cwc	Cold subtropical highland	La Paz, Bolivia	17.0	0.7
Dfa	Hot-summer humid continental	Bucharest, Romania	32.7	10.3
Dfb	Warm-summer humid continental	Oslo, Norway	24.8	8.5
Dfc	Subarctic	Fairbanks, Alaska	25.6	7.0
Dfd	Extremely cold subarctic	Yakutsk, Russia	28.4	9.0
Dsa	Hot, dry-summer continental	Bishkek, Kirgizstan	33.7	7.6
Dsb	Warm, dry-summer continental	Sivas, Turkey	30.7	6.9
Dsc	Dry-summer subarctic	Homer, Alaska, USA	17.3	7.7
Dwa	M-i, warm-summer humid continental	Beijing, China	33.5	12.5
Dwb	M-i, subarctic	Irkutsk, Russia	26.9	8.7
Dwc	M-i, extremely cold subarctic	Mohe County, China	28.3	9.5

Based on the conducted simulation it is stated that in two cases (Cwc, Dsc), the IEC heat and mass exchanger is not recommended as a heat recovery unit for cooling purposes due to the low temperature of the outdoor airflow. The IEC exchanger may be rather considered as a cooling unit that operates on the outdoor air instead on the exhaust airflow. In other cases, IEC exchanger can operate as a heat recovery device even if outdoor air temperature is lower than temperature of the working airflow (Cfc, Csc, Dfb). Under these climatic conditions, $\overline{Q}_1^T$ of the IEC exchanger is the lowest ($\overline{Q}_1^T$ for Cfc zone is equal to 1.1kW for the counter- and cross-flow configuration), nevertheless it is possible to decrease the temperature of the outdoor airflow on average by 6.1°C (Table 3). Among all considered cases, the highest $\Delta\overline{t}_1$ is achieved for Cwa climate zone (22.7°C for the counter- and 22.2°C for the cross-flow configuration). The lowest $\Delta\overline{t}_1$ is achieved for Csc climate zone (5.1°C for the counter- and 4.9°C for the cross-flow configuration).

The water vapor condensation from the product airflow occurs in four tropical climate zones (Af, Am, As, Aw). In these cases, the $\Delta\overline{x}_1$ is equal to 4.1g/kg for the counter-flow configuration and 3.5g/kg for the cross-flow configuration. The highest $\Delta\overline{x}_1$ is achieved for Aw climate zone (6.7g/kg for the counter- and 6.0g/kg for the cross-flow configuration). For the Aw climate zone, also $\overline{Q}_1^T$ of the of the IEC exchanger is the highest (5.1kW for the counter- and 4.6kW for the cross-flow configuration). It is due the fact that with intense condensation process in product air channels, additional latent heat is released, and in this case it accounts for up to 56% of total cooling capacity.

Table 3. Performance indicators of the counter- and cross-flow configuration of the IEC heat and mass exchanger for different climatic conditions (VCC - vapor compression cooling unit; DEC - direct evaporative cooling unit).

Climate zone	Counter-flow	Cross-flow	Counter-flow	Cross-flow	Counter-flow	Cross-flow	AC system
	$\Delta \bar{t}_1$, °C		$\Delta \bar{x}_1$, g/kg		$\bar{Q}_1^S / \bar{Q}_1^L / \bar{Q}_1^T$, kW		
Af	13.9	13.3	4.3	3.7	2.3/1.9/4.2	2.2/1.6/3.8	IEC + VCC
Am	13.4	12.9	3.2	2.6	2.2/1.4/3.6	2.2/1.2/3.4	IEC + VCC
As	10.3	9.9	2.1	1.7	1.7/0.9/2.6	1.6/0.8/2.4	IEC + VCC
Aw	12.9	12.2	6.7	6.0	2.2/2.9/5.1	2.0/2.6/4.6	IEC + VCC
BWk	20.1	19.6	0.0	0.0	3.4/0.0/3.4	3.3/0.0/3.3	IEC + DEC
BWh	18.8	18.4	0.0	0.0	3.2/0.0/3.2	3.1/0.0/3.1	IEC + VCC or IEC + DEC
BSk	14.9	14.6	0.0	0.0	2.5/0.0/2.5	2.4/0.0/2.4	IEC + DEC
BSh	19.7	19.2	0.0	0.0	3.3/0.0/3.3	3.2/0.0/3.2	IEC + DEC
Cfa	12.1	11.8	0.0	0.0	2.0/0.0/2.0	2.0/0.0/2.0	IEC + VCC
Cfb	7.7	7.5	0.0	0.0	1.3/0.0/1.3	1.3/0.0/1.3	IEC + VCC or IEC + DEC
Cfc	6.5	6.3	0.0	0.0	1.1/0.0/1.1	1.1/0.0/1.1	IEC + VCC
Csa	17.8	17.4	0.0	0.0	3.0/0.0/3.0	2.8/0.0/2.8	IEC + VCC or IEC + DEC
Csb	11.9	11.6	0.0	0.0	2.0/0.0/2.0	1.9/0.0/1.9	IEC + VCC or IEC + DEC
Csc	5.1	4.9	0.0	0.0	0.8/0.0/0.8	0.8/0.0/0.8	IEC + DEC
Cwa	22.7	22.2	0.0	0.0	3.8/0.0/3.8	3.7/0.0/3.7	IEC + VCC or IEC + DEC
Cwb	10.2	9.9	0.0	0.0	1.7/0.0/1.7	1.7/0.0/1.7	IEC + DEC
Cwc	-0.7	-0.7	0.0	0.0	n/a	n/a	NOT APPLICABLE
Dfa	14.7	14.4	0.0	0.0	2.5/0.0/2.5	2.4/0.0/2.4	IEC + VCC or IEC + DEC
Dfb	6.9	6.8	0.0	0.0	1.2/0.0/1.2	1.1/0.0/1.1	IEC + VCC or IEC + DEC
Dfc	7.7	7.5	0.0	0.0	1.3/0.0/1.3	1.3/0.0/1.3	IEC + DEC
Dfd	10.5	10.2	0.0	0.0	1.8/0.0/1.8	1.7/0.0/1.7	IEC + VCC or IEC + DEC
Dsa	15.7	15.3	0.0	0.0	2.6/0.0/2.6	2.6/0.0/2.6	IEC + VCC or IEC + DEC
Dsb	12.8	12.4	0.0	0.0	2.1/0.0/2.1	2.1/0.0/2.1	IEC + DEC
Dsc	-0.4	-0.4	0.0	0.0	n/a	n/a	NOT APPLICABLE
Dwa	15.5	15.1	0.0	0.0	2.6/0.0/2.6	2.5/0.0/2.5	IEC + VCC
Dwb	9.0	8.8	0.0	0.0	1.5/0.0/1.5	1.5/0.0/1.5	IEC + VCC or IEC + DEC
Dwc	10.4	10.1	0.0	0.0	1.7/0.0/1.7	1.7/0.0/1.7	IEC + VCC or IEC + DEC

It could be stated that the counter-flow configuration of the IEC exchanger allows to achieve better performance indicators than the cross-flow configuration under various climate conditions. However, the counter-flow unit requires additional inlet and outlet branches that are not involved in heat exchange (Fig. 2). It results in higher material costs to produce the IEC exchanger without increasing it performance. The required sheet surface of the counter-flow exchanger is 1.5 times higher than in the cross-flow configuration (Table 1) and therefore it can be assumed that investment cost will be 1.5 higher too. In no case addressed in this study, total cooling capacity of the counter-flow IEC exchanger does not exceeds the total cooling capacity of the cross-flow configuration high enough to compensate the high investment cost of the exchanger.

3.2 AC system applicability under different climate conditions

In this section, the AC system is proposed for each climate zone considered in this study. First system consists of the IEC heat and mass exchanger used for heat recovery, followed by the VCC unit for additional cooling and dehumidification. The exemplary processes in the product and working air channels of the IEC exchanger and in the VCC unit are presented on the psychrometric chart in Figure 3(a). In the second system, the IEC exchanger is followed by the DEC unit, where additional adiabatic process of cooling and humidifying is realized (Fig. 3(b)). The appropriate AC system is selected based on its ability to provide supply airflow parameters that allow to meet comfort parameters in the conditioned space (analysis performed for design conditions).

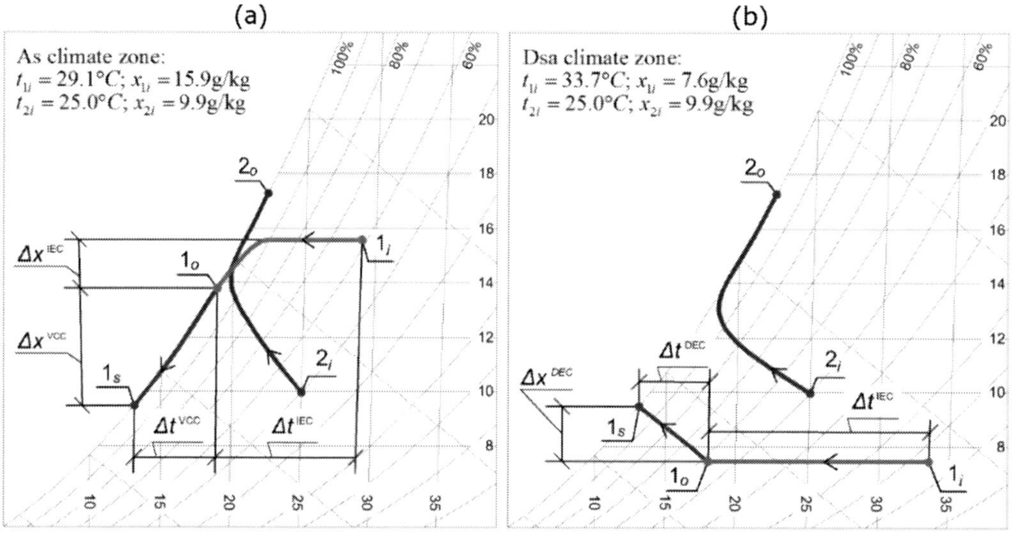

Fig. 3. Psychrometric chart for the AC system with the counter-flow IEC heat recovery unit followed by: (a) vapor-compression cooling unit; (b) direct evaporative cooling unit (1 – product channel; 2 – working channel; *i* – inlet parameters; *o* – outlet parameters; *s* – related to supply airflow).

In this study, comfort parameters in the conditioned space in summer period cover the range of temperature between 23 and 26°C and relative humidity within the range of 40 to 60%. Based on this assumption, it was stated that if the airflow that leaves IEC exchanger has the humidity ratio higher than 11.0g/kg (all tropical and some of the temperate climates), the VCC unit is required. When the humidity ratio is lower than 7.5g/kg (arid and some of the temperate climates) then the AC system should be equipped with DEC unit that follows the IEC exchanger.

With the DEC unit applied in the AC system, achieving needed parameters of the supply airflow may require partial reduction of the cooling capacity of the IEC exchanger to meet the requirements of the adiabatic process in the DEC unit. It was also concluded that when the airflow at the outlet of the IEC exchanger has the humidity ratio within the range of 7.5 and 11.0g/kg (most of the continental and some of the temperate climates) it is possible to apply either the system with the VCC unit or with the DEC unit. In these cases the cooling capacity of the exchanger does not have to be adjust to the requirement of the DEC unit.

4 Summary and conclusions

In this paper, performance studies of the counter- and cross-flow IEC heat and mass exchanger applied as a heat recovery device in the AC system are presented. Studies are carried out for inlet product airflow parameters characteristic for different climate conditions. Numerical simulations are performed with validated analytical ε–NTU model.

It was concluded that the IEC exchanger in both, the counter- and cross-flow configuration has a high potential to be employed as a heat recovery device under wide range of climate parameters. While the counter-flow configuration of the IEC exchanger is characterized by higher performance, its cost significantly exceeds the price of the cross-flow configuration. It is also stated that depending on the climate zone, the AC system consisted of IEC exchanger and VCC unit is suitable for tropical, temperate and some of the continental climates while the AC system consisted of IEC exchanger and DEC unit is suitable for arid and continental climate zones.

One of the co-authors, Demis Pandelidis, received financial support for his research from resources for scientific work for years 2016–2019 from Polish Ministry of Science and High Education (program "Iuventus Plus"), project number IP2015 058274 and from Polish National Agency for Academic Exchange (Bekker program), agreement number PPN/BEK/2018/1/00347/U/00001.

References

1. X. Cao, X. Dai, J. Liu, Energ. Buildings **128**, 198–213 (2016)
2. P. M.Cuce, S.Riffat, Renew. Sust. Energ. Rev. **54**, 1240–1249 (2016)
3. B. Porumb, P. Ungureşan, L. F. Tutunaru, A. Şerban, M. Bălan, Energy Procedia **85**, 452–460 (2016)
4. Y. Chen, Y.Luo, H. Yang, Energ. Buildings **108**, 387–400 (2015)
5. Cui, K.J. Chua, M.R. Islam, K.C. Ng, Energ. Convers. Manage. **102**, 140–150 (2015)
6. W.-Y. Li, Y.-C. Li, L.-Y. Zeng, J. Lu, Int. J. Heat Mass Tran. **124**, 1245–1261 (2018)
7. D.Meng, J.Lv, Y. Chen, H. Li, X. Ma, App. Therm. Eng. **145**, 165–173 (2018)
8. D.Pandelidis, A.Cichoń, A.Pacak, S.Anisimov, P.Drąg, Int. J. Heat Mass Tran. **130**, 757–777 (2019)
9. B. Riangvilaikul, S. Kumar, Energ. Buildings **42**, 5, 637–644 (2010)
10. American Society of Heating, Refrigerating and Air-conditioning Engineers Inc., ASHRAE Handbook-Fundamentals, Atlanta, GA (2017)
11. M.C. Peel, B.L. Finlayson, T.A. Mcmahon, Hydrol. Earth Syst. Sci. Discuss **4**, 439–473 (2007)

Daily variability of air pollution in locations of different population density

Robert Cichowicz[1,*], and *Artur* Stelęgowski[1]

[1]Lodz University of Technology, Faculty of Civil Engineering, Architecture and Environmental Engineering, al. Politechniki 6, 90-924 Lodz, Poland

Abstract. The air quality levels vary during a day, especially in inhabited areas. Therefore, it seems reasonable to observe and analyze the occurrence of daily maximum and minimum level of air pollution. In this article, data obtained from automatic air quality monitoring stations located in 5 large, 5 small and medium cities and 5 villages in Poland was analyzed in 2012–2016. Those locations vary, inter alia, depending on number of inhabitants and population density, and for this reason also due to the presence of air contaminants. As an indicator of daily variability air pollution it was determined the ratio of maximum to minimum concentrations of selected air pollutants (NO_2 and NO_x, and O_3, SO_2, CO, PM_{10} and $PM_{2.5}$, and benzene) in urban and agricultural areas. In winter, the daily changes were bigger in cities than in villages. While in summer, the level of daily variability was similar, irrespective of size of the settlement unit. The biggest daily changes concerned nitrogen oxides, the lowest – sulfur dioxide and dusts.

1 Introduction

Air pollution affects, in different ways, the state of the environment. This impacts the state of fauna and flora, and pose a threat to human health in exposed areas [1–3]. Nitrogen oxides, sulfur oxides, tropospheric ozone and dusts are the so-called "classic air pollutants". Primarily, they have a harmful effect on the human respiratory system [4–6]. Other air pollutants can be toxic, such as carbon monoxide, which achieves low concentrations in atmospheric air, or carcinogenic, such as benzene [7–10]. Today, steps are being taken to significantly reduce the air pollutant emissions [11], and also to better understand the processes of their formation and dispersion in the air [12–16]. Anthropogenic sources of air pollution in Europe are mainly: intensity of road transport, activity of the power sector, and other types of burning fossil fuels [17].

Due to variable activity of emission sources, and changes in meteorological conditions, the concentration of air pollutants varies during a day. However, the nature of changes depends on the density of emission sources, and hence indirectly on the density of population in the considered areas. In large cities there are usually higher concentrations of air pollutants than in less urbanized areas, and unfavorable air quality conditions deteriorate

* Corresponding author: robert.cichowicz@p.lodz.pl

health of its citizens [18]. However, the level of air pollution vary from hour to hour. Therefore, it was decided to carry out an analysis of daily variability in concentrations of air pollutants, in fifteen selected areas of different settlement size and density, in Poland, Europe.

2 Materials and methods

The analysis was based on the measurements of average hourly concentrations from automatic air monitoring stations. The monitoring data came from the years 2012–2016, and the monitoring stations belonged to the Provincial Inspectorates of Environmental Protection in Poland. For the purpose of the analysis, fifteen different locations were selected in central Poland (Table 1). From those were distinguished cities, towns and villages. Settlements with over 50,000 inhabitants and density of at least 1,500 inhabitants/km^2 were treated as cities [19]. Towns (small and medium sized towns) were defined as settlements with population of 25,000–100,000 inhabitants and up to 1500 inhabitants/km^2, or 5,000–25,000 inhabitants and over 300 inhabitants/km^2 [20]. Villages were defined as settlements that did not meet population criteria for very small towns. This means that villages had up to 5,000 inhabitants, did not have city rights, and had up to 300 inhabitants/km^2.

Table 1. Selected locations in Poland.

Settlement type	Name	Population, (inh)	Density, (inh/km^2)
City	Wrocław	639,000	2181
	Poznań	539,000	2057
	Łódź	690,00	2354
	Warsaw	1,765,000	3412
	Lublin	340,000	2310
Town	Kłodzko	27,000	1098
	Konin	75,000	919
	Piotrków Tryb.	74,000	1114
	Piastów	23,000	3977
	Biała Podlaska	57,000	1159
Village	Osieczów	345	< 300
	Piaski	124	< 300
	Gajew	114	< 300
	Belsk Duży	790	< 300
	Obrocz	525	< 300

Air quality monitoring stations were located in areas of different characteristics, inter alia, with different assumed density of pollution emission sources and population (Table 2). The air pollution (considering PM and NO_x) is usually the highest in the high-traffic sites, lower in the highly-populated areas, and the lowest in the rural areas [21–23]. This is mainly due to conjestion of non-industrial and traffic emissions. In Poland, which has a coal-based economy, these pollution sources contribute to 30% of SO_2, 42% of NO_x, 55% of $PM_{2.5}$, and 81% of CO emissions [24]. Power industry contributes to 45% of SO_2, 25% of NO_x, 9% of $PM_{2.5}$, and 2% of CO emissions. However, the pollution from power generation usually wide-spreads and dillutes across large area.

From analyzed locations in cities, there were distuinguished areas of: urban traffic, city background, and suburban background. In towns – town background areas, and in villages – rural background areas. However, the urban traffic and suburban background areas were defined only in the following large cities: Wrocław, Łódź, Warsaw. The UT areas were affected mostly by road traffic emissions. In CB, SB and TB areas occurred mostly traffic and non-industrial emission sources. And in RB areas the air pollution came mostly non-industrial combustion and agriculture. Also, in each location there was some background pollution, originated, inter alia, from power industry.

Table 2. Assumed population density in the areas of air quality monitoring stations.

Station type	Assumed density of air pollution emissions	Assumed population density
Urban traffic (UT)	Very high	Very high
City background (CB)	High	High
Suburban background (SB)	Medium	Medium
Town background (TB)	Low	Low
Rural background (RB)	Very low	Very low

To evaluate the occurrence / intensity of daily variability in the concentration of pollutants, the Maximum to Minimum Concentrations Ratio (*MMR*) was determined. *MMR* was calculated for the analyzed measurement stations as the maximum average hourly concentration of a given pollutant on a "typical" day divided by the minimum concentration value on a "typical" day. The "typical" day included averaged hourly values of concentrations from the analyzed periods of winter (from October to March), and summer (from May to August). Thus, the *MMR* determined the level of changes between extreme values throughout a day. Additionally, for the selected database, the coefficient of variation of concentrations during the day was determined. It was calculated as the standard deviation of hourly concentrations of pollutants, divided by the arithmetic mean of concentrations. Therefore, coefficient of variation determined the level of hourly changes in pollutant concentrations relative to the daily average.

3 Results and discussions

Daily variability in concentrations of analyzed pollutants was expressed by means of the *MMR* coefficient. The highest daily variability in winter periods occurred in the areas characterized by the highest population density (Table 3), i.e. in cities in urban traffic areas (UT). And the lowest occurred in areas with the lowest population density, i.e. in rural areas (RB). This could lead to conclusion that daily variability of air pollution in winter depend on the population density in the analyzed area. However, the highest *MMR* values were observed for NO_2 and NO_x, and the smallest for $PM_{2.5}$. Level of nitrogen oxides increased and decreased significantly ($MMR > 2.0$) during a day. However, in the case of dusts, the ratio between average extreme daily values was only 1.28–1.71. This indicated the occurrence of periods with significantly elevated and reduced levels of nitrogen oxides, while the level of dust remained similar throughout a day. In summer periods the daily variability of air pollution seemed to be independent from the population density of the analyzed areas (Table 4). For example, changes in the concentration of nitrogen oxides and ozone were the highest in towns, while the changes in CO and PM_{10} concentrations were the highest in cities. However, the lowest *MMR* values occurred, depending on the pollutant type, in urban areas ($PM_{2.5}$ and benzene in UT, and NO_2 and O_3 in CB), towns (SO_2 in TB) and rural areas (NO_x, CO and PM_{10} in RB). Interestingly, daily changes in nitrogen dioxide concentration were greater in suburban areas (SB) than in city centers (UT and CB). The highest daily variability of air pollution in summer, i.e. the highest *MMR* values, occurred in all areas for NO_2, NO_x and O_3, as well as for benzene in SB areas. However, the smallest daily variability of air pollution was for dusts (PM_{10} and $PM_{2.5}$), for which the *MMR* ratio was 1.31–1.70. This meant that the level of dust remained similar throughout a day in summer, like in winter.

Table 3. Average hourly Maximum to Minimum Concentration Ratio (*MMR*) for the analyzed areas in winter (dark grey = the highest values, light grey = the lowest values).

Air pollutant	*MMR* in winter period				
	UT	**CB**	**SB**	**TB**	**RB**
NO_2	2.59	2.23	1.87	1.90	1.46
NO_x	3.82	2.98	1.96	2.11	1.42
O_3	n/a	1.81	1.85	1.76	1.60
SO_2	1.58	1.52	1.24	1.57	1.36
CO	1.86	1.70	1.48	1.67	1.32
PM_{10}	1.71	1.61	1.40	1.63	1.29
$PM_{2.5}$	1.28	1.59	1.41	1.63	n/a
C_6H_6	1.71	1.97	1.86	n/a	n/a

Table 4. Average hourly Maximum to Minimum Concentration Ratio (*MMR*) for the analyzed areas in summer (dark grey = the highest values, light grey = the lowest values).

Air pollutant	*MMR* in summer period				
	UT	**CB**	**SB**	**TB**	**RB**
NO_2	2.59	2.85	3.29	3.62	3.11
NO_x	3.13	3.14	2.93	3.24	2.76
O_3	n/a	2.60	2.64	3.15	2.97
SO_2	1.83	1.85	1.85	1.63	1.72
CO	2.18	1.67	1.65	1.51	1.17
PM_{10}	1.57	1.47	1.52	1.49	1.43
$PM_{2.5}$	1.31	1.68	1.63	1.70	n/a
C_6H_6	1.97	2.34	3.87	n/a	n/a

In addition, the *MMR* calculated for selected locations were compared with data derived from the literature (Fig. 1), and a wide variety of results was visible. The most striking is the difference between the model, which used temporal profiles of emission factors of selected economy sectors [25], and the results from experimental measurements in European cities [26–28]. This could be due to the large size of the calculation grid in the model, while experimental measurements were made at specific points. The results of experimental measurements and numerical calculations differ especially in the case of NO_2 concentrations. However the biggest daily fluctuations occurred for O_3 concentrations. Research by Zheng et al. in Pearl River Delta in China showed that daily changes in O_3 concentrations were significantly higher in urban areas (CB) than in suburban (SB) and rural areas (RB) [29]. However, from the analysis performed in Poland, it was clear that O_3 differed only slightly between all locations (UT, CB, SB, TB, RB). Also, a coefficient of variation [31] of air pollutant concentrations was determined for the analyzed locations. The coefficient was calculated as the standard deviation of hourly concentrations of pollutants, divided by the arithmetic mean of concentrations. In this case, the greatest variation in pollution levels occurred mostly in the summer (Table 5). Coefficient of variations for NO_2 and benzene generally decreased with increasing assumed density of emissions. But increased for CO and PM_{10}. However, in the case of other pollutants, no clear dependencies were observed. Therefore it could be concluded that the deviation of NO_x, O_3, SO_2 and $PM_{2.5}$ levels from the daily average does not depend on the assumed density of emissions.

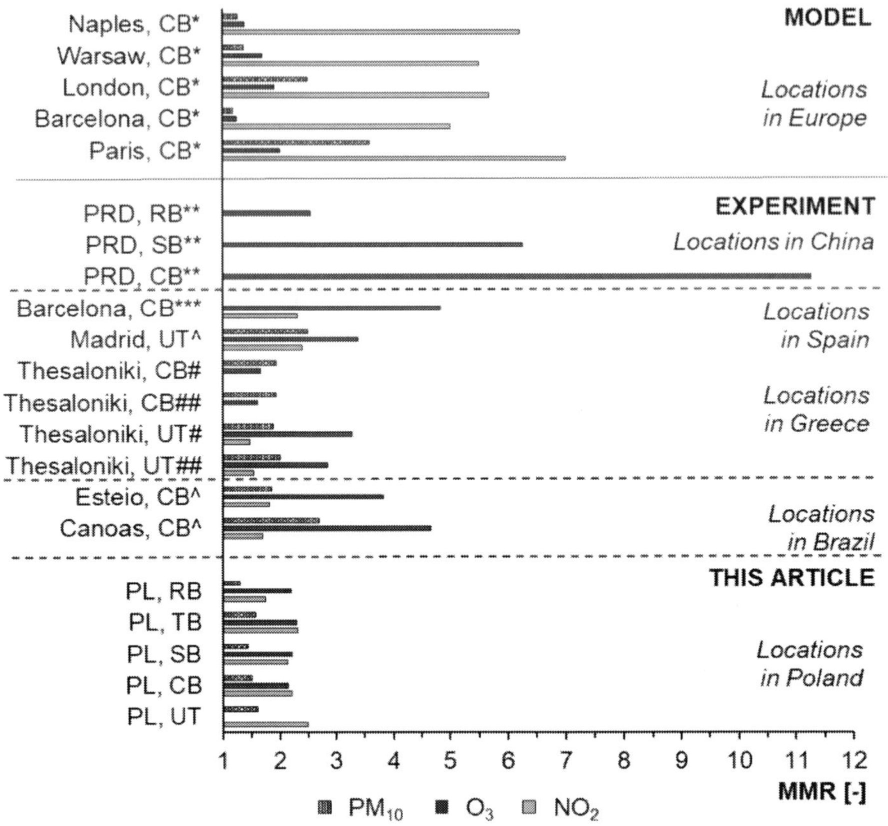

Fig. 1. Average dialy variability of air pollutants. Own study based on : *Menut et al. [25], **Zheng et al. [29], ***Perez et al., summer period [26], #Vouitsis et al., summer period [27], ##Vouitsis et al., winter period [27], ^Moreno et al., winter period [28], ^^Agudelo-Castaneda and Teixeira [30].

Table 5. Coefficient of variation (%) of average hourly concentrations in a day, calculated as the standard deviation divided by the arithmetic mean of concentrations (dark grey = the highest values, light grey = the lowest values).

Air pollutant	Coefficient of variation (%) Winter periods					Coefficient of variation (%) Summer periods				
	UT	CB	SB	TB	RB	UT	CB	SB	TB	RB
NO_2	37	38	37	39	39	41	53	55	60	62
NO_x	51	52	45	50	39	50	61	57	61	58
O_3	n/a	43	44	43	35	n/a	39	40	44	43
SO_2	n/a	37	39	39	40	n/a	51	48	42	43
CO	32	32	27	32	26	33	29	28	28	19
PM_{10}	39	36	35	34	36	37	36	35	36	33
$PM_{2.5}$	32	36	33	35	n/a	29	37	34	30	n/a
benzene	36	44	57	n/a	n/a	45	62	106	n/a	n/a

4 Conclusions

In the winter, the character of changes of air pollution depended on the analyzed area. Level of this changes (*MMR*) increased along with assumed emission density. The exceptions were concentrations of finer dust particles ($PM_{2.5}$), which showed the opposite pattern. However, in summer there was no clear dependency between daily variability of concentrations and the assumed emission density. In winter, when the pollution level was usually high, the daily variability was greater in more congested areas. But in summer, when air pollution was usually lower, there was no such dependency. However, greater variability of pollutant concentrations occurred in summer than in winter. This could be related to the lower "background" level of air pollution in the summer. Also, the highest daily variability concerned nitrogen oxides, and the smallest concerned dusts. This indicated the occurrence of significant differences in air quality in the "most" and "least favorable" periods during a day. Which is important in terms of protecting the health of people living in the areas of cities, towns and villages. Also, the analysis of literature [21–26] showed similar, or larger, daily variability of air pollution in urban areas in other countries around the world. However, when considering the coefficient of variation, the daily changes in air quality were not unequivocally dependent on the density of the analyzed cities. In addition, this was comparable in winter and in summer. However, variability in concentrations of carbon monoxide and particulate matter (PM_{10}) were decreasing along with the decrease in density of emissions (and also population). The coefficient of variation was the highest for nitrogen oxides and for benzene, which indicated large deviations of these values from the daily average, and the lowest coefficient was for CO, PM_{10} and $PM_{2.5}$. Nevertheless, in the winter generally there was the highest daily variability of air pollution in the most congested areas (cities), and the lowest in the least populated places (villages). However, in the summer, the daily variability was different in all locations. Knowledge about the daily changes in concentrations of air pollutants in areas of various characteristics, may help in better planning of air protection strategies in modern cities, and thus may positively affect the protection of health of people living in those areas.

References

1. M. Baciak, K. Warmiński, A. Bęś, Forest Research Papers **76**, 401–409 (2015)
2. R. Afroz, M. N. Hassan, N. A. Ibrahim, Environmental Research **92**, 71–77 (2003)
3. B. Brunekreef, S. T. Holtage, Lancet **360**, 1233–1242 (2002)
4. World Health Organisation, *Air Quality Quidelines for Europe* (WHO Regional Publications, Copenhagen, 2000)
5. World Health Organisation, *Air Quality Quidelines for Europe – Global Update 2005*. (WHO Regional Publications, Copenhagen, 2005)
6. R. Bascom, P. A. Bromberg, D. A. Costa, R. Devlin, D. W. Dockery, M. W. Frampton, W. Lambert, J. M. Samet, F. E. Speizer, M. Utell, American Journal of Respiratory and Critical Care Medicine **153**, 3–50 (1996)
7. C. Reboul, J. Boissière, L. André, G. Meyer, P. Bideaux, G. Fouret, C. Feillet-Coudray, P. Obert, A. Lacampagne, J. Thireau, O. Cazorla, S. Richard, Scientific Reports **7**, 39715 (2017)
8. H. Lawin, L. Ayi Fanou, V. Hinson, J. Wanjiku, N. K. Ukwaja, S. B. Gordon, B. Fayomi, J. R. Balmes, P. Houngbegnon, E. Avokpaho, A. Sanni, BMC Public Health **17**, 320 (2017)

9. D. F. Costa, M. Goldbaum, Ciencia & Saude Coletiva **22**, 2681–2692 (2017)
10. C. B. B. Guerreiro, V. Foltescu, F. de Leeuw, Atmos. Environ. **98**, 376–384 (2014)
11. Directive 2008/50/EC of the European Parliament and of the Council of 21 May 2008, O*n Ambient Air Quality and Cleaner Air for Europe*, OJ L 152, 11.6.2008, 1–44.
12. R. Cichowicz, E3S Web Conf. **28**, 01007 (2018)
13. R. Cichowicz, A. Stelęgowski, Chemical Papers (2018)
14. R. Cichowicz, G. Wielgosiński, A. Targaszewska, Ecological Chemistry and Engineering S **23**, 49–60 (2016)
15. R. Cichowicz, G. Wielgosiński, W. Fetter, Environ. Monit. Assess. **12**, 189–605 (2017)
16. G. Wielgosiński, J. Czerwińska, O. Namiecińska, R. Cichowicz, E3S Web Conf. **28**, 01039 (2018)
17. https://www.eea.europa.eu/publications/2599XXX/page010.html
18. K. N. Priftis, M. B. Anthracopoulos, A. G. Paliatsos, G. Tzavelas, A. Nikolaou-Papanagiotou, P. Douridas, P. Nicolaidou, Mantzouranis E, Respiratory Medicine **101**, 98–106 (2007)
19. L. Dijkstra, H. Poelman, *Cities in Europe – The new OECD-EU definition* (European Comission, Brussels, 2012)
20. L. Servillo, R. Atkinson, I. Smith, A. Russo, L. Sykora, C. Demaziere, A. Hamdouch, *TOWN, small and medium sized towns in their functional territorial context – Final Report* (Espon, Luxembourg, 2014)
21. J. Putaud, F. Raes, R. van Dingenen, E. Brüggemann, M. Facchini, S. Decesari, A. Wiedensohler, Atmos. Environ. **38**, 2579–95 (2004)
22. A. Hagenbjörk, E. Malmqvist, K. Mattisson, N.J. Sommar, L. Modig, Environ Monit. Assess. **189**, 161 (2017)
23. I. Vouitsis, S. Amanatidis, L. Ntziachristos, A. Kelessis, M. Petrakakis, I. Stamos, Z. Samaras, Atmos. Environ. **122**, 577–587 (2015)
24. B. Dębski, A. Olecka, K. Bebkiewicz, I. Kargulewicz, J. Rutkowski, D. Zasina, M. Zimakowska-Laskowska, M. Żaczek, *National balance of SO_2, NO_x, CO, NH_3, NMLZO, dust, heavy metals and TZO in the SNAP and NFR classification system. Basic report* (Warsaw, The National Centre for Emissions Management, 2015)
25. L. Menut, A. Goussebaile, B. Bessagnet, D. Khvorostiyanov, A. Ung, Atmos. Environ. **49**, 233–244 (2012)
26. N. Pérez, J. Pey, M. Cusack, C. Reche, X. Querol, A. Alastuey, M. Viana, Aerosol Science and Technology **44**, 487–499 (2010)
27. I. Vouitsis, S. Amanatidis, L. Ntziachristos, A. Kelessis, M. Petrakakis, I. Stamos, E. Mitsakis, Z. Samaras, Atmos. Environ. **122**, 577–587 (2015)
28. T. Moreno, A. Karanasiou, F. Amato, F. Lucarelli, S. Nava, G. Calzolai, M. Chiari, E. Coz, B. Artinano, J. Lumbreras, R. Borge, C. Linares, A. Alastuey, X. Querol, W. Gibbons, Atmos. Environ. **68**, 33–44 (2013)
29. J. Zheng, L. Zhong, T. Wang, P. K. K. Louie, Z. Li, Atmos. Environ. **44**, 814–823 (2010)
30. D. M. Agudelo-Castaneda, E. C. Teixeira, F. N. Pereira, Atmos. Pollut. Res. **5**, 411–420 (2014)
31. C. E. Brown, *Coefficient of Variation*, In: *Applied Multivariate Statistics in Geohydrology and Related Sciences* (Springer, Berlin-Heidelberg, 1998)

Experimental analysis of waste tyres as a sustainable source of energy

Dina Czajczyńska[1,*], *Krzysztof M.* Czajka[2], *Renata* Krzyżyńska[1], and *Hussam* Jouhara[3]

[1]Wroclaw University of Science and Technology, Faculty of Environmental Engineering, Wyb. Wyspianskiego 27, 50-370 Wroclaw, Poland
[2]Wroclaw University of Science and Technology, Faculty of Mechanical and Power Engineering, Wyb. Wyspianskiego 27, 50-370 Wroclaw, Poland
[3]College of Engineering, Design and Physical Sciences, Brunel University London, Uxbridge, Middlesex UB8 3PH, UK

Abstract. Nowadays, the stability of energy supply that additionally should be sustainable is one of the most important global issues. Thus, many new potential energy sources are being investigated. Since automobile industry is growing, a huge amount of waste tyres (WT) occur each year. Pyrolysis of scrap tyres can be considered as a sustainable way to recover significant amounts of energy as well as the valuable materials. Potential of waste tyres in the energy sector is studied in this work. Proximate, ultimate and thermogravimetric (TG) analysis of mechanically grounded WT sample was done. Waste tyres feature high heating value combined with carbon content as high as 87.90 % (on dry ash-free basis). Additionally, TG analysis allows to choose optimal operating temperature for pyrolysis process which is between 350 and 500°C. However, the sulphur content is also relatively high – around 2 wt.% - and it is the most important challenge for utilizing this waste in a thermochemical way.

1 Introduction

It is evident that the number of vehicles is growing each year. According to European Tyre & Rubber Manufacturers' Association (ETRMA) there was 1.022 billion passenger cars in 2016 and production of tyres had reached 4.94 million tonnes. Moreover, it is estimated that the number of passenger cars will exceed 1.3 billion worldwide until 2024 [1]. Unfortunately for waste managers but luckily for drivers, all produced tyres eventually will become a difficult-to-utilize waste since they are generally very durable. There are several widely available methods of utilization of Waste Tyres (WT). For example, in Poland majority of them is combusted in cement plant where the construction of kiln and temperature approximately 2000°C ensures complete and safe disposal [2, 3]. Another popular way of WT processing is mechanical shredding combined with further application in asphalt, sport fields or wipers production.

* Corresponding author: dina.czajczynska@pwr.edu.pl

Among sustainable energy sources, waste utilization is one of the most prominent research topics of recent years. Scrap tyres are often considered as a raw material for thermochemical processes different from combustion in order to recover energy as well as valuable chemicals. This can be achieved by pyrolysis which has been known since ancient times. It is a process that causes thermal decomposition of almost any kind of organic matter under high temperature and different pressure conditions but in the absence of oxygen. For instance, Choi et al. [4] obtained pyrolysis oil from waste tyres which contains sulphur content of less than 1 wt.% together with high concentration of precious limonene and xylenes. Additionally, char treated with acid and then activated gained specific surface area as high as 437 m^2/g [4]. It can be concluded, that good quality product with industrial potential were produced. Literature review concerning production of pyrolytic oils form waste tyres was made by Hita et al. [5]. In turn, Antoniou et al. [6] critically analysed activated carbon production form char. It is important to note that when comparing with traditional waste management of tyres such as combustion, pyrolysis provides lower emissions of pollutants such as polycyclic aromatic hydrocarbons, volatile organic compounds and CO_2, thus it can be considered as an environmentally respectable solution [7–9]. On the other hand, a large variety of equipment possible to use in waste pyrolysis process may confused researchers and potential investors [10]. Furthermore, multiple process adjustment options also need to be studied. In this paper several analysis was investigated in order to describe and evaluate the waste tyres as a pyrolysis substrate. Many researchers investigated pyrolysis of tyre rubber, but limited information is available about pyrolysis of whole tyre samples. Thus obtained results will provide very useful information for possible application of the process in an industrial scale where probably whole WT will be utilized.

Firstly, the most basics properties that must be identified before planning waste pyrolysis are fuel parameters. Among them proximate and ultimate analysis as well as heating value may be listed. It is reported that they vary significantly depending on waste type. Some of them are presented in the Table 1. Additionally, fuel ratio (FR) defined as the content of fixed carbon (FC) divided by the content of volatile matter (V) may be investigated to highlight the quality of feedstock for thermochemical processes. The ratio of FC to V provides an indication of the ease of solid fuel ignition and subsequently gasification, or oxidation, depending on how the fuel is to be converted.

Table 1. Fuel parameters of different feedstocks for thermochemical processes.

Material	**Proximate analysis,** *wt. %*				**Ultimate analysis,** *wt. %*					**Heating value,** *MJ/kg*	**Ref.**
	M	**V**	**FC**	**A**	**C**	**H**	**O**	**N**	**S**		
MSW	5.1	59.3	8.4	27.2	48.6	6.0	12.5	3.9	0.3	18.6^l	[11]
WEEE	-	25.5	21.3	53.2	19.8	1.8	77.6	0.7	0.2	-	[12]
Cardboard	5.4	79.1	10.0	10.9	38.0	4.6	45.6	0.3	0.3	13.7^l	[13]
Spruce Wood	8.4	83.5	16.1	0.4	48.3	6.3	44.6	0.3		19.5^h	[14]
Biomass	14.7	64.4	16.0	4.9	51.1	6.2	41.4	1.1	0.2	18.0^h	[15]
Coal	5.5	30.8	43.9	19.8	78.2	5.2	13.6	1.3	1.7	25.0^h	[15]
PET	0.6	86.8	13.2	0.0	62.5	4.2	33.30	0.0	0.0	22.1^l	[13]

M – moisture, A – ash, MSW – municipal solid waste, WEEE – Waste of Electrical and Electronic Equipment, ^{l} – lower, ^{h} – higher

Secondly, also kinetics of the process, determined with the use of the thermogravimetry (TG) coupled with differential scanning calorimetry (DSC), are important for improvement of any process where waste tyres are a part of the chemical system and to determine how the waste decomposition takes place [11]. The knowledge about fuel behaviour is essential for good prediction of the ignition and burnout, the temperature profile, the product distribution of tar, char and evolved gases and the overall process efficiency.

2 Methodology

2.1 Samples preparation

Samples of waste tyres was provided by recycling company "Recykl" located in Greater Poland, Poland. Three different-size fractions were obtained by grinding whole tyres in special mills. The smallest fraction with particle size about 0.5 cm was chosen for further pyrolysis experiments. It was reported in the literature that using of particle sizes below 5 mm does not influence on the process rate and allows to avoid temperature gradient leading to heat transfer limitations [16]. Because the whole tyre was shredded, the sample contains rubber matter as well as fibre and steel pieces.

The sample was placed indoor for a few days to reduce the moisture since the grinded tyres were kept outdoor at the company's premises, but no additional drying was applied. However, there was a necessity to carryout extra mechanical fragmentation to investigate the potential of waste tyres as a sustainable source of energy by determination of fuel parameters and TG analysis. Analytical equipment need a feedstock with particle size of up to 0.25 mm which was achieved by using cutting mill RETSH SM-2000.

2.2 Determination of fuel parameters

The proximate analysis was performed according to following standards:

1. *PN-G-04511:1980* – determination of M content;
2. *PN-G-04512:1980/Az1:2002* – determination of A content;
3. *PN-G-04516:1998* – determination of volatile matter content.

The fixed carbon was accounted from a mass balance to 100 wt.% based on M, V, A share.
The ultimate analysis was done in accordance with:

1. *PN-G-04571* – determination of carbon, nitrogen and hydrogen content by using automatic analyser Leco TruSpec CHNS, and
2. *PN-ISO 35* – determination of total sulphur content.

The share of oxygen was defined indirectly by difference.
Additionally, the heating value was investigated according to *PN-C-04375-2:2013-07*. The higher heating value (HHV) was measured in calorimeter IKA C2000 and the lower heating value (LHV) was accounted based on Equation (1):

$$LHV = HHV - 24{,}42 \cdot (M + 8.9 \cdot H) \qquad (1)$$

2.3 Thermogravimetric analysis

The analysis was performed by using thermogravimetry (TG) coupled with differential scanning calorimetry (DSC). The simultaneous TG/DSC measurement enables determination of a mass loss and a heat flow as a function of temperature and time. Experimental tests were carried out by means of Setaram Setsys Evolution 18 thermogravimeter. Figure 1 shows the TGA/DSC used for the analysis.

 https://doi.org/10.1051/e3sconf/201910000012

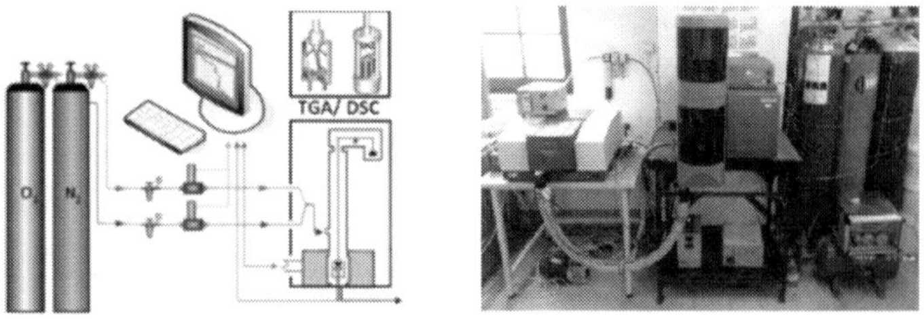

Fig. 1. The scheme and the photo of the TGA/DSC stand.

The thermogravimetric analyser comes equipped with a cylindrical, electrically heated furnace, offering heating rates up to about 100°C/min and a maximum operation temperature up to 1500°C. Inside the furnace, a microbalance is located with a sample pan connected to programmable control temperature. A precise control of the atmosphere ensures a gas control panel equipped with mass flow controllers.

During operation, the thermogravimeter measures the change of sample mass and the heat flow over time and temperature. For the purpose of this study, an isothermal approach is applied, due to inability to reach the high temperature without release of the most reactive volatiles [17]. The experimental procedure consist of heating about 5 mg samples with a flow of 32 ml/min of a carrier gas. Samples are held at 120°C for 50 min in nitrogen, to remove moisture. Afterwards, to simulate pyrolysis process, samples are heated at a rate of 20°C/min, from 120 to 900°C.

In order to represent the physical phenomenon properly under the investigated conditions, two models are applied, namely the reaction rate constant model, which assumes that the fuel is decomposed with the reaction rate k (model B) and the multi-component model, which assumes that the fuel is decomposed through several parallel first order reactions (model A).

3 Results and discussion

3.1 Determination of fuel parameters

Waste tyres can be considered as a fuel with very good heating parameters. In contrast with other types of waste, such as biomass or MSW, waste tyres have a noticeably high heating value in-comparison with traditional solid bitumen. The lowest heating value for raw sample of waste tyres is 31.88 MJ/kg – two times higher than that for firewood (see Table 1.). Additionally, the higher and the lower heating value for dry ash-free basis was 36.83 MJ/kg, and 35.31, respectively.

Moreover, the composition of waste tyres also makes them a suitable material for pyrolysis. The results of the tyres proximate and ultimate analysis are listed in Table 2. For dry ash-free tyre sample, volatiles and fixed carbon content are 73.31 and 26.69 wt.%, respectively. Fuel ratio for waste tyres is 0.36 which is below the fuel ratio range of 1 to 10 that is normally used for commercial coal directed for combustion. This means that tyres become volatile much easier than traditional solid bitumen and present perfect feedstock for pyrolysis or gasification. Contrary to biomass, which is considered as a renewable energy source too, moisture content in analysed sample was only 2.3 % thus tyres do not need high energy-consuming drying devices before thermochemical processes, such as pyrolysis, to take place.

Table 2. Proximate and ultimate analysis of waste tyres.

state of the sample	Proximate analysis				Ultimate analysis				
	M	A	V	FC	C	H	N	S	O
	%				%				
raw	2.30	7.26	66.31	24.14	77.67	6.15	0.59	1.99	4.04
dry		7.43	67.87	24.70	79.50	6.29	0.61	2.04	4.13
dry ash-free			73.31	26.69	87.90	6.95	0.67	2.26	2.22

Furthermore, it is clearly indicated in the Table 2. that tyres consist mainly of carbon, which is a component of rubber matter as well as carbon black which both are used for tyres production. The percentage of carbon in dry free-ash basis exceed 85%. Additionally, relatively low content of nitrogen and oxygen allows to avoid high concentration of oxides in evolved pyrolysis gases. Significant concentration of carbon monoxides have been reported in literature [18, 19]. However, sulphur content in tyres is approximately 2% since it is used as a vulcanization agent during their production process. The presence of sulphur in feedstock for pyrolysis is unfavourable, because under high temperature and oxygen-free condition it is converted to H_2S and thiols which have strong unpleasant odours and are toxic. Their removal from pyrolytic gas is one of the biggest challenges for researchers working on the commercialisation of the process [19, 20].

The obtained results from the waste tyre analysis are comparable with literature data, thus it is possible to conclude that the tyre sample is representative and can be used for further experiments. For instance, Kandasamy & Gökalp [21] investigated heating value of passenger and truck tyres achieving results of 28.9 and 30 MJ/kg, respectively. These are slightly lower than that presented in this paper. There was a slight higher amount of water noticed in the sample from "Recykl", but the fraction of volatiles were higher. Thus, the heating value also was higher. In turn, ultimate analysis showed the following percentage composition: C – 79.1, H – 6.7, N – 0.6, S – 2.2, O – 2.8 and it was almost the same as presented here. On the other hand, Gonzalez *et al.* [22] obtained higher heating value of analysed sample, namely 36.3 MJ/kg. It contained: 0.7 wt.% of moisture, 8.0 wt.% of ash, 61.9 wt.% of volatiles and 29.5 wt% of fixed carbon. Elemental composition was as follows: C – 86.7 %, H – 8.1, N – 0.4%, S – 1.4%, O – 1.3%, and it was a bit diverse than presented here. Some differences depending on type of tyres may appear and they were reported for passenger cars, trucks, motorbike [23] or bike tyres [24]. Obviously, also each tyre producer has its own unique composition [25], but it is still very comparable. Generally, it can be assumed that tyres compositions are fairly similar and they can be considered as a homogenous waste as well as source of sustainable energy.

3.2 Thermogravimetric analysis

Figure 2 depicts a heating curve and an exemplary loss of sample mass as a function of time. The obtained results indicate that the most rapid evolution of volatiles was observed between 55 and 64 minute of the experiment, which correspond to the temperature range between 350 and 500°C (see Fig. 3.). It was proven in further experiments that evolution of gases and vapours were most intensive in this temperature range.

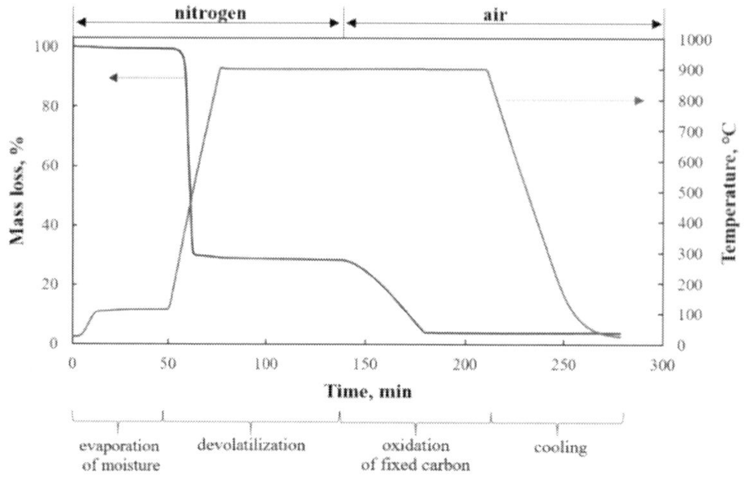

Fig. 2. Heating curve and exemplary loss of sample mass as a function of time.

The derivation of mass loss over time, known as a reactivity, indicates at least two-stage character of pyrolysis, where each stage can be attributed to the decomposition of different fractions of the investigated fuel. In order to calculate the kinetic parameters describing the evolution of each fraction, the reactivity curve was deconvoluted. Obtained curves are depicted in Fig. 4. Performed deconvolution of TG signal indicates the existence of at least one more stage (1'), confirming observations made based on the DSC measurement depicted in Figure 5. According to the DSC analysis, peak 1' may be attributed to the evolution of highly reactive volatiles, being the early stage of pyrolysis (ca. 350°C). Additionally, from Figure 5 a highly exothermic reaction (peak 3) not associated with the loss of mass is observed. This can be attributed to the change of carbon structure or ash transformation.

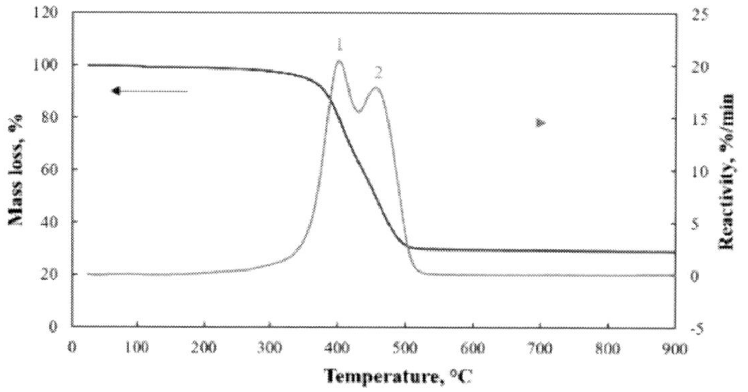

Fig. 3. Mass loss and reactivity curves presented as a function of time.

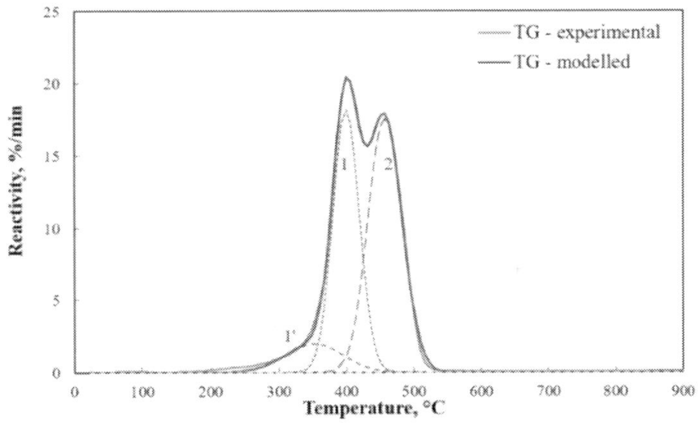

Fig. 4. Deconvolution of reactivity curve.

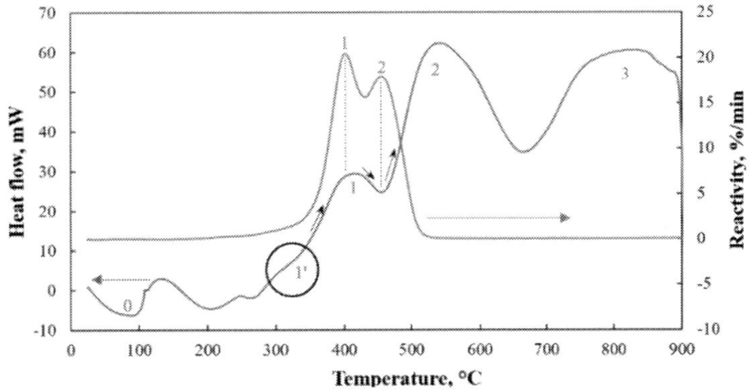

Fig. 5. The heat flow and reactivity as a function of temperature.

Based on the deconvoluted profiles of reactivity, for each constituent a set of kinetic parameters was calculated according to models A and B and summarized in Table 3.

Table 3. Sets of kinetic parameters describing pyrolysis of waste tyres sample (model B and A).

	Temperature range	Activation energy, E	Pre-exponential factor, k_0	Fraction of the ith component, Y	
	°C	kJ/mol	1/s	%	
Peak 1'	260–480	68.6	4.04×10^3	9.2	A
Peak 1	360–460	268.9	3.00×10^{18}	40.5	A
Peak 2	400–570	149.2	4.75×10^8	50.3	A
Total	370–500	41.1	4.62	100	B

In the case of model A, the activation energies determined for the three constituents of waste tyre are 68.6, 268.9 and 149.2 kJ/mol, while the pre-exponential factors vary from 4.04×10^3 to 3.00×10^{18}. Obtained values are similar to these presented in literature, e.g. in paper of Chen et al. [26] for industrial non-tyre rubber waste. According to [26], the activation energy of reaction, reaching its maximum at 360°C, is about 69 kJ/mol and it is almost the same as that obtained for peak 1' (reaching its maximum at 370 °C). In both papers, the peak achieves its maximum rate at 400–420 °C and it has the highest activation energy of 268.9 kJ/mol and about 166 kJ/mol respectively. Chen et al. [26] discovers three constituents being evolved with maximum rate in temperatures from 470 to 510°C, while in this paper only one reaction was observed (the peak 2) in this range. Nevertheless, the activation energies presented in [26] are from about 106 to 128 kJ/mol, while in the present study E of peak 2 is 149.2 kJ/mol. In case of the pre-exponential factors, the discrepancies between the obtained results and the data presented in [26] are higher. Nevertheless, some correlations are still visible. The kinetic parameters determined with the use of model B are also close to these found in the literature. The obtained activation energy of 41.1 kJ/mol is close to values determined by Singh et al. [27] for the automotive tyre waste (53.2–64.6 kJ/mol). Some discrepancies between obtained and cited results may be attributed to differences in process conditions (e.g. the heating rate), sample composition and variations of kinetic parameters, known as compensation effect [28]. Thermochemical processes such as pyrolysis is being extensively developed by researchers [9, 29] as it seems to be a promising mechanism to deal with a variety of waste and the energy content of the samples tested in this study support the potential even further.

4 Conclusion

The experimental investigation carried out in this paper has proven the potential of waste tyres to be used as a fuel for thermochemical processes.

The heating value of the sample is as high as 36.83 MJ/kg, the fuel ratio is 0.36 and C content is around 80–90%. These parameters clearly indicate the feasibility of scrap tyres as a fuel for pyrolysis. Additionally, the low moisture content makes them even more attractive since an additional drying does not have to be implemented. However, a relatively high sulphur content is present. TG/DSC analysis have provided valuable information to design the laboratory or potentially industrial scale process parameters. It was observed that temperature range 350–500°C featured the most rapid volatilisation. Thus, to ensure the entire decomposition of organic matter in tyres, a temperature of 600°C seems to be the most suitable for the pyrolysis of waste tyres. The presence of several process stages was proven by the TG/DSC analysis and the obtained kinetic parameters can be considered as reference for experiments design.

The performed analysis and the comparison of the obtained results with those reported in literature allow to conclude that whole WT – not only the rubber – can be considered as a suitable source of sustainable energy through thermochemical processes such as pyrolysis.

The paper was financial supported by Department of Air Conditioning, Heating, Gas Engineering and Air Protection; Wroclaw University of Science and Technology (No. 0401/0055/18).

References

1. ETRMA, *European Tyre & Rubber Industry. Statistics Edition 2017 N.9.* (Brussels, Belgium, 2017)
2. A. Wojciechowski, R. Michalski, E. Kamińska, Polimery **57**, 656–60 (2012)
3. J. Malinauskaite, H. Jouhara, D. Czajczyńska, P. Stanchev, E. Katsou, P. Rostkowski, Energy **141**, 2013–44 (2017)
4. G. G. Choi, S. H. Jung, S. J. Oh, J. S. Kim, Fuel Process Technol. **123**, 57–64 (2014)
5. I. Hita, M. Arabiourrutia, M. Olazar, J. Bilbao, J. M. Arandes, P. Castano, Renew. Sustain. Energy Rev. **56**, 745–59 (2016)
6. N. Antoniou, G. Stavropoulos, A. Zabaniotou, Renew. Sustain. Energy Rev. **39**, 1053–73 (2014)
7. J. Zeaiter, F. Azizi, M. Lameh, D. Milani, H. Y. Ismail, A. Abbas, Renew. Energy **123**, 44–51 (2018)

8. H. Jouhara, D. Ahmad, D. Czajczyńska, H. Ghazal, L. Anguilano, A. Reynolds, Therm. Sci. Eng. Prog. **5**, 579–88 (2018)
9. H. Jouhara, D. Czajczyńska, H. Ghazal, R. Krzyżyńska, L. Anguilano, A.J. Reynolds, Energy **139**, 485–506 (2017)
10. D. Czajczyńska, L. Anguilano, H. Ghazal, R. Krzyżyńska, A.J. Reynolds, N. Spencer, Therm. Sci. Eng. Prog. **3**, 171–97 (2017)
11. X. Zheng, Z. Ying, B. Wang, C. Chen, Appl. Therm. Eng. **144**, 242–7 (2018)
12. Y. Shen, X. Chen, X. Ge, M. Chen, J. Environ. Manage. **214**, 94–103 (2018)
13. J. M. Heikkinen, J. C. Hordijk, W. De Jong, H. Spliethoff, J. Anal. Appl. Pyrolysis **71**, 883–900 (2004)
14. A. M. Azeez, D. Meier, J. Odermatt, T. Willner, Energy & Fuels **24**, 2078–85 (2010)
15. S. V. Vassilev, C. G. Vassileva, V. S. Vassilev, Fuel **158**, 330–50 (2015)
16. J. M. Encinar, J. F. González, J. González, Fuel Process Technol. **68**, 209–22 (2000)
17. R. Moliner, I. Suelves, J. Anal. Appl. Pyrolysis **47**, 111–25 (1998)
18. P. T. Williams, Waste Manag. **33**, 1714–28 (2013)
19. D. Czajczyńska, R. Krzyżyńska, H. Jouhara, N. Spencer, Energy **134**, 1121–31 (2017)
20. D. Czajczyńska, R. Krzyżyńska, H. Jouhara, Proceedings **2**, 1455 (2018)
21. J. Kandasamy, I. Gökalp, Energy & Fuels **29**, 346–54 (2014)
22. J. F. Gonzalez, J. M. Encinar, J. L. Canito, J. J. Rodriguez, Inter. Symp. Anal. App. Pyrolysis **59**, 667–83 (2001)
23. M. I. Rofiqul, H. Haniu, A. B. M. Rafiqul, Fuel **87**, 3112–22 (2008)
24. M. R. Islam, H. Haniu, J. Fardoushi, Waste Manag. **29**, 668–77 (2009)
25. M. Kyari, A. Cunliffe, P. T. Williams, Energy & Fuels **19**, 1165–73 (2005)
26. R. Chen, Q. Li, Y. Zhang, X. Xu, D. Zhang, Fuel **235**, 1224–37 (2019)
27. R. Kumar Singh, B. Ruj, A. Jana, S. Mondal, B. Jana, A. Kumar Sadhukhan, J. Anal. Appl. Pyrolysis **135**, 379–89 (2018)
28. K. M. Czajka, J. Anal. Appl. Pyrolysis **133**, 82–90 (2018)
29. H. Jouhara, T. K. Nannou, L. Anguilano, H. Ghazal, N. Spencer, Energy **139**, 1210–30 (2017)

Life support system in the mature Martian colony for 1000 people

Natalia Ćwilichowska[1,*], *Joanna* Kuźma[2], *Anna* Jurga[3], *Paweł* Piszko[4], and *Thien* Nguyen[5]

[1]Wroclaw University of Science and Technology, Faculty of Chemistry, ul. Norwida 4/6, 50-373 Wrocław, Poland
[2]Wroclaw University of Science and Technology, Faculty of Mechanical and Power Engineering, Wyb. Wyspianskiego 27, 50-370, Wroclaw, Poland
[3]Wroclaw University of Science and Technology, Faculty of Environmental Engineering, Wyb. Wyspianskiego 27, 50-370, Wroclaw, Poland
[4]Jagiellonian University, Faculty of Chemistry, ul. Gronostajowa 2, 30-387, Kraków, Poland
[5]University of Adelaide, Faculty of Engineering, Computer and Mathematical Sciences, Adelaide, South Australia 5005, Australia

Abstract. The development in the field of space engineering is aimed at achieving another milestone, which is the creation of an extraterrestrial colony on, for example, Mars. Missions of this kind will be long-term and long-distance, which results in the need to design systems with a high degree of self-sufficiency. It is synonymous with high recovery resources. This article is a concept proposed for the needs of the Mar Colony Prize competition – Design The First Human Settlement On Mars, created by The Mars Society. This competition assumed designing highly-autonomous base for 1000 people. This means, that all environmental systems must be connected with each other and work in closed loops. In this paper four main environmental systems are proposed and characterized. The state of art of each subsystems technology is presented.

1 Introduction

It is projected that Mars will be inhabited before 2050 [1]. It is a great challenge for technology aimed at supporting long-term life support of humans in extraterrestrial habitats. Successful habitation of Mars requires life support systems than can be autonomous for periods of 1000 days or more [2]. However, Mars gravitational acceleration (0.38 g), atmosphere composition (about 95.3% CO_2, 2.7% dinitrogen (N_2), 1.6% argon, and only 0.13% O_2), pressure (less than 1% the Earth's one), temperature (-150 to +20°C), and low solar radiations (approximately 40% less than on Earth) creates very limited conditions compared to Earth. To create permanent base is not a simple task. It requires design of many systems responsible for all environmental parameters that will enable human life in hostile conditions. This set of subsystems is called the Life Support System (LSS) [3] and is responsible for maintaining appropriate environmental conditions, as well as managing the

* Corresponding author: n.cwilichowska94@gmail.com

waste generated in the space facility. The LSS in the International Space Station is the only one currently functioning in space conditions. While, short distance to the Earth enables supply of required onboard resources. This wouldn't be applicable for long term missions placed on Mars. In this case LSS is required to be self-sufficient and to recover each of essential resources in-situ, which are primarily oxygen, water and nutrients [3]. After processing and recovery they will be able to return to the cycle, providing the needs of the crew, but also serving as a fertilizer for grown crops, or various different processes. Due to lack on Martian surface easily accessible sources of life support sustainable compounds, LSS needs to characterised as closed system for mater circulation.

This paper is focused on presenting of our LSS concept. This LSS system is a part of 1000-man Mars base, designed for the *Mars Colony Prize competition – Design The First Human Settlement On Mars,* organised by The Mars Society. It was assumed to create the concept of a 1000-man Mars base, characterised as highly-autonomous, and capable of producing its own goods. This challenge is not trivial and to tackle with it we based on comprehensive literature study. The Figure 1 presents a summary of the subsystems of the LSS and their main functions covered by this study.

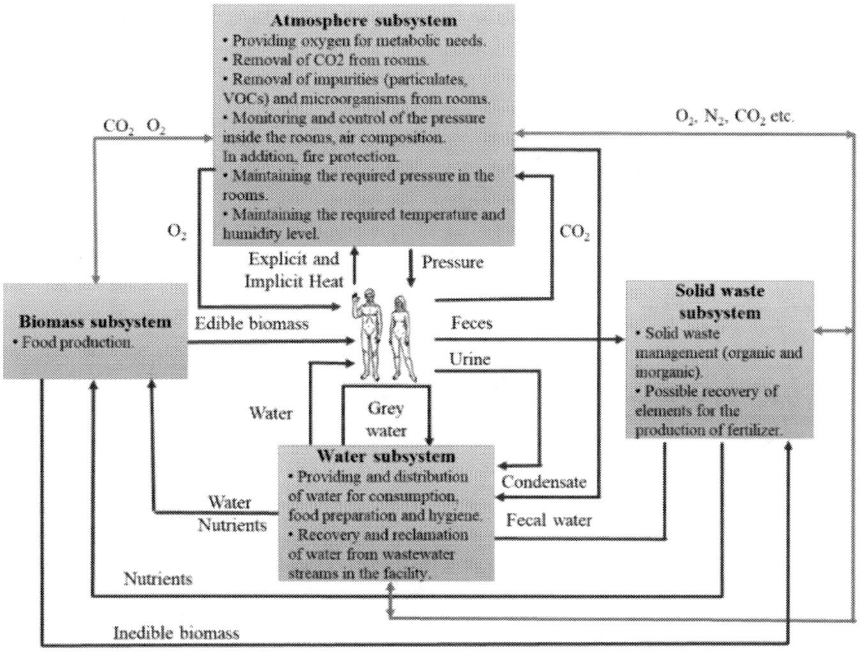

Fig. 1. Martian Life Support System scheme and each subsystem function.

2 Atmosphere subsystem

The major issue which is ought to be tackled in the concept of the colony is the atmosphere and the possibility of the inhabitants to genuinely breath. In order to increase the life comfort of the colony's inhabitants it is more preferable to fabricate a gas ambience rather than equip 1000 colonists with personal breathing apparatus in addition to presence of multiple depressurizing compartments. The key to achieving this goal is the constant levelling of pressure, temperature and most importantly atmosphere compositions as they are on Earth: 78% N_2, 21% O_2 and ppm values of other gases such as noble gases.

There are many variables in the equation of forming and maintaining such atmosphere. Those include: fluctuation on Mars conditions affecting its atmosphere, volume of the space which will be filled with gas and e.g. oxygen usage and CO_2 emission to name a few. Average crewmember is consuming 0.816 kg oxygen per day [3]. In this concept biomass cultivation is producing a sufficient amount of oxygen for crew purposes (see chapter 5). However, since an emergency system must be also provided it is decided to utilize the oxygen generating system MOXIE – currently developed by NASA. It is the most sustainable way of in-situ oxygen generating system available in remote conditions and from carbon dioxide as substrate which is ideal for Mars conditions [4]. This system will be validated onboard the rover during Mars 2020 mission, thus, by the time the extraterrestrial colony will stand, this process will be examined precisely. This device is capable of generating oxygen from carbon dioxide similar to the photosynthesis. However, the idea behind this system is nothing alike the plant-based process mentioned. The operation principle is electrochemically based. It utilizes solid oxide electrolysis cell doped with yttria-stabilized zirconia in elevated temperature (800°C) according to equation $2CO_2 \rightarrow 2CO + O_2$. The scheme of MOXIE is depicted below in the Figure 2. At the cathode oxygen molecules are generated and at the anode - carbon monoxide. The process takes place in high temperature and thus it requires sophisticated thermal isolation which can be hard to achieve while installing MOXIE devices onto the scaffolds of the Colony's buildings.

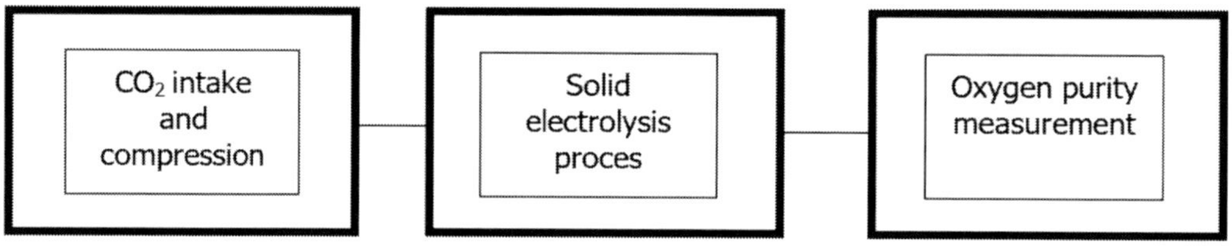

Fig. 2. MOXIE scheme [5].

The current prototype of MOXIE has dimensions of: 23.9 x 23.9 x 30.9 cm, weights approximately 15 kg and has 300 W of operational power. It is capable of generating about 10 g of pure oxygen which is equivalent of 7.52 dm^3 in NTP (293.15 K, 1013 hPa). Devices will be paired in racks and installed to provide clean oxygen for the colonists in significant places of the colony. Oxygen from the MOXIE system will be additionally supplied to the wastewater and waste system in order to avoid the situation where oxygen will not be available for proper implementation of these processes.

The nitrogen will be generated in larger quantities while treating the biological waste in the Colony in the manner described in the Solid waste and Biomass subsystems chapters of this publication. The remaining elements such as noble gases will be dosed from the pressurized bottles. In creating the micro-atmosphere, the exact same stratification as Earth's has not to be sustained. The pressure of the gas will be artificially pumped to maintain the atmospheric pressure (~1013 hPa). The two-way selective filters of the dome will evacuate excess carbon dioxide as well as other components of the atmosphere and its contaminants. Thanks to the system of O_2, CO_2, CO and N_2 sensors and heat exchangers the environmental conditions could be equalized. The system will constantly monitor pressure and temperature to adjust it to the desired level and to detect anomalies. For MOXIE system CO_2 will be sucked in directly from the Mars atmosphere of which it constitutes almost all (95%).

3 Water and wastewater subsystem

Currently, at the International Space Station wastewater processing contains several physicochemical steps [3, 6], which includes stabilization by acid dosing, distillation and multifiltration (active carbon and ion-exchange resin beds). In general, the physicochemical methods includes: distillation, reverse osmosis, electrodialysis, multifiltration (mechanical filtration, purification in an active carbon bed, purification using ion exchange resins). In case of extraterrestrial colony this methods would not be sufficient, due to unreliability (e.g. flogging) and the need for spare parts. However, the most important aspect is to recover water and nutrients, which can be implemented in well-known terrestrially biological methods. Process might be conducted in activated sludge or trickling filters [7, 8]. Both methods are currently used in advanced space research. Thus, in this concept biological wastewater treatment is used. With regard to chosen biomass production type, the nitrification process is favorable. Nitrification rate in the range of 90% is assumed. The issue with this process is acquiring the alkalinity to carry out the nitrification. It is assumed, that it will be extracted from Mars surface.

Table 1 presents the quantity and quality of generated in the colony wastewater (based on [3]), composition of reactor inlet and outlet. There are three main wastewater streams: urine and grey water with condensate. Since, the use of the diversion toilet is more than certain, faeces will be separated in the source and transported directly to the solid waste management system. Only fecal water will be treated in the biological reactor. Urine is human origin stream, rich in phosphorus and nitrogen [9, 10] and will be used as a main source of nutrients for biomass cultivation. Assuming 1.2 g $N\cdot m^{-2}$ is average demand on nitrate for most cultivated plants, outlet composition ensure 10505 m^2 of soilless cultivation, which would ensure 15% of daily energy demand for habitants. The next type of wastewater is grey water, which is for instance hygiene water, water for meals preparation, water after dish washing, water after laundry. After biological treatment water not used in cultivation will be directed into physicochemical unit, where water for human needs will be recovered. It should be noted, that the water production probably will be assisted by water mining from Mars surface, with use of ISRU (in situ resource utilization). The method is based on separating frozen water by evaporating it from regolith during pre-processing, which is extracted in open-cast mines The last stream is condensate. It is water recovered from the ventilation system. More precisely it is perspiration and respiration water (e.g. sweat, tears). This stream is not included in the Table 1, since in the colony, this stream would also contain water coming from the food production (transpiration water) and will be recirculated back to the food production system. Assuming average transpiration rate in the range of 91–99%, volume of this stream will be 243623.4 $kg\cdot day^{-1}$.

It should be noted water also might be consumed during various processes, which might occur in the colony. This stream should be determined after assumption of each branches of industry, which colony would have.

Table 1. Quantity and quality of wastewater streams created in the conceptual colony (based on [3]).

Basic parameters				**Reactor inlet (for 1000 crewmembers)**		**Reactor outlet (for 1000 crewmembers)**	
-	Urine	Grey water	Fecal water	Load	Concentratio n	Load	Concentration
Constituent	$g \cdot crewmember^{-1} \cdot day^{-1}$ (except for flow: $L \cdot crewmember^{-1} \cdot day^{-1}$)			$g \cdot day^{-1}$	$g \cdot L^{-1}$	$g \cdot day^{-1}$	$g \cdot L^{-1}$
Flow	1.53	24.45	0.087	26067	-	26067	-
Total Nitrogen	11.2	2.81	1.09	15096	0.58	15096	0.58
$N\text{-}NO_3$	0	0	0	0	0	13586	0.52
$N\text{-}NH_4$	4.36	0	0	4360	0.17	1510	0.06
P	0.85	1.28	0.91	3036	0.12	3036	0.12
PO_4	0.81	1.21	0.3	2322	0.089	2322	0.09
K	3.02	1.06	0.41	4489	0.17	4487	0.17
Ca	0.24	0.096	0.79	1126	0.043	1126	0.04
Mg	0.12	0.24	0.15	510	0.020	510	0.02

4 Solid waste subsystem

The concept of the Martian base proposed in this paper assumes large-area plant cultivation, as crops will be the main source of food for astronauts. However, such crops are associated with the generation of a large amount of solid waste – mostly non edible parts of plants, like: leaves, stems or rotten plants. In our conception, each day the amount of waste biomass will be equal to 28 989 kg. Such wastes is rich in water and various other compounds, that can be successfully separated, modified and used for the need of colony. These compounds include: proteins, sugars, vitamins, fatty acids, cellulose, lignin, hemicellulose. Those two polymers – cellulose and hemicellulose, due to the lack of solubility in water, are a problematic waste also on the Earth. The percentage content of carbohydrates in the dry mass of deciduous stems is quite high: 45–50% cellulose, 25–35% hemicellulose and 25–35% lignin [11–14]. According to the amount of daily biomass, which is need to be utilized, we propose implementation of efficient processing technology, so called biorefinery, which will play an essential role in our concept.

Biorefineries allow the use of biomass for the production of chemicals, substrates for further syntheses or fuels, to name a few. Due to the nature of the colony conception, the techniques used in biorefineries, can be also used in Mars colony.

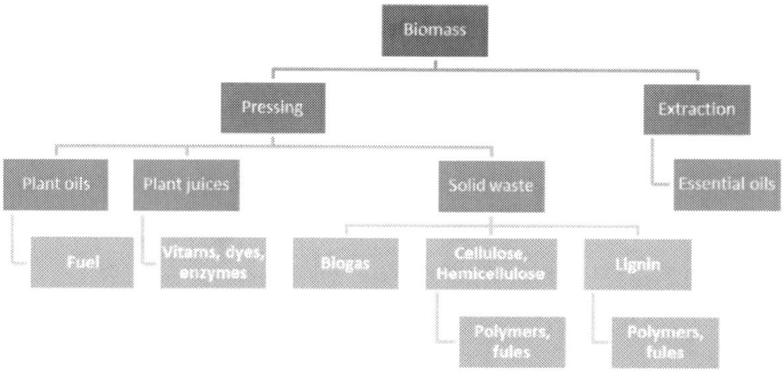

Fig. 3. The overview of the biorefinery concept [11–14].

The concept of the Martian biorefinery assumes the use of the most universal devices, such as filters, bioreactors or extractor, so that the installation could be easily adapted to current needs and produce a range of different compounds [11–14].

The entire biomass, will be collected in aerobic composter, equipped with a biogas discharging system. This gas could be used as a fuel source for the entire installation [11–14]. The residence time of the biomass in the composter will depend on the process being carried out. The biomass from the composter will be firstly transported to the macerator, which allows grinding of wet plant material, and then to the pressure filter. Three main products are achieved: green juice, oil and biomass (filter cake) [11–14]. Obtained biomass could be subjected to a supercritical extraction process with carbon dioxide. Such a solution brings with it a number of advantages. The atmosphere of Mars is rich in CO_2, the solvent after the gas expansion is removed from the product, and the sensitive compounds contained therein do not decompose. Depending on the process parameters: pressure and temperature, various compounds will be extracted [15]. The extract after quantitative and qualitative analysis, could be used as a dietary supplement for astronauts, enriching the nutritional value of food. Before use the extract, will also be further purified, for example using chromatographic columns or filtration technique [14, 16].

The oil fraction could then be used as a substrate for the production of fuel for the installation's operation. The water fraction, rich in such compounds like vitamins, dyes, proteins could be purified on chromatographic columns in order to get rid of undesirable compounds and also be an astronaut dietary supplement [11].

A filter cake, rich in water-insoluble compounds – cellulose, hemicellulose, will be transported to a bioreactor, where suitable microorganisms will use it as a source of energy and produce biogas in anerobic conditions, for energy purposes [11]. The polymers contained in the filter cake could also be used to produce useful raw materials for further syntheses. The cellulose can be subjected to the two different processes: enzymatic hydrolysis to obtain a simple sugar, which can then be used, for example, for the production of ethanol [17, 18]. Cellulose will also be used to obtain cellulose fibers: lyocell, modal or viscose [19]. These fibers are now widespread, and wood cellulose is produced due to the lower cost of the process.

Biomass is rich in numerous of compounds, which will be extracted and use for many another purposes. However there is need to develop a processing technology, which does not require much chemicals due to lack of many compound or Martian surface. One of the solutions is to use biotechnology – microorganism and enzymes for it, but for it more advance research on influence of microgravity and radiation for living organisms must be performed.

5 Biomass subsystem

For permanent settlement of other celestial bodies, it is necessary to create a food production system. Dietary requirements for one crewmember is defined by Anderson et al. [3] as the crewmember mass range from a 95th percentile American male, with a total body mass of 99 kg, to a 5th percentile Japanese female, with a total mass of 53.0 kg for whom daily calories consumption is 3072 kcal for one crewmember a day. Which means that the total amount of produced calories for 1000 people should be approximately 3,800,000 kcal a day. Three types of systems are taken into account: Hydroponic, Aeroponic and Aquaponic. Hydroponics and aeroponics are soilless, plant cultivation techniques, which provides the nutrient solution directly to the plants and have been proven successful to produce crops for relatively long time periods. Hydroponics and aeroponics do not use soil, because of that, the mass of the necessary equipment is much lower compare to the soil techniques. Additionally they can be highly automated to control nutrient and water dosage [20]. The diet requirements of humans dictates that up to 15 species of plants must be provided to complete menu [21]. These crops were be selected for small size, fast growth, nutrition, short plant cycles, and high harvest indices [20]. To the most-studied plants in space environmental belong: lettuce, mizuna, wheat, tomato, onion, radish, Chinese cabbage, cucumber, barley and pea [23]. However plants have relatively low calorific value. Because of that, for a long term mission additional source of calories should be found. In our case aquaponics seems to be the solution to this problem. It is a system that combines conventional aquaculture, raising aquatic animals with hydroponics. Including fish into the LSS gives two benefits food and waste treatment sub-process. Fish are the best choice among aquatic animals for LSS for several reasons. First of all they have relatively low energy requirements, 5- to 20-fold less than terrestrial vertebrates [24]. Fish are 10- to 13-fold more efficient than other animals. This efficiency is due to a lack of thermoregulation expenditures, lower sodium pump activity, neutral buoyancy which negating the effects of gravity, and nitrogen excretion mechanisms [25]. It is proved that fish nutrition has a direct impact on the nutritional composition of fish. Lots of studies showed differences between wild caught and farm raised fish [26]. On the example of Nil Tilapia has been proved that the incorporation of Spirulina in the diet of those fish will result in an acceptable and nutritious dietary source for human consumption [25]. That is why we propose to use spirulina in our aqua system. Additionally spirulina will be a part of astronauts diet. It is a great source of vitamins, carotenoid pigments, proteins, lipids and polysaccharides and spirulina can be easily adsorbed by humans [27]. Taking those two facts into consideration. Spirulina will be cultivated in the colony like this. To provide half of the caloric demand it should be produced 2000 kg of edible biomass a day which means that the cultivation area must be about 40 000 m^2. Cultivation of such sizes will need 1700 kg of CO_2 and will produce 1200 kg of Oxygen. 1000 people will consume 816 kg of O_2 and will produce 1040 kg of CO_2 a day [3]. The surplus of oxygen produced in the crop will be directed to the oxygen tanks used for EVA (Extravehicular activity). Additional source of CO_2 is sustained with atmosphere subsystem by uptake from Martian atmosphere.

6 Conclusions

The concept of LSS system proposed in this paper is based on comprehensive literature background. What is more all technologies used in this project are already implemented. Our study shows that the level of current technology for each of the four subsystems is sufficient to fulfil its task in the colony covered by this concept. Martian environment is unfavourable for human kind, due to lack of easily accessible sources of elements, essential for life. All this goods need to be produce in artificial way or getting them is very energy or time consuming. To overcome this challenge, we assumed LSS system as a closed system for material circulation, what was shown in each subsystem description. All waste and outputs are treated as substrates for another process. We shown that all units work together, and depend on each other. Nevertheless, the main problem is implementation of proposed solutions and analysing in experimental way if all individual processes are compatible with each other. Nevertheless, the main problem is the implementation of these solutions, not as individual processes but acting together as a unity. This unification was tested during the previously conducted projects investigating self-sufficient systems such as BIOS, Biosphere or MELiSSA. However each of these experiments was conducted on a smaller scale or even in pilot scale. Together with the increase to such a large area, there are a number of problems related to operation, monitoring of installations and conditions. Thus, to create an extraterrestrial colony that can accommodate 1,000 people, a lot of research must be carried out during earlier stages of colonisation. During our work we came across a lot of technological problems. To sum up main research problems regarding the described systems as well as our literature study, that should be examined in the coming years are as follows:

- atmosphere subsystem: reducing the weight, efficiency improvement, reduction of energy demand
- water subsystem: inoculation of biological reactors, the effect of reduced gravity on the aeration system and the collection of excessive sludge, recovery of elements from brine from physicochemical devices.
- solid waste subsystem: significant improving the efficiency, recovering of valuable chemical compounds and development of technology to process them
- biomass subsystem: clogging of aeroponics, precipitation in each system, collecting harvests by the crew.

References

1. M. Yamashita, Y. Ishikawa, Y. Kitaya, E. Goto, M. Arai, H. Hashimoto, K. Tomita-Yokotani, M. Hirafuji, K. Omori, A. Shiraishi, A. Tani, K. Toki, H. Yokota, O. Fujita, Ann. N. Y. Acad. Sci. **1077**, 8, 232–243 (2006)
2. Society of Automotive Engineers (SAE), *Regenerative life support systems & processes* (1991)
3. M. S. Anderson, M. K. Ewert, J. F. Keener, *Life Support Baseline Values and Assumptions Document* (National Aeronautics and Space Administration Washington, D.C. Report No. NASA/TP-2015–218570, 2018)
4. https://mars.nasa.gov/mars2020/mission/instruments/moxie/for-scientists/
5. https://mars.nasa.gov/resources/7391/moxie-functional-block-diagram/
6. R. Zubrin, *The Case for Mars* (Free Press, Revised edition, New York, 2011)
7. D. J. Barta, K. D. Pickering, C. Meyer, S. Pensinger, L. Vega, M. Flynn, A. Jackson, R. Wheeler, *A Biologically-Based Alternative Water Processor for Long Duration Space Missions* (NASA Johnson Space Center; Houston, TX, United States, Report No. NASA/JSC-CN-33488, 2015)
8. G. Bornemann, K. Waßer, J. Hauslage, Life Sci. Space Res. **18**, 12–20(2018)
9. W. Pronk, D. Kone, Desalination **248**, 1–3, 360–368 (2009)
10. H. W. Jones, M. H. Kliss, Adv. Space Res. **45**, 917–928 (2010)
11. W. Pronk, D. Kone, Desalination **248**, 1–3, 360–368 (2009)
12. H. W. Jones, M. H. Kliss, Adv. Space Res. **45**, 917–928(2010)
13. B. Burczyk, Wiadomości chemiczne **63**, 759–760, 755, 771 (2009)
14. G. A. Płaza, D. Wandzich, Management Systems in Production Engineering **23**, 3 (2016)
15. https://bit.ly/2uBYfPa
16. S. Ji, B. Orlikova, M. Diederich, J. Cancer Prev. **19**, 1 (2014)
17. V. Mićić, D. Novaković, Ž. Lepojević, M. Jotanović, B. Pejović, P. Dugić, Z. Petrović, Contemporary Materials **2**, 1 (2011)
18. M. Krauze-Baranowska, I. Malinowska, LAB Laboratoria, Aparatura, Badania **22**, 6 (2010)
19. I. Szwach, R. Kulesza, Chemik **68**, 10, 894 (2014)

20. G. A. Płaza, D. Wandzich, Management Systems in Production Engineering **23**, 3, 150–155 (2016)
21. https://bit.ly/2UmMESF
22. F. Maggi, C. Pallud, Adv. Space Res. **46**, 1257–1265 (2010)
23. R. L. Olson, M. W. Oleson, T. J. Slavin, HortScience **23**, 275–286 (1988)
24. O. Monje, G. W. Stutte, G. D. Goins, D. M. Porterfield, G. E. Bingham, Adv. Space Res. **31**, 151–167 (2003)
25. P. Zabel, M. Bamsey, D. Schubert, M. Tajmar, Life Sci Space Res. **10**, 1–16 (2016)
26. D. Bureau, S. Kaushik, C. Cho, *Bioenergetics* (Fish Nutrition, third ed. Academic Press, New York, NY, 1–50, 2002)
27. J. M. Gonzales Jr., Adv. Space Res. **43**, 1250–1255 (2009)

Inequality of water consumption for hygienic and sanitary purposes in production halls

Edyta Dudkiewicz[1,*], and *Marta* Laska[1]

[1]Wroclaw University of Science and Technology, Faculty of Environmental Engineering, Wybrzeże Wyspiańskiego 27, 50–370 Wroclaw, Poland

Abstract. Water consumption in hall buildings is substantial and characterised by variability over the time: yearly, seasonality, monthly, weekly, daily, hourly and second. Both, water consumption and their irregularity, influence on many processes and technical issues. Purposes for which water is consumed are various. Beyond the technological processes, one of the most water-intensive demand is need for water for hygienic and sanitary purposes, especially in production buildings with high degree of soiling work. The paper presents the study of hot water (DHW) consumption and its profile for sanitary and hygienic purposes in two production halls. The halls vary in the degree of soiling work, sanitary equipment and numbers of employees. The aim of this article is to define DHW consumption profiles and hourly water-consumption irregularity coefficient applicable in design process of hot water preparation for hygienic purposes in production buildings. The average daily hot water consumption varies significantly and amounts 7 or 16 m^3 respectively to both halls. Despite the discrepancy of sanitary equipment and number of workers the hourly water-consumption irregularity coefficients for both halls are in the similar range, about 2, and daily domestic hot water profiles present typically three main consumption peaks that occur at the end of work shifts.

1 Introduction

The large-cubage buildings belong to the group in which the level of energy consumption is significant and reaches about 400 kWh/(m^2·year) [1]. Beyond the demand of large amount of energy, industrial buildings are characterized of high water consumption, particularly freshwater [2]. In year 2011 industrial plants in Poland itself consumed about 651.6 hm^3 of water [3]. The publications [1, 4] present the examples and calculations of water usage and energy needed to prepare domestic hot water in such buildings. These demands depend on the building purposes, the operating schedule, number of people and sanitary equipment. However the mentioned literature point out the additional difficulty in energy and water usage calculations caused by the discrepancy in guidelines for domestic water calculations and thus problems with the estimation of the real water and energy demand.

However each branch of industry requires an individual approach to the analysis of energy and water consumption [2, 3, 5] depending on the purposes, for instant:

* Corresponding author: edyta.dudkiewicz@pwr.edu.pl

- hygienic/sanitary (cold and DHW),
- fire-fighting, sprinkler installation,
- technological processes,
- production of steam,
- additional purposes (e.g. irrigation/watering/cleaning).

The most water-intensive purposes in production halls are technological processes [5]. In production buildings with high degree of soiling work and a large amount of workers, where water is used for hygienic and sanitary purposes, domestic hot water (DHW) is also accounted as large percentage of water consumption and generates significant energy consumption. In such buildings the showers and washbasins are basic equipment. Thus the water consumption depends on the number and type of hygienic equipment and sanitary facilities, degree of soiling work, type of work and sex of workers [6, 7].

Acquisition of knowledge about the amount and profile of water consumption and their irregularity allows to evaluate the development of technology in water supply networks and installations as well as water supply and sewage systems [2, 5, 8]. Additionally water load pattern gives the possibility to understand and forecast the daily sanitary needs of workers. Further the forecast can be integrated with a control strategy to optimize the energy costs of DHW preparation and possibility of implementation of renewables and waste sources [7]. Furthermore it is also worth to emphasize the fact that, especially from the engineering point of view, neglecting knowledge concerning the characteristics of hot water consumption in a building would result in oversizing or undersizing the installation or the whole system and thus problems with exploitation [6, 9, 10]. Additionally wrongly sized installation may cause an occurrence of too low or too high velocities of the flow and thus may have negative effect on the system operation [11, 12]. Too low velocities may lead to water stagnation in the water network, the occurrence of chemical and biological processes in pipes, the intensive sedimentation on the internal surfaces of water pipes and the development of biofilms [13]. The occurrence of biofilm, temperature increase and water stagnation are the main reasons of occurrence and multiplication of Legionella pneumophila [15, 18]. All these aforementioned factors have a negative effect on decreasing water quality which can be dangerous for human life.

In all buildings, also in the industrial halls, water consumption is characterised by variability over the time. Concept of annual, seasonal, monthly, weekly, daily, hourly and second fluctuations can be distinguished [15, 16]. This inequality of water consumption influences on various processes and issues aforementioned.

Reviewing the literature in the field of hot water consumption profiles and its irregularity it was found that researchers have been widely studied domestic hot water consumption profiles in residential buildings or public buildings such as hospitals, hotels or office buildings [10, 16–18]. However the subject related to production halls has been omitted and in the literature there is a the lack of the described research in such buildings. The daily hot water demand profiles have an inevitable and crucial influence on hot water system design [18, 19] therefore this issue related to production halls is the main goal of this paper.

The study presents the results of domestic hot water consumption research for sanitary purposes in two production buildings with different degree of soiling work. The aim of this study is to analyse daily changes in water consumption in industrial halls and investigation the irregularity coefficients of hourly water consumption.

2 Description of measurements

The measurements of the hot water consumption were carried out in two production halls with different demand of hot water, degrees of soiling work and sanitary equipment, amount of workers and their sex. The halls are localized on Lower Silesia region in Poland.

The production hall with the high degree of soiling works will be referred as the industrial hall no.1 in the next part of this paper. The measurements were carried out over 30 days in October and November in 2017. The industrial hall consists of few departments with sanitary units in each of them. People work on 3 shifts, however on each shift different number of people is working. Total number of workers is 235. The biggest group of people work on first shift, from 6 am to 2 pm. All workers are male. The work in the hall is organized 5 days per week, from Monday to Friday. Due to high soiling work the bathrooms are equipped with 24 showers and 22 washbasins.

The second hall will be referred in this paper as the industrial hall no. 2. The work in the hall is of low soiling potential. The measurements were performed over 3 months from March to June in 2018. People work in the hall in a three-shift system, six days per week. About 350 people work during the morning shift from 6 am to 2 pm, about 250 people during the day shift from 2 pm to 10 pm, and about 100 people during the night shift from 10 pm to 6 am. Most workers in the industrial hall no.2 are women. The bathrooms are equipped in 22 washbasins and no showers are installed.

The Table 1 presents summarised information about the characteristics of the industrial halls where the measurements were conducted.

To measure water flow rate and water velocity the portable ultrasonic flow meter (PORTAFLOW C) was installed on the pipe delivering water to hot water tank (the storage unit). Data was sampled and archived every 10 seconds.

3 Results of the research

The significant parameter in water installation design and assessment of water consumption in buildings is the average daily water consumption. The average daily hot water consumption in industrial hall no.1 was 16 m^3 while average daily hot water consumption per person equals 68 l. The average temperature in hot water tank was 46°C. This value is very close to the temperature reported in many publications [20]. The calculated value of the water consumption for one bath in hall no. 1 is greater than the described in literature value of 50 l per person and temperature 35°C [20].

The hot water consumption in industrial hall no.2 also varied. Daily values fluctuated between 1.9 m^3 in Saturday to 7.3 m^3 during a week. Average daily hot water consumption per person was estimated as 8.5 l what is more than 7 l/person and day recommended in guidelines and handbooks, e.g. [21]. However it has to be highlighted that the recommended value assumes tap water temperature of 55°C while the hot water temperature of the measuring period was 45°C. The average weekly hot water consumption in production hall no. 2 was almost half lower than in hall no.1 and equalled 45 m^3.

The second important factor in designing process of hot water systems is the energy demand for hot water preparation. It is crucial when implementation of renewable sources of energy are considered, especially utilization of solar collectors [7, 22]. It is strongly related to the number of people using hot water and the schedule of the usage. Table 1 presents the average amount of water consumption and estimated energy demand for shifts, day, week and a month. Monthly heat demand for production of hot tap water was calculated from the formula:

$$Q_{htw} = [(x/7)\cdot l_d\cdot m_d\cdot c_w\cdot(t_{hw}-t_{cw})\cdot 10^{-3}]/3600 \tag{1}$$

where: l_d – number of days in a month (-), m_d – daily demand of the hot tap water (kg/day), c_w – specific heat of water (kJ/(kg·K)), t_{hw} – temperature in the hot water tank (°C), t_{cw} – temperature of the cold water (°C). The $x/7$ coefficient in the formula (1) was introduced to take into account the water draw-off over 5 or 6 days per week in both halls.

Table 1. The data related to working conditions, water and energy consumption in both halls.

	Industrial hall no. 1	**Industrial hall no. 2**
Number of work shifts	3	3
Total amount of people per day	235	750
Working days per week	5	6
Degrees of soiling work	high soiling	low soiling
Sanitary equipment (of hot water):	Yes	No
- Shower - Washbasin	Yes	Yes
Average temperature in hot water tank (°C)	46	45
Average daily hot water consumption G_d (m^3)	16	7
Daily energy demand (kWh)	763	297
Energy demand per person (Wh/person)	3250	396
Average daily hot water consumption (l/person)	68	8.5
Average weekly DHW consumption (m^3)	80	45
Weekly energy demand (kWh)	3818	2095
Monthly energy demand DHW purposes (kWh)	16361	7642

3.1 Hot water consumption profiles

In generally in production building hot water demand has periodical peaks at the end of each work shift when showers and washbasins are used. Duration of draw-off periods lasts for about 30 minutes. The periods of installation usage are dependent on the number of working shifts and working/holiday days during a year. The general rule in the DHW designing process assumes the constant value of hot water demand during the year [7]. Fig. 1 presents profile of dynamic of hot water consumption for hall no. 2 during seven days. As is shown distribution of hot water consumption is very similar in weekdays.

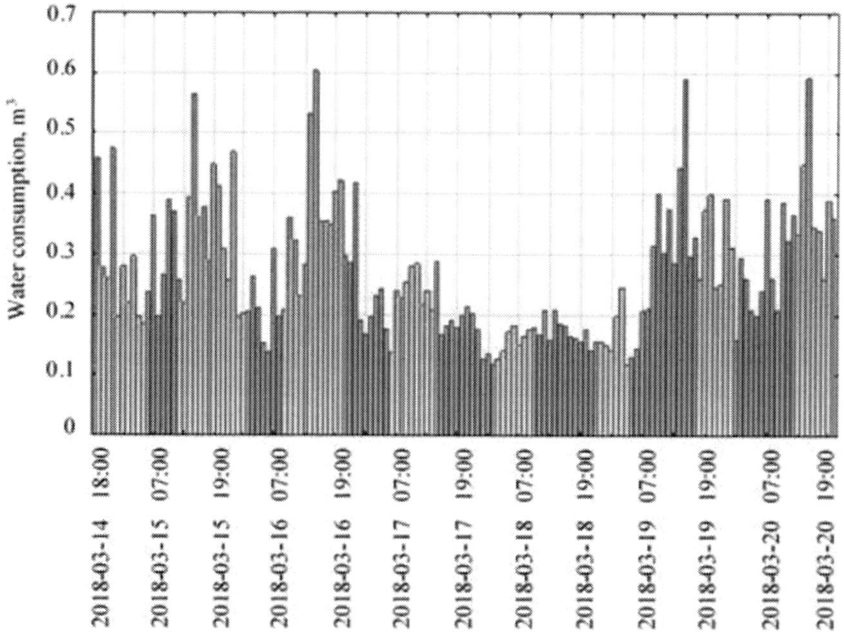

Fig. 1. Profile of dynamic of hot water consumption for hall no. 2 for one week [23].

The studied industrial halls have different amount of water consumption, but as shown in Fig. 2 daily water load patterns are quite similar. A considerably larger water demand is observed after the first shift between 6 am to 2 pm during the weekdays. The peak of water consumption occurs at 2 pm and hourly domestic water consumption fluctuates between 8.4–8.9% of daily water consumption. After the second shift also the second peak is observed – in both halls the percentage of hourly domestic water consumption is in a range of 5.6–6.1%. Analysing Fig. 2, in hall no.1 with a high soiling work, also the third peak, after night shift at 6 am is visible. During the remaining working hours discrepancies in hourly domestic water consumption between two industrial halls are substantial, but still, the highest consumption for both buildings is after the first shift. The actual distributions of water consumption based on the measurements in both halls are significantly different from the theoretical ones described in the literature [23, 24]. Better understanding of the reductions in average daily peak hour water demand impacts on design process.

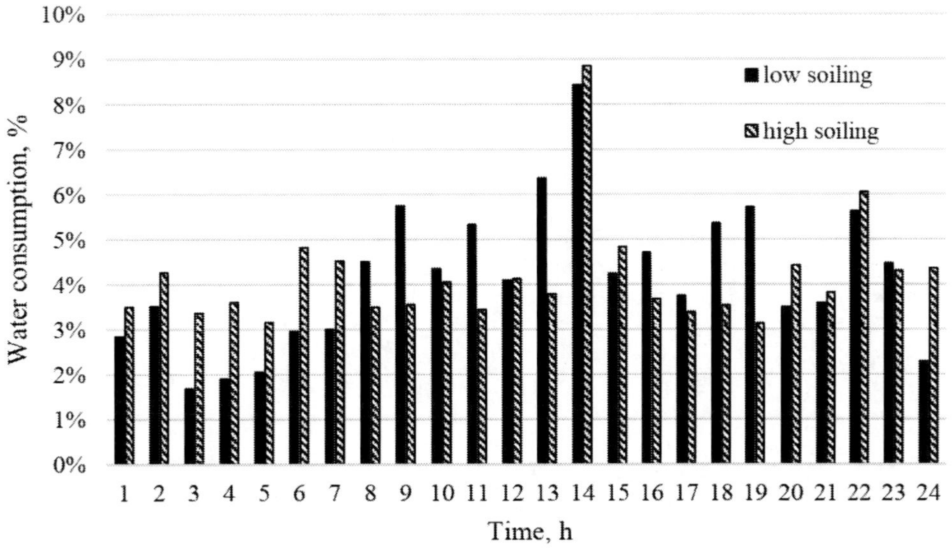

Fig. 2. Profile of percentage distribution of hot water consumption for both halls (right – industrial hall no. 1, left – industrial hall no. 2).

Analysing the hot water consumption profiles the conclusion arises that the consumption of hot water is irregular throughout the day [14]. Irregularity of water consumption in buildings is characterized by various coefficients. One of the most important is the hourly water irregularity coefficient N_h. It is defined by the quotient of maximum hourly demand G_{hmax} and average hourly water demand G_{hav}:

$$N_h = G_{hmax}/G_{hav} \tag{2}$$

The coefficient is dimensionless and its value defines upper limits of variability ranges of hourly water consumption. The hourly water-consumption irregularity coefficients for both halls are in similar range and oscillate around the value of 2. For all measured days the irregularity coefficients fluctuate between 1.74 and 2.36 for hall no. 1, and 1.24 and 2.56 for hall no. 2.

In residential buildings the value of hourly water-consumption irregularity coefficients decrease with the rising number of residents [9]. This is due to the fact that for all residential buildings, the same amount of hot water consumption per person is assumed. Whereas in production halls this value depends on equipment and number of sanitary facilities, and rises with the degree of soiling work. It was estimated, that for hall no.1 the average daily hot water consumption per person is 68 l and for hall no. 2 – 8.5 l. In production halls the values of hourly water-consumption irregularity coefficients decrease with the increase of the water consumption in the buildings during the day. Fig. 3 presents the relationship between hourly water-consumption irregularity coefficients and daily water demands G_d. On the left hand site the relationship for the hall no.2 is presented, on the right – for hall no. 1. Following the measurements, the equations describing the relationship between the daily demand of water and hourly water consumption for particular hall were derived and presented in Fig. 3. The formulas are as follow:

- for industrial hall no. 1:

$$N_h^1=0,44\ G_d^1+1,315 \tag{3}$$

- for industrial hall no. 2:

$$N_h^2=-0,079\ G_d^2+1,9076 \tag{4}$$

The values of the hourly irregularity coefficient of hot water is higher than the hourly cold water irregularity coefficient, which in turn is higher than the coefficient of hourly irregularity distribution of the total water (together cold and hot):

$$N_h^{DHW}>N_h^{cw}>N_h^{tw} \tag{5}$$

It is crucial to pay particular attention for above relationship (5) when dimensioning the water supply installations. It happens that designers accept the value of the irregularity coefficient, not paying the proper attention to which water it is dedicated.

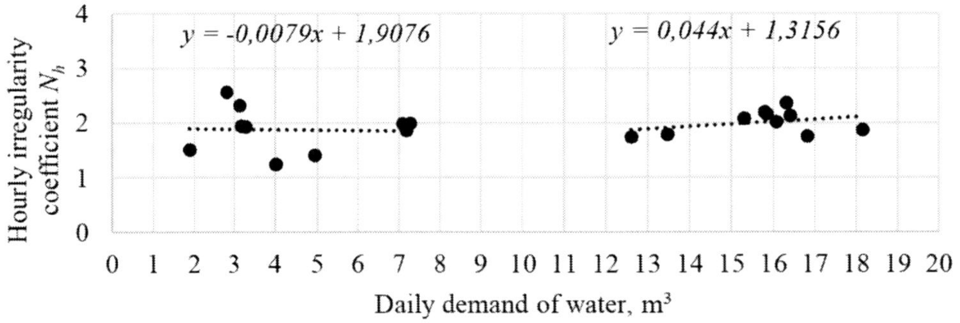

Fig. 3. The functional dependence of daily demand of water and hourly water consumption irregularity coefficient N_h (left- for industrial hall no.2, right- states for industrial hall no. 1).

3.2 Energy consumption

In the technical design process, on the stage of the sizing the DHW system, the assumption of water storage (water accumulation) option is on the great importance. Characterization of domestic hot water profiles would allow for designing innovative control strategies for DHW systems [16]. The Table 2 produces an overall relationship in methodological approach between the heat load of the heat exchanger and the volume of the storage tank [19]. From the engineering point of view the relation between the value of heat load reduction factor ψ and accumulation coefficient φ plays the crucial role.

Table 2. Dependence between the heat load of heat exchanger and volume of water storage tank.

Accumulation	Heat load of heat exchanger	Volume of water storage tank
Non-accumulation	Q_{hmax}*	$V = 0$
Partial accumulation	$\Psi \cdot Q_{hmax}$	$V = \varphi \cdot V_z$
Full-accumulation	Q_h**	$V = V_z$

* $Q_{hmax} = G_{hmax} \cdot c_w \cdot (t_{dhw} - t_{cw})$, ** $Q_h = G_{hav} \cdot c_w \cdot (t_{dhw} - t_{cw})$

Volume of water storage tank V_z is calculated following on standardized values of hot water demand per person and day (l/person·day) or energy consumption (Wh) during the utilization of the sanitary equipment. Following the literature energy consumption for the washbasin and shower varies between 500 and 870 (Wh), and from 1450 to 2300 (Wh) respectively. Energy demand per person in industrial hall no.1 was calculated to 3250 (Wh/person) and 396 (Wh/person) in hall no. 2 (see Table 1). Whereas the value for hall no.2 is in the recommended range aforementioned, the energy demand for the hall no. 1 with high soiling work is significant higher.

4 Conclusion

The acquisition of knowledge about the ranges of water consumption of each equipment and person, consumption profiles and peak times occurrence is helpful especially at early stage of the technical design, namely for selection of a measuring device (flow meters), sizing hot water preparation system, pipe dimensioning, defining the volume of the storage tank, etc. Furthermore the domestic hot water consumption is an important part of the building energy demand. The forecast of consumed DHW volume allows to control and reduce the costs of energy needed for hot water preparation.

The paper presented inequality of water consumption for hygienic and sanitary purposes calculated on the basis of the measurements conducted in two production halls located in Poland. Despite the substantial differences in the operating parameters of the considered buildings, the distribution profiles of hot water consumption are similar. The halls differ with the sanitary equipment, the degree of soiling work, number and sex of workers. However, the three-shift work schedule, common for both halls, was the reason of the occurrence of the typical water peaks in the water consumption profiles after each completed shift.

Despite the discrepancy of sanitary equipment and number of workers the hourly water-consumption irregularity coefficients for both halls are in the similar range. In our study this value oscillates around 2. Due to the lack of information in the literature about values of the hourly water-consumption irregularity coefficient N_h for sanitary purposes in production halls, and following the results of the research, the authors assume that the aforementioned value could be suitable also for other industrial halls.

The work was partially financed by The Faculty of Environmental Engineering, Wroclaw University of Science and Technology, Poland. Ref. no. 0401/0055/18.

References

1. Financing of buildings' energy efficiency improvement in Poland, BPIE (2016)
2. E. Ahmetović, I. E. Grossmann, AIChE Journ. 57, **2** (2010)
3. J. Wojdalski, B. Drózdz, J. Piechocki, Polish J. Chem. Technol. 15, **2** (2013)
4. M. Małetka, M. Laska, E3S Web of Conferences, **22** (2017)
5. A. Reynaud, Environ. Resour. Econ. **25** (2003)
6. E. Dudkiewicz, A. Zabnieńska-Góra, E3S Web of Conferences, **45** (2018)

7. N. Fidorów-Kaprawy, E. Dudkiewicz, E3S Web Conf. **22** (2017)
8. M. C. Amiri, H. Mohammadifard, G. Ghaffari, AIP Conf. Proc. **1758** (2016)
9. P. Podwójci, Inżynieria Ekol. **26** (2011)
10. E. J. Pieterse-Quirijns, A. H. Van Loon, H. Beverloo, Procedia Eng. **70** (2014)
11. M. Świderska-Bróż, M. Wolska, Ochr. Środowiska **26**, 4 (2004)
12. B. Toczyłowska, Technologia Wody **1**, 45 (2016)
13. N. Lin, H. Arabnejad, S. A. Shirazi, B. S. McLaury, Powder Technol. **336** (2018)
14. L. Zhang, J. Xia, J. E. Thorsen, O. Gudmundsson, S. Svendsen, Energy, **116** (2016)
15. R. Pasela, M. Goraczko, Rocz. Ochr. Sr. **15**, 1 (2013)
16. E. Fuentes, L. Arce, J. Salom, Renew. Sustain. Energy Rev. **81** (2018)
17. J. García-Sanz-Calcedo, F. López-Rodriguez, T. Yusaf, Energies **10**, 4 (2017)
18. R. Rankin, P. G. Rousseau, Energy Convers. Manag. **47** (2006)
19. B. Nowak, G. Bartnicki, COW, **3** (2004)
20. Compedium of knowledge Recknagel ER, OmniScala (2008)
21. A guide to the designer of boiler rooms with innovative solutions from Brötje (2014)
22. J. C. Evarts, L. G., Swan, Energy Build. **58** (2013)
23. E. Dudkiewicz, A. Żabnieńska-Góra, Rynek Instal. **7/8** (2018)
24. S. Mańkowski, *Designing Domestic Hot Water Installations* (Arkady, 1981)

Environmental aspects of heat recovery systems of boiler plants

Nataliia Fialko[1,*], *Raisa* Navrodska[1], *Malgorzata* Ulewicz[2], *Georgii* Gnedash[1], *Sergii* Alioshko[1], and *Svitlana* Shevcuk[1]

[1]The Department of Thermophysics of Energy Efficient Heat Technologies, Institute of Engineering Thermophysics of National Academy of Sciences of Ukraine, Kyiv, Ukraine
[2]Faculty of Civil Engineering, Czestochowa University of Technology, Czestochowa, Poland

Abstract. The results of studies on improving the environmental characteristics of boiler plants of municipal heat-power engineering in the application of technologies for the deep recovery of heat from exhaust-gases of gas-fired boilers are presented. The data on the reduction of environment by reducing the amount and temperature of harmful emissions resulting from fuel combustion are given. The data are given for boiler plants equipped with complex heat-recovery systems characterized by cooling the exhaust-gases below the dew point of the water vapor contained in them. It is shown that the decrease in emissions is due to a decrease in fuel consumption in boilers due to the beneficial use of the heat of exhaust-gases in these systems and the dissolution of nitrogen and carbon oxides in the condensate, formed in the heat-recovery equipments. The analysis of improving the environmental safety of boiler plants when used in heat-recovery technologies of corrosion protection systems for chimneys has been performed. Anticorrosion protection is provided by preventing condensate formation in the exhaust-gas ducts of boiler plants when using technologies for the deep recovery of exhaust-gas heat. To prevent condensate formation, the method of pre-drying cooled exhaust-gases in a heat exchanger-preheater installed after heat-recovery equipment is used. It is also shown that the use of complex heat-recovery systems provides in the boiler plant additional water in the form of condensate formed during condensation of moisture from exhaust-gases. Receipt of this condensate is another ecological effect of heat-recovery, which allows reducing the consumption of natural water resources for supply municipal heat networks.

1 Introduction

At present, the problems of rational use of fuel and energy resources and improvement of the ecological situation are priorities in the world energy practice. The relevance of scientific tasks in these areas is increasing due to the constant increase in the cost of fuel and the increasing requirements to reduce environmental pollution. The main ways to solve

* Corresponding author: nmfialko@ukr.net

these important problems in the production of heat energy in the municipal sector is to improve the environmental performance of gas-fired boilers and increase the efficiency of their use of fuel in the application of technologies the deep recovery-heat of the exhaust-gases [1–6] or other companion waste heat [7]. It should be noted that the use of these technologies is characterized by obtaining an environmental effect in several directions.

2 Prerequisites and means for solving the problem

These technologies provide a reduction in fuel consumption in the boiler due to a decrease in the exhaust-gas temperature through the beneficial use of their heat. At the same time, the coefficient the use heat of fuel (CUHF) of boiler or its efficiency significantly increases. In this case, the environmental performance of the boiler is also improved by reducing temperature pollution of the environment and reducing harmful emissions resulting from fuel combustion.

The reduction in fuel consumption and, consequently, the reduction of harmful emissions of ΔC into the environment are determined by the depth of cooling of the exhaust-gases in a boiler plant containing a boiler equipped with a heat-recovery system. In certain modes of the boiler in this system, the exhaust-gas temperature decreases below the dew point of the water vapor contained in them. There comes the so-called condensation mode of operation of heat-recovery equipment, when in addition to the so-called clear heat of exhaust-gases, the latent heat of condensation of this vapor is used.

In fig. 1 shows data on the magnitude of the reduction in emissions of ΔC due to an increase in boiler efficiency and a decrease in fuel consumption depending on the exhaust-gas temperature at the outlet of the boiler plant.

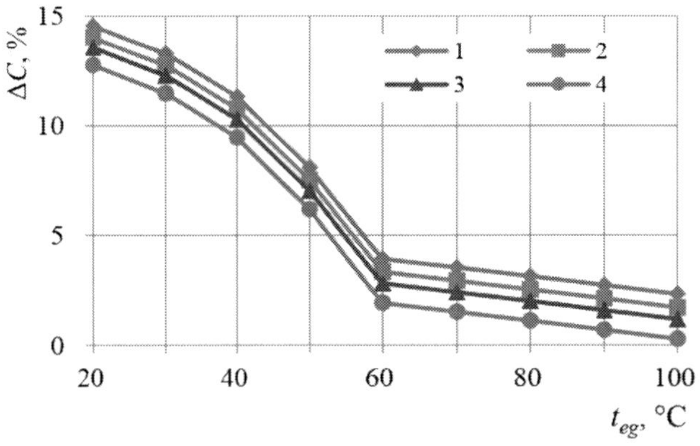

Fig. 1. The dependence of the relative reduction of harmful emissions ΔC of the boiler plant from the temperature of exhaust-gases at the outlet t_{eg} for various heat loads of the boiler:
1 – Q_b/Q_{nom} = 100%; 2 – 70%; 3 – 50%; 4 – 30%.

The data are given for a gas-fired water-heating boiler, type KSVA-2.0G with a nominal heating capacity of 2 MW when calculated from the highest calorific value of fuel. The nominal temperature t_{eg} of the boiler exhaust-gases and the coefficient of air excess α in them are 170°C and 1.1, respectively. With this value of α, the dew point t_{dp} of water vapour contained in the exhaust-gases is close to 58°C.

As can be seen from the presented results, in all modes of operation of the boiler, there is a tendency of a significant increase in the value of ΔC when cooling the exhaust-gases below the dew point t_{dp}, that is, when operating in the condensation mode. So, with a nominal load of the boiler and a decrease in the exhaust-gas temperature from 170°C to the temperature at which water vapor starts to condense on the heat exchange surfaces (i.e., when heat-recovery is carried out in the dry mode), the maximum efficiency of the boiler and the corresponding reduction in harmful emissions is 4.3%. Under the conditions of operation of the boiler plant with this load in the condensation mode, a reduction in harmful emissions of ΔC may be as the exhaust-gas temperature drops to 50°C – 8.5%; up to 40°C – 11.8% and up to 30°C – 13.7%. In the event of a decrease in the boiler load below the nominal, the relative amount of harmful emissions of ΔC decreases slightly due to a decrease in fuel consumption in the boiler.

In traditional heat-recovery systems used in municipal heat and power engineering mainly for heating boilers and intended only for heating the return heat-network water entering the boiler, this condensation mode is usually implemented only in the autumn-spring period of the heating season with relatively minor loads of the boiler. In order to ensure this mode throughout the entire heating period, it is necessary to use the recovery heat also for heating heat-transfer agents colder than return heat-network water. Such heat-transfer agents can be cold water of the chemical water-purification system of the boiler plant, technological water, combustion air, etc.

3 Solution of the examined problem

The complex heat-recovery system using the received heat to heat the return heat-network water and the combustion air is shown in fig. 2.

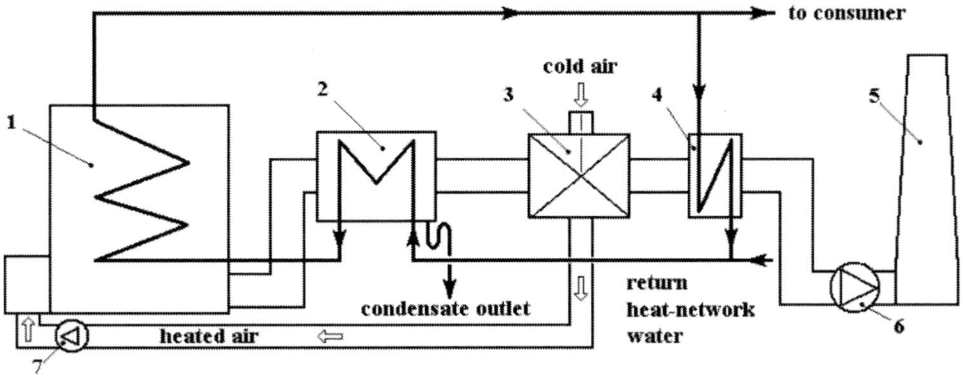

Fig. 2. Basic scheme the heat-recovery installation with the complex use water-heating and air-heating heat-recovery exchangers: 1 – boiler; 2 – heater of return heat-network water; 3 – air-heater; 4 – gas-heater; 5 – chimney; 6 – exhauster; 7 – fan.

In this system, the exhaust-gases of the boiler are cooled first in the water-exchanger, heating the return heat-network water before it is supplied to the boiler, and then cooled in the air heat exchanger, which heats the combustion air. Both heat-recovery exchangers can operate in condensation mode. This mode is carried out in the cold season mainly in the air-heating heat-recovery exchanger due to low air temperature, and in the autumn-spring heating season in the water-exchanger when the heated water temperature going down below the dew point t_{dp} of exhaust-gases.

When calculating the thermal and increasing the efficiency of the boiler, the characteristics of the above-mentioned KSVA-2.0G boiler, calculated from the highest calorific value of fuel, were used. The air-heating heat exchanger of the system was made up of packages of stainless steel plates, and the water-heating one was made of a bundle of transverse-finned bimetallic pipes (carbon steel base, aluminium fins).

4 Results and discussion

In fig. 3 shows the results of computational studies of the levels of reduction of harmful emissions ΔC by increasing the boiler efficiency when using the heat-recovery system during the heating period. Wherein, the heat load of the boiler changed depending on the environment air temperature with the adopted temperature difference in the heat-network of 70–115°C.

The results show that the use of air and water heat-recovery exchangers provides a level of reduction of harmful emissions during the heating period in the ranges of 2.1–4.0% and 2.8–5.9 %, respectively. At the same time, the reduction of emissions of the entire system is 6.0–7.5%, which is a significantly higher level compared to the use of single heat exchangers.

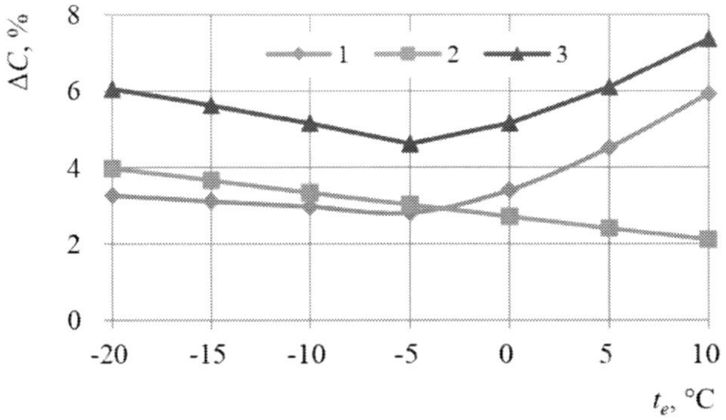

Fig. 3. Dependence on the outdoor temperature the relative reduction of harmful emissions ΔC boilers with a complex heat-recovery system through the use of a heat exchanger for heating return heat-network water (1), combustion air (2) and when they are used together (3).

It should be noted that the realized of the condensation mode in the complex heat-recovery systems of boiler plants in addition to reducing harmful emissions by increasing the efficiency boiler is characterized by an additional effect. This effect is to reduce harmful emissions due to partial dissolution of the nitrogen and carbon oxides contained in the exhaust-gases in the condensate obtained during the deep cooling of gases. This effect is much smaller in magnitude compared with the effect associated with reduced fuel consumption. Its value depends on the composition of the fuel and the environmental performance of burners, and for modern boilers it is estimated to be less than 1%.

The next, very important environmental aspect of the use of modern complex heat-recovery systems is to ensure the environmental safety of municipal facilities through the use of anti-corrosion protection of boiler chimneys in these systems. Chimneys, as is well known, must fulfil strict environmental requirements regarding the maintenance of regulatory dispersion of pollutants contained in the flue gases of boiler plants, and high demands are placed on them for their reliability.

The use of heat-recovery technologies without protection systems for gas exhaust-ducts leads to corrosion destruction of the exhaust-ducts and chimneys [8–10]. This is due to the deep cooling of exhaust-gases in heat-recovery systems. In this case, further cooling of the exhaust-gases and condensation of the water vapor remaining in them occurs in the gas-exhaust ducts of the boiler plant, including the chimney, and a problem called "chimneys are crying" occurs. The condensate formed is acidic; its pH in gaseous fuel is 5–6. This circumstance enhances the corrosion destruction of exhaust-ducts. At the same time, the most vulnerable section of the gas exhaust-duct is the mouth of the flue, the last section of this duct along the exhaust-gas path.

It should be noted that even when heat-recovery installations are operating in the dry mode (without reducing the exhaust-gas temperature below the dew point), there is an increased condensate formation in the chimney due to the subsequent cooling of these gases in the exhaust-gas duct. When using heat-recovery technologies with deep cooling of exhaust-gases, a significant amount of moisture from them is condensed in the heat-recovery equipment. This reduces the dew point of the exhaust-gases, which significantly improves the conditions for ensuring the prevention of condensate formation in the exhaust ducts.

To prevent the formation of condensate in the exhaust-gas ducts of boiler plants in modern heat-recovery systems, thermal methods are used to prevent condensate formation [4, 9, 10]. The proposed heat-recovery system (see fig. 2) provides for thermal protection of the exhaust ducts from condensation, which is provided by applying the thermal method of drying exhaust-gases after the heat-recovery system in the heat exchanger gas-heater.

The effectiveness of this method essentially depends on the type of chimney used, which determines the insulating properties of its body. In fig. 4 shows data on the heat and humidity characteristics at the mouth of chimneys of various types (brick and metal with thermal insulation) during the heating period when used in systems for heat recovery of gas-heaters.

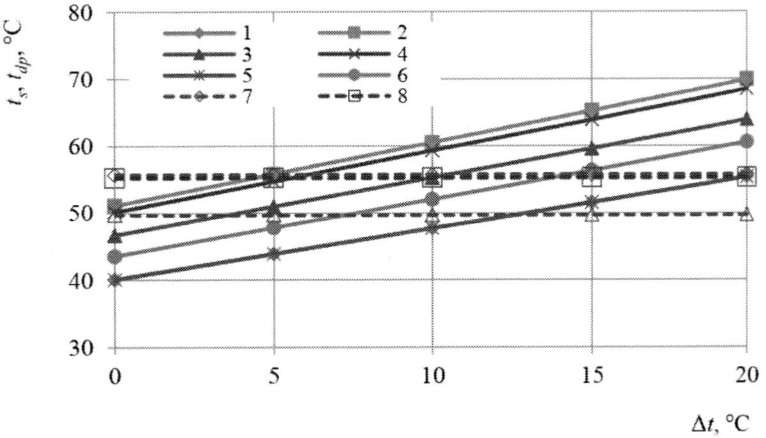

Fig. 4. The dependence of the temperature of the inner surface t_s at the mouth of chimneys (1–6) and the dew point t_{dp} (7-9) from the levels of heating Δt of exhaust-gases at different environment temperatures t_e: 1, 3, 5 – brick; 2, 4, 6 – metal with heat insulation (λ = 0.065 W/(m·K), δ = 50 mm) chimneys: 1, 2, 7 – t_e = -20°C; 3, 4, 8 – t_e = -5°C; 5, 6, 9 – t_e = 10°C.

The data obtained indicate that for the absence of condensation at the mouth of the chimney in these modes of operation of the boilers it is necessary to heat the flue gases to values Δt = 10°C and 17°C for brick and metal pipes, respectively. That is, for the environmental safety of boiler plants of municipal heat-power engineering facilities, it is advisable to apply heat-recovery technologies with anti-corrosion protection of the gas exhaust ducts.

The environmental aspects of heat-recovery systems also include the possibility of condensate in them. In fig. 5 shows the dependencies regarding the amount of condensate heat generated in the complex heat-recovery system also for the gas-fired boiler type KSVa-2.0 with a nominal heating capacity of 2 MW. Data is based on 1 MW of installed boiler capacity.

As can be seen from the presented results, during the heating period, the output of condensate in the complex heat-recovery system is from 21 to 40 kg/h per 1 MW. When operating a boiler plant with an installed capacity of 30 MW, for example, up to 29 tons of condensate can be formed per day.

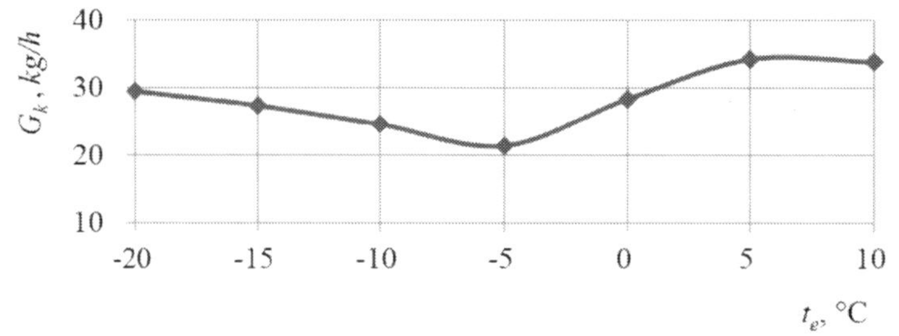

Fig. 5. The formation of condensate at 1 MW of installed power of the boiler depending on the environment temperature t_e during the heating period when using a complex heat-recovery system.

The resulting condensate can be used, in particular, in chemical water-purification systems of boiler plants to supply heating-networks. This circumstance ensures a reduction in the consumption of natural water resources in the heat supply systems of municipal heat and power engineering.

5 Conclusions

In conclusion, it should be noted that the conducted studies have shown significant environmental efficiency of modern heat-recovery technologies for boiler plants. So the application of these technologies in the municipal heat-power engineering provides:

- a reduction of up to 7.5% of the harmful emissions generated by the combustion of natural gas in boilers, by reducing its consumption with the beneficial use of the heat of exhaust-gases;
- a reduction of up to 1% of the emissions of oxides of nitrogen and carbon contained in the flue gases by partially dissolving them in the condensate obtained during the deep cooling of these gases;
- ecological safety of communal heat and power engineering objects when using heat-protection technologies of heat-recovery systems of exhaust-gas ducts against condensate formation;
- saving natural water resources due to the beneficial use of condensate formed in heat-recovery systems.

References

1. A. V. Efimov, A. L. Goncharenko, A. B. Goncharenko, Bulletin of the National Technical University "KhPI" **13**, 987, 73–80 (2013)
2. I. N. Artemov, E. A. Artemova, *Most complete utilizing the heat of exhaust gases boilers* (Energy efficient and resource-saving technologies and systems, 126–129, 2014)
3. N. M. Fialko, R. A. Navrodskaya, G. A. Presich, M. A. Novakovskiy, Technological systems **77**, 4 (2016)
4. N. Fialko, R. Navrodskaya, G. Gnedash, G. A. Presich, A. I. Stepanova, S. I. Shevchuk, International Scientific Journal for Alternative Energy and Ecology **15**, 126–129 (2014)
5. A. A. Dolinskiy, N. M. Fialko, R. A. Navrodskaya, G. A. Gnedash, Industrial Heat Engineering **36**, 4, 27–35 (2014)

6. N. M. Fialko, G. A. Presich, G. A., Gnedash, S. I. Shevchuk, I. L. Dashkovska, Industrial Heat Engineering **40**, 3, 38–45 (2018)
7. N. M. Fialko, L. B. Zimin, S. V. Dubovskoy, Promyshlennaya teplotekhnika **22**, 1, 90–93 (2000)
8. A. A. Kudinov, K. A. Ziganshina, A.V. Fedotenkova, Energetic **8**, 60–62 (2014)
9. N. M. Fialko, R. O. Navrodska, S. I. Shevchuk, G. O. Presich, G. O. Gnedash, Scientific Bulletin of UNFU **27**, 6, 125–130 (2017)
10. N. M. Fialko, R. A. Navrodskaya, S. I. Shevchuk, G. A. Presich, G. A. Gnedash, O. Glushak, Modern Science: Researches, Ideas, Results, Technologies **2**, 15, 13–17 (2014)

The use of solar radiation for generating heat in a solar air collector in northern Poland

Jolanta Fieducik[1,*]

[1]University of Warmia and Mazury in Olsztyn, Faculty of Technical Sciences, 10-900 Olsztyn, Poland

Abstract. This article analyzes the use of solar radiation for generating heat in a simple solar air collector built in a laboratory. The structure of the collector and the equipment for measuring physical parameters were described. The aim of the study was to analyze the operating parameters and the thermal efficiency of a solar air collector in northern Poland. Various applications of solar air collectors were discussed. Solar air collectors can be used for heating, ventilating and drying indoor premises and for heating water.

1 Introduction

Global fossil fuel depletion has increased the interest in solar energy for generating electricity and heat [1, 2]. This is a particularly important consideration in small buildings whose energy needs can be fully met by solar energy. Single-family homes meet the above definition. Cheap in the construction and operation of air collectors can be used for heating, drying, ventilation and hot water heating. New solutions in the construction of air collectors are introduced based on new research. Numerical analyzes [13] and modeling using the Green function [14] are carried out for research in the field of flow, heat transfer and flat plate constructions. Collectors are tested with heat absorbing plates [15], or with an absorber consisting of containers filled with phase change material [16]. The purpose of these tests is to ensure the highest possible heat transfer efficiency and minimize the costs of building air collectors. In the new type of solar collector with a parabolic concentrator (CPC) (SAC) and a flat microthermal tube system (FMHPA), an average efficiency of 61% was achieved, and the instantaneous efficiency was 68% [18]. The aim of this study was to analyze the applicability of solar air collectors for generating heat in northern Poland. In Poland the average annual irradiation on a horizontal plane is in the range 950–1150 kWh/m^2. In Olsztyn, this irradiation is around 950–1000 kWh / m^2. The analysis was performed in the city of Olsztyn with the use of a simple solar air collector built in a laboratory.

* Corresponding author: jolanta.fieducik@uwm.edu.pl

2 Conversion of solar radiation

The Sun is the main source of energy reaching the Earth. The electromagnetic radiation emitted by the Sun has a broad wavelength spectrum of around 0.2–2.5 μm, and the transmitted energy varies subject to wavelength. The radiation emitted by an ideal blackbody at a given frequency of the electromagnetic wave is described by Planck's law and can be expressed by the following formula [4]:

$$I(f) = \frac{2hf^3}{c^3} \cdot 1/(e^{\frac{hf}{kT}} - 1) \quad (1)$$

where:
$I(f)$ – spectral radiance (radiance of a surface per unit frequency)
h – Planck constant of $6.626 \cdot 10^{-34}$ Js
T – absolute temperature (K) of a blackbody
c – speed of light (in vacuum)
k –Boltzmann constant
f – radiation frequency (for which irradiance is calculated)

The dependency between solar irradiance and wavelength is presented in Figure 1.

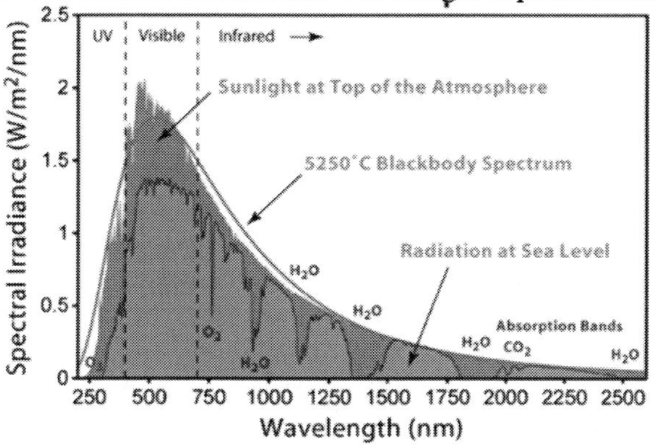

Fig. 1. Distribution of solar irradiance subject to wavelength [5].

Solar spectral irradiance reaching the upper layer of the Earth's atmosphere and the radiation reaching the surface of the Earth are presented in Figure 1. Solar irradiance in the visible range occupies the wavelength of 350 to 750 nm. Solar energy that reaches the ground is around 46% visible light, 7% ultraviolet light and 47% infrared light.

Total solar irradiance can be determined with the use of the Stefan-Boltzmann law which defines the total electromagnetic energy radiated per unit surface area of an ideal blackbody at a given temperature.

The Stefan-Boltzmann law is expressed by the below formula [6]:

$$\Phi = \sigma T^4 \quad (2)$$

where:
Φ – energy radiated per unit area of a body [W/m²]
σ –Stefan-Boltzmann constant
T – temperature (K)

The overall energy carried by solar radiation to the upper layer of the Earth's atmosphere is estimated at 1.39 kW/m², and it is referred to as the solar constant. Solar radiation is reflected, absorbed and diffused in the atmosphere, and only 1 kW/m² of that energy reaches the surface of the Earth on a cloudless day. The loss of solar radiation in the atmosphere is presented in Figure 2.

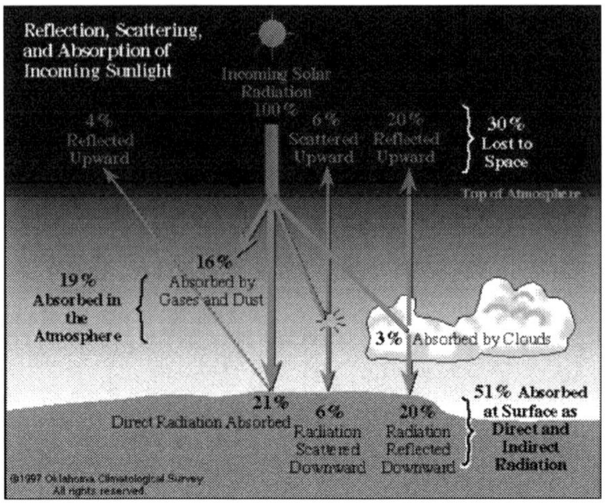

Fig. 2. Distribution of solar radiation reaching the surface of the Earth [8].

Solar irradiance on the surface of the Earth is also determined by the Sun's position relative to the zenith and the thickness of the atmospheric layer which absorbs solar radiation. The loss of solar radiation in the atmosphere resulting from differences in the solar zenith angle is defined by the air mass (AM) coefficient. In Figure 3, coefficient AM is defined by the quotient of the optical path length L of solar radiation passing the Earth's atmosphere and path length L_0 at the zenith. Coefficient AM can be used to describe the solar spectrum after sunlight has crossed the atmosphere, and it is widely applied to calculate the efficiency of solar cells under normalized conditions. Air mass is expressed by appending its value to the acronym AM, for example AM1.5 [9].

$$AM = L / L_0 = 1 / \cos(\Theta z) \qquad (3)$$

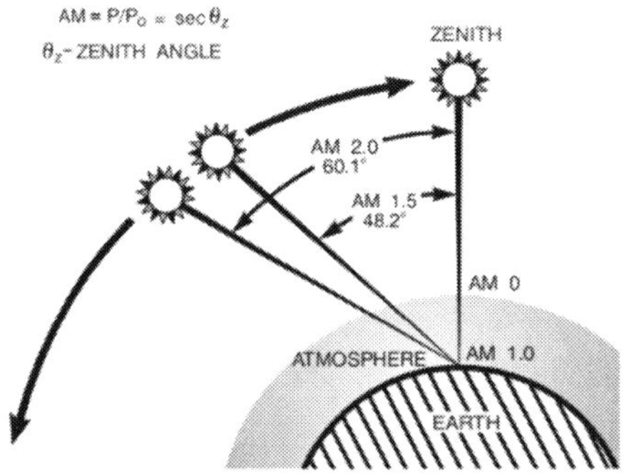

Fig. 3. Dependency between coefficient AM (Air Mass) and the zenith angle [9].

Solar irradiance is also significantly influenced by weather conditions in a given location and time. Solar irradiance can reach 1000 W/m^2 on a cloudless day, and it decreases to around 50 W/m^2 on a cloudy winter day. Solar irradiance is determined by season and sunshine duration which is estimated at 3500 h/year in locations with high solar exposure. In Poland, sunshine duration ranges from 1300 to 1900 h/year, and it is estimated at 1550 h/year in north-eastern Poland on average.

3 Solar air collectors – structure and operation

Several types of solar air collectors are in use. They differ in structure, but operate on a similar principle. A solar air collector consists of an absorber plate, a transparent cover plate and, in some cases, a heat insulating layer. The absorber plate is the most important part of a collector. Solar energy is transferred to the absorber which heats the flowing air stream. Solar air collectors have numerous advantages. The absorbing medium is air which does not freeze, heats up rapidly and can be directly used to heat indoor premises.

Solar air collectors can be uncovered or covered. In uncovered collectors, sunlight falls directly on the absorber plate, whereas in covered collectors, the absorber plate is mounted under a transparent cover plate. Depending on the solution, the air stream flows above or under the absorber, and in more complex structures – above and under the absorber or through the absorber. In the most advanced collectors, air flows through specially profiled channels. Collector channels have rectangular, triangular or oval cross-section. The absorber plate can be flat, rough or porous to increase the absorbing area and maximize the efficiency of heat transfer between the absorber and the air stream. In collectors with perforated absorbers, air is sucked in by the openings, which maximizes contact between the air stream and the absorber and increases the efficiency of heat transfer.

The operating principle of a solar air collector is presented in a schematic diagram in Figure 4.

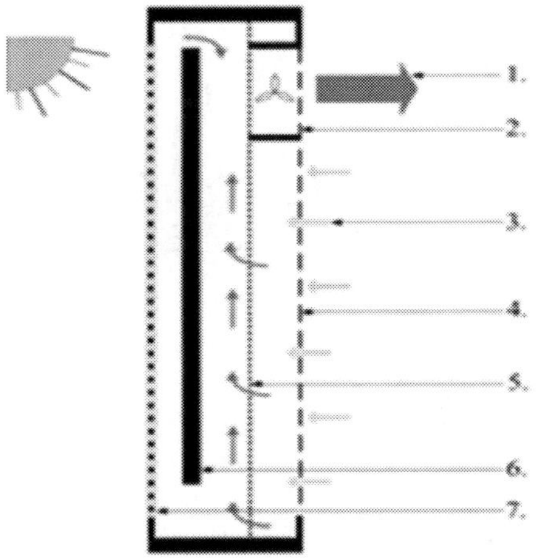

1. Fresh and heated air is blown inside
2. Fan
3. Air is sucked by the perforated aluminum plate across the entire surface
4. Perforated aluminum plate
5. Black felt mat
6. Cover
7. Photovoltaic module

Fig. 4. The operating principle of a solar air collector [10, 11].

In solar air collectors, air is the heat absorbing medium, and solar energy is efficiently converted to heat [1–3]. In the temperature range characteristic of the Earth's climate, air is not susceptible to phase change; therefore, changes in temperature do not disrupt the operation of a solar air collector. Air heated inside the collector can be directly used to heat indoor premises without the involvement of additional media.

A solar air collector can also be used for indoor ventilation. Air flow and air temperature are easily controlled, and the thermodynamic status of the premises connected to a solar air collector can be easily monitored.

4 Parameters of the analyzed solar air collector

Below is presented geographical location and climatic data of Olsztyn. Olsztyn lies on the northern latitude: 53.78 and east longitude: 20.48. The height of the position from 87.7 to 154.4 m. The climate in Olsztyn is moderately cold, with an average temperature of 7.9°C. In Olsztyn the average annual rainfall is 635 mm. The solar radiation is 900–950 kWh / m^2. The average exposure to the Sun is about 1,600 hours a year. The average wind speed during the year reaches the level of 3.0 m/s. The analyzed solar air collector was built in a laboratory. Collector dimensions were 2700 x 1400 x 150 mm, the dimensions of the aluminum sheet absorber were 2500 x 1250 mm, and total absorption area was 3.125 m^2. The absorber plate was covered with matte black paint on both sides. The absorber was mounted inside a wooden box whose rear and side walls were insulated with a layer of felt on the inside and styrofoam on the outside. The collector was covered with a 4-cm-thick glass pane. The aluminum absorber plate was mounted at a distance of 25 mm from the rear wall and 30 mm from the glass cover to enable free air flow and heat collection on both sides of the absorber. Two 9G180 NMB fans (12 V, 1.34 A) supplied by a 25 W photovoltaic module (12 V) were used for forced air circulation inside the collector.

The solar air collector was used to heat a residential building. The cold air inlet and the heated air outlet were positioned on the same side of the collector based on the building's layout and the assumption that cold air drawn from the building would be heated and returned to the building. A schematic diagram of the collector and airflow are presented in Figure 5.

Fig. 5. Schematic diagram of the collector and airflow inside the collector.

The collector was positioned around 5 m above ground level at 205° to the south (25° to the west). The angle of inclination between the collector and the ground was 45° to capture more sunlight in autumn and winter. Solar irradiance was measured with the 6450 DAVIS solar radiation sensor which measures both direct and diffuse components of solar irradiance. Air temperature at the collector inlet was measured with the VA18B multimeter with a temperature converter, and the signal was supplied to the RS232 terminal and registered by a computer. Air flow rate, temperature and humidity at the collector outlet were measured with the MS6252B MASTECH digital anemometer.

5 Results and discussion

The research was carried out from February 2015 to March 2016. The data collected during the period 27- 30 September 2015 selected for the analysis of the collector efficiency were presented in Fig. 6. Solar irradiance, air temperature at the collector outlet and the energy flux measured between 27 and 30 September 2015 are presented graphically in Figure 7.

Cool air flowing through the collector is heated by solar radiation, and its temperature and enthalpy increase. The process of heating air with humidity X from temperature T_1 to T_2 inside the collector is associated with changes in enthalpy Δh and an increase in internal energy Δe:

$$\Delta h = \Delta e = h_2(1+X) - h_1(1+X) = 1.006 \cdot (T_2 - T_1) + 1.86 \cdot X \cdot (T_2 - T_1) \quad (4)$$

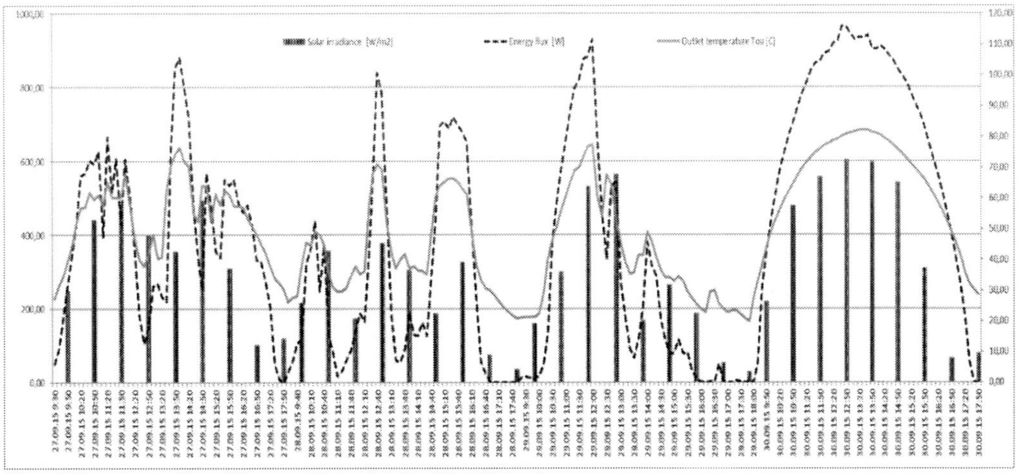

Fig. 6. Data registered on 27-30 September 2015.

The mass flux m of air with density ρ is calculated based on the velocity v of air leaving the collector with cross-section S_w:

$$m = \rho \cdot v \cdot S_w \quad (5)$$

The energy flux e [W] transferred by heated air in the collector is:

$$e = m \cdot \Delta h \quad (6)$$

The collector's thermal efficiency η can be determined based on energy flux e, the collector's surface area S_k and solar irradiance I:

$$\eta = e / I \cdot S_k \quad (7)$$

The collector's thermal efficiency was calculated based on the data registered on 30 September 2015, a relatively cloudless day, to minimize the effects of external factors.

 https://doi.org/10.1051/e3sconf/201910000016

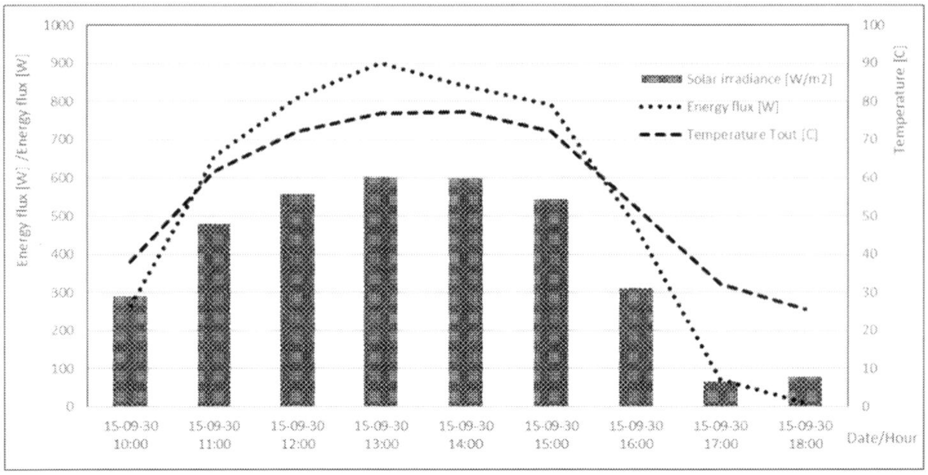

Fig. 7. Data registered on 30 September 2015.

The data for calculating the collector's thermal efficiency are presented graphically in Figure 7, and the results are presented in Table 1.

Table 1. Collector's thermal efficiency based on the data registered on 30 September 2015.

Date/hour	**Temper ature Tout [C]**	**Humi dity [%]**	**Air veloc ity [m/s]**	**ho [kJ/ kg]**	**h [kJ/ kg]**	**Δh [kJ/ kg]**	**Ener gy flux [W]**	**Solar irradia nce [W/m^2]**	**Efficie ncy [%]**
15-09-30 10:00	38.02	31.1%	3.91	62.0 5	80.3 9	18.3 5	264. 96	291	38.7%
15-09-30 11:00	61056	9.3%	4.19	62.4 9	104. 69	42.4 0	653. 89	479	43.7%
15-09-30 12:00	72.10	4.6%	4.14	62.7 2	115. 58	52.8 6	809. 27	558	46.4%
15-09-30 13:00	76.79	2.4%	4.18	62.2 7	120. 41	58.1 5	898. 91	603	47.7%
15-09-30 14:00	77.35	1.4%	4.17	66.6 5	120. 98	54.3 4	838. 69	598	44.9%
15-09-30 15:00	72.21	1.4%	4.17	64.5 1	115. 69	51.1 8	789. 47	543	46.5%
15-09-30 16:00	52.10	2.7%	4.18	64.5 5	94.9 3	30.3 8	469. 15	310	48.4%
15-09-30 17:00	32.12	6.0%	1.94	64.1 6	74.3 0	10.1 5	72.7 7	65	35.8%
15-09-30 18:00	25.65	10.3%	0.80	64.0 8	67.6 2	3.55	10.4 9	78	4.3%

Based on the data registered on 30 September 2015, the collector's maximum thermal efficiency was determined at 48.4%, and average efficiency was estimated at 40%.

6 Conclusions

One of the greatest advantages of solar air collectors is that they rely on air as a working medium that never freezes. Solar air collectors can be used for indoor heating, ventilation, drying and water heating [12]. The amount of heat supplied by a collector is determined by solar irradiance in a given time and location. Therefore, to introduce solar energy technologies it is necessary to promote them as modern options for energy conservation in buildings that should be implemented when energy-efficient measures have already been applied. Thus both options can reduce total energy consumption significantly and, in consequence, reduce environmental pollution [17]. Solar air collectors can also be deployed as backup heat sources in buildings equipped with other heating systems, such as gas heating. In October and November and from February to May, the thermal energy supplied by a collector can decrease indoor heating costs. Solar air collectors can also be used for ventilation, both independently and in conjunction with the main ventilation system. Collectors used for ventilation purposes are equipped with a fan for forced air circulation. The fan can be powered by a photovoltaic module, which makes the ventilation system independent of the power grid. In systems powered by a photovoltaic module, the collector is activated automatically by sunlight. Air flow and temperature can be controlled with the use of additional devices. Solar air collectors supply buildings with fresh and heated air which can be directly used for indoor heating without the involvement of other media. Air flow and air temperature are easily controlled, and the thermodynamic status of premises connected to a solar air collector can be easily monitored. Solar air collectors are an ideal solution for summer cottages to ensure that indoor premises are adequately ventilated and heated outside the season to prevent fungal growth and eliminate musty smell [7].

The maximum thermal efficiency of the analyzed solar air collector approximated 50%, which is a satisfactory result. The maximum thermal efficiency of collectors supplied by renowned manufacturers, such as Solar Venti Ltd., exceeds 70% (model SV30 Air). According to the latest research conducted by TestLab Solar Thermal Systems of the Fraunhofer Institute for Solar Energy Systems in Germany, the thermal efficiency of innovative Solhotair air collectors exceeds 80%. The results of the present study indicate that a simple and inexpensive solar air collector is characterized by satisfactory thermal efficiency in northern Poland. In November and January, the performance of solar air collectors is significantly compromised by cloudy and foggy weather, which can be attributed to a large number of lakes in the Olsztyn area.

References

1. W. Smolec, *Fototermiczna konwersja energii słonecznej* (PWN, Warsaw, Poland, 2000)
2. G. R.Timilsina, L. Kurdgelashvili, P. A. Narb, *A review of solar energy; Markets, economics and policie, Polisy Research Working Paper 5845* (Soc. Sci. Res. Netw., 2011)
3. M. Maier, GetSolar software. version 10.1.1. (2009)
4. https://pl.wikipedia.org/wiki/Prawo_Plancka
5. https://upload.wikimedia.org/wikipedia/commons/4/4c/Solar_Spectrum.png
6. https://pl.wikipedia.org/wiki/Prawo_Stefana-Boltzmanna
7. http://www.instsani.pl/401/zasoby-energii-slonecznej
8. https://en.wikipedia.org/wiki/Air_mass_(solar_energy)
9. https://www.newport.com/t/introduction-to-solar-radiation
10. J. Fieducik, J. Godlewski, Czasopismo Inżynierii Lądowej, Środowiska i Architektury, **61** (2014)
11. J. Fieducik, J. Godlewski, Czasopismo Techniczne. Budownictwo, **3-B** (2014)

12. J. Fieducik. Czasopismo Inżynierii Lądowej, Środowiska i Architektury, **63** (2016)
13. T. C. Hung, T. J. Huang, D. S. Lee, C. H. Lin, B. S.Pei, Z. Y. Li, Applied Thermal Engineering, **111** (2017)
14. A. Grine, A. Radjouh, S.Harmand, Solar Energy, **105** (2014)
15. S. Xu, H. Dong, *Heat transfer characteristics study on solar air collector of heat-absorbing plate with holes* (World Automation Congress Proceedings, 6321398, 2012)
16. P. Charvat, O. Pech, J. Hejcik, EPJ Web of Conferences **45**, 01127 (2013)
17. D. Chwieduk, Polish J. of Environ. Stud. **19**, 3 (2010)
18. T. Zhu, T. Y. Diao, Y. H. Zhao, F. F. Li, Applied Thermal Engineering, **98** (2016)

Performance of special type heat sink with an integrated synthetic jet actuator

Paweł Gil[1,*]

[1]Rzeszow University of Technology, The Faculty of Mechanical Engineering and Aeronautics, al. Powstańców Warszawy 8, Rzeszów, 35-959, Poland

Abstract. The performance of special type heat sink with integrated synthetic jet actuator has been presented in this work. Synthetic jet is a flow technique which synthesizes stagnant air to a form of jet. Synthetic jet produces high turbulent flow and thus high heat transfer coefficient can be achieved. Standard heat sink with fan have limited applications in particular in a dusty industrial environment. Therefore, the use of new flow technique is optimistic. The paper presents preliminary results of heat sink thermal power and characteristic temperatures during synthetic jet switched on and off. The results show that under synthetic jet switched on, the dissipated heat is 3.7 times higher than when synthetic jet is switched off.

1 Introduction

A synthetic jet is produces by a diaphragm oscillating in a cavity with one or more orifices, which produces a volume changing and thus pressure variation. As the diaphragm oscillate, the air is periodically sucked inside and pushed out of the cavity through openings [1]. Under a certain condition [2–3] synthetic jet is formed, which is a train of vortex ring issuing from orifice. The synthetic jet actuators can utilize electroacoustic transducer [4–5] or piezoelectric diaphragm [6]. Since the end of the last century, the synthetic jet has been the subject of interest especially in the field of heat transfer enhancement [7] and flow control [8]. Synthetic jet is widely used for heat removal especially in two geometrical configurations: impinging cooling and with laminar cross flow. Synthetic jet impinging cooling required some distance between synthetic jet actuator orifice and heated plate for effective cooling because for small jet to surface distance, hot air is sucked in synthetic jet actuator cavity reducing thermal power dissipation [9]. One of the unique design solutions of the heat sink is proposed by Mahalingam and Glezer [10] who experimentally investigated performance of synthetic jet ejector heat sink. For natural convection, heat sink thermal resistance was 3.15°C/W and for synthetic jet forced convection was 0.76°C/W.

The aim of this paper is to propose a new design of heat sink with an integrated synthetic jet actuator. The synthetic jet cavity and orifices is integrated with heat sink, enabling more compact solutions. In such a geometrical configuration the problem with hot air suction process by synthetic jet actuator is solved. The thermal performance and

* Corresponding author: gilpawel@prz.edu.pl

characteristic temperature of heat sink with an integrated synthetic jet actuator was also presented in paper.

2 Experimental set-up

The heat sink with integrated synthetic jet actuator designed and manufactured for the purpose of current measurement is presented in Fig. 1. It consist of an loudspeaker (electroacoustic transducer) actuator STX W.18.200.8.FGX of 180 mm external diameter mounted to heat sink made from aluminium alloy. The heat sink has external diameter of 180 mm and 30 mm high, inside cavity is 32 fins and in the side wall there is 16 axisymmetric orifices. The heat transfer area was 0.10 m^2. The loudspeaker has nominal impedance 8 Ω and resonance frequency 38 Hz. The audio amplifier Auna CD708 was used to amplify sinusoidal signal from Rigol DG4062 function generation and to power synthetic jet actuator. The working fluid is air. The heat sink with integrated synthetic jet actuator is mounted on circular electric heater powered by DC power supply Sorensen XFR 300 V and 2 A. Temperature measurement was perform with thermocouple type K. Three thermocouple were fixed to heat sink base and seven thermocouple were used to measurement ambient temperature. The cold junction compensation was realize with Kaye 170 ice point reference. Signals from thermocouple were measure with Keithley 2700 digital multimeter (6.5 digit, 22-Bit) with 7707 all in one I/O module (10 analog input). The resolution of temperature measurement was 0.001°C and accuracy better than 0.25°C. The root mean square voltage, current and frequency of signal feeding electroacoustic transducer and DC voltage and DC current feeding electric heater was measure with Keithley 2700 digital multimeter with 7702 all in one I/O module (20 analog input) (Fig. 2). The accuracy of voltage and current measurement were better than 0.2%. LabVIEW program was utilize for data acquisition. Measurements of heat sink thermal resistance were carried out in a steady state. Each measurement series was 3,600 seconds (1 hour) and included 100 measurements of each parameter.

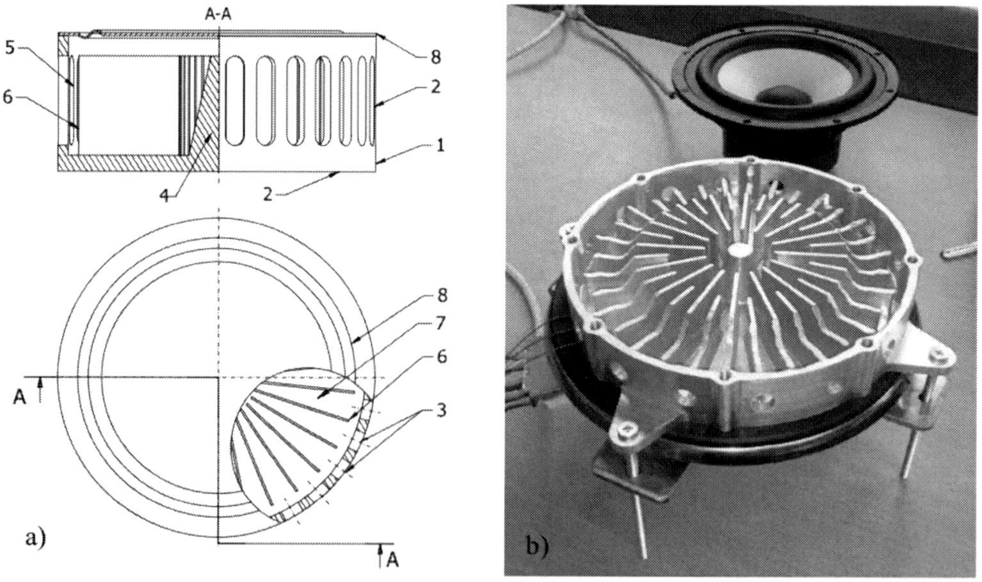

Fig. 1. a) Heat sink with an integrated synthetic jet actuator: 1 – heat sink, 2 – base, 3 – orifices, 4 – inflow steering, 5 – gap, 6 – fins, 7 – duct, 8 – actuator, b) photo of milled heat sink and actuator.

The synthetic jet actuator was powered by sinusoidal signal with frequency 38 Hz and root mean square voltage equal 5 V for each case. The electric power consumed by the loudspeaker was 1.3 W. The heat sink thermal resistance was calculated as:

$$R = \frac{T - T_{amb}}{Q} \tag{1}$$

and

$$Q = U \cdot I \tag{2}$$

where:
T – is average heat sink base temperature, °C
T_{amb} – is average ambient temperature, °C
Q – thermal power, W
U – DC voltage drop on electric heater, V
I – DC electric current feeding electric heater, A.

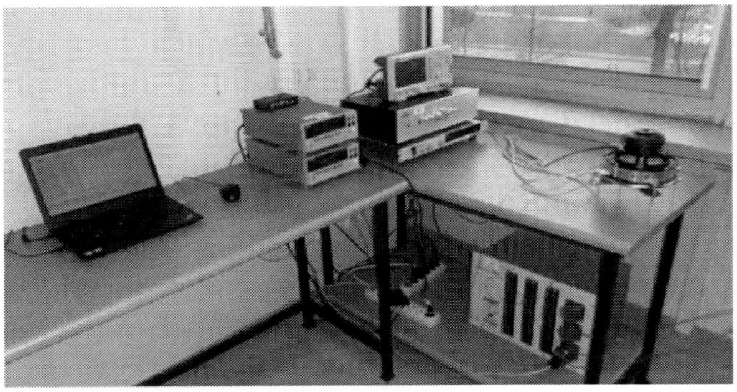

Fig. 2. Experimental set-up.

3 Results

The heat sink base temperature variation with time is present in Fig. 3. For the first 10 minutes the DC power supply was switch off, after which was set up at Q = 50 W. The ambient temperature was almost constant during all measurement $T_{amb} \approx 20°C$. After about 2h 30min the steady state for free convection was established, reaching $T \approx 80°C$, then a synthetic jet was switched on and rapid cooling capability can be seen. After next hour the heat sink base temperature was reduced to $T \approx 38°C$.

The heat sink temperatures for free and forced convection at varies thermal power is presented in Fig. 4. The temperature in the first 20 minutes is for free convection after which synthetic jet actuator is switched on (forced convection). For Q = 50 W heat sink base temperature reduction was 42°C, for 40 W and 20 W the heat sink base temperature reduction were 35°C and 19°C respectively.

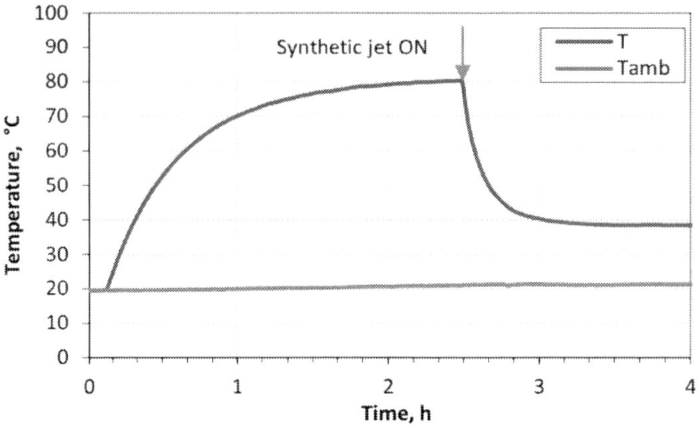

Fig. 3. The heat sink base temperature variation with time.

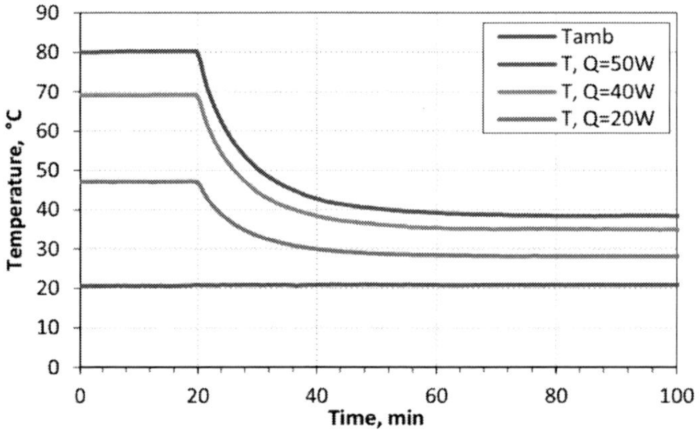

Fig. 4. The heat sink base temperature variation with time for free and forced convection.

The heat sink with integrated synthetic jet actuator thermal performance is presented in Fig. 5. Heat sink thermal resistance for free convection is R = 1.21°C/W while for forced convection is R = 0.33°C/W. As can be seen from Fig. 5 for temperature difference ΔT = 50°C (heat sink base temperature $T \approx$ 70°C) synthetic jet can dissipate more than Q = 152 W while for free convection Q = 41 W.

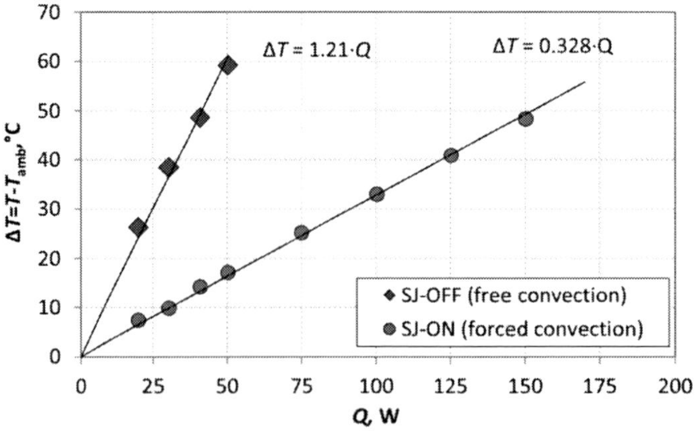

Fig. 5. Heat sink thermal resistance for free and forced convection.

The new type of heat sink can be used for heat dissipation from industrial electronics components. Especially this design is dedicated for industrial LED lighting. The LED diodes changes about 35% of the supplied electricity to the usable light and the remaining 65% is converted into thermal power. Also the temperature of the diode junction also has a significant effect on the parameters of the produced light. The proposed heat dissipation device can be relatively simple continuous controlled for providing heat sink base isothermal operation by changing the amplitude or frequency of signal feeding the actuator. In the industrial environment the dust is also significant factor which reduces a reliability of heat dissipation devices. The device in the presented geometric configuration is not sensitive to dusty environment because the fins are located inside the synthetic jet actuator cavity. In addition, electroacoustic transducers are resistant to dust due to their reciprocating movement.

The energy balance of the device can be represented by coefficient of performance COP. The COP can be defined as the ratio of increased thermal power dissipation to electrical power feeding device. The increased thermal power dissipation can be defined as the difference in power dissipation under forced (synthetic jet on) and free convection:

$$COP = \frac{Q_{Forced} - Q_{Free}}{P} \tag{3}$$

where:

Q_{Forced} – is thermal power under forced convection, W
Q_{Free} – is thermal power under free convection, W
P – electrical power feeding synthetic jet actuator, W.

The highest COP = 78 was achieved for the Q_{Forced} = 150W.

4 Summary

The thermal performance and new design of special type heat sink with an integrated synthetic jet actuator was presented and discussed in the paper. The heat sink base temperature reduction was achieved for synthetic jet operation. For thermal power 50 W the temperature drop from about 80°C for free convection to 38°C for forced convection was achieved. Significantly lower temperatures of the heat sink base result in greater reliability and durability of the electronics. The heat sink thermal resistance was reduced from R = 1.21°C/W to R = 0.33°C/W with the use of synthetic jet. The highest COP = 78 was achieved for the tested heat dissipating devices.

References

1. B. L. Smith, A. Glezer, Phys. Fluids **10**, 2281–2297 (1998)
2. R. Holman, Y. Utturkar, R. Mittal, B. L. Smith, L. Cattafesta, AIAA Journal **43**, 10, 2110–2116 (2005)
3. P. Gil, J. Fluids Struct. **81**, 466–478 (2018)
4. P. Gil, P. Strzelczyk, Exp. Therm Fluid Sci. **76**, 163–174 (2016)
5. P. Strzelczyk, P. Gil, J. Phys Conf Ser, **760**(1), p. 012029 (2016)
6. Q. Gallas, R. Holman, T. Nishida, B. Carroll, M. Sheplak, L. Cattafesta, AIAA Journal **41**, 2, 240–247 (2003)

7. M. Chaudhari, B. Puranik, A. Agrawal, Int. J. Heat Mass Transfer, **53**, 5–6, 057–1069 (2010)
8. P. Gil, J. Appl. Fluid Mech. **12**, 1, 293–302 (2019)
9. P. Valiorgue, T. Persoons, A. McGuinn, D. B. Murray, Exp. Therm Fluid Sci. **33**, 4, 597–603 (2009)
10. R. Mahalingam, A. Glezer, J. Electron. Packaging **127**, 2, 172–177 (2005)

Energy analysis of charging infrastructure for electric vehicles on the TEN-T road network

Maciej Gis[1,*], *Mateusz* Bednarski[2], and *Piotr* Orliński[2]

[1]Motor Transport Institute, ul. Jagiellonska 80, 03-301 Warsaw, Poland
[2]Warsaw University of Technology, Faculty of Automotive and Construction Machinery Engineering, ul. Narbutta 84, 02-524 Warsaw, Poland

Abstract. Electromobility is a European vision of future motorization. In Poland, there are plans to introduce a million vehicles of this type by 2030. Currently, their share is marginal (about 1 percent). This vision is to be made real. This is due to the fact that vehicle manufacturers are developing newer EV vehicle constructions. Increasing the number of electric vehicles requires the development of their charging infrastructure. Based on the work of the authors regarding the EV vehicle charging network on the Trans-European Transport Network road network, it was possible to extend this issue with energy calculations related to energy demand for supplying the EV vehicle charging network. This is an important topic from the point of view of the State's energy needs. The calculations made in the article present the problem of the need to increase the production of electricity, which in the case of Poland is associated with increased emissions of harmful substances and the possibility of periodic interruptions in the supply of electricity. Due to excessive domestic consumption with too little production. The second issue is the need to supply electricity to the charging station (infrastructure), as well as transmission losses, which limit the possibility of building multi-station fast charging stations. The issue presented by the authors in this article is one of the key problems relating to the introduction of electromobility in Poland. The key is to determine how large the demand for electricity in the country will be if a greater number of electric vehicles is put into operation. Considering that there are power shortages during a hot summer, this may affect the possibility of using electric vehicles in the country.

1 Introduction

The first attempts to work on electric vehicles began around 1800. Their popularity was significant until 1900. In later years they were supplanted by vehicles fueled with conventional fuels [10].

* Corresponding author: maciej.gis@its.waw.pl

Electric vehicles have been considered as vehicles for everyday use for about 5–6 years. This is the result of an increasing number of vehicles in the world, and thus a significant increase in emission of harmful exhaust gas, which in the end has negative impact on the natural environment and human health. These aspects combined with shrinking resources of oil-derived fuels have contributed to the intensification of works on alternative drives and fuels in transport [11–13, 17, 18].

According to the IEA report, the number of electric vehicles (BEV), hydrogen vehicles (FCEV) and plug-in hybrid (PHEV) on global roads in 2017 reached 3.1 million copies (an increase of 54% y/y). In 2017, more than 1 million electric cars were sold in total (54% more than in 2016), half of which (580 thousand) in China. The second place was taken by the United States with a result of 280.000 vehicles [14].

On the roads of the Middle Kingdom, there are currently 1.23 million electric vehicles (40% of the global EV fleet), in Europe 0.82 million, in the US 0.76 million, while in the rest of the world – 0.3 million. In 2018, the largest EV share in the new vehicle market was recorded in Norway (39.2%), Iceland (11.7%) and Sweden (6.3%). China (2.2%), Germany (1.6%), USA (1.2%) and Japan (1.0%) ranked next [14].

It is assumed that electric vehicles will be one of the most important pillars of the automotive industry. The development of this type of vehicle now allows the assumption that in the medium-term, i.e. until 2030, they are to account for around 30–35% of all vehicle sales in the world. For this plan to be accomplished their market share must be increased. In addition to convincing customers to purchase vehicles, it is also necessary to provide the right infrastructure. In the case of these vehicles, a network of vehicle charging points must be built to ensure that the vehicles' batteries are charged to 80% within 30 minutes or faster [9]. In Poland, there are currently over 300 generally available charging stations for electric cars – according to the estimates of the Polish Alternative Fuels Association (there are about 100.000 in Europe). The largest operator in Poland is Greenway. At the end of 2017, the company had 33 own charging stations (32 fast and one accelerated AC) and 4 partner stations. Now, 78 own and 15 partner stations operate in the Greenway network, which in total gives 83 [15, 19, 20, 21].

2 Electric vehicles range

Several models of electric cars are available on the market. Currently in Poland, despite an extensive promotion of electric vehicles, their sale is still insignificant. It represents less than 1% of the sales of all vehicles on the market. The number of purchased copies in the years 2015-2018 is presented in Table 1.

The vehicles offered so far have a fairly diverse range. It is largely dependent on the capacity of the batteries used. Based on the range of available electric vehicles, it is possible to determine the average mileage that vehicles can pass in a mixed cycle with fully charged batteries (table 2) [2–5]. It should be noted that the values given are those declared by the producers. In real traffic conditions, they can be much smaller.

When calculating the average range of electric vehicles available in Poland, vehicles such as Tesla Model S, Hyundai Kona Electric (extended range) and Nissan LEAF with 62 kWh battery are not included. These vehicles were rejected due to the much wider range, which would affect the overall average range of these vehicles included in the analysis. If the above vehicles had not been rejected, the average range for other available electric vehicles would increase significantly.

Based on the collected data, it has been determined that the average mileage of electric vehicles available in Poland is about 175 km. On this basis, further calculations presented in this work were performed.

Table 1. Number of electric vehicle registrations in the country in the years 2015–2018 [1].

Model	**2015**	**2016**	**2017**	**2018**	**Summary**
BMW i3	67	32	117	164	380
Nissan e-NV200	9	3	21	10	43
Nissan LEAF	24	8	239	269	540
Tesla Model S	21	4	40	20	85
Tesla Model X	–	–	31	33	64
Volkswagen Golf	1	3	13	41	58
Volkswagen up!	1	3	2	1	7
Renault ZOE	–	–	11	49	60
Other	–	–	10	50	60
Summary on the market	**123**	**53**	**484**	**637**	**1297**

3 TEN-T transport routes

The Trans-European Transport Network (TEN-T) is an instrument for coordinating and ensuring coherence and complementarity of infrastructure investments [6].

As a result of the revision of the TEN-T guidelines completed in 2013, a new agreement has been established in the territory of the EU Member States. This system includes a core network that underpins the development of the transport network on which EU activities are to focus, in particular on cross-border sections, missing links, multimodal connections and major bottlenecks, and a comprehensive network ensuring the availability and connectivity of all EU regions [6].

The trans-European TEN-T transport network includes [6]:

- road, rail, air, sea and river routes constituting the most important connections from the point of view of the development of the European Union,
- point elements of infrastructure in the form of sea, air, inland and road-rail terminals.

Its integral component is also intelligent transport systems, the implementation of which contributes to the improvement of network capacity, traffic safety and reduction of environmental pollution by transport [6].

The aim of developing the TEN-T network is to ensure territorial cohesion of the EU and to improve free movement of people and goods. An efficiently functioning transport system within the European Union is to contribute to the improvement of the operation of the single internal market, to stimulate the region's economic growth and to increase the competitiveness of individual Member States and the entire EU on a global scale [6].

The objective of EU policy in the context of the development of the TEN-T network is to create a coherent and interoperable, multimodal transport network with uniform, high technical parameters throughout the European Union [6].

Because of the nature of the work, the article focuses only on road transport. According to [7] for the TEN-T base network in Poland in 2010, roads in the Baltic-Adriatic corridor, the average traffic volume was 13,284 vehicles /day, while roads located in the North Sea-Baltic corridor 10,876 vehicle per day. On national international roads, by far the largest traffic load, amounting to over 35 thousand vehicles/day (including almost 24,000 passenger cars daily), was noted in the Śląskie Voivodship, followed by 26,513 vehicles/day (including 17,764 passenger cars/day) in the Opolskie Voivodeship. High traffic load, amounting to more than 20,000 vehicles/day (including over 13,000 passenger cars/day), was also recorded in the following provinces: Małopolskie and Mazowieckie [7].

On the remaining national roads, by far the largest traffic, with over 11,000 vehicles/day (including 8,214 passenger cars/day), was noted in the Śląskie Province. Very high load on the network of other national roads, amounting to more than 9,000 vehicles/day (including over 6,000 passenger cars/day), also occurred in the following provinces: Małopolskie and Wielkopolskie [7].

The traffic forecast for 2020 was developed for the state of the national road network in 2015. The assumption was made that no changes will take place therein and there will be no significant factors that may affect the changes in traffic behavior [7, 8].

According to [8], the highest road load according to the adopted forecast will take place around the largest urban agglomerations in Poland (including Warsaw, Katowice and Kraków, Poznań, Wrocław, Trójmiasto), where the average daily traffic volume is expected to exceed 20 thousand vehicles [7].

4 Proposal to depoy charging stations for electric vehicles at the TEN-T transport routes

On the basis of TEN-T transport routes and traffic forecasts on national roads in 2020, developed by GDDKiA, it was possible to determine road sections of the largest and most significant importance. These roads include, among others national road no. 7 (Tricity–Warsaw–Krakow), A1 motorway (Tricity–Łódź), A2 motorway (Warsaw–Frankfurt/Oder), A4 motorway (Świecko–Kraków), express road S8 – A1 motorway – national road 1 (Warsaw–Katowice) or national road 17 (Warsaw–Hrebenne).

Based on the above it is proposed to deploy fast charging stations for electric vehicles. The distance between the charging points was calculated based on the average range of electric vehicles and the possibility of charging electric vehicles. The exact calculation of the distance between the charging stations is shown in Table 2.

Table 2. Calculating the distance between electric vehicle charging stations.

	Range
The average range of electric vehicles when the batteries are 100% charged	175 km
The average range of electric vehicles when the batteries are 80% charged	140 km
Current reserve in 15% batteries	26 km
5% safety factor	~9 km
Distance between electric vehicle charging stations	**~110 km**

Based on the calculated distance between the charging points, it was possible to determine the minimum number of charging stations for electric vehicles on routes within TEN-T.

The suggested proposal comprises the construction of a quick charging station on the forecasted main transport routes in the country. Considering the assumptions regarding the distance between fast loading points and analyzing the possibility of their location and considering the current road infrastructure in the country (intersections, interchanges, road type), it was possible to determine the minimum number of charging points on individual routes (Świecko–Suwałki –14; Trójmiasto–Kraków – 12 ; Trójmiasto–Łódź – 6; Łódź–Wrocław – 6; Katowice–Rzeszów – 4; Warszawa–Hrebenne – 6; Warszawa–Katowice – 6; Warszawa–Katowice – 6; Warszawa–Kraków – 6; Overall – 60).

Taking into account such variables as the location and infrastructure of given road sections, it was possible to determine the number of fast charging points for electric vehicles. Their minimum number is 60 points. With such infrastructure it will be possible to use electric vehicles on roads within the TEN-T transport routes and constituting road sections that will be heavily loaded in the perspective of 2020, without worrying about the lack of space for charging vehicles.

5 Energy demand of an electric vehicle

The demand for electricity is growing in Poland. According to the available information, in 2017 more than 170 TWh of energy was produced in the country. It is 2.2% more than in 2016. Table 3 presents the division of electricity production from various sources for years 2016–2017.

Table 3. Electricity production in Poland [16].

Specification	**2016**	**2017**	**Change**
	GWh		%
Total production	**166 597**	**170 335**	**+2.2**
Thermal power stations	140 378	140 259	–0.1
- brown coal	51 082	52 281	+2.3
- coal	80 173	79 265	–1.1
- gas	5 604	6 161	+9.9
- biomass/biogas	3 519	2 552	–17.5
Hydroelectric power stations	2 335	2 719	+16.4
- including pumped storage	482	474	–1,6
Wind power stations	2 981	3 485	+16.9
Industrial power plants	9 897	11 417	+15.4
Independent power plants – renewable energy installations	11 006	12 454	+13.2

On the basis of own tests of an electric vehicle in real traffic conditions, it was possible to determine its average energy consumption. Then it was determined with the assumption of daily runs (about 60 km), what would be the energy demand of one vehicle and energy costs related to it (Table 4).

Table 4. Annual energy consumption needed to recharge the batteries of one vehicle.

Specification	
Daily mileage on business days	60 km
Daily mileage on Sundays and holidays	60 km
Annual mileage on business days	15 060 km
Aaily mileage on Sundays and holidays	6840 km
Total mileage	**21 900 km**
Price 1 kWh for a household (G11 tariff)	0,55 zł/kWh
The cost of charging one vehicle a year	**2529.45 zl**
Annual energy consumption needed to recharge the batteries of one vehicle	**4599 kWh**

Next, the results of the annual electricity consumption needed to charge the batteries of one electric vehicle were referred to the assumed introduction in Poland of a million electric cars. On this basis, the annual energy consumption needed to recharge the batteries of one million electric vehicles could be determined (Table 5). Next, the percentage of energy demand in the country will be estimated.

Table 5. Energy demand for powering electric vehicles.

Specification	
Annual energy consumption needed to recharge the batteries of one electric vehicle	4599 kWh
Annual energy consumption needed to recharge the batteries of a million electric vehicles	4,599 TWh
A change in energy demand while supplying a million electric vehicles	+2.7%

Calculations show that with one million vehicles the energy demand of the country will increase by 2.7%. This is not a significant change with regard to the energy demand in the country in 2017, which amounted to over 170 TWh. One can say that the only problem is the time of charging electric vehicles. It is best for financial reasons that this process takes place at night. The tariff adopted for the purpose of the calculation is a tariff applicable to households (G11) and amounts to 0.55 PLN/kWh. With this assumption, the annual costs relating to the charging of an electric vehicle amount to PLN 2529.45. With regard to conventional vehicles, this is a significant difference at the moment.

6 Infrastructure for charging electric vehicles on the TEN-T road network

On TEN-T routes, the so-called quick chargers for electric vehicles are logical. There are currently 55 kWh variants. On this basis, it is possible to calculate the energy demand assuming that all specified stations are simultaneously used (the so-called black scenario). The above calculations showed that a sufficient number of stations on the TEN-T road network is 60 points. Considering the number of electric vehicles registered in the country (table 1), at each point there should be four stations for simultaneous charging of electric vehicles. This gives an overall ability to load 240 vehicles simultaneously in the country (table 6).

Table 6. The number of charging points for electric vehicles.

Specification	
The number of charging points for electric vehicles on TEN-T transport routes	60
Number of stands at each charging point	4
The total number of electric vehicles that can be recharged	**240**

With such a large number of stations and the assumption that electric vehicles on routes will be loaded within 30 minutes (battery charging up to 80%), the energy demand to power fast chargers can be determined. Once these values were determined, it was possible to determine daily and annual energy demand (Table 7).

Table 7. Energy demand for supplying fast chargers (55 kWh) for powering the batteries of electric vehicles.

Specification	Change
Hourly energy demand of fast chargers	55 kW/h
Energy demand of fast chargers for charging electric vehicle batteries in 80%	27.5 kW/h
Individual energy demand	**6.6 MWh**
Daily energy demand	**0.3168 GWh**
Annual energy demand	**115.632 GWh**

7 Conclusions

The proposal to build fast-charging points for electric vehicles on roads belonging to the TEN-T transport routes and constituting road sections that will be heavily loaded in the perspective of 2020, allowed to determine that to move the above-mentioned it is necessary to build 60 charging points for the type of vehicle. Other conclusions are:

- the proposal to build fast-loading points for vehicles would already allow the operation of available electric vehicles on the main transport routes,
- taking into account the proposal to build charging points on the main transport routes and the assumed development of electric vehicles (increasing their range) will enable further development of the network of charging points beyond the main transport routes,
- the development of the network of charging points can have a significant impact on the popularization of electric vehicles, and thus reduce the emission of harmful substances from transport,
- the increase in the energy load associated with the increase in the number of electric cars amounting to 2.7% will have insignificant environmental importance, considering lower consumption of petrol due to the limitation of the use of cars with an internal combustion engine.

References

1. www.samar.pl
2. www.bmw.pl
3. www.nissan.pl
4. www.tesla.com

5. www.volkswagen.pl
6. www.mib.gov.pl
7. Praca zbiorowa: „Przesłanki narodowego planu wodoryzacji transportu samochodowego w Polsce". Projekt HIT-2-Corridors. Sprawozdanie z realizacji pracy nr 5502/ITS. 25.09.2015
8. www.gddkia.gov.pl
9. K. Qian, Ch. Zhou, M. Allan, Y. Yuan, *Modeling of Load Demand Due to EV Battery Charging in Distribution Systems* (IEEE Power & Energy Society, 2010)
10. K. Rajashekara, *Present Status and Future Trends in Electric Vehicle Propulsion Technologies* (IEEE Power & Energy Society, 2013)
11. Biała Księga. Plan utworzenia jednolitego obszaru transportu – dążenie do osiągnięcia konkurencyjnego i zasobooszczędnego systemu transportu. KOM (2011)
12. J. Merkisz, I. Pielecha, *Alternatywne paliwa i układy napędowe pojazdów* (WPP, Poznań, 2004)
13. S. W. Kruczyński, M. Ślęzak, W. Gis, P. Orliński, Eksploatacja i Niezawodność, **18**, 3, 342–347 (2016)
14. http://www.elektro.info.pl/aktualnosc/id8599,iea-liczba-pojazdow-elektrycznych-na-swiecie-przekroczyla-3-mln-sztuk
15. https://wysokienapiecie.pl/13819-stacje-ladowania-powstaja-przy-trasach-i-w-miastach/
16. https://rynek-energii-elektrycznej.cire.pl/st,33,207,tr,75,0,0,0,0,0,podstawowe-dane.html
17. J. Merkisz, J. Pielecha, P. Lijewski, *Exhaust emissions from vehicles in real traffic conditions in the Poznan agglomeration* (21st International Conference on Modelling, Monitoring and Management of Air Pollution, Siena, Italy, 2013)
18. W. Gis, J. Pielecha, J. Waśkiewicz, *Use of certain alternative fuels in road transport in Poland* (Scientific Conference on Automotive Vehicles and Combustion Engines KONMOT, Krakow, Poland, 2016)
19. R. Jasinski, *Mass and number analysis of particles emitted during aircraft landing*, E3S **44**, 00057
20. R. Jasinski, J. Pielecha, J. Markowski, *Emission of particulate matter during aircraft landing operation*, E3S, **10** (2016)
21. Z. Chłopek, J. Lasocki, P. Wójcik A. J. Badyda, International Journal of Green Energy, **15**, 13, 773–779

Characteristics of water sludge from Cracow Water Treatment Plants – case study

Renata Gmurkowska[1,*]

[1]University of Technology, Department of Environmental Engineering, ul. Warszawska 24, Cracow 31-155, Poland

Abstract. During water treatment a large amount of sludge is created – in the form of sewage and sediments. The largest amounts of sludge are produced during coagulation, ozonation and backwashing rapid filters. The quality and quantity of treated water, the type and dose of used coagulants are factors affecting the quantity, composition and properties of the sludge. Sludge produced during processing of drinking water is important problem and their quantity has been increasing. The study focuses on characteristics of water treatment sludge from four water treatment plants in Cracow. It includes theoretical and experimental part. The first part is based on analysis of literature and information obtained from MPWiK [3]. The second experimental part, concerns the analysis of dry matter, organic dry matter, capillary suction time and visual parameters: the color and consistency of the sludge. Result shows that every sludge contains organic matter. The highest concentrations of organic compounds and the largest diversity has been observed in the sludge collected in the Water Treatment Plant Raba, reaching even up to 70% of organic compounds in the dry mass of sludge.

1 Introduction

Water treatment plants, apart from the main product – a potable water, generate also some by-products (waste) such as e.g. hydrated sediments formed during the coagulation, sedimentation and ion exchange processes. The waste include also technological wastewater i.e. water resulting from backwashing of rapid filters, iron and manganese filters or ion exchangers. The amount of waste is closely related to the raw water quality, the assumed process efficiency and the type and doses of reagents [1]. Due to the more stringent quality standards applied to drinking water (the Regulation of the Minister of Health of December 7, 2017 (Journal of Laws, 2017, item 2229) the amount of sludge is constantly increasing, as well. The sludge produced during a water treatment processes (water sludge) can be polluted with heavy metals; treatment a surface water may also generate waste with polycyclic aromatic hydrocarbons (PAH), pesticides, chlorinated hydrocarbons and polyelectrolytes dosed during water treatment. These compounds are considered hazardous and therefore have to be monitored [2]. They can strongly limit a possible sludge reuse and management. The biological composition of water sludge

* Corresponding author : renata.gmurkowska@gmail.com

depends primarily on quality of a treated water. Sludge resulting from a surface water treatment is definitely more biologically contaminated then sludge resulting from a groundwater treatment; the former contains a large number of bacteria (including Escherichia coli, Salmonella, *Pseudomonas aerubinosa*, Legionella), viruses and intestinal parasitic cysts. The number of bacteria found in the coagulation sludge (moisture 99%) is 2300/100 cm^3 of sludge [2].

All these issues contributed to changes in the legislation, which in turn forced some changes in sludge management. The new techniques of sludge processing and final handling are investigated. Currently, the legislation tries to limit the amount of by-products. According to the Act of December 14, 2012 on waste (Journal of Laws of 2013, No. 0, item 21, article 22, chapter 5), sludge is considered a hazardous waste and the waste management costs are borne by the original producer or by the holder. It means that the sludge cannot be utilized, as it was before, without any treatment and its characteristics should be closely monitored. In addition, the Regulation of the Minister of Economy of July 16, 2015 on the admission of waste for landfill entered into force on January 1, 2016; it prohibits a storage of waste (including sewage sludge) with a significant energy potential (over 6 MJ/ kg of dry solids). The new law also limits the possibility of storage and agricultural use of waste considered hazardous and non-hazardous, e.g. water sludge and bottom sediments. Due to a growing pollution of the natural environment, including surface waters, a water sludge may contain and most often contains heavy metals and organic and toxic substances harmful to human health and the environment. However, once organic parts are removed from the sludge, the mineral residue can be disposed as a mineral raw material or safely stored. The following parameters should be specified in such case:

- physic-chemical characteristic of sludge,
- hazardous and toxic substances, including heavy metals,
- pollutants leaching and how it is affected by environmental factors and a deposition time.

For this reason, the physic-chemical characteristic of waste becomes an important issue. Often, it is the waste characteristic and leaching of pollutants that are the basic criteria that help to define: class of waste, the method of a final sludge handling, the process design and efficient operation.

2 Methods

The material for the study was collected at three Cracow water treatment plants: ZUW Raba, ZUW Rudawa and ZUW Dłubnia; infiltration ponds at the ZUW Bielany were not included into analysis. Grab sludge samples were collected in clean and properly prepared containers in summer, fall and winter. At the ZUWs Raba and Rudawa, the sludge was collected from sludge drying beds, whereas at the ZUW Dłubnia from a sludge well, located ahead of the dewatering press. The samples were analyzed on the same day in the Laboratory of Water and Wastewater Technology, the Faculty of Environmental Engineering, the Cracow University of Technology. Each sample was tested in triplicate. Analysis of dry solids (DS) and dry volatile solids (DVS) were done according to PN-EN12879:2004, while a capillary suction time (CST) was tested according to PN-EN14701-1:2007. In addition, pH, color and consistency of each sludge sample was determined.

3 A research area – technological characteristics of the Cracow water treatment plants

3.1 ZUW Raba

The largest treatment plant in Cracow takes water from the Dobczycki reservoir, constructed as a result of partition of the Raba river bed. Water is collected from three different levels at the tower intake, located on the bank of the reservoir. Water from the intake is pumped to the treatment plant where it enters the ozone contact chamber and then the mixing chambers where the reagents are dosed. Next, water splits into two technological lines: Raba I and Raba II. In the Raba I line, the water with coagulant flows through mixers I, II and III and then goes to the vortex chambers, where the coagulation process begins. Next, water passes to clarifiers and rapid filters. In the Raba II line, water from the mixing chambers is directed to four accelerator clarifiers, where coagulation and sedimentation take place and then to sand-anthracite filters. Finally, water from both lines is collected in clear water tanks and pumped via disinfection chambers to the three Gorzków reservoirs. Water is disinfected with UV lamps and sodium hypochlorite produced with electrolytic method [3].

During water treatment, sludge is produced, which contains both pollutants removed from water and chemicals added to water during treatment (coagulant, carbon, lime). The sludge is discharged from the production line at the following locations: a contact tank (common for both lines), accelerators (RII line), clarifiers (RI line) and filters (lines RI and RII, separately). Sludge from the RII line is transported to six sludge thickeners. On the other hand, in the RI line sludge settles in gravity settlers (4 units), from which is discharged to sludge drying beds (two sets). Sludge drying beds of the RI line comprise two units of area 160 m^2 each, the sludge level is 0.65 m and sludge drying beds of the RII line comprise 14 covered beds (8 units of area of 290 m^2 and 6 units of area of 490 m^2); the maximum sludge level 0.85 m [3].

3.2 ZUW Dłubnia

The Dłubnia water treatment plant supplies water for the metallurgical plant and residents of Nowa Huta. The raw water from the Dłubnia river is coagulated in a coagulation system comprising: coagulant dosing units, rapid mixers, slow mixers and clarifiers. Water from the clarifiers passes through the overflow channel to ten rapid filters. After filtration and disinfection with chlorine dioxide water goes to the "lower" tanks [3].

The sludge from sedimentation processes as well as filter backwash waters flow to a sludge thickener. The supernatant, after settling, is discharged into the sewage system while remaining sludge is pumped to the thickened sludge tank. The sludge moisture is 99%. Further processing involves flocculant dosing and sludge dewatering at a belt press. After serious dewatering sludge moisture is lowered to 35% [3].

3.3 ZUW Rudawa

Water treatment at the ZUW Rudawa involves coagulation with aluminum sulphate and filtration on rapid filters. Raw water is stored in the dam located in the village of Szczyglice and from there it flows into two sedimentation ponds and then to retention reservoirs. Next, water passes to the collection wells located at the plant site and then is pumped successively to four high-speed mixing chambers and slow-speed mixing chambers. After sedimentation in clarifiers water is transported to 12 rapid sand filters (filter bed thickness of 1.1 m) and then to 6 carbon filters. The filter effluent is disinfected with chlorine dioxide (IV) produced on site [3].

Sludge generated in the course of the treatment process is transported to sludge thickeners and then to a filter press for further dewatering [3].

3.4 ZUW Bielany

The oldest water treatment plant in Cracow has been put into operation in 1901. Currently, its production (12–18 thousand m^3/day) is based solely on water from the Sanka river. The water, once stored in the dam, flows through bar screens and the grit chamber to the collecting well. In the well, there are pumps that pump water to the clarifier or on slow filters. In slow filters, raw water passes through a natural filter bed. There are 96 wells near the filtration basins to collect water from the aquifer. The water is drained with siphons to the two collecting wells and then disinfected with sodium hypochlorite generated with a electrolytic method. Finally, water free of microorganisms is discharged to three drinking water tanks [3].

4 Disscusion of results

The paper presents analyzes of sludge produced at the ZUWs Raba, Dłubnia and Rudawa, as a technological process by-product. The results include the visual sludge characteristic, i.e. color and consistency as well as other measurable parameters, i.e. pH, dry solids (DS), dry volatile solids (DVS) and capillary suction time (CST).

At the ZUW Raba, both color and consistency of sludge varied with seasons. The sludge produced in summer had the highest density (a solid consistency) and therefore a sample of 10 g of sludge was needed for DS and DVS analysis; in other tests 50 ml of liquid sludge was required. The sludge pH turned out to be the most stable parameter and it stayed at approx. 6.7. The highest pH value as well as the highest CST (268.10 s) was found in the sample collected in July. The lowest CST was detected in the last measurement series, i.e. in December. The CST value can be directly linked to the sludge consistency. The sludge with a higher moisture and a lower CST releases water more easily.

The lowest content of organic compounds was found in sludge from the ZUW Dłubnia. The maximum value was recorded in summer while the minimum one in fall. The sludge had also high pH, if compared to the sludge produced during treatment of water from the Dobczycki reservoir. The average pH value was 7.48. The CST was much lower than at the ZUW Raba; the maximum value was recorded in August, while the minimum one in winter.

The mass loss after heating the sample to 550°C determines a relative content of organic matter in a water sludge. Higher mass losses (a higher content of organic compounds) are observed during treatment of strongly polluted water as well as water where powdered activated carbon (PAC) was used during coagulation. The PAC content in the sludge increases the amount of organic material, while on the other hand, PAC stabilizes the sludge biologically, reducing its ability to decay [4].

Sludge from surface water coagulation usually has black and gray color due to the presence of humus, clay particles, and in some cases also powdered organic carbon. This color predominated in the sludge from the ZUW Dłubnia and the ZUW Rudawa, both in fall and in winter.

Table 1. Sludge from the Cracow water treatment plants.

Season	Colour	Consistency Liquid/semi-liquid /solid	pH	Dry solids, g/dm^3	Dry volatile solids, g /dm^3	CST, s
ZUW Raba						
Summer	gray-brown	solid	6.9	117.6 g /kg	32.91 g /kg	268.10
Fall	black and gray	semi-liquid	6.79	9.52	6.58	106.98
Winter	gray and green	liquid	6.52	17.37	4.42	68.66
ZUW Dłubnia						
Summer	grey and brown	liquid	7.23	45.91	6.31	55.79
Fall	black and gray	liquid	7.62	9.51	1.65	39.73
Winter	black and gray	liquid	7.59	10.82	3.02	31.12
ZUW Rudawa						
Summer	grey and brown	liquid	7.26	60.53	13.88	54.06
Winter	black and gray	liquid	7.56	39.96	7.936	25.14
Fall	black and gray	semi-liquid	6.96	56.54	10.75	92.05

Table 2. DS, DVS and organic content (%) in sludge at three Cracow water treatment plants.

	Season	Dry solids, g /dm^3	Dry volatile solids, g/dm^3	Organic content, %
ZUW Raba	Summer	117.64 g/kg	32.91 g/kg	28.0
	Fall	9.52	6.58	69.1
	Winter	17.37	4.42	25.5
ZUW Dłubnia	Summer	45.91	6.31	13.7
	Fall	9.51	1.65	16.4
	Winter	10.82	3.02	27.9
ZUW Rudawa	Summer	60.53	13.88	22.9
	Fall	39.96	7.94	19.1
	Winter	56.54	10.75	19.0

Table 2 summarizes the results of DS, DVS and organic content (%) in sludge for three water treatment plants at the Cracow agglomeration. The highest organic content was recorded in sludge samples collected at the ZUW Raba. The maximum value over 69%was observed in fall and it was much higher than the other ones. This may be caused by some accumulation of organic compounds in summer, which resulted in a significantly higher content of an organic material in sludge in November.

The Dobczycki reservoir is located in heavily urbanized areas that may pose a threat to water quality. Also, the basin of the Raba river, the main source of water for the reservoir, still requires some efforts in terms of sewage management control. The inflow of polluted water is one of the reasons for the unsatisfactory water quality in the reservoir. The sources of pollution include: nutrients from municipal sewage, mineral fertilizers or surface runoff from the catchment. A significant input of nutrients to surface waters results in escalation of eutrophication processes. Pawełek and Spytek [5] in their research in the years 2003–2005 found that the maximum daily loads of the analyzed nutrients (NO_3, NO_2, NH_4, PO_4andP_{total}) were recorded in 2005. The main nutrient that stimulates eutrophication is phosphorus. The authors pointed out that in 2005 the phosphorus load exceeded more than twenty times the hazardous load and over forty times the acceptable load. The eutrophication process is manifested at first by the ecosystem imbalance and following accumulation of excess organic matter in surface waters [6]. A continuous input of nutrients to the Dobczyce reservoir may therefore have negative effects on quality of water supplied to the Cracow water supply system; nutrients may also be responsible for a higher organic content in the sludge.

Also, the other two plants: ZUW Dłubnia and ZUW Rudawa showed the highest sludge concentrations (DS) in summer (Tab.1); the sludge concentration at the ZUW Rudawa was the highest and reached 60.53 g/kg. Such high values were also accompanied by the highest concentrations of DVS. The lowest DS and DVS values for both plants were found in fall and can be attributed to weather conditions, i.e. heavy rainfall. In addition, the lower sludge concentrations may result from the particular technology used. At the ZUW Dłubnia, the sludge concentration is kept at a certain level so it can be easily pumped to a belt press for dewatering. At the other two plants, the sludge moisture largely depends on the atmospheric conditions. At the ZUW Rudawa, the sludge is deposited in uncovered thickeners, while at the ZUW Raba on roofed sludge beds, which significantly reduce the weather impact, though do not eliminate it.

Another important parameter defining the water sludge characteristic and possibility of its safe management is a heavy metals content. In Poland, there are no regulations regarding acceptable levels of toxic materials, i.e. heavy metals and other substances. However, according to the American standards, waste is classified as hazardous or toxic based on leaching tests. The tests are carried out according to two procedures EP and TCLP. Monitoring of dangerous substances required for agricultural sludge management depends on the decisions of state bodies. However, the Federal Rules for the Control of Pollutants and Waste Management Rules specify the maximum allowable concentrations of heavy metals in mg/kg of sludge, which determine its classification. In Table 1 the maximum allowable concentration of heavy metals in sludge suitable for agricultural use were listed, according to the American standards. The seven heavy metals were specified, which are usually analyzed by the Central Laboratory of Cracow Water Works. Comparing the values with data provided by the Central Laboratory for the sludge from the ZUW Raba, it can be noticed that none of the values exceeds the maximum concentration. Moreover, for six heavy metals the concentrations are much lower than the maximum allowable concentrations [7].

Table 3. Maximum allowable concentrations of heavy metals in sludge used for agriculture, as specified by the Federal Water Pollution Control Act and the Solid Waste Management Act [7].

Metal	Maximum allowable concentrations, mg/kg dry solids
Zinc	7500
Lead	840
Copper	4300
Chromium	1200
Cadmium	85
Mercury	57
Nickel	420

The capillary suction time tests were carried out to evaluate how easily sludge releases water. The method measures the time of crossing of the frontal boundary layer of filtrate between the concentric circles 32 and 45 mm in diameter, on the filter paper. The lower capillary sucking time (CST) the more easily (sooner) sludge releases the liquid [8].

Table 4. CST for three water treatment plants and different sludge consistencies.

	Season	Consistency solid/semi liquid/liquid	CST, s
ZUW Raba	summer	solid	268,10
	fall	semi liquid	106,98
	winter	liquid	68,66
ZUW Rudawa	summer	liquid	54,06
	fall	liquid	25,14
	winter	semi liquid	92,05
ZUW Dłubnia	summer	liquid	55,79
	fall	liquid	39,73
	winter	liquid	31,12

Table 4 illustrates sludge consistency and sludge CST, as observed in the samples collected at all three water treatment plants. For the ZUW Raba, it was clear that CST varied with a sludge consistency. In summer, when the collected sludge had a solid consistency, the highest CST was recorded; it means that the samples did not release water easily. In the next sample (November 10) the CST was more than a half of the summer sample. The consistency of the sludge also changed; it was semi-liquid in fall. The lowest CST equal to 68.66 s was recorded for the last liquid sample, taken in December. The sludge collected in winter dewatered easily. The highest CST was recorded in the semi-liquid sludge samples at the ZUW Rudawa, in winter. The results obtained for the ZUW Dłubnia and the ZUW Raba were similar; they showed the lower CST values for the lower air temperatures. In summer, CST was 55.79 s and it was close to CST measured in summer for sludge from sludge drying beds at the ZUW Rudawa.

5 Summary

According to current regulations, water sludge poses a significant environmental problem. The ways of its utilization are mostly determined by its properties and physic-chemical characteristic, while the strong trend of the closed-circuit economy forces the scientists to look for new ways of sludge handling. Water sludge may be utilized as a substrate in some technological processes but its by-products should be significantly reduced. In addition, heavy metals in sludge limits a possibility of its use in agriculture, which until now was a popular practice. Therefore, it is necessary to look for new options for sludge management, that are possible to use in a technological scale and consider all restrictions resulting from the legal provisions.

References

1. A. Nawrocka, M. Włodarczyk-Makuła, Tech. Wod. **34–39** (2014)
2. A. L. Kowal, M. Świderska-Bróż, *Oczyszczanie wody. Podstawy teoretyczne i technologiczne procesy i urządzenia* (2009)

3. MPWiK S.A. Informacje udostępnione przez Miejskie Przedsiębiorstwo Wodociągów i Kanalizacji Spółka Akcyjna w Krakowie (2016)
4. A. Florkiewicz, *Technologia usuwania i unieszkodliwiania osadów z uzdatniania wody* (1999).
5. J. Pawełek, M Spytek, Infrastruktura i Ekologia Terenów Wiejskich **3**, 1, 107–116 (2006)
6. E. Neverova-Dziopak, Z. Kowalewski, Czas. Inż. Ląd., Środ. i Arch. **7–9**, 47–58 (2013)
7. M. Leszczyńska, M. M. Sozański, Ochr. Środ. i Zas. Nat (2009)
8. A. Piotrowska-Cyplik, Z. Czarnecki, Journ. of Res. and Appl. in Agr. Eng. **50**, 1 (2005)
9. I. Płonka, *Met. cięż. w os. pow. przy uzd. wod.* (2012)

Experimental investigation of thermoelectric cooling system with heat recovery

Justyna Gołębiowska[1,*], and *Agnieszka* Żelazna[1]

[1]Lublin University of Technology, Faculty of Environmental Engineering, Institute of Renewable Energy Engineering, ul. Nadbystrzycka 40B, 20-618 Lublin, Poland

Abstract. Thermoelectric modules are implements used as power generators and cooling devices. In case of TE module application for space cooling, it is highly recommended to reuse energy generated on the hot side of the module for example for domestic water heating. In this paper, an example of a small scale thermoelectric cooling system with heat recovery was presented. In the tested system, the performance of TE module was supported by two heat exchangers. On the cold side of the TE module the forced convection was provided by the implementation of aluminium heat sink integrated with fan. On the hot side a water heat exchanger was built in order to reuse waste energy to heat water in the water tank. Different values of current were applied to the TE module to observe the differences in the module performance. The correlation between the air temperature inside the experimental room and the water temperature in the tank were determined for all conducted series. In general, the best cooling effects were achieved when the water temperature in the tank was not higher than 26.4°C. The maximum reduction of temperature inside the experimental room was noted for the amperage 7 A and it was 11.7°C.

1 Introduction

1.1 Construction and operation of TE module

Thermoelectric module (TE module) is a semiconductor-based electronic component consisted of n- and p-type doped semiconductor elements placed between two ceramic substrates. Thermoelectric elements in the module are connected electrically in series and thermally in parallel [1]. The typical construction of the TE module is shown in Fig. 1.

As a result of doping, the p-type semiconductor has fewer electrons in a given density compared to n-type. When a DC voltage is applied to the module, the electrons movement occurs. The transition of electrons from the lower energy state to the higher one requires energy supply. Therefore, on the p-n junction, the energy from the environment is absorbed (this is the cold side of the module). The opposite situation occurs on the n-p junction where due to the transition of electrons the excess energy is released into the environment (this is

* Corresponding author: j.golebiowska@pollub.pl

the hot side). This phenomenon is known as the Peltier effect and it is used in cooling process [2].

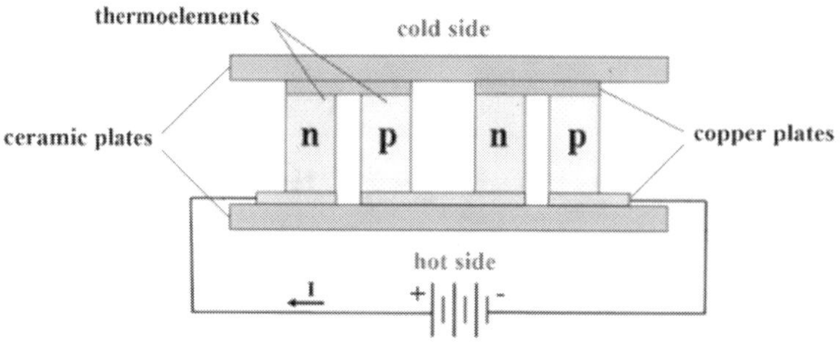

Fig. 1. Schematic construction of TE module consisted of n-type and p-type semiconductor elements [1].

Nevertheless, according to the Seebeck effect, TE modules can be also used as power generators (as a result of temperature difference between the hot and cold sides of the module) [3]. There are also other physical phenomena, such as Joule and Thomson effects as well as heat conduction, affecting the TE module performance [1, 4–6].

The cooling capacity of the TE module, understood as its ability to remove heat, depends on the electric current applied to the module, the temperatures obtained at the hot and cold sides of the module and properties of thermoelectric material. A good thermoelectric material should be characterised by high Seebeck coefficient, high electrical conductivity and low thermal conductivity. A lot of studies were conducted on this issue [7–11]. Currently, the most popular thermoelectric material is bismuth telluride (Bi_2Te_3) characterized by ZT (dimensionless figure of merit) value of 1 – with this value of ZT the TE module can achieve only 10% of Carnot efficiency [7].

Refrigeration based on thermoelectric technology has advantages of simple use, compact size, lack of moving parts, lover levels of noise and vibration, no working fluid and small cooling capacities. The main disadvantages of thermoelectric cooling are the high cost and low energy efficiency.

1.2 Application of TE modules for cooling

Thermoelectric modules are often used for cooling electronic devices such i.e. processors or graphics cards, as well as in laboratory equipment due to high precision temperature control [4, 12, 13]. They are used in tourist refrigerators (also those powered by solar energy) [14].

There is also a possibility to use TE modules as components of building walls or windows in *Active Building Envelopes* systems, to support cooling and heating processes in the buildings [15, 16].

Cosnier et al. [17] presented an example of TE modules implementation in ventilation system (TE modules were integrated with the ventilation duct). Reported studies showed that a *Coefficient of Performance* (COP) of the system solution, understood as a ratio of useful energy gained from the system to the energy input to the system, in air cooling mode was more than 1.5, while in heating mode was up to 2.0.

He et al. [18] presented theoretical and experimental investigations of a small-scale thermoelectric cooling system, powered directly from photovoltaic panels. In the studies heat-pipe exchanger was used to reuse waste heat for applied modules for water heating purposes. The obtained COP of the system reached values above 0.45.

Liu et al. [19] presented theoretical and experimental research of a thermoelectric solar air conditioner with the possibility of heat recovery for domestic water heating. The system was tested in laboratory conditions. The conducted studies showed that the designed system, operating in the cooling mode with the option of heat recovery achieved a relatively high COP value – up to 4.51 (at water temperature in the tank of 20°C).

2 Materials and methods

2.1 Experimental setup

The thermoelectric cooling system (used for microscale room cooling and water heating under the laboratory conditions) was applied to an experimental room of cubature 0.125 m^3. The experimental room consisted of polystyrene boards (5 cm of thickness) characterized by thermal conductivity coefficient of 0.034 W/(m·K). In the conducted research study, the cooling-type thermoelectric module was used. The basic parameters of the TE module are as following: maximum amperage 8.5 A, maximum voltage 15.5 V and the maximum temperature difference between the hot and cold sides 70°C.

The direction of the current applied to the TE module ensured the cooling effect inside the experimental room. On the cold side of the module (inside the room) the aluminum heat sinks integrated with fans was installed. On the hot side of the module the water heat exchanger was installed. Water exchanger consisted of water cooling block Thermaltake Pacific W1, insulated water tank (5 dm^3 of water volume), PVC pipes (inside diameter of 9 mm) and universal water pump New Jet NJ 600.

To measure air and water temperature, voltage and intensity of current applied to the module and water flow the following devices were used:

- temperature sensors Pt500 (measuring accuracy ± 0.1°C) within the water exchanger,
- temperature sensors Pt500 (measuring accuracy ± 0.1°C) inside the experimental room,
- APEK AL 154 recorder connected to temperature sensors,
- digital multimeters (PeakTech),
- 1/2" single-jet dry-running water meter with pulse transmitter (BMETERS).

The scheme of applied laboratory installation allowing tests of thermoelectric module is presented in Fig. 2.

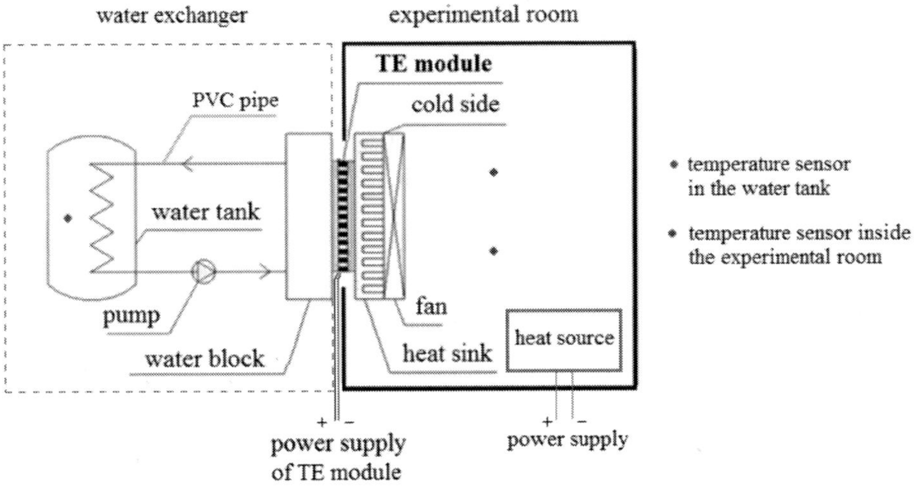

Fig. 2. Schematic construction of air cooling and water heating system with TE module.

2.2 Experiment assumptions

The single measurements series lasted until the water temperature in the tank reached 55°C (4 A series was an exception). The mentioned above assumed value (55°C) was not exceeded during all performed measurement series, for all applied values of current, because of the strength parameters of the materials used in the system and required tightness of connections of water exchanger elements.

A heat source supplied from the power grid was placed inside the experimental room in order to create an internal heat gains (see Fig. 2).

3 Results and discussion

3.1 Experiment results

The changes of air temperature inside the experimental room for different current values are presented in Fig. 3. The initial temperature of air in all cases was approx. 20.5°C. In all cases the decrease in the air temperature to the minimal value was achieved after approx. 40 minutes of cooling process duration. The minimum measured temperature values were as follow: 13.1°C (4 A), 11.3°C (5 A), 9.5°C (6 A) and 8.55°C (7 A). Afterwards, the measured temperature values increased significantly. The following values of the final air temperature were observed: 20.5°C (4 A), 19.1°C (5 A), 16.7°C (6 A) and 15.1°C (7 A). Thus, the highest values of applied current the better cooling effects, understood as the maximum decrease in air temperature, were observed.

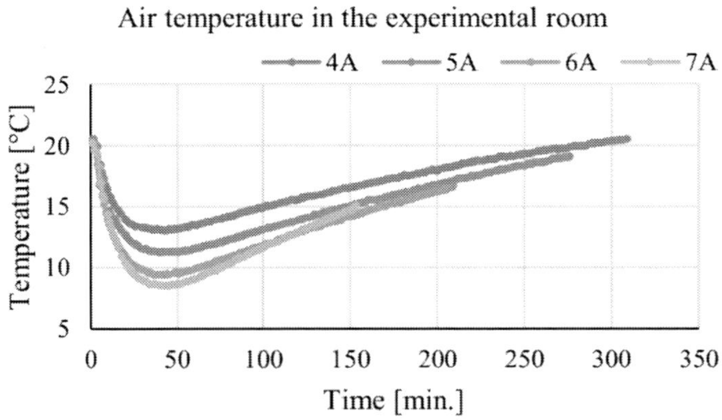

Fig. 3. Air temperature changes inside the experimental room during the tests.

However, there is a maximum possible temperature difference between the cold and the hot side of the TE module. During the experiments, the continuous increase in water temperature inside the tank (due to zero water demand) triggered the eventual increase in the air temperature inside the experimental room which can be noticed after 40 minutes of cooling process duration (see Fig. 4).

In the Fig. 4 the changes of water temperature in the tank for different currents are presented. The initial temperature of water in all analyzed series was approx.10.5°C. In the case of amperage 4 A, the maximum temperature of water at the end of process was 50.5°C (for this value of water temperature, temperature of the air inside the experimental room was equal to the initial one). In the other cases, for amperage values 5 A, 6 A and 7 A, the temperature of water at the end of series was 55°C. The higher value of applied current, the faster process of water heating was noted. It results from the additional heat, described by Joule's first law, generated by current flowing through the electrical conductor of non-zero resistance (TE module) [1].

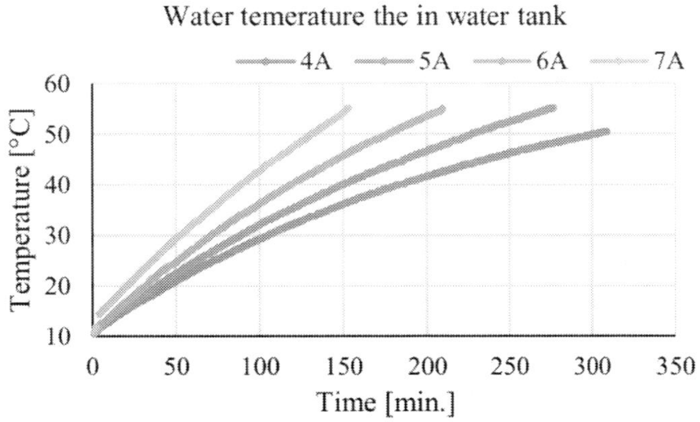

Fig. 4. Water temperature changes in the water thank during the tests

The performed statistical analysis of the obtained results of laboratory experiment showed very strong significant correlations among values of water temperature as well as air temperature for all of the applied currents (see Table 1). So, the time-related changes in measured air and water temperature values are similar in all tested cases of different currents. The distributions of studied variables tested by the Shapiro-Wilk test of normality were in all studied cases different than normal. The applied Kruskal-Wallis one-way analysis of variance supported by post-hoc multiple comparisons for both cases, i.e. temperature of water and temperature of air in the experimental room allowed to obtain different results. The Kruskal-Wallis test for temperature of water showed that in most of the studied cases (except pair 4A and 5A) distribution functions do not differ significantly. On the other hand, the same test performed for values of temperature of air inside the experimental room demonstrated that distribution functions for all the tested currents differ significantly.

Table 1. Correlation matrix for water and air temperature measured during the experiment for different current values.

	Water temperature in water tank					**Air temperature in experimental room**			
	4A	5A	6A	7A		4A	5A	6A	7A
4A	-	0.99990	0.99990	0.99974		-	0.9922	0.9970	0.9949
5A	0.99990	-	0.99991	0.99974		0.9922	-	0.9979	0.9870
6A	0.99990	0.99991	-	0.99966		0.9970	0.9979	-	0.9929
7A	0.99974	0.99974	0.99966	-		0.9949	0.9870	0.9929	-

During the experiment, correlations between the temperature values in monitored points (in the experimental room and in the water tank) were observed and presented in Fig. 5. It is visible that in all tested cases the correlations between studied values of temperature were successfully fitted to fifth degree polynomial with very high values of R^2 (0.971–0.997).

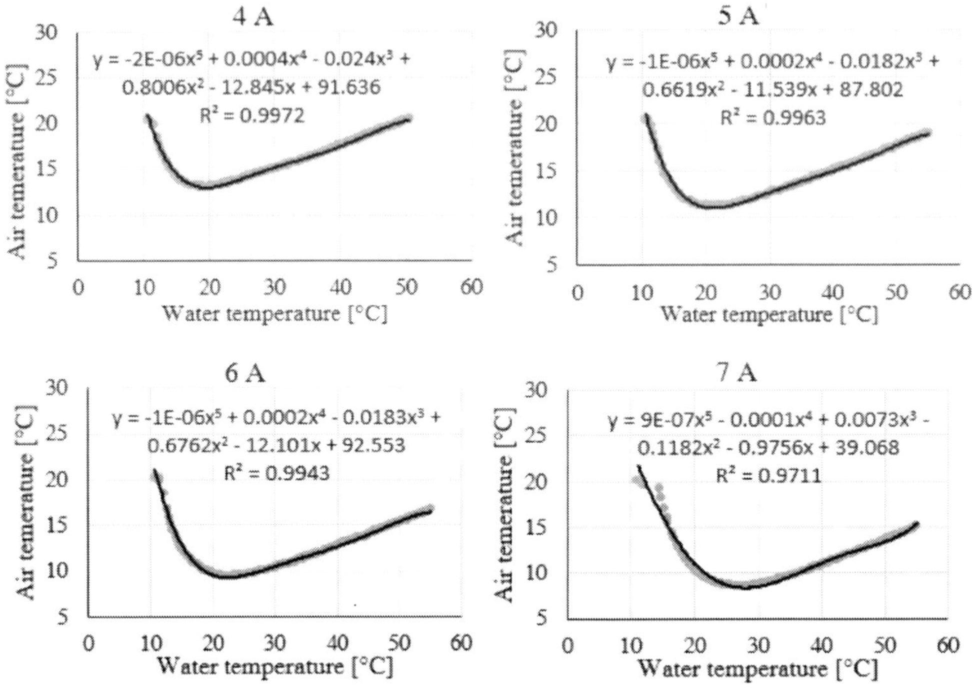

Fig. 5. Correlations between the water temperature in the water tank and air temperature inside the experimental room.

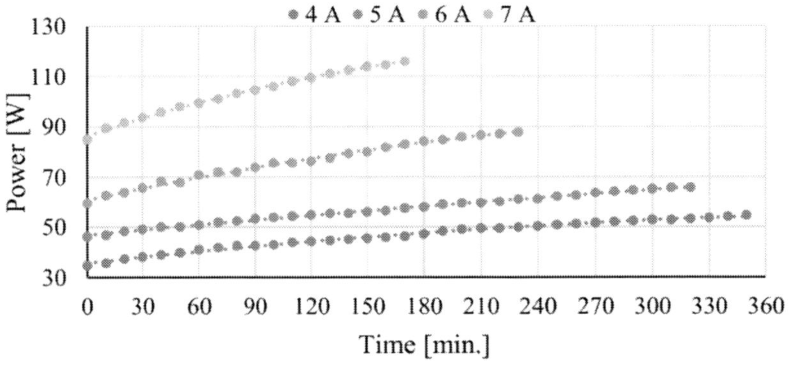

Fig. 6. Power applied to the TE module during the different series of the experiment.

During the performed experiments the paramagnets of the electric current (voltage and amperage) were monitored. These parameters were used to determine the values of power applied to the TE module (Fig. 6). There is a visible relation between cooling effects of TE module resultant from applied amperage and required power demand. Application of higher values of current allowed to obtain lower values of air temperature inside the experimental room determining better cooling effects of tested system, nevertheless the increase in power consumption was observed.

3.2 Coefficient of performance

On the basis of achieved results the values of the system COP were estimated for two points of experiment series: for the lowest air temperature in the room, and at the end of measuring series, when temperature of water was 55°C (for 4 A 50.5°C). To determine the COP for the entire system, the following formula, suggested by the literature reports for studied type of system utilising TE module for air conditioning and water heating [18–20], was used:

$$COP = (Q_c + Q_h) / (P_{TE} + P_{fan} + P_{pump})^{-1} \quad (1)$$

where:
Q_c – energy absorbed from the experimental room, Wh,
Q_h – energy generated on the hot side of the TE module, Wh,
P_{TE} – energy demand for TE module powering, Wh,
P_{fan} – energy demand for fan powering, Wh,
P_{pump} – energy demand for pump powering, Wh.

The results of COP value calculation are presented in Table 2, as well as the energy demand for powering all devices in the system (TE module, fan and pump) and energy obtained from the system for space cooling and water heating. According to the obtained results, in all cases, the highest values of COP in the range of 1.3–1.58 were obtained after approx. 40 min. of experiment, when the lowest values of temperature of air inside the experimental room were noted. Thus, basing on the obtained results of COP values, the tested system with TE module could be recommended to be used for preliminary water heating.

Table 2. Energy delivered to the system, energy obtained from the system and COP coefficient of the system for a given current value.

Current applied to the TE module	**Min. air temp. inside the experimental room**					**Water temp. in the water tank 55 °C**			
	Process duration [min.]	Energy delivered to the system [Wh]	Energy obtained from the system [Wh]	COP		Process duration [min.]	Energy delivered to the system [Wh]	Energy obtained from the system [Wh]	COP
4 A	36	31.69	45.61	1.44		309	331.62	232.64	0.7
5 A	39	35.53	56.09	1.58		276	345.75	261.82	0.76
6 A	36	44.01	63.7	1.45		209	331.87	258.99	0.78
7 A	40	68.56	90.57	1.32		153	321.65	256.72	0.8

4 Conclusions

In this paper the operation of a thermoelectric cooling system with heat recovery, powered by four applied values of current was studied. Based on the conducted experimental analysis, the following conclusions can be formulated:

- In all analyzed cases, temperature of water in the tank at the end of experiment series exceeded 50°C. The higher value of the current applied to the TE module, the shorter the water heating time was observed.
- The observed cooling effects depended on the current applied to the TE module. In all studied cases, the best cooing results were achieved after approx. 40 min. of the experiment, so when the temperature in the water tank was between 18.3°C and 26.4°C, depended on applied current, so the tested system with TE module could be successfully used for preliminary water heating.
- In general, application of the proposed system with TE module allowed to decrease the temperature inside the experimental room by approx. 7.4–11.7°C regarding the minimum temperatures obtained in the experimental room and by approx. 1.5–5.1°C regarding to the results achieved at the end of experiment series. In all of the studied cases the observed

differences in measured temperature for various values of applied currents were statistically significant.

This research work was supported by Polish National Science Center under the project UMO-2015/17/N/ST8/02824.

References

1. D. Enescu, E. O. Virjoghe, Renew. Sust. Energ. Rev. **38** (2014)
2. H. J. Goldsmid, *Introduction to thermoelectricity* (Springer, Berlin, Heidelberg, 2016)
3. D. Champier, Energy Conv. Manage. **140** (2017)
4. M. J. Huang, R. H. Yen, A. B. Wang, Int. J. Heat Mass Transfer **48**, 2 (2005)
5. W. H. Chen, C. Y. Liao, C. I. Hung, Appl. Energy **89**, 1 (2012)
6. H. S. Kim, W. Liu, Z. J. Ren, Appl. Phys. **118**, 11 (2015)
7. D. Zhao, G. Tan, App. Therm. Eng. **66**, 1 (2014)
8. B. Liao, B. Qiu, J. Zhou, S. Huberman, K. Esfarjani, G. Chen, Phys. Rev. Lett. **114**, 11 (2015)
9. S. Hébert, D. Berthebaud, R. Daou, Y. Bréard, D. Pelloquin, E. Guilmeau, A. Maignan, J. Phys. Condens. Matter **28**, 1 (2015)
10. O. Bubnova, Z. U. Khan, A. Malti, S. Braun, M. Fahlman, M. Berggren, X. Crispin, Nat. Rev. Mater. **10**, 6 (2011)
11. N. Toshima, Synth. Met. **225** (2017)
12. V. Novak, B. Podobnik, J. Možina, Appl. Thermal Eng. **57**, 1–2 (2013)
13. D. Liu, F. Y. Zhao, H. X. Yang, G. F. Tang, Energy, **83** (2015)
14. S. Su, X. Chen, J. Wang, J. Chen, Int. J. Refrig. **60** (2015)
15. R. A. Khire, A. Messac, S. Van Dessel, Int. J. Heat Mass Transfer **48**, 19 (2005)
16. X. Xu, S. Van Dessel, A. Messac, Build. Environ. **42**, 3 (2007)
17. M. Cosnier, G. Fraisse, L. Luo, Int. J. Refrig. **31**, 6 (2008)
18. W. He, J. Zhou, J. Hou, C. Chen, J. Ji, Appl. Energy, **107** (2013)
19. Z. B. Liu, L. Zhang, G. Gong, Y. Luo, F. Meng, Energy and Build. **86** (2015)
20. Z. Liu, L. Zhang, G. Gong, Energy Convers. Manage. **87** (2014)

The use of various species of fungi in biofiltration of air contaminated with odorous volatile organic compounds

Milena Gospodarek[1,*], *Piotr* Rybarczyk[1], *Anna* Brillowska-Dąbrowska[2], and *Jacek* Gębicki[1]

[1]Department of Process Engineering and Chemical Technology, Chemical Faculty, Gdańsk University of Technology, ul. G. Narutowicza 11/12, 80-233 Gdansk, Poland
[2]Department of Molecular Biotechnology and Microbiology, Chemical Faculty, Gdańsk University of Technology, ul. G. Narutowicza 11/12, 80-233 Gdańsk, Poland

Abstract. Air pollution with odorous compounds is a significant social and environmental problem. This paper presents biological deodorization methods. The attention is focused on the application of biotrickling filters for air deodorization. Principles of their operation are discussed, indicating the key role of the selection of microorganisms responsible for the degradation of odorous compounds. A literature overview of the used fungal species is presented and the advantages of using fungi in comparison with bacteria are indicated. The results of experimental studies on the n-butanol removal in biotrickling filter are presented.

1 Introduction

Air pollution with odorous compounds is becoming a growing problem, especially for residents of large agglomerations and areas adjacent to industrial plants [1]. In order to reduce the emission of odorous substances into the atmosphere, various deodorization techniques are used, including physical methods (combustion), physicochemical methods (absorption, adsorption) and biological methods. Relatively low operating costs, low waste generation and the ability to clean large volumes of gases, containing low concentrations of odorous compounds, make the use of biological methods particularly beneficial [2, 3].

The biofiltration process consists in passing a polluted stream of gas through a filter bed, inhabited by microorganisms of various species (bacteria, fungi) [4]. The impurities diffuse from the gas phase to the so-called biofilm, being formed on the surface of the elements of the packed bed. The compounds adsorbed on the surface or absorbed in the biofilm are biodegraded and the air leaves the biofilter cleansed and free of unpleasant odours. The biofiltration process can be carried out in two main types of apparatus, i.e. in a conventional biofilter or in a biofilter with a trickled bed. In a conventional biofilter, the contaminated gas is initially humidified in a separate chamber, and only then it is passed through the biofilter bed. The packing of such a biofilter is most often made of natural

* Corresponding author: milena.gospodarek@pg.edu.pl

materials, e.g. bark, cones or peat. The construction of a biotrickling filter, on the other hand, enables the process to be carried out in one apparatus. The filter bed, made of natural or synthetic materials, is regularly trickled with a liquid enriched with minerals necessary for the growth of microorganisms. The spraying liquid may circulate in a closed system and is subjected to periodic regeneration. Biofilters with a trickled bed allow easier control and regulation of the process, compared to conventional biofilters [5, 6].

Microorganisms are basic elements of biological processes of air purification [3]. Considering the deodorization mechanism described above, the selection of microorganisms colonizing the biofilter bed is crucial for the effective degradation of specific groups of chemical compounds. Depending on the type of air pollution, biofilters can be inoculated with selected strains or mixed consortia of microorganisms [7].

In recent years, many efforts have been made to develop optimal methods of air deodorisation using microorganisms. Fungi and bacteria deserve the most attention. The undoubted advantage of fungi used in biofilters, compared to other organisms, is a much higher resistance to drying out and acidification [8, 9]. This makes it possible to carry out the process using fungi in relatively dry conditions, and thus it is possible to significantly reduce the thickness of the liquid layer adjacent to the biofilm that inhabits the microorganisms. Jin et al. [10] proved that fungi can be successfully used to remove hydrophobic compounds during biofiltration. Fungal biofilms are highly resistant to acidic conditions with low humidity, and they are able to absorb hydrophobic compounds from the gas phase faster and more effectively than in the case of bacterial biofilms.

In the case of filamentous fungi, compared to other fungal species, high pressure losses are more common. This is directly related to the problems of clogging and channelling of biofilter packed beds. The solution to these problems may be the use of mites in the biofiltration process, in addition to fungal organisms. Application of higher organisms allow to prevent a rapid drop in pressure, and thus result in a significant reduction in energy requirements. It has been proven that the mites used in biofilters are relatively easy to maintain during the biofiltration process, and may be successfully used as a means to control fungal biomass [10].

The advantage of fungi, as compared to bacteria, is that they have an aerial mycelium that significantly increases the surface area of the biofilm in the gas phase, thanks to which volatile odour compounds may be more efficiently captured. The key role in the growth and development of filamentous fungi is played by hydrophobins. These proteins secreted by fungi are involved in the formation of surface structures, as well as the formation of hyphae and their attachment to various types of hydrophobic surfaces. The removal of hydrophobic compounds is a current problem in biofiltration processes [11]. Hydrophobins are able to fulfill these functions due to their production by fungi on hydrophobic-hydrophilic surfaces. Thanks to these properties, fungi in biofilters can be used with high efficiency to remove impurities directly from the gas phase, and thus it is possible to overcome the resistance of mass transfer of hydrophobic compounds to aqueous liquids [12].

Table 1. Studies on the RE, and fungal community of current biofiltration systems for removing VOCs and odor waste gases.

Predominate strain	Contaminant	Packing material	RE [%]	Reference
Sporothrix variecibatus	Styrene	Ceramic monolith	95	[15]
Sporothrix variecibatus, Sporothix sp.	Acetone	Perlite	75.6	[16]
Ascomycota strain	Trichloroethane	Wood chips	59.2	[17]
Fusarium solani	Hexane	Perlite	-	[18]
Ophiostoma sp.	Alpha-pinene	Perlite	89	[19]
Cladophialophora sp.	Toluene	Perlite	99	[20]
Fusarium oxysporum	Nitric oxide	Porous silicate pellets	93	[21]

The hydrophobicity of the mycelium surface can increase consistently as the amount of hydrophobic substrates increases. This phenomenon can be explained by the higher uptake of hydrophobic gas substances by fungi used in biofilters. Thanks to this, fungi are able to adapt to hydrophobic conditions [14]. Table 1 presents different studies where a fungus, as a dominant organism, was used for pollutant degradation.

As object of the study, n-butanol was chosen. It is due to its presence in gases emitted from municipal wastewater treatment plants or landfills [22, 23]. Additionally, the removal of n-butanol in biotrickling filters is rarely studied [24].

2 Aim of the work

In the experimental part of this paper, the effectiveness of n-butanol removal from air in a biotrickling filter as well as a preliminary assessment of the microbiological composition of the applied filter packing were investigated. The n-butanol removal efficiency was evaluated for the first twenty days of a biofilter operation, starting from introducing the n-butanol-air mixture flow through an activated filter bed. In the fifth day of the process, samples of a packed bed material for microbiological tests were taken. Selection of a period of 20 days of biofilter performance evaluation aims at verifying whether it is a sufficient time to reach steady-state conditions i.e. if the start-up period for the system is achieved.

3 Calculations

The effectiveness of n-butanol removal from air was assessed on the basis of the removal efficiency (RE), calculated according to equation (1):

$$RE = (C_{in} - C_{out})/C_{in}) \times 100\% \quad (1)$$

where C_{in} and C_{out} stand for concentrations of n-butanol in the inlet and outlet gas streams, respectively. The value of empty bed residence time (EBRT) was calculated from formula (2):

$$EBRT = V_G/Q \quad (2)$$

where V_G is the volume of packing in the biofilter and Q is the volumetric flow rate of a gas phase.

4 Experimental set-up

The research was carried out in a two-section biotrickling filter, shown schematically in Figure 1. The biofilter was made of organic glass and packed with a commercially available mixture of peat and perlite (COMPO SANA, Compo, Germany). Dimensions and process parameters of a biofilter are given in Table 2.

In order to reduce the pressure drop across the packing, in each of the biofilter sections was placed loosely a layer of 10 × 2.4 mm Raschig ceramic rings. A layer of 6 x 1.5 mm Raschig rings was also placed over the peat layer to ensure uniform distribution of the trickling liquid in the biofilter packed bed. Trickling liquid was sprayed over the filter packing for 1 min per each 60 min.

A stream of n-butanol and air from the gas mixture generator (1) was introduced via a pipe (2), equipped with an inlet gas sampling port (3), to the base of the biofilter (4). The purified gas leaves the biofilter with a pipe (5), equipped with an outlet gas sampling port

(3). In countercurrent to the gas phase, at the top of the biofilter, a trickling liquid (7) was fed in via the pipe (6) taken from the trickling liquid reservoir (8) via a peristaltic pump (7). The trickling liquid circulates in a closed system, flowing from the biofilter through the pipe (9) to the reservoir (8). The trickling liquid was replaced by means of valves (10) and (11). The measurement of the gas pressure drop in the biofilter was made using a differential manometer (12).

Table 2. Basic dimensions and process parameters of a biotrickling filter.

Parameter	Unit	Value
Internal diameter of a biofilter	dm	0.8
Total height of a two-section biofilter	dm	6.8
Total volume of biofilter packing	dm^3	2.5
Volumetric flow rate of gas phase	$dm^3\ min^{-1}$	2.5
EBRT	min	1
C_{in}	ppmv	200
Height of a Rasching rings layer: 10 × 2.4 mm 6 × 1.5 mm	dm	 0.3 0.2
Volumetric flow rate of trickling liquid	$dm^3\ min^{-1}$	0.2
Pressure drop across the biofilter packing during the experiment	Pa	549.4–902.5

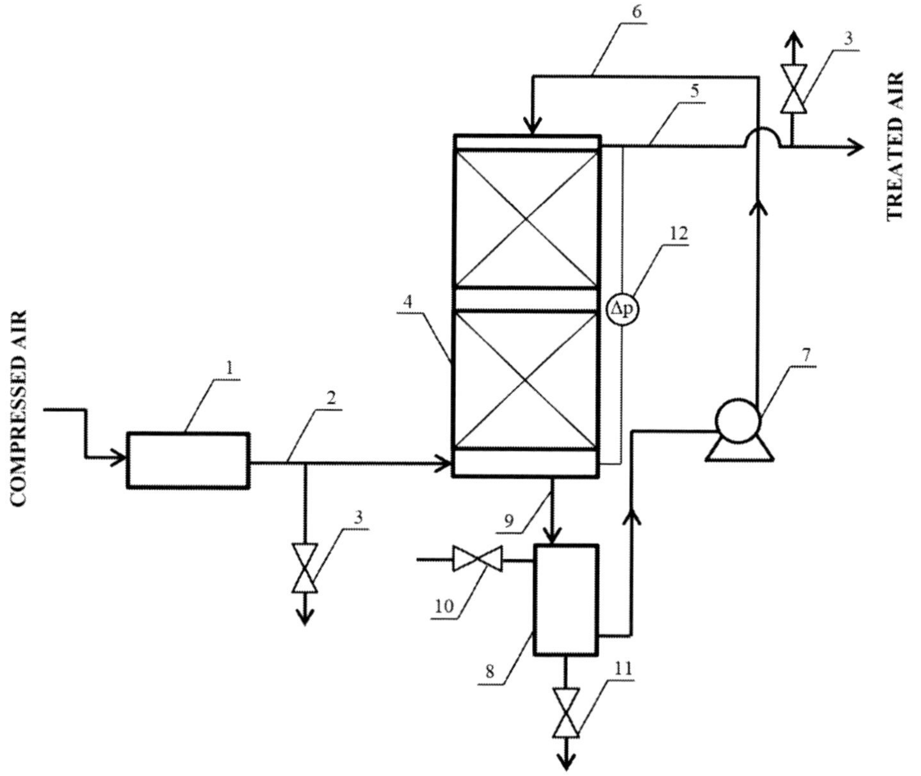

Fig. 1. Schematic diagram of the experimental set-up.

4.1 Biofiltration – process initiation and operation

After placing a filter packing bed in the individual sections and having the bioreactor twisted, a flow of a clean air as well as a trickling of a liquid were introduced through a packed bed. The intensity of air and trickling liquid flows are as shown in Table 1. The activation of the biofilter packing lasted 4 days, and as a trickling liquid a Buffered Peptone Water medium was used. After activation of the bed, the flow of n-butanol and air mixture was initiated, further using a trickling liquid as described above. After 4 days of operation of the system, the liquid used was replaced with a fresh medium. The composition of the medium was changed and it is was as follows: Na_2HPO_4 x $2H_2O$, KH_2PO_4, NaCl and NH_4Cl. A trickling liquid of such a composition was changed every 4 days, up to day 20 of the process. During the tests, at least once a day, gas samples were taken from the inlet and outlet streams. The flow of the gas mixture stream at the inlet to the biofilter and the pressure drop on the filter bed were regularly controlled.

4.2 Determination of n-butanol concentration

Concentrations of n-butanol in the gas samples at the inlet and outlet of the biofilter were determined using a VARIAN CP-3800 gas chromatograph with a flame ionization detector (FID). A 30 m long HP-5MS capillary column was used (inner diameter 0.25 mm, film thickness 0.25 µm). Conditions of chromatographic analysis were as follows: carrier gas: nitrogen; carrier gas flow: 2 ml/min; split: 3; oven temperature: 150°C; injector temperature: 150°C; FID temperature: 200°C; total time of a single analysis: 5 min. The analyzes were performed in triplicate and the results are presented for mean values.

4.3 Microbiological methods

To isolate the DNA of fungal strains, cultivation was carried out on Sabouraud medium. The plates were incubated at room temperature for about 2 days. In order to isolate the strains DNA, a short method developed by Brillowska-Dąbrowska et al. was used [25].

To obtain PCR products, ITS1 and ITS4primers were applied [26].The electrophoretic separation was carried out for 30 minute in a 2% agarose gel placed in 1xTAE buffer and at 100 V.

10 µl of the mixture obtained after the PCR reaction was applied to each well. In order to compare the size of the products obtained, the DNA M100-500 marker was used. Electrophoresis results were recorded using the ChemiDOC apparatus.

5 Results and discussion

5.1 Performance of a biotrickling filter

Data presenting the changes of n-butanol removal efficiency with time of biofiltration is presented in Figure 2. Immediately after introducing the n-butanol-air mixture to the biofilter, RE about 37% is achieved. When the biofiltration is continued, an increase of n-butanol removal efficiency is observed. This may be a result of the adaptation of a filter medium to the composition of treated gas stream. In the second day of the process, values of RE reach the level of about 85%. Further continuation of the process results in the stabilization of RE values in the range between about 85–92%. During this period, the biofilter packing is further accommodated to the treated gas and the microbes present in the peat structure consume n-butanol as a source of carbon. Obtained results show that after 20 days of biofiltration, removal efficiency of n-butanol exceeds 90%. This may indicate that the start-up period, characterized with a stable values of RE and assumed steady-state conditions in the filter bed, is achieved and a system is ready for changes of process parameters, including the variable inlet loading or EBRT[27]. Such investigations are currently performed in our research team.

The obtained removal of n-butanol is quite high (RE > 90%) and it is in accordance with literature data [24]. Because n-butanol is a hydrophilic volatile organic compounds, its high removal efficiency when applying aqueous trickling liquid is not surprising. Further investigations of the applied biotrickling filter system i.e. for the system employing natural packing materials should concentrate on the removal of hydrophobic compounds like toluene or cyclohexane. As excellent performance of natural packings in conventional biofilters is well known [6], further evaluation of the durability of the applied packing to biotrickling filters seem to be reasonable.

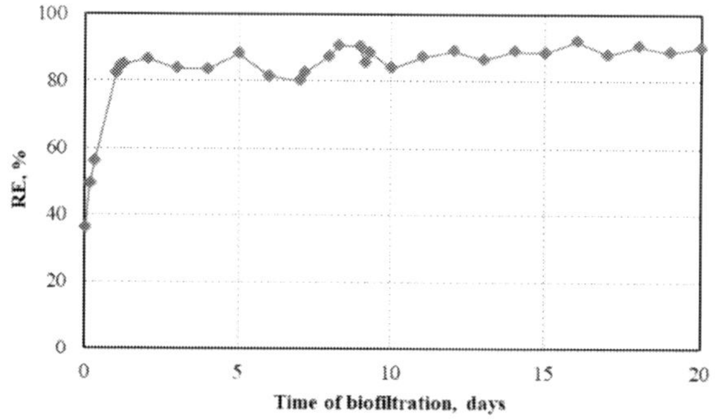

Fig. 2. Performance of a biotrickling filter packed with a mixture of peat and perlite.

5.2 Evaluation of microbial composition of a biofilter packed bed

Peat mixed with perlite was used as a biofilter packing material in this study. Peat is characterized by high wetland biodiversity. To date, 601 species of fungi from peatlands have been identified worldwide. The most popular are ascomycetes with 276 species (46%), followed by basidiomycetes (40%, 243 species), zygomycetes (9%, 55 species) and chytridiomycetes (4%, 26 species). However, the most species-rich species are *Penicillium* (48 species), *Galerina* (41 species) and *Mortierella* (20 species). It should be noted that the 20 most common types of fungi constitute 42% of all fungi in peat bogs (252 from 601 species). Most fungi on peat bogs are saprobic organisms, so they participate in the decomposition of organic matter. Often this process is accompanied by a deodorization process[28].

For the taxonomic identification of fungi from the analyzed peat, cultivation was carried out on Sabouraud medium, which is a selective medium for the cultivation of fungi. Then a PCR reaction was carried out. A pair of primers with the sequences given in Table 3 were used for this purpose. The PCR reaction was carried out using the temperature-time profile described in Table 4.

Table 3. Primer sequences used in the PCR reaction.

Name	Primer Sequence
ITS1	TCCGTAGGTGAACCTGCGG
ITS4	TCCTCCGCTTATTGATATGC

Table 4. Temperature-time profile of the PCR reaction.

Response stage	Temperature [°C]	Time [s]	Number of cycles
Initial denaturation	95	180	1
Specific denaturation	95	30	
Attachment of starter	57	30	35
Elongation	72	45	
Final elongation	72	300	1
Cooling	4	∞	1

The Figure 3 presents the results of the electrophoretic separation of the amplified DNA fragments using the ITS1 and ITS4 primers for the isolates obtained. The authors of the manuscript anticipated obtaining single bands, which would indicate the isolation of DNA from a single microorganism. Unfortunately, in the case of three out of five tested samples (3-5), double bands where received, which indicates a failed attempt to obtain pure culture. Double bands indicate the presence of at least two species.

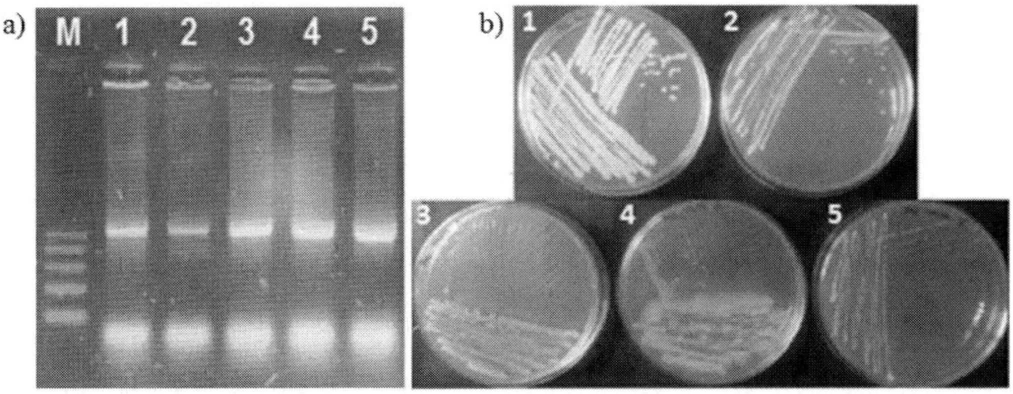

Fig. 3. (a) Result of electrophoretic separation of PCR reaction products for unidentified isolates. Agarose gel 2%, separation time: 30 minutes, voltage 100V, M - marker size 100 - 500 bp. (b) Fungi from the analyzed peat.

In the next step, a streak method was applied for the obtained cultures to confirm the presence of different species. The colonies found on plates 1 and 2 represent yeast, however cultures on plates 3, 4, 5 consist of yeast and molds (Figure 4). It confirmed the necessity of application of several streak plate cultures to obtain pure cultures.

6 Conclusions

Experimental results presented in this study indicate that n-butanol vapors may be effectively eliminated from air in a biotrickling filter packed with a mixture of peat and perlite. Removal efficiency of n-butanol exceeding 90% has been achieved after a period of 20 days of biofiltration. Such a period may be taken as a start-up period, indicating the stable conditions in the biofilter bed. Based on the literature review and the results of microbiological investigations it can be concluded that peat is inhabited by many species of microorganisms, including fungi. Further studies on the efficiency of removal of hydrophobic volatile organic compounds as well as the microbial compositions of a peat-packed biotrickling filter are suggested.

This work was financially supported by the project "INTERPHD2" No. POWR.03.02.00-IP.08-00-DOC/16.

References

1. P. Lewkowska, B. Cieślik, T. Dymerski, P. Konieczka, J. Namieśnik, Environ. Res. **151** (2016)
2. Q. Zhou, L. Zhang, J. Chen, B. Xu, G. Chu, J. Chen, Chem. Eng. J. **284** (2016)
3. K. Barbusinski, K. Kalemba, D. Kasperczyk, K. Urbaniec, V. J. Kozik, Clean. Prod. **152** (2017)
4. D. McNevin, J. Barford, Biochem. Eng. J. **5** (2000)
5. R. Lebrero, L. Hernández, R. Pérez, J. M. Estrada, R. J. Muñoz, Environ. Manage. **151** (2015)
6. P. Rybarczyk, B. Szulczyński, J. Gębicki, J. Hupka, Biochem. Eng. J. **141** (2019)
7. S. Mudliar, B. Giri, K. Padoley, D. Satpute, R. Dixit, P. Bhatt, R. Pandey, A. Juwarkar, A. J. Vaidya, Environ. Manage. **91** (2010)
8. H. Cox, Rijksuniversiteit te Groningen (1995)
9. J. W. van Groenestijn,W. N. van Heininge, N. J. Kraakm, Water Sci. Technol. **44** (2001)
10. J. R. Woertz, K. A. Kinney, N. J. R. Kraakman, W. N. M. van Heiningen, M. H. A. Eekert, Exp Appl Acarol. **27** (2003)
11. Y. Cheng, H. He, C. Yang, G. Zeng, X. Li, H. Chen, G. Yu, Biotechnol. Adv. **34** (2016)
12. H. A. B. Wösten, M.-A. van Wetter, L. G. Lugones, H. C. van der Mei, H. J. Busscher, J. G. H. Wessels, Curr. Biol. **9** (1999)
13. H. A. B. Wösten, Annu. Rev. Microbiol. **55** (2001)
14. A. Vergara-Fernández, B. van Haaren, S. Revah, Biotechnol. Lett. **28** (2006)
15. E. R. Rene, M. E. López, M. C. Veiga, C. Kennes, Bioresour. Technol. **101** (2010)
16. E. R. Rene, R. Spačková, M. C. Veiga, C. J. Kennes, Hazard. Mater. **184** (2010)
17. Y. Quan, H. Wu, C. Guo, Y. Han, C. Yin, Bioresour. Technol. **259** (2018)
18. S. Arriaga, S. Revah, Can. J. Civ. Eng. **36** (2009)
19. Y. Jin, L. Guo, M. C. Veiga, C. Kennes, Biotechnol. Bioeng. **96** (2007)
20. W. Woertz, M. van Heiningen, J. van Ee, W. van Heiningen, M. van Eekert, N. Kraakman, K. Kinney, J. van Groenestijn, Appl. Microbiol. Biotechnol. **58** (2002)
21. J. Woertz, K. Kinney, P. J. Szaniszlo, Air Waste Manage. Assoc. **51** (2001)
22. B. Szulczyński, P. Rybarczyk, J.Gębicki, Monatshefte für Chemie - Chem. Mon. **149** (2018)
23. B. Szulczyński, J. Gębicki, J. Namieśnik, Chemical Papers, **72** (2018)
24. T. Schmidt, W. Anderson, Environments, **4** (2017)
25. A. Brillowska-Dąbrowska, A. Świerkowska, D. M. Lindhardt Saunte, M. C. Med. Mycol. **48** (2010)
26. T. White, T. Bruns, S. Lee, Mycologia, **64** (1990)
27. H. Qian, Y. Cheng, C. Yang, S. Wu, G. Zeng, J. Xi, Environ. Sci. Pollut. Res. **25** (2018)
28. M. N. Thormann, A. V. Rice, Fungal Diversity, **24** (2007)

The use of pulsed flushing to improve the efficiency of polymer floating media filters

Zhanna Govorova[1], and *Alexey* Andrianov[1,*]

[1]Department of Water Supply and Sewerage, Moscow State University of Civil Engineering (National Research University), Yaroslavskoe shosse 26, Moscow, 129337, Russia

Abstract. Surface runoff from urban area is a potential source of pollution of water bodies. The composition of pollutants in surface runoff depends on various factors and seasons. The main pollutants are oil products, suspended solids, COD, chlorides and other mineral salts. Traditional wastewater treatment plants include the mechanical removal of large items, debris and leaves; sand removal; water clarification in accumulating settling tanks; chemical dosing, contact filtration and deep purification on sorption filters. Filters with inert floating polystyrene media of various types are widely used as first stage filters. The operation of industrial filters in the technological schemes of wastewater treatment strongly depends on the backwash efficiency. Floating media flushing is carried out in the direction from top to down using the water accumulated in the over-filter space. The article presents the results of the study of pulsed flushing on a model plant. Experimental study was conducted on the real wastewater treatment plant. The kinetics of suspended solids and oil products removal from filter media was studied. The operational parameters and mode of pulsed flushing were determined. The research results were used during the reconstruction of Moscow surface runoff water treatment plants.

1 Introduction

The main contaminants of the surface runoff from urban areas are suspended solids, petroleum products, COD, chlorides, etc. Furthermore, recently there has been a significant pollution of surface runoff with heavy metals, deicing agents and surface-active agents [1–10].

The design of treatment facilities depends on their capacity, quality of surface wastewater from urban areas and requirements for its purification level [11–14]. Wastewater treatment plants generally include: mechanical stage using bar or mesh screens and grit chambers, water clarification in settling tanks which also work as flow equalization basins; removal of the bulk of organic and mineral contaminants using sedimentation, flotation or contact filtration with preliminary reagent treatment; advanced treatment of residual mechanical impurities with petroleum products and organic matter adsorbed on

* Corresponding author: alexei-waterlab@yandex.ru

them using mechanical filtration on standard deep bed filters; advanced treatment using activated carbon filters to remove residual dissolved petroleum products and other organic compounds; disinfection of treated wastewater to be discharged back into water bodies or to be reused for industrial water supply.

Various types of polymer floating media filters are used for contact filtration [12–15]; some of them are presented in the Table 1.

Table 1. Types of filters with floating polymer loading.

Type	Name	Specification
Contact filter with floating media	CFPZ-1, (FPZ-1)	Upward filtration through a heterogeneous floating media
Hydraulic automatic floating media filter	AFPZ-4 FPZ-4	Downward filtration through a heterogeneous floating media (A – automatic flushing)
Clarifying and sorption contact filter with floating media	CFPZ (CS)	Two compartment filter. Downward filtration through polystyrene floating media, followed by downward filtration through a sorption media

Flushing of floating media filters is carried out by a downward water stream from the space above the filter bed. The granules of filter media remain motionless at low speed of water movement. The head of water passing through pores of granular bed is partially spent on overcoming the friction force at the surface of the granules. When the value of hydrodynamic pressure becomes equal to the mass of the filtration media in the water, the filter layer expands, its porosity increases, the granules switch from the stationary state to the condition of irregular motion in a definite volume [15].

Flushing is an important procedure of the filtration process. The key factors affecting the efficiency of granular media flushing are: flushing intensity; the duration of the washing of granules surface from adhering suspended solids and sediment; media expansion ratio; the degree of uniformity of distribution and collection of flush water. The main goal of the filter flushing is adjusting flushing intensity and expansion ratio which guarantee almost complete washing of the granules from adhering contaminants. The uneven collection and distribution of flush water over the filter surface area may cause a local expansion of the media and washout of polystyrene granules with the flush water. Excessive expansion of the floating media during flushing leads to necessity for an increase of the construction height of the filter body and capital investment.

Progressive clogging and oiling of the upper layer may occur due to reagentless filtration of wastewater that contains petroleum products and suspended solids and due to insufficiently effective flushing. The consequences are a sharp increase of the initial headloss in the filter media, a reduction of the filtration duration time and a decrease in the overall efficiency of the filter installation.

To intensify the flushing process a series of short pulses is used that are aimed at more efficient washing of the granular media from suspended solids and adsorbed petroleum products. The investigation of this flushing procedure is an urgent engineering problem and was the objective of the experiments described below.

2 Materials and methods

The flushing efficiency of industrial filters FPZ-4 and FPZ-1, which were a part of surface runoff treatment facility (Moscow), was studied. At all treatment facilities the water used for flushing was either treated (FPZ-1) or clarified in intermittent settling tanks (FPZ-4).

A pilot plant was installed in the filtering chamber of the treatment facility (Fig. 1) to investigate impulsive flushing of the floating filter media. It included pilot filter with polystyrene floating media, equipped with pipes, shutoff and control valves, and samplers. Pipe fitting allowed carrying out upward (FPZ-1) and downward (FPZ-4) filtration. Characteristics of floating media are given in Table 2.

Fig. 1. General view and fragments of the pilot plant.

The pilot plant was supplied with water from the last section of the settling tank. Flow rate did not exceed 10 m/h. Filtration time ranged from 10 to 14 hours. Pulsed water flushing included a short-time supply of treated water from the space above the filter bed with the intensity of 35–40 $L/m^2 \cdot s$ during the first 30–60 seconds and then flushing with reduced intensity down to 7–10 $L/m^2 \cdot s$ for the next 3–5 minutes. A criterion for the evaluation of flushing effectiveness is the rate of washing out of contaminants – suspended solids and petroleum products. Flushing effectiveness was evaluated separately for three different layers of filter media, comparing the contamination degree (dirt-holding capacity) of these layers throughout the filtration cycle.

Table 2. Filter media characteristics.

Layer number	Grain size, mm	Layer thickness, m
1	1–1.5	0.05
2	1.5–3	0.65
3	4–8	0.1

Flush water was sampled at regular intervals of 0.5–1 minute. The amount of suspended solids was determined by the gravimetric method (PND F 14.1:2:4.254-2009). The mass concentration of petroleum products was determined by IR spectroscopy with pre-extraction of emulsified and dissolved petroleum products from water with carbon tetrachloride and its separation from the related organic compounds of other classes in a column filled by aluminum oxide (PND F 14.1:2.116-97).

3 Results and discussion

At the first stage of experiments, the analysis of water flushing efficiency of industrial filters operating in reagentless mode was carried out. Influent wastewater entering the FPZ-1 and FPZ-4 filters has the concentration of suspended solids up to 250 mg/L, petroleum products – up to 5.5 mg/L. Grain size of filter media was 2–4 mm, layer thickness – 1.5–1.6 m. The flow rate in nominal operating mode did not exceed 8 m/h. Flushing intensity was 12–15 L/m^2·s.

The results showed that at the appointed washing rate contaminants were discharged too slowly over time (Table 3).

Table 3. The concentration of suspended solids (SS) and petroleum products (PP) in the flush water after industrial filters.

Washing time, min.	FPZ-1		FPZ-4	
	SS, mg/L	PP, mg/L	SS, mg/L	PP, mg/L
0.5	1100	9.4	150	9.8
1	1500	7.1	720	7.6
2	1800	3.4	2250	4.7
3	1320	2.5	1740	3.4
4	640	2.1	1050	3.0
5	230	1.8	300	2.0
6	150	0.9	160	1.8
7	80	0.8	150	1.5

With a washing time of 7 minutes, a sufficiently large amount of suspended solids (50–150 mg/L) and petroleum products (0.8–1.5 mg/L) remained in the flush water.

Insufficient intensity of floating media flushing leads to the accumulation of residual contamination in the layer of large polystyrene granules. Performing a flushing with constant high intensity (35–40 L/sec·m^2) leads to 2.5–3 times excessive consumption of flush water and to an increase of the pipe diameters of the underdrainage.

Therefore, an alternative method of flushing the filter bed media was chosen for further research – this is pulsed downward flushing that uses purified water, accumulated in the space above the filter bed. The duration of filtration cycle was set at 14 hours. The concentration of suspended solids in wastewater entering the pilot plant was 35–90 mg/L, petroleum products – 2.6–3.2 mg/L. Dirt-holding capacity per filter cycle did not exceed 4.14 kg/m^3 and 0.073 kg/m^3 for SS and PP respectively.

The impulsive flushing mode was as follows: for the 1st minute the intensity was 39.5 L/m^2·s, and then for the next 5 minutes the intensity did not exceed 9 L/m^2·s. The results of experiments, during which the mode of accumulation and distribution of the forming sediment and petroleum products in the interporous space of filter media was changing, are shown in Fig. 2 and 3.

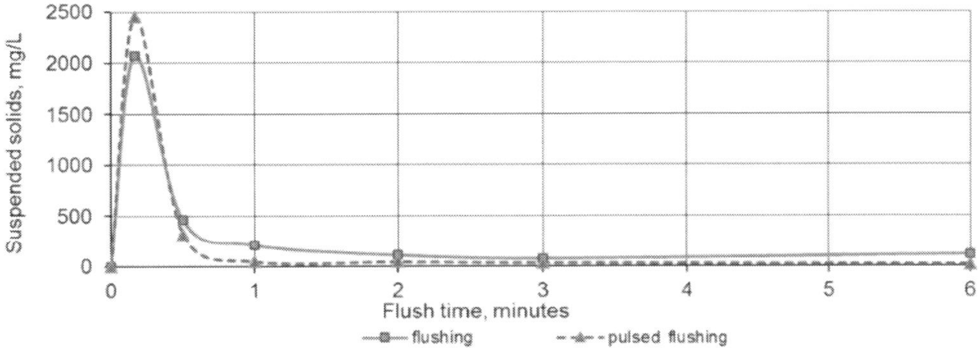

Fig. 2. The kinetics of the washing out of suspended solids during the flushing of FPZ-1 filter.

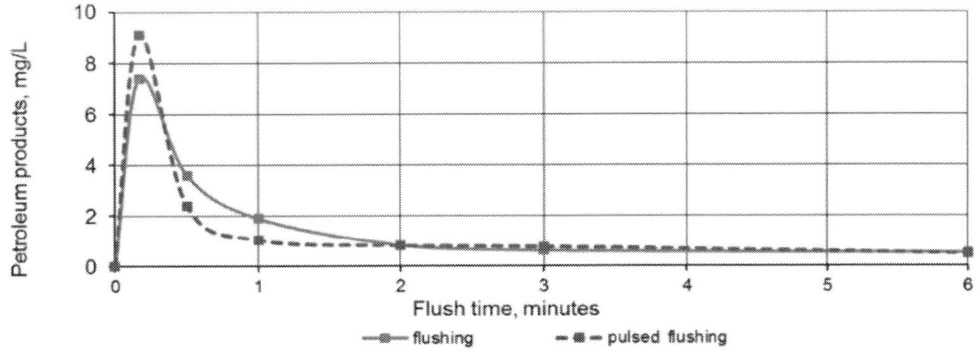

Fig. 3. The kinetics of the washing out of petroleum products during the flushing of FPZ-1 filter.

As can be seen from the graphs, during the pulsed flushing with the intensity of 39.5–9 L/m^2·s compared to flushing with constant intensity of 15 L/m^2·s for 6 minutes, the amount of contaminants washed out increases by 18–25% in the first 20 sec.

Pulsed flushing with a short-time intensity of 39.5 L/m^2·s provides an increase of the expansion ratio of large-granular layer with a density of 60–80 kg/m^3 within 30–40% and mixing of the granules. Therefore, contaminants that are washed out from large-granular layer then transported through the expanded layers of smaller granules during next 3–5 minutes of flushing and then washed out from the model filter.

The results of conducted research have been introduced into industrial treatment facility with a capacity of 4000 m^3/day, its general view is shown in Fig. 4.

Fig. 4. Industrial filters FPZ-1 (Moscow).

4 Conclusions

The analysis of the flushing process of industrial filters with floating polystyrene media revealed that after flushing with constant intensity up to 15 L/m^2·s for 7 minutes a large amount of suspended solids (50–150 mg/L) and petroleum products (0.8–1.5 mg/L) remained in a flush water. It demonstrates that granules of the load are not clean enough and might lead to accumulation of contaminants and reduction of the filter cycle.

As an alternative method, impulsive flushing, with the intensity of 35–40 L/m^2·s for the first 30–60 sec and reduced intensity of 7–10 L/m^2·s for the next 3–5 minutes, was studied. It was established that in the first 30–60 sec the amount of contaminants carried out with the flush water increases and by the end of the flushing concentration of suspended solids and petroleum products left in water is no more than 10 mg/L and 1.0 mg/L, respectively.

The results of conducted research were adopted for the reconstruction of Moscow surface runoff treatment plants.

References

1. M. A. Alekseyev, A.M. Kurganov, *Organizatsiya otvedeniya poverkhnostnogo (dozhdevogo i talogo stoka) s urbanizirovannymi territoriyami* (ASV, Moscow, 2000)
2. E. D. Palagin, A. K. Strelkov, P. G. Bykova, A. V. Tsypin, D. V. Vdovin, Water Supply and Sanitary Tech. **12**, 41–46 (2016)
3. E. D. Palagin, M. A. Gridneva, P. G. Bykova, A. A. Pavlukhin, Water Supply and Sanitary Tech. **12**, 28–35 (2018)
4. G. Ignatavičius, V. Valskys, I. Bulskaya, D. Paliulis, A. Zigmontienė, J. Satkūnas, Estonian J. of Earth Sci. **66**, 1, 13–20 (2017)
5. R. Buzier, M. H. Tusseau-Vuillemin, M. Keirsbulck, J. M. Mouchel, Phys. and Chem. of the Earth **36**, 500–505 (2011)
6. S. Marshall, V. Pettigrove, M. Carew, A. Hoffmann, Envir. Pollut. **158**, 1716–1725 (2010)
7. M. Scholz, S. Kazemi Yazdi, Water Air Soil Pollut. **198**, 55–64 (2009)
8. M. Zhang, H. Chen, J. Wang, G. Pan, J. of Environ. Sci. **22**, 1, 40–46 (2010)
9. R. H. Kim, S. Lee, Y. M. Kim, J. H. Lee, S.K. Kim, S. G. Kim, Environ. Technol. **26**, 411–420 (2005)
10. J. W. Zhao, B. Q. Shan, C. Q. Yin, Techniques and Equip. for Environ. Pollut. Control **7**, 12, 14–17 (2006)
11. *Rekomendatsii po raschetu sistem sbora, otvedeniya i ochistki poverkhnostnogo stoka s selitebnykh territoriy, ploshchadok predpriyatiy i opredeleniyu usloviy vypuska yego v vodnyye ob"yekty* (NII VODGEO, Moscow, 2014)
12. M. G. Zhurba, Zh. M. Govorova, O. B. Govorov, Water Supply and Sanitary Tech. **3**, 25–29 (2003)
13. Zh. M. Govorova, *Regulirovaniye i ochistka poverkhnostnykh stochnykh vod: problemnyy doklad* (VNIINTPI, Moscow, 2005)
14. M. G. Zhurba, Zh. M. Govorova, O. B. Govorov, Vodoochistka. Vodopodgotovka. Vodosnabzheniye **1**, 29–34 (2008)
15. M. G. Zhurba, *Vodoochistnyye fil'try s plavayushchey zagruzkoy* (Moscow, 2011)
16. C. Weidemann, S. Vogt, H. Nirschl, J. of Food Eng. **132**, 29–38 (2014)

Regional strategy of energy and CO_2 emissions management – base study on the example of Gliwice region

Paulina Grelich[1], *Marta* Moszczyńska[1], *Grzegorz* Stańczyk[1], and *Krystian* Szczerbak[1,*]

[1]Wrocław University of Science and Technology, Environmental Engineering Department, Wybrzeże Wyspiańskiego 27, Wrocław 50-370, Poland

Abstract. The European Union has set the requirements for year 2030, aimed to increase the energy production from the renewable sources and to lower the CO_2 emissions. Currently the percentage of renewable energy production in Poland is too low. The actual data of Gliwice County from Study of Conditions and Directions of Spatial Development (SUIKZP), Plan of Low-Emission Management (PONE/PGN) and Predictions of Impact on the Environment (POŚ) documents was used to calculate the regional energy consumption for heating and electricity. The data has been used in the simulation of the energy consumption and the CO_2 emission for year 2030, including the production of the energy using region's potential renewable resources. Based on that, the authors have calculated how much more energy must be produced from the renewable sources to meet the European Union requirements. The analysis includes the investment cost in the new energy sources and its comparison to the financial penalties the region would have to pay for not meeting the requirements.

1 Introduction

Powering buildings in electricity and heating states for 40% of global energy consumption and 30% of CO_2 emission. In central and northern Europe 80% of total energy for buildings is used for heating. High level of heating energy is caused by low insulation standard of majority of buildings [1]. Most of Poland's energy is produced from fossil fuels, hard and brown coal in particular. Rising mining costs, emission problems, European regulations are making conventional energy sources more and more expensive and cumbersome. On the other hand the potential for renewable and sustainable energy sources is very big. It is speculated that by year 2020 emissions will be reduced by 20% and by year 2050 all of the buildings will be neutral to the environment, due to the usage of renewable resources in 100% [1]. Poland stands behind western countries in climate matter, but there is possibility to catch up and increase the usage of clean energy. Gliwice region is popular for its great coal potential, and it stands for 82% of its energy demand [4]. Although the region has also very great potential for renewable technologies. The investment cost in new energy sources is big, but the costs of waste emissions is also big, and will rise in the future, and soon the local

* Corresponding author: 220282@student.pwr.edu.pl

authorities will not be able to pay for it. The main point of this article is to calculate the costs of new energy sources including actual region's potential and comparison to emission penalties.

2 Methodology

All of the data used for this study was collected from Gierałtowice, Knurów, Pilchowice, Pyskowice, Rudziniec, Sośnicowice, Toszek, Wielowieś documents of: Study of Conditions and Directions of Spatial Development (SUIKZP) [2], Plan of Low-Emission Management (PONE/PGN) [3] and Predictions of Impact on the Environment (POŚ) [4]. The data contains actual energy consumption for buildings of all sectors (residential, commercial, industrial, public utilities) and energy sources. Based on that and the specification of the region the simulation for 2030 was made including various new, renewable and sustainable energy sources in an exact number to reduce the CO_2 emissions by 30% to current level. The European Councils decision demands reducing emission by 40% below 1990 by 2030 [6]. Due to lack of data from that period the emission condition was changed. The investment cost was evaluated using the referenced prices established by the Ministry of Energy [7]. In the end the combined costs were compared to the value of penalties the region will pay for emissions, if they would keep the current level by 2030.

3 Region's specification

The energy distribution by sector is almost balanced in the region as shown in figure 1. The majority of energy is used in residential (40%) and industrial (35%) sectors. The minor part is commercial (15%) and public utilities (10%) sectors. Most of the residential buildings in the region are single-family and small multi-family buildings, the only exception is the city of Knurów, where majority of buildings are blocks of flats. Commercial and public utilities sector are both well insulated buildings powered by regulations-fulfilling energy sources. Industrial sector is consist mainly of mines, coal processing facilities, woodworking, farming and food production.

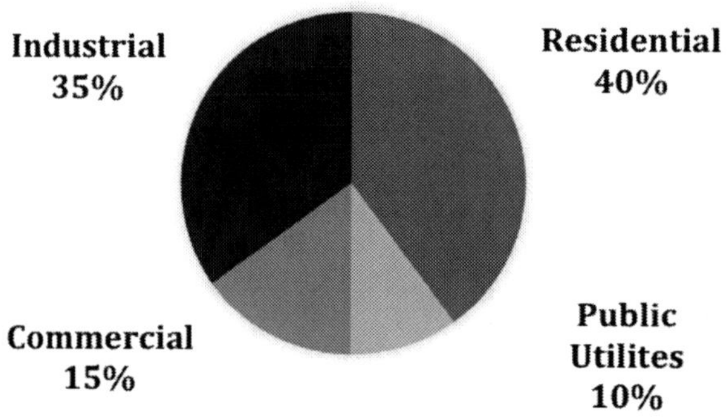

Fig. 1. Energy consumption percentage distribution by building sectors [2].

Figure 2 shows 3 maps defining the regions energy and emission specification basing on actual data. The firs shows the imbalance of energy usage for heating (pink) and electricity (red). The reason for this is poor insulation of old buildings in the village parts of the region. The imbalance is the smallest in Knurów and Gierałtowice, because of recent building insulation programmes, which eliminated the problem in all sectors except for industrial.

Most of industrial sector in Knurów are mines, coal processing and water treatment buildings that are not insulated at all. On the second map energy sources are shown. Energy mainly comes from coal (grey), and depending on local politics they are also other sources such as natural gas (yellow) and renewable (teal). Cities of Knurów and Pyskowice are the only parts of the region with heat network powered by cogeneration (pink) and providing 1/3 of needed heat energy. The last map shows emission level, which is the biggest in Knurów, because of concentration of majority of region's industry, mining facilities and power plants.

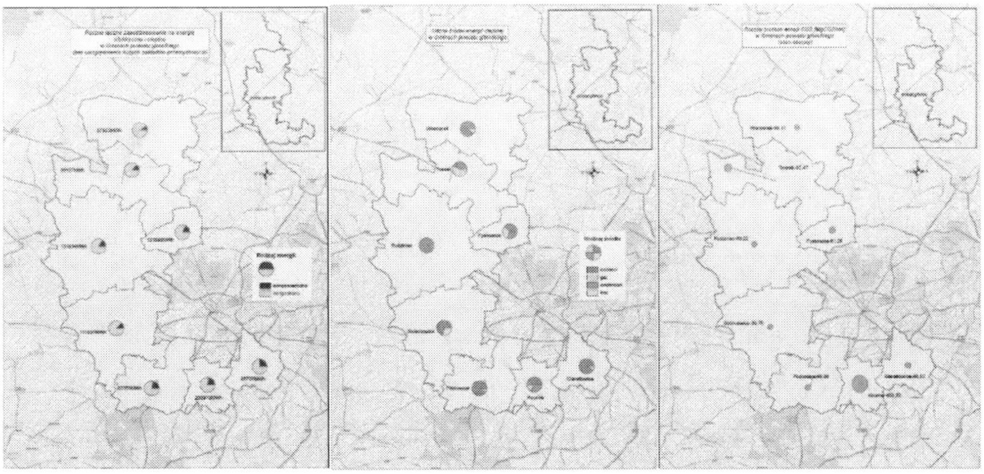

Fig. 2. Region's energy consumption, energy sources and CO_2 emission [4].

4 Energy production

4.1 Current energy production

Currently the region's power safety lies in coal industry. Renewable energy stands for 4.20% of energy production, while coal (heating + electricity) stands for 72.65% of total 996 630 MWh energy produced in 2018. Figure 3 shows shares of energy production by source. Coal column on the chart shows the usage for heating, while electricity was separated. Domination of coal is determined by region's traditions, local mines and high number of old buildings. Natural gas and cogeneration heat networks are built only in more populated parts of region, thus its share is overall small and cannot be applied in all of the region. Renewable energy consists of solar panels used in residential sector and heat pumps, used mostly in commercial and public utilities buildings.

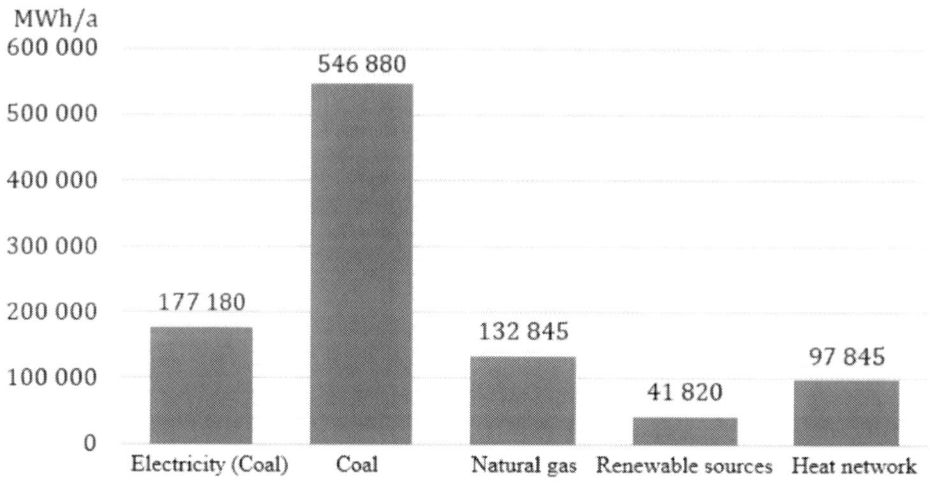

Fig. 3. Current energy production by sources [2].

4.2 Renewable and sustainable energy potential in the region

The region has many buildings appropriate for installing photovoltaic installations and a lot of unused land around its dense road network for photovoltaic or solar power plants. Especially Wielowieś, which declare willingness for such energy source in its development plans. Farming and food production facilities cover 52% of region's surface and produce tons of biological waste, which is currently not used, or used inefficiently. This makes a great source for biogas plant and biomass power plants. More settlements are ready for connection to natural gas or cogeneration heat networks. Buildings in less populated areas can be powered in heat by air heat pumps, because although Gliwice region lies in a 3rd climate zone of Poland, with temperatures reaching -18 Celsius degrees, in last 30 years winters are less harsh, and temperatures drop below 10 Celsius degrees less than 100 hours a year. Such temperature regime is optimal for air heat pumps and rarely require additional electrical heating. The region is also rich in rivers, moist soil, abandoned mining shafts. These can be effectively used as source for ground and water heat pumps for adjacent areas and buildings.

4.3 Predicted energy production in 2030

Considered actions to reduce usage of coal and conventional energy sources were based on the regions unused potential and technologies that would bring the biggest result, with the lowest cost. Main objective is to reduce CO_2 emissions by 30% and raise renewable sources usage to 30% (287 805 MWh in 2030) of total 959 350 MWh used speculated required energy. In case analysis following in-vestments were used, separately for all sectors:

- Residential (32.8% of total renewable energy): heat pumps, photovoltaics, solar panels
- Commercial (18.4%): heat pumps, photovoltaics, solar panels
- Industrial (42.5%): heat pumps, photovoltaics, solar panels, biogas
- Public utilities (6.3%): heat pumps, photovoltaics, solar panels

Figure 4 shows the energy production profile in 2030 with applied new investments and technologies, cutting off the coal usage.

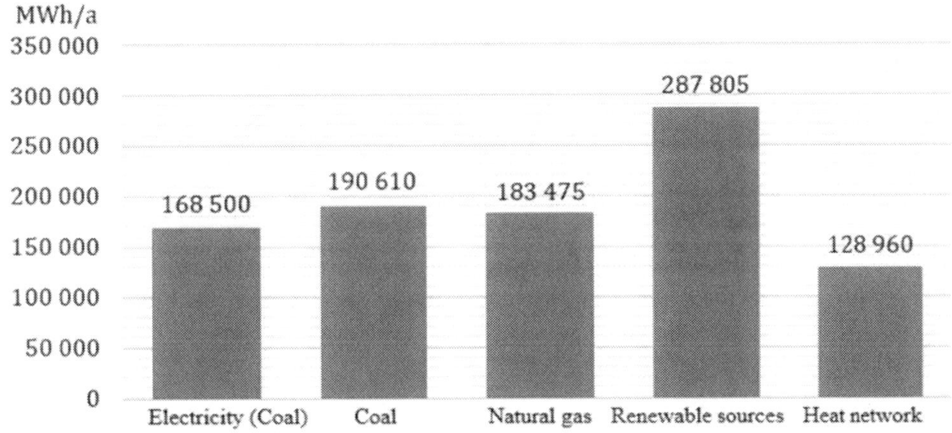

Fig. 4. Speculated 2030 energy production by sources.

In order to fulfil renewability and emissions requirements the analysis adopted changes that included: decentralisation of energy sources, reducing of coal usage, expansion of natural gas and cogeneration heat networks, biogas production and various renewable energy sources appropriate for the region.

5 CO_2 emission

Actual emission level is very high in the region. However usage of different than coal energy sources in this simulation makes it possible to lower CO2 emissions by 30%.

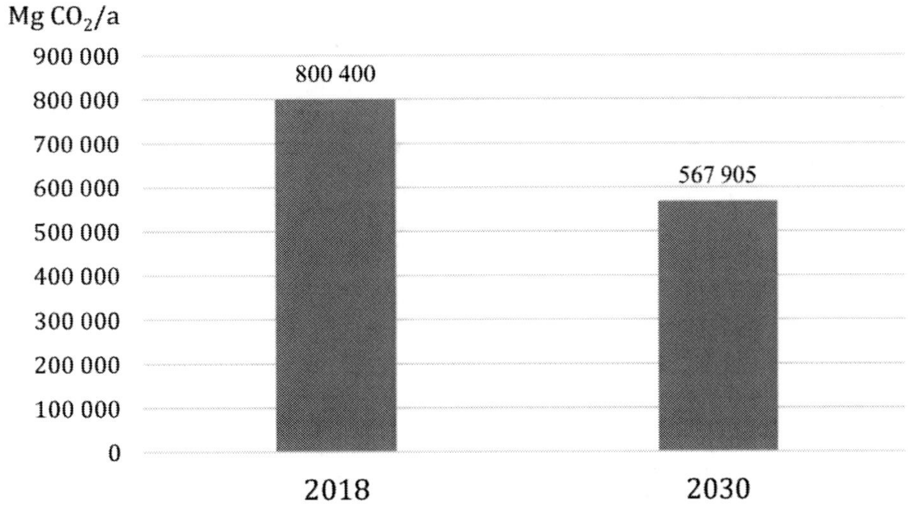

Fig. 5. CO_2 emission comparison.

6 Financial case

Investing in new, environmental energy sources is not cheap. But in a long-term view it will be not only more ecologic for the region but also more profitable. It's because the financial penalties that the region will have to pay every year for emissions higher than those accepted by European Union. Calculated cost of the new energy sources in the analysis would be 125 million PLN, according to reference prices. However the emission penalties would reach

297 million PLN if the emission level would not change by 2030. The penalties can be even higher, because the analysis is based only on CO2 emissions, and penalties can be applied to other pollution too.

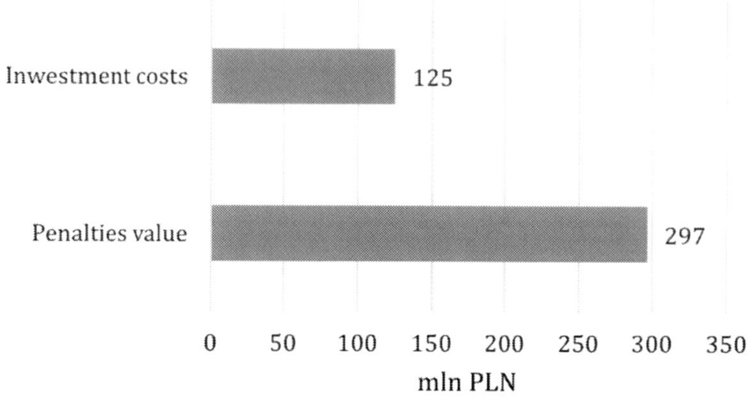

Fig. 6. Investment costs and CO2 emission penalties comparison.

7 Conclusions

It is not an easy task to meet up with requirements of European Union for countries like Poland, with its energy economy centralised around coal. The Gliwice region has serious problem with current emissions level. Basing on the newest technologies and complex, actual data of buildings energy experts are creating a program of funding new energy sources in the whole region. The program will include all sectors and will be used as an example for the rest of Poland. It shows that there is still possibility and desire for executing European requirements. Lowering the emissions by 30% and increasing the usage of renewable resources by 30% is the first step that the region, as well as all of Poland, needs to make to eventually decrease energy production emissions to 0% and make buildings neutral to environment by 2050 (or later in the future). Every year of delay will result in higher emission penalties and local environment degradation. It is crucial to make a difference, because the more outdated and conventional energy sources will be, the more costly requiring the modernizations will be.

References

1. E. Ursula, *Enegry Efficient Buildings with Solar and Geothermal Resources* (Wiley, Chichester, 2014)
2. Studium uwarunkowań i kierunków zagospodarowania przestrzennego gmin: Gierałtowice, Knurów, Pilchowice, Pyskowice, Rudziniec, Sośnicowice, Toszek, Wielowieś
3. Plan gospodarki niskoemisyjnej gmin: Gierałtowice, Knurów, Pilchowice, Pyskowice, Rudziniec, Sośnicowice, Toszek, Wielowieś
4. Prognoza oddziaływań na środowisko gmin: Gierałtowice, Knurów, Pilchowice, Pyskowice, Rudziniec, Sośnicowice, Toszek, Wielowieś

5. Dz.U. 2002 nr 75 poz. 690, Rozporządzenie Ministra Infrastruktury z dnia 12 kwietnia 2002 r. w sprawie warunków technicznych, jakim powinny odpowiadać budynki i ich usytuowanie.
6. Directive of the European Parliament and of the Concil amending Directive 2003/87/EC to enhance cost-effective emission reductions and low-carbon investments and Decision (EU) 2015/1814
7. Reference prices for renewable and sustainable energy sources established by the Ministry of Energy, http://gramwzielone.pl/trendy/25180/ceny-referencyjne-w-aukcjach-dla-oze-na-2017-r

Odour nuisance assessment of the food industry wastewater treatment plant

Agnieszka Grzelka[1,*], *Elżbieta* Romanik[1], and *Urszula* Miller[1]

[1]Wroclaw University of Science and Technology, Faculty of Environmental Engineering, Wybrzeże Wyspiańskiego 27, 50-370 Wrocław, Poland

Abstract. Odour nuisance generated by food industry objects is a significant problem, the result of which is often numerous complaints from local residents about the quality of air. Apart from the production process, this problem often concerns industrial wastewater treatment processes, which contribute to a large extent to the emissions of odorants from the sulphur group, including hydrogen sulphide and mercaptans. In order to estimate the odorous air quality around these types of objects, the modelling of dispersion of pollutants: odours or odorants is often used. The paper presents the results of tests on odour emission as well as hydrogen sulphide and mercaptans emission as compounds typical for waste gases from the sewage treatment process from the food industry. In order to estimate the range of odour impact, model calculations using the Polish reference model for the aforementioned substances were made. The values of annual average concentrations and frequency of odour detection thresholds for odorants and odour concentration of 1 $ou_{E/m}{}^3$ for odours as a mixture in a computational grid of 1400 m x 1400 m with a step of 50 m were calculated. The results showed the highest range of odour impact of mercaptans, whose value of the average annual concentration exceeded the odour detection threshold up to 700 m in the east and west direction from the emission source.

1 Introduction

The emission of gases contaminated with odorants into the atmosphere is a significant environmental problem. Among the anthropogenic sources of odour emission, the group of industrial sources is the most diversified. The most frequently represented group of them are objects of agro-food processing [1, 2]. In many cases, thermal processing of substrates takes place in the technological process. This is caused by the fact that the raw materials used in the food industry are characterized by low durability and complex organic structure, so they are partially decomposed, and thus results in the formation of volatile chemical compounds.

Both in the context of odours and individual odorants, two types of detection thresholds can be distinguished. The lower limit of the perceived odour intensity range that can be detected is the odour detection threshold. This is the minimum concentration of a substance that can trigger a sensory response in the olfactory receptor of a specific percentage of a given population, usually 50% of the cases in which the odour is detected. In this concentration,

* Corresponding author: agnieszka.grzelka@pwr.edu.pl

the odour is not necessarily correctly identified. The second characteristic threshold is the odour threshold, which refers to the lowest concentration at which the sensory effect can be correctly combined with a specific substance that is sensed by 50% of cases in the representative group [3]. For example, for hydrogen sulphide, the average geometric recognition threshold, at which the odour is correctly identified as hydrogen sulphide is 6.3 $\mu g/m^3$ [3], and the detection threshold is 0.57–1.42 $\mu g/m^3$ [4-8], hence, it is nearly 5–10 times lower.

One of the most commonly used techniques to assess the effects of odour emissions is the simulation of the topographical and meteorological data of the object in order to estimate the concentration of odours and odorants in the air using dispersion modelling tools. This method allows forecasting of distances from the emission source that plume of pollution can reach [1, 2]. In Poland, in accordance with the methodology contained in the Regulation of the Minister of the Environment of 26 January 2010 on reference values for some substances in the air, such calculations are carried out with a reference model based on the Gaussian plume model [9].

The paper presents the results of calculations of the model dispersion of odours and odorants: hydrogen sulphide and mercaptans emitted from deodorization plant at food industry wastewater treatment plant. The evaluation of the odorous impact of WWTP was made using the Polish reference model in the OPERAT FB software for the grid size of 1400 m x 1400 m and the calculation step of 50 m. The obtained results allowed to compare the odour nuisance assessment of the object due to the emission of odours and selected odorants.

2 Odours at WWTPs

Odours produced in wastewater treatment plants are the result of biological degradation of organic matter under anaerobic conditions. The odour emission may result from the fact that odorants are present in the sewage supplied to WWTPs and are released into the air during the treatment plant operation. Another possibility is the formation of odorants as a result of biological activity in wastewater. An anaerobic activity leads to the production of chemical compounds such as hydrogen sulphide, methane, ammonia, and organic compounds: mercaptans, amine, indole, skatol. Sulphurous compounds are a typical group for WWTPs, including hydrogen sulphide characterized by the smell of rotten eggs and mercaptans: methyl and ethyl mercaptans with a decaying cabbage smell, as well as propyl and butyl mercaptan with a generally unpleasant smell [10].

Table 1 summarizes the literature data on odour detection thresholds for hydrogen sulphide and mercaptans typical for emissions from WWTPs: methyl mercaptan, ethyl mercaptan, propyl mercaptan, butyl mercaptan, benzyl mercaptan. The mean value of the odour detection threshold for hydrogen sulphide and for the sum of mercaptans was used for calculations performed as part of the work.

The value of odour detection threshold determined for hydrogen sulphide in research works performed in the last twenty years ranged from 0.491 to 0.97 $\mu g/m^3$. The average value from the values in the table below was used for further analyses. In the case of mercaptans, chemical compounds typical of wastewater treatment processes have different odour thresholds (from 0.01 $\mu g/m^3$ for butyl mercaptan up to 2.16 $\mu g/m^3$ for methyl mercaptan). In order to carry out analyses in the further part of the work, the arithmetic mean from detection threshold values determined for methyl mercaptan, propyl mercaptan, butyl mercaptan, and benzyl mercaptan were used as the detection threshold value for all mercaptan mixture.

Table 1. Comparison of odour detection thresholds for hydrogen sulphide and mercaptans typical for emission from WWTPs.

References	Hydrogen sulphide [µg/m³]	Mercaptans				
		Methyl mercaptan [µg/m³]	Ethyl mercaptan [µg/m³]	Propyl mercaptan [µg/m³]	Butyl mercaptan [µg/m³]	Benzyl mercaptan [µg/m³]
[4]	0.65	2.16	0.48	0.404	0.01	0.96
[5]	0.97	0.98	-	-	-	-
[6]	0.57–1.42	-	-	-	-	-
[7]	0.57	0.14	0.022	0.404	0.01	0.96
[8]	0.491–0.946	-	-	-	-	-
Average	**0.802**	**1.093**	**0.251**	**0.404**	**0.010**	**0.96**
		0.546				

3 Materials and methods

3.1 Study site

The analysed object is an anaerobic biological wastewater treatment plant that treats wastewater from the food industry. The air from the sewage treatment plant is sucked in through the pipeline system transporting to the deodorization plant – trickle bed bioreactor. The outlet from the deodorization installation is located at an altitude of 6.600 m, and its diameter is 0.80 m. The facility is located in the southern part of Poland on the outskirts of the city above 100,000 inhabitants. In the area surrounding the plant, at a distance of approx. 100 m single- and multi-family housing estates are located, as well as industrial and agricultural areas.

Table 2 summarizes the parameters of the emitter (the outlet channel of the trickle bed bioreactor).

Table 2. Parameters of emitter.

Parameter	Unit	Value
Emitter diameter	[m]	0.8
Emitter height	[m]	6.60
Exhaust gas velocity	[m/s]	2.76
Exhaust gas flow	[m³/s]	1.39
Exhaust gas humidity	[%]	77.5
Exhaust gas temperature	[K]	303.8

3.2 Input data for modelling

In order to determine the odour concentration in the outlet channel from the trickle bed bioreactor, gas samples were taken in accordance with the methodology described in EN 13725:2007 [11]. Sampler and PET bags characterized by the lack of absorption and odour extraction were used for sampling. The other equipment used for sampling is also made of odourless materials. PET bags were conditioned prior to takeingsamples. Samples were taken at an average time of 30 s using a standard sampler CSD30.

Immediately after the collection, the samples were transferred to the Olfactometric Laboratory, where odour concentrations were determined. The measurement of the odour concentration was made using the dynamic olfactometry method in accordance with the procedures described in the EN 13725: 2007 standard [11]. A TO8 four-station olfactometer was used for the measurement together with the necessary equipment. The unit of the odour concentration is reported as the European odour unit per cubic meter (ou_E/m^3).

Measurements of the concentration of hydrogen sulphide and mercaptans were made using the dynamic dosimetry method using Dräger indicator tubes used for instantaneous measurements.

Table 3 lists the concentrations of odorants - hydrogen sulphide and mercaptans, the odour concentration, expressed in European odour unit per cubic meter. On the basis of the determined concentrations of pollutants in the off-gas and gas flow rate, the emission value for the above mentioned substances was determined.

Table 3. Data on the emission of odour and odorants (hydrogen sulphide and mercaptans) from the analysed object.

Name of pollutant	Pollutant concentration [ou_E/m^3] - odours [mg/m^3] – odorants	Pollutant emission [ou_E/s] – odours [mg/s] - odorants
Odours	228393	317466
Hydrogen sulphide	69.53	96.65
Mercaptans	196.32	272.44

3.3 Model studies

The calculations were carried out with the Polish reference model according to the methodology contained in the Regulation of the Minister of the Environment of 26 January 2010 on reference values for certain substances in the air in OPERAT FB software. The location and parameters of odour and odorant emission sources were determined, the aerodynamic roughness of the terrain was determined and the meteorological parameters for the designated area were selected.

Using the model, for a computational grid of 1400 m x 1400 m and a computational step of 50 m, the average annual concentration of odour and average annual concentration of analysed odorants (mercaptans and hydrogen sulphide) to the reference value of 1 ou_E/m^3 for odours and to odour detection threshold for odorants calculations were made. For the same assumptions, calculations of the frequency of exceedances of the odour concentration of 1 ou_E/m^3 and odour detection thresholds for the analysed odorants during the year were carried out.

4 Results and discussion

The calculations carried out in the OPERAT FB software using the Polish reference model have made it possible to estimate the odour nuisance impact of the analysed object. For each pollutant, calculations of the average concentrations and frequency of exceeding one-hour concentrations were made. Based on the obtained results, it can be concluded that the object may be a cause of odour nuisance for residents of nearby buildings. Figure 1–3 shows the results of calculations of pollutants – odours (Fig. 1), hydrogen sulphide (Fig. 2) and mercaptans (Fig. 3) dispersion.

In order to perform model calculations for hydrogen sulphide and mercaptans, according to data included in Table 1, the odour detection threshold was assumed at the level of 0.802 µg/m^3 and 0.546 µg m^3, respectively.

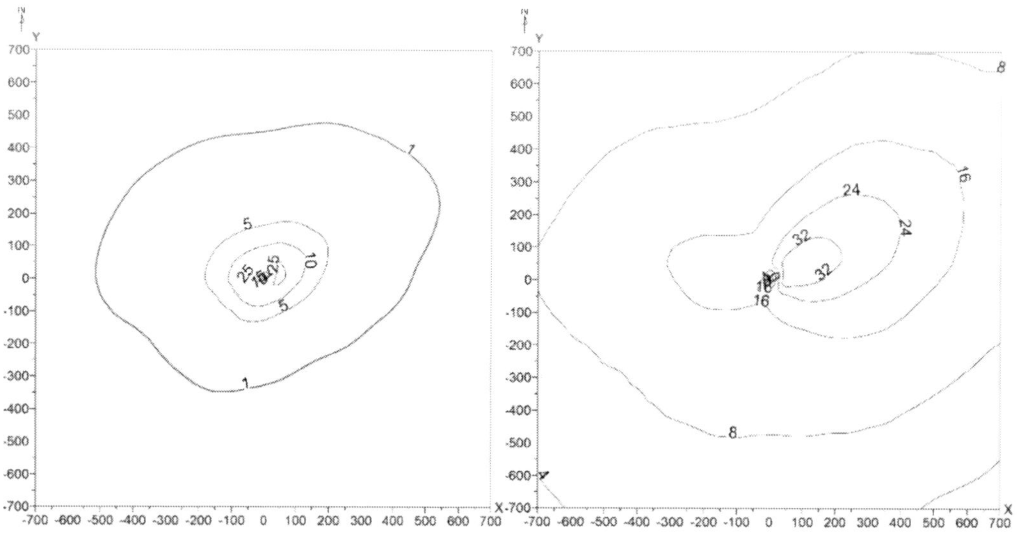

Fig. 1. Isolines of an average annual odour concentration (exceedance threshold of 1 ou$_E$/m^3) – the left side. Isolines of the frequency of exceeding the one-hour odour concentrations (limit value 1 ou$_E$/m^3) – the right side.

The model calculations of the average annual odour concentration (Figure 1) showed the highest range of the odour impact in the eastern direction – 550 m from the emission source. The range of impact in the northern, western and southern directions was 450 m, 500 m, and 350 m, respectively. Also, the frequency of exceeding the one-hour odour concentrations reached the highest values in the east direction and amounted to about 14% at a distance of 700 m from the emission source. In the north, west and east directions, the frequency of exceedances at a distance of 700 m from the emission source was 6%, 8%, and 6%, respectively.

Modeling the dispersion of hydrogen sulphide (Figure 2) around the emitter showed the least impact on the odour nuisance compared to the remaining analysed pollutants. Calculations of the average annual hydrogen sulphide concentration showed the highest range of the odour impact in the eastern direction – 350 m from the emission source. The range of impact in the northern, western and southern directions was 250 m, 275 m, and 200 m, respectively. Also, the frequency of exceeding the one-hour hydrogen sulphide concentrations at a level of odour detection threshold reached the highest values in the east direction and amounted to about 5% at a distance of 700 m from the emission source. In the north, west and east directions, the frequency of exceedances at a distance of 700 m from the emission source was 3%, 4%, and 3%, respectively.

https://doi.org/10.1051/e3sconf/201910000024

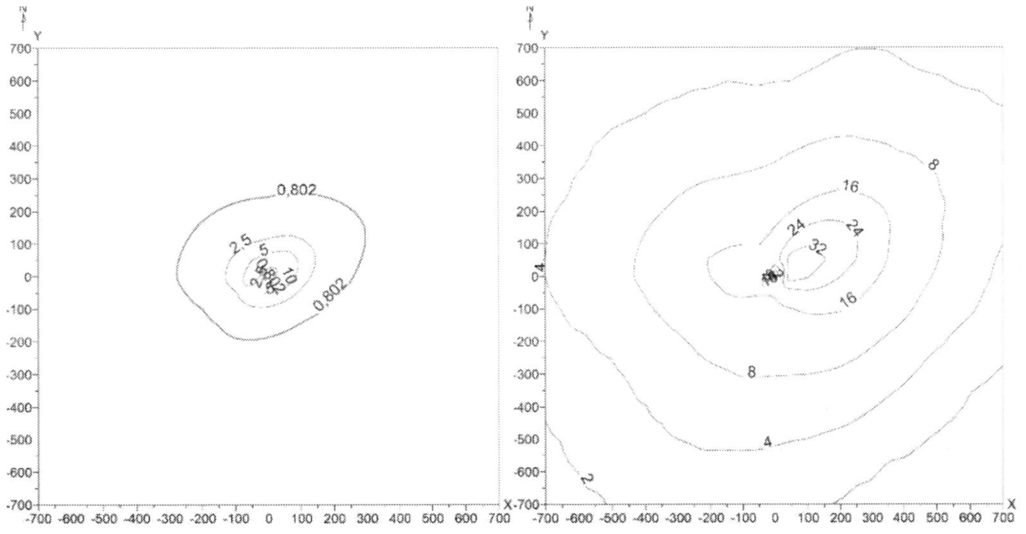

Fig. 2. Isolines of an average annual concentration of hydrogen sulphide (up to odour detection threshold for hydrogen sulphide) – the left side. Isolines of the frequency of exceeding the one-hour concentrations at the level of odour detection threshold for hydrogen sulphie – the right side.

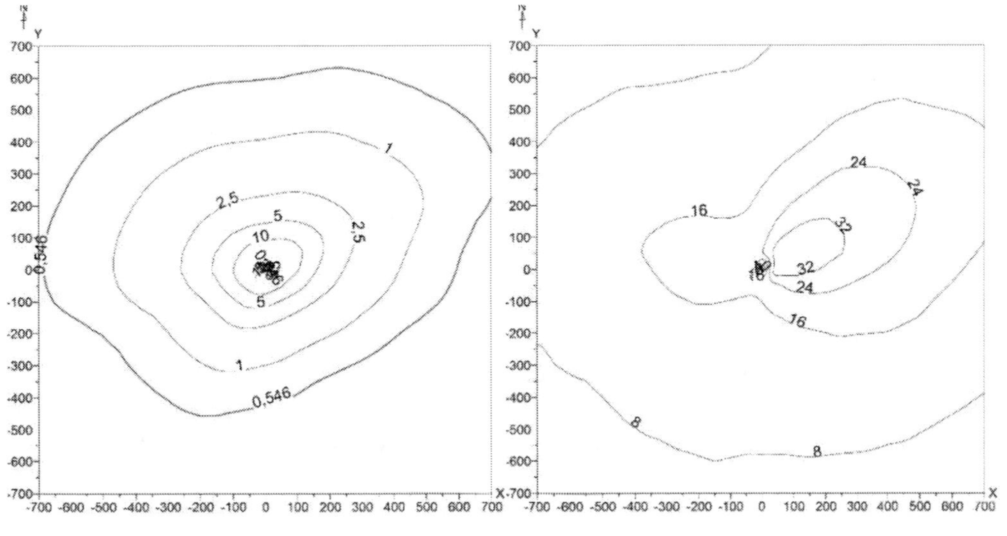

Fig. 3. Isolines of an average annual concentration of mercaptans (up to odour detection threshold for mercaptans) – the left side. Isolines of the frequency of exceeding the one-hour concentrations at the level of odour detection threshold for mercaptans – the right side.

Model calculations carried out showed that among the analysed pollutants, mercaptans have the largest range of the odour impact Calculations of the average annual mercaptans concentration showed the highest range of the odour impact in the eastern direction – 350 m from the emission source. The range of impact in the northern, western and southern directions was 250 m, 275 m, and 200 m, respectively. Also, the frequency of exceeding the one-hour mercaptans concentrations at a level of odour detection threshold reached the highest values in the east direction and amounted to about 5% at a distance of 700 m from the emission source. In the north, west and east directions, the frequency of exceedances at a distance of 700 m from the emission source was 3%, 4%, and 3%, respectively.

5 Summary and conclusions

Water and sewage management is one of the main sources of odour nuisance of food industry facilities. The main pollutants that may affect the odour impact of these objects include sulphurous compounds: hydrogen sulphide and mercaptans. The range of the odour impact depends not only on the odour detection thresholds of the individual compounds contained in the gases but also on the overall composition of the mixture, in which some compounds can affect the aroma parameters (intensity, hedonic quality) of others. For this reason, such an important element of the odour impact assessment is the use of methods allowing the determination of the odour concentration and the inclusion of odours as a mixture.

The study was co-financed within the framework of the project No. 0402/0092/18 with the specific subsidy granted for the Faculty of Environmental Engineering Wroclaw University of Science and Technology (W7) by the Minister of Science and Higher Education to conduct research and development work and related tasks contributing to the development of young scientists and doctoral students in 2018/2019.

References

1. M. Szklarczyk, J. Zwoździak, I. Sówka, *Industrial sources of odour emissions, Współczesna problematyka odorów* (praca zbiorowa pod redakcją Szynkowskiej i Zwoździaka, 54–85, Wydawnictwo Naukowo-Techniczne, Warszawa, 2010)
2. I. Sówka, *Metody identyfikacji odorotwórczych gazów emitowanych z obiektów przemysłowych* (Oficyna Wydawnicza Politechniki Wrocławskiej, Wrocław, 2011)
3. W. van Niekerk, M. Fourie, *Odour Thresholds and Odour Descriptors of Substances. Measured at the EnviroServ Shongweni Landfill Site* (INFOTOX (Pty) Ltd, Report No. 011-2017, 2017)
4. A. Talaiekhozani, M. Bagheri, A. Goli, M. R. T. Khoozani, J. Environ. Manage. **170**, 186–206 (2016)
5. D. Glindemann, J. Novak, J. Witherspoon, Environ. Sci. Technol. **40**, 202–07 (2006)
6. M. A. McGinley, C. M. McGinley, Water Sci. Technol. **50**, 4, 75–82 (2004)
7. Y. Nagata, *Measurement of Odor Threshold by Triangle Odor Bag Method In Odor Measurement Review* (Japan: Office of Odor, Noise and Vibration Environmental Management Bureau, Ministry of the Environment, Government of Japan, 118–12, 2003)
8. D. Mannebeck, H. Mannebeck, Gefahrstoffe Reinh. Luft. **62**, 135–40 (2002)
9. Regulation of the Minister of Environment dated 26 January 2010 on reference values for some substances in the air. (Journal of Laws 2010, no. 16, item 87).
10. B. Rajbansi, *Control of specific odorants from Sewage Treatment Plants STP development and validation of the control scheme* (Ph.D. thesis, Jadavpur University, Kolkata, 2013)
11. PN-EN 13725:2007 standard: Air quality. Determination of odour concentration, with the use of dynamic olfactometry method

Failure frequency of water supply networks in the Podlaskie voivodeship

Joanna Gwoździej-Mazur[1,*], and *Kamil* Świętochowski[1]

[1]Department of Environmental of Engineering Systems, Bialystok University of Technology, ul. Wiejska 45E, 15-351 Bialystok, Poland

Abstract. This article describes the subject matter of the failure frequency of water supply pipes in Poland. The introduction to the article describes the causes of failures in the water supply network as well as a tool for assessment of the failure frequency of water supply pipes, i.e. damage intensity index. The status of Polish literature concerning the problem of failure frequency of water supply networks is also presented, and attention is paid to the absence of studies within the area of the Podlaskie voivodeship. The research part of the article presents the results of calculations of failure frequency indexes for water supply networks in the Podlaskie voivodeship. Materials from the Central Statistical Office for the period of 2015–2017 served as the basis for studies. Damage intensity indexes throughout the entire Podlaskie voivodeship were analyzed. The status of failure frequency of water supply networks was assessed in poviats of the Podlaskie voivodeship during the years 2015–2017. The mean level of failure frequency of water supply networks was also compared between the Podlaskie voivodeship, other voviodeships and the national average. As a result of calculations that were carried out, reference material was obtained for current and future studies concerning failure frequency of water supply networks in the Podlaskie voivodeship as well as other voivodeships of northern Poland.

1 Introduction

Failure repair (e.g. water leaks) in water supply pipes is part of a water supply system's operation[1, 2]. The occurrence of failures is a direct cause of water leakage from a water supply system[3]. Failures may be responsible for up to 75% of non-revenue water in a water company. There are many causes of failures [4–8]:
- soil subsidence, tectonic movements and other geological causes,
- material and design defects of pipes, fittings, connectors, etc.,
- excessive loading of the overburden, particularly dynamic loading,
- overloading with many years of exploitation,
- frost penetration to soil in the vicinity of pipe,
- hydraulic shocks,
- high pressure in the network and high daily pressure fluctuations,

* Corresponding author: j.mazur@pb.edu.pl

- corrosiveness of soil and groundwater, as well as of water flowing inside pipes,
- stray voltage accompanying electrical traction,
- mining exploitation of mineral deposits.

Despite the fact that pipe failures are a normal phenomenon over the course of exploitation, their number, and thus the amount of water that is irreversibly lost, should be minimized [9, 10]. The condition of water supply infrastructure as one of many factors affects the safety of water consumers [11–15]. Damage intensity indexes λ are determined in the operational practice of a water supply system [16]. According to Kwietniewski [17], the value of damage intensity index λ should not exceed:

- for mains $\lambda < 0.3$,
- for distribution pipes $\lambda < 0.5$,
- for service lines $\lambda < 1.0$.

The main area of studies on the failure frequency of water supply pipes in Poland are the southern voivodeships of the country, i.e. the Małopolskie, Śląskie and Podkarpackie voivodeships. The area of southern Poland is characterized by mountainous and piedmont terrain. Meanwhile, Śląsk (Silesia) is an area of intensive mining operations, which translate directly to the failure frequency of water supply networks. Studies by Bergel [18] based on 374 water supply systems in Poland provided important information about the failure frequency of water pipes. Bergel determined the mean level of a Polish water supply system's failure frequency at the level of $\lambda = 0.27$ failures/km/year [18]. This study is an important point of reference for assessment of other water supply systems. Works [6, 19–24] provide further and increasingly more recent results with regard to the failure frequency of water supply pipes in southern Poland. Poland is lacking studies on the areas of central and northern Poland. Central and northern Poland are characterized by lowland areas. The problem of failure frequency of water supply pipes in this area is touched upon in the works of Gwoździej-Mazur and Świętochowski [7, 25]. Comparing the water supply systems of central Poland with values for areas where mining damage occurs may result in an excessive tolerance for failures occurring in water supply systems in the northern part of the country. Therefore it is justified to conduct studies intended to provide information about the current level of failure frequency of water supply networks in the northern regions of the country.

2 Material and methods

The Podlaskie voivodeship is administratively divided into 17 poviats (including 3 cities under poviat rights). 118 municipalities, including 13 urban municipalities, 27 urban-rural municipalities and 78 rural municipalities, are distinguished within the entire voivodeship.

At the end of 2017, over 13,544 km of water supply distribution network was functioning within the Podlaskie voivodeship. Water supply networks supplied water to consumers through 197,128 service lines, which translates to a mean density of 14.6 service lines/km of distribution network.

Analyses of failures in water supply networks in the Podlaskie voivodeship were carried out based on data (the length of the distribution network pipes, the number of failures) from the Central Statistical Office for the years 2015–2017 [26]. The presented results describe the failure of whole systems without division into main, distribution and connection pipes. The damage intensity index and data from the literature were used to assess and determine failure frequency and its level.

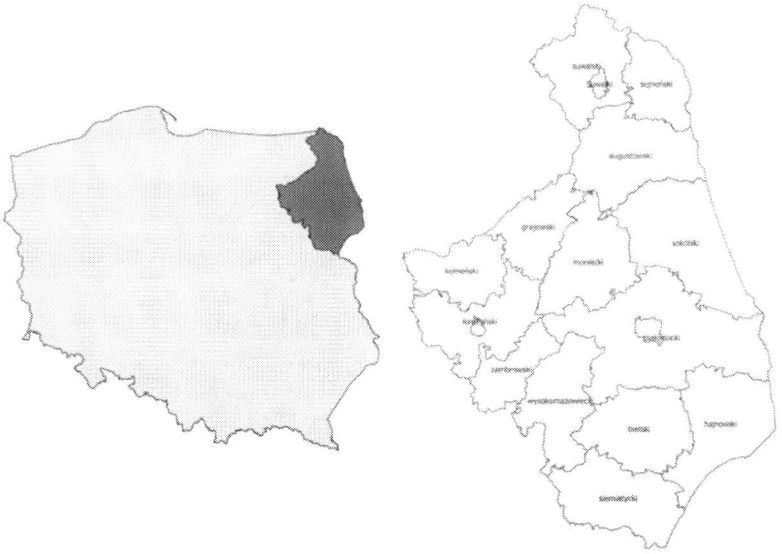

Fig. 1. Location of the Podlaskie voivodeship and division of the voivodeship into poviats.

Failure frequency index [27]:

$$\lambda(\Delta t) = n(\Delta t) / L \cdot \Delta t \ (1) \qquad (1)$$

where:
$\lambda(\Delta t)$ – unit failure intensity, fail/(km·a)],
$n(\Delta t)$ – number of failures within time interval Δt,
L – length of pipes tested within time interval Δt (mean length within this interval), km,
Δt – considered time interval, year.

The ranges described by Kwietniewski served as the basis for assessment of obtained values of the damage intensity index [7]:
- low failure frequency, when $\lambda \leq 0.1$,
- medium failure frequency, when $0.1 < \lambda \leq 0.5$,
- high failure frequency, when $\lambda > 0.5$.

3 Results

Failure frequency of water supply networks in Poland decreased from the value of 0.30 in 2015 to 0.27 in 2017. The lowest mean failure frequency indexes (λ, fail/(km·a)) for voivodeships were obtained in the Lubelskie voivodeship in the years 2015–2016 and in the Lubelskie and Wielkopolskie voivodeships in 2017. The highest failure frequency was noted in the Śląskie voivodeship. In the Podlaskie voivodeships, the mean failure frequency of water supply pipes decreased from the value of 0.26 in 2015 to 0.23 in 2017. Failure frequency of water supply pipes during the period of observation was lower than the mean value for the entire country.

In 2015, the highest failure frequency indexes (λ_{2015}, fail/(km·a)) of water supply pipes were noted in the Śląskie, Dolnośląskie, and Małopolskie voivodeships, and the lowest in the Lubelskie, Wielkopolskie and Mazowieckie voivodeships. In 2016, the highest failure frequency indexes (λ_{2016}, fail/(km·a)) of water supply pipes were noted in the Śląskie, Dolnośląskie, and Małopolskie voivodeships, and the lowest in the Lubelskie, Kujawsko-Pomorskie, Lubuskie, Łódzkie and Mazowieckie voivodeships. In 2017, the highest failure frequency indexes (λ_{2017}, fail/(km·a)) of water supply pipes were noted in the Śląskie,

Małopolskie and Dolnośląskie voivodeships, and the lowest in the Lubelskie, Wielkopolskie and Łódzkie voivodeships.

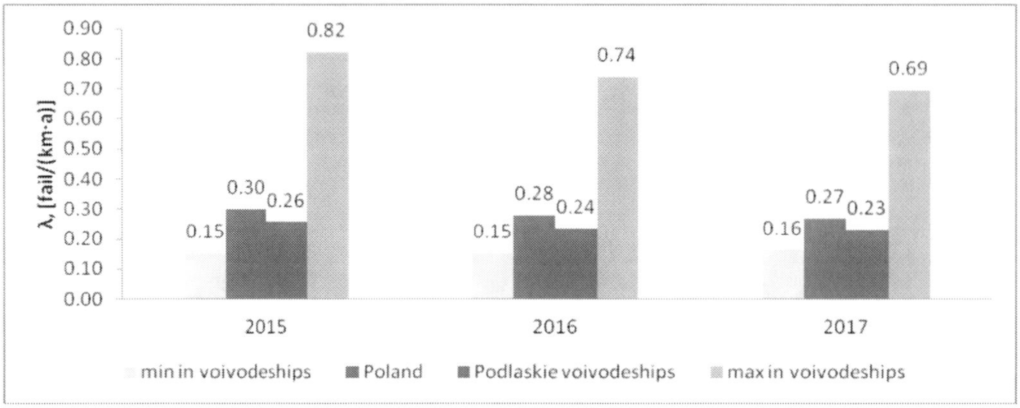

Fig. 2. Mean failure index of a water supply network in the years 2015–2017.

Table 1. Mean failure indexes of a water supply network in voivodeships.

Voivodeship	**λ, fail/(km·a)**	**λ, fail/(km·a)**	**λ, fail/(km·a)**
	2015	**2016**	**2017**
Dolnośląskie	0.46	0.38	0.36
Kujawsko-pomorskie	0.22	0.20	0.19
Lubelskie	0.15	0.15	0.16
Lubuskie	0.22	0.20	0.22
Łódzkie	0.20	0.20	0.19
Małopolskie	0.43	0.40	0.42
Mazowieckie	0.19	0.20	0.17
Opolskie	0.37	0.25	0.30
Podkarpackie	0.27	0.23	0.23
Podlaskie	**0.26**	**0.24**	**0.23**
Poland	**0.30**	**0.28**	**0.27**
Pomorskie	0.35	0.31	0.29
Śląskie	0.82	0.74	0.69
Świętokrzyskie	0.26	0.28	0.28
Warmińsko-mazurskie	0.28	0.28	0.27
Wielkopolskie	0.18	0.18	0.16
Zachodniopomorskie	0.35	0.35	0.34

Failure frequency indexes were calculated for every municipality of the Podlaskie voivodeship in the years 2015–2017. In 2015, the lowest minimum index, $\lambda = 0$ fail./km/year, was reached in the Hajnowski, Kolneński and Siemiatycki poviats. In 2016, the lowest index, $\lambda = 0$ fail./km/year, was reached in the Białostocki, Grajewski, Siemiatycki and Sokólski poviats. In 2017, the lowest index, $\lambda = 0$ fail./km/year, was reached in the Białostocki, Grajewski, Hajnowski, Kolneński, Łomżyński and Sokólski poviats. The results of $\lambda = 0$ fail./km/year occurred in various communes in poviats in different years. Impact on such a result $\lambda = 0$ fail./km/year could be incomplete data in companies associated with poor care for the failure logs or lack of them. The highest minimum index in the years 2015-2017 was noted in the poviat of the city of Białystok (2015 – 0.34 fail./km/year, 2016 – 0.31 fail./km/year, 2017 – 0.37 fail./km/year). Based on the lowest values of the damage intensity index for municipalities, it can be stated that, during the studied period, 5 poviats had low failure frequency of water supply pipes, and 12 poviats were characterized by medium failure frequency of water supply pipes.

Table 2. Failure frequency indexes of the water supply network in poviats of the Podlaskie voivodeship.

Poviat	λ_{2015}, **fail/(km·a)**			λ_{2016}, **fail/(km·a)**			λ_{2017}, **fail/(km·a)**		
	mean in poviat	min in poviat	max in poviat	mean in poviat	min in poviat	max in poviat	mean in poviat	min in poviat	max in poviat
Augustowski	0.27	0.03	1.70	0.24	0.03	1.43	0.20	0.02	0.97
Białostocki	0.39	0.02	3.57	0.32	0.00	3.15	0.26	0.00	3.22
Bielski	0.59	0.09	1.52	0.52	0.11	1.32	0.54	0.09	1.37
Grajewski	0.13	0.01	0.38	0.13	0.00	0.40	0.11	0.00	0.34
Hajnowski	0.28	0.00	1.75	0.20	0.00	0.87	0.19	0.00	0.83
Kolneński	0.16	0.00	0.94	0.15	0.05	1.00	0.15	0.00	0.73
Łomżyński	0.18	0.02	1.35	0.18	0.02	0.85	0.23	0.00	1.27
Poviat of the city of Białystok	0.33	0.33	0.33	0.31	0.31	0.31	0.37	0.37	0.37
Poviat of the city of Łomża	0.18	0.18	0.18	0.26	0.26	0.26	0.24	0.24	0.24
Poviat of the city of Suwałki	0.34	0.34	0.34	0.24	0.24	0.24	0.25	0.25	0.25
Moniecki	0.15	0.02	0.46	0.13	0.03	0.57	0.09	0.02	0.28
Sejneński	0.22	0.04	0.86	0.22	0.04	0.71	0.24	0.03	0.86
Siemiatycki	0.29	0.00	1.35	0.20	0.00	1.12	0.22	0.04	0.84
Sokólski	0.17	0.02	0.93	0.16	0.00	1.04	0.15	0.00	0.98
Suwalski	0.12	0.01	0.34	0.15	0.05	0.37	0.21	0.03	0.51
Wysokomazowiecki	0.31	0.02	2.44	0.34	0.02	2.59	0.34	0.02	3.28
Zambrowski	0.22	0.15	0.48	0.25	0.02	0.78	0.16	0.08	0.44

In 2015, the lowest mean index, λ = 0.12 fail./km/year, was reached in the Moniecki poviat. In 2016, the lowest index, λ = 0.13 fail./km/year, was reached in the Grajewski and Moniecki poviats. In 2017, the lowest index, λ = 0.11 fail./km/year, was reached in the Grajewski poviat. The highest mean index in the years 2015–2017 was noted in the Bielski poviat (2015 – 0.59 fail./km/year, 2016 – 0.52 fail./km/year, 2017 – 0.54 fail./km/year). Based on the mean values of the damage intensity index for municipalities, it can be stated that, during the studied period, 1 poviat had high failure frequency of water supply pipes, and the other poviats were characterized by medium failure frequency of water supply pipes.

In 2015, the lowest maximum index, λ = 0.18 fail./km/year, was reached in the poviat of the city of Łomża. In 2016, the lowest maximum index, λ = 0 fail./km/year, was reached in the Białostocki, Grajewski, Siemiatycki and Sokólski poviats. In 2017, the lowest maximum index, λ = 0.24 fail./km/year, was reached in the poviat of the city of Suwałki. The highest maximum index in the years 2015-2017 was noted in the Białostocki poviat (2015 – 3.57 fail./km/year, 2016 – 3.15 fail./km/year, 2017 – 3.22 fail./km/year). Based on the lowest values of the damage intensity index for municipalities, it can be stated that, during the studied period, 13 poviats had high failure frequency of water supply pipes, and 4 poviats were characterized by medium failure frequency of water supply pipes.

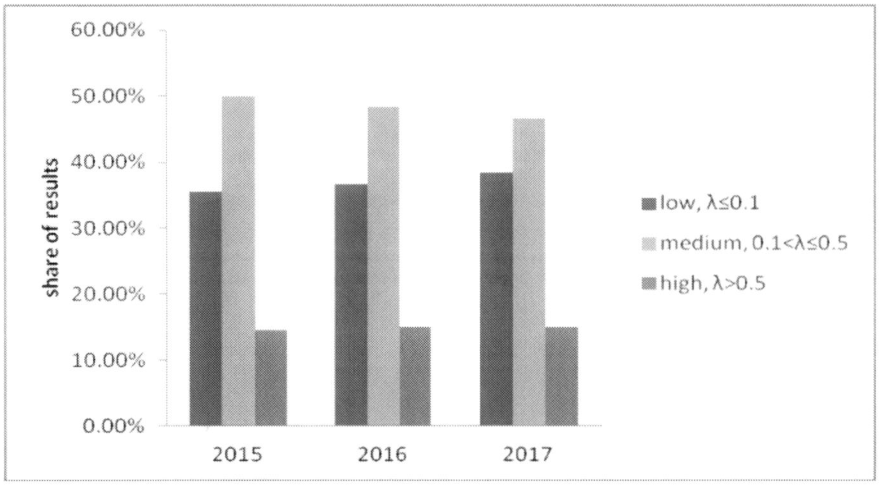

Fig. 3. Failure assessment of water supply in Podlasie voivodship in the years 2015–2017.

The largest number of water supply (individual municipalities) in the Podlasie voivodeship in 2015–2017 was characterized by average failure rates. Low failure rate ranged from 35.5% to 38.4% of water supply. The high failure rate was characterized by between 14.5% and 15.1% of water supply. In 2015-2017, the number of high and low failure water supply increased. The number of medium failure rate water supply is decreasing.

4 Summary

During the period of study, 2015–2017, the Podlaskie voivodeship was characterized by a general failure frequency index of waters supply pipes below the national average. During the period of study, the failure frequency index of water supply pipes in the Podlaskie voivodeship was not one of the three lowest values in the country. Considering the mean values of damage intensity indexes of water supply pipes in poviats of the Podlaskie voivodeship, it can be assessed that the Podlaskie voivodeship is characterized by medium failure frequency of water supply pipes (over 94% are characterized by medium failure frequency). Conducted studies and analyses constitute a good foundation and point of reference for the characterization of individual water supply systems in the Podlaskie voivodeship during the years 2015–2017, and also serve as a point of reference for assessment of the failure frequency of water supply systems in the Podlaskie voivodeship in future years.

Basic data on water supply collected from the Central Statistical Office do not contain information on breakdowns, on the number of failures in individual months and on the age or material on which the failure occurred. Considering the purpose for which research and analysis were carried out, they are sufficient. According to the authors, the results presented in the article are a good reference point for comparing the results used in studies and analyzes of individual water systems (in which you can focus, for example, on the age of pipes, material, type of pipes). Completed calculations and presented results supplement the basic gap in the Polish literature on the subject of waterline failure regarding information on the failure frequency of the lowland water supply in Poland. Comparing the failure rate in water supply in the Podlasie voivodship to the water supply of the mountain areas or mining damage areas is incorrect and would show the Podlaskie water supply as an almost perfect condition. This article is a good starting point for further, detailed research on the failure of water supply in the Podlasie voivodship.

This study supplements the studies of Polish researchers of the subject matter of failure frequency of water supply systems in Poland, which until now have mainly been focused on the area of southern Poland.

References

1. S. J. F. Gould, Water Scien. Techn. **63** (2011)
2. A. Pagano, Water Res. Manag. **32** (2018)
3. J. B. Martínez-Rodríguez, I. Montalvo, J. Izquierdo, Water Resour. Manag. **25** (2011)
4. H. Hotloś, Env. Protection, **1** (2003)
5. S. Jeewon, Procedia Engin. **119**, (2015)
6. A. Kuliczkowski, A. Parka, G. Stefański, Instal, **2** (2013)
7. J. Gwoździej-Mazur, K. Świetochowski, B. Kaźmierczak, Econ. and Environ. **2**, (2018)
8. L. Berardi, R. Ugarelli, J. Rostum, O. Giustolisi, Water Resour. Res. (2014)
9. Y. Kleiner, B. J. Adams, J. S. Rogers, Journal of Comp. Civil Engin. **15** (2001)
10. R. Sadiq, Y. Kleiner, B. Rajani, Jour. of Wat. Sup. Res. and Techn.-AQUA, **53** (2004)
11. W. Balcerzak, S. M. Rybicki, Jour. of Wat. Sup. Res. and Techn.-AQUA, **60** (2011)
12. R. Jafar, I. Shahrour, I. Juran, Math. and Comp. Mod. **51** (2010)
13. D. Szpak, B. Cecilia, E3S Web of Conf. **44** (2018)
14. B. Tchórzewska-Cieślak, K. Boryczko, M. Eid, Adv. in Saf., Reliab. and Risk Manag. (Taylor & Francis Group, 2492–2499, London, 2012)
15. B. Tchorzewska-Cieslak, J. Rak, Environ. Engin. (Taylor & Francis Group, 521–526, London, 2010)
16. K. Shahata, T. Zayed, Struct. Infrast. **8** (2012)
17. M. Kwietniewski, *Failure of water supply and wastewater infrastructure in Poland based on the field* tests (In Proceedings of the XXV Scientific-Technical Conference, Międzyzdroje, Poland, 2011)
18. T. Bergel, G. Kaczor, P. Bugajski, *Technical condition of water supply networks in small water supply pipes of the Małopolskie and Podkarpackie voivodeships* (Infrastructure and ecology of rural areas, Polish Academy of Sciences, Kraków Branch, 2013)
19. K. Pietrucha-Urbanik, A. Studziński, Env. Prot. **18**, (2007)
20. J. Rak, Env. Prot. **29**, (2016)
21. J. Rak, B. Tchórzewska Cieślak *Failure frequency of waters supply networks in the main cities of the San* Valley (3rd "Blue San" Scientific and Technical Conference, Dubiecko, 2006)
22. D. Szpak, B. Tchórzewska, Chemik **68**, 10 (2014)
23. B. Tchórzewska-Cieślak, K. Boryczko, *Analysis of the exploitation of the water supply system of the city of Mielec* (Polish environmental engineering five years after accession to the European Union, PAN, **58**, 1, 2009)
24. K. Świętochowski, Water San. Techn. **5** (2018)

25. I. Vloerbergh, M. Blokker, Water Asset Manag. Intern. **6** (2010)
26. https://www.stat.gov.pl
27. M. Kwietniewski, J. Rak, *Reliability of water supply and wastewater infrastructure in Poland* (Institute of Fundamental Techno-logical Research, Warsaw, 2010)

Human health risk assessment of heavy metals in road dust collected in Cracow

Gabriela Hajduga[1,*], *Agnieszka* Generowicz[1], and *Małgorzata* Kryłów[1]

[1]Cracow University of Technology, Department of Environmental Engineering, Department of Environmental Technologies, ul. Warszawska 24, 31-155 Cracow, Poland

Abstract. Road dust is viewed as one of the major contributors for metal pollution in urban environment and long-term exposure can cause chronic damage through ways of inhalation, ingestion, and dermal contact so they pose a great threat on human health. The article presents a study conducted to determine the concentrations of seven heavy metals in road dust from a chosen street in Cracow, and the impact of heavy metals contamination in surface street dust on human health using Health Risk Assessment. The health risk was assessed using Hazard Quotient (HQ), Health Index (HI) and Carcinogenic Risk (RI).

1 Introduction

Dust on urban impervious surface which also contain heavy metals, has become one of the most important issues in urban environmental management and it's considered as toxic [1–3]. It can be easily re-suspended under certain outside dynamic condition, pollutants adsorbed on them enter human body by the pathways of respiratory inhalation and direct skin contact [1, 4, 5]. Pollutants attached to surface dust can be also transferred to the surrounding aquatic environment with runoff.

In the face of the harmful effects of heavy metals, more attention is devoted to issues of health risk assessment, which relies on qualitative and quantitative characteristics of the probability of negative health effects as a result of exposure to a given factor. As different types of chemicals bring different influences on human bodies, risk assessments of these chemicals can be characterized by non-carcinogenic risk (HI) and carcinogenic risk (RI), Carcinogenic risk refers to the incremental probability of an individual developing any kind of cancer as a result of exposure to carcinogens [7, 8].

The purpose of this work is to assess the health exposure of the inhabitants of Halszki Street in Krakow, resulting from the presence of heavy metals such as cadmium, nickel copper, lead zinc, chromium and manganese in the street dust of one of the streets [9–12]. In addition, the results were referred to the results of the Life Cycle Assessment (LCA) carried out on a similar subject and compared with result of Health Risk Assessment carried out for another major street in Kraków [11, 12].

* Corresponding author: gabriela.hajduga@gmail.com

2 Materials and methods

The basis for this research were the results of the qualitative composition collected material as a result of sweeping processes and washing up of streets - waste and sewage. Street sweeping activities were carried out in cooperation with the following entities: Miejskie Przedsiębiorstwo Oczyszczania in Cracow Sp. z o.o. (Municipal Cleaning Company), Provincial Inspectorate for Environmental Protection and the Faculty of Environmental Engineering of the Cracow University of Technology.

The results of the composition of collected and washed waste comes from Halszki Street in Cracow. The segment on which the research was carried out was about 1.1 km long. This street has been qualified for roads with traffic exceeding 1000 cars per hour. Halszki Street is the main access road to the nearest housing estates with a variable width of 9 to 10 m. There are no parking spaces or green lanes along the route. On both sides of the road, along its entire length, there is a footpath for pedestrians with a width of about 2 m. The street, is mainly surrounded by residential blocks and urban greenery at sports facilities.

The sampling were collected within 7 consecutive days of the week in the month of April. To avoid impediments related to car traffic, samplings were carried out at night, starting from 23:00. First, the street was swept along both street sides and across its entire width. After this stage, each day samples of solid waste material were collected from the sweeper tank for the needs of laboratory tests. Each day they were collected to 500 ml polyethylene containers as in similar studies carried out by [4]. Next, using a water tanker, the street section was washed, and at selected points (at the beginning of the route, in the middle and at the end), samples of sewage were collected, which were also subjected to separate laboratory analysis for the purposes of Life Cycle Assessment.

Laboratory tests were carried out at the Provincial Inspectorate for Environmental Protection in Cracow. The samples were tested for the concentration of such elements as Zinc, Manganese, Copper, Nickel, Lead, Cadmium and Total Chrome. The concentrations of heavy metals were determined according to the following standards: PN-EN 13657:2006, PN-ISO 8288:2002, PN-EN 1233:2000 p. 3. The first one determined the method of extraction of soluble pearl substitutes from waste by solution with royal water. From samples taken each day, 3 samples were prepared, which were further prepared for analysis. This included drying and reduction of grain size below 250 μm particle size, and docking after drying, the material was sieved through a 1 mm mesh sieve and the fraction < 63 μm was separated as potentially posing the greatest threat to health). The weight of a single sample prepared according to the recommendations always exceeded 20 mg and amounted to 0.50 g (weighed to the nearest 0.0001 g), then 6 ml of HCl and 2 ml of HNO_3 were added separately and the dilution procedure according to the program 1) 2 min, 250 W; 2) 2 min, 0 W; 3) 5 min, 250 W; 4) 5 min 400 W; 5) 5 min, 500 W, was started. Then the contents of vessels were quantitatively transferred to the measuring flask, supplemented to the mark and then the analysis was performed. Concentrations of heavy metals were determined by atomic absorption spectrometry with flame atomization (PN-ISO 8288:2002, PN-EN 1233:2000 p.3.). The accuracy of the method was evaluated using the standard addition method. Table 1 presents average concentrations of heavy metals in solid waste.

Table 1. Average concentrations of heavy metals in collected solid waste.

Chemical element	Unit	Result	Chemical element	Unit	Result
Cadmium	mg/kg	5.13	Zinc	mg/kg	333.73
Nickel	mg/kg	24.93	Chrome	mg/kg	122.17
Copper	mg/kg	171.33	Manganese	mg/kg	788.97
Lead	mg/kg	46.77			

2.1 Health risk assessment

In this study, the model that was used to assess health risk of Halszki Street residents was derived from a model advanced by the United States Environmental Protection Agency and it was widely used by other authors in the literature [4, 14–18]. It was assumed that heavy metals penetrate the body in an unintentional manner through oral, inhalation and dermal routes (exposure paths) [1, 4, 5, 7, 9, 11]. Another assumption proposed by EPA was the duration of life, during which residents would be exposed to heavy metals. In the case of adults, it EPA assumes that it would be 70 years, and in the case of children – 6 years [8, 9, 11]. Environmental exposure was estimated on the basis of the average dose, which determines the amount of harmful substance taken by the studied population per day and per 1 kg of body weight. Dose values for each route of exposure have been calculated from equation from 1 to 3 [9, 12–14].

$$D_{ing} = \frac{c \cdot R_{ing} \cdot CF \cdot EF \cdot ED}{BW \cdot AT} \quad (1)$$

$$D_{inh} = \frac{c \cdot R_{inh} \cdot EF \cdot ED}{PEF \cdot BW \cdot AT} \quad (2)$$

$$D_{derm} = \frac{c \cdot SA \cdot CF \cdot SL \cdot ABS \cdot EF \cdot ED}{BW \cdot AT} \quad (3)$$

For carcinogens, the Lifetime Average Daily Dose (LADD) (inhalation exposure route for nickel, and cadmium; inhalation, ingestion exposure route for chromium) was utilized in the evaluation of cancer risk then calculated using equation (4) [7, 11, 12, 14–18]. The meanings and values of the parameters are listed in Table 2.

$$LADD = \frac{c \cdot EF}{PEF \cdot AT} \cdot \left(\frac{CR_{child} \cdot ED_{child}}{BW_{child}} + \frac{CR_{adult} \cdot ED_{adult}}{BW_{adult}} \right) \quad (4)$$

Table 2. Parameters and their values [1, 4, 5, 9, 11, 12, 14–18].

Factor	Definition	Unit	Value	
			Children	Adult
c	concentration of the contaminant in dusts	mg/kg	Table 1	
R_{ing}	ingestion rate	mg/day	200	100
R_{inh}	inhalation rate	m^3/day	7.6	20
EF	exposure frequency	days/year	330	
ED	exposure duration	years	6	70
BW	average body weight	kg	15	70
AT	average time	days	365·ED 365·70 for carcinogens	
CF	conversion factor	kg/mg	$1 \cdot 10^{-6}$	
PEF	particle emission factor	m^3/kg	$1 \cdot 10^{9}$	
SA	surface area of the skin that contacts the dust	cm^2	2800	5700
SL	skin adherence factor for dust	mg/cm^2	0,2	0,7
ABS	dermal absorption factor	-	0.001	

Health risk, which is a function of human exposure to toxic substances, was assessed using the Hazard Quotient (HQ) which was calculated according to equation 5 [9].

$$HQ = \frac{D}{RfD} \quad (5)$$

D –absorbed dose $\left[\frac{mg}{kg \cdot d}\right]$; RfD –reference dose$\left[\frac{mg}{kg \cdot d}\right]$.

The reference dose (RfD) is an estimation of maximum permissible risk on human population through daily exposure during a lifetime. The threshold of RfD value can be used to indicate whether there would be an adverse health effect during a life time. If an average daily dose (D) value is lower than the reference dose, there would not be any adverse health effect; otherwise it is likely that the exposure pathway will cause adverse human health effect [1, 4, 5, 9, 11, 12, 14–18, 19] (table 3).

The HQs can be added to generate Hazard Index (HI) (equation 6) which allows to estimate the risk of mixed heavy metal contaminates. HI presents total risk of non-carcinogenic metals and if the value of $HI \leq 1$ there is no significant risk, but if $HI > 1$, there is a great chance of non-carcinogenic effects to human health [1, 4, 5, 9, 11, 12, 14–18, 20–23]. The same rule applies to individual HQ values.

Table 3. Reference doses for heavy metals in accordance do exposure pathways [11].

Chemical element	Unit	RfD		
		RfD_{ing}	RfD_{inh}	RfD_{derm}
Cadmium	mg/kg·d	1.00E-03	1.00E-03	1.00E-03
Nickel		2.00E-02	2.06E-02	5.40E-03
Copper		4.00E-02	4.02E-02	1.02E-02
Lead		3.50E-03	3.52E-03	5.25E-04
Zinc		3.00E-01	6.00E-02	3.00E-01
Chrome		3.00E-03	2.86E-05	6.00E-05
Manganese		1.40E-01	5.00E-05	5.60E-03

$$HI = HQ_{Cd} + HQ_{Cu} + HQ_{Pb} + HQ_{Zn} + HQ_{Mn} + HQ_{Ni} + HQ_{Cr} \qquad (6)$$

It is also possible to calculate the total exposure to the considered heavy metals for each exposure path (7).

$$HI = HQ_{ing} + HQ_{derm} + HQ_{inh} \qquad (7)$$

For carcinogens the index of RI is used and it can be calculated by the following equations:

$$RI = LADD \cdot SF \qquad (8)$$

$$R = RI_{ing} + RI_{derm} + RI_{inh} \qquad (9)$$

The slope factor (SF) converts estimated daily intake averaged over a lifetime of exposure directly to incremental risk of an individual. The cancer risk is sum of carcinogenic risk of individual carcinogens in the possible exposure pathways (R) [9, 11, 16, 18, 21]. According to the U.S. EPA, the value of cancer risk in the range of 10^{-6} to 10^{-4} is an acceptable or tolerable risk, a risk of less than 10^{-6} can be ignored, and a risk exceeding 10^{-4} is considered to unacceptable. In calculations for carcinogenic exposure, the US EPA guidelines do not make calculations in the distinction between adults and children [1, 9, 11, 22, 23].

3 Results

3.1 Comparison with Aleje Krasińskiego

On the exact same date, samples was also taken from Aleje Krasińskiego located in the city center. This route characterizes with heavy traffics over 3000 vehicles per hour which is 3 times more than on Halszki Street. Along of this road we have old tenements, commercial buildings and public buildings which use gas or electrical heating. For samples taken from that site, it was also possible to calculate Health Risk for citizens from that area. Result for both location are presented in tables from 4 to 7. During the time of sampling the nearest air quality measurement station to Halszki Street registered max. concentration of PM10 – 80 μm/m3, min. 34 μm/m3 (three days when, the concentration was below the alarm level – 50 μm/m3), whereas on Krasińskiego max. concentration of PM10 – 160 μm/m3, min. 39 μm/m3 (only during one day, the concentration was below the alarm level).

Table 4. Comparison of the Health Risk Assessment results for Halszki Street and Aleje Krasińskiego.

	Adults Halszki Street				Adults Al. Krasińskiego			
Chemical element	HQ_{ing}	HQ_{inhal}	HQ_{derm}	HI_{adults}	HQ_{ing}	HQ_{inhal}	HQ_{derm}	HI_{adults}
Cadmium	6.63E-03	9.74E-07	2.64E-04	7.70E-03	7.40E-03	1.09E-06	2.95E-04	6.89E-03
Nickel	1.61E-03	2.30E-07	2.37E-04	2.18E-03	1.90E-03	2.70E-07	2.80E-04	1.85E-03
Copper	5.53E-03	8.08E-07	8.66E-04	8.48E-03	7.33E-03	1.07E-06	1.15E-03	6.40E-03
Lead	1.73E-02	2.52E-06	4.59E-03	3.94E-02	3.11E-02	4.55E-06	8.25E-03	2.19E-02
Zinc	1.44E-03	1.06E-06	5.73E-05	2.05E-03	1.97E-03	1.45E-06	7.87E-05	1.50E-03
Chrome	5.27E-02	8.11E-04	1.05E-01	1.69E-01	5.60E-02	8.64E-04	1.12E-01	1.59E-01
Manganese	7.29E-03	3.00E-03	7.27E-03	2.22E-02	9.21E-03	3.80E-03	9.20E-03	1.76E-02

Table 5. The Results of Health Risk Assessment results for Aleje Krasińskiego.

Cancer Risk Al. Krasińskiego					
Chemical element	SF_{inh}	$LADD_{inh}$	SF_{ing}	$LADD_{ing}$	Cancer risk (R)
Cadmium	6.3	1.3E-09			8.19E-09
Nickel	0.84	6.4E-09			5.38E-09
Chrome	42	2.9E-08	0.5	2.20E-07	1.33E-06

Table 6. The Results of Health Risk Assessment results for Halszki Street.

Cancer Risk Halszki Street					
Chemical element	SF_{inh}	$LADD_{inh}$	SF_{ing}	$LADD_{ing}$	Cancer risk (R)
Cadmium	6.30	1.1E-09			6.93E-09
Nickel	0.84	5.5E-09			4.62E-09
Chrome	42.00	2.7E-08	5.0E-01	2.1E-07	1.24E-06

Table 7. Comparison of the Health Risk Assessment results for Halszki Street and Aleje Krasińskiego.

	Children ul. Halszki				**Children Al. Krasińskiego**			
Chemical element	**HQ_{ing}**	**HQ_{inhal}**	**HQ_{derm}**	**$HI_{children}$**	**HQ_{ing}**	**HQ_{inhal}**	**HQ_{derm}**	**$HI_{children}$**
Cadmium	6.18E-02	1.73E-06	1.73E-04	6.20E-02	6.91E-02	1.93E-06	1.93E-04	6.93E-02
Nickel	1.51E-02	4.08E-07	1.56E-04	1.53E-02	1.77E-02	4.80E-07	1.84E-04	1.79E-02
Copper	5.18E-02	1.44E-06	5.67E-04	5.24E-02	6.85E-02	1.90E-06	7.52E-04	6.93E-02
Lead	1.61E-01	4.49E-06	3.01E-03	1.64E-01	2.89E-01	8.04E-06	5.41E-03	2.94E-01
Zinc	1.34E-02	1.87E-06	3.77E-05	1.34E-02	1.84E-02	2.57E-06	5.17E-05	1.85E-02
Chrome	4.90E-01	1.44E-03	6.87E-02	5.60E-01	5.23E-01	1.53E-03	7.33E-02	5.98E-01
Manganese	6.79E-02	5.32E-03	4.75E-03	7.80E-02	8.64E-02	6.74E-03	6.02E-03	9.92E-02

3.1. Comparison with LCA

Life cycle assessment according to the official definition given by the European Commission is the process of collecting and evaluating the input and output data of the product as well as assessing the potential impact on the environment throughout its life cycle (production, use and utilization). In used LCA model, there were two categories of influence relating to humans. One is "Toxic effect on humans – carcinogenic effect" and the other "Toxic effect for humans - action other than carcinogenic". The unit for these categories of human impact is the Comparative Toxic Unit for Humans (CTUh) and expresses the estimated increase in mortality in the total human population, i.e. the number of cases per 1 kg of emitted substance (Table 8). In case of Halszki Street, LCA analysis was carried out on the basis of developed various emission variants, assuming different distribution of pollutant emissions contained in street dust to the environment. The most unfavorable in terms of impact on humans turned out to be a variant where 70% of pollutants are emissions to air, another 15% goes to water and 15% to soil. For adopted allocation of emission, the Health Risk Assessment was carried out, with the assumption that 70% of the emission goes to air and puts people at risk through accidental ingestion, dermal contact and inhalation (variant 1); 15% of emission goes to water and puts people at risk only through ingestion (variant 2); 15% of the emission goes to the soil and puts people at risk through ingestion and dermal contact (variant 3), then the results were referred to the results of the LCA analysis.

Table 8. The results of the LCA analysis in relation to categories concerning human health.

Category of influence	**Unit**	**Value**
Toxic effect for humans - action other than carcinogenic	CTUh	0.0052
Toxic effect on humans - carcinogenic effect	CTUh	0.0004

Due to the change in concentrations, the risk values decreased accordingly. There were no changes in the range of elements from the least to the most toxic.

4 Disscusion and conclusions

Although these areas are different from each other, in many respects mentioned above, the analysis indicates that zinc and chromium are the greatest threats to the health of adults. More interestingly, the results show an almost twofold risk pattern for lead closely related to car traffic. For none of the above cases, the risk does not exceed the limit value. In the case of children, again the smallest health hazard in Halszki Street is characterized by zinc, and the largest chrome. This time the results are not completely transferred to Aleje Krasińskiego, where 'the less risky' of metals turned out to be nickel. Again, the biggest difference is for lead where the risk value increased by 80%. The risk values for chromium for children, compared to adults in both cases are almost 3.5 times higher.

The results obtained for each case confirm the information from the LCA analysis. The risk analysis shows that when it comes to carcinogenic activity, chromium is the most important, and the LCA analysis showed that chromium is in more than 90% responsible for the result in the Toxic effect on humans category, which gives us 0,0004 cancer cases per 1 kg emission of a given substance. In turn, in the case of Toxic effect for humans – action other than carcinogenic in 94.8%, for the result of the category, responsible were lead, zinc and chromium. That gives us 0.0052 cases per 1 kg of emission of a given substance.

In an ideal scenario the analysis would be carried out based on average year load of heavy metals, but at this moment, it's not possible due to the lack of access and the possibility to carry out regular field studies. Because of that, the Health Risk Assessment was calculated for a given composition of street dust, converted into exposures to that composition throughout the year. At this moment, although the method was developed by the U.S. Agency, it can be considered transferable on favourable terms. A handbook with examples for analysis does not say anything to develop it for a particular community or area. It is based on a breakdown into exposure pathways and dose limits that may endanger humans. These have been mainly determined by laboratory tests on animals. In addition, elements such as exposure time, can be assumes. In this case, 330 days were assumed assuming 26 days of leave for the average person and additional 9 special days where an adult with a family can leave the city. The EPA guidelines are also used by authors conducting research in China, Thailand or Pakistan. In Poland, risk is not a popular indicator, but if there was a greater interest in the subject, environmental agencies maybe would undertake to prepare their own guidelines.

As mentioned above, the topic needs to be further developed. In the nearest future, research with different assumptions will start to be carried out with a plan of keeping regularity in sampling and for closed area.

Acknowledgements

We would like to thank the Management Board and Employees of Cracow MPO Sp. z o.o. for the opportunity to undertake research and analysis.

In a special way we would like to thank the Presidents of MPO Sp. z o.o. Kraków: Henryk Kultys and Andrzej Natkaniec for indicating the area of research and their technical support, Director Jacek Sobczyk for his goodwill, time and professional help

References

1. X. Liu, Q. Song, Sci. Total. Environ. **1**, 463–464 (2013)
2. A. Generowicz, K. Gaska, G. Hajduga, E3S Web. Conf. **44** (2018)
3. G. Hajduga, A. Generowicz, E3S Web. Conf. **17** (2017)
4. X. Han, X. Lu, Int. J. Env. Res. Pub. He. **14**, 8, 886–898 (2017)
5. Y. Liang, X. Yi, Int. J. Env. Res. Pub. He. **14**, 12, 1557–1574 (2017)
6. K. Gaska, A. J. Wandrasz, Waste. Manage. **28**, 6, 973–985 (2007)

7. L. He, B. Gao, X. Luo, J. Jiao, H. Qin, C. Zhang, Y. Dong, Sustainability-Basel. **10**, 4, 1–12 (2018)
8. G. Sun G, Z. Li, J. Chen, T. Wu, X. Feng, Int. J. Env. Res. Pub. He. **14**, 3, 261–272 (2017)
9. M. Trojanowska, R. Świetlik, *Bezpieczeństwo i Ekologia* **12**, 476–477 (2016)
10. A. Mahmood, R. N. Malik, Pakistan. Arab J. Chem. **7**, 1, 91–99 (2014)
11. http://www.epa.gov/ncea/efh.
12. J. Ma, W. Singhirunnusorn, Proc. Soc Behv. **50**, 1, 280–293 (2012)
13. L. Ferreira-Baptista, E. De Miguel, Atmos. Environ. **39**, 4501–4512 (2005)
14. G. Shi, Z. Chen, C. Bi, L. Wang, J. Teng, Y. Li, S. Xu, Atmos. Environ. **45**, 764–771 (2011)
15. H. Chen, X. W. Lu, L. Y. Li, T. N. Gao, Y. Y. Chang, Sci. Total Environ. **484**, 27–53 (2014)
16. A. A. Jennings, Z. Li, J. Environ. Manag. **160**, 16–29 (2015)
17. A. A. Jennings, Z. Li, J. Environ. Manag. **160**, 226–240 (2015)
18. M. Alghamdi, M. Khoder, Int. Env. Res. Pub. He. **15**, 11, 2397–2417 (2018)
19. J. Mohmand, S. A. Eqani, M. Fasola, A. Alamdar, I. Mustafa, N. Ali, L. Liu, S. Peng, H. Shen, Chemosphere **132**, 142–151 (2015)
20. Y. Du, B. Gao, H. Zhou, H. Hao, S. Yin, Procedia. Environ. Sci. **18**, 299–309 (2013)
21. C. Zhang, Q. Qiao, E. Appel, B. Huang, J. Geochem. Explor. **119–120**, 60–75 (2012)
22. J. A. Acosta, M. Gabarrón, A. Faz, S. Martínez-Martínez, R. Zornoza, J. M. Arocena, Chemosphere **134**, 328–337 (2015)
23. G. F. Birch, A. Scollen, Aust. J. Soil. Res. **41**, 1329–1342 (2003)

The sorption of organic compounds from water during technological start-up of carbon filters

Dorota Holc[1,*], *Alina* Pruss[2], *Małgorzata* Komorowska-Kaufman[2], *Beata* Mądrecka[2], and *Zefiryn* Cybulski[3]

[1]Poznan University of Technology, Faculty of Civil and Environmental Engineering, Institute of Environmental Engineering, ul. Berdychowo 4, 60-965 Poznan, Poland; Envirotech Ltd. Co., ul. Jana Kochanowskiego 7, 60-845 Poznan, Poland
[2]Poznan University of Technology, Faculty of Civil and Environmental Engineering, Institute of Environmental Engineering, ul. Berdychowo 4, 60-965 Poznan, Poland
[3]Greater Poland Cancer, Microbiology Laboratory, ul. Garbary 15, 61-860 Poznan, Poland

Abstract. Natural waters contain organic substances that have a major impact on the quality of water, in particular on its organoleptic properties. They can also change its physical and chemical properties and have a negative impact on the technological processes used during water treatment. There is no doubt that organic substances contained in natural waters have to be removed during treatment processes in order to achieve concentrations consistent with the requirements concerning the quality of water intended for human consumption. A very effective method of organic compounds elimination from water is to use of biologically active filters (BAF). This article presents the results of pilot scale research on the effectiveness of removal of organic compounds from water using filtration through a bed of activated carbon filters, during their technological start-up. The obtained results showed high, more than 91% efficiency of reducing the content of organic compounds expressed by TOC, COD ($KMnO_4$) and UV_{254} absorbance. The predominant process at the start of the use of carbon filters has been sorption. During the research, the development of biofilm in the filter bed was also controlled. The identification of microorganisms showed the presence of Gram-negative bacteria in the beds of both filter columns.

1 Introduction

Organic compounds found in natural waters can be divided into two categories: the first one includes natural organic matter, while the other organic compounds of anthropogenic origin. These compounds have a fundamental impact on the quality of water, in particular on its organoleptic properties, causing deterioration of water colour, odour and taste. They can also change the physicochemical properties of water and have a negative impact on technological processes during water treatment. In addition, their inadequate removal

* Corresponding author: dorota_holc@op.pl

from water entering the water supply system may cause secondary pollution development, which in turn leads to poor water quality at the tap. It is therefore necessary to remove organic substances from water to achieve concentrations consistent with the requirements for the quality of water intended for human consumption [1–5].

The use of biologically active filters (BAF) to remove organic compounds from water gives very good results [6–8]. Biofiltration is also a way to ensure biological stability of tap water and reduce the doses of disinfectants [3, 9, 10, 11].

Removal of contaminants on the bed of biologically active filters takes place as a result of two parallel processes: adsorption and biodegradation [9, 12, 13]. The BAFs are filled with a filter bed characterized by a well-developed specific surface of grains and high porosity, which is a very good substrate for the proliferation of microorganisms. The most popular filling of such filters is activated carbon, which additionally has the ability to adsorb pollutants contained in the water. A constant supply of oxygen and nutrients is essential for the development of microorganisms in the deposit [9, 13]. The biofilm formed is capable of processing and elimination (biodegradation) of pollutants adsorbed in the pores and on the surface of the bed grains. At the beginning of the filters operation, when the biofilm-forming bacteria are in the adaptation period, dissolved organic matter is removed by way of physical adsorption. This start-up period may take from 2 to 3 months [9, 14]. Due to continuous development of microorganisms, at a later stage the degree of dissolved organic matter removed through adsorption is gradually decreasing. The active sites of the bed grains are saturated and the rate of biological degradation increases considerably. With time, the biodegradation process taking place in the filter prevails over adsorption, which is also related to the exhaustion of the adsorptive capacity of activated carbon [9, 15, 16]. The microorganisms responsible for the metabolism of contaminants also extend the working time of the filter bed and limit the development of pathogenic microflora competing with it for food [17].

In this article, the authors present the results of their own research which is a continuation of the research initiated in April 2015. In April 2018, the authors having 3 years of experience in operating the pilot BAF installation, once again launched the research stand. The filter columns were filled again with fresh WG-12 activated carbon. This time, the filtration media were not activated through recirculation of backwashings from carbon filters operated on a technical scale [6, 7, 16]. From the beginning of the installation operation, tap water and water modified with humic compounds was filtered. The article will present the results obtained during the first stage of research, directly after the commissioning of BAFs, when the sorption process prevailed in the filters.

2 Materials and methods

The research was conducted on a pilot scale on the physical model of rapid filters at the Institute of Environmental Engineering of the Poznan University of Technology. The model consisted of two filter columns with an inner diameter of 100 mm, filled with fresh WG-12 activated carbon. Constant temperature was maintained across the entire filter bed height thanks to the use of a water jacket – a pipe with an inner diameter of 140 mm filled with water. The water flowing through the pipe had a temperature equal to the temperature of the water subject to filtration. The total height of the filter bed was 2.10 m. The research installation is shown in Figure 1. Water was supplied to both filtering columns at a flow rate of 20 dm^3/h. Constant rate of filtration was provided by rotameters. The filter column 1 was supplied with tap water. The filter column 2 was flooded with modelled water, i.e. contaminated with biohumus (100 ml of biohumus/1000 ml of tap water). Biohumus is an organic fertilizer, rich in nutrients and microorganisms, derived from processed manure. Filter columns have been equipped with taps for collecting water samples evenly distributed throughout the entire vertical section of the bed. The samples were taken at the inflow of water to the filters and from the outflow, and at the depths of 45 cm, 85 cm, 125 cm, 165 cm and 205 cm. Samples from the filtration columns' inflow and outflow were collected daily at the same time, from day 3 to day 40 of the system operation. Water samples from the vertical cross-section of the filtration columns were collected once a week. The effectiveness of removing organic compounds from water was evaluated on the basis of three main indicators of organic pollutants: COD ($KMnO_4$), total organic carbon (TOC) and UV_{254} absorbance. In addition, the presence of psychrophilic bacteria in the bed was investigated using the heterotrophic plate count method (HPC). Water samples for microbiological analyses were sterilely collected once a week from the vertical section of the filtration columns. The identification of microorganisms was carried out using the automated system Vitek 2 Compact (bioMerieux). Identification was performed using bacterial identification charts (GN).

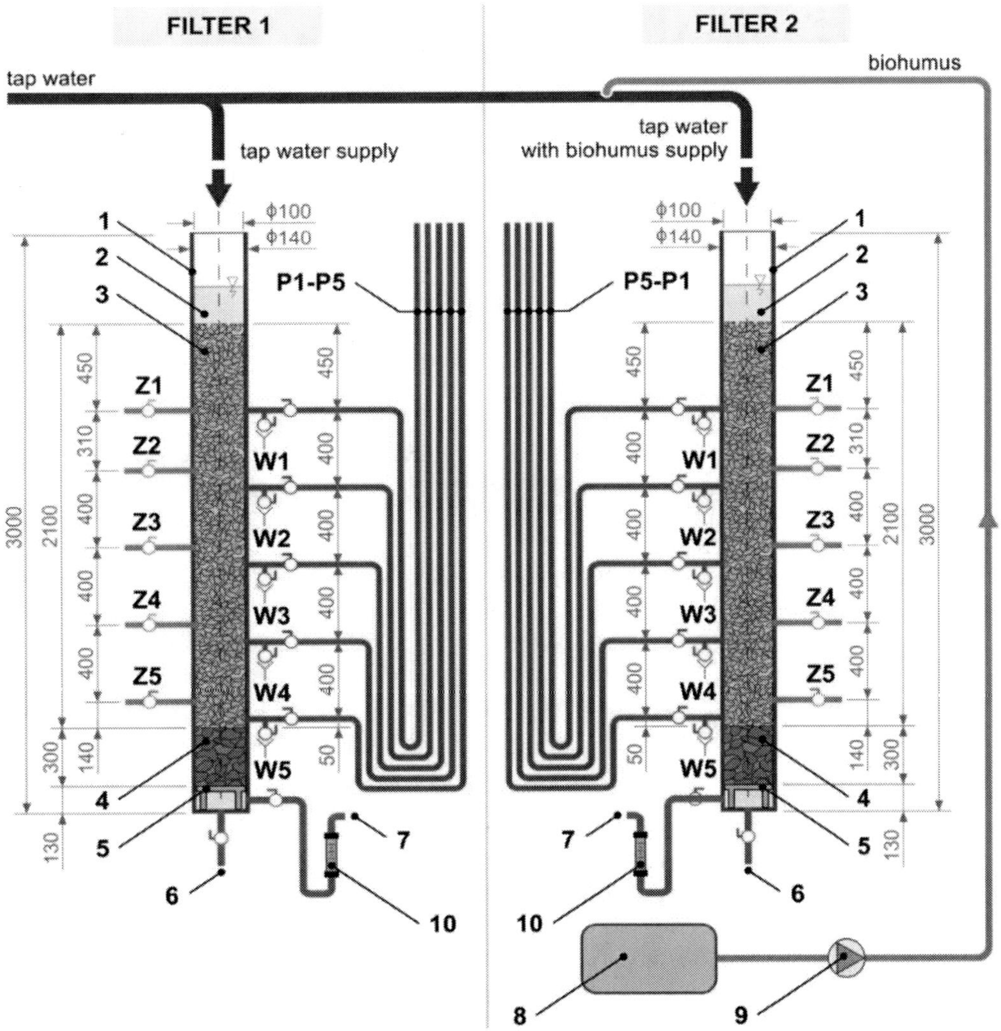

Fig. 1. Pilot plant with two identical filtration columns:
1 – water jacket, 2 – filtration column, 3 – filter bed, 4 – gravel supporting layer, 5 – drainage, 6 – backwash water, 7 – outlet, 8 – biohumus tank, 9 – diaphragm pump, 10 – rotameter, P1–P5 – piezometers, W1–W5 water sampling points, Z1–Z5 filter bed sampling points.

3 Results and discussion

In order to check which process is predominant in the entire filter bed, the Eberhardt, Madsen and Sontheimer index (EMS) was determined. This index is defined as the ratio of COD change (ΔCOD) to the change of the dissolved oxygen concentration (ΔO_2) during the flow of water through the bed [17, 18]. The results of the EMS test for the technological start-up of both filters are shown in Figure 2.

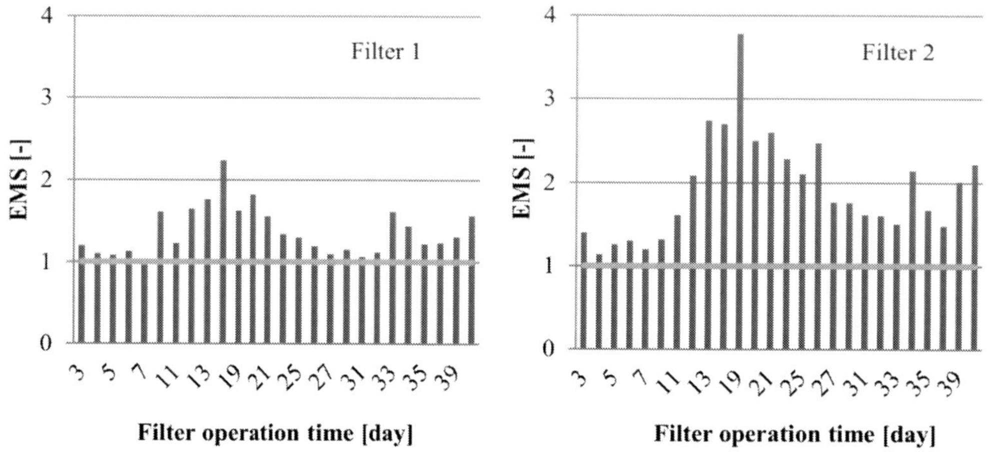

Fig. 2. Relationship between the sorption and biodegradation processes in Filter 1 and Filter 2.

The value of the EMS > 1 confirms that the sorption process prevailed in both filters throughout the entire forty-day study period. Over time, when the adsorbed pollutants deplete the sorption capacity of the filter bed and a sufficient number of microorganisms develop in the filter, biodegradation will become the predominant process. The development of microorganisms in the filter bed is possible when appropriate conditions are ensured, i.e. the optimum temperature, oxygen access and sufficient nutrient content [6, 7, 16]. During the tests, the abovementioned parameters were controlled. The water temperature ranged from 14.1°C to 21.1°C,the concentration of oxygen dissolved in water measured in the inflow was 4.72 mg O_2/dm^3 and pH was 7.5, on average. The average value of total organic carbon in the water flowing into Filter 1 was 5.7 mg C/dm^3, while in the modelled water 5.9 mg C/dm^3. The measured values confirm that the conditions in both filter beds were conducive to colonization by heterotrophic bacteria. During the research, microbiological analyses were carried out to detect the presence of psychrophilic bacteria. The obtained results are shown in Figure 3. Already on the 5th day of the system operation, an increase in the number of bacteria in the samples of water collected from the filter profile, in relation to their number in the inflow was found. With time, the amount of bacteria living in the filter bed increased. In Filter 1, the amount of bacteria increased in relation to the inflow throughout the entire study period, while in Filter 2, an increase occurred in the initial period of research (on days 5 and 12). Later on, more bacteria were detected in the inflow than in the samples taken from the bed and from the outflow, which may indicate that the biofilm is very well attached to the bed. In total, more microorganisms were found in the bed of Filter 2, supplied with biohumus modelled water. The solution dosed was not sterile, therefore more bacteria capable of colonizing the deposit were found in the water supplying Filter 2 than in the case of Filter 1. The presence and growth of microorganisms indicates proper conditions of the process, mainly a sufficient content of organic compounds being nutrients for the bacteria.

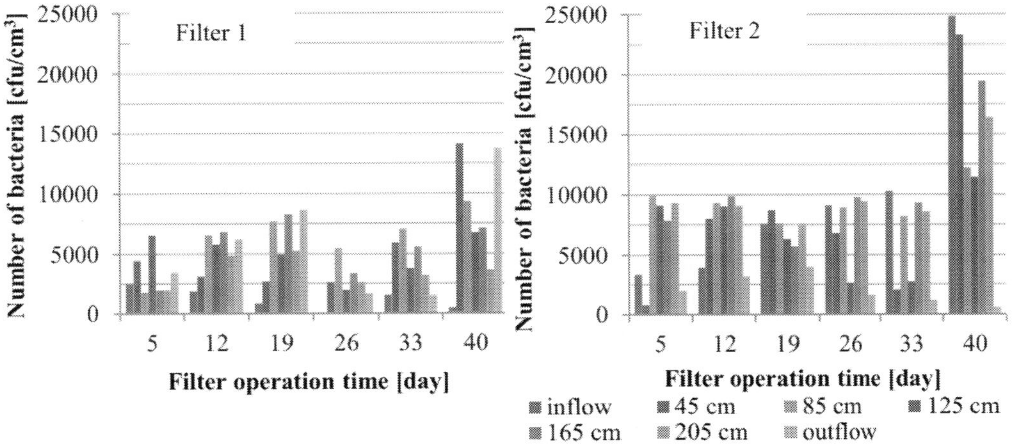

Fig. 3. Changes in the number of psychrophilic bacteria at different depths of Filter 1 and Filter 2.

The identification of the microorganisms showed the presence of Gram-negative bacteria in the beds of both filter columns. In both filters the *Sphingomonas paucimobilis*, *Stenotrophomonas maltophilia*, *Acinetobacter lwoffii*, *Aeromonas salmonicida* strains were identified. In Filter 1 the *Achromobacter denitrificans* were also detected. In Filter 2 fed with water modified with biohumus, other strains were detected, including *Pseudomonas fluorescens*. The differences in microorganisms inhabiting the two beds are undoubtedly related to the different bacteriological composition of the water feeding the filtration columns. No specimens from the *Enterobacteriaceae* family which are hazardous to human health were isolated from the collected samples. The identified strains are different from the ones found in the filter bed during the research in 2015 [6].

The study was carried out for a period of 40 days which turned out to be too short for the development of biofilm capable of biodegradation across the filter section. The earlier inoculation of the bed with backwashings from the biologically active filters has a positive effect on the development of biofilm in the filter bed. The backwashings contain large amounts of biofilm bacteria which produce a layer of organic substances - extracellular polymeric substances (EPS) (polysaccharides, proteins, nucleic acids, lipids) facilitating the adhesion of bacteria to the bed [9]. In filters in which the bed had been previously inoculated, the biodegradation process took place from the first day of the system operation, i.e. after cutting off the supply of recirculated backwashings [6, 7]. The backwash water recirculation with optimal nutrient composition lasted for 16 days.

The waters supplied to the test system i.e. tap water from the Poznan waterworks supplied Filter 1 and water contaminated with the biohumus solution supplied to Filter 2, had variable COD ($KMnO_4$). In the inflow to Filter 1, the COD ($KMnO_4$) of water ranged from 4.6 to 6.3 mg O_2/dm^3, while in the case of Filter 2, the values of the same parameter in the inflow were slightly higher and ranged from 4.7 to 6.6 mg O_2/dm^3. Minor differences in the content of organic substances in the inflow to both filters, despite the dosing of biohumus into the water feeding the filtration column 2, were associated with problems with biohumus mixing in the modelled water tank. The problems were solved in the further course of the pilot installation operation. The COD ($KMnO_4$) of water, due to the flow through the WG-12 carbon bed, decreased in both analysed cases. The COD ($KMnO_4$) measured in the outflow from Filter 1 ranged from 0.02 to 0.40 mg O_2/dm^3, while from Filter 2, from 0.08 to 0.50 mg O_2/dm^3 (Fig. 4). The parameter decreased significantly after filtration through a 45 cm thick layer of the bed. The efficiency of COD reduction down to the depth of 45 cm versus the total efficiency ranged from 97% to 67% for Filter 1 and from 91% to 64% for Filter 2. In both cases the upper bed layer efficiency decreased over time. The recorded COD ($KMnO_4$) reduction efficiency was very high for both filters, in excess of 95% on average. Throughout the entire research period, the efficiency was never below 93%. Based on that it was concluded that the fresh WG-12 activated carbon was highly effective in reducing the content of organic compounds in water through the sorption process.

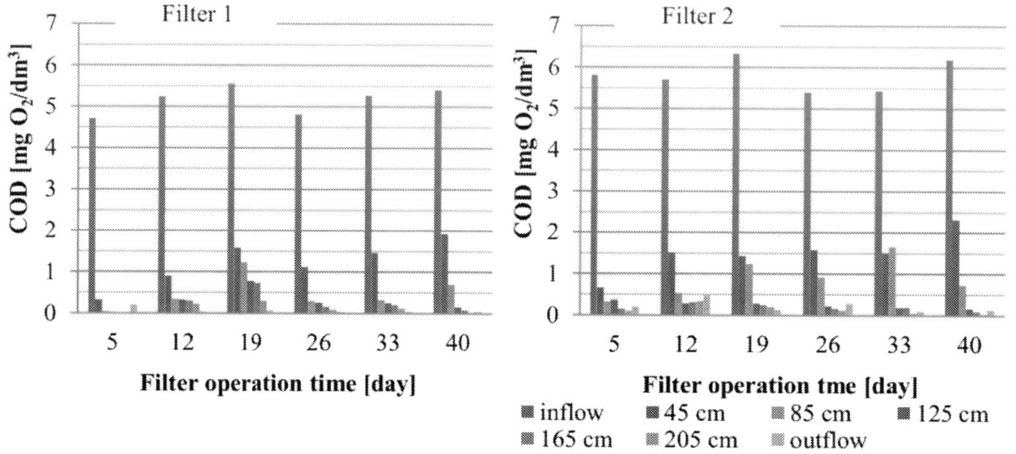

Fig. 4. COD of water at different depths of Filter 1 and Filter 2.

The TOC values in the inflow, outflow and in the vertical cross-section of filters are presented in Figure 5. The average TOC value in water supplied to Filter 1 was 5.7 mg C/dm^3, while in the case of water treated with the solution of humic compounds amounted to 5.9 mg C/dm^3. Biohumus caused a slight increase in that parameter, which was associated with its poor mixing in the biohumus tank. On the outflow from both filters, the value of TOC was the same and amounted to 0.5 mg C/dm^3. This indicates the high efficiency of the sorption process during the filtration period.

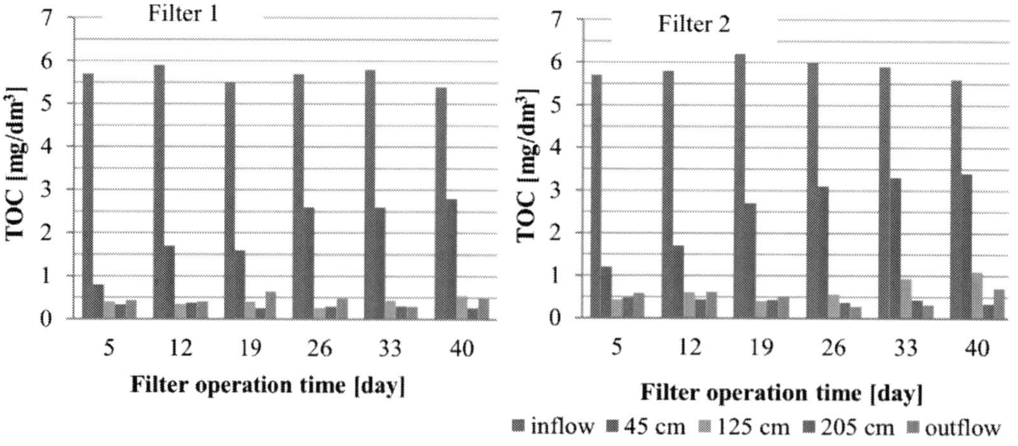

Fig. 5. TOC in the water at different depths of Filter 1 and Filter 2.

The UV absorbance at 254 nm at different bed depths is shown in Figure 6. The mean value of this parameter at Filter 1 inlet was 0.22 m^{-1}. The flow through the WG-12 active carbon bed caused a decrease in UV_{254} absorbance at the outlet to the value of 0.00 m^{-1}. This result was already obtained at a bed depth of 45 cm. For Filter 2, the average UV_{254} value in the inflow was 0.29 m^{-1}. As a result of the flow through the filter bed, the water was purified and reached an average value in the outflow of 0.04 m^{-1}. An exception to that rule was the 19th day of the filter operation. The increase in UV_{254} absorbance on that day may indicate the detachment of fragments of the biological film formed on the activated carbon surface.

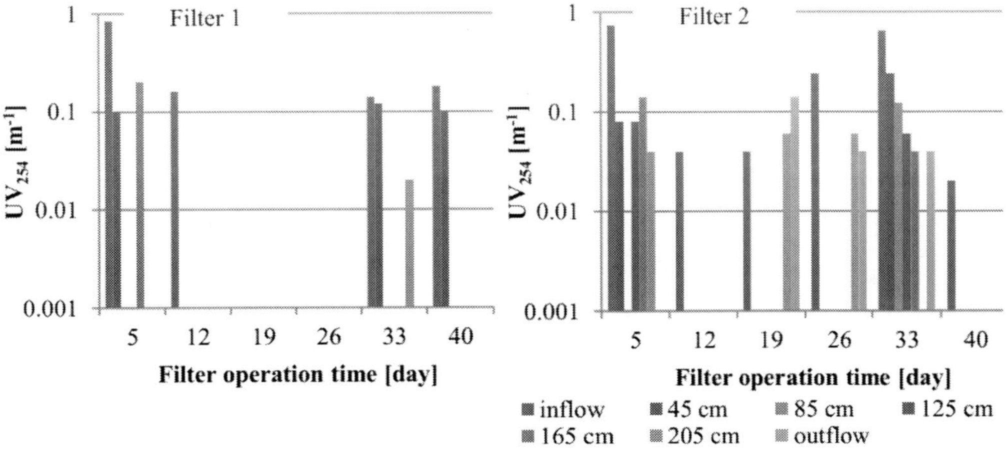

Fig. 6. Changes in UV_{254} absorbance at different depths of Filter 1 and Filter 2.

4 Summary

Removal of organic compounds from water during the technological start-up of rapid carbon filters operated on a pilot scale showed that during the first 40 days of the system operation, the sorption process prevailed in the filter bed.

The efficiency of reducing the organic matter content through a sorption process in the fresh WG-12 activated carbon filter bed during the first 40 days of the filter operation was as follows: for COD ($KMnO_4$) 96–97%, for TOC 91–92% and for UV_{254} absorbance 100%.

The identification of microorganisms showed the presence of Gram-negative bacteria in the beds of both filter columns. Differences in the strains inhabiting both deposits were undoubtedly related to the different bacteriological composition of the water feeding the filtration columns.

When comparing the results of current research, to the research carried out in 2015 [6], it can be concluded that:

- the strains of bacteria identified in the filter bed are different,
- there is a significant difference (17 days) in the length of time during which sorption prevailed over biodegradation.

Such differences occurred, despite the use of the same commercial carbon WG-12, with the same declared parameters.

References

1. S. A. Baghoth, *Characterizing natural organic matter in drinking water treatment process and trans*, UNESCO-IHE PhD Th. (2012)
2. I. Krupińska, Zeszyty Naukowe Uniwersytetu Zielonogórskiego. Inżynieria Środowiska **28**, 148, 55–72 (2012)
3. O. Gibert, B. Lefevre, A. Teuler, X. Bernat, J. Tobella, Journal of Water Process Engineering **6**, 64–71 (2015)
4. M. Świderska-Bróż, M. Wolska, Rocz. Ochr. Sr. **33**, 4, 77–80 (2011)
5. W. Adamski, M. Szlachta, Technologia Wody **1**, 17–21 (2011)
6. D. Holc, A. Pruss, M. Michałkiewicz, Z. Cybulski, Rocz. Ochr. Sr. **18**, 2, 235–246 (2016)
7. D. Holc, A. Pruss, M. Michałkiewicz, Z. Cybulski, *Zaopatrzenie w wodę jakość i ochrona wód* (Wyd. PZITS, Poznań, 2016)
8. P. Kołaski, A. Wysocka, A. Pruss, I. Lasocka-Gomuła, M. Michałkiewicz, Z. Cybulski, *Dezynfekcja wody, zagrożenia, wyzwania, nowe technologie* (Wyd. AGH, Kraków, 2017)
9. D. R. Simpson, Water Res. **42**, 2839–2848 (2008)
10. L. Han, W. Liu, M. Chen, M. Zhang, S. Liu, R. Sun, X. Fei, Water Res. **47**, 4861–4868 (2013)
11. A. Lis, Ł. Pasoń, L. Stępniak, Inżynieria i Ochrona Środowiska **19**, 3, 4113–425 (2016)
12. D. Papciak, A. Pietrzyk, *Dezynfekcja wody, zagrożenia, wyzwania, nowe technologie* (Wyd. AGH, Kraków, 2017)
13. O. Gibert, B. Lefevre, M. Fernandez, X. Bernat, M. Paraira, M. Calderer, X. Martinez-Llado, Water Res. **47**, 1101–1110 (2013)
14. H. Xiang, X. Lu, D. Zhang, F. Yang, L. Yin, G. Zhu, W. Liu, Chinese Journal of Applied and Environmental Biology **18**, 4, 642–646 (2012)
15. M. Wolny, E. Kudlek, M. Dudziak, *Inżynieria środowiska – młodym okiem. T. 30, Wody powierzchniowe i podziemne* (Wyd. Oficyna Wydawnicza Politechniki Białostockiej, Białystok, 2017)
16. A. Pruss, M. Komorowska-Kaufman, B. Mądrecka, *Water Supply and Wastewater Disposal* (Wyd. Lublin University of Technology, Lublin, 2018)
17. P. Olesiak, L. Stępniak, *Interdyscyplinarne zagadnienia w inżynierii i ochronie środowiska* (Wyd. Oficyna Wydawnicza Politechniki Wrocławskiej, Wrocław, 2014)
18. D. Papciak, J. Kaleta, A. Puszkarewicz, B. Tchórzewska-Cieślak, Journal of Ecological Engineering **17**, 3, 119–124 (2016)

Cemetery waste management in Poland. A case study for Wrocław city

Anna Janda[1,*], and *Tadeusz* Marcinkowski[1]

[1]Wrocław University of Science and Technology, Faculty of Environmental Engineering, Department of Waste Technology and Land Remediation, Plac Grunwaldzki 9, 50-370 Wrocław, Poland

Abstract: The annual amount of cemetery waste in Wrocław does not exceed 1% of generated municipal waste. The largest amount of cemetery waste is generated in autumn months. For the Osobowicki cemetery, the total volume of containers for selective waste collection has doubled since 2013 and now constitutes 49% of the total volume of containers. Research of the content of selected waste containers revealed the segregation of glass, plastics and biodegradable waste in the range from 81% to 92%. For the St. Lawrence cemetery the volume of containers for selective waste collection constitutes only 6% of the total volume of containers there. Despite these difficulties, the degree of separation of glass and plastics is very high (from 89% to 94%). Increase of the volume of containers for biodegradable waste would reduce the high share of green fraction (from 22% to 38%) in non-biodegradable waste containers.

1 Introduction

Cemeteries are places of worship and remembrance of the deceased. Due to the strongly religious and reflective nature of these objects, it is desirable that there are places neat and well-kept. Cemeteries are one of the elements of spatial development in which waste is generated.

Two basic sources of cemetery waste (CW) can be distinguished. The first of them, which generates about 10% of the waste, is associated with the care and renovation activities in the cemetery. As a result, biodegradable waste (leaves, grass, branches), sand, stones, soils and debris are formed. Second and most dominant source of CW are items brought to the cemetery by visitors in order to decorate and clean-up tombstones This group includes: live and artificial bunches, flowers, candles and cartridges, plastic pots, various types of packaging (plastic, cardboard), as well as textile waste [1–5]. The above-mentioned waste is very often included in the group of multi-material waste. After use, decorative elements are transformed into a stream of waste, which visitors want to get rid of quickly and "on the spot".

Due to the fact of CW generation, it is necessary to create a waste collection system. According to the Act [6], cemeteries are divided into municipal and denominational ones.

* Corresponding author: anna.janda@pwr.edu.pl

Maintenance of municipal cemeteries (MC) belongs to the city's self-government authorities and denominantional cemeteries (DC)- to the parish. According to Polish legislation [7, 8], waste from cemeteries is classified as a municipal waste group, a subgroup of waste from gardens and parks (including cemeteries) – **20 02**. Packaging waste is also generated in the cemeteries – group **15**.

Due to the fact that CW belong to the municipal waste (MW) group, their management is specified by statute [9], which imposes an obligation to conduct selective waste collection (SC). With regard to CW, a SC system should include glass, plastic and biodegradable waste. The problem in CW collecting may result in the variability of the amount of waste generated over time. In the post-holidays periods (Christmas, Easter, All Saints Day) the amount of waste generated at cemeteries increases significantly [1]. In addition, the collected waste is difficult to process. Literature data show that even a selectively collected biodegradable waste stream, due to impurities, cannot be subjected to composting [10]. There are also no data about effectiveness of the SC of CW.

The following article aims to show the system of CW collection on selected cemeteries located in Wrocław (Poland). Based on the analysis of the research carried out as part of the master's diploma theses at the Faculty of Environmental Engineering of the Wrocław University of Science and Technology in 2013–2018 [2–5], the effectiveness of waste SC system was estimated.

The problem of CW management is rarely mentioned in scientific publications. Therefore it was almost impossible to base the obtained research results on other literature reports. For this reason, this publication seems to be purposeful, as one of the pioneering issues related to the management of CW in Poland, on the example of the Wrocław city.

2 Case study for Wrocław city

Wrocław city is located in the south-western part of Poland and is the capital of the Lower Silesian Voivodship. The city has over 670,000 inhabitants and covers area of 293 km^2. There are 6 MCs and 15 DCs in the city. MCs are characterized by a much larger area (1.27 km^2) than DCs (0.25 km^2). The largest MCs are: Osobowicki cemetery (0.53 km^2), Psie Pole cemetery (0.40 km^2) and Grabiszyński cemetery (0.28 km^2). The largest DCs in Wrocław are: St. Lawrence cemetery (0.04 km^2) and St. Family cemetery.

According to the Polish law [6] the MCs are supervised by the president of Wrocław city. In practice, all Wrocław MCs are managed by the Board of Municipal Cemeteries. All of these facilities are served by WPO ALBA S.A. Wrocław. From season 2012/2013, the SC of CW fractions at MCs has been introduced. In the implementation year, the total volume of biodegradable SC containers, located at 6 MCs, was 158.4m^3 . The volume of glass and plastic fraction containers was 73.4 m^3 and 19.5 m^3 respectively [1]. The volume of containers for non-biodegradable wastes was 393.8 m^3, which accounted for 61% of the total container volume. With the increase of cemeteries 'users' awareness about waste segregation, the number of containers for SC is gradually increasing.

For DCs, where the denominational community is responsible for managing cemeteries, the situation is different. Due to the separate governing bodies of each of the DC, the waste collection system is dispersed and run by various enterprises. Currently, on most DCs at least the fraction of biodegradable waste is released.

Research [1] showed that the total amount of CW generated in Wrocław constitutes about 1% of the total MW stream. Stejskal [11] obtained similar results for 11 cemeteries in Brno (Czech Republic). The largest amount of waste is generated in the following months: October, November and December (successively 32.9%, 19.5% and 13.7% of the total CW stream) [1]. Increased waste stream is associated with the cleaning and preparing tombstones in pre- and post-holiday periods. A similar trend in the variability of the amount of CW produced over time was noticed in research for Wodzisław Śląski city (Poland) [3]. The estimated research for Wrocław [12] also retains the trend of changes in the amount of waste generated in the annual cycle. According to calculations, 39.5% of the CW stream is generated in the autumn season.

2.1 System of selective waste collection at the Osobowicki municipal cemetery

Osobowicki Cemetery is the largest MC of Wroclaw city with an area of 0.53 km^2. There are currently around 100,000 graves on its site, and burials are still carried out [2]. In 2012 SC of CW, including the fraction of biodegradable waste, glass and plastics, was introduced. Table 1 presents the number and volume of containers for CW in the implementation year 2013 and in 2018.

Table 1. Quantity and total volume of waste containers at the Osobowicki Cemetery in 2013 and 2018 [1, 3].

Type of waste container	2013			2018		
	Number, pcs.	Volume, m^3	Total volume, m^3	Number, pcs.	Volume, m^3	Total volume, m^3
Non-biodegradable waste	218	1.10	239.80	189	1.10	207.90
Biodegradable waste	61	1.10	67.10	151	1.10	166.10
Green waste	-	-	0.00	12	0.24	2.88
Plastic	50	0.12	6.00	71	0.24	17.04
Glass	50	0.12	6.00	59	0.24	14.16
TOTAL	**379**	-	**318.90**	**482**	-	**408.08**
Including:						
for SC waste	**161**		**79.10**	**293**		**200.18**

As shown in Table 1, the quantity and volume of containers for SC in 2018 was much higher than in 2013. Year 2013 was the implementation period of SC system. In this time public awareness about the correct waste fractions distribution was very low. In 2013 the share of SC containers constituted only 25%. In season 2013/2014 only about 15% of the CW stream was collected selectively [1].

Over the 5 years period, the total number of containers for waste collection has increased by over 100. The quantity and volume of containers for non-biodegradable waste decreased by 29 pieces and 32 m^3, which is a decrease by 13% compared to 2013. After 5 years since the introduction of the SC system, the overall demand for the volume of SC containers has increased from 79 m^3 to 200 m^3. The volume of containers for biodegradable waste increased almost 2.5 times. Very similar results were also reported for glass and plastic fraction containers. In 2018, the volume of SC containers accounted for 49% of the total volume of the CW containers A sufficient number of containers and their good deployment in the cemetery ensure a well-functioning waste collecting system.

The overall increase in the total volume of containers for waste collection by 27.5% undoubtedly indicates an increase in the amount of CW generated over the analyzed period of time. This trend is consistent with the simultaneously increasing total MW stream. However, it is positive, that from 2013 only the number and volume of containers for biodegradable waste, glass and plastic fraction increased (from 25% to 49%).

The SC waste system is not limited to setting containers for waste fractions. The actual degree of segregation is closely related to the users of the cemetery. Paper [2] includes research about content of individual waste fractions in SC containers. The research methodology included mass analysis of the content of selected containers for non-biodegradable and biodegradable waste as well as for glass and plastics. Obtained results allowed for calculating the percentage composition of each analyzed waste fraction in SC containers. The analyzed waste fractions included: green waste, soil, glass, plastic and other waste, i.e., paper, textiles, metals, wood, stones and other waste groups. The research was carried out in spring (12/05/2018 and 05/06/2018). The results are shown in Table 2.

Table 2. Percentage composition of individual waste fractions in selected containers for non-biodegradable and selectively collected waste at the Osobowicki Cemetery on 12/05/2018 (I) and 05/06/2018 (II) [2].

Waste fraction	**Non-biodegradable waste container composition, %**				**Biodegradable waste container composition, %**		**Glass waste container composition, %**		**Plastics waste container composition, %**	
	I		II		I	II	I	II	I	II
Green	17		13		**81**	**85**	0	0	0	0
Soil	3	**∑=83%**	6	**∑=87%**	**2**	**7**	0	0	0	0
Glass	32		20		8	0	**90**	**92**	0	0
Plastic	18		42		2	3	5	7	**81**	**92**
Other	30		19		7	5	5	1	18	8

As shown in Table 2, the share of selectively collected fractions in the SC containers ranged from 81% to 92%. These results show a high degree of CW segregation. In the best way the fraction of glass was separated. The main contamination of the glass fraction was made of plastics (from 5% to 7%), including mainly unseparated plastic cartridges. Selectively separated plastic fraction consisted mainly of paraffin contaminated cartridges as well as pots and bags. In containers for plastic fraction, the only hazardous CW-eternal candles- was found. They include batteries, which are a source of power. For the biodegradable waste container, the largest contamination was glass and plastic fraction. In addition, it was observed that a greater share of the green fraction in each container had green conifer waste. For biodegradable waste containers , the total share of this fraction was 50% and 47% in series 1 and 2 respectively.

The content of biodegradable waste in containers for non-biodegradable waste was 13% and 17%. In addition, the significant content of analyzed containers was SC fractions of glass and plastics: 32–20% and 18–42% respectively. As shown in Table 1, the number of containers for non-biodegradable waste at Osobowicki Cemetery is 189, while the number of containers for glass, plastic and biodegradable waste is, in sequence, 59, 71 and 151. These data suggest that containers for non-biodegradable waste are still the most accessible for the cemetery users. Increasing the number of containers for SC waste and locating them near containers for non-biodegradable waste could increase the separation degree of glass, plastic and green fraction.

2.2 System of selective waste collection at the St. Lawrence denominational cemetery

St. Lawrence Cemetery is one of the largest Roman Catholic cemeteries in Wrocław, with area of 0,04 km^2. Still in 2015, there was no full system of selective waste collection. Table 3 presents the number and volume of containers for CW in period 2015–2018.

Table 3. Quantity and total volume of waste containers at the St. Lawrence cemetery in 2015 and 2018 [4, 5].

Type of waste container	**2015**			**2018**		
	Number, pcs.	Volume, m^3	Total volume, m^3	Number, pcs.	Volume, m^3	Total volume, m^3
Non-biodegradable waste	15–30	1.10	16.50–33.00	27	1.10	29.70
Biodegradable waste	-	-	-	1	1.10	1.10
Plastic	-	-	-	2	0.24	0.48
Glass	-	-	-	4	0.24	0.24
TOTAL			**33.00**			**31.52**
Including						
for SC waste:	-	-	-	**7**	-	**1.82**

Results presented in table 3 show, that the number and volume of containers for CW in 2015 and 2018 are similar. However, because of changes in the total available volume of waste containers throughout the year, data from 2015 are hard to compare. In 2018, containers for selective collection of biodegradable fraction, glass and plastics were used. However, volume of these containers was only about 6% of the total volume of all available containers.

In analysis [4] it was noted, that no information about SC waste system was posted on the cemetery area. Another problem is bad location of SC containers. Containers for glass and plastics were located in three places: at the main entrance gate to the cemetery, at the main intersection of cemetery alleys and at the back of the cemetery area. The only container for biodegradable waste was placed at the back of the cemetery. Most of the graves are too far away from this container. Therefore, cemetery users throw green waste into non-biodegradable waste containers, which are located along the entire length of the main cemetery paths. The only solution of this problem may be to increase the number of containers for SC and to place them at the main intersections of internal roads of the cemetery.

As in the case of the Osobowicki cemetery, for the St. Lawrence cemetery research about effectiveness of SC were also conducted. In tests containers for non-biodegradable waste (1, 2 and 3), glass waste (4 and 5) and plastic waste (6) were used. The research methodology and division of waste into fractions was the same as in point 2.1. Tests were carried out after Easter, on 24/04/2018. For research at St. Lawrence cemetery, the key stage was appropriate selection of analyzed containers. For testing the composition of non-biodegradable waste containers, intentionally containers located at different distances from biodegradable, glass and plastic fraction containers were used. The container 1 was in the largest and container 3 in the shortest distance from the container for the biodegradable fraction. The results are summarized in tables 4 and 5.

The obtained results indicate that despite the largest distance of container 1 from the biodegradable waste container, the percentage content of green fraction was the lowest (20.6%). The highest share of green fraction in the analyzed containers was recorded for the container 2 located in the middle part of the cemetery. A high content of plastics fraction was observed for container 3, which results from the lack of SC containers in this area of cemetery. Interestingly, for the container 3 placed close to the container for plastic, content of this fraction was almost 45%. That could be caused by a high filling degree of the container for plastic fraction (about 90%) during the research.

Table 4. Percentage composition of individual waste fractions in selected containers for non-biodegradable waste at St. Lawrence cemetery on 24/04/2018 [4].

Waste fraction	Biodegradable waste container (1) composition, %		Biodegradable waste container (2) composition, %		Biodegradable waste container (3) composition, %	
Green	20.6		38.4		22.1	
Soil	4.9	**Σ= 79.4%**	2.4	**Σ= 61,6%**	3.1	**Σ= 77,9%**
Glass	18.3		11.9		9.2	
Plastic	39.2		9.6		44.5	
Other	17.0		37.7		21.1	

Table 5. Percentage composition of individual waste fractions in SC containers at the St. Lawrence cemetery on 24/04/2018 [4].

Waste fraction	Glass waste container (4) composition, %	Glass waste container (5) composition, %	Plastic waste container (6) composition, %
Green	0.1	0.0	0.0
Soil	0.0	0.0	0.0
Glass	**93.7**	**91.8**	8.2
Plastic	5.8	6.5	**89.0**
Other	0.5	1.7	2.8

The results presented in Table 5 shows the SC degree of glass and plastic fractions ranging from 89% to 94%. The largest contamination for the glass fraction were plastics and vice versa – for plastics – glass. Similar results for glass and plastic fraction containers were recorded for research at the Osobowicki Cemetery (table 2). The contamination of the glass fraction results mainly from multi-material construction of candles. Their SC should include separating the glass part from the plastic cartridge. This, however, requires additional work from the cemetery users.

The results obtained and summarized in Tables 4 and 5 show that the users of the cemetery are the most decisive in determining the degree and efficiency of selective waste collection. Even the correct placement of CW containers does not guarantee compliance with the rules of selective waste collection system.

3 Conclusion

Cemeteries are public facilities on which waste is generated. CW collection should be carried out in a selective way, including the fraction of biodegradable waste, glass waste and plastics. The impact on the SC efficiency has the number of waste containers as well as their appropriate location in the cemetery area. The human factor is also decisive.

There are 6 MCs and 15 DCs in Wrocław city. The annual amount of emerging cemetery waste accounts for about 1% of the total weight of generated MW. The amount of CW varies throughout the year. The largest amount of generated CW is observed in autumn months – October, November and December. This is mainly related to the celebration of Catholic holidays.

On MCs in the years 2012–2013, SC systems were implemented. For the biggest MC-Osobowicki cemetery- the volume of containers for selective waste collection is currently almost 50% of the total volume of containers and is twice as high as in the implementation year 2013. Research of the content of waste fractions in SC containers revealed the segregation of glass, plastic and biodegradable waste at the level of 80–90%. Reducing the number of containers for non-biodegradable waste in favour of containers for SC of biodegradable, glass and plastic waste would allow to increase the degree of waste segregation.

The situation in DCs is different. Results clearly indicate that at the St. Lawrence cemetery the number of SC containers is too small. Mass analysis of containers for collection of glass and plastic waste showed the content of these fractions in the range from 89% to 94%. Unfortunately, too small volume of biodegradable waste container and its bad location caused a high content of green fraction (from 20% to 38%) in non-biodegradable waste containers. However, the high separation degree of glass and plastic fractions suggests that increasing the volume of containers for SC, including containers for biodegradable waste, would enable a more efficient separation of waste fractions.

References

1. A. Jaworska-Szott, T. A. Marcinkowski, Wyd. Uczelniane Politechniki Koszalińskiej, **T/10**, 67 (2014)
2. F. Ziemiński, MA thesis (not published), Faculty of Environ. Eng., Wrocław University of Science and Technology (2018)
3. A. Bobrzyk, MA thesis (not published), Faculty of Environ. Eng., Wrocław University of Science and Technology (2015)
4. K. Prędkiewicz, MA thesis (not published), Faculty of Environ. Eng., Wrocław University of Science and Technology (2018)
5. A. Chmielewska, MA thesis (not published), Faculty of Environ. Eng., Wrocław University of Science and Technology (2015)
6. Ustawa z dnia 31 stycznia 1959 r. o cmentarzach i chowaniu zmarłych
7. Ustawa z dnia 14 grudnia 2012 r. o odpadach
8. Rozporządzenie Ministra Środowiska z dnia 9 grudnia 2014 r. w sprawie katalogu odpadów
9. Ustawa z dnia 1 lipca 2011 r. o zmianie ustawy o utrzymaniu czystości i porządku w gminach oraz niektórych innych ustaw
10. B. Stejskal, PAN Kraków **4**, 1217 (2014)
11. B. Stejskal, PAN Kraków **2**, 121 (2011)
12. E. den Boer, R. Szpadt, J. den Boer, *Analiza stanu gospodarki odpadami komunalnymi na terenie gminy Wrocław za rok 2015 w celu weryfikacji możliwości technicznych i organizacyjnych gminy w zakresie gospodarowania odpadami komunalnymi, zgodnie z wymogami ustawy o utrzymaniu czystości i porządku w gminach* (2016)

Estimation of particles emissions from a jet engine in real flight

Remigiusz Jasinski[1,*]

[1]Poznan University of Technology, Institute of Combustion Engine and Transport, 3 Piotrowo Street, Poznan 60-965, Poland

Abstract. The article presents the results of particles measurements from a jet engine. Engine parameters were analysed during the actual flight and an attempt was made to map it in stationary tests. Research results were supported by correlation analysis. Based on measurements of particulate matter it was found that the most favourable conditions for particle formation occur during taxiing. Mass distributions were presented and emission factors were determined for three basic phases: taxiing, take-off and approach. The applied research methodology was based on the latest guidelines and was expanded to include propositions related to the analysis of engine operation in real flight conditions.

1 Introduction

Particles emitted by aircraft engines adversely affect human health and contribute to deterioration of air quality. Particles with a diameter of 10 μm or less can cause various diseases and related deaths. The severity of diseases is combined with the long-term effects of particles in the environment. They contribute to the occurrence of diseases such as asthma and bronchitis. They are also one of the causes of cardiac arrhythmia and heart attacks. The most important problems arise from the interaction of fine particles. The lowest resistance to the negative effect of particles is demonstrated by people with heart and lung diseases, the elderly and children [1, 2]. It is assumed that the diameters of particles emitted by aircraft engines are not larger than 2.5 μm and can be divided into primary and secondary particles. Primary particles consist of a non-volatile fraction of carbon (mainly black carbon) and other components that accumulate on the carbon core, using it as the nucleus of condensation. Primary particles include nitric and sulfuric acid, water and heavy hydrocarbons containing up to 30 carbon atoms. The diameters of the discussed particles reach values of several dozen nanometers.

The secondary particles are formed as a reaction result of the primary particles and other exhaust toxic components, such as nitrogen oxides, sulfur oxides and light hydrocarbons. The products of reactions occurring on particles that can be found in the composition of secondary particles are: ammonium sulphate, ammonium nitrate and other compounds, mainly hydrocarbons. Research shows that as much as 70% of the particles emitted by aircraft engines are associated with the emission of nitrogen oxides, 14% are primary

* Corresponding author: remigiusz.jasinski@put.poznan.pl

particles, 12% of particles are associated with sulfur oxides, and 4% are formed with hydrocarbons [3–5]. Due to complicated processes of particle formation, the measurement of their emissions is very difficult. The approval process for jet engines with thrusts greater than 26.7 kN is based on the LTO (Landing and Take-Off) test but it doesn't include particles measurements. The main assumption of the LTO test is to reduce exhaust emissions in areas adjacent to airports, which is why it includes operations performed by the aircraft at an altitude of less than 3000 feet. The individual phases of the test (taxiing, takeoff, climb and approach) correspond to the constants of thrust force, so the procedure can be described as a four-point test. There are procedures for estimating particles emissions, but they are based on the LTO test, which is a clear simplification of engine operation conditions during air operations. Due to the above fact, the search for a methodology allowing to carry out reliable measurements of particles emission while maintaining the actual engine operation parameters is underway.

2 Methodology

2.1 Purpose and scope of the research

The research was divided into three stages. The first was a research flight made to get information about the engine's operation parameters during taxiing, takeoff and approach. Then, the engine test on the dynamometer and the measurement of particles number (PN) was made. Engine parameters obtained during the flight and in stationary tests were subjected to correlation analysis. In the last step, the most important parameters of PN emission were determined.

2.2 Research objects

The research flight was made by a multi-task F-16 fighter (Fig. 1). On-board equipment allowed the registration of drive system parameters. The aircraft was equipped with the Pratt & Whitney F100-PW-229 engine. The research object was Pratt & Whitney engine, F100-PW-229. It is a turbofan, twin-shaft engine with three-stage low pressure compressor and a ten-stage high pressure compressor. More detailed specification of the engine is shown in Table 1.

Fig. 1. The view of F-16 fighter.

Table 1. Technical data of the F100-PW-229 engine.

Parameter	Value
Maximum diameter	1080 mm
Length	4855 mm
Engine weight	1370 kg
Maximum thrust	79.13 kN

2.3 Measuring apparatus

Measurement of particle diameters was performed with an EEPS 3090 (engine exhaust particle sizer™) (Fig. 2a). It enabled the measurement of a discrete range of particle diameters (from 5.6 nm to 560 nm) on the basis of their differing speeds. The degree of electric mobility of particulate matter is changed exponentially, and measurement of their size is carried out at a frequency of 10 Hz. In addition, the Semtech DS gas analyzer (Fig. 2a) was used. The above apparatus is used to measure the concentration of exhaust gas compounds. It allows, however, to additionally determine the excess air coefficient, which allowed to perform additional calculations to determine the exhaust mass flow. Measurements of carbon dioxide allowed to determine the dilution coefficient. Diagram of the measurement system used for testing was shown in Fig. 2b.

a) b)

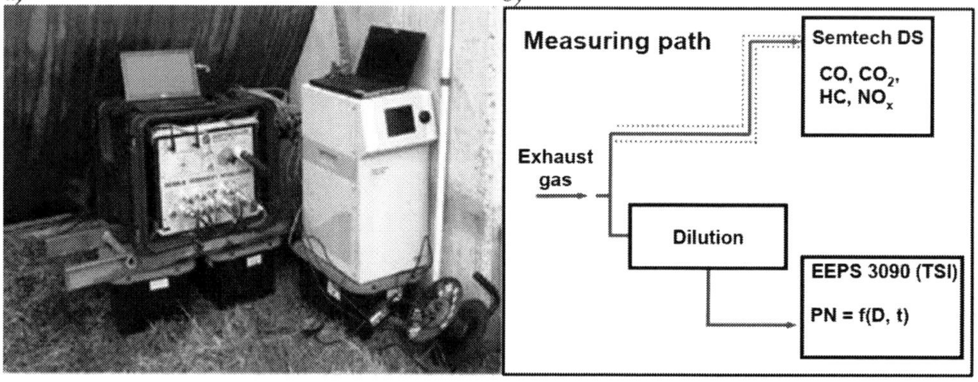

Fig. 2. The view of the apparatus during measurements (a) and diagram of the measurement system.

According to the literature, particles forming behind the engine nozzle have a significant contribution to the total particulate matter emission [6–9]. This is due to the presence in the exhaust gases chemical compounds containing sulfur, hydrocarbons and nitrogen, which in contact with air and other particles cause the formation of the so-called of secondary particles. Due to the above fact, it is necessary to measure the PN at an appropriate distance from the engine outlet nozzle, which is necessary to ensure a specific contact time of the exhaust gas with the air. Based on the literature analysis, it was assumed that the PN measurement should be carried out approximately 25 m behind the engine [8–10]. Figure 3 shows the probe 25 m behind the engine outlet nozzle.

Fig. 3. Location of the measuring probe.

Due to the mixing of exhaust gases with air during the PN concentration measurement, there is a significant dilution of the exhaust gases. In order to determine the dilution rate, it was necessary to measure the carbon dioxide immediately after the engine outlet nozzle and at a distance of 25 m. Additionally, to correctly determine the number of particles created due to the engine's operation, it is necessary to determine the PN concentration in the air.

3 Results and analysis

3.1 Research flight

The record of the propulsion parameters of the combat aircraft was carried out during a training flight. The data recording was carried out by on-board recorders, being a standard equipment of the F-16 fighter. The data from the recorders was divided into three phases: taxiing, take-off and approach. The taxiing phase refers to the time in which the aircraft is traveling on or before the runaway. In the case of a research flight, the total time of taxiing before take-off and after landing was 964 s, which accounted for 81% of the total test flight time. The take-off phase is defined as the period in which the aircraft increases speed in order to achieve the minimum lift force value needed to tear off the aircraft and initial climb to 1000 m above sea-level. The take-off phase accounted for 7% of the total flight time of the test flight and lasted 78 seconds. The approach is a period in which the aircraft reduces its height (from 1000 m above sea-level) in order to touch down and start taxiing. The approach took 150 seconds, which was 12% of the duration of the research flight. The average speed of descent during the maneuver was equal to 5.3 m/s.

The parameter relative to which the drive system is directly controlled is the position of the power lever. Due to the above fact, a further analysis of the operating parameters of the propulsion system registered in the research flight was carried out in the field of the power lever position (Z). The most important parameters of the engine operation were fuel consumption (FC) and fan speed (N_1), which are shown in Figure 4.

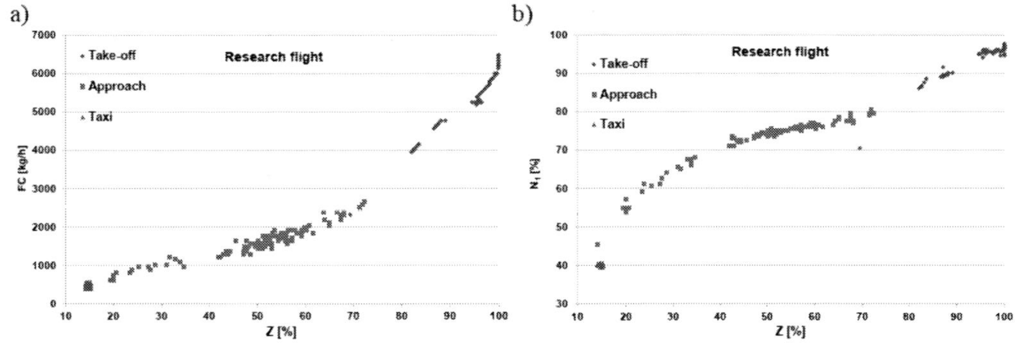

Fig. 4. Fuel consumption (a) and fan speed (b) obtained during individual phases of research flight.

Recorded parameters of the propulsion system have been grouped into phases (taxi, takeoff, approach) based on the time of each maneuver. The above fact allowed assigning ranges of power lever position values to individual phases according to Table 2. Additionally for the purpose of further analyzes, the above intervals were divided into smaller ones in order to determine the time share of particular groups of points in the given phases of the research flight. The taxi phase due to the small range of changes in the power lever position has not been further divided.

Table 2. Division of engine work points relative to the power lever position.

<table>
<tr><th colspan="6">Taxi</th></tr>
<tr><td>Interval (Z)</td><td colspan="5"><14;20)%</td></tr>
<tr><td>Time share</td><td colspan="5">100%</td></tr>
<tr><th colspan="6">Approach</th></tr>
<tr><td>Interval (Z)</td><td><20;30)%</td><td><30;40)%</td><td><40;50)%</td><td><50;60)%</td><td><61;75>%</td></tr>
<tr><td>Time share</td><td>6%</td><td>6%</td><td>28%</td><td>51%</td><td>9%</td></tr>
<tr><th colspan="6">Take-off</th></tr>
<tr><td>Interval (Z)</td><td colspan="2"><80;90)%</td><td colspan="3"><90;100>%</td></tr>
<tr><td>Time share</td><td colspan="2">27%</td><td colspan="3">73%</td></tr>
</table>

3.2 Stationary tests

3.2.1 Engine parameters

Measurements of the engine operation parameters were carried out as part of a trial after standard engine service. The engine test lasted 87 min and included an engine test at various work points. For the purpose of analysis of engine operating parameters during stationary tests, only those work points were selected which were considered useful due to the mapping of engine operation during the research flight. The selected points have been classified in relation to the position of the power lever in accordance with Table 2. Fuel consumption (FC) and fan speed (N_1) measured on stationary test are shown in Figure 5.

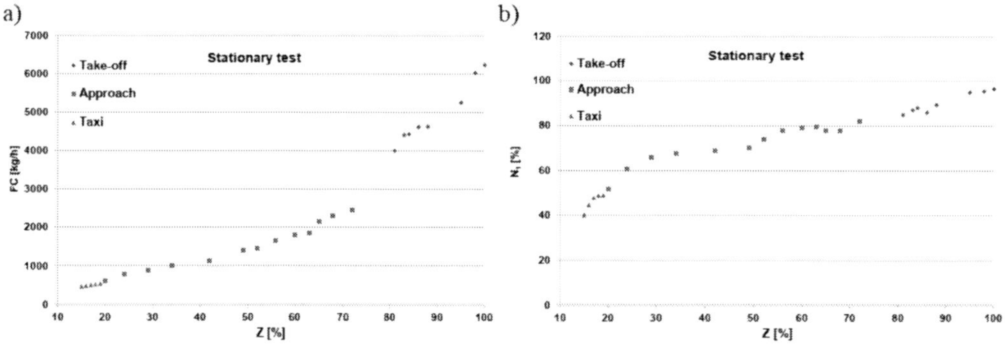

Fig. 5. Fuel consumption (a) and fan speed (b) obtained during stationary test.

3.2.2 Correlation of engine parameters

In order to show the dependence between engine operation in flight conditions and in stationary tests, a correlation analysis of obtained parameters in relation to the position of the power levers was performed. Parameters analyzed were fuel consumption (FC) and fan speed (N1). Figure 6 presents the results of the correlation according to Pearson's correlation coefficient. Obtained data in both cases show very strong dependence of parameters obtained in flight conditions and stationary tests. The above fact indicates very good mapping in the stationary test of the engine's operating conditions during the actual flight. This provides the basis for analyzing the results of particles measurement and relating them to individual operations performed by the aircraft in real operating conditions.

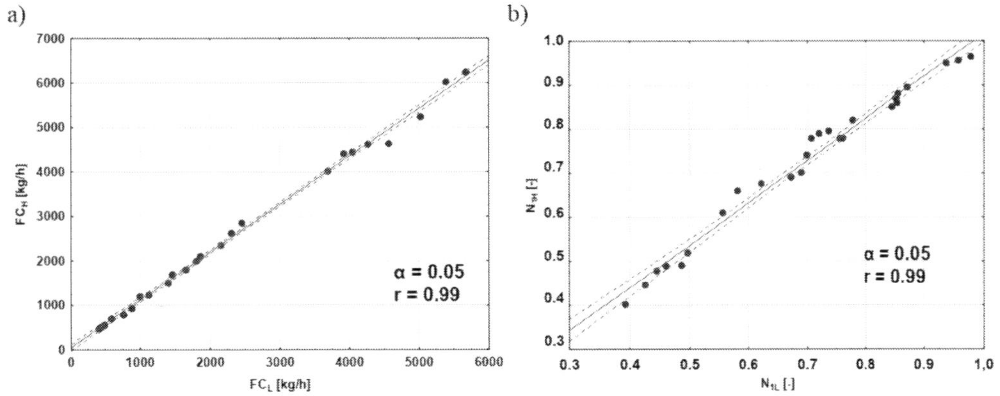

Fig. 6. Correlation analysis results of engine parameters obtained during flight (FC_L, N_{1L}) and stationary test (FC_H, N_{1H}).

3.2.3 Particles measurement results

At individual points of the engine operation, the particles mass distributions were measured. In order to present the particles mass concentration results (PM) depending on the diameter, it was necessary to average the obtained dimensional distributions in the individual phases of the research. Due to the fact that the time share of individual work points differed, it was necessary to assign weights. The weights were determined on the basis of the individual work points timeshare presented in Table 2. On this basis, the equation 1 was determined, which allows to determine the average dimensional distribution in a given phase on a weighted average basis.

$$PM_p(D) = \frac{\sum_{i=1}^{n} a_i PM_i(D)}{\sum_{i=1}^{n} a_i} \tag{1}$$

where:
$PM_p(D)$ – average mass distribution of particles emitted in each phase,
$PM_i(D)$ – mass distribution of particles in each engine work point,
a_i – weights of each mass distribution based on time share of each work point in whole phase,
n – number of engine work points in each phase,
D – diameter of particles.

During determining the average particle mass distributions (Fig. 7), the degree of exhaust gas dilution resulting from the sampling of exhaust gases at a certain distance from the outlet nozzle was taken into account. The degree of exhaust gases dilution was determined on the basis of the measurement of the carbon dioxide concentration immediately after the outlet nozzle and at the place of analyzed gas sample collection. All obtained mass distributions are bimodal with distinct characteristic diameters. In the case of taxiing, the main part of emitted particulate matter corresponds to particles with the size of 8–40 nm. Solid particles with dimensions greater than 100 nm correspond to approximately 20% of the total mass emitted. Landing is characterized by a more balanced distribution. Most of particulate matter emission is caused by the particles with diameters greater than

50 nm. The take-off operation in terms of particles mass emission was dominated by particles with diameters in the range of 6–40 nm. This is due to the fact that at a high combustion temperature nucleating fraction of the particles predominates.

a) b)

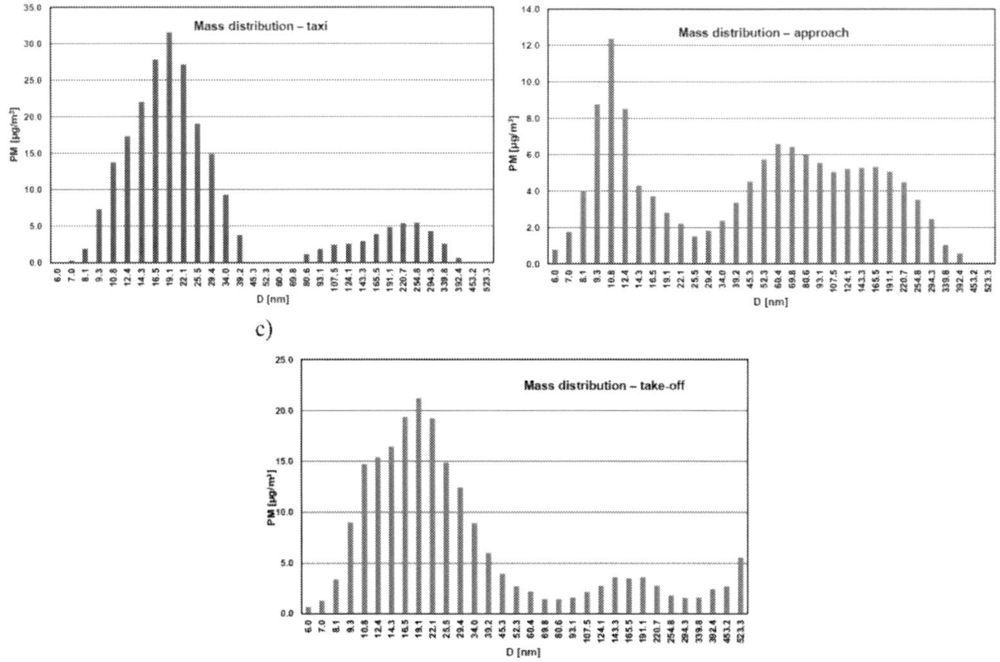

Fig. 7. Particulate matter distribution in each phase: taxi (a), approach (b) and take-off (c).

On the basis of the excess air coefficient and fuel consumption measurements, the emission factor of the particulate mass per 1 kg of fuel was determined according to the formula:

$$WPM = \frac{m_f \cdot TPM}{G_e} \left[\frac{\mu g}{kg_{fuel}} \right] \quad (2)$$

where:
m_f – exhaust mass flow,
TPM – total concentration of particles in averaged mass distribution for each phase,
G_e – fuel mass flow.

Figure 8 shows the values of the WPM coefficient for individual phases. The largest mass of particles is generated during taxiing. From one kilogram of fuel the engine produces $2.7 \cdot 10^5$ μg. The above fact results from the dimensional distribution of the emitted particles. During taxiing, larger particles are emitted than in case of landing or taking-off. Taxiing is an operation that lasts the longest and is directly affecting the quality of air, which is breathed by airport employees and passengers.

https://doi.org/10.1051/e3sconf/201910000029

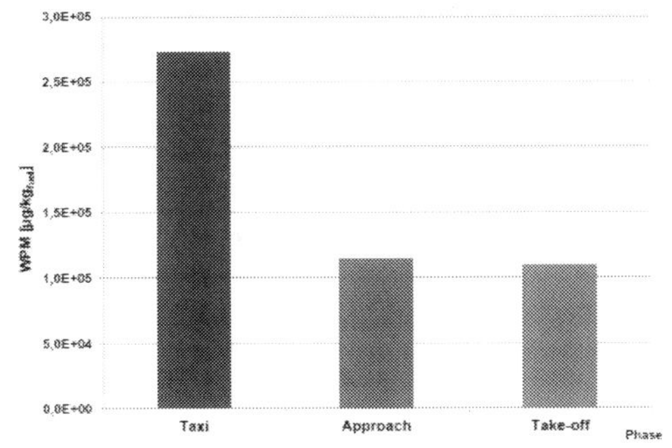

Fig. 8. Value of WPM coefficient in each phase.

4 Conclusions

The research has been carried out in accordance with the latest guidelines for the measurement of particles emissions from jet engines. The extension of the methodology with the analysis of engine operation parameters in real flight and their correlation analysis with the parameters in stationary tests is the author's contribution. The results of particulate matter emission tests showed significant differences between individual aircraft operations. The presented test results indicate that during taxiing operation from one kilogram of fuel the biggest mass of particulate matter is produced. This is extremely disadvantageous from the point of view of air pollution, because the emission takes place on the ground, therefore the dispersion of particulate matter is more difficult in comparison to the other air operations. Particles emission from air transport contribute to environmental pollution mainly in the local aspect. Cities are increasingly closer to airports, and because of that air traffic contributes to increase the concentration of particles in urban areas. This mainly concerns particles of the smallest size (up to 500 nm), whose mass is relatively small and therefore they do not increase the value of the mass concentration to which the assessment of air quality is carried out. However, it should be noticed that the smallest particles (mainly from combustion engines) are the most dangerous for human health.

References

1. R. Jasinski, E3S Web of Conferences **44**, 00057 (2018)
2. R. Jasinski, J. Markowski, J. Pielecha, Procedia Engineering **192**, 381–386, (2017)
3. R. Jasinski, J. Pielecha, J. Markowski, E3S Web of Conferences **10**, UNSP, 00030 (2018)
4. J. Merkisz, M. Gis, E3S Web of Conferences **44**, 00109 (2018)
5. W. Gis, J. Pielecha, J. Waskiewicz, M. Gis, M. Menes, IOP Conference Series-Materials Science and Engineering **148**, UNSP, 012040 (2016)
6. R. Wayson, G. Fleming, R. Iovinelli, J. Air Waste Manage. **59**, 91–100
7. J. Markowski, J. Pielecha, R. Jasinski, Procedia Engineering **192**, 557–562 (2017)
8. J. Kinsey, U.S. Environmental Protection Agency, Research Triangle Park, NC. Personal communication (2006)
9. Aircraft Particle Emissions experiment, APEX1; Particulate Data Provided by National Aeronautics and Space Administration, University of Missouri–Rolla, Aerodyne Research Inc., and Wright Patterson Air Force Base

10. J. Merkisz, P. Lijewski, P. Fuc, M. Siedlecki, A. Ziolkowski, IOP Conference Series-Materials Science and Engineering **148**, UNSP, 012077 (2016)

Energy self-consumption from PV systems: estimations for two office buildings in Krakow (Poland)

Magdalena Jurasz[1,*], and *Jerzy* Mikulik[1]

[1]AGH University, Krakow, Poland

Abstract. The analysis presented in this paper focuses on the energetic aspects of the use of photovoltaic (PV) systems in office buildings. Energy generation from PV system has been simulated on an hourly time scale and compared with the energy demand of two office buildings located in Krakow (southern Poland). The buildings' annual load exceeded 1 GWh in both cases. The analysis dealt with estimating how much energy generated by a PV system can be utilized on-site (self-consumption) and how big the energy surpluses will be (energy generation greater than demand). Capacities of PV systems ranging from 0.1 to 1.0 MW have been considered. Also, the impact of PV generation on the residual load parameters and changes in the maximal monthly and hourly energy demand were investigated. The results show that although the building energy demand is similar (in terms of annual volume of energy consumed) the potential of PV systems to cover it is different. The 100-kW PV system can reduce the observed hourly peak energy demand by 1% in December but by over 30% in June (respectively, from 171 kWh to 169 kWh, and from 333 kWh to 255 kWh). Considering the annual patterns of the office buildings' energy demand and PV generation (both have their respective peaks in summer), the application of a PV system changes the office building energy demand pattern significantly from the perspective of the power system operator. After installing 500 kW PV in an office building consuming 1 GWh annually, the months with highest demand are no longer in summer but in winter.

1 Introduction

Power systems across the globe are undergoing dramatic transformation in the search for the ultimate goal of sustainability and minimal environment impact [1]. The changes being seen constitute an increasing share of renewable energy sources in individual countries' energy mixes. The world population is moving from rural to urban areas, which are regions of ever-increasing energy consumption. A recent paper by Jurasz et al. [2] investigating the larger scale renewables-based energy system for Lower Silesia in Poland has shown that there is a significant spatial mismatch between wind energy potential and energy demand; thus, if

* Corresponding author: magda.krzywda01@gmail.com

wind energy is to be the energy source of choice the distribution network will have to be improved. However, on a spatial scale, wind generation can be replaced by photovoltaics, which can be placed on building roofs. A paper by Dabek and Jurasz [3] indicates that a city of close to 300 km^2 has a potential to install PV systems on a roof area exceeding 11 km^2 (which translates to exactly 1.1 GW of PV capacity if 10 m^2 are required per 1 kW of capacity).

Since cities are becoming the major centres of human activity and energy consumption, it is mandatory to ensure that their energy needs are satisfied in a sustainable way [4]. As indicated in the previous paragraph, a huge potential lies in the utilization of PV systems. A recent paper by Mikulik [5] shows that office buildings in Poland are starting to play a significant role as energy consumers. Even more importantly, their energy demand is strongly dependent on meteorological parameters, and positively correlated with irradiation (a driving factor of PV energy generation) [5]. The need to investigate the role of PV systems in office buildings is even higher, as a paper by Jurasz and Campana [6] shows that, in case of energy tariffs in Poland, PV systems can significantly reduce the energy cost in office buildings. Considering the above, the objective of this paper is to investigate the potential of a PV system to cover the electricity load in two office buildings in Krakow (Southern Poland). The analysis focuses on aspects such as self-consumption, peak load reduction and impact on the energy demand curve. The paper provides evidence on the role of PV systems in covering office building loads. The remainder of this paper is organized as follows: the following section presents used data and methods; section 3 discusses and presents the results; and the paper ends with conclusions and some potential future research directions.

2 Materials and methods

For the purpose of this study, hourly measurements of electricity demand for 2017 in two office buildings have been used. The measurements are a part of the OBEMS research grant realized at AGH University (Faculty of Management) by Prof. Mikulik and his research group. The details of the office buildings cannot be disclosed for reasons of confidentiality. The solar irradiation and temperature data has been downloaded from http://www.soda-pro.com/. Figure 1 presents the time series.

The energy demand from the first building amounted to 1.034 GWh in 2017, whereas in the second building it was only 3 MWh more. The mean hourly energy demand in both buildings is almost the same and amounts to 118 kWh. However, the energy demand standard deviation is higher in the first building, being 49.8 kWh compared to 41.7 kWh. In both buildings the maximal energy demand exceeds 400 kWh but the minimal demand is no greater than 40 kWh. The majority of daily energy consumption occurs from 8 AM to 5 PM (over 54%) and, over the year, occurs in summer (40% of energy being consumed from May to September).

To estimate the energy generated by a PV installation on an hourly basis the following formulas were applied.

$$E^{PV} = P^{PV}\frac{H}{H^{STC}}[1-\mu(T^C - T^{STC})]\eta^{PV} \tag{1}$$

$$T^C = T_a + H\left(\frac{T^{NOCT}-20}{8}\right) \tag{2}$$

where: E^{PV} – energy generated by PV system [kWh], P^{PV} – installed capacity if PV system [kW], H – irradiation [kWh], H^{STC} – irradiation in standard testing conditions [kWh], μ – temperature-dependent efficiency reducing coefficient [-], T^{C} – operating temperature of PV modules [°C], T^{STC} – PV modules operation temperature in standard testing conditions [°C], η^{PV} – overall efficiency of the PV system [%], T_a – air temperature [°C], T^{NOCT} – temperature in normal operating conditions [°C].

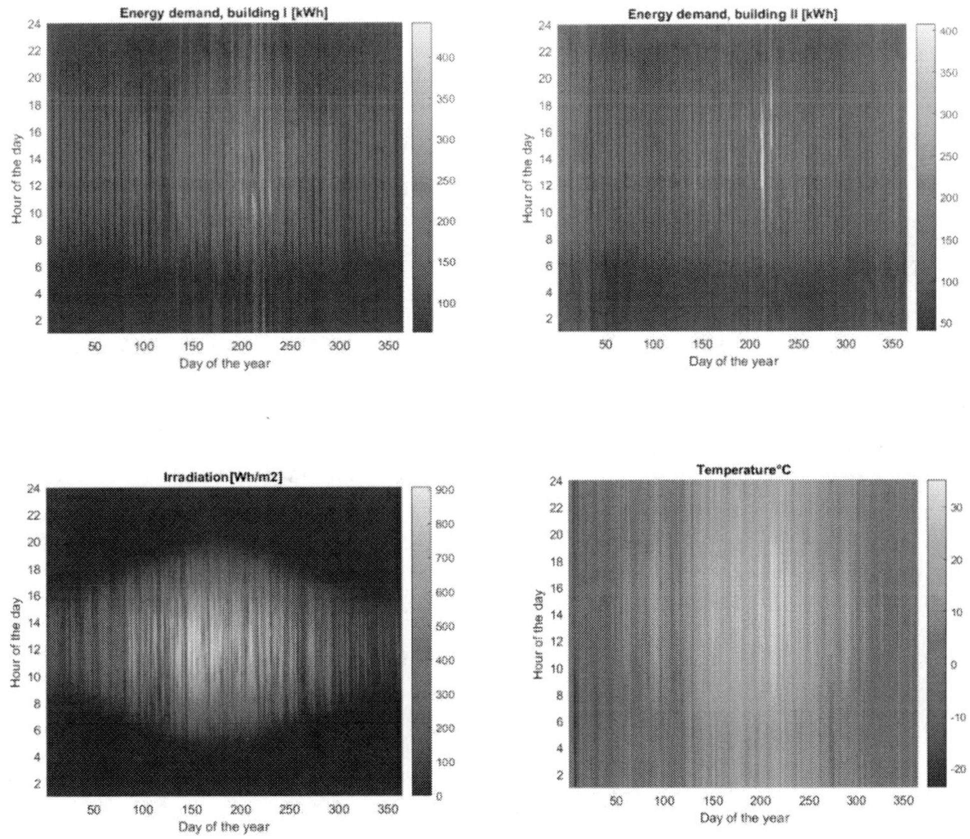

Fig. 1. Spectrum chart of input data for the year 2017.

It was assumed that the modules will be optimally mounted (from the energy generation perspective), namely, facing south at a 30° inclination. This approach was selected since the objective of this paper was not to investigate the potential of specific buildings to accommodate PV installation. Also, it was not the intention of the authors to investigate the potential of different PV module orientations, as indicated in the paper by Chattopadhyay et al. [7]. The overall efficiency of the PV system was set to $\eta^{PV} = 90\%$ (this included the losses in the inverter, wiring, etc.), whereas the temperature-dependent efficiency reducing coefficient was $\mu = 0.35\%/°C$. The equations presented above are in line with the recommendation of Duffie and Beckman [8].

Based on Eq. 1 and Eq. 2 it was found that, on a yearly basis, 1 kW of capacity installed in a PV system will generate 1,060 kWh, which results in a capacity factor of 12.1%. Such values are common for this part of Europe [9]. The way how residual load and self-consumption were calculated are provided in section 3.1. The differences between office buildings load profiles results from the variation of the number of employees, buildings orientation or type of activities. The factors driving those differences were not subject of this analysis.

3 Results and discussion

This section presents and discusses the results of the conducted analysis. The subsections present the following: self-consumption from PV generation; impact of PV on residual load; changes in annual energy demand patterns; PV potential to reduce monthly peak load.

3.1 Self-consumption and residual load

The first part of the analysis focuses on self-consumption and residual load. The former is defined as the ratio of energy generated by a given energy source (PV) to the volume of that energy utilized by the energy consumer (office building). If the energy generation is greater than the energy demand, an energy surplus will occur (which in the case of a grid-connected system will have to be absorbed by the grid, impacting the national load balance [10]. The latter (residual load) is understood as a difference between energy demand and energy supply from PV generator, but if supply is greater than demand the residual load is set to "0".

The values of the parameters presented above are shown in Fig. 2. As can be observed, both have a non-linear relationship with the capacity of the PV system. For PV capacities of up to 150 kW, over 99% of the energy generated is used to cover the energy demand of the building. When the capacity of the PV system exceeds 0.95 MW and 0.8 MW (for buildings I and II, respectively) the energy self-consumption drops below 50%.

The residual load parameter coincides with self-consumption. As can be seen on the right-hand chart in Fig. 2, in both buildings, with a PV capacity of up to 300 kW, the PV systems have a similar impact on the residual load. However, at a certain point each additional kW installed in PV systems in building I lowers the residual load. A decrease in the residual load means that the building needs less energy from the national power system.

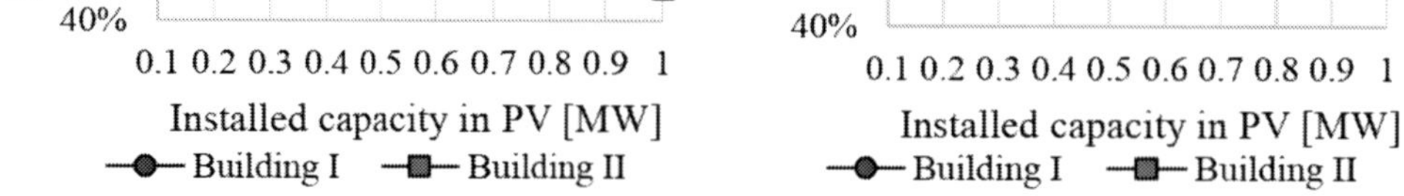

Fig. 2. Energy generated from PV used by the building (left), impact of PV on the building total load (right).

3.2 Covering the energy demand

The previous subsection has shown how changing the capacity of the PV system affects self-consumption and residual load. The results were, however, presented on an annual time scale. Considering the load and irradiation variability, it is important to investigate this on a shorter, monthly time scale.

Figure 3 shows the sum of monthly electricity demand in both buildings. It shows a clear annual energy demand pattern, with the highest energy consumption from May to August. In the first building the monthly demand ranges from 67 to 112 MWh, but from 70 MWh to 104 MWh in the second building. For this specific analysis, a 500 kW PV system was selected in both cases. As can be observed, it significantly changes the annual demand pattern from peak demand being observed in the summer to peak load in winter. Such a change might also be important from the perspective of the operation of the national power system, especially the Polish one, which is dominated by thermal generation and is beginning to face some power generation capacity constraints during hot summer days. Not surprisingly (considering the irradiation patterns in Fig. 1), the highest reduction in monthly electricity consumption is observed in June. In buildings I and II the energy demand was reduced by 169% and 158%, respectively, from 99 MWh to 36 MWh and from 89 MWh to 34 MWh.

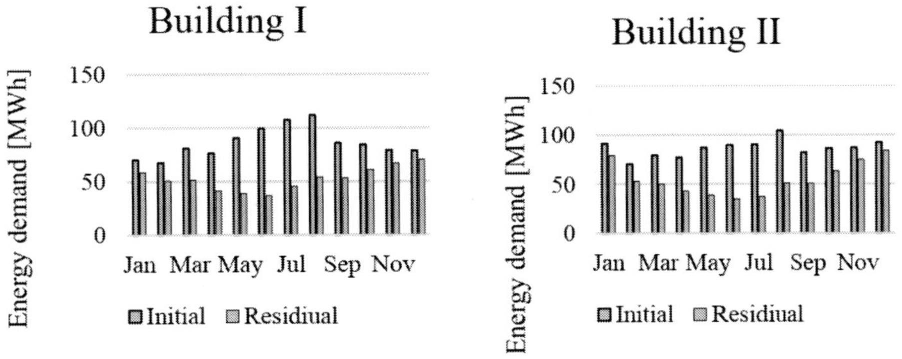

Fig. 3. Energy demand reduction for a 500-kW PV system in both buildings.

3.3 Changes in monthly maximal loads

Last but not least, the analysis now focuses on the potential of the PV system to reduce the maximal hourly energy demand observed during individual months. As shown by Jurasz and Campana [6] peak demand plays an important role in the electricity bill in the case of tariffs for office buildings. Jurasz and Campana estimated that PV systems can significantly reduce the peak load and thereby reduce the electricity bill. Accordingly, for Polish electricity tariffs, the owner can declare his peak load for each month of the year, and, if exceeded, a penalty will be imposed. Figures 5 and 6 present the current (blue continues line) peak load during each month and the impact of adding 0.1, 0.2 and so on of PV capacity on its reduction.

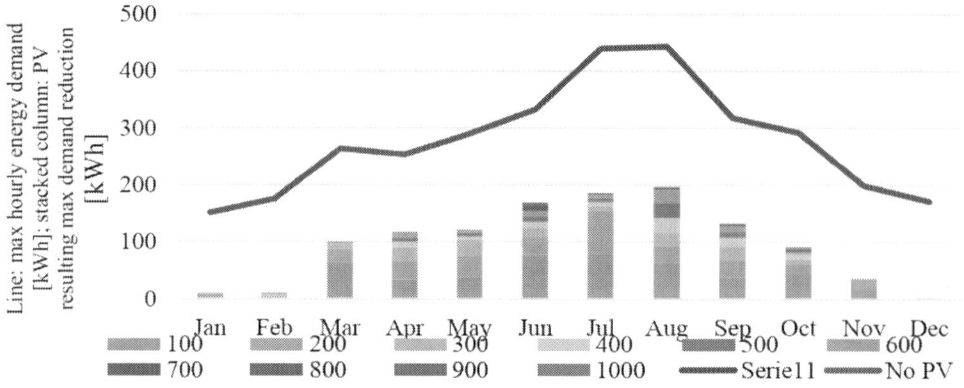

Fig. 4. Impact of adding PV capacity on the reduction of observed hourly maximal energy demand during individual months (Building I).

As can be seen in Fig. 4 and Fig. 5, the buildings exhibit different patterns in terms of maximal hourly energy demand. The patterns differ, but in both cases the maximal values are observed in August, and exceed 400 kWh (significantly in building I, and slightly in building II). Naturally, the potential of the PV system to reduce peak load depends on the match between solar energy availability and the time when the maximal load is observed. As shown in Fig. 4 and Fig. 5, for months like November–February (varies between the two buildings) the additional capacity in PV system does not significantly lower the maximal energy demand. For example, in building I in January, adding 100 kW of PV reduces the peak load from 151.25 kWh to 143.2 kWh, but adding 900 kW of PV capacity only reduces it by a further 0.7 kWh to 142.5 kWh. This clearly shows that the peak load is observed in hours different from the availability/operation of the PV generator. A different situation is observed in (especially) summer months, however. Then, adding 100 kW and 200 kW in PV capacity has an immense impact on the peak load. A reduction of as much as 154 kWh (July, building I) can be observed. It must be noted that this potential of PV systems to reduce peak demand is strongly dependent on the match between electricity demand and supply. It is naturally recommended that office buildings with big air-conditioning units should be equipped with PV systems to reduce their impact on the load curve on the national level [5].

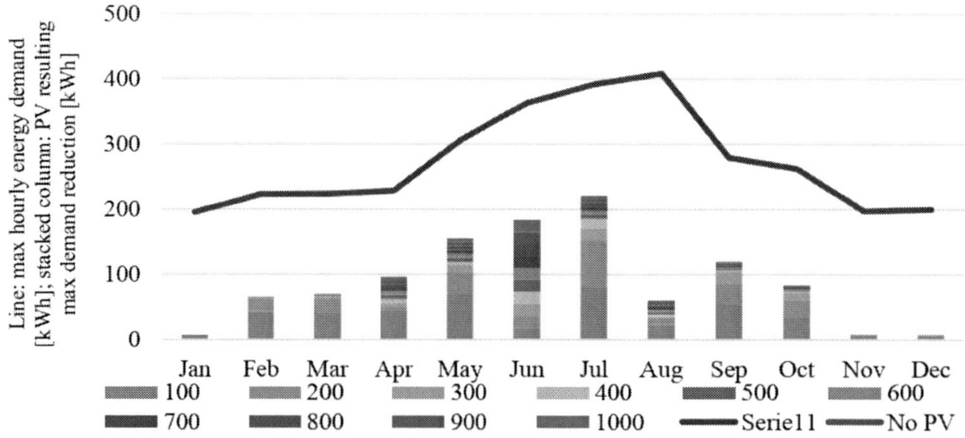

Fig. 5. Impact of adding PV capacity on reducing observed hourly maximal energy demand during individual months (Building II).

3.4 Energy flow

The final part of the analysis briefly presents the issue of the energy flow between an office building with a PV system and the national power system. The PV generator is a non-dispatchable energy source and, as such, it has a significant impact on the operation of the national and regional power system. Considering their non-dispatchability and temporal variability of power output, their integration into the power system is a significant challenge [11].

The self-consumption and residual demand analysis presented in Fig. 2 indicates that: a) the PV system is not able to completely cover the electricity demand, and b) not all energy generated by the PV system is consumed by the building/used to cover its demand. If the building is operated in an on-grid mode it means that there is a significant variability in the energy flow direction. Figure 6 presents this issue for both buildings and 500 kW installed capacity in PV. As can be observed, there is a significant energy surplus (flow from the PV system to the grid) during daytime hours and from the grid to the building during night- time hours. As indicated before, adding a PV system to the building significantly changes the building energy demand pattern. This opportunity should be seriously considered by governmental organizations and power system operators, as PV systems can clearly support the operation of the power system. It is, however, important to note that electricity tariffs may significantly affect the self-consumption rate [12] and also that some new conceptual business models might be needed for such situations [13]. On the other hand, building owners may consider not only a PV system but also, bearing in mind the energy flow, the potential application of energy storage technology [14, 15].

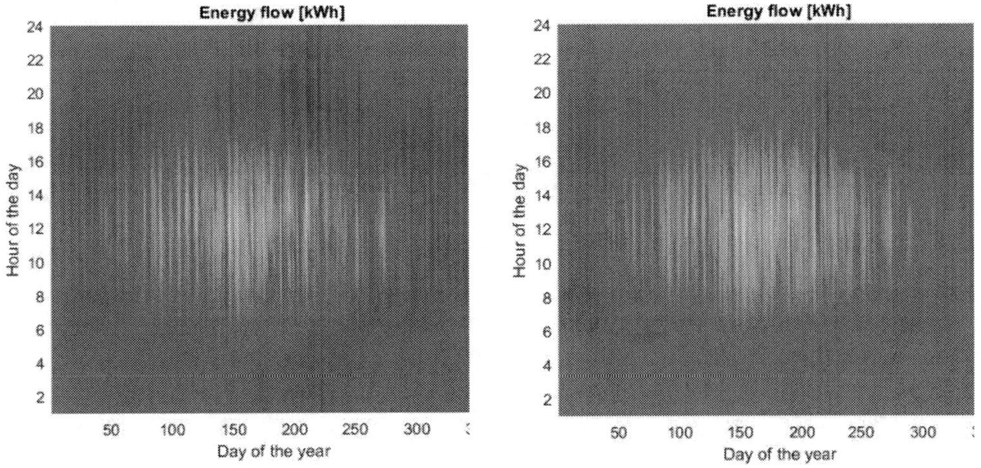

Fig. 6. Energy flow between office buildings (left building I, right building II) with 500 kW PV installations. Positive values show energy flow to the grid (surplus energy from PV generation), negative values show energy demand of the building being covered from the grid

4 Conclusions

This paper reports an outcome of research intending to analyse the impact of PV generation on energy demand in office buildings. Based on real data on hourly energy demand and simulated energy generation from a PV system it was found that, for the considered office buildings, PV systems with capacities up to 200 kW have very high self-consumption factors (meaning that all the generated energy is consumed onsite). Also, adding PV systems to support the office buildings in covering their electricity needs reduces the observed maximal energy demand and significantly alters the annual energy demand curve from peaking in summer to peaking in winter. The results of this research present important evidence that PV systems may not only potentially reduce electricity bills in office buildings (by reducing the peak load, as also shown by Jurasz and Campana [6] but that their impact on the energy demand might be beneficial from the perspective of the national power system. In future research we would like to focus on the economic aspects of PV utilization in office buildings, their hybridization with wind energy and application with energy storage. Our intention is also to increase a portfolio of considered office buildings in order to work on a more representative sample.

References

1. S. C. Parkinson, M. Makowski, V. Krey, K. Sedraoui, A. H. Almasoud, N. Djilali, Appl. Energ. **210**, 477–486 (2018)
2. J. Jurasz, P. B. Dąbek, B. Kaźmierczak, A. Kies, M. Wdowikowski, Energy **161**, 183–192 (2018)
3. P. B. Dąbek, J. Jurasz, E3S Web. Conf. **45**, 00014 (2018)
4. M. Z. Jacobson, M. A. Cameron, E. M. Hennessy, I. Petkov, C. B. Meyer, T. K. Gambhir, M. L. Miccioli, Sustain. Cities and Soc. **42**, 22–37 (2018)
5. J. Mikulik, Sustainability-Basel. **10** (2018)
6. J. Jurasz, P. E. Campana, Sustain. Cities Soc. **44**, 871–879 (2019)
7. K. Chattopadhyay, A. Kies, E. Lorenz, L. von Bremen, D. Heinemann, Renew. Energ. **113**, 176–189 (2017)
8. J. Duffie, W. A. Beckman, *Solar engineering of thermal processes* (John Wiley & Sons, 2013)
9. A. Bugała, K. Bednarek, L. Kasprzyk, A. Tomczewski, E3S Web. Conf. **19**, 01002 (2017)
10. J. Jurasz, M. Krzywda, J. Mikulik, E3S Web. Conf. **10**, 00059 (2016)
11. J. Jurasz, B. Ciapała, Sol. Energy **173**, 675–690 (2018)
12. J. C. Solano, M. C. Brito, E. Caamaño-Martín, Energy Pol. **122**, 322–331 (2018)
13. Y. Yang, Y. Zhang, P. E. Campana, J. Yan, Energy Proced. **142**, 3182–3193 (2017)
14. P. E., Campana, L. Cioccolanti, B. François, J. Jurasz, Y. Zhang, B. Stridh, Y. Jinyue, *10th International Conference on Applied Energy* (ICAE 2018)
15. J. C. Solano, L. Olivieri, E. Caamaño-Martín, Appl. Energy. **206**, 249–266 (2017)

Experiment design to investigate the possibility of using grey water in aeroponic cultivation of various crops for future long-term space missions

Anna Jurga[1,*], *Joanna* Kuźma[2], *Kamil* Janiak[1], and *Tobias* Niederwieser[3]

[1]Wroclaw University of Science and Technology, Faculty of Environmental Engineering, Wyb. Wyspianskiego 27, 50-370, Wroclaw, Poland
[2]Wroclaw University of Science and Technology, Faculty of Mechanical and Power Engineering, Wyb. Wyspianskiego 27, 50-370, Wroclaw, Poland
[3]University of Colorado Boulder, Ann and H.J. Smead Department of Aerospace Engineering Sciences, 429 UCB, Boulder, CO, 80309, USA

Abstract. One of the crucial steps in the development of self-sufficient extra-terrestrial human colonies is water and biomass management. One of the attempts to simplify the treatment scheme is to directly implement greywater into soil-less cultivation as no greywater treatment unit would be necessary. In this case, the most essential key factor is the influence of surfactants. In this paper, an analysis of various experiment designs is investigated and shown. While full factorial design of experiment would require more than 500 000 tests, with duration of at least 30 days each, an approach to reduce this number to a total of 5 experiments is shown in this paper for initial experiments.

1 Introduction

Water management and food production are one of the most important key factors for human habitation, either on Earth or beyond. Creating self-sufficient Life Support Systems (LSS) with each subsystem working in cooperation is essential. LSSs include four basic subsystems: biomass production, atmosphere revitalization, as well as waste and water treatment, shown in Figure 1. As it can be seen, each subsystem by-product has to be used to facilitate at least one other subsystem. Thus, recovery of essentials such as water and nutrients is required. It is rather certain, that hybrid (simultaneous use of biological and physiochemical wastewater treatment) wastewater treatment will be considered for future, long-term space missions according to NASA Space Technology Roadmap [1]. It is still undetermined whether it is possible to use untreated waste streams (such as grey water) directly for plant cultivation. Cultivation methods might be divided into two main groups: classical cultivation (in soil) and soilless cultivation. Trends, however, show that the latter type should be considered for spaceflight applications due to the decreased system mass

* Corresponding author: anna.jurga@pwr.edu.pl

 https://doi.org/10.1051/e3sconf/201910000031

[2]. The long-term goal of this project is to close a loop between the water and biomass subsystem, by suppling biomass production with biologically treated wastewater. The first milestone is to study whether plants cultivated with soilless methods (more precisely aeroponics) could be supplied with untreated greywater. Overall assumptions, aims, and beneficial aspects of discussed system design are shown in next chapter.

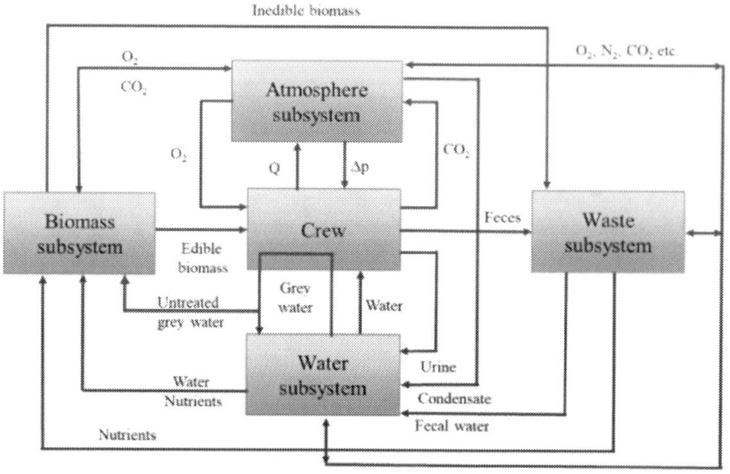

Fig. 1. Self-sufficient Life Support System [own source].

1.1 Why aeroponics?

Typical soil cultivation is not applicable in extraterrestrial conditions as it requires more water and nutrients as well as space than soilless types (aeroponics or hydroponics). In hydroponics, roots are submerged in nutrients solution (water and nutrients), while in aeroponics, roots are suspended in air, and nutrient solution is supplied in form of mist. In both soilless systems, roots have better access to water and nutrients and therefore plant growth is faster. Water that is not taken up by roots returns to the tank, and therefore very limited losses occur [3]. Aeroponics however require less water and have a higher yield. Lower water consumption is due to lower evaporation, while higher yield is due to better oxygenation of roots [4]. Aeroponic systems are therefore intensively studied as part of future life support systems, as this type of cultivation has potential superior parameters over soil and other soilless types of cultivation. With application of aeroponics, future colonies will require less water and nutrients to produce food, as well as less space. These savings in materials will lead to decrease in cost of transport from Earth to further destinations such as Moon or Mars. An example of an aeroponic system is shown in Figure 2. Besides many beneficial features of aeroponics, it has to be noted that a system might be exposed to failures, mainly in regard to the small capillaries and nozzles, which are delivering nutrient solution to the roots and are susceptible to clogging. As grey water is rich in solids and surfactants, enhanced clogging and biofilm forming may occur. While the system does not use any cultivated substrate, the risk of plant withering is greater in the event of a hydraulic system failure. Thus, experiments conducted on grey water contaminated with surfactants might also reveal hydraulic responses that can provide valuable lessons learned for future system designs.

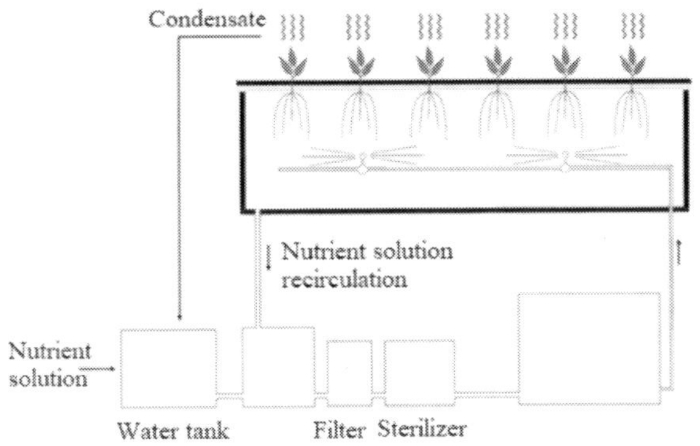

Fig. 2. Aeroponic system scheme [4].

1.2 Why grey water?

Wastewater originating from showers, laundry, bathing, and washing the dishes [5] is called grey water. Grey water or total hygiene water [2] is a main source of water in all waste streams. The composition of grey water is provided for extra-terrestrial condition by Schubert et al. [6], Wydeven and Golub [7] and most important by Anderson et al. [2]. However, it should be noted that data are not obtained directly from space mission conditions. Despite lack of data, hygiene water is a major stream from which water must be recovered and major stream that should be considered as water source in regenerative LSSs.

2 Design of experiment

2.1 Full factorial design

Utilization of untreated grey water for irrigation requires extensive research. List of variables that have influence on results of irrigation with grey water are summarized in Table 1. Grey water contains more contaminants than just surfactants but presence of surfactants is a main obstacle to use grey water. Variable 1 and 2 are connected directly with surfactant presence. Variables 3–10 are variables connected with parameters of aeroponic cultivation or species of plant. While presence of surfactant poses threat to plants, probably some or all negative effects can be diminished by maximization of plant strength with selection of optimal parameters. Therefore, the results of exposure of plants to surfactants will be highly dependent on overall performance of cultivation and must be included in experimental design. Decisive variables can be tested on different levels. In Table 1, data about the number of levels as well as possible ranges of values are obtained from literature review and presented.

Only anionic surfactants (Igepon TC-42: sodium N-coconut acid-N-methyl taurate; Sodium methyl cocoyl taurate (SMCT); Sodium dodecylbenzenesulfonate (SDBS); Sodium laureth sulfate (SLES)) are used during space missions or in research related to space technologies. The detergent concentrations differ, depending on the type of use of water (hygiene water, laundry water, dishwash water). One to three concentrations can be distinguished for each surfactant. In soil cultivation the influence of anionic surfactant was measured and results showed negative effect on wheat growth [27]. It is possible that it was related to the decrease in microbial activity, which in turn reduced solubilizing effects. Studies performed in hydroponic systems while using Igepon TC-42 shows that concentration above 0.25 $g{\cdot}L^{-1}$ exhibit phytotoxic effects [28]. On board the International

Space Station concentrations obtained in grey water are 55% higher [2]. Additionally, in most space references surfactants concentrations tend to be higher than stated in Bubenheim et al. [28]. Only Wydeven and Golub [7] state lower concentration of SDBS for dishwasher and laundry stream, 0.22 and 0.056 $g \cdot L^{-1}$ respectively. Plant growth stages consist of 4 main stages: sprouting, seedling, vegetative, and reproductive phase [29].

Table 1. List of variables in experiment design process of surfactants impact on aeroponically cultivated plants.

Number	Variable	Number of levels	References
1	Surfactant concentration	1–3 (0.056–0.94 $g \cdot L^{-1}$)	[2, 7–9]
2	Surfactant type	4	
3	Growth stage of plant	2 (seedling stage/vegetative stage)	[10, 11]
4	Duration of sprinkling	3 (1–5min)	[12, 13]
5	Sprinkling interval	3 (5–15min)	
6	Nutrients solution composition	4	[14–17]
7	CO_2 concentration	4 (360–50 000 ppm)	[18]
8	Plant species	14–23	[19, 20]
9	Illumination intensity	6 (150–930 $\mu mol \cdot m^{-2}\ s^{-1}$)	[21–26]
10	Illumination time	7 (10–24h)	
11	Total number of combinations	677376–1112832	

The age of the plant can affect its response to the stress factor. Lutts et al. [30] compered salt resistance in various development stages of plants. It appeared that the young seedling stage tends to be the most sensitive to NaCl. Impact of two different surfactants in two different concentrations on lettuce (*Lactuca Sativa L.*) in two stages were conducted [11]. Results showed, that for each case the young seedling stage was more susceptible to surfactants.

Sprinkling time and intervals between them depend on the size of the plant roots. The main rule is that the roots should not be allowed to dry between cycles. At the early stage of development of plants, it should be done often, but shorter, and in later stages longer, but less frequent [3]. During each cycle the whole root should be well moisturized and not dry out before the next cycle. The times for sprinkling are from 1 to 3 min, and intervals from 5 to 15 minutes [12, 13].

A nutrient solution for aeroponic systems is an aqueous solution which contains mainly inorganic ions from soluble salts of necessary elements for plant growth. Eventually, some organic compounds such as iron chelates may be present [31]. All of those elements have a clear physiological role and they are necessary to complete a plant's life cycle [32]. 17 elements are considered essential for most plants: carbon, hydrogen, oxygen, nitrogen, phosphorus, potassium, calcium, magnesium, sulphur, iron, copper, zinc, manganese, molybdenum, boron, chlorine and nickel [3]. Carbon and oxygen, can be provided from the atmosphere. The rest of the macro and microelements are obtained from the growth medium [33]. The most basic nutrient solutions include nitrogen, phosphorus, potassium, calcium, magnesium and sulphur. The nutrient composition determines electrical conductivity and osmotic potential of the solution. The nutrient solutions must contain the chemical form of ions that can be absorbed by plants, so in aeroponic systems the plant productivity depends on nutrient uptake and pH [33]. Analyzing the impact of macro and micro elements on the growth and development of plants, it can be seen that each nutrient solution must be properly balanced [3, 4]. Nitrogen has a strong influence on the growth and development of plants, but its excess reduces the resistance to damage of the plant and delays blooming. Higher levels of magnesium and calcium cause plants to produce thicker shoots that are more resistant to external factors, however, providing too much of these elements can cause stunted growth. Potassium positively affects the growth of plants and their resistance to external factors, but its excess lowers the content of nutritional elements in the yield. Phosphorus affects the growth rate of the root system and yield. Phosphorus is not a toxic component for plants, but its excess may cause reduced uptake of other elements. Boron has a direct impact on the formation of the cell wall in plants, however, its excess negatively affects the size and quality of the crop.

The lighting is affecting plant growth via three main factors: light cycle, light intensity, and light spectrum. There are 6 most popular configurations for plant illumination cycles (constant light – plants are illuminated for 24 h; 20 h light/4h dark photoperiod; 17 h light/7 h dark; 16 h light/ 8 h dark; 13 h light/11 h dark; 12 h light/ 12 h dark; 10 h light, 14 h dark). An increase in lighting duration might facilitate compensation for low light intensity. Light intensity is responsible for photosynthesis rate. In general, the higher intensity is, more photosynthesis would occur in the plants. In terms of light spectrum, both red and blue spectrum is required for undisturbed growth. However, Sago [26] stated that rapid plant growth under favourable environmental conditions might lead to some adverse effects (*e.g.* decreased calcium concentrations in leaves, tipburns). Thus, selection of appropriate light conditions is essential. Nam et al. [25] studied effects of supplementary lighting intensity and duration on hydroponically grown plants. Five different plants were tested in four configurations (3 h additional lighting with 4000 lux and 8000 lux; 6 h additional lighting with 4000 lux and 8000 lux). Additional lighting did not support plant growth in one case. The height and diameter were smaller than in reference. The best configuration varied for different plants. Thus, it cannot be stated conclusively, which option would provide the most effective growth, as it is strongly dependent on the type of plant.

The number of plant species required to provide a nutritious diet for colonists differs from study to study, however it is certain that many plants have to be cultivated to deliver attractive and sustainable diet for colonists for long time. Therefore, it can be assumed that the number of cultivated plant species will be approximately 20 and every species may have different response to surfactants present in grey water. Increase in CO_2 concentration from ambient concentration to a maximum level of 50 000 ppm results in a higher yield of lettuce. It seems that plants exposed to an abundant CO_2 source prevail better, and that phenomena can be used to limit negative consequences of exposure to surfactants. It is however important to consider that creation of high concentrations of CO_2 require resources and energy, and costs of such modification may outperform costs of surfactant removal from grey water. All mentioned above parameters have influence on plants growth and strength and therefore when applied at proper levels might lead to success in cultivation on water containing surfactants. However, full factorial design in case of that problem is composed of at least 500 000 combinations of tests (calculated in Table 1) with each test requiring time equal to the full growth period of selected plant. Mostly this is due to large number of plant species that has to be cultivated.

2.2 Optimized approach and proposed design

The overall number of possible combinations leads to an immense experiment if full factorial design would be applied. However, the size of the experiment can be diminished for initial pilot studies. From a large number of cultivated plants, mostly consumed crops are those that deliver energy such as rice, wheat, soybean, or peanut Liu et al. [19]. Those crops require most of the cultivation area as well as water. For example, in studies conducted by Tako et al. [20] rice, soybean, and peanut covered 120 m^2 of cultivation area, while all other 19 crops covered only 18 m^2. Therefore, delivery of grey water to dominant crops is the only solution that will provide complete utilization of grey water. The number of tested species can be decreased to 3 or 4 without important loss of quality and usefulness. However, at initial stage of research only one species should be chosen. In this experiment design rice is chosen. It is also necessary to introduce cost analysis into the experiment design to eliminate those combinations that are more expensive, require more mass to be transported to colonies, or are difficult to obtain in extraterrestrial conditions than treatment of grey water. Take high CO_2 concentrations as an example as it requires heavy and high energy consuming equipment such as membranes and pumps. Additionally, high CO_2 concentrations are dangerous for humans. Therefore, even if increase in CO_2 would be beneficial for plants resistance to surfactants, costs of implementation may overcome positive aspects. The CO_2 concentration would be also dependent on chosen location. On Mars for example it might be easier to have high CO_2 concentrations in a green house because there are very high CO_2 concentrations available in the Martian atmosphere. Overall analysis led to the assumption, that first experiment should be conducted in lab scale. Our proposed experiment design includes five separate aeroponic modules, where rice is cultivated. A single module is shown in Figure 3.

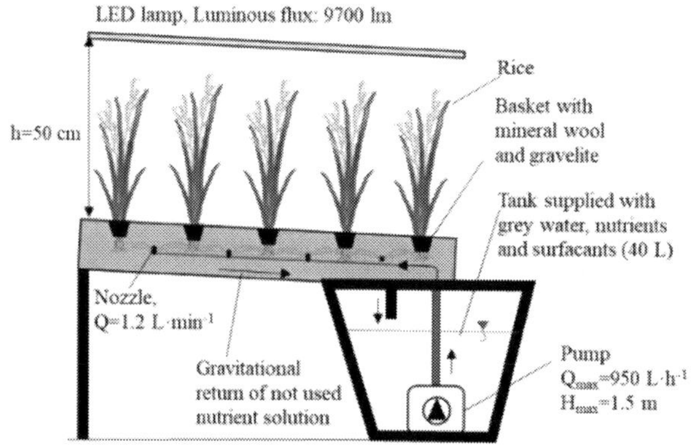

Fig. 3. Aeroponic module scheme [own source].

Four modules would be supplied with different surfactants used for space missions (SMCT, SDBS), as well as SLES, which is used worldwide. The chosen concentrations for SMCT and SDBS are 0.25 $g \cdot L^{-1}$, since this value is reported to be the boundary between affecting plant growth [28]. For SLES two concentrations are considered: 0.78 $g \cdot L^{-1}$ (corresponding average consumption of washing agents at the volume of water consumed at an international space station) and 0.11 $g \cdot L^{-1}$ (corresponding average consumption of washing agents at the volume of consumed terrestrial grey water). The fifth module is used as a reference would be fed with pure diluted nutrient solution, based on Zanão Junior et al. [34]. This solution was formulated with 1.25 $mmol \cdot L^{-1}$ K, 0.25 $mmol \cdot L^{-1}$ P, 3.75 $mmol \cdot L^{-1}$ N, 1.0 $mmol \cdot L^{-1}$ Ca, 0.5 $mmol \cdot L^{-1}$ Mg, 0.5 $mmol \cdot L^{-1}$ S, 20 $mmol \cdot L^{-1}$ Fe, 0.3 $mmol \cdot L^{-1}$ Cu, 0.33 $mol \cdot L^{-1}$ Zn, 11.5 $mmol \cdot L^{-1}$ B and 0.1 $mmol \cdot L^{-1}$ Mo. The duration of the experiment will be determined by the duration of the rice growing cycle (105–150 days), with standard illumination period of 16h.

3 Conclusions and future work

Conclusions can be derived from presented studies:

1. For untreated grey water to be applicable for aeroponics, influence of surfactants on plants growth must be determined.
2. Full factorial design of experiment aimed at determination of influence of surfactants would requires more than 500 000 tests with every test lasting at last 3 months. Large number of tests is due to large number of factors that influence plants response to surfactants.
3. It is advised to optimize approach and to limit test to crops that are cultivated to deliver energy to astronauts as these crops will consume most of the water.
4. It is also advised to develop cost analysis to rule out those combinations of parameters that would be too expensive to implement.
5. First design proposed by the authors consists of five separate modules, where rice is cultivated aeroponically. Four modules would include surfactants (SMCT, SDBS, SLES in different concentrations) and the fifth module would be a reference.

References

1. D. J. Barta, K. D. Pickering, C. Meyer, S. Pensinger, L. Vega, M. Flynn, A. Jackson, R. Wheeler, *A Biologically-Based Alternative Water Processor for Long Duration Space Missions* (NASA Johnson Space Center; Houston, TX, United States, Report No. NASA/JSC-CN-33488, 2015)
2. M. S. Anderson, M. K. Ewert, J. F. Keener, *Life Support Baseline Values and Assumptions Document* (National Aeronautics and Space Administration Washington, D.C. Report No. NASA/TP-2015–218570, 2018)
3. W. Breś, A. Golcz, A. Komosa, E. Kozik, W. Tyksiński, *Żywienie roślin ogrodniczych* (Wydawnictwo Uniwersytetu Przyrodniczego, Poznań, 2016)
4. A. Komosa, *Aeroponiczna uprawa roślin ogrodniczych* (X Ogólnopolskie Sympozjum Naukowe: Efektywność stosowania nawozów w uprawach ogrodniczych, Kraków, Poland, 17–18 June, 2004)
5. WHO, *Guidelines for the safe use of wastewater, excreta and greywater* (Volume 4, WHO Press, Geneva, Switzerland, 2006)
6. E. H. Schubert, R. A. Wynveen, P. D. Quattrone, Adv Space Res. **4**, 279–288 (1984)
7. T. Wydeven, M. A. Golub, *Generation rates and chemical compositions of waste streams in a typical crewed space habitat* (National Aeronautics and Space Administration, Ames Research Center; Moffett Field, CA. Report No. NASA-TM-102799, 1990)
8. D. L. Carter, *Waste Water Characterization for the ISS Water Processor* (SAE Technical Paper 981616, 28th International Conference on Environmental Systems, Society of Automotive Engineers, Warrendale, PA, 1998)
9. J. Silverstein, G. M. Brion, R. Barkley, A. Dunham, C. Hurst, P. Todd, J. Schulz, Acta Astronaut. **33**, 317–338 (1994)
10. S. Lutts, J. M. Kinet, J. Bouharmont, J. Exp. Bot. **46**, 12, 1843–1852 (1995)

11. K. Janiak, A. Jurga, W. Breś, J. Kuźma, Surfactant impact on lettuce aeroponic cultivation, [in preparation] (2019)
12. S. Leoni, B. Pisanu, R. Grudina, *A new system of tomato greenhouse cultivation: high density aeroponic system (HDAS)* (In International Symposium on New Cultivation Systems in Greenhouse **361**, 210–217 (1993)
13. J. Osvald, N. Petrovic, J. Demsar, Acta Aliment. **30**, 1, 53–61 (2001)
14. K. Yamazaki, *Nutrient solution culture* (Pak-kyo Co., Tokyo, Japan, 41, 1982)
15. D. S. Domingues, H. W. Takahashi, C. A. Camara, S. L. Nixdorf, Comput. Electron. Agric, **84**, 53–61 (2012)
16. E.A. Maier, R.D. Matthews, J.A. McDowell, R.R. Walden, B.A. Ahner, J. Environ. Qual., **32(4)** (2003) 1356–1364.
17. H. M. Resh, *Hydroponic food production: a definitive guidebook for the advanced home gardener and the commercial hydroponic grower* (CRC Press, 2016)
18. J. He, P. T. Austin, S. K. Lee, J. Exp. Bot. **61**, 14, 3959–3969 (2010)
19. H. Liu, C. Y. Yu, N. S. Manukovsky, V. S. Kovalev, Y. L. Gurevich, J. Wang, Adv. Space Res. **42**, 6, 1080–1088 (2008)
20. Y. Tako, S. Tsuga, T. Tani, R. Arai, O. Komatsubara, M. Shinohara, Adv. Space Res. **41**, 5, 714–724 (2008)
21. R. M. Wheeler, C. L. Mackowiak, G. W. Stutte, N. C. Yorio, L. M. Ruffe, J. C. Sager, R. P. Prince, W. M. Knott, Adv. Space Res. **41**, 5, 706–713 (2008)
22. F. Galangau, F. Daniel-Vedele, T. Moureaux, M. F. Dorbe, M. T. Leydecker, M. Caboche, Plant Physiol. **88**, 2, 383–388 (1988)
23. K. Herdel, P. Schmidt, R. Feil, A. Mohr, U. Schurr, Plant Cell Environ. **24**, 1, 41–52 (2001)
24. N. S. Mattson, E. D. Harwood, *Effect of light regimen on yield and flavonoid content of warehouse grown aeroponic Eruca sativa* (In VII International Symposium on Light in Horticultural Systems 956, 417–422, 2012)
25. S. Y. Nam, H. S. Lee, S. Y. Soh, R. A. M. Cabahug, Flower Res. J. **24**, 1–9 (2016)
26. Y. Sago, HortScience **51**, 9, 1087–1091 (2016)
27. A. Mohammad, A. Moheman, (In Chemistry of Phytopotentials: Health, Energy and Environmental Perspectives, Springer, Berlin, Heidelberg, 261–264, 2012)
28. D. Bubenheim, K. Wignarajah, W. Berry, T. Wydeven, J. Am. Soc. Hortic Sci. **122**, 6, 792–796 (1997)
29. A. A. Schneiter, J. F. Miller, Crop Sci. **21** 901–903 (1981)
30. S. Lutts, J. M. Kinet, J. Bouharmont, J. Exp. Bot. **46**, 12, 1843–1852 (1995)
31. A. A. Steiner, *Soilless Culture* (Proceedings of the IPI 1968 6th Colloquium of the Internacional Potash Institute, 324–341, Florence, Italy, 1968)
32. L. Taiz, E. Zeiger, *Plant Physiology* (Sinauer Associates, Inc. Publishers. Sunderland, ISBN : 0878938311, Massachusetts, USA, 1998)
33. L. I. Trejo-Téllez, F. C. Gómez-Merino, *Nutrient solutions for hydroponic systems* (In Hydroponics-A Standard Methodology for Plant Biological Researches. InTech, 2012) [Accessed online 28.01.2019]
34. L. A. Zanao Junior, F. Á. Rodrigues, R. L. F. Fontes, G. H. Korndörfer, J. C. L. Neves, J. Phytopatho. **157**, 2, 73–78 (2009)

Methodology and impact categories of environmental life cycle assessment in geothermal energy sector

Michał Kaczmarczyk[1,*]

[1]AGH University of Science and Technology, Faculty of Geology, Geophysics and Environmental Protection, Department of Fossil Fuels, 30-059 Krakow, Poland

Abstract. The development of renewable energy sources, including geothermal energy, causes that more and more attention should be devoted to their environmental impact in a holistic approach. Currently, the impact of geothermal installations on the environment is analyzed in the context of emissions of pollutants into the atmosphere at the operational stage. Additionally, sustainable development and creating a responsible energy mix requires the use of tools that will provide precise information about the effectiveness of the proposed technological solutions and their impact on the environment. The LCA lifecycle assessment can be such a tool. The complexity of the investment process, which is the construction of a heating plant, a combined heat and power plant or power plant based on geothermal energy, results mainly from changing geological, topographical and environmental conditions. In such a case, the precise determination of the methodology and the definition of impact categories are crucial. The article is a review of the applied methodologies in the world, taking into account impact categories and databases used in conducting the LCA in the geothermal energy sector.

1 Introduction

Sustainable development that respects the environment, requires a responsible approach to creating an optimal energy mix. Natural elements of that mix are currently renewable energy sources which percentage share in energy production is systematically increasing. Therefore, there is a situation where a wider view of renewable energy sources and its impact on the environment becomes a necessity. Geothermal energy, being a renewable energy source, is perceived as almost completely emission-free during working time [1]. In terms of emission of combustion products into the atmosphere, this is a correct approach, but in order to determine the real impact on the environment, the proper way is the implementation of life cycle assessment (LCA) to geothermal energy sector.

The main categories (Fig. 1) that need to be defined at the stage of lifecycle assessment include land use, geological threats, emissions to the atmosphere, emissions to water,

* Corresponding author: mkz@agh.edu.pl

emissions to rock mass, water consumption, impact on biodiversity, noise and light emissions, and waste heat [2, 3]. The scale of the undertaking which is the life cycle assessment, taking into account the above information, requires a great interdisciplinary and consideration of each case in an individual and as precise a manner as possible.

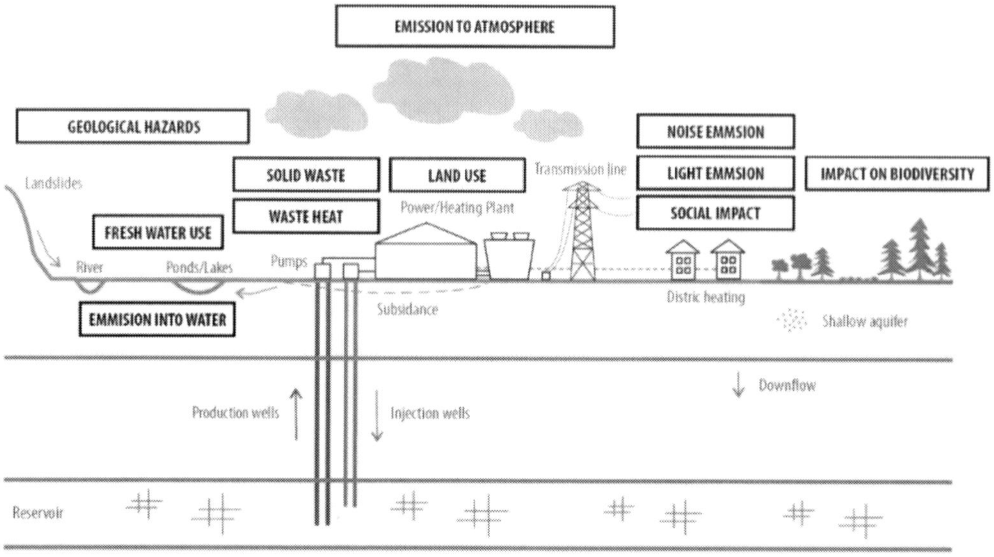

Fig. 1. Environmental impact of geothermal power plant/district heating (based on [3, 4] – modified).

2 Life Cycle Assessment methodology

Life cycle assessment is a standardized method of conduct for the analysis of environmental hazards associated with the product/process throughout its lifetime. The principles and structure of the LCA are defined in the PN-EN ISO 14040 standard "Environmental management. Life cycle assessment. Principles and framework" [5], while the requirements and guidelines for conducting LCA are contained in the PN-EN ISO 14044 standard "Environmental management. Life cycle assessment. Requirements and guidelines" [6]. Additionally, standards PKN-ISO/TR 14047 "Environmental management. Life cycle impact assessment. Examples of applications of ISO 14042", ISO/TS 14048 "Environmental management. Life cycle assessment. Data documentation format" [8] and PN-EN ISO 14001" Environmental management systems. Requirements with guidance for use" [9] are helpful in the development of LCA.

LCA allows to proper assessment of all environmental aspects at the stage of extraction and processing of mineral resources, production, distribution, operation, recycling and final disposal of waste [10–13]. For the basic tasks resulting from the implementation of the LCA, it is important to consider the documentation of the impact of the product/process on the environment at every stage of its life and to analyze the possibility of interrelated environmental impacts resulting from the product/process improvement.

The first stage of the LCA is to determine the purpose and scope of the research. At this stage, it is important to justify why life cycle assessment is carried out, which will determine the level of detail that it will be performed. To make this possible, it is necessary to define the product system and functional unit. Product system is defined as a set of material and energetically connected unit processes [14]. The functional unit is defined as the quantitative effect of the product system (the reference unit) in the LCA [5]. In the considered case of geothermal energy, the scope of the research should be the environmental impact of the electricity/heat production, while the functional unit should constitute as a unit of electricity – i.e. MWe, or heat – i.e. TJ. The next step in the LCA is to determine the set of inputs and outputs (LCI, life cycle inventory), on the input side – raw materials and energy, on the output side – products, waste and emissions to the environment. This stage, in which the product is creating, determines the boundaries of the system. Life cycle impact analysis (LCIA) is the third stage. It consists of assigning particular items in the developed catalog of input and output data of relevant impact categories, which allows for a reliable assessment of the environmental impact for the analyzed product/process. The last stage is the interpretation of the results. This stage should be considered as the most difficult and the riskiest for error.

Despite the normalization of the LCA method by the applicable ISO standards, it is important to calculate indicators for individual impact categories. For this purpose, various methods are used (Tab. 1), of which in the context of application to renewable energy sources, including geothermal energy, the following should be mentioned: Eco-indicator, CML, ILCD, Monte Carlo and ReCiPE. In the case of geothermal energy, the analyses performed so far [15–25], all methods except ReCiPE were used. Each of these methods is an attempt to resolve two major problems – difficulty of interpreting the obtained results in a comprehensive approach and the diligence of collecting all necessary environmental data.

In summary, the mandatory elements during LCA implementation are: selection of impact categories, category indicators and characterization models, LCI results assignment (classification) and calculation of the category index value (characterization). The optional elements are: calculation of the category index value relative to reference information (normalization), grouping, weighing and data quality analysis.

3 Database and impact categories

Conducting LCA remains rare in geothermal energy. The vast majority of publications concern geothermal power plants [3, 4, 15, 17, 19, 20, 22–25]. From the point of view of the usefulness of the current experience from the use of the LCA method in heat engineering using geothermal energy, the publications Nitkiewicz and Sekret [18] and Bartolozzi et al. [21] could be helpful. They do not analyze geothermal heat plants, but the definition of a functional unit or analysis of a heating network could be useful. In the case of the functional unit, it is clear that most mentioned researchers used kWh. Only Nitkiewicz and Sekret [18] defined functional unit as the total amount of heat that should be delivered to customers connected to the heating network.

Considering the above, any guidelines resulting from world experience in the development of LCA for geothermal energy should be seen in publications on renewable energy sources or conventional systems. It is due to the fact that certain aspects of LCA for geothermal district heating and geothermal power plants have characteristics in common with the generally perceived energy industry. In addition, in the context of the assessment of district heating systems, the geothermal energy production sector, to which the above mentioned publications relate, may constitute a significant complement to knowledge for district heating.

Table 1. Short description of LCA methods.

Method	Description	Used by
Eco-indicator 99	Significant emphasis is placed on the assessment of damage. The LCA method in this approach has been extended to the weighing method. It allows calculating a single result for the total environmental impact. The method includes the assessment of three groups of impact categories: human health (number and duration of diseases, life years lost due to premature death), ecosystem quality (species diversity) and resource consumption (surplus energy needed in the future to extract lower quality mineral and fossil resources) [26].	Nitkiewicz and Sekret (2014)
CML	CML allows characterizing the coefficients in order to quantify the share of impact categories and normalization factors. The method limits quantitative modeling to the early stages of the cause-and-effect chain to reduce uncertainty as much as possible. The results are grouped into categories of midpoints according to common mechanisms (i.e. climate change) or commonly accepted groups (i.e. ecotoxicity). The normalization rates are calculated on the basis of total substance emissions and characterizing factors per substance [27].	Buonocore et al. (2015)
ILCD	Method of midpoints recommended by the European Commission, which complies with the applicable PN-EN ISO 14040 and PN-EN ISO 14044 standards. The ISO standard is an important framework for LCA, however these frameworks leave a number of important choices. These choices may affect the credibility and comparability of the LCA results. The method was developed to provide guidelines to increase the consistency of the methodology and to ensure the quality of the results obtained [28].	Pratiwi et al. (2018), Bartolozzi et al. (2017)
Monte Carlo	Simulation based on modeling systems, the course of which depends on random factors. The method assumes that all analyzed data are to some extent uncertain. This is particularly true when LCA is performed for the comparison of two products. In practice, simulation based on the assumption of random values for each uncertain input. The procedure is repeated many times, and its results are recorded, creating a distribution of uncertainty for the final result [19, 22].	Buonocore et al. (2015); Hanbury and Vasquez (2017)

In general, the issues of databases and impact category are related to the specific methodology used. However, this is not a rule. The most popular databases are: Ecoinvent, Probas, Agri-BALYSE, Blue Mountain, JRC's European Reference Life Cycle Data System (ELCD), NREL's U.S. Life-Cycle Inventory Database (USLCI), EXIOBASE and CML's database of impact assessment factors (CML-IA). Among these, the most popular in geothermal energy is Ecoinvent, widely recognized as the most comprehensive and complete database for the implementation of environmental life cycle assessment. At the same time, attention should be paid to the development of LCA research in Europe. The European Reference Life Cycle Data System (ELCD) may be an equally popular solution as Ecoinvent in the near future. Due to the small number of LCA studies in the field of geothermal energy, it is difficult to state clearly which of the databases is the best for use. It is worth noting, that it is not necessary to choose one base, which confirms Heberle et al. [20] and Pratiwi et al.

[25]. In the first case, the authors used Ecoinvent and Probas bases, while in the second case Ecoinvent, USLCI and Agri-BALYSE. It seems that more important than choose of database is which method will be implemented.

The commonly used impact categories include climate change, stratospheric ozone depletion, photooxidation, acidification, eutrophication, toxicity to humans and ecotoxicity [7]. More specifically, impact categories can be divided into global, regional and local. The first group includes global warming, stratospheric ozone depletion and resource consumption. The regional categories include photochemical oxidant formation, acidification, nutrient enrichment, ecotoxity, human toxity and resource consumption. The third group – local – are photochemical oxidant formation, acidification, nutrient enrichment, effects of waste heat water, ecotoxicity, human toxicity, odor, noise, resource consumption, land use and waste [30]. An example of defining impact categories is presented in Table 2.

Table 2. A simplified example of defining 4 impact categories and defining category indicators together with the characterization parameter, indicator value and end points of category [based on 30].

Impact category	LCI results	Category indicator	Parameter characterization	Indicator value	Category endpoints
Depletion of primary energy resources	Extraction of fossil fuels resources	Energy resources in primary fuels	Lower heating value per unit of mass	The total lower calorific value	Energy
Depletion of mineral resources	Extraction of resources, expressed using useful materials	Extraction of ore material	Current material extraction	The total mass of material used from the ore	Availability of resources
Climate change	Greenhouse gas emissions	Increasing the intensity of infrared radiation	Global warming potential (i.e. GWP100)	A kilogram equivalent to CO_2	Lost years of life
Destruction of stratospheric ozone	Emission of gases destroying the ozone layer	Increases the stratosphericity of the ozone hole	Ozone depletion potential (ODP)	Kilograms of equivalent CFC-11	Days of illness

4 Geothermal energy sector in the world

The development of renewable energy sources in the world, causes that more and more attention should be devoted to their environmental impact in a holistic approach. How big is the potential of geothermal energy shows the data published periodically during World Geothermal Congresses, which indicate that year-on-year the installed capacity in geothermal sector increases.

In the case of geothermal power plants, it was noted that in 1995 the total installed capacity was 6 832 MWe, while twenty years later (2015) it was almost 12 635 MWe (Fig. 2), which is an increase of 185%. In the case of electricity production, it was a growth form 136 926 TJ/yr in 1995 to 264 776 TJ/yr in 2015 (193%). In addition, it is estimated that in 2020, 21 443 MWe will be achieved [31]. Among the countries in which installed capacity grew the most in the last five years, Kenya stands out – an increase of 392 MWe (194%), the USA – 352 MWe (11%) and Turkey – 306 MWe (336%). In addition, Germany should be distinguished as a percentage increase in relation to the installed capacity in 2010 – 280% (20 MWe) [31, 32]. The dynamics of the geothermal sector in Europe shows that in 2017, installed capacity in 117 geothermal power plants amounted to 2 800 MWe (16 power plants with a total capacity of 22 MWe were put into operation in 2017). It is worth noting that there are another 30 projects under implementation, mainly in Turkey, France, Hungary, Germany, the UK, Iceland, Croatia and Belgium. Moreover, works are in progress on 276 new projects [33].

In the opposition to the electricity production, direct use of geothermal energy is possible on a much larger scale, which results from the available geothermal resources, determined primarily by the temperature of the geothermal water at the outflow. World data indicates strong increases in installed capacity and thermal energy production in the years 1995–2015. In the case of installed capacity, it was a growth from 8 664 MWth in 1995 to 70 885 MWth in 2015 (818%). In the case of thermal energy production, it was respectively 112 441 TJ/yr and 592 638 TJ/yr, which is an increase in annual production of 527% [34]. The largest share of installed capacity belongs to heat pumps (55.15%) and to bathing and swimming (20.18%). However the third result belong to district heating (14.96%).

Analyzing the data above and trends on the energy market, further growth resulting primarily from the development of the use of geothermal energy for agricultural purposes (greenhouse heating), should be expected (in Europe – 10% per year). In addition, according to EGEC data from 2017 [33], 294 geothermal heat plants with a total capacity of 4.9 GWth was operating in Europe. Only in 2017, nine geothermal heat plants with a total capacity of 75 MWth (France, Italy, the Netherlands) were launched. In addition, 35 new projects are currently being under realization.

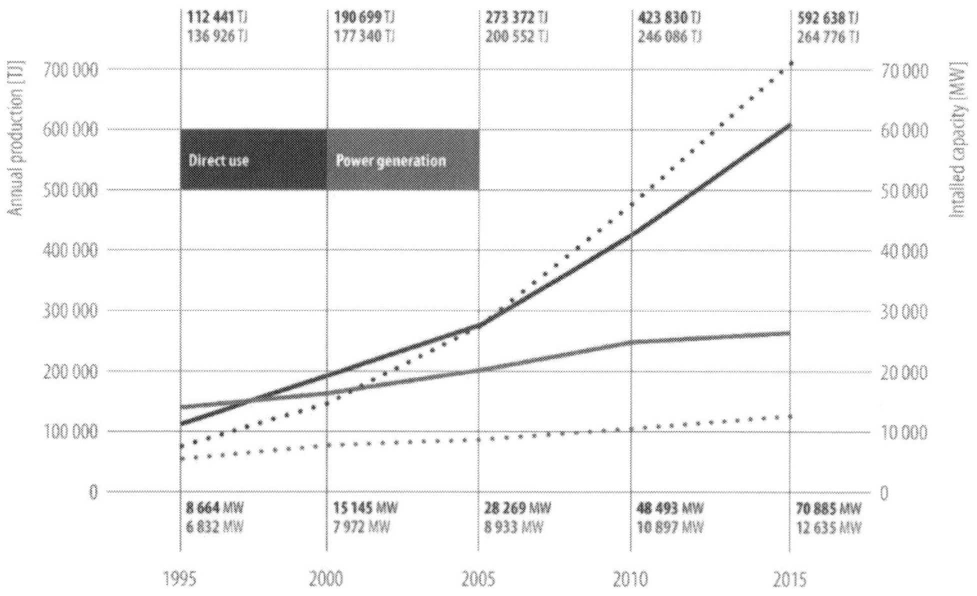

Fig. 2. Geothermal power generation and direct use in 1995–2015 (based on [31, 34] – modified).

5 Summary and conclusions

Modern energy, based more and more on the use of renewable energy sources, aims to reduce emissions of pollutants into the environment. However, actions aimed at protecting the environment are mainly focused on the work stage of power/heat plants and do not take into account the impact on the environment throughout the entire life cycle. Such an approach may turn out to be short-sighted, hence the need to implement methods and use of tools that will provide a precise answer to the question about the impact of individual energy projects on the environment.

The analysis of the literature data indicates that various tools (methodologies) are used in the conduct of environmental life cycle assessment in geothermal energy. The situation is similar in the case of databases. What is interesting, the ILCD method promoted by the European Commission is not dominant in previously published works. A similar freedom of choice takes place when the impact category is determined. In result, the surveys that have been thoroughly verified are very difficult to compare in a reliable way.

Considering the above, the more justified it seems to use the environmental life cycle assessment not only in geothermal energy, but energy in general. In another way, we will not be able to answer the question to what extent individual technologies of electricity or heat production affect the natural environment. At the same time, it should be remembered that LCA is not a "negative" method, but a tool that seeks to answer the question of how to improve products/processes and minimize their negative impact on the environment.

The paper has been prepared under the AGH-UST statutory research grant No. 11.11.140.031.

References

1. M. Kaczmarczyk, E3S Web of Conferences **44**, 00062 (2018)
2. M. Kaczmarczyk, Technika Poszukiwań Geologicznych. Geotermia, Zrównoważony Rozwój, **2** (2018)
3. P. Bayer, L. Rybach, P. Blum, R. Brauchler, Renewable and Sustainable Energy Reviews, **26** (2013)
4. K. Menberg, P. Blum, S. Pfister, L. Rybach, P. Bayer, *Life cycle assessment of geothermal power generation* (European Geothermal Congress, 2016)
5. PN-EN ISO 14040-06, Zarządzanie środowiskowe. Ocena cyklu życia. Zasady i struktura (2009)
6. PN-EN ISO 14044-06, Zarządzanie środowiskowe. Ocena cyklu życia. Wymagania i wytyczne (2009)
7. PKN-ISO/TR 14047, Zarządzanie środowiskowe. Ocena wpływu cyklu życia. Przykłady stosowania ISO 14043 (2006)
8. ISO/TS 14048, Zarządzanie środowiskowe. Ocena cyklu życia. Format dokumentowania danych (2003)
9. PN-EN ISO 14001, Systemy zarządzania środowiskowego. Wymagania i wytyczne stosowania (2016)
10. G. Rebitzer, T. Ekvall, R. Frischknecht, D. Hunkelerd, G. Norrise, T. Rydbergf, W. P. Schmidtg, S. Suhh, B. P. Weidemai, D. W. Penningtonf, *Life cycle assessment. Part 1: Framework, goal and scope definition, inventory analysis, and applications* (Environmental International, **30**, 2004)

11. K. Grzesik, Inżynieria Środowiska **11**, 1 (2006)
12. Z. Kowalski, J. Kulczycka, M. Góralczyk, *Ekologiczna ocena cyklu życia procesów wytwórczych* (2007)
13. R. Dąbrowski, M. Dzikuć, PAK, **58** (2012)
14. J. Kulczycka, M. Góralczyk, K. Koneczny, P. Przewrocki, A. Wąsik, *Ekologiczna ocena cyklu życia (LCA) nową techniką zarządzania środowiskowego* (2001)
15. S. Frick, M. Kaltschmitt, G. Schroeder, Energy, **35** (2010)
16. C. Chiavetta, F. Tinti, A. Bonoli, Procedia Engineering, **21** (2011)
17. M. Lacirignola, I. Blanc, Renewable Energy, **50** (2013)
18. A. Nitkiewicz, R. Sekret, Energy Conversion and Management, **87** (2014)
19. E. Buonocore, L. Vanoli, A. Carotenuto, S. Ulgiati, Energy, **86** (2015)
20. F. Heberle, C. Schifflechner, D. Bruggemann, Geothermics, **64** (2016)
21. I. Bartolozzi, F. Rizzi, M. Frey, Renewable and Sustainable Energy Reviews, **80** (2017)
22. O. Hanbury, V. R. Vasquez, Renewable Energy, **115** (2017)
23. J. I. Martinez-Corona, T. Gibon, E. G. Hertwich, R. Parra-Saldivar, Journal of Cleaner Production, **142** (2017)
24. C. Tomasini-Montenegro, E. Santoyo-Castelazo, H. Gujba, R. J. Romero, E. Santoyo, Applied Thermal Engineering, **114** (2017)
25. A. Pratiwi, G. Ravier, A. Genter, Geothermics, **75** (2018)
26. Ministry of Housing, Spatial Planning and the Environment, *Eco-indicator 99. Manual for Designers. A damage oriented method for Life Cycle Impact Assessment* (2000)
27. J. B. Guinée, M. Gorrée, R. Heijungs, G. Huppes, R. Kleijn, A. Koning, L. Oers, A. Wegener Sleeswijk, S. Suh, H. A. Udo de Haes, H. Bruijn, R. Duin, M. A. J. Huijbregts, *Handbook on life cycle assessment. Operational guide to the ISO standards. I: LCA in perspective. IIa: Guide. IIb: Operational annex. III: Scientific background. Kluwer Academic Publishers* (2002)
28. M. A. Wolf, R. Pant, K. Chomkhamsri, S. Sala, D. Pennington, *The International Reference Life Cycle Data System (ILCD) Handbook. Towards more sustainable production and consumption for a resource efficient Europe* (2012)
29. M. Bogacka, K. Pikon, M. Landrat, Waste Management, **70** (2017)
30. H. K. Stranddorf, L. Hoffmann, A. Schmidt, Environmental News, **78** (2005)
31. R. Bertani, *Geothermal Power Generation in the World 2010–2014 Update Report* (Proceedings World Geothermal Congress, 2015)
32. B. Matek, *2015 Annual U.S. & Global Geothermal Power Production Report* (2015)
33. P. Dumas, T. Garabetian, V. Pinzuti, C. Marchitelli, *2017 EGEC Geothermal Market Report. Key Findings* (2018)
34. J. W. Lund, T. L. Boyd, Geotermics, **60** (2016)

Assessment of the Ortachala HPP low head dam abutment deformation and seepage regime

Mirian Kalabegishvili[1], *Levan* Klimiashvili[1], *David* Gurgenidze[1], and *Adam* Ujma[2,*]

[1]Georgian Technical University, 68 M.Kostava St. 0175 Tbilisi, Georgia
[2]Czestochowa University of Technology, Faculty of Civil Engineering, ul. Akademicka 3, 42-200 Czestochowa

Abstract. During Ortachala HPP operation period cracks were formed in the right-bank of foundation, where is located underground fish-way sections. Right bank dam-foundation contact opened only in the horizontal direction, that caused 5.5 cm gap development, while there were no changes in the vertical direction. Visual researches showed that cracks (size is 4–5 cm) developed in fish-way construction. Mathematical models: fish-drainage-foundation, retaining wall-drainage-foundation, with the use of the finite element scheme, allowed us to finalize the suffusion-deformation process developed during the operation, namely: in the cracking zones of fish-way developed high gradients of seepage pressure, causing suffusion; under the influence of the weight of the soil and transport, shrinkage of the soil developed, which for the foundation of the right bank, created bending deformations causing the rotation of the soil in the vertical plane; bending deformation of the foundation at the contact of the concrete structure and the embankment caused a horizontal opening, spread out – at a depth of 4m. In connection with this, the retaining wall, leaning against the ground, loses support and, under the influence of hydrostatic pressure, tensile stress develops in it. In the case of crack formation in the wall, concentrated seepage will develop, which will cause groundwater rise in the main road and various unfavourable processes. Having established the reason for opening the contact, the fish was closed, as a result of which above mentioned process was stopped. Thus, the waterproofing of the fish-way is a necessary measure for the normal operation of the waterworks.

1 Introduction

Ortachala HPP is the law head waterworks facility situated in central part of Tbilisi. Station is being operated since 1955. Reservoir, placed in river canyon, is surrounded by on both sides, with underground draining galleries behind them. The latter is purposed (by design) to avoid rising of groundwater table level on banks that may cause flooding of the city communication lines (public systems, cable-lines and etc.), deformation of the road surface and development of other subsequent unfavorable processes. On the right-bank is also situated under ground fish-way sections (Fig. 1, photos 1, 2).

*Corresponding author: aujma55@wp.pl

During operation period cracks were formed in the right-bank foundation in HPP and under ground fish-way sections. Surprisengly, right bank dam-foundation contact opened only in the horizontal direction, that caused 5.5 cm gap development, while there were no changes in the vertical direction. Visual reaseraches showed that cracks developed in fish-way construction. Gap size is 4–5 cm. Some results of deformation and seepage developed analysis in right bank foundation is given below.

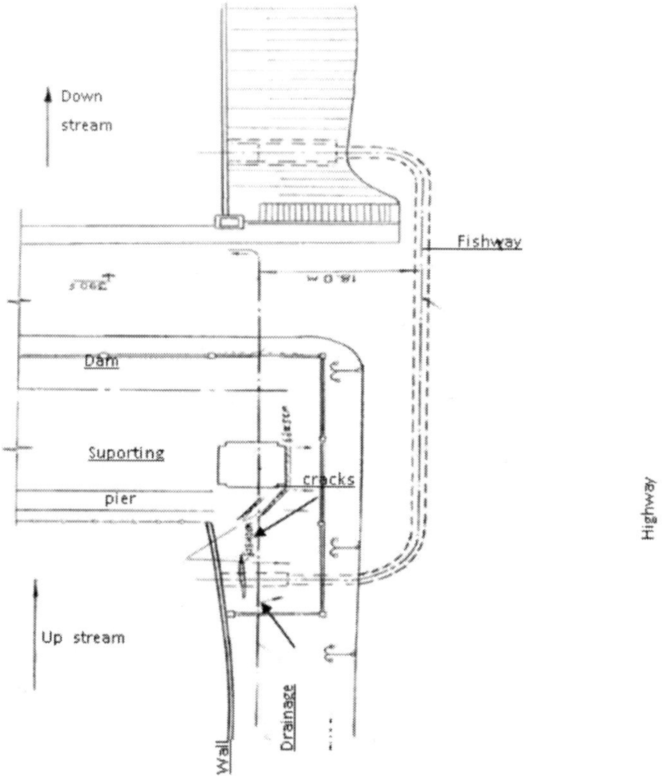

Fig. 1. Cracks developed in Ortachala HPP right bank.

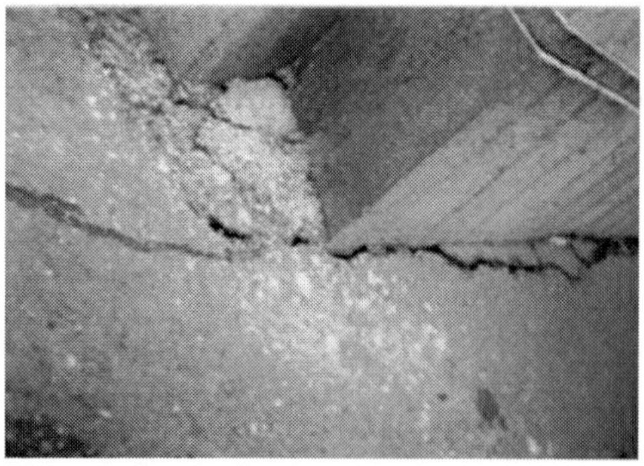

Photo 1. Cracks developed in Ortachala HPP right bank.

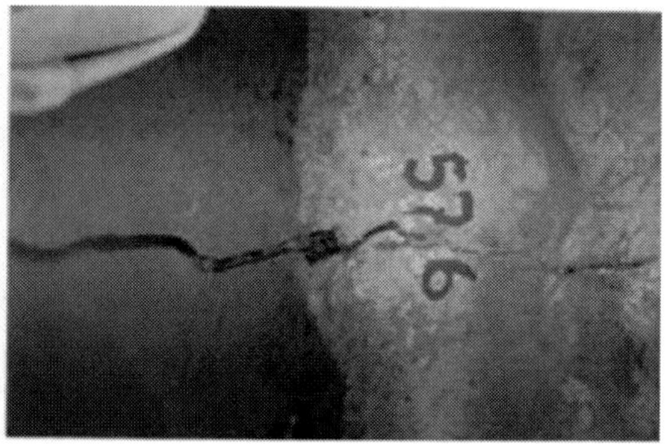

Photo 2. Cracks developed in under ground fish-way.

2 Mathematical model of seepage-deformation proses

Seepage regime and deforamtions, developed in water bank, are interconnected processes. Exact estimation of the task requires seepage and static calculation review of the "fish-way – drainage – foundation" system.

Generally, pressure-free seepage task is reviewed. Above mentioned considers depression surface unknown outline and seepage flow parameters definition. Calculations are based on second row differential equation solution considering anisotropic massive:

$$\frac{\partial}{\partial x}\left(K_x\frac{\partial H}{\partial x}\right)+\frac{\partial}{\partial y}\left(K_y\frac{\partial H}{\partial y}\right)+Q=0 \quad (1)$$

where: H – is a pressure, K_x and K_y are hydraulic conductivity in the direction of x and y axis; accordingly, Q – is seepage flow discharge.

On base of presented differential equation, with finite element method usage, in condition of system seepage flow total potential energy minimization, task shall be solved. Calculations were held with iteration scheme (Fig. 2). By choosing step, high degree of convergence shall be achieved.

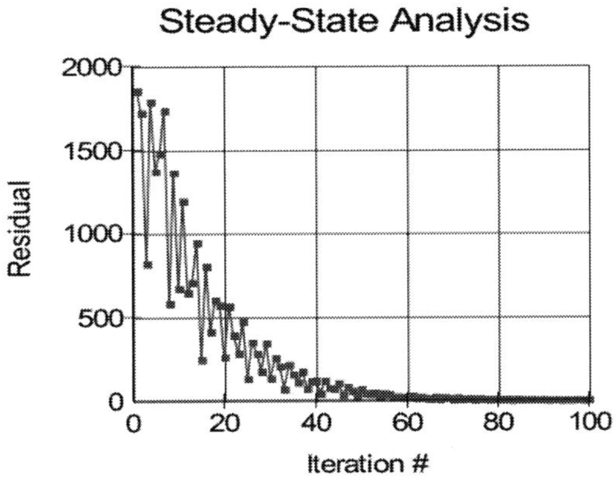

Fig. 2. Cycle - iteration scheme of calculation.

Fish-way, that is totally located in the embankment, in the process of operation was compacted with soil. In the result, along it's longitudinal axis, cross cracks were developed in the construction (especially in the bottom), that itself caused seepage. Generally, soil seepage strength condition has it's faces:

$$J < J_{cr}$$

where: J and J_{cr} are seepage gradients calculation and critical meanings.

According to the calculation results, it was acknowledged that maximum meaning of the seepage flow gradients developed from cracks, is 1.5 (Fig. 3). From this point starts disturbance of the seepage steadiness and beginning of suffusion.

Suffusion development causes pore increase in soil. Suffusion development model in time was accepted. Connection of the deformation modulus (E) with pores (n) function is presented with using decrease regularity curve $E_y = f(n)$ (Fig. 4).

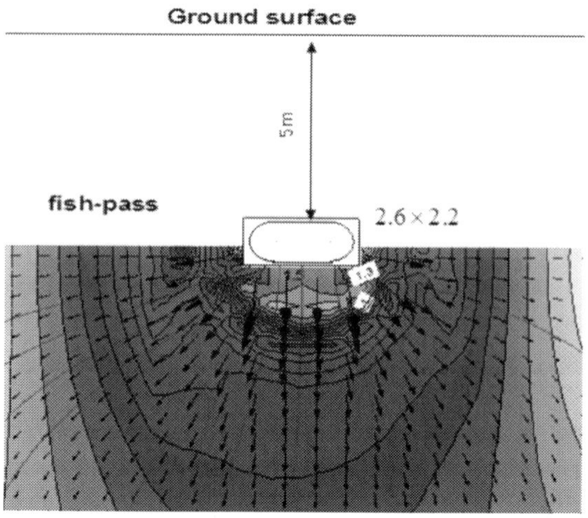

Fig. 3. Seepage gradients in fish-way foundation.

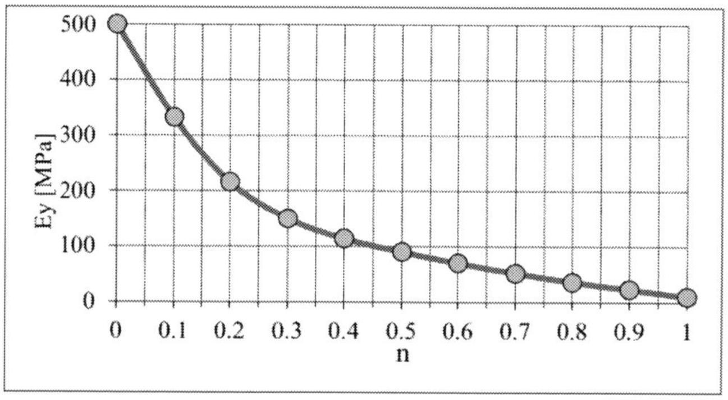

Fig. 4. $E_y = f(n)$ - reletionship between elastice modulus and pores.

3 Deformation prosess in right bank foundation

Estimation of deformation development processes in time in bank foundation was held with statistic calculation (cyclic-iteration sequence) using soil operation non-liner model [2]. Square is parametrical elements was used as an finite element mesh measurer. In zone of tensile zone and high compression, also in high seepage gradient zone element gains anisotropic properties. Anisotropic axis in element are orientated according to the main stress (Fig. 5).

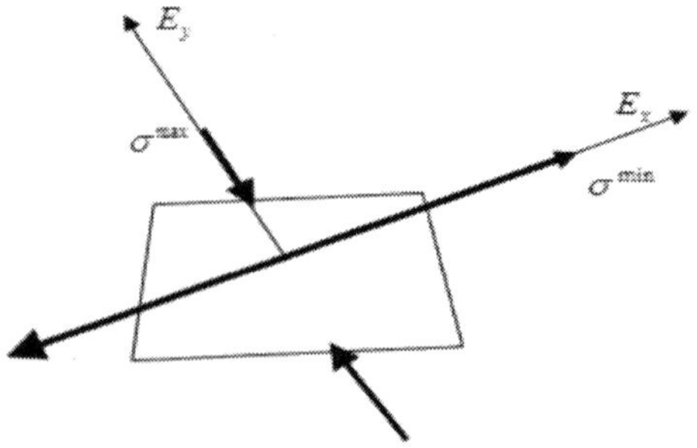

Fig. 5. Finite element anisotropic model.

Deformation modulus values are corrected in the following way:

- in compressions direction - $\sigma = f(\varepsilon)$ according to the curve;
- in zone of high seepage gradients, according to the $E_y = f(n)$ curve (in the fish-way foundation)
- towards tensile stress direction - $E_x \approx 0$, when $\sigma \succ R$, where R – generally, soil strength on expansion (zone of cracking behind concrete wall).

Calculation results showed that change of deformation modulus in fish-way foundation influences deformation development of bank (Fig. 6, 7). Under fish-way, in zone of high gradients, developed suffusion increased its settlement and cracks opening in fish-way construction. This causes increase suffusion. This means that crack development and suffusion causes inter-progressive development.

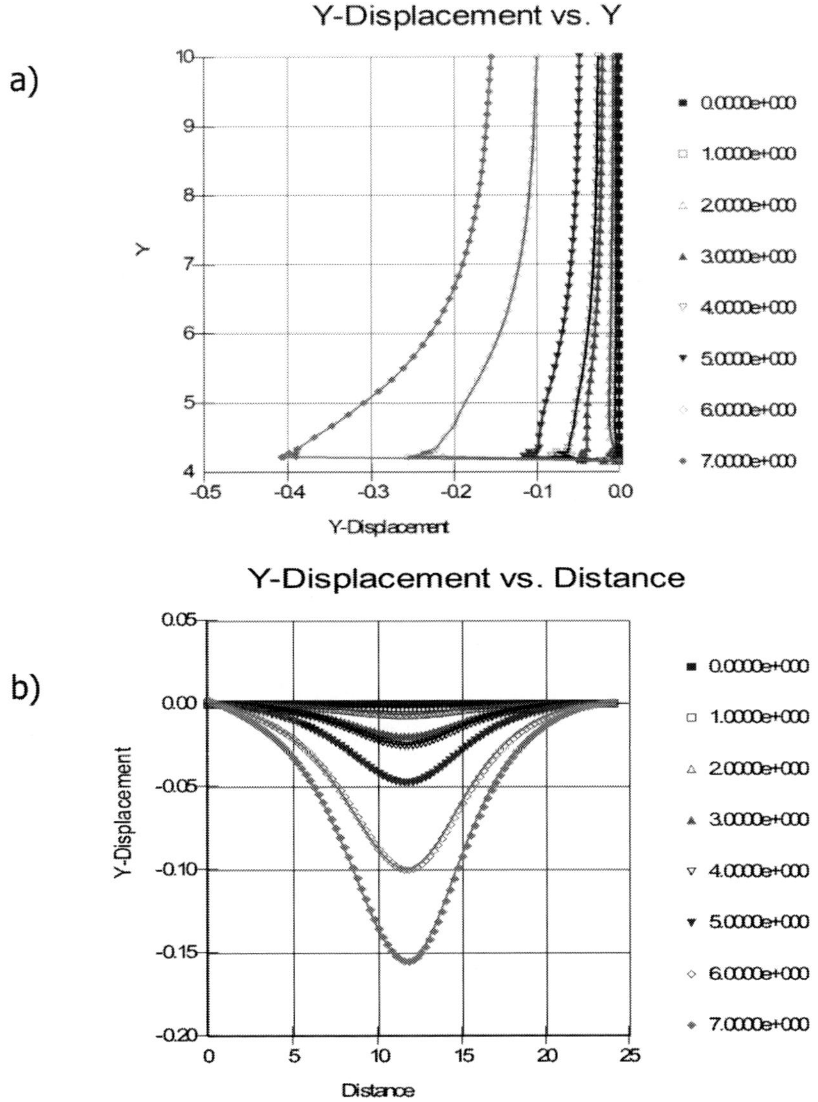

Fig. 6. Settlement developed in fish-way structures cross section:
a) in depth of embankment; b) embankment surface.

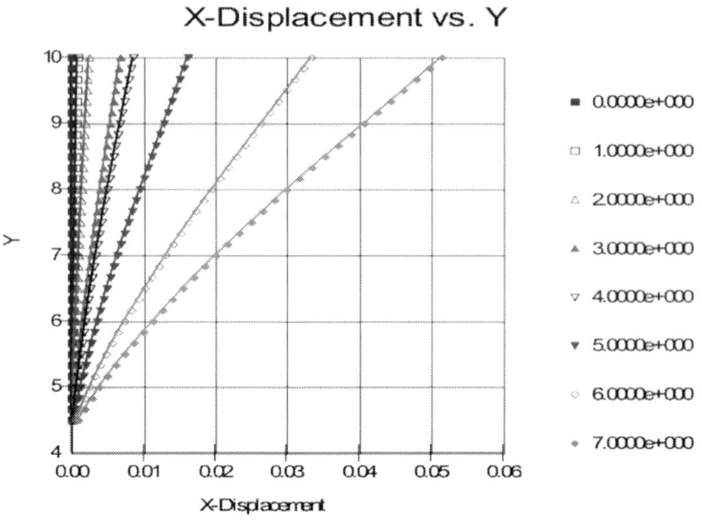

Fig. 7. Behind the wall crack development dynamic in depth of embankment.

Fish-way settlement (Fig. 6), developed caused soil replacement and gap open development in horizontal direction next cracking in bank foundation (it spread in depth down to 5.4 m, Fig. 7).

Within the cracks developed depth behind the concrete wall is loosing support, that causes increase tensile stress on its pressure surface. This itself causes lost of strength bearing capacity. Currently, bearing capacity is almost out. Increase of depth crack in soil will cause crack development in the wall herewith, hydraulic break-through and groundwater table rise

4 Seepage parameters analasys in the right bank foundation

Seepage parameters study were carried out in bank foundation using "fish-way – drainage – foundation" system. Basic rock mass is presented with tuphogenic sand, but the bank foundation with 12 m capacity embankments with fish-way and drainage located there.

Hydraulic conductivity normative values (using the reference books and analogs) are:

- basic massive $K_m \approx 10^{-4}$ cm/sec;
- embankment $K_E \approx 1$ cm/sec.

Hydraulic conductivity values was corrected during the iteration calculations. It was accepted that groundwater table elevation, in correspondence with piezometer measuring results, is satisfied for the $K_E / K_m \approx 4.5 \bullet 10^5$ value (Fig. 8).

Ratio K_0 / K_m^{5} is quite big, consequently, basic massive and embankment contact is presented practically as a seepage outlet section. Groundwater table reaches to the highest elevation near the right bank wall, farther it slowly goes down.

On the first stage of the calculation, drainage wasn't foreseen. Drainage operation was studied separately, using classical solutions and finite elements method [3];

Lowered pressure in drainage gallery cross, made according to the classical solution, is defined with the formula [4]:

$$\varphi_b = \frac{1}{1 + \frac{N \cdot r_c}{\pi}\left(\frac{1}{R+b} - \frac{1}{2t}\right)}$$

where: N – quantity of bores in the drainage galleries, r_c – bore radius, R – radius of drainage galleries, b – half distance between existing bores, t – distance from the drainage center to the groundwater table elevation. $\varphi_b = 0.96m$ as calculated.

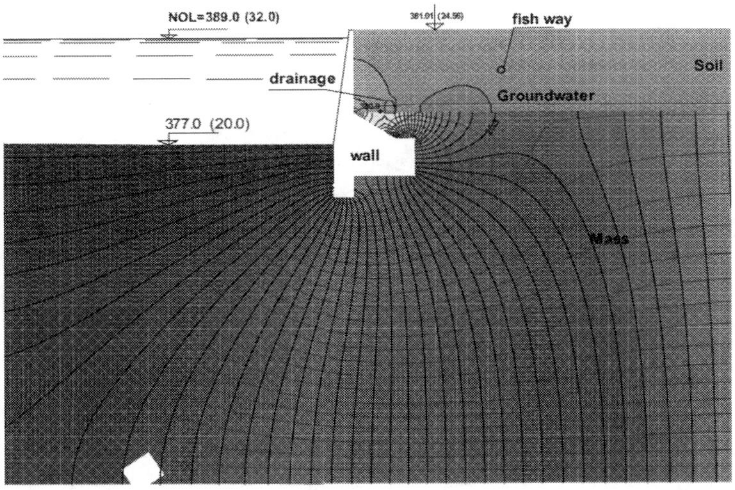

Fig. 8. Seepage head distribution in the "Fish-way - drainage - foundation" system.

By using finite element model, was defined that depression surface lowers till 100 cm in drainage cross (Fig. 9) that is in good compliance with the classic solution results.

Hence seepage is provided within the basic mass (tuphogenic sand rocks) and determines groundwater table level of the bank, and drainage is incapable to change seepage regime. It's maximum effectiveness can be estimated by the decrease of the seepage pressure on the bank wall not exceeding 1 m. So drainage is actually ineffective.

Seepage gradient distribution along the wall tooth and its vertical surface characterizes significantly unequal and maximum meaning does not exceed 0.6 sm.

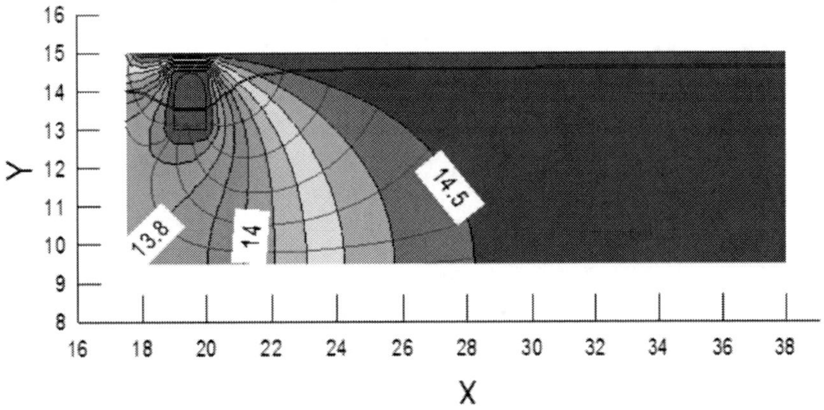

Fig. 9. Seepage head distribution in drainage.

5 Conclusions

1. Mathematical models – "fish-way foundation", "retaining wall - drainage - foundation" was elaborated by the connected static and seepage scheme, that showed developed deformations and seepage regime in right bank foundation considering drainage in Ortachala HPP.
2. Because of the foundation basic mass and embankment hydraulic conductivity ratio high value, their contact surface appears as a seepage outlet section. So rock mass - embankment contact surface defines the groundwater table elevation in bank foundation.
3. Drainage can't influence seepage regime (groundwater table elevation) development in the bank foundation. Drainage influence is spread only near the wall and decreases pressure on it within 1.0 m. Thus, drainage can't be considered as a right decision for projecting.
4. Fish-way, that is located in the embankment undergo settlement, that was followed by crack development and seepage. Suffusion developed under the fish-way, in the zone of high gradients, increased its settlement and opened the cracks. This means, crack development and suffusion causes inter-progressive development.
5. Settlement, developed in fish-way structure lining, caused soil replacement towards horizontal direction relative to concrete retaining wall and depth crack development near the wall. Above mentioned increases tensile stress in wall and causes lost of strength bearing capacity. In the result, crack development in wall, will causes hydraulic break-through and groundwater table elevation rise in embankment - all inconvenient results connected to it.
6. Having established the cause of these processes, the fish way was stopped, and by result of which the fractures were stopped. So fish-way internal surface hydro isolation is necessary as a rehabilitation measure.

References

1. L. J. Segerlind, *Applied Finite Element Analysis* (2nd Ed. John Wiley & Sons Inc., New York, 1984)
2. O. C. Zienkiewicz, R. L Taylor, *The Finite Element Method* (5th Ed. Vol. 1. Butterworth-Heinemann, Oxford, 2000)
3. P. Novak, A. I. B. Moffat, C. Nalluri, R. Narayanan, *Hydraulic Structures* (4th Ed. Taylor & Francis, London/New York, 2007)
4. *Hydraulic Structures* (Reference book, Moscow, 1983)

The influence of pipe diameter selection on operating costs of heating installation in the context of the anticipated increase in electricity prices

Maciej Knapik[1,*]

[1]Institute of Thermal Engineering and Air Protection, Faculty of Environmental Engineering, Cracow University of Technology

Abstract. The article presents an analysis of the influence of selection a pipe diameter on the operating costs of a heating installation in the context of the expected increase in electricity charges required to motive the main circulation pump of the installation. As a result of a combination of several factors affecting the Polish energy sector based on coal, energy price increases seem inevitable. Polish government's response to the forecasted increases in charges for electricity is a proposal to introduce subsidies for electricity bills. The article presents the results of the analysis of pipe selection from several standard types, including investment and operating costs, and a comparison of results before and after the increase in electricity charges and including the influence of limescale precipitation on the internal surfaces of pipe walls. The results analyzes show that the increase in charges for electricity affects the selection of the pipe diameter, like the problem of the formation of limescale can significantly reduce the efficiency of the installation and in extreme cases, can lead to a complete stop of installation.

1 Introduction

In 2018, in Poland through media reports [1–3], speculations have started on the rise in electricity prices in 2019, which according to various sources can range from 10 to 80%. The Polish economy and in particular professional energy branch is based on coal. In 2015, 85% of electricity generated in Poland came from coal. Therefore, as a reason for the increase in electricity prices in Poland are the price increase in the global coal market, the increase in prices of CO_2 emission allowances, and also the costs of buying certificates related to the obligation to support renewable energy sources.

The increase in electricity prices [4–6] will effect on every area of the economy as well as daily life, including industry and residential installations powered by electricity. In order to reduce the influence on daily life of electricity price increases, the Polish government is analyzing and considering introducing a system of subsidies for electricity bills.

*Corresponding author: maciek.knapik@gmail.com

On the example of a heating installation, where the main circulation pump of the installation, powered by electricity, during the heating season (usually from September to April), works almost continuously in this time, which results in the amount of electricity collected from the network, it achieves significant amounts, and in the context of increased operating costs, it can have a significant influence on the design process of a given system.

On the amount of electricity collected from the network by the circulation pump has several factors such as: flow of the heating medium, resistance of the installation, size of the pump, efficiency and the working time of the device. The main factor that generates operational costs is the flow resistance, which depends on the type of pipes used in the project, their diameters, fittings and the flow resistance of the receivers.

The article presents an analysis of the selection of pipes for heating installations [7, 8] in the context of the expected price increases for electricity, taking into account possible complications (precipitation of limescale on the internal walls of pipes) created during the operation of a heating installation.

2 Assumptions for the case analysis

In order to analyze the selection of heating pipes in the context of the increase in electricity prices, should be adopted certain assumptions, which are presented in Table 1.

Table 1. The basic assumptions of the case study (study of Author).

Assumptions	Numerical data	Units
Installation power	150	kW
Temperature parameters of the heating medium	80/60	^{0}C
Design flow	6.6	m^3/h
The duration of the heating season	200	days
Analyzed length of the installation (summary)	400	m
Cost of kWh before the increase electricity prices	0,55	PLN
The assumed increase in electricity prices	40	%
Cost of kWh after the increase electricity prices	0.77	PLN

According to the Polish standard [9], the designer of the heating installation in the selection of heating installation pipes should be guided by three criteria: the flow rate of the heating medium in the pipe, which should to be in the range from 0.1 to 1.0 m/s, linear pressure drops, no greater than 150 Pa/m and also an economic issue. For the assumed heating power of 150 kW, with the parameters 80/60°C, the design flow is 6.6 m^3/h.

Copper and PP (polypropylene) pipes are characterized by the highest investment costs, but from other side, steel pipes are characterized by a lower coefficient of roughness, thanks to which these, copper and PP pipes have greater resistance to limescale deposits. Steel pipes are cheaper, but these pipes that have a greater roughness, which supports the deposition of limescale and reduce the inner diameter of the pipe. As a consequence of this process, during the heating installation operation, the pressure losses increases and it also increases the total costs of system operation. The results are summarized in Fig. 1.

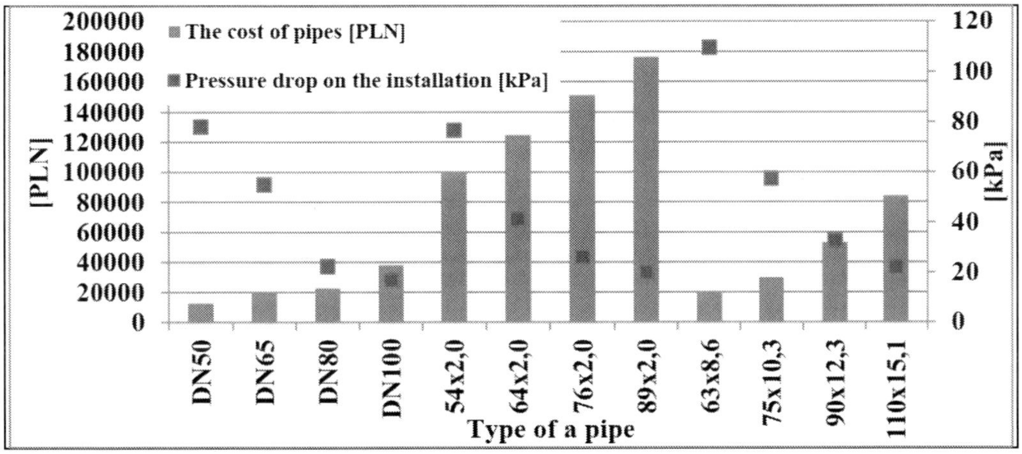

Fig. 1. Summary of investment costs with the total pressure drop of installation, designed on the type and diameter of the pipe (study of Author).

3 Analysis of the pipe selection in the context of rising the electricity prices

The operating costs of a heating installation depends on the type of the main circulation pump and the type of boiler room and used fuel [10, 11]. The process of selecting the pump is based on the design flow of the heating medium and the pressure drops on the critical cycle. The higher values reach the indicated parameters, the higher amounts of the electric energy is needed to motive the pump and the size of the pump increases as well as the investment costs. Based on the initial assumptions from the Table 1 as the calculation flow of the heating medium of 6.6 m^3/h, and pressure drops on the system depending on the type of pipe used, several variants of pumps were selected, based on the Grundfos technical catalog [12] (for the time of operation the manufacturer gives a period of 15 years). The results are summarized in Fig. 2.

The obtained results of the selection of pumps and their nominal power (Fig. 3) coincide with the results of pressure drop in the installation (Fig. 1). The higher the pressure drop, the higher the pump power.

After taking into account an increases in electricity prices (Table 2), the type of selected pipe has not changed. It is still a steel pipe, but its diameter has changed.

In this case, the most profitable in economic terms, there has been a change to a pipe with two larger dimensions (DN80 steel pipe), where the total cost amounts 30880.22 PLN. The results are summarized in Fig. 3.

Fig. 3 presents the expected increase in the cost of operating a heating installation in the case of a 40% increase in electricity prices. The largest increases in costs were observed on an ex ample of smallest diameters of pipes of a given type.

The obtained results may be influenced by such factors as a constant increase in the costs of the energy, which was not taken into account in this analysis, as well as losses resulting from the duration of operation, of system like the precipitation of limescale on the inner wall surfaces of pipes.

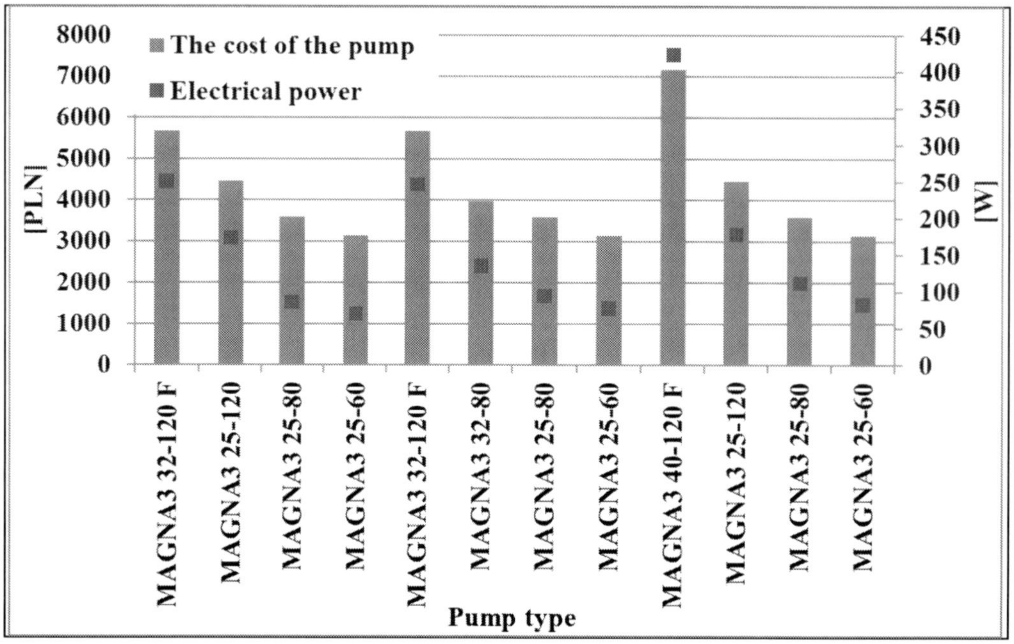

Fig. 2. Summary of the costs of the pump and the power of the device under nominal conditions (study of Author).

Table 2. Summary of the amount of energy consumed by pumps during the heating season, operating costs and total costs before and after the increase of electricity prices.

Type of a pipe	Electricity consumption [kWh]	Operating costs I [PLN]	Operating costs II [PLN]	Total costs I [PLN]	Total costs II [PLN]
Steel DN50	1204.8	9939.6	13915.44	28237.21	32213.05
Steel DN65	830.4	6850.8	9591.12	31068.78	33809.1
Steel DN80	408	3366	4712.4	29461.82	30808.22
Steel DN100	336	2772	3880.8	44122.77	45231.57
Copper 54x2.0	1180.8	9741.6	13638.24	115115.20	119011.9
Copper 64x2.0	648	5346	7484.4	134087.70	136226.1
Copper 76x2.0	456	3762	5266.8	158493.80	159998.6
Copper 89x2.0	374.4	3088.8	4324.32	182755.60	183991.1
PP 63x8.6	2030.4	16750.8	23451.12	43625.09	50325.41
PP 75x10.3	859.2	7088.4	9923.76	41438.38	44273.74
PP 90x12.3	537.6	4435.2	6209.28	61139.02	62913.1
PP 110x15.1	398.4	3286.8	4601.52	90349.57	91664.29

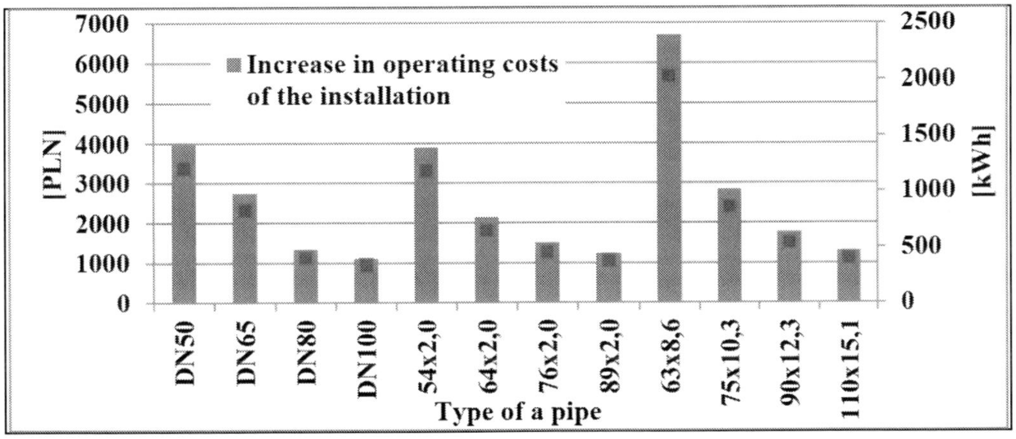

Fig. 3. Increase in operating costs of the heating installation in relation to the base state and the amount of energy consumed by the pump depending on the selected diameter and type of pipe (study of Author).

4 Influence of limescale on the operation of pipes

The steel pipes are characterized by the highest roughness factor in the scope of the analyzed pipes. Roughness is the mechanically perceptible surface irregularities resulting not due to its shape but the nature of their processing. The greater the roughness factor of the pipe, the greater it is their potential to precipitate on the inner wall surface of the pipe of the limescale. According to the principle of electrostatic interaction, particles that have already deposited on the inner wall surface, it will attract the next particles leading to sediment growth. The deposition of limescale decreases the inner diameter of the pipe, and consequently increases the pressure losses and operating costs of the heating installation.

The limescale [13, 14] is a layer of sediment of calcium carbonates ($CaCO_3$) and magnesium ($MgCO_3$), formed as a result of the thermal decomposition of calcium and magnesium bicarbonates contained in water, especially in hard water. As a result of the heating process, the bicarbonate ions HCO^{3-} to carbonate CO_3^{2-} decompose. At the time of decomposition, the most difficult soluble carbonates then precipitate in the form of a precipitate (limescale) containing mainly $CaCO_3$ (solubility 14 mg/l) and $MgCO_3$ (solubility 106 mg /l) as shown by the reaction equation:

$$Ca(HCO_3)_{2(aq)} \rightarrow CaCO_{3(s)} \downarrow + H_2O_{(l)} + CO_{2(g)} \uparrow \quad (1)$$

The higher the hardness of water [10] (it is a water feature as a function of the concentration of calcium cations (Ca^{2+}) and magnesium (Mg^{2+}), and optionally iron (II) (Fe^{2+}) and manganese (II), (Mn^{2+}), and the higher the temperature parameters of the heating medium, the phenomenon of the precipitation of limescale becomes more intense.

Preventing the formation of limescale in water systems usually consists of the use of water softeners or demineralization, but not everywhere these treatments are used [15–18].

The growth of the limescale on the inner surface of the pipe reduces the internal diameter of the pipe by twice the thickness of this layer, so that the thickness of the scale layer of 1 mm will reduce the internal diameter of the pipe by a total of 2 mm. Fig. 4 presents the results of changes in total and linear pressure drops depending on the thickness of the limescale layer in the analyzed steel pipes.

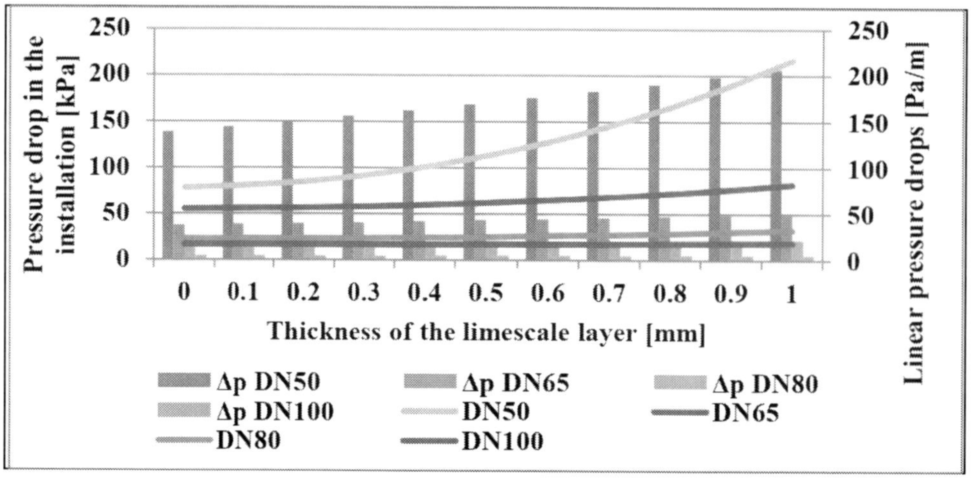

Fig. 4. The value of linear and total pressure drops of steel pipes depending on the thickness of the limescale in heating installation (study of Author).

In the case of a steel pipe with a diameter of DN50, thickness of the scale of 1 mm, results in a reduction of inner pipe diameter by 2 mm and in consequence linear pressure drop are increased by almost 50%. The effect of limescale growth on pressure drops in the analyzed installation is presented on Fig. 4.

The growth of the limescale will increase the operating costs of the heating installation, due to the increase in total pressure drops in the installation. In several cases, the installation will not work properly, because the selected pump (Fig. 2) is not able to meet the specified requirements (the flow can be increased at the expense of the pump pressure, however the nominal power of the heat receiver will be significantly reduced).

On Fig. 5 for diameters DN50 and DN65, for the limescale layer thickness of 0.5 mm, for the MAGNA3 pumps, which were initially selected, the pressure drop exceeds the pump characteristic, which means a drop in efficiency, therefore the installation becomes ineffective, The nominal heat power of installation is reduced. In the case of the phenomenon of limescale deposition, the initial selection of the pipe, for economic reasons – the diameter of the DN50 steel pipe, during the operations turns out to be the wrong selection, the designer should be guided by a larger number of decision criteria. For diameters DN80 and DN100, differences in pump power values are minimal, therefore in further analysis only DN80 and DN100 steel pipes will be taken into consideration, because as mentioned above, for DN50 and DN65 diameters, pressure drops are too high, as a consequence of which the installation becomes ineffective.

Fig. 6 presents the results of operation costs of the installation based on DN80 and DN100 steel pipes and nominal power of the pump, taking into account the limescale deposits on the inner walls of the pipes.

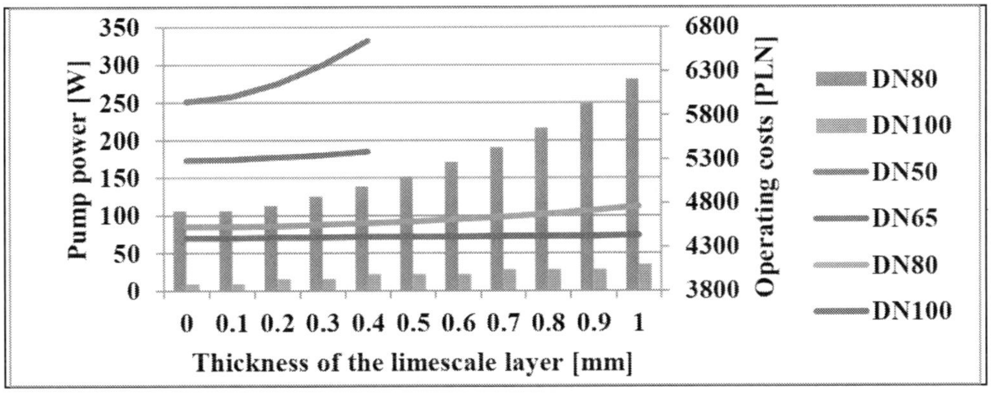

Fig. 5. The nominal power of the pump and operating costs of a heating installation designed on steel pipes depending on the thickness of the limescale (study of Author).

For a DN80 steel pipe, operating costs, over 15 years of operation of the heating installation, under influence of limescale, will increase by about 32%, and in the case of a DN100 steel pipe costs will increase by about 6%.

Taking into account all the analyzed aspects such as the increase in electricity prices and inclusion of limescale in steel pipes, obtained results of the analysis are presented on Fig. 6.

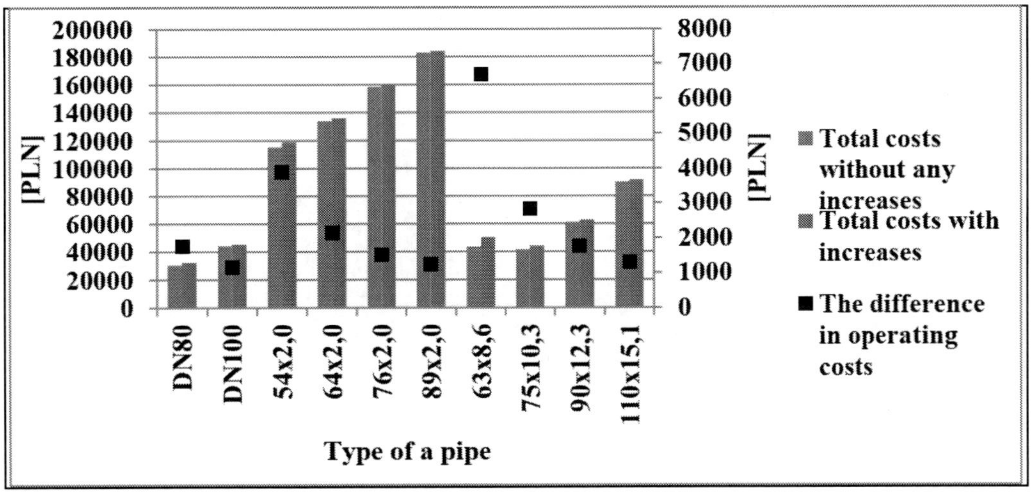

Fig. 6. Comparison of operating and total costs of heating installation before and after the prices increase for electricity (study of Author).

The DN50 and DN65 steel pipes were omitted, because taking into account the limescale, a new pump should be purchased or the pipes replaced to a new ones, which generates additional investment costs.

The smallest total costs, despite the expected price increases for electricity, are characterized by DN80 steel pipes, taking into account the formation of limescale, which amount depends on the water hardness. The largest total costs are characterized by copper pipes with a diameter of 89 x 2.0.

The smallest difference in operating costs before and after the increase in electricity charges was obtained for the DN100 steel pipes and the largest for PP pipes with a diameter of 63 x 8.6.

5 Summary and conclusions

In connection with the planned increase in electricity costs, branch designers should take a broader look at the selection of pipes in the installation. As indicated by the analysis, not always following the guidelines indicated in the standard [9], leads to economically advantageous solutions. Lower investment costs do not always mean that the installation will be economically viable, by increasing operating costs.

On the example of the analysis of the selection of steel, copper and PP pipes, which allow to estimate that the price increase for electricity, which is used to motive a pump for heating installations, causes pipes that will be characterized by lower linear pressure drops over 15 years operation will be more economically (Fig. 6).

It should be known that the selection of pipes depends on the user with the support of the designer. Copper and PP pipes are more expensive (Fig. 1), but they have a lower coefficient of roughness, thanks to which they are more resistant to the formation of limescale on the surface of the inner walls of pipes. This phenomenon is more common in the analyzed steel pipes, which causes the reduction of the pipe inner diameter and as consequence it will increase installation pressure drops and the increase in operating costs of the system (Fig. 5). In extreme cases it will reduce nominal power of installation or it will completely stop the installation.

On the example of this analysis for installations characterized by a flow of 6.6 m^3/h (150 kW), for a 400 m pipe (sum of supply and return), the increase in electricity costs increases the operating costs of the system over 15 years from around 1200 PLN up to 6700 PLN, depending on the type of selected pipe (Fig. 6).

The influence of limescale on the operation of the installation is presented on the background of steel pipes. The precipitation of a 1 mm thick layer of limescale reduces the pipe inner diameter by 2 mm. Such a state of affairs based on the example of DN50 and DN65 steel pipes shows that the installation does not meet its requirements, works inadequately, inefficiently and uneconomically (Fig. 5).

For steel pipes DN80 and DN100, the precipitation of the limescale will accordingly increase the pressure loss on the system by 32% and 6% (Fig. 5).

The current trend in construction and installation branches is to reduce the energy demand of buildings by receiving eco-friendly certificates, obtaining passive house standards or low energy demand [19–25]. Therefore, the situation related to the anticipated increase in electricity prices is a good moment to verify project guidelines aimed at finding optimal solutions that ensure proper system operation, while maintaining the economic criterion during the operation of a given system and as well as reducing energy demand.

References

1. Ceny prądu w 2019 r. Jak duże podwyżki cen energii elektrycznej czekają nas w przyszłym roku? [online], Available at: https://gazetalubuska.pl/ceny-pradu-w-2019-r-jak-duze-podwyzki-cen-energii-elektrycznej-czekaja-nas-w-przyszlym-roku/ar/13596866, (Accessed: 30th January 2019)
2. Ceny prądu 2019. Podwyżki cen prądu w 2019 roku. Ile zapłacimy za prąd w 2019? [online], Available at: https://polskatimes.pl/ceny-pradu-2019-podwyzki-cen-pradu-w-2019-roku-ile-zaplacimy-za-prad-w-2019/ar/13721250, (Accessed: 30th January 2019)
3. Ceny prądu w 2019 w roku - będą podwyżki? [online], Available at: https://enerad.pl/aktualnosci/ceny-pradu-2019/, (Accessed: 30th January 2019)
4. J. Liang Zhang, Y. Jun Zhang, D. Zhi Li, Z. Fu Tan, J. Fei Ji, Int. J. Elec. Power **105**, 541–548 (2019)

5. G. Müller, A. Seibert, Energy Econ. **78**, 267–277 (2019)
6. H. Shavandi, M. Pirnia, J. D. Fuller, Appl. Energ, **240**, 251–264 (2019)
7. O. Ibrahim, F. Fardoun, R. Younes, H. Louahlia-Gualousc, Build Environ. **72**, 259–286 (2014)
8. K. C. Kavvadias, A. P. Tosios, Z. B. Maroulis, Energy Convers. Manag. **51**, 833–845 (2010)
9. Polish standard, PN-EN 12:831 – 2006: *Instalacje ogrzewcze w budynkach – Metoda obliczania projektowego obciążenia cieplnego*
10. G. Martinopoulos, K. T. Papakostas, A. M. Papadopoulos, Energ. Buildings **124**, 79–87 (2016)
11. M. Knapik, RI **9**, 36–38 (2016)
12. Grundfos | The full range supplier of pumps and pump solutions. [onlnie], Available at: https://www.grundfos.com/, (Accessed: 30th January 2019)
13. Kamień kotłowy – Wikipedia, wolna encyklopedia [online], Available at: https://pl.wikipedia.org/wiki/Kamień_kotłowy, (Accessed: 30th January 2019)
14. D. Dobersek, D. Goricanec, Math. Models Methods Appl. Sci. **1**, 55–61 (2007)
15. Twardość wody. [online], Available at: https://pl.wikipedia.org/wiki/Twardość_wody, (Accessed: 30th January 2019)
16. K. Konieczny, S. Tanżyna, J. Marjanowski, M. Kosorz, Instal **12**, 18–21 (2006)
17. J. Listewnik, O. Bezioukov, A. Nowik, *Ultradźwiękowa metoda walki z tworzeniem się kamienia kotłowego lub innych osadów w urządzeniach cieplno-energetycznych* (Prace Naukowe Instytutu Techniki Cieplnej i Mechaniki Płynów Politechniki Wrocławskiej, Konferencje, **10**, 120–127, 2000)
18. M. Mulik, M. Bożym, RI **5**, 84–88 (2017)
19. M. Knapik, E3S Web Conf. **17**, 00039 (2017)
20. M. Knapik, *Budownictwo proekologiczne – zrównoważony rozwój* (I Ogólnopolska Konferencja Naukowa OSA – Odpady, Środowisko, Atmosfera, 5–6 czerwca 2014, materiały pokonferencyjne, 87–92, 2014)
21. M. Knapik, Czas. Techn. **9**, 137–150 (2017)
22. M. Knapik, E3S Web Conf. **44**, 00069 (2018)
23. M. Knapik, RI **11**, 22–30 (2017)
24. C. Thormark, Build. Environ. **37**, 429–435 (2002)
25. I. Sartori, G. Hestnes, Energ. Buildings **39**, 249–257 (2007)

Critical analysis of decay rate measurement methodology

Łukasz Kokurewicz[1], *Mateusz* Muszyński-Huhajło[1], *Stanisław* Miodoński[1,*], and *Maciej* Leoniak[1]

[1]Faculty of Environmental Engineering, Wroclaw University of Science and Technology, Plac Grunwaldzki 9, 50-377 Wroclaw, Poland

Abstract. In modeling studies, estimation of microorganisms kinetic parameters set is a key element for proper model operation and predictability. Nitrification process is very often, a crucial element of the wastewater treatment systems as bacteria responsible for ammonium and nitrite oxidation are slow growing microorganisms, making whole nitrification process vulnerable to external factors i.e. temperature, inhibition and load fluctuations. Growth and decay rate of nitrifiers decide about amount nitrifying biomass in the wastewater treatment plants, thus the nitrification efficiency. Paper presents analysis of the decay rate (b_a) estimation methodology based on respirometric assays measuring the oxygen uptake rate (OUR). Evaluation of this simple and cheap method was made based on decay estimation tests performed on sludge samples from side-stream partial nitritation reactor treating reject water from digested sludge dewatering. Database obtained from these tests were analyzed to evaluate the impact of respirometric assay duration on calculated decay rate values. 11 time ranges were selected for the performed analysis. Calculated b_a values were compared showing the optimal test duration between 5–6 hours, while test shorter than 2 hours resulted in unsatisfactory b_a outcome.

1 Introduction

1.1 The idea of decay rate

In activated sludge models (ASMs), decay rate (b_a, d^{-1}) is a part of the function describing net microbial growth rate presented in equation 1 [1]. In case of autotrophic, nitrifying biomass, a simplified version (omitting inhibition and operational conditions) of this function describes the relationship between the biomass yield coefficient (Y_A, g COD/g NH_4-N), decay rate, solids residence time (SRT, d) and the oxidized ammonium load ($NH_{4,ox}$, g NH_4-N/d) and relates those parameters to amount of biomass produced daily:

$$X_A = Y_A \cdot \frac{1}{1+b_A \cdot SRT} \cdot NH_{4,OX}, \text{ [g COD/d]} \quad (1)$$

* Corresponding author: stanislaw.miodonski@gmail.com

where:
Y_A – nitrifiers yield coefficient, g COD/g NH_4-N
b_A – nitrifiers decay rate, d^{-1}
SRT – solids residence time, d
$NH_{4,OX}$ – oxidized ammonium load, g NH_4-N/d

Early versions of ASM models assumed the decay rate as integrated coefficient representing many different processes. Recent versions of these models (ASM3) was developed including number of components as presented in Figure 1. Main process included in this coefficient represents cell death in time due to natural life cycle. Other processes included in this parameter are: endogenous respiration, cell lysis, motility and predation due to protozoa activity [2].

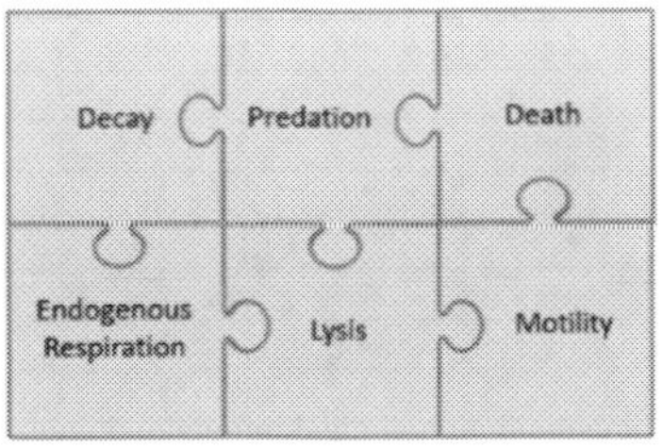

Fig. 1. Processes included in decay processes in oxic conditions.

Determination of the decay rate coefficient requires proper modification in the activated sludge operational regime. To measure the rate of nitrifying biomass loss, its growth must be minimalized. Due to these requirements, during the decay rate estimation test, activated sludge sample is kept in starvation conditions without substrate addition. The only source of substrate available for the bacterial growth are products obtained due to biomass decay processes. Next, the change of process kinetics under starvation conditions is investigated in specified time intervals to evaluate the nitrifiers loss.

Duration of the starvation period may vary depending on reports and was in range from 1 to 30 days [3, 4]. Characteristics of the decay rate coefficient, which is related to daily change, induces required intervals between successive process rate tests at 24 hours. Literature reports considering decay rate in aerobic conditions indicates that nitrifiers mass drops to 20% and 17% of the initial mass after 10 and 17 days of starvation, respectively [4, 5].

Based on available reports, analysis of published data about nitrifiers decay rate tests in aerobic conditions was performed and summarized in Table 1. In each of reviewed studies, decay rate was determined for both groups of nitrifiers or separately for each group of nitrifiers separately (ammonia oxidizers, AOB or nitrite oxidizers, NOB) if such data was available. Except operational conditions of each test (pH and temperature) also evaluation method of process kinetics and starvation period length was highlighted.

The lowest decay rate was significantly different from other reports and was equal 0.02 d^{-1} [6]. However some explanations can be found in the methodology of that test which was performed on pure cultures of nitrifiers feed with synthetic wastewater. Such conditions are significantly different from observed in activated sludge thus observed decay rate can be lower due to no protozoa grazing or neglected impact of toxic substances causing bacterial death or enhancing cell lysis. Maximal coefficient rate was 0.43 d^{-1} observed in *Nitrosomonas europea* (AOB) also cultivated in laboratory conditions [7]. Range of decay rate coefficient observed in conditions similar to full-scale activated sludge systems (mixed-cultures) was between 0.13 and 0.4 d^{-1}. Selected method of measuring biomass activity probably also had impact on the final decay rate result, however in reports where oxygen uptake rate was used the spread of the results was smaller.

Table 1. Values of decay rate in the literature (measurement method: OUR – respirometric assay, NB – nitrification rate test, Model – mathematical model calibration). Test performed on pure cultures signed with (*).

AOB	(AOB+NOB)	NOB	pH	Temp	Method	Starv. period	Literature
d^{-1}	d^{-1}	d^{-1}	-	°C	-	d	
	0.19 ± 0.05		7.3–7.6	12	OUR	7 day	Siegrist et al. (1999)
	0.21 ± 0.05		7.3–7.6	20	OUR	7 day	Siegrist et al. (1999)
	0.153 ± 0.02		7.7 ± 0.2	20	NB	9 day	Lee and Oleszkiewicz (2003)
	0.43		7.5–7.8	28	OUR		Nowak et al. (1994)
	0.40			30			Slazer (1992)
0.43*			7.4	30	OUR	14 day	Leenen et al. (2000)
	0.05			10	Model		Henze et al. (2000)
0.02 ± 0.002*		0.08 ± 0.000*	7.5	20	NB	30 day	Salem et al. (2005)
0.20 ± 0.016		0.21 ± 0.024	7.5	20	NB	30 day	Salem et al. (2005)
0.35			7.3	20	NB	14 day	Oleszkiewicz (2011a)
0.15			7.5	20	Model	9 day	Siegrist et al. (2006)
	0.175 ± 0.002		7.0–8.0	20	NB	6 day	Zhou et al. (2015)
	0.13			10	NB		Choubert (2008)
0.28		0.27	7.3	20	NB	3 day	Oleszkiewicz (2011b)
	0.19		7.2	20	NB	6 day	Dold (2005)
	0.06		7.3	20	NB	5 day	Katehis (2002)
	0.218			25	NB	1 day	Lesouef (1992)
	0.13			20	Model	1 day	Lesouef (1992)
	0.24			20	OUR	30 day	Julián Carrera (2011)
	0.08–0.36		7.6–7.8	30	OUR	5 day	Martinage (2000)
0.26		0.07			NB		Volcke E.I.P. (2006)
0.24 ± 0.02		0.35±0.02	5.6	29	NB	7 day	Wenlong Liu (2017)

No detailed information is provided about taking into account the protozoa activity influencing the overall observed oxygen uptake rate and additionally removing nitrifiers from the system. The same mechanism can be related to heterotrophic processes consuming oxygen, thus affecting observed OUR. These phenomena should be considered during calculations of decay coefficient by use of correction factors including influence of estimated non-nitrifying biomass activity. As the internal substrate source production is strictly connected with the processes responsible for cell destruction, its rate is relatively stable throughout whole test. To include this phenomenon in the final results, rate of endogenous respiration of sludge sample should be determined in each test and subtracted from the measured OUR. Considering other parameters, like pH and temperature, no significant differences between presented studies was noticed. In most of presented reports selected temperature and pH were 20°C and about 7.5, respectively.

1.2 Decay rate methodology

There is a number of methods to determine the decay rate for a activated sludge. Regardless of the used method, decay rate is determined by a comparative analysis in a time period. Most of the laboratory experiments (Table 1) determine this coefficient based on the isolation of the activated sludge sample from the substrates necessary for the multiplication of individual groups of bacteria in certain process conditions: aerobic, anoxic or anaerobic.

The most common methods in this regard are primarily measurements of the process kinetics, such as OUR or substrate utilization rate by microorganisms (i.e. Ammonia Uptake Rate, AUR). In the case of autotrophic organisms such as nitrifiers, it will be ammonium nitrogen for AOB or nitrite nitrogen for NOB. In the case of heterotrophic organisms, the monitored substrate will be organic carbon. The nitrification rate test is based on the determination of the dynamics of decreasing or increasing in individual nitrogen forms during the experiment. Based on the dynamics of changes, the maximal bacterial activity is determined in optimal conditions, without substrate limitation and potential inhibition by too high substrate concentration [8–10].

The OUR test is based on oxygen concentration depletion measurements due to bacterial respiration. Oxygen is, alongside ammonium and nitrite, the main substrate for the metabolic processes of nitrifiers. To observe the maximal dynamics of process, and thus the oxygen consumption, measurements have to be performed in a significantly higher range of dissolved oxygen concentration than the nitrifiers oxygen affinity, which in ASM models is set default at 0.5 g O_2/m^3 [1]. At the moment of aeration there is a situation where the half-saturation constants is much lower than the substrate actual oxygen concentration, hence the mathematical description of the reaction can be represented by the zero order equation, where the reaction rate is directly determined based on the decrease in oxygen concentration over time. In addition, it should be ensured that the system does not provide organic carbon, which would result in measuring the activity of several groups of bacteria, including heterotrophs.

Except the experimental methods for determining the decay rate highlighted previously, there are also molecular analysis techniques based on the recognition of RNA or DNA of specific microorganisms, by the selective determination of genes or enzymes. These techniques allows to determine the amount of bacteria in sample. In this group of methods, the main ones are those based on fluorescence, where the reaction to the introduction of specific antibodies give a visual response. The most commonly used molecular analysis techniques are the FISH (fluorescent in situ hybridization) method based on visual fluorescence identification and the PCR (polymerase chain reaction) method, which uses the amplification of available genetic material by introducing an enzyme catalysing the synthesis of cell genetic information carriers (DNA/RNA) [11]. It should be emphasized that the molecular methods, at the current state of knowledge, are not significantly developed and are economically disadvantageous due to the cost of conducting such analyses.

Another group of methods for determining the decay rate is the use of existing mathematical models. Based on the principle of mass conservation, mathematical models describing transformation processes and removal of biogenic compounds have been created. The first advanced model was published in 1985 under the name ASM1. It gained general approval and over the years it was developed with additional elements (ASM2, ASM2d, ASM3). In subsequent versions, there is a distinction and separation of processes carried out by individual groups of bacteria. In the ASM2, ASM2d and ASM3 models, the default value for decay rate for nitrifiers is $b_a = 0.15$ d^{-1} [1]. The effectiveness of the model depends on the input values. Determination of the constant of decay rate with this method is usually based on prior calibration of the model, which will allow a meaningful reference of the simulation to the actual conditions, which also requires the use of laboratory methods.

This study focused on measuring OUR in aerobic conditions for nitrifiers. OUR tests are characterized by the relatively lowest costs of all the aforementioned methods. In addition, it does not require highly specialized equipment contrary to the molecular techniques. Also, the OUR method significantly minimizes the use of chemical reagents, limiting them to the dosing of the substrate. The test itself is relatively simple, both regarding the methodology and the subsequent analysis of the collected data. Based on these premises, respirometric method was used during the described experiment.

1.3 Decay rate in side-stream treatment

Contrary to mainstream nitrogen removal systems, there are only few reports about determination of the decay rate of nitrifiers present in side-stream systems treating high-strength ammonium wastewater (i.e. landfill leachate or digester supernatant) [12, 13]. Available data about decay coefficient in such systems is gained due to model calibration rather than experimental determination. However, observed decay rates might be higher than in the conventional nitrification systems due to higher temperature, especially considering membrane bioreactors with long SRT [14]. Surprisingly, in side-stream attached biomass systems, decay rate is presented as equal or even lower than in conventional systems, both for AOB and NOB [15]. The performed literature analysis revealed no attempt for experimental determination of nitrifiers decay rate coefficient in side-stream systems so far.

2 Materials and methods

2.1 Starvation Reactor parameters

A CSTR reactor (150dm^3) was used as the starvation reactor. Reactor was equipped with a mechanical mixer, aeration system controlled by on-line dissolved oxygen (DO) probe (Hach LDO101) to continuously maintain DO level between 5–7 g O_2/m^3. Temperature was kept above 25°C using a single heating segment, thus while ambient temperature was higher than this value, also the reactor content temperature was increasing. During reactor operation alkalinity was kept at non-limiting level, above 20 mval/dm^3.

2.2 Inoculum

Biomass used in this study was an activated sludge from a partial nitritation (PN) SBR treating reject water from digested sludge dewatering to obtain NO_2/NH_4 ratio suitable for an Anammox reactor. PN reactor was operated under following conditions: SRT ~3 days, temperature 25°C, low DO level (~1 g O_2/m^3) and pH 6.4–6.6. Ammonium and nitrite concentration level in this reactor varied between 300–350 mg NH_4-N/dm^3 and 350–400 NO_2-N/dm^3, respectively.

2.3 OUR test methodology

Two decay rate tests (Series 1 & Series 2) were performed using sludge from PN reactor. Sludge samples from the starvation reactors were transferred to a lab-scale reactors (4 dm^3) with mixing, aeration, temperature and pH control to measure OUR in following days. 5 respirometric assays in 5 days were completed during Series 1, while Series 2 took 6 days with 5 respirometric assays. OUR tests were performed daily in 26°C and 30°C for Series 1 and Series 2, respectively.

Each respirometric assay started from the initial substrate concentration (50 mg NH_4-N/dm^3 added as NH_4Cl) and alkalinity as $NaHCO_3$. pH was controlled at 7.5 and monitored on-line by a pH-electrode (Hach, PHC101). To determine OUR, oxygen oscillated between 4–7 g O_2/m^3. Providing such conditions, nitrification process was not limited or inhibited by any factors and allowed to determine the nitrifiers mass loss in following days.

2.5 Data analysis

The values of the OUR's recorded in subsequent days were lower and on this basis the value of decay rate was determined. The analysis of each batch test allowed to determine the reliable instantaneous values of the oxygen uptake of the process. For the Series 1 a total of 160 instantaneous OUR values were determined, and 450 in the Series 2. Data analysis was carried out in two consecutive stages. The first stage was based on the analysis of established time ranges in terms of instantaneous OURs. Ranges were determined on the basis of the dynamics of instantaneous rate variation and are presented in Table 2 (ranges 1–9). Further for individual ranges, the analysis of cumulative oxygen consumption from the beginning of the experiment to the moment determined by the range was performed. The last step (stage 2) was the elimination of a specific time interval (1 h and 2 h from the beginning of the test), thus creating successive time intervals (ranges 10 and 11) based on the ranges selected from stage 1.

Table 2. Accepted time ranges for stage 1 and stage 2.

	Series I	**Series II**
Stage I		
Range 1	0.5 h	0.5 h
Range 2	1 h	1 h
Range 3	2 h	2 h
Range 4	3 h	3 h
Range 5	5 h	5 h
Range 6	-	7 h
Range 7	-	9 h
Range 8	-	11 h
Range 9 (all)	All (6.6 h)	All (12.8 h)
Stage II		
Range 10 (cut time 1 h)	5.6 h	8 h
Range 11 (cut time 2 h)	4.6 h	7 h

The range 9 was determined to include all 86 and 180 OUR values for Series 1 and Series 2, respectively. For each time range, the total oxygen consumption was determined at that time and the value decreased relative to the first test. Due to the use of the longest range for the first series of measurements, it was decided to extend the second measurement series to about 12 hours in order to search for a period for which time extension will not cause a significant change in observed values. Time of approx. 6–12 h is too short and potential effect of autotrophic biomass growth will not affect the observed OUR values. The obtained results were subjected to the least-squares estimation using the non-linear Gauss-Newton regression method.

3 Results

The first stage of data analysis mentioned in section 2.2 was to determine OUR's for selected ranges depending in each series. The results of the measured values are presented in the diagrams below (figure 2 and 3).

 https://doi.org/10.1051/e3sconf/201910000035

Measured values clearly shows the decrease of the observed OUR value in following days, obviously due to nitrifiers decay process. During the initial 2 hours of each respirometric assay, very low values of OUR were noticed probably due to a bacterial lag phase after starvation period. This phenomenon led to the second stage of performed analysis where first one or two hours were excluded from the database (range 10 and 11, respectively).

Cumulative values for each day was compared to first day and percentage values was determined. Result are summarized in Table 3.

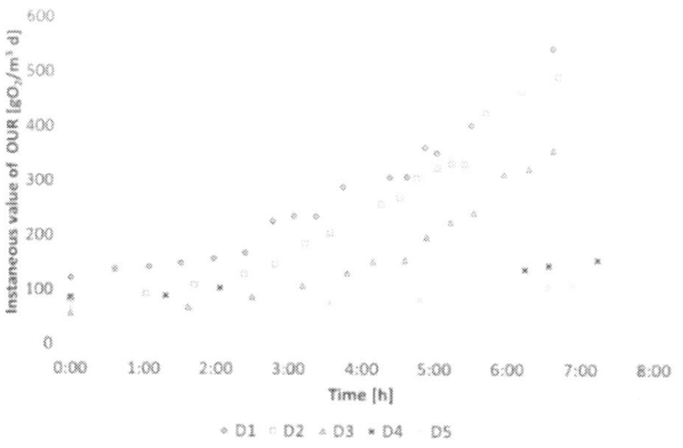

Fig. 2. OUR's during Series 1 in days 1–5.

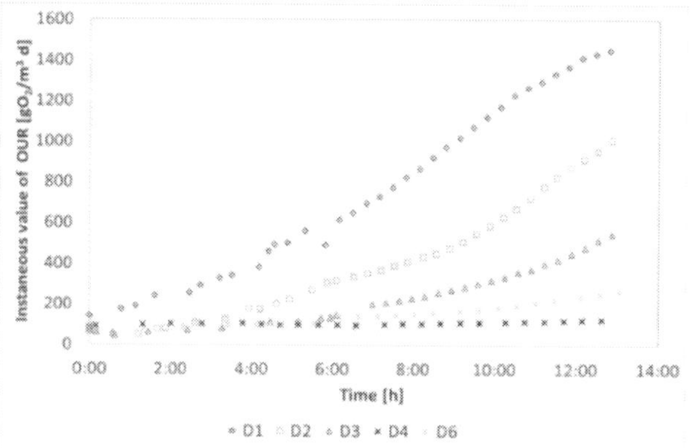

Fig. 3. OUR's for series no. 2 from day 1 to day 6.

The collected data was used to calculate the decay rate by fitting the experimental data to model. Calculated decay coefficient was corrected due to temperature as proposed in previous studies using equation 2 [11, 16]. Empirical coefficient of for (θ) temperature changes was 1.072. The b_a^t calculated for Series 1 & 2 (26°C and 30°C, respectively) and values corrected to 20°C are presented in Table 4.

For Series I the best model fit was for range 9 (6.6 h – full time of experiment) while for Series 2 there was no significant difference in curve fitting for ranges from 7 to 9. Estimation results was show in fig. 4.

Table 3. Values of cumulative oxygen consumption (Stage 1 of the analysis) compared to day 1.

	Day 2	**Day 3**	**Day 4**	**Day 5**	**Day 6**
			Series I		
Range 1	66.6%	46.0%	67.2%	51.2%	-
Range 2	66.0%	45.5%	65.1%	49.0%	-
Range 3	68.1%	45.7%	63.8%	46.1%	-
Range 4	69.2%	45.9%	60.6%	42.2%	-
Range 5	76.7%	47.8%	50.1%	33.9%	-
Range 9	83.2%	54.1%	42.2%	29.0%	-
			Series II		
Range 1	40.2%	40.6%	59.8%	-	48.6%
Range 2	17.2%	34.1%	57.2%	-	45.4%
Range 3	32.3%	30.1%	49.8%	-	38.8%
Range 4	34.3%	28.8%	45.0%	-	35.9%
Range 5	38.9%	26.5%	34.3%	-	29.3%
Range 6	44.6%	25.8%	26.3%	-	25.4%
Range 7	45.5%	28.8%	21.1%	-	22.8%
Range 8	47.7%	28.0%	17.0%	-	21.0%
Range 9	51.4%	28.9%	14.9%	-	21.0%

Table 4. Decay coefficient values observed in experimental conditions and corrected to 20°C.

	Range 1	**Range 2**	**Range 3**	**Range 4**	**Range 5**	**Range 6**	**Range 7**	**Range 8**	**Range 9**
					Series I				
b^{26}	0.166	0.179	0.194	0.217	0.273	-	-	-	0.299
b^{20}	0.110	0.118	0.128	0.143	0.180	-	-	-	0.197
					Series II				
b^{30}	0.147	0.154	0.258	0.315	0.457	0.531	0.557	0.591	0.580
b^{20}	0.073	0.077	0.129	0.157	0.228	0.265	0.278	0.295	0.289

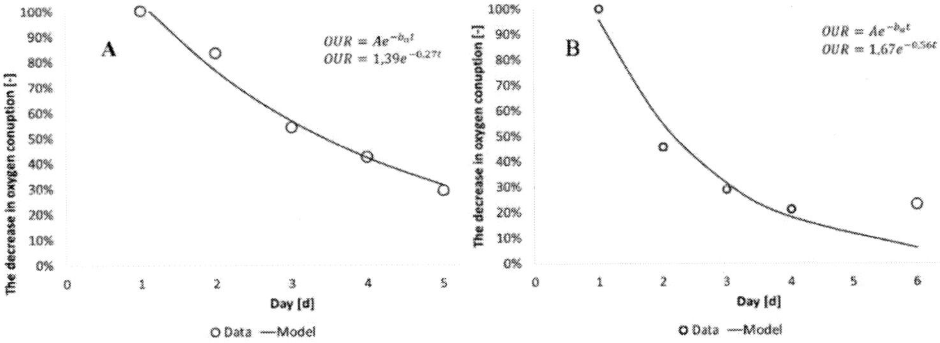

Fig. 4. Estimation results for A – range 9 (6.6 h) for Series 1, B – range 7 (9 h) for Series 2.

Analyzing the short test duration (ranges 1–3) substantial differences between experimental values and model predictions were observed. As mentioned before, such situation can be caused by slow adaptation of nitrifiers after starvation period, however no similar observation has been discussed in the literature so far. When test duration was extended (higher ranges in this analysis), calculated values was gradually approaching to the model predictions (fig. 5).

The best range for Series 1 & 2 (range 6 and range 7 respectively) was used in stage 2 of the analysis. In this stage first on or two first hours was cut from the database and new range were created: 10 and 11. Estimation results for new ranges were compared with previous ranges (Table 5).

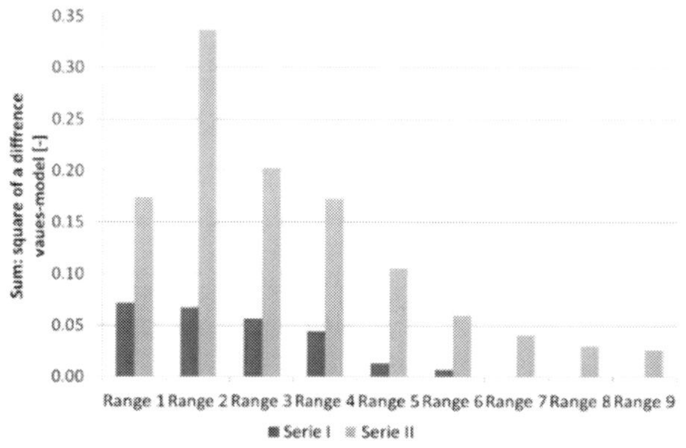

Fig. 5. Results of the estimation for all ranges in stage 1 of the analysis.

Table 5. Obtained results of decay rate for stage 2 of the analysis.

	Range 6	**Range 7**	**Range 10**	**Range 11**
Series I				
b^{26}	0.299	-	0.301	0.298
b^{20}	0.197	-	0.199	0.196
Difference between model and test (lower is better)	0.0067	-	0.0086	0.0152
Series II				
b^{30}	-	0.557	0.555	0.565
b^{20}	-	0.278	0.277	0.282
Difference between model and test (lower is better)	-	0.0408	0.0377	0.0356

Database modification performed in stage 2 (ranges 10 and 11) had no significant impact on decay coefficient calculations in comparison to previous ranges. That clearly indicates that respirometric assays used for decay coefficient estimation should be longer than several hours (c.a. 7 h or longer) and initial lag phase and low respirometric activity of the examined biomass has no relevant impact on the final result.

4 Conclusions

- Decay rate estimation using different methods is widely known topic. Decay rates determined for nitrifiers presented in literature are in the wide range of values, depending on biomass characteristics, experimental conditions and selected method of measurement
- Respirometric assays based on OUR measurements are an advantageous alternative for methods based on intensive chemical analysis (process kinetics observations) or molecular analysis (FISH, PCR) without significant decrease in the obtained results quality
- Performed data analysis revealed that short respirometric tests (up to 2 hours) resulted in unsatisfactory results comparing to model calculations. Low

respirometric activity, probably due to the lag phase after starvation period, affected the quality of experimental curve fitting to the existing decay model

- The best results were observed in long tests, which duration varied between 5–9 hours for both performed experimental series.
- Omitting the data from the observed lag phase (first 1–2 hours of respirometric assay) had no impact on the final results.

The authors gratefully acknowledge co-funding from the National Centre for Research and Development (grant no LIDER/16/0172/L–7/15/NCBR/2016).

References

1. M. Henze, W. Gujer, T. Mino, M. Loosdrecht, *Activated sludge models ASM1, ASM2, ASM2d and ASM3* (IWA, London, 2000)
2. B. Petersen, K. Gernaey, M. Henze, P. A. Vanrolleghem, Focus Biotechnol. **3C**, 101–186 (2003)
3. A. Lesouef, M. Payraudeau, F. Rogalla, B. Kleiber, Water Sci. Technol. **25**, 105–123 (1992)
4. J. Carrera, J. A. Torŕ, J. Lafuente, J. A. Baeza, Bioresource Technol. **102**, 9870–9875 (2003)
5. H. Yu, B. Ni, G. Sheng, Chem. Eng. Sci. **66**, 747–754 (2011)
6. S. Salem, M. S. Moussa, M. C. M. van Loosdrecht, Biotechnol. Bioeng. **94**, 252–262 (2006)
7. O. Nowak. K. Svardal, P. Schweighofer, Water Sci. Technol. **31**, 115–124 (1994)
8. H. Siegrist, R. Manser, W. Gujer, Water Res. **40**, 2416–2426 (2006)
9. J. A. Oleszkiewicz, G. Munz, C. Lubello, Water Res. **45**, 557–564 (2011)
10. J. A. Oleszkiewicz, G. Munz, C. Lubello, Chemosphere **83**, 720–725 (2011)
11. G. Tchobanoglous, F. L. Burton, H.D. Stensel, *Wastewater engineering: Treatment and Reuse / Metcalf & Eddy 4th edition* (McGraw Hill, New York, 2003)
12. S. Ge, S. Wang, X. Yang, S. Qiu, B. Li, Y. Peng, Chemosphere **140**, 85–98 (2015)
13. Y. Ahn, Process Biochem. **41**, 1709–1721 (2006)
14. S. Wyffels, S. van Hulle, P. Boeckx, E. Volcke, Biotechnol. Bioeng. **86**, 531–542 (2004)
15. N. Hubaux, G.Wells, E.Morgenroth, Water Res. **68**, 127–139 (2015)
16. J. A. Oleszkiewicz, S. A. Berquist, Water Res. **22**, 1163–1171 (1988)
17. H. Siegrist, I. Brunner, G. Koch, L. C. Phan, V. C. Le, Water Sci. Technol. **39**, 129–137 (1992)
18. J. A. Oleszkiewicz, Y. Lee, Water Res. **37**, 4202–4210 (2003)
19. R. Salzer, *A contribution to the determination of kinetic parameters for nitrification in activated sludge plants treating industrial wastewaters* (Institute for Water Quality, Viena, 1992).
20. E. J. T. M. Leenen, A. A. Boogert, A. A. M. van Lammeren, J. Tramper, R. H. Wijffels, Biotechnol. Bioeng. **55**, 630–641 (2000)
21. Z. Zhou, D. Hu, L. Jiang, C. Xing, Y. Zhu, M. Jiang, W. Qiao, Z. Li, Desalin. Water Treat. **56**, 2046–2054 (2015)

22. J. M. Choubert, A. Marquot, A. Stricker, S. Gillot, Y. Racault, A. Heduit, Water SA, **34** (2008)
23. P. L. Dold, R. M. Jones, C. M. Bye, Water Sci. Technol. **52**, 469–477 (2005)
24. D. Katehis, J. Fillos, L. A. Carrio, Water Sci. Technol. **46**, 289–295 (2002)
25. V. Martinage, E. Paul, Environ. Technol. **21**, 31–41 (2000)
26. E. I. P. Volcke *Modelling, analysis and control of partial nitritation in a SHARON reactor* (PhD thesis, Ghent University, Belgium 2006)
27. W. Liu, Q. Yang, B. Ma, J. Li, L. Ma, S. Wang, Y. Peng, Environ. Sci. Technol. **51**, 4001–4008 (2017)

Preliminary ecotoxicity assessment of selected flavour and fragrance compounds using Microtox assay

Wojciech Kołodyński[1,*], *Katarzyna* Piekarska[1], and *Daniel* Strub[2,3]

[1]Wrocław University of Science and Technology, Faculty of Environmental Engineering, Wyb. Wyspiańskiego 27, 50-370 Wrocław, Poland
[2]Wrocław University of Science and Technology, Faculty of Chemistry, Wyb. Wyspiańskiego 27, 50-370 Wrocław, Poland
[3]Liquid Technologies sp. z o.o., ul. Chełmońskiego 12, 51-630 Wrocław, Poland

Abstract. The bioluminescence inhibition bioassay using marine bacteria *Vibrio fischeri* is widely used as a tool to assess the toxicity of chemical compounds, because of the many benefits comprising cost and time effectiveness, rapidness, sensitivity, and ease of operation. In addition, the test is used for various types of organic and inorganic compounds. Due to the insolubility of organic compounds in water, a suitable solubilizer should be applied. The ecotoxicity of the solvent should be negligible in relation to marine bacteria. On account of superior human activities the synthesis of new chemical substances is inextricably linked to the emergence of chemical compounds that are a potential threat on environment. A Microtox system utilizing the 81.9% Basic Test with 14 dilutions was applied to pre-evaluate the ecotoxicity of β-cyclocitral oxime, citronellal oxime and perillaldehyde oxime. Substances solutions with an initial concentration of 0.036% in DMSO were prepared. The preliminary results show that the studied scent compounds are characterized by quite high toxicity. In order to confirm the received ecotoxicity values, it is necessary to carry out additional surveys using another range of concentrations and biotests on organisms at each trophic level.

1 Introduction

Over the last years, a constantly growing request for synthesis of new fragrance components has been observed [1]. The olfactory potential depends mainly on the structure of the designed molecules and an osmophore (functional group) [2, 3]. The chemical groups responsible for a pleasant smell include: ester, ketone, aldehyde, ether and hydroxyl groups, whereas the presence of thiocarbonyl, thioformyl, thioether, thiol and amine groups result most often in unobjectionable smell. Fragrances with slight changes in the chemical structure may differ significantly in aroma [4]. A small change in the spatial structure caused by the removal or addition of one atom or chemical group in the molecule may be sufficient to

* Corresponding author: wojciech.kolodynski@pwr.edu.pl

increase the intensity of the smell. Additionally the concentration of the substance plays a key role in the perception of the fragrance, because it can affect the quality of the scent[5]. Development of synthetic methods for the preparation of natural compounds was possible by unravelling their structures. Moreover, modern chemistry manners have allowed both the chemical synthesis of compounds existing in nature as components of essential oils or animal products, as well as the discovery of fragrant molecules that do not exist in nature. Fragrance ingredients are divided into natural and synthetic origin compounds. Due to the high cost of scents isolated from natural sources, more often aroma compounds are prepared via synthesis [4]. Upwards of 3000 scent-bearing compounds are applied in perfumery industry [3]. The largest amount of fragrance compounds is present in cosmetics, perfumes and household chemicals [6].

Chemical synthesis of new products is associated with the conduct of many tests to check the quality, density, purity, structure, physical-chemical properties, as well as toxicity and ecotoxicity in order to estimate the potential risk to the environment [7]. Microbiotests based on bioindicators have been available on the market for several years. They allow to determine whether a given substance may pose a threat to the environment. The toxikit microbiotests use test organisms in immobilized or dormant form, which allows them to be applied on demand if required [22]. In addition, they are an alternative to classical analytical methods [23]. This allows to avoid the step of cultivating test organisms in laboratory conditions [22]. Toxicity action of the tested chemicals is manifested by metabolic disturbances resulting from physiological and conservative changes of test organisms [10]. In these studies, the toxicological hazard assessment of the three fragrances was investigated. Test compounds, namely β-cyclocitral oxime, citronellal oxime, perillaldehyde oxime were supplied by the Department of Bioorganic Chemistry, Faculty of Chemistry, Wrocław University of Science and Technology. This fragrances can be further used in household chemicals, cosmetics and perfumes or can become a substrate for the synthesis of another odour components [8]. Due to the increasingly frequent withdrawal of fragrances from the market and use, it is necessary to search for new fragrances. The scents or their derivatives may be delivered to the environment, therefore their impact on living organisms must be known, as it was in the case of camphor oxime and citral oxime ethers.

Many substances used in the cosmetic industries have an allergenic effect. In addition, they may accumulate in living organisms [8]. Fragrance components present in cosmetics commonly lead to the allergic response of the human body [9]. For example, polycyclic and nitro musk are one of the substances applied in the cosmetic industry that have a negative impact on people. These compounds and their derivatives are characterized by high toxicity and bioaccumulation [8].

Fast and reliable assessment of the acute ecotoxicity of a new substance can be carried out by employing Microtox test. This biotest is widely applied to monitor toxicity due to many benefits encompassing cost and time effectiveness, rapidness, sensitivity, and ease of operation.. Biotest based on the inhibition of luminescence in marine bacteria *Vibrio fischeri* is utilized to all types of matrices, both organic and inorganic compounds. Furthermore, conducting a complete ecotoxicity assessment requires the usage of bioassays based on organisms from each trophic level. In addition, surveys based on marine luminescent bacteria are frequently first of the bioassays utilized in the creation of biotest battery on account of the speed and cost of the procedure for its execution [11, 12].

The Microtox Acute Toxicity Test utilize a clonal strain of bioluminescent bacteria supplied in lyophilized vial format. During the toxicological evaluation of organic substances that do not dissolve in the aquatic environment, solubilization is used. For this purpose, carrier solvents such as DMSO, acetone or ethanol may be necessary to solubilize chemicals. Finding the right solubilizer is the main challenge when assessing the ecotoxicity of organic compounds. However, solvents possess their own ecotoxicity and should be selected so that they do not affect the results of the test substance. Therefore, an important step during the assessment of the ecotoxicity of chemical substances is to eliminate the toxic effects of the solubilizer. Therefore, an analysis based on a comparison of the ecotoxic effect of the solvent and the solvent with dissolved substance is performed.

In the paper we present preliminary studies on the ecotoxicity of three flavour and fragrance substances, namely β-cyclocitral oxime, citronellal oxime and perillaldehyde oxime. The chemical structures of β-cyclocitral oxime, citronellal oxime and perillaldehyde oxime are presented in Figure 1. The aim of the research was to determine the level of ecotoxicity of the compounds listed above. Microtox system was chosen for ease of operation and high sensitivity.

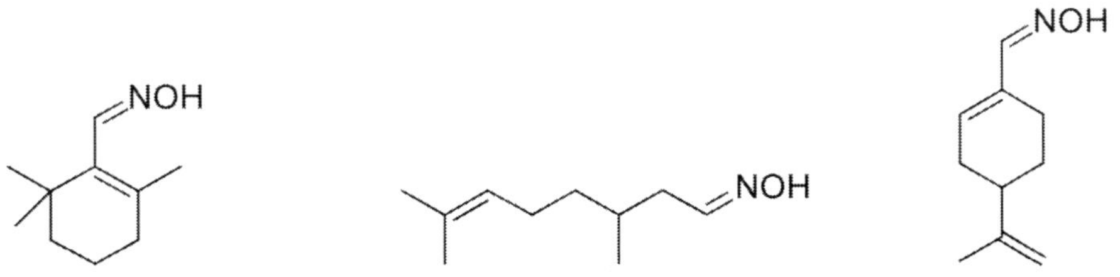

1 2 3

Fig 1. β-cyclocitral oxime (1), citronellal oxime (2), perillaldehyde oxime (3) chemical structures.

2 Materials and methods

2.1 Microtox toxicity analysis

The Microtox toxicity analysis is based on a specific clonal strain *Vibrio fischeri* marine bacteria, earlier referred as *Photobacterium phosphoreum* [11]. Light production is directly related to the metabolic activity of bioluminescent microorganisms [12]. The test relies on disrupting the emission of light in optimal environmental conditions as a result of toxic substances acting on the marine bacteria *Vibrio fischeri*, which lead to the reduction of light emission. The reaction mechanism is that marine luminous bacteria contain luciferase that catalyses the oxidation of a reduced flavin mononucleotide and aliphatic long-chain aldehyde by oxygen. As a result of this reaction, an unstable complex is formed which in next step emits photons during decomposition with its highest intensity at wavelength 490 nm. Bioluminescence is directly related to respiration through the electron transport chain, therefore the bacteria are used to evaluate the toxicity of the test substances [10, 13]. The Microtox 81,9% Basic Test with 14 dilutions was chosen to provide a wider range of tested concentrations in view of the fact that the ecotoxicity of tested substances was completely unknown. The ecotoxicity of fragrance substances dissolved in DMSO was measured. Possible toxicity carrier solvent DMSO, a substance that facilitates the dissolution of organic compounds in water, has been tested in previous studies to assess the potentially ecotoxicity effects on bioluminescent bacteria [7]. For this reason, the re-evaluation of the ecotoxicity of carrier solvent DMSO is omitted.

2.2 Calculations

Luminescence intensity was measured by means of a Microtox ™ Model 500 Analyser after 0-, 5-, 15-minute exposure time. For the calculations of percent inhibition of light emission, the following equations were applied:

$$R_t = \frac{I_{Ct}}{I_{C0}} \quad (1)$$

$$\Gamma_t = \frac{R_t \cdot I_{s0}}{I_{st}} - 1 \quad (2)$$

$$\%inhibition = \frac{\Gamma_t}{1+\Gamma_t} \cdot 100 \quad (3)$$

where R_t is a correction for any inhibition induced by a negative control sample (blank cuvette) and I_{C0}, I_{Ct} are absolute light intensities produced by negative control at time t and at initial time 0, respectively. I_{s0} and I_{st} are light intensities produced by a sample at time t and at initial time 0, respectively. Γ_t is the ratio of light lost at time t to light remaining at time t, which can be convent to %inhibition. The final expression of the toxic response of the samples is presented as EC_{50} which is the Effective Concentration of the samples reducing light emission by 50% and estimated from linear regression of the log of each concentration level of the contaminant versus percent inhibition (%inhibition), as described in [7, 14].

2.3 Samples preparation

Samples of β-cyclocitral oxime, citronellal oxime and perillaldehyde oxime were supplied by the Department of Bioorganic Chemistry, Faculty of Chemistry, Wrocław University of Science and Technology. On account of insolubility in water of each of substances, their solubility in DMSO was conducted. It was found that each of the substances easily dissolved in DMSO. According to the article [7], a ecotoxicity comparison of two solubilizers was carried out using Microtox 81.9% Basic Test, namely acetone and DMSO. Due to the lower toxicity of DMSO than acetone to marine bacteria *Vibrio fischeri*, DMSO was chosen as the carrier solvent. Due to the lack of EC_{50} data, solutions 0.036% of fragrances in DMSO were prepared and tested.

The concentrations of the samples used in the calculations took into account the initial dissolving of the sample in DMSO. Therefore, the initial concentration was:

$$\frac{81{,}9}{100} \cdot 0{,}036 = 0{,}0295\% \quad (4)$$

3 Results and discussion

In the first stage of the studies, a screening test was carried out to check whether the test substances inhibit the bioluminescence of marine bacteria *Vibrio fischeri*. All of the tested fragrances showed a toxic effect on bioluminescent bacteria. In the second stage of the study, a series of 14 dilutions was made to find the concentration of the substance causing 50% inhibition of light produced by the bacteria. At the moment of dissolving the fragrances in DMSO and subsequently in water, cloudy milky solutions were formed during the test. At the time the test fragrance was added to the cuvette containing the high concentration of NaCl (derived from 2% in Diluent and 22% in Osmotic Adjusting Solution), no precipitate formed in the solution, therefore a test was carried out according to the standard Microtox protocol (81.9% Basic Test).

The analysis results of acute ecotoxicity measurements are shown in Table 1 as concentration EC_{50} and ratio of $1/EC_{50}$.

Table 1. EC_{50} parameter values of the test substances dissolved in DMSO after 5 and 15 minutes of exhibition to bioluminescent marine bacteria.

Sample	**EC_{50} [%]**		**$1/EC_{50}$**	
	t = 5 min	**t = 15 min**	**t = 5 min**	**t = 15 min**
perillaldehyde oxime	0.000653	0.000684	1531.394	1461.988
β-cyclocitral oxime	0.000279	0.000301	3584.229	3322.259
citronellal oxime	0.000231	0.000222	4329.004	4504.505

The results of ecotoxicity measurements of studies scent compounds are quite high. Comparing them with the toxicity values of the pure solvent, the toxicity of the carrier solvent can be considered negligible. Referring to [7], the $1/EC_{50}$ value after the 5 and 15 minute exposure time is 0.259 and 0,380 for solutions 5%, respectively. This suggests that DMSO is a good solubilizer for further studies of organic compounds as mentioned in the Microtox Acute Toxicity Test, and due to its solubility range, low vapor pressure and low freezing point [11]. However, the results obtained need confirmation with a different range of concentrations using the Microtox test, as well as other ecotoxicity bioassays that use living organisms from different trophic levels. In the interim the conducted research can be considered as an initial assessment of the toxicity of selected investigated scent compounds.

Perillaldehyde oxime, also called perillartine, is a compound 2000 times sweeter than sucrose [15]. Perillaldehyde found in *Pertilla frutescens* (Lamiaceae) is precursor in synthesis perillartine, which is an oxime derivative [16]. May cause skin irritation, serious eye damage or respiratory irritation [17, 18]. Due to the lack of literature reports on perillartine, its ecotoxicity cannot be compared with other studies. However, the perillaldehyde substrate used for the synthesis of perillartine has been tested for genotoxicity. According to the authors of the study, perillaldehyde has no genotoxic potential [24]. β-cyclocitral is used as a flavoring ingredient. β-cyclocitral is produced by Microcystis species. So far, it has not been demonstrated whether β-cyclocitral may have a negative impact on the environment [19]. In literature, the amount of available data about β-cyclocitral oxime is modest, it is a compound that has not been fully understood yet. Citronellal oxime is a toxic chemical substance to aquatic organisms with long lasting effects, it can also cause eye irritation [20]. Citronellal mutagenicity was tested using Salmonella strains. Citronellal was found to be not a mutagenic compound. In addition, it was proven that citronellal inhibited the embryonic development of mosquito *Aedes aegypti* eggs responsible for causing the yellow fever [25]. Citronellal oxime proved to be the most toxic of the substances tested. As the only one of the tested compounds, it showed an increase in ecotoxicity during a longer time of exposure to bioluminescent bacteria *Vibro fischeri*. Marine bacteria were found to be less sensitive to perillaldehyde oxime and β-cyclocitral oxime. Effective concentrations that inhibits 50% of the light produced by the bacterial culture is much higher than concentration citronellal oxime. For all substances tested no data on their ecotoxicity effects were established before. Only carbonyl substrates were assessed for their toxicity on humans [17–20].

In order to evaluate the potential impact of the studied fragrances on the environment, other ecotoxicity biotests using organisms from different trophic levels should be used. The main problem of tested chemical compounds is their insolubility in water. Most of the ecotoxicological biotest uses aquatic organisms such as *Thamnocephalus platyurus*, *Daphnia magna*, *Chlorella vulgaris* and so on [21]. The lipophilic components must be solubilized in organic solvents. A suitable carrier solvent should have low or no toxicity [11].

4 Conclusions

Biological tools, such as bioindicators, are applied to assess the toxic effects of newly formed chemicals. The synthesis of new fragrances and their introduction on the market involves the assessment of ecotoxicity in order to eliminate the potentially negative impact on the environment. The assessment of acute toxicity utilizing the Microtox system revealed quite greatly high EC_{50} values of β-cyclocitral oxime, citronellal oxime and perillaldehyde oxime to relation marine bioluminescent bacteria *Vibrio fischeri*. The most ecotoxicity of all three substances tested was citronellal oxime, while the lowest acute toxicity was demonstrated by the perillaldehyde oxime, although the results necessitate confirmation. Other available ecotoxicity bioassays involving organisms at each trophic level are necessary to confirm the obtained results. Moreover, the insolubility properties of the studied fragrance compounds is a challenge in selecting the right conditions when performing the ecotoxicity assessment on aquatic organisms.

This work was supported by the project "Synthesis of new fragrances from raw materials of a natural origin with application in perfumery, cosmetics and household chemistry" (SYNFRA); grant no. LIDER/4/0099/L-7/15/NCBR/2016; financed by the National Centre for Research and Development within the LIDER Programme.

References:

1. J. Krzyczkowska, E. Białecka-Florjańczyk, I. Stolarzewicz, ZNTJ **16**, 5–18 (2009)
2. M. A. Amerine, R. M. Pangborn, E. B. Roessler, Trends Food Sci. Technol. 153–154 (1965)
3. D. J. Strub, J. Kula, M. Sikora, J. Gibka, S. Lochyński, Flavour Frag. J. **31**, 81 (2016)
4. K. Mitka, J. Staryńska, Tech. Trans. **17**, 135–148 (2012)
5. I. Wojciechowska, A. Wojciechowska, K. Wieszczycka, **1**, 39–43 (2017)
6. M. Kieć-Świerczyńska, B. Kręcisz, D. Świerczyńska-Machura, Med. Pr. **55**, 1, 203–206 (2004)
7. E. Kielka, A. Siedlecka, M. Wolf, S. Stróżak, K. Piekarska, D. Strub, E3S Web of Conferences **44**, 1–5 (2018)
8. J. Bogacki, P. Marcinowski, P. Wiliński, J. Naumczyk, Gaz, Woda i Tech. Sanit. 8–13 (2016)
9. M. Kieć-Świerczyńska, **3**, 2, 61–65 (1998)
10. M. Abbas, M. Adil, S. Ehtisham-ul-Haque, B. Munir, M. Yameen, A. Ghaffar, G. Abbas Shar, M. A. Tahir, M. Iqbal, Sci. of the Tot. Environ. **626**, 1295–1309 (2018)
11. B. T. Johnson, Environ. Micro. **1**, 69–105 (2005)
12. S. Parvez, C. Venkataraman, S. Mukherji, Environ, Int. **32**, 265–268 (2006)
13. A. Trusz-Zdybek, A. Szymczycha-Madeja, T. M. Traczewska, K. Piekarska, Kos. **61**, 417–423 (2012)
14. M. Pogorzelec, K. Piekarska, E3S Web of Conferences **17**, 1–8 (2017)
15. R. Kalinowski, E. Chrzanowska, M. Brytan, 1 (2012)

16. A. Tavares, B. C. Arruda, E. S. Boes, V. Stefani, H. K. Stassen, L. F. Campo, I. H. Bechtold, A. Aa. Merlo, J. Braz. Chem. Soc. **23**, 5, 880–888 (2012)
17. https://echa.europa.eu/information-on-chemicals/cl-inventory-database/-/discli/details/5046 (access 25.01.2018)
18. https://pubchem.ncbi.nlm.nih.gov/compound/Perillartine#section=Cellular-Locations (access 25.01.2018)
19. https://pubchem.ncbi.nlm.nih.gov/compound/beta-Cyclocitral#section=Pharmacology-and-Biochemistry (access 25.01.2018)
20. https://pubchem.ncbi.nlm.nih.gov/compound/6299203#section=Canonical-SMILES (access 25.01.2018)
21. J. Rybak, *Ecotoxicology: theory and laboratory practice: course in English* (2011)
22. P. Jakubowicz, T. Steliga, D. Kluk, Nafta-Gaz **69**, 5, 409–417 (2013)
23. M. Wieczerzak, J. Namieśnik, B. Kudłak, Environ. Int. **94**, 341–361 (2016)
24. C. A. Hobbs, S. V. Taylor, C. Beevers, M. Lloyd, R. Bowen, L. Lillford, R. Maronpot, S. Hayashi, Food and Chem. Toxicol. **97**, 232–242 (2016)
25. https://pubchem.ncbi.nlm.nih.gov/compound/citronellal#section=Non-Human-Toxicity-Excerpts (access 30.01.2019)

Active torque distribution strategy for electric vehicle independent axle drive

Artur Kopczyński[1,*]

[1]Warsaw University of Technology, Department of Multisource Propulsion Systems, ul. Narbutta 84, 02-524 Warsaw, Poland

Abstract. Vehicles equipped with an electric independent axle drive have different properties compared to conventional vehicles. Distribution of driving or braking torque can be achieved by the proper control of the operation of electric machines without applying additional mechanisms. In this paper a method of active torque distribution between front and rear axles is presented. The method allows to use the maximum tyres adhesion and minimize slips. The results of simulation tests are presented, in which the active torque distribution with evenly torque distribution were compared.

1 Introduction

Electric cars, different in many respects compared to conventional vehicles powered by fossil fuels, are characterised by distinctive features widely discussed in all types of media sources nowadays. Some of them like obvious issues related to driving range [1–3], energy consumption [4, 5], charge or refuel time [6] or impact on the local environment [7] are most often repeated topics. However, there are also other aspects scrutinized more deeply by scientists. These include for example the possibility of energy recovery from the braking process [8, 9] or the fact that the electric drive is much more compact, which creates the possibility of greater diversity in the structure of the propulsion system. In an electric vehicle, the motors can for instance be placed in wheel hubs, or two machines can independently drive the front and rear axles of wheels [10, 11]. The problem of the distribution of drive torques in multimotor electric powertrain is a normal consequence of the development of technology. In articles [12, 13], the issue of controlling the electric independent axle drive of the vehicle in the event of loss of adhesion by one of the wheels was analysed. However, relatively few research papers address the issue of how to distribute the drive torque between axles or wheels to minimize the risk of losing adhesion by tyres.

2 Electric vehicle independent axle drive

The electric vehicle independent front and rear axle drive scheme is shown in Figure 1. The vehicle is driven by electric motors (EM) and powered by an electrochemical battery (Bat). EM transmit the driving/braking torque to the wheels via the drive shafts and differential mechanism (X). The prototype Italian car P-Mob and commercially available Tesla Model S

* Corresponding author: artur.kopczynski@pw.edu.pl

All-wheel Drive Dual Motor [14] are examples of such a construction. The correct operation of independent front and rear electric drive is handled by the Central Control Unit (CCU). CCU decides about actions such as torque distribution between axles or defines the mode of EMs operation (motor for accelerating and drive with constant speed and generator in case of regenerative braking). In Figure 1 the continuous (——) and dashed (– – –) lines denote the mechanical and electrical connections, respectively. The arrows indicate the possible direction of the energy flow depending on the EM operating mode.

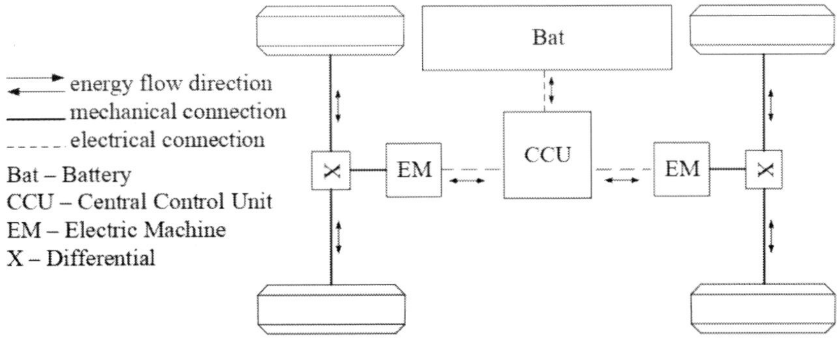

Fig. 1. Electric vehicle independent front rear axle drive.

3 Modelling

The preparation of appropriate simulation model is crucial to analyse vehicle's dynamic properties. One can refer to [15] in order to get acquainted with this topic. However, the most important elements that should be taken into consideration while creating such a simulation model adapted to the independent front and rear axle drive are presented below.

3.1 Motion equation of vehicle equipped with independent axle drive

The Equations (1-3) of motion of the vehicle equipped with independent axle drive are derived from the second law of Newton's dynamics.

$$d^2x/dt^2 = 1/m\ [\{[\tau_F + \tau_R]/r_d\} - F_{mr}] \quad (1)$$

$$\tau_F = J_F d\omega_F/dt + F_{xF} r_d \quad (2)$$

$$\tau_R = J_R d\omega_R/dt + F_{xR} r_d \quad (3)$$

where: x – means vehicle longitudinal displacement, m – vehicle mass, r_d – tyre dynamic radius, F_{mr} – sum of motion resistance forces, τ_F, τ_R – front, rear wheels torque, J_F, J_R – front and rear wheels moment of inertia, ω_F, ω_R – front, rear wheels angular speed, F_{xF}, F_{xR} – front, rear tyre longitudinal force.

3.2 Tyre

A basic classic tyre model was adopted, taking into account only progressive (longitudinal) movement. The Equation (4) describes tyre longitudinal slip s_x during vehicle acceleration due to tyre elasticity [16]. Whereas, the slip during the vehicle braking is expressed by the Equation (5).

$$s_x = [r_d\omega - v_x]/\, r_d\omega \tag{4}$$

$$s_x = [v_x - r_d\omega]/\, v_x \tag{5}$$

When driving or braking torque is applied to the wheels, tyre deformation occurs. In such a situation tyre wheel has less ability to transfer power. An example of the characteristics of the coefficient of adhesion μ_x as a function of tyre slip s_x is presented in Figure 2.

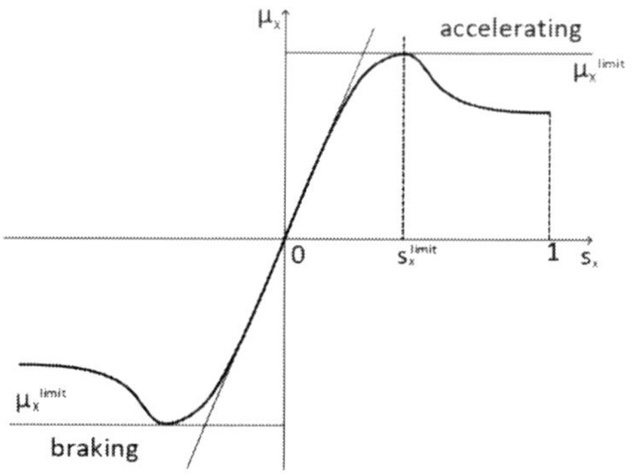

Fig. 2. Tyre adhesion coefficient characteristic.

Taking into account the slip, the force transmitted by the tyre is expressed by Equation (6).

$$F_x = \mu_x F_z \tag{6}$$

where: F_x – means tyre longitudinal force and F_z – tyre vertical load. In order to analyse the motion of the vehicle cornering, it is necessary to extend the model with lateral deformation of the tyre [16].

3.3 Wheel vertical load caused by pitch

The method of determination of vertical load caused by pitch (during accelerating or braking) was presented in the work [8]. Forces of vertical load of a given axle during vehicle accelerating are expressed by Equations (7) and (8), where a – means vehicle acceleration or deceleration, h – height of COG (Centre of Gravity) from ground and L – wheelbase.

$$F_{zf} = \{m[gl_r - ah]\}/L \tag{7}$$

$$F_{zr} = \{m[gl_f + ah]\}/L \tag{8}$$

In case of vehicle braking the forces F_{zf} and F_{zr} are defined by similar Equations to (7, 8) but the signs in numerators are opposite.

4 Torque distribution strategy

In conventional cars, a differential mechanism with two degrees of freedom is used to compensate the difference in wheels speed. In 4x4 vehicles, this differential mechanism is also located between the front and rear axles. The relation between speeds of the differential mechanism is described by Equation (9).

$$[\omega_F + \omega_R]/2 = \omega_{engine} \tag{9}$$

While the torque on each exit shafts T_1, T_2 from differential mechanism is described by Equation (10) and (11) [16].

$$T_1 = 0.5T_0 + 0.5T_f \tag{10}$$

$$T_2 = 0.5T_0 + 0.5T_f \tag{11}$$

If the value of internal friction torque T_f is small (as it is for most common differential mechanisms), then the torque is distributed evenly.

In the case of independent front and rear electric powertrain, any torque on both motors can be set (limited by ME parameters). Equation (12) describes the active torque distribution method between the front T_F and the rear T_R, based on the maximum usage of tyre adhesion.

$$T_F / T_R = [gl_r - ah] / [gl_r + ah] \tag{12}$$

The Equation (12) results directly from the relation of vertical loads occurring on both wheel axles of the vehicle, which is described by Equations (7) and (8).

Figure 3 shows the percentage torque distribution during acceleration, while Figure 4 refers to the braking of the vehicle. Both graphs present results for a vehicle with the parameters given in Table 1. The acceleration plot has been limited to $a = 4\ m/s^2$, because it is the highest acceleration that the analysed vehicle can reach. For both cases (acceleration/braking) the initial difference (for $a = 0\ m/s^2$) results from the position of the centre of gravity. As the acceleration / deceleration value changes, the relation between torques on front and rear wheels also differentiates.

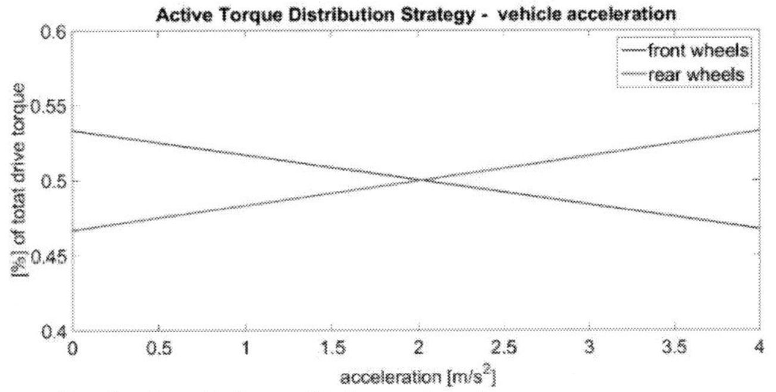

Fig. 3. Active torque distribution during vehicle accelerating.

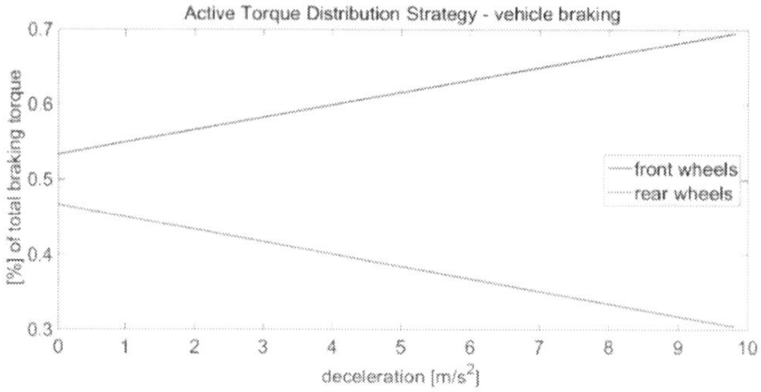

Fig. 4. Active torque distribution during vehicle braking.

5 Simulation study

Simulation tests were carried out for a small electric vehicle, the parameters of which were defined in Table 1.

Table 1. Input data.

Vehicle Parameter	Value
Mass	m = 500 kg
Frontal area	A = 1.35 m^2
Drag coefficient	Cd = 0.45
Rolling resistance coefficient	ft = 0.01
Tyre dynamic radius	rd = 0.27 m
Front axle to COG	lf = 1.0 m
Rear axle to COG	lr = 1.5 m
Centre of gravity height	h = 0.34 m
Rotor inertia	J = 0.128 kgm^2
Wheel inertia	Jk = 0.7 kgm^2

Acceleration of the vehicle on a straight road was simulated. The results of these tests are presented in Figures 5–10. In the graphs in Figures 5 and 6 the waveforms of vehicle's velocity and acceleration for the two analysed strategies were compared, i.e. evenly and active torque distribution. In both cases, the waveforms are very similar to each other, which allows a good comparison of subsequent parameters.

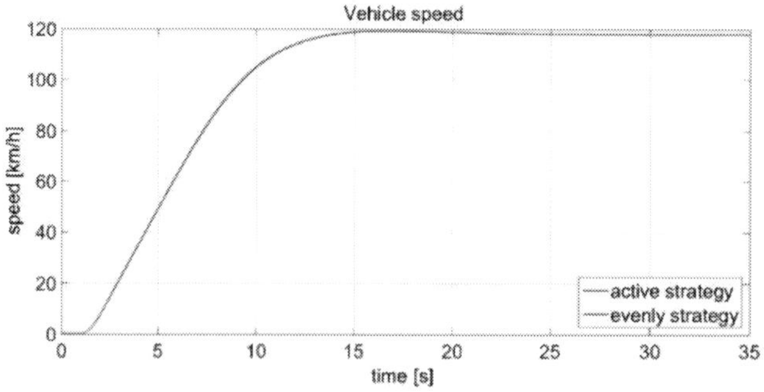

Fig. 5. Vehicle longitudinal speed in tested scenario.

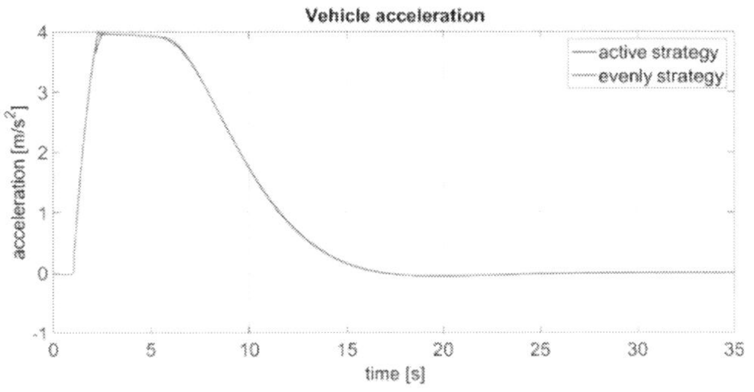

Fig. 6. Vehicle longitudinal acceleration in tested scenario.

The torque on wheels waveforms on the front and rear axle for evenly strategy and the waveforms of tyres slips are shown in Figures 7 and Figure 8, respectively. The torque graph for both axles is almost identical, which is in line with the assumptions of this strategy. In the case of slips, there is a significant difference in the time range when the vehicle achieves the greatest acceleration. The values of slip on the front axle are higher because this axis is less loaded during acceleration (according to Equations 7 and 8) and their ability to transfer torque is lower.

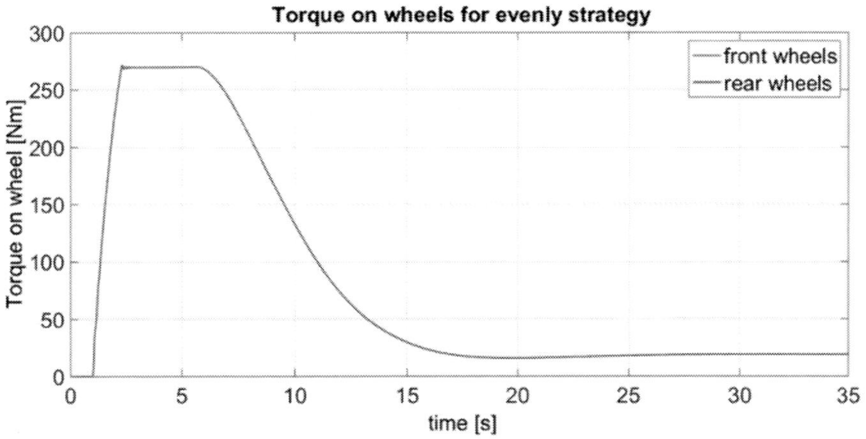

Fig. 7. Front and rear wheels torques for evenly torque distribution strategy.

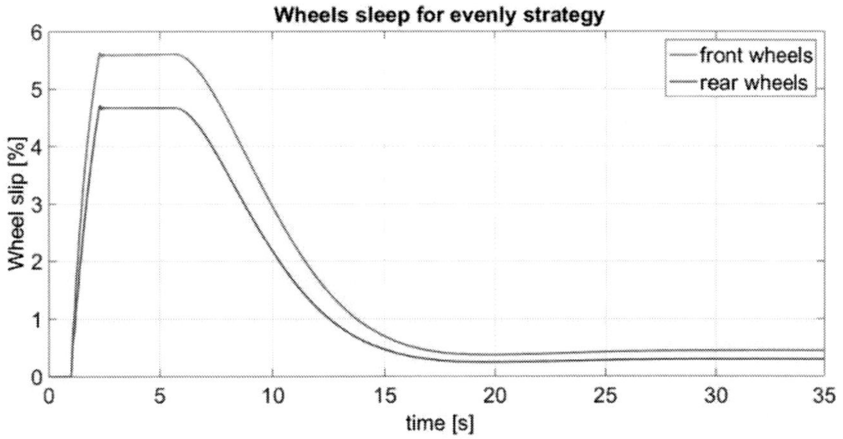

Fig. 8. Front and rear wheels tyres slip for evenly torque distribution strategy.

The simulation results of active torque distribution, are shown in Figures 9 and 10. Greater drive torque was delivered to the rear wheels of the vehicle, because these wheels are loaded during acceleration, and thus tyres have better adhesion with ground. The torque of the front axle was proportionally lower. Due to active distribution, slips on both axes adopted similar values which means that the vehicle has a greater stability margin. Potential loss of stability of vehicle movement occurs at the moment of maximum use of adhesion by all wheels. With this method, all wheels approach their adhesion limits before losing adhesion on one of them. As a consequence, the risk of losing adhesion by one of the wheels is lower than in the evenly distribution strategy.

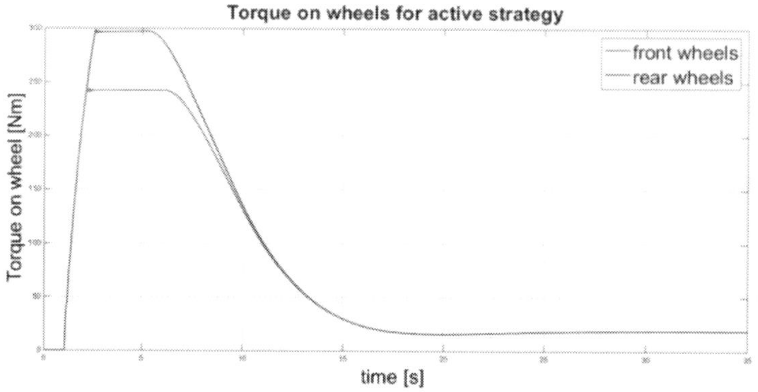

Fig. 9. Front and rear wheels torques for active torque distribution strategy.

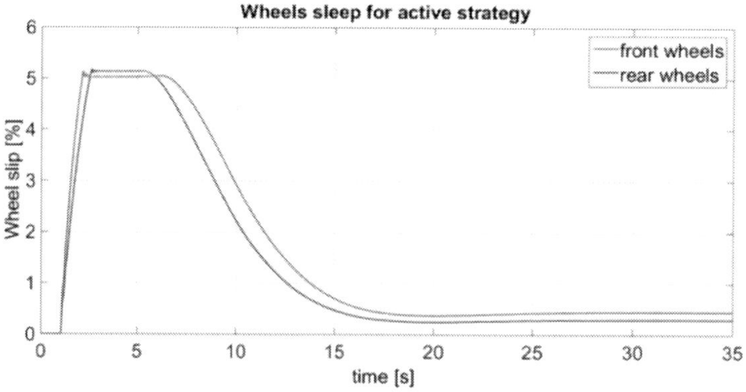

Fig. 10. Front and rear wheels tyres slip for active torque distribution strategy.

6 Conclusion

The presented method of active torque distribution allows to better use of tyre adhesion, and thus positively affects vehicle stability. In comparison to the evenly distribution strategy, slips on all wheels reach similar values. For a vehicle with a higher EM power, it would be possible to approach the limit values of wheel slip. Therefore, the differences in the indicated parameters (discussed in chapter 5) would be even greater, and of course in favour of the active method. The position of COG has also a significant influence on the differences in the results. The higher COG is located from the ground and the closer it is to one of the wheel axles, the advantage of the active distribution strategy will be greater. A slight drawback of the indicated method is the operation of torque distribution algorithm depending on vehicle's momentary longitudinal acceleration. In a real car, it is possible to define the function of the dependence of the current work parameters of EMs and the position of the accelerator pedal and determine the theoretical relations on the acceleration of the vehicle. On the other hand, in modern vehicles equipped with ESP (Electronic Stability Program) systems, acceleration sensors in all three directions are widely used, so this should not be a problem. Another weak point of the indicated method as well as other such systems is sensitivity to the position of COG, which depending on whether only the driver, or e.g. 5 passengers and luggage are travelling, creates a significant difference. However, taking into account all the above-mentioned aspects and the results of simulation tests, the presented method of active torque distribution is significantly better than the evenly distribution method.

References

1. A. Hajduga, Proceedings of 23rd International Conference on Methods & Models in Automation & Robotics (MMAR), 843–848 (2018)
2. A. Kopczyński, P. Krawczyk, J. Lasocki, E3S Web of Conferences **44**, 00073 (2018)
3. A. Hajduga, Proceedings of 23rd International Conference on Methods & Models in Automation & Robotics (MMAR), 622–627 (2018)
4. A. Szumanowski, Y. Chang, Z. Liu, P. Krawczyk, Int. J. Vehicle Des. **76**, 1–19 (2018)
5. A. Kopczyński, P. Piórkowski, P. Roszczyk, IOP Conf. Ser. Mater. Sci. Eng. **421**, 022016 (2018)
6. A. Kopczyński, Z. Liu, P. Krawczyk, E3S Web of Conferences **44**, 00074 (2018)
7. Z. Chłopek, J. Lasocki, P. Wójcik, A.J. Badyda, Int. J. Green Energy **15**, 773–779 (2018)
8. A. Kopczyński, Logistyka: czasopismo dla profesjonalistów **3**, 2269 (2015)
9. N. Mutoh, IEEE Trans. Ind. Electron. **59**, 10 (2012)
10. N. Mutoh, K. Takita, Conference Record of the 2004 IEEE Industry Applications Conference, 39th IAS Annual Meeting, 8245720 (2004)
11. L. Wei, H. Hongwen, Y. P. Yang, C. P. Lo, Control Eng. Pract. **16**, 1285–1292 (2008)
12. H. Hongwen, P. Jiankun, X. Rui, F. Hao, Energies **7**, 3748–3763 (2014)
13. W. Liu, H. He, J. Peng, Math. Probl. Eng. **2013**, 408965 (2013)
14. https://www.tesla.com/sites/default/files/tesla-model-s.pdf
15. A. Kopczyński, P. Krawczyk, Proceedings of the Institute of Vehicles **112**, 49 (2017)
16. D. Schramm, M. Hiller, *Vehicle Dynamics Modeling and Simulation* (Springer, 2014)

Power distribution in multi-motor (AWD) powertrain of electric vehicle

Artur Kopczyński[1,*], and *Paweł* Roszczyk[1]

[1]Warsaw University of Technology, Department of Multisource Propulsion Systems, ul. Narbutta 84, 02-524 Warsaw, Poland

Abstract. This article presents the results of analysis of power distribution in an electric vehicle independent all-wheel drive. The utilized method of velocity distribution takes into account the change in the motion resistance occurring on particular wheel. Moreover, the method of determining the change of vertical loads on traction wheels is also described. Theoretical considerations were verified on a dedicated laboratory stand that allows to perform real time simulation for analysed powertrain structure. The results of two different scenarios of vehicle driving in curvilinear motion are presented.

1 Introduction

In the last decade, a significant increase in the popularity of electric road vehicles has been observed. Therefore, the intensification of scientific research devoted to this subject is an indispensable process. Most of these studies raise issues related to energy analysis of the propulsion system [1, 2], component modelling [3, 4] and the possibility to reduce energy consumption [5, 6]. While other studies consider ecological aspects and the impact of transport on the environment [7, 8]. The variety of power trains structures and the increasing safety requirements pose new problems for designers and researchers. In this paper, the aspects of motion kinematics of a four-wheeled electric vehicle equipped with independent drive of all wheels will be elaborated. These considerations constitute a continuation and extension of the studies presented in [9–11]. In contrast to results described in [12–15], the loss of vehicle stability will not be scrutinised but the power distribution between vehicle wheels will be analysed depending on the parameters set by the driver, in order to meet the kinematic conditions and avoid possible disturbances in the movement trajectory.

2 Electric vehicle independent All-Wheel Drive (AWD)

According to the scheme presented in Figure 1, the vehicle is driven by an electric motors (EM) controlled by a motor controllers (MC) and powered by an electrochemical battery (Bat). EM transmit the drive/braking torque to the wheels via drive shafts. A transmission with a fixed ratio can be mounted between the motors and the traction wheels. The prototype vehicle Fiat Phylla and the commercially available Mercedes SLS AMG Electric Drive are

* Corresponding author: artur.kopczynski@pw.edu.pl

examples of such construction. EM can also be mounted directly in the wheel hubs which save space inside a car. This solution allows for the bus floor lowering and was utilized in the buses of the Chinese company BYD. However, in typical passenger cars the dynamics of movement is definitely bigger. The hub motors cause an unsprang mass increase, which significantly reduces the stability of vehicle movement and performance.

The Central Control Unit (CCU) which is responsible for the proper functioning of the propulsion system, determines the mode and parameters of individual EM on the basis of parameters set by the driver (steering angle, acceleration pedal position and brake pedal position). In Figure 1 the mechanical and electrical connections are presented with continuous (——) and dashed (– – –) lines, respectively. While the arrows indicate the possible direction of the energy flow depending on the EM operating mode (acceleration/braking).

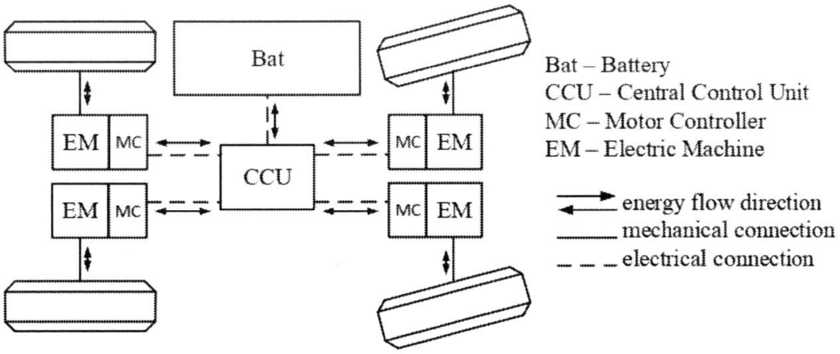

Fig. 1. Power train configuration of electric vehicle multi-motor independent AWD.

3 Modelling

In a curvilinear motion, each wheel of a four-wheeled vehicle moves on a different trajectory, which means that each wheel has a different speed. As shown in Figure 2, the speed of each wheel depends on the longitudinal velocity of the centre of gravity (COG) of the vehicle and steering wheel angle (set by the driver) as well as on the geometric parameters of the vehicle (front and rear wheels track – d_f, d_r, wheelbase – L). In the conventional cars, a differential mechanism with two degrees of freedom is responsible for the difference in speeds compensation. In an independent drive this speed compensation must be carried out by controlling the operation of particular EM.

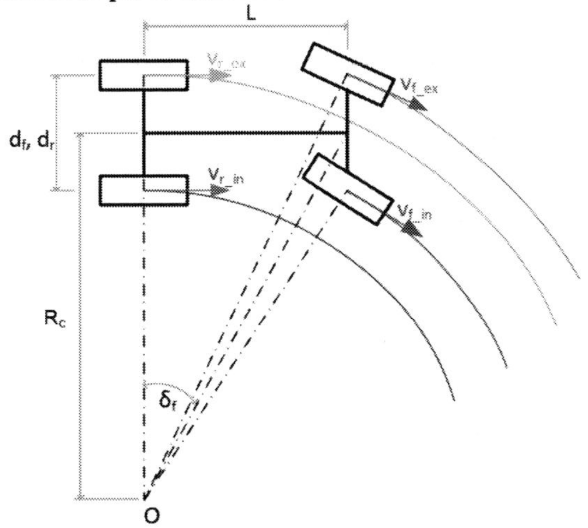

Fig. 2. Kinematics of four-wheeled vehicle during cornering.

Equations allowing to determine the wheel speed of a four-wheeled vehicle with the Ackerman steering mechanism of front wheels are described and derived in the literature [e.g. 9, 16] and are also presented below.

The angular speed of the front external wheel ω_{f_ex} is described by the equation:

$$\omega_{f\_ex}(t) = [v_x(t)/r_d]\{1 + tan(\delta_f)d_f/L + tan^2(\delta_f)[1+d_f^2/L^2]\}^{1/2} \quad (1)$$

The following equation was obtained for the front internal wheel ω_{f_in}:

$$\omega_{f\_in}(t) = [v_x(t)/r_d]\{1 - tan\ (\delta_f)d_f/L + tan^2(\delta_f)[1 + d_f^2/L^2]\}\ ^{1/2} \quad (2)$$

The angular speed of the rear external wheel ω_{r_ex} is described by the equation:

$$\omega_{r\_ex}(t) = [v_x(t)/r_d]\{1 + [d_r /2L]tan(\delta_f)\} \quad (3)$$

And the internal wheel of the rear axis ω_{r_in} using the equation:

$$\omega_{r\_in}(t) = [v_x(t)/r_d]\{1 - [d_r /2L]tan(\delta_f)\} \quad (4)$$

where v_x means vehicle longitudinal speed, δ_f – steering angle of virtual wheel located in the middle of front axis, r_d - tyre dynamic radius, d_f, d_r – front and rear wheel track, L – wheelbase.

3.1 Forces acting on vehicle's wheels

The vehicle moving on a flat surface is mainly influenced by the forces of aerodynamic drag and rolling resistance. If there is also a change of vehicle speed (e.g. acceleration), the inertia resistance force is also present. Curvilinear movement is additionally influenced by the resistance of the cornering. The rolling F_R and cornering F_C resistance, presented by Equations (5) and (6), are directly associated with traction wheels [16].

$$F_R = F_z f_t \quad (5)$$

The rolling resistance depends on the type of tyre and the surface on which the vehicle moves. This is determined by rolling resistance coefficient $f_t = f_0 + v_x(t)^2 10^{-5}$[–], where f_0 depends on tyre properties and type of surface .

$$F_C = F_z[v_x(t)^2/R_c g]\{[l_r /L]sin(\theta_f) + [l_f /L]sin(\theta_r)\} \quad (6)$$

The cornering resistance depends on many factors and is described by more complicated mathematical equation. Lateral deformability of the tyre, defined by the side sleep angles (for front wheels θ_f and for rear wheels θ_r), is the main reason of occurrence of the cornering resistance. Moreover, the cornering resistance depends on relation of centrifugal acceleration ($v_x(t)^2/R_c$, where R_c means cornering radius), to acceleration due to gravity g. In addition, the value of cornering resistance depends on the location of the centre of gravity (COG), as specified: l_f, l_r – front and rear axis to COG, and L means wheelbase ($L = l_f + l_r$). Units of all variables are compatible with the SI system.

Both resistance forces depend on the vertical load F_{zi} on the given wheel. Due to this fact in the next chapter a universal method of determination of the instantaneous load on particular wheels of the vehicle will be presented.

3.2 Wheel vertical load: pitch and roll

The method of determination of vertical load caused by pitch (during accelerating or braking) was presented in the work [10]. Forces of vertical load of a given axis during vehicle accelerating are expressed by Equations (7) and (8), where a means vehicle acceleration or deceleration and h – height of COG from ground.

$$F_{zf} = \{m[gl_r - ah]\}/L \quad (7)$$

$$F_{zr} = \{m[gl_f + ah]\}/L \quad (8)$$

In case of vehicle braking the forces F_{zf} and F_{zr} are defined by similar Equations to (7, 8) but the signs in numerators are opposite.

Determination of the load on each of the wheels caused by roll (during vehicle cornering) is troublesome because a change in the vertical load depends on many factors and constitutes the individual feature of the particular car and its suspension. Whereas the simplified universal model presented below, takes into account the following assumptions that will not significantly affect the results:
- The considered system is an isolated system on which no external forces act;
- Coefficient of tyre adhesion is constant in time and it is the same for all of the wheels;
- There are no slippages.

Another important assumption can be formulated as follows:
- Internal wheels of the vehicle during cornering are unloaded by the same value by which the external wheels are loaded.

This assumption can be described as follows:

$$mg = F_{z\_in} + F_{z\_ex} \quad (9)$$

For further considerations, please refer to Figure 3.

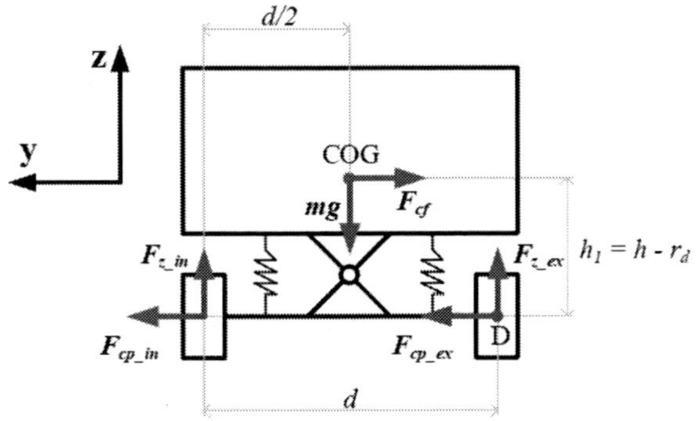

Fig. 3. Forces acting on a vehicle during cornering.

The centrifugal force F_{cf} of such a setting equals to:

$$F_{cf} = mv_x^2/R_c \quad (10)$$

The sum of moments in terms of point D:

$$\Sigma M_D = 0 \quad (11)$$

$$F_{cf}h_1 + F_{z\_in} - mgd/2 = 0$$

From (10, 11) the following is obtained:

$$F_{z\_in} = m[g/2 - v_x^2 h_1/R_c d] \tag{12}$$

The sum of forces acting on the vehicle vertically (along with axis z):

$$\Sigma F_z = 0 \tag{13}$$

$$F_{z\_ex} = mg - F_{z\_in} = m[g/2 + v_x^2 h_1/R_c d]$$

4 Study

In order to analyse the control of a multi-motor drive, a special laboratory stand was prepared. The block diagram of the stand is presented in Figure 4. The main components are two sets consisting of two electric machines (EM-1 with EM-3 and EM-2 with EM-4) with rotor shafts connected with each other. The inertia discs J_1 and J_2 are located on shafts that connect the motors. In this set EM-1 works as a drive motor, and EM-3 simulates the motion resistance (generator mode). The situation for the second set of machines looks the same: EM-2 acts as a driving motor and EM-4 as load. Each motor is connected to its controller MC-i equipped with current and voltage probes. Two encoders (E) are used for measuring the rotational speed, one for each set. Power for motors is provided by lead-acid electrochemical batteries that also accumulate energy from EMs operating in generator mode. The verification of the operation of the drive of one axle (front or rear) is possible during one test. The tested sets represent the right and left wheels.

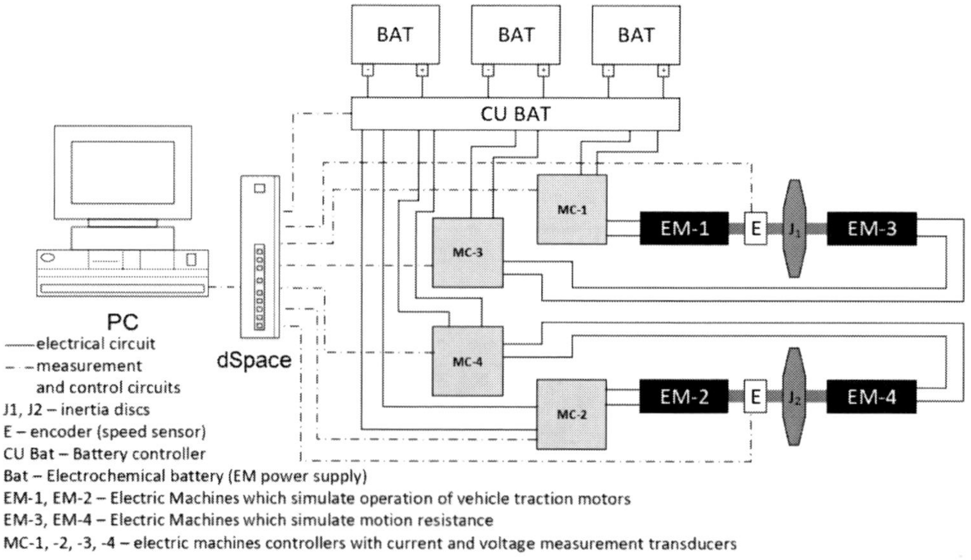

Fig. 4. Bloc diagram of laboratory stand.

The system is managed by rapid control prototyping hardware with dSpace software which allows to perform real time simulation. On the basis of the equations from Chapter 3, a model was developed in the Matlab Simulink environment, to which the elements responsible for cooperation with the real object (controllers, sensors) were attached. The model developed in Simulink has been automatically converted to the dSpace environment. The instantaneous parameters determined from the simulation model i.e. speeds for drive motors (EM-1 and EM 2) and load current for motors (EM-3 and EM-4), were taken as set parameters during tests. The determined values have been scaled to the capabilities of the stand. After taking into account own non-linear resistances of the stand, the calculated resistance forces were scaled to the load current values.

The image of the stand and the measuring and control panel prepared in the dSpace software are presented in Figures 5a and 5b, respectively. The panel can be used to select the reference vehicle speed and the reference steering angle. During the test, the waveforms of given and measured parameters are displayed in plots.

It has been defined that each test is carried out in the following procedure: Initially, the system accelerates to the reference speed assuming a straight line movement and after the speed is reached, the steering angle changes to the reference value. Then, when the vehicle is cornering with a constant radius, braking begins. Set values are determined for each instantaneous condition. In this study, two scenarios were verified for a reference speed of 90 km/h and a steering angle of 20° and a speed of 50 km/h and a steering angle of 45°. The tests were carried out for the simulation of the operation of the front and rear drive for the same parameters.

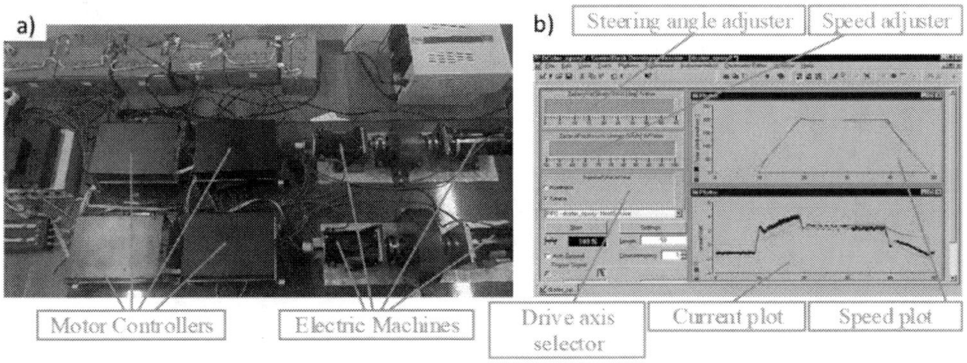

Fig. 5. a) Photo of laboratory stand, b) dSpace control panel.

Table 1 contains the parameters used in the simulation.

Table 1. Input data.

Vehicle mass	**Frontal area**	**Aerodynamic drag**	**Rolling resistance coefficient**	**Gear total ratio**	**Steering column ratio**	**Centre of gravity height**
m = 600 kg	A = 1.3 m^2	C_d = 0.45 [-]	f_0 = 0.01 [-]	i_c = 4	i_{SC} = 15.7	h = 0.45 m
Wheelbase	Front axis to COG	Rear axis to COG	Front wheel track	Rear wheel track	Wheel dynamic radius	Laboratory stand scale
L = 2.3 m	l_f = 1.3 m	l_r = 1 m	df = 1.42 m	dr = 1.4 m	r_d = 0.285 m	1 : 100

4.1 Real-time simulation results

The results for the tests are shown in Figures 6 and 7. A comparison of the set and measured signal is included in each plot. In order to show better the waveform in time when the steering angle changes only the relevant part of each plot has been presented. The acceleration and braking phases have been omitted because they are not relevant to the considerations undertaken in this work.

In the case of speed waveforms, it is visible that the measured signals follow the set values. When there is a change in speed, the differences appear in dynamic states, which is related to the mechanical inertia of the elements. A good correlation of the results was also obtained for the current waveforms. However, in this case the calculated load (from the motion resistance) is compared to the actual current waveform of the drive motors. By this, the correct operation of the stand, which satisfactorily reflects the tested powertrain system together with the application of instantaneous motion resistance on particular wheels, has been confirmed. A better reflection of the given speeds was obtained for the test carried out at a speed of 50 km/h, however, worse was obtained in the case of currents. The improvement in reaching the set parameter can be obtained by improving the settings of the regulators. Unfortunately, the tested object is nonlinear and it is impossible to select these settings of regulators based on the classical theory, simultaneously ensuring correct operation in the full range of speeds for both analysed parameters (speed and current).

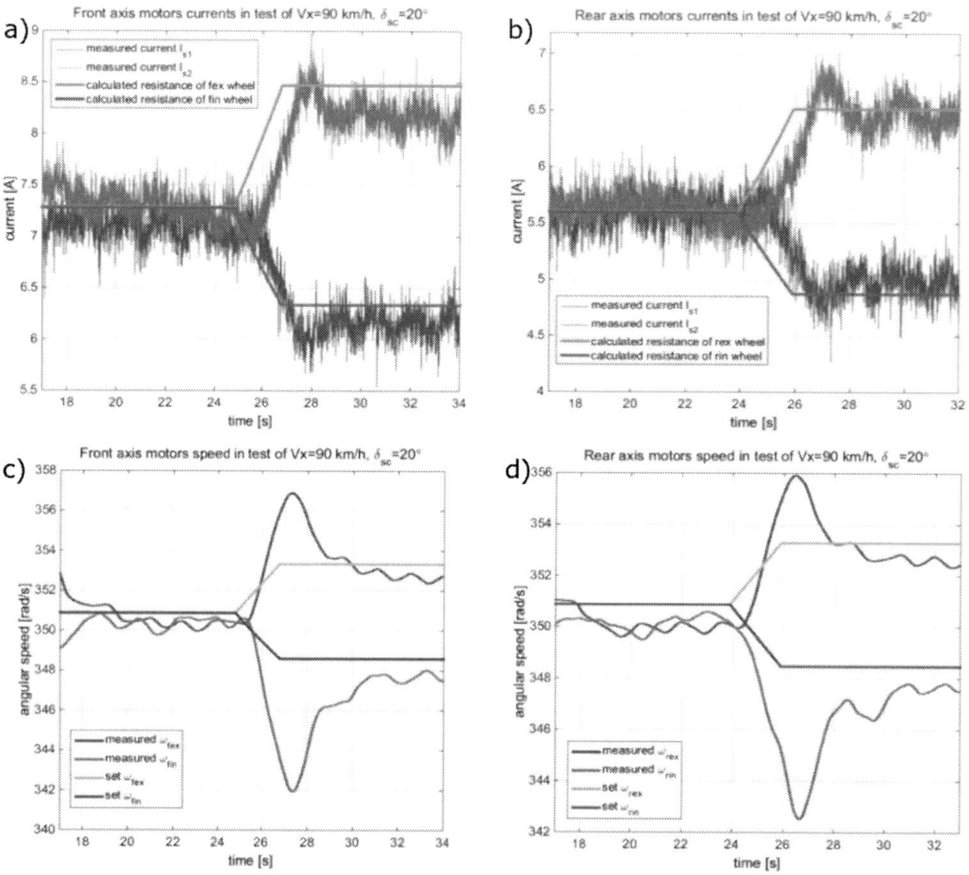

Fig. 6. Results of real time simulation of the test of vehicle speed equal to 90 km/h and steering angle δsc equal to 20° a) front axis currents, b) rear axis currents, c) front axis wheels angular speeds, d) rear axle wheel angular speeds.

https://doi.org/10.1051/e3sconf/201910000038

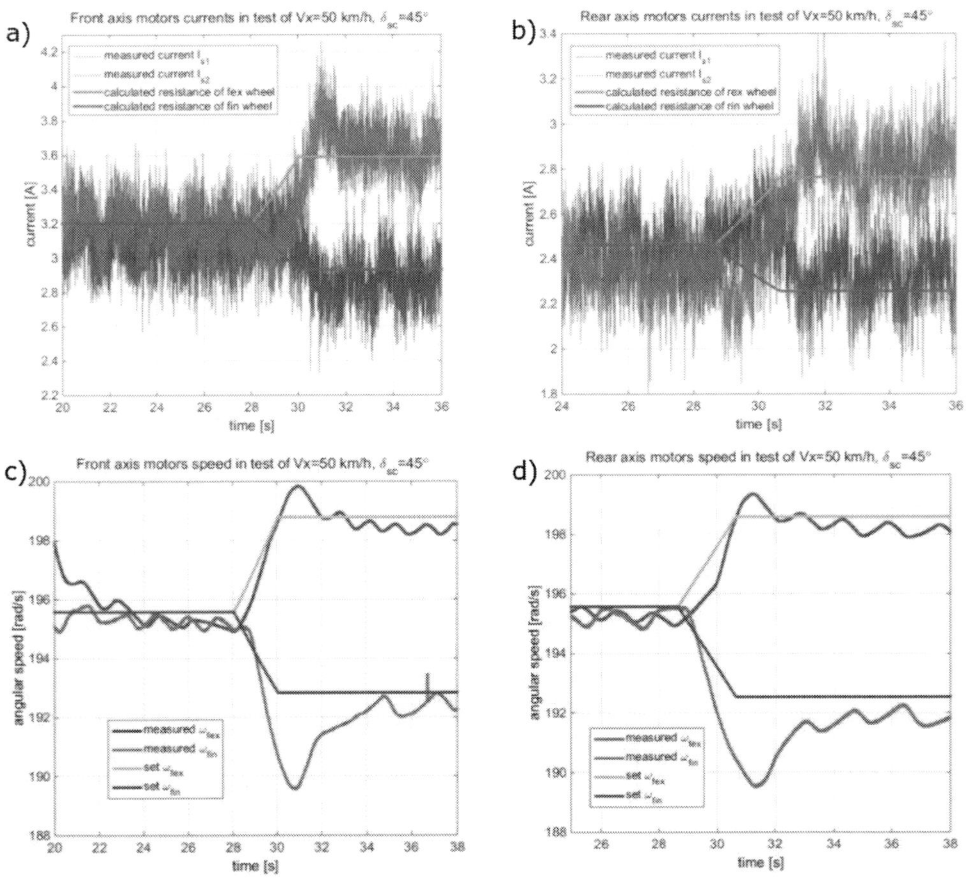

Fig. 7. Results of real time simulation of the test of vehicle speed equal to 50 km/h and steering angle δsc equal to 45° a) front axis currents, b) rear axis currents, c) front axis wheels angular speeds, d) rear axle wheel angular speeds.

The current percentage error for the worst case (test front drive 90 km/h) is presented in Figure 8a while the speed error (for the worst case test – rear drive with the speed of 50 km/h) in Figure 8b. The speed error is below 2% which is satisfactory, while the mean value of the current error amounts to approx. 10 % (excluding the peaks). Given the small values of the measured current and the possible accuracy of the measurement and the relatively large own stand resistances, this error should also be considered as acceptable.

https://doi.org/10.1051/e3sconf/201910000038

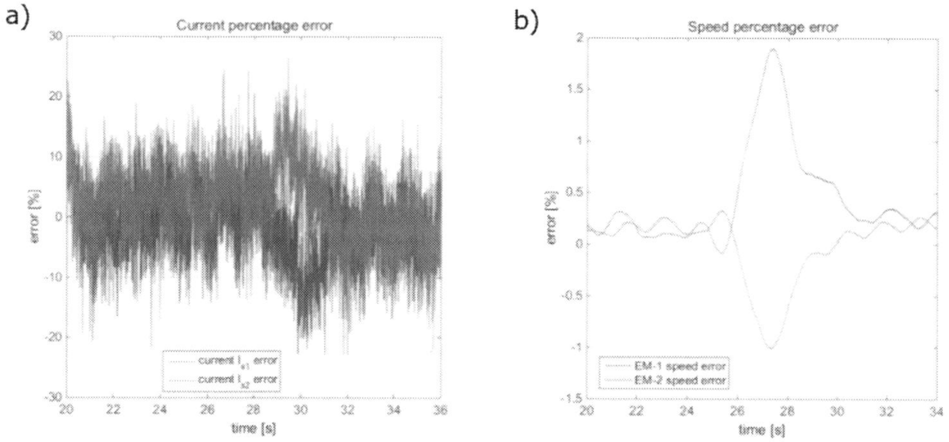

Fig. 8. a) example of current percentage error, b) example of speed percentage error.

5 Conclusion

The article presents the method of speed distribution for electric vehicle independent all-wheel drive. The change in motion resistance on each wheel was taken into account depending on the instantaneous values of the parameters determining the change in vertical loads on each wheel. The simplifications used in the determination of vertical loads have been made intentionally to create a more versatile model. However, it is expected that these simplifications will not significantly affect the obtained results. For the verification of theoretical considerations, tests were carried out on a laboratory stand which reflects the work of the analysed propulsion system. The correctness of the operation of the laboratory stand and thus of the analysed method of multi-motor drive control was demonstrated. A drawback constitutes the high value of the stand's own resistance, which also hampered the adjustment of the settings of the regulators, which translated into the accuracy of the results. Nevertheless, the accuracy is sufficient to draw valuable conclusions. Even though in a real environment there is a number of other factors that have been omitted in these considerations such as the occurrence of tyre slip, road gradient, weather condition, etc., the presented research results take into account these that have a main, fundamental impact.

References

1. Z. Chłopek, J. Lasocki, P. Wójcik, A. J. Badyda, Int. J. Green Energy **15**, 773–779 (2018)
2. A. Kopczyński, Z. Liu, P. Krawczyk, E3S Web of Conferences **44**, 00074 (2018)
3. E. Jankowska, K. Kopciuch, M. Błażejczak, W. Majchrzycki, P. Piórkowski, A. Chmielewski, K.J. Bogdziński, Advances in Automation, Robotics and Measurement Techniques **743**, 339–349 (2018)
4. A. Hajduga, Proceedings of 23rd International Conference on Methods & Models in Automation & Robotics (MMAR), 843–848 (2018)
5. A. Hajduga, Proceedings of 23rd International Conference on Methods & Models in Automation & Robotics (MMAR), 622–627 (2018)
6. A. Kopczyński, P. Piórkowski, P. Roszczyk, IOP Conf. Ser. Mater. Sci. Eng. **421**, 022016 (2018)

7. Z. Chłopek, J. Biedrzycki, J. Lasocki, P. Wójcik, Eksploatacja i Niezawodnosc – Maintenance and Reliability **17**, 35–41 (2015)
8. A. Kopczyński, P. Krawczyk, J. Lasocki, E3S Web of Conferences **44**, 00073 (2018)
9. A. Szumanowski, Ed., *Designing of electronic differentials of electric road vehicles* (WN ITeE-PIB, Radom, 2007)
10. A. Kopczyński, Logistyka: czasopismo dla profesjonalistów **3**, 2269 (2015)
11. A. Kopczyński, P. Krawczyk, Proceedings of the Institute of Vehicles **112**, 49 (2017)
12. W. Cao, Z. Liu, Y. Chang, A. Szumanowski, Math. Probl. Eng. **2017**, 5170492 (2017)
13. W. Cao, H. Liu, Y. Chang, Z. Liu, A. Szumanowski, Energies **10**, 1641 (2017)
14. W. Liu, H. He, J. Peng, Math. Probl. Eng. **2013**, 408965 (2013)
15. H. He, J. Peng, R. Xiong, H. Fan, Energies **7**, 3748–3763 (2014)
16. D. Schramm, M. Hiller. *Vehicle Dynamics Modeling and Simulation* (Springer, 2014)
17. A. Szumanowski, Y. Chang, Z. Liu, P. Krawczyk, Int. J. Vehicle Des. **76**, 1–19 (2018)

Economic efficiency of biogas generation from food product waste

Viktor Koval[1], *Inesa* Mikhno[2], *Gabriela* Hajduga[3,*], and *Krzysztof Gaska*[4]

[1]Department of Applied Economics, Odessa Institute of Trade and Economics of Kyiv National University of Trade and Economics Odessa, 19 Kyoto Street, 02156 Kyiv, Ukraine
[2]Institute of Water Problems and Land Reclamation NAAS, 37 Vasylkivska Street, 03022 Kyiv, Ukraine.
[3]Cracow University of Technology, Department of Environmental Engineering, Department of Environmental Technologies, ul. Warszawska 24, 31-155 Kraków, Poland
[4]Silesian University of Technology, Faculty of Energy and Environmental Engineering, Institute of Water and Wastewater Engineering, ul. Stanisława Konarskiego 18, 44-100 Gliwice, Poland

Abstract. The issue of waste accumulation has become one of the global problems of humanity. In Ukraine, the main method of waste management is landfill depositing, which is unproductive and affects the ecosystem negatively. However, large landfill sites should be used to produce biogas, thereby reducing the environmental burden and the earned revenue should be used to introduce recycling and a gradual transition to a European waste management policy. The aim of our study was to investigate the state of waste recycling and the possibility of using waste landfills and food with lost consumptive qualities for biogas generation. To analyze the economic efficiency of the installation of biogas equipment in landfills and to investigate the current state of waste management in Ukraine. The research demonstrates that at present in Ukraine the problem of waste accumulation becomes actual. At the same time, the number of large waste landfills is increasing. It is proved that at landfills with an area of more than 50 hectares, the installation of biogas equipment is cost-effective, and the raw material can be distributed at the feed-in tariff, thus solving the energy problem.

1 Introduction

Currently, the problem of resource saving is becoming relevant in energy economy. One of the most promising sources of energy in recent years is the biogas produced as a result of decomposition of solid household waste organic part in anaerobic conditions that shortly arise after its burial. Biogas is considered as an alternative source of energy, but the different structural composition of the landfills and their specifics should be taken into account.

At present, there is growth of public awareness of the limited nature of environmental resources, which promotes the development and implementation of new economic models that can effectively manage resources and recycle them [1–3]. The amount of different wastes

* Corresponding author: gabriela.hajduga@gmail.com

increases in relation to the speed of urbanization, but inappropriate treatment of wastes leads to health hazards and urban environment degradation [1, 4, 5].

However, at the time of global demand for energy escalation, the importance of ecologically clean environment is also growing. Due to the extremely high energy potential, along with the safety and purity of its operation, a biogas plant, as one of the methods of utilization of food waste and energy complex, can be considered as a major investment in the future [5].

The development of biogas plants has a potential for growth and it can contribute to the development of the economy [4, 6], since biogas can reduce the greenhouse effect by diminishing the amount of carbon dioxide and methane generated during the process of waste biogas generation [2, 4, 7].

2 Materials and methods

Theoretical and methodological basis of the research consists in general scientific and special methods. The general scientific methods such as monographic and dialectic are used for the process of collecting, systematizing and processing information for the research, as well as theoretical generalizations; the methods of synthesis and comparison are implemented for processing the statistical data necessary for assessing the state, dynamics and comparison of the studied indicators [8–10]. Analytical methods and analogical methods for studying waste management in supermarkets of different countries of the world were used. In addition to the general scientific methods, method of economic and mathematical modeling is applied [11]. The basis of the study is not only economic analysis, but also the consideration of the problem from the environmental point of view, which makes it possible to make this study systemic and complete. The information base of the study was legislative and regulatory acts of the European countries, analytical reviews and reports as well as statistical data on ecology and natural resources.

3 Results

An increase in the amount of waste in Ukraine and their irrational distribution and use have created huge mountains of garbage and trash in the territory of about ten percent of the country's total area. The dominant direction of energy use of biogas is the generation of electricity with the subsequent supply to the electricity grid. Over the last decade, biomethane production projects have been intensively developed, followed by pumped into the gas supply network.

The profitability of biogas production increases with the organic waste percentage in the structure of utility waste. The disadvantage of creating biogas production sites is the increase of self-inflammation risk and explosion hazard. Biogas produced at landfills is used for power generation mostly. At the Municipal solid waste-5 landfill, 1 m. 3 of biogas can generate with 1.5 kWh of electricity.

Currently, in Ukraine, the amount of waste is increasing due to its accumulation at landfills that could be used for biogas production (Table 1). From Table 1, we see there is not any growth in the rate of waste formation in some categories of municipal solid waste, this process occurs due to the economic crisis and inflation in the country, but we can observe constant accumulation of the total storage weight.

Table 1. Waste generation according to the categories of materials that could be used for biogas production in thousands of tons [12].

	Generated (thousands of tons per year)					**Utilized (thousands of tons per year)**			**Burnt to produce energy (thousands of tons per year)**		
	2012	2013	2014	2015	2016	2014	2015	2016	2014	2015	2016
Animal waste and mixed food waste	12850,0	9864,4	954,2	897,0	996,0	485,6	220,0	315,1	2,1	2,0	0,0
Waste of vegetable origin	267,5	287,2	9061,4	7742,3	8606,0	3020,4	2674,4	3158,4	396,9	462,0	392,3
Animal droppings, urine and manure	5062,0	4736,1	4037,2	4938,0	4288,7	3082,0	3231,5	3146,9	0,0	0,0	0,0
Domestic household waste and similar waste	9713,3	10804	7125,7	6789,2	6946,2	3,8	4,0	6,5	152,8	256,4	257,3

However, the industrial usage of biogas is possible after a landfill creation; it will become economically efficient if more than 1 million tons of solid domestic waste is reached. Since today, there is a large number of waste deposits with such characteristics on the territory of Ukraine, an interest in the task of their geographical and production identification for the purpose of cost-effective extraction of biogas is increasing (Figure 1).

Fig. 1. Interactive map of waste disposal sites register in Ukraine as of 2018 [13].

During the process of creating a complex for biogas production, it is important to calculate the efficiency of its extraction from household waste for a specific waste landfill (Equation 1). The peculiarities of this method are the installation of the equipment on the landfill site, which could also have negative sides and affect the ecosystem around (it takes into account the indicators A, B) [8, 9–12]. Fractional waste treatment is considered to be more environmentally friend as it leads to reduce of residues amount (infinitesimal) and refusal of landfills as a form of waste storage as well as gradual processing of existing landfills [10–13]. However, in the current situation in Ukraine, it is expedient to install biogas plants until the process of recycling process throughout the country is established.

The efficiency of biogas extraction can be calculated according to the following formula:

$$E = \frac{\sum_{i=1}^{n}(C_i \cdot Q_i) - \left(\sum_{j=1}^{m} A_j + \sum_{k=1}^{l} B_k\right)}{\sum_{i=1}^{n}(D_i \cdot Q_i)} \quad (1)$$

E – the efficiency of introducing biogas production by installing equipment at landfills for solid domestic waste, C_i – the rime cost of the products sold, Q_i – amount of sold products for a certain period of time (horsepower of equipment), D – the price of biogas or energy unit sold, A – the sum of expenses spent (m) on the amount of environmental protection measures and the restoration of the environment near the waste landfills, B – the sum of compensation expenses spent (l) for the population living in the vicinity of the landfill's impact.

It should be noted, that the investment project with discounted net revenue must be higher than the contribution of equipment and costs associated with the activity of the complex biogas production (E> NPV). For European countries, the E index should reach at least 2–3%, while for countries with small planning horizons, that Ukraine belongs to it should be more than 15%, which complicates the implementation of biogas collection systems, but taking into account the large raw material base and availability landfills more than 50 hectares, the process of biogas production becomes attractive.

With increasing landfill volume increases the amount of biogas production (R).

$$R = k \cdot \Delta s \cdot \Delta h \quad (2)$$

k – the index of biogas production growth with the increase of landfill waste area size of and height garbage mountain, h – (from the zero mark – the bottom of the ditch).

It should be noted that with the growth and fall of the volume of landfill there is different effect on the nature of the indicators of the investment project of installing equipment in landfills solid household waste. It depends on the influence on their life cycle of external factors; therefore, it is advisable to apply the coefficient of elasticity in the stages of the life cycle [5, 11, 14–18] to determine the degree of influence of external factors:

$$\mathrm{KE} = \left| \frac{\frac{P_{(x)t} - P_{(x)t-1}}{P_{(x)t-1}}}{\frac{X_t - X_{t-1}}{X_{t-1}}} \right| \quad (3)$$

KE – elasticity coefficient, $P_{(x)t}$, $P_{(x)t-1}$ – profitability indicator x enterprise activity at stages of the life cycle in the periods t and t-1 [%], X, X_{t-1} – external factor of the environment of the enterprise operation in periods t and t-1 [%], $P_{(x)}$, $X > 0$.

According to the analysis the increase in landfill area of 10 hectares and garbage mountain's height of 10 meters, R increases by 0.5, that increases and profitability and proves the efficiency of large landfills and generation of 120 m^3 gas from 1 ton of food waste can increase biogas production up to 200 m^3 in mixed food waste.

Nowadays in Ukraine, there is growth in biofuel production by a method of crop products processing and by biogas production at landfills (Figure 2). At the garbage landfill # 5 in the village of Pidhirtsi, Obukhiv district, Kyiv region, There is a line of five biogas engines with an installed capacity of 177 kW each, and the landfill gas company also extracts at Municipal solid waste- landfill in the villages of Gliboke (capacity 1.06 MW) and Rozhivka (capacity – 1.06 MW). The same biogas plants operate on landfills in the Zhytomyr and Mykolaiv regions. In a year, LLC LNK plans to launch a biogas recycling plant to produce electricity in Nikolaev. Biogas plants for power generation operate on two landfills of solid waste at Mariupol and at a municipal landfill near Vinnitsa. Lviv Yeast Plant, as well as Kyiv and Zapizhzhia Brewery plants receive biogas as a result of in-depth sewage treatment [17, 19].

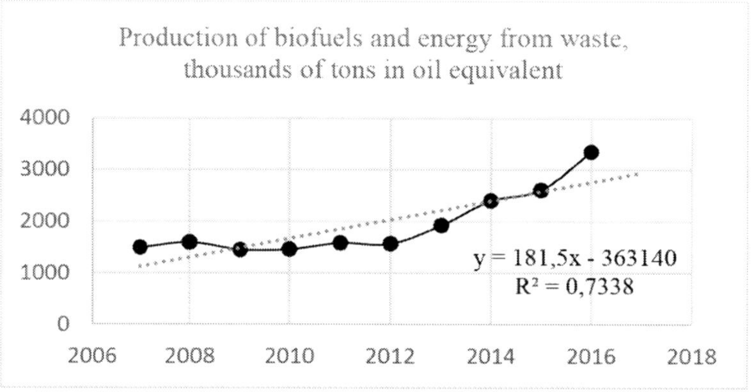

Fig. 2. Dynamics of production of biofuel and energy from waste in Ukraine for 2007–2016.

A large part of household waste is food. This component has a decomposition term of 1–6 months, and therefore they require rapid disposal. For example, in Germany, companies that have equipment for biogas production are involved in the recycling of food waste. Depending on the structure of food waste, from 1 ton of the food waste from 30 m^3 to 800 m^3 of biogas (lactose) can be obtained, which is an average of 130 m^3 per 1 ton of food waste. From 1 m^3 of biogas it is possible to produce 1.5–2.2 kW of electricity depending on the weight of methane. Biogas derived from biomass can be sold at gas stations for the needs of cars and agricultural machinery, or users of the gas network. According to the Law of Ukraine "On Electric Power", enterprises that produce electricity from alternative fuels can sell it, and the state must implement green tariffs purchasing the energy, which is 2.6 times more than wholesale tariffs [7, 20–22].

Currently, "green tariff" in Ukraine is set for 132 enterprises; seven of them produce biogas [3, 23–27]. According to the resolution of the Cabinet of Ministers of Ukraine dated on 02.02.2015, #220, electricity is retailed to the population from March 1, 2017, at a price of 0.9 UAH (green tariff is 1.1 UAH that equals $ 0.04) (up to 100 kWh per month). Each year there are 6,9 million tons food waste (some of the waste is not recycled using different techniques), it is possible to get 1.2 billion kW. electricity ($ 0.05 billion per year) [1, 27–29]. This indicator can be substantially increased in case of introduction of European legislation in relation to the consumption of out of products by population [28, 29].

In case of introduction of equipment for biogas production on the territories of large garbage landfills (more than 50 hectares), there are economic as well as social and environmental benefits (Figure 3).

https://doi.org/10.1051/e3sconf/201910000039

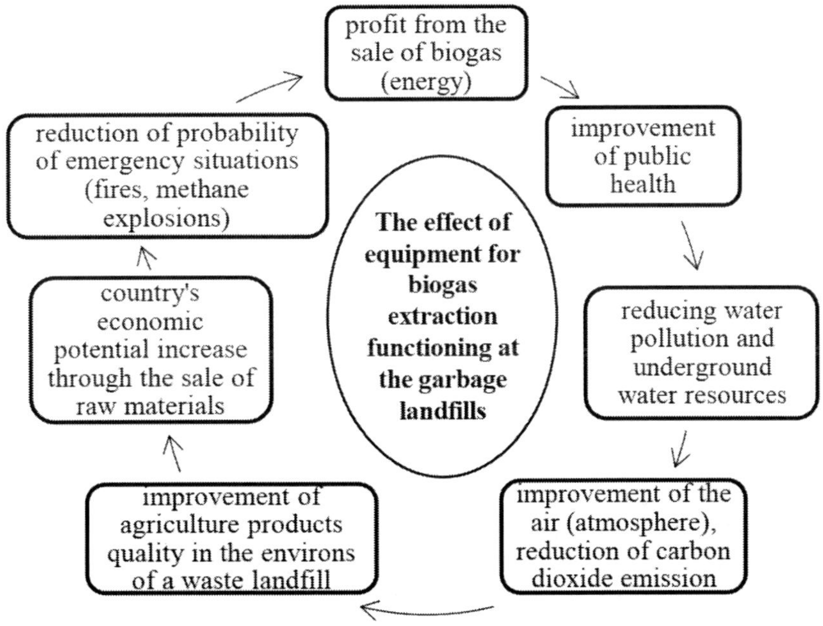

Fig. 3 The effect of equipment for biogas extraction functioning at the garbage landfills.

In Ukraine, there are not any regulations on food waste disposal that has not been merchandised, the life cycle of food products is increased by its processing and repackaging.

Figure 4 describes the life circle of food products in Ukraine, it depicts the desirable patterns of products processing, and that society should strive for. In particular, the role of entrepreneurs grows, as they are the basis for goods quality shaping process. Therefore, only a small amount of the food products is utilized.

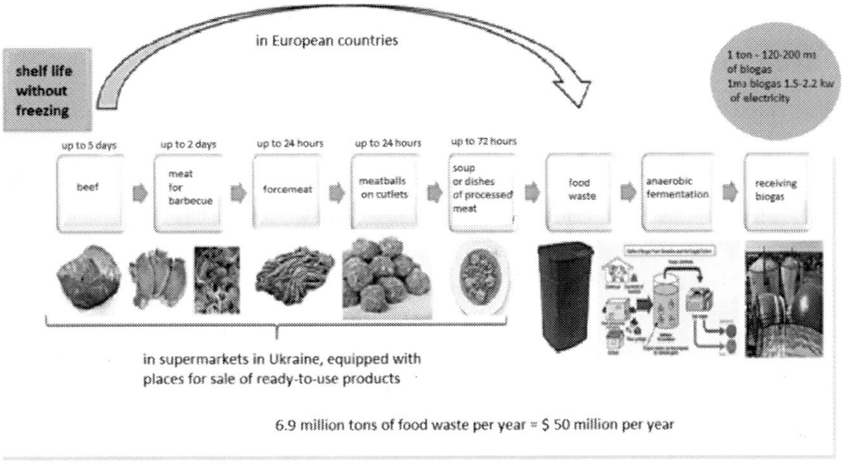

Fig. 4. Life circle of food products (that were not merchandised) and desirable patterns of products processing, on the example of beef (source: created by the authors)

In Germany, organic waste is recycled through recycling plants for biogas production. Currently, Germany has more than 100 waste recycling facilities, with the capacity of 18 million tons, although its annual production is 16 million tons. The government in Germany sets new tougher conditions for recycling for the society. Last year, a new law was passed on increasing the quota of supermarket processing of overdue products from 36 to 65%, and starting from 1 January 2021 it would be 70% [21, 30–32]. In France, since 2016, it is prohibited for supermarkets to recycle of food products. There are high penalties for the violation of the regulation that equal 3750 euros. The food products reached 75 % of their life circle, must be transferred to special charity food banks.

Creation of specially equipped centers for food processing, requires appropriate investments with the economic reliability justification and assessment of probabilistic nature of parameters of the environment [15, 33, 34].

4 Discussion

In Ukraine, it is advisable to strengthen the system of control over food products sale and improve the system of penalties. These activities will improve the quality of products; create a system of discounts on goods, which life circle has reached 50%. These changes can improve the work of transportation companies involved in the transport of waste to the points of garbage processing and introduce a waste sorting system. These actions would improve the quality of services in companies that sell food products, reduce the probability of poisoning among the population, and will lode to usage of overdue products for secondary production of biogas and change the existing waste management system. Taking into account the large number of landfills that operate in Ukraine, biogas equipment should be installed on the territories of landfills that there should be the installation of on their territory, the profitability of which is confirmed on similar sites in other countries of the world [16, 29]. At present, such technology is already in use in Ukraine, however, it is not massive. A number of private companies processing biogas are processing less than 1% of waste that can be used for biogas production. We are invited to make adjustments to the state budget for allocating funds for the construction of biogas complexes at all large landfills in Ukraine at public funds, which would help solve the energy problem and reduce heating and electricity prices for the population, as only systemic actions on the territory of the country and control over the implementation of all necessary measures can solve the existing problem and effectively use internal resources.

5 Conclusions and further research prospects

Nowadays, the problem of non-utilized waste accumulation is increasing worldwide. At the same time, raw materials and energy problems remain urgent. Biogas production, using a waste disposal method, becomes more widespread and cost-effective at landfills with the area of over 50 hectares. Since the main method of waste management is landfill in Ukraine, it is recommended to install biogas equipment on the territory of all large waste landfills. These actions can solve the energy problem as well as other problems, and in general, it can increases the potential of the country and affects the ecosystem positively. It is proposed to improve the control system at the points of food product sale, and to send expired goods to waste recycling points for biogas production. In connection with the signing of the Association Agreement with the European Union it is proposed to use the experience of European countries in relation to the strategy of waste management.

References

1. N. M. Bocken, I. de Pauw, C. Bakker, B. van der Grinten, Journal of Industrial and Production Engineering **33**, 303–308 (2006)
2. S. Begum, G. Anupoju, S. Sundergopal, S. Bhargava, Detritus **1**, 75–82 (2018)

3. Z. Kremljak, Annals of DAAAM & Proceedings, **28** (2017)
4. L. Deng, Renewable and Sustainable Energy Reviews **70**, 845–851 (2017)
5. M. Franchetti, J. Environ. Manage. **123**, 42–48 (2013)
6. H. Sun-Kee, S. Hang-Sik, Japca. J. Air. Waste. Ma. **54**, 2, 242–249 (2004)
7. J. Morrissey, J. Browne, Waste Manage. **24**, 297–308 (2004)
8. P. Aragonés-Beltrána, J. A. Mendoza-Rocab, A. Bes-Piáa, M. García-Melónb, E. Parra-Ruizb, J. Hazard. Mater. **164**, 1, 288–295 (2009)
9. G. Hajduga, A. Generowicz, E3S Web. Conf., **17** (2017)
10. K. A. Weitz, S. A. Thorneloe, S. R. Nishtala, S. Yarkosky, M. Zannes, Japca. J. Air. Waste. Ma. **52**, 9, 1000–1011 (2002)
11. K. Gaska, A. Generowicz, Architecture, Civil Engineering, Environment **10**, 117–126 (2017).
12. https://ukrstat.org.
13. https://ecomapa.gov.ua.
14. K. Gaska, A. J. Wandrasz, Waste Manage. **28**, 973–985 (2008)
15. I. S. Mikhno, Archivarius **4**, 49–54 (2015)
16. H. Hu, X. Li, A. Dung Nguyen, P. Kavan, Int. J. Environ. Res. Publ. Health **12**, 7593–7614 (2015)
17. K. Govindan, S. Rajendran, J. P.Sarkis, P. Murugesan, J. Clean. Prod. **98**, 1, 66–83 (2015)
18. O. Eriksson, M. Bisaillon, M. Haraldsson, J. Sundberg, J. Environ. Manage. **175**, 33–39 (2016)
19. http://www.cewep.eu/2018/09/25/congress-2018/
20. A. Generowicz, K. Gaska, G. Hajduga, E3S Web. Conf., **44** (2018)
21. https://eur-lex.europa.eu/eli/dir/2004/12/oj.
22. M. Marttunen, V. Belton, Eur. J. Oper. Res. **265**, 1, 178–194 (2018)
23. Klang, P. A. Vikman, H. Btattebř, Conserv. Recycling **38**, 317–334 (2003)
24. M. Garfe, S. Tondelli, A. Bonoli, Waste Manage. **29**, 2729–2739 (2009)
25. E. den Boer, A. Jędrczak, Z. Kowalski, J. Kulczycka, R. Szpadt, Waste Manage. **30**, 369–377 (2010)
26. http://aggeek.net/ru/technology/id/12-zanimatelnyh-faktov-o-biogaze-059
27. Generowicz, Pol. J. Environ. Stud. **23**, 2, 57–63 (2014)
28. http://www.biteco-energy.com/vyhod-biogaza-iz-razlichnogo-syrya.
29. W. Balcerza, A. Generowicz, Z. Mucha, Pol. J. Environ. Stud. **23**, 3, 983–987 (2014)
30. K. Jóźwiakowski, K. Mucha, A. Generowicz, Arch. Environ. Prot. **41**, 3, 76–82 (2015)
31. Y. P. Li, G. H. Huang, J. Environ. Manage. **81**, 188–209 (2006)
32. A. Generowicz, J. Kulczycka, Z. Kowalski, M. Banach, J. Environ. Manage. **92**, 4, 1314–1320 (2011)
33. V. Koval, Y. Prymush, V. Popova, Baltic Journal of Economic Studies **3**, 183–187 (2008)
34. M. Czop, 17th International Multidisciplinary Scientific GeoConference SGEM 2017, **17**, 52, 153–160 (2017)

The influence of the dye adsorption time on the DSSC performance

Ewelina Krawczak[1,*], and *Agata* Zdyb[1]

[1]Faculty of Environmental Engineering, Lublin University of Technology, ul. Nadbystrzycka 38, 20-618 Lublin, Poland

Abstract. Dye-sensitized solar cells (DSSC) of third generation photovoltaic technology, are nowadays one of the most investigated due to possibility to apply ecological and natural materials (dyes) such as alizarin. This paper reports the influence of electrode immersion time on alizarin-based dye-sensitized solar cells performance. The absorption spectra of alizarin dye were measured in the range of 300-800 nm. Fully structured dye-sensitized cells of working area equal to 0.8 cm^2 have been fabricated in the sandwich way using four different immersion times of the TiO_2 electrodes: 10', 40', 1 h, 24 h. The high-performance EL- HPE electrolyte was instilled into the space between electrodes. Current-voltage (I-V) dark and illuminated characteristics have been measured using solar light simulator. Solar cells characterization was carried out under standard test conditions. The solar irradiance was set at 100 mW/cm^2 and temperature of the module was maintained at 25°C. Characteristic parameters of the fabricated cells were determined on the basis of measured I-V curves. Series resistances were extracted from I-V characteristics at an open circuit voltage using first order derivatives. It was found that 60 minutes of electrode immersion in dye solution is sufficient to obtain appropriate stage of the dye adsorption.

1 Introduction

In recent years, the intensive development of renewable energy sources have been noticed all over the world. It is motivated by depletion of fossil fuel resources, increasing of their prices and still growing global energy demands. It is estimated that taking into account the current demand for electricity, the reserves of fossil fuels will last only 40, 60 and 200 years for oil, natural gas, and coal respectively [1]. The climatic consequences of fossil fuels combustion are not without significance because of greenhouse gases emission to the atmosphere. Preservation of the environment during the process of energy production is crucial. According to Directive 2009/28/EC of the European Parliament and of the Council of 23 April 2009, the target for European countries to be achieved by the end of 2030 is 32% of EU's final energy consumption [2]. The above mentioned factors are the most significant reasons to explore alternative energy sources, especially renewable ones. Many research

* Corresponding author: e.krawczak@pollub.pl

groups focus on solar energy because it is considered as environmentally friendly source with huge potential, which is able to meet the current and subsequent generations energy requirements. Direct conversion of the energy from the Sun into electricity can be achieved by photovoltaic (PV) devices. Photovoltaic cells are divided into three categories called generations, which are related directly to different materials and have various photoelectric conversion efficiencies. The first generation of solar cells is based on crystalline silicon material. It is relatively expensive which causes high cost of producing. However, silicon wafer-based solar cells are the most effective ones. Yoshikawa et al. [3] obtained highest noticed silicon cell efficiency of 26.7%. The second generation of solar cells is characterized by less material usage [4], which allows to produce cells of high efficiency. Materials that are typically used in thin film solar cells are cadmium sulfide (CdS), cadmium telluride (CdTe) [5], amorphous silicon (a-Si) [6] and copper-indium-diselenide/copper-indium-gallium-diselenide (CIS/CIGS) [7, 8]. Recent highest noticed efficiency of the laboratory CIGS cell equals 22.4% and is close to the efficiency of the silicon devices, which makes the possibility of replacing Si crystalline solar cells with the CIGS [9]. The third generation of photovoltaic cells is a promising alternative to conventional cells based on the p-n junction. Dye-sensitized solar cell (DSSC) technology can be considered as an economical substitute with relatively high conversion efficiency. Typical DSSC device consists of photoanode, counter electrode with platinum catalyst layer and electrolyte, made of mesoporous semiconductor layer and dye sensitizer. Both electrodes are based on transparent conductive oxide materials (TCOs), especially on indium tin oxide (ITO) [10] or fluorine-doped tin oxide (FTO) [11]. Interesting alternative to ITO is also ZnO doped with trivalent elements such as Al [12–15], Ga [12, 16] or B [17] which is a material that can even improve its electrical properties, like conductivity, during annealing process required in DSSC.

The dye sensitizer is considered as a crucial component that strongly affects the performance of the working cell. The phenomenon which plays an important role in the performance of dye-sensitized solar cells is electron transfer from the excited state of the dye molecule to the conduction band of titanium dioxide nanoparticle [18–20]. The course of the light-induced transfer of electrons from the sensitizer to a semiconductor depends mainly on the type of the dye i.e. the shape of its molecule and energy levels positions. The best performing dyes are made of Ru complexes (e.g. N3, N719, and black dyes) [21, 22]. Mathew et al. [23] obtained the record efficiency of 13% for Ru-based DSSC. However, the high production cost, complicated synthesis and harmful effect on the environment, made the researchers search alternative to Ru-based compounds [24]. Variety of natural resources, which are environmentally friendly, can be used as a origin of dyes sensitizers, such as plants, flowers, fruits [25, 26]. Nevertheless, the efficiency of natural dye-based cells should be improved for the extensive application of DSSC. Various groups [27–29] have been working on implementing natural dyes, such as alizarin, quercetin or luteolin. Alizarin is a red dye extracted from the root of *Rubia tinctorum*. However, this dye can be also obtained by laboratory methods from anthraquinone. Alizarin is characterized by rapid injection photoexcited electrons into the semiconductor conduction band.

The amount of the adsorbed dye also influences the efficiency of the DSSC device and can be adjusted by the change of the concentration of the dye solution or the time of the photoelectrode dipping in the solution. This paper presents the time of electrode immersion influence on alizarin-based dye-sensitized solar cells performance. DSSC devices were prepared by the use of prefabricated components and dyed with alizarin dye.

2 Materials and methods

In this research, the dye-sensitized solar cells were prepared with the use of alizarin dye (Fig. 1) purchased by Sigma-Aldrich. Initially, the 2 mM solution of alizarin dye was prepared in ethanol (99.8%). The absorption spectra were measured in the range of 300–800 nm by Shimadzu UV–vis spectrophotometer. Addition of dye resulted in colour progress, from transparent to dark purple. In order to improve solubility, the dye solution was ultrasonically mixed for 10 minutes. Afterwards, the photoanodes coated with TiO_2 (Greatcell Solar) were placed in a dye solution. The immersion time was different for each series of experiments: 10 minutes, 40 minutes, 1 hour and 24 hours. The container of dye solution was left without access to light, owing to the fact that particles are extremely photosensitive, especially in a solution form. After that, the electrodes were taken out and rinsed with the ethanol in order to eliminate excessive dye molecules. The photoelectrodes were dried for 30 minutes in the ambient environment and then were immediately used in the DSSC structures. DSSCs were assembled in the sandwich way. Counter electrode with a platinum catalyst layer (Greatcell Solar) was used to face the titania surface deposited onto photoanode. DSSC devices were sealed by a thermoplastic sealant (Dye Sol). The whole assembly was heated in high-temperature titan hotplate. The heat was supplied to the structure from the top and the bottom. Heating was carried out in three steps. Firstly, the structure was heated up to 90°C for 2 minutes, secondly, the temperature was raised to 110°C and held by 2 minutes. Finally, the temperature of 115°C was set and kept by the time of 2 minutes. After this procedure, the DSSC assemblies were left to cool down. Then, the high-performance EL- HPE electrolyte by Greatcell Solar was instilled into the space between electrodes through a drilled hole in the counter electrode. The working area of the DSSC structure was fixed as 0.8 cm^2. Then, the contacts were made by the use of silver tape. A scheme of the dye-sensitized solar cell is presented in Fig. 2.

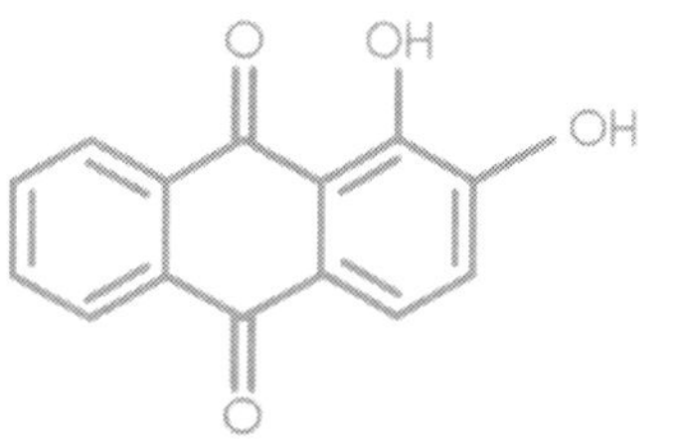

Fig. 1. The molecular structure of alizarin dye.

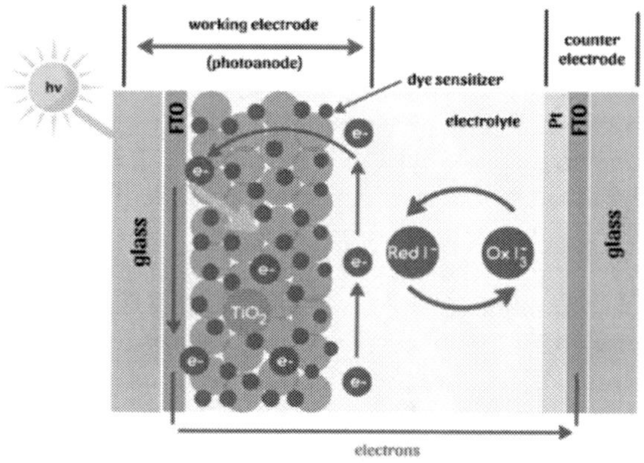

Fig. 2. A scheme of the dye-sensitized solar cell.

The current-voltage characteristics were immediately measured in dark and in illuminated conditions by SUN 3000 solar light simulator of Abet Technologies, class AAA. The simulator is equipped with Xe lamp (450 W). The solar irradiation was set at 100 mW/cm^2. The data were collected by a Keithley Instruments, type 2440, electrometer with integrated power supply. The ReRa Tracer software was used to measure and calculate various performance parameters.

3 Results and discussion

Initially, alizarin dye was characterized in order to obtain absorption spectrum by the use of spectrophotometer (Fig. 3). The dye exhibits the wide absorption spectrum in the range of 400 to 600 nm (in the visible range) with a major peak at 432.5 nm. It can be seen that characterized dye has a strong absorption in the violet region. On the basis of the absorption data, the value of optical band gap, HOMO-LUMO distance was estimated to be equal to 2.13 eV.

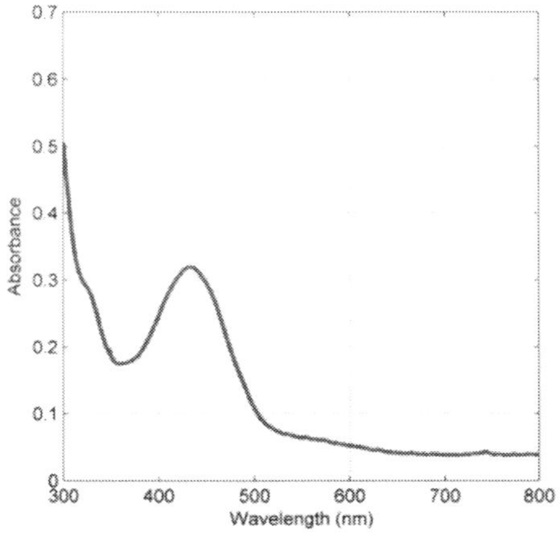

Fig. 3. UV-vis absorption spectrum of alizarin dye.

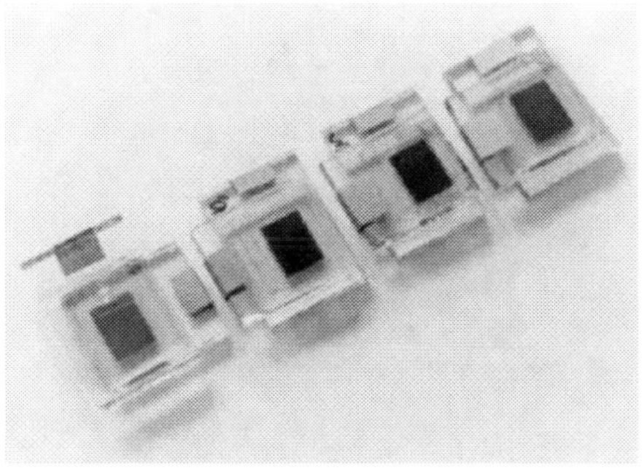

Fig. 4. Alizarin-based dye-sensitized solar cells for different immersion times, from left: 10 min, 40 min, 60 min, 24 h.

The dye sensitized solar cells were prepared in the way described in Experimental section. The obtained cells sensitized with alizarin dye as a sensitizer are shown in Fig.4. The change in colour from orangish to dark purple was noticed for different times of photoanode immersion in dye solution (10 min, 40 min, 60 min and 24 h). As can be seen in Figure 4, the time of the immersion has a strong influence on the colour of the semiconductor layer. Electrodes immersed for short time, such as 10 minutes and 40 minutes are characterized with a brighter colour (near to orange) while longer immersion times results in the dark purple colour of the electrodes, which means that the adsorption process proceeds in time.

The I-V characteristics of the cells based on a TiO_2 substrate using alizarin as a sensitizer were measured under the standard light intensity of 100 mW/cm^2 by a solar simulator. On the basis of obtained parameters, i.e. short circuit current (I_{sc}), short-circuit current density (J_{sc}), open circuit voltage (V_{oc}), maximum power point current (I_{MPP}) and maximum power point voltage (V_{MPP}), the value of fill factor was calculated. The fill factor is a parameter which defines the quality of the solar cell and can be defined as the ratio of the maximum power to the short-circuit current density and open circuit voltage values. The more the characteristic resembled a rectangle, the higher value is reached by fill factor. The fill factor value can be calculated on the basis of equation (1) [26].

$$FF = \frac{V_{MPP} \cdot I_{MPP}}{V_{oc} \cdot I_{sc}} \tag{1}$$

Fill factor is used to assess the power loss of electric current resulting from the internal resistance of the cell, the losses caused by recombination and other losses.

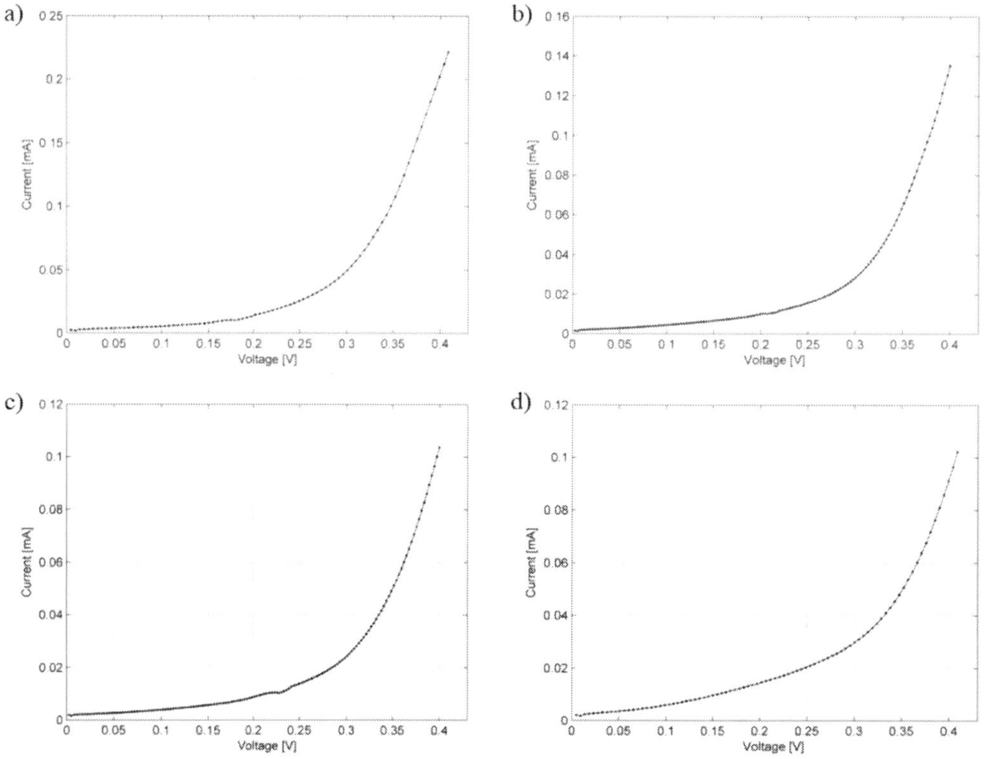

Fig. 5. I-V dark characteristics for alizarin dye-based DSSC for different immersion time: a) 10 min, b) 40 min, c) 60 min, d) 24 h.

Dark characteristics of the prepared dye-sensitized cells are shown in Fig. 5. Their shape is typical for semiconductor diode. The comparison with the I-V curves obtained for illuminated devices, which are presented in Fig. 6., depicts that introducing the light results in the less bent shape of the curves. Thus the role of series resistance is visible. The values of the series resistance were obtained as a reverse shape of the straight line fitted near I_{sc}=0. Series resistance, *Rs*, is a parameter that represents the series resistive losses which occur inside the solar cell during the current flow through the solar cell volume. Current generated inside the solar cell travels through resistive semiconductor layer of FTO/TiO_2. There are also resistances on TCO interconnections and cell metallization. Although the open circuit voltage is independent of the series resistance *Rs*, the short circuit current and especially the fill factor strongly depends on *Rs* and decreases with the increase of series resistance of the cell. The value for *Rs* can be calculated from the current-voltage (I-V) curve of the solar cell [30]. The slope at open circuit voltage V_{oc} provides the negative inverse value for $R_{s,0}$ according to the formula (2).

$$\left.\frac{dI}{dV}\right|_{V=V_{oc}} = -\frac{1}{R_{s,0}} \tag{2}$$

Using the extracted value of $R_{s,0}$ from I-V curve the series resistance can be calculated on the basis of the formula (3).

$$R_s = R_{s,0} - \frac{nV_{th}}{I_0} e^{\frac{-V_{oc}}{nV_{th}}} \tag{3}$$

where $n = 1$, V_{th} is the thermal voltage (kT/q, $k = 1.381\cdot10^{-23}$ J/K, $q = 1.602\cdot10^{-19}$ C and $T = 297$ K). Reverse saturation current, I_0 can be calculated by equation (4).

$$I_0 = I_{sc} \cdot e^{-\frac{V_{OC}}{nV_{th}}} \quad (4)$$

On the basis on the above-mentioned method, the Rs values for immersion times equal to 10 min, 40 min, 60 min, and 24 h were estimated to be 2.11 kΩ, 2.27 kΩ, 2.57 kΩ, and 2.74 kΩ respectively. Figure 6 shows the current-voltage characteristics for the devices with the various immersion times. Usage of the electrode immersed for 10 minutes leads to the worse properties of DSSC devices. Adsorption of the dye molecules onto the TiO_2 layer, in such a short time, was insufficient which resulted in low parameters values.

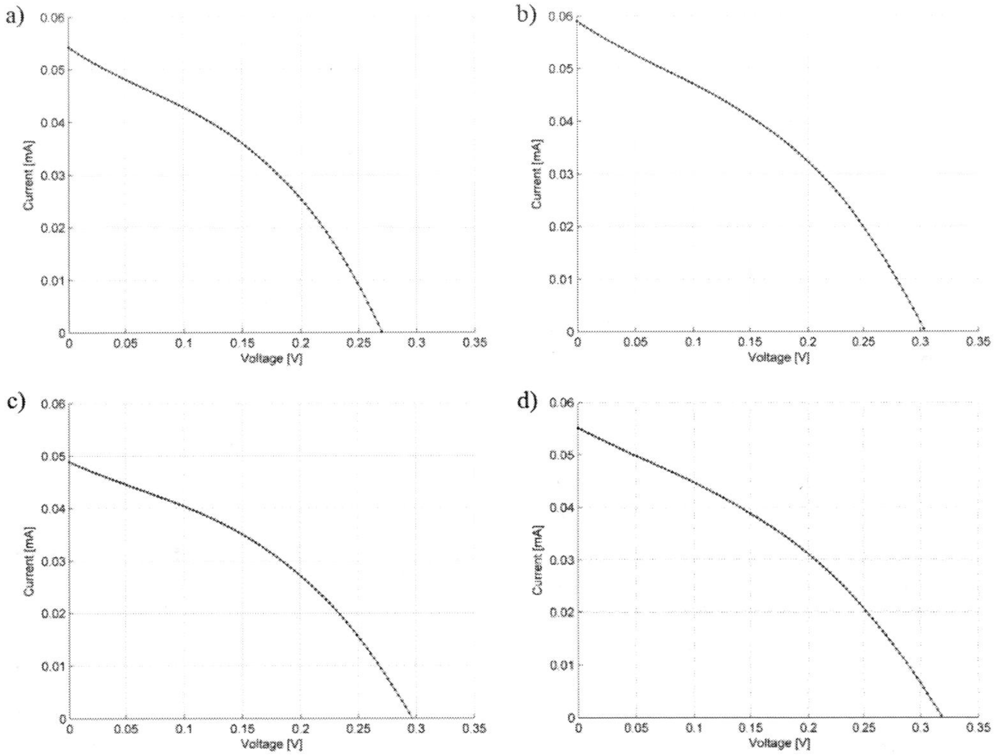

Fig. 6. I-V characteristics for alizarin dye-based DSSC for different immersion time: a) 10 min, b) 40 min, c) 60 min, d) 24 h.

A comparison of various performance parameters for different immersion times are presented in Table 1. As can be seen, the highest short-circuit current density value was found for an electrode submerged for 10 minutes. Increasing of the immersion time led to decreasing J_{sc} value in the range of 0.095–0.058 mA/cm^2. Open circuit voltage increases for longer absorption times and its value was found to be the best for cell 4 and equal to 318.82 mV, whilst the lowest was obtained for cell 1 (270.52 mV). The open circuit voltage is the parameter depending on the energy of LUMO level of the dye and its distance from to the conduction band of a semiconductor used in the device. The main limitation of V_{oc} value is the recombination of electrons present in conduction band with the ions from the electrolyte. Comparing all the parameters, it can be summarized that 60 minutes immersed electrode-based cells present the best performance parameters, especially fill factor, with 24 times shorter time, than cell 4.

Table 1. Performance parameters for different immersion times.

N° cell	Immersion time	J_{sc}	I_{sc}	V_{oc}	I_{mpp}	V_{mpp}	P_{mpp}	*FF*	R_S
		[mA/cm²]	[mA]	[mV]	[mA]	[mV]	[mW]	[%]	[kΩ]
cell 1	10 min	0.095	0.054	270.52	0.031	158.34	0.005	33.47	2.11
cell 2	40 min	0.075	0.060	304.19	0.034	189.21	0.006	35.82	2.27
cell 3	**60 min**	**0.063**	**0.048**	**296.60**	**0.030**	**183.88**	**0.006**	**38.06**	**2.57**
cell 4	24 h	0.058	0.055	318.82	0.028	183.58	0.005	29.22	2.73

The fill factor parameter was found to increase with increasing immersion time up to 60 min. Further increase of the immersion time have not led to better performance of the device. What is more, the slightly worse value of fill factor has been noticed for the immersion time of 24 h (cell 4).

4 Conclusion

The influence of the immersion time of the photoelectrode in the dye solution on dye-sensitized solar cells performance was studied. DSSC devices were obtained on the basis of alizarin dye with various immersion time equal to 10 min, 40 min, 60 min and 24 hours. The performed investigation leads to the conclusion that alizarin dye is appropriate to be used as a sensitizer in dye-sensitized solar cells because it is characterized by a wide absorption spectrum in the visible range of 400 to 600 nm, which can be compared with others dyes reported in the literature. The performance of obtained devices was studied by employ illuminated and dark current-voltage characteristics. It was found that the shape of dark I-V curves is typical for semiconductor diode. The influence of the series resistance on the illuminated characteristics is visible. *Rs* values have a strong influence on short circuit current values, which is very low. However, the cells were characterized by V_{oc} values in the range from 270.52 to 318.82 mV, which are appropriate for dye-sensitized assemblies. The fill factor values changed in the range of 29.22%–38.06%. The cell 3 characterized by 60 minutes of immersion time shows maximum FF value, while the lowest value was found for immersion time equals 24 hours. It can be summarised that DSSC structure immersed for 60 minutes reveals the best performance parameters and this time can be considered as optimal for alizarin dye. Further increasing of the immersion time does not lead to better parameters, fill factor especially. Further investigation is necessary in order to improve the values of all significant parameters of working dye-sensitized solar cells prepared by the use of natural dye, such as alizarin.

References

1. B. Li, L. Wang, B. Kang, P. Wang, Y. Qiu, Sol. Energ. Mat. Sol. C **90**, 549–573 (2006)
2. Eurepean Commisssion Statement, http://europa.eu/rapid/press-release_STATEMENT-18-4155_en.htm
3. K. Yoshikawa, H. Kawasaki, W. Yoshida, T. Irie, K. Konishi, K. Nakano, T. Uto, D. Adachi, M. Kanematsu, H. Uzu, K. Yamamoto, Nat. Energy **2**, 17032 (2017)

4. J. M. Olchowik, K. Cieslak, S. Gulkowski, A. Kaminski, A. Fave, *2010 35th IEEE Photovoltaic Specialists Conference*, 003546-003548 (2010)
5. J. Sites, J. Pan, Thin Solid Films **515**, (15), 6099–6102 (2007)
6. T. Matsui, H. Sai, A. Bidiville, H.-J. Hsu, K. Matsubara, Sol. Energy **170**, 486–498 (2018)
7. V. Bermudez, Sol. Energy **175**, 2–8 (2018)
8. S. Gulkowski, E. Krawczak, E3S Web of Conferences **19**, 03006 (2017)
9. P. Jackson, R. Wuerz. D. Hariskos, E. Lotter, W. Witte, M. Powalla, Phys. Status Solidi RRL **10**, 583–586 (2016)
10. A. S. Shikoh, Z. Ahmad, F. Touati, R. A. Shakoor, S. A. Al-Muhtaseb, Ceram. Int. **43**, 10540–10545 (2017)
11. W. Li, F. Lv, T. Shu, X. Tan, L. Jiang, T. Xiao, P. Xiang, Mater. Lett. **243**, 108–111 (2019)
12. K.- M. Kang, Y. Wang, M. Kim, H.-H. Park, Thin Solid Films **660**, 913–919 (2018)
13. A. Zdyb, E. Krawczak, P. Lichograj, Opt. Appl. **46**, 181–185 (2016)
14. E. Krawczak, A. Zdyb, S. Gulkowski, A. Fave, E. Fourmond, E3S Web of Conferences **22**, 00090 (2017)
15. A. Zdyb, E. Krawczak, S. Gulkowski, Opto-Electron Rev. **26**, 247–251 (2018)
16. C. Bhoomanee, S. Nilphai, S. Sutthana, P. Ruankham, S. Choopun, D. Wongratanaphisan, Integr Ferroelectr. **165**, 121–130 (2015)
17. L.- H. Wong, Y.-S. Lai, Thin Solid Films **583**, 205–211 (2015)
18. A. Zdyb, S. Krawczyk, Appl. Surf. Sci. **256**, 4854–4858 (2010)
19. A. Zdyb, S. Krawczyk, Chem. Phys. **443**, 61–66 (2014)
20. A. Zdyb, S. Krawczyk, Spectrochim. Acta A Mol. Biomol. Spectrosc. **157**, 197–203 (2016)
21. A. S. Shikoh, Z. Ahmad, F. Touati, S. A. Al-Muhtaseb, Opt. Mater. **78**, 201–2016 (2018)
22. P. Naik, I. M. Abdellah, M. Abdel-Shakour, R. Su, K. S. Keremane, A. El-Shafei, A. V. Adhikari, Sol. Energy **174**, 999–1007 (2018)
23. S. Mathew, A. Yella, P. Gao, R. Humphry-Baker, B. F. E. Curchod, N. Ashari-Astani, I. Tavernelli, U. Rothlisberger, Md. K. Nazeeruddin, M. Grätzel, Nat. Chem. **6**, 242–247 (2014)
24. D. Zhang, S. M. Lanier, J. A. Downing, J. L. Avent, J. Lum, J. L. McHale, J. Photoch. Photobio. A **195**, 72–80 (2008)
25. H. Zhou, L. Wu, Y. Gao, T. Ma, J. Photoch. Photobio. A **219**, 188–194 (2011)
26. W. A. Ayalew, D. W. Ayele, Journal of Science: Advanced Materials and Devices **1**, 488–494 (2016)
27. R. Sánchez-de-Armas, M. A. San-Miguel, J. Oviedo, J. Fdez. Sanz, Comput. Theor. Chem. **975**, 99–105 (2011)
28. S. Çakar, M. Özacar, J. Photoch. Photobio. A **346**, 512–522 (2017)
29. S. Shalini, N. Prabavathy, R. Balasundaraprabhu, T. S. Kumar, D. Velauthapillai, P. Balraju, S. Prasanna, Optik **155**, 334–343 (2018)
30. F. Ghani, G. Rosengarten, M. Duke, J. K. Carson, Renew. Energ. **72**, 105–112 (2014)

Comparison of daily flows simulated for the year 2060 on the Kaczawa River for various scenarios of climate change by simple time series analysis

Leszek Kuchar[1,*], *Ewa* Broszkiewicz-Suwaj[1], *Slawomir* Iwanski[1], and *Leszek* Jelonek[2]

[1]Wroclaw University of Environmental and Life Sciences, Department of Mathematics, ul. Grunwaldzka 55, 50-357 Wroclaw, Poland
[2]Institute of Meteorology and Water Management, ul. Parkowa 30, 51-616 Wroclaw, Poland

Abstract. In this paper a time series analysis for daily flow simulations according three climate change scenario for Kaczawa River a left side tributary of the Odra River in south-west Poland is presented. The flow sequences were simulated using the hydrological model MIKE SHE and the spatial SWGEN meteorological data generator. Meteorological data for the hydrological model were generated based on data from 24 meteorological stations and 35-year daily data from the Institute of Meteorology and Water Management of the National Research Institute (IMGW). Data were generated for future climate condition for 2060 according GISS Model E, HadCM3, and GFDL R15 scenarios as well for the present conditions. The year 2000 was used as a reference year. The results obtained on the basis of a simple time series analysis point to small changes in flows for current and simulated conditions for 2060 for the Kaczawa River.

1 Introduction

This work is a continuation of earlier studies on changes in the Kaczawa River flows simulated for various scenarios of climate change [14–17]. The following studies compared the time series of daily flows in a period of the year in contrast to previous studies on the probability distributions of flows, e.g. maximum and minimum flows. The aim of this work is to answer the question whether in simulated flows for the year 2060 there are significant changes in the periodicity of the phenomenon, amplitude, time shift or others resulting from the analysis of the time series [2, 9, 21, 26]. For example, floods in Poland, which are accompanied by high flows, appeared mainly in the summer. However, in the last 20 years, as a results of changes in the course of winters, mainly higher temperatures, meltwater thaw floods began to appear more frequently and on a much larger scale [15, 19, 23]. In the studies, flows were simulated for the year 2060 according to the three most likely scenarios GISS Model E, HadCM3, and GFDL R15 which are considered to comply with the new scenarios of Representative Concentration Routes (RCPs) 4.5 and 6.0. Three selected SRES scenarios were adopted in the present study due to the extensive experience in generating meteorological data for these scenarios for the condition of Poland, as well as due to their high convergence with the RCP4.5 and RCP6.0 scenarios [7, 8, 15]. It has been assumed that

[*] Corresponding author: leszek.kuchar@upwr.edu.pl

the background in the research, i.e. the reference period, are the years at the turn of the 20th and 21st centuries. Similarly to previous studies, the spatial SWGEN meteorological data generator was used, the MIKE SHE rainfall-outflow hydrological model and the operating scheme were consistent with earlier studies and described in [10, 13, 20, 16, 17]. Simulations were performed for the small tributary of the Odra (south-west region of Poland) due to the high quality of observational data and previous experience in hydrological modelling [17, 18, 25, 31].

2 River flow simulation procedure

As in previous studies, the simulation of river flows for the future climate is carried out in three stages [14–17]. Climate description of all meteorological stations are prepared using an observed data. Above characteristics are modified according to climate change scenario and it is required for the weather generator producing synthetic data [3, 12, 13, 16]. In the next step, synthetic meteorological data are generated for all stations by the SWGEN weather generator. These data are used as an input for the hydrological rainfall-runoff model MIKE SHE to simulate river flows. This is step 3. The described procedure is presented in Figure 1 originally described in the papers [16, 17] and has been used many times to simulate processes in river catchment [14–17]. The daily flows sequences obtained are the basis for analysis illustrating the hydrological effect of climate changes in rivers [5, 16, 17, 22, 24].

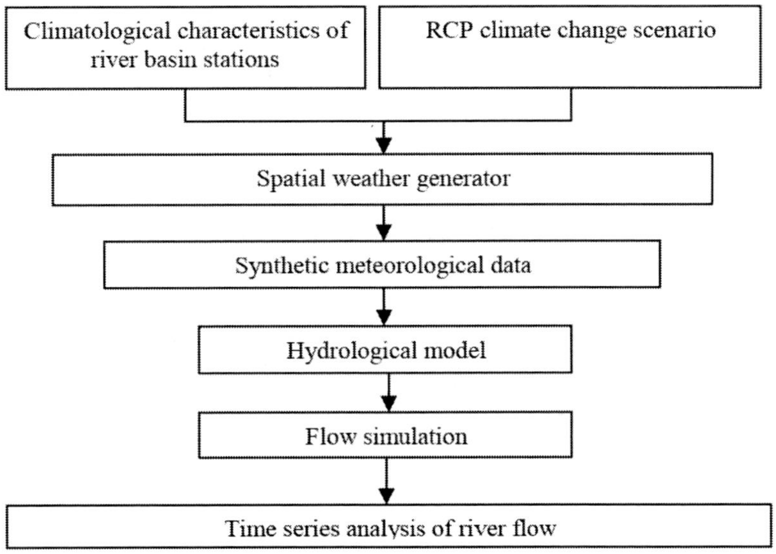

Fig. 1. Diagram of river flow simulation for future climate using synthetic meteorological data and climate change scenario.

3 Catchment and data

Our research has been conducted for many years in south-western Poland in the Kaczawa river basin (Fig. 2) [16, 17]. This choice is related to recommendation of the Institute of Meteorology and Water Management National Research Institute (IMGW) due to the first order meteorological and hydrological data. A 35-year dataset (1981–2015 series) for daily solar radiation, maximum and minimum air temperature and total rainfall were obtained for 24 stations from the Institute network and used for presented research [13, 16].

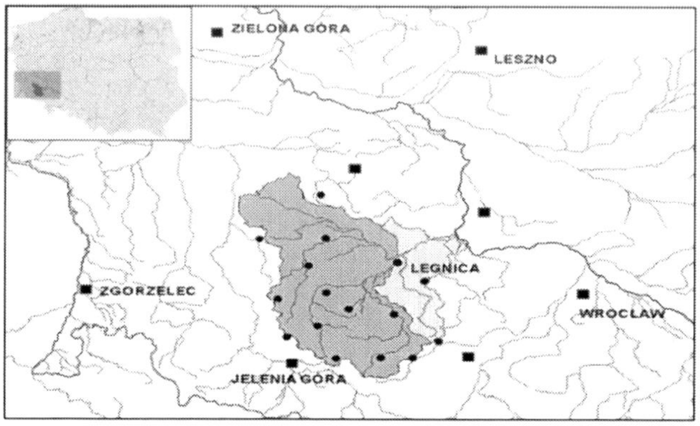

Fig. 2. The Kaczawa River catchment (left, area of 1807 km2, main sit: Legnica 51°13'N, 16°14'E) with meteorological (•) and first order (▪) meteorological stations.

An information about climate change, required for the simulation were derived from the Special Report on Emissions Scenarios (SRES) as well from the Representative Concentration Routes (RCP) scenarios [7, 8]. Used from 2014 new RCP scenarios are considered for simulations, but due to the extensive experience in generating meteorological data for SRES scenarios for the condition of Poland, SRES A1B scenarios finally were selected.

For the study three typical SRES A1B scenarios (GISS Model E, HadCM3, and GFDL R15) were used, which correspond to the RCP 4.5 and 6.0 scenarios assumes a changes in CO_2 concentrations in Poland, amounting to 538 ppm and 670 ppm respectively up to 2100 (with additional consideration of changes in CH_4 and N_2O emissions, the total concentration would be equivalent to CO_2 values of 630 ppm and 800 ppm) [7, 8, 15].

4 Spatial weather generator and rainfall runoff model

The weather generator SWGEN is used to generate daily data for n years and for k stations as described in earlier papers [10, 11, 29, 30]. The model generates total precipitation by means of the first-order Markov chain to determine the occurrence of wet/dry days, and then for the amount of precipitation the multidimensional two-parameter gamma distribution is used [4, 16]:

$$\left(\Gamma_m(\alpha_1, \beta_1), \dots, \Gamma_m(\alpha_k, \beta_k)\right) \tag{1}$$

where m is the month number (m = 1, ..., 12, i.e. January = 1, February = 2, ..., December = 12) and k is the location number. Daily values of solar radiation (SR), temperature maximum (T_{max}) and minimum (T_{min}) are treated as a multidimensional time series AR(1) in the following form:

$$\mathbf{X_t} = \mathbf{\Phi_m} \cdot \mathbf{X_{t-1}} + \mathbf{\varepsilon_t} \tag{2}$$

where $\mathbf{X_t}$ and $\mathbf{X_{t-1}}$ are vectors (3k × 1) of standardized values for all three variables for day t and t-1, ε_t is a vector (3k × 1) of independent random components normally distributed with vector of means equal to zero and matrix of covariance $\mathbf{\Sigma_m}$, and $\mathbf{\Phi_m}$ (for m = 1, …, 12) is a matrix of parameters [10, 18, 27]. For this study, the SWGEN was used to produce long series of 400 years of synthetic data for 24 stations as an input for the rainfall runoff model. Generated data are applied to the MIKE SHE hydrological model [6, 14, 28] to simulate daily flows for closing water-gauges. In these studies, 2000 was selected as the reference year to identify potential changes in river flows [1, 16, 17].

5 Results

As in our previous papers [14, 15, 16, 17] the simulations of daily runoff in the Kaczawa River catchment were done at discharge point in Piatnica. The simulations were done for the time horizon 2060 with 2000 as a reference year. Three typical scenarios (GISS Model E, HadCM3, and GFDL R15) were considered. The number of generated years (400) for each case, with total of 1200 simulations for the flow. The rainfall-runoff MIKE SHE model was used in each year for the simulation a daily flow at discharge point (basic information about the average daily flows are included in table 1).

Table 1. Means, standard deviations, maximum and minimum of daily flow over 400 simulated flows, and different scenarios.

Time horizon	Scenario	Mean, m³/s	Std Dev	Maximum, m³/s	Minimum, m³/s
2000	Present	7.07	1.76	12.94	3.10
2060	GISS Model E	8.47	1.80	14.83	3.78
	HadCM3	6.85	1.55	13.17	3.52
	GFDL R15	7.87	1.73	14.48	3.48

The first step of data analysis is the investigation of yearly mean and standard deviation shape for all scenarios. All plots are given in the Figure 3.

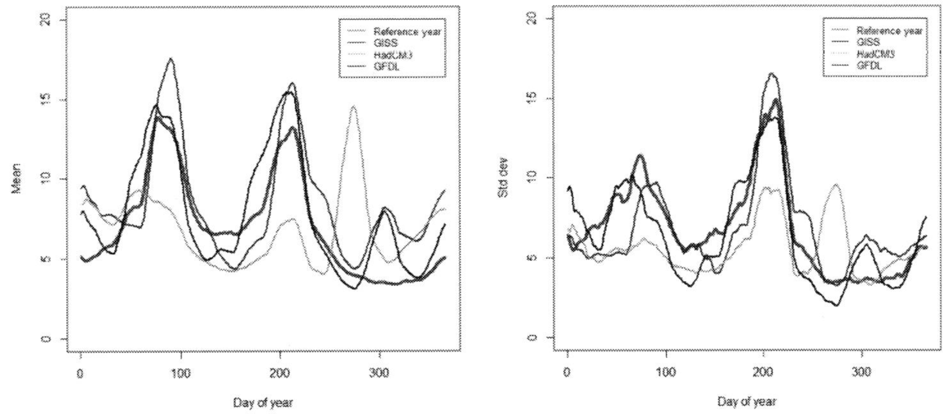

Fig. 3. Plot of yearly smoothed (moving average with window size of 15 days) mean and standard deviation for four scenarios.

For all data types based on the yearly mean shape we can distinguish the regularity of flows within the year. For the present conditions two periods with high flows are identified during March–April and second during July–August. Simulated flows for the GISS Model E and GFDL R15 scenarios for the years 2060 show a high compliance with flows for 2000, however, in the period October-November, the third larger flow appears clearly (on a much smaller scale). In the case of simulated flows for the HadCM3 scenario there are also three periods with larger flows, however, with significantly different values. In the March–April and July–August periods, these are small values, while in the autumn period they are on a level comparable to those simulated for the GISS Model E and GFDL R15 scenarios for the spring and summer. Additionally, in the case of the HadCM3 scenario, high flows appear at the turn of September and October. It should also be noted that for the scenarios considered, the maximum flows may increase by up to 25% and the minimum flows may be reduced by 20% (mainly in May) compared to the current conditions. Already these simulated changes in flows are important information not only in the context of flood safety, but also proper water management. The variance of simulated flows (expressed by standard deviation) does not indicate large changes, in particular their increase in periods of high or low flows. Only in the case of the GISS Model E scenario, an increase in the standard deviation by about 10% along with higher flows by 20% in the July-August period may determine a higher flood risk.

Simulated flows over the year were compared to the reference period (2000), assessing the periodicity of time series. As a standard tool for periodicity detection the periodogram was applied as an asymptotically unbiased estimator of the spectral density. On the Figure 4 the plots of periodograms of flows for reference year and for climate scenarios are presented. Peaks at points d, 2d, 3d etc. indicate periodicity with period of length $T = N/d$ where N is the length of given time series. The analysed data have length $N=400\cdot365$. For all scenarios we observe peaks (Fig. 4) in points $d = 400$, $2d = 800$ and $3d = 1200$ which means that periodogram detected the yearly periodicity. From the above computations, it can be concluded that in the simulated flows for the climate scenarios considered, the periodicity in relation to the reference period (year 2000) will be at a similar level.

In the next step the structure of autocorrelation functions of given flow series were compared. On the graph (Fig. 5), there are vertical lines (a "spikes") corresponding to each lag. The height of each spike shows the value of the autocorrelation function for the lag. Each spike that rises above the dashed line is considered to be statistically significant. In all scenarios, the spikes are statistically significant for lags up to 20. This means that the daily flows are highly correlated with each other. In other words, when the flow rises, it tends to continue rising. When the flow falls, it tends to continue falling. However, the most important issue is that the presented autocorrelation functions for flows simulated for given climate change scenarios do not differ from the autocorrelation function for the reference period. This suggests that flows for the reference period and for three simulations consistent with the adopted climate change scenarios maintain similar relationships and may confirm the absence of major changes in the flows of the Kaczawa River for 2060.

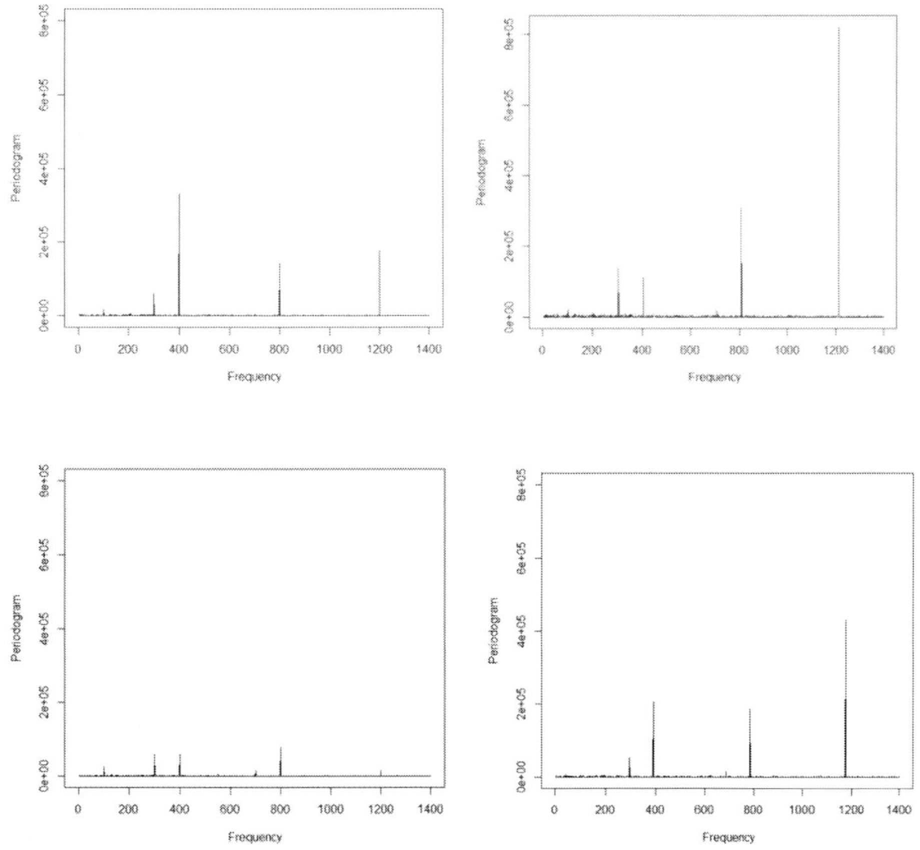

Fig. 4. Plot of periodogram for reference year (left top panel), GISS scenario (right top panel), HadCM3(left bottom panel), GFDL (right bottom panel).

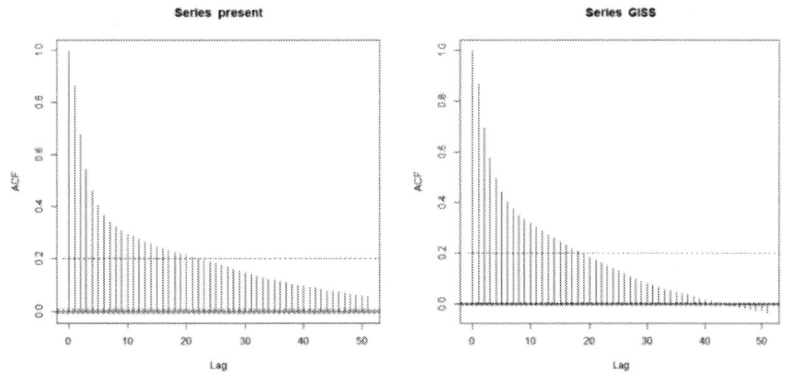

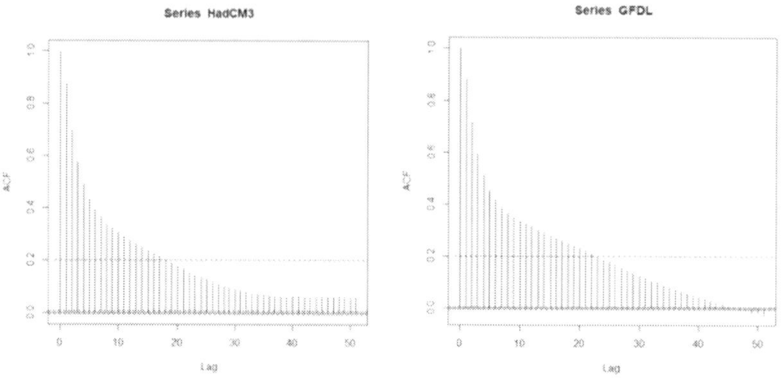

Fig. 5. Plot of autocorrelation function for reference year (left top panel), GISS scenario (right top panel), HadCM3(left bottom panel), GFDL (right bottom panel).

6 Conclusions

The maximum daily flow during the March-April and July-August periods simulated for the GISS scenarios Model E and GFDL R15 increase by 20–25% in relation to the reference period.

In the case of the HadCM3 scenario, only one period of higher flows was detected shifted for the period October–November but with similar values in relation to the reference period.

The course of flow variances in the year for the GISS Model E and GFDL R15 scenarios shows compliance with the reference period. In the case of the HadCM3 scenario, the values of flow variances show a significant drop except for the period October–November.

No changes were detected in long-term flows as well as in the structure of the autocorrelation function.

References

1. T. Barnett, T. Malone, W. Pennell, D. Stammer, B. Semtner, W. Washington, Climate Change **62**, 1–11 (2004)
2. S. Bergstrom, B. Carlsson, M. Gardelin, G. Lindstrom, A. Pettersson, M. Rummukainen, Climate Research **16**, 101–112 (2001)
3. F. Brissette, M. Khalili, R. Leconte, Journal of Hydrology **345**, 3–4, 121–133 (2007)
4. J. Chen, F. P. Brissette, X. C. Zhang, Transactions of the ASABE **57**, 5, 1375–1397 (2014)
5. N. S. Christensen, A. W. Wood, N. Voisin, D. P. Lettenmaier, R.N. Palmer, Climate Change **62**, 337–363 (2004)
6. D. N. Graham, M. B. Butts, [In:] Watershed Models, V.P. Singh, D.K. Frevert (Eds.). CRC Press, 245–272 (2005)
7. IPCC-SRES-SPM. IPCC Special Report (2000) https://www.ipcc.ch/pdf/special-reports/spm/sres-en
8. IPCC. Climate Change 2014: Synthesis Report. Contribution of Working Groups I, II and III to the Fifth Assessment Report of the IPCC, pp.151 (2014) http://www.ipcc.ch/pdf/assessment- report/ar5/syr/SYR_AR5_FINAL_full_wcover.pdf

9. S. Goswami, S. C. Kar, Meteorology Hydrology and Water Management **6**, 1, 13–25 (2018)
10. S. Iwanski, L. Kuchar, Acta Scientiarum Polonorum-Formatio Circumiectus **2**, 1, 113–121 (2003)
11. M. Khalili, R. Leconte, F. Brissette, Journal of Hydrometeorol. **8**, 3, 396–412 (2007)
12. L. Kuchar, Mathematics and Computers in Simulation **65**, 69–75 (2004)
13. L. Kuchar, S. Iwański, EKO-DOK 2018, E3S Web of Conferences **44**, 00083 (2018)
14. L. Kuchar, S. Iwański, L. Jelonek, E3S Web of Conferences **17**, 00046 (2017)
15. L. Kuchar, S. Iwański, L. Jelonek, XLVIII Seminar of Applied Mathematics, ITM Web of Conferences **23**, 00021 (2018)
16. L. Kuchar, S. Iwański, L. Jelonek, W. Szalinska, Geografie **119**, 1, 1–25 (2014)
17. L. Kuchar, S. Iwański, L. Jelonek, W. Szalinska, Meteorology Hydrology and Water Management **2**, 2, 49–63 (2014)
18. L. Kuchar, A. Tiukało, Meteorol. Hydrol. and Water Management **6**, 2, 79–83(2018)
19. A. Michalski, Meteorology Hydrology and Water Management **4**, 1, 41–46 (2016)
20. MIKE 11. A modelling system for Rivers and Channels, User Guide, DHI Water and Environment (2003)
21. H. T. Mitosek, Journal of Hydrology **228**, 3–4, 188–205 (2000)
22. D. I. Müller-Wohlfeil, G. Bürger, W. Lahmer, Climate Change **47**, 61–89 (2000)
23. B. Ozga-Zieliński, M. Ciupak, J. Adamowski, B. Khalil, J. Malard, Journal of Hydrology Regional Studies **6**, 26–51 (2016)
24. C. Prudhomme, N. Reynard, S. Crooks, Hydrol. Processes **16**, 1137–1150 (2002)
25. W. Szalinska, I. Otop, T. Tokarczyk, Meteorology Hydrology and Water Management **2**, 1, 13–20 (2014)
26. J. R. Thompson, Hydrol. Res. **43**, 4, 507–530 (2012)
27. R. E. Walpole, R. H. Myers, S. L. Myers, K. Ye, *Probability and statistics for engineers and scientists* (Prentice Hall, 7th Ed., NJ, 2002)
28. T. Vansteenkiste, M. Tavakoli, V. Ntegeka, P. Willems, F. De Smedt, O. Batelaan, Hydrological Processes **27**, 25, 3649–3662 (2012)
29. R. L. Wilby, Environmental Modelling & Software **22**, 12, 1705–1719 (2007)
30. D. S. Wilks, Climate Change **1**, 898–907 (2010)
31. M. Wdowikowski, B. Kaźmierczak, O. Ledvinka, Meteorology Hydrology and Water Management **4**, 1, 53–63 (2016)

Biochar compared with activated granular carbon for landfill leachate treatment

Anna Kwarciak-Kozłowska[1,*], *Renata* Włodarczyk[2], and *Katarzyna* Wystalska[1]

[1]Czestochowa University of Technology, Institute of Environmental Engineering, Faculty of Infrastructure and Environment, ul. J. H. Dąbrowskiego 73, 42-200 Czestochowa, Poland
[2]Czestochowa University of Technology, Institute of Advanced Energy Technologies, Faculty of Infrastructure and Environment, ul. J. H. Dąbrowskiego 73, 42-200 Czestochowa, Poland

Abstract. The aim of the research was to compare the effectiveness of sorption of pollutants from landfill leachate on two granulated activated carbons (AG and BA-10) and biochar (BC). The sorption process of wastewater components on activated carbons was carried out under static conditions for 72 hours. The doses of adsorbents used varied from 2–8 $g\cdot dm^{-3}$. Results of removing of COD were estimated on the base of adsorption capacity and COD removal efficiency. After 72 h of the process at the highest doses (8 $g\cdot dm^{-3}$), the COD removal rate ranged from 40% (BA-10) to 60% (AG). In the case of biochar, the COD adsorption rate was 53% The state of sorption equilibrium takes a similar character for BC and AG and is determined after 4 h. The degree of sorption of selected metals from landfill leachate using the analyzed range ranged from 84% to 96% depending on its type.

1 Introduction

At municipal landfills, the waste undergoes constant physicochemical and biological changes. In the results of rainwater infiltration through the waste deposit, organic and inorganic compounds are extracted. These compounds are often toxic and are a reason for the strong contamination of waters called landfill leachate [1–3]. Direct discharge of leachate into the natural environment could cause irreparable damage, especially to the surface and underground water systems [2]. Many factors influence the amount and composition of leachate, among others topography of the site, waste storage technology, type of waste and degree of fragmentation, hydrogeological and meteorological conditions or the age of the landfill [3–6]. It is estimated that from a densely compacted landfill with an annual rainfall of 0.7 m per 1 m^2 of waste, within a day, 0.278 to 0.487 dm^3 of leachate is generated. For landfills with low density, this number is even three times higher [4]. The amount of leachate generated is also limited by the age of the landfill. On the young landfills, the production of leachate is slightly lower compared to the old ones (over 10 years of operation). This is related to the decreasing water retention capacity in waste and

* Corresponding author: akwarciak@is.pcz.czest.pl

the increasing organic stabilization over time [6, 7]. According to the landfill age, leachate can be classified into three types: young, middle-aged and stabilized (Table 1). The amount of pollution in one m^3 of leachate from the landfill is equivalent to the amount of pollution found in 100 tonnes of municipal wastewater [1]. To date, over 200 organic substances have been identified as impurities in leachates, 35 of which belong to the so-called "Priority pollutants" [4]. Landfill leachate contains large amounts of organic and inorganic material, including a number of refractory organics such as aromatic compounds and humus; inorganic salts such as ammonia nitrogen, carbonate, and sulfate; and metal ions such as chromium, lead, and copper because the composition of the leachate is very complex and variable over time, there is no single universal method of purification [2–4].

Table 1. The effect of municipal landfill age on physicochemical characteristics of landfill leachate and efficiency of its treatment [1, 8, 9].

Characteristics of the landfill leachate	**Leachate type**		
	< 5 years	**5–10 years**	**> 10 years**
pH (-)	< 6.5	6.5–8.0	> 8.0
COD (mg/dm^3)	> 10000	< 10000	< 3000
BOD/COD (-)	0.5–0.7	0.3–0.5	< 0.13
COD/NH_4^+-N (-)	5–10	3–4	< 3
NH_4^+-N (mg/dm^3)	500–1000	800–2000	1000–3000
Treatment methods	Process efficiency		
biological proces	+++	++	+
chemical oxidation	+	++	++
chemical precipitation	+	++	+
adsorption	+	++	+++
reverse osmosis	++	+++	+++
ozonation	+	++	++
ion exchange	+	++	++

+ low, ++ average, +++ beneficial

One of the decisive factors in the selection of leachate treatment technology is the economics of the process (investment and operating costs) and the effectiveness of a given technology in removing contaminants.

It is believed that the adsorption process is the most suitable technique for removing contaminants including metals from stabilized effluents [8–10]. Granular activated carbon (GAC) is a safe product with a porous structure and extremely large specific surface area. Its properties depend on the properties of the raw material and the technology used. Currently, activated carbons are obtained in the process of thermal and/or chemical decomposition of substances containing significant amounts of fixed carbon. On an industrial scale as sorbents, hard coal, lignite, peat, half-coke or coke are used, but also the seeds of apricot and almond shell, the shell of the Italian coconut as well as nut shell [11–13]. Common GAC materials are comprised primarily of micropores (> 1 nm) which are highly effective at adsorbing dissolved contaminants, but are easily blocked by suspended solids [14, 15]. In recent years, research has been carried out to replace GAC with cheaper material. Biochar (BC) is often used directly with little post treatment modification and is generally less resource, energy, and capital intensive to manufacture compared to highly refined GAC [16]. It is assumed that the amount of energy needed to produce GAC from various sorbents ranges from 44 to 170 MJ·kg^{-1}, while for producing biochar it requires much smaller amounts in the range from 1.1 to 16 MJ·kg^{-1} (i.e. up to

90% lower demand). The biochar changes the pH of soil, improves its fertility and yields crops [14–16]. Using spent BC material after wastewater/landfill leachate treatment as an agricultural amendment could further increase its life-cycle benefits. Table 2 depicted raw materials for the biochar's production, temperatures of pyrolysis process used during biochar production and impurities removed using a biocarbon adsorbent. The carbonization temperature affects the porosity of materials and the same on the adsorption capacity. As it appears from the literature, the process of removing contaminants using biochar proceed according to various mechanisms [19–27].

Table 2. The possibility of using BC for adsorption of inorganic compounds from liquid solutions.

Raw material	Process temperature	Pollution	Type and mechanism of adsorption	Refs.
manure	200°C	Pb	precipitation of phosphates	[19]
straw from sugar cane	300°C	Cr	reduction of Cr(VI) to Cr(III)	[20, 21]
oak wood	450°C		sorption	[22]
wheat straw	400°C	Cu	adsorption by surface complexation	[23]
nut shells	800°C		sorption of humic acids at pH 6	[24]
soybean stalk	700°C	Hg	precipitation, complexation and reduction	[25]
broiler litter manure	250°C	Ni	adsorption and partitioning mechanisms	[26]
wheat straw and rapeseed	100°C	Cd/Pb	adsorption using aromatic structure	[27]

To complement the performance of the adsorption process, the contact time (from several minutes to several hours), the temperature of the adsorption process (from 19°C to 46°C), the pH of the solution from which the purification process was carried out should be taken into account. The adsorption effect is strongly influenced by the physical and chemical properties of sorbent-biochar (S-BC) [17]. The key parameters controlling its biochar as sorbent properties include pyrolysis temperatures, residence time in the reactor, heating speed and type of raw material [28–31]. The effectiveness of biochar in capturing contaminants depends on its specific surface area (SSA). SSA for biocarbon may range from approx. 0.77 $m^2 \cdot g^{-1}$ to 1160 $m^2 \cdot g^{-1}$ [18]. The high temperature of the pyrolysis process gives a product that is an effective adsorbent for the sorption of organic pollutants from the soil due to the increased surface area, microporosity and hydrophobicity. Biochars obtained at low temperatures are more suitable for removing inorganic/polar organic pollutants containing oxygen functional groups by electrostatic attraction. Sorption of pollutants from solutions is complex and the indication of the dominant process requires systematic analyzes taking into account the type of pollution and its origin, and biochar characteristics as a sorbent. This paper compares the efficiency of removing contaminants from leachate generated at an old landfill by using two types of granular activated carbons and biochar.

2 Materials and methods

2.1 Substrate

The leachate was derived from the regional municipal waste landfill in Silesian Province (Poland). The landfill has been in operation since 1987, therefore it should be included in the group of "old landfills". This is evident from the values of leachate pollution indicators and the low BOD/COD ratio (0.059) indicating their low biodegradability (Table 3).

Table 3 Characteristics of landfill leachate used.

Parameter	Unit	Value
pH	-	8.2–8.6
COD	$mg \cdot dm^{-3}$	3098±20
BOD	$mg \cdot dm^{-3}$	185±5
TOC	$mg \cdot dm^{-3}$	550±10
Ni	$mg \cdot dm^{-3}$	0.23
Zn	$mg \cdot dm^{-3}$	1.5
Cu	$mg \cdot dm^{-3}$	0.32
Cd	$mg \cdot dm^{-3}$	0.00303
Pb	$mg \cdot dm^{-3}$	0.01396

The landfill is conducted by the prismatic method. The area designated for the landfill is 128.4 ha (plus a protection zone of 342 ha). As a result of infiltration of rainwater and technological sprinkling, waste solids are created there by approx. 60 $m^3 \cdot d^{-1}$ of leachate. The purification process is carried out in a two-stage reverse osmosis process in a ROAD 9142 PALL-ROCHEM installation.

2.2 Activated granular carbon

The analysis was based on two granular activated carbons (GAC) with symbols: AG and BA-10. Both granular activated carbons were of Polish production (AG- Gryfskand Sp. z.o.o and BA-10 ELbar Katowice Sp. z.o.o.). Characterization of the activated carbons is presented in Table 4.

Table 4. Characteristic of studied activated carbons (PN-90/C-97555).

Parameters	AG	BA-10
Specific surface, $m^2 \cdot g^{-1}$	750–850	min.1000
Bulk mass, $g \cdot dm^{-3}$	420–480	490±30
Typical grain size, mm	1.5	3
Resistance mechanical, %	92–96	97
The ash content, %	5–20	5.94

2.3 Biochar

Biochar (BC) was obtained in the autothermic process of carbonizing o the miscantus at 450°C during 60 min. Technical analysis of biochar was made according to Polish Standard PN-G-04560:1998P (Table 5). The analysis of the content of carbon (C), hydrogen (H), nitrogen (N) and sulphur (S) was carried out in elemental analyzer (LECO). Heat of combustion was determined using IKA Basic C 2000 calorimeter using the isoperibolic method. Philips XL30/LaB6 scanning microscope was used for structural analysis of biochar. Porosity of biochar was determined using PoreMaster 33 mercury porosimeter with Quantachrome Instruments software. The open porosity of BC was estimated to 24.5% and specific surface area about 12.16 m^2 g^{-1}.

Table 5. Proximate and elemental ultimate analysis of biochar sample (dry): WT – total water, VM – total volatile fractions, A_d – ash content (575°C- sample combustion temperature), FC – fixed carbon, C_d- carbon content, H_d- hydrogen content, N_d- nitrogen content, O_d-oxygen content.

WT, %	VM_d, %	A_{d575}, %	FC, %	C_d, -	H_d, -	N_d, -	S_d, -	O_d, -
7.9	10.00	15.80	74.20	74.20	2.82	0.95	0.23	18.99

3 Research methodology

Adsorption from landfill leachates was carried out under static conditions for jar tests of 250 cm^3 with addition respectively: 0.5 g (dose – 2 $g\cdot dm^{-3}$), 1 g (dose – 2 $g\cdot dm^{-3}$), 1.5 g (dose – 6 $g\cdot dm^{-3}$) and 2 g (dose – 8 $g\cdot dm^{-3}$) adsorbents (GAC and BC). The contact time was 72 h with the mixing intensity was set at 200 rpm min^{-1}. The effectiveness of the adsorption process at the assumed doses and the contact time, was examined based on the degree of removal of impurities designated as COD and concentrations of selected metals (Ni, Zn, Cu, Cd, and Pb). Chemical Oxidation Demand (COD) was analyzed using colorimetric tests on Hach-DR 4000 photometer. Heavy metals were analysed inductively coupled plasma mass spectrometry (ICP-MS) analyser. The sorption capacity of the sorbents tested was calculated from the formula (1):

$$q = \frac{(C_0 - C_t)\cdot v}{m} T \tag{1}$$

where q is the adsorbent removal capacity of the given contaminant; C_0 is the initial and C_t is the equilibrium concentration of the given contaminant ($mg\cdot dm^{-3}$); v is the volume of solution (dm^3) and m is the weight of the adsorbent (g).

4 Results and discussion

4.1 Analysis of the adsorbent dose effect on sorption capacity and on removal pollutions degree from landfill leachates

In the first step, the time of sorption equilibrium was determined in static conditions on the adsorbents tested. The initial concentration of impurities designated as COD in the leachate from the landfill amounted to 3098 $mg\cdot dm^{-3}$. It was found that irrespective of the adsorbent used, the sorption proceeded most intensely in the first 2 hours of the process. The time of the sorption equilibrium was similar to the biocarbon and active carbon AG and was established after 4 hours. The sorption equilibrium took place on the BA-10 granulated carbon (6 h at the dose of 2 $g\cdot dm^{-3}$). The discussed changes are presented in Fig.1. The obtained test results in this stage show that extending the contact time above 10 h slightly affects the efficiency of the sorption process. In the case of biochar as well as activated carbon AG, already for 0.5 h the sorption process at the dose of 2 $g\cdot dm^{-3}$ resulted in an average of 20% removal of impurities designated as COD. Increasing the dose of sorbents from 2 $g\cdot dm^{-3}$ to 4 $g\cdot dm^{-3}$ for all tested adsorbents resulted in an increase in the COD removal rate on AG, BA-10 and biochar, respectively by 9%, 12% and 6%. After 72 h of the process at the highest doses (8 g·dm^{-3}), the COD removal rate ranged from 40% (BA-10) to 60% (AG). In the case of biochar, the COD adsorption rate was 53% (Figure 2).

The research carried out showed that with increasing doses of adsorbents there was a decrease in the sorption capacity. It was noted that increasing the dose from 6 g·dm^{-3} to 8 g·dm^{-3} in the case of biochar and active carbon-AG did not affect its significant reduction anymore. The highest sorption capacity (152.6 mg·g^{-1}) was recorded for active carbon AG at the dose of 2 g·dm^{-3}. Comparably at this dose, the sorption capacity for activated carbon BA-10 and biochar was 102.62 mg·g^{-1} and 147.92 mg·g^{-1}, respectively (Figure 2).

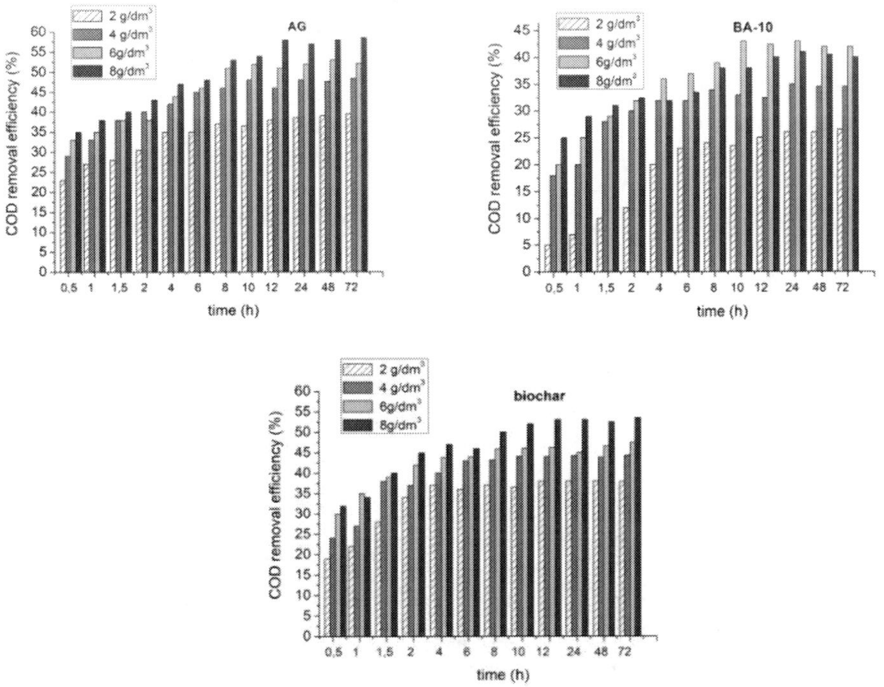

Fig. 1. Time determination of the sorption equilibrium for studied activated carbons and biochar.

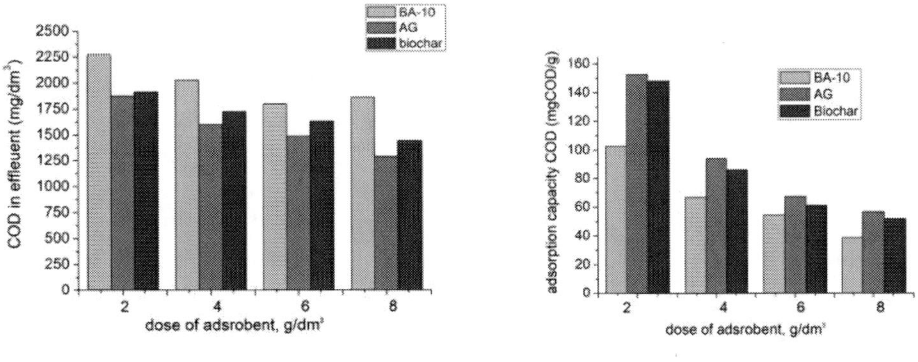

Fig. 2. The effect of adsorbent dose on sorption capacity and COD in effluent.

The Freundlich equation was used to describe the sorption isotherms. Using the logarithmic form of this equation, KF and 1/n constants were determined for the studied activated carbons and biochar, whose values are shown in Table 6.

Table 6. The Freundlich isotherm constans from COD.

Carbons	**K_F, mg/g**	**1/n**	**R^2**
AG	6.94	0.6098	0.931
BA-10	10.30	0.6502	0.928
Biochar	5.80	0.5182	0.879

Numerous studies confirm that the sorption of pollutants on biochars can proceed with similar effectiveness as on granular activated carbons [11, 13–15, 32]. The degree of removal of COD may vary from 45% to 95% and depends on the type of biocarbon used, its activation or absence as well as on the concentration and type of pollutants in the wastewater. In the future, it is planned to carry out a modification of biochar (thermal or chemical) in order to improve the efficiency of sorption of contaminants on its surface. Huggins and el. 2016 [15] have observed even higher COD (raw wastewater – 1243 mg·dm^{-3}) removal rate by over 10% on the biochar compared to granular activated carbon. The biochar used in their studies was characterized by a 30% higher sorption capacity (70 mg COD g^{-1}) compared to GAC (49.3 mg COD g^{-1}). Unlike the authors of the work, Huggins and el. 2016 [15] determined the dependence of sorption capacity on the equilibrium concentration and not on the amount of adsorbents used.

After analyzing the obtained test results, it is stated that in the case of activated carbon AG as well as biochar, it is most advantageous to conduct the sorption process of contaminants designated as COD from landfill leachate at their dose level of 6 g·dm^{-3}. Therefore, in the above attempts of treated landfill leachates changes in concentrations of selected metals were analyzed. The degree of metal sorption has the adsorbents used and was similar and ranged from 84% to 96% depending on its type (Table 7).

Table 7. Intial and final concentration for AG, BA-10 and biochar.

Metal	**Intial metal concentration**	**Final metal concentration**		
		AG	**BA-10**	**Biochar**
Ni	0.23	0.0253	0.0207	0.0368
Zn	1.5	0.033	0.03	0.03
Cu	0.32	0.0576	0.0672	0.064
Cd	0.00303	0.00048	0.00033	0.00039
Pb	0.01396	0.0011	0.00055	0.000695

As results from the research and reports, biochar is an attractive adsorbent of organic pollutants and heavy metals [19–27]. The use of porous biochar in the aeration chamber during the biological treatment of leachate as a carrier of activated sludge flocs will allow simultaneous biodegradation and adsorption of pollutants.

5 Conclusions

In the work, the adsorption of pollutants from landfill leachates was analyzed in static conditions. The sorption properties of two granular activated carbons (GAC) with symbols: AG and BA-10 with biochar made from miscantus in pyrolysis process were compared. On the basis of the conducted research, the following statements and conclusions were formulated:

– the state of sorption equilibrium takes a similar character for BC and AG and is determined after 4 h. The sorption equilibrium was observed on BA-10 granulated coal (after 6 h at the dose of 2 g·dm^{-3}),

- 30 min of conducting sorption process from sewage effluents with the inclusion of BC and AC gives a 20% effect of removing impurities designated as COD, and extending the duration of the process to 72 hours allows the removal of COD up to 60%,
- the sorption capacity AG at a dose of 2 $g \cdot dm^{-3}$ was established at 152.6 $mg \cdot g^{-1}$, comparable with this dose, the sorption capacity for activated carbon BA-10 and biochar was 102.62 $mg \cdot g^{-1}$ and 147.92 $mg \cdot g^{-1}$,
- the degree of sorption of selected metals from landfill leachate using the analyzed range ranged from 84% to 96% depending on its type.

The study has been funded by BS/PB–401–301/11.

References

1. K. Wang, L. Li, F. Tan, D. Wu, Hindawi Archea, **2018** (2018)
2. F. N. Ahmed, C. Q. Lan, Desalination **287**, 41–54 (2012)
3. G. Li, W. Wang, Q. Du, Journal of Applied Polymer Science **116**, 2343–2347 (2010)
4. B. Zadroga, K. Olańczuk-Neyman, Wydawnictwo Politechniki Gdańskiej (2001)
5. J. Bohdziewicz, E. Neczaj, A. Kwarciak, Desalination **221**, 559–565 (2008)
6. S. M. Joshi, P. R. Gogate, Separation and Purification Technology **211**, 10–18 (2019)
7. L. Jurczyk, J. Koc-Jurczyk, M. Balawejder, Journal of Ecological Engineering **20**, 204–213 (2019)
8. J. Bohdziewicz, A. Kwarciak, Desalination **222**, 128–134 (2008)
9. A. Kwarciak-Kozłowska, L. Sławik-Dembiczak, Inżynieria i Ochrona Środowiska **19**, 561–575 (2016)
10. A. Kwarciak-Kozłowska, Journal of Ecological Engineering **19**, 186–193 (2018)
11. X. Tan, Y. Liu, G. Zeng, X. Wang, X. Hu, Y. Gu, Z. Yang, Chemosphere **125**, 70–85 (2015)
12. L. Dąbek, Annual Set The Environment Protection **17**, 616–645 (2015)
13. D. Mohan, A. Sarswat, Y. S. Ok, Ch. U. Pittman Jr., Bioresource Technology **160**, 191–202 (2014)
14. S. S. Dalahmeh, SLU Report, **090** (2016)
15. T. Huggins, A. Haeger, J. C. Biffinger, Z. J. Ren, Water Research **94**, 225–232 (2016)
16. fingerlakesbiochar.com
17. B. Wang, B. Gao, J. Fang, Critical Reviews In Environmental Science And Technology **0**, 1–50 (2017)
18. A. D. Lewis, E. G. Fletcher, T. H. Fletcher, *Pyrolysis and CO_2 Gasification Rates of Biomass at High Heating Rate Conditions* (Western States Section of the Combustion Institute and Hosted by the University of Utah May 19–22, 2013)
19. A. B. Duwiejuah, S. J. Cobbina, N. Bakobie, International Journal of Environmental Bioremediation & Biodegradation **5**, 27–40 (2017)
20. R. A. Wuana, F. E. Okieimen, Communicat. Soil Sci. Plant Anal. **42**, 111–122 (2011)
21. M. Uchimiya, I. M. Lima, K. T. Klasson, S. Chang, L. H. Wartelle, J. E. Rodgers, J. Agric. and Food Chem. **58**, 5538–5544 (2010)

22. A. Dowie, A. Crosky, P. Muroe, *Physical properties of biochar* (in BC for environmental management, J. Lehmann and S. Joseph, Eds., 47–82, Earthscan, London, UK, 2009)
23. R. Qiu, H. Lu, W. Zhang, Y. Yang, X. Huang, S. Wang, Wat. Res. **46**, 854–862 (2012)
24. S. Babel, T. A. Kurniawan, Chemosph. **54**, 7, 951–967 (2004)
25. Z. Song, F. Lian, Z. Yu, L. Zhu, B. Xing, W. Qiu, Chemical Engineering Journal, **242**, 36–42 (2014)
26. D. Mohan, S. Rajput, V. K. Singh, P. H. Steele, C. U. Pittman Jr., Journal of Hazardous Materials **188**, 319–333 (2011)
27. X. Tong, J. Li, J. Yuan, R. Xu, Chemical Engineering Journal **172**, 828–834 (2011)
28. S. De Gisi, G. Lofrano, M. Grassi, M. Notarnicola, Sustainable Materials and Technologies **9**, 10–40 (2016)
29. H. Kong, J. He, Y. Gao, H. Wu, X. Zhu, J. Agric. Food Chem. **59**, 12116–12123 (2011)
30. M. Uchimiya, I. M. Lima, K. T. Klasson, S. Chang, L. H. Wartelle, J. E. Rodgers, J. Agric. and Food Chem. **58**, 5538–5544 (2010)
31. K. Štefušová, M. Lovás, A. Zubrik, M. Matik, M. Václavíková, Nova Biotechnologica Et Chimica **11**, 139–146 (2012)
32. S. B. Chavda, M. J. Pandya, IJIRT **1**, 101472, 1–5 (2014)

Environmental effects of driving style: impact on fuel consumption

Jakub Lasocki[1,*], and *Karol* Boguszewski[1]

[1]Warsaw University of Technology, Faculty of Automotive and Construction Machinery Engineering, ul. Narbutta 84, 02-524 Warsaw, Poland

Abstract. From an environmental point of view, the fuel consumption of vehicles with combustion engines is directly related to the depletion of non-renewable crude oil resources and pollutant emission. The aim of this paper is to evaluate the effect of driving style on fuel consumption of light-duty vehicles. The study considered five metrics used for quantitative description of driving style: Dynamic Performance Index (DPI), Aggressiveness Factor (AF), Vehicle Aggressivity (VA), Total Aggressivity (TA), based upon the previous works of other researchers, and a newly proposed metric named Driving Style Indicator (DSI). All metrics were applied to the results of chassis dynamometer tests of two light-duty vehicles with spark-ignition and compression-ignition combustion engines. The values of metrics were plotted against corresponding values of fuel consumption to create dependences. Their analysis revealed that AF metric has strong correlation with fuel consumption, but is mathematically complex and requires numerous input data. DSI metric has simple mathematical form and is based only on the speed profile of the vehicle, and yet is characterized by a strong correlation with fuel consumption. DSI metric was further employed to investigate the influence of driving style on greenhouse gas (GHG) emissions from the Well-to-Wheel (WtW) perspective.

1 Introduction

Fuel consumption is one of the most important features of modern vehicles. First of all, it is clear that a decrease in fuel consumption is beneficial for the environment. The less fuel is used by the internal combustion engine (the more efficient it is), the less exhaust gases are produced, and the less is the consumption of crude oil – a non-renewable resource [1]. In some areas of the world, fuel consumption of vehicles is limited by law (directly or indirectly – in the form of a carbon dioxide emission limit) [2]. Besides, from the consumer's point of view, fuel consumption is directly related to the costs of vehicle use borne by the driver, which is often a decisive factor considered when buying a vehicle. For these reasons, a lot of effort is put into investigating the determinants of the fuel use of vehicles, both at the academic level and in the R&D centres of the automotive industry [2–7].

Detailed literature review reveals that although there are many factors affecting fuel consumption of vehicles, the most important include: the use of air conditioning devices,

* Corresponding author: jakub.lasocki@pw.edu.pl

ambient temperature and environmental conditions, roof add-ons, driving style, tyre pressure and vehicle weight [1, 8, 9]. This study focuses on a driving style, which is usually characterised by vehicle speed profile and choice of gears [4, 8]. Aggressive driving is known to increase fuel consumption by up to 24% [1, 2], while eco-driving can lead to certain gains of around 4–10% [3, 10, 11] (with some sources claiming even 30% [1]), compared to standard driving. Therefore driving style optimization is a relatively low-cost and immediate measure to reduce fuel consumption significantly [12]. At the same time it is very difficult to evaluate a personal driving style, as it is one of the most unstable and psychological-related factor [13, 14]. The complexity of driver's behaviour and its connection with vehicle operating parameters is a timely research problem being approached by more and more scientists working in the field of road transport.

One of the main descriptors of personal driving style is the choice of vehicle speed, which has direct implication on fuel consumption [15]. Not only the average value of speed, but also the way the speed changes is of a great importance [16]. For example, intensive acceleration and braking, causing higher fuel consumption [17, 18], are often regarded as indicators of driving aggressiveness. Therefore thorough investigation of speed data obtained from real road measurements is the basic approach in the development of driving cycles for testing of pollutant emission and fuel consumption of vehicles in laboratory conditions.

The aim of this paper is to quantify the impact of driving style on fuel consumption of light-duty vehicles and to assess the related environmental effects. Assuming the course of vehicle speed as a representative characteristic of a driving style, different metrics were determined based on the results of chassis dynamometer tests of two vehicles with different type of combustion engine, i.e. spark-ignition (SI) and compression-ignition (CI). Calculated metrics were matched with corresponding values of fuel consumption to create dependencies. Evaluation of those dependencies allowed to choose metric with the best correlation to fuel consumption. This metric was selected to examine the influence of driving style on greenhouse gas (GHG) emission from the Well-to-Wheel (WtW) perspective, which includes not only fuel combustion in the engine, but also the entire chain of fuel production.

2 Driving style metrics

A wide range of researchers have studied how fuel consumption is affected by the way a vehicle is driven. Some of these studies made an attempt at defining a metric to be used in the quantitative assessment of this relation. Five such metrics, selected after extensive research, have been considered in the current study.

The first study [19, 20], by Casanova, Fonseca and Espinosa, concerned the link between parameters of driving pattern determined based on signals recorded in a vehicle and pollutant emissions produced over a certain urban route. The Dynamic Performance Index (DPI) has been established in order to classify driving into one of three categories: aggressive, normal or calm. DPI has two different formulas, depending on the type of vehicle's engine. For SI engines the formula is [19]:

$$DPI_{SI} = -0.17 + 0.55a_{AV} + 0.006t_{n>2000} \tag{1}$$

while for CI engines [20]:

$$DPI_{CI} = -0.00158t_{g1} + 0.00205t_{g2} + 0.0582a_{AV} + 0.00148t_{n>1850} \tag{2}$$

where: a_{AV} – average acceleration, $t_{n>2000}/t_{n>1850}$ – time share of driving with engine speed higher than 2000 rpm/1850 rpm, t_{g1}/t_{g2} – time share of driving spent in the first/second gear.

In the next study considered [21] Berry proposed to evaluate driving style using Aggressiveness Factor (AF) that incorporates coast down coefficients used to model resistance which acts on the vehicle in operation, e.g. tire rolling resistance and aerodynamic drag. To achieve the best correlation between aggressiveness factors and fuel consumption, AF is normalised to vehicle mass and customised to suit different traffic conditions defined by the average vehicle speed. The resulting formulas were named: 'neighbourhood driving' for speed below 32 km/h, 'city driving' for speed between 32 and 72 km/h and 'highway driving' for speed above 72 km/h. They all have a similar mathematical form, differing only in details. For example, aggressiveness factor used for city driving is given by equation:

$$AF_C = \frac{1}{\mathrm{M}}\left[\frac{\int (\mathrm{A}v + \mathrm{B}v^2 + \mathrm{C}v^3 + \mathrm{M}av)^+ \mathrm{d}t}{\int v\mathrm{d}t} - F_r(v_{AV})\right] \tag{3}$$

where: M – vehicle mass, A, B, C – coast down coefficients, v – speed, v_{AV} – average speed, $F_r(v) = \mathrm{A} + \mathrm{B}v + \mathrm{C}v^2$ – the force of resistance, a – acceleration, t – time.

The third metric considered in this paper is Vehicle Aggressivity (VA) taken from the study by Ford Scientific Research Laboratory [22]. It was found to have a very good correlation with pollutant emission, however did not show as significant match with fuel economy as would be preferred. The advantage of this metric is its simple mathematical form and a limited need for input data:

$$VA = \sqrt{\frac{1}{N}\sum_{1}^{N}(2v_i a_i)^2} \tag{4}$$

where: N – the number of events that occur in a drive cycle, v – speed, a – acceleration.

Ford's original concept of VA was further extended by MAHLE Powertrain Ltd. Pedal Aggressivity (PA), a factor related to the position of the accelerator pedal, was added [23]:

$$PA = \sqrt{\frac{1}{N}\sum_{1}^{N}(2s_i s'_i)^2} \tag{5}$$

where: s – relative position of the accelerator pedal for given acceleration profile, s' – the average rate of the relative position of the accelerator pedal for this profile.

In contrast to Ford's metrics, MAHLE's VA considers only positive vehicle acceleration. Another unique feature of this method is averaging speed and acceleration over an acceleration profile (that goes between adjacent minimum and maximum) and then their normalization by the number of events (acceleration profiles considered). This same technique is used for the PA as well. VA is combined with PA into one metric called Total Aggressivity (TA):

$$TA = \sqrt{VA} + \sqrt[4]{PA} \tag{6}$$

The goal of TA metric is to alleviate the differences between vehicles, which characteristics (e.g. power output) affect the driving dynamics, and thus to focus primarily on the impact of the driver. In other words, TA helps normalize between vehicles.

In search of a simplified way of assessing the impact of driver's driving style on fuel consumption, a new metric has been developed in this study. The starting point was driving aggressivity metric proposed by Stichter [24], which combines average positive and negative vehicle acceleration with vehicle speed – three parameters that are known to correlate well with fuel consumption. The original idea of Stichter has been implemented in current research with an emphasis on obtaining a maximally simplified mathematical form. In this way, the Driving Style Indicator (DSI) was proposed as follows:

$$DSI = \frac{a^{+}_{AV} - a^{-}_{AV}}{v_{AV}} \tag{7}$$

where: a^{+}_{AV} – average positive acceleration of a vehicle, a^{-}_{AV} – average negative acceleration (deceleration) of a vehicle.

Each metric outlined in this section was used for the investigation of the effect of driving style on fuel consumption of light-duty vehicles.

3 Materials and methods

The general concept of the study was to determine the dependences of fuel consumption on the driving style metrics using results of testing vehicles in standard driving cycles, on a chassis dynamometer. It is therefore assumed that the driver's driving style is expressed exclusively by the speed of a vehicle, or specifically by the speed profile imposed by a given driving cycle. In the next step, the environmental impact of driving style was quantified by linking the dependencies obtained to total GHG emission, in line with the WtW approach.

The study was conducted based on previously published data gathered at the Advanced Powertrain Research Facility of Argonne National Laboratory in the USA [25]. The tests were carried out using a chassis dynamometer with rollers of 1.22 m diameter, able to provide variable equivalent inertia weight of 136–5443 kg. Test cell conditions, i.e. temperature, humidity, and solar load, were fully controlled. Detailed specifications of testing equipment can be found in [25].

The testing included two vehicles: Dodge RAM with SI combustion engine (hereinafter referred to as vehicle 1) and Chevrolet Cruise with CI combustion engine (hereinafter referred to as vehicle 2). Their main technical specifications are shown in Table 1.

Table 1. Specifications of the tested vehicles.

Parameter	Unit	Vehicle 1	Vehicle 2
Make and model		Dodge RAM	Chevrolet Cruise
Model year		2013	2014
Curb/test mass	kg	2074/2245	1587/1727
Engine type		SI	CI, turbocharged
Engine displacement	cm^3	3604	1956
Maximum power (at engine speed)	kW (rpm)	227 (6400)	112 (4000)
Maximum torque (at engine speed)	N·m	365 (4175)	358 (2600)
Transmission type		8-speed automatic	6-speed automatic
Other features		Start-stop system, variable valve timing	DPF, cooled EGR, urea injection after-treatment

Seven different driving cycles were performed on a chassis dynamometer, including FTP-72, HWFET, US06, WLTC, NEDC, NYCC and LA92. Vehicle 1 was subjected to all of them, while in the case of vehicle 2 the last two cycles were omitted. In order to achieve bigger accuracy of dependencies determined, WLTC and NEDC cycles were split into phases of which they are composed. The values of driving style metrics described in section 2 of this paper were calculated for the actual speed course of the vehicle, which is always slightly different from the nominal one defined by the driving cycle.

The DPI_{CI} metric, intended to be used for vehicles with CI engines, was not applied for vehicle 2 due to the lack of data on the time share of driving in individual gears. AF metric was calculated separately for the three ranges of vehicle average speed, according to the original concept of Berry, although this does not agree with the formal classification of some driving cycles (e.g. the EUDC simulating 'extra-urban' conditions was classified as city driving). TA metric was considered only for vehicle 1, due to the lack of data on the position of the accelerator pedal for vehicle 2. Besides, this metric was slightly modified in accordance with the Stichter proposal [24], i.e. instead of looking at an acceleration profile, speed points for each individual time interval were considered.

In order to evaluate the relationship of driving style with the environmental effects of vehicle use, data from the WtW analysis were employed. This approach allows to quantify environmental load of the production, transport and distribution of fuels (Well-to-Tank, WtT), as well as of fuel combustion in the vehicle's engine (Tank-to-Wheel, TtW). The methodology considered in this paper is presented in detail in the report of JRC-Eucar-Concawe consortium (JEC) [26]. The most important assumptions for the calculations made in the current study include:

- WtT GHG emission factors: gasoline – 13.8 g_{CO2eq}/MJ, diesel oil – 15.4 g_{CO2eq}/MJ [26],
- TtW GHG emission factors: gasoline – 73.4 g_{CO2eq}/MJ, diesel oil – 73.2 g_{CO2eq}/MJ [26],
- global warming potentials (in time horizon of 100 years): CO_2 – 1, CH_4 – 23, N_2O – 296,
- heating value of fuels used in tests: gasoline – 42.64 MJ/kg, diesel oil – 42.80 MJ/kg [25],
- density of fuels used in tests: gasoline – 0.741 kg/dm^3, diesel oil – 0.851 kg/dm^3 [25].

4 Results and discussion

Fuel consumption of vehicles, related to the distance travelled, was plotted against the values of driving style metrics. The resulting dependences are presented in Figures 1–5.

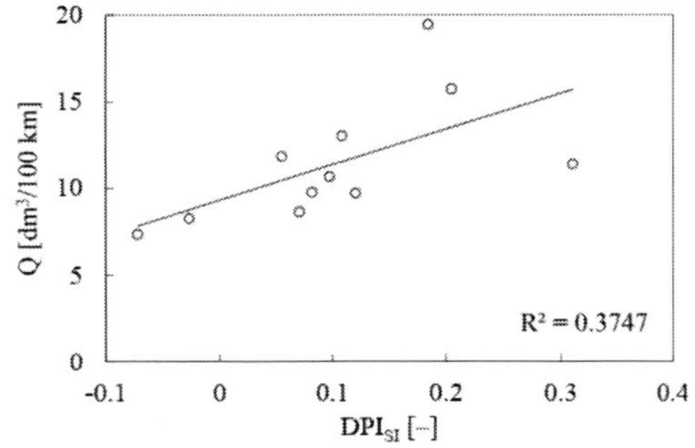

Fig. 1. The dependence of fuel consumption on DP_{SI} for vehicle 1.

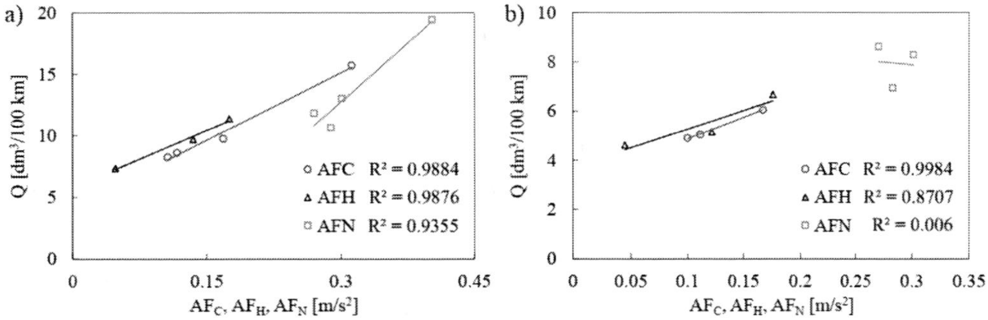

Fig. 2. The dependence of fuel consumption on AF for: a) vehicle 1, b) vehicle 2.

https://doi.org/10.1051/e3sconf/201910000043

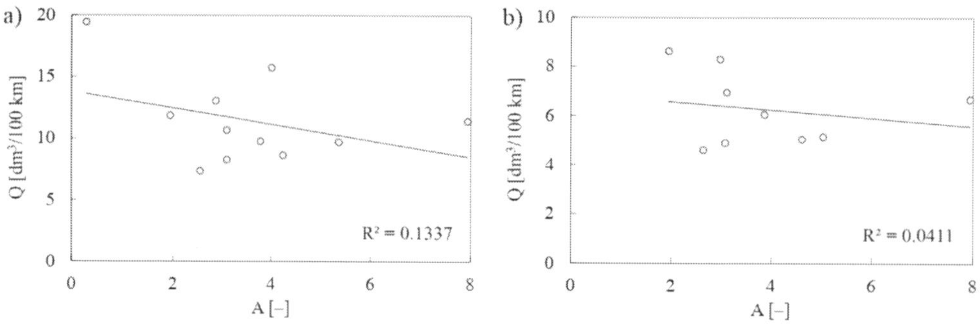

Fig. 3. The dependence of fuel consumption on A for: a) vehicle 1, b) vehicle 2.

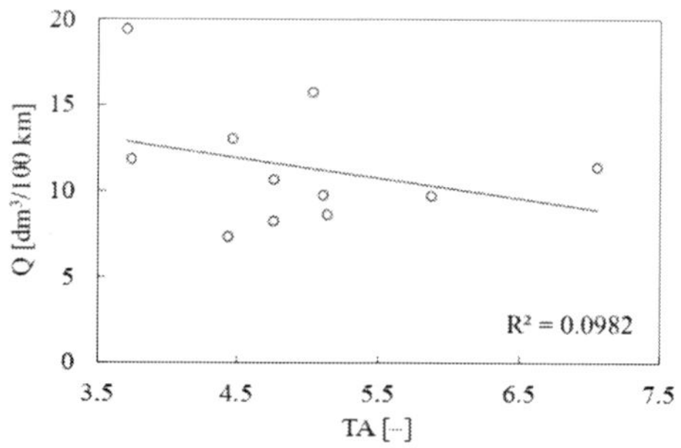

Fig. 4. The dependence of fuel consumption on TA for vehicle 1.

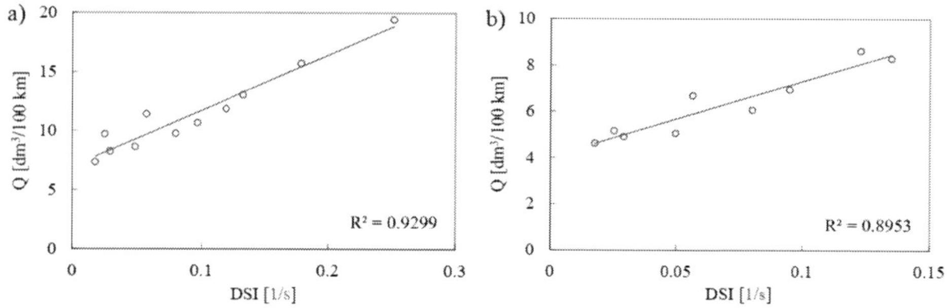

Fig. 5. The dependence of fuel consumption on DSI for: a) vehicle 1, b) vehicle 2.

Analysing the obtained results, it can be seen that metrics A and TA have definitely the worst correlation with fuel consumption. The R^2 values of the linear trend line drawn are around 0.1 or lower for both vehicles. It should be noted, however, that metrics A and TA were intended by their authors to assess pollutant emissions and not fuel consumption. DP_{SI} metric is somewhat better correlated with fuel consumption, but still it is not a sufficient dependence, because the R^2 value is only about 0.37. In contrast, AF was found to have very strong correlation to fuel consumption, having R^2 values that exceed 0.9. Being specific, the best results were obtained for urban driving conditions to which the AF_C metric is applied. This is true for both vehicles. Similarly, for high speed driving, fuel consumption is strongly correlated with AF_H metric. In the case of low-speed driving (AF_N), called 'neighbourhood driving' by Berry, this relationship is rather ambiguous. On the one hand, a strong dependence was obtained for vehicle 1, with R^2 at the level of 0.94, while on the other hand a very weak dependence, with R^2 being much less than 0.01 was achieved for vehicle 2. Despite this, the AF is actually the only driving style metric derived from other authors' research that showed close dependency with fuel consumption. The new metric – DSI, proposed in this study, allows to obtain comparatively good results. This is confirmed by the R^2 values of the linear trend lines, which are 0.93 and 0.90 for vehicle 1 and vehicle 2, respectively. It must be emphasized that these results were obtained using data from all driving cycles, regardless of the vehicle's average speed. It shows the universality of the new metric. Another advantage of DSI over AF is a small number of input data necessary for calculations and no need to determine the vehicle's coast down coefficients. DSI metric has been selected for further research into the impact of driving style on GHG emission. The reasons behind this choice were twofold: strong correlation with fuel consumption and its almost linear character. Figure 6 shows the resultant dependence of GHG emission on DSI metric, obtained for both vehicles.

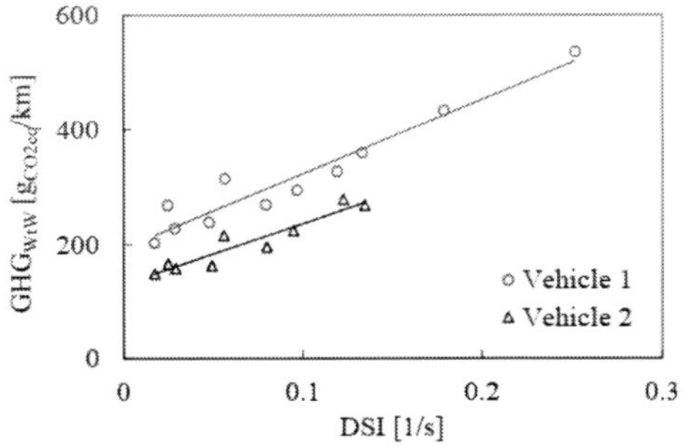

Fig. 6. The dependence of Well-to-Wheel GHG emission on DSI for vehicles 1 and 2.

For obvious reasons, the dependencies determined in this way are strongly related to the properties of the vehicle and fuel, but for a specific case, this approach can be recommended for predicting the approximate effects of driving style in terms of GHG emission. When assessing the results presented, the limitations of this study should be kept in mind. First of all, a relatively small number of driving cycles has been included. This is especially evident in the case of AF metric, where sometimes three points had to suffice to infer dependence on fuel consumption. Secondly, the driving cycles performed on a chassis dynamometer do not fully reflect the actual driving conditions of vehicles in road traffic. It would therefore be beneficial to carry out further studies in road traffic. This can be a future direction of development of this work.

5 Conclusions

Based on the research described in this paper, the following conclusions can be drawn:

- To evaluate the impact of a driving style on fuel consumption, certain metrics can be used, which combine parameters characterizing the course of vehicle speed, gear changes, engine speed and technical specifications of the vehicle.
- The dependence of fuel consumption on driving style metrics known from the literature is approximately linear.
- Driving style metrics known from the literature can be divided into two groups: those strongly correlating with fuel consumption, but requiring numerous input data, sometimes difficult to access, and those to which data is readily available, but giving results with a relatively weaker correlation.
- The new metric, Diving Style Indicator (DSI), proposed in this study is not mathematically complicated and is based only on the data regarding the speed of the vehicle, and yet is characterized by a strong correlation with fuel consumption.
- It is possible to employ selected metrics for predicting the approximate environmental effects of driving style in terms of GHG emission associated with the use of fuel.

References

1. N. Zacharof, G. Fontaras, B. Ciuffo, S. Tsiakmakis, K. Anagnostopoulos, A. Marotta, J. Pavlovic, *Review of in use factors affecting the fuel consumption and CO_2 emissions of passenger cars* (Publications Office of the European Union, Luxembourg, 2016)
2. L. Schipper, Transport Policy **18**, 358–372 (2011)
3. B. Beusen, S. Broekx, T. Denys, C. Beckx, B. Degraeuwe, M. Gijsbers, K. Scheepers, L. Govaerts, R. Torfs, L.I. Panis, Transport. Res. D-Tr. E. **14**, 514–520 (2009)
4. K. Brundell-Freij, E. Ericsson, Transport. Res. D-Tr. E. **10**, 213–229 (2005)
5. A. Kopczyński, Z. Liu, P. Krawczyk, E3S Web of Conferences **44**, 00074 (2018)
6. W. Gis, J. Pielecha, J. Waśkiewicz, M. Gis, M. Menes. IOP Conf. Ser.-Mat. Sci. **148** (2016)
7. A. Kopczyński, P. Piórkowski, P. Roszczyk, IOP Conf. Ser.-Mat. Sci. **421**, 022016 (2018)
8. E. Ericsson, Transport. Res. D-Tr. E. **5**, 213–229 (2005)
9. A. Kopczyński, P. Krawczyk, J. Lasocki, E3S Web of Conferences **44**, 00073 (2018)
10. J. N. Barkenbus, Energ. Policy **38**, 762–769 (2010)
11. I. Jeffreys, G. Graves, M. Roth, Transport. Res. D-Tr. E. **60**, 85–91 (2018)
12. Y. Huang, E. C. Y. Ng, J. L. Zhou, N. C. Surawski, E. F. C. Chan, G. Hong, Renew. Sust. Energ. Rev. **93**, 596–609 (2018)
13. Y. Lei, K. Liu, Y. Fu, X Li, Z. Liu, S. Sun, Research on driving style recognition method based on driver's dynamic demand, Adv. Mech. Eng. **8**, 1–14 (2016)
14. Y. Wang, W. Qu, Y. Ge, X. Sun, K. Zhang, Plos One **13**, e0202126 (2018)
15. Z. Chłopek, J. Lasocki, P. Wójcik, A. J. Badyda, Int. J. Green Energy **15**, 773–779 (2018)
16. Z. Chłopek, J. Biedrzycki, J. Lasocki, P. Wójcik, Eksploatacja i Niezawodnosc – Maintenance and Reliability **17**, 35–41 (2015)
17. A. Szumanowski, Y. Chang, Z. Liu, P. Krawczyk, Int. J. Vehicle Des. **76**, 1–19 (2018)
18. J. Merkisz, M. Gis. E3S Web of Conferences **44**, 00109 (2018)
19. J. Casanova, N. Fonseca, F. Espinosa, Proc. 17th Transport and Air Pollution Symp. and 3rd Environment and Transport Symp., 2–4 June 2009, Toulouse, France (2009)
20. N. Fonseca, J. Casanova, F. Espinosa, Proc. 18th International Symp. Transport and Air Pollution, 18–19 May 2010, Dübendorf, Switzerland (2010)
21. I. Berry, *The effects of driving style and vehicle performance on the real-world fuel consumption of U.S. light-duty vehicles* (PhD thesis, Massachusetts Institute of Technology, Massachusetts, 2010)
22. E. K. Nam, C. A. Gierczak, J. W. Butler, *A comparison of real-world and modeled emissions under conditions of variable driver aggressiveness* (82nd Annual Meeting of the Transportation Research Board, 12–16 January 2003, Washington DC, 2003)
23. R. Daniel, T. Brooks, D. Pates, SAE Technical Paper, 2009-01-0236 (2009)
24. J. S. Stichter, *Investigation of vehicle and driver aggressivity and relation to fuel economy testing* (PhD thesis, University of Iowa, Iowa, 2012)
25. Argonne National Laboratory, https://www.anl.gov/ [online] [accessed 24.01.2019]
26. JEC (JRC-Eucar-Concawe), *Well-To-Wheels Report Version 4.a: Well-to Wheels Analysis of Future Automotive Fuels and Powertrains in the European Context* (Publications Office of the European Union, Luxembourg, 2014)

Evaluation of different methods supporting swimming pool water disinfection in the aspect of removing organic micropollutants

Anna Lempart[1,*], *Edyta* Kudlek[1], and *Mariusz* Dudziak[1]

[1]Silesian University of Technology, Institute of Water and Wastewater Engineering, Gliwice, Poland

Abstract. As a result of the intensive development of analytical techniques that allow to reduce the detection limits of tested compounds, the latest scientific research alerts the growing number of organic micro-pollutants identified in the swimming pool water environment. These compounds can both react with the disinfectant, causing the formation of highly toxic disinfection by-products or constitute a serious health risk to swimmers due to their biological activity. The aim of the research presented in this paper is to evaluate the modern methods supporting disinfection used in swimming pool systems, in the aspect of removing compounds from the group of organic microorganisms. The total content of organic matter in systems using UV radiation and ozonation was compared. The identification of organic micropollutants was also carried out. The presence of two phthalates in the tested pools has been documented. It has been shown that their concentration depends on the disinfection method used. Photocatalytic degradation of phthalates in the pool water system allowed to obtain over 35% removal rate, while the efficiency of ozonation in the most favourable case was just over 16%.

1 Introduction

In the light of current legal regulations and technical guidelines [1–3], water in public swimming pools should meet a number of microbiological and physico-chemical requirements that guarantee it will not endanger swimmers' health. Such properties are ensured by a properly designed pool water treatment system, which consists of basic unit processes, i.e. coagulation, filtration, pH correction and disinfection. The choice of pool water treatment method depends on many factors. At the first stage, it is extremely difficult to predict which kind of system will be the most beneficial in particular case. Investment and operating costs are often very important when deciding in this regard.

The basic process of water treatment is filtration, which allows removal of impurities with a diameter > 0.1 µm. In order to increase efficiency, filters are often preceded by coagulation. The essence of this process is to reduce the dispersion of the colloidal system as a result of combining individual particles of the dispersed phase into larger clusters that can be effectively retained on the filter beds. To destroy the living and spore

* Corresponding author: anna.lempart@polsl.pl

forms of pathogenic organisms and to prevent their secondary development, water must be disinfected. Mainly chemical methods involved the dosing of strong oxidants such as chlorine, sodium hypochlorite or chlorine dioxide are used for disinfection of swimming pool water. One of the strongest oxidants is ozone - almost twice stronger than chlorine. An important disadvantage of ozone as a disinfectant, is its low durability, and the danger of secondary bacterial growth. In the absence of an active disinfectant in water, the growth of bacteria can occur even in the presence of ozone products. Therefore, in the swimming pool water it is necessary to provide the remaining concentration of the disinfectant, and ozonation can only be used as a method of disinfection support.

During chemical oxidation processes, new impurities may be formed in the water, as by-products of the oxidation reaction with organic components of the treated water. One of the ways to reduce the efficiency of disinfection by-product formation is to reduce the dose of disinfectant. It is possible only as a result of chemical disinfection preceded by the physical method of destroying microorganisms, e.g. with UV irradiation [4]. The most bactericidal properties are UV-C radiation. Microorganisms have the highest sensitivity to radiation with a wavelength of 254 nm. This radiation causes rapid photochemical reactions in the DNA, thus microorganisms are destroyed or they lose the ability to reproduce [5].

As a result of the intensive development of analytical techniques that allow to reduce the detection limits of tested compounds, the latest scientific research alerts the growing number of organic micro-pollutants identified in the swimming pool water environment. This group includes a number of chemical compounds, e.g. Pharmaceuticals and Personal Care Products (PPCPs), flame retardants (FRs) or insecticide compounds [6]. All mentioned compounds can react with the disinfection agent introduced into the pool water, contributing to the formation of toxic or irritating products, leading to a decrease in the comfort of use of swimming pool facilities or even a health risk to swimmers [7]. Some of micropollutants recently identified in swimming pool water are biologically active even in trace concentrations. Among the health effects of human exposure to these compounds the most frequent are the occurrence of reproductive system diseases, hormonal disorders and even cancer [8, 9]. The classical methods used to treat swimming pool water are not effective in removing most of the organic micro-pollutants [10]. The closed-circuit system may even contribute to the accumulation and increase their concentration level [11]. The contamination level of pool water with micropollutants depends on many factors [12], among which the technology of water treatment is mentioned.

The aim of the research presented in this paper is to evaluate different disinfection methods used in swimming pool systems in the aspect of organic micro-contaminates removal. The content of organic matter, a precursor of dangerous disinfection by-products in systems, using UV radiation and ozonation was compared. A qualitative and quantitative analysis of the organic micropollutants profile was also carried out.

2 Materials and methods

2.1 The research subject

Sampling took place in accordance with the guidelines of the Polish standard PN-ISO 5667-5:2003. Swimming pool water samples were collected from the aqua park equipped with four water circuits, independent of each other. All of them are equipped with the same pressure filters, before which the aluminum sulfate by Kemipol of 0.1–0.8 mm granulation is dispensed as a coagulant, containing 17% Al_2O_3.

Filters contain a multi-layer filter bed consisting of:
- 15cm of quartz gravel, fraction 3–5mm,
- 15 cm of quartz gravel, fraction 1–2mm,
- 80 cm of quartz sand,
- 20 cm of activated carbon Sorbotech LGCO100,
- 20 cm of Hydro-flit bed, fraction 1.5–2mm.

The entire aquapark is equipped with a homogeneous system of disinfection by chlorination with Chlorox (Sodium Sodium Chloride Specification S) from NTCE, which is dosed for installation before entering each pool basin. In three cycles (P1, P2, P3), disinfection is supported by the ozonation using the Prominent Bono Zon ozone generator (Fig. 1), while one of the circuits (P4), which provides water treatment in the swimming pool for learning to swim, is an exception. Instead of the ozone generator, the UV lamp by Lifetech (Figure 2) is used. This is the only difference in the technological system of analyzed circuits.

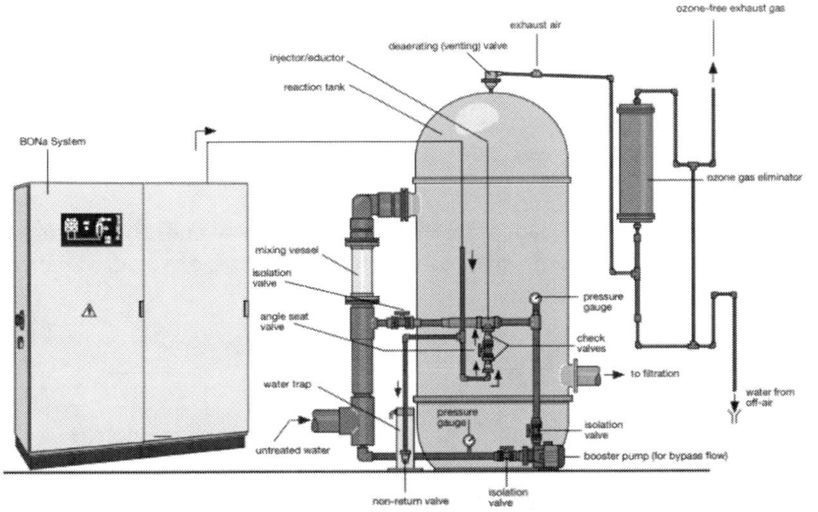

Fig. 1. The technological scheme of the Bono Zon system applied in the technological circuits of the analyzed swimming pool (*source: Prominent's catalog*).

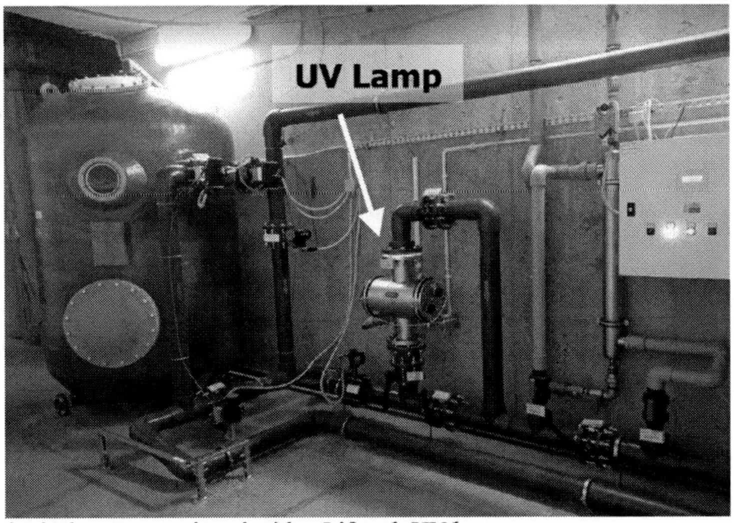

Fig. 2. Technological system equipped with a Lifetech UV lamp.

2.2 Analysis of selected indicators of organic swimming pool water contamination

The assessment of water quality before and after the treatment process was done on the basis of selected parameters related to the presence of organic matter in water:
- ultraviolet absorbance (UV_{254}) measured at a wavelength of 254 nm in a cuvette with a luminal length of 1 cm, using the Analytik UV VIS Cecil 1000,
- the content of total organic carbon (TOC) measured by Shimadzu's TOC-L analyzer.

2.3 Methodology for the determination of organic micropollutants in pool water

Determination of micropollutants in water was carried out in two stages:
1. Extraction of compounds from water by solid phase extraction (SPE).
2. Qualitative and quantitative analysis of extracts by gas chromatography coupled with mass spectrometry (GC-MS).

Organic solvents (methanol and acetonitrile) with analytical purity from Avantor (Gliwice, Poland) and columns for SPE, ENVI-18 by Sigma Aldrich (Poznań, Poland) were used. Details of the applied analytical method and its validation are presented in paper [13].

2.4 Evaluation of research and results analysis

The results presented in the paper are arithmetic mean values from four replications of individual tests. In all cases, the error estimated on the basis of standard deviation did not exceed 5%.

3 Results and discussion

Both indicators of water pollution (TOC and UV_{254}) showed an increase in the organic matter content after water treatment in relation to the measurement before treatment in all circuits using ozonation as a disinfection method (Fig. 3). The results of research [14] indicated that ozonation increases the share of hydrophilic fractions in sewage organic matter, especially in hydrophilic acids, which leads to the increase of the potential of formation the specific haloacetic acids in chlorination. In the same studies [14], the potential of creating trihalomethanes decreased after the initial ozonation but increased with further ozonation. Probably these two phenomena caused the increase in the organic matter content in swimming pool installations using ozone. In system equipped with the UV lamp, after irradiation of water, there was a marked decrease in organic matter content in comparison to the level of pollution at the inlet to the treatment system (Fig. 3). The TOC removal rate was 19.5%, and absorbance decreased by over 33%. Among the known methods for reducing the concentration of disinfection by-products, the removal of precursors present in water in the form of organic matter is considered as the most effective [5]. Significantly higher results of absorbance in one of ozonation system compared to the others, result from the specific physico-chemical properties of brine water.

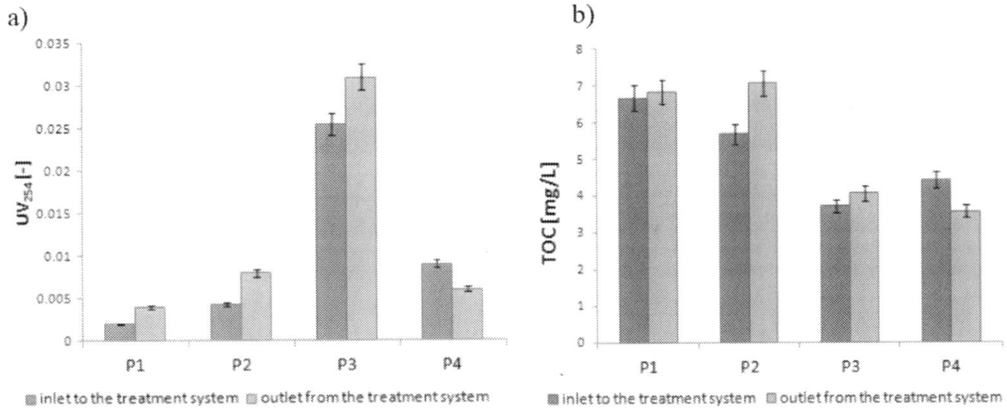

Fig. 3. Change of the organic matter content before and after ozonation (P1, P2, P3) and UV radiation processes (P4), expressed by pollution indicators: a) UV_{254} b) TOC.

Chromatographic analysis showed the presence of many different organic micro-pollutants in the tested water samples. The content of three compounds was clearly distinguishable (Figure 4).

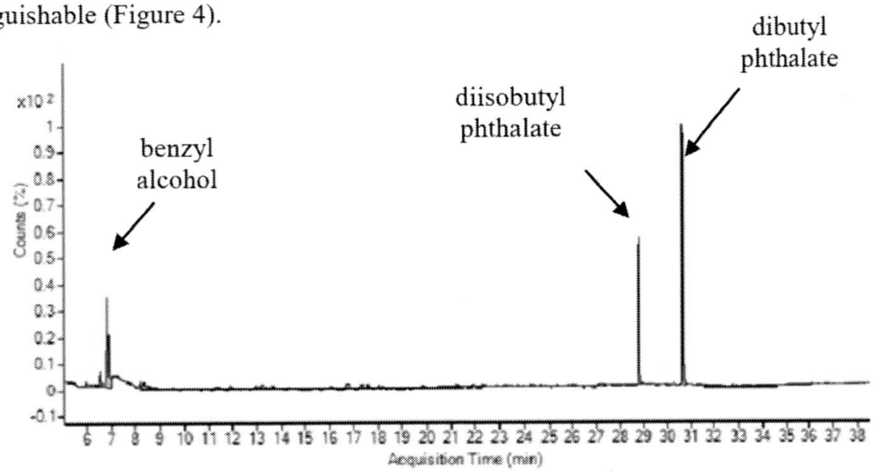

Fig. 4. An example of a chromatogram obtained as a result of testing one of the pool water samples.

Benzyl alcohol, due to its pleasant smell, is a component of many cosmetics, it is found among others in body butter, creams, lotions or balms. It can also be found in basic cosmetics for daily care, such as shampoos and foot creams. It is also used as a regulator of the product consistency. Moreover, it prevents the growth of microorganisms during storage, protecting the cosmetic against bacterial infection. This compound can also be used in the production of paints, varnishes, waxes and resins, while its derivatives are used in the pharmaceutical industry as a preservative.

Phthalates are synthetic compounds mainly used for the production of plastics, paints and cosmetics. This results in their wide application in many everyday products. They are an additive that increases flexibility, durability and strength of materials, mainly containing PVC. Both dibutyl phthalate (DBP) and diisobutyl phthalate (DIBP) are often found in personal care products, e.g. for cleaning nails. They can also be used as an additive to printing inks or glues. They are widely used as plasticizers in various products, ie floor coverings, balls, shower curtains. Due to their widespread use, phthalates can potentially occur even in drinking water sources. For example, the study of water samples from the

Mopanshan reservoir as a source of drinking water for the city of Harbin in north-east China [15] showed the presence of DBP in the concentration range from 52.5 ng/L to 4498.2 ng/L and DIBP from 40.0 ng/L to 658.8 ng/L. The fate of phthalates in two water supply networks capturing water from the Mopanshan reservoir was also analyzed. There were no significant differences in the level of concentrations of the subject compounds in the purification of water in a conventional purification station in relation to the concentration level in raw water [15].

The presence of DBP and DIBP in swimming pools seems to be particularly worrying, as they have been included in the list of high-risk substances REACH as compounds suspected of having a negative impact on the human endocrine system. The harmfulness is that they affect the human endocrine system, disrupting its proper functioning, due to its chemical structure. This may lead to fertility problems, cancer of the reproductive system, abnormal fetal development, early maturation of girls, genital damage or decreased production of semen, metabolic disorders, obesity and diabetes [16]. They may also increase the risk of allergies, asthma or cancer. Pursuant to Commission Delegated Directive (EU) 2015/863 of 31 March 2015 amending Annex II to Directive 2011/65 / EU of the European Parliament and of the Council, these compounds will be subject to a restriction on application from 22 July 2019.

The comparative analysis showed the smallest content of identified phthalates in the pool equipped with the UV lamp (Figure 5). Lau et al. [17] showed in laboratory tests that under the influence of UV radiation, up to 90% DBP can degrade during one hour of irradiation. The actual exposure time in working swimming pool system is much shorter and amounts to only a few to several seconds, which has reduced the content of diisobutyl phthalate 37.3% in water and 35.1% dibutyl phthalate (Figure 6).

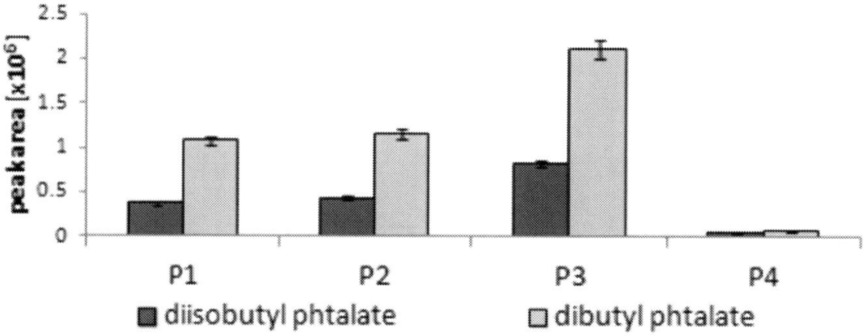

Fig. 5. Comparison of concentration levels of phthalates depending on the disinfection method used – ozonation (P1, P2, P3) and UV radiation (P4).

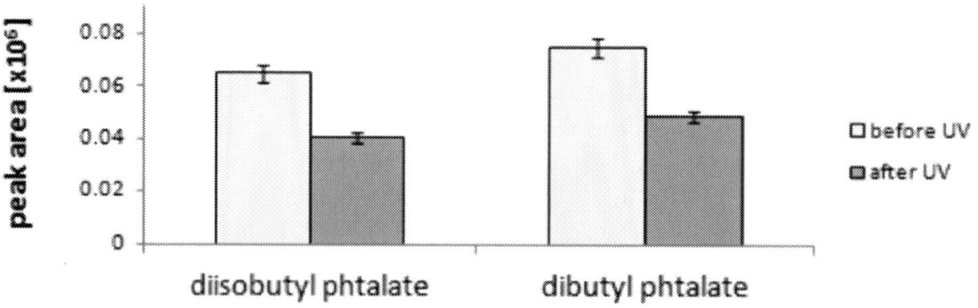

Fig. 6. The content of phthalates in swimming pool water before and after applying the UV lamp (P4).

The efficiency of ozonation in the aspect of removing the subject compounds was much lower, it was 6.96%–16.39% for diisobutyl phthalate (Figure 7a) and 5.32%–13.46% for dibutyl phthalate (Figure 7b).

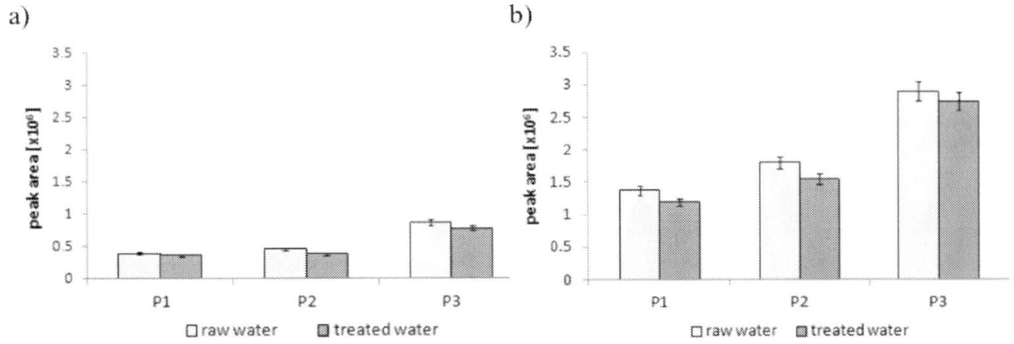

Fig. 7. The content of a) diisobutyl phthalate b) dibutyl phthalate in swimming pool water before and after ozonation (P1, P2, P3).

Research [17] documented six major intermediates of final DBP photodegradation products, including: butyl benzoate, benzoic acid, and phthalic acid, and butyl mono phthalate (MBP) and its derivatives: alcoholic and ketone or aldehyde. These compounds affect human health. For example, monobutyl phthalate may adversely affect the thyroid, reproductive and respiratory systems, resulting in lung function decline, allergies and asthma [18-20]. It can also reduce both sperm concentration and motility [21, 22]. However, the effect of long-term chronic exposure to benzoic acid in animal studies resulted in decreased appetite and reduced growth. This compound also caused an increased number of resorptions and malformations in hamsters. Clinical data indicate that it may produce non-immune contact urticaria and non-immune direct contact reactions characterized by the appearance of blisters, erythema and pruritus [23].

In paper [24] it was shown that in the initial period of photocatalysis there is an increase in water toxicity as a result of formation by-products of oxidation. For this reason, it has been recommended to use a long exposure time. In pool installations, this solution has not yet been applied. UV lamps work in these systems as flow devices, so the exposure time for their rays is very short.

4 Conclusions

- In order to minimize the concentration of disinfection by-products in swimming pool water, disinfectants should be dosed after the removal of organic compounds. The use of UV radiation is an effective method in this regard.

- In the assessed pools, the presence of micropollutants, among others dibutyl phthalate and diisobutyl phthalate was documented. It has been shown that their concentration depends on the disinfection support method used in the technological system.

- Photocatalytic degradation of phthalates in the pool water system allowed to obtain over 35% removal rate, while the efficiency of ozonation in the most favorable case was just over 16%.

- UV radiation is a more beneficial method of supporting disinfection of pool water than ozonation in the aspect of removing organic micropollutants from swimming pool water. However, this process often leads to the formation of by-products, that can significantly degrade the quality of the treated water streams. Therefore, the key role in assessing the effectiveness of advanced oxidation processes should not only be the assessment of the removal rate of parent compounds, but also the identification and evaluation of toxicological by-products of their degradation.

This work was supported by National Science Centre, Poland (No. 2018/29/N/ST8/01352) and Ministry of Science and Higher Education Republic of Poland within funds for young scientists BKM-508/RIE-4/2018 awarded to the Institute of Water and Wastewater Engineering Silesian University of Technology in 2018.

References

1. Polish Ordinance of the Minister of Health, Official Journal 2015, item. 2016 (2015)
2. DIN 19643, *Water Treatment for Swimming and Bathing Pools* (2012)
3. Guidelines for Safe Recreational Water Environments. Vol. **2**: *Swimming Pools and Similar Environments*, WHO (2006)
4. J. Wyczarska-Kokot, Ochr. Środ. **36**, 37–42 (2014)
5. A. L. Kowal, M. Świderska-Bróż, *Oczyszczanie wody. Podstawy teoretyczne i technologiczne, procesy i urządzenia* (PWN, 2007)
6. T. L. L. Teo, H. M. Coleman, S. J. Khan, Environ. Int. **76**, 16–31 (2015)
7. J. Wyczarska-Kokot, Desalin. Water Treat. **134**, 7–14 (2018)
8. L. Xing, H. Liu, J. P. Giesy, X. Zhang, H. Yu, J. Environ. Sci. **24**, 239–334 (2012)
9. P. Westerhoff, Y. Yoon, S. Snyder S., E. Wert, Environ. Sci. Technol. **39**, 6649–6663 (2005)
10. M. Włodarczyk-Makuła, *Physical and chemical fates of organic micropollutants* (2018)
11. A. Lempart, E. Kudlek, M. Dudziak, A. Szyguła, Inż. Ekol. **19**, 23–31 (2018)
12. A. Lempart, E. Kudlek, M. Lempart, M. Dudziak, J. Ecol. Eng. **19**, 29–37 (2018)
13. A. Lempart, E. Kudlek, M. Dudziak, Water **10**, 10 (2018)
14. W. Qi, H. Zhang, C. Hu, H. Liu, J. Qu, Sci Total Environ. **610–611**, 1057–1064 (2018)
15. Y. Liu, Z. Chen, J. Shen, J. Anal. Methods Chem. **2**, 419349 (2013)
16. J. D. Meeker, S. Sathyanarayana, S. H. Swan, Philos. Trans. R. Soc. Lond B. Biol. Sci. **364**, 2097–2113 (2009)
17. T. K. Lau, W. Chu, N. Graham, Chemosphere **60**, 1045–53 (2005)
18. M. Ema, E. Miyawaki, Reprod. Toxicol. **15**, 189–94 (2001)
19. J. D. Meeker, A. M. Calafat, R. Hauser, Environ. Health Perspect. **115**, 1029–1034 (2007)
20. J. A. Hoppin, R. Ulmer, S. J. London, Environ. Health Perspect. **112**, 571–574 (2004)
21. S. M. Duty, M. J. Silva, D. B. Barr, J. W. Brock, L. Ryan, Z. Chen, R. F. Herrick, D. C. Christiani, R. Hauser, Epidemiology **14**, 269–277 (2003)
22. R. Hauser, J. D. Meeker, S. Duty, M. J. Silva, A. M. Calafat, Epidemiology **17**, 682–691 (2006)
23. B. Nair, Int. J. Toxicol. **20**, 23–50 (2001)
24. J. Bohdziewicz, E. Kudlek, M. Dudziak, Proceedings of ECOpole **8**, 135–140 (2014)

Minor pressure losses for different connections of PP-R and PEX/Al/PEX installation pipes

Kinga Ligaj[1], *Marcin K.* Widomski[1,*], and *Anna* Musz-Pomorska[1]

[1]Faculty of Environmental Engineering, Lublin University of Technology, ul. Nadbystrzycka 40B, 20-618 Lublin, Poland

Abstract. This paper presents results of laboratory and numerical research concerning determination of water flow resistance through three types of two-way connection of polymer installation pipes: PP-R 20x3.4 mm and PEX/Al/PEX 16x2.0 mm. The following fittings were applied: the direct connection, pipe union and coupler, allowing to test six measurement variants. The laboratory measurements of pressure loss for the tested pipes connections were performed for variable Reynolds number, from approx. 5000 to 50000. The numerical modeling allowing to assess the distributions of velocity of flow and turbulence intensity were performed using FLUENT, Ansys Inc. modelling software. The relations between determined values of minor pressure loss and coefficients of local pressure losses and type of pipes connection, direction of flow as well as the value of Reynolds number were observed. The applied nonparametric statistics, combined with multi comparison, showed that in most cases of analyzed connections, besides the pipe union, the observed differences in pressure losses for various directions of flows are statistically significant for $p = 0.05$.

1 Introduction

The proper calculations of losses pressure losses in modern plastic pipes, including polyethylene (PE), cross-linked polyethylene (PEX), polypropylene (PP) and polyvinyl chloride (PVC) pipelines, which are nowadays very popular in various domestic installations, for which minor pressure losses may reach even the significant level of 40–60% of total pressure drop, is the important and rather difficult issue [1–4]. The modern pipelines may be connected in various combinations and in different manners, using several possible fittings, which may not fit to the values of coefficients of minor pressure losses presented in the former PN-M-34034:1976 [5], developed for different materials and fittings. Thus, the final calculated values of minor pressure drop were often reported as different than the real, measured values [3, 6–10]. Moreover, the possible connections (such as pipe unions and couplers) may be located very close to the other elements installed on the plastic pipelines, which may additionally influence the possible pressure drop [11]. Finally, the assumed popular methodology of minor pressure drop calculations does not include the possible effect of Reynolds number [12, 13].

* Corresponding author: M.Widomski@pollub.pl

Taking into account that values of minor pressure loss coefficients presented in PN-M-34034:1976 [5] for direct connections of pipes as well as for pipe union and coupler are in the range between 0.25–0.5, the determination of real coefficients of minor pressure losses for selected connections of different pipelines is required.

This paper presents results of laboratory measurements and numerical studies of pressure loss during water flow through three types of two-way connection of polymer installation pipes: PP-R 20x3.4 mm and PEX/Al/PEX 16x2.0 mm. Three different possible connections of the tested pipelines were tested: the direct connection, pipe union and coupler.

2 Materials and methods

The performed research covered laboratory and numerical studies of water flow through three types of bidirectional connections of two popular plastic domestic installation pipes, PP-R 20x3.4 mm PN 20, according to PN-EN ISO 15874-2:2005 [14] and PEX/Al/PEX 16x2.0 mm according to DIN 4726 [15]. The following fittings were applied: the direct connection, pipe union and coupler.

The laboratory measurements of minor pressure losses on tested connections were performed on installation presented in Fig. 1. The applied range of water volumetric flow rate was in range approx. 100–1500 dm^3/h, reflected by the dimensionless Reynolds number between approx. 5000 and 50000. Measurements were conducted for two directions of flow and three selected connections of pipes, thus six variants of research were developed.

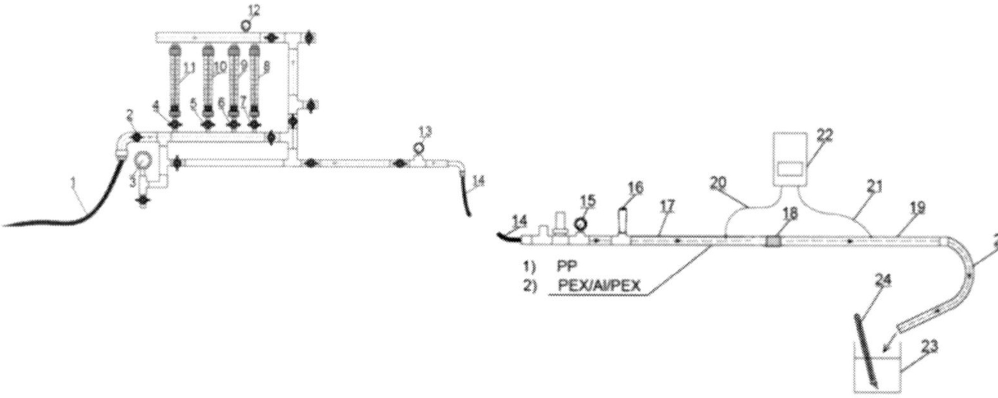

Fig. 1. Scheme of laboratory installation: 1– water supply pipe, 2, 4, 5, 6, 7 – valves, 3, 12, 13, 15 – manometers, 8, 9, 10, 11 – rotameters, 14 – elastic water pipe, 16 – vent valve, 17, 19 PP-R and PEX/Al/PEX pipes, 18 – studied fitting, 20, 21 – pressure impulse pipes, 22 – electronic differential manometer, 23 – water reservoir, 24 – thermometer.

The performed experiments were based on measurements of pressure difference by electronic laboratory differential manometer by Lutron Electronic, Taiwan with 2% accuracy, while volumetric flow rate of water was determined by set of rotameters produced by Meister Strömungstechnik, Germany. Location of points of pressure measurements met requirements of PN-EN 1267:2012 [16]. The pressure sampling point before the connection was established at distance > 10 d, while the point after the tested local loss was located at distance > 60 d.

All laboratory measurements were repeated three times for each tested connection of pipes and applied volumetric flow rate of water. The laboratory installation was supplied with tap water with constant pressure obtained due to application of elevated water reservoir.

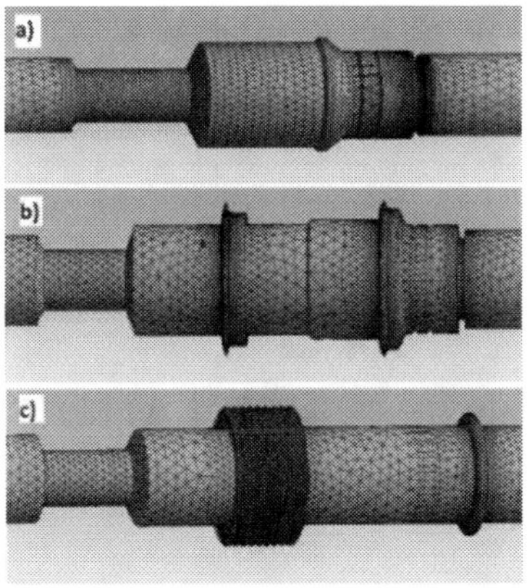

Fig. 2. Developed FEM models of tested connections: a) direct connection, b) pipe union, c) coupler.

Numerical modeling of water flow through tested pipes connections for two tested piping materials was performed in Fluent, ANSYS Inc. computing software. Three different 3D models in 1:1 scale, reflecting the real dimensions and shapes of waterbody for tested connections of pipes were developed. The variable direction of flow was obtained due to the assigned boundary conditions. The developed models consisted of the following numbers of nodes and finite elements: 47637 and 222010 for direct connection, 44016 and 203868 for pipe union and 48683 and 221004 for coupler, respectively. The developed models are presented in Fig. 2. Our calculations of viscous liquid flow were based on the standard two-equation k-epsilon turbulence model [17].

The required input data for numerical modeling covered characteristics of materials and boundary conditions. The assumed characteristics of water were directly based on measurements performed in laboratory and covered temperature (16.5–21.5°C), dynamic viscosity (0.00094–0.00107 kg/(m·s)) and density (997.45–998.82 kg/m^3). The inlet velocity boundary condition, covering velocity of flow, gauge pressure, hydraulic radius and turbulence intensity, was assumed at the inlet surface of the developed models. The initial percentage turbulence intensity I was calculated for each of the computational variants according to the following formula [11]:

$$I = 0.16 \cdot Re^{-\frac{1}{8}} \tag{1}$$

The wall boundary condition was assumed in Fluent as pipe material's roughness equal $k = 7 \cdot 10^{-6}$ m.

The resultant values of coefficient of minor losses for the tested pipe connections were calculated according to the transformed Bernoulli's equation:

$$\zeta = \frac{2g \cdot \left(\Delta h - \lambda_1 \frac{l_1 \cdot v_1^2}{d_1 \cdot 2g} - \lambda_2 \frac{l_2 \cdot v_2^2}{d_2 \cdot 2g} - \frac{v_2^2 - v_1^2}{2g}\right)}{v_2^2} \tag{2}$$

where: ζ – coefficient of minor (local) losses [-], g – gravity [m/s^2], Δh – total pressure loss read from the differential manometer [mH$_2$O], v_i – mean velocity of flow through pipeline before and after the fitting [m/s], λ – coefficient of friction losses [-], l_1, l_2 – lengths of pipelines before and after the connection, measured to locations of pressure impulse pipes [m], d_1, d_2 – diameters of pipes before and after the connection [m].

The required values of λ coefficient of friction pressure losses were calculated according to the popular Colebrook-White's formula:

$$\frac{1}{\sqrt{\lambda}} = -2\log\left(\frac{2.51}{Re\sqrt{\lambda}} + \frac{k}{3.71d}\right) \tag{3}$$

where: Re – dimensionless Reynolds number, k – roughness of pipe material [mm].

The obtained results of laboratory research were statistically analyzed to assess the statistical significance of the observed differences between pressure losses and coefficients of minor pressure loss for tested connections and variable direction of flow. The Shapiro-Wilk test of normality and Wilcoxon signed-rank test as well as non-parametric Kruskal-Wallis' one-way analysis of variance were applied to our studies. Results of numerical calculations were validated by R^2, the root mean square error (RMSE), ratio of the root mean square error to the standard deviation of measured data (RSR) and Nash-Sutcliffe coefficient of model efficiency (NSE).

3 Results and discussion

The values of local pressure drop and coefficients of minor pressure loss for all tested connections of pipe, variable Reynolds number and two directions of flow PP – PEX/Al/PEX and PEX/Al/PEX – PP, respectively, determined during laboratory studies, are presented in Fig. 3.

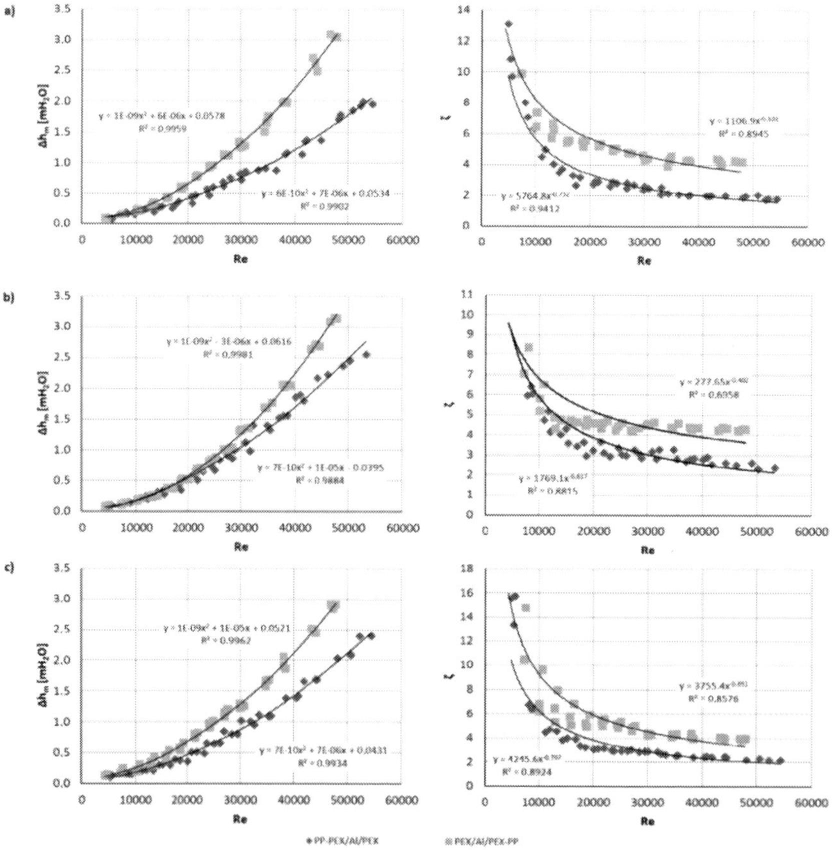

Fig. 3. Laboratory measurements of local pressure losses and coefficients of minor pressure loss for studied pipe connections.

It is visible that all tested connections for different directions of flow generated various values of minor pressure drop and coefficients of minor pressure loss. Moreover, it should be underlined that in all the tested cases, for the three studied connections, the greater resistance of flow for the same Reynolds number was generated for connection of pipelines in direction PEX/Al/PEX – PP. Additionally, the dependence of minor loss coefficient to Reynolds number was also observed. The obtained results were summarized in Tab. 1.

Table 1. Range of variability of determined Re number, confidents of minor pressure loss and local pressure losses for tested connections.

Type of connection	PP – PEX/Al/PEX				PEX/Al/PEX – PP			
	Re range	ζ range	ζ mean	Δh_m [-] range	*Re* range	ζ range	ζ mean	Δh_m [-] range
Direct	4881–54303	1.77–13.16	3.57	0.08–2.00	4352–47942	3.91–16.79	6.24	0.10–3.10
Pipe union	4741 – 53248	2.31–13.41	4.1	0.10–2.56	2470–27942	4.23–15.63	5.51	0.07–3.16
Coupler	4881–54303	2.20 – 15.76	4.09	0.11–2.41	4352–47942	3.91–24.63	6.84	0.12–2.92

The determined values of coefficients of minor pressure loss for all tested connections are clearly higher than values presented in former Polish national standard PN-76/M-34034:1976 [5], which suggested constant values of ζ for all discussed fittings from range 0.25–0.5. Moreover, the obtained results of our laboratory tests are generally higher than most of values of minor pressure loss coefficients suggested by nowadays producers, which commonly are in range 0.5–4.5 for direct connection, 0.6–0.7 for pipe union and 0.1–0.73 for coupler, respectively.

The performed statistical analyses showed that according to Shapiro-Wilk test of normality all the tested variables showed distributions different than normal. The Wilcoxon test applied for pairs of series of results obtained for the same connection but different direction of flow showed statistically significant differences between determined medians of coefficients of minor pressure loss. The Kruskal-Wallis ANOVA test supplemented by multiple comparisons applied to all six series of laboratory measurements showed that the observed differences in distribution functions were statistically significant. The obtained results of post-hoc test of multiple comparisons for values of coefficients of minor pressure losses determined in laboratory are presented in Tab. 2.

Table 2. Results of p for post-hoc multiple comparisons test for obtained values of minor loss coefficients for tested fittings and two directions of flow.

Type of connection	Direct PEX-PP	Direct PP-PEX	Pipe union PP-PEX	Pipe union PEX-PP	Coupler PP-PEX	Coupler PEX-PP
Direct PEX-PP	-	0.00057	0.01419	1	0.00779	1
Direct PP-PEX	0.00057	-	1	0.00493	1	0.00032
Pipe union PP-PEX	0.01419	1	-	0.08202	1	0.00886
Pipe union PEX-PP	1	0.00493	0.08202	-	0.0488	1
Coupler PP-PEX	0.00779	1	1	0.0488	-	0.00477
Coupler PEX-PP	1	0.00032	0.00886	1	0.00477	-

The numerical simulation in the performed studies was generally used to determine the factors affecting generation of resistance of flow for the three tested fittings and two directions of flow. The above assessment was based on determined distributions of

 https://doi.org/10.1051/e3sconf/201910000045

magnitude of flow velocity and turbulence intensity for each tested case. The exemplary graphs showing distribution of v and I for each tested case and selected volumetric flow rate are presented in Figs 4 and 5.

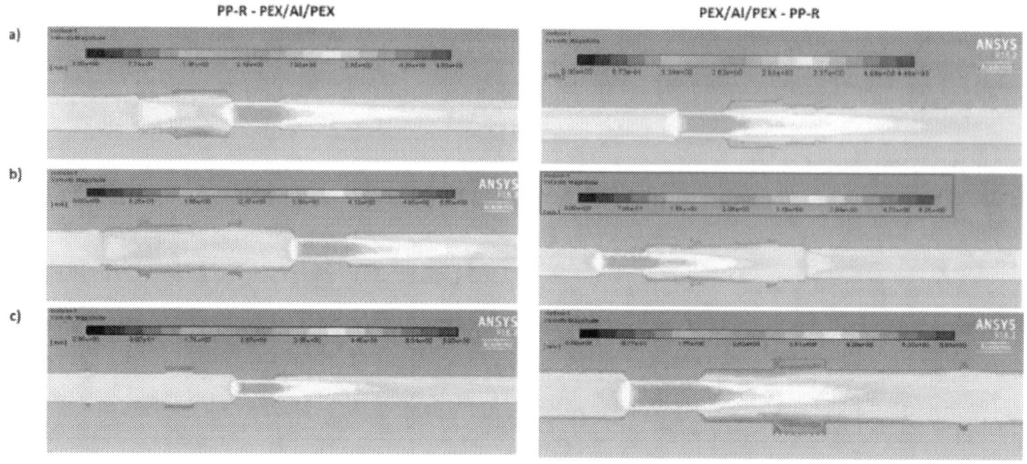

Fig. 4. Modelled distributions of velocity magnitude for selected flow rate (800 dm^3/hr): a) direct connection, b) pipe union, c) coupler.

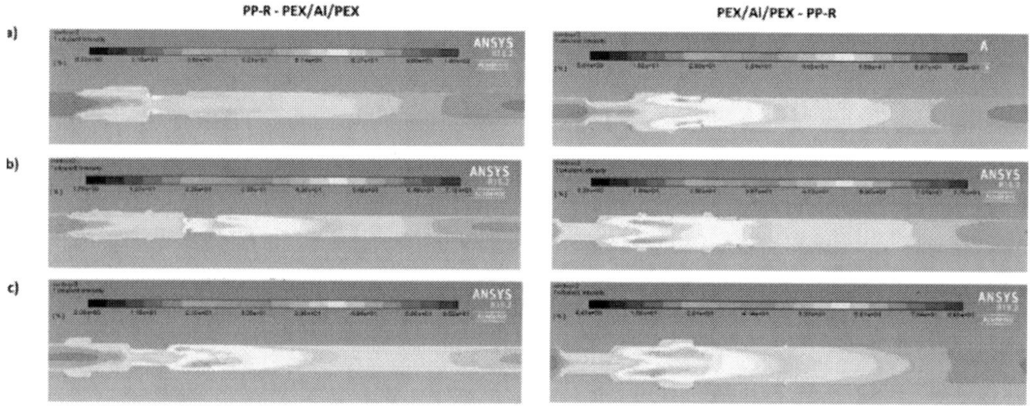

Fig. 5. Modelled distributions of turbulence intensity for selected flow rate (800 dm^3/hr): a) direct connection, b) pipe union, c) coupler.

The clear differences between distributions of flow velocity and turbulence intensity caused by different shape of waterbodies are visible. The highest local velocity of flow was detected for both direction of flow through the tested coupler. The highest local turbulence intensity in direction PP-PEX/Al/PEX was observed for the direct connection, while, in the opposite direction, PEX/Al/PEX-PP, the highest level of turbulence intensity was reached in waterbody of the coupler. Generally, connection of tested domestic installation pipes by all studied fittings resulted in significant increase in turbulence intensity, from approx. 5% typical for undisturbed turbulent flow inside the pipeline to level of approx. 65–108%. Similarly, installation of the tested fittings resulted in increase in the local velocity of flow from approx. 1.7 m/s to the values close to 4.86–5.93 m/s.

The determined results of model validation covering R^2, RMSE, RSR and Nash-Sutcliffe coefficient of model efficiency presented in Tab. 3 showed the satisfactory performance of the developed model [18, 19].

Table 3. Results of model validation for all tested fittings and flow directions.

Type of connection	Direction of flow	R^2	RSME	RSR	NSE
Dirtect	PP-PEX/AL/PEX	0.9967	0.086	0.135	0.982
	PEX/AL/PEX-PP	0.9983	0.383	0.425	0.819
Pipe union	PP-PEX/AL/PEX	0.9967	0.624	0.881	0.224
	PEX/AL/PEX-PP	0.9983	0.700	0.828	0.314
Coupler	PP-PEX/AL/PEX	0.9972	0.358	0.471	0.778
	PEX/AL/PEX-PP	0.9984	0.514	0.552	0.695

4 Conclusions

The performed laboratory measurements of local pressure losses for three different connections of PP-R and PEX/Al/PEX domestic installation pipelines and two applied directions of flow showed that all tested fittings generated different local pressure drop, also in relation to the applied direction of flow. The greater resistance of flow, related to the shape of waterbody of each connection, leading to higher local pressure losses and expressed by higher values of coefficients of minor pressure losses were in all cases observed for direction of flow PEX/Al/PEX – PP-R. Additionally, in all of the tested cases (different connections and variable direction of flow) the determined values of minor pressure losses coefficients were greater than values suggested by the standards and technical guidelines. The performed numerical simulation of water flow through the tested fitting showed that the developed shape of waterbody for each of the tested connections and each applied Reynolds number significantly influenced the resultant pressure drop, due to different values and spatial distributions of flow velocity and turbulence intensity. In our opinion the presented problem should be carefully studied, because underestimated minor pressure losses inside various domestic installations may significantly decrease their functionality.

References

1. M. Annan, E. A. Gooda, A. E. J. **57**, 4299–4305 (2018)
2. I. Cisowska, A. Kotowski, Fundations of Civil and Environ. Eng. **8**, 37–57 (2006)
3. P. Grajper, J. Smołka, Gaz, Woda i Technika Sanitarna, **7–8** (2010)
4. K. Weinerowska-Bords, Instal **6**, 42–49 (2014)
5. PN-M-34034 Pipelines – Rules for the calculation of pressure losses (1976)
6. T. Siwiec, D. Morawski, G. Karaban, Gaz, Woda i Technika Sanitarna **2** (2002)
7. I. Cisowska, A. Kotowski, Gaz, Woda i Technika Sanitarna **10** (2004)
8. K. Strzelecka, K. Jeżowiecka-Kabsch, Ochrona Środowiska **30**, 2 (2008)
9. K. Strzelecka, K. Jeżowiecka-Kabsch, Ochrona Środowiska **32**, 1 (2010)
10. M. K. Widomski, A. Musz, M. Iwanek, Gaz, Woda i Technika Sanitarna **9**, 380–385 (2012)
11. Ansys. Inc., Ansys Fluent User's Guide. Release 19.0 (2018)
12. Y. Bae, Y. In Kim, Ann. Nucl. Energy **73**, 33–38 (2014)

13. N. Moallemi, J. R. Brinkerhoff, Comput. Fluids. **140**, 209–221 (2016)
14. PN-EN ISO 15874-2 *Plastic pipe systems for the installation of hot and cold water-Polypropylene (PP) – Part 2: Pipes* (2005)
15. DIN 4726 *Warm water surface heating systems and radiator connecting systems – Plastics piping systems and multilayer piping systems* (German National Standard, 2017)
16. PN-EN 1267 Industrial fittings – Testing the resistance of water flow (2012)
17. B. E. Launder, D. B. Spalding, J. Heat Mass Transfer **1**, 2, 131–138 (1974)
18. D. N. Moriasi, J. G. Arnold, M. W. Van Liew, R. L. Bingner, R. D. Harmel, T. L. Veith, ASABE **50**, 885–900 (2007)
19. G. Golmohammadi, S. Prasher, A. Madani, R. Rudra, Hydrology **1**, 20–39 (2014)

Effect on louver dampers configuration on gas maldistribution in a heat exchanger for waste heat recovery

Grzegorz Ligus[1,*], *Daniel* Zając[2], and *Jatinkumar* Panchal[3]

[1]Faculty of Mechanical Engineering, Opole University of Technology, Opole, Poland
[2]Kelvion Central Europe Sp. z o. o., Opole, Poland
[3]Laser Microjet Solution, Gujarat, India

Abstract. This paper reports the results of visualization research involving a heat exchanger applied for waste heat recovery using of Digital Particle Image Velocimetry (DPIV) method. Five louver dampers configurations responsible for typical operation modes were tested. As a consequence, airflow maldistribution in the tube bundle section and bypass section were identified and described. Based on the measurements and analysis carried out, the authors have developed a way to implement design changes to optimize gas distribution inside the unit.

1 Introduction

Waste heat recovery unit (WHRU) are found in various process parts such as exhaust gas streams or used drying process air. In the case of process boilers that are commonly applied in the industry practice, WHRUs are called economizers. The task of WHRU is to use the enthalpy of the exhaust gases to heat up another medium and transfer heat into another process area. The application of such equipment leads to an increase in the energy efficiency of the installation in which WHRU is installed [1]. To ensure the required performance in variable operating conditions or balance WHRU with other heat exchange equipment, WHRUs comprise regulating mechanisms in the form of louver dampers of various types and bypass sections. These components are responsible for the distribution of the exhaust gases to individual heat-receiving elements as well as enable complete shutdown of the equipment by initiating bypass flow. Irrespectively of the dimensions of WHRUs, louver dampers are commonly used for regulating. The variations in the configuration of the louver dampers determine the variations in the gas velocity value and its profile [2]. In addition, the location of the louver dampers plays a significant role of pressure drops generation in the apparatus [8]. The total of these changes considerably alters the aerodynamic characteristics of the equipment in relation to the simplifications adopted during the design phase. The objective of the current research involved the development of a correlation between the configuration of louver dampers and variations in the direction and velocity of gas flow in a heat recovery

* Corresponding author: g.ligus@po.opole.pl

heat exchanger. Usually, Computational Fluid Dynamics (CFD) is used to evaluate changes of geometrical and flow parameters in industrial equipment [3, 4]. The idea of reconstructing gas flow patterns as a result of identification of undesired phenomena like vortices or flow maldistribution can provide assistance in redesign industrial apparatus [5]. In the field of experimental research, the DPIV method is mentioned as one of the most excellent tools for assessing gas behaviour in flow devices [6, 7].

2 Experimental setup and methodology

DPIV method was applied in the experimental research. An industrial scale transparent model of WHRU was built for this purpose. The external dimensions of the device were: length 900 mm, width 456 mm and height 450 mm. The pitch of the tube arrangement in the bundle was equal to 24 mm. The inlet and outlet with a diameter of 200 mm were placed in the central section of the front and the back panels. The louver dampers configurations in the WHRU model correspond to three principal modes of operation: full gas flow through the tube bundle section, full gas flow through the bypass section and gas flow separated into two parts: through the tube bundle section and through bypass section, resulting from the selection of an operating mode or controlling of the heat reception intensity. The inclination angles of the louver dampers during transient operation modes were selected to generate pressure drops at similar levels at each stage of opening or closing of both sections. The configuration parameters of the tested louver dampers locations are summarized in Table 1 and their visualizations are given in Figure 1.

Table 1. Louver dampers configuration.

	Angle of bypass louver damper	Angle of main louver dampers	Pressure drop ΔP, Pa	Description of the WHRU operating mode
Position 1	90°	0°	44	Full gas flow through a tube bundle section
Position 2	70°	20°	42	Closing / opening
Position 3	50°	30°	42	Closing / opening
Position 4	30°	40°	41	Closing / opening
Position 5	0°	90°	61	Full gas flow through a bypass section

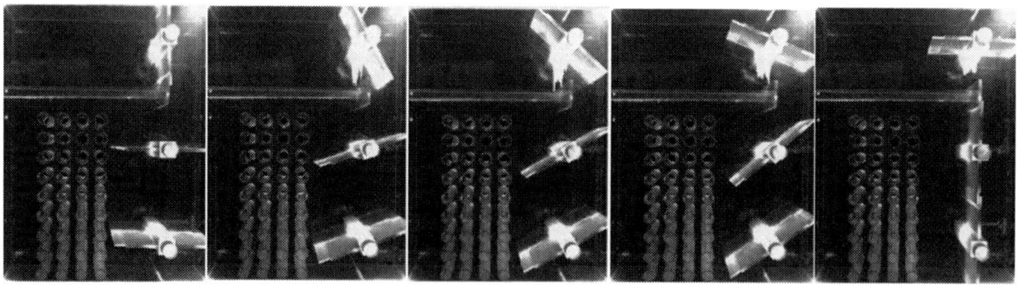

Fig. 1. Visualization of the tested louver dampers configurations.

The study applied flow through WHRU using atmospheric air with a suspended droplets of Di-Ethyl-Hexyl-Sebacat (DEHS) as seeding particles. Airflow was generated by the Helios MVZ315 fan unit. The calculated airflow velocity in the tube bundle section was equal to 4 m/s and was within the airflow velocity range recommended by the manufacturer (from

2 m/s to 6 m/s). The region that was selected for observations was illuminated by application of the laser sheet technique. The laser sheet was generated by the Dantec Dynamics DualPower TR laser and directed with the use of a mirror, to the bottom of the transparent WHRU model, perpendicular to the vertical symmetry line of the inlet. The airflow images (100 images for each measurement series) were recorded using the Dantec Dynamics FlowSense EO 4M camera in the double frame mode. The computer images processing and velocity vector fields analysis were carried out using the Dynamic Studio 2015 software. In a table 2 the technical parameters of the DPIV system were summarized. The pressure drop measurement was performed for each louver dampers configuration regardless of the applied DPIV method. The sketch and the actual view of the experimental stand are presented in Figure 2 and Figure 3.

Table 2. Technical parameters of the DPIV system.

Parameter	Description
Type of the devices	Camera: Dantec Dynamics FlowSense EO 4M, Laser: Dantec Dynamics DualPower TR, Pulse generator: BNC 575-8, Seeding particles generator: Flow Visualization Components Flow Tracker 700CE
Time between pulses, μs	56
Interrogation area	Size 32 x 32 (overlap 50% in both directions), Central difference offset
DPIV algorithm	Adaptive Correlation with Local Neighbourhood Validation
Vector validation	Universal Outlier Detection method (normalized median validation)

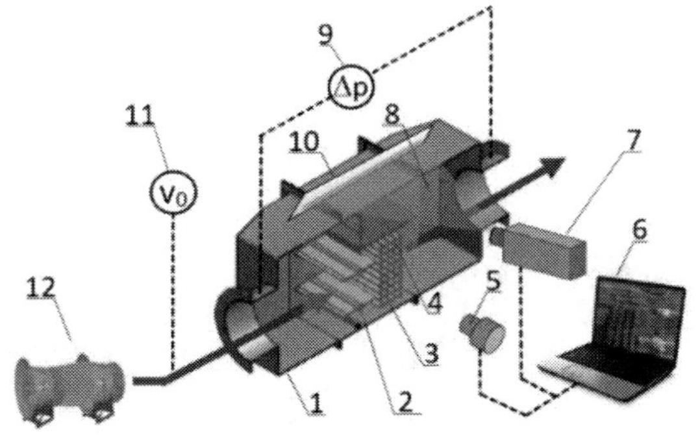

Fig. 2. Experimental stand and equipment: 1) transparent WHRU model, 2) main louver dampers, 3) tube bundle section, 4) bypass section, 5) CCD camera, 6) pulse generator and data acquisition, 7) double pulse Nd:YAG laser, 8) laser sheet, 9) gas differential manometer, 10) mirror, 11) thermo-anemometer, 12) fan unit.

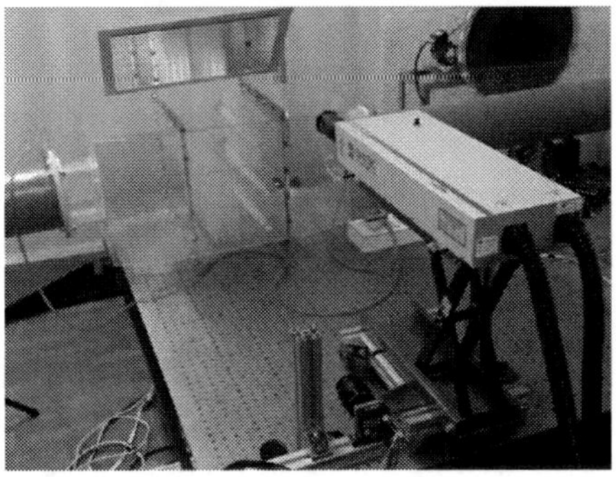

Fig. 3. View of experimental setup.

3 Results and discussions

The main focus of the research reported here involved the determination of the velocity vector fields and streamlines, which make it possible to assess the velocity distribution and identification of vortex patterns in the WHRU. Due to the occurrence of the perspective phenomenon (covering the spacing between the tubes), a comprehensive analysis of the resulting visualizations was unfeasible. A similar problem was noted in the work [9]. However, the above optical effect did not apply to the region before the first vertical tube row, which became the basis for assessing the gas fraction. Considering the main mode of the operation of WHRU with full gas flow through a tube bundle section (position 1, fig. 4), flow maldistribution in the tube bundle section were recorded. At the level of the three horizontal lower tube rows, the forecasted value of the gas flow velocity derived from computations could not be achieved. An additional factor that negatively affected the performance of the device was associated with the occurrence of a maximum flow velocity of 5 m/s only in the region located between the tube bundle and the baffle separating the tube bundle section from the bypass section. Despite the fact that the main louver dampers were completely opened, a vortex structure emerged directly behind the vertical structural element extending beyond the baffle separating both sections of the WHRU (Fig. 5a). In this operating mode, secondary air flows into the closed bypass section from the outlet side start to occur, which leads to the development of a vortex in the entire section volume. This phenomenon, in the case of the flow of a dusty agent, will lead to the deposition of contamination in the bypass section and the uncontrolled outflow when the section is opened.

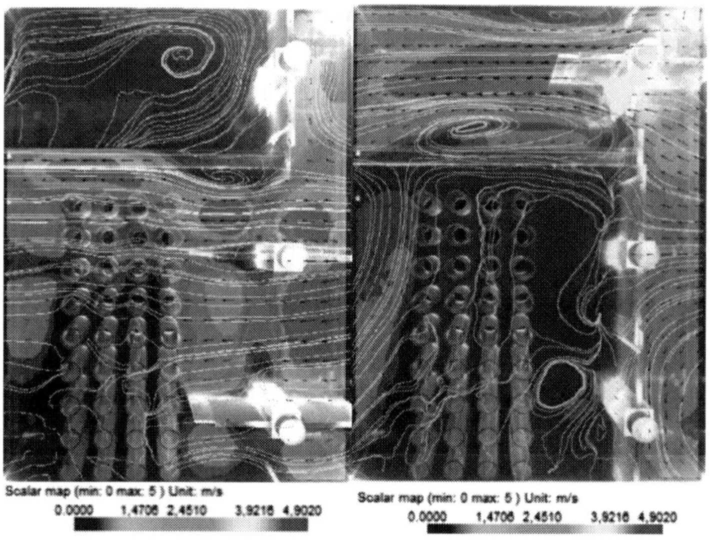

Fig. 4. Gas velocity vector field with a streamline in the WHRU during position 1 (on the left) and position 5 (on the right).

In the case of the analysis of the full gas flow through a bypass section, a similar relation was established between the occurrence of secondary air in this bypass (position 5 on Fig. 4). As a result of closing the main louver dampers, secondary air starts to flow into the tube bundle section. This is particularly undesirable due to potential contamination of the tube bundle and potential difficulties associated with removal of the resulting residue. In addition, leakage streams were identified as a consequence of the technological clearances. The total opening of the bypass louver damper with the main louver dampers closed also causes a vortex behind the extending structural element above baffle between the bypass section and the tube bundle section (Fig. 5b). Therefore, there is a need to redesign this element completely.

Fig. 5. The influence of the extending structural element (in a white frame) on the formation of vortex zones (vector velocity in m/s): a) position 1, b) position 5.

Figure 6 shows gas behaviour in transient states (positions 2, 3 and 4), which are primarily responsible for the direction of gas flow from the one section to another. The analysis of these operating modes is also particularly important because the control of the heat reception can be realized by selecting the appropriate configuration of louver dampers [16]. The combination of simultaneous gas flow in both sections and the blocking of gas flow through the upper main louver damper results in a decrease of the mean gas velocity in the tube bundle section. As a consequence, the amount of heat reception decreases as well. In positions 2, 3 and 4, maldistribution is intentional and expedient. The effect of these operating modes is associated with an increased intensity of the vortex phenomena. The vortex located behind the structural element extending beyond the baffle separating the two sections develops so as to occupy the space between this baffle and the upper main louver damper. In addition, an increase of the angle of main louver dampers leads to the development of another vortex region in the upper inlet slot to the tube bundle section. In turn, a huge vortex region is formed in the lower part of the bypass section irrespectively from the location of the bypass louver damper. It can be only reduced as a consequence of the full opening of the bypass section. The size of this vortex is also determined by the height of the structural element extending beyond the boundary of the partition between the sections. The limitation for the flow maldistribution observed in the above cases may be to relocate the inlet from the axis of symmetry of the front panel to the axis of symmetry of the tube bundle. Such operating mode should ensure an increase in the gas flow velocity. In such circumstances, gas flows directly into the tube bundle section without a negative effect of changes of the air flow direction and intensive vortices structures development.

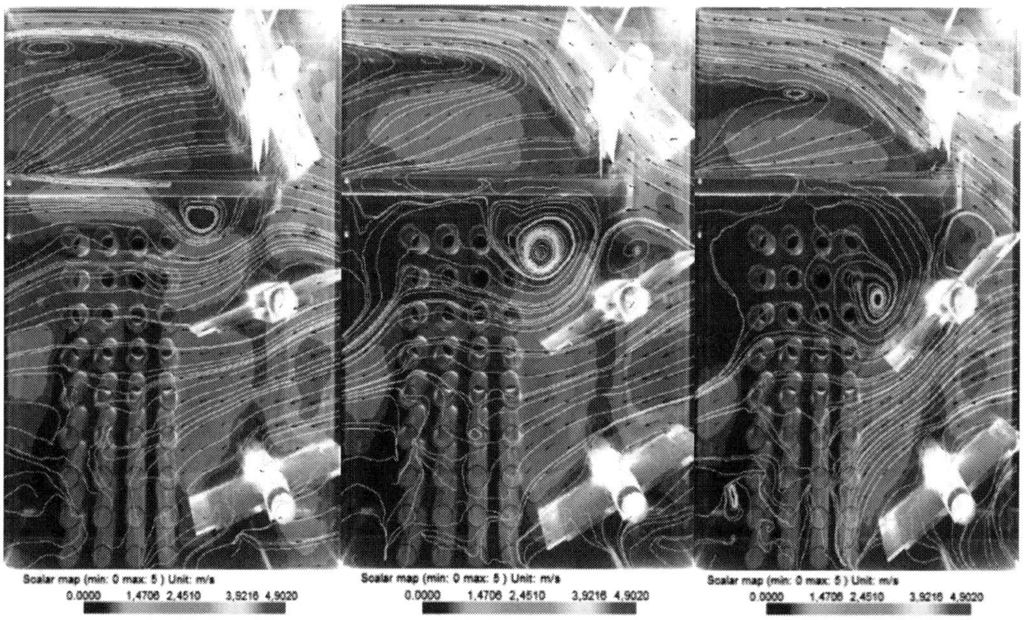

Fig. 6. Gas velocity vector field with a streamline in the WHRU during opening/closing the louver damper: position 2 (on the left), position 3 (in the middle) and position 4 (on the right).

A quantitative assessment was carried out into flow maldistribution depending on the configuration of louver dampers by comparing gas velocity profiles before the first vertical tube row (Fig. 7) for all tested locations. However, due to the fact that position 5 does not include the flow of the agent through the tube bundle, it is not taken into account in the quantitative analysis. The expected, calculated value of gas flow velocity (4 m/s) was achieved only for position 1 and only at the height of the two upper tube rows. In contrast, at the height of the two bottom tube rows, the gas flow velocity greater than 0.5 of the calculated value could not be reached for any of the tested positions. By analyzing velocity vector fields, it was observed that along with an increase in the incline angle of the main louver dampers, the area of increased flow velocity is displaced downwards, at the expense of the top rows of tubes. On the basis of these observations, four regions of gas flow intensity in the tube bundle section were determined:

- high flow velocity zone ($Re > 2.02 \times 10^5$)
- medium flow velocity zone ($1.86 \times 10^5 < Re < 2.02 \times 10^5$)
- low flow velocity zone ($Re < 1.86 \times 10^5$)

- the aerodynamic shadow region, which is characterized by a rapid drop of the velocity followed by a subsequent increase in its value. This region can occur irrespectively of the gas velocity.

In the case of positions 1 and 2, the zone in the height range of tube bundle section from 130 mm to 260 mm was identified. These areas are separated by a distinct aerodynamic shadow region developed by the upper main louver damper. The area of average gas velocity, that is characteristic of all tested positions, is located at a height range from 50 mm to 130 mm (except position 1, where due to the presence of the aerodynamic shadow zone, its lower limit must be raised up to 70 mm). Below the height of tube bundle of 50 mm, a low gas flow zone was determined for all positions. The occurrence of such a clear flow maldistribution in the tube bundle may create conditions for selective reception of the agent from particular horizontal rows of pipes, which leads to an improvement of the performance in terms of flexibility.

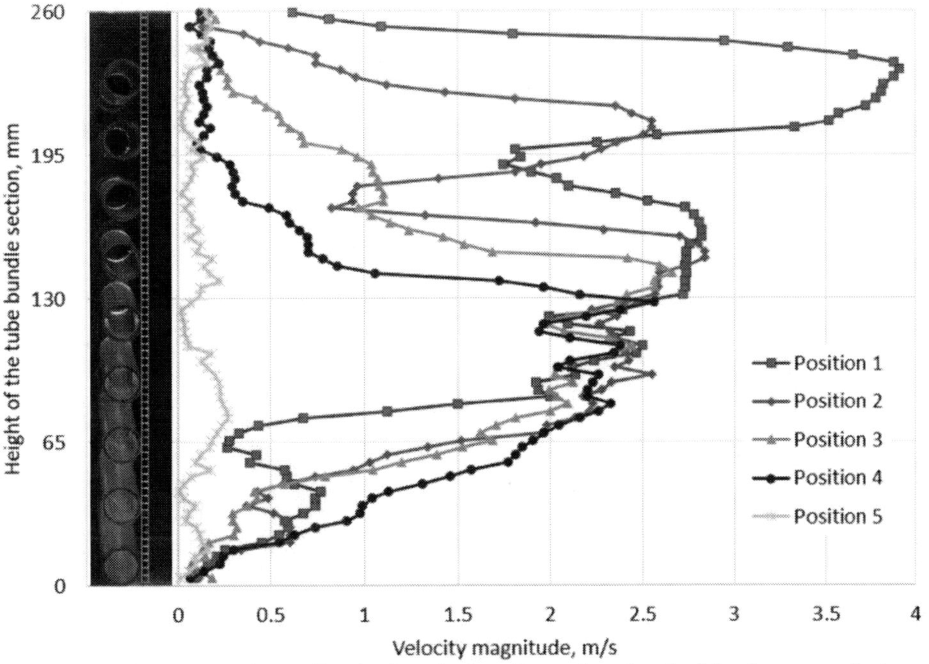

Fig. 7. Velocity magnitude profiles in the tube bundle section ahead of the first row of tubes with marked location of the measuring plane.

The uncertainty of the measurements of velocity vector fields depends on correctly recorded images of the phenomena. Assuming the correct type and size of the seeding particles as well as the optimal optical conditions of the equipment and identifying an average of 10 seeding particles in each interrogation area, the relative error of the DPIV method, as recommended by [10], was estimated to be equal to 10%. To a lesser extent, the uncertainty of measuring is also affected by the uncertainty of estimating the displacement of seeding particles in the applied DPIV algorithm. Usually, it is in the range of 0.1–0.2 pixel unit [10]. According to the principle of displacement seeding particles by 0.25 linear dimension of the interrogation area, displacement at the level of 8 pixels was obtained. The studies in [11–13] determined that for these conditions the uncertainty in the measurement of seeding particles displacement is about 1–2%. It is also necessary to take into account the occurrence of random uncertainty. On the basis of the study presented in [15], it was demonstrated in [14] that for the 95% confidence level (that is, a typical value that is assumed in experimental studies regarding gas phase flows), the random error could be determined in the range of 5% of the mean velocity that is derived by the DPIV method.

4 Conclusions

The following conclusions can be stated on the basis of the experimental research into gas behavior in the heat exchanger applied for waste heat recovery:

- the DPIV method offers an effective diagnostic tool that enables the evaluation of the WHRU operating mode by reconstruction of flow structures and in particular the identification of vortices and maldistribution of gas,
- in order to eliminate partial maldistribution, it is recommended to reduce the distance of the tube bundle from the baffle that separates the tube bundle section and the bypass section,
- asymmetric location of the gas inlet relative to the tube bundle plays a significant role in the development of maldistribution of the gas. It is recommended to adjust the displacement of the gas inlet to a center position to be relative to the tube bundle,
- due to the fact that gas flow disturbance is generated, a suggestion is made to replace the vertical structural element extending beyond the baffle located between the sections of WHRU by application of another solution that can lead to stabilization of the louver dampers in the system.

References

1. M. Ghazi, P. Ahmadi, A. Sotoodeh, A. Taherkhani, Energy Convers. Manage. **58**, 43–156 (2012)
2. F. Haug, J. Zheng, M. J. Beleynaud, J. Lu, J. Energy Inst. **90**, 951–961 (2017)
3. M. Wasilewski, J. Duda, Powder Technol. **289**, 143–158 (2016)
4. W. Ludwig, D. Zając, G. Ligus, P. Korman, Chem. Eng. Sci. **197**, 386–403 (2019)
5. M. Wasilewski, Sep. Purif. Technol. **168**, 114–123 (2016)
6. R. Goyal, B. K. Gandhi, M. J. Cervantes, Renew. Sust. Energ. Rev. **81**, 2976–2991 (2018)
7. F. C. Li, K. Hishida, Adv. Chem. Eng. **37**, 87–147 (2009)
8. P. Montgomery, *Fundamentals of HVAC control systems* (Elsevier, Amsterdam, 2008)
9. I. Sunghyuk, T..K. Hyoung, W..R. Bo, J..S. Hyung, Ann. Nucl. Energy **98**, 1–11 (2016)
10. R. J. Adrian, J. Westerweel, *Particle Image Velocymetry* (Cambridge University Press, New York, 2011)
11. C. E. Willert, Appl. Sci. Res. **56**, 79–102 (1996)
12. J. M. Foucaut, B. Miliat, N. Perenne, M. Stanislas, *Particle Image Velocimetry: Recent Improvements* (Springer, Berlin, 2008)
13. J. Westerweel, Exp. Fluids **26**, 3–12 (2006)
14. B. P. Husted, P. Petersson, I. Lund, G. Holmstedt, Fire Safety J. **44**, 1030–1045 (2009)
15. H. W. Coleman, W. G. Steele, *Experimentation and Uncertainty Analysis for Engineers* (John Wiley & Sons, New York, 1999)
16. V. Vinod, S. Sathishkumar, R. K. Mourya, S. Chandramouli, G. Padmakumar, B. K. Nashine, P. Selvaraj, Nucl. Eng. Des. **318**, 174–181 (2017)

The impact of operating conditions on exhaust emissions from a two-wheeled urban vehicle

Piotr Lijewski[1], *Natalia* Szymlet[1,*], *Łukasz* Rymaniak[1], *Barbara* Sokolnicka[1], and *Andrzej* Domowicz[1]

[1]Politechnika Poznańska, Instytut Silników Spalinowych i Transportu

Abstract. The subject of the article is the exhaust emission analysis of harmful gases from an two-wheeled urban motor vehicle in real operating conditions. For this purpose, characteristics of ecological indicators of compounds such as hydrocarbons (HC), carbon monoxide (CO), carbon dioxide (CO_2) and nitrogen oxides (NO_x) were presented. The tests were carried out on a two-wheeled urban vehicle, equipped with an engine with a stroke volume of 49 cm^3 and a maximum power of 2.5 kW. What's more, the operating parameters of a moped registered during the test were also analyzed. The measurements were performed in real operating conditions in the Poznań agglomeration. The designated route included streets in the city center, and the maximum permissible speed did not exceed 50 km/h. The AXION R/S+ mobile apparatus belonging to the PEMS (*Portable Emissions Measurement Systems*) group was used for the research. The device is one of the most modern measuring instruments, enabling the measurement of concentrations and emissions of both gaseous and solid exhaust components. What is more, the obtained results of road emissions of harmful compounds were referred to the limit value of the exhaust emission norms, which the tested vehicle is expected to meet (Euro 3). The results indicated that the limit value for carbon monoxide was exceeded by 52%.

1 Introduction

In accordance with the Polish traffic law of June 20, 1998, a moped is a one- or two-axle vehicle, equipped with a combustion engine with a stroke volume not exceeding 50 cm^3, whose design limits its travel speed to 45 km/h. [1]. Such vehicle characteristics make it preferable for driving in urban traffic, where road vehicle users and city authorities face the problem of congestion or lack of parking spaces. Therefore, in order to limit the time and costs two-wheeled vehicles are being used more and more often, this is particularly true for mopeds. This phenomenon is visible primarily in densely populated urban agglomerations such as in India and China, where the total share of two-wheelers constitutes 70% of the entire fleet of road vehicles [2]. Extensive literature studies have shown that the most extensive research works are carried out in Asian research and development centers, aimed at assessing the emission of harmful and toxic compounds of exhaust gases from this type

* Corresponding author: natalia.r.szymlet@doctorate.put.poznan.pl

of vehicles. These tests focus primarily on the modification of type approval tests based on the operating parameters obtained in real driving conditions [3, 4]. The lack of regulations concerning the RDE (Real Driving Emissions) procedure for two-wheeled vehicles, as well as the lack of available equipment for exhaust emission tests from this group of vehicles, resulted in them only being tested in laboratory conditions thus far.

2 Research methodology

2.1 Research object

The tests in real driving conditions were performed on a moped whose four-stroke engine was characterized by a 50 cm^3 displacement volume and a power of 2.5 kW. The vehicle was manufactured in 2013, so it is subject to the Euro 3 emission norm. The vehicle was checked for any technical defects before the test. Table 1 presents the technical data of the test vehicle, while Figure 1 shows the tested vehicle together with the measuring apparatus.

Table 1. Tested vehicle technical parameters.

Engine type	**SI**
Number and arrangement of cylinders, number of valves	single-cylinder, 2 valves
Displacement	49 cm^3
Bore/stroke	37,8 mm / 44 mm
Maximum power	2.5 kW/7000 rpm
Maximum torque	3.5 Nm/6750 rpm
Cooling	liquid
Valve timing	ohc

Fig. 1. Test vehicle with Axion R/S+ device.

2.2 Research methodology

The equipment used for real operating conditions tests was Axion R/S+, belonging to the PEMS (*Portable Emissions Measurement System*) group [5–7]. The apparatus allows the measurement of both gaseous (CO, CO_2, NO, HC) and solid (PM) emission compounds. To determine compounds such as carbon monoxide, carbon dioxide or hydrocarbons, an NDIR (*Nondispersive Infrared Sensor*) analyzer was used, while an electrochemical analyzer was used to measure nitric oxide concentration. The method based on Laser Scatter was used for the measurement of solid particles. Figure 2 presents the Axion R/S+ measuring apparatus.

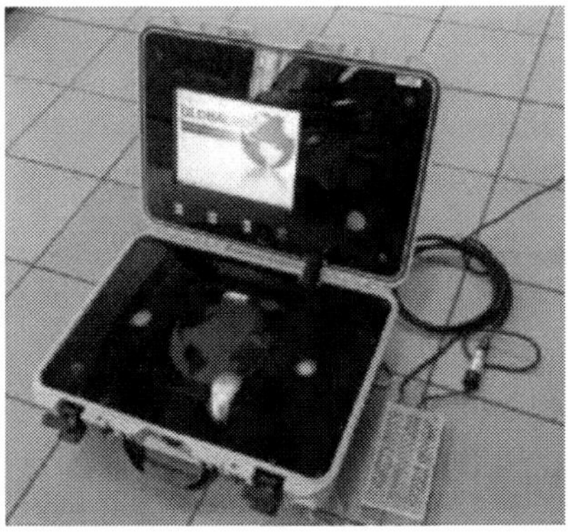

Fig. 2. The AxionR/S+ device.

Axion R/S+ was equipped with a meteorological station, a GPS as well as a module enabling registration of data from the on-board diagnostic system – OBD. The device measured and acquired data at a frequency of 1 Hz. Based on the recorded data, corrections were made to the obtained results, and then the road/specific exhaust emission values from the test were calculated [8].

In the aspect of exhaust emissions testing from two-wheeled vehicles in actual conditions of operation, this apparatus provides unparalleled flexibility and ease of use. Its small size and weight (only 18 kg), allow measuring the emissions in real driving conditions even with the smallest two-wheeled vehicles. Considering the fact that this measuring device has become available for purchase worldwide only recently, this publication is one of the first research works on the subject of pollutant and toxic compounds emission tests of two-wheeled vehicles in real operating conditions. Researchers from Austria [9] measured the emissions of gaseous compounds from a vehicle in real conditions for a larger motorcycle, however, this was done using equipment dedicated to passenger cars, which excludes measurements of emissions from urban two-wheeled vehicles.

2.3 Test route

The test route was the city center of the Poznań agglomeration. The aim of the research was to determine the real emission from two-wheeled vehicles in real conditions of their operation. Based on their design and construction, mopeds are adapted for urban driving, hence the route included only roads with a maximum speed of up to 50 km/h. The route started at the Poznań University of Technology, which greatly facilitated the research process, and its total length was 7.1 km. The route, determined on the basis of data obtained from GPS, is shown in Figure 3.

Fig. 3. The test toure used for the drive cycle [10].

3 Analysis and results

3.1 Analysis of engine operating conditions

The data obtained during the moped tests in real operating conditions allowed to obtain the engine operation time density characteristics, which are a graphical representation of the share of engine operating time (compared to the duration of the whole test) for individual vehicle speed and acceleration ranges. The characteristic showed that the vehicle most often worked at a speed of 0 km/h and at accelerations in the range of -0.5–0 m/s^2, i.e. idling. The share of operating time in this area amounted to 24% (Figure 4), which results from the characteristics of urban traffic (road congestion, a large number of traffic lights). A significant share of operating time was also found for the speed range of 10-30 km/h and accelerations of 0–1 m/s^2, which amounted to 21% in total. This is confirmed by the average vehicle speed and acceleration values of 21 km/h and 0.57 m/s^2, respectively.

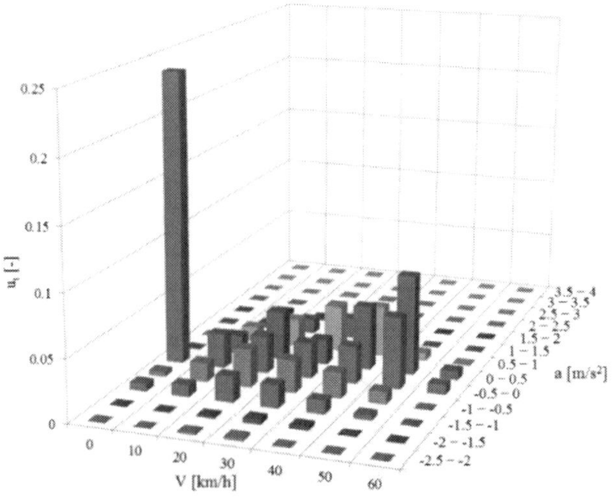

Fig. 4. Time density characteristics of the tested vehicle driving in real operating conditions.

3.2 Analysis of the engine ecological indicators

The environmental performance characteristics for the tested vehicle were also presented in the function of individual vehicle speed and acceleration ranges, as was the case with the engine time density characteristics. The impact of these vehicle performance parameters on emissions has been assessed for each toxic exhaust component (HC, CO, NO_x) limited under Euro 3 norm as well as for the harmful CO_2 which is not limited by the emission norms.

CO_2 emissions as a function of time depend on both vehicle speed and acceleration (Figure 5a). An even distribution of carbon dioxide emission values was recorded in all acceleration ranges. However, the maximum value of carbon dioxide emissions (0.52 g/s) for the tested moped was recorded for the speed range of 20–30 km/h and accelerations of 2.5–3 m/s^2, i.e. in the range of medium speeds and high accelerations. The average emission for this compound was 0.23 g/s.

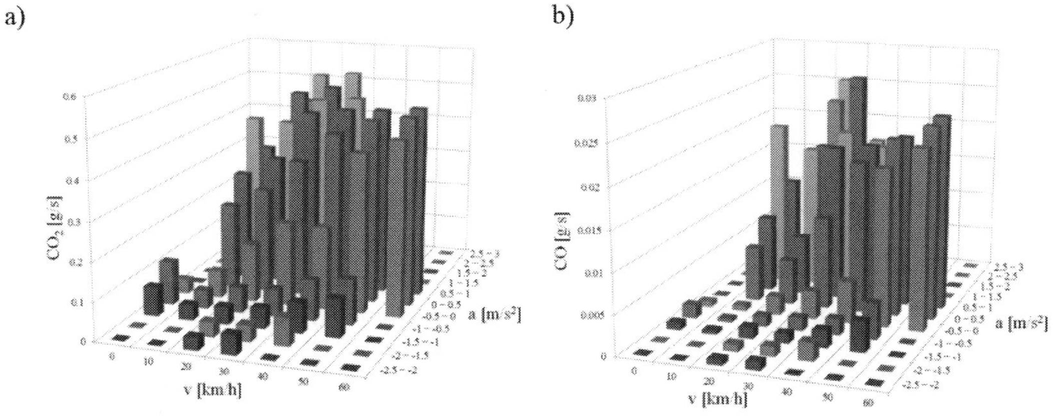

Fig. 5. Emission value per second as a function of vehicle speed and accelerations intervals for a) CO_2, b) CO.

The characteristics of the per second carbon monoxide emission as a function of vehicle speed and accelerations (Fig. 5b) showed that the highest emission values of this compound were in the ranges of medium speeds and high accelerations (10–30 km/h and 1–2 m/s^2). High engine loads favor global and local oxygen deficiency, which is closely related to the formation of carbon monoxide. In addition, a high temperature in the cylinder also contributes to the formation CO through the decomposition of CO_2. The highest emission value, 0.027 g/s, was recorded for speeds in the range of 20–30 km/h and accelerations of 1.5–2 m/s^2.

The characteristics of hydrocarbon emissions as a function of vehicle speed and accelerations are shown in Figure 6a. The obtained data indicates that the highest emission values were in the area of medium speeds in the range of 10–30 km/h and accelerations in the range of 1–2.5 m/s^2. For such load values, the engine operates in transient states, subjected to inertia resistances, resulting in temporary enrichment of the fuel-air mixture, which is a factor favoring the formation of hydrocarbons. The highest value, 0.004 g/s, was recorded for speeds in the range of 10–20 km/h and accelerations of 1.5–2 m/s^2.

Analysis of the nitrogen oxides ecological indicators characteristics showed that the largest share in the total emission can be found the high engine speed ranges ranging between 50-60 km/h and medium accelerations of 0–1.5 m/s^2 (Fig. 6b). The highest value, 0.0021 g/s, was recorded for vehicle speed range 50–60 km/h and acceleration in the range 0.5–1 m/s^2, while the average emission value of this compound was 0.00081 g/s. The high speed of the vehicle generates a higher temperature in the cylinders, which directly promotes the formation of nitrogen oxides. Such a high vehicle speed value for a moped (> 50 km/h) results from the local terrain features, which lead to rolling downhill.

a) b)

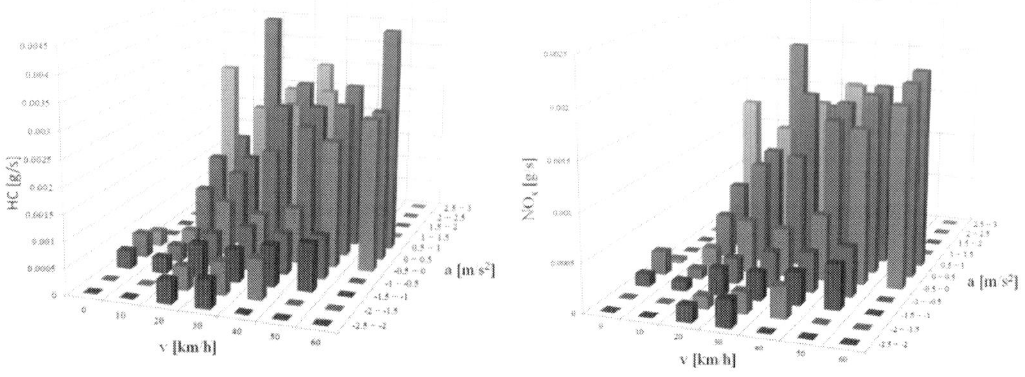

Fig. 6. Emission per second value as a function of vehicle speed and acceleration ranges for a) HC, b) NO_x.

The road emission values of all exhaust components limited by the Euro 3 norm have been referred to the limit values of the said norm and carbon monoxide emissions have exceeded the norm by 52% (Figure 7). This phenomenon results from more diversified engine operating conditions under real operating conditions than in the case of a type approval tests, where the engine loads are limited. For the test in real conditions, therefore, we have an engine load characteristic, and not a full engine power characteristic as in the case of a laboratory test.

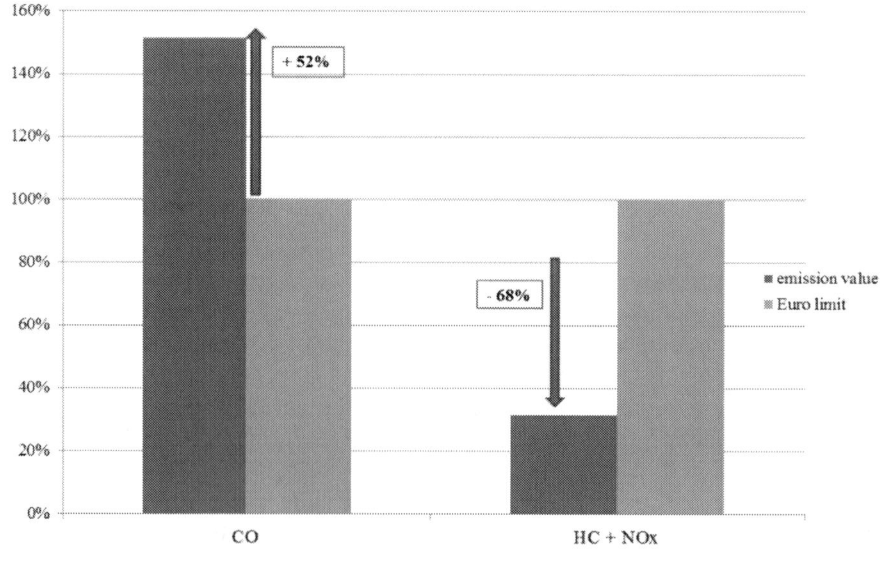

Fig. 7. Comparison of the exhaust emission results with the Euro 3 norm limits.

4 Conclusions

The vehicle group to which the tested vehicle belonged is characterized by a relatively simple construction (simplified power systems, limited number of electronic systems), which results in the users of such two-wheeled vehicles often making slight adjustments or repairs by themselves, rather than at authorized service centers. In part because of this these vehicles are characterized by significant emission factors, primarily hydrocarbons and carbon monoxide, as shown in this article. The continuous population increase, and hence also the number of vehicles used (including two-wheeled vehicles), means that inhabitants of densely populated cities are particularly exposed to threats resulting from the environmentally harmful operation of two-wheeled vehicles.

What's more, exceeding the emission limit values specified in the Euro 3 norm results from the more diversified operating conditions that occur in the case of actual vehicle operation. Many scientific and research works have shown significant differences in the results of exhaust emission tests performed in laboratory conditions and in real operating conditions [11]. This fact confirms the legitimacy of conducting emission tests in real driving conditions for this group of vehicles, because only this method can provide full and reliable results.

References

1. Ustawa z dnia 20 czerwca 1997 r. – Prawo o ruchu drogowym
2. J. H. Tsai, P. H. Hunag, H. L. Chiang, Atm. Env. **153**, 126–134 (2017)
3. H. Y. Tong, H. D. Tung, W. T. Nguyen, Atm. Env. **45**, 5191–5199 (2011)
4. J. H. Tsai, H. L. Chiang, Y. C. Hsu, B. J. Peng, R. F. Hung, Atm. Env. **39**, 6631–6641 (2005)
5. J. Merkisz, P. Lijewski, P. Fuć, M. Siedlecki, Appl. Eng. In Agr. **31**, 875–879 (2015)
6. J. Merkisz, P. Lijewski, P. Fuć, S. Weymann, Ekspl.i Niez. **15**, 364–368 (2013)
7. J. Merkisz, J. Pielecha, P. Fuć, P. Lijewski, VPPC, 903–907 (2013)
8. www. globalmrv.com
9. J. Hiesmayr, S. Schmidt, S. Hausberger, R. Kirchberger, C. Zinner, P. Filips, R. Wanker, H. Friedl, JSAE **32** (2017)
10. www.gpsvisualizer.com
11. W. Saleh, R. Kumar, H. Kirby, P. Kumar, Trans. Research **14**, 326–333 (2009)

The quality of the microclimate in educational buildings subjected to thermal modernization

Anna Lis[1,*], and *Nadiia* Spodyniuk[2]

[1]Czestochowa University of Technology, Faculty of Civil Engineering, ul. Akademicka 3, 42-201 Czestochowa, Poland
[2]National Technical University of Ukraine "Igor Sikorsky Kyiv Polytechnic Institute", Metal Science and Heat Treatment Department, 35 Politechnichna str., Kyiv, 03056, Ukraine

Abstract. Realization and exploitation of buildings involves in European Union about 40% of total energy consumption [1]. One of the elements of rationalization of energy consumption in buildings are the undertakings related with thermal modernization of buildings. The actions related with reducing the energy intensity of buildings are not always correlated with improvement of microclimate conditions in the rooms. Errors in the implementation of the energy efficiency program led to the phenomenon of sick building syndrome. The paper presents the results of the research conducted in a few educational buildings before and after thermal modernization. The research includes energy consumption for the heating of building and selected parameters of the interior microclimate. This analysis was carried out to evaluate the influence of energy saving activities on microclimate interior conditions. It was found that in many cases commonly used gravitational ventilation is not able to ensure the proper conditions of the interior microclimate, and the concentrations of carbon dioxide recorded in the tested rooms exceeded the applicable standards.

1 Introduction

The issues of thermal protection of buildings are directly related with energy savings but also with other aspects. The purpose of thermal protection is also to ensure appropriate indoor microclimate conditions and thermal comfort of occupants. The rationalization of energy consumption is a widely addressed issue in strategic documents determining the development direction of the European Union [2, 3]. All members of the Union committed to elaborate national, long-term strategies supporting the modernization investments covering the specification of optimal ways for improving energy efficiency of buildings and specification of instruments mobilising the investments. Starting from 2021 all newly constructed buildings will be required to have very low energy consumption, covered mainly by the renewable energy resources.

Presently building stock of European Union covers about 200 million buildings, out of which 6 million are located in Poland. Buildings existing in Poland, both residential and public are characterized by a great potential in terms of energy efficiency. The main element

* Corresponding author: alis@bud.pcz.pl

of improving their energy efficiency are the thermal modernization activities, which adapts existing buildings to present and future energy requirements. It has been estimated that through the modernization of an average building in accordance with the currently valid regulations up to 40–50% of energy may be saved [5]. However, according to the European Union Directive on the energy performance of buildings [4], it is important not only to improve the energy efficiency of buildings, but also to define the conditions for the classification of objects in terms of the indoor microclimate requirements.

The energy saving activities focusing only on the insulation of building envelope and air tightness of fenestration joinery often results in worsening of the microclimate quality. Excessive airtightness of buildings often leads to improper operation of natural ventilation systems and impaired air quality. High concentration of pollutants in the air inside the rooms is usually the reason for occurring a sick building syndrome. Requirements relating to ensuring the proper state of the indoor environment in buildings should be therefore given priority in relation to the requirements of energy saving due to the need to ensure first of all the right conditions for people to stay in closed rooms.

The evaluation of influence of thermal modernization activity on indoor microclimate conditions was conducted for an educational buildings. The lack of thermal insulation in building envelope caused mainly that the buildings failed to comply with thermal protection requirements and were not energy efficient therefore generating considerable energy losses. Thermal modernization of buildings consisted of insulation of building envelope and modernization of the heating system and hot water preparation.

2 The formation of microclimate conditions in buildings

The elements that affect the human are both thermal parameters of the microclimate as well as chemical, microbiological pollutants and air ionization and also ionizing and non-ionizing radiation, lighting and noise [6]. Long-term influence of disadvantageous of environmental conditions may cause or exacerbate many symptoms associated with abnormal functioning of the organism, lead to the depletion of mental and physical efficiency as well as to weakness or illness. As people spend a considerable amount of time indoors, buildings should be a structure that will not have a negative impact on them. However, the World Health Organization already in 1983 found that many of the disease symptoms to which users complain relate to the quality of the indoor environment [6]. The attention was paid to the relationship between indoor environmental quality and increasing sickness absence and its high economic cost [7, 8]. This phenomenon is called the sick building syndrome (SBS), and its symptoms appear to be closely linked to time spent in the building and decrease or disappear after leaving it [9]. Symptoms of the syndrome are primarily eye, nose or throat irritation, itching and dry skin, hoarseness and dry cough, cold and flu-like symptoms, dizziness and headache, nausea, sensitivity to odours, allergies and increased incidence of asthma attacks or fatigue, irritability and difficulty in concentration. The World Health Organization presented guidelines for the protection of health against many substances commonly found in indoor air [10–12].

The amount of air exchanged in the ventilation process has a large impact on the quality of the indoor environment, but also on the heat consumption, especially at high thermal insulation of the building envelope [13]. Errors in the implementation of the energy efficiency program, increasing the tightness of buildings and inadequate ventilation, but also the state of the external environment often results in worsening of the microclimate quality and the intensification of the syndrome phenomenon in modernized and new energy-saving buildings [14–16]. Particularly noticeable is deterioration of air quality in the rooms, the increase of its relative humidity and the occurrence of different kinds of moulds and fungus.

Keeping the required values of microclimate parameters has obviously a significant influence on the building energy needs. Their level is mainly affected by the air temperature maintained in the rooms. Requirements regarding air temperatures to be maintained in individual rooms are included in the technical conditions [17]. Maintaining a lower temperature value promotes the energy efficiency. Lowering the value of air temperature from 20 to 18 degrees causes the reduction in heat demand by about 12%, but it is not favourable for people working in a sitting position [18]. The ambient radiation temperature is related mainly to the temperature of the building envelope, which is affected by their thermal insulation. In the perfectly clean air, the low relative humidity is not burdensome for users of the rooms. Signalized health problems at low relative humidity are associated with indoor air pollution. High humidity is unfavourable due to energy saving and the possibility of mould and fungus growth.

At an air temperature of 20–22°C, the recommended air velocity is 0.15–0.20 m/s, for people working in a sitting position. Research conducted by Fanger [19] showed that these values are too high. About 20% of respondents reported dissatisfaction with existing environmental conditions in the rooms.

3 Improving the energy efficiency of educational buildings

The buildings, in which the research was conducted, failed to comply with thermal protection requirements and were not energy efficient. In the years 2006–2012, they were subjected to thermal modernization. As part of the thermal modernization the insulation of building envelope and replacement of windows and external doors were made. Various changes in the heating and hot water preparation system in individual buildings were made too. First of all, the boiler room was modernized, coal or gas boilers were exchanged for devices with better efficiency provided with weather automation system. In the buildings after thermal modernization, the gravity ventilation system was left.

The heat transfer coefficients for walls, roofs, ceilings above basements, floors on the ground, windows and doors were calculated based on project data. The values of coefficients before and after thermal modernization are presented in Table 1.

Table 1. Heat transfer coefficients before and after thermal modernization.

Heat transfer coefficient, W/(m^2 K)	**Before thermal modernization**		**After thermal modernization**	
	Harmonic mean	Standard deviation	Harmonic mean	Standard deviation
Walls	1.35	0.22	0.22	0.02
Roofs	0.96	0.23	0.16	0.03
Ceilings above basements	1.21	0.15	0.33	0.06
Floors on the ground	0.88	0.10	0.35	0.04
Windows	2.6	0	1.3	0.2
Doors	3.1	1.8	1.8	0.3

Before the thermal modernization, the building envelope of the buildings had low thermal insulation. This resulted in excessive heat losses and was not conducive to energy saving. After insulation, the thermal parameters of the building envelope has significantly improved and the thermal insulation requirements applicable then were met.

Analyzing the energy parameters of buildings, the indicator of the annual demand for usable energy (EU), final energy (EK) and primary energy (EP) as well as the energy consumption index in relation to surface with adjustable temperature (Q/A_f) were determined. The energy parameters of buildings before and after thermal modernization are presented in Table 2.

Table 2. Energy parameters of buildings before and after thermal modernization.

Indicator, kWh/(m^{2}year)	Before thermal modernization		After thermal modernization	
	Harmonic mean	Standard deviation	Harmonic mean	Standard deviation
EU	184.6	43.4	64.3	12.5
EK	281.4	54.4	76.9	13.3
EP	312.6	53.6	83.1	14.1
Q/A$_f$	274.5	49.6	79.6	17.4

The calculated final energy demand decreased by 72%. A slightly higher energy consumption after thermal modernization was observed in relation to the determined final energy demand. The previous inspection of thermal insulation works did not show errors, also the temperatures prevailing during the research were not significantly different from those used for calculations. Perhaps the cause of excessive energy consumption was frequent airing of rooms due to poor air quality.

4 Evaluation of the indoor environment conditions after thermal modernization

The research on the condition of the indoor microclimate was carried out in buildings before and after thermal modernization in the same period of the heating season (during two months December, January) in similar temperature and humidity conditions outside. The research were conducted only in typical classrooms in which work was carried out during the research. The research included most of the classrooms in individual buildings. The average size of the room was 62 m^2, there were about 19 persons in the room.

Many factors influencing indoor environmental quality as well as the occupants' perception of thermal comfort were considered in the evaluation. First of all, the basic thermal parameters of the microclimate were measured: air temperature (t_a), average ambient radiation temperature (t_{ra}), relative air humidity (φ_a) and air velocity (v_a). In the Table 3 are shown average values of thermal parameters of the microclimate with standard deviation value.

Table 3. Basic thermal parameters of the microclimate before and after thermal modernization.

Parameters of the microclimate	Before thermal modernization		After thermal modernization	
	Arithmetic mean	Standard deviation	Arithmetic mean	Standard deviation
t_a, °C	19.2	2.4	22.4	1.8
t_{ra}, °C	20.7	2.8	23.8	1.5
φ_a, %	48	12	32	4
	Harmonic mean	Standard deviation	Harmonic mean	Standard deviation
v_a, m/s	0.25	0.11	0.14	0.03

The air temperature in the rooms before thermal modernization was on average 19.2°C with the standard deviation equal to 2.4°C. A lower temperature value was noted at the windows, which were leaky. After thermal modernization in the rooms an increase in air temperature was noted (table 3). There was no significant temperature drop near the windows area. The temperature difference between the level of head and feet was bigger before thermal modernization and it was around 2.4°C. After thermal modernization this temperature difference did not exceed 1.0°C. The temperature near the floor did not differ by more than about 1.2°C from the average air temperature designated for a given room. The outside temperature during the study was on average between -9.5°C and -0.5°C.

The ambient radiation temperature before thermal modernization was about 1.5°C higher than the air temperature, just like after thermal modernization. After thermal modernization works average value of ambient radiation temperature was on level 23.8°C. The outside relative humidity during the study was on average between 69% and 86%. Before thermal modernization a higher air relative humidity level was found. The value of the air humidity ranged from 30 to 60% and amounted to an average of 48%. During the day there was a gradual increase in the humidity value by about 10%. The value of the air humidity noticeably decreased after thermal modernization. A considerable air humidity level in the building in connection with low thermal insulation of building envelope is related with the growth of microorganisms on their internal surfaces. Growth of moulds on the surfaces of walls within the area of window occurring sometimes before thermal modernization was completely eliminated after thermal modernization. The air velocity in the rooms was slightly higher before thermal modernization, especially near leaky windows. After thermal modernization air velocity was maintained at a relatively low level around 0.1 m/s and did not exceed 0.2 m/s.

High carbon dioxide concentrations have a negative impact on the human well-being and health [20]. The correct level is 450–1000 ppm. By the level 2000 ppm poor air quality and sleepiness is already observed. At values 2000–5000 ppm the headaches are possible. Value over 5000 ppm leads to discomfort and rapid heart rate, over 15,000 ppm to problems with breathing and over 30,000 ppm to dizziness and indisposition. Concentration above 60000 ppm may cause a blackout [21]. The highest permissible concentration CO_2 in the work environment is 90.00 mg/m^3 i.e. about 5000 ppm.

The mainly source of indoor carbon dioxide pollution was the rooms users breathing. During the day classes in educational buildings are periodically used and during the breaks, intensive airing of rooms occurs by opening windows. The carbon dioxide concentration in the classrooms changed dynamically depending on the number of people, the size of the rooms and the frequency of their airing. The measurement of carbon dioxide concentration was made only in two buildings. A disturbing phenomenon was observed. Although before the thermal modernization the CO_2 level did not exceed 1400 ppm, after thermomodrization a maximum of 2307 ppm was noted at the end of the work.

Fig. 1. shows the change in the concentration of carbon dioxide (in ppm) in the air in the classroom during the working day, in the educational building in which the thermal modernization was carried out.

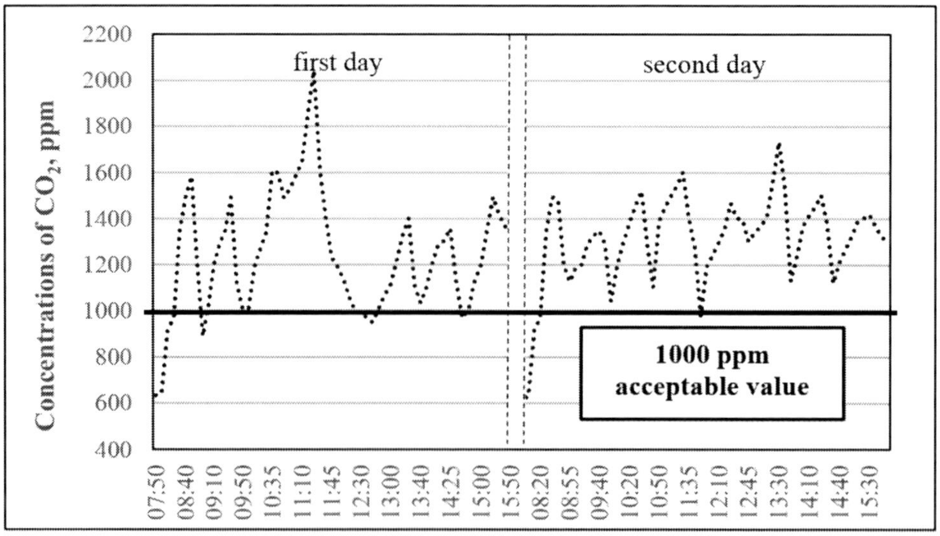

Fig. 1. Distribution of concentrations of carbon dioxide in the air.

The initial carbon dioxide content in the air was about 600–800 ppm when the room were used daily. When the room were not used for a longer period, the concentration of carbon dioxide dropped below 500 ppm. After the start of the lesson, intensive growth of carbon dioxide was observed. Within an hour, this increase was on average 500–600 ppm and even up to 1000 ppm. The largest measured CO_2 value in this room was 2043 ppm, which significantly exceeded the proper acceptable value. Intensive airing of the rooms caused a rapid drop in CO_2 levels. Increasing the number of air exchanges in rooms will reduce the level of CO_2 pollution [22]. During the study the value of outside carbon dioxide fluctuated in the range from 438 ppm to 456 ppm.

During the research, the sick building syndrome symptoms were noted in people staying in the rooms, which they identified with long-term staying in a given environment. The study was based on questionnaires filled out by 189 persons before thermal modernization and 207 after thermal modernization. The occurrence of symptoms in the studied population of people before and after thermal modernization is presented in Table 4.

Table 4. Sick building syndrome symptoms before and after thermal modernization.

Sick building syndrome symptoms	**Percent symptom occurrence**	
	Before	**After**
Headache	32.1	33.8
Throat irritation, cough	26.1	38.8
Malaise	24.6	52.1
Fatigue and sleepiness	21.3	45.5
Nose irritation, rhinitis	24.8	11.3
Difficulty in concentration	17.7	43.8
Eye irritation	26.4	17.8
Irritability	7.3	6.2
Flu-like symptoms	41.5	14.4
Itching and dry skin	3.4	1.2

Some of the sick building syndrome symptoms such as above all flu-like symptoms or nose irritation and rhinitis partly receded after thermal modernization. Increased fatigue, sleepiness, malaise and difficulty in concentration which may be related with the growth of level of carbon dioxide. Generally poor air quality in the room have been reported on the about 90% people staying in rooms. The feeling of lack of fresh air created the necessity of additional airing of rooms by opening windows. During the heating season, this was the reason for the increase in heat loss and temporary deterioration of thermal comfort conditions, due to the lowering of the room temperature and the increased speed of the cold air flow.

5 Conclusions

The realization and exploitation of buildings consumes about 40% of total energy consumption in the European Union [5]. Thermal protection of buildings is directly related to, but not only to, energy saving. The purpose of thermal protection should also be to ensure proper conditions for the interior microclimate and thermal comfort of people staying in in closed rooms. However, attention is paid to the deterioration of indoor air quality in rooms in relation to the tightening of the thermal protection requirements of buildings associated with the energy efficiency program.

Increasing thermal insulation of buildings envelope and modernization of the heating system and hot water preparation significantly improved the energy efficiency of educational buildings. Thermal modernization activities partly improved also indoor microclimate conditions and thermal comfort perception of occupants as well as contributed to the elimination of some health problems related with sick building syndrome. The air relative humidity decreased, even down to the value unacceptable fully by respondents. Furthermore, the moulds growth in the area of window occurring before thermal modernization, was totally eliminated.

However after thermal modernization, the intensification of some symptoms of sick building syndrome was found, which the occupants related with worsening of air quality in the rooms. A significant increase in carbon dioxide concentration after prolonged use has also been noted. Also recorded a significant increase in the concentration of carbon dioxide after a longer time of rooms use. Probably the ventilation of rooms was not sufficient. It would be beneficial to introduce a constant, gradual inflow of fresh air in the amount necessary to ensure adequate hygienic and health conditions. This would eliminate the need for frequent ventilation by opening windows, contributed to reducing heat losses and improving the thermal comfort conditions of people staying in the rooms. But this is not possible with gravity ventilation.

In order to get a more complete picture of the initially observed phenomena, a long-term monitoring of the conditions prevailing in the rooms should be carried out.

References

1. Energy, Statistical Information and Elaborations, Central Statistical Office, Warsaw (2018)
2. EUROPE 2020. A strategy for smart, sustainable and inclusive growth, European Commission, Brussels (2010)
3. Energy 2020. A strategy for competitive, sustainable and secure energy, European Commission, Luxembourg (2011)
4. Directive 2018/844/EU of The European Parliament and of the Council of 30 May 2018 amending Directive 2010/31/EU *on the energy performance of buildings* and Directive 2012/27/EU *on energy efficiency* (2018)
5. A. Guła, *The strategy of thermal modernization of buildings. Roadmap 2050* (Institute of Environmental Economics, Cracow, 2014)
6. Y. Al Horr, M. Arif, M. Katafygiotou, A. Mazroei, A. Kaushik, E. Elsarrag, Int. Jour. of Sus. Bu. Env. **5**, 1, 1–11 (2016)
7. Indoor air pollutants: exposure and health effects, EURO Reports and Studies, World Health Organization (1984)
8. Economic cost of the health impact of air pollution in Europe: Clean air, health and wealth, World Health Organization (2015)
9. Sick building syndrome, World Health Organization Regional Office for Europe (1998)
10. WHO guidelines for indoor air quality: dampness and mould, World Health Organization (2009)
11. WHO guidelines for indoor air quality: selected pollutants, World Health Organization (2010)
12. Evolution of WHO air quality guidelines: past, present and future, World Health Organization (2017)
13. P. Kapalo, N. Spodyniuk, *Effect of the variable air volume on energy consumption – case study* (IOP Conf. Series: Materials Science and Engineering 415, 2018)
14. J. Sowa, Izol. **23**, 2, 76–82 (2018)

15. J. Sowa, *Jakość powietrza w budynkach energooszczędnych* ([W]: Budownictwo energooszczędne w Polsce – stan i perspektywy, Red.: M. Wesołowska, Wydawnictwo Uczelniane Uniwersytetu Technologiczno-Przyrodniczego, Bydgoszcz, 155–166, 2015)
16. K. Januszkiewicz, Budownictwo o zoptymalizowanym potencjale energetycznym **2**, 20, 43–50 (2017)
17. Regulation of the Minister of Infrastructure of 12th April 2002 on technical conditions, which should correspond to the buildings and their location (consolidated version: OJ 2015 item. 1422, as amended OJ 2017 item. 2285)
18. A. Lis, A. Ujma, *Aspekty poprawy charakterystyki energetycznej budynków* ([W]: Współczesne problemy budownictwa. Rozwiązania praktyczne i numeryczne. Monografie nr 330. Red.: M. Major, J. Selejdak. Częstochowa Wydawnictwo Politechniki Częstochowskiej, 213–222, 2017)
19. P. O. Fanger, Z. Popiołek, P. Wargocki, *Środowisko wewnętrzne. Wpływ na zdrowie, komfort i wydajność pracy* (Politechnika Śląska, Gliwice, 2003)
20. P. Kapalo, F. Domnița, C. Bacoțiu, N. Spodyniuk, Jour. of Appl. Eng. Scien. **8**, 21, 61–66 (2018)
21. P. Kapalo, A. Sedláková, D. Košicanová, O. Voznyak, J. Lojkovics, P. Siroczki, *Effect of ventilation on indoor environmental quality in buildings* (The 9th International Conference on Environmental Engineering Selected Papers, Vilnius, VGTU Press, 1–6, 2014)
22. S. Vilcekova, P. Kapalo, L. Meciarova, E. Krídlová Burdová, V. Proc. Eng. **190**, 496–503 (2017)

Lighting of urban green areas – the case of Grabiszyn Park in Wrocław. Searching for the balance between light and darkness through social and technical issues

Agata Łopuszyńska[1,*], and *Małgorzata Bartyna-Zielińska*[1]

[1]Wrocław University of Science and Technology, Faculty of Architecture, Chair of Urban Planning and Settlement Processes, ul. B. Prusa 53/55, 50-317 Wrocław, Poland

Abstract. Illuminating the urban green spaces could lead to conflicts of spatial, technical and social nature. This relatively new, though already global, problem is expected to grow bigger with the further increase of urban areas artificial brightness. The case of Grabiszyn Park in Wrocław is an example of how difficult it is to find a balance between big-city lights and a natural darkness. The situation is even more difficult if the light is not legally recognized as a significant source of air pollution and direct nuisance at the legal level. The aim of the paper is to recognize the broader perspective of urban greenery lighting issues, global recommendations basis and the local awareness. The authors also made an attempt to analyze and assess the project implementation, as well as to characterize the components of the quality of an urban green areas lighting.

1 Introduction

Illumination of urban area is a field of growing interest not only for designers and researchers, but for its users as well. Defining the components of the high quality of the city's lighting after dark, and consequently creating the proper guidelines, is a very complex and changing process though. Outdoor lighting, as a public good, is expected to be functional, economical and respond to the conditions of places and users' needs within a diversified urban space. However, the situation becomes complicated when not only people are exposed to artificial light but the lively elements of cities' ecosystems, as well as in the case of urban green areas.

1.1 Lighting regulations and recommendations

Current guidelines related to lighting are based on legal regulations and corresponding norms, as well as a lighting culture of the society and the awareness and choices of designers. In Poland, there are no specific requirements dedicated to the lighting of green areas. The existing schematic technical standards refer mainly to roads [1], outdoor work spaces and sports areas [2]. Although some of these elements could be found in urban

* Corresponding author: agata.lopuszynska@pwr.edu.pl

greenery, they are not always applicable. Park lighting in Poland is not a subject to standards as it is in the case of streets, often creating more nuisance than road lighting [3].

Internationally recognized guidelines referring to green areas focus mostly on the prevention of intrusive light, suggesting zoning broken down into, for example, wild nature reserves, villages, suburbs, city's downtowns and centers [4, 5]. Illuminating Engineering Society draws attention to the environmental issue of urban lighting, especially the light pollution, light trespass and glare associated with ill-considered use of lighting techniques [6]. Environmental Zones are also proposed, determining the brightness levels. The more detailed recommendations due to the quality of outdoor lighting refer to the light pollution reduction publications [7, 8]. However, they are often questioned whether they are right for urban areas due to insufficient flexibility and rigorous approach to lighting. This is also connected with the erroneous fear of switching off the city lights, more than with paying attention to their distribution direction and proper parameters.

In the specialist and scientific literature, lighting of green areas is still an unresolved topic, presenting many, often contradictory, approaches. Most publications refer to the impact of light on fauna and flora [9], especially to the blue-rich LED lighting [10], energy efficiency, visuality and lighting engineering [11], aesthetics and a landscape beautification [12, 13]. In the case of parks, the attention is focused on illumination in relation to seasonal changes, plant vegetation periods. A lot of attention is also devoted to digital control and new technologies used in greenery areas lighting, for example, adaptation to moonlight or the number of its users [14]. The difficulty of green areas illumination is also associated with its wild, dynamic character, difficult to measure with parameters. However, the art of illumination of objects differs from the lighting of larger urban green spaces with a specific functional and spatial program. There are also voices raised that the naturalness of the place is important in the perception of space after darkness [15]. In line with such concept, the lighting of the greenery should not be restricted to certain lighting level but to mimic its daytime character. On the other hand, the ecological approach to illumination recommends preserving the darkness due to the harmfulness of exposure to light during the night-time and maintaining its warmest color – low color temperature [16].

1.2 Urban green areas issues

Urban greenery areas vary in terms of the functional sphere or level of development. In the downtown conditions single isolated trees are preferred, but they are treated more as elements of city furniture than living objects. Lighting them is not a biologically beneficial, but extremely popular, seen in many cities' lighting plans [17]. Parks, squares and green areas retain their natural function to a greater extent in the vicinity of the buildings. There are also city green areas that areas that make up extensive ecosystems bordering with agricultural areas. Indeed, all of the presented types of urban greenery are used by residents and tourists. When planning to illuminate it, one should consider whether the space of the park is a typical 24-hour, accessible public space subordinated to people's expectations and global trends. In the conditions of today's permanent brightness of urban areas, the schematic and normative lighting model seems to not work anymore. Urban greeneries are often the last remaining oasis of wild, undeveloped space. On the other hand, they provide an attractive, and deficit, space for outdoor recreation, which result in the desire to adapt it to the users' needs over time.

Currently, the main goals of green area lighting are to increase the sense of safety of the users, create attractive landscape with the light and make people dare to stay out more. The most frequently illuminated elements of green areas are paths and squares, street furniture, buildings and characteristic objects that help to orientate after dark [18]. However, evolutionally every living element of the environment, including people, requires nighttime darkness conditions, so that circadian rhythm function properly on the psychophysical level [19]. Planning the green spaces lighting is therefore the result of seeking a balance between the natural and urban environment, artificial brightness and the darkness. An example of such process is the lighting of Grabiszyn Park in the city of Wroclaw.

2 Methods and material

2.1 The aim and scope of the study

The aim of the paper is to present the issues of urban green areas illumination, which represent an example of spatial conflict associated with artificial light. According to the authors' assessment, in the future there will be more and more of them due to the dynamic increase of brightness in cities and the growing level of knowledge on the negative consequences of excessive or inappropriate lighting.

The analysis of the lighting process of Grabiszyn Park in Wroclaw was conducted, which allowed to learn the real arguments for and against brightness in parks. The implemented lighting project was evaluated by the authors in terms of functionality, social and light pollution related issues. The timeframe of the research covers the latest events (last two years) and focuses on the park itself, however, the literature study mainly refers to the last decade. The analysis of sources is the basis for assessing whether it is possible to reach a compromise solution and considerations on the components that constitute to the quality of urban green areas lighting.

2.2 Research hypothesis and methods

An interdisciplinary study of contemporary literature was adopted in the introduction section of the paper, involving fields of lighting engineering, legal issues, composition and visual side of light, ecology and astronomy. A practical part of research is the case study of Grabiszyn Park illumination process. The analysis of documents, formal and press reports as well as interviews and observations have been adopted to examine mentioned issues. For the purpose of the paper, an authorial analysis and assessment of park lighting project implementation were also carried out.

Considering the low social awareness, the authors hypothesized that currently it is not possible to reach a compromise solution regarding the park lighting socially, but it is technically. However it is difficult due to the lack of national and local legal regulations in the field of lighting and light pollution.

3 Grabiszyn Park – the urban greenery lighting case study

The Grabiszyn Park is a vast area of urban greenery (~50 ha) of a natural character, with an extensive network of pedestrian routes. The Ślęza River, the cemetery complex and the allotment gardens that limit the park, affect its indirect isolation in the city tissue. In recent years, along with the multi-family housing development, the Park is gaining attention due to its recreational potential.

3.1 Participation and masterplanning

Lighting of Grabiszyn Park lanes has been the one of the winning projects in 2016 Wroclaw Participatory Budget edition. The main aim of the project was a new lighting installation to make jogging and walking safe and accessible after dark. The construction works had started at the end of 2017. Before then, only about dozen park lamps have been located in the park, situated around the entrance zone from Grabiszyńska Street.

The lighting project was one of the factors leading to protests regarding development and further maintenance of Grabiszyn Park. At the beginning of 2018 people started to protest against the city's maintenance works in park that included cutting down over 500 dead trees. Around the same time a project of Grabiszyn Park Masterplan [20] was presented to the citizens. The public consultation of Masterplan had started the discussion concerning the maintenance of the park and its future development. One of the topics was new lighting of alleys, which was meeting citizens' need of safety but influencing fauna and the trees roots. During the time of discussion, the participatory budget lighting had been installed, and for some of neighborhood citizens is was considered as park devastation. In the meantime 10 lamps were set on fire, and during the next months, from January to July, fastening and control elements had been stolen and cables cut of another lamps located at the main alley (Fig. 1.). Because it had not happened before in any other park in the city, it was assumed to be not a hooligan act, but rather a deliberate action [21].

Fig. 1. Burnt lamps, January 2018, [photography courtesy of Tomasz Pajączek/Onet].

Due to many issues and conflicts it was decided to extend the time of the public consultation till June 2018. In view of such a big change, the residents' reservations and fears seem to be natural, even though the lighting of the park lanes was a bottom-up initiative. The maintenance works interfering with the structure of the park, the Masterplan project and the above events turned out to be so crucial for the local community that the civic committees were established in order to protect the park from changing its natural and forest-like character.

The second, so far suspended, stage of lighting in Masterplan assumed a significant increase in the number of planned lamps and additional illumination. During the consultation, it was possible to learn about arguments for and against artificial brightness [22]. The main „for" arguments were such as: an improvement of security and maintaining the continuity of communication corridors after dark, eliminating the fear of darkness, facilitating access to public transport sites and the possibility of longer activity and recreation in the park. However, far more requests concerned objections to lighting or its significant limitation. It was claimed that lighting at night generates too high costs in relation to a small number of its users, as an expression of excessive 'convenience', that also can encourage pathological behavior. Many considered nature as the primary function of the park and the protection of its wild nature as a priority (mainly protection of trees, plots and nocturnal animal species). It was proposed to completely turn off the lights at night, beyond the trams running time. There were also votes for the dark-sky protection and the possibilities of amateur astronomical observations. For many citizens, enough would be the use of flashlights at night and repairing the surface of paths for security purposes.

The public consultation resulted in, among other issues, conclusions that the lighting as part of participatory budget has been implemented in a very limited extent in relation to Masterplan (Fig. 2.). It was determined that the fixtures must prevent the upper light emission and recommended to create a lighting curfew. Traffic sensors could not be implemented in order to protect nature.

Fig. 2. The brown dots show existing lamps, orange dots represent winning participatory budget project and the smallest red ones lamps planned in Masterplan [20]. Spatial distribution of the new luminaires is indicated by the blue line [graphic based on ZZM Wrocław data].

3.2 Analysis and assessment of the project implementation

Although not much time has passed since the changes in the park area brightness after dark, the implementation can be assessed on several levels. At the same time, it should be noted that access to the public data and technical parameters of light is still difficult, which limits the assessment methods for observations and simple tools, without specialized measurement being necessary.

Public consultations, protests and the devastation of lamps, led to reduction of the Masterplan version, simultaneously developing the vision from the Participatory project and postponing further plans for the park area modernization. In the end, only the main communication park paths were illuminated and by arranging the lamps the bypasses for the roots of nearby trees were included or the works with the use trenchless method were conducted. Energy-efficient LED luminaires have been used with four times less energy consumption than already existing ones, and with an estimated life expectancy of 25 years. Lamps are equipped with a remote control system and the possibility of gradual brightening, which is especially important in the case of such sensitive area and with thoughtful usage can serve as a part of the compromise solution.

From a public space user's point of view, lighting can be evaluated in terms of its functionality and usage comfort. Selected communication routes have been illuminated continuously, allowing one to walk through the whole area of the park in the most popular directions. The light of the lamps does not affect the housing zones, and its interference with the plant tissue is limited only to the closest surroundings of the paths. However, the discomfort associated with the glare is appreciable. It involves the appearance of excessive brightness or contrast within the visual field [23], which in the case of Grabiszyn park is caused by the choice of luminaires shape. Contrary to the good practice rule, that proper lighting ensures the visibility of light and illuminated space, not its direct source, the flat-lens luminaire exposes strong, unshielded LED diodes. It can disturb the user or limits the ability to recognize objects in space, which also may affect one's sense of security.

The impact of light on the environment is another aspect of assessment. The new full cut-off fixtures do not emit light above the horizon, which minimizes light pollution (Fig. 3.). Although it is visibly bright within the paths, it is possible to recognize the main stellar constellations. Currently, the night-sky quality of the park centre corresponds to 6 "bright suburban sky" in the Bortle scale (with 1 as the darkest and 9 as the brightest) [24]. The park is not big enough to curb the surrounding city lights and an ubiquitous sky glow, but still remains a stargazing potential.

The color temperature of luminaires is 4000 K, which is a neutral-cool light limit. The recommended range for green areas is approximately 1900–2500 K, to reduce glare and insect attraction or to minimize ecological effects and impact to human dark adaptation [4, 6, 7]. Therefore, the color could be more nature-friendly, warmer, especially since with a quality luminaires it is possible to obtain an adequate color rendering index. There are some doubts, as well, regarding the lamp height (5 m) that in some points violates the level of tree crowns, but at the same time it limits the needed lamps quantity.

Fig. 3. New lighting of Grabiszyn Park at night: distinct lack of upward light spill and visible distant visual glare from the luminaires, [A. Łopuszyńska].

3.3 Discussion: a quality lighting

While the applicable procedures, technical standards and the legal situation of outdoor lighting in Poland are regulated by the government, and their possible changes and updates are long-lasting, the bottom-up initiative had influenced local authorities and their actions towards using new, more individualized, solutions. Certainly, this is a factor introducing the diversity of opinions and an element of discussion in the sphere of shaping public space, and in consequence it may result in faster reactions to globally changing needs. However, it is worth noting that the 'loudest' activist groups are not always representing the interests of entire local community; there is a risk of pushing their agenda at the expense of citizens who tend not to participate in public consultations.

It is the quality that could become the platform for a common compromise and the multi-level perception of security after dark. On the one hand, it is to ensure a democratic sense of security for people after dark through public lighting; but also safe for the environment, i.e. minimize the impact on the environment and light pollution as well as safe for the public budget in terms of reducing energy consumption and preserving the environment in the best possible condition, in accordance with the rule of sustainable city. At the same time, only an interdisciplinary approach to the issue could guarantee this. Nowadays, technical-economic and visual premises dominate, such as caring about low lighting maintenance costs, striving to meet the minimum required technical standards and beautify the space after dark. Due to the lack of regulations, it might result in chaos and visual nuisance for its users.

Future changes in lighting regulation are unavoidable considering the need to care for the quality of increasing brightness in cities. It is necessary to adopt an urban lighting zones that will fill the gap between areas of protected darkness and the bright city centers. Currently, the use of globally recognized recommendations is not obligatory, and there are no tools to enforce lighting policy. The current transformation into energy-saving lighting makes perhaps the greatest opportunity to focus on its quality.

4 Conclusion

The devised solution that has not evolved towards lighting most of the Grabiszyn Park alleys, but focused on the main ones is a kind of technical compromise. However, there can be no question of finding a reconciliation of involved social groups. Such reconcilement would require prior determining and social acceptance of the real quality of lighting. Social awareness and general knowledge about the modern possibilities that lighting technologies bring are still too small. After all, a sense of security and recognition in space after dark is possible without major losses for the night landscape and ecosystems. Emphasizing the importance of light pollution issues, it must be said that they are not only related to the night-sky darkness, but thoughtful and functional lighting usage.

It seems right to assume, that the urban greenery, should be considered special, but still a part of a city space. Therefore, green areas, as well as its illumination, should serve the people in contrast to agricultural or protected areas. Shutting off the lights seems to be inappropriate and too radical for urban areas, but this does not mean, however, disrespect for the conditions and surroundings of such a place. It is rational to strengthen the sense of security for park users after dark by lighting: creating its continuity, uniformity and clear communication routes. The same aim is served also by minimizing the light trespass and glare. Pedestrian-bicycle paths and selected squares, as well, could remain the character of a fully-fledged public space, but with keeping the rules of safe lighting. The color and intensity of light should harmonize with the natural circadian rhythm of people, fauna and flora, so it is reasonable to eliminate the blue light. It is worth noting that the use of energy-saving fixtures and fulfillment of color rendering index is not necessarily connected with the 'cool' color temperature.

In order to define the criteria for urban greenery high quality lighting design, the individual approach is recommended, meeting the requirements of a specific area and its users. The key is to choose the right parameters of light, depending on the urban functional zoning, as well as the correct selection, assembly and placement of the fixtures. This allows to keep the light within the desired boundaries, so that, for example, the paths will be illuminated instead of the tree crowns. In case of not yet developed 'culture' of lighting and possible change of needs over time, it is a good solution to use controllable lighting devices, including the intensity (reducers), the usage time of lighting (timers), as well as manipulation of the light direction.

Due to light pollution prevention, that becomes a part of urban planning in many places in the world, limiting the upward light spill is crucial. Minimizing the negative effects of ill-considered and excessive lighting, thus increasing the care for its quality could create the basis for a sustainable lighting environment. The issues presented in the paper are gaining attention, which gives hope for further discussion and a more humanistic approach to artificial brightness.

References

1. Polska Norma PN-EN 13201: Oświetlenie dróg (2007)
2. P. Pracki, Spektrum **3–4**, I–V (2014)
3. P. Nawalkowski https://www.ciemneniebo.pl/pl/przyklady-oswietlen/parkowe (2013)
4. CIE, Technical Report 150:2017
5. ILP, Guidance Notes for the Reduction of Obtrusive Light, GN01:2011
6. IES, Lighting for Exterior Environments, SKU:RP-33-14 (2014)
7. IDA, Model Lighting Ordinance (MLO) with User's Guide (2011)
8. CIE, Technical Report 001-1980 (Joint Publication IAU/CIE (1980)
9. T. Longcore, C. Rich, Front. Ecol. Environ. **2**, 191–198 (2004)
10. IDA, *Visibility, Environmental and Astronomical Issues Associated with Blue-Rich Outdoor Lighting* (2010)
11. W. Żagań, *Iluminacja obiektów* (Oficyna Wydawnicza PW, 2003)
12. R. Narboni, *Landscape Lighting* (Design Media Publishing UK Ltd., 2016)
13. K. Michalak, A. Nawrowski, Architecture Technical Transactions **19**, 313–326 (2012)
14. Arup Foresight, *Cities alive: Rethinking the Shades of Night* (ARUP, London, 2005)
15. L. J. Nasar, *The Evaluative Image of the City* (Sage Publications, USA, 1998)
16. Ł. Kowalewska, A. Mostowska, Kosmos. Problemy Nauk Biologicznych **64**, 471–483 (2015)
17. I. Vassileva, H. Kollhof, *Night and Day Image of Public Space, Urban Nightscape* (Conference Proceedings, 2006)
18. H. Łakomy, K. Hodor, Czasopismo Techniczne Wyd. Polit. Krak. **4–A**, 131–138 (2007)
19. R. Stevens, Y. Zhu, Phil. Trans. R. Soc. B **370**, 20140120 (2015)
20. https://www.wroclaw.pl/rozmawia/park-grabiszynski-konsultacje-do-pobrania (2018)
21. https://wroclaw.onet.pl/ktos-celowo-niszczy-nowe-latarnie-w-parkugrabiszynskim/xwg42ee
22. UM Wrocław, *Park Grabiszyński. Planuj! Konsultacje Parku Grabiszyńskiego* (2018)
23. I. Lewin, *Light Trespass Research. Final Report to the Lighting Research Office of EPRI: TR-114914* (Lighting Research Institute, EPRI Palo Alto, CA, 2000)
24. J. E. Bortle, Sky and Telescope **101**, 2, 126–129 (2001)

Risk assessment in municipal wastewater treatment plant

Magdalena Łój-Pilch[1,*], *Anita* Zakrzewska[1], and *Ewa* Zielewicz[1]

[1]Institute of Water and Wastewater Engineering, Silesian University of Technology, Gliwice, Poland

Abstract. Each technical facility is exposed to a risk that will hinder or prevent its proper functioning. Risk assessment is a component of the risk management process in municipal wastewater treatment plants, the purpose of which is to reduce the occurrence of events as well as their resulting effects, which adversely affect the operation of the treatment plant. The authors, having based their work on the identification of risk-related phenomena, estimated the value of risk and determined its acceptability. The result of the work was to obtain a risk matrix – in other words, a risk map with a designated hierarchy. This could then be used as a basis to determine the level of risk that an event triggered. This allows for faster response to risky situations and an optimal selection of remedies.

1 Introduction

The concept of risk is inseparably related to any system or technical device. Complicated technical facilities, such as municipal wastewater treatment plants, are particularly susceptible to risk. They consist of many connected devices working with variable loads of pollutants, hydraulic load as well as changing and difficult weather conditions, especially in winter. During the operation of such specific facilities, there are a number of hazards that may lead to emergency events and losses, therefore it seems necessary to implement a continuously verified risk management process at sewage treatment plants [1–3].

Risk management consists in undertaking of activities that aim to make an assessment, control the risk and monitor the actions taken. The first stage of risk assessment – the identification of risk – leads to the determination of risk factors and the determination of the type of risk and factors that cause it [4, 5]. In the case of sewage treatment plants, at an earlier stage of research, the following types of risk were identified: qualitative, operational, ecological and, related to each of them, financial risk. The factors that cause them were determined as: internal, external, latent, explicit, ordinary and extraordinary [6–8]. The subject of this article is the presentation of subsequent stages of risk assessment – estimation and determination of its acceptability. Risk management has been well

* Corresponding author: magdalena.loj-pilch@polsl.pl

researched and described in the literature in the field of water supply systems, for e.g. in [9–12]. In the case of municipal sewage treatment plants, have been made attempts of modelling risk, e.g. [13], and few studies on its management also can be found [14, 15]. On this basis, the authors of the paper attempt to analyze and assess the risk of municipal sewage treatment plants based on real operational data.

2 Risk assessment

Risk management in an enterprise should lead to a situation in which the management of the facility will be aware of the presence of a risk during the operation of the plant and its size, while its level in the current operation of the plant does not exceed the contractually accepted risk limit [4, 5]. Risk assessment is an element of this process [6–8]. The result of risk assessment of a technical facility should be a risk hierarchy, which is best represented graphically, creating the so-called risk map. An illustrative picture of the risk hierarchy is determined on the basis of an estimation of the risk value and the determination of its acceptability [16].

2.1 Risk identification

The main purpose of risk identification is to get acquainted with the functioning of the facility, to determine the risk factors and the types of risks that cause them. Risk recognition can be divided into the following stages [6–8]:

- becoming acquainted with the characteristics of the tested facility,
- conversation with the manager and employees of the facility,
- analysis of historical data regarding events that occurred on the site,
- verification of the results obtained.

The risk-related situations identified in this way should be subjected to further stages of risk assessment.

2.2 Risk estimation

Risk estimation consists in the adoption of its measures and on the basis of the determination of the risk value. The choice of measure depends primarily on the availability of input data, their number and reliability, selected for estimating risk parameters and expected output values. Therefore, you can distinguish [14, 15, 17, 18]:

- quantitative numbers – determine the exact numerical value of risk, e.g. probability,
- qualitative variables – they determine the risk value as a relative quantity
- on a conventionally agreed scale, different for each case,
- single numbers – they focus on only one parameter,
- comprehensive resolutions – they consider several parameters, most often the probability and amount of losses,
- focused numbers – the result obtained is a single numerical value,
- spread numbers – the result obtained is a function of the distribution of a random variable, usually presented in graphic form.

Figure 1 presents an example of a risk matrix of a so-called risk map resulting from an undertaken estimation.

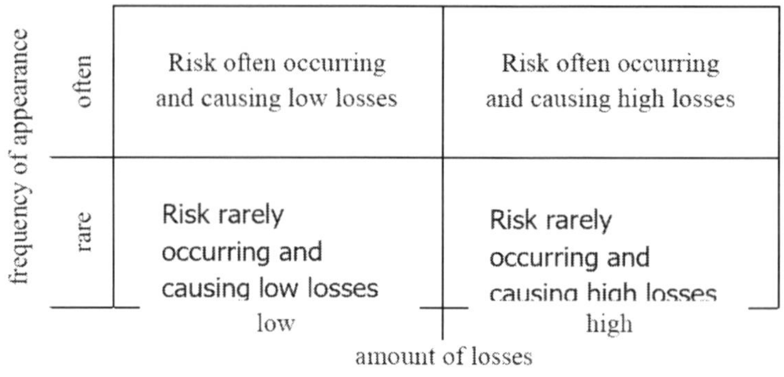

Fig. 1. Risk map [9].

2.3 Risk admissibility

The admissibility of risk is determined on the basis of existing legal acts and standards. In the case of municipal wastewater treatment plants, the basic document is the Regulation of the Minister of Environment from November 18, 2014 [19]. In reference to an individual facility, it could also be considered to issue a permit for specific water treatment. The mentioned act defines the conditions of sewage disposal - the highest acceptable values of selected indicators or the minimum percentage of contamination reduction and the maximum number of samples that do not meet these requirements. These are immission criteria, the fulfilment of which ensures that discharged sewage will not cause contamination of the receiver [14, 15].

The following risk tolerance levels are distinguished [14, 15, 18]:

- low risk (acceptable) – is perceived as not relevant to the overall functioning of the facility, does not require special security measures,
- medium (tolerated) risk – conditionally acceptable risk, when the costs of risk reduction are disproportionately large to the possibility of reducing losses,
- high risk (unacceptable) – is directly related to the occurrence of threats to the environment, health and human life, requires immediate action and reduction of risk regardless of the cost.

The risk acceptability should be applied to the previously obtained risk map (Fig. 1). In this way, graphical interpretations of the enterprise risk hierarchy are obtained (Fig. 2).

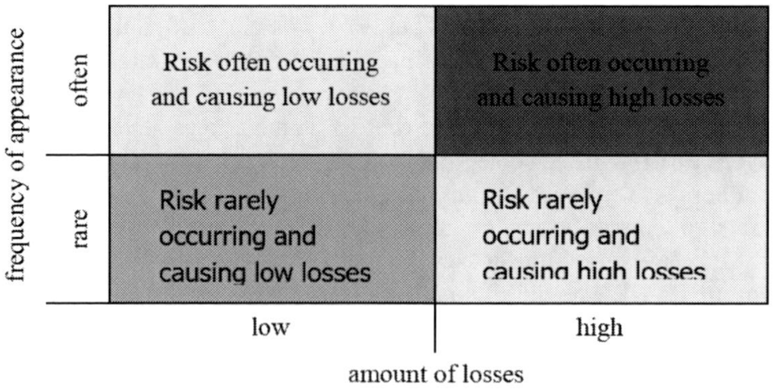

Fig. 2. Risk hierarchy.

3 Obtained results

The functioning of a municipal sewage treatment plant using activated sludge technology located in the Upper Silesian Industrial District (GOP) was considered. The sewage plant serves 42 thousand inhabitants, which in addition to municipal sewage also receives industrial sewage from meat processing and electroplating. The risks assessed were identified at an earlier stage of research work. A three-year operational period of the treatment plant was analyzed, during which 36 different risk-related situations occurred with varying frequency [6]. Estimating the risk of municipal sewage treatment plants consisted in determining the incidence rate (I) of an adverse event and determining the amount of losses (L) resulting from its consequences. As the basic risk parameters, the authors accepted the values of F and L. Events, depending on the incidence, were grouped as rare, common and very common, and weighed from 1 to 3 (Table 1.)

Table 1. Incidence rate (I).

Occurring events	Incidence rate	
	[1/year]	(I)
rare	≤ 2	1
common	2–4	2
very common	≥ 4	3

The amount of losses (L) was determined on the basis of the type of risk assigned to each event and numerical weights were also assigned to them (Table 2). Table 3 presents examples of events together with the incidence rate (I) and assigned loss (L). On the basis of the determined values I and L, a risk map was prepared taking into account its hierarchisation (Fig. 3).

All analyzed events were in the area of acceptable and tolerated risk, with the advantage of the former, which proves the proper functioning of the treatment plant. In order to better illustrate the risk occurring in the normal operation of the sewage treatment plant, a risk map was prepared taking into account events occurring at individual devices of the technological sewage treatment plant (Fig. 4).

Table 2. Amount of losses (L).

Type of risk	Amount of losses (L)
qualitative	1
qualitative, operational	2
qualitative, operational, financial	3
qualitative, operational, ecological, financial	4

Table 3. Examples of events together with the incidence rate (I) and amount of losses (S).

Device	Event	Incidence rate		Type of risk	Amount of losses (L)
		[1/year]	(I)		
Bar screen	clogging of bars	4.33	3	qualitative	1
Grit chamber	large dump of greasy sewage	2.67	2	qualitative	1
Activated sludge chamber	emergence of filamentous bacteria	0.33	1	qualitative, operational	2

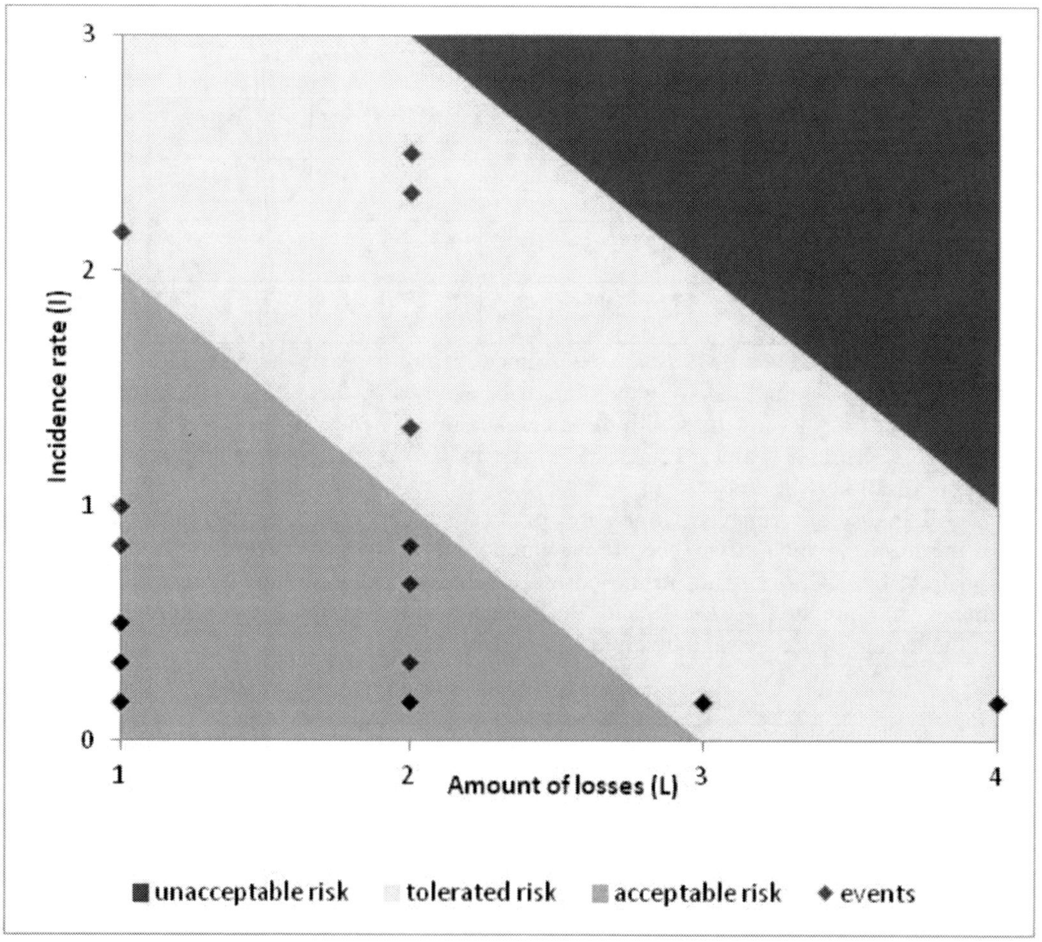

Fig. 3. A risk map taking into account its hierarchy.

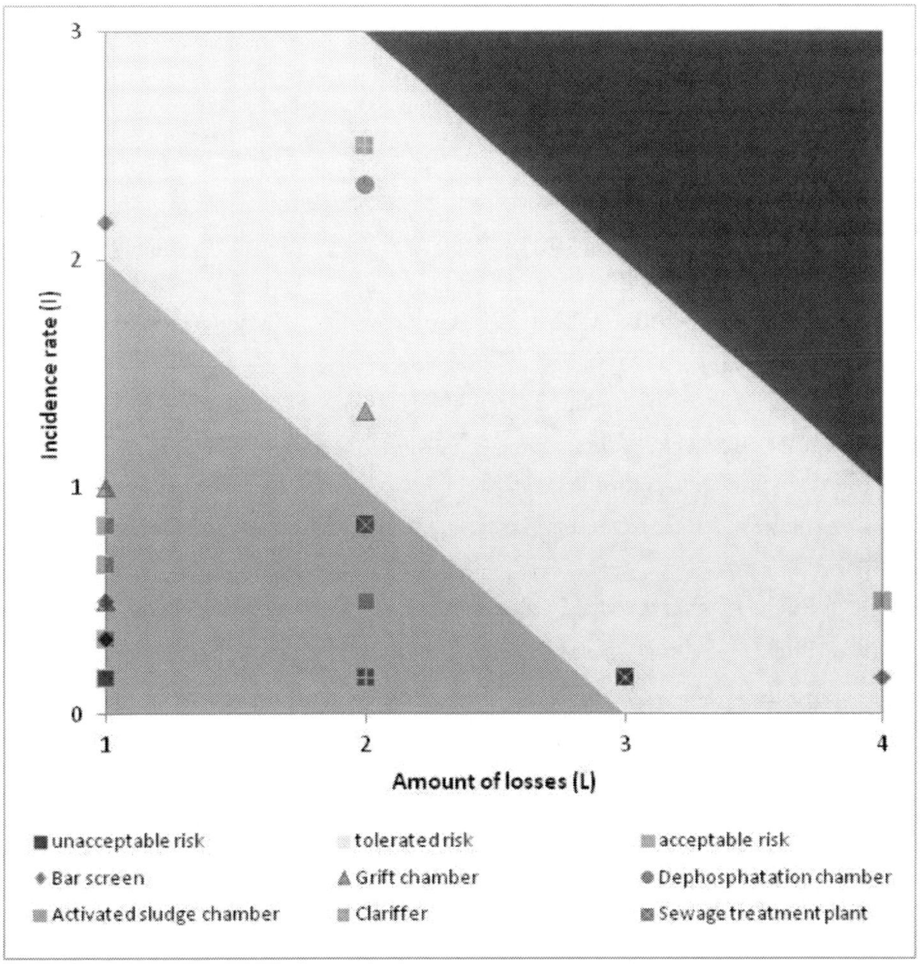

Fig. 4. A risk map taking into account its hierarchy with division into technological line devices

The highest risk level was assigned to events located in the activated sludge chamber, dephosphatation chamber, clarifier and on the bar screen. The first three of these devices are responsible for biological - chemical treatment. The last - the bar screen, is the first device in the technological process and is responsible for mechanical cleaning.

4 Conclusions

Analyzing the results of the risk map, it can be noted that the largest density of points representing events in the treatment plant is in the area of acceptable risk (I = <0,1>; L = {1}). The largest losses (L = 4) were attributed to events that occurred at the bar screen and in the activated sludge chamber. These are the devices most exposed to risk occurrence. The bar screen is the first in the technological process to come into contact with sewage, so it is exposed to damage resulting from the physical properties of the sewage. The circulating activated sludge chamber is a device in which biological wastewater treatment takes place, so it is on the one hand exposed to the risk associated with the change in the composition of wastewater and on the other hand, its influence is affected by external conditions, such as ambient temperature. There were no events in the area of unacceptable risk. It provides information about proper functioning of the treatment plant, good training of the crew and conscious decision making by the management of the facility.

This work was supported by Research Funds For Young Researchers awarded to the Institute of Water and Wastewater Engineering of the Silesian University of Technology.

References

1. D. Andraka, L. Dzienis, Zeszyty Naukowe Politechniki Białostockiej. Inżynieria Środowiska **16**, 24–28 (2003)
2. T. Żaba, A. Królikowski, J. Królikowska, Inżynieria Ekologiczna **24**, 215–225 (2011)
3. E. Kulińska, Logistyka – nauka **2**, 385–409 (2011)
4. J. Zawiła-Niedźwiecki, *Zarządzanie ryzykiem operacyjnym w zapewnianiu ciągłości działania organizacji* (edu-Libri, 2013)
5. Y. Y. Haimes, *Risk Modeling, Assessment, and Menagement* (John Wiley & Sons, 2009)
6. M. Łój-Pilch, A. Zakrzewska, E. Zielewicz, E3S Web of Conferences, **44** (2018)
7. M. Łój-Pilch, A. Zakrzewska, E. Zielewicz, Instal **4**, 61–63 (2018)
8. M. Łój-Pilch, A. Zakrzewska, E. Zielewicz, Proceedings of Ecopole **12**, 195–201 (2018)
9. J. Rak, *Istota ryzyka w funkcjonowaniu systemu zaopatrzenia w wodę* (Oficyna Wydawnicza Politechniki Rzeszowskiej, 2004)
10. J. Rak, B. Tchórzewska-Cieślak, *Czynniki ryzyka w eksploatacji systemów zaopatrzenia w wodę* (Oficyna Wydawnicza Politechniki Rzeszowskiej, 2007)
11. J. Rak, *Bezpieczna woda wodociągowa. Zarządzanie ryzykiem w systemie zaopatrzenia w wodę* (Oficyna Wydawnicza Politechniki Rzeszowskiej, 2009)
12. R. Iwanejko, Technical Transactions. Environment Engineering **102**, 8, 113–127 (2005)
13. D. Andraka, L. Dzienis, Middle Pomeranian Scientific Society Of The Environment Protection **15**, 1111–1125 (2013)
14. R. Iwanejko, S. M. Rybicki, Gaz, woda i technika sanitarna **2**, 10–13 (2008)
15. R. Iwanejko, S. M. Rybicki, Gaz, woda i technika sanitarna **3**, 34–38 (2008)
16. J. Zawarska, Zarządzanie i Finanse **1**, 65–75 (2012)
17. R. Iwanejko, Czasopismo Techniczne. Środowisko **Z8-S3**, 113–127 (2006)
18. J. Rak, Monografia Komitetu Inżyierii Środowiska PAN, **28** (2005)
19. Rozporządzenie Ministra Środowiska z dnia 18 listopada 2014 r. w sprawie warunków, jakie należy spełnić przy wprowadzaniu ścieków do wód lub do ziemi, oraz w sprawie substancji szczególnie szkodliwych dla środowiska wodnego (Dz.U. 2014 poz. 1800)

Radioactivity of waste materials coming from the largest power plants in Poland

Aneta Łukaszek-Chmielewska[1], *Martin* Girard[1], *Barbara* Piotrowska[2], *Karol* Wojtkowski[2,*], *Krzysztof* Isajenko[2], *Olga* Stawarz[2], and *Robert* Godlewski[3]

[1]The Main School of Fire Service, ul. Słowackiego 52/54, 01-629 Warszawa, Poland
[2]Central Laboratory for Radiological Protection, Department of Dosimetry, ul. Konwaliowa 7, 03-194 Warszawa, Poland
[3]Warsaw University of Technology, Faculty of Physics, ul. Koszykowa 75, 00-662 Warszawa, Poland

Abstract. The Central Laboratory of Radiological Protection together with 30 other laboratories in Poland is involved in the analysis of radioactivity among indirect products of combustion (ashes and slags) coming from polish power plants and thermal power station. Systematic studies are conducted on the concentrations of natural radioactive isotopes in waste materials obtained from the power industry, and their results are recorded in a database. This database is supervised by the Central Laboratory for Radiological Protection. The database already contains 46300 by now and the oldest results come from the eighties. In this article the results of studies on radioactive concentration of natural radioisotopes ^{40}K, ^{226}Ra and ^{228}Th in waste materials coming from the largest power plants in our country are presented. Furthermore an analysis was carried out on the possibility to use the selected waste materials (ashes and slags) in housing construction, road construction and in public facilities buildings, considering the values of activity indexes f_1 and f_2.

1 Introduction

The largest power plants active in Poland are: Bełchatów (5354 MW), Kozienice (2193 MW), Turów (2106 MW), Połaniec (1800 MW), Rybnik (1775 MW), Jaworzno III (1535 MW), Opole (1492 MW), Dolna Odra (1362 MW) and Łaziska (1155 MW) [1].

The main task of those power plants is the production of electric and thermal energy through the combustion of hard coal or lignite.

As a result of the coal combustion, energetic waste called indirect products of combustion (UPS) are produced such as: slags, fly ashes, ash-slag mixtures, microspheres, etc. About 24 million tons of such waste is produced annually [2]. Despite the fact, that for years we have been seeking renewable sources of energy, forecasts aren't optimistic for the coming years, as they predict a coal consumption reaching around 10 560 million tons in 2030. If we take into account, that 60% of this coal is dedicated to the production of electric

* Corresponding author: karol_wojtkowski@go2.pl

energy, it is estimated that there will be an increase in the amount of UPS to approximately 950 million tons annually [3].

In the eighties of 20th century, the level of economic use was only 30%, while it reaches now 70%, but it isn't identical for all groups of waste. In the case of ash-slag mixtures the level of economic use is around 50%, for slags it's over 88% and for fly ashes it reaches even 96% [4]. Therefore, at the present time, energy waste is not treated as troublesome production waste anymore, but as a valuable raw material used in various industries such as construction, mining or agriculture.

One of the main industries using ashes is the cement industry. Ashes act as aggregates and binders in the production of cellular concrete and is partly a replacement to cement. In addition, ashes are used as an active additive – ground together with Portland clinker, it affects the change of usable characteristics of the binder and leads to the formation of pozzolanic cement. It should also be mentioned that ashes increase the weight of cement, and improve also the resistance of concrete to aggressive factors, which prevents their corrosion and reduces their shrinkage. Siliceous fly ashes introduced in the amount of 20% of the cement mass results in a significant improvement of concrete resistance to cracking [5]. Fly ashes are currently used as an additive to multicomponent cements, drilling cements and conventional, high-value and self-compacting cements [6–12].

Ashes and slags are also used for road embankments, soil stabilization and to improve the base and sub-base layers [2, 5, 13].

In addition to the construction industry, fly ashes are also used in mining. They are used to fill underground cavities by injecting them under the pressure of 0.05–0.2 MPa. Additionally, ashes are used to "shield" coal heaps from fire by covering them with a sealing layer [2, 14].

Furthermore, fly ashes are used in agriculture, where it plays a role of calcium and magnesium fertilizer. Fly ashes obtained from lignite accelerate the alkalization process, as well as increase the absorption capacity and absorbency to water of the soil and also reduced its density [2, 15].

In the plastics and paint industry, fly ashes play the role of fillers, which significantly reduce the costs of plastic production, improve their mechanical properties and reduced the flammability of the materials obtained [2].

Bearing in mind the widespread use of indirect products of combustion in various branches of the industry, the question of where UPS are safe materials for people arises. It must be emphasized, that ash as a waste material with mineral origins contains in its composition natural radioactive elements such as: potassium ^{40}K, radium ^{226}Ra and thorium ^{228}Th. In order to take a decision to allow the use of building materials in selected types of constructions, measurements of their radioactivity level are necessary.

The basic legal act defining the requirements for raw material and construction materials used in various types of construction is the regulation from the Council of Ministers of January 2, 2007 (Dz. U. nr 4/2007, pos. 29) [16].

This regulation classifies the raw materials and building materials considering their use in various types of construction by determining two parameters:

1. Activity index f_1, which takes into account the content in natural radioactive isotopes, is defined by the equation:

$$f_1 = \frac{S_K}{3000} + \frac{S_{Ra}}{300} + \frac{S_{Th}}{200} \quad (1)$$

where: S_K, S_{Ra}, S_{Th} are the radioactive concentrations of the following natural elements: potassium (^{40}K), radium (^{226}Ra) and thorium (^{228}Th), in Bq/kg.

The value of activity index f_1 gives information on the exposure of the human body to gamma radiation emitted by natural radionuclides.

2. Activity index f_2, which determines the content of radium (^{226}Ra), being the parent isotope of radon. Therefore, the value of this index determines the degree exposure to alpha radiation coming from radon (it is the index of lung epithelium exposure to alpha radiation emitted by radon decay products inhaled with air by humans):

$$f_2 = S_{Ra} \quad (2)$$

2 Method of measurement

The studies of radioactivity of waste materials were performed using the MAZAR type and AZAR type analyzers combined with NaI(Tl) scintillation probe. These analyzers work in three measuring ranges and let determine the radioactive concentrations of natural elements such as: ^{40}K, ^{226}Ra and ^{228}Th. Based on the determined concentrations, the activity indexes f_1 and f_2 can be calculated. The measuring ranges of the analyzers are contained in the following energy ranges: range for ^{40}K – from 1.26 MeV to 1.65 MeV, range for ^{226}Ra – from 1.65 MeV to 2.30 MeV and range for ^{228}Th – from 2.30 MeV to 2.85 MeV.

Calibration of the detector efficiency is based on measurements of three volume calibration references: ^{40}K, ^{226}Ra and ^{228}Th and measurement of the matrix reference (background measurement). The geometry of the reference sources are Marinelli beakers with a capacity of 1.5 dm^3. The density of the reference sources equals 1.6 g/cm^3, however the density of the measured samples of raw and building materials is in the range from 0.6 g/cm^3 to 2.0 g/cm^3.

In order to minimize the external background, the detector is placed in a protective casing made of lead with a wall thickness of 50 mm. Shredded and sieved through a sieve with a mesh thickness of 2 mm, the samples are packed into Marinelli beakers with a capacity of 1.5 dm^3 and then sealed tightly. Samples are studied only once, after minimum 5 days, when the radioactive equilibrium has been reached between radium ^{226}Ra and bismuth ^{214}Bi as well as between thorium ^{228}Th and thallium ^{208}Tl [16]. Measurements were made in series of multiple repetitions (from 5 to 32) [17].

3 Results of radioactivity studies

In tables 1 and 2, we present the mean values and ranges of radioactive concentration in ash and slag samples coming from the largest power plants in Poland during the years 1980–2015 for the following radioactive natural isotopes: potassium (^{40}K), radium (^{226}Ra) and thorium (^{228}Th).

Mean values of radioactive concentrations of natural isotopes ^{40}K, ^{226}Ra and ^{228}Th for ashes for the whole Poland are: 703 Bq/kg, 120 Bq/kg and 91 Bq/kg, respectively [18]. Therefore, in the case of potassium isotope the mean values of concentration in Poland were not exceeded for the following power plants: Jaworzno III, Turów and Bełchatów. However when comparing the mean values of radium concentration for Poland with the studied power plants, it can be concluded that lower values than the mean value of ^{226}Ra were only obtained in the Łaziska power plant. Considering the mean value of thorium isotope concentration, which is 91 Bq/kg for Poland, it can be stated, that it wasn't exceeded only in the Bełchatów power plant.

The highest mean values of radioactive concentration of natural isotopes were obtained in ash samples coming from Rybnik for ^{40}K, Turów for ^{226}Ra and Połaniec for ^{228}Th.

Table 1. Radioactive concentrations of ^{40}K, ^{226}Ra and ^{228}Th in ash samples coming from the largest power plants in Poland.

Power plant	**Radioactive concentration [Bq/kg]**					
	^{40}K		**^{226}Ra**		**^{228}Th**	
	Mean	Range	Mean	Range	Mean	Range
Łaziska	788	594–954	107	16.9–183.3	103	7.2–128.8
Dolna Odra	834	692–916	131	11.0–178.9	98.3	92–116.8
Opole	780	158–1053	136	7.8–237.5	113	5.5–141.5
Jaworzno III	693	242–869	140	23.0–182.3	9.2	40.0–130.0
Rybnik	904	830–1043	128	98.0–154.6	113	104–119
Połaniec	767	470–911	153	79.0–203.4	117	72.0–137.0
Turów	450	378–529	202	185–216	106	95.0–115.0
Kozienice	799	468–1024	127	89.5–189.0	90.9	65.7–130.0
Bełchatów	74	70–80	150	83.0–209.0	53.3	37.0–71.0

The lowest mean values of radioactive concentration of ^{40}K and ^{228}Th were obtained for ash samples coming from the Bełchatów power plant, and the lowest value of radioactive concentration of ^{226}Ra was noticed in ashes from the Łaziska power plant [19].

Table 2. Radioactive concentrations of ^{40}K, ^{226}Ra and ^{228}Th slag samples coming from the largest power plants in Poland.

Power plant	**Radioactive concentration [Bq/kg]**					
	^{40}K		**^{226}Ra**		**^{228}Th**	
	Mean	Range	Mean	Range	Mean	Range
Łaziska	560	118–807	87.1	32.0–130.0	78.8	16.0–107.0
Dolna Odra	861	680–1260	89.9	52.1–113.0	94.0	76.1–161.0
Opole	552	339–939	115	32.3–183.1	82.7	43.2–105.9
Jaworzno III	505	191–878	88.3	38.0–156.9	73.8	20.4–144.4
Rybnik	743	669–828	117	104–124	87.4	74.1–101.1
Połaniec	763	592–883	101	78.0–117.0	91.1	65.0–106.0
Turów	385	161–695	86.1	37.0–127.0	60.4	34.0–98.0
Kozienice	706	542–806	115	77.7–148.2	79.7	56.9–92.0
Bełchatów	52	39–64	161	102–219	59.0	41.0–77.0

The highest mean values of radioactive concentrations of ^{40}K and ^{228}Th were obtained for slag samples coming from the Dolna Odra power plant. The highest mean value of radioactive concentration of ^{226}Ra was obtained for slags from the Bełchatów power plant.

The lowest mean values of radioactive concentrations of ^{40}K and ^{228}Th were obtained for slags delivered by the Bełchatów power plant, while the lowest mean value of radioactive concentration of ^{226}Ra was obtained for the Turów power plant.

4 Assessment of possibilities to use ashes and slags in building construction

In order to assess the possibility to use waste materials in a given type of construction, the activity indexes f_1 and f_2, must be calculated and then the studied materials must be classified in accordance to the guidelines contained in the regulation of the Council of Ministers of January 2, 2007.
Mean values of activity index f_1 for ash samples coming from the largest power plants in Poland are presented in Figure 1. The red line shows the limit of activity index f_1 = 1.2 determined for construction intended for the stay of people and livestock.

The values of the activity index f_1 describing the exposure of body to gamma radiation coming from natural radionuclide, are contained in the range from 0.79 to 1.36. Most of them exceed the accepted values from the regulation [16] concerning raw materials used in building intended for human stays and livestock by more than 20 %.

The lowest mean value of activity index f_1 = 0.79 was obtained for ashes coming from the Bełchatów power plant and was about 34 % lower than the limit of activity index (f_1 = 1.2). However the highest mean values of activity index f_1 = 1.36 was obtained in the ashes delivered by the Turów power plant and was about 13.3 % higher than the limit value. For four power plants (Opole, Rybnik, Połaniec and Turów) the mean value of activity index f_1 was noticed to be higher than the limit value determined for the housing construction of 6.6 %, 7.5 %, 12.5 % and 13.3 %, respectively. For comparison, the mean value of activity index in Poland is 1.08 [19].

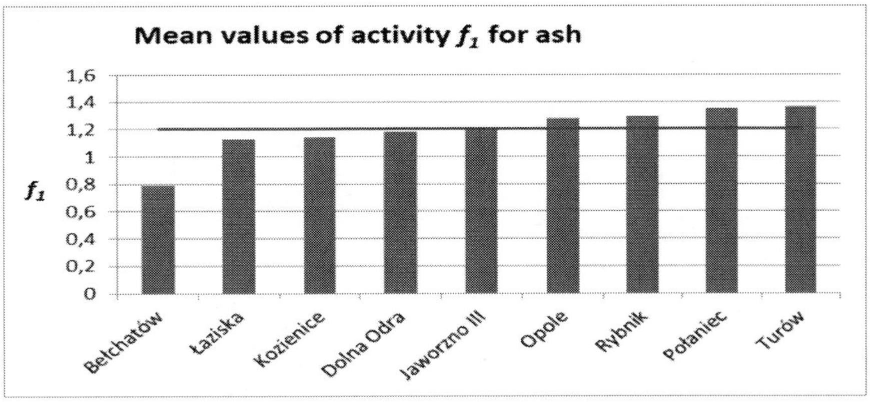

Fig. 1. Mean value for the activity index f_1 for ash samples coming from the largest power plants in Poland.

Mean values of activity index f_2 for ashes coming from the largest power plants in Poland are presented in Figure 2. The red line shows the limit of activity index f_2 = 240 Bq/kg determined for housing construction.

Mean values of the activity index f_2 for ashes ranges from 107 Bq/kg (Łaziska power plant) to 202 Bq/kg (Turów power plant). For all the ash samples, the mean value of activity index f_2 obtained are below the limit determined for the housing construction (Fig. 3). The lowest mean value of activity index f_2 was obtained for the ash samples coming from Łaziska power plant and was about 55.5% lower than the limit value of activity index f_2 = 240 Bq/kg. Mean values of the activity index f_1 for slag samples coming from the largest power plants in Poland are presented in Figure 3. The red line shows the limit for activity index f_1 = 1.2 determined for construction intended for human stays and livestock.

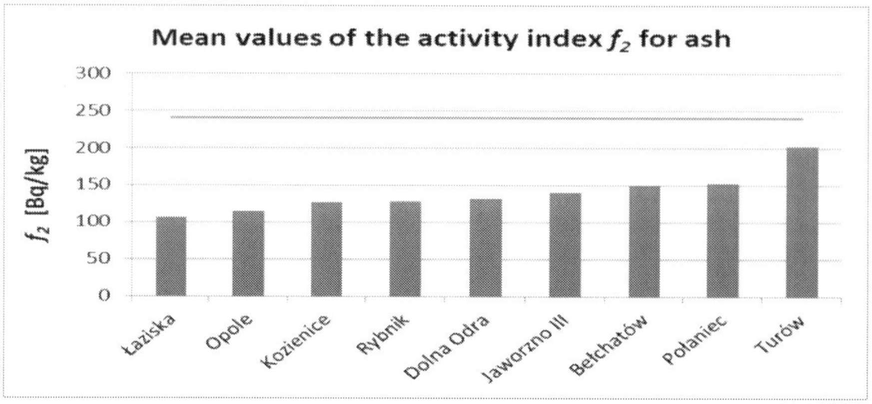

Fig. 2. Mean values of the activity index f_2 for ash samples coming from the largest power plants in Poland.

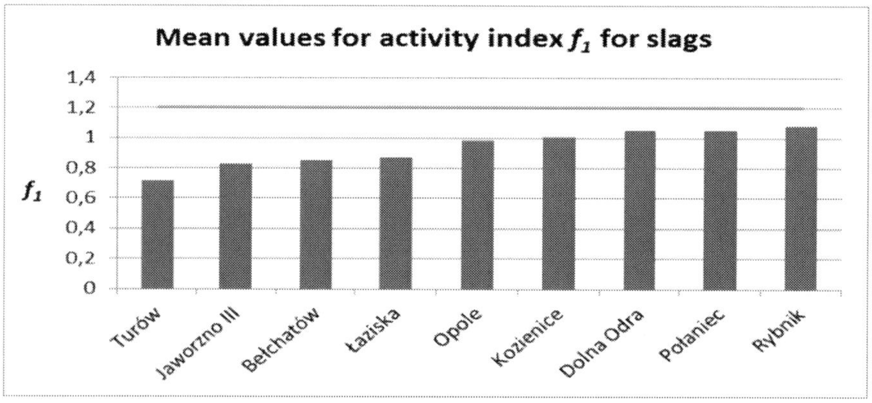

Fig. 3. Mean values of activity index f_1 for slag samples coming from the largest power plants in Poland.

The values of activity index f_1 are contained in the range from 0.72 to 1.08 and do not exceed by more than 20 % the values accepted in the regulation [16] for raw materials used to produce building materials intended to the housing or livestock construction.

The lowest mean value of the activity index f_1 for the slag samples was obtained for the Turów power plant (f_1 = 0.72), while the highest mean value of the activity index f_1 was measured for the Rybnik power plant (f_1 = 1.08). In the case of slag samples, it can be seen that the limit value of activity index f_1 = 1.2 wasn't exceed, therefore waste can be used in the housing construction.

Mean values of activity index f_2 for slag samples coming from the largest power plants in Poland are presented in Figure 4. The red line shows the limit for activity index f_2 = 240 Bq/kg. determined for housing construction.

The range of values of activity index f_2 for the studied slag samples is between 86.1 Bq/kg (Turów power plant) and 161 Bq/kg (Bełchatów power plant). As can be seen from the data presented in Fig. 5, the values of activity index f_2, giving information about the content in radium in the studied materials, are below the limit value determined for the housing construction. Therefore, in this case as well all the studied samples meet the requirements presented in the regulation from the Council of Ministers of January 2, 2007 "on the requirements for the content of natural potassium K-40, radium Ra-226 and thorium Th-228 in raw materials used in buildings intended for human stays and livestock, as well as the content in industrial waste used in construction and control of the content of these isotopes."

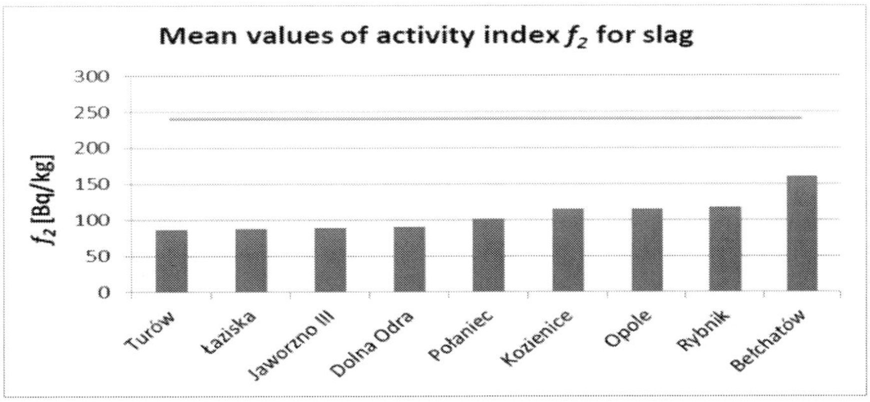

Fig. 4. Mean values of activity index f_2 for slag samples coming from the largest power plants in Poland.

5 Conclusions

Based on the results of measurements taken in the years 1980 – 2015 on ashes and slags samples, it can be observed that in most of the cases the concentration of potassium was a few times higher than the concentrations of radium and thorium (Tables 1 and 2) [20].

Analyzing the values of all indexes, both the ones giving information about the exposure of human body to gamma radiation emitted by natural radionuclide like ^{40}K, ^{226}Ra and ^{228}Th, defined by f_1, and the ones describing the degree of exposure to alpha radiation coming from radon, defined by f_2, it can be stated that the slag samples are characterized by much lower values of activity indexes f_1 and f_2 comparing to ash samples.

Based on the determined activity indexes f_1 and f_2, it can be observed, that all the slag samples meet the requirements described in the regulation of the Council of Ministers of January 2, 2007 and can be used in all types of constructions. In the case of ashes, the limit value of activity index f_1 for the housing construction was exceeded in most of the samples, but ashes can still be used in other types of constructions.

The lowest mean value of activity index f_1 was obtained for ashes coming from the Bełchatów power plant. However for the slag samples delivered by the largest power plants in Poland, the lowest radioactive concentrations of natural isotopes was found for the Turów power plant. Analyzing the content of radium defined by the activity index f_2 in ashes and slag samples, the lowest mean values were observed in Łaziska power plants for ashes and in Turów power plant for slags. All mean values of activity index f_2 are below the limit determined for buildings intended for human stays and livestock.

A permanent control of building materials coming from the industry insures, that the use in construction of combustion waste such as ashes and slags is safe in regard to the radiological effects to humans health.

References

1. http://geoland.pl/dodatek/energia-xxv/efektywnosc-konkurencyjnosc-ekologia/ (access: 27.09.2016)
2. G. Borowski, IE **22**, 52–62 (2010)
3. J. J. Hycnar, *EUROCOALASH weryfikacja popiołów ze spalania węgla* (ENP, 2009)
4. K. Galos, A. Uliasz-Bocheńczyk, GSM **1**, 21, 23–42 (2005)
5. G. L. Golewski, CWB **2**, 106–114 (2015)
6. E. Tkaczewska, J. Małolepszy, CWB **1**, 26 (2009)
7. W. Kurdowski, A. Bochenek, CWB **6**, 434 (2012)
8. W. Kurdowski, *Dodatki mineralne do cementu a trwałość betonu* (Mon. **106**, WPK, 1990)
9. G. L. Golewski, P. Golewski, T. Sadowski, CMS **62**, 75 (2012)
10. Z. Giergiczny, J. Małolepszy, J. Szwabowski, J. Śliwiński, *Cementy z dodatkami mineralnymi w technologii betonów nowej generacji* (Górażdże Cement, 2002)
11. J. Bensted, J.R. Smith, CWB **1**, 17 (2008)
12. J. Filipiak, ROŚ **15**, 1153 (2013)
13. J. Kukiełka, *Wykorzystanie popiołów lotnych do budowy dróg*, E **4**, 8–11 (1998)
14. Z. Chrzanowski, *Formalno-prawne zasady wykorzystania popiołów do makroniwelacji i rekultywacji wyrobisk* (KNT, Rekultywacja terenów zdegradowanych w województwie szczecińskim, Nowe Czarnowo, 1995)
15. Z. Piotrowski, A. Uliasz-Bocheńczyk, GSM **24**, 2/1 (2008)
16. Rozporządzenie Rady Ministrów z dnia 2 stycznia 2007 roku w sprawie wymagań dotyczących zawartości naturalnych izotopów promieniotwórczych potasu K-40, radu Ra-226 i toru Th-228 w surowcach i materiałach stosowanych w budynkach przeznaczonych na pobyt ludzi i inwentarza żywego, a także w odpadach przemysłowych stosowanych w budownictwie oraz kontroli zawartości tych izotopów, Dz. U. **4/2007**, poz. 29.
17. K. Isajenko, B. Piotrowska, O. Stawarz, Ł. Obigowski, A. Łukaszek-Chmielewska, ZN SGSP **60**, 4 (2016)
18. A. Żak, K. Isajenko, B. Piotrowska, M. Kuczbajska, A. Ząbek, T. Szczygielski, Nukl. **55**, 3, 387–391 (2008)
19. B. Piotrowska, M. Fujak, K. Isajenko, S. Krawczyńska, BiTP **44**, 4, 127–134 (2016)
20. K. Kovler, CBM **29**, 158 (2012)

Measurements of natural radioactivity in soil samples collected in the Kampinoski National Park

Aneta Łukaszek-Chmielewska[1], *Martin* Girard[1], *Olga* Stawarz[2], *Barbara* Piotrowska[2], *Karol* Wojtkowski[2,*], and *Krzysztof* Isajenko[2]

[1]The Main School of Fire Service, ul. Słowackiego 52/54, 01-629 Warszawa, Poland
[2]Central Laboratory for Radiological Protection, Department of Dosimetry, ul. Konwaliowa 7, 03-194 Warszawa, Poland

Abstract. In this article are presented the results of measurements of natural radioactivity in twelve soil samples from the Kampinoski National Park by using high purity germanium detector (HPGe). The average values of the natural radionuclides ^{40}K, ^{226}Ra and ^{228}Ac concentrations in the soil samples are 8.54, 6.65 and 206 Bq/kg respectively. Additionally radiation hazard indices were calculated to evaluate the radiological risk for the public and environment. The results show that the mean values of radium equivalent activity (Raeq) and gamma absorbed dose rate (D) in the air, outdoor annual effective dose equivalent (E_{out}), and representative level index (I_γ) for analyzed samples were lower than the limit recommended by the United Nations Scientific Committee on the Effect of Atomic Radiation (UNSCEAR). This study is the first to evaluate the radiological impacts in the investigation area.

1 Introduction

Ionizing radiation has always been present in human life. It is an integral part of the natural environment. At first, ionizing radiation was coming exclusively from natural sources (described in the literature on natural background radiation) including: cosmic radiation, terrestrial radiation and internal radiation [1].

The intensity of cosmic radiation, mainly coming from the Sun, depends on the height above sea level. The average radiation dose for the general population coming from cosmic rays is 0.33 mSv/year [1].

The main sources of terrestrial radiation are radioactive elements presents in the earth's crust. Among most important ones, we can list uranium (^{238}U), actinium (^{228}Ac) and potassium (^{40}K) and their decay products. The intensity of terrestrial radiation isn't homogenous around the world. It depends on the geological structure of the region and ranges from 0.3 mSv/y to 1.5 mSv/y (average 0.55 mSv/y) [1]. There are places on earth

* Corresponding author: karol_wojtkowski@go2.pl

where the terrestrial radiation is much more intense due to the presence of monazite rocks, which contain high concentration of uranium. The highest natural radiation dose was measured in the mountains of Brazil and reached 120 mSv/year. High natural radiation background doses are also present in the southwest coast of India (due to the presence of monazite sand), in Iran (in Ramsar and Mahallat) as a result of a high concentration of radium in rocks and soils as well as radium in the air, in the US, in Canada and in other countries [1–3].

The next component generating natural radiation background is the internal radiation, whose main source comes from radioactive elements introduced into the body through inhalation(radium and its decay product) or food. Mean value of internal radiation is 0.17 mSv/y [1].

Scientific studies revealed, that around 99% of the radioactive precipitations gets stocked in soils [8] because soils act as a buffer accumulating radioactive substances coming from natural sources as well as anthropogenic ones [4]. Studies show, that the mean concentration of radium (^{226}Ra) in polish soils is 25 Bq/lg while the worldwide mean value of ^{226}Ra concentration reaches 35 Bq/kg. Nevertheless some areas of Poland are also rich in radium including Sudety, Jakuszyce, Szklarska Poręba, Karpacz. The maximum concentration of ^{226}Ra in these areas is 132 Bq/kg. In the case of potassium (^{40}K) isotope, mean value in polish soils and worldwide mean value are very similar, reaching 408 Bq/kg for Poland and 400 Bq/kg worldwide. The concentration of actinium isotope (^{228}Ac) is lower in soils of Poland. Mean concentration of this isotope is 23.3 Bq/kg for Poland while worldwide mean concentration is 30 Bq/kg [5].

Kampinoski National Park (KPN) is a popular region among inhabitants of the Warsaw agglomeration and the localities nearby due to its valuable natural values, numerous patriotic memorials and touristic tracks. No radioactive studies of these soils has been carried out yet. The knowledge of radioactivity level in soils samples will provide people living nearby and visitors of KPN relevant information concerning radiological effects for their health.

2 Materials and methods

2.1 Study area

Kampinoski National Park (KPN) is of unique natural value, not only for Poland but for the whole Europe. It covers an area of 38544 ha and is situated in the region of the Vistula river valley terraces, in the basin of the Łasica river in the central part of the Warsaw valley extending from Góra Kalwaria to Wyszogród. The area of the Park is made up of: forest land (28219 ha), agricultural land (7805 ha), water (154 ha), ecological land (1484 ha) and other land occupying an area of 886 ha. In 2000, the park together with the protected zone was recognized as biosphere reserve "Puszcza Kampinoska". In 2004, however, the park was classified as a Natura 2000 site. The area of the Park is characterized by the occurrence of parallel bands of dunes areas separated by marshy depressions. It is a typical ice-marginal valley area, which was created by waters flowing into the sea through the wide riverbed of the Prawisła in post-glacial period. KPN surface is composed of quaternary sediments, including: preglacial sands and gravels, boulder clays and water-glacial sediments of three glaciations and river sediments of two interglacial periods, with a thickness of 20 m to 50 m. Pliocene deposits are formed by compact clays, sandy and silty muds, fine-grained sands and dusty sands that usually form lenses with a small range and thickness [6–12].

2.2 Field sampling

Soil samples were collected between June and July 2017 from 12 collection points situated in the Kampinoski National Park area. They were collected from a depth of 0–10 cm using a hand probe. Geographical coordinates of collection points for materials used in further analysis are presented in Table 1.

Table 1. Geographical coordinates of soil samples collection points in Kampinoski National Park.

Location of samples collection	Longitude	Lattitude
Wilcze Śladowskie	20° 18,447'	52° 21,549'
Bór Kazuński	20° 37,986'	52° 24,148'
Palmiry - Łomna Las	20° 46,894'	52° 21,820'
Brzozówka	20° 39,702'	52° 20,668'
Dziekanów Leśny	20° 50,421'	52° 20,407'
Famułki Brochowskie	20° 21,469'	52° 18,627'
Górki	20° 31,347'	52° 18,700'
Posada Sieraków	20° 47,784'	52° 19,102'
Granica	20° 26,911'	52° 17,385'
Julinek	20° 35,720'	52° 16,676'
Mariewskie Łąki	20° 43,981'	52° 17,728'
Opaleń	20° 52,267'	52° 18,179'

2.3 Laboratory works

Soil samples were collected with special cylindrical probes with a diameter of 6.9 cm and a height of 10 cm as follows: 6 portions from points situated on the perimeter of a circle with a radius of 1 m and 1 portion from the center of this circle. Then the sample portions were placed in a bag to mix them in order to obtain a homogeneous representation of the given area. In the laboratory, soil samples were dried first at room temperature, then during 16 h at 105°C. They were then weighed to get the total mass. Once dried, samples were crushed, sieved through 2 mm sieves and placed in Marinelli Beakers tightly sealed. The prepared material was then set aside for one month in a safe place to obtain radioactive equilibrium between ^{238}U and ^{228}Ac and their product decays.

Research on soil samples were conducted on a Canberra gamma ray spectrometer. The set consisted of HPGe coaxial detector with an efficiency of 25% working together with an analyzer, using special software. The measurement time of tested samples was 80 000 s. The spectrometer provided a spectrum analysis for gamma ray for photons with energies in the range from 10 to 1800 keV. The spectral resolution for ^{60}Co ($E\gamma$ = 1330 keV) is 1.8 keV. Detector was placed in a lead protective casing ensuring the reduction by two orders of magnitude of the background coming from external gamma radiation [13].

3 Measurements results and analysis

Measurements of natural radionuclides concentrations in the soils samples were done based on the observed spectral lines and for each sample the concentrations of radionuclide were determined in Bq/kg (Table 2).

Table 2. Activity concentrations of natural radionuclides (^{40}K, ^{226}Ra and ^{228}Ac) [Bq/kg] in Kampinoski National Park soil samples.

Location of samples collection	**Activity concentrations [Bq/kg]**		
	^{40}**K**	^{226}**Ra**	^{228}**Ac**
Wilcze Śladowskie	188±7	4.7±0.5	4.5±0.2
Bór Kazuński	154±6	7.4±0.5	5.0±0.2
Palmiry - Łomna Las	140±5	5.7±0.5	3.6±0.2
Brzozówka	230±9	12.6±0.7	12.0±0.4
Dziekanów Leśny	154±6	6.9±0.5	5.1±0.2
Famułki Brochowskie	196±8	4.8±0.4	4.2±0.2
Górki	174±7	6.9±0.5	4.5±0.2
Posada Sieraków	230±9	8.3±0.6	7.0±0.2
Granica	205±8	9.0±0.6	6.8±0.2
Julinek	252±10	10.5±0.6	8.5±0.3
Mariewskie Łąki	269±10	9.6±0.6	7.3±0.3
Opaleń	275±11	16.1±0.7	11.2±0.2
Mean	**206±13**	**8.5±1.0**	**6.7±0.8**
UNSCEAR 2000 [14]	**400**	**30.0**	**35.00**

The highest value of activity of potassium (^{40}K) was in Opaleń – tourist glade, being equal to 275 Bq/kg, while the lowest value of activity of ^{40}K was in Palmiry-Łomna Las, equal to 140 Bq/kg with mean value of 206 Bq/kg. The range of ^{40}K concentration in Poland is from 45 Bq/kg to 1065 Bq/kg. The present results have shown that values of specific activity of potassium in KPN were less than its recommended value given by UNSCEAR 2000 (equal to 400 Bq/kg) [14, 15]. Mean value of natural potassium in polish soils is equal to 0.68%, while maximum is 2.8%. The highest concentration of this isotope is in places reach in loess and granite like Sudets, Carpathians and North – East Poland [16].

The highest value of concentration of radium (^{226}Ra) was in Opaleń – tourist glade, being equal to 16.06 Bq/kg, while the lowest value of activity of ^{226}Ra was in Wilcze Śladowskie, equal to 4.74 Bq/kg with mean value of 8.54 Bq/kg. The present results have shown that values of activity of radium in KPN were less than its recommended value given by UNSCEAR 2000 (equal to 30 Bq/kg) [14, 15]. Radium ^{226}Ra concentration in soil in Poland is in the range of 4.6 Bq/kg to 128.0 Bq/kg [16]. I normal conditions the content of radium in environment is little and closely related with the content of uranium ^{238}U (parent) Extraction of uranium ores and other minerals, as well as the application of phosphate fertilizers with an increased amount of radium, is the reason for the increase this isotope in soil, in ground and surface water as well as in flora and fauna [16].

The highest value of activity of ^{228}Ac was in Brzozówka being equal to 12.03 Bq/kg, while the lowest value of activity of ^{228}Ac was in Palmiry-Łomna Las, equal to 3.62 Bq/kg, with mean value of 6.65 Bq/kg. The present results have shown that values of activity of ^{228}Ac in KPN were less than its recommended value given by UNSCEAR 2000 (equal to 35 Bq/kg) [14, 15].

Furthermore radiological hazards indexes were defined for all the studied soil samples in order to determine the changes in ionizing radiation level and the exposure for humans health resulting from the presence of natural radionuclide in soils.

Table 3 shows values of radium equivalent (Ra_{eq}), level index (I_γ) for soil samples in Kampinoski National Park, absorbed gamma dose rate (D) and outdoor annual effective close equivalent (E_{out}).

Table 3. Values of radium equivalent (Ra_{eq}), level index (I_γ) for soil samples in Kampinoski National Park, absorbed gamma dose rate (D) and outdoor annual effective dose equivalent (E_{out}).

Location of samples collection	Radiological hazards indexes			
	Ra_{eq} [Bq/kg]	**$I\gamma$ [Bq/kg]**	**D [nGy/h]**	**E_{out} [mSv/y]**
Wilcze Śladowskie	25.65	0.10	12.91	0.016
Bór Kazuński	26.38	0.10	12.88	0.016
Palmiry - Łomna Las	21.64	0.08	10.70	0.013
Brzozówka	47.48	0.18	22.75	0.028
Dziekanów Leśny	26.06	0.10	12.75	0.016
Famułki Brochowskie	25.91	0.10	13.10	0.016
Górki	26.76	0.10	13.24	0.016
Posada Sieraków	36.07	0.14	17.82	0.022
Granica	34.49	0.13	16.89	0.021
Julinek	42.03	0.16	20.60	0.025
Mariewskie Łąki	40.76	0.16	20.22	0.025
Opaleń	53.25	0.20	25.66	0.031
Mean	**33.87**	**0.13**	**16.63**	**0.020**
UNSCEAR 2000 [14]	**≤370**	**≤1**	**55**	**≤1**

On this basis, the obtained values of radiological hazard indexes were analyzed with the reference values determined by the UNSCEAR scientific committee.

Radium equivalent activity index is the most widely used as a radiation hazard index associated with the three radionuclides of ^{40}K, ^{226}Ra and ^{228}Ac. Ra_{eq} is a weighted sum of activities of these radionuclides based on the estimation that 4810 Bq/kg of ^{40}K, 370 Bq/kg of ^{226}Ra and 259 Bq/kg of ^{228}Ac produce the same γ-ray dose rates. Ra_{eq} is given by the following equation [17]:

$$Ra_{eq}\left[\frac{Bq}{kg}\right] = A_{Ra} + 1.43A_{Ac} + 0.077A_K \qquad (1)$$

where: A_K, A_{Ra} and A_{Ac} are the activity concentrations (Bq/kg) of ^{40}K, ^{226}Ra and ^{228}Ac. When the value of Ra_{eq} is 370 Bq/kg, this corresponds to 1 mSv/y [14, 17]. The radium equivalent concept allows a single index or number, which is a widely used hazard index to describe the gamma output from different mixtures of potassium, radium and actinium in the soil samples from different locations. The calculated values vary from 21.64 Bq/kg (Palmiry-Łomna Las) to 53.25 Bq/kg (Opaleń) with mean 33.84 Bq/kg (Tab 3.). These values are lower than the permissible maximum value of 370 Bq/kg [14, 17].

The second one was representative level index (I_γ) which is used to estimate the level of gamma radiation hazard .The representative level index must be less than unity to keep the radiation hazard insignificant for the public. The following equation was applied to calculate I_γ for soil samples under investigation [18]:

$$I_\gamma\left[\frac{Bq}{kg}\right] = \frac{A_{Ra}}{300} + \frac{A_{Ac}}{200} + \frac{A_K}{3000} \tag{2}$$

where: A_K, A_{Ra} and A_{Ac} are the specific activities of ^{40}K, ^{226}Ra and ^{228}Ac in Bq/kg, respectively. I_γ varies from 0.08 Bq/kg (Palmiry-Łomna Las) to 0.20 Bq/kg (Opaleń) with a mean value of 0.13 Bq/kg as shown in Table 3. The values of representative level index for all soil samples was lower than unity [14, 18].

The index of the absorbed gamma dose rate D [nGy/h] in air at 1 m above the ground surface for the uniform distribution of radionuclides was calculated based on guidelines provided by [14, 19]:

$$D\left[\frac{nGy}{h}\right] = 0.427A_{Ra} + 0.623A_{Ac} + 0.043A_K \tag{3}$$

where: A_K, A_{Ra} and A_{Ac} are the activity concentrations [Bq/kg] of ^{40}K, ^{226}Ra and ^{228}Ac, respectively, in the samples. The absorbed dose rate expresses the received dose in the open air from the radiation emitted from radionuclide concentrations in environmental materials. It is observed that the total absorbed dose rate resulting from ^{40}K, ^{226}Ra and ^{228}Ac ranges between 10.70 Bq/kg (Palmiry-Łomna Las) to 25.66 Bq/kg (Opaleń), and the mean value was 16.63 Bq/kg, which was lower than the limits recommended value of 55 nGy/h given by the International Commission on Radiological Protection [20, 21].

The outdoor annual effective dose equivalent (E_{out}) was calculated from the absorbed dose by applying the dose conversion factor of 0.7 Sv/y with an occupancy factor of 0.2 for outdoor [14, 19]:

$$E_{out}\left[\frac{mSv}{y}\right] = D\left[\frac{nGy}{h}\right] \cdot 8760h \cdot 0.7\left[\frac{Sv}{y}\right] \cdot 0.2 \cdot 10^{-6} \tag{4}$$

To estimate annual effective doses, conversion coefficient should be taken into account from absorber dose in air to effective dose and the indoor occupancy factor. Based on the present results, values of outdoor annual effective dose equivalent in Kampinoski National Park were less than the recommended value of 1 mSv/y given by UNSCEAR 2000 [14].

4 Conclusion

The natural radioactivity levels of ^{40}K, ^{226}Ra and ^{228}Ac have been measured in soil in Kampinoski National Park using gamma ray spectroscopy. The mean activity concentrations of ^{40}K, ^{226}Ra, ^{228}Ac and in soil are 8.54 Bq/kg, 6.65 Bq/kg and 206 Bq/kg.

The results of the determination parameters such as the radium equivalent activity (Ra_{eq}), representative level index (I_{γ}), absorbed gamma dose rate (D) and outdoor annual effective dose equivalent (E_{out}) have been found to be lower than their corresponding allowed limits, hence they will pose relatively no serious health risk.

This work has established background guideline on the natural radioactivity levels in the study area, which will provide a future reference for other studies.

References

1. A. Kuliczkowski, R. Pluta, D. Zwierzchowski, MŚRŚ **1**, 47–50 (1993)
2. L. A. Najam, H. L. Mansour, N. F. Tawfiq, M. S. Karim, JRN, Appl. **1**, 25–30 (2016)
3. A. G. E. Abbady, M. A. Adel, A.El-.Taher, JER **84**, 1, 65–78 (2005)
4. S. Wołkowicz, R. Strzelecki, PG **50**, 10/2 (2002)
5. Mirosław Janik, Shinji Tokonami, Jpn. JHP **44**, 1, 116–121 (2009)
6. A. Andrzejewska, *Monitoring wód powierzchniowych i podziemnych KPN* ([w:] M. Sosnkowski, B. Gołecki (red.). *Raport o stanie środowiska w województwie stołecznym warszawskim w 1996 roku*, BMŚ, 1996)
7. A. Andrzejewska, *Klimat Puszczy Kampinoskiej* ([w:] R. Andrzejewski (red.), KPN, **1** , 41–68, 2003)
8. A. Andrzejewska, PK **3**, 8–11 (2008)
9. R. Andrzejewski, PK 2, 231–239 (2004)
10. S. Borek, *Niektóre właściwości fizyczne gleb Rezerwatu Granic w Kampinoskim Parku Narodowym*, (1983)
11. K. Konecka-Betley, *Wpływ działalności człowieka na środowisko glebowe w Kampinoskim Parku Narodowym* (IG SGGW-AR, KGiChR PAN, Warszawa, 1983)
12. T. J. Chmielewski, *Operat generalny planu ochrony Kampinoskiego Parku Narodowego* ([w:] *Plan Ochrony Kampinoskiego Parku Narodowego*, NFOŚ, Warszawa, 1996)
13. K. Konecka-Betley, *Geneza gleb Kampinoskiego Parku Narodowego* ([w:] B. Dobrzyński, K. Konecka-Betley (red.) *Wpływ działalności człowieka na środowisko glebowe w Kampinoskim Parku Narodowym*, Instytut Gleboznawstwa SGGW-AR, KGiChR PAN, Warszawa, 1983)
14. K. Isajenko, B. Piotrowska, O. Stawarz, A. Łukaszek-Chmielewska, S. Krawczyńska, BiTP **46**, 2 (2017)
15. United Nations Scientific Committee on the Effects of Atomic Radiation (UNSCEAR) Sources, ERIR, New York, United Nations (2000)
16. M. Abusini, K. Al-ayasreh, J. Al-Jundi, RPD **128**, 2, 213–216 (2003)
17. K. Isajenko, M. Fujak, B. Piotrowska, *Monitoring stężenia* ^{137}*Cs w glebie w latach 2012–2013* (2014)
18. H. M Dia, S. A. Nouh, A. Hamdy, S. A. EL-Fiki, NRP **3**, 1, 53–62 (2008)
19. A. Al-Taher, S. Makhluf, Upper Egypt, IJP & AP **48**, 697–702 (2010)
20. S. Sherif, D. Al-Othmany, H. Q. Hamidalddin, J. H. Al-Zahrani, W. R. Alharbi, H. M. Barashed, JGEP **5**, 66–75 (2017)
21. A. M. El-Arabi, IJP & AP **43**, 422–426 (2005)

Exhaust emissions of an LPG powered vehicle in real operating conditions

Jerzy Merkisz[1], *Dawid* Gallas[1,*], *Maciej* Siedlecki[1], *Natalia* Szymlet[1], and *Barbara* Sokolnicka[1]

[1]Poznan University of Technology, Department of Transport Engineering, ul. Piotrowo 3, 61-138 Poznań, Poland

Abstract. The paper contains exhaust emission results for an RDE test drive of a passenger vehicle with a spark ignition bi-fuel engine fuelled with LPG. The tests were performed in accordance with the RDE test requirements, and their results were compared to the emission limits of the applicable EURO norm. The aim of the paper is to investigate the discrepancies in the real exhaust emissions between the gasoline-specific emission norms and real values from vehicle use when powered by LPG.

1 Introduction

Due to the repeated problems of detecting and enforcing the exhaust emission limits on vehicle manufacturers changes to the test methods are introduced continuously. The exhaust emission norms for passenger cars have adopted emission tests in real driving conditions, known as RDE (*Real Driving Emissions*) tests in the recent years. These tests are to increase the reliability of emission data obtained for tested vehicles undergoing the type approval process, especially considering the increasingly more popular direct injection methods and emission problems that come along with them [1]. They are a part of a WLTP (*Worldwide Harmonized Light Vehicle Test Procedure*) drive cycle, which replaced the much simpler NEDC (*New European Drive Cycle*) test. These tests differ significantly, as the newer WLTP test is designed to best mimic emissions in real driving conditions, which is greatly improved by including RDE drive cycle as a part of the emission test. The main differences can be seen in instantaneous acceleration values and the vehicle velocity, which are designed to better reflect the real conditions when driving. Despite that change there are still several aspects of exhaust emission testing that have not been addressed, specifically in the new limits and testing procedures, and are thus not normally measured or tested [2]. One example of such an oversight are bi-fuel vehicles, such as vehicles equipped with an additional fuel supply system for LPG (*Liquified Petroleum Gas*). While not particularly common in vehicles worldwide, this matter is especially significant for the polish transport sector, due to the popularity of adding LPG systems to gasoline vehicles, which significantly reduces the fuel costs for the user [3, 4]. These vehicles are subject to the same exhaust emission tests as all

* Corresponding author: dawid.p.gallas@doctorate.put.poznan.pl

other new vehicles but the tests are only performed for drive cycle when the engine is powered by gasoline, even though majority of vehicle operating time in reality would normally be spent powered by the LPG fuel. And while emissions from LPG powered engines have been subject of research to a reasonable degree [5, 6], the emission limits have not been adjusted to account for it. Thus the lack of exhaust emission data for driving when powered by LPG remains a significant problem in consistently assessing vehicle exhaust emissions. This is particularly important for countries with large numbers of LPG powered vehicles, such as Turkey (at about 2 400 000), South Korea (2 300 000), Poland (2 300 000) and Italy (1 700 000) (based on data from WLPGA, 2011).

The goal of the paper is to establish a wider understanding of the environmental impact of LPG powered vehicles, in relation to the emission limits and norms they are subject to. Thus ensuring that powering regular gasoline vehicles with LPG using a non-factory fuel supply system does not result in the vehicle significantly exceeding the emission limits set for it. This information is particularly important when considering solutions aimed at reducing air pollution in urban agglomerations and the fact that the use of non-standard fuels can significantly impact a vehicle's exhaust emissions [7]. Even more so because some of the components of vehicle exhaust emissions can be very harmful to human health as well as the environment. The paper is part of a larger study in testing the real driving emissions of LPG vehicles subject to different emission norms (manufactured in different years). The focus of this paper is the RDE tests of a Euro 5 vehicle with a spark ignition engine when powered with LPG.

2 Test vehicle, method and equipment

The tested vehicle was a Chevrolet Aveo (Fig. 1) equipped with a 1.2 dm^3 displacement spark ignition engine and a rated power of 62 kW. The vehicle was produced in 2011, thus it is expected to meet the Euro 5 emission limits.

Fig. 1. The vehicle used for emission tests.

The RDE test has been performed using a PEMS (*Portable Emission Measuring System*) device SEMTECH DS (Fig. 2), produced by Sensors Inc. The device analysers measured the emission of CO_2 (*Carbon Dioxide*), CO (*Carbon Monoxide*), HC (*Hydrocarbons*) and NO_x

(*Nitrogen Oxides*). Detailed description of the PEMS equipment, its use and its measuring ranges can be found in [8, 9]. A flowmeter was used to determine the volume of exhaust gasses and to enable the calculation of emission values. These devices were mounted on the tested vehicle and fastened firmly to ensure safety (Fig. 3).

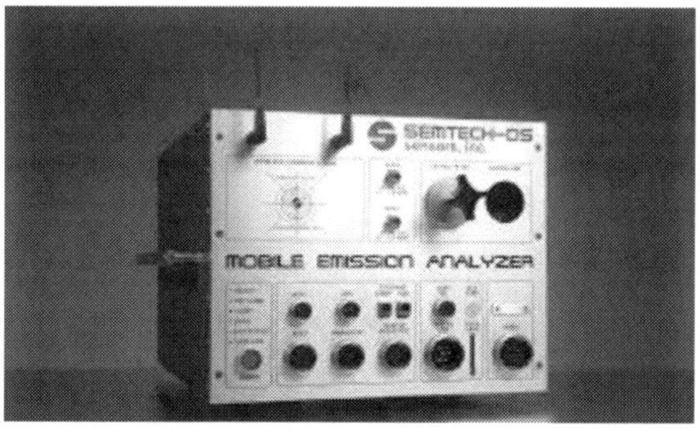

Fig. 2. The SEMTECH DS mobile emission analyser.

Fig. 3. The SEMTECH DS measurement equipment with the flowmeter mounted on the vehicle.

The road emission measurement was conducted in line with the RDE test requirements. The drive cycle comprised of three sections: urban, rural, and motorway, defined by their different speed profiles. In accordance with the procedure each section constitutes approx. 23–43% of the total drive distance, and each section needs to be longer than 16 km. The average vehicle speed ranges for each section are 15–40 km/h, 60–90 km/h, and above 90 km/h for the urban, rural and motorway sections respectively.

3 Results

The results of RDE test cycles have been analysed and presented as diagrams representing the share of the vehicle engine operating time based on its emission value of a given exhaust component as a function of velocity and acceleration intervals. The total share of driving time spent in each velocity-acceleration interval has been shown (Fig. 4). The obtained results do not deviate from expectations, especially the notable share of idling for when the vehicle is stationary (v = 0 km/h and a = ⟨-0.5 m/s², 0 m/s²⟩), constituting the highest individual share of drive time at approx. 17%. The other three highest shares all fall in the same acceleration range but for higher velocities, indicating situation of idle engine speed when slowing down, mostly before the brakes are applied, or engine braking.

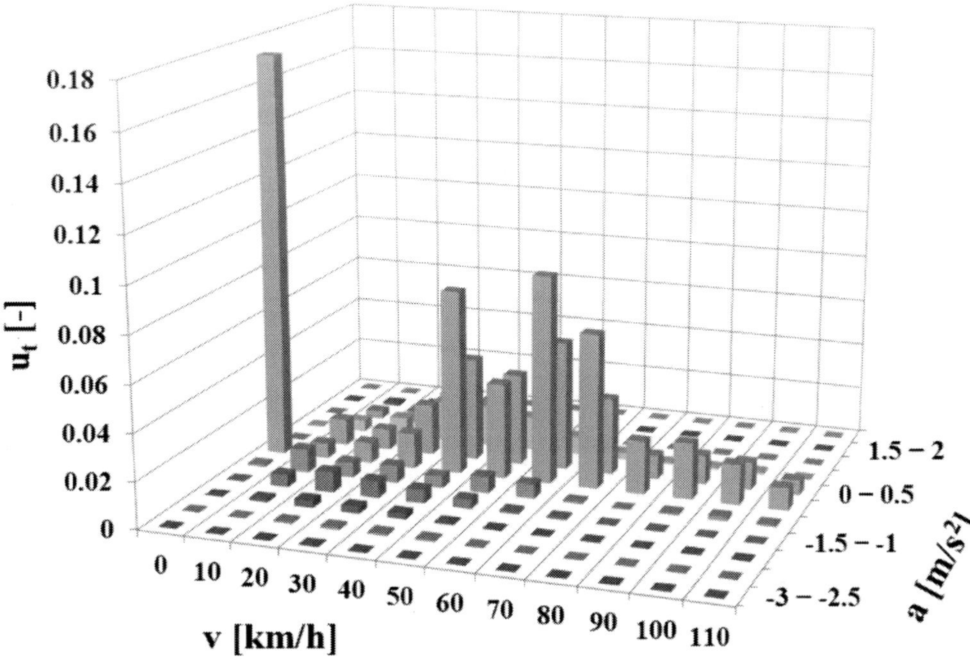

Fig. 4. Share of engine operating time in velocity-acceleration intervals.

The CO_2 emission (Fig. 5) could further be used as a reference for other types of exhaust emissions for a set of results that could be comparable between different vehicle types, as described in [10]. The results indicate a stable level of CO_2 emission for the range of low velocity values, which can be used to establish the carbon dioxide emission for the engine at idle speed. In this case these emissions reach about 0.5 to 1 g/s of CO_2 as measured by the PEMS device. Similar values for the higher velocity ranges with negative net acceleration reflect the driver's driving style, which consisted of using idle engine speed and brakes rather than engine braking for deceleration. Hence the total measured CO_2 emission could have been lower with slight variation to the driving style, as indicated in the eco-driving principles described in [11]. The highest recorded values of CO_2 emission were observed for the highest positive acceleration intervals of ⟨2 m/s², 2.5 m/s²⟩. Emission of carbon dioxide is strictly related to the vehicle fuel consumption, thus a reduction in this parameter would also result in a decreased fuel consumption and thus a better fuel economy.

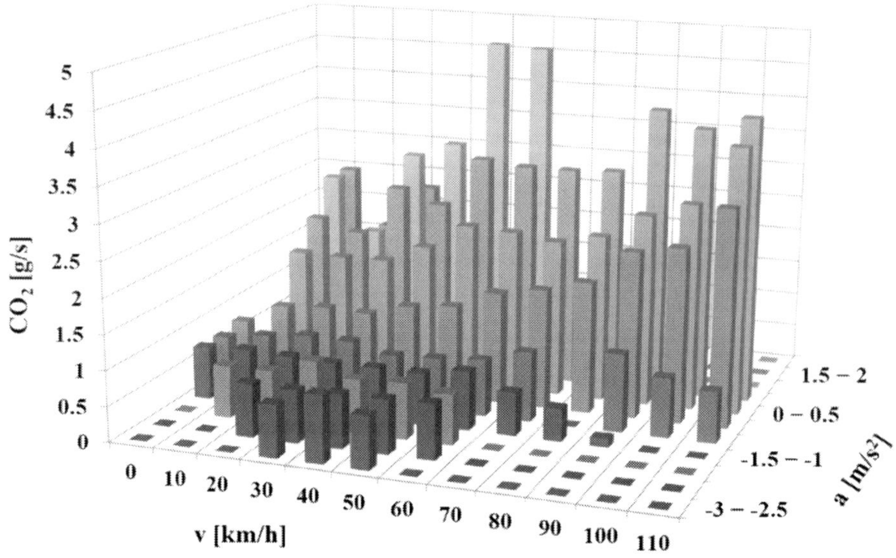

Fig. 5. CO_2 emission per second characteristics in velocity-acceleration intervals.

Emission results of carbon monoxide (Fig. 6) indicate high values for all positive acceleration intervals, largely regardless of vehicle velocity. The highest emission values were observed for acceleration intervals between (0.5 m/s^2, 1.5 m/s^2⟩. Equally high values were reached for the interval (1.5 m/s^2, 2 m/s^2⟩ but only for low velocity ranges (0 km/h, 30 km/h⟩. This is mostly the result of the vehicle power limit when travelling at high speeds. CO emission is strongly influenced by acceleration, as a result of worsening combustion conditions in the cylinder.

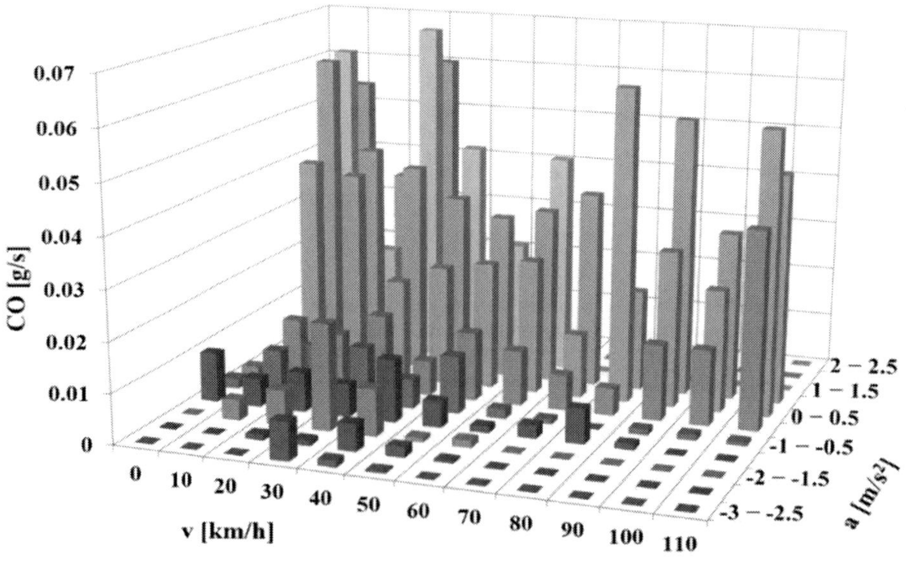

Fig. 6. CO emission per second characteristics in velocity-acceleration intervals.

The emission values for hydrocarbons (Fig. 7) indicate an increase in emission values for higher velocities, mainly in velocity ranges (80 km/h, 110 km/h⟩. This steady increase in HC emissions per second indicate that for these intervals the engine was running on a rich mixture rather than, as it would be expected for LPG vehicles, a lean mixture. The highest emission value of 0.0042 g/s, however, was found for the interval a = (1.5 m/s^2, 2 m/s^2⟩ and v = (20 km/h, 30 km/h⟩. It is an operating point at a large acceleration value but relatively low vehicle velocity, implying a situation of dynamic acceleration.

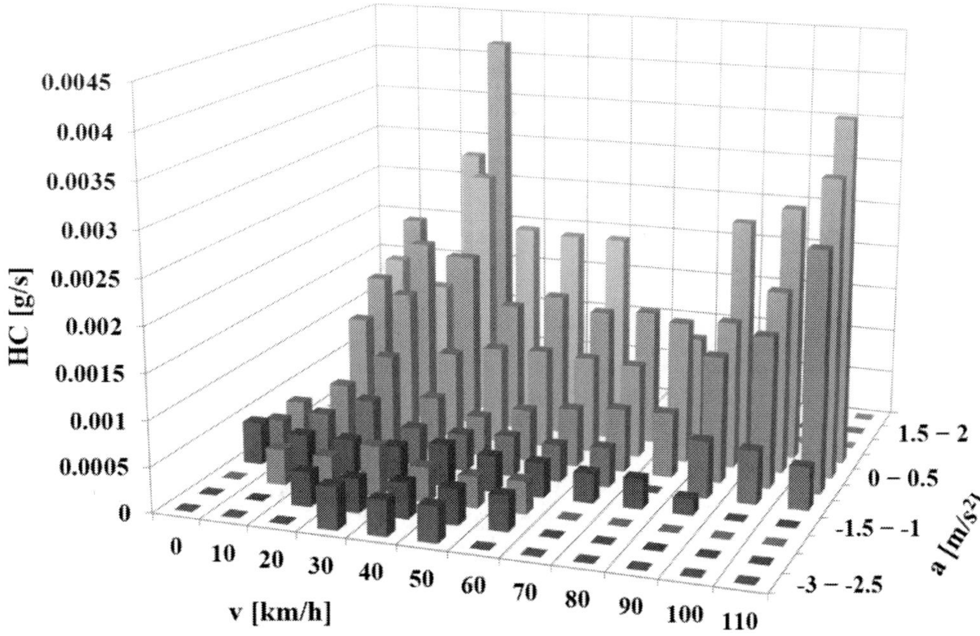

Fig. 7. HC emission per second characteristics in velocity-acceleration intervals.

Emission of nitrogen oxides indicates a strong impact of velocity and acceleration (Fig. 8). The highest NO_x emission values (of up to 0.0028 g/s) were observed in the intervals v = (80 km/h, 110 km/h⟩ and a = (0 m/s^2, 1 m/s^2⟩. This indicates high temperature during the combustion process, as resulting from the richer air-fuel mixtures. Unlike for gasoline or diesel oil, which require thermal energy from the combustion chamber to evaporate, LPG fuel does not decrease the thermal energy of the combustion chamber as significantly (which is also reflected by the problems of cylinder head overheating in such systems). Hence leading to increased NO_x emissions.

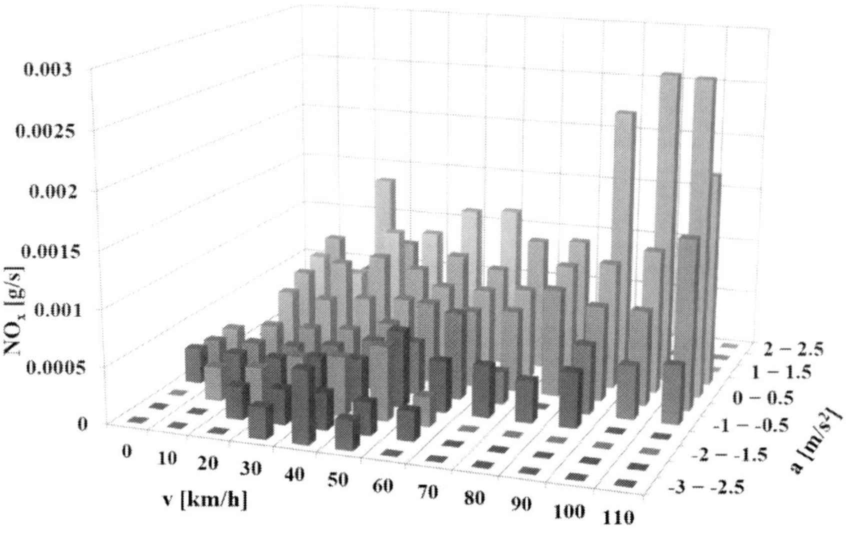

Fig. 8. NO_x emission per second characteristics in velocity-acceleration intervals.

Other results include the cumulative CO_2 emission for the entire RDE test cycle. It amounted to just over 10 kg, which averages out at approx. 150 g/km. This value exceeds the EU passenger cars 2015 CO_2 target of 130 g/km. The highest rate of cumulative CO_2 increase can be seen around the 1500–2500 s time period. This period corresponds to the vehicle switching from rural drive section to the motorway section, thus requiring a significant increase in speed to match the motorway cruising speed of other vehicles. Especially since for the purposes of the RDE testing motorway section is assessed based on the average vehicle speed being over 90 km/h. Highest recorded values of CO_2 emission reached 5 g/s, while the average value for the whole drive was determined to be around 2 g/s.

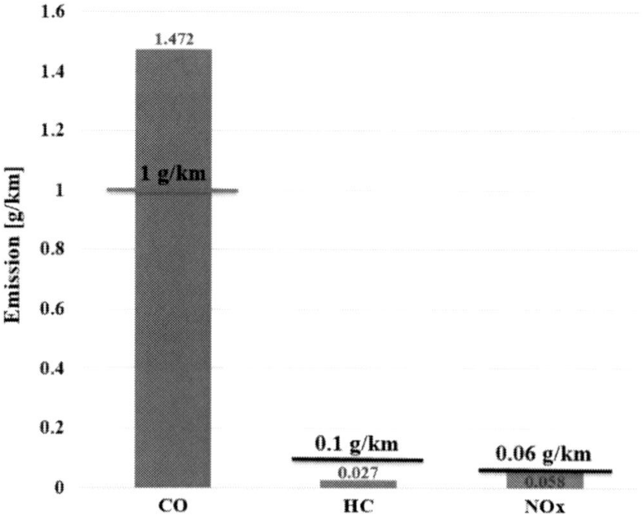

Fig. 10. Measured road emission values for CO, HC and NO_x in relation to the Euro 5 limits.

The test results indicate a relatively similar set of emission characteristics of an LPG powered vehicle to the emission limits set for regular gasoline powered vehicles (Fig. 10). These results have been compared to the Euro 5 emission norms.

4 Conclusions

As a general conclusion it should be noted that despite not being subject to testing, the emissions of vehicles powered by LPG fuel are reasonably within the limit values, even in the case of CO the RDE test emission results are permitted to reach 210% of the limit value due to the introduction of "conformity factor" equal to CF = 2.1. Although the conformity factor was introduced only recently to RDE tests in newer emission norms (Euro 6), it provides a good reference for older vehicles that RDE testing doesn't legally apply to. Use of RDE tests are also intended to compliment rather than supplement stationary dynamometer tests, however, their improved accuracy in representing real emissions does impact the use and utility of stationary tests [12].

The specific conclusions for the tested vehicle include:

1. The vehicle NO_x emissions at their highest points were higher than expected. The total amount of NO_x emitted in the test drive was nearly 4 grams, and the peak emission value reached was just under 0.01 g/s. Despite that the emission limit value was not exceeded (Fig. 10). Problems in meeting the NO_x emission limits tend to be a more common issue for vehicles equipped with compression ignition engines, as shown clearly in a comprehensive study by O'Driscoll et. al [13].
2. The HC emission was significantly below the limit value, despite the mixture being richer than expected for high velocity and acceleration intervals. Average road HC emissions were recorded at only 0.027 g/km compared to the limit value of 0.1 g/km.
3. The CO emission exceeded the limit value by approx. 47%. Thus a conformity factor of 1.5 would be needed for the test to have yielded acceptable values [14].
4. High CO emission indicates that either the air intake was insufficient (mixture too rich) or the aftertreatment systems failed to adequately reduce the emitted CO.

References

1. M. Siedlecki, M. Galant, L. Rymaniak, A. Ziolkowski, Autobusy **12**, 404–409 (2017)
2. A. Merkisz-Guranowska, Arch. Of Trans. **31**, 47–59 (2014)
3. M. Buczaj, A. Sumorek, Econtechmod **6**, 47–54 (2017)
4. E. Liu, S.Y. Yue, J. Lee, Research & Library Services Division Legislative Council Secretariat RP05/96-97 (1997)
5. S. Oh, S. Lee, Y. Choi, K-Y. Kang, SAE International **1461** (2010)
6. H. Bayraktar, O. Durgun, Energy Conv. and Management **46**, 2317–2333 (2005)
7. C. Y. Choi, R. D. Reitz, Fuel **78**, 1303–1317 (1999)
8. Sensors Inc., Emissions Measurement Solutions. *SEMTECH® DS On Board In-Use Emissions Analyzer* (Erkrath 2010)
9. P. Lijewski, J. Merkisz, P. Fuć, Croat. J. of Forest Eng. **34**, 113–122 (2013)
10. J. Merkisz, Ł. Rymaniak, Maint. & Reliab. **4**, 522–529 (2017)
11. J. Merkisz, M. Andrzejewski, J. Pielecha, CE **52**, 66–74 (2013)
12. M. Schroder, N. Baltes, J. Danzer, MTZ Worldwide **78**, 28–35 (2017)
13. R. O'Driscoll, H. M. ApSimon, T. Oxley, N. Molden, M. E. J. Stettler, A. Thiyagarajah, Atm. Env. **145**, 81–91 (2016)
14. J. Pielecha, J. Merkisz, J. Markowski, R. Jasiński, E3S Web Conf. **10**, 00073 (2016)

Modeling chlorine distribution in water supply system utilizing empirically determined chlorine decay rate

Anna Musz-Pomorska[1,*], *Marcin K.* Widomski[1], *Agnieszka* Matczuk[2], and *Konrad* Sadura[2]

[1]Faculty of Environmental Engineering, Lublin University of Technology, ul. Nadbystrzycka 40B, 20-618 Lublin, Poland
[2]Graduate of the Lublin University of Technology

Abstract. Analysis of free chlorine propagation in water supply network has a significant meaning for the process of water distribution. Results of numerical studies allow the proper selection of disinfectant or suitable monitoring of pipelines endangered by stagnation of water. The first-order reaction of chlorine decay in pipe boundary layer and inside the waterbody is commonly successfully assumed in numerical modeling. The aim of this studies was to analyze transport of chlorine inside the rural water supply system. The calculations were performed with application of Epanet 2.0 with assumed the first-order re action of chlorine decay and empirically determined chlorine decay rate in the mass of waterbody. The periodical disinfection of water in the network with the constant chlorine concentration 0.3 mg·dm^{-3} introduced during the whole time duration of simulation was assumed to calculations. The obtained results of chlorine distribution showed that even after 4 days there were available pipelines in which concentration of free chlorine was lower than 0.2 mg·dm^{-3}. Thus, the microbiological protection of water quality is unavailable in these pipelines.

1 Introduction

Concentration of free chlorine in water is one of the most important factors directly affecting the quality of water delivered to consumers. Doses of chlorine used in practice are result of the required need of assuring the microbiological protection of water during transport to the customers as well as limiting development of by-products of disinfection (DBPs) threatening the humans health [1–5]. According to guidelines by WHO, the concentration of free chlorine in sampling points should be in range 0.2–0.5 mg·dm^{-3} [6]

Two parallel reactions occur in the process of chlorine decay. One in the boundary layer, the second in mass of waterbody. According to the complexity and numerous factors affecting ratio of these reactions, modeling of chlorine transport in water distribution systems requires the separate determination of values of chlorine decay constants. The total chlorine decay constant (k) (1) is commonly expressed as sum of constant of chlorine decay

* Corresponding author: a.musz-pomorska@pollub.pl

in mass of water (k_b) (dependent to the initial characteristics of water: temperature, initial chlorine concentration and content of organic components) [7–9] and constant of chlorine decay in the boundary layer (k_w) (related to pipe diameter, velocity of flow and pipe material) [10].

$$k = k_b + k_w \quad (1)$$

The constant of chlorine decay in mass of water is commonly determined empirically in numerical modeling [11, 12] while the constant of chlorine decay in the boundary layer is obtained by the trial and errors method during calibration of qualitative model [13]. The k_w constant is less frequently determined empirically [14].

Nowadays, several models allowing simulations of water quality inside the water supply systems are available [15–17]. One of the most popular is Epanet 2.0 [18], allowing calculations of chlorine transport, studies of increase in DBPs and trihalomethanes concentrations as well as determination of water age [19].The kinetics of chlorine decay in Epanet 2.0 is described by the first-order reaction (2), in relation to which, the final concentration is dependent only to the initial concentration of one component. In case of chlorine transport, its final concentration is related only to its initial concentration.

$$c = c_0 e^{-kt} \quad (2)$$

where: c – chlorine concentration in time t [mg·dm^{-3}], c_0 – initial chlorine concentration [mg·dm^{-3}], k – constant of chlorine decay rate [min^{-1}], t – time [min].

The studies reported by Ozdemir and Ucak (2002) [20] based on comparison of modeling results obtained in Epanet and DYNAQ to results of measurements of chlorine concentration in nodes of water supply system showed satisfactory agreement between results of calculations and measurements.

The aim of this paper was to analyze chlorine transport in rural water supply system. Our studies were performed in Epanet 2.0 modeling software with applied hydraulic model of the tested network. The first-order reaction of chlorine decay and empirically determined value of constant of chlorine decay in water mass were assumed to our calculations. The periodical disinfection of water in the network with the constant chlorine concentration 0.3 mg·dm^{-3} introduced during the whole time duration of simulation was assumed to calculations.

2 Materials and methods

Numerical calculations of chlorine transport inside the water distribution network were performed in Epanet 2.0 computational software, distributed by EPA, US. (United States Environmental Protection Agency), with the applied hydraulic model of the tested system. The applied hydraulic model was earlier calibrated in relation to measured pressure and values of volumetric flow rate of water inside the pipelines. The studied network had annular-branched structure and was supplied with water by single water supply station. The network consisted of polyvinyl chloride (PVC) pipes of total length equal to 16.4 km. Diameters of pipes were in range 80 – 200 mm, but the greatest share had pipelines of diameter 150 mm – approx. 37 %, while the lowest part was represented by pipes of 200 mm – 0.9 %.

2.1 Hydraulic model

The developed hydraulic model representing, in a simplified manner, the studied water supply network, consisted of 200 nodes, 203 lines and one reservoir. The computational water demands in the developed model were based on readouts of domestic watermeters while hourly water demand pattern was based on readouts of the main watermeter located in the water supply station (Fig. 1).

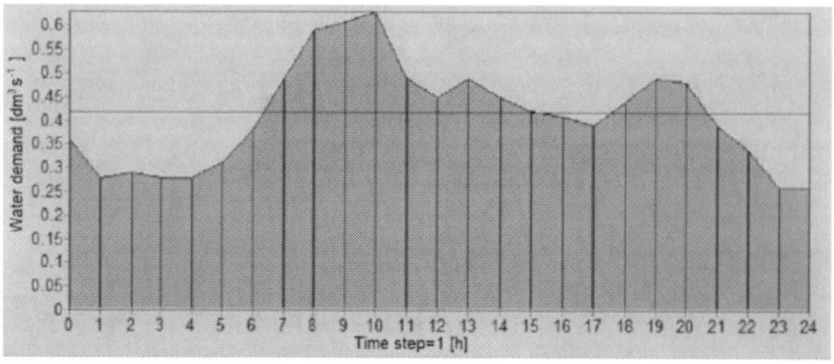

Fig. 1. Water demand pattern.

The determined value of unit water demand for an individual water supply network user was equal 0.012 $dm^3 \cdot s^{-1}$. The value of pipe material roughness was assumed as 0.02 mm for all of the modeled pipelines [21]. Pressure at the exit of water supply station was sustained at the level of approx. 35 m H_2O. Volumetric flow rates of water during hours of minimal and maximal water demand inside the main water supply pipeline were equal 4.77 $dm^3 \cdot s^{-1}$ and 1.97 $dm^3 \cdot s^{-1}$, respectively.

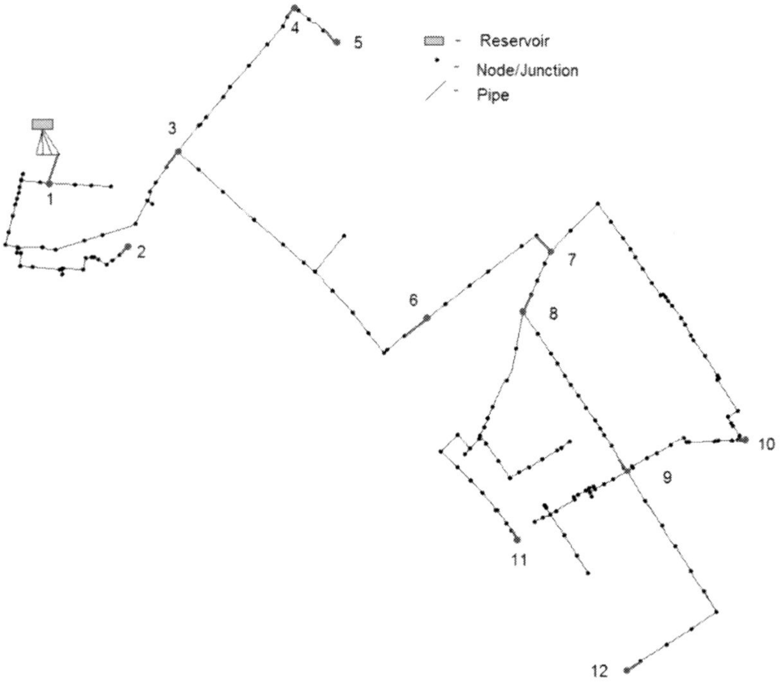

Fig. 2. Network scheme with marked characteristic nodes and pipelines.

In order to determine the chlorine distribution inside the tested water supply network 12 characteristic nodes and 12 pipelines were selected (Fig. 2). The choice was determined by the distance to pumping station, value of daily changes in values of hydraulic parameters and required coverage of the whole network. Description of the selected nodes and pipelines as well as their labels in the model are presented in Table 1.

Table 1. Description of the characteristic nodes and pipes.

Description	1	2	3	4	5	6	7	8	9	10	11	12
Nodes	w2	J9	w128	J3	J1	w18	w149	w151	w157	W85	J53	J69
Pipelines	R2	R45	R10	R64	R17	R190	R80	R142	R178	R105	R159	R185

2.2 Quality model

Numerical calculations of chlorine transport inside pipelines of the studied network were based on the first-order model of chlorine decay, according to equation (2) [14, 22, 23] with value of constant of chlorine decay rate in the boundary layer of the pipeline accepted after literature as $k_w = 0.04$ h^{-1} [7, 8] and empirically determined constant of chlorine decay in the mass of water k_b. The chlorine concentration equal 0.3 mg·dm^{-3} was assumed at the output of the pumping station. Dosing of chlorine was assumed as constant during the whole time duration of simulation – 4 days (96 hr). Time duration of water chlorination process was assumed basing directly on information obtained from the local water supply company. Taking into account that water in the studied network is being chlorinated periodically after failures or servicing, the initial concentration of chlorine (c_0) was assumed in our research as 0 mg·dm^{-3}. The assumed time duration for determination of water age was accepted as equal to time of water chlorination in the network. This time duration is in accordance with so called "long" time of water retention inside the pipelines, according to literature data, greater than 3 days [24].

2.2.1 Determination of the chlorine decay rate in water body

The constant of chlorine decay rate in mass of water k_b was determined with application of the bottle test described by Vieira et al. (2004) [12] on the laboratory set-up presented in Fig. 3.

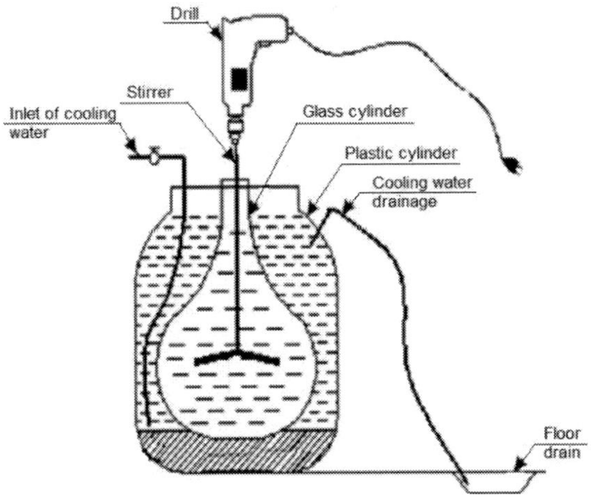

Fig. 3. Laboratory set-up for k_b determination.

Water for the research was sampled in water supply station and immediately transported to the laboratory. Then, the glass balloon of 54 dm^3 (Fig. 3) was filled with the tested water. Next, sodium hypochlorite containing 200 mg·dm^{-3} of the active chlorine was added to water. Temperature of water inside the glass reservoir was sustained at the level of approx. 16°C. After exhaustive mixing of water with stirrer, samples were taken according to the assumed schedule. The chlorine content in the obtained samples was determined with PC Mulit Direct meter using the colorimetric method DPD, according to PN-EN ISO 7393-2:2011 [25].

3 Results

The results of laboratory changes of chlorine concentration in water sampled from the tested water distribution network are presented in Figs. 4 and 5. The obtained results show a very quick consumption of chlorine during the opening phase of the experiment (the first 3 hours of measurements) (Fig.4). During this phase the initial concentration of free chlorine decreased by 0.4 mg·dm^{-3} (from 2.24 mg·dm^{-3} to 1.84 mg·dm^{-3}). During the second phase, from 3rd to 24th hour of the experiment the significantly slower decrease in free chlorine concentration was observed, from 1.84 mg·dm^{-3} to 1.40 mg·dm^{-3}. The observed dynamics of decrease in chlorine concentration is being supported by previously reported results by Vieira et al. (2004) [12]. The constant of chlorine decay in mass of water was determined basing on the observed changes in chlorine concentration and the assumed first-order equation of chlorine decay. In case of the first-order decay the k_b constant is determined as tangent of inclination angle of the straight line obtained by the natural logarithm of c_t/c_0 in time t [18] (Fig.5). The determined value of constant k_b was equal 0.0182 h^{-1}.

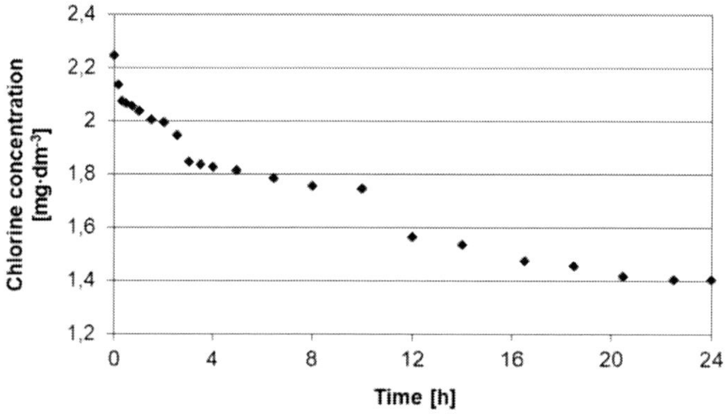

Fig. 4. Chlorine concentration in water during laboratory measurements.

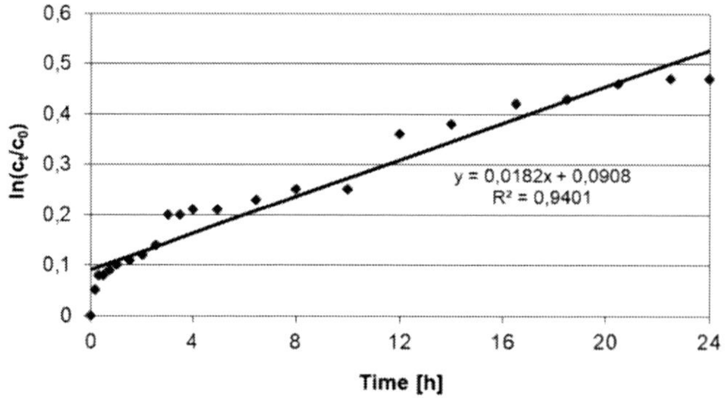

Fig. 5. Changes of chlorine concentration in water.

The calculated numerical results of chlorine transport obtained for the selected nodes are presented in Table 2. Table 3 presents computed results covering selected characteristics of flow in the chosen pipelines obtained for the 96th hour of the assumed time duration of simulation.

The performed numerical calculations showed low values of water flow velocity inside the modeled pipelines (from 0.06 $m \cdot s^{-1}$ to 0.15 $m \cdot s^{-1}$ in the pipeline supplying the studied network). The required value of velocity equal 0.5 $m \cdot s^{-1}$ was not achieved in none of the pipelines [26], which may negatively affect the qualitative characteristics of water and may result in deposition of sediments inside the pipelines and increase in roughness as well as development of bio-film [27]. The maximum water demand (q = 4.77 $dm^3 \cdot s^{-1}$) was observed at 10.00 while the minimal at 23.00 (q = 1.97 $dm^3 \cdot s^{-1}$).

Table 2. Chlorine concentration on selected nodes of the network.

Time [hour] / Number of node	Chlorine concentration [$mg \cdot dm^{-3}$]											
	1	2	3	4	5	6	7	8	9	10	11	12
12	0.30	0.00	0.25	0.00	0.00	0.00	0.00	0.00	0.00	0.00	0.00	0.00
24	0.30	0.16	0.24	0.00	0.00	0.18	0.16	0.15	0.00	0.00	0.00	0.00
48	0.30	0.16	0.24	0.17	0.00	0.18	0.16	0.15	0.02	0.07	0.00	0.00
72	0.30	0.16	0.24	0.17	0.12	0.18	0.16	0.15	0.02	0.07	0.00	0.03
96	0.30	0.16	0.24	0.17	0.12	0.18	0.16	0.15	0.02	0.07	0.04	0.04

*Grey cells – chlorine concentration higher than 0.2 $mg \cdot dm^{-3}$, minimal concentration required by WHO for drinking water

Table 3. Parameters of flow in selected pipes at 96 hours of modeling.

Parameter	Unit	Number of pipe											
		1	2	3	4	5	6	7	8	9	10	11	12
Diameter	[mm]	200	80	150	80	100	150	150	100	100	80	80	80
Flow	[$dm^3 \cdot s^{-1}$]	2.73	0.03	1.30	0.09	0.05	1.05	0.97	0.52	0.19	0.25	0.00	0.03
Velocity	[$m \cdot s^{-1}$]	0.09	0.01	0.07	0.01	0.01	0.06	0.06	0.07	0.02	0.03	0.00	0.01
Age	[h]	1.64	15.8	5.6	41.4	53.1	12.6	15.4	16.4	26.6	32.0	81.0	65.3

Figure 6 presents scheme of isolines of chlorine concentration in water and age of water inside the pipelines.

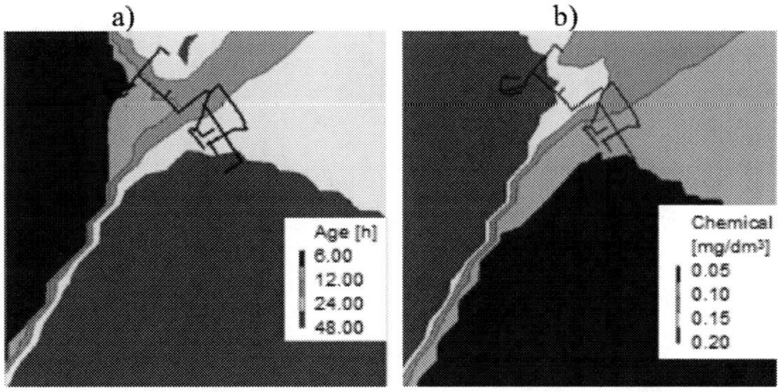

Fig. 6. Scheme of counter lines dividing: a) the water age in pipelines, b) chlorine concentration in water.

The obtained computational results of chlorine transport inside the studied rural water supply system show that the periodical chlorination of water in the network with the assumed dose of disinfectant through the selected time does not assure the required minimal concentration of free chlorine (according to WHO recommendations, minimum 0.2 mg $Cl_2 \cdot dm^{-3}$). Simultaneously, the acceptable concentration of chlorine in water (0.3 $mg \cdot dm^{-3}$) was not exceeded in any node of the network [28]. The highest chlorine concentration was observed in nodes close to the water supply station (nodes 1 and 3), located on the main conduit supplying the whole water distribution system (Fig. 6).

The modeled disinfectant reaches the final nodes in the network with the significant delay since the start of chlorination process (Table 2). Pipelines supplying water to final nodes of the network are characterized by very low values of volumetric flow rate and velocity of flow as well as very long age of water in the pipelines (Table 3). The age of water in pipeline supplying water to node No 11 exceeds 3 days. Figure 5 shows that the shortest water age (less than 6 h) was observed in pipelines close to water supply station, while the longest in the final parts of the system. The significant water age, accompanied by lack of the required concentration of disinfectant may result in development of microorganisms, deposition of sediments and deterioration of water quality [29].

4 Conclusions

Availability of computer simulation models allows analysis of qualitative characteristics inside water supply systems, including chlorine transport. The quality of obtained results depends on the assumed input data, selection of the proper model of disinfectant transport or determined values of constants of disinfectant decay rate. Our calculations of chlorine transport in rural water distribution system, performed in Epanet 2.0 with assumption of first-order equation of chlorine decay and determined empirically constant of chlorine decay rate exposed very low concentrations of free chlorine in modeled tap water, below 0.2 $mg \cdot dm^{-3}$, which combined with low values of flow velocities and extended water age do not assure the microbiological protection of water quality. In relation to periodic disinfection of water inside the studied supply network, measurements of chlorine concentration in selected points of system were unavailable and indisposed calibration of developed qualitative model. Taking into account lack of qualitative model calibration, the obtained results should be treated as preliminary and approximate. However, they pose an important information about operation conditions of the tested network. In future, the developed qualitative model should be calibrated basing on measured values of free chlorine concentrations in the selected nodes of the studied system. The presented study will be also extended to allow the selection of the proper chlorine dose due to results of numerical studies of the influence of various doses of disinfectant on water quality in the network.

References

1. Y. Zhao, Y. J. Yang, Y. Shao, J. Neal, T. Zhang, Water Res. **141**, 32–45 (2018)
2. R. V. Goyal, H. M. Patel, Appl. Water Sci. **5**, 311–319 (2015)
3. D. L. Boccelli, M. E. Tryby, J. G. Uber, R. S. Summers, Water Res. **37**, 2654–2666 (2003)
4. Y. Bixiong, W. Wuyi, Y. Linsheng, W. Jianrong, E. Xueli, J. Hazard. Mater. **171**, 1–3, 147–152 (2009)
5. Z. K. Chowdhury, L. Passantino, R. S. Summers, L. Work, N. Smith, L. Rossman, *Assessment of Chloramine and Chlorine Residual Decay in the Distribution System* (AWWA Research Foundation, Denver, Colorado, 2006)
6. WHO Guidelines for drinking water quality, 4th ed., (WHO, Geneva, 2011)

7. J. C. Powell, N. B. Hallam, J. R.West, C. F. Forster, J. Simms, Water Res. **34**, 1, 117–126 (2000)
8. R. M. Clark, R. C. Haught, J. Water Res. Plan. Man. **131**, 3, 208–217 (2005)
9. F. Shang, G. Uber, J. Water Res. Plan. Man., ASCE **133**, 4, 363–371 (2007)
10. N. B. Hallamn, J. R.West, C. F. Forster, J. C. Powell, I. Spencer, Water Res. **36**, 3479–3488 (2002)
11. L. Monteiro, D. Figueiredo, S. Dias, R. Freitas, D. Covas, J. Menaia, Proc Eng. **70**, 1192–1200 (2014)
12. P. Vieira, S. T. Coelho, D. Loureiro, J. of Water Supply Res. T. – AQUA, **53**, 7, 453–467 (2004)
13. S. H. Maier, R. S. Powell, C. A. Woodward, Water Resour. Res. **34**, 8, 2301–2309 (2000)
14. A. O. Al-Jasser, Water Res. **41**, 387–396 (2007)
15. I. E. Karadirek, S. Soyupak, H. Muhammetoglu, Desalin. Water Treat. **57**, 25, 11592–11598 (2016)
16. A. G. Seyoum, T. T. Tanyimboh, C. Siew, Water Sci. Tech.-W. Sup. **13**, 4, 1161–1166 (2013)
17. A. Musz, B. Kowalska, M. K. Widomski, Ecol. Chem. Engineer. S, **16**, 2, 175–184 (2009)
18. L. A. Rossman, *EPANET 2 Users Manual. Water Supply and Water Resources Division, National Risk Management Research Laboratory, Office of Research and Development* (US Environmental Protection Agency, Cincinnati, Ohio, 2000)
19. F. Shang, J. G. Uber, L. A. Rossman, J. Environ. Sci. Technol. **42**, 3, 808–814 (2008)
20. O. N. Ozdemir, A. Ucak, J. Environ. Eng. ASCE, **128**, 1, 31–39 (2002)
21. W. W. Sharp, T. M. Walski, J. Am. Water Works. Ass. **80**, 11, (1988)
22. B. Kowalska, D. Kowalski, A. Musz, Environ. Prot. Eng. **32**, 2, 5–16 (2006)
23. H. Kim, S. Kim, J. Koo, Procedia Eng. **119**, 370–378 (2015)
24. S. Shu, S. Liu, X. Wang, L. Yu, S. Shu, D. Zhang, International Conference on MACE. 1918–1921 (2010)
25. PN-EN ISO 7393-2 Water quality - Determination of free chlorine and general chlorine – Part 2: Colorimetric method with N, N-diethyl-1,4-phenylenediamine, for routine control purposes (2011)
26. M. Kwietniewski, W. Olszewski, E. Osuch-Pajdzińska, *Projektowanie elementów systemu zaopatrzenia w wodę* (Oficyna Wydawnicza Politechniki Warszawskiej, Warszawa, 2009)
27. D. Kowalski, M. Kwietniewski, A. Musz, M. K.Widomski, Ochr Środ. **30**, 1, 27–30 (2008)
28. Rozporządzenie Ministra Zdrowia z 2015 r. w sprawie jakości wody przeznaczonej do spożycia przez ludzi (Dz.U. 2015, poz. 1989)
29. S. C. Tinker, C. L. Moe, M. Klein, W. D. Flanders, J. Uber, A. Amirtharajah, J. Water Health. **7**, 2, 332–343 (2009)

Reducing nitrogen oxide emissions in boilers at moistening of blowing air in heat recovery systems

Raisa Navrodska[1], *Nataliia* Fialko[1,*], *Georgii* Presich[1], *Georgii* Gnedash[1], *Sergii* Alioshko[1], and *Svitlana* Shevcuk[1]

[1]The Department of Thermophysics of Energy Efficient Heat Technologies, Institute of Engineering Thermophysics of National Academy of Sciences of Ukraine, Kyiv, Ukraine

Abstract. Considered indicators of reducing harmful emissions into the environment by gas-fired water-heating and steam boilers of the municipal heat-power engineering, which are equipped with heat-recovery systems of boiler exhaust-gases. The results of studies on the relative decrease of emissions of nitrogen oxides and other harmful substances generated during fuel combustion in boilers when used for boilers complex heat-recovery plants for heating heat-network water and water for a chemical water-purification system of a boiler plant, as well as for heating and humidifying combustion air are presented. It is shown that the use of such complex heat-recovery systems, due to the useful heat of exhaust-gases from boilers, provides a reduction in fuel consumption and, consequently, the volumes of harmful emissions generated during its combustion, by 8–12%, depending on the mode of operation of boilers during the heating period. The efficiency of using combustion-air humidifying in these installations to decrease the level of formation of nitrogen oxides in the combustion area of boilers is considered. It is also shown that the use of these systems provides a decrease emissions of nitrogen oxides to the environment by up to 60% by reducing the combustion temperature in the combustion area at moisture introduction with the combustion air and, as a result, suppression of formation these oxides during fuel combustion in boilers. The substantiation of the accuracy of the obtained calculated values of the levels of relative decrease of emissions of nitrogen oxides with the moisture introduction into the boiler by comparing them with experimental data is given. Comparative of the results shows a good agreement between the calculated and experimental data.

1 Introduction

A significant share in pollution of environment is made by fuel-fired installations of various purposes. Among them, heat-generating boiler plants are especially distinguished, they are intended for heat supply of residential and municipal facilities. The reduction of emissions

*Corresponding author: nmfialko@ukr.net

to the environment by such installations is currently being solved through the use of heat-recovery technologies that decrease fuel consumption and, as a result, reduce harmful emissions generated during its combustion. The greatest thermal and ecological effect is achieved in this case at the use of complex heat-recovery systems [1–6] in which deep cooling of the exhaust-gases of boiler plants is realized (below the dew point of the water vapor contained in them).

2 Prerequisites and means for solving the problem

This paper is devoted to research on the effect-use of complex heat-recovery systems for gas-fired boilers designed to heat and humidity combustion air and to heat cold water for chemical water-purification system. These systems are characterized the additional environmental effect associated with a decrease in the formation of nitrogen oxides during fuel combustion in boilers. In these systems, the heating of combustion air and water of the chemical water-purification system (CWP) is intended to increase the fuel efficiency in the boiler, and its humidifying is reduced the formation of nitrogen oxides by lowering the combustion temperature in the boiler [7–9].

The aim of the study is to assess the levels of decrease of nitrogen oxide emissions in water-heating and steam gas-fired boilers in different modes of their operation at humidifying the combustion air.

3 Solution of the examined problem

In fig. 1 shows a schematic circuit of a heat-recovery plant with complex use of heat of recovery for heating and humidifying the combustion air and heating cold water of CWP system.

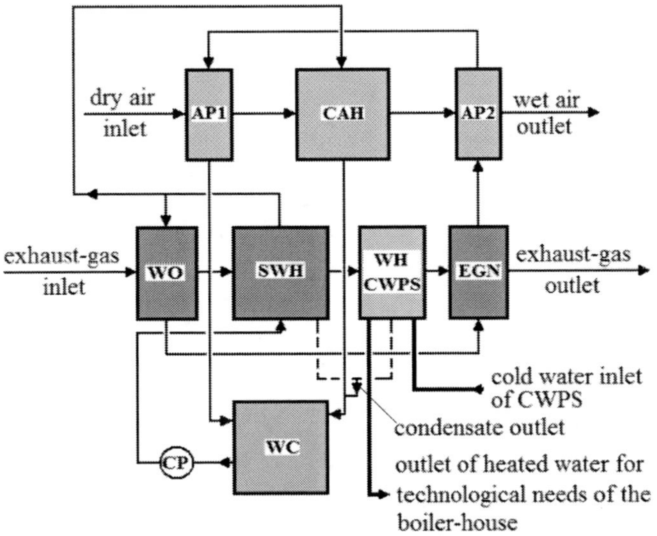

Fig. 1. Schematic circuit of a complex heat-recovery system with heating and humidifying the combustion air and heating the cold water of the chemical water-purification system: AP1 – first air preheater; CAH – contact air heater; AP2 – second air preheater; WO – water-overheater; SWH – surface water heater; WH CWPS – water heater of chemical water-purification system; EGH – exhaust-gas heater; WC – water collector; CP – the centrifugal pump.

The heat-recovery installation works as following way. Exhaust-gases of the boiler with high moisture content are first cooled in the water-overheater WO, and then come to the surface water heater SWH where the cooling takes place below the dew point of the water vapor contained in these gases, that is, under conditions of condensation of steam from the exhaust-gases. Further cooling of the exhaust-gases with the implementation of the condensation regime is carried out in a water heater of the chemical water purification system WH CWPS. In condensing mode of operation of the water heaters WO and WH CWPS, the heat of condensation of water vapor contained in the exhaust-gases is used. After these water heaters, the exhaust-gases are dried in the exhaust-gas heater EGH to a level that avoids the formation of condensate in the exhaust ducts of the boiler plant [10], and leave the heat-recovery system.

Water from the water collector WC is taken by the pump and first pumped through the SWH, where it heats up, and then the circulation circuit splits. Part of the water enters the contact air heater CAH, where it flows through the heat exchange surface, is cooled with partial evaporation and enters the WC. The rest of the water passes through WO, EGH, second air preheater AP2 and first air preheater AP1, being cooled successively, and also enters the WC. Then the cycle repeats.

Cold dry air outside the boiler house first enters the first air preheater AP1, where it is heated, then sent to the contact air heater CAH, where it is heated and humidified, and through the second air preheater the AP2 is removed from the heat recovery system. Then goes into the air duct and enters the gas burner device the boiler.

An increase in air temperature provides a corresponding reduction in fuel consumption, and the humidifying of this air helps to decrease the formation of nitrogen oxides by lowering the combustion temperature by using part of the thermal potential of the flame to heat the introduced moisture.

Studies of the thermal and environmental performance of the heat-recovery installation were carried out at the load of the water-heating boiler in accordance with the heat boiler schedule shown in fig. 2, and for the steam boiler in two modes of operation, namely: 100% and 50%. The environment air temperature t_e for heating and humidifying in a heat-recovery system during the heating period was taken for two types of boilers in accordance with fig. 2

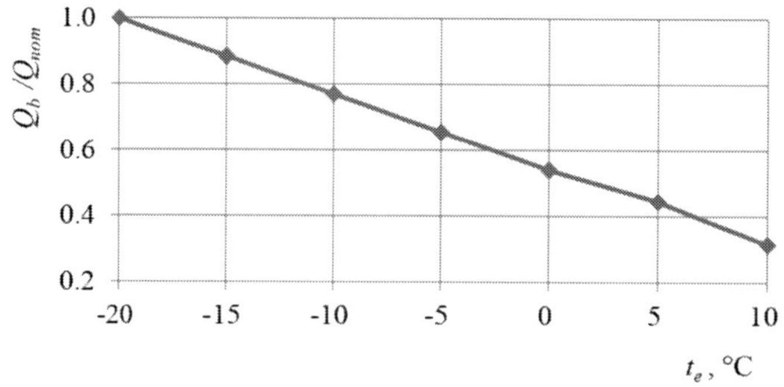

Fig. 2. Dependence of the load of the water-heating boiler on the environment temperature.

Thermal calculations of the elements of the complex heat-recovery system to determine the levels of fuel consumption reduction were performed according to well-known methods [11] using the calculated experimental dependencies obtained at the Institute of Technical Thermophysics of the National Academy of Sciences of Ukraine [12] in condensation modes.

The volumes of NO_x reduction were dependent on the water-fuel ratio β in the boiler furnace according to the data of [9]. Namely, the relative decrease in nitrogen oxides in combustion products due to the humidification of the blown air was calculated by formulas:

$$\Delta C_{NOx} = 1 - (0{,}947 \cdot e^{-0{,}995\beta} + 0{,}066) \tag{1}$$

$$\beta = G_w / G_{ng}, \tag{2}$$

where β is water-fuel ratio, kg/kg; G_{ng}, G_w – mass flows of natural gas and moisture introduced into the boiler furnace with combustion air.

4 Results and discussion

The calculated data of the decrease of harmful emissions ΔC_f into the environment due to the reduction of fuel consumption to the boiler when using the complex heat-recovery system under consideration for the two types of boilers are shown in Fig. 3.

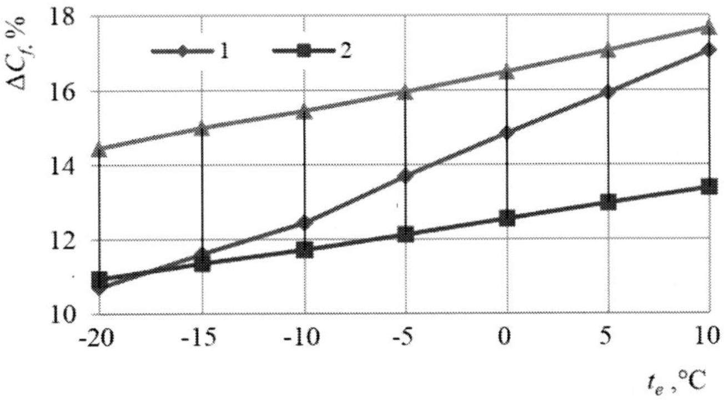

Fig. 3. The change from the environment temperature t_e relative levels of decreasing the formation of harmful emissions in the boiler ΔC_f for water-heating and steam boilers when using the proposed heat-recovery system, ensuring a reduction in fuel consumption: 1 – water-heating boiler with a load in accordance with the heat schedule of Fig.2; 2 and 3 – steam boiler at a load of 100 and 50%, respectively.

The results obtained indicate that the use of the heat-recovery system with the complex use of recycled heat provides for the considered boilers the significant decrease in the formation of harmful emissions in accordance with a reduction in fuel consumption. Namely: the value of ΔC_f f is 10.8–16.6% for a water-heating boiler, and 11.1–17.5% for a steam boiler, depending on the operation modes of the boilers.

The results of studies on the decrease of the formation of nitrogen oxides ΔC_{NOx} are shown in Fig. 4.

As can be seen from the obtained results, when using the proposed complex system for a water-heating boiler, there is a relative decrease in ΔC_{NOx} of the formation of nitrogen oxides, from 22% to 63% depending on the mode of operation of the boiler in accordance with the environment temperature t_e. In this case, the maximum values of ΔC_{NOx} correspond to the highest temperature of the heating period. With a higher air temperature at the inlet to the heat-recovery system, and therefore in the contact air heater CAH, its consumption is reduced in accordance with the mode of operation of the boiler. This makes it possible to humidify the combustion air to a higher level and, accordingly, increase the water-fuel ratio

β for fuel combustion. This ensures a decrease in the formation of nitrogen oxides in combustion area of boiler.

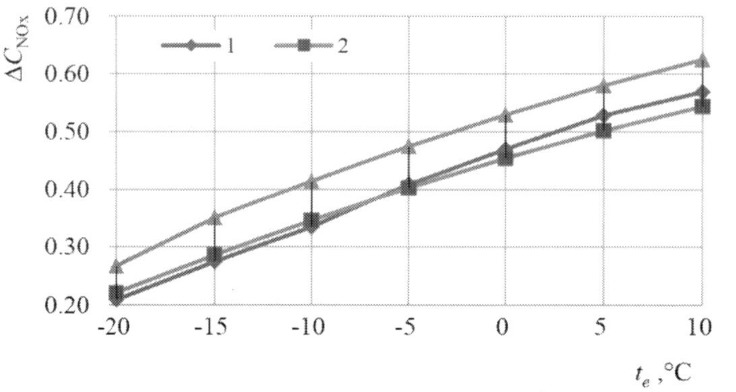

Fig. 4. The change at the application of the complex heat-recovery system of the relative decrease of nitrogen oxides ΔC_{NOx} in the combustion area, depending on the environment temperature t_e for water-heating and steam boilers in different modes of their operation at: 1 – water-heating boiler with a load in accordance with the heat schedule of Fig.2; 2 and 3 – steam boiler at a load of 100 and 50%, respectively.

The same tendency is observed for the steam boiler, that is, the ΔC_{NOx} values also depend on the mode of its operation and the input temperature of the heated air. At the same time, a greater decrease in the ΔC_{NOx} ratio corresponds to 50% of the boiler load, which is associated with a decrease in the air consumption for combustion. So with this load, the relative levels of reduction of the formation of nitrogen oxides ΔC_{NOx} vary from 27% to 63%, and under a load of 100% – only from 21% to 54%.

The proposed heat-recovery system is applied in the boiler house of PSC "KYIVENERGO" (Kyiv, Ukraine) (see fig. 5) for the steam boiler (type E-1/9) with a thermal capacity of 0.63 MW.

Fig. 5. Complex heat-recovery installation with heating and humidification of combustion air in the boiler house of PSC "KYIVENERGO" (Kyiv, Ukraine).

To assess the accuracy of the calculations obtained when determining the values of the decrease of emissions of nitrogen oxides C_{NOx} with the introduction of moisture from the combustion air into the boiler furnace, a comparison was made of these calculations and experiments. The corresponding results are presented in fig.6.

The experimental values near curve 5 of fig. 6 correspond to personal measurements of the NO_x content in the exhaust-gases at the outlet from the complex heat-recovery system, as well as the results of other researchers [9]. The calculated data for the decrease of NO_x are determined using formula (1). Comparative of the results shows a good agreement between the calculated and experimental data.

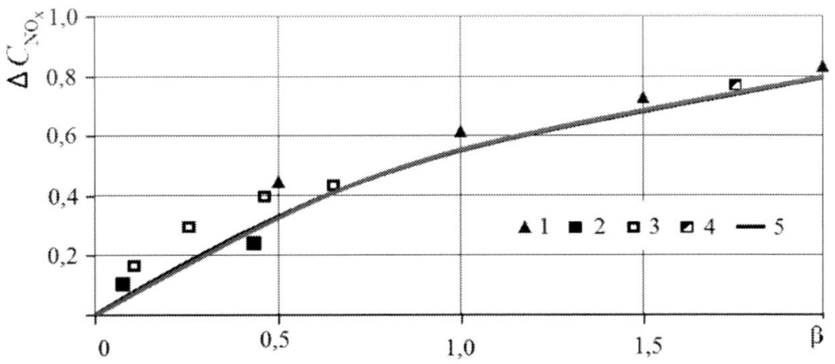

Fig. 6. Comparative of calculated and experimental data of the relative ΔC_{NOx} concentration in the exhaust-gases from the boiler with air humidification depending on the water-fuel ratio β: 1, 2, 3 – experimental data of other researchers [9]; 4 – experimental data of personal tests; 5 – theoretical curve.

5 Conclusions

The results of the conducted studies show that the use of modern heat-recovery technologies for gas-fired boiler plants with the complex use of recycled heat for heating boiler water, water of CWP systems and combustion air decrease the formation of nitrogen oxides ΔC_{NOx} to 60% due to the suppression of their formation in the combustion area the boiler. The use of this technology also ensures the reduction of harmful emissions from boilers by 10.8–17.5% due to a corresponding decrease in fuel consumption due to the beneficial use of the recovery heat of the exhaust-gases of the boiler.

References

1. N. Fialko, R. Navrodskaya, G. Gnedash, G. A. Presich, A. I. Stepanova, S. I. Shevchuk, International Scientific Journal for Alternative Energy and Ecology **15**, 126–129 (2014)
2. R. Navrodskaya, N. Fialko, G. Gnedash, G. Sbrodova, Industrial Heat Engineering **39**, 4, 69–75 (2017)
3. N. Fialko, A. Stepanova, R. Navrodska, J. Sherenkovsky, Promyshlennaya teplotekhnika **36**, 1 ,41–47 (2014)
4. N. Fialko, A. Stepanova, G. Presich, G. Gnedash, Promyshlennaya teplotekhnika **37**, 4, 71–79 (2015)

5. L. Jia, X. F. Peng, Y. Yan, Int. J. Heat Mass Transfer **44**, 4257–4265 (2001)
6. P. Rączka, K. Wójs, Rynek Energii **111**, 87–92 (2014)
7. A. I. Sigal, Thermal Engineering **12**, 34–37 (2004)
8. E. N. Bukharkin, Industrial Power Engineering **6**, 40–42 (2004)
9. Y. M. Dashevskiy, B. Novikov, L. R. Khasanov-Agayev, Thermal Engineering **4**, 61–63 (1991)
10. N. M. Fialko, R. A. Navrodskaya, S. I. Shevchuk, G. A. Presich, G. A. Gnedash, O. Glushak, Modern Science: Researches, Ideas, Results, Technologies **2**, 15, 13–17 (2014)
11. V. V. Mitor, *Thermal calculation of boiler units* (Normative method, Ekolit, 296, 2011)
12. N. M. Fialko, V. I. Gomon, R. A. Navrodskaia, V. G. Prokopov, G. A. Presich, Promyshlennaia teplotekhnika **22**, 2, 49–53 (200)

Cross-building cooling-to-heating energy transfer

Elżbieta Niemierka[1,*], and *Piotr* Jadwiszczak[1]

[1]Wroclaw University of Science and Technology, Faculty of Environmental Engineering, pl. Grunwaldzki 13, 50-377 Wrocław, Poland

Abstract. Nowadays office buildings are faced with high and long-term cooling demand with grate heat recovery potential. In low heating demand office buildings not all of recoverable excess heat can be utilised, so it forces to search the consumers beyond the energetic boundary of office building. One of more promising way is supplying residential building by excess heat to meet the space heating and domestic hot water demands. Proposed cross-building cooling-to-heating energy flow allows transferring and utilizing excess heat from office building in residential as a useful heat. This solution creates the flexible and sustainable environment and meets the energy challenges of the future, in line with current energy trends and policy.

1 Introduction

Nowadays office buildings are characterized by a high cooling load. During the operation hours, HVAC systems remove large amounts of heat from office spaces, technical rooms and data centres [1], losing energy by rejection into the environment. Huge amount of rejected excess heat causes surrounding thermal pollution [2] and increases the inlet air temperature, secondarily increasing the cooling demand [3], especially in urban areas.

Re-use of excess heat from office cooling is an attractive and favourable option in energetic, economic and environmental approach. Heat recovery technologies in HVAC are well known and developed in various application. The heat recovery from SC systems is limited by available heat sinks. Profitable solutions of heat recovery and usage are commonly applicate in hospitals [4], industry [5] or commercial buildings [6]. In office buildings, with low heating demand, not all of large amount of recoverable excess heat can be consumed. The excess heat overcapacity forces to search the future consumers beyond the energetic boundary of office building. Residential neighbourhood seem to be an extensive [7] and profitable consumer.

Proposed cross-building cooling-to-heating energy transfer will integrate office as a donor and residential as an acceptor building to take advantages excess heat utilization. Heat recovered in office building will be transferred to cover residential building demands: space heating (SH) and domestic hot water (DHW). Office buildings as a heat sources are usually located in high heat demand city centre and can generate excess heat in temperatures ranges suitable for residential heat supply [8]. Cross-building cooling-to-heating energy

[*] Corresponding author: elzbieta.niemierka@pwr.edu.pl

transfer offers high flexibility on supply and demand site. Office excess heat can supply the residential as a single heat source (covers full demand), as a part of multi heat source mix (covers part of demand) or can be rejected in overproduction periods (disadvantageous energy lost).

General overview of excess heat generation and residential heat demand profiles shows the energy potential of cross-building heat transfer (Fig. 1). The excess heat profile for typical office building is characterized by generation peaks during the operation hours (as green). Residential building profile consists the steady demand for storage DHW (as dark grey) and notable demand for SH in heating season (as grey). There are overlapped and time shifted areas of both building types energy profiles. Overlapping areas allow direct covering of residential demands by transferred excess heat. To utilize the non-overlapping heat, the energy management and balancing solutions need to be considered.

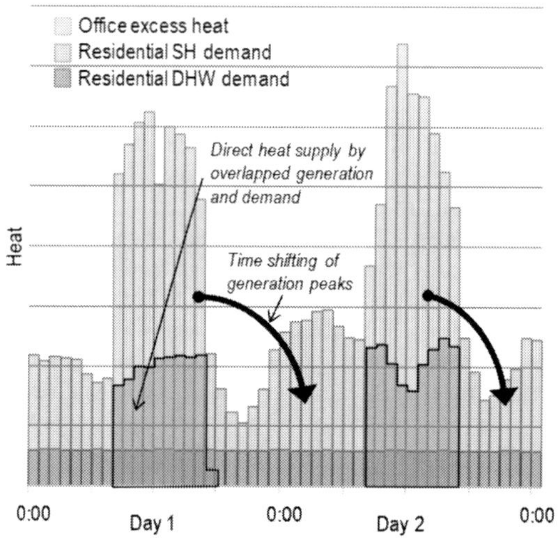

Fig. 1. Office building excess heat and residential building demand profiles for two specific days.

There are two main issues in cross-building cooling-to-heating energy transfer: 1) maximize the office excess heat transferring and consumption in residential building and 2) avoid the disadvantageous excess heat rejection. Based on defined assessment indicators specific cross-building cooling-to-heating connection modes will be developed and investigated to reach above mentioned energy targets.

2 Methods

The aim of the investigation was to test the energy potential of cross-building cooling-to-heating energy transfer based on the standardised buildings profiles. The sites were represented by the office building with excess heat from SC as a producer and by residential building with SH and DHW heat demands as a consumer. Three heat flow indicators were defined to assess the energy potential of specific cross-building connection modes: 1) the excess heat consumed in residential, 2) the excess heat needed to be rejected from office and 3) the complementary heat from local HS to cover residential demand.

To define the cross-building cooling-to-heating heat flows the yearly profiles of excess and demand heat were hourly simulated using building thermal modelling EDSL TAS software. Two variants of residential building energy standard were considered in office to residential heat transfer. First variant includes standard office building and standard residential building with SH and DHW demand. In second variant residential building was modified to low-energy standard. Buildings were simulated in accordance with the ASHRAE 90.1 [9] and ISO [10, 11] relevant specification, with local climatic conditions based on Wrocław city (Poland) data, as a central Europe location. The envelope and glazing of simulated building are shown as 3D models on Fig. 2. The annual energy performance of residential and office buildings was detailed and shown in Table 1. Due to the same net floor area the analysis results could be related to the unit area. The annual heat demand for SH and DHW of standard residential building equals the total office SC excess heat generation and amounted 133.7 kWh/(m²·a). Low-energy residential building has reduced the annual heat demand for SH and DHW by 73.8 kWh/(m²·a) which was nearly half of the standard building load.

Table 1. The annual excess heat and heat demands by end-use for all analysed buildings.

Building	Useful floor area, m²	Space cooling excess heat, kWh/(m²·a)	Heat for SH demand, kWh/(m²·a)	Heat for DHW demand kWh/(m²·a)
Office building	16 250	133.7	1.0	-
Standard residential building	16 250	no cooling	58.8	74.9
Low-energy residential building	16 250	no cooling	28.8	45.0
Office building	16 250	133.7	1.0	-

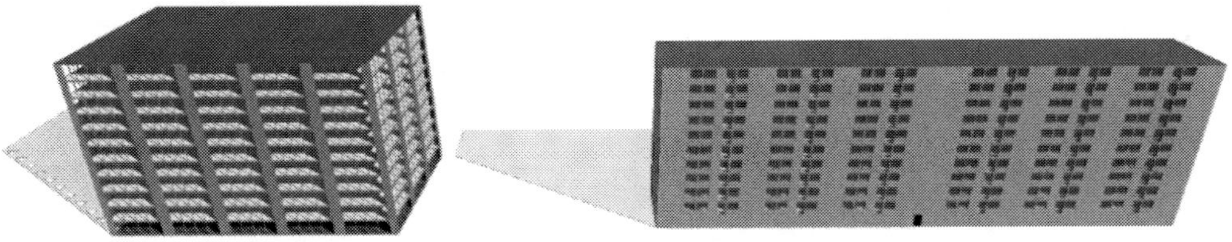

Fig. 2. Office (left) and residential (right) buildings 3D simulation models.

The essential connection modes for cross-building cooling-to-heating energy transferring was identified and analysed. Six investigated connection modes contain further improvements to maximize excess heat consumption and to avoid heat rejection. The following figures provide an overview of office to residential heat flows as a base for simulation to determine assessment indicators in specific connection modes.

In base mode analysed buildings are deprived of cross-building connection (Fig. 3a). The office building needs to reject all excess heat from SC while the residential uses the local HS to cover SH and DHW needs. The total energy consumption includes the sum of energy demands of both buildings.

Direct connection mode for cross-building cooling-to-heating energy transfer is shown in Fig. 3b. Excess heat from office building SC supplies the substation (SS) in residential building to meet SH and DHW demands. For residential building the excess heat plays the role of the primary HS. Excess heat from overlapped profiles areas could only be transferred and consumed. Unused heat from office SC must be rejected. The remaining heat demand must be covered by local HS in periods of waste energy deficit. Total energy consumption is less than the of sum energy demands in both buildings.

Fig. 3c presents semi-direct connection mode of cross-building cooling-to-heating energy transfer with short term thermal energy storage (TES) and local HS. The TES collects the excess heat to balance the office excess and residential demand. The TES capacity was defined to stored excess heat generation of the largest 24 subsequent hours. Due to the TES time shifting of heat loads more excess heat could be consumed and less of energy will be rejected. The local HS operation will be reduced. Limited storage capacity restricts the excess and demand heat balancing. The total energy consumption is lower both: than the sum of energy demands in residential and office buildings and then in cross-building direct connection mode.

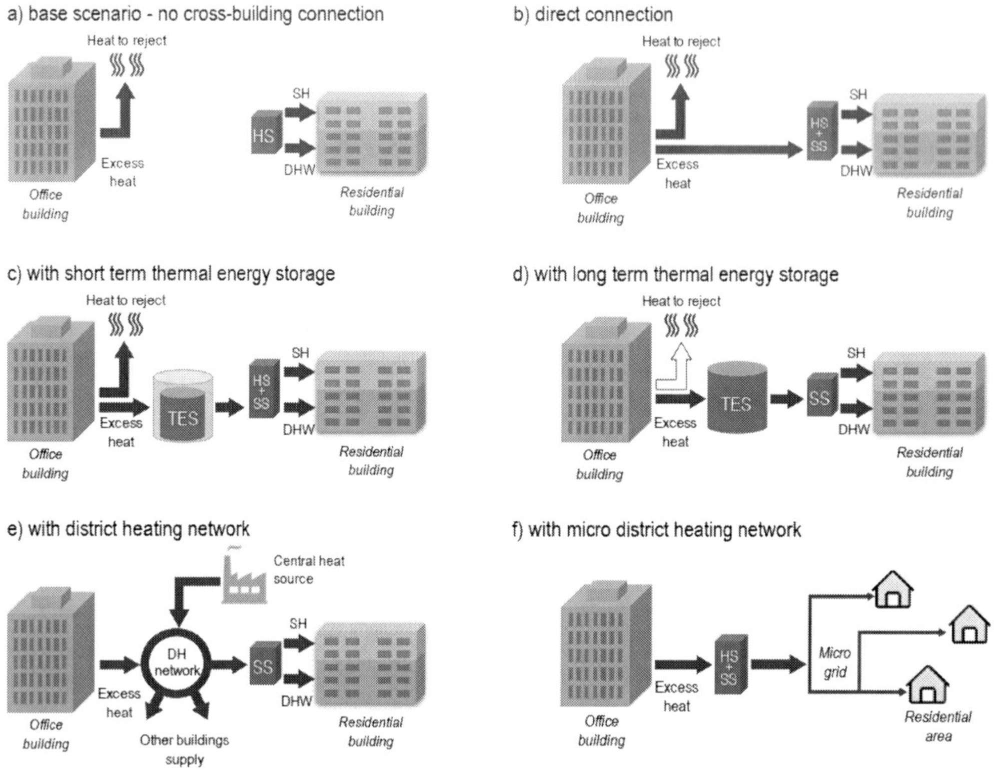

Fig. 3. Investigated cross-building connection modes.

Cross-building heat transfer shown on Fig. 3d illustrates semi-direct connection mode with seasonal TES. The TES capacity is sized to annual balance the office excess heat generation and the residential heat demand. For building with similar annual performance TES connection mode allows transferring, storing and consuming of all excess heat. A heat rejection and local HS operation are no longer necessary. In this mode the both buildings require as much energy as only one building in a no-cross-building connection mode.

Cross-building connection through district heating (DH) network mode is shown in Fig. 3e. Integration into a fourth-generation DH network provides fully usage of excess heat as a supported by the central HS. Existing DH infrastructure is suitable to take over and distribute the excess heat. The DH system can balance the excess and demand in multiple office and residential buildings, connected into DH, at the district, city or even regional scale.

The office building can be a heat source also for small-scale customers in areas with no DH infrastructure. Cooling-to-heating energy transfer will be an in-situ heat source for low-temperature micro DH network as shown in Fig. 3f. In idealized system whole office SC excess heat is transferred into DH as a base supply for SH and DHW demand. Operation of local HS in SS complements the residential heat demands. In this connection-mode the total energy use will be locally decreased in scale of community.

3 Results

The yearly excess and demand heat profiles were defined based on EDSL TAS thermal simulation data. Dynamic simulation with hourly accounting step enables assessing the energy profiles for three investigated buildings: office, standard residential and low-energy residential. Daily data aggregation clearly presents the annual heat profiles of office and residential buildings, as is shown in Fig. 4. The A-shape excess heat profile covers only part of the U-shape SH and DHW demands directly. Thus, the direct excess heat supply is limited to overlapped areas. To maximize the excess heat consumption in residential building and to avoid the disadvantageous waste heat rejection the load time shifting is needed.

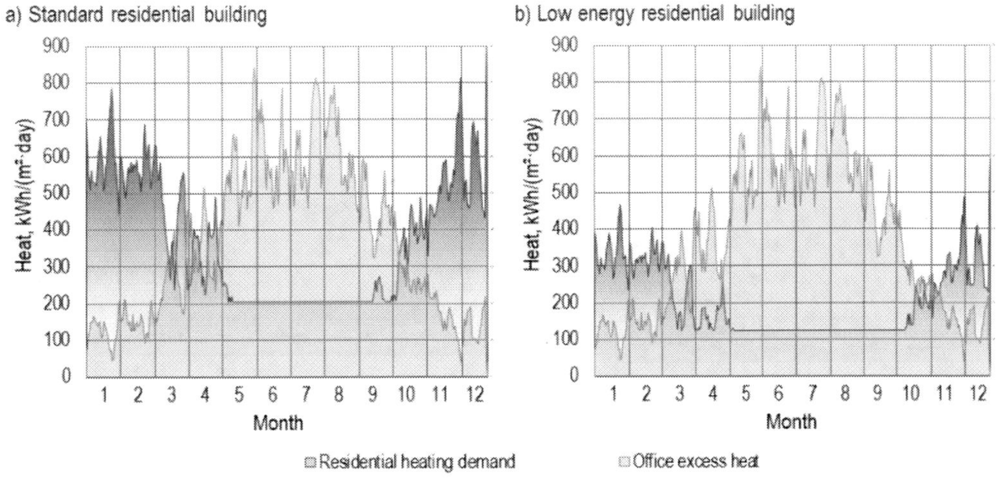

Fig. 4. Annual profiles of office excess heat generation (green) and residential heat demand (grey) daily distribution in standard (a) or in low-energy (b) residential buildings.

To illustrate the investigated cross-building heat transfers in six developed connection modes the EDSL TAS results were daily integrated. Fig. 5 shows the three assessment indicators of cross-building heat transfer: 1) excess heat consumed in residential building (blue), 2) rejected unused excess heat from office building SC (green) and 3) complementary heat from local HS in residential buildings (red).

In the Fig. 5a and 5b are shown the results for base no cross-building connection mode for two variants of residential. The no office to residential heat transfer scenario illustrates the annual profiles of excess heat generation (green area) and residential heat demand (red area). The SH and DHW loads of standard and low-energy residential buildings differ almost twice, however have analogical shapes and trends. Standard residential building load peaks to almost 813 kWh/(m^2·d) and has nearly the same annual heat demand as the excess heat generated in office. The low-energy building riches the maximum of 496 kWh/(m^2·d).

Simultaneously appearance of excess from office and demand in residential entail the application of cross-building direct connection. The amount of cooling-to-heating energy is clearly visible and maintain during the year almost at the same level (blue area on Fig. 5c). The mismatching of residential and office energy profiles causing the fluctuates. The daily aggregation of data allows to clearly visualisation of the trends, however affect the hourly episodes flatting. In direct connection mode the excess heat covers up to 100% of hourly residential demand. The large green area of rejected excess heat illustrates the unused potential of office cooling system as a heat source. In low-energy residential the trends are analogous, thus the next connection modes are plotted for standard residential building only.

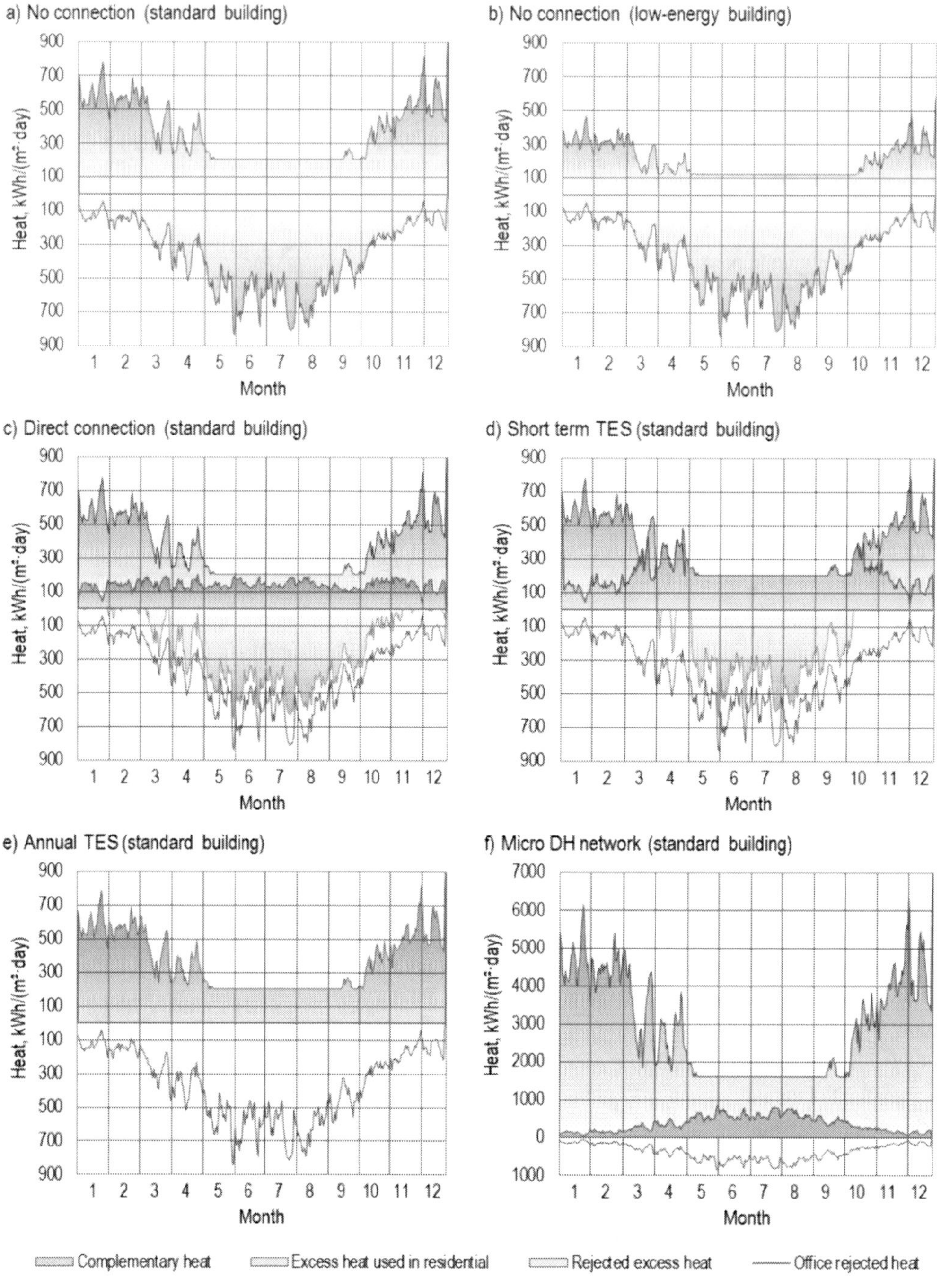

Fig. 5. Annual profiles of office excess heat generation (green) and residential heat demand (grey) daily distribution in standard (a) or in low-energy (b) residential buildings.

Thermal storage smoothing influence of generation and demand sites profiles. Short term TES (Fig. 5d) increases the excess heat utilisation especially in spring and autumn. The 24-hour heat storage is beneficial for the system, but limited TES capacity causes incomplete balancing of daily excess and demand peaks. The rejected excess heat was noticeably limited (green area), covering residential demands in summer (blue area). The annual TES (Fig. 5e) allows to fully use the SC excess heat in residential SH and DHW. Green area of rejected heat disappears and annual residential demand was represented by blue filling. There is no need to reject excess heat and complementary HS operation. Whole amount of waste heat was transferred and consumed in standard residential building as a single heat source.

In DH cross-building cooling-to-heating connection mode district network takes over and distributes all excess heat generated in connected office building. The integrated office buildings become a part of multi-source district heating system.

Micro-grid (Fig. 5f) according to assumptions are based on excess heat as a primary energy source. The high of residential demand profile will overlap each of the hourly excess peak. On daily scale chart the residential heat demand seems to be much lower than the SC excess. All generated excess heat from office SC will be locally transferred and consumed in-situ. The office SC excess heat (blue area) was complemented by the local HS operation (red area) eliminating heat rejection (no visible green area).

The three assessed indicators allow evaluating the energy effect of cross-building cooling-to-heating energy transfer both on demand and supply site. On demand site, based on simulation data, the annual proportion of consumed excess and complementary heat were defined and assessed (Table 2).

Table 2. The annual excess heat and heat demands by end-use for all analysed buildings.

Indicator	Residential building	No connection	Direct connection	Short term TES	Seasonal TES
Excess heat in residential demand	SB	0.0%	38.9%	57.4%	100.0%
	LEB	0.0%	46.3%	73.4%	100.0%
Complementary heat in residential demand	SB	100.0%	61.1%	42.6%	0.0%
	LEB	100.0%	53.7%	26.6%	0.0%

In no connection mode whole excess heat was rejected and wasted. In direct connection mode the simulation data provides the annual share of excess heat in residential demand at the level of 38.9% and 46.3% for standard (SB) and low-energy (LEB) residential building respectively. The rest of uncovered demand was complemented by local HS operation, supplying 61.1% (SB) and 53.7% (LEB) of annual SH and DHW. The hourly simulation shows up to 100% coverage of residential demand by excess heat in overlapping periods. In remaining time, the SH and DHW were supplied by local HS or by excess and HS mix. Short term TES increases share of excess heat in residential demand. The 57.4% (SB) and 73.4% (LEB) of annual heat demand was covered by excess heat from office building SC. The rejected heat was favourable limited from one-third (SB) to almost twice (LEB). Simulation of annual TES connection mode shows full covering of SH and DHW demand by excess heat. Due to time shifting of long term storage the excess heat covers 100% of residential demand both in SB and LEB. The district heating and micro-grid connection modes allow transferring and utilizing all excess heat generated in office building SC for both residential building variants.

Application of simple direct connection mode contribute to 38.9% (SB) and 46.3% (LEB) energy benefit. Implementation of short-term TES intensifies the excess heat utilisation by 18.5% (SB) and 27.1% (LEB) in relation to the direct connection. The TES extension to the annual accumulation level provides the independence of additional heat source required in SB and LEB. Accomplishment these requirements have much greater involvement with comparable profits than transition from standard to direct or from direct to short-term TES modes.

Despite almost two-fold annual demand reduction changing from standard to low-energy building not increase the excess energy share of twice. In the direct connection mode, the share of waste heat growth only 7.4% and in short term TES 16.0%. In annual TES connection mode, the whole SH and DHW demands were covered by excess heat with the waste heat overproduction in low-energy building variant.

The percentages of rejected excess heat on supply site are shown in Table 3. No connection mode precludes the cross-building heat transfer and whole generated excess heat needed to be rejected and wasted. In direct connection mode, due to non-overlapping profiles, up to 61.2% (SB) and 74.7% (LEB) of generated excess heat was unused and lost in rejection process. Short term TES application decreases the untransferred heat rejection on generation site for 43.1% (SB) and for 60.1% (LEB). In office to standard residential building connection mode the annual TES allows overtaking, storing and transferring 100% of excess heat generated in office building. In low-energy building supply until 44.8% of available excess heat was rejected, due to the small SH and DHW needs regarding to available excess heat value. The surplus heat can be transferred to supply next low-energy residential buildings.

Table 3. Assessment indicator for supply site.

Indicator	Residential building	No connection	Direct connection	Short term TES	Seasonal TES
Unused and rejected excess heat	SB	100.0%	61.2%	43.1%	0.0%
	LEB	100.0%	74.7%	60.1%	44.8%

Presented results forms the connection-modes ranking in energetic approach. The developed assessment methods, based on annual simulations, can be easily extend by next indicators related to environment, economic, building services, district infrastructure or other local factors and conditions. The specific set of assessment indicators determines the ranking and best solution of cross-building connection modes.

4 Conclusions

The study revealed large energetic potential of cross-building cooling-to-heating energy transfer, as a favourable heat source for residential buildings. In excess heat abundant cities cross-building connection will integrate office donor and residential acceptor buildings to beneficial waste heat transferring and consumption beyond the donor building energy boundaries. Office buildings are an excess heat sources usually located in high heat demand city centre.

Simulation results show energy profits in every cross-building connection mode. The share of excess energy in residential demand varies according to applied connection mode, energy standard and energy profiles of integrated buildings. The basic direct cross-building connection is already beneficial and technical developments in heat transferring improve the energy effect. Energy standard of supplied building affects the supply and demand site. On demand site the low-energy standard of residential building increases the share of excess heat in annual demand for all investigated connection modes. On supply site lower residential demand limits the excess heat utilisation and forces energy losses in waste heat rejection. On other site it creates opportunity to supply the next residential buildings.

Based on presented assessment indicators the most favourable connection mode can be determined depending on expected energy effects and local conditions. The indicators methodology could be also extended to include ecological or economic issues. Energy and environmental benefits of excess heat utilisation are convergent. Cross-building heat transferring reduces or even eliminates operation of conventional heat sources for residential SH and DHW supply. The environmental impact will be double-reduced in the emission mitigation and no-thermal pollution.

The proposed connection modes for cross-building cooling-to-heating energy transferring are high applicable. Cross-building cooling-to-heating transfer can be applied in existing, refurbished and new constructed buildings. Cross-building connection creates the flexible and sustainable environment for excess heat transferring and utilization in integrated buildings of different function, as a local heat linking and neighbourhood energy community.

Proposed energy linking can be applied as an island energy system or as a part of small or large scale energy system including multi-source and smart grid technologies. This solution meets the energy challenges of the future, in line with current energy trends and policy. This concept should be submitted by holistic framework of multi-functional masterplaning and should be considered in all excess energy abundant locations, especially in urban and DH areas.

References

1. M.Wahlroos, M. Pärssinen, J. Manner, S. Syri, Energy, **140** (2017)
2. X. Xu, F. Chen, S. Shen, S. Miao, M. Barlage, W. Guo, A. Mahalov, J. Geophys. Res. Atmos. **123**, 5 (2018)
3. Y. Wang, Y. Li, S. Di Sabatino, A. Martilli, P. W. Chan, Environ. Res. Lett. **13**, 3 (2018)
4. https://www.hfmmagazine.com/articles/3494-new-boiler-and-chiller-advances-for-hospitals
5. S. Brückner, H. Schäfers, I. Peters, E. Lävemann, Sustain. Cities Soc. **13** (2014)
6. W. Loibl, R. Stollnberger, D. Österreicher, Sustainability **9**, 2 (2017)
7. International Energy Agency. Energy balances (2016) https://www.iea.org/Sankey
8. K. Ebrahimi, G. F. Jones, A. S. Fleischer, Renew. Sust. Energ. Rev. **31** (2014)
9. ANSI/ASHRAE/IES Standard 90.1-2016 - Energy Standard for Buildings Except Low-Rise Residential Buildings
10. ISO 17772-1:2017 Energy performance of buildings – Indoor environmental quality – Part 1: Indoor environmental input parameters for the design and assessment of energy performance of buildings
11. ISO 17772-2:2017 Energy performance of buildings – Overall energy performance assessment procedures – Part 2: Guideline for using indoor environmental input parameters for the design and assessment of energy performance of buildings

Potential of individual heat pumps for renewable energy storage in Smart Grid

Elżbieta Niemierka[1,*], and *Piotr* Jadwiszczak[1]

[1]Wroclaw University of Science and Technology, Faculty of Environmental Engineering, pl. Grunwaldzki 13, 50-377 Wrocław, Poland

Abstract. Ever-increasing power market and environmental policy enforce growth of renewable power sources. Renewables inflexibility and dependency on weather condition causes periodically imbalance in power system due to the green power overproduction. With the increase of renewable sources, the balancing problems in power system will be increasingly significance issue. It is proposed to use individual heat pumps as a next tool for energy system adjustment support. Power system adjustment will be carried out by active demand side management by intended domestic hot water tanks overheating. The smart grid individual heat pumps setpoints will be switched at community or even country scale. The strategy allows shaving the overproduction peaks through short-term increase of electricity consumption in remote controlled heat pumps and to lowering power demand during green power deficits using the thermal energy stored in overheated domestic hot water. The dynamic mathematical simulations were made to define the operation and limitation of active control strategy of heat pumps integrated into smart grid. The results allow testing and assessing the potential of individual heat pumps as a next tool for balancing the power system with large scale of renewable power.

1 Introduction

Worldwide power market steadily growing driven by ever-increasing electricity demand. Maintaining the economic growth and national power security decoupled from the associated environmental pressures are the nowadays issue which are mainly deployed and implemented through increasing share of renewable energy sources. At present, wind power is the leading source of renewable energy [1]. In sustainable power system the renewables are complemented with conventional combustion power, CHP or nuclear plants. Such energy policies lead to the challenge of balancing electricity demand and supply from specific power sources. Increase of green power will indicate overwriting the control and management strategies. Applying the smart grid technologies allows maintaining balance in community, country or even region power system.

The electricity overproduction risk mainly depends on energy power sources mix. The higher share of renewables and lower operational flexibility of conventional sources, the greater risk of overproduction. Despite the growth of world homogeneity each country

* Corresponding author: elzbieta.niemierka@pwr.edu.pl

is characterised by different electricity generation technologies mix. On the Fig. 1 are shown electricity production structure by different fuels in 1990 and in 2016, for overall world, European Union Member States and four distinct countries. Germany, Poland, Spain and Sweden represent different fuel mixes and different rates of renewable development between 1990 and 2016. A significant increase in renewables in electricity production is visible, both at the national, regional and worldwide level. On global scale the green power grew of 4.3% amounted more than 3.6 million GWh. At the European Union level, the regional legislation promotes and forced increasing renewables of 20% final energy consumption by 2020 [2].

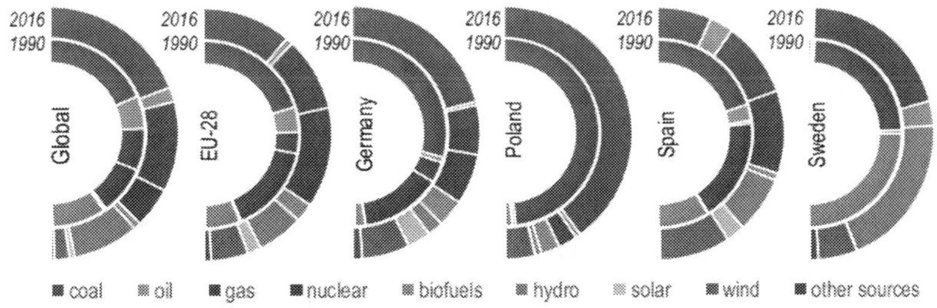

Fig. 1. Electricity production by fuel in 1990 and 2016 [3].

High share of renewable power sources is favourable for mitigation on climate change and power system safety, however, increases the unbalance risk. Renewable sources are majority dependent on meteorological condition and in mostly are complemented with non fast-respond conventional sources (Fig. 2). The long hot start-up time and limited ramping rate of common conventional power generation technologies in combination with inflexibility of renewable power sources require dynamic management and increase electricity overproduction risk.

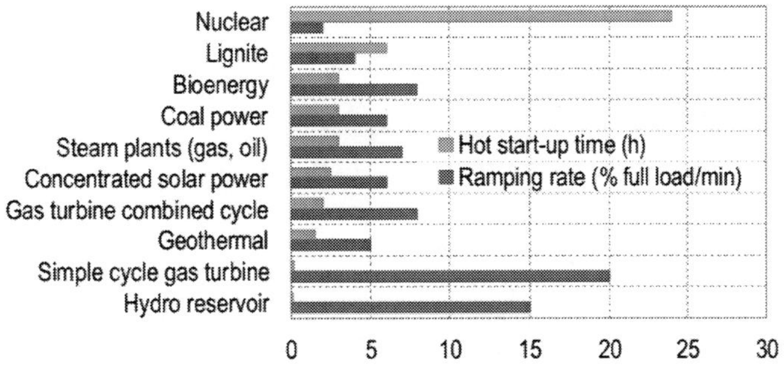

Fig. 2. Flexibility and dynamics of common power generation technologies by fuel [4].

The main goal of current research and development is to solve the problems of excess green energy in power systems. The search areas are focused around smart grid solution in supply and demand management, energy storage and energy conversion. Due to high environmental impact [5], investment and life-cycle costs the effective surplus electricity storage or conversion technologies are still at research and development stage.

Heat pumps (HP) in combination with domestic hot water storage tanks emerge as a favourable and flexible power-to-heat technology for excess electricity storage. Increasing the individual HP market runs parallel to renewable power sources. As shown on Fig. 3 the number of installed HP units in European Union achieved most than 10.6 million in 2017 and is characterised by steady upward trend. More than three quarters of the installed HP are working in standalone installation [6] and a large number of them are fitted with thermal energy storage, mostly for domestic hot water demand.

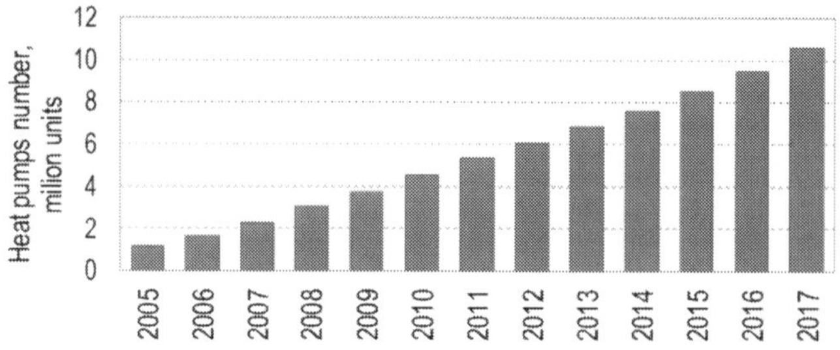

Fig. 3. The number of heat pumps installed in European Union [7].

Millions of installed HP units create a large-scale potential for smart and controlled storage of excess green power in individual water tanks. The smart grid technology will involve the large group of existing and future individual HP in renewable power system active balancing. HP integration into a smart grid creates a new opportunity to remote demand and power-to-heat management in individual HP. Driving the HP units with green power is double-environmentally friendly, zero emission and helps to increase the green power share. In smart grid with green power pricing the heat pumps are profitable both for demand and supply site [8]. In overproduction periods the individual HP units help to balance the power system by consuming excess power in low-price or free energy tariffs. Additionally, the individual HP units are smart grid ready in plug and play technology and are financed by the users. For these reasons the smart grid HP units should be considered as a next tool for active balancing the power system with large share of renewables.

2 Method

The aim of the investigation is to assess the power peaks shaving potential of individual domestic hot water storages supplied by individual HP units integrated into smart grid. The electricity overproduction will be storage as a water sensible heat and utilise during the subsequent domestic hot water demand. The advanced fast-response Smart Grid Control (SGC) strategy remotely changes the setpoint of water storage temperature in periods with electricity overproduction. The SGC changes the setpoint from the user value to the maximum technically or economically justified temperature for analysed HP units. The number of overheated HP is dynamically changed along with the amount of excess energy, considering technical limitations and user safety.

The investigated power system is based on mixed conventional and green power generation on the supply side and large number of individual heat pumps for thermal energy storage in domestic hot water tanks on the demand side. All components are integrated into smart grid to enable supply and demand side communication, remote setpoint control and energy management. The demand on conventional and green power were investigated by numerical simulation for two HP operation scenarios (Fig. 4). Base scenario where each HP unit operates separately based on the own digital controller setpoints. Developed SGC scenario activates an additional power-to-heat storage capacity by smart grid setpoint control, as an additional field of power demand management. The SGC strategy creates scalable power-to-heat storage capacity by integration of individual HP units, for remote switching of stored water temperature setpoints.

For the purpose of the assessment investigated HP units are divided into five groups (from HP1 to HP5) with individual SGC for setpoints control. Small number of groups allows to clearly illustrate the SGC operation principles and assessing the energy storage potential. In the real condition denser division heat pumps on SGC group, the greater adaptation to supply site.

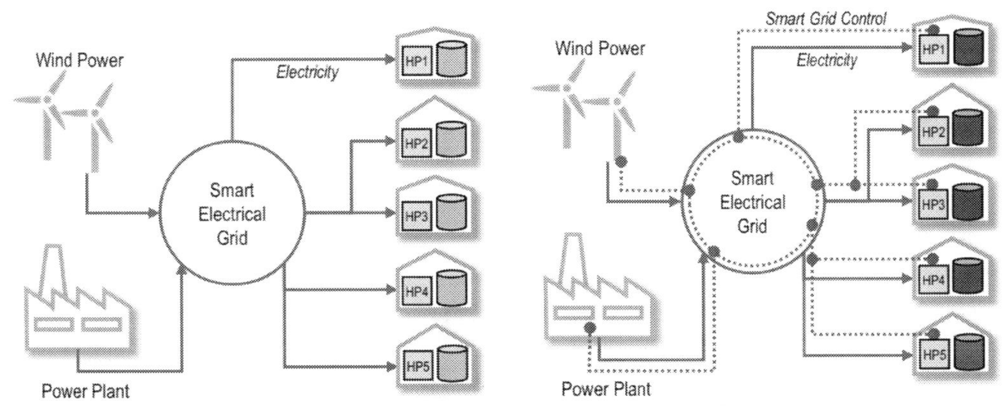

Fig. 4. Investigated power system in base scenario (left) and smart grid control scenario (right).

Assessing the potential of power-to-heat SGC for overproduction limitation in power system requires embed in quasi-real boundaries condition. All system components used to create system reflecting system idea are detailed with appropriate mathematical models expressed by differential and algebraic equation simulated in TRNSYS software [9]. Prepared simulation layout is shown on Fig. 5.

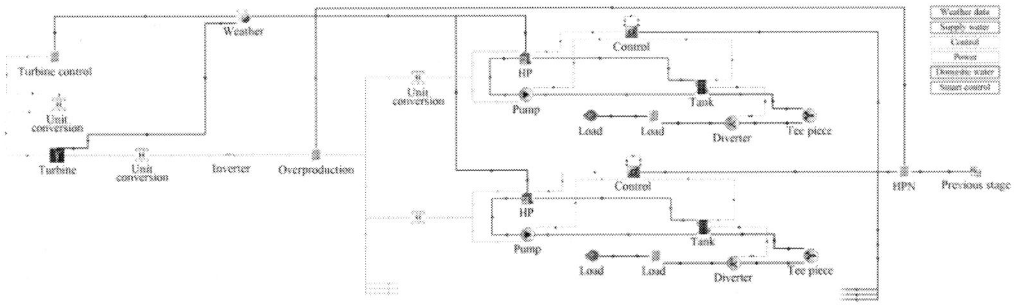

Fig. 5. TRNSYS simulation model.

The production (power supply) site was represented by the Type90 model parameterized with the Nordex N27/150 wind turbine power curve and required technical data [10]. The demand (power consumption) site consist of the individual HP units (Type917) with the thermal energy storage (Type4) for the domestic hot water (Type14) with ASHRAE profile [11]. The temperature of tap water is stabilising independent on time by mixing in Type11 with cold water to ASHRAE requirements [11]. Type2 model reflecting the control system delivered by the HP manufacturer with the default setpoint (DSP) temperature. The weather data came from the Poland database for Kołobrzeg city. TRNSYS flexible model was built to allow using technical and climatic data specific for other locations and systems in further research. Despite the established power source and meteorological condition, the conclusions will be proper to other renewable power sources or them groups.

In simulated systems the overproduction is monitored with very short time-step (5 seconds) in annual scope. For annual assessment the four groups of simulation output data were collected: renewable and conventional power consumptions, power system overproduction and individual HP demand site condition (temperatures and operation mode).

SGC strategy involves individual HP to shaving the overproduction peaks. While the system is balanced, the HP units operate in DSP mode (45°C) Every HP unit will automatically switch on and off keeping DSP by the basic manufacturer's controller according specific profile of domestic hot water consumption. In the case of excess power (overproduction), the SGC will switch the appropriate number of HP groups to overheating setpoint (OSP) temperature 70°C for power-to-heat store mode. The aim is to achieve power system balance by active SGC of HP power consumption, according to the current electricity overproduction. Specific groups of HP units will operate as long as the overproduction exist. The overproduction was limited by the amount of controlled and ready to OSP work HP units.

3 Preliminary results

The dynamic mathematical model of investigated power system, build in TRNSYS software, includes and simulates all devices, controls, operation modes and energy flows. The whole year simulation results enable interpretation and assessment both the input and output data. To illustrate the acting and potential of SGC strategy the four most characteristic operation patterns in power system are shown. Each pattern includes three sections illustrating the HP operation, generated power variability and water temperatures in storage tanks. The top section shows the operation modes of five HP groups. Particular HP group is characterised by three operating modes: OFF (switched off), DSP (operation in default domestic hot water setpoint) and OSP (operation in overheating domestic hot water setpoint). The middle section shows the generated wind power and power overproduction in the system. Dashed line represents the minimum overproduction power to activate SDC strategy. The bottom section presents the hot water temperature changes. Temperature in storage tank increases by HP charging and decreases by domestic hot water consumption.

The pattern on Fig. 6 illustrates the operation mode of five HP groups in case of negligible overproduction in power system. In well balanced power system SGC sets the DSP in all HP units. Almost all generated power is consumed, which indicates the high system efficiency in relation to the electricity unit. While overproduction is lower than the minimum OSP switch level in SGC strategy (dashed line), the SGC did not interfere with the HP control. This pattern defines the base operation mode to assess the SGC effect and actions in the following cases.

The second pattern (Fig. 7) illustrates the long-term and large overproduction case in power system. In green overproduced, unbalanced system the SGC automatically sequentially switches HP groups into OSP, to consume and store increased excess power. Finally, all controlled HP units operate in OSP mode and all domestic hot water tanks will be fully load. This case illustrates availability of fast-respond SGC strategy to power system balancing support by active short-term electricity consumption increase in controlled HP. The SGC stores the excess green energy in overheated tanks to utilise during green power deficits.

Fig. 8 illustrates the full potential of SGC strategy in dynamic conditions. Two specific periods in energy supply and demand are visible: green power overproduction period and green power deficit period. Excess power forces the SGC to OSP sequentially switch and HP fully thermal energy storage. The accumulated heat will be consumed in green energy deficit period with no conventional power demand. Two advantages of SGC strategy are noticeable: the reduction of overproduction problem and no conventional power demand for almost one day.

The SGC achieves beneficial results both in energy and environment approach. The SGC covering overproduction could also cause in electricity price reduction and minimise battery power accumulation with higher environmental and economical costs.

In real conditions the excess green power periods can occur frequently, more than once a day. The fourth pattern (Fig. 9) includes more than one period of overproduction with too short interval among causes non-fully thermal energy utilisation. First overproduction forces the SGC to OSP switch. The surplus green power storages in overheated domestic hot water tanks will not be fully consumed in the DST mode. During the second overproduction period SGC switches HP into OST, although the tanks low discharged-levels enables power utilisation. This issue is connected with mismatched amount of connected HP to SGC and overproduction peak size. To fully load the potential the SGC strategy should be implemented on area with large number of smart grid consumers.

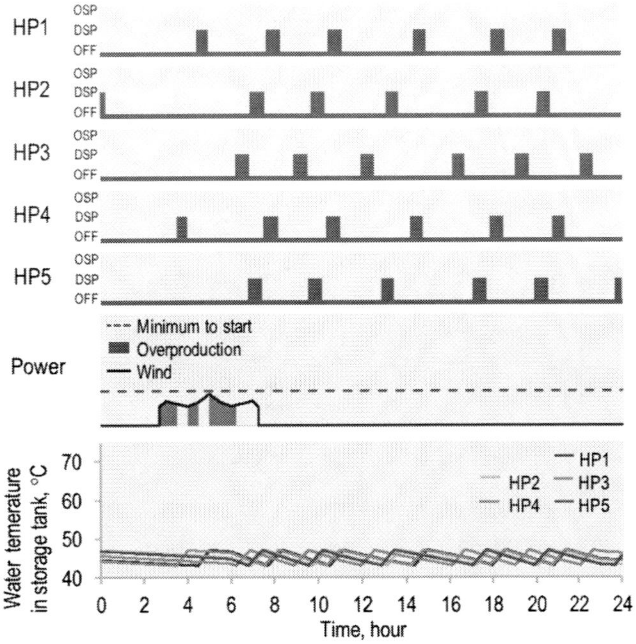

Fig. 6. Smart grid control in default setpoints conditions.

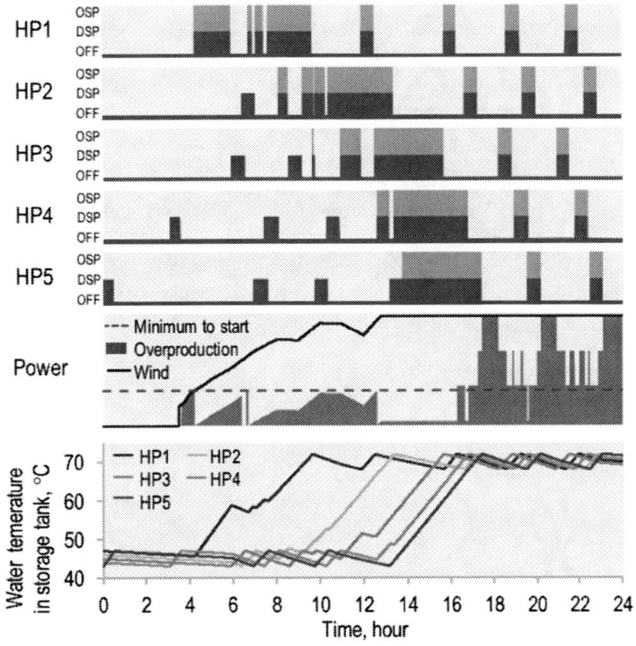

Fig. 7. Smart grid control in case of long-term overproduction.

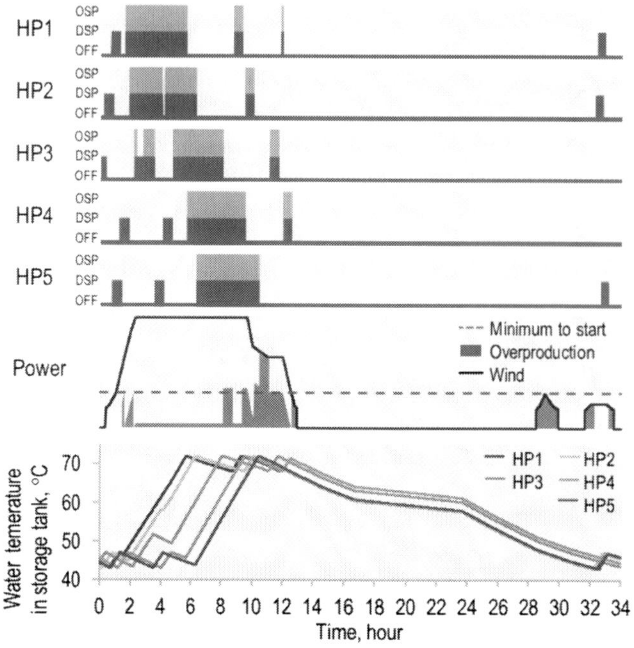

Fig. 8. Smart grid control in case of short-term overproduction and fully thermal energy utilisation.

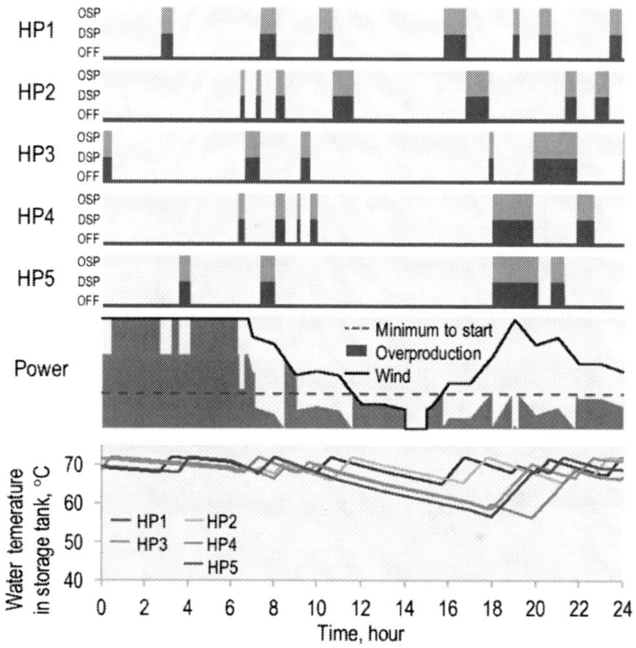

Fig. 9. Smart grid control in case of multiple overproduction day.

4 Conclusions

Ever-increasing renewable power sources caused by legislation and environmental impact limitation, generates pros and cons on power market level. Unbalance and power system blackouts caused by green power overproduction will be the major challenges of going more and more renewable. Nowadays, technology development aim is to extend and overwrite control procedures to overproduction minimisation. Purposed Smart Grid Control (SGC) is investment and environmental favourable. The SGC strategy is based on existing generation, transmission and demand infrastructure. Applying flexible and fast-respond tool for active balancing power system with large scale of renewables, will be achieved by using individual heat pumps with thermal energy storage. Using SGC strategy to green power overproduction limitation creates double-environmentally friendly system. Firstly, by using renewable power to supply HP units as a renewable heat sources, secondly by utilising excess green power to overheating and using accumulated thermal energy while lack of green power. SGC on the global level could: mitigate environmental impact of energy generation through them more fully usage, on country level increase the power production efficiency and ensure energy security, and on user level reduce HP operating costs.

Dynamic mathematical models enabled to simulate and tracking the operation modes and energy flows in whole year period. Based on simulation results it can be seen that investigated strategy caused beneficial both by fully and partly overproduction utilisation and is promising solution to balance power system with cheap and easy to implement smart grid strategy. The potential of implementation in all countries are large enough to further develop this control strategy. In this paper SCD was based on domestic hot water storage but could be also extend on HP space heating by active using the buildings thermal capacity. Proposed solution uses modern technologies and supports development of green power sector moving them into a low-carbon future.

References

1. L. Freris, D. Infield, *Renewable energy in power systems* (John Wiley & Sons, Chichester, 2008)
2. Directive 2009/28/EC of the European Parliament and of the Council of 23 April 2009 on the promotion of the use of energy from renewable sources and amending and subsequently repealing Directives 2001/77/EC and 2003/30/EC
3. International Energy Agency Statistics (2018) Statistics data browser. https://www.iea.org/statistics. Accessed 15 January 2019
4. M. A. Gonzalez-Salazar, T. Kirsten, L. Prchlik, Renew. Sust. Energ. Rev. **82** (2018)
5. J. F. Peters, M. Baumann, B. Zimmermann, J. Braun, M. Weil, Renew. Sust. Energ. Rev. **67** (2017)
6. K. Dawson, *Latest trend in the World Traditional & Renewable Heating Market* (BSRIA, 2015)
7. A. Rathi, *A 19th-century solution to heat homes is helping the world cut emissions* (Quartz, 2018)
8. L. Hirth, Energ Econ. **38** (2013)
9. University of Wisconsin-Madison. *Solar Energy Laboratory* (1975) TRNSYS, a transient simulation program. Madison. Wis. The Laboratory
10. Nordex SR Technical data. http://www.nordex-online.com
11. ASHRAE Standing Standard Project Committee 90.2 (2007) *Energy-Efficient Design of Low-Rise Residential Buildings*, ASHRAE

Isolation of anaerobic bacteria of the *Bacteroides fragilis* group from environmental samples

Sebastian Niestępski[1,*], *Monika* Harnisz[1], *Ewa* Korzeniewska[1], and *Adriana* Osińska[1]

[1]University of Warmia and Mazury in Olsztyn, Faculty of Environmental Sciences, Department of Environmental Microbiology, ul. Prawocheńskiego 1, 10-720 Olsztyn, Poland

Abstract. The aim of this study was to analyze the effectiveness of identification of *Bacteroides fragilis* group (BFG) strains isolated from human and rat feces, hospital wastewater, and untreated and treated sewage from a wastewater treatment plant (WWTP). BFG strains were plated on Bacteroides Bile Esculin (BBE) agar. Characteristic colonies were isolated from the culture medium, tested for antibiotic resistance and identified by PCR and MALDI-TOF MS. A total of 319 strains that formed characteristic colonies were isolated from BBE agar, of which 250 were resistant to kanamycin, colistin and vancomycin. PCR revealed that only 135 strains harbored the *bfr* gene specific to the BFG. In MALDI-TOF MS analysis, only 123 isolates were classified as members of the BFG. The most frequently isolated species was *Parabacteroides distasonis* (51.22% of strains). *B. fragilis, B. ovatus* and *B. thetaiotaomicron* were less frequently encountered in the analyzed environmental samples (30.01%, 8.13% and 5.69%, respectively). The strains isolated from human and rat feces on BBE agar were most reliably identified, and 100% of the isolated strains were classified as members of the BFG. The effectiveness of isolation of BFG strains from hospital wastewater and untreated and treated sewage from the WWTP was relatively low (20%, 19%, 40%, respectively). The results of this study suggest that the method for isolating BFG strains from environmental samples on BBE agar requires modification. Additional methods, such as PCR and MALDI-TOF MS, can be used to more effectively identify BFG strains isolated from different environmental samples.

1 Introduction

Bacteria of the genus *Bacteroides* and the closely related genus *Parabacteroides* are Gram-negative obligate anaerobes. These immobile and non-sporulating bacteria are part of normal gut microbiota in humans and animals [1]. Species of the genera *Bacteroides* and *Parabacteroides* include the *Bacteroides fragilis* group (BFG) of 44 potentially pathogenic species which are the most frequently isolated anaerobes in human infections [1, 2]. BFG strains are the most antibiotic-resistant anaerobic pathogens [3,4]. Clinical BFG strains are presumptively identified on Bacteroides Bile Esculin (BBE) agar, a selective and differential culture medium [5, 6]. The medium is used to differentiate bacteria capable of growing in the

* Corresponding author: sebastian.niestepski@uwm.edu.pl

presence of 20% bile and gentamycin, bacteria hydrolyzing esculin and bacteria capable of growing in the presence of hemin, which can be used in the catalase test. Vitamin K is added to the medium to stimulate the growth of many *Bacteroides* spp. [6]. Due to its specific composition, BBE agar stimulates the growth of BFG strains only which are identified based on the blackening of the medium around the formed colonies. Most BFG species are resistant not only to gentamycin but also to colistin, kanamycin and vancomycin. This trait is used to differentiate BFG strains from other anaerobic bacteria. However, BFG species can differ in resistance to the above antibiotics. Isolates have to be rapidly and accurately identified for effective treatment of bacterial infections. BFG strains are characterized by high phenotypic similarity, which is why they are often incorrectly identified in routine biochemical tests [2]. In addition, the existing methods for the identification of BFG strains based on phenotypic traits do not account for recent taxonomic changes and new species of the genus *Bacteroides*.

The progress in molecular biology and the availability of genotyping techniques have contributed to the development of new methods for the identification of the BFG. Bacterial species, including BFG strains, can be identified in hybridization tests with DNA and RNA probes [7–9], and in polymerase chain reaction (PCR) and quantitative PCR (qPCR) assays with the use of specific primers [10, 11]. 16S rRNA gene sequencing is regarded as the most effective method of bacterial identification [2]. However, this method is costly and time consuming, and it is not widely used. Genotyping techniques support the identification of bacterial strains based on the presence of conserved nucleotide sequences that are specific to a given genus, species or a group of species, mostly in the region of 16S rRNA and 23S rRNA genes. Alternative methods of microbial identification involve spectroscopic techniques [12]. Raman spectroscopy and Matrix Assisted Laser Desorption and Ionization Time-of-Flight Mass Spectrometry (MALDI-TOF MS) are regarded as the most promising techniques. Microbiological laboratories rely on MALDI-TOF MS to identify bacterial and fungal species [13–15]. This method has also been effectively deployed to identify BFG strains [12, 14]. MALDI-TOF MS generates characteristic mass spectral fingerprints that are unique for every bacterial genus or species. However, reliable identification of bacteria requires an expanded database [12].

The aim of this study was to analyze the effectiveness of isolation of BFG strains from different environmental samples, including human feces, laboratory rat feces, hospital wastewater, and untreated and treated sewage from a municipal wastewater treatment plant (WWTP) in Olsztyn. The analysis was carried out in four consecutive stages: bacteria were cultured on selective and differential BBE agar, the isolated strains were tested for resistance to colistin, kanamycin and vancomycin, the presence of the *bfr* gene specific to BFG was determined by PCR, and the strains were identified to the species level by MALDI-TOF MS.

2 Materials and methods

2.1 Sample collection

Bacteroides fragilis group strains were isolated from human and lab rat feces, from wastewater from a hospital in Olsztyn and from untreated and treated sewage from the WWTP in Olsztyn. The sampling sites in the WWTP and the hospital were described in detail in our previous studies [16, 17]. Thirteen grab samples from hospital wastewater and untreated and treated sewage were acquired in the spring and summer of 2016. Bacterial strains from nine samples of human feces from healthy volunteers were isolated by a private microbiological laboratory in Olsztyn. Three samples of laboratory rat feces were obtained from the Department of Industrial and Food Microbiology of the University of Warmia and

Mazury in Olsztyn. Immediately after collection, all samples were transported on ice to the laboratory and analyzed.

2.2 Isolation of *Bacteroides fragilis* strains

Bacteroides fragilis group bacteria were isolated from rat feces, hospital wastewater, and untreated and treated sewage from the WWTP by serial dilution in saline (1 g of feces or 1 mL of wastewater in 9 mL of 0.85% NaCl). Then, 0.1 mL of each solution was transferred to BBE agar and incubated for 48 h at 37°C under anaerobic conditions. BBE agar was used for the initial identification of BFG strains [5]. Bacteria were cultured under anaerobic conditions in the BACTRON I-2 chamber (Sheldon Manufacturing Inc., USA). BFG strains formed small, regularly spherical, gray colonies on BBE agar where esculin hydrolysis resulted in the blackening of the medium.

All of presumptive *Bacteroides fragilis* group strains isolated from HWW, UWW, TWW and RF samples as well as the strains isolated from HF samples in a private laboratory were streaked onto Brucella Blood agar (BD) (Merck) and incubated for 24 hours at 37°C under anaerobic conditions. Subsequently, the strains were tested for resistance to kanamycin (1000 μg), colistin (10 μg) and vancomycin (5 μg) by the disk diffusion method on Brucella Blood agar (24 h/37°C, anaerobic conditions). BFG strains are resistant to these antibiotics. All strains presumptively identified as members of the BFG were stored in LB-Broth (Miller, Merck) with 10% addition of glycerol, at a temperature of -80°C.

2.3 Genomic DNA extraction

Genomic DNA was extracted by suspending a loopful of bacterial colonies from Brucella Blood agar in 0.5 mL of sterile water. The suspension was heated at 95°C for 10 min and centrifuged at 5000 rpm for 5 min at 4°C according Harnisz et al. [18]. The concentration and quality of extracted DNA was determined with the NanoDrop spectrophotometer (NanoDrop® ND-1000, NanoDrop Technologies, Willmington, DE). DNA was extracted in duplicate and stored at -20°C for further analysis.

2.4 Identification of *Bacteroides fragilis* group strains by PCR

Isolates were identified by PCR using an oligonucleotide primer pair flanking conserved regions of the 16S rRNA *bfr* gene (5'-CTGAACCAGCCAAGTAGCG-3' and 5'CCGCAAACTTTCACAACTGACTTA3') based on the method described by Liu et al. [10]. PCR amplification was carried out in a total reaction volume of 15 μl containing 1.0 U of Run polymerase DNA (A&A Biotechnology, Poland), 1.5 μl of 10 x PCR buffer, 0.5 mM of each primer, 0.2 mM of dNTPs and 1.5 μL of template DNA. The reaction protocol was as follows: initial denaturation at 95°C for 5 min, 35 denaturation cycles at 95°C for 20 s, annealing at 52°C for 1 min, extension at 72°C for 30 s. A 5 min cycle at 72°C was added to the final extension step.

After amplification, DNA was separated by electrophoresis in agarose gel stained with ethidium bromide (0.5 μg/mL). A DNA marker (100-1000 bp, A&A Biotechnology, Poland) was used to determine the size of amplification products. Electrophoresis was performed in a buffer (1x TBE) for 60 minutes at 100 V. After electrophoresis, gels were visualized with Gel DocTM EZ Imager and Image LabTM v.6.0.0 2017 software (Bio-Rad Laboratories Inc).

2.5 Identification of *Bacteroides fragilis* group strains by matrix-assisted laser desorption-ionization time of flight mass spectrometry (MALDI-TOF MS)

Each isolate was cultured overnight on Brucella Blood agar, and it was identified in the Vitek MS Plus system (bioMérieux, Marcy l'Etoile, France). The isolates were plated and incubated in a BD GasPak™ EZ container under anaerobic conditions generated with the use of BD GasPak EZ anaerobe container system sachets. The samples were prepared on barcode-labeled target slides (Vitek MS-DS; bioMérieux, Inc.). An overnight culture was transferred and smeared directly onto the plate to inoculate the spots. The preparations were overlaid with 1 μl of α-cyano-4-hydroxycinnamic acid matrix solution (Vitek MS-CHCA; bioMérieux, Inc.) and were air dried for 1-2 min at room temperature. The calibration spots of each acquisition group were inoculated with the Escherichia coli ATCC 8739 strain as a calibrator and internal ID control, according to the manufacturer's recommendations.

The mass spectrum (2,000-20,000 Da) was generated with the Vitek MS Plus mass spectrometer (bioMerieux, Marcy l'Etoile, France). For each bacterial sample, mass fingerprints were processed by Compute Engine and the advanced spectrum classifier (ASC) algorithm of the Vitek MS system. The strains were identified to the species level by comparing the obtained spectra against the spectra typical of each claimed species (Vitek MS IVD version 3.0.0). The RUO mode (Vitek MS Saramis database) was used when an isolate could not be identified by Vitek MS IVD. In line with the manufacturer's instructions, a confidence interval of 60-99% was considered acceptable for species identification (ID).

3 Results and discussion

A total of 319 bacterial strains were isolated from selective and differential BBE agar. The appearance of all colonies was characteristic of the BFG: they were small (⌀ = 1–2 mm), gray and spherical, with visible blackening of the medium around the colonies. The highest number of 157 strains were isolated from untreated wastewater flowing into the WWTP in Olsztyn, followed by hospital wastewater (56 strains), treated wastewater evacuated from the WWTP (40 strains), and human and rat feces (33 strains each). All strains isolated from BBE agar tested positive for catalase.

During preliminary identification, the isolated strains were tested for resistance to colistin, kanamycin and vancomycin on Brucella agar. All strains isolated from samples of human and rat feces were resistant to the above antibiotics (100%, 33 strains each). Lower levels of resistance to the tested antibiotics were noted in the strains isolated from hospital wastewater (83.93%, 47 out of 56 strains) and treated sewage (77.50%, 31 out of 40 strains), whereas the lowest percentage of resistant strains was noted in raw sewage (67.52%, 106 out of 157 strains). Strains resistant to colistin, kanamycin and vancomycin were presumptively identified as members of the BFG and were subjected to further analyses.

In the following two stages, bacterial strains were analyzed with the use of molecular methods. In the first step, the species identity of the isolated strains was resolved by PCR. The presence of the *bfr* gene specific to the BFG was analyzed in the DNA from the studied isolates. Only 135 out of the 250 strains identified as members of the BFG with the use of the culture-dependent method harbored the *bfr* gene (PCR product with a size of 126 bp). The *bfr* gene was detected in all strains isolated from human and rat feces (100%, 33 strains each). The above gene was also identified in 17 out of 31 strains (54.84%) from treated sewage evacuated by the WWTP, in 21 out of 47 strains (44.68%) from hospital wastewater, and in 31 out of 106 strains (29.25%) from untreated sewage. Liu et al. [10] demonstrated that PCR was a highly effective technique for identifying BFG strains. The application of additional primers, which were not described in this study, could facilitate rapid and reliable identification of ten species belonging to the BFG: *B. caccae*, *B. distasonis*, *B. eggerthii*,

B. fragilis, *B. merdae*, *B. ovatus*, *B. stercoris*, *B. thetaiotaomicron*, *B. uniformis* and *B. vulgatus* [10].

Strains that presented a characteristic growth profile on BBE agar, were resistant to colistin, kanamycin and vancomycin and harbored the *bfr* gene were identified to the species level by MALDI-TOF MS. Ultimately, 119 out of 135 strains were classified as members of the BFG, and another 4 isolates, despite failed attempts to resolve their species identity, were also classified as members of the BFG. The percentage of strains from environmental samples which were identified as BFG species is presented in Figure 1. Hong et al. [9] demonstrated that *B. fragilis*, *B. thetaiotaomicron*, *B. caccae*, *B. eggerthii*, *B. vulgatus*, *B. uniformis* and *B. massiliensis* were frequently encountered in samples of human feces. In the present study, the species composition of the strains isolated from human feces differed from published data. In human feces isolates, the predominant BFG species were *B. fragilis* (61%) and *P. distasonis* (21%). The remaining strains were identified as *B. thetaiotaomicron* (9%), *B. ovatus* (6%) and *B. caccae* (3%). In the clinical trials described by Snydman et al. [19, 20], Eitel et al. [21] and Nagy et al [22], *B. fragilis* was also most frequently isolated from human feces (up to 80% isolates).

Parabacteroides distasonis (70%) was the predominant species in rat feces isolates. The remaining species, including *B. ovatus*, *B. fragilis* and *B. stercoris*, accounted for 3-15% of BFG strains. In a study analyzing the presence of *Bacteroides* spp. in pig and cow feces, Hong et al. [9] did not detect the above BFG species, but identified *P. distasonis, P. goldsteinii* and *P. merdae.*

All isolates from hospital wastewater were classified as *P. distasonis* (100%). The strains isolated from raw sewage in the WWTP belonged to three species, mostly *P. distasonis* (60%), as well as *B. fragilis* (37%) and *B. thetaiotaomicron* (3%). *Bacteroides fragilis* was the predominant species (31%) in the group of strains isolated from treated wastewater which also contained *P. distasonis* (25%), *B. ovatus* (19%) and *B. thetaiotaomicron* (19%). One BFG strain (6%) was not identified to the species level. The remaining strains from hospital wastewater and untreated and treated sewage in the WWTP did not belong to the BFG and were identified by MALDI-TOF MS as *Serratia marcescens* (10 isolates) and *Raoultella ornithinolytica* (2 isolates) of the family *Enterobacteriaceae*. These isolates where phenotypically similar to BFG strains. However, the PCR test screening for the presence of the *bfr* gene produced additional bands on agarose gel, which probably led to the misclassification of the above 12 strains as members of the BFG. The occurrence of the BFG in wastewater has been rarely investigated. Hong et al. [9] relied on fluorescence in situ hybridization (FISH) to detect *Bacteroides* spp. in raw sewage in a municipal WWTP. They identified the following BFG species in sewage samples: *B. fragilis*, *B. caccae*, *B. uniformis, B. vulgatus, B. massiliensis* and *B. thetaiotaomicron*. Their results are partially consistent with our findings.

In the current study, anaerobic BFG strains were detected in treated wastewater despite the presence of an aeration step in the treatment process. It can be probably attributed to the presence of catalase which enables anaerobes to survive under aerobic conditions [23]. This applies particularly to *B. fragilis* and *P. distasonis*, as well as selected strains of *B. thetaiotaomicron, B. ovatus* and *B. eggerthii* [24]. In this study, the above species were detected in all environmental samples, and they were predominant in samples of treated wastewater.

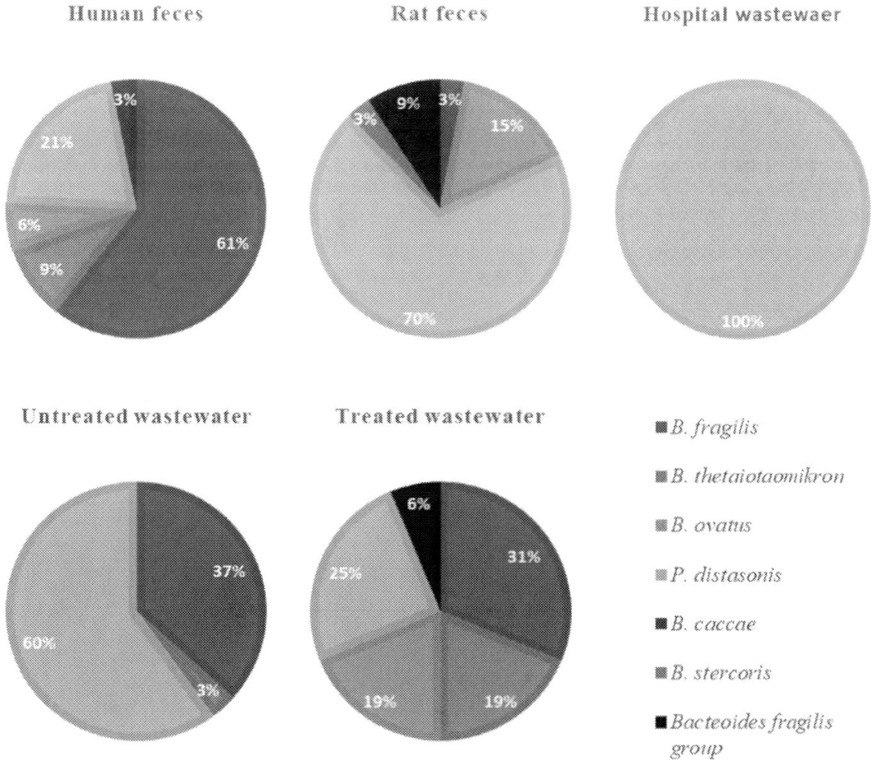

Fig. 1. Percentage composition of *Bacteroides fragilis* group in environmental samples.

The effectiveness of isolation of BFG strains in successive stages of the study varied across the analyzed environmental samples (Fig. 2). All strains isolated from human and rat feces on BBE agar belonged to the BFG (100% accuracy). The percentage of BFG strains isolated from hospital wastewater and untreated and treated sewage which demonstrated traits characteristic of the BFG continued to decrease in successive stages of analysis. BFG strains isolated from treated sewage were identified with 40% accuracy, and those from hospital wastewater and raw sewage – with 19–20% accuracy.

Culebras et al. [2] and Nagy et al. [25] identified *Bacteroides* spp. by MALDI-TOF MS with the use of reference strains and clinical isolates identified by 16S rRNA gene sequencing. Their results revealed that MALDI-TOF MS was an easy, rapid, low-cost and highly effective (87–98%) technique for the routine identification of *Bacteroides* spp. However, the MALDI-TOF MS database is incomplete, and new species such as *B. dorei*, *B. xylanisolvens* and *B. faecis* can be erroneously classified as *B. vulgatus, B. ovatus* and *B. thetaiotaomicron*, respectively [2, 26]. The MALDI-TOF MS database should be expanded to increase the accuracy of bacterial species identification.

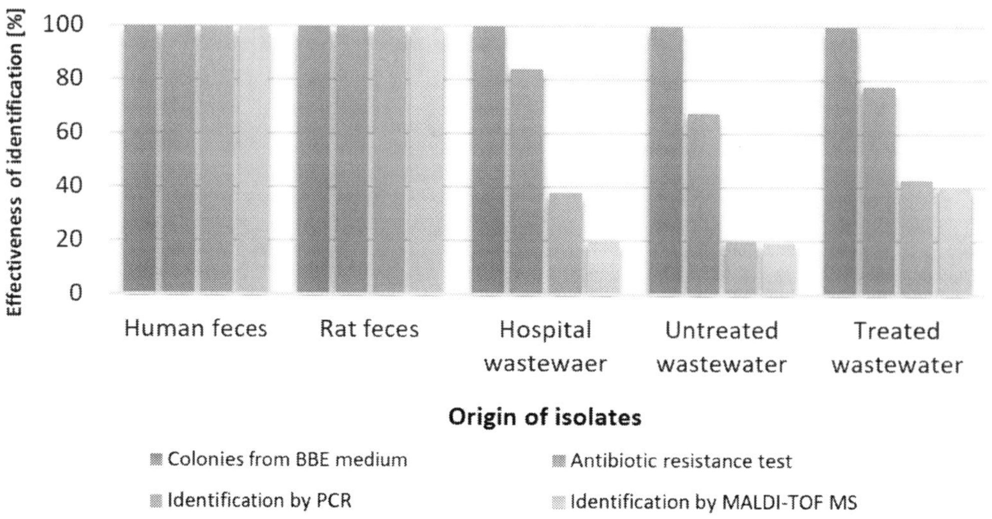

Fig. 2. Effectiveness of isolation of *Bacteroides fragilis* group from environmental samples.

4 Conclusions

All strains isolated from samples of human and rat feces on BBE agar were identified as members of the BFG. The evaluated samples were probably similar to clinical samples for which diagnostic methods have been developed and optimized. The isolates from hospital wastewater and untreated and treated sewage from the WWTP were identified with low accuracy, which can probably be attributed to the different character of those samples, greater bacterial diversity and the presence of bacteria with modified phenotypic traits in these environments. The results of this study suggest that the method for isolating BFG strains from environmental samples on BBE agar requires modification. Additional methods, such as PCR and MALDI-TOF MS, can be used to more effectively identify BFG strains isolated from different environmental samples However, the MALDI-TOF MS database should be expanded to increase the accuracy of identification.

Human feces were characterized by the highest diversity of BFG strains. Anaerobic BFG strains were detected in treated sewage despite the presence of an aeration step in the treatment process. Treated wastewater is evacuated to surface water bodies, and it can pose a health risk for humans and animals.

This work was supported by the National Science Centre (Poland) [grant No. 2016/23/N/NZ9/02167].

References

1. H. M. Wexler, Clin. Microbiol. Rev. **20**, 4, (2007)
2. E. Culebras, I. Rodriguez-Avial, C. Betriu, M. Gomez, J. J. Picazo, Anaerobe **18**, 1, 163–165 (2012)
3. D. R. Snydman, N. V. Jacobus, L. A. McDermott, R. Ruthazer, Y. Golan, E. J. C. Goldstein, S. M. Finegold, L. J. Harrell, D. W. Hecht, S. G. Jenkins, C. Pierson, R. Venezia, V. Yu, J. Rihs, S. L. Gorbach, J. Antimicrob. Agents Chemother. **51**, 5, 1649–1655 (2007)
4. J. Soki, M. Hedberg, S. Patrick, B. Balint, R. Herczeg, I. Nagy, D. W. Hecht, E. Nagy, E. Urban, J. Antimicrob. Chemother. **71**, 9, 2441–2448 (2016)

5. S. J. Livingston, S. D. Kominos, R. B. Yee, J. Clin. Microbiol. **7**, 5, 448–453 (1978)
6. N. R. Krieg, W. Ludwig, J. P. Euze´by, W. B. Whitman, Springer, 25–469 (2011)
7. A. H. Franks, H. J. M. Harmsen, G. C. Raangs, G. J. Jansen, F. Schut, G. W. Welling, Appl. Environ. Microbiol. **64**, 9, 3336–3345 (1998)
8. L. Rigottier-Gois, V. Rochet, N. Garrec, A. Suau, J. Dore, Syst. Appl. Microbiol. **26**, 1, 110–118 (2003)
9. P. Y. Hong, J. H. Wu, W. T. Liu, Appl. Environ. Microbiol. **74**, 9 2882–2893 (2008)
10. C. X. Liu, Y. L. Song, M. McTeague, A. W. Vu, H. Wexler, S. M. Finegold, FEMS Microbiol. Lett. **222**, 1, 9–16 (2003)
11. R. A. Haugland, M. Varma, M. Sivaganesan, C. Kelty, L. Peed, O. C. Shanks, Syst. Appl. Microbiol. **33**, 6, 348–357 (2010)
12. A. C. M. Veloo, G. W. Welling, J. E. Degener, Anaerobe **17**, 4, 211–212 (2011)
13. Cherkaoui, J. Hibbs, S. Emonet, M. Tangomo, M. Girard, P. Francois, J. Schrenzel, J. Clin. Microbiol. **48**, 4, 1169–1175 (2010)
14. P. Seng, M. Drancourt, F. Gouriet, B. La Scola, P. E. Fournier, J. M. Rolain, D. Raoult, Clin. Infect. Dis. **49**, 4, 543–551 (2009)
15. S. Q. van Veen, E. C. J. Claas, E. J. Kuijper, J. Clin. Microbiol. **48**, 3, 900–907 (2010)
16. A. Osinska, M. Harnisz, E. Korzeniewska, Environ. Sci. Pollut. Res. Int. **23**, 11, 10818–10831, (2016)
17. E. Korzeniewska, M. Harnisz, J. Environ. Manage. **128**, 904–911 (2013)
18. M. Harnisz, E. Korzeniewska, S. Ciesielski, I. Golas, Sci. Total Environ **505**, 704–711, (2015)
19. D. R. Snydman, N. V. Jacobus, L. A. McDermott, Y. Golan, E. J. C. Goldstein, L. Harrell, S. Jenkins, D. Newton, C. Pierson, J. Rosenblatt, R. Venezia, S. L. Gorbach, A. M. Queenan, D. W. Hecht, Anaerobe **17**, 4, 147–151 (2011)
20. D. R. Snydman, N. V. Jacobus, L. A. McDermott, E. J. C. Goldstein, L. Harrell, S. G. Jenkins, D. Newton, R. Patel, D. W. Hecht, Anaerobe **43**, 21–26 (2017)
21. Z. Eitel, J. Soki, E. Urban, E. Nagy, E. S. G. Anaerobic, Anaerobe **21**, 43–49 (2013)
22. E. Nagy, E. Urban, C. E. Nord, E. S. G. A. Res, Clin. Microbiol. Infect. **17**, 3, 371–379, (2011)
23. B. M. Meehan, M. H. Malamy, Microbiology-Sgm. **158**, 539–546 (2012)
24. S. Patrick, Molecular Medical Microbiology, 1921–1948 (2002)
25. E. Nagy, T. Maier, E. Urban, G. Terhes, M. Kostrzewa, E. S. G. A. Res, Clin. Microbiol. Infect. **15**, 8, 796–802 (2009)
26. A. C. M. Veloo, H. Jean-Pierre, U. S. Justesen, T. Morris, E. Urban, I. Wybo, M. Kostrzewa, A. W. Friedrich, E. Workgrp, Anaerobe **54**, 224–230 (2018)

The separation of maximum amounts of precipitation for the Polish Atlas of Rains Intensities (PANDa)

Michał Oktawiec[1,3,*], *Marcin* Wdowikowski[2], *Bartosz* Kaźmierczak[1], *Janusz* Zaleski[2], and *Paweł* Licznar[1,3]

[1]Wrocław University of Science and Technology, Faculty of Environmental Engineering, Plac Grunwaldzki 13, 50-377 Wrocław, Poland
[2]Institute of Meteorology and Water Management – National Research Institute, Podleśna Street 61, 01-673 Warsaw, Poland
[3]RETENCJAPL Sp. z o.o., ul. Szymanowskiego 4, 80-280 Gdańsk, Poland

Abstract. In this paper selection of maximal precipitation amounts for Polish Atlas of Rains Intensities (PANDa) has been presented. PANDa supposes to be the source of actual and indisputable information about reliable rainfall intensities for designing of storm water drainage and retention systems in Poland. During the realization of the maximum amounts separation procedure, for chosen 100 meteorological stations with the use of peak over threshold (POT) method, a number of problems has been found, for which proceeding algorithms have been described.

1 Introduction

In Poland, the dimensioning of storm water drainage meets the basic difficulty resulting from the lack of a reliable precipitation model necessary to determine the authoritative rainfall in the national, regional or specified point range [1]. The most commonly used in engineering practice (since 1954), Blaszczyk's rainfall model underestimates the results of rainfall calculation streams by 40% [2]. This is caused by the non-stationarity (variability over the years) of rainfall – nowadays in Poland, higher sums of intense rainfalls are recorded than in the past [3–6].

Reliable precipitation models developed on the basis of current measurement data were derived only for individual towns [7]. The lack of an up-to-date atlas of rain intensities that are decisive for Poland (similar to the KOSTRA atlas in Germany) results in underestimating the required hydraulic capacity of the channels. This has its consequences when dimensioning drainage areas according to the recommendations of the European standard PN-EN 752 - directly affecting the higher incidence of urban floods in Poland [8].

The necessity of developing reliable dependencies of IDF (intensity-duration-frequency) and DDF (depth-duration-frequency) for the whole country for the purposes of designbroadly understood sanitary and water-drainage infrastructure as well as for flood

* Corresponding author: michal.oktawiec@pwr.edu.pl

protection needs has been indicated for many years by many countries. In some countries, rainfall frequency atlas exists since the 1960s. KOSTRA atlas in Germany has already a history exceeding 30 years. Therefore, Poland, using these experiences has a chance to create a tool for the 21st century [9–12].

Polish Company Retencjapl in cooperation with the Institute of Meteorology and Water Management - National Research Institute, applied to the National Center for Research and Development with a request for co-financing of the project entitled: *Development and implementation of the Polish Atlas of Rains Intensities* (PANDa) – the Smart Growth Operational Programme 2014-2020 POIR.01.01.01-00-1428/15 [13]. This article describes one of the stages of the project, which consists in separating the maximum rainfall amounts from the previously prepared (majorly digitalized from archival meteorological paper reports) rainfall time series as a partial duration series (PDS) of 16 durations from $t_1 = 5$ to $t_{16} = 4320$ min. Preparation of such an accurate numerical database of pluviographs allowed in subsequent stages to develop reliable rainfall models, which in the final stage will enable to construct a long-awaited atlas of intensive rainfall in Poland in the form of IDF and DDF dependencies. Moreover uniformly methodically developed rainfall database covering the whole country will allow to supplement the lack of knowledge about intensive rainfall in the area of interdisciplinary activities of science, climate change and civil engineering [13].

2 Materials and methods

2.1 Research area

As part of the PANDa project, a digital base of rainfall series was developed with a one-minute time step. This database, covering the years 1986–2015, consists of digitized pluviographic paper strips records (mostly rainfall data from 1986 do 2005) and registration from electronic rain gauges available from 2000 to 2015 (RG-50 SEBA working at synoptic stations and Met One Instruments 60030 - unheated and 60030H - heated, working at lower-graded IMWM-NRI stations). Rainfall data comes from a network of 100 rain gauges of the national hydrological and meteorological service, whose location is shown in Figure 1 [13].

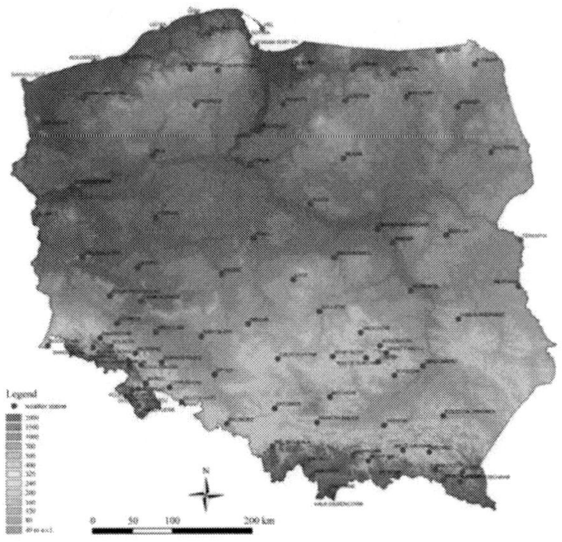

Fig. 1. Locations of IMWM-NRI rain gauges included in the digital database developed under the PANDa project (adopted from [13]).

The meteorological stations were selected for the project in such a way as to spatially cover the whole country - with the highest density in mountainous areas in the south of Poland (to better capture the orographic effect). As a result, the maximum rainfall values registered at selected 100 stations will allow to develop a rainfall atlas, which will be applied to the whole country [13].

2.2 Data selection algorithm

Input data for the calculations were one minute precipitation heights time series from 1986–2015. Primary database consisted 3000 files (30 files with a yearly series prepared as a data-value column structure, where date were stored in numerical [year-month-day hour-minute] format, for each of the 100 station) [13]. The aim of the assumed task was to isolate 30 sets of independent extreme rainfall values h_j (h_1 – the largest, …, h_{30} – the smallest) for the 16 rainfall durations t_i: $t_1 = 5$, $t_2 = 10$, $t_3 = 15$, $t_4 = 30$, $t_5 = 45$, $t_6 = 60$, $t_7 = 90$, $t_8 = 120$, $t_9 = 180$, $t_{10} = 360$, $t_{11} = 720$, $t_{12} = 1080$, $t_{13} = 1440$, $t_{14} = 2160$, $t_{15} = 2880$ and $t_{16} = 4320$ minutes. As a result matrix of maximum rainfall partial duration series was obtained. In order to preserve the independence of maximum rainfall random variables, necessary to perform statistical analysis in the next stage of the project, measured data was isolated using complete review method. As a result 30 largest values h_j in each year (from the 30 years analyzed) were selected for each time period t_i. This approach enabled further data selection in two recommended in the literature and commonly used directions: determination of annual maximum precipitation (AMP or sometimes called annual maximum series AMS) and determination of the absolute highest precipitation values appearing above the peak-over threshold (POT).

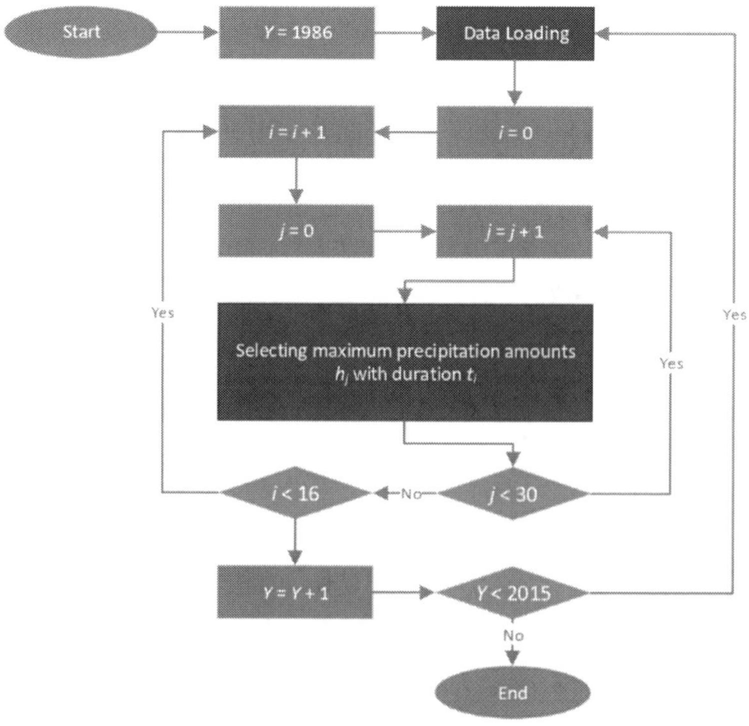

Fig. 2. Algorithm for determining the maximum rainfall partial duration series in the PANDa project.

Due to the relatively large size of the database (because of the yearly data series length, every year was in separate files), repeatability of the calculation procedure for each of the 100 stations and the obvious need to automate the calculation process, in the project self-made software based on Microsoft Excel Visual Basic for Applications (MS Excel VBA) scripts was developed. The calculation scheme was shown in the form of block diagram (Fig. 2).

The proposed algorithm (Fig. 2) for determining the maximum rainfall for a selected station in simplification presents as followed:

- step I: assigning start values ($Y = 1986, j = 1, i = 1$) and data upload to a workspace
- step II: searching for the highest independent value h_j for duration t_i then saving the result to a separate matrix and removing the workspace within the searched data. Step II internal procedure sets value $j = j + 1$, and repeats it in the loop until $j < 30$ is reached.
- step III: setting value $i = i + 1$ and return to step II. This part is repeated in the loop until $i < 16$ is reached.
- step VI: setting values $Y = Y + 1, j = 1, i = 1$ and return to step II. This part of calculations is repeated in the loop until $Y < 2015$ is reached, that is the end of the procedure at the same time.

The final results of the conducted procedure are tables containing the largest precipitation heights registered at the selected station in the years 1986–2015. For each of the 16 analyzed rainfall durations, the matrixes of the largest values (precipitation depth) in particular years as well as the matrix of absolutely the largest values in whole analyzed period (regardless to the year of occurrence) were obtained. The procedure is repeated for each of the 100 stations analyzed in the project, while the automation of the calculation process applies only to one measurement station. The following operatioins: summation of one-minute values in the assumed intervals of the rainfall duration, filtering and extracting 30 independent maximum values in a year, wer implemented in the form of a cascade of internal procedures. However, at each stage, it was possible to run the internal procedure for one duration, e.g. only 5 or 1440 minutes.

As mentioned at the beginning of the chapter 2.2, the basic data files with one-minute rainfall amounts for each year were prepared in a time series, therefore part of the internal procedures of the program concerned calculations on time units. Due to the methodically required need to ensure the independence of individual precipitation heights with respect to each other, special attention have been paid to the mechanism identifying that kind of situation, especially for short rainfall durations, for t from 5 to 30 minutes, when during one rainfall event, e.g. 2 hours, values that fulfill the selection criteria could be from a few to a dozen. In the case of data selection using the annual maxima method, this is of no great importance, because only one value from a given year is used for further statistical analyzes, while the POT method requires caution in the data selection [14, 15].

Regarding to the independent maximum rainfall values searching procedures, it is worth to mention that a number of individual annual series records was zero, especially in years with deep meteorological drought. Such a situation did not cause methodical problems, however, it influenced the time of the automatic data search procedure. These and other problems were described in the section below.

3 Specific problems and solution proposals

During the implementation of the separation of maximum values procedure, a few of specific problems were encountered, among which the most important were:

- What to do if at least two time-dependent precipitation depths had an equal value? Which one or maybe both choose for further analysis?

- What to do if the next largest value of rainfall was between previously selected rainfall with duration t_i, however, the length of the selected series was smaller than t_i. Should such a value be taken for further analysis?
- How to maintain the independence of rainfall data?

The first problem was exemplified in Figure 3 which presents that the largest precipitation depth $h_1 = 2.3$ mm with duration $t_1 = 5$ min occurred from 24 to 28 minutes inclusive. Another maximum height of $h_2 = 2.2$ mm, occurred from 1 to 5 minute (Fig 3 – case I), or from 2 to 6 minute (Fig. 3 – case II). Because $h_2 = 2.2$ mm in both cases was time-dependent, one of it should be chosen for further analysis. Choosing the first option, we obtained the following results: $h_1 = 2.3$ mm, $h_2 = 2.2$ mm, $h_3 = 2.0$ mm and $h_4 = 2.0$ mm, while for the second option was: $h_1 = 2.3$ mm, $h_2 = 2.2$ mm, $h_3 = 2.0$ mm and $h_4 = 1.9$ mm.

Case I

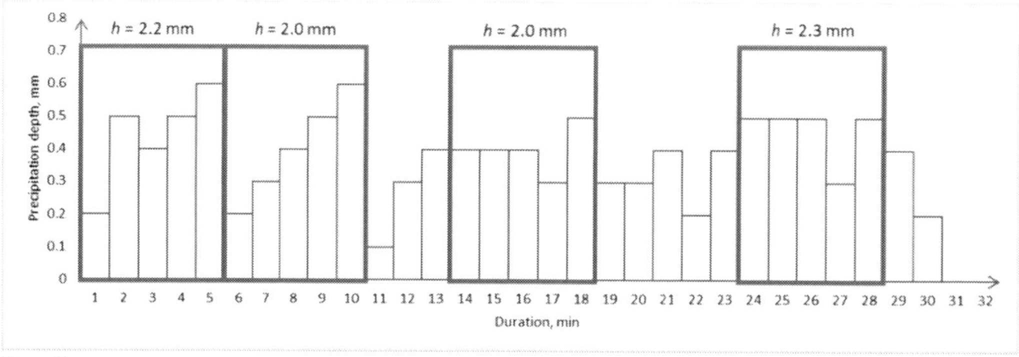

Case II

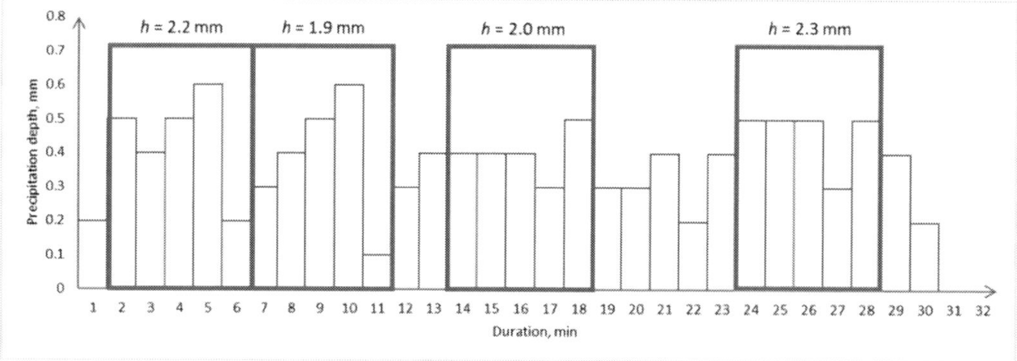

Fig. 3. Determination of maximum values with duration $t_1 = 5$ min.

Because the task is to select the maximum values, we should choose the first option. The solution to this problem required programming intervention, which consisted of:

- In the first step, the precipitation values beginning as the earliest (from time-dependent precipitation of the same height) was selected, and then the remaining precipitation sums (time-independent) were counted and ranked not descending.
- In the second step, the precipitation beginning as the latest (from time-dependent precipitation of the same height) was selected, and then the remaining precipitation sums (time-independent) were counted and ranked descending.
- In the third (final) step, both of the maximum value sequences determined in the first and second steps were compared and selected for further analysis as data string with larger values.

The second problem was exemplified in Fig. 4. The largest rainfall amounts with duration t_2 = 10 min were h_1 = 4.2 mm and h_2 = 4.0 mm and occurred respectively from 1 to 10 and from 20 to 29 minutes (Fig. 4 – case I). Another maximum height, amounting to h_3 = 3.1 mm, occurred from 11 to 19 minute – a 9-minute long string (Fig. 4 – case II).

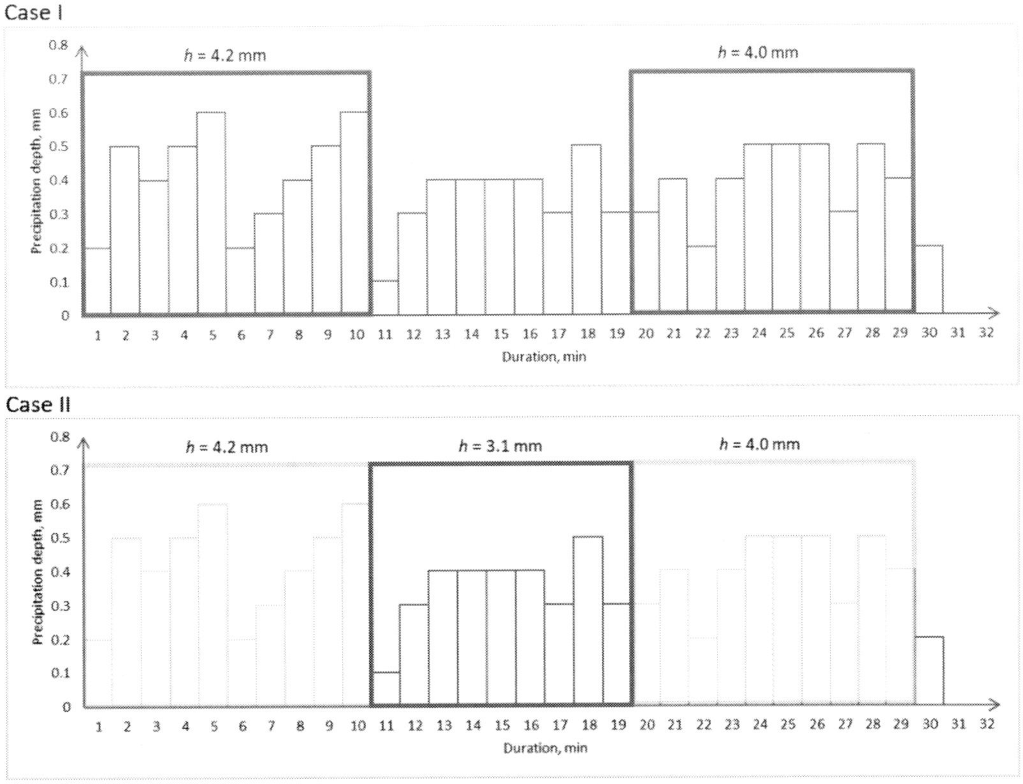

Fig. 4. Determination of precipitation maxima with duration t_2 = 10 min.

The second problem required making a decision whether to include rainfall in the analysis described above. After careful analysis, it was decided that a safer solution would be taken into account it in the development of rainfall formulas (rainfall height dependent on rainfall duration *t* and probability of exceedance *p*), because it was included in the sum of the 30 largest values with a given duration *t*.

The issue of independence of the separated interval values of precipitation in the examined cases t = 5 minutes (Fig. 3) and t = 10 minutes (Fig. 4) was also the subject of major problems. The literature does not provide clear guidelines on how to maintain the independence of rainfall maxima selecting partial duration series from 5 to 4320 minute, indicating that the safest way is to use the method of annual maxima. At the beginning, the calculation procedure assumed extracting one interval maximum value from one precipitation episode, assuming that the interval between successive precipitation is 1 day. It quickly turned out that for the rainfall durations from three hours to three days, it was not possible to complete the 30 largest values in a year that fulfilled the accepted POT criterion. Large precipitation values of short duration were also omitted, which occurred for example at the beginning and then at end of the precipitation episode. In the next approach, an analysis was performed in which the interval between the values might have no exceeded the length of the analyzed range of precipitation duration, as in Figure 4. The results were much better but for several durations still insufficient, that is why in view of the time-consuming calculations, it was considered to accept data selection from successive intervals as a final approach. The final control of data independence was implemented during the final selection of the 30 largest values from the entire multiannual period to statistical analyzes. If in the sequence of consecutive rainfall depths the selected values would come from almost the same dates, such record should have been removed, and in this place, another separated value was passed. Despite the use of quite mild criteria for the selection of maximum rainfall, a 30-year series of measurements allowed to create such a large data set that in the end all maxima were independent in time and most often came from various genetic rainfall episodes.

4 Instead of a conclusions

Both global progressive urbanization and climate change [6] have a negative impact on the efficiency of the functioning of sewerage systems, causing more and more frequent overloads leading to local flooding or urban flooding [16, 17]. In Poland, an atlas of rains intensities (PANDa) will be created soon. It will surely become a reliable source of the knowledge used in design flood control facilities and in development of risk maps, and also for sizing stormwater drainage systems whose task is to provide the required land drainage standard [13]. The rigorous requirements for the choice of random variables, the quality of which consequently determines the quality of probabilistic models, obliges the researcher to act carefully and precisely. The aim of this paper was to present the calculation algorithms for maximum rainfall interval sums selection, which were used for the project purposes. In addition, several main technical and methodological problems were presented, which the research team encountered during the development of rainfall data, for further statistical analysis. During the preparation and development of rainfall data, it was confirmed that such analyzes should be performed using the complete review method (preferably based on one-minute data). This method is much more time-consuming but it allows to identify much larger values than in case of data processing in rigid time intervals, e.g. for $t = 5$ minutes, between 10:00 and 10:04, and also between 10:05 and 10:09 etc. In the calculation procedures, very restrictive method of selecting time-independent data was not adopted for the project's needs, whereas in the final result, no values were dependent on each other.

The key problems have been solved empirically, with expert methods, whenever the accessible Polish and foreign literature sources did not indicate transparent answers.

Acknowledgements

The project entitled: Development and implementation of the Polish Atlas of Rains Intensities (PANDa) – POIR.01.01.01-001428/15, is financed by the National Centre for Research and Development under the Smart Growth Operational Programme 2014-2020, Priority axis: Support for R&D in enterprises; Measure: R&D projects in enterprises; Sub-measure: Industrial research and development conducted by enterprises

References

1. A. Kotowski, *Podstawy bezpiecznego wymiarowania odwodnień terenów* (Seidel-Przywecki, Warsaw, 2015)
2. P. Licznar, K. Siekanowicz-Grochowina, M. Oktawiec, A. Kotowski, E. Burszta-Adamiak, Ochrona Środowiska **40**, No. 2, 17–22 (2018)
3. A. Dai, WIREs Climate Chang **2**, No. 1, 45–65 (2011)

4. A.K. Fleig, L.M. Tallaksen, P. James, H. Hisdal, K. Stahl, Hydrology and Earth System Sciences **19**, 3093–3107 (2015)
5. Z.W. Kundzewicz, S. Kanae, S.I. Seneviratne, J. Handmer, N. Nicholls, P. Peduzzi, R. Mechler, L.M. Bouwer, N. Arnell, K. Mach, R. Muir-Wood, R. Brakenridge, W. Kron, G. Benito, Y. Honda, K. Takahashi, B. Sherstyukov, Hydrological Sciences Journal **59**, No. 1 (2012)
6. Q. Schiermeier, Nature **470**, No. 7334, 316 (2011)
7. A. Kotowski, B. Kaźmierczak, Journal of Hydrometeorology **14**, 1958–1965 (2013)
8. B. Kaźmierczak, A. Kotowski, Theoretical and Applied Climatology **118**, No. 1, 285–296 (2014)
9. S. Perica, D. Martin, S. Pavlovic, I. Roy, M.St. Laurent, C. Trypaluk, D. Unruh, M. Yekta, G. Bonnin, *NOAA Precipitation-Frequency Atlas of the United States* (Silver Spring, Maryland, 2013)
10. W.H. Asquith, M.C. Roussel, *Atlas of Depth-Duration Frequency of Precipitation Annual Maxima for Texas* (U.S. Geological Survey, Reston, Virginia, 2004)
11. G. Malitz, H. Ertel, *KOSTRA-DWD-2010, Starkniederschlagshöhen für Deutschland (Bezugszeitraum 1951 bis 2010)*, (Deutscher Wetterdienst, Offenbach am Main, 2015)
12. A.V. Dyrrdal, *Estimation of extreme precipitation in Norway and a summary of the state-of-the-art*, (Norwegian Meteorological Institute, Oslo, 2012
13. E. Burszta-Adamiak, P. Licznar, J. Zaleski, Meteorology, Hydrology and Water Management-Research and Operational Applications **7** (1), 3–13 (2019)
14. S.G. Coles, *An Introduction to Statistical Modeling of Extreme values* (Springer Series in Statistics, London, 2001)
15. D. Koutsoyiannis, Hydrological Sciences, **49** (4), 575–590 (2004)
16. N. Bezak, M. Šraj, M. Mikoš, Journal of Hydrology **541**, 272–284 (2016)
17. M.A.M. Saboia, F.A. Souza Filho, L.M. Araujo Junior, C.S. Silveira, Brazilian Journal of Water Resources **22**, No. 21, 1–15 (2017)

Fulvic acids generated in municipal waste landfills in Promnik

Tomasz Orliński[1,*], and *Anna M.* Anielak[1]

[1]Department of Environmental Engineering, Institute of Water Supply and Environmental Protection, Cracow University of Technology, ul. Warszawska 24, 31-155 Cracow, Poland

Abstract. Landfill leachates are resistant to biodegradation, contain substances harmful to the environment, which is why they are troublesome in the water and sewage management. The study tested the leachate from the municipal waste landfill in Promnik, whose age of waste disposal is 33 years. The research consisted of extracting from average leachates (with a volume of 800 m^3) fulvic acids resistant to biodegradation. The extraction of acids was carried out using a hydrophobic and hydrophilic ionite in a hydrogen form. The extracted fulvic acids (FAs) was subjected to qualitative analysis on the content of micro-pollutants and the content of elementary elements forming the structure of the FAs molecule, their infrared and UV-VIS spectra were examined. Then, a full analysis of the obtained values was carried out in order to determine the properties of the obtained extractant and demonstrate the convergence of its features with humus substances that are common in the environment, which are an important element of soils, affect their quality and participate in the transport of plant nutrients. It was demonstrated that the concentration of FAs in the effluent is 905.6 mg/l.

1 Introduction

According to the definition contained in art. 3 par. 1 of Directive 2008/98/EC, "waste means any substance or object which the holder discards or intends or is required to discard" [1]. In 2017, the average Pole generated 307 kg of municipal waste, which is 7.3% more than in 2015. With regard to the inhabitants of European countries, Poland has one of the lowest rates of waste generated. In EU countries, each resident generates approximately 482 kg of municipal waste [2]. Each year, the amount of waste generated increases, creating a huge problem with waste management.

Waste is subjected to disposal processes, i.e. biological, physical and chemical processes in order to bring them to a condition that does not pose a threat to human life or health and the environment. These processes include: landfilling, incineration, gasification and pyrolysis and processing into solid fuel. The oldest and most popular method of waste

* Corresponding author: t.orlinski1990@gmail.com

disposal is landfilling. In Poland in 2016, 37% of the total amount of waste generated was allocated for landfilling (i.e. 4.3 million tonnes) [3].

Landfilling of waste generates leachate containing many hazardous substances in high concentrations. The composition and properties of leachate depend on the age of the landfill, the type of waste, climatic conditions and the operating mode of the landfill [4]. A significant part of the waste deposited undergoes so-called humification. The humification process stabilizes the organic matter in the waste deposit and the formation of humic substances that penetrate the leachate [5].

It is generally known that the leachates contain different contaminants [6], including: biodegradable and non-biodegradable organic substances, microorganisms, heavy metals, chlorinated organic and inorganic salts, which are mostly dangerous to the environment and human health. The results of tests carried out on the leachate from eight landfills in France indicate that they react with fulvic acids to form macromolecular complexes with them [7]. The leachates of landfills, as a result of leaching of pollutants from the landfills and infiltration into the soil, is a source of potential contamination of underground and surface waters. Some types of organic leaching substances interact with heavy metals and become their carriers [8–10]. Studies have shown that the concentration and composition of leached pollutants depend not only on the type of waste being deposited but also on the time of landfilling. The leachate can be divided into: young, medium and old. The old leachates is characterized by a high content of organic substances with considerable resistance to biodegradation. These are high molecular weight substances. The humic acids of the old leachate are characterized by a higher fluorescence intensity [11]. The leachate from landfills are directed to biological municipal sewage treatment plants. The humic substances contained in them do not undergo biochemical degradation, they are partially adsorbed on the activated sludge, they are partially subject to further humification and finally they are discharged with the treated wastewater to the ground and surface waters.

2 Characteristics of municipal waste landfill in Promnik

The municipal waste landfill is located approximately 17 km north-west of the center of Kielce, in the village of Promnik, in the Strawczym municipality, in the Świętokrzyskie province.

In the landfill, processed municipal waste and non-hazardous waste are deposited, from group 19 05, 19 06, 19 08, 19 12, 20 02 and 20 03 according to the waste catalog contained in the Regulation of the Minister of Environment of December 9, 2014 on the catalog waste (Journal of Laws 2014 item 1923), in an amount not exceeding 122,000 Mg per year [12]. Wastes are deposited in layers, compacted with compactors to a layer of about 1.35–1.50 m and insulated with a mineral layer with a thickness of 0.15–0.20 m. After the exploitation, the area will be reclaimed [12].

The landfill is degassed with a system of degassing wells, and the captured biogas is directed to a small power plant in order to obtain energy [12].

Waters and leachates are captured through a drainage system consisting of a perforated pipes, a discharge pipeline and an underground pumping station. The leachates from the communal waste landfill are collected in two connected tanks, recirculated to the currently operated landfill part or exported to the Sewage Treatment Plant in Sitkówka. The annual amount of leachate produced is approximately 40,000 m^3 [12].

3 Material and methods

Samples were collected permanently, in the month of December from two connected tanks for leachate, with a total capacity of approx. 800 m^3. The tank is located near the active third sector of the landfill. It can be assumed that the collected sample represented leachate collected for at least 2 months. The total volume of the collected and averaged sample was 20 L.

Extraction of fulvic acids

A sample of leachate was acidified to pH ≤ 2 with HCl. After 24 hours, the filtrate was filtered through a medium thickness filter with glass balls < 0.5 cm in diameter and passed through the ionic bed at a rate of about 1 L/hr. The volume of leachate directed to the ionic bed was 20 L.

Then NaOH solution was directed onto the ionite to desorb adsorbed humic acids. To remove NaOH from the solution and obtain pure FAs in the hydrogen form, the effluent (eluent) was directed to the cation exchange bed in the hydrogen form. The resulting leach was concentrated in a vacuum evaporator at a vacuum of 350 bar and a temperature of 85°C. The leakage was then evaporated on a water bath at 100°C to give a dry weight of FAs. FAs were obtained according to the method accented by the International Humic Substances Society (IHSS).

Analysis of carbon, hydrogen and nitrogen content was done using a chromatographic detection technique (elemental analysis). For the analysis used from 5 to 10 mg of sample. The analysis of the remaining elements was done using X-ray fluorescence (XRF). The analysis was carried out in a powder analysis vessel on 4μm thick Proline film. Brune's ED-XRF spectrometer, Bruker's S8 Tiger spectrometer and Thermo's EA Flash element analyzer were used. The ash designation was done by thermogravimetric method using the SDT Q600 Thermogravimeter from TA Instruments. The heating rate was 10°C/min to 700°C in airflow of 100 ml/min. For the analysis used from 5 to 10 mg of sample. Identification of compound was made using FT-IR with ATR. The spectral range was 65–4000 cm^{-1} at 1 cm^{-1} resolution and 32 scans. The Thermo Nicolet FT-IR iS10 spectrophotometer was used.

4 Results and discussion

Leachates from the Promnik municipal landfill were characterized by dark brown color and unpleasant odor. After filtration and ion exchange, the leachates partially lost coloration, turning the color to light brown. From 905.6 mg of fulvic acid with 1 L of leachate were obtained. The extracted FAs formed a clear brown solution with water. Dried FAs were subjected to qualitative analysis. The elemental composition of the obtained FAs is shown in Table 1, the atomic quotient in Table 2, and the micropollutants in Table 3.

Table 1. Elemental composition of FAs coming from leachate [%].

Elemental composition	Content [%]	Literature Content [%]
C	37.07	45.6–55.7
H	3.31	6.2–9.6
N	3.08	1.0–12.0
S	4.09	1.53–1.86
O	52.45	29.6–50.0

[a] Based on 6 different landfills [13]

The extracted FAs were slightly different in composition compared to the literature data. They contained less carbon but more sulfur and oxygen. The data may indicate a low degree of humification and landfilling of waste containing a significant amount of sulfur, which creates the structure of the FA molecule, and may additionally be their trace contaminations. The content of hydrogen and nitrogen was within the range specified by the authors. Comparison with other values [13] in literature is presented in Table 1.

For detailed information on FA, molar ratios H/C, O/H, O/C and C/N were calculated. The atomic ratios are convergent with the research of other authors, and slight differences result from the creation of different elementary compositions related to the environmental conditions in which the humification process takes place [14]. The results are shown in Table 2. H/C atomic ratio allows to determine the aromaticity and maturity of acids [15–17]. With the age of the landfill, the degree of maturity and aromatic condensation of FAs increases [5]. The H/C ratio is 1.07 and is within the range of literature data. This value suggests that the obtained FAs is characterized by moderate maturity [15, 18].

The studies of Kukkonen J. [19], and Perminowa and others [20] on FAs, confirmed the crucial importance of the aromatic core in the binding of impurities. A large number of aromatic groups in a molecule usually means high adsorption capacity of hydrophobic compounds. This explains why the acids obtained easily adsorb pollutants present in the leachate.

Table 2. Atomic ratios of the fulvic substances from leachate.

Atomic ratio	Atomic ratios	Literature[a]
	Calculated as mole fraction	
H/C	1.07	0.90–2.53
O/H	0.99	-
O/C	1.06	0.40–0.78
C/N	14.04	4.35–33.33

[a] Based on 6 different landfills [13, 14, 15]

The molar O/H quotient is used as an indicator oxidation of acids [21]. Therefore, the O/H ratio for tested FAs of 0.99 indicates a low degree of humification. For comparison, the average values of the ratio of FAs extracted from river sediments and lakes in China range from 5.4 to 9.1 [14]. The high difference results from the duration of the decomposition processes of organic compounds. Bottom sediments are characterized by a high content of organic compounds, deposited for many years on the bottoms of rivers and lakes, creating optimal conditions for the humification process. The low O/H ratio extracted from FAs is caused by several factors. Firstly, the waste has been stored for only 33 years, and the leachate collected was a homogenized solution from 3 landfill parts. It can be assumed that the largest amounts of leachates came from the quarters no. II of the second stage, i.e. the currently exploited part, where waste decomposition processes only begin to take shape.

The O/C ratio is an indicator of the carbohydrate and carboxylic acid content in the humic substance molecule [14]. According to literature data, the O/C ratio ranges from 0.5 to 0.9. FAs from the landfill showed a very high O/C ratio, demonstrating a high O-alkyl and carboxylic acid composition, as well as low melting of aromatic condensation and the aliphatic nature of the molecule. In addition, the indicator determines the degree of humification, which should be interpreted as follows: increased humification causes an increase in organic carbon and oxygen decrease, which translates into a decrease in the O/C ratio [21]. The O/C ratio is confirmed by the low degree of humification of FAs. Research on changes in O/C quotients in FAs from landfill leachates confirms the correlation of the landfilling age with the number of functional groups and the degree of aromatic condensation. With the age of waste landfilling, the number of functional groups decreases, and the degree of aromatic condensation increases [5].

Analyzing the quotient of carbon to nitrogen, can determine the directions and intensity of transformation processes of organic matter and nitrogen compounds [22]. In most soils, the C/N ratio ranges from 10 to 15 and it means the balance of organic matter decomposition with the synthesis and accumulation of new organic matter. As the rate or degree of humufication increases, the C/N ratio will decrease. Atomic ratios between 10 and 15 are commonly regarded as an index characterizing well-developed FAs [23]. The resulting C/N ratio of 14.04 clearly shows that the organic material in the landfill is largely stabilized.

Table 3. Micropollutants analysis of FAs from leachate of Promnik landfill.

Element	Symbol	Content, %		Element	Symbol	Content, %
Chlorine	Cl	4,47		Iron	Fe	0,07
Sodium	Na	1,66		Praseodymium	Pr	0,05
Silicon	Si	0,91		Scandium	Sc	0,05
Bromine	Br	0,85		Aluminium	Al	0,01
Phosphorus	P	0,65		Sickel	Ni	< 0,01
Calcium	Ca	0,36		Potassium	K	< 0,01
Magnesium	Mg	0,11		Barium	Ba	< 0,01

In the analyzed elements constituting FAs micropollutants, two leading ones can be distinguished – chlorine (4.47%) and sodium (1.66%). Silicon, bromine and phosphorus (< 1%) had much smaller share. The remaining elements studied constituted a small fraction of the total sample content (< 0.5%).

Ash content in extracted FAs accounted for 27.15% of the total mass, which indicates high contamination with inorganic substances, the value of which depends on the extraction of acids. In the Table 3 shows some micro-pollutants. In addition, in the elemental composition of FAs, there were many elements commonly regarded as toxic. Negative functional groups, found in parts of aliphatic FAs, easily adsorb heavy metals and other toxic substances.

Infrared spectrum

The FT-IR spectra of FAs extracted from leachate were analyzed.

The signal at the 2942 cm^{-1}, adsorption band was attributed to the symmetric stretching of CH in $-CH_3$ i $-CH_2$ aliphatic chains and indicates a high proportion of aliphatic structures in the construction of fulvic acids [13, 24].

A clearly marked adsorption in the band of approximately 1713 cm^{-1}, is associated with the occurrence of C = O carbonyl groups in peptides, aldehydes and ketones, C = O double bonds, C = O stretching of carbonyl groups in ketones and quinones. The occurrence of C = O bonds confirms the moderate and high maturity of FAs [13, 25, 26].

The band in the 1326 cm^{-1} range is associated with C-H deformation in the CH_2 and CH_3 groups, the asymmetric COO- stretch and the C-N stretch [4, 25].

The highest intensity of adsorptive bands was found in the range of 1158 cm^{-1} and 1038 cm^{-1}, this proves the vibration of alcohol functions, stretching of C-O-C carbohydrates, stretching of C-O in polysaccharides, shocking vibrations of C-H methyl linked to the S atom [13, 27].

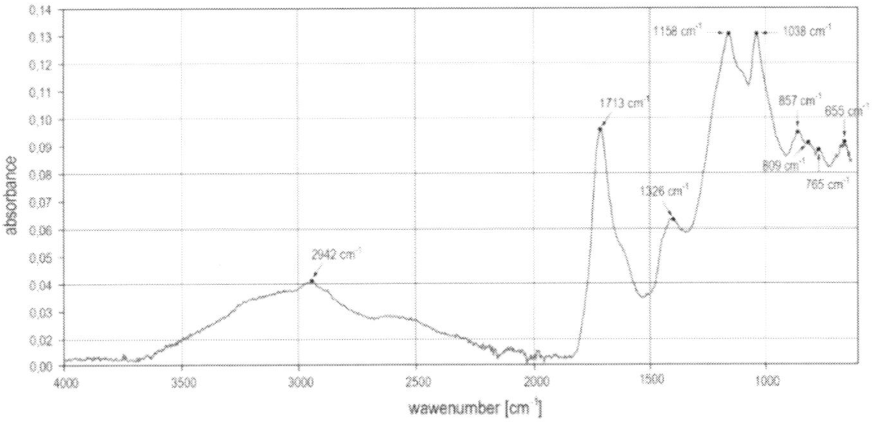

Fig. 1. Infrared spectra of FAs from leachate of Kielce landfills.

FAs showed adsorption in the 857–765 cm^{-1}, range due to N-O bonds with nitrogen esters, aromatic C-H bending, N-H vibration and vibration NH_2 out of plane [13, 15]. Bands at 655 cm^{-1} are associated with the presence of inorganic sulphates (S-O bends and S-O stretch, respectively) and nitrite esters [15, 25].

5 Conclusions

1. The leachate from the Promnik landfill was characterized by the presence of fulvic acids in an amount 905 mg/L.
2. Sample analysis showed significant inorganic contaminants. Dry matter content after roasting was 27.15%. This demonstrates the high chemical activity of the tested acids, which become carriers of toxic micro-contaminants.
3. H/C, O/H, O/C and C/N atomic ratios showed:
   - moderate FAs maturity and low degree of humification,
   - high carboxylic acid O-alkyl composition and low degree of aromatic condensation,
   - an aliphatic nature of a molecule consisting mainly of an aromatic structure,
   - a high degree of stabilization of organic material.
4. Infrared spectra showed a high proportion of aliphatic and aromatic structures in the study of fulvic acids and moderate maturity of FAs.

References

1. Dyrektywa Parlamentu Europejskiego i Rady 2008/98/WE z dnia 19 listopada 2008 r. w sprawie odpadów oraz uchylająca niektóre dyrektywy (Tekst mający znaczenie dla EOG) (z. Urz. UE z 22 listopada 2008 r. L 312/3 – dyrektywa ramowa). Strasburg: Dz. Urz. UE (2008)
2. D. Rozkrut, E. Adach-Stankiewicz, R. Bielak, M. Bieniek, M. Błażej, M. Jeznach, E. Kamińska-Gawryluk, L. Kursa, M. Mojsiewicz, D. Rogalińska, J. Stańczal, W. Szuchta, G. Szydłowska, W. Tkaczyk, K. Walkowska, M. Węgłowska, H. Woźniak, *Mały Rocznik Statystyczny Polski* (GUS, Warszawa, 448, 2018)
3. D. Bochenek, E. Gorzkowska, A. Górska, R. Józwicka, M. Klimkiewicz, A. Kulasza, B. Nowakowska, T. Pawłowska, J. Sulik, M. Wojciechowska, A. Wrzosek, *Ochrona środowiska 2017* (GUS, Warszawa, 40–42, 2017)

4. H. Wang, YN. Wang, X. Li, Y. Sun, H. Wu, D. Chen, Waste Manage. **56**, 271–279 (2016)
5. C. Xiaoli, T. Shimaoka, G. Qiang, Z. Youcai, Waste Manage. **28**, 5, 896–903 (2008)
6. P. Ghosh, I. S. Thakur, A. Kaushik, Ecotoxicol. Environ. Saf. **141**, 259–270 (2017)
7. F. Clareta, C. Tournassata, C. Crouzeta, E. C. Gauchera, T. Schäferbc G. Braibanta, D. Guyonneta, Waste Manage. **31**, 9–10, 2036–2045 (2011)
8. P. Kjeldsen, M. A. Barlaz , A. P. Rooker , A. Baun , A. Ledin , T. H. Christensen, Crit. Rev. Env. Sci. Tec. **32**, 4, 297–336 (2002)
9. Y. Zhou, Y. Zhang, G. Li, T. Jiang, Powder Technol. **302**, 90–99 (2016)
10. A. M. Anielak, M. Wojnicz, Arch. Environ Prot. **36**, 3, 27–37 (2010)
11. C. Xiaoli, L. Guixiang, Z. Xin, H. Yongxia, Z. Youcai, Waste Manage. **32**, 3, 438–447 (2012)
12. A. Siemieniec, M. Siemieniec, *Instrukcja prowadzenia składowiska odpadów komunalnych w Promniku* (Kielce, 2015)
13. Y. S. Han, J. Y. Lee, C. J. Miller, L. Franklin, Waste Manage. Res. **27**, 3, 233–241 (2009)
14. M. He, Y. Shi, C. Lin, J. Environ. Sci.-China. **20**, 11, 1294–1299 (2008)
15. G. Qi, D. Yue, Y. Nie, Front Env. Sci. Eng. **6**, 5, 711–716 (2012)
16. K. H. Kang, H. S. Shin, H. Park, Water Res. **36**, 16, 4023–4032 (2002)
17. C. Xiaoli, J. Rong, W. Jun, T. Huanhuan, Z. Youcai, Chemosphere **78**, 11, 1362–1367 (2010)
18. F. Delarue, S. Derenne, K. Sugitani, F. Baudin, F. Westall, B. Kremer, R. Tartese, A. Gonzales, F. Robert, Org. Geochem. **122**, 140–146 (2018)
19. F. De Paolis, J. Kukkonen, Chemosphere **34**, 8, 1693–1704 (1997)
20. I. V. Perminova, N. Y. Grechishcheva, V. S. Petrosyan, Environ. Sci. Technol. **33**, 21, 3781–3787 (1999)
21. A. M. Gonach, M. H. El-Halafawi, T. M. El-Essawi, Alex. J. Agric. Res. **26**, 3, 747–754 (1978)
22. K. Walenczak, *Charakterystyka gleb naturalnej i wschodniej części Wrocławia* Rozprawa doktorska (Uniwersytet Przyrodniczy we Wrocławiu, Wrocław, 2011)
23. K. T. Tan, *Humic Matter in Soil and the Environment: Principles and Controversies* (Second Edition, **5**, 90–91, 2014)
24. M. Linczar, Zesz. Nauk. AR we Wrocławiu **48**, 1–79 (1985)
25. S. Huo, B. Xi, H. Yu, L. He, S. Fan, H. Liu, J. Environ. Sci.-China. **20**, 4, 492–498 (2008)
26. S. Zhang, L. Yuan, W. Li, Z. Lin, Y. Li, S. Hu, B. Zhao, Chemosphere **166**, 334–342 (2017)
27. M. A. Nanny, N. Ratasuk, Water Res. **36**, 6, 1572–1584 (2002)

The occurrence of antibiotic-resistant bacteria, including *Escherichia coli*, in municipal wastewater and river water

Adriana Osińska[1,*], *Ewa* Korzeniewska[1], *Monika* Harnisz[1], *Sebastian* Niestępski[1], and *Piotr* Jachimowicz[1]

[1]Department of Environmental Microbiology, University of Warmia and Mazury in Olsztyn, Poland

Abstract. Wastewater treatment plants (WWTPs) are major reservoirs of antibiotic-resistant bacteria (ARB) which are transported to the natural environment with discharged effluents. Samples of untreated wastewater (UWW) and treated wastewater (TWW) from four municipal WWTPs and samples of river water collected upstream (URW) and downstream (DRW) from the effluent discharge point were analyzed in the study. The total counts of bacteria resistant to β-lactams and tetracyclines and the counts of antibiotic-resistant *Escherichia coli* were determined. Antibiotic-resistant bacteria, including antibiotic-resistant *E. coli,* were removed with up to 99.9% efficiency in the evaluated WWTPs. Despite the above, ARB counts in TWW samples were high at up to 1.25×10^5 CFU/mL in winter and 1.25×10^3 CFU/mL in summer. Antibiotic-resistant bacteria were also abundant (up to 10^3 CFU/ml) in URW and DRW samples collected in winter and summer. In both UWW and TWW samples, the counts of ARB and antibiotic-resistant *E. coli* were at least one order of magnitude lower in summer than in winter. The study revealed that despite the high efficiency of bacterial removal in the wastewater treatment processes, considerable amounts of ARB are released into the environment with TWW and that the percentage of ARB in total bacterial counts increases after wastewater treatment.

1 Introduction

Bacterial resistance to antibiotics poses a considerable challenge in medicine, and it is not limited to pathogenic microorganisms. Research studies investigating the antibiotic resistance of environmental strains revealed that the natural environment, including soil and aquatic ecosystems, are often reservoirs of clinical bacterial strains with increased resistance to antibiotics [1]. Antibiotic-resistant environmental bacteria can transfer antibiotic resistance genes (ARGs) to human pathogens, and the resulting infections are increasingly difficult or even impossible to treat with the available antibiotics [2]. Antibiotic resistance increases the prevalence of infections, often with lethal consequences.

* Corresponding author: adriana.osinska@uwm.edu.pl

By limiting therapeutic options, antibiotic resistance constitutes one of the most urgent threats to public health.

Beta-lactams and tetracyclines account for approximately 95% of all antibiotics prescribed in the world. Antibiotic metabolism processes in humans and animals are considerably different, and antibiotic residues are evacuated with urine and feces into the sewage system and wastewater treatment plants (WWTPs). Prolonged exposure to even low concentrations of antibiotics in the sewage system contributes to the transfer of ARGs, exerts selective pressure on sewage-borne microorganisms [3] and leads to the spread of antibiotic resistance among bacteria in WWTPs' wastewater and in the natural environment [4].

WWTPs are significant reservoirs of bacteria, in particular gut bacteria harboring ARGs that can be transferred to other microorganisms, including environmental bacteria. *Escherichia coli* colonizes the gastrointestinal tract of humans and animals, and it is an important indicator of fecal contamination. Most *E. coli* strains are commensal bacteria, but some pathogenic strains produce enterotoxins that cause acute diarrhea and infections [5]. The presence of gut bacteria, including pathogenic and potentially pathogenic strains, in these ecosystems can exert negative effects on human and animal health. The relevant risks are heightened when these bacteria are resistant to many groups of antibiotics. Drug-resistant fecal bacteria can also confer antibiotic resistance to environmental microorganisms through horizontal gene transfer.

Activated sludge bioreactors are characterized by high concentrations of dissolved oxygen and high availability of nutrients, and they offer a supportive environment for microbial proliferation and transfer of ARGs [6]. Wastewater contains antibiotics, heavy metals, biocides and disinfectants which contribute to selective pressure and dissemination of antibiotic resistance in bacteria [7]. The effectiveness of wastewater treatment is also determined by temperature which affects bioconversion, the composition and structure of activated sludge [8, 9]. Bacterial decomposition of wastewater contaminants is less effective or completely ineffective at lower temperatures. For this reason, the concentration of ammonium nitrate and the counts of nitrifying bacteria are significantly lower in winter than in summer [10]. Differences in temperature can also induce changes in the concentration of genes encoding antibiotic resistance and the efficiency of their removal [9]. Temperature influences the expression of microbial functional genes, which can promote the spread of ARGs among bacteria. The Polish climate is characterized by a broad annual temperature range (up to 60 degrees), with temperatures up to 35°C in summer and -30°C in winter.

Antibiotic-resistant bacteria and ARGs are discharged with treated sewage and reach soils, surface water and sources of potable water. The above poses considerable health risks because ARB and ARGs evacuated to the environment can be transferred back to humans and animals [11, 12]. Aquatic ecosystems are often subjected to strong anthropogenic pressure, which creates ideal conditions for the acquisition and spread of ARGs. Treated effluents can increase antibiotic concentrations in the aquatic environment and expose bacteria to selective pressure from antibiotics [13]. Antibiotic-resistant bacteria and ARGs can compromise the quality of drinking water, agricultural water (crop cultivation, irrigation) and water used for recreational purposes [14, 15].

The aim of this study was to compare the occurrence and counts of antibiotic-resistant bacteria, including *Escherichia coli,* in wastewater of municipal WWTPs in summer and winter and in samples of river water collected upstream and downstream from the effluent discharge point.

2 Materials and methods

2.1 Sampling sites and sample collection

Samples of untreated (UWW) and treated sewage (TWW) and samples of upstream (URW) and downstream (DRW) river water were collected in four municipal WWTPs located in the Warmia and Mazury district (in total number samples: 8 of UWW, 8 of TWW, 8 of URW, 8 of DRW). The evaluated plants deploy different wastewater treatment systems based on activated sludge. Samples were collected in February 2018 and July 2018 when the consumption of antibiotics in outpatient care is highest and lowest, respectively. Samples of wastewater were collected into sterile bottles, transported to the laboratory at a temperature of 4°C and processed on the day of collection.

2.2 Total bacterial counts and the counts of antibiotic-resistant bacteria

Samples of downstream (DRW) and upstream river water (URW), untreated wastewater (UWW) and treated wastewater (TWW) were diluted with 0.85% NaCl to obtain 30–300 colony forming units (CFU) per plate. Where low bacterial counts were expected, selected URW and DRW samples were passed through a cellulose filter (filter diameter 47 mM, pore diameter 0.45μM, Millipore) to obtain 8–80 CFU per filter. Greater accuracy was achieved by plating triplicates. Total bacterial counts, the counts of bacteria resistant to β-lactams (ampicillin, cefuroxime) and tetracyclines (oxytetracycline, doxycycline), total *Escherichia coli* counts, and the counts of *Escherichia coli* resistant to the above drugs were determined on plates containing TSA (Oxoid) and mFc agar (Merck) growth media, with or without ampicillin (8 μg/mL), cefuroxime (8 μg/mL), oxytetracycline (16 μg/mL) and doxycycline (16 μg/mL). Antimicrobial dosing was based on EUCAST guidelines [16]. The plates for the determination of total microbial counts were incubated at 30°C for 48 h. *E. coli* were cultured at 44.5±0.2°C for 24 h, and *E. coli* colonies were counted based on the number of dark blue colonies formed on mFc agar.

3 Results and discussion

In winter, the average total bacterial counts ranged from 2.30 to 8.30×10^6 CFU/mL in UWW samples and from 1.40×10^4 to 9.60×10^5 CFU/mL in TWW samples (Tab. 1). In summer, the average total bacterial counts were lower by at least one order of magnitude in the range of 3.50×10^4 to 6.85×10^5 CFU/mL in UWW samples and 8.80×10^2 to 8.60×10^3 CFU/mL in TWW samples. The counts of ARB were generally lower in summer than in winter, and average ARB counts in UWW samples were determined at 3.39×10^3 to 6.49×10^4 CFU/mL in summer and at 6.70×10^4 to 3.5×10^6 CFU/mL in winter. In TWW samples, ARB counts ranged from 8.7×10^1 to 1.25×10^3 CFU/mL in summer and from 3.10×10^3 to 1.25×10^5 CFU/mL in winter.

In a study by Ziembińska-Buczyńska et al., [17] total bacterial counts in activated sludge fluctuated between 12×10^9 CFU/mL in samples collected in summer and 42×10^9 CFU/mL in winter, and they increased steadily between winter and spring months. The cited authors also reported higher counts of bacteria resistant to erythromycin and sulfamethoxazole/trimethoprim in samples from winter. In winter, the predominant ARB were ampicillin-resistant bacteria that accounted for 26–91% of total bacterial counts in UWW samples, whereas summer samples of UWW were characterized by a predominance of cefuroxime-resistant bacteria (6–28%). Summer samples of TWW were also dominated by ampicillin-resistant bacteria (15–64%), whereas cefuroxime-resistant bacteria accounted for 8–18% of total bacterial counts in winter samples. It should be noted that the percentage of ARB in total bacterial counts after wastewater treatment decreased in winter, but increased significantly in summer. Osińska et al. [18] analyzed various wastewater treatment systems based on activated sludge and reported an increase in the percentage of ARB in total bacterial counts during the treatment process. An increase in the proportions of ARB in treated wastewater can be attributed to the presence of mobile genetic elements in bacterial cells, including integrons, plasmids and transposons, which play a key role in the transmission of ARGs [19–21]. The spread of ARGs among sewage-borne bacteria can also be induced by antibiotics and antibiotic degradation products which increase selective pressure on microorganisms [22]. According to Jutkina et al., [23] sublethal antibiotic concentrations not only exacerbate selective pressure, but also directly contribute to the dissemination of ARGs through horizontal gene transfer.

Table 1. Total counts of bacteria, counts of bacteria resistant to beta-lactams and tetracycline, percent of reduction and the share of antibiotic resistant bacteria in total counts of microorganisms in wastewater.

	Summer			Winter		
	UWW (CFU/mL)	TWW (CFU/mL)	% of reduction	UWW (CFU/mL)	TWW (CFU/mL)	% of reduction
TB	0.35–6.85x10^5	0.88–8.6x10^3	97–99.99	2.3–8.3x10^6	0.14–9.6x10^5	88–99.99
TBA	0.634–5.67x10^4	0.416–1.25x10^3	86–99.99	1.65–3.5x10^6	0.165–1.25x10^5	94–99.99
%ARB	1–18%	15–64%		26–91%	12–17%	
TBC	0.956–5.68x10^4	0.087–1.25x10^3	96–99.65	0.53–1.1x10^6	0.175–1.03x10^5	90–99.77
%ARB	6–28%	10–20%		8–43%	8–18%	
TBO	0.829–6.49x10^4	2.3–4.56x10^2	94–99.47	0.67–9.0x10^5	0.545–1.1x10^4	83–99.94
%ARB	2–25%	3–52%		1–14%	1–8%	
TBD	3.39–6.5x10^3	1.2–4.6x10^2	92–96.46	0.93–1.26x10^6	0.31–1.97x10^4	98–99.74
%ARB	1–15%	5–43%		1–15%	1–3%	

CFU – colony forming unit, UWW – untreated wastewater, TWW – treated wastewater, TB – counts of total bacteria, TBA – counts of bacteria resistant to ampicillin, TBC – counts of bacteria resistant to cefuroxime, TBO – counts of bacteria resistant to oxytetracycline, TBD – counts of bacteria resistant to doxycycline, %ARB- the share of antibiotic resistant bacteria in total counts of microorganism

Total bacterial counts and ARB counts were reduced and characterized by similar proportions in TWW samples collected both in summer and winter. Total bacterial counts and ARB counts were reduced by 86.06–99.9% in summer and by 83.58–99.99% in winter. In the work of Czekalski et al. [11] and Korzeniewska and Harnisz [24], bacterial counts were reduced in TWW, but the the presence of multidrug resistant strains and an increase in the counts of ARGs were observed in discharged effluents. Proia et al., [25] also reported a significant reduction in the abundance of ARB, but did not note significant differences in ARB counts after wastewater treatment. In a study by Bodnarczuk et al., [26] wastewater treatment led to a significant decrease in the counts of heterotrophic bacteria (by up to two orders of magnitude). Despite the above, the counts of ampicillin-resistant bacteria increased in lake sediments, which could point to the spread of ARB in the environment.

In winter, the average total bacterial counts ranged from 2.33x10^3 to 1.10x10^4 CFU/mL in URW samples and from 2.20 to 9.20x10^3 CFU/mL in DRW samples (Tab. 2). In summer, the average total bacterial counts were determined in the range of 3.60x10^3–1.24x10^4 CFU/mL in URW samples and 4.20x10^3–1.23x10^4 CFU/mL in DRW samples.

Table 2. Total counts of bacteria, counts of bacteria resistant to beta-lactams and tetracycline, percent of reduction and the share of antibiotic resistant bacteria in total counts of microorganisms in river water/

	Summer		**Winter**	
	URW	DRW	URW	DRW
TB	0.36–1.24x10⁴	0.42–1.23x10⁴	0.233–1.1x10⁴	2.2–9.2x10³
TBA	0.28–4.56x10³	0.28–4.9x10³	1.4–3.76x10³	1.8–6.7x10³
%ARB	8–47%	7–41%	30–67%	1–89%
TBC	0.18–3.2x10³	0.19–3.4x10³	1.74–2.89x10³	1.45–4.78x10³
%ARB	5–33%	5–28%	24–75%	1–72%
TBO	0.09–1.8x10³	0.095–1.4x10³	0.003–2.86x10³	0.065–1.27x10³
%ARB	3–18%	2–12%	0.13–26%	0.007–32%
TBD	0.15–1.4x10²	0.3–2.0x10¹	0.05–1.6x10²	0.5–6.5x10¹
%ARB	0.2–1%	0.1–0.2%	0.21–1%	0.047–0.38%

URW – upstream river water, DRW – downstream river water, remainder explanation of abbreviations under Table1

The absence of a reduction in bacterial counts in summer samples of river water relative to winter samples could be associated with the specific character of sampling sites and their surroundings (allotment gardens, forests and urban areas). In URW samples, ARB counts ranged from $1.5x10^1$ to $4.56x10^3$ CFU/mL in summer and from $3x10^0$ to $3.76x10^3$ CFU/mL in winter. In DRW samples, ARB counts were determined in a range of $3x10^0$–$4.90x10^3$ CFU/mL in summer and $5x10^0$–$6.7x10^3$ CFU/mL in winter. Samples of URW collected in both winter and summer were characterized by a predominance of ampicillin-resistant bacteria (up to 67% and 47% of total bacterial counts, respectively) and cefuroxime-resistant bacteria (up to 75% and 33%, respectively). A predominance of ampicillin- and cefuroxime-resistant bacteria was also noted in DRW samples where these microorganisms accounted up to 89% and 72% of total bacterial counts, respectively, in winter, and 41% and 28% of total bacterial counts, respectively, in summer. Doxycycline-resistant bacteria were least prevalent in URW and DRW samples (less than 1% of total bacterial counts) in both seasons. In most DRW samples collected in winter, the percentage of bacteria resistant to ampicillin, cefuroxime and oxytetracycline increased relative to URW samples. In contrast, the proportions of ARB in most DRW samples collected in summer decreased relative to URW samples collected in the same season. Similar results were reported in a study by Harnisz et al., [27] total bacterial counts were determined at up to $1.5x10^3$ CFU/mL in URW samples and $3.1x10^3$ CFU/mL in DRW samples. They observed that the counts of bacteria resistant to oxytetracycline and doxycycline ranged from $3x10^1$ to $1.0x10^2$ CFU/mL in URW samples and from $3.0x10^1$ to $2.0x10^2$ CFU/mL in DRW samples. Proia et al., [25] reported an increase in *E. coli* counts in DRW samples where total bacterial counts were determined at 10^8 CFU/L. In the cited study, the abundance of ARGs was relatively high in URW samples, and it increased significantly in DRW samples, which indicates that urban activities contribute to the dissemination of antibiotic resistance. Xu et al., [13] also reported a high percentage of ARB in river water and bottom sediments. Samples of river water were abundant not only in ARB and ARGs, but also in antibiotics and antibiotic degradation products, which suggests that effluents discharged to rivers can substantially promote antibiotic resistance. Osińska et al., [28] isolated multidrug resistant *E. coli* from DRW samples (32% of isolates), but not from URW samples.

In winter, the average total *E. coli* counts ranged from $1.85x10^4$ to $6.45x10^5$ CFU/mL in UWW samples and from $2.5\ x10^1$ to $1.55x10^3$ CFU/mL in TWW samples (Tab. 3).

Table 3. Total counts of *Escherichia coli*, counts of *Escherichia coli* resistant to beta-lactams and tetracycline, percent of reduction and the share of antibiotic resistant bacteria in total counts of microorganisms in wastewater.

	Summer			Winter		
	UWW (CFU/mL)	TWW (CFU/mL)	% of reduction	UWW (CFU/mL)	TWW (CFU/mL)	% of reduction
EC	2.2–7.59x10^3	0.11–4.5x10^2	82–99.86	0.185–6.45x10^5	0.025–1.55x10^3	9–99.99
ECA	2.75–9.95x10^2	0.65–8.7x10^1	75–99.35	0.351–2.75x10^4	0.065–2.4x10^2	93–99.93
%ARB	9–27%	19–59%		2–56%	15–39%	
ECC	0.19–1.35x10^3	0.2–5.6x10^1	95–99.7	0.190–1.35x10^3	0.2–8.75x10^1	93–99.67
%ARB	3–61%	2–22%		0.03–7%	1–10%	
ECO	2.2–3.75x10^2	0.4–1.2x10^1	94–98.67	0.29–2.20x10^4	4–1.45x10^1	99.7–99.93
%ARB	4–14%	1–67%		0.45–63%	0.48–34%	
ECD	1.5–5.25x10^2	0.35–3.5x10^1	88–99.33	1.5–9.25x10^3	0.035–1.55x10^2	95–99.96
%ARB	7–13%	8–56%		1–18%	10–42%	

CFU – colony forming unit, UWW – untreated wastewater, TWW – treated wastewater, EC – counts of total *E. coli*, ECA – counts of *E. coli* resistant to ampicillin, ECC – counts of *E. coli* resistant to cefuroxime, ECO – counts of *E. coli* resistant to oxytetracycline, ECD – counts of *E. coli* resistant to doxycycline, %ARB – the share of antibiotic resistant *E. coli* in total counts of *E.coli*

Similarly to total bacterial counts, the average total counts of *E. coli* were at least one order of magnitude smaller in summer than in winter, ranging from 2.2 x10^3 to 7.59 x10^3 CFU/mL in UWW samples collected in summer and from 1.85x10^4 to 6.45 x10^5 CFU/mL in UWW samples collected in winter. In TWW samples, the counts of antibiotic-resistant *E. coli* were determined in the range of 2x10^0 to 8.7x10^1 CFU/mL in summer and 2x10^0 to 2.4x10^2 CFU/mL in winter. In a study investigating the prevalence of bacteria of the family *Enterobacteriaceae,* Korzeniewska et al., [29] found that despite a 99% reduction in these bacterial counts during wastewater treatment, bacterial abundance in discharged wastewater was high at 0.6–3.5x10^4 CFU/mL. Proia et al., [25] also reported a decrease in *E. coli* counts from $6.0 \pm 4.4 \cdot 10^4$ CFU/mL to $7.1 \pm 7 \cdot 10^2$ CFU/ mL and from $8.7 \pm 3.0 \cdot 10^4$ CFU/mL to $7.1 \pm 7.0 \cdot 10^2$ CFU/ mL in treatment of wastewater in two WWTPs, whereas the reduction in the counts of antibiotic-resistant *E. coli* was significantly lower at around 2 log. Despite the observed decrease in the abundance of *E. coli*, antibiotic resistance was not diminished, and the counts of antibiotic-resistant *E. coli* were high (approx. 10^2 CFU/mL) in TWW samples. In our study observed that in winter, UWW samples were characterized by a predominance of oxytetracycline-resistant *E. coli* strains which accounted for 0.45–63% of total *E. coli* counts, whereas in TWW samples ampicillin-resistant bacteria represented 15–39% of *E. coli* strains and doxycycline- resistant bacteria represented 10–42% of *E. coli* strains. In summer, cefuroxime-resistant bacteria accounted for 3–61% of total *E. coli* counts in UWW samples, whereas ampicillin-resistant and oxytetracycline-resistant *E. coli* were predominant (respectively 19–59% and 1–67%) in TWW samples. Total *E. coli* counts and the counts of antibiotic-resistant *E. coli* were reduced during wastewater treatment in all analyzed plants. Total *E. coli* counts and the abundance of antibiotic-resistant *E. coli* were reduced by 75-99.86% in summer and 91–99.99% in winter. In a study by Mokracka et al., [30] the counts of ARB and *E. coli* decreased by 59-99% after treatment. In contrast, Osińska et al., [18] reported an increase in the counts of antibiotic-resistant *E. coli* after wastewater treatment in selected plants. The increase in the counts of antibiotic-resistant *E. coli* could be attributed to horizontal gene transfer between bacterial strains [29, 31].

The discharge of wastewater after the treatment process to aquatic ecosystems is an important problem from both the economic point of view and also from environmental safety. Although the applied wastewater treatment processes allow reducing the number of bacteria up to 99%, in discharged wastewater from WWTPs, there may still be up to several thousand to one million coliform bacteria cells in 100 ml of treated wastewater [24, 25, 27]. Similarly, the results presented in this article indicate significant amounts of the total number of bacteria and ARB in wastewater flowing out of WWTPs. It is important to underline that, in this article were analyzed WWTPs which took only municipal wastewater. Nevertheless, significant amounts of bacteria resistant to both older generation antibiotics (ampicillin, oxytetracycline) and newer generation antibiotics (cefuroxime, doxycycline) were found in WWTPs' effluents. Moreover, obtained results of the study indicate significant amounts of the total number of *Escherichia coli* and antibiotic-resistant *E. coli* in both untreated wastewater as well as in treated wastewater. *E.coli* belongs to a group of bacteria that, because of its plastic genome, are often used to monitor the spread of antibiotic resistance in various environments. The presence in the aquatic ecosystems of intestinal bacteria, originating from WWTPs, may have a negative impact on public health because of these bacteria can be characterized by multi-drug resistance and they frequently possess ARGs or virulence genes [11, 28]. Additionally, these ARB can transmit resistance to native bacteria from the environment, including both pathogenic or non-pathogenic bacteria, mediated by mobile genetic elements, such as plasmids (conjugation), bacteriophages (transduction), or free DNA from the environment (transformation) [11, 17]. The obtained results indicate the presence of ARB also in river water before and after the inflow of WWTPs, which may pose a threat to human health and animals using water resources. Thus, antibiotic-resistant environmental bacteria can be a natural reservoir for the further spread of ARGs in the environment. The results of other authors also suggest that wastewater flowing into surface water is the main source of ARB and ARGs in the environment [11, 25]. It is also worth emphasizing that in WWTPs in Poland, the processes of wastewater disinfection prior to their discharge into the environment are not commonly used. Wastewater disinfection processes could be used as potential tools to reduce the spread of ARB and ARGs to the environment. Therefore, it seems to be particularly important to monitor the effectiveness of wastewater treatment processes and the microbiological quality of wastewater discharged from WWTPs to the natural environment.

4 Conclusions

The results of this study indicate that wastewater treatment plants are major reservoirs of antibiotic-resistant bacteria which are evacuated to water bodies with treated effluents, thus contributing to the spread of antibiotic resistance in the natural environment. Antibiotic-resistant bacteria were effectively removed in all evaluated WWTPs. Despite the above, ARB counts continued to be high in discharged wastewater. In UWW and TWW samples, the counts of ARB and antibiotic-resistant *E. coli* were at least one order of magnitude lower in summer than in winter, but the percentage of ARB in total bacterial counts increased after treatment. Antibiotic-resistant bacteria were also highly abundant (up to 10^3 CFU/ml) in river water sampled upstream and downstream from the effluent discharge point in winter and in summer. These findings indicate that wastewater evacuated from wastewater treatment plants should be closely monitored to minimize the spread of ARB in the environment.

We would like to thank the staff of the WWTPs in Białuty, Dobre Miasto,Gietrzwałd and Dobre Miasto for the possibility of samples collecting. This study was supported by grant No. UMO-2016/23/N/NZ9/02150 from the National Science Center (Poland).

References

1. R. L. Finley, P. Collignon, D. G. Larsson, S. A. McEwen, X. L. Li, W. H. Gaze, R. Reid-Smith, M. Timinouni, D. W. Graham, E. Topp, Clin. Inf. Dis. **57**, 5 (2013)
2. World Health Organization. Antimicrobial Resistance: Global Report on Surveillance 2014. Available online: http://www.who.int/en/ (accessed on 6 February 2017)
3. A. Novo, S. André, P. Viana, O. Nunes, C. Manaia, Water Res. **47**, 5 (2013)
4. X-X. Zhang, T. Zhang, H. H. P. Fang, Appl. Microbiol. Biotechnol. **82**, (2009)
5. L. J. Dubreuil, S. Mahieux, C. Neut, C. Miossec, J. Pace, Int. J. Antimicrob. Ag. **39**, 6 (2012)
6. A. Luczkiewicz, K. Jankowska, R. Bray, E. Kulbat, B. Quant, A. Sokolowska, K. Olanczuk-Neyman, Water Sci. Technol. **64**, 12 (2011)
7. L. Rizzo, C. Manaia, C. Merlin, T. Schwartz, C. Dagot, M. C. Ploy, I. Michael, D. Fatta Kassinos, Sci. Total Environ. **447** (2013)
8. A. Karkman, T. A. Johnson, C. Lyra, R. D. Stedtfeld, M. Tamminen, J. M. Tiedje, M. Virta, FEMS Microbiol. Ecol. **92**, 3 (2016)
9. Y. N. Jiao, Z. C. Zhou, T. Chen, Y. Y. Wei, J. Zheng, R. X. Gao, Environ. Pollut. **234** (2018)
10. T. F. Ducey, M. B. Vanotti, A. D. Shriner, A. A. Szogi, A. Q. Ellison, Bioresource Technol. **101**, 2 (2010)
11. N. Czekalski, T. Berthold, S. Caucci, A. Egli, H. Bürgmann, Front. Microbiol. **3** (2012)
12. Y.-G. Zhu, T. A. Johnson, J.-Q. Su, M. Qiao, G.-X. Guo, R. D. Stedtfeld, S. A. Hashsham, J. M. Tiedje, PNAS, **110** (2013)
13. Y. Xu, C. Guo, Y. Luo, J. Lv, Y. Zhang, H. Lin, L. Wang, J. Xu, Environ. Pollut. **213** (2016)
14. J. Xu, Y. Xu, H. Wang, C. Guo, H. Qiu, Y. He, Y. Zhang, X. Li, W. Meng, Chem. **119** (2015)
15. Y. Yang, B. Li, S. C. Zou, H. H. P. Fang, T. Zhang, Water Res. **62** (2014)
16. EUCAST, European Committee on antimicrobial susceptibility testing. Breakpoints tables for interpretation of MICs and zones diameters. Version 4.0, http://www.eucast.org (2014)
17. A. Ziembińska-Buczyńska, E. Felis, J. Folkert, A. Meresta, D. Stawicka, A. Gnida, J. Surmacz-Górska, Arch. Environ. Protect. **41**, 4 (2015)
18. A. Osińska, E. Korzeniewska, M. Harnisz, S. Niestępski, E3S Web Conf. **17** (2017)
19. S. Xia, L. Duan, Y. Song, J. Li, Y. M. Piceno, G. L. Andersen, L. Alvarez-Cohen, I. Moreno-Andrade, C.-L. Huang, S. W. Hermanowicz, Environ. Sci.Technol. **44**, 19 (2010)
20. J. Mokracka, R. Koczura, A. Kaznowski, Water Res. **46**, 10 (2012)
21. W. Xu, G. Zhang, X. Li, S. Zou, P. Li, Z. Hu, J. Li, Water Res. **41**, 19 (2007)
22. Y. Y. Zhang, J. J. Geng, H. J. Ma, H. Q. Ren, K. Xu, L. L. Ding, Sci. Total Environ. **571** (2016)

23. J. Jutkina, N. P. Marathe, C.-F. Flach, D. G. J. Larsson, Sci. Total Environ. **616–617** (2018)
24. E. Korzeniewska, M. Harnisz, Sci. Total Environ. **639** (2018)
25. L. Proia, A. Anzil, J. Subirats, C. Borrego, M. Farrè, M. Llorca, J. L. Balcázar, P. Servais, Sci. Total Environ. **628–629** (2018)
26. K. Bondarczuk, Z. Piotrowska-Seget, Sci. Total Environ. **650** (2019)
27. M. Harnisz, E. Korzeniewska, I. Gołaś, Chem. **128** (2015)
28. A. Osińska, E. Korzeniewska, M. Harnisz, S. Niestępski, Sci. Total Environ. **577** (2017)
29. E. Korzeniewska, M. Harnisz, J. Environ. Manage. **128** (2013)
30. J. Mokracka, R. Koczura, A. Kaznowski, Water Res. **46**, 10 (2012)
31. A. Osińska, M. Harnisz, E. Korzeniewska, Environ. Sci. Pollut. Res. **23**, 11 (2016)

Analysis of the multi-stage desiccant cooling system performance in Wrocław (Poland)

Anna Pacak[1,*], *Aleksandra* Cichoń[1], *Demis* Pandelidis[1], and *Sergey* Anisimov[1]

[1]Wrocław University of Science and Technology, Faculty of Environmental Engineering, Department of Air Conditioning, Heating, Gas Engineering and Air Protection, ul. Norwida 4/6, 50-373 Wroclaw, Poland

Abstract. In this study, the application potential of desiccant cooling system with indirect evaporative precooling is analyzed for moderate climatic conditions. Wrocław (Poland) was selected as a representative city. On the base of summer season weather parameters different system running modes were proposed such as cooling and dehumidification mode (Mode 1) and only cooling mode (Mode 2). The average month Thermal and Electrical COP were established for selected climatic conditions. It was concluded that proposed system allows to cool the air without occurrence of condensation in Mode 2 with high average monthly electrical COP (COP_{el} = 62.2 in August). On the other hand when dehumidification is needed, system operates in Mode 1 and obtains relatively high average Thermal COP values (COP_{th} = 2.2).

1 Introduction

Indirect evaporative cooling (IEC) is currently one of the most promising solutions for air conditioning. These devices provide the airflow temperature decrease, by sensible cooling, without adding any moisture to the air. One of the best methods of IEC approaching the low supply air temperatures such as dew point temperature, is the novel thermodynamic cycle, known as the Maisotsenko cycle (M-cycle). On the other hand, in typical direct evaporative coolers, humidity ratio of the treated airflow increases during cooling process, and a wet-bulb temperature is a limitation of this process. M-Cycle heat and mass exchangers (HMX) are already produced for air conditioning systems by Coolerado Corporation [1]. The main advantage of the M-Cycle units over typical vapor-compression systems is that it utilizes significantly less electricity [2]. It is the reason why these devices are treated as an environmental friendly solution for cooling energy production. Individual devices based on the M-Cycle technology are implemented for systems which operate especially in hot and dry area for now [3]. In humid and moderate climatic conditions there is a necessity to control the humidity level of the supply airflow. It is possible to use M-Cycle IEC devices in those areas also. In such a climatic conditions they need to cooperate with some devices which provide the outdoor air dehumidification. The most representative solid dehumidification units are the thermally driven desiccant wheels [4]. There are two main advantages of solid desiccant systems confirming its cost saving potential. They may use

* Corresponding author: anna.pacak@pwr.edu.pl

https://doi.org/10.1051/e3sconf/201910000062

solar or waste thermal energy for heating the regeneration air which is used for desorption of desiccant wheel, and they allow to control the temperature and humidity of the supply airflow independently [5, 6]. Either solid desiccant which usually consists of desiccant wheel and cooling device, or M-cycle dew point evaporative coolers has been extensively reported in the literature. Chaudhary et. al. [7] proposed a solid desiccant cooling system equipped with a cross-flow M-Cycle unit and solar power as a heat source. They analyzed its operation experimentally considering the performance indicators such as dehumidification effectiveness, dew point effectiveness, Thermal COP, and cooling capacity. The results show that this system reaches an average Thermal COP equal to 0.91 for analyzed hot and humid climatic conditions. Comino et. al. [8] established the annual energy consumption of a hybrid system with a desiccant wheel activated at low temperatures and an IEC for 6 climate zones. The system was compared to the direct expansion (DX) system and it represents a significant energy savings, up to 46.8%. The authors concluded that this system has a great potential to reduce energy costs and ensure proper thermal comfort air conditioning spaces concurrently. Shahzad et. al. [9] also performed an experiment in which the performance investigation of a solid desiccant dehumidifier integrated with Maisotsenko cycle (M-Cycle) was presented. The comparison of proposed system to the typical desiccant air conditioning system shows that it is around 60–65% more efficient in terms of Thermal COP.

2 Proposed system description

The interest in desiccant cooling systems is increasing due to its ability to produce cooling utilizing thermal energy. Moreover, the energy required to regenerate the desiccant wheel in the summer may come from many energy sources like solar collectors, heat pumps or waste heat. In the case of a conventional desiccant cooling system, the end-point air treatment is limited by the humidification of the airflow in direct evaporative coolers. In this study, a novel system which is a simple way of improving the M-Cycle desiccant system performance is proposed. The new system uses precooling of the air stream before the dehumidification process. In this system, the ambient air is cooled two times, both before the dehumidification process and after. It allows lower supply air temperature and humidity ratios to be obtained [10]. The proposed system consists of a combination of two indirect evaporative air coolers combined with a desiccant wheel (see Fig. 1).

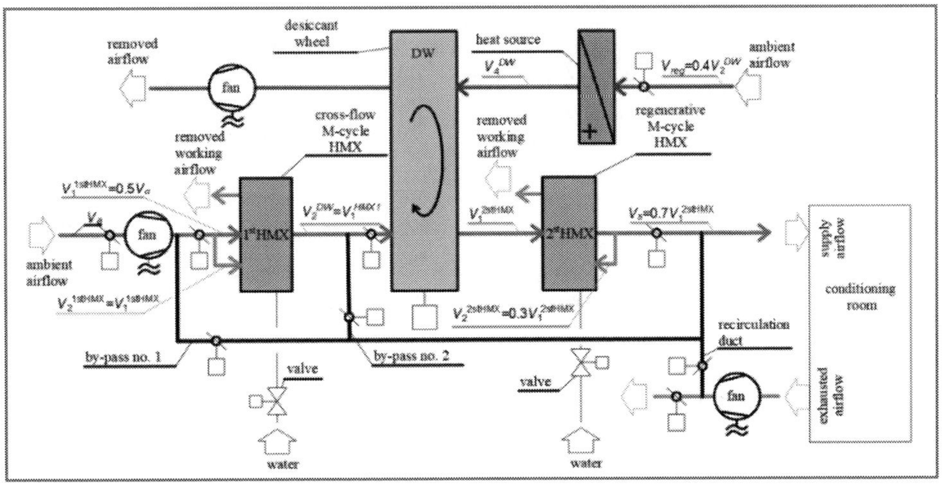

Fig. 1. Analyzed system scheme.

The first cross-flow M-Cycle unit is located before the desiccant wheel, to initially pre-cool the outdoor air, before it enters the dehumidification section. The second regenerative M-Cycle unit is located after the desiccant wheel and it realized the post cooling of the process airflow. The aim of this study is to establish the system performance in moderate climatic conditions numerically, on the base of the real outdoor air conditions. To do that two key performance indicators are proposed those are: Thermal Coefficient of Performance and Electric Coefficient of Performance.

3 Methods

Performance of M-Cycle regenerative air cooler, cross-flow M-cycle air cooler and desiccant wheel were simulated using original ε-NTU models. These models are created with an assumption that the airflow is treated as gaseous fluid, which has temperature, velocity and mass transfer potential equal to bulk average values in sections normal to the plate surface of heat exchanger. Models equations, validation of these models are presented in author's previous works [10, 11]. Main system components characteristics are presented in Table 1.

Table 1. Main system components operating characteristics [10].

Indirect evaporative air coolers parameters		**Desiccant wheel parameters**	
Cross-flow M-Cycle HMX		Specific heat capacity (silica gel), J/(kg K)	750
Working to intake air ratio	1.0	Channel type	Sinusoidal
NTU_1	8	Channel height, mm	1.24
NTU_2	8	Channel width, mm	2.21
		Wall thickness, mm	0.30
		Channel length, m	0.1
Working to intake air ratio (regenerative M-Cycle HMX)	0.3	Wheel diameter, m	0.5
NTU_1	10	Percent of wheel used for regeneration airflow, %	40
NTU_2	16.7	Rotational speed, cycles/hour	8

4 Selected outdoor air parameters

Authors dedicate the system especially for moderate climatic conditions. Wrocław (Poland) is a representative city for which a selected summer season outdoor air parameters has been analyzed in this study. Each system configuration can work in 3 different operation modes. The mode selection depends on the ambient air parameters which are presented in Table 3. The limits were established with an assumption that the comfort indoor air parameters are in temperature range between 20°C and 25°C and humidity ratio below 12 g/kg. The maximum humidity ratio is established on the base of an assumption that the share of sensible heat loads in total heat loads is about 0.8. That is why, 11 g/kg is the maximum humidity ratio of the supply airflow which enables to maintain a typical indoor thermal comfort.

Table 2. Conditions for operation modes.

Mode 1	Mode 2	Mode 3
Cooling and dehumidification	Cooling	Ventilation mode
$t_a \geq 20°C \pm 0.5°C$ and $x_a \geq 11 g/kg \pm 0.5 g/kg$	$t_a \geq 20°C \pm 0.5°C$ and $x_a < 11 g/kg \pm 0.5 g/kg$	$t_a < 20°C \pm 0.5°C$

After the hourly analysis of the outdoor air parameters such as air temperature and humidity ratio, the amount of mode working hours was specified (See. Fig. 2). It may be observed that during selected months in year 2018, System would work in Mode 2 for most of the summer season. It means that, the air is cooled down only in 1[st]HMX (see Fig. 1) and next it is bypassed using the bypass no. 2. Mode 1 uses each device of the system (1[st]HMX, desiccant wheel and 2[st]HMX) and its operation hours makes up about 30% of total working hours in August, about 20% in July and the least 13% in June. In polish climatic conditions there are the days and hours when the air needs to be delivered to the room but the temperature and humidity ratio are low enough to cover sensible and latent heat loads. In such operating conditions the bypass no 3 is used and system works in Mode 3. What is more, in Mode 3, the recirculation duct may be used. When the ambient air temperature is to low, the air may be mixed with the exhausted air and delivered to the air conditioning space (see Fig. 1). The amount of mode working hours is specified in Fig. 2. It has to be underlined that authors analyzed 920 hours in total.

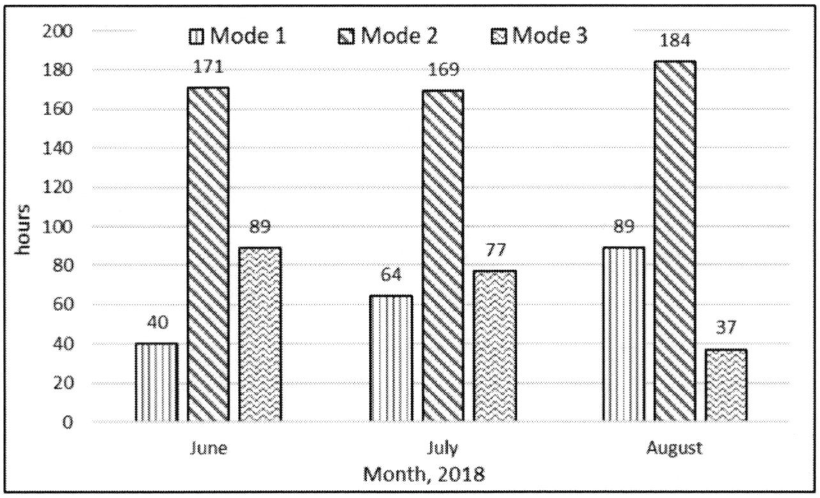

Fig. 2. System operation modes for selected summer operating conditions for Wrocław (Poland).

For considered climatic conditions, proposed system operates in Cooling mode (Mode 2) mostly. There is a significant advantage of such a solution. It enables to cool the air sensibly down without occurrence the condensation process. It is a main difference of such a solution comparing to the typical vapor-compression systems in which usually decreasing temperature is connected with decreasing the humidity ratio of the treated airflow. It pushes up the energy demand for air conditioning system. It is the fact that M-Cycle evaporative coolers produce the cooling energy with high energy efficiency. To establish the system efficiency for selected climatic zone three month average COP_{el} factor is determined for Cooling mode (Mode 2). In moderate climates there are days where the dehumidification of the ambient airflow is necessary. In such a conditions, the proposed device is working in Mode 1 in which the cooling and dehumidification processes run separately. To establish the effectiveness of this mode, two average key performance indicators COP_{el} and COP_{th} (Thermal Coefficient of Performance) were calculated.

5 System performance

In this section, the results of the system performance analysis based on the computational simulations are presented. Below, the basic equations which were used for calculation of chosen key performance indicators. System cooling capacity is calculated on the base of the outdoor air temperature and humidity ratio measured in Wroclaw (Poland) which is located in moderate climatic conditions.

The Thermal Coefficient of Performance was calculated using the following formula:

$$COP_{th} = \frac{\Phi}{P} \quad (1)$$

Where the cooling capacity of the supply airflow was calculated using the formula:

$$\Phi = V_{1o}^{2^{st}HMX}\rho(i_E - i_{1o}^{2^{st}HMX}) \quad (2)$$

Energy for regeneration purposes was calculated from the equation:

$$P = V_4\rho c_p(t_{reg} - t_E) \quad (3)$$

The Electrical Coefficient of Performance was calculated using the equation:

$$COP_{el} = \frac{\Phi}{N} \quad (4)$$

Where the energy consumption by fan was calculated using the equation:

$$N = \Delta p V_{1i}^{1^{st}HMX}/\eta \quad (5)$$

A fan efficiency of $\eta = 0.4$ was used.

In Table 3 there are considered ranges of ambient, exhausted and regeneration air parameters for which the simulations of system performance were conducted.

Table 3. Air parameters for simulations.

Ambient airflow:	$V_a = 1000 m^3/h$	**Minimum value**	**Maximum value**
Assumed exhausted air parameters			
Temperature	t_{CR}	20°C	25°C
Humidity ratio	RH_{CR}	40%	60%
Assumed regeneration air parameters			
Temperature	t_4^{DW}	45°C	45°C
Humidity ratio	$x_4^{DW} = x_a$	10.5 g/kg	16.2 g/kg
Ambient air parameters during selected period of time			
Temperature	t_a	12.9	33.4
Humidity ratio	x_a	10.5 g/kg	16.2 g/kg

Table 4. Obtained COP_{th} and COP_{el} for system operation modes.

	Mode 1			**Mode 2**		
running hours	193			524		
Month	June	July	August	June	July	August
average COP_{th}	1.5	1.8	2.2	-	-	-
average COP_{el}	18.0	18.4	19.1	59.4	62.0	62.2
maximum COP_{th}	2.7	2.4	3.7			

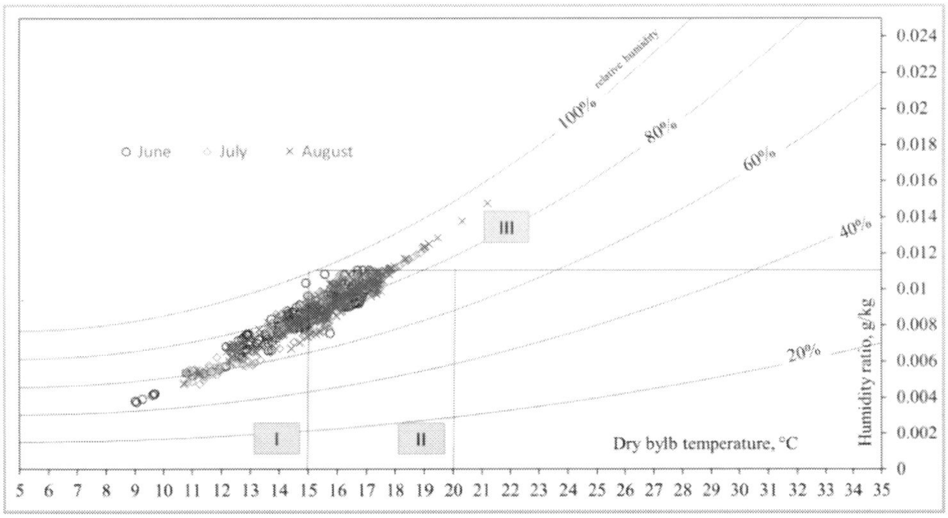

Fig. 3. Supply airflow parameters on psychometric chart.

According to the data presented in Table 4 proposed system represents high electrical performance in Cooling Mode 2 because only 1stHMX is in operation. On the other hand, the month average monthly thermal Coefficient of Performance is above 1.0 and it equals 1.5 in June, 1.8 in July and 2.2 in August. The higher outdoor air temperature the higher system COP_{th} is observed. It is because of the fact that the desiccant wheel is regenerated by the heated ambient airflow. The calculations allowed to establish the COP values for each working hour during selected summer season. In Table 5 the appearance of COP_{th} is presented and it may be noticed that the COP_{th} is higher than 3 during 11% of operation hours (in Mode 1), it is between 2 and 3 during 28% of operation hours and between 1 and 2 during 55% of operation hours. The maximum hourly COP_{th} value equals 3.73.

Table 5. COP_{th} appearance.

COP_{th}	Percent of hours appearance in working Mode 1
$[1-2)$	55%
$[2-3)$	28%
≥ 3	11%

In Fig. 3 obtained supply airflow parameters for selected months are presented. It may be seen that analyzed system for considered regeneration air temperature for most cases is able to assimilate latent and sensible heat loads (see Fig. 3 zone II). The supply airflow parameters included in zone III in psychometric chart would not provide the comfort indoor conditions. In such situations a higher regeneration temperature is necessary to decrease the humidity ratio of the supply airflow. When the supply air temperature is too low (see zone I) the system performance regulation is necessary. It can be realized by mixing the air after 2stHMX with the hot air after the desiccant wheel in Mode 1. In Mode 2 when the supply air temperature is below 15°C the 1st HMX may be bypassed by the part of the airflow and mixed after the heat and mass exchanger. Second possibility for temperature increase in Mode 2 is to recirculate the exhausted air.

6 Conclusions

This paper presents a performance analysis of a novel desiccant air conditioning system with indirect evaporative precooling for selected real moderate climatic conditions in summer season. Authors proposed 3 different system running modes. Mode 1 provides cooling and dehumidification of the ambient airflow. Mode 2 provides only cooling the ambient airflow and Mode 3 is a ventilation mode in which the air treatment such as cooling or dehumidification is unnecessary. For each mode the average key performance indicators were calculated on the base of simulation results. Those are: Thermal Coefficient of Performance (COP_{th}) and Electrical Coefficient of Performance (COPel). During 55% of Mode 1 operational hours COPth is between 1.0 and 2.0, during 28% between 2.0 and 3.0 and during 11% the COPth is above 3.0. The maximum hour COPth equals 3.73. For most of the ambient airflow parameters during summer season in Wroclaw (Poland) the system is able to provide the sufficient supply airflow parameters. Nevertheless, the in some cases the regeneration temperature equal to 45°C is inadequate and it needs an increment to provide proper supply airflow humidity ratio.

One of the co-authors, Demis Pandelidis, received financial support for his research from resources for scientific work for years 2016-2019 from Polish Ministry of Science and High Education (program "Iuventus Plus"), project number IP2015 058274 and from Polish National Agency for Academic Exchange (Bekker program), agreement number PPN/BEK/2018/1/00347/U/00001.

COP_{el} – Electrical Coefficient of Performance, COP_{th} – Thermal Coefficient of Performance, i – enthalpy; kJ/kg; N – energy consumption by fan, W; P – energy for regeneration purposes, W; RH – relative humidity; t – temperature, °C; V – airflow, m^3/h; x – humidity ratio, g/kg; Φ – cooling capacity, W; Δp – Pressure, Pa; ρ – Air density, kg/m^3, Subscripts: i – inlet; o – outlet; reg – regeneration. Acronyms: 1^{st}HMX – cross-flow heat and mass exchanger; 2^{st}HMX – regenerative heat and mass exchanger; a – ambient; CR – conditioning room; DW – desiccant wheel, s – supply NTU – Number of Transfer Units; 1 – product; 2 – working; 3 – process; 4 – regeneration.

References

[1] B. Weerts, *CU Scholar NSIDC Green Data Center Project: Coolerado and Modeling an Application of the Maisotsenko Cycle* (2017)

[2] Inernational Energy Agency, *The Future of Cooling* (2018)

[3] W. Gao, W. Worek, V. Konduru, K. Adensin, Energy Build. **86** (2015)

[4] H. Sadighi Dizaji, E. J. Hu, L. Chen, Energy, **156** (2018)

[5] D. B. Jani, M. Mishra, P. K. Sahoo, Renew. Sustain. Energy Rev. **60** (2016)

[6] D. La, Y. J. Dai, Y. Li, R. Z. Wang, T. S. Ge, Renew. Sustain. Energy Rev. **14** (2010)

[7] G. Qadar Chaudhary, M. Ali, N. A. Sheikh, S. I. ul H. Gilani, S. Khushnood, Appl. Therm. Eng. **140** (2018)

[8] F. Comino, M. Ruiz de Adana, F. Peci, Appl. Therm. Eng. **131** (2018)

[9] M. Kashif Shahzad, M. Ali, N. Ahmed Sheikh, G. Qadar Chaudhary, M. Shahid Khalil, T. U. Rashid, Appl. Therm. Eng. **128** (2018)

[10] D. Pandelidis, A. Pacak, A. Cichoń, S. Anisimov, P. Drąg, B. Vager, V. Vasilijev, Energy Convers. Manag. **177** (2018)

[11] D. Pandelidis, S. Anisimov, W. M. Worek, P. Drąg, Energy Build. **123** (2016)

Application of mathematical modelling in evaluation of odour nuisance from selected waste management plant

Marcin Pawnuk[1,*], and *Izabela* Sówka[1]

[1]Wroclaw University of Science and Technology, Faculty of Environmental Engineering, pl. Grunwaldzki 9, 50-377 Wrocław, Poland

Abstract. A wide range of tools is required to comprehensively assess the impact of odours on people and the environment. Among them, mathematical tools for odour dispersion modelling are widely used. We can distinguish models based on Gaussian, Lagrangian and Eulerian approach. In Poland, the reference model, Operat FB, based on Gaussian way is commonly used for odour dispersion modelling. Modelling was conducted for two scenarios: first including area and point emission sources, while second one included only point sources. Comparison of both scenarios shows significant improvement of odour impact range of selected Plant in case when area sources are excluded.

1 Introduction

A wide range of tools is required to comprehensively assess the impact of odours on people and the environment. Effective evaluation of a specific odour emission source includes activities such as inventory of odour emission sources, determination of the sampling method, determination of odour concentration from the source, determination of emissions and odour concentrations at receptor points [1]. It is possible due to the use of a wide range of measuring tools, which include sensory methods (e.g. dynamic olfactometry), analytical methods using techniques from classical analytical chemistry and sensory-instrumental methods (e.g. electronic noses) [1–3]. The final assessment of the impact of odorous substances emitted into the atmosphere on people and the environment should be based on one of two approaches. The first is the assessment based on the characteristics of odour emission sources and prediction of concentrations at receptor points. The second approach concerns direct measurements in areas adjacent to the emission source. These approaches can be used in combination or individually [4, 5]. Different types of mathematical models are used to predict odour concentrations. As shown in the published literature [6], the emission source characteristics alone is not sufficient to determine the impact of odorous pollution on humans and the environment, therefore, mathematical models are used to determine how odorous pollution spreads in the atmosphere.

* Corresponding author: marcin.pawnuk@pwr.edu.pl

The purpose of the work was to characterize the available tools for modelling the dispersion of odours, together with a case study of the Polish reference model for analyses and calculations of the dispersion of odours emitted from a selected waste management plant including point and area sources.

2 Review of models used in the odour dispersion modelling

There are many approaches to mathematical modelling of pollutants dispersion (including odorous pollutants) in the atmosphere. Due to the mathematical basis, we can distinguish three main types of models used in pollution modelling. These models follow Gaussian, Lagrangian or Eulerian approaches.

In modelling the dispersion of odour pollution into the atmosphere, Gaussian models are considered the simplest and cheapest tools [6]. These are models based on the assumption that the spreading plume of pollution emitted from the source does not change over time (the flow is homogeneous and constant) [1, 7], they are so-called stationary plume models. These are the fastest calculation models used in the atmospheric simulation of pollution transport. This is possible due to the fact that only one equation, which can be successfully calculated on most computers, is solved to compute the concentrations at the receptor points. The most important disadvantage of these models is that in stationary conditions or low wind speeds they give unfavourable results. Another critical parameter for these models is the complex topography of the area for which modelling is carried out [6, 7]. In the case of the most advanced Gaussian models, in order to obtain the most accurate simulation results, the impact of the complex topography of the terrain and the influence of turbulence are parameterized with appropriate coefficients [7]. The Gaussian models initially served to simulate the dispersion of pollutants from point sources, however, they were quickly applied also to surface, linear and volume sources [8]. Due to limitations resulting from the fact that the plume emitted from the source as well as meteorological conditions are not variable over time, they are used for modelling average annual and hourly pollution concentrations [8]. There are many different models using the Gaussian approach to predict the concentration of odorous contaminants at receptor points. These are models such as SCREEN3, ISCST3, ISC-PRIME, TAPM, AODM. The most commonly used are AERMOD, ADMS [4, 5, 7]. Gaussian models can be used in simulating odour concentrations for a wide range of different facilities. Among them, we can distinguish waste management facilities such as landfills, wastewater treatment plants or industrial facilities.

In Poland, the Gaussian model based on the Pasquill formula is mainly used for predicting odour concentration at receptor points. It is considered as a reference model based on the guidelines contained in the *Regulation of the Minister of Environment dated 26 January 2010 on reference values for some substances in the air (Journal of Laws 2010, no. 16, item 87)*. Pasquill's equation is as follows [9]:

$$S_{xyz} = \frac{E_g}{2\pi\bar{u}\sigma_y\sigma_z} \cdot \exp\left(-\frac{y^2}{2\sigma_y^2}\right) \cdot \left\{exp\left[-\frac{(z-H)^2}{2\sigma_z^2}\right] + exp\left[-\frac{(z+H)^2}{2\sigma_z^2}\right]\right\} \cdot 1000 \quad (1)$$

where:

S_{xyz} – concentration of the substance in the air averaged over 1 hour (μg/m^3);

E_g – maximum emission of gaseous substance (mg/s);

σ_y – horizontal atmospheric diffusion coefficient (m);

σ_z – vertical atmospheric diffusion coefficient (m);

$\bar{u}$ – average wind velocity in the layer from the geometric height of the emitter h to the effective height of the emitter H (m/s);

y – component of the emitter distance from the point at which the calculation is made perpendicular to the direction of the wind (m);
z – the height for which the concentration of the substance in the air is calculated (m);

In order to carry out modelling using the aforementioned model, it is necessary to collect data such as meteorological data, information on land cover, parameters of individual emitters and emissions from particular emitters [9]. This model can be used to predict the concentrations of a wide range of pollutants, including odours.

The Lagrangian models use a more complex approach than the previous model type. They are known as particle models. Forecasting the spread of pollutants in the air involves calculating the trajectory of computational particles at small time intervals in a given wind field and three-dimensional turbulence field. These particles represent a specific mass of pollutants. The dispersion of pollutants in Lagrangian models uses both stochastic (turbulence) and deterministic (wind field and buoyancy) effects. The movement of computational particles is considered to be random what makes the Lagrangian models more expensive in terms of computational costs [6–8, 10]. Lagrangian models can be used under low wind and calm conditions [11]. There are models that can use both the Lagrangian and Gaussian approach. An example of such models are the so-called puff models. These models to represent the spread of pollutants use clouds/puffs of a defined volume. Concentrations inside puffs are driven by gaussian approach (e.g. equation (1)), while trajectories are defined by Lagrangian approach [1, 7]. An example of such a solution is the CALPUFF tool, which was used worldwide for environmental and scientific purposes. CALPUFF model can be used to simulate odour dispersion from point, area and volume sources. To simulate odour dispersion it is necessary to obtain data such as emissions from sources, meteorological and topographical data [11, 12]. Similar to the gaussian models, CALPUFF can be used in evaluation of odour impact range from many sources such as animal by-product utilization plants [12], landfills [13], waste and residue treatment incineration/utilization plants [14] or composting food wastes [15].

The most advanced models in terms of mathematical approach are the Eulerian models. They are known as grid models or 3-D models. These terms derive from the fact that the calculations of the dispersion of pollutants take place in a three-dimensional domain divided into specific cells / grids. Eulerian models numerically solve the dispersion equations of wind-generated turbulent flow and calculate the average concentration of pollutants in a specific 3D domain [8]. The mathematical equation is solved by means of specific initial and boundary conditions. These models allow for a better and more accurate spatio-temporal representation of simulated contaminants. Eulerian models can be used for both small scale and global scale. Pollution concentrations are calculated for individual spatial cells. They can be successfully used in unstable conditions. The biggest disadvantage of these models is that they need extremely high computer power for calculations, which generates significant costs in their use [6–8]. Mainly, these models are used to simulate dispersion of different types of pollutants in a global scale (e.g. ozone). But literature shows that they can be easily adapted to local scale use. The tool named ModOdour can be used for simulation of odour dispersion from waste landfills [8] and other waste management facilities.

3 Evaluation of the odour impact range of the selected waste management plant using the Polish reference model

3.1 Plant characteristic

The subject of the research was the Waste Management Plant located on the outskirts of a chosen city in Poland. The operations of the plant include activities related to the management of non-hazardous and inert waste, i.e. municipal waste. Its activity includes handling of local systems for selective waste collection, sorting and recycling of waste. The plant has its own waste sorting line. During the research time, the plant had two landfill sites. The first landfill site was closed and reclaimed, the second was under current exploitation. The exploited landfill site had the necessary systems for leachate and gases collecting. The plant does not directly border with residential development. The nearest houses are located at a distance of about 1,100 meters to the west and about 1,700 meters to the east. The plant is surrounded by a strip of greenery and is adjacent to the agricultural areas.

3.2 Purpose and methodology of research

In order to determine the odour impact of the Waste Management Plant, at the beginning, the odour emission sources were identified. Two types of odour emission sources have been identified: point sources and area source. The area of the landfill site being under current exploitation was identified as the area emission source. As point sources: 4 of them located above the waste reception hall and 5 directly above the sorting building were found.

The odour sampling from identified sources has been carried out in accordance with the methodology described in the PN-EN 13725:2007 "*Air quality. Determination of odour concentration, with the use of dynamic olfactometry method*" [16] and the VDI 3880: 2011 [17] guidelines. In the case of point sources, the sampling was carried out using a CSD30 vacuum probe, while a ventilated sampling hood was used to collect samples from the area source. In both cases, PTFE plastic bags were used. All components of the sampling kit are made of materials that do not emit or absorb odours. According to the methodology mentioned above, the bags and the remaining equipment were conditioned and cleaned before sampling. Meteorological parameters such as temperature, humidity and pressure were also measured during sampling. During the transport of samples, a constant ambient temperature was maintained to prevent potential chemical reactions that could interfere with later results.

Measurement of odour concentration was performed in the Olfactometric Laboratory. Measurements of odour concentrations have been carried out in accordance with the PN-EN 13725:2007 guidelines [16]. In determining the odour concentrations, a four-station olfactometer TO 8 with necessary equipment was used. The measurement team consisted of 5 people (4 investigators and 1 operator) which were selected in accordance with the mentioned guidelines [16, 17]. The obtained data of the odour concentration determining tests were calculated as the geometric mean of all measurement series. The odour concentrations from tests are shown as European odour unit per cubic meter (ou_E/m^3).

3.3 Emissions calculations and input data for modelling

Table 1 shows the results of odour concentrations measurements and odour emission rate calculations for identified nine point and one area sources. Odour concentrations are presented as an average concentration from collected samples. Table 1 shows also necessary parameters for odour dispersion modelling, i.e.: emitter height, a diameter of the emitter (in case of area source the area of landfill site is presented in the table) and gas velocity for point sources. The temperature of emitted gases was determined at 293 Kelvin degrees.

The highest odour emission rate was obtained for the area source (E-10): 699317,0 ou_E/s. It represents emissions from landfill of area equal of 27500 square meters. Summing up the obtained emission values, the surface source can be responsible for about 99.95% of odour emission from Waste Management Plant under study.

Table 1. Odour emission sources parameters, odour concentration measurements and estimated emission values results.

Name of point	Type of source	Localization of emission source	Emitter height [m]	Diameter of the emitter [m]	Avg. odour conc. [ou_E/m^3]	Odour emission rate [ou_E/s]	Gas velocity [m/s]
E-1	point	waste reception hall	10.5	0.345	60.0	12.8	2.3
E-2	point	waste reception hall	11	0.345	60.0	12.8	2.3
E-3	point	waste reception hall	12.5	0.345	60.0	12.8	2.3
E-4	point	waste reception hall	13	0.345	60.0	12.8	2.3
E-5	point	waste sorting building	13	0.345	265.5	56.8	2.3
E-6	point	waste sorting building	11	0.345	265.5	56.8	2.3
E-7	point	waste sorting building	13	0.345	265.5	56.8	2.3
E-8	point	waste sorting building	11	0.345	265.5	56.8	2.3
E-9	point	waste sorting building	10.5	0.345	265.5	56.8	2.3
E-10	area	landfill site	3	27500 m^2 (landfill area)	2333.0	699317.0	-

Obtained results were used as a input data for odour dispersion modelling from selected Waste Management Plant. Software called "Operat Fb" was used in the assessment of odour impact range. It follows Gaussian approach (equation (1)) to simulate pollution transport in the air. The software follows the guidelines contained in the *Regulation of the Minister of Environment dated 26 January 2010 on reference values for some substances in the air (Journal of Laws 2010, no. 16, item 87)* [9] and is considered as a reference model in pollution dispersion modelling. Table 1 shows the necessary emitter parameters used for modelling study. The aerodynamic roughness coefficient of the terrain was defined as z = 0.55. The location of the nearest residential buildings was determined (receptor points B1–B7). In accordance with the Draft law on preventing odour nuisance from 27.02.2009 (Poland) [18] the environmental detection threshold for odours pollution was established at a range of 1 ou_E/m^3. The maximum frequency of exceedances of the odour concentration at a level of 1 ou_E/m^3 was determined at 3% during the year [19]. The grid size was determined to be equal to 5800 m x 4500 m with a grid step at 100 m.

Odour dispersion modelling was carried out in two different scenarios. The first one included both point and area sources as input data for modelling, while the second scenario included only point sources. These simulations could help assess the impact odour range of selected Waste Management Plant under two different operating conditions, when the landfill is being under exploitation (first scenario) and when its closed (second scenario). According to Polish law [19], waste deposition on landfills is one of the least desirable options included in the rules of waste management handling. And therefore, the amount of operating municipal landfills should be reduced, with the intention of reducing their number to zero. As mentioned before, the Waste Management Plant activity includes not only waste deposition on landfill, so modelling in these conditions could help assess the odour impact range in the future of the Plant.

3.4 Results and discussion

Figure 1 shows the maximum odour concentrations for both scenarios in a defined computational grid. Under the first scenario, the maximum odour concentration obtained from the modelling was found at point X = 2700, Y = 2400 and equals to 564.2 ou_E/m^3. While under the second scenario it was found at point X = 2800 Y = 2100 and equals to 0.2 ou_E/m^3.

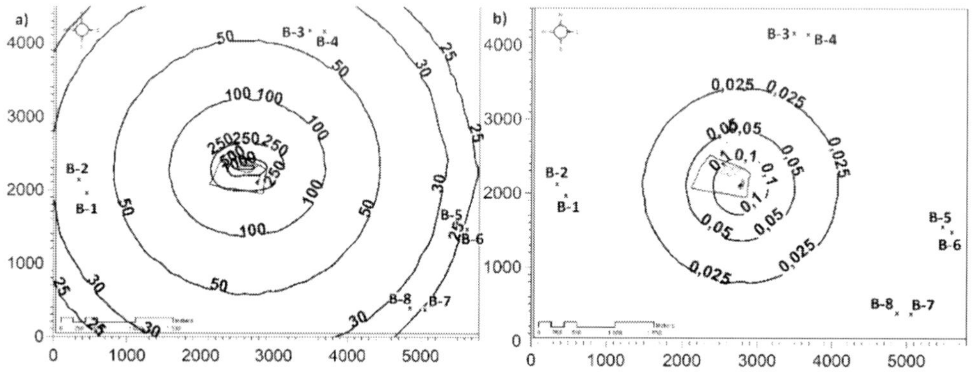

Fig. 1. The maximum odour concentrations obtained for: point and area sources (a) and point sources (b), ou_E/m^3.

The frequency of exceeding the maximum odour concentration is shown in figure 2. When the point and area sources were examined together, the highest frequency of exceedances was reported at point X = 2700, Y = 2400 and equals to 50.9 % and exceed the determined threshold level of 3% (for 1 ou_E/m^3). In the case of the second scenario, no exceedings were obtained.

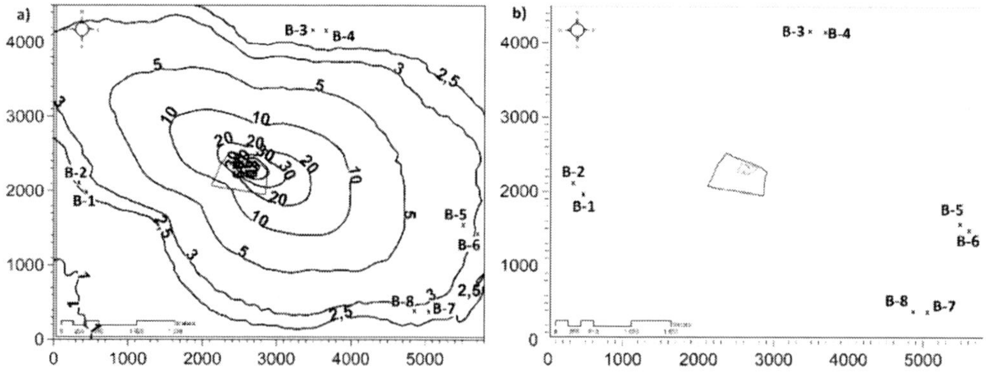

Fig. 2. The frequency of exceeding the maximum odour concentrations for: : point and area sources (a) and point sources (b), %.

In the case of average annual odour concentration for both scenarios, the results are shown in figure 3. The highest average annual odour concentration was calculated at the same point as maximum odour concentration (first scenario) and was equal to 27.0 ou_E/m^3 (while the determined threshold level is at 1 ou_E/m^3). Results obtained from the second modelling does not exceed determined threshold level and the highest value equals to 0.01 ou_E/m^3 (X = 2800, Y = 2100).

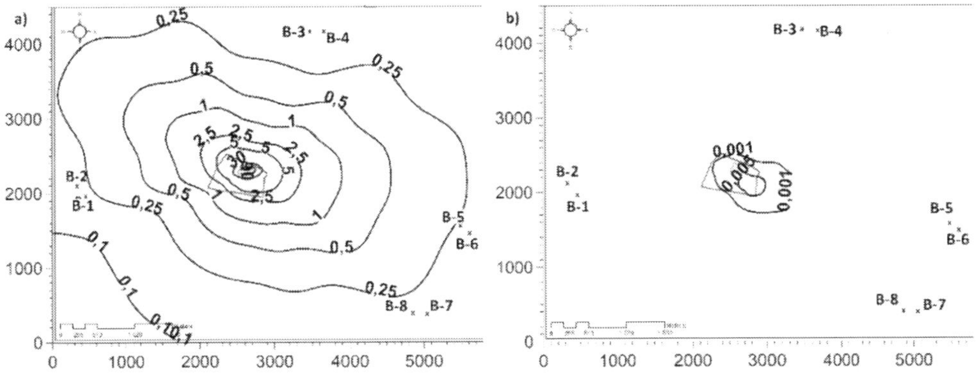

Fig. 3. The average annual odour concentrations for: : point and area sources (a) and point sources (b), ou_E/m^3.

The results of the odour dispersion modelling for additional points outside the Waste Management Plant are shown in table 2. These points (B-1 to B-8) represent the nearest residential buildings. The highest maximum odour concentration for the first scenario was calculated at point B-3 (40.5 ou_E/m^3) approx. 1500 m from boundary of Plant. The frequency of exceedances at this point equals to 2.1 %. The highest frequency of exceedances for residential buildings was reported at point B-5 and equals to 3.2% of permissible threshold level of 3% per year. These values shows that the emission from the Plant could have potential impact on humans living in the neighbourhood of point B-5. The annual average odour concentration does not exceed the 1 ou_E/m^3 in any additional point. In case of the second scenario, there is a significant improvement of odour impact range obtained from modelling. There is no exceedances of threshold level at any points related to residential buildings, when maximum and average annual odour concentration stays at relatively low levels.

Table 2. Comparison of odour modelling for both scenarios for additional points at calculation grid.

Name of point	Max. odour Conc. [ou_E/m^3]	Annual Avg. Conc. [ou_E/m^3]	Freq. of. exc. [%]	Max. odour Conc. [ou_E/m^3]	Annual Avg. Conc. [ou_E/m^3]	Freq. of. exc. [%]
-	First scenario			Second scenario		
B-1	38.5	0.2	2.5	0.0	0.0	0.0
B-2	36.6	0.2	2.6	0.0	0.0	0.0
B-3	40.5	0.1	2.1	0.0	0.0	0.0
B-4	39.7	0.1	1.9	0.0	0.0	0.0
B-5	26.2	0.2	3.2	0.0	0.0	0.0
B-6	25.1	0.2	2.9	0.0	0.0	0.0
B-7	24.6	0.2	2.8	0.0	0.0	0.0
B-8	26.3	0.2	2.9	0.0	0.0	0.0

4 Summary and conclusion

Many of different tools can be used in odour dispersion modelling, for example, Gaussian, Lagrangian and Eulerian models. Among them, Polish reference model "Operat Fb" that follows Gaussian approach, can be distinguished as one of the useful tools for predicting odour concentration at receptor network, especially under averaged meteorological parameters and for area and point sources. In this paperwork, with help of the polish reference model it was possible to obtain valid data about the odour impact range from selected Waste Management Plant under two different scenarios. Obtained results for the first scenario shows that the odour impact range of selected Plant is significant. The Plant could have potential impact on humans and environment. The exceeding of permissible threshold level occurs at residential area, approx. 2100 m from plant boundary. Therefore, it is necessary to take certain actions aimed at reducing the influence of the odour impact of selected Waste Management Plant. In the case of the second scenario the odour impact range is significantly lower, results did not show any impact near residential buildings. No actions are needed. The conducted research shows that the most of odour emission comes from landfill site and it is responsible for almost all the odour nuisance caused by the Plant. Closing the landfill site could improve the air quality near the Plant. The research shows that polish reference model "Operat Fb" could be used as a valid tools in odour impact assessment.

European Funds
Knowledge Education Development

European Union
European Social Fund

References

1. T. Kiełbasa, P. Sobczyński, M. Skrętowicz, I. Sówka, *Mathematical tools in odour impact assessment of the municipal facilities* (Oficyna Wydawnicza Politechniki Wrocławskiej, Wrocław, 2013) (in Polish)
2. R. Barczak, *Comparative Analysis of Field Olfactometry with Other Research Methods in the Odour Impact Assessment of Wastewater Treatment Plants* (PhD dissertation, 2014) (in Polish)
3. J. Gębicki, H. Bylinski, J. Namieśnik, Environ. Monit. Assess. **188**, 1 (2016)
4. I. Sówka, *Methods of identification of odour gases emitted from industrial plants* (Oficyna Wydawnicza Politechniki Wrocławskiej, Wrocław, 2011) (in Polish)
5. A. J. Nicell, Atm. Env. **43** (2009)
6. L. Capelli, S. Sironi, R. Del Rosso, J. M. Guillot, Atm. Env. **73** (2013)
7. A. Leelossy, F. Molnár, F. Izsák, A. Havasi, I. Lagzi. R. Mészáros, Cent. Eur. J. Geosci. **6**, 3 (2014)
8. L. Yanjun, Z. Yan, L. Wenjing, W. Hongtao, H. Qifei, Env. Mod. & Soft. **113** (2019)
9. Regulation of the Minister of Environment dated 26 January 2010 on reference values for some substances in the air. (Journal of Laws 2010, no. 16, item 87)
10. J. Nielinger, R. Röckle, H. C. Hölf, W. J Kost *Lagrange versus Eulerian dispersion modeling comprasion for investigations concerning air pollution caused by traffic* (2004)
11. L. Capelli, S. Sironi, Atm. Env. **191** (2018)
12. S. Sironi, L. Capelli, P. Centola, R. Del Rosso, S. Pierucci, Atm. Env. **44** (2010)
13. J. Badach, P. Kolasińska, M. Paciorek, W. Wojnowski, T. Dymerski, J. Gębicki, M. Dymnicka, J. Namieśnik, J. Environ. Manage. **213** (2018)
14. S. C. Doğruparmak, H. Pekey, D. Arslanbaş, Env. For. **19**, 1 (2018)
15. L. Dentoni, L. Capelli, S. Sironi, J.M. Guillot, Water. Sci. Technol. **68**, 8 (2013)
16. PN-EN 13725:2007: *Air quality. Determination of odour concentration, with the use of dynamic olfactometry method*
17. VDI 3883: *Olfactometry. Static sampling* (Verein Deutscher Ingenieure, 2011)
18. *Waste management Act* dated 14 December 2012 (Journal of Laws 2013, item 21)
19. The Draft law on preventing odour nuisance from 27.02.2009 (Poland)

Forecasting water consumption in towns of different sizes

Adam Piasecki[1*], *Agnieszka* Pilarska[2], and *Radosław* Golba[2]

[1]AGH University, 30-001 Cracow, Poland
[2]Nicolaus Copernicus University, 87-100 Toruń, Poland

Abstract. The aim of the work was to compare water consumption forecasting in two towns of different sizes. The objects of research were the town of Toruń and the town of Żnin in central Poland. Two models were built for each. The models were constructed using the multiple regression method. In constructing the models, explanatory variables determined by Principal Component Analysis (PCA) were used. The set of explanatory variables identified to construct each individual model differed. The models for Toruń obtained better forecast quality assessment criteria values. This was mainly due to the water supply system in the small town (Żnin) being less resilient to sudden, short-term changes in consumers' water use. At the same time, the importance of the location of the meteorological stations from which data was taken to build the model was emphasised.

1 Introduction

The infrastructure for supplying water to urban residents is classed as critical. Therefore, even at the design and construction stage, it is necessary to consider all conditions that may negatively impact its operation. One basic requirement is to quantify water demand in a prospective period, taking into account the needs of all end-users. In practice, this is done using guidelines that provide a normative measure of water demand. Research results show that real water consumption in many towns deviates significantly from the assumed design values [1, 2]. Thus, water consumption forecasts are an important additional element for facilitating the rational operation of water supply systems. The literature has presented many methods and solutions in this domain [3–5]. At the same time, it is impossible to indicate the best method, due to limitations in the ability to compare forecast results (the use of diverse measures of forecast accuracy). Local conditions and the unique characters of objects of study (including size of water supply system) can be decisive in determining results, so are also of significant importance. The aim of the work is to compare water consumption forecasts in two towns of different sizes. This was done using a multiple regression model. A set of explanatory variables was first defined using Principal Component Analysis (PCA).

*Corresponding author: adm.piasecki@gmail.com

2 Research area

The objects of study were the town of Toruń and the town of Żnin in the Kujawsko-Pomorskie voivodeship in central Poland. Toruń has over 200,000 inhabitants and a very well developed water and sewage infrastructure. The length of the water supply system is currently 621.2 km, and the sewerage network is 654.2 km long. At present, almost all the town's residents are connected to the municipal water and sewage system [6]. Żnin is much smaller – a town of only 14,000 residents. In recent years, a number of investments have been made in water and sewage infrastructure in the town. This has increased the availability to this infrastructure. Currently, approximately 97% of the town's population is connected to the water supply and 86% to the sewage system. In both towns, industrial use of mains water is proportionally small.

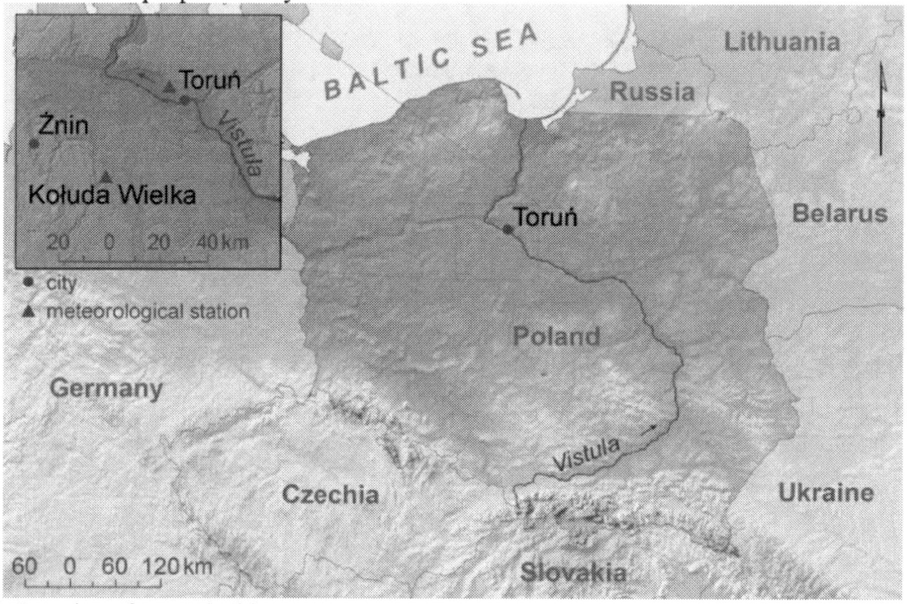

Fig. 1. Location of research objects

3 Materials and methods

The study uses daily water consumption values for the towns being studied. The data was made available by the companies Torun Waterworks Company Ltd and Zakład Wodociągów i Kanalizacji "WIK" Sp. z o. o. The study also used daily values of selected meteorological parameters recorded at the Toruń-Wrzosy and Kołuda Wielka stations. These meteorological stations belong to the Institute of Meteorology and Water Management - National Research Institute (IMWM-NRI). Data was used for the period 2011–2017.

The research progressed in three main stages:

1. determining explanatory variables,
2. building the models,
3. forecasting, and evaluating the models.

A set of explanatory variables was determined using Principal Component Analysis (PCA) based on the procedure presented by M. M. Haque et al. [7]. To determine the explanatory variables, data from 2011–2016 were used. The analysis adopted a correlation matrix. A Bartlett sphericity test was conducted and the Kaiser–Mayer–Olkin coefficient (KMO) was determined, which confirmed the validity of using the PCA method for the whole period and for warm periods (April–September). The KMO coefficient of data from the cold periods (Jan–March and October–December) for Żnin was below the threshold value of 0.500(defined by Hutcheson and Sofroniou [8]), and only slightly exceeded it for Toruń (0.502). For this reason cold periods were not subjected to a separate detailed analysis.

After performing the calculations, it was decided that only two components would be taken into account in determining the explanatory variables: PC1 and PC2. Principal component 3 (PC3) was omitted despite having an eigenvalue above 1.000 (the criterion of H. F. Kaiser [9]). A. Balicki [10] emphasises the importance of the ability to interpret components in determining their number. In this case, including PC3 would have prevented the results from being properly interpreted.

According to the procedure presented by M. M. Haque et al. [7], the set of explanatory variables was determined based on the following parameters:
– selecting the variables with the highest correlation coefficients for a given component (variable loadings).
– attempting to avoid the multicollinearity problem in the regression model by selecting mutually uncorrelated variables.
– attempting to ensure diversity in the character of the selected variables, e.g. "previous-day water consumption" is not a meteorological variable, and so was included in the construction of the explanatory models.

The models were constructed using the multiple regression method. In constructing the models, explanatory variables determined by Principal Component Analysis (PCA) were used. Two models were built for each. The first model, "Year", was built using daily data for 2011–2016. The second model, "Hot Periods", was built using daily data for the warm months (April–September) of 2011–2016. The models were then used to predict water consumption in 2017. The following indices were used to assess the quality of forecasts:

$$E = 1 - \frac{\sum_{i=1}^{n}(A_t - F_t)^2}{\sum_{i=1}^{n}(A_t - A_{mean})^2}$$

$$\text{RMSE} = \sqrt{\frac{1}{n}\sum_{i=1}^{n}(A_t - F_t)^2}$$

$$\text{NRMSE} = \frac{RMSE}{\frac{1}{n}\sum_{i=1}^{n} A_t}$$

$$MAPE = \frac{1}{n}\sum_{i=1}^{n}\left|\frac{A_t - F_t}{A_t}\right|$$

where: A_t – observed values, F_t – predicted values.

Statistical analyses were performed in PS IMAGO 5 software using the IBM SPSS Statistics analytical engine and in Statistica.

4 Results

The Principal Component Analysis results are presented in Tables 1 and 2. In the case of both the town and the town, the two principal components identified for both the entire analysed period and the hot half-year explained a total of 60% of variance. It should also be emphasised that while both components combined explained a similar percentage of the variance in each town, the amount of the variance explained by PC1 was more than double that of PC2 (Table 1).

Table 1. Eigenvalues and percentage of explained variance.

Town	Żnin				Toruń			
Model	Year		Hot Periods		Year		Hot Periods	
PC No.	PC1	PC2	PC1	PC2	PC1	PC2	PC1	PC2
Eigenvalue	3.333	1.151	2.947	1.347	3.357	1.186	2.994	1.441
Explained variance [%]	47.62	16.45	42.11	19.24	47.96	16.94	42.77	20.58

For the entire period, the highest variable loadings of the PC1 component were attained by variables representing thermal conditions. A high loading was also attained by the "previous-day water consumption" variable. For PC2, the highest loadings were for "daily sum of precipitation" and "average daily relative humidity" (Table 2). Considering all the mentioned criteria for selecting explanatory variables, the following were selected:
– for Żnin: "average daily temperature", "previous-day water consumption", and "daily sum of precipitation".
– for Toruń: "maximum daily temperature", "previous-day water consumption", and "daily sum of precipitation".

Table 2. Correlation of the two distinguished components with the analysed variables in 2011–2016 and in hot periods of 2011–2016.

Variable	Model Year				Model Hot Periods			
	Żnin		Toruń		Żnin		Toruń	
	PC1	PC2	PC1	PC2	PC1	PC2	PC1	PC2
Minimum daily temperature [°C]	0.934	0.184	0.900	0.257	0.900	0.132	0.827	0.397
Maximum daily temperature [°C]	0.972	0.074	**0.978**	0.078	0.953	-0.034	0.952	0.042
Average daily temperature [°C]	**0.976**	0.120	0.976	0.137	**0.984**	0.030	**0.969**	0.136
Average daily relative humidity [%]	-0.561	0.603	-0.651	0.552	-0.182	**0.833**	-0.405	**0.785**
Daily sum of precipitation [mm]	0.161	**0.695**	0.109	**0.684**	0.196	0.464	0.086	0.635
Average daily wind speed [m/s]	-0.080	-0.359	-0.100	-0.357	-0.218	-0.483	0.084	-0.294
Water use (previous day)	**0.465**	-0.351	**0.438**	-0.440	**0.376**	-0.429	**0.535**	-0.396

For warm periods, the variables that met the adopted criteria for PC1 were "average daily temperature" and "previous-day water consumption", while only "mean daily relative humidity" met the criteria for PC2 (Table 2). As a result, for the construction of the explanatory model using the multiple regression method, the same variables were adopted for both towns: "average daily temperature", "previous-day water consumption" and "average daily relative humidity".

Table 3 presents the water consumption models that were constructed and the selected statistics characterising them. The highest R^2 coefficients were obtained by model II, and the lowest by model IV. In all models the most important variable was variable x_1 ("previous-day water consumption"). It explained 54–66% of the variance of the dependent variable. The remaining variables were of minor importance, except variable x_4 in models II and IV.

Table 3. Models of water consumption and their selected statistics (x_1 – previous-day water consumption, x_2 – maximum daily temperature, x_3 – daily sum of precipitation, x_4 – average daily relative humidity, x_5 – average daily temperature).

Model	Variable	Statistics				
		R^2variance	Partial correlation	Semipartial correlation	Tolerance	p
M1 (Toruń – year)		$W_{T\_year} = 7640.52+0.779x_1+31.307x_2-79.938x_3$				
	x_1	0.650	0.142	0.083	0.892	0.000
	x_2	0.007	-0.158	-0.093	0.991	0.000
	x_3	0.007	0.787	0.740	0.899	0.000
M2 (Toruń – hot period)		$W_{T\_hot} = 18984.14+0.65x_1-94.94x_2+45.18x_3$				
	x_1	0.606	0.715	0.575	0.788	0.000
	x_4	0.074	-0.415	-0.257	0.821	0.000
	x_5	0.003	0.096	0.054	0.848	0.001
M3 (Żnin – year)		$W_{Z\_year} = 329.94+0.792x_1+1.956x_2-3.748x_3$				
	x_1	0.661	0.793	0.749	0.896	0.000
	x_5	0.003	0.115	0.067	0.877	0.000
	x_3	0.004	-0.103	-0.059	0.977	0.000
M4 (Żnin – hot period)		$W_{Z\_hot}= 749.60+0.687x_1-3.103x_2+2.187x_3$				
	x_1	0.542	0.704	0.650	0.896	0.000
	x_4	0.025	-0.229	-0.155	0.938	0.000
	x_5	0.002	0.071	0.047	0.927	0.018

Models M1 and M2 built for Toruń have the best forecast accuracy according to the selected evaluation criteria (Table 4). M2 was assessed as having forecast quality better than M1,butonly according to criterion E. At the same time, a slight difference in value (between M1 and M2) for the other criteria should be noted. All the criteria indicated much weaker forecast quality in the M3 and M4 models.

Table 4. Forecast quality indices according to the selected criteria.

Model	Criterion			
	E	RMSE	NRMSE	MAPE [%]
M1 (Toruń – year)	0.418	1748.295	0.050	3.542
M2 (Toruń – hot period)	0.513	1886.063	0.053	4.099
M3 (Żnin – year)	0.289	141.115	0.099	6.958
M4 (Żnin – hot period)	-0.085	166.968	0.111	7.979

5 Discussion

The set of explanatory variables identified to construct each individual model differed. The differences between the variables in the year model and the warm-period model seem understandable. In models for the warm period, relative humidity is more important, while atmospheric precipitation is less. The reasons can be found in the high variability of the second of these parameters. This applies particularly to warm periods, when intense rainfall usually occurs on one, or a few, days a month. The relative humidity parameter indirectly provides information on both rainfall (high values) and evaporation (low values). To some extent it indicates plants' water needs. As shown by numerous studies, the watering of gardens and lawns is the main source of increased water consumption in the warm period [11–12].

The differences between the variables in the year models for the selected towns appear to be interesting. This may be due to the sizes of the two towns, and to their local conditions. However, this is somewhat countered by the compatibility in the sets of variables for their warm-period models. This is thus most likely a result of the location of the meteorological station from which data for Żnin town was obtained. Despite the station's proximity (about 20 km from Żnin), some of the parameters it records may not entirely reflect the town's prevailing conditions. Such parameters may include the sum of precipitation (a very spatially variable parameter, especially in the warm period) and maximum temperature (which often depends on local conditions).

The models constructed differed in forecast accuracy. The models for Toruń obtained much better forecast quality assessment criteria values. This may be in part due to the aforementioned problem with the meteorological data for Żnin. However, what seems more important is the specificity of local conditions and associated water consumption characteristics in a small town. Smaller water supply systems are more susceptible to variability in water consumption, as it can be caused by a small group of consumers. This means that forecasting water consumption is more difficult in a small town than in a large one.

One problematic issues in the built models is the very large significance of the x_1 variable ("previous-day water consumption"). It results in a clear shift in the graph showing forecasted values in relation to observed values (Fig. 2). In all the constructed models, the largest forecasting errors were found on days of sudden changes in water consumption. As a result, the models are ineffective for forecasting extreme situations. To improve forecast quality, it appears necessary to expand the model with other variables that will balance out that weighting. This is confirmed by the rather low determination index value. Research results [13–14] indicate that one such variable could be a "day of the week" variable. The variation in amount of water consumed on particular days of the week stems from the normal cycle on which society functions and the activity of economic entities [15]. The main difference is between water consumption on Monday to Saturday versus consumption on Sundays and public holidays. With the "day of the week" variable being qualitative, the information is usually coded binarily. This work did not use that variable due to the PCA method's limitation in taking into account binary coded data.

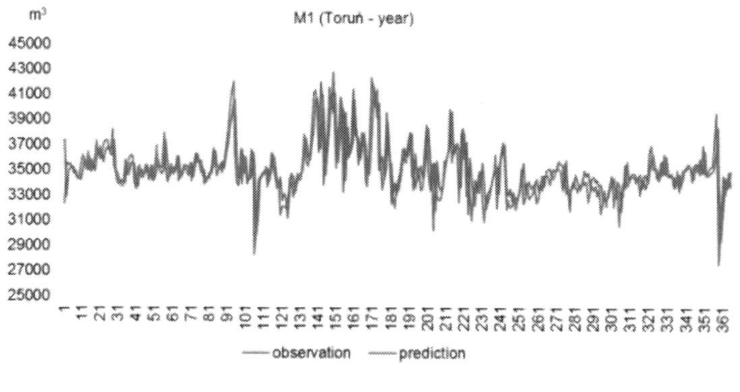

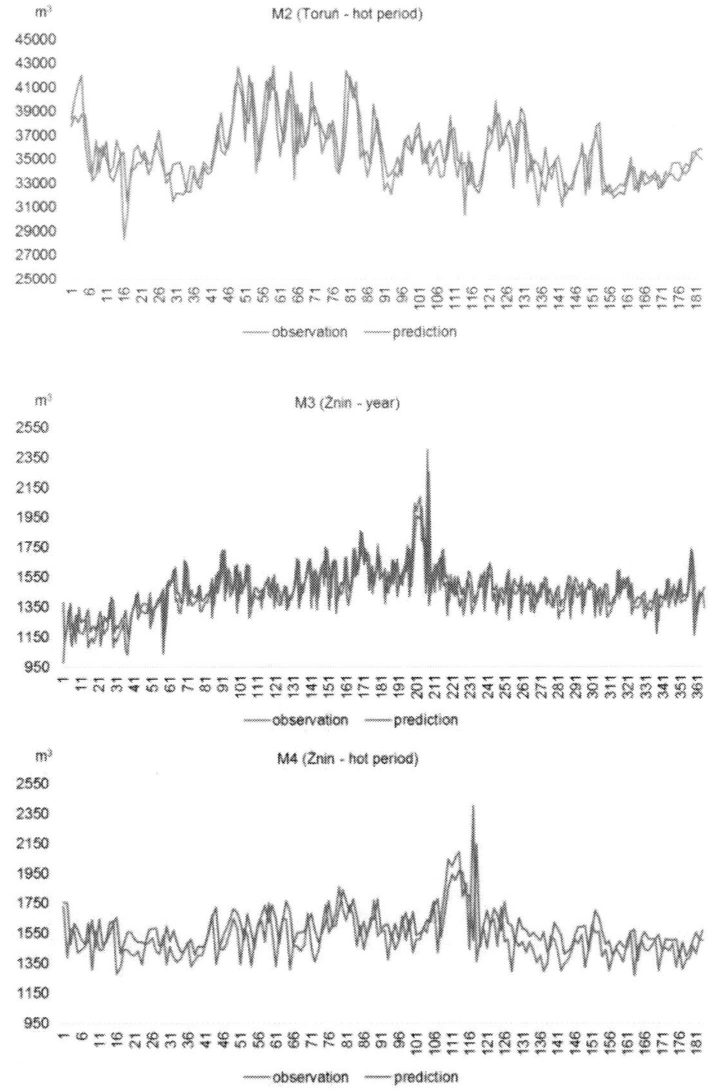

Fig. 2. Matching of forecasted values to observed water consumption in 2017 in Toruń and Żnin.

6 Summary

Comparative analysis of the forecasting models for water consumption in the towns of different sizes showed that the models built for the larger of the two (Toruń) had better forecast accuracy. This was mainly due to the water supply system in the small town (Żnin) being less resilient to sudden, short-term changes in consumers' water use. In addition, the meteorological data source used in constructing the prognostic model was shown to have been important. Using meteorological data from a station located outside the town can significantly affect model quality. In the case of towns without meteorological stations, it is in the interest of water and sewage companies to have their own measurement system. Modern technical solutions make building and maintaining such a system relatively inexpensive. This is especially true when we consider the benefits of well constructed forecasting models, which apply particularly to supporting decision-making in the design, development and maintenance of water supply networks, and in the implementation of procedures to optimise the operation of pumping stations, water treatment plants and sewage treatment plants [16].

The work demonstrated the effectiveness of the PCA method in determining meteorological variables for water consumption forecasting models. At the same time, the method was shown to be limited in terms of its inability to handle qualitative data. As a result, the constructed models have fairly low forecasting effectiveness for days with sudden, large changes in water consumption. Their practical use is thus limited. The quality of forecasts would probably be improved by including a "day of the week" variable.

References

1. Z. Hiedrich, J. Jędrzejkiewicz, Environment Protection **29(4)**, 29–34 (2007).
2. H. Hotloś, Environment Protection, **32(3)**, 39–42 (2010)
3. G. Bárdossy, G. Halász, J. Winter, Journal of Water Supply: Research and Technology - AQUA, **58(3)**, 203–211 (2009).
4. J. Adamowski, H. Fung Chan, S.O. Prasher, B. Ozga-Zielinski, A. Sliusarieva, Water Resources Research, **48(1)**, (2012).
5. A. Piasecki, J. Jurasz, B. Kaźmierczak, Periodica Polytechnica Civil Engineering, **62(3)**, 818-824 (2018).
6. A. Piasecki, J. Jurasz, W. Marszelewski, Environment Protection, **38(2)**, 17-22 (2016).
7. M.M. Haque, A. Rahman, D. Hagare, R.K. Chowdhury, Water, **10**, 419 (2018).
8. G. D. Hutcheson, N. Sofroniou, The Multivariate Social Scientist : an introduction to generalized linear models (SAGE Publications, London-Thousand-Oaks-New Delhi, 1999)
9. H. F. Kaiser, EPM, **20**, 1 (1960)
10. A. Balicki, Statystyczna analiza wielowymiarowa i jej zastosowania społeczno-ekonomiczne (Wydawnictwo Uniwersytetu Gdańskiego, Gdańsk, 2009)
11. U. Kępa, L. Stępniak, E. Stańczyk-Mazanek, Annual Set The Environment Protection, **15(3)**, 2546-2562 (2013).
12. W. Żuchowicki, R. Gawin, Annual Set The Environment Protection, **15(1)**, 924-929 (2013).
13. S.L. Zhou, T.A McMahon, A. Walton, J. Lewis, Journal of hydrology, **259(1-4)**, 189-202 (2002).
14. A.Studziński, K. Pietrucha-Urbanik M. Dąbek, Journal of Civil Engineering, Environment and Architecture, **61(1)**, 323-332 (2014).
15. A. Piasecki, Ł. Górski, Infrastructure and Ecology of Rural Areas, **4(1)**, 973-984 (2018).
16. J. Stańczyk, J. Kajewska-Szkudlarek, J. Łomotowski, P. Lipiński, P. Rychlikowski, T. Konieczny, Gaz, Woda i Technika Sanitarna, **10**, 372-377 (2018).

Quasi-dynamic model of the energy efficiency for an air-to-water heat pump

Krzysztof Piechurski[1,*], and *Małgorzata* Szulgowska-Zgrzywa[1]

[1]Wrocław University of Science and Technology, Wybrzeże Stanisława Wyspiańskiego 27, 50-370 Wrocław, Poland

Abstract. The energy efficiency of air-to-water heat pump operating in an actual heating installation depends on many factors. In order to create a reliable model of the unit, it is necessary to include as many of them as possible. Unfortunately, the most common data provided by heat pump manufacturers are based on tests performed in accordance with the EN 14511 standard [1]. These tests are performed in steady-state conditions and do not provide reliable information on the impact of dynamic effects on the energy efficiency of the device. The solution may be the tests in quasi-dynamic conditions. The article presents the possibility of creating the characteristics of an air-to-water heat pumps based on operational data. The accuracy of the created model has been compared with the characteristics resulting from measurements in steady state conditions. It has been confirmed that dynamic test data, after proper selection, will allow to determine the characteristics of repeatable parameters and this can be an alternative to tests performed in fixed conditions.

1 Introduction

The heating power and the coefficient of performance (*COP*) of air-to-water heat pumps are influenced by many factors. The most important of these are the parameters of the lower and upper sources, primarily the ambient temperature (T_a) and the supply temperature of the heating system (T_{in}). Correct determination of the actual *COP* of the heat pump in various operating conditions is crucial to determine its seasonal coefficient of performance (*SCOP*). Typically, the basis for the calculation of the instantaneous *COP* value and the heating power (Q_{HP}) is data provided by the heat pump manufacturer, which is based on tests performed in accordance with the EN 14511 [1] standard. At the moment, this is the most accurate and common source of information for the purpose of air-to-water heat pumps work simulations. The tests proposed in the standard [1] take into account changes of the heating system inlet temperature (upper source), changes in the ambient temperature (lower source) and the process of defrosting the heat pump's evaporator. Values of the declared coefficient of performance (COP_d) resulting from tests in accordance with the [1] standard also include the electrical energy supplied to the compressor, fans, circulation

* Corresponding author: krzysztof.piechurski@pwr.edu.pl

pump (if installed in the heat pump) and automation during the active cycle of the device. The tests are conducted at several measuring points. Most often, the data set includes three upper-source temperatures: 35°C, 45°C, 55°C and five to six lower source temperatures, e.g. -15°C, -10°C, -7°C, 2°C, 7°C, 20°C. Based on these points, the COP_d and the device heating power (Q_{HP}) calculation models should be developed. The model proposed by T. Afjei is a commonly used method for this purpose [2, 3]. A significant problem, however, is the extrapolation beyond the range of measurement points. The model assumes that outside this range the characteristic will be linear. The results obtained may differ significantly from the actual ones, which was observed in the works [3, 4]. It should be emphasized, that the transformation from single points measurements into a one-function characteristic is an important advantage of this solution. In the original method, it is a polynomial of the second degree with two variables. Regardless of the *COP* and the heating power model used, the most important issue is the quality of the input data. In general, the tests performed in accordance with EN 14511 [1] are not entirely appropriate for this purpose. The main problems concern: insufficient number of measuring points (e.g. no data for inlet temperature below 35°C), constant temperature difference on the condenser (5 K), no evaluation of the frequency inverter influence on the COP_d and heating power of the device, as well as the lack of evaluation of the partial load effect and the way of the device's power adjustment to its efficiency. The COP_d value determined on the basis of these tests concerns only the situation when the heat pump works at full heating load in the active cycle. In more precise simulations, corrections to take into account the effect of these dynamic conditions on the efficiency of the heat pump are taken into account [5–8].

The article presents the possibility of creating the characteristics of an air-to-water heat pump based on operational data. As mentioned above, tests in steady state conditions do not provide reliable information about the impact of dynamic effects on the energy efficiency of these devices. The solution may be the tests in quasi-dynamic conditions. In order to check this possibility, preliminary works described in this article are aimed at preparing the energy efficiency model of the device based on the collected measurement data. The accuracy of such model was compared with the characteristics resulting from measurements in steady state conditions.

2 Description of the measurements

The measurement data of the heating installation's operation was used to calibrate the efficiency model of the heat pump. In this case, the heat is distributed to the offices of approx. 300 m^2 situated on the 1st floor and to the air handling unit room in which the air-to-water heat pump with a compressor (split type) of the nominal power of 10.6 kW (A2/W35°C) is installed. The heat pump performance data: the heating capacity (Q_{HP}) and the declared coefficient of performance (COP_d), measured according to EN 14511 standard, are shown in Table 1.

Table 1. Declared heating capacity (Q_{HP}) and COP_d of the unit according to EN 14511 [1].

	T_a, °C	-15	-7	2	7	10	12	20
Q_{HP} (kW)	T_{in} = 35°C	5.80	8.47	10.60	14.60	14.80	15.82	17.90
	T_{in} = 45°C	5.20	7.50	10.00	13.10	14.10	14.70	16.80
COP_d	T_{in} = 35°C	1.89	2.62	3.25	4.29	4.40	4.63	5.29
	T_{in} = 45°C	1.50	2.10	2.80	3.10	3.40	3.50	4.10

The unit works in the monoenergetic mode. In the periods of insufficient heating capacity or the ambient temperature below -15°C, the built-in electric heater delivers auxiliary energy. The installation does not have any heat buffer and is not used for domestic hot water. For further analysis, it is also important that the unit has a frequency inverter installed. It allows the capacity control of the heat pump up to the value of 4.4 kW.

The basic results of the heating system measurements in the months from IX to V are summarized in Figures 1 and 2. Figure 1 shows the heating demand of the building (Q_B) and the electricity consumption of the heat pump system (Q_{EL}) in a monthly step.

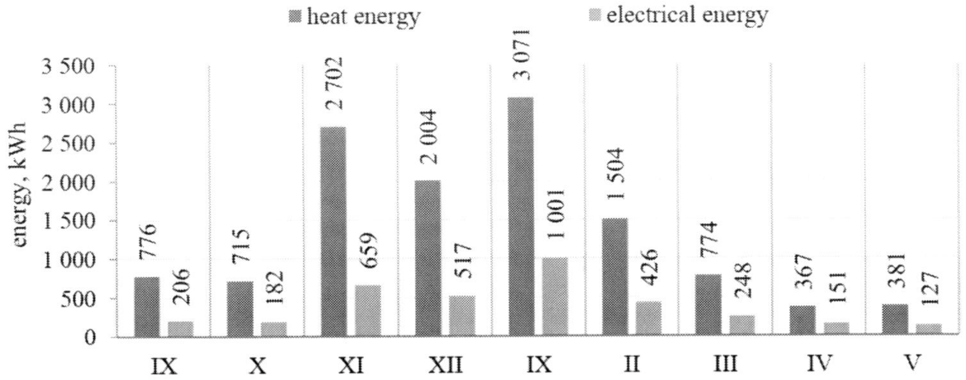

Fig. 1. Thermal energy delivered to the building and electricity consumption of the heat pump in the months from IX to V (measurement data).

Figure 2 contains the monthly *COP* (calculated on the basis of measurement data shown in Figure 1), COP_d and the average ambient temperature ($T_{a,avg}$) data. It is worth noting that the dependence of the monthly *COP* to the ambient temperature is not as it appears directly from the COP_d model under full load. In the transitional periods (e.g. IX, III) the decrease in the *COP* value relative to the declared value (COP_d) is the result of operating in the part load conditions. This fact was discussed in previous research regarding the impact of work under partial load on the energy efficiency of an air-to-water heat pump [7, 8].

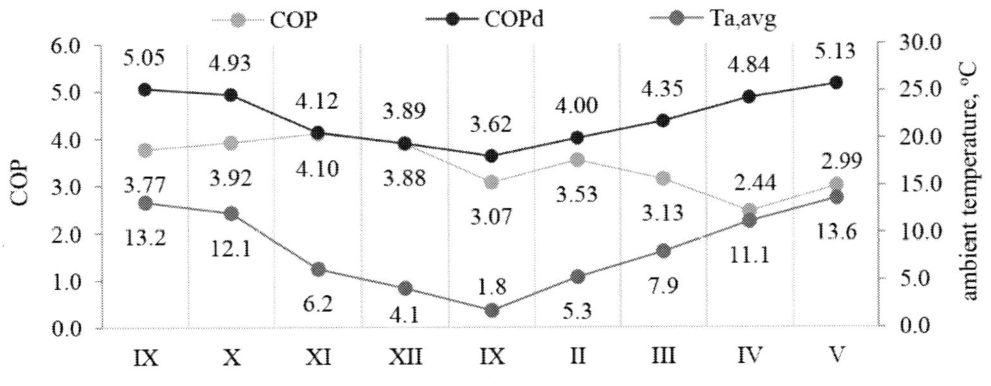

Fig. 2. Average monthly values of the average ambient temperature ($T_{a,avg}$), average monthly COP_d and *COP* (measurement data).

3 The model development

3.1 Model of the energy efficiency for an air-to-water heat pump

In this article, the model of T. Afjei [2] was used to describe the energy efficiency of the heat pump. The independent variables for COP model are the inlet temperature (T_{in}) and the ambient temperature (T_a). In the original method, the parameters A-F of the function described by equation (1) are determined by statistical analysis (by multiple regression) of measurement data obtained in steady state test conditions performed in accordance with the EN 14511 [1] standard.

$$COP_d = A + B \cdot T_a + C \cdot T_{in} + D \cdot T_a \cdot T_{in} + E \cdot T_a{}^2 + F \cdot T_{in}{}^2 \quad (1)$$

Additionally, in order to include the influence of dynamic effects on the change of the energy efficiency of the heat pump, the correction factor described in the article [7] was applied. Correction of the COP_d value allows taking into account the effect of the partial load of the heat pump (described by the *PLR* parameter) based on the part load factor (*PLF*) value. The reduced energy efficiency value is called COP_{pl} (*COP* in part load conditions) and is determined in accordance with the equation (2).

$$COP_{pl} = COP_d \cdot PLF \quad (2)$$

where the *COP* reduction factor (*PLF*) is calculated according to equation (3). The *PLR* value was calculated as the ratio of hourly average building heating demand (Q_B) to the hourly average heating power of the heat pump (Q_{HP}).

$$PLF = 1 + a \cdot ln\left(PLR + e^{(-1/a)}\right) \quad (3)$$

The energy efficiency model described by equation (1) was developed on the data presented in Table 1. The calculations performed for the heat pump tested proved that the T_a^2 and T_{in}^2 variables are not statistically significant for this function (the p value is higher than 0.05). The coefficient of determination (R^2) for the estimated function is 0.99. Parameter a in function (3) is also determined on the basis of statistical analysis. A detailed description of how to determine this value was presented by the authors in the article [7]. Its value determined in relation to the COP_d function developed for measurement data under steady state operating conditions is 0.28. The proportion of the dependent variable translated by the model R is 0.81. The final form of the model developed on the basis of test data under steady state working conditions according to the standard EN 14511 [1] is therefore (4).

$$COP_{pl} = (5.67 + 0.18 \cdot T_a - 0.067 \cdot T_{in} - 0.0023 \cdot T_a \cdot T_{in}) \cdot$$
$$\cdot \left(1 + 0.28 \cdot ln\left(PLR + e^{(-1/0.28)}\right)\right) \quad (4)$$

In the case of determining the energy efficiency characteristics of a heat pump operating in dynamic conditions, the same equations (1), (2) and (3) were used. However, the statistical analysis is based on the hourly measurement data of the work of the heating installation described in section 2. Function (1) applies to the operation of the heat pump under a significant heating load, therefore hourly average measurement data with a PLR value higher than 0.8 were selected to determine the A-F parameters. As in the case of previous calculations performed for working in steady state conditions, the T_a^2 and T_{in}^2 variables proved to be irrelevant to this function (the p value is higher than 0.05). The coefficient of determination (R^2) for the estimated function is 0.90. The correction of this characteristic related to work under partial load conditions was determined from equation (3). In this case, the key range of the inlet temperature (T_{in}) varied from 0°C to 10°C and the PLR from 0.17 to 0.80 were selected. The proportion of the dependent variable translated by the model R is 0.86. In this case, the a parameter is 0.31. The difference between this value and the previously determined for the measurements in steady state working conditions (0.28) results from different *PLF* values used for parameter estimation. The change in the PLF value is related to the different COP_d value determined from function (1). The final form of the model developed on the basis of measurement data of a heat pump operating in dynamic conditions is therefore (5).

$$COP_{pl} = (2.64 + 0.49 \cdot T_a + 0.031 \cdot T_{in} - 0.010 \cdot T_a \cdot T_{in}) \cdot$$
$$\cdot \left(1 + 0.31 \cdot ln\left(PLR + +e^{(-1/0.31)}\right)\right) \quad (5)$$

3.2 Scope of simulations

The purpose of the calculations was to precisely determine the SCOP value and the electricity consumption of the heat pump system. For the comparison, the commonly used methods of taking into account the partial load were used: according to EN 14825 [5] (Simulation 1) and EN 15316 [6] (Simulation 2). The key parameters of these functions have been determined on the basis of measurement data. The results obtained from four simulations were compared with the results of measurements. The characteristics of each simulation are described below:

- Simulation 1: calculated for a heat pump model determined on steady state test data and for the reduction coefficient Cc = 0.8 (Cc value was determined based on measurements of heat pump operation [7]).
- Simulation 2: calculated for a heat pump model determined on steady state test data and for $Q_{el,s\text{-}by}$ = 91 W ($Q_{el,s\text{-}by}$ value was calculated based on measurements of heat pump operation [7]).
- Simulation 3: calculated for a heat pump model determined on steady state test data and for a reduction coefficient a = 0.28 (a value was determined based on measurements of heat pump operation [7]).
- Simulation 4: calculated for a heat pump model determined on measurement in dynamic conditions data and for a reduction coefficient a = 0.31 (a value was determined based on the operational characteristic of the heat pump).

In order to compare the results of the simulation (s) with the measurements (m), the errors of estimating the monthly values of $SCOP$ ($\delta SCOP^m$) and hourly COP values (δCOP^h) were calculated according to the equations (6) and (7).

$$\delta_{SCOP}{}^{m} = \frac{SCOP_m - SCOP_s}{SCOP_m} \quad (6)$$

$$\delta_{COP}{}^{h} = \frac{\sum_1^i \left|\frac{COP_m - COP_s}{COP_m}\right|}{i} \quad (7)$$

3.3 Results and discussion

The results of simulations and measurements for individual months and the entire year are summarized in Table 2. Error values are presented in Figure 3 ($\delta SCOP^m$) and Figure 4 (δCOP^h).

Comparing the results of the monthly values of $SCOP$ for simulations 1, 2 and 3 (that is for the COP_d model determined on steady state test data) shows that simulation 3 allowed the most precise description of the heat pump's operation. With two methods proposed in the standards, the method based on the Cc reduction coefficient (simulation 1) allows to achieve more precise results. In addition, it should be noted that all models describing the impact of working in part load conditions have been calibrated with measurement data. Using the manufacturer test data instead of operating data can bring significant overstatement, especially during transient periods. It is worth noting that the *Cc* coefficient given by manufacturers oscillate in the range of 0.96–0.99. Authors of the publications [7, 8, 10] suggest that such an approach to the overall impact of operating in part load conditions is inappropriate and leads to errors in simulation results. For this reason, research has been carried out and the model described in equation (3) has been developed. The model allows to obtain the best fit among the respondents to the operational data. In addition, a more accurate characteristic of a heat pump operating under full load was developed by replacing the test data in steady state conditions by dynamic, measurement data. This allowed for further improvement of the results. Errors in monthly *SCOP* values, obtained with this modelling method do not exceed 4.5%. The average errors of hourly COP^h values in particular months are shown in Figure 4. The average errors for simulations 1 and 2 are respectively 13.5% and 12.3%. When analyzing the results for individual months, it should be noted that the highest errors relate to the transitional period (e.g. III, IV). Simulation 3 is characterized by slightly lower errors of hourly COP^h values, with the annual average of 11%. Simulation 4 generates the lowest errors, the annual average is 8.7%.

Table 2. Simulation results and measurements for the heat pump system.

Month	Measurement			Simulation 1		Simulation 2		Simulation 3		Simulation 4	
	Q_B	*SCOP*	Q_{EL}	*SCOP*	Q_{EL}	*SCOP*	Q_{EL}	*SCOP*	Q_{EL}	*SCOP*	Q_{EL}
IX	772	3.77	205	3.40	227	3.90	198	3.48	222	3.93	196
X	715	3.92	182	3.69	194	4.02	178	3.69	194	4.07	176
XI	2 699	4.10	658	4.00	675	4.05	666	3.98	677	4.20	642
XII	2 004	3.88	516	3.82	525	3.92	511	3.80	528	3.97	505
I	3 071	3.07	1 002	3.32	926	3.38	909	3.30	931	3.02	1 018
II	1 504	3.53	426	3.54	425	3.84	391	3.48	432	3.69	408
III	772	3.13	247	2.92	265	3.50	221	2.89	267	3.09	250
IV	366	2.48	148	2.19	167	2.93	125	2.31	158	2.54	144
V	381	3.00	127	2.60	146	3.30	115	2.75	139	3.05	125
year	12284	3.50	3 511	3.46	3 550	3.71	3 315	3.46	3547	3.62	3 391

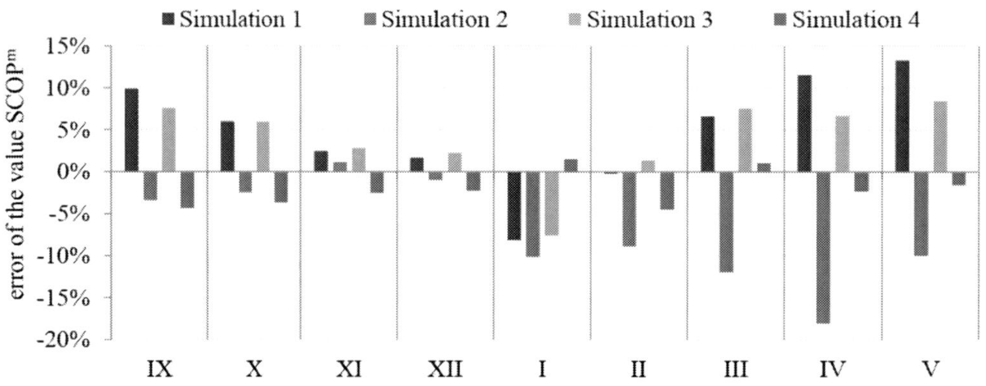

Fig. 3. Errors of monthly $SCOP^m$ values (simulation results in relation to measurement data).

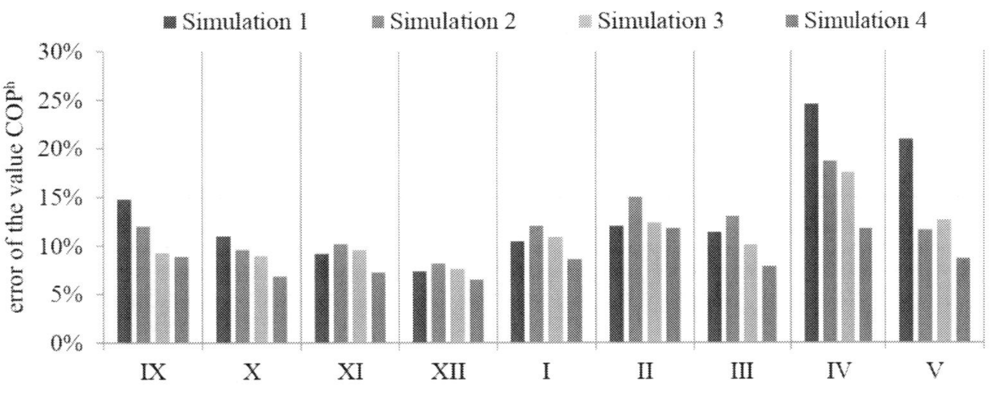

Fig. 4. Errors of hourly COP^h values in individual months (simulation results in relation to measurement data).

Figure 5 presents the hourly COP_h values for measurement data and simulations 1–4 for two selected months: January and March. Plots allow to observe the effects of model calibration. Simulations 1, 2 and 3 are based on static characteristic of COP_d and three different models of operating in part load conditions calibrated with measurement data. Correct results are visible for simulations 1 and 2. A detailed analysis of errors showed that simulation 2 has slightly better accuracy. Results of COP^h values based on simulation 3 are overestimated. Simulation 4, determined on the basis of operating characteristics of the *COP*, better describes the actual work of the air-to-water heat pump. The model takes into account the influence of changes in air humidity, a different defrost cycles frequency and the effect of inverter frequency change on the *COP*. It should be noted that this impact has not been demonstrated directly due to the insufficient scope of changes of these parameters during the operation of the device. Under the influence of these factors the parameters of variables T_{in} and T_a have been corrected.

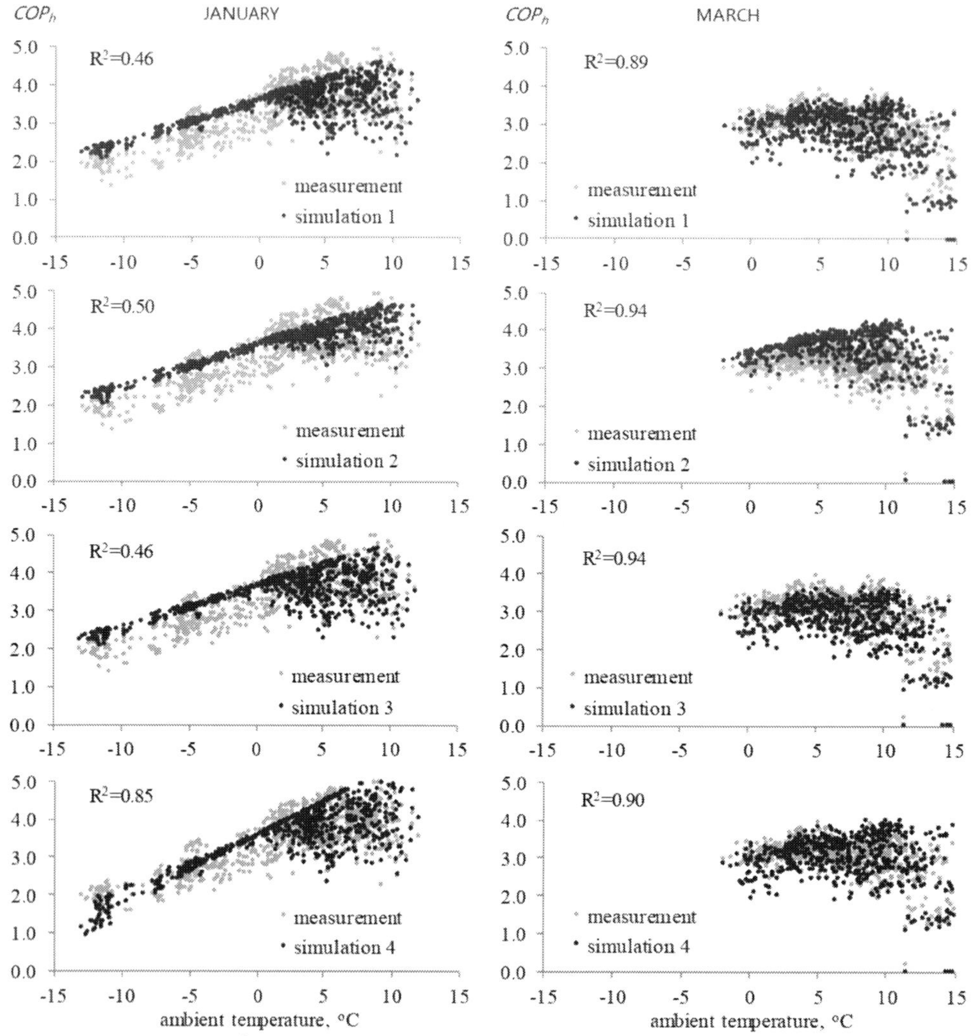

Fig. 5. The hourly COP^h values in January and March (simulation results in relation to measurement data).

4 Conclusion

The article discusses the possibility of creating an air-to-water heat pump energy efficiency model based on operational data. Tests in steady-state conditions do not provide much information about dynamic effects affecting the energy efficiency of these devices, because they are conducted in conditions that rarely exist in heating installation. Tests under quasi-dynamic conditions can be an alternative to these procedures. In this article, it was confirmed that dynamic test data, after proper selection, will allow to determine the parameters of the *COP* characteristics. Of course, the research for procedure of such a test is necessary. For example, in this study, the influence of air humidity and compressor rotation reduction is not directly determined. However, comparable device models were developed on the basis of dynamic measurement data and data from test in steady-state conditions. The simulation results carried out on both characteristics were compared. The agreement between the results of the dynamic model and the measurement results was better than for the data developed in static conditions.

References

1. Air conditioners, liquid chilling packages and heat pumps with electrically driven compressors for space heating and cooling, EN 14511:2013 (2013)
2. T. Afjei, M. Wetter, Compressor heat pump including frost and cycle losses (1997)
3. J. Ruschenburg, T. Ćutić, S. Herkel, Energy and Buildings **84**, 506–515 (2014)
4. F. Simon, J. Ordoñez, T. A. Reddy, A. Girard, T. Muneer, Renewable Energy **95**, 413–421 (2016)
5. EN 14825:2013, Air conditioners, liquid chilling packages and heat pumps, with electrically driven compressors, for space heating and cooling – Testing and rating at part load conditions and calculation of seasonal performance (2013)
6. EN 15316-4-2, Heating systems in buildings - Method for calculation of system energy requirements and system efficiencies – Part 4-2: Space heating generation systems, heat pump systems (2008)
7. K. Piechurski, M. Szulgowska-Zgrzywa, J. Danielewicz, EDP Sciences, E3S Web of Conferences **44**, 00139 (2018)
8. E. Fuentes, D. Waddicor, J. Salom, REHVA Journal, 45–49 (2016)
9. K. Piechurski, M. Szulgowska-Zgrzywa, J. Danielewicz, EDP Sciences, E3S Web of Conferences **17**, 00072 (2017)
10. M. Dongellini, C. Naldi, G. L. Morini, Applied Thermal Engineering **90**, 1072–1081 (2015)

Drug resistance in airborne bacteria isolated from waste management and wastewater treatment plants in Olsztyn

Jacek Potorski[1,*], *Izabela* Koniuszewska[1], *Małgorzata* Czatzkowska[1], and *Monika* Harnisz[1]

[1]Department of Environmental Microbiology, University of Warmia and Mazury in Olsztyn, Poland

Abstract. Wastewater treatment plants (WWTPs) and municipal waste management plants (MWMPs) emit bioaerosols containing potentially pathogenic biological components which post a threat for human health. Microbiological monitoring supports evaluations of the antibiotic resistance (AR) of airborne microorganisms and the relevant health risks. The aim of this study was to analyze the microbiological quality of air sampled in a WWTP and MWMP in Olsztyn based on total bacterial counts, the presence of bacteria resistant to three antibiotic classes (beta-lactams, tetracyclines and chloramphenicol) and genes encoding resistance to these antibiotics (bla_{TEM}, bla_{SHV}, bla_{CMY-2}, bla_{AmpC}, *tet*(M), *tet*(A), *tet*(X), *tet*(B), *cmlA*, *floR*, *fexA*, *fexB* and *catA1*). Bacterial counts were higher in air samples collected from the MWMP (~10^4 CFU/m^3) than from the WWTP (10^1–10^3 CFU/m^3). A similar trend was noted in the counts of antibiotic resistant bacteria (ARB). The abundance of ARB did not exceed 1.7 x 10^2 CFU/m^3 in WWTP samples, but was higher at up to 4.2 x 10^3 CFU/m^3 in MWMP samples. Bacteria resistant to doxycycline were least prevalent in the analyzed ARB. In the group of 49 tested bacterial strains, 44 harbored at least one of the analyzed antibiotic resistance genes (ARGs). A comparison of ARGs in all bacterial strains isolated from WWTP and MWMP air samples revealed the highest diversity and prevalence of ARGs in the samples collected in the mechanical segment of the waste processing line in MWMP and the biological segment of the wastewater processing line in WWTP. The results of this study point to high microbiological contamination of air in MWMPs and WWTPs which are reservoirs of ARB and ARGs and potential sources of AR.

1 Introduction

Wastewater treatment plants (WWTPs) and municipal waste management plants (MWMPs) emit odors and biological aerosols which are potential sources of microbiological contamination in atmospheric air [1–3]. Plant employees are exposed to class 2 biological factors (Regulation of the Minister of Health of 2005) that are potentially hazardous for human health [4]. Bioaerosols contain bacteria, viruses, fungi, yeasts and yeast-like fungi.

* Corresponding author: jacek.potorski@uwm.edu.pl

Potentially pathogenic airborne microorganisms can cause many infectious diseases in humans and animals [5, 6]. The concentration and range of bioaerosol emissions are determined by the size of the processing plant, the applied technology and conditions in every segment of the plant (mechanical/biological waste treatment). Weather conditions, including strong winds, also contribute to the spread of airborne microorganisms [3, 6]. The type and reach of processing plants environmental impact is influenced by initial microbial counts in solid waste and wastewater, microbial growth stages, the applied processing technology, weather and environmental conditions [7]. According to the invalid, but the only available microbial contamination reference standards for atmospheric air which expired in 2015 [8], bacterial counts of 10^3 CFU/m^3 are indicative of strongly contaminated air. Highly contaminated air can also contain antibiotic resistant bacteria (ARB) which are transferred from processed waste in plants, many of which are large open areas.

Antibiotic resistant bacteria (ARB) pose a significant threat to public health because they progressively less sensitive to pharmacological treatment. These pathogens have highly specialized molecular mechanisms which enable bacteria to rapidly adapt to and proliferate in environments with high antibiotic concentrations, such as WWTPs and MWMPs where various sources of waste are mixed. Growing levels of antibiotic resistance and the increasing biodiversity of antibiotic resistance genes (ARGs) pose a considerable problem. Microbiological monitoring systems in WWTPs and MWMPs should also be able to detect ARB strains [9–11].

The aim of this study was to analyze the microbiological quality of air sampled from a WWTP and a MWMP in Olsztyn based on total microbial counts, the presence of bacteria resistant to three antibiotic classes (beta-lactams, tetracyclines and chloramphenicol) and genes encoding resistance to these antibiotics (bla_{TEM}, bla_{SHV}, $bla_{\mathrm{CMY\text{-}2}}$, bla_{AmpC}, *tet*(M), *tet*(A), *tet*(X), *tet*(B), *cmlA*, *floR*, *fexA*, *fexB* and *catA1*).

2 Sampling sites

Air was sampled from the Municipal Waste Management Plant (MWMP) in Olsztyn and the Łyna Municipal Wastewater Treatment Plant (WWTP) in Olsztyn. In both plants, air samples were collected in the mechanical and biological segments of solid waste and wastewater processing lines.

3 Materials and methods

3.1 Determination of total bacterial counts and counts of drug resistant bacteria

Air samples were collected by aspiration with the MAS-100 Eco air sampler (Merck). The total counts of airborne bacteria were determined on Miller's LB agar (A&A Biotechnology) with peptone. Antibiotic resistant strains were isolated on Miller's LB Agar (A&A Biotechnology) with ampicillin (8 mg/L), doxycycline (2 mg/L) or chloramphenicol (8 mg/L). Antibiotic concentrations in the culture medium were determined based on EUCAST 2018 guidelines [12]. Air samples were collected once in the summer of 2018. The results were presented as CFU/m^3.

3.2 Isolation and amplification of dsDNA

Up to five morphologically different bacterial colonies were isolated from each ARB culture. Every culture was suspended in 0.5 mL of distilled water, and genomic DNA was isolated from 22 ampicillin-resistant strains, 15 chloramphenicol-resistant strains and 12 doxycycline-resistant strains. DNA was isolated by thermal lysis [14]. The presence of genes encoding resistance to β-lactams (*bla*$_{TEM}$, *bla*$_{SHV}$, *bla*$_{AmpC}$, *bla*$_{CMY-2}$), tetracyclines (*tet*(A), *tet*(B), *tet*(M), *tet*(X)), chloramphenicols and florfenicols (*cmlA*, *floR*, *fexA*, *fexB*, *catA1*) was determined by standard PCR, according to previous described parameters [15–21].

4 Results and discussion

4.1 Determination of microbiological air quality based on total bacterial counts and the counts of drug resistant bacteria

The environmental impacts of waste and wastewater treatment plants are determined by biological factors and local conditions, including initial microbial counts in solid waste and wastewater, bacterial growth stages, the applied processing technology, weather and environmental conditions [22]. In this study, the total bacterial counts were lowest (8.75×10^1 CFU/m^3) in air samples collected in the mechanical segment of the WWTP and highest (4.51×10^4 CFU/m^3) in air samples collected in the biological segment of the MWMP. Total bacterial counts were 24-times higher in air samples from the MWMP than from the WWTP. Air samples from the mechanical segments of both plants were more highly contaminated than the samples from biological segments. The counts of all analyzed microbial groups in air samples from the evaluated sites are presented in Figure 1.

In a study by Korzeniewska et al. [1], the counts of heterotrophic mesophilic bacteria in untreated wastewater ranged from 10^5 to 10^7 CFU/mL. In the work of Osińska et al. [23], the average total bacterial counts were determined at 10^7 CFU/mL in untreated sewage and at 10^4 to 10^5 CFU/mL in treated sewage. WWTPs can emit bioaerosols which contribute to the contamination of atmospheric air and pose a public health risk [24]. The microbial contamination of air in the WWTP evaluated by Vítězová et al. [25] ranged from 10^1 to 10^4 CFU/m^3 for psychrophilic bacteria and from 10^1 to 10^4 CFU/m^3 for mesophilic bacteria. Air contamination was highest in the proximity of the activated sludge chamber, dried sludge storage and the mechanical segment of the wastewater treatment line. In general, aeration chambers in WWTPs are most exposed to biological aerosols [26], but in this study, total bacterial counts were 35-times higher in the mechanical than in the biological segment of the wastewater processing line.

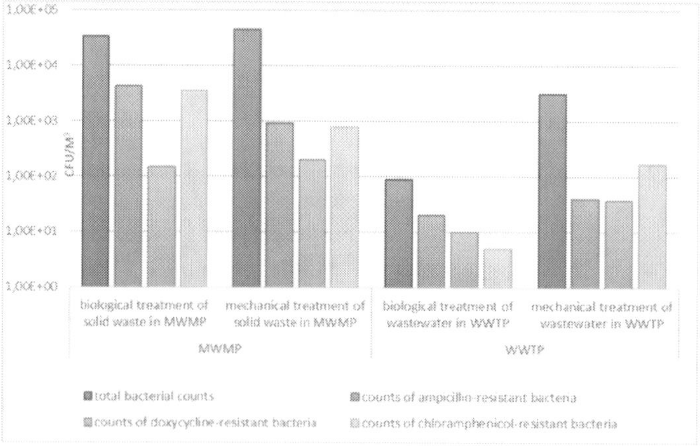

Fig. 1. Counts of all analyzed microbial groups in air samples from the evaluated sites.

Breza-Boruta [27] analyzed the microbiological quality of bioaerosols in municipal waste management plants and reported the highest total bacterial counts in the waste sorting station (10^4 CFU/m^3). In the current study, the total counts of airborne bacteria also approximated 10^4 CFU/m^3 in the mechanical and biological segments of the MWMP in Olsztyn. Chmielowiec-Korzeniowska et al. [28] determined total bacterial counts at 10^6 CFU/m^3 in a study evaluating the microbiological contamination of air in the vicinity of an animal waste management plant.

At present, there are no valid microbiological contamination standards for atmospheric air in Poland. However, in view of the standards that had expired in 2015 [8], air samples collected from the MWMP and from the mechanical segment of the WWTP were characterized by high levels of microbial contamination.

The counts of ampicillin-resistant bacteria were highest in the biological segment of the MWMP (4.2 x 10^3 CFU/m^3) and lowest in the biological segment of the WWTP (2 x 10^1 CFU/m^3). The counts of ampicillin-resistant bacteria were twice higher in the mechanical segment than in the biological segment of the WWTP. In the MWMP, the counts of ampicillin-resistant bacteria were 4-times higher in the biological segment than in the mechanical segment of the waste processing line.

The counts of doxycycline-resistant bacteria were lowest in the biological segment of the WWTP (1 x 10^1 CFU/m^3) and highest in the mechanical segment of the MWMP (2 x 10^2 CFU/m^3). Air contamination with doxycycline-resistant bacteria was higher in the mechanical segments of both analyzed plants. Odeyemi [29] evaluated the microbiological contamination of air near a landfill site in Nigeria and reported that 95.9% of all isolated bacteria were resistant to amoxicillin (amoxicillin and ampicillin belong to the group aminopenicillin antibiotics), whereas 84.6% of the isolated strains were resistant to tetracyclines. In a study by Reinthaler et al. [30], *E. coli* isolated from wastewater, a source of toxic bioaerosols, were most frequently resistant to penicillins, including ampicillin (18%), as well as tetracyclines (57%).

In the present study, the counts of chloramphenicol-resistant bacteria were highest in the biological segment of the MWMP (3.5 x 10^3 CFU/m^3) and lowest in the biological segment of the WWTP. In the MWMP, the counts of chloramphenicol-resistant bacteria were more than 4-times higher in the biological segment than in the mechanical segment of the processing line. In the WWTP, the prevalence of this microbial group was more than 30-times higher in air samples collected from the mechanical segment.

Total microbial counts were highest in the mechanical segments of both analyzed plants. Published data are consistent only with the results noted in the MWMP.

4.2 Drug resistance genes in the isolated bacterial strains

The bacterial strains isolated from air samples collected in the MWMP and the WWTP harbored mainly genes encoding resistance to tetracyclines. The percentage of genes responsible for resistance to beta-lactams, chloramphenicol and florfenicol was lower. None of the bacteria isolated from the WWTP or the MWMP harbored the *fexB* gene encoding resistance to florfenicol. A comparison of all airborne bacteria isolated from both plants revealed the highest prevalence of antibiotic resistance genes in the strains isolated from the MWMP (Fig. 2).

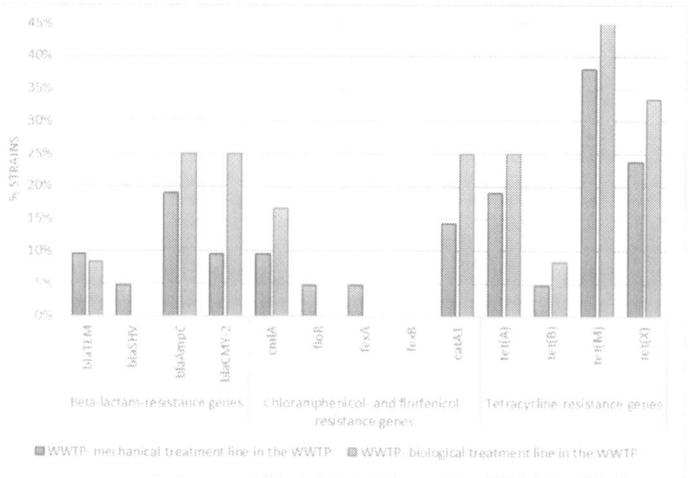

Fig. 2. Prevalence of antibiotic resistance genes [%] in airborne bacteria isolated from the MWMP and the WWTP.

The prevalence of ARGs was also compared in 24 bacterial strains isolated from air in the mechanical and biological segments of the WWTP. The strains isolated from the biological segment of the WWTP were characterized by the highest prevalence of ARGs, and 8 out of the 9 isolated strains harbored at least 2 genes encoding resistance to antibiotics, mostly tetracyclines (data not shown). Interestingly, *tet*(M) was most prevalent gene in wastewater, whereas in a study of the Olsztyn WWTP conducted by Harnisz et al. [14], the most ubiquitous gene was *tet*(A). However, Zhang and Zhang [31] reported that both genes were most prevalent in a WWTP. Genes encoding resistance to beta-lactams and chloramphenicol were also noted in our study. Around 25% of the tested isolates harbored both bla_{AmpC} and $bla_{\mathrm{CMY\text{-}2}}$ genes which are responsible for resistance to all beta-lactams, excluding fourth-generation drugs and monobactams, which poses a significant threat for public health.

All of the 15 bacterial strains isolated from the mechanical segment of the WWTP harbored at least one ARG. This part of the WWTP was colonized by significantly fewer bacterial strains with ARGs than the biological segment. Most of the isolated strains harbored genes encoding resistance to tetracyclines and beta-lactams, but their prevalence was somewhat lower than in the biological segment of the plant. The *cmlA* gene was observed in isolated cases. None of the isolated strains harbored *floR*, *fexA* and *fexB* genes which encode resistance to florfenicol (Fig. 3).

A comparison of 24 airborne bacterial strains isolated from the mechanical and biological segments of the MWMP (Fig. 3) revealed a significantly higher prevalence of ARGs in the strains isolated from mechanical part of the plant. The most prevalent ARGs in the strains from the mechanical segment encoded resistance to tetracyclines. Genes responsible for resistance to tetracyclines were also most prevalent in the biological segment of the MWMP. The lower prevalence of ARGs in the strains isolated from the biological part of the plant could be attributed to the type of processed waste and the design of the waste processing line. Similar results were reported by Li et al., McEachran et al. and Xie et al. who identified ARGs in airborne bacteria isolated from WWTPs in urban and rural areas, including tetracycline resistance genes (*tet*(A), *tet*(C), *tet*(M), *tet*(O), *tet*(Q), *tet*(W), *tet*(B), *tet*(L)) [32-34]. In our study, the isolates from the mechanical part of the MWMP also harbored genes encoding resistance to beta-lactams and chloramphenicol. It should be noted that more than 35% of the isolated strains harbored the bla_{AmpC} gene which is responsible for resistance to broad-spectrum beta-lactams that are widely used in clinical practice. Bacteria with mutations in regulator genes are able to produce AmpC beta-lactamases in the absence of an antibiotic inductor. The removal of the mechanism which represses the production of the AmpC enzyme is referred to as derepression. Isolates with derepressed AmpC become resistant to all beta-lactams, including all penicillins, cephalosporins (excluding fourth-generation drugs) and monobactams, and they are not sensitive or are largely insensitive to beta-lactamase inhibitors: clavulanic acid, sulbactam and tazobactam. This process has a tremendous impact on the clinical environment because it drives the search for new and stronger antibiotics. bla_{AmpC} genes are transferred by mobile genetic elements (ISEcp1, ISCR1, IS26) in the genomes of bacteria such as *C. freundii*, *Enterobacter* spp. and *M. morganii*. Enzyme CMY-2 is a derivative of enzyme AmpC. Beta-lactamases of the CMY-2 family are the most ubiquitous derivatives of AmpC in *C. freundii* [35-36]. In the present study, the bla_{CMY-2} gene was noted in more than 20% of the strains isolated from the mechanical segment of the MWMP, whereas the bla_{TEM} gene was detected in more than 10% of the tested isolates. The bla_{SHV} gene was not identified in any of the analyzed strains. In a study by Li et al. (2018), genes encoding resistance to beta-lactams were highly prevalent in airborne bacteria sampled from a rural area, and bla_{TEM} was most frequently noted. The above authors concluded that airborne ARGs contributed to the emergence of new bacterial genotypes. They demonstrated that urban air was highly contaminated with ARGs which pose a serious threat to public health [32]. Considerable air pollution in the mechanical segment of the MWMP could be attributed to high proportions of organic matter in the processed waste, which offers an ideal environment for microbial growth and development. Contaminated air is released outside the MWMP, and ARB and ARGs are transferred to urban areas. Airborne pollutants and microorganisms can overcome geographic barriers, and this mode of transmission significantly contributes to the spread of ARB and ARGs across large distances [33, 34]. The results of this study confirm that airborne transmission of ARGs poses a threat in urban areas and that air quality standards should be revised to protect public health.

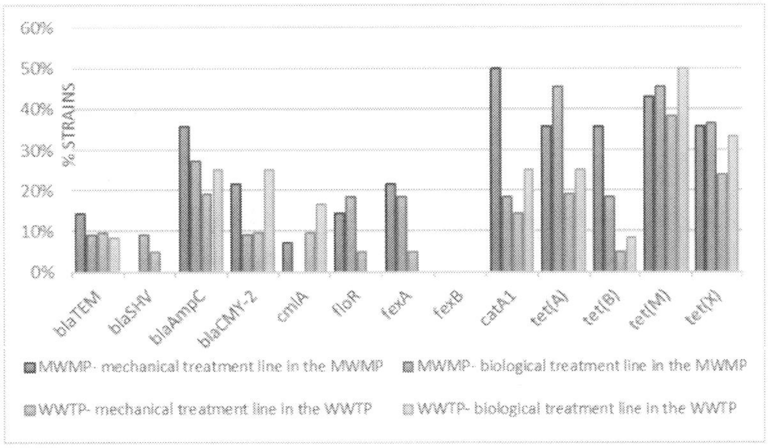

Fig. 3. Prevalence of antibiotic resistance genes [%] in airborne bacteria isolated from the mechanical and biological segments of the WWTP and MWMP.

5 Conclusions

This study demonstrated that WWTPs and MWMPs are reservoirs of airborne ARB and ARGs. Antibiotic resistance genes were most diverse in the mechanical part of the MWMP and the biological segment of the WWTP in Olsztyn. In both plants, employees who come into contact with bioaerosols should handle contaminated waste and wastewater with utmost caution, should strictly observe occupational health and safety regulations and wear protective equipment and clothing.

This project was funded by a Student Grant awarded by the Rector of the University of Warmia and Mazury in Olsztyn.

References

1. E. Korzeniewska, Z. Filipkowska, A. Gotkowska-Płachta, W. Janczukowicz, B. Dixon, M. Czułowska, Water Res. **43** (2009)
2. E. Korzeniewska, Front Biosci. (Schol Ed) **1**, 3 (2011)
3. E. Korzeniewska, M. Harnisz, J. Environ. Manage. **128** (2013)
4. Regulation of the Minister of Health of 22 April 2005 on biological factors that are hazardous for health in the work environment and health protection measures for employees who are exposed to hazardous factors in the work environment; J. Law **81** (2005)
5. R. I. Aminov, Environ. Microbiol. **11** (2009)
6. M. Szyłak-Szydłowski, A. Kulig, E. Miaśkiewicz-Pęska, Int. Biodeter. Biodegr. **115** (2016)
7. O. Golovko, V. Kumar, G. Fedorova, T. Randak, R. Grabic, Chemosphere **111** (2014)
8. PN-89/Z-04111/02 (2002) Air quality protection. Microbiological analyses. Determination of bacterial counts in atmospheric air (emission) sampled by aspiration and sedimentation
9. F. Barancheshme, M. Munir, Front Microbiol. **8** (2018)
10. M. N. Alekshun, S. B. Levy, Cell **128**, 6 (2007)
11. K. Yoo, T. L. Lee, E. J. Choi, J. Yang, S. K. Shukla, S. Hwang, J. Park, J. Environ. Sci. **51** (2017)
12. EUCAST (2014) European Committee on antimicrobial susceptibility testing. Breakpoints tables for interpretation of MICs and zones diameters. Version 4.0. http://www.eucast.org
13. M. Gołofit-Szymczak, J. Skowroń, Bezp. Pr. Nauk Prakt. **3** (2005)
14. M. Harnisz, E. Korzeniewska, S. Ciesielski, I. Gołaś, Sci. Total Environ. **505** (2015)
15. J. Kim, S. Jeon, H. Rhie, B. Lee, M. Park, H. Lee, J. Lee, S. Kim, J. Infect. Chemother. **41**, 3 (2009)
16. E. Ruppé, S. Hem, S. Lath, V. Gautier, F. Ariey, J. Sarthou, D. Monchy, G. Arlet, Emerg. Infect. **15**, 5 (2009)
17. J. Li, B. Shao, J. Shen, S. Wang, Y. Wu, Environ. Sci. Technol. **47**, 6 (2013)
18. H. Maidhof, B. Guerra, S. Abbas, H. M. Elsheikha, T. S. Whittam, L. Beutin, Appl. Environ. Microbiol. **68** (2002)
19. M. Nawaz, K. Sung, S. A. Khan, A. A. Khan, R. Steele, Appl. Environ. Microbiol. **72** (2006)
20. L. K. Ng, I. Martin, M. Alfa, M. Mulvey, Mol Cell Probes **15** (2001)
21. Z. Eitel, J. Soki, E. Urban, E. Nagy, Anaerobe **21** (2013)
22. N. Patentalakis, A. Pantidou, N. Kalogerakis, Water Air Soil Pollut. **8** (2008)
23. A. Osińska, E. Korzeniewska, M. Harnisz, S. Niestępski, E3S Web Conf. **17** (2017)

24. L. Fracchia, S. Pietronave, M. Rinaldi, M. G. Martinotti, Water Res. **40** (2006)
25. M. Vítězová, T. Vítěz, H. Mlejnková, T. Lošák, Acta Univ. Agric. Silvic Mendel Brun. LX **3** (2012)
26. Z. Filipkowska, W. Janczukowicz, M. Krzemieniewski, J. Pesta Biul. Nauk UWM **15**, 217 (2002)
27. B. Breza-Boruta, Ecol. Chem. Eng. A **19**, 8 (2012)
28. A. Chmielowiec-Korzeniowska, L. Tymczyna, A. Drabik, Ł. Krzosek, Ann. Agric. Environ. Med. **23**, 1 (2016)
29. A. T. Odeyemi, Microbiol. Res. **2**, 2 (2012)
30. F. F. Reinthaler, J. Posch, G. Feierl, G. Wust, D. Haas, G. Ruckenbauer, F. Mascher, E. Marth, Water Res. **37** (2003)
31. X. Zhang, T. Zhang, Environ. Sci. Technol. **45**, 7 (2011)
32. J. Li, J. Cao, Y. Zhu, Q. Chen, F. Shen, Y. Wu, S. Xu, H. Fan, G. Da, R. Huang, J. Wang, A. Lorelei de Jesus, L. Morawska, C. K. Chan, J. Peccia, M. Yao, Environ. Sci. Technol. **52**, 19 (2018)
33. A. D. McEachran, B. R. Blackwell, J. D. Hanson, K. J. Wooten, G. D. Mayer, S. B. Cox, P. N. Smith, Environ. Health Perspect. **123** (2015)
34. J. Xie, L. Jin, X. Luo, Z. Zhao, X. Li, Environ. Sci. Technol. Lett. **5** (2018)
35. M. Barlow, B. G. Hall, Agents Chemother. **46** (2002)
36. K. Bush, J. F. Fisher, Ann. Rev. Microbiol. **65** (2011)

Constructed wetlands aided with bio-preparation to neutralize wastewater from a small fruit and vegetable processing plant with high load pollutants

Monika Puchlik[1,*]

[1]Bialystok Technical University, Faculty of Civil and Environmental Engineering, Chemical Laboratory, Wiejska Street 45E, 15-351 Bialystok, Poland

Abstract. The research was carried out for real wastewater collected directly from a fruit and vegetable plant located in north-eastern Poland. Effectiveness of organic matter metabolism expressed as BOD_5, COD, total suspended solids depending on the load of pollutants and bio-preparation, was determined. The use of bio-preparation in the constructed wetland allowed to increase the efficiency of wastewater treatment throughout the research period: for BOD_5 by 94% in deposit with bio-preparation and by 87% in deposit without the addition of bio-preparation; however, for COD – by 93% in deposit with the addition of bio-preparation and by 87% in deposit without bio-preparation added.

1 Introduction

In rural areas, as well as with dispersed buildings, the problem is the treatment of sewage discharged from individual houses, their small clusters and small industrial plants that are not and will not be connected to collective sewage systems. Although the amount of wastewater discharged is small, they require treatment to the same extent as larger settling units [1, 2, 17]. Solution to the problem of wastewater treatment in small fruit and vegetable processing plants in non-urbanized areas may be the application of a proven, low-cost method of wastewater treatment using constructed wetlands. Properly designed and manufactured, they guarantee the required degree of treatment with sewage disposal to the ground or water. The constructed wetland method has found the widest application for the treatment of domestic wastewater. It can be used in agricultural areas where, due to the dispersion of buildings, it is more economical to build individual treatment plants than collective ones. It works perfectly in mountainous areas, valuable in nature, as well as in seasonally used facilities [2, 3]. In the case of household wastewater treatment, it is most often used after preliminary mechanical treatment of sewage. The constructed wetland method can be applied in the second or third degree in sewage treatment plants operating

* Corresponding author: m.puchlik@pb.edu.pl

using the activated sludge method or biological deposits. In the United Kingdom, a number of municipal wastewater treatment plants use rotary beds, and then constructed wetland with horizontal or vertical flow for wastewater treatment [4]. Such solutions can be applied in operating municipal and industrial wastewater treatment plants in order to increase the efficiency of their operation [5, 6]. Their efficiency is high while ensuring proper operation and control of the treatment process, even for the smallest objects. Due to the possibility of removing various impurities, the constructed wetland systems are increasingly used for the treatment of industrial wastewater [4, 5]. The reason for undertaking the research was the problems with the discharge and treatment of sewage from the small food industry plant in north-eastern Poland located in a scattered development area. The innovative nature of own research results from the lack of domestic and foreign experience in the field of treatment of wastewater originating from small plants of the fruit and vegetable industry using the constructed wetland method with subsurface vertical flow, grown with common reed, and additionally fed with bio-preparation to increase its operational efficiency. Obtained dependences concerning the influence of the analyzed factors on the amount of the removed load of selected pollutants in the constructed wetland deposit were confronted with general knowledge of this method of wastewater treatment. In the available literature, the assessment of significance of factors affecting the efficiency of contaminants removal in constructed wetlands is limited to the analysis carried out on the basis of percentage reduction of pollution. In this study, it was recognized that development of every dependence is possible only while taking into account the influence of factors such as pollution load, hydraulic load and mutual interactions between them [8].

2 Application of constructed wetland and bio-preparation

Constructed wetlands systems are most often used for the treatment of household sewage from individual settlements, but also rainwater, surface runoff, flowing waters, leachate from municipal waste landfills, and for drainage of sewage sludge [2, 3, 42]. They are also used for wastewater treatment from agro-food processing, which contains contaminations susceptible to biochemical degradation processes [7]. This solution was applied in the USA to wastewater from the vegetable industry and from sugar factories, as well as in New Zealand and Great Britain to wastewater from meat plants [8, 9]. Interest in this convenient and pro-ecological solution is high enough that it is aimed at that they can work on any plot, regardless of ground conditions and the property size [10, 11]. Undoubtedly, by opting for a household treatment plant, ecological benefits are gained. This definitely improves the cleanliness of waters in a given area. It is also worth paying attention to the aesthetic value of the constructed wetland treatment plant modeled on natural ecosystems providing shelter for birds, amphibians and insects associated with the aquatic environment. Thus, they allow to reduce the differences between natural and anthropogenic elements and a well-integrated constructed wetland treatment plant, which may increase the aesthetic values of a landscape. Vegetation growing in such treatment plant has a beneficial effect on the microclimate, reduces the wind speed and effects of heavy rainfalls [1, 12]. There are known applications of constructed wetland systems for treatment of wastewater from farms, especially in the United States and Japan [13–15]. Constructed wetland systems can be used for treatment of wastewater from the cellulose industry [26]. In the United States and China, there are constructed wetland systems in refineries and mines. Other common applications of constructed wetland systems include cleaning the rain runoffs from roads and streets, bridges, petrol stations or airports [5]. Among sewage treated in vertical (VF) and horizontal (HF) flow systems, there are also wastewater from agro-food processing, which contain contaminants susceptible to biochemical degradation processes [7, 11]. This solution was used in the United States for wastewater from sugar factories, as well as in

New Zealand and the United Kingdom for wastewater from meat plants [8, 9]. Interest in this convenient and pro-ecological solution is so big that it is aimed at that they can work on any plot, regardless of ground conditions and the property size [10, 11].

Bio-preparations are preparations, the main component of which are microorganisms or enzymes that influence the acceleration or support of the decomposition processes of organic compounds contained in wastewater, liquefaction of solid waste, preventing the formation of deposits, formation of so-called *skin* on the surface of the sewage. Biological wastewater treatment process consists in the decomposition of organic pollutants by microorganisms specially selected for this purpose. Bio-preparations are used both in municipal wastewater treatment plants as well as in individual farms, e.g. in cesspools or household wastewater treatment plants. A bio-preparation is composed of a starter conditioner and suitably selected composition of non-pathogenic microorganisms with strictly targeted effect, causing decomposition and degradation of harmful substances, like ammonia, nitrite, hydrogen sulfide, indole, skatole, mercaptans, phosphorus compounds and other mono- and multi-carbon organic compounds. The microbiological composite has the ability to bind heavy metals, direct the desired fermentation processes, change the pH of sewage and liquid manure, and strongly inhibit the development of pathogenic microorganisms, as well as enzymatic destruction of eggs and spores of insects and endoparasites [18]. Bio-preparations are manufactured in the form of tablets, powder, granules, emulsions and liquid. They usually contain selected strains of saprophytic bacteria, fungi and enzymes of natural origin. All microorganisms must belong to the first category of harmfulness, including microorganisms with low probability of causing diseases in humans and not posing any threat to the environment. In addition, bio-preparations include, among others, carriers and nutrients for microorganisms and fragrances.

3 Material and methods

The paper presents the research upon the treatment of wastewater from the fruit and vegetable industry in constructed wetlands supported by a bio-preparation, which were carried out in 2014–2016. The deposit was fed with sewage from a small processing plant, that was created in the period from V to X (2014 and 2016) during washing, sorting and production of fruit and vegetable juices. The hydraulic load was 0.02 m^3/m^2d. One of the deposits was additionally supported by the bio-preparation (every 7 days in the amount of 4 grams per 1 m^3 of raw sewage) that ensures the maintenance and reactivation of biological wastewater treatment plants. It is intended for wastewater with high COD parameters containing cellulose. It enriches biological deposits with microorganisms to improve their degradation properties. Samples of raw and treated wastewater were collected every 14 days from May 2014 to October 2016. The sewage treatment plant consists of a retention reservoir and two parallel systems consisting of ground and plant deposits with common reed (*Phragmites australis*). Ground and plant deposits have the dimensions of a square – 75 × 75 centimeters and depth of 80 centimeters. They were built in a system with the subsurface vertical flow of sewage. Deposits have four layers of filling, from the top: layer I (sand 0–2 mm, 0.15 m), layer II (gravel 2–8 mm, 0.15 m), layer III (fine gravel 8–20 mm, 0.20 m), layer IV (stones 20–80 mm, 0.15 m), on which the common reed (*Phragmites australis*) was planted. In the raw and treated sewage samples taken, in accordance with mandatory methodology, the following parameters were determined: COD – dichromate method according to: PN-74/C-04578.03 [37], BOD_5 – manometric method using Oxi-Top Standard system, total suspended solids (TSS) – gravimetric method PN-EN 872:2007+Ap1:2007 [38].

4 Results and discussion

When analyzing the results of own research, a significant dispersion of the values for basic pollutants was noted (characteristic for sewage from the food industry) (Table 1). The highest variability was demonstrated by organic substance parameters expressed as BOD_5 and COD as well as total suspended solids. According to the guide of the United States Environmental Protection Agency (US EPA 1994) [27], such a wide diversity of raw sewage quality creates the necessity of conducting local research to design the purification process. Figures 1–6 show the relationships between the quantities of removed ($L_{Rem.}$) and delivered ($L_{Del.}$) loads vs. hydraulic load $q_h = 0.02$ m^3/m^{2}d. The amount of the removed load (L_{Rem}) for BOD_5 in the deposit with the addition of bio-preparation changed from 8.20 to 86.20 g O_2/m^{2}d, while in deposit without bio-preparation – from 7.80 to 74.80 g O_2/m^{2}d. Volume of the removed load ($L_{Rem.}$) for COD in deposit with the addition of bio-preparation changed from 32.40 to 155.36 g O_2/m^{2}d and in deposit without bio-preparation, it changed from 29.42 to 154.74 g O_2/m^{2}d. For total suspensions, the amount of charge removed ($L_{Rem.}$) in deposit with the addition of bio-preparation changed from 22.18 to 56.80 g/m^{2}d. However, in deposit without bio-preparation, it changed from 21.74 to 55.80 g/m^{2}d. It was found that more organic substance expressed as BOD_5 and COD as well as total suspended solids is removed in a bio-preparation-supported deposit, and thus a smaller pollutant load is discharged in treated wastewater ($L_{Dis.}$).

Table 1. Values of parameters in raw wastewater from the fruit and vegetable industry delivered to and discharged from the bio-preparation supported constructed wetland deposit and deposit without the addition of bio-preparation for the entire research period.

Parameters	Unit	Raw wastewater*	Wastewater treated with bio-preparation addition*	Wastewater treated without bio-preparation addition*
BOD_5	$mgO_2 \cdot dm^{-3}$	2650 / 500–4400	112.5 / 10–440	332.5 / 95–850
COD	$mgO_2 \cdot dm^{-3}$	4162.7 / 1760–7909	255.1 / 127–539	483.3 / 172–1037
TSS	$mg \cdot dm^{-3}$	1403.6 / 1340–1490	704 / 320–950	887 / 300–840

* numerator: the arithmetic mean; denominator: minimum-maximum, source-own research.

The effectiveness of organic substance removal expressed as BOD_5 in the process of treatment of wastewater from the fruit and vegetable industry in the constructed wetland deposit during the whole research period with hydraulic load $q_h = 0.02$ m^3/m^{2}d was 94.84% with added bio-preparation and 87.87% in the deposit without bio-preparation. Effectiveness of organic substance removal expressed as COD in the process of treatment of wastewater from the fruit and vegetable industry in the constructed wetland with medium deposit load $L_{Del.} = 83.26$ g O_2/m^{2}d and hydraulic load $q_h = 0.02$ m^3/m^{2}d, was 93.05% for the deposit with added bio-preparation and 87.34% in the deposit without the addition of bio-preparation. Based on them, dependencies have been developed, which show that the higher the load of organic matter, the higher load is removed, and thus the higher efficiency of its removal. The removed load decreases slightly when the hydraulic load increases. Due to the lack of previous studies on the treatment of wastewater from the fruit and vegetable industry in bio-preparation supported constructed wetlands, achieved results can be confronted with the results obtained for the treatment of household sewage, municipal wastewater and treatment of leachate in constructed wetlands. Obtained results of research by Dąbrowski [3] in studies upon the treatment of leachate in vertical deposits showed lower efficiency of BOD_5 removal (87.3% to 89.8%) than it was achieved in own research in vertical deposit with the addition of bio-preparation, while higher than in deposit without bio-preparation. Lower efficiencies than those obtained in own research in deposit with and without the addition of bio-preparation, were reached by Liénard and Payrastre [21] for COD with a value of 83% in research conducted in the VF-CW deposit during household wastewater treatment, under load of COD ($L_{Del.}$ = 280 g O_2/m^2d). Results of research carried out by Troesch et al. showed the same efficiency – 93% decrease in COD values in relation to those obtained in the own research in the constructed wetland with the addition of bio-preparation [36]. Higher (98.5%, on average) efficiency in reducing BOD_5 and COD values was obtained in studies conducted at a 6 times higher BOD_5 load ($L_{Del.}$ = 329.5 g O_2/m^2d) and a comparable COD load ($L_{Del.}$ = 79.6 g O_2/m^2d) in municipal wastewater treatment in constructed wetland deposit [20]. In both cases, preservation of high loads with a simultaneous low hydraulic load was possible due to very high concentrations of organic matter, even several times higher than the average level observed in own research. Comparable results were obtained by Kengne et al. and Vincent et al. in the constructed wetland with the addition of bio-preparation during the treatment of household wastewater [23, 28]. Very high (95%) efficiency of organic substance removal expressed as COD was obtained by Koottatep et al. [39] in tests carried out with varying loads of total suspended solids, considerably higher than the load of TSS than in the own research. Kengne et al. reported the same effectiveness [23], carrying out tests with a load of suspended solids $100 \leq L_{Del.} \leq 300$ kg/m^{2}year. High (88%, on average) effectiveness of organic substance removal expressed as COD was demonstrated by Vincent et al. [28] in the research carried out with the deposit load of total suspended solids $L_{Del.}$ = 30 and $L_{Del.}$ = 50 kg/m^{2}year during the treatment of sewage in constructed wetlands. In all cases, the average value of COD in raw sewage delivered into the deposit was significantly higher than that observed in own research (respectively: 16 000, 31 100 and 42 000 mg $O_2 \cdot dm^{-3}$, own study: 4 400 mg $O_2 \cdot dm^{-3}$). It can therefore be assumed that the organic matter load on the deposit was also higher. Observed regularity in the increase of organic substance removal efficiency along with the increase in the load of the constructed wetland was also confirmed by the results of studies carried out by Tanner [24] and Gajewska and Obarska-Pempkowiak [6]. This dependence is explained by the linear increase in the constant reaction rate along with the increase in the organic matter load of the deposit. According to Makowska et al. [30], the constructed wetland is a kind of biological deposit, in which development of a biomass is the more intense, the more loaded with pollution is the bioreactor. According to Sekułowicz et al. [29], high gravel deposit load with organic matter does not reduce their efficiency over time due to colmatation, but it causes its growth. This is explained by development of a biological membrane and plants inhabiting the deposit. In the case of deposits filled with fine-grained soil, overloading with the organic substance worsens the hydraulic conditions, which creates a high probability of colmatation [8, 31]. As a result, after some time, the efficiency of removing impurities deteriorates [28]. The high efficiency of organic substance removal is determined by good aeration of the VF-CW deposits, which is achieved by means of their cyclic flooding [40]. Lower than in own research (both in deposit with the addition of bio-preparation and without it), the average treatment efficiency of dairy leachate delivered to the deposit was achieved by Dąbrowski et al. [32]. It changed for BOD_5 by 81.8% on deposit with subsurface vertical sewage flow and by 72.7% on deposit with subsurface horizontal sewage flow during vegetation, as well as by 64.4% (SS VF) and 45.2% (SS HF), out of vegetation period [32]. Giuseppe et al. obtained lower values than those achieved in own research (COD 81%, BOD_5 79%) of average efficiency in the HF-VF1-VF2 hybrid system (CW) for municipal wastewater [33]. Masi et al. studies have shown higher efficiency of municipal sewage treatment in two-stage CW systems in Moldova for COD 96%, while lower for BOD_5 66% [34] than in own research (in the deposit with and without bio-preparation). Milani et al. in multi-stage systems (CW) for the treatment of wastewater produced by a small winery in southern Italy (Sicily) showed average removal of COD in 79% [35], which is the lower value obtained in the research upon small fruit and vegetable processing plant on vertical deposit (both in deposit with and without the addition of bio-preparation). Results of studies conducted in Poland on household wastewater confirm a few percent reduction in the effectiveness of organic substance removal in the out-of-growing period [6]. Dependence of increasing the effect of hydraulic load increase on the reduction of the load removed in the out-of-growing period observed in the case of COD can be explained by lower intensity of biochemical processes in the out-of-growing period, which along with higher raw sewage flow rate through the constructed wetland and consequently shorter contact time with the biological membrane, contributes to lowering the efficiency of the process. Many authors emphasize that in the case of wastewater treatment with the use of constructed wetland, it is necessary to preliminarily remove the total suspended solids [25]. Brix and Arias [25] reported the efficiency of removing the TSS in 91% to 97% in vertical flow deposits. In turn, Vincent et al. [28] determined the effectiveness of removing total suspended solids based on the research carried out in France in eighty facilities. On average, it was 85%. However, Obarska-Pempkowiak [6] during the treatment of leachates from anaerobic stabilization of sediments obtained a reduction in the concentration of total suspended solids from 452.10 to 56.10 $mg{\cdot}dm^{-3}$ and 368.10 to 59.70 $mg{\cdot}dm^{-3}$, using a sequence of two vertical flow deposits. Lower efficiency of removing the TSS in wastewater from the fruit and vegetable industry in relation to Obarska-Pempkowiak [6], Brix and Arias [25], was determined during the period of basic research. The results of own research showed similar tendency in removing total suspended solids (72.25%) in the constructed wetland with the addition of bio-preparation and 49.80% in the constructed wetland deposit without the bio-preparation with medium load $L_{Del.} = 28.08$ g/m²d and hydraulic load $q_h = 0.02$ m³/m²d. Troesch et al. [36] reported higher value of the efficiency of the total suspension removal, while lower than Brix and Arias [25]. Efficiency in the deposit without the addition of bio-preparation was lower than that obtained by other authors (74.50%). On their basis, a relationship was developed, which shows that the load of total suspended solids increases as the delivered load is increased and decreases as the hydraulic load increases. This dependence is confirmed by the results of own research. Higher efficiency of removing the total suspended solids (99%) was achieved by Kinsley and Crolla [22] in vertical deposit tests. Paing and Voisin [20] reached equally high efficiency in the treatment of municipal wastewater in vertical constructed wetland deposits. The deposit was operated with a load of total suspended solids $L_{Del.} = 46$ kg/m²year and a comparable hydraulic load as that used in own research. Higher efficiency of removing the TSS than that obtained in the tests (90%) was obtained by Kengne et al. [23] conducting research in a vertical deposit. Higher (81%) effectiveness of removing total suspended solids was reported by Koottatep et al. [39] in research carried out upon the household wastewater as compared to own research, but lower than in the off-campaign period for a deposit with bio-preparation. The dehydrated solid fraction accumulated on the surface reduced the hydraulic capacity of deposit, reducing its effectiveness in the area of removing total suspended solids. Lower value (53.2%) than in own research, was recorded by Soroko [7] for the efficiency of removing the TSS during industrial wastewater treatment in horizontal flow (HF) deposits with load $L_{Del.} = 626.00$ g/m²d. Relationship between the amount of load delivered and removed was also noticed by Gajewska and Obarska-Pempkowiak [6]. As the hydraulic load increases, the concentration in the outflow increases, thus the removed load decreases. However, in studies conducted by Vincent et al. [28], with a lower load of the total suspended solids $L_{Del.} = 30$ and $L_{Del.} = 50$ kg/m²year, no statistically significant effect of load with the total suspended solids on the efficiency of their removal was observed. Although the efficiency of removing the TSS increases as the load of the constructed wetland increases, the majority of authors suggest the use of a preliminary mechanical wastewater treatment, which is one of the factors protecting the deposit against colmatation [8, 20].

5 Conclusions

Analysis of the measurements made allows the following statements:

1. Constructed wetlands with vertical flow assisted with bio-preparation purify wastewater from a small fruit and vegetable processing plant with high efficiency. In the constructed wetland with the addition of bio-preparation, there was no reduction in the efficiency of organic substance removal along with the increase in hydraulic load, which should result from shorter sewage contact time with the biological membrane developed in the deposit, and responsible for biochemical purification processes.
2. Enrichment of the biological membrane with nutrients by dosing bio-preparation in the constructed wetland allowed to increase the efficiency of wastewater treatment for organic substances known as BOD_5 by 6.97, COD by 5.71%, and total suspended solids by 22.45 %, in relation to the deposit without the addition of bio-preparation.
3. Based on the obtained results of the treatment efficiency of wastewater from the fruit and vegetable industry in the constructed wetland, the relations between quantities of removed and delivered loads and hydraulic load were formulated, as well as their mutual interactions, which show that for all analyzed pollution indicators (BOD_5, COD, TSS), amount of the removed load increases with the increase of the delivered load in the examined range.

Acknowledgements The research has been carried out in the framework of project No. MB/WBiIŚ/13/2015 in Bialystok Technical University and financed from the funds for science MNiSW in Poland.

References

1. E. Burszta-Adamiak, M. Kęszycka, B. Ryglewska, Landscape Architecture, **2** (2008)
2. A. Jucherski., A. Walczewski, IBMER GCB (2009)
3. W. Dąbrowski, *Purification of leachate from dairy plants in hydrophyte systems* (Monography, 2014)
4. P. F. Cooper, *Fundacão Calouste Gulbenkian* (Lisbona, Portugal, 2003)
5. M. Puchlik, E3S Web of Conf. **44** (2018)
6. H. Obarska-Pempkowiak, M. Gajewska, E. Wojciechowska, (PWN, 2010)
7. M. Soroko, Water - Environment - Rural Areas **11**, 1 (2011)
8. M. Puchlik, *The effect of bio-preparation on the efficiency of wastewater treatment from the fruit and vegetable industry in constructed wetlands* (PhD Thesis, Poland, 2018)
9. C. M. Weedon, 9th International Conference of Wetlands System for Water Pollution Control, France (2004)
10. Z. Dymaczewski, PZIiTS (Poznań, Poland, 2011)

11. J. Ryńska, *Household sewage treatment plants* (Guide, Log InMedia, 2006)
12. M. Wolska, http://ladnydom.pl/budowa/1,113332,13137657 (2014)
13. J. A. Schafsma, C. Baldwin, C. Streb, Ecol. Eng. **14** (2000)
14. K. Kato, T. Koba, H. Ietsugu, T. Saigusa, S. Yanagijya, K. Kitagawa, S. Kobayashi, W CCin Japan (2007)
15. M. G. Haley, M. Rodgers, J. Mulqueen, Biores. Techn. **98** (2007)
16. B. Adrados, C. A Arias, L. M. Pérez, F. Codony, E. Becares, H. Brix, J. Morató, Ecol. Eng. **124** (2018)
17. M. Puchlik, J. Ecol. Eng. **17**, 1 (2016)
18. M. Puchlik, K. Ignatowicz, W. Dąbrowski, J. Ecol. Eng. **16**, 1 (2015)
19. M. Soroko, WERA **11**, 1 (2011)
20. J. Paing, J. Voisin, Wat. Sci. Tech. **51**, 9 (2005)
21. A. Liénard, E. Payrastre, Water Pollution Control (1996)
22. Ch. Kinsley, A. Crolla, AOWMA Annual Conference (2013)
23. I. M. Kengne, P.-H. Dodane, A. Amougou, D. Koné, Des. **248** (2009)
24. C. Tanner, Wat. Sci. Tech. **44**, 11–12 (2001)
25. H. Brix, C. A. Arias, Ecol. Eng. **25**, 5 (2005)
26. A. Hammer, B. P. Pullin, D. K Mc Murry, J. W. Lee, CWWQI (1993)
27. U.S. Env. Prot. Agency, Guide to Septage Treatment and Disposal, EPA Office of Research and Development, Washington (1994)
28. J. Vincent, P. Molle, C. Wisniewski, A. Liénard, IWA Water Pollution Control, Venice (2010)
29. J. Sekułowicz, A. Karczmarczyk, J. Mosiej, WSOW **2**,18 (2006)
30. M. Makowska, M. Spychała, R. Błażejewski, J. Borowski, IWA Kraków, Poland (2005)
31. A. Karczmarczyk, J. Mosiej, Agricult. Univ. Land Recl. **35** (2004)
32. W. Dąbrowski, B. Karolinczak, D. Boruszko, P. Malinowski, IWA Water Pollution Control, Gdańsk, Poland (2016)
33. L. Giuseppe, Cirelli, S. Barbagallo, G. O. Conti, M. Ferrante, A. Marzo, M. Milani, IWA Water Pollution Control, Gdańsk, Poland (2016)
34. F. Masi, R. Bresciani, N. Martinuzzi, G. Cigarini, A. Rizzo, IWA Water Pollution Control, Gdańsk, Poland (2016)
35. M. Milani, S. Barbagallo, G. L. Cirelli, S. Consoli, A. Marzo, IWA Water Pollution Control, Gdańsk, Poland (2016)
36. S. Troesch, A. Liénard, P. Molle, G. Merlin, D. Esser, Wat. Sci. Tech. **60**, 3 (2009)
37. PN-74/C-04578.03
38. PN-EN 872:2007+Ap1:2007
39. T. Koottatep, Ch. Polprasert, N. T. K. Oanh, U. Heinss, A. Montangero, M.Strauss, Elsevier (2001)
40. S. C. Reed, E. J. Middlebrooks, R. W. Crites, 2nd ed. McGraw-Hill, NY (1995)
41. P. S. Rajendra, D. F. Fu, D. N. Fu, H. Juan, Arab. J Sci. Eng. (2014)
42. Y. Ogata, T. Ishigaki, Y. Ebie, N. Sutthasil, Ch. Witthayaphirom, Ch. Chiemchaisri, M. Yamada, Springer Japan KK, part of Springer Nature (2018)

Utilization of waste forest biomass: pelletization studies of torrefied sawmill wood chips

Michał Rejdak[1,*], *Agata* Czardybon[1], *Karina* Ignasiak[1], and *Jolanta* Robak[1]

[1]Institute for Chemical Processing of Coal, ul. Zamkowa 1, 41-803 Zabrze, Poland

Abstract. The paper presents the effects pelletization process of torrefied waste forest biomass – sawmill wastes. The sawmill waste are generated in significant quantities and has significant energy potential. The use of torrefaction and pelletization process enables to produce mechanically durable fuel pellets. Aim of the studies was to evaluate the effect of different kind of additives (binders) to torrefied waste forest biomass on pelletization process characteristic (production capacity, energy consumption) and properties of produced pellets (i.e. density, mechanical durability). The most effective binder among the tested was modified wheat starch. Pellets made with this addition were characterized by the highest mechanical strength.

1 Introduction

The development of biomass conversion processes have been an important issue for the past several decades and seems to be attractive in the future due to its clean, renewable, and carbon–neutral properties. The most important sources of biomass acquisition in Poland are currently: forestry, agriculture, industry and municipal management [1, 2]. Among biomass with potential for energy use, a significant amount of it is generated by the wood industry (c.a. 4.06 million m^3) [3]. It is estimated that about 70% is sawmill waste (c.a. 2.83 million m^3). However, above mentioned waste biomass is characterized some drawbacks in terms of its utilization. Its heterogeneity, low energy density and high moisture content have become major problems in its efficient transport, handling, storage and conversion into energy products.

To overcome the above drawbacks and improve quality of biomass, the thermal pretreatment e.g. pyrolysis might be used [4]. Due to the temperatures used, pyrolisis processes can be divided into: low temperature (< 350°C), medium temperature (350–800°C), and high temperature pyrolysis (> 800°C) [5]. Medium- and high temperature pyrolysis is commonly used to biochar production, which is currently tested in the processes of enriching degraded soils, water cleaning, in construction, in electronics industry, in metallurgy, in medicine, and in cosmetics [6]. Low temperature pyrolysis, also known as torrefaction, is often used to thermal valorization of biomass. It is carried out in

* Corresponding author: mrejdak@ichpw.pl

temperatures between 200–350°C ussually in an inert atmosphere [7–11], sometimes in hot compressed subcritical water (hydrothermal carbonization, HTC) [12–14] or saturated steam presence (steam explosion treatment) [15–17].

As result of aforementioned processes, the chemical structure of biomass undergoes certain transformations, which in turn contribute to the change in the physical and chemical properties of the obtained product. In comparison to raw biomass, the produced fuel is characterized by lower volatile matter content, increased calorific value (higher energy density), improved grindability (reduced costs of grinding) and hydrophobic properties (less risk of environmental degradation – bacterial and mycological), which makes that thermally treated biomass have more favorable usable properties in term of such processes as combustion, co-combustion and gasification applications [18–27]. Despite a number of these advantages, low bulk density of the pyrolysis products, their porous structure and excessive fragility may cause logistical problems (high transport costs, large storage areas), as well as excessive dust emissions, resulting in an increased explosion hazard [21]. However, these problems can be eliminated by the pressure agglomeration, the most commonly used method for compaction of powdery materials including biomass [21, 26–31].

Recently, the issue of densification of thermally treated biomass has been the subject of research by many authors [11, 20, 21, 26, 27, 32–39].

In general, most of them found that strength of pellets produced from thermally treated biomass decreases with increasing final temperature of biomass treatment. In most cases, pellets made from biomass processed at a temperature above 300°C were characterized by negligible mechanical strength. Therefore, to produce mechanically durable pellets from biomass processed at temperatures higher than 300°C it is necessary to use binding agents.

At the Institute for Chemical Processing of Coal (IChPW) in Zabrze for several last years, research on development of autothermal biomass carbonization technology, production of pelleted fuels from torrefied biomass, combustion test, as well as environmental aspects of biomass torrefaction technology have been performed [18, 21, 24–26, 40].

Aim of our recent studies was to evaluate the effect of different kind binders to torrefied waste forest biomass on pelletization process characteristic (production capacity, energy consumption) and properties of produced pellets (i.e. density, mechanical durability).

2 Experimental

2.1 Biomass torrefaction

Sawmill waste wood chips (mixtures of pine, spruce and alder) were torrefied in an Chemical Processing of Coal demonstration installation (capacity of up to 120 kg/h) under autothermal conditions at 350°C [21]. Properties of raw and torrefied biomass were presented in Table 1. Volatile matter content (V) was determined according to PN-EN ISO 18123:2016-01. Ash content (A) was determined according to PN-EN ISO 18122:2016-01. Ultimate analysis (C, H, N) was determined according to PN-EN ISO 16948:2015. Total sulfur content (S) was determined according to PN-EN ISO 16994:2016. Heating value (Q_i) was determined according to PN-EN ISO 18125:2017-07.

As a result of torrefaction process, the material with higher carbon content and calorific value, reduced content of volatile components, and an increased grindability (assessed by modified Hardgrove method – HGI_{eqv}) in comparison with raw wood chips was obtained.

Table 1. Properties of raw and torrefied waste chips (dry basis).

Parameter	Proximate Analysis				Ultimate Analysis				
Fuel	FC^d %	V^d %	A^d %	Q_i^d MJ/kg	C^d %	H^d %	N^d %	S^d %	HGI_{eqv}
Wood chips	14.20	85.15	0.65	19.57	52.17	5.73	0.14	0.02	16
Torrefied wood chips	33.20	64.83	0.97	22.65	62.30	5.02	0.24	0.02	68

2.2 Pelletization tests of torrefied wood chips

The compacting tests were carried out using a pellet mill with a flat die with a nominal capacity of 150 kg/h and a power of 3 kW (Fig. 1). The pellet mill was equipped with 24 and 30 mm die with a diameter hole of 6 mm. Before the compacting process, the torrefied biomass was ground in a hammer mill for grains below 3 mm. Then, the different binders in the amount of 5% by mass were added to the raw material and the mixture was humidified to the appropriate level. The following types of binder were tested: modified wheat starch, waste glycerol, molasses, sodium lignosulphonate and two binders being a mixture of the mentioned above: binder 1 (modified wheat starch 60%, molasses 15%, glycerol 15%, sodium lignosulfonate 10%), and binder 2 (wheat starch 60%, molasses 30%, sodium lignosulfonate 10%). After homogenizing in a planetary mixer the mixture was pelletized. After densification process was completed, pellets produced were dried to moisture content below 5%. Pellet density (apparent), mechanical strength according to PN-EN 15210-1:2010 and its bulk density were determined. The production capacity (yield) and energy consumption in the pelleting process was also assessed. Prior to the actual pelleting tests, preliminary tests were carried out to select the right amount of moisture in the feed to the process. The general view of the pellets produced is illustrated on Fig. 2.

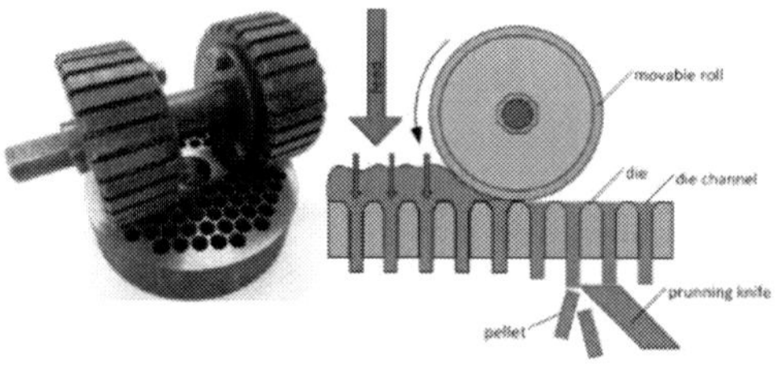

Fig. 1. The principle of flat die pellet mill operation.

Fig. 2. General view of pellets produced from torrefied waste forest biomass.

3 Results and Discussion

3.1 Preliminary tests

Preliminary tests of torrefied biomass pelletizing were carried out for 5% wheat starch addition. The aim was to evaluate proper level of moisture content for further tests. As can be seen from the data presented in Fig. 4 and Fig. 5 an increase in moisture content in the material directed to the process results in an increase in the production capacity of the pelletizer, and thus a decrease in the unit energy consumption, with a decrease in mechanical strength and density of the pellets produced. Pellet density for 30%, 35%, and 40% moisture content was respectively: 740, 720, 680 kg/m^3.

The value of moisture content at the level of 30% is the limit value at which a very high level of energy consumption was observed. At the moisture content of 25%, the die of the pellet mill was blocked. For further tests, the moisture content in the sample directed to the compacting process was set at 35%.

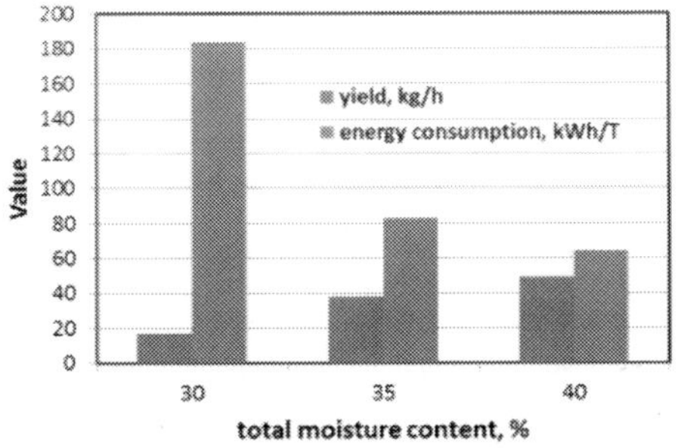

Fig. 4. The effect of moisture content on the efficiency and energy consumption during pellet production (24 mm die).

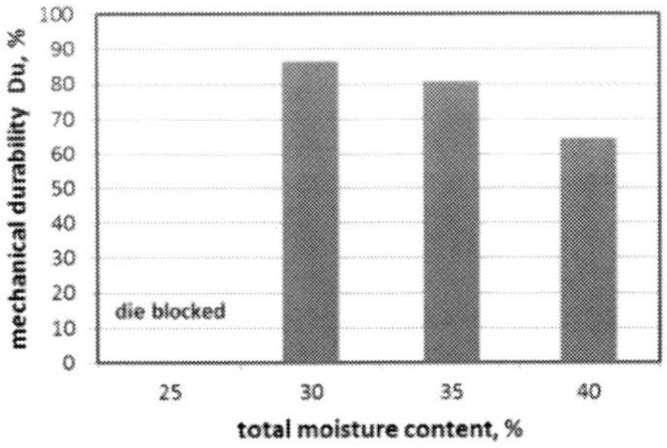

Fig. 5. The effect of moisture on pellets mechanical strength (24 mm die).

3.2 Pellet production capacity

As is apparent from the data presented in Fig.6 the type of binder used affects the pellet production efficiency. It is caused by the influence of the binder addition on the frictional resistance of the material passing through the die holes.

For 24 mm die, the lowest efficiency (30 kg/h) for waste glycerol was obtained. Slightly higher efficiency was obtained for the addition of sodium lignosulphonate and binder 2, which amounted 34 and 35 kg/h, respectively. The highest value was obtained with the addition of molasses – the pellet production efficiency was reached 40 kg/h. Slightly lower yields were noted for wheat starch and binder 1 addition.

Different results were obtained for pelleting tests using a 30 mm thick die (Fig. 6–9). In general, thicker die – longer die channels – generates higher resistance during material transfer through channels, which in turn causes a decrease in efficiency (yield), and an increase in energy consumption. On the other hand, it positively affects the mechanical durability and density of pellets.

In case of production capacity (yield), the best results for addition of binder 2 were obtained (31 kg/h). Slightly lower efficiency values: about 25 kg/h were obtained using wheat starch and binder 1. With the addition of sodium lignosulphonate, yield of 22 kg/h was obtained. The lowest yield was obtained (12 kg/h) which was surprising, due to the beneficial results obtained for the 24 mm matrix. In case of waste glycerol addition, the die of pellet mill was blocked after about 20 seconds from the start of work.

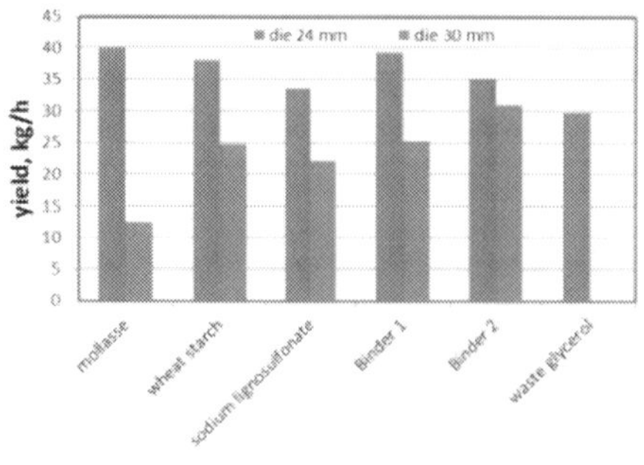

Fig. 6. The influence of the type of binder on the pellet production efficiency.

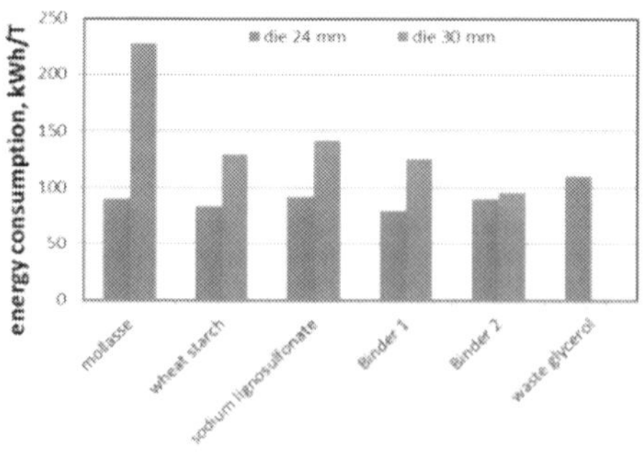

Fig. 7. The influence of the type of binder on energy consumption during pellet production.

3.3 Energy consumption

The energy consumption during the pelleting process of torrefied biomass due to the resistance of the material moving through the die holes is closely dependent on the binder used to pellet production. The influence of the type of binder used in pellet production on energy for two different dies: 24 mm and 30 mm is presented in Fig. 7.

In the case of 24 mm die, the highest energy consumption in pellet process production was noted for the waste glycerol addition, which was 110 kWh/T, and the lowest for binder 1 – 79 kWh/T. A similar result (83 kWh/T) was observed for pelleting process with wheat starch addition. Slightly higher energy consumption was achieved for: binder 2, molasses and lignosulphonate, which amounted to 90, 90, 92 kWh/T, respectively.

If a thicker 30 mm die was used, a significant increase in energy consumption was observed. The highest energy consumption level was recorded for pellet production with molasses addition – 227 kWh/T, and the lowest for binder 2 – 96 kWh/T. In the case of the use of starch, binder 1 and lignosulphonate, the energy consumption was 130, 125, 142 kWh/T, respectively.

In general achieved results are consistent with the results obtained for production capacity evaluation. For mixtures that generated higher resistance during pelletizing process, higher energy consumption were noted and thus, lower yield were obtained.

3.4 Pellet apparent density

Resistance accompanying the pelleting process also affect the density of the product obtained. In the case of a die with a thickness of 24 mm, the density of the pellets produced was in the range of 675–768 kg/m^3, the highest value was obtained for waste glycerol addition and the lowest for use of lignosulphonate. Densities above 700 kg/m^3 were noted for the addition of molasses, wheat starch and binder 2 and amount respectively: 747, 720, 707 kg/m^3.

The use of a 30 mm thick die resulted in a slight increase in the density of the pellets produced. All of the pellets produced were characterized by a density above value of 700 kg/m^3. The highest density was obtained for the wheat starch additive – 781 kg/m^3, and the lowest for molasses – 716 kg/m^3. The influence of the type of binder on pellets density for different dies is presented on Fig. 8.

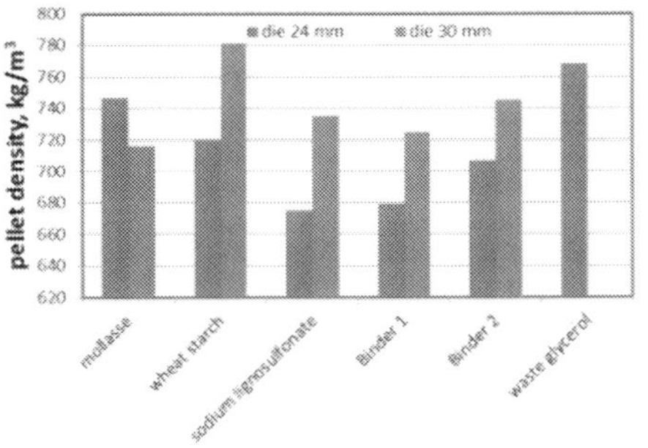

Fig. 8. The influence of the type of binder on the pellets density.

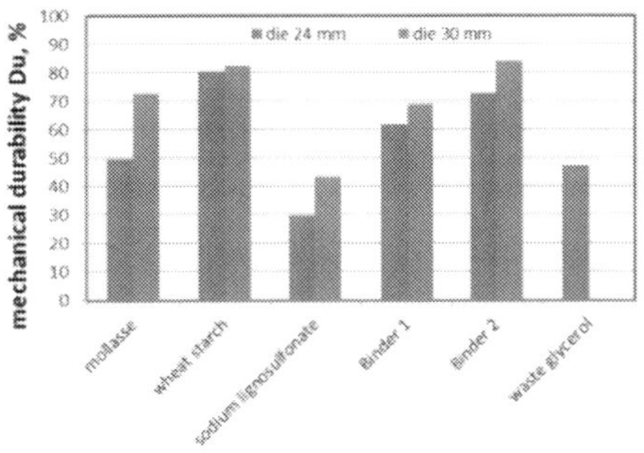

Fig. 9. The influence of the type of binder on the pellets strength.

3.5 Mechanical durability of pellets

Both: type of the binding additive and the die size has an impact on obtained mechanical durability of pellets produced. The influence of the type of binder on pellets mechanical strength is presented on Fig. 9.

The highest mechanical durability was obtained with the addition of wheat starch – 80.5% and binder 2 – 72.7%. The strength of pellets produced with the use of other additives was at a much lower level. For the lignosulphonate additive the strength was 29.5%. Increasing the thickness of the applied die had a positive effect on the mechanical strength of the pellets produced. The highest strength was obtained for the binder 2 (83.9%) and starch (82.5%).

The type of binder and the thickness of the die affect the amount of the obtained undersize (below 3.15mm) (Fig. 10.). Too much amount of it reduces the efficiency of pellet production due to the necessity of turning the undersize back to re-pelleting.

The best results were obtained in the case of wheat starch and binder 2, for which in both cases (die thickness 24 and 30 mm) the undersize below 3.15 mm content did not exceed 5%. For the 24 mm die, the worst results were recorded for waste glycerol – 38.5%. The use of a thicker die resulted in a significant decrease in the undersize content in case of molasses additive and a slight decrease for lignosulphonate. For the other additives (starch, binder 1 and 2) the content of the undersize has practically remained unchanged.

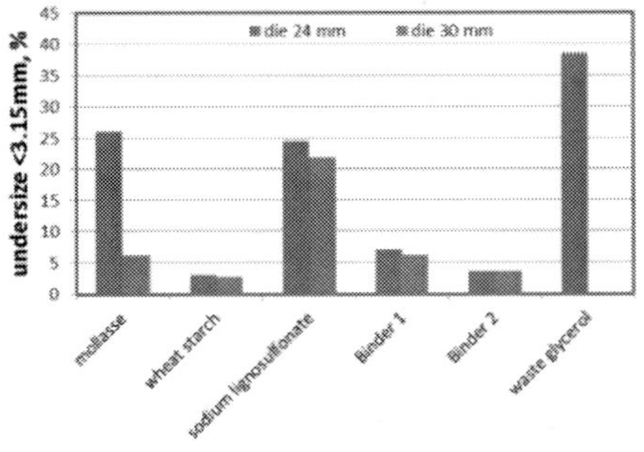

Fig. 10. The influence of the type of binder on the amount of undersize.

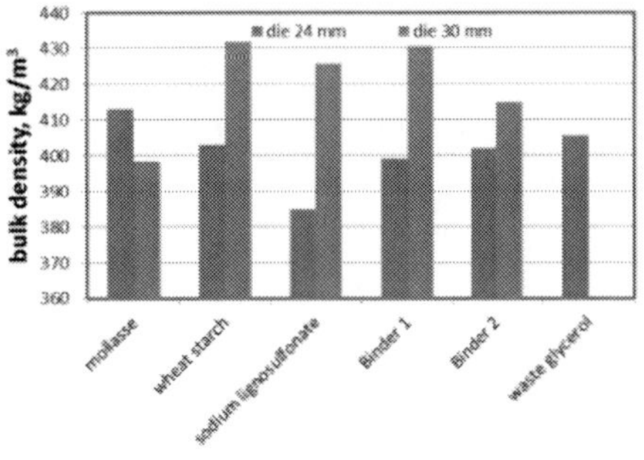

Fig. 11. The influence of the type of binder on the pellets bulk density.

3.6 Bulk density of pellets

The influence of the type of binder on bulk density of pellets is presented on Fig. 11. In the case of a 24 mm die for all samples tested, with the exception of pellets with the addition of lignosulphonate, a bulk density of approx. 400 kg/m^3 was obtained. The highest density was obtained for molasses – 413 kg/m^3 and the lowest for lignosulphonate – 385 kg/m^3. The use of a thicker 30 mm die positively influenced the achieved bulk density, with the exception of pellets with the addition of molasses, whose lower bulk density results from a slightly lower apparent density of pellets.

4 Conclusions

On the basis of the conducted studies, it was found that it is possible for tested material – torrefied waste forest chips to produce pellets with a mechanical strength of approx. 80% using a pelleting machine with a flat die with a power of 3 kW. The moisture content is the key parameter affecting the effectiveness of the pelleting process. The higher the moisture content in the material, the higher process efficiency and lower energy consumption were observed. An increase in the moisture content, however, causes a decrease in the density of the compacted material, decrease in mechanical strength and require more energy for drying of produced pellets. Lowering the moisture content below a certain limit level results in a drastic increase in resistance to material compacting and blocking of the die. For the investigated case, the optimum moisture level was 35%. But it should be taken into consideration that it might depend on the origin and properties of the material, as well as the type of pelleting equipment.

The most effective binder among the tested was modified wheat starch. Pellets made with this addition were characterized by the highest mechanical strength. Another additive that gave satisfactory results was a binder 2, which consisted of: wheat starch 60%, molasses 30%, and sodium lignosulfonate 10%. The pellets obtained with it had slightly lower mechanical strength than those made with starch. The advantage of binder 2 is approx. 18–20% lower than the price of modified wheat starch.

Work carried out as part of the project titled „Biocoal for Power Generation" (BIOPOGEN), financed by the European Institute of Innovation and Technology (EIT).

References

1. E. Pietras, Inżynieria i Ochrona Środowiska **20**, 1, 59–70 (2017)
2. S. Parzych, Leśne Prace Badawcze **76**, 3, 256–264 (2015)
3. P. Gołos, A. Kaliszewski, Instytut Badawczy Leśnictwa, 59–76 (2013)
4. M. J. C. Stelt, H. Gerhauser, J. H. A. Kiel, K. J. Ptasinski, Biomass Bioenergy **35**, 9, 3748–3762 (2011)
5. J. Xiao, F. Li, Q. Zhong, J. Huang, B. Wang, Y. Zhang, Journal of Analytical and Applied Pyrolysis **117**, 64–77 (2016)
6. H. P. Schmidt, Ithaka Journal **1**, 286–289 (2012)
7. J. Peng, W. H. Chen, X. T. Bi, Renewable and Sustainable Energy Reviews **44**, 847–866 (2015)

8. M. J. Prins, PhD Thesis,Technische Universiteit Eindhoven ,The Netherlands (2005)
9. J. H. A. Kiel, *BUS final meeting* (Wageningen, The Netherlands, 2007)
10. T. G. Bridgeman, J. M. Jones, I. Shield., S. P. T. William, Fuel **87**, 6, 844–856 (2008)
11. M. J. Prins, K. J. Ptasinski, J. J. Janssen, Journal of Analytical and Applied Pyrolysis **77**, 1, 28–34 (2006)
12. M. T. Reza, M. Uddin, J. G. Lynam, C. J. Coronella, Biomass and Bioenergy **63**, 229–238 (2014)
13. S. K. Hoekman, A. Broch, C. Robbins, Energy Fuel **25**, 1802–1810 (2011)
14. H. S. Kambo, A. Dutta, Applied Energy **135**, 182–191 (2014)
15. P. S. Lam, S. Sokhansanj, X. T. Bi, C. J. Lim, S. H. Larsson, Bioresource Technology **116**, 396–402 (2012)
16. P. S. Lam, P. Y. Lam, S Sokhansanj, X. T. Bi, C. J. Lim, Biomass Bioenergy **56**, 116–126 (2013)
17. Z. Tooyserkani, S. Sokhansanj, X. Bi, C. J. Lim, J. Saddler, A. Lau, Transactions of the ASABE **55**, 6, 1–8 (2012)
18. M. Kopczyński, J. Zuwała, Energy Policy Journal **16**, 4, 271–284 (2013)
19. M. J. Prins, K. J. Ptasinski, F. J. Janssen, Energy **31**, 3458–3470 (2006)
20. W. Stelte, A. R. Sanadi, L. Shang, J. K. Holm, J. Ahrenfeldt, U. B. Henriksen, BioResources **7**, 3, 4451–4490 (2012)
21. A. Czardybon, K. Ignasiak, M. Rejdak, A. Sobolewski, Przemysł Chemiczny **96**, 11, 2265–2271 (2017)
22. P. C. A. Bergman, *Combined torrefaction and pelletisation: the TOP process* (ECN publication, 2005)
23. H. Pawlak-Kruczek, M. Czerep, Nowa Energia **1**, 31–38 (2016)
24. M. Kopczyński, J. Zuwała, Chemik **6**, 540, 540–545 (2013)
25. J. Lasek, M. Kopczyński, M. Janusz, A. Iluk, J. Zuwała, Energy **119**, 362 (2017)
26. M. Rejdak, A. Czardybon, K. Ignasiak, A. Sobolewski, J. Robak, Journal of Ecological Engineering **18**, 1, 183–192 (2017)
27. W. Stelte, C. Clemons, J. K. Holm, A. R. Sanadi, J. Ahrenfeld, L. Shang, U. B. Henriksen, Biomass and Bioenergy **35**, 4690–4698 (2011)
28. A. Bazagran, S. L Rough, G. Mckay, Biomass and Bioenergy **70**, 489–497 (2014)
29. M. Rejdak, R. Winkler, K. Supernok, K. Ignasiak, Inżynieria Ekologiczna **49**, 100–106 (2016)
30. S. Obidziński, M. Joka, O. Fijoł, International Agrophysics **31**, 515–523 (2017)
31. G. Borowski, Environment Protection Engineering **33**, 2, 79–86 (2007)
32. M. Phanphanich, MSc Thesis,University of Georgia, USA (2010)
33. J. H. Peng, X. T Bi, S. Sokhansanj, C. J. Lim, Energy Fuels **26**, 3826–3939 (2012)
34. J. Peng, H. Bi, C. Lim, S. Sokhansanj, Energy Fuels **27**, 967–974 (2013)
35. H. Li, X. Liu, R. Legros, T. Xiaotao, C. Bi, C. J. Lim, S. Sokhansanj, Applied Energy **93**, 680–685 (2012)

36. W. Stelte, N. P. K Nielsen, H. O. Hansen, J. Dahl, L. Shang, A. R. Sanadi, Biomass and Bioenergy **53**, 105–112 (2013)
37. H. Li, X. Liu, R. Legros, T. Xiaotao, C. Bi, C. J. Lim, S. Sokhansanj, Applied Energy **93**, 680–685 (2012)
38. L. Shang, N. P. Nielsen, J. Dahl, W. Stelte, J. Ahrenfeldt, J. K. Holm, T Thomsen, U. B. Henriksen, Fuel Processing Technology **101**, 23–28 (2012)
39. P. Nanou, W. Huijgen, M. Carbo, J. Kiel, Biomass and Bioenergy **111**, 248–262 (2018)
40. M. Bałazińska, Polish Journal of Environmental Studies **26**, 6, 2471–2477 (2017)

URANS and ACM for determining the aerodynamic performance of vertical-axis wind turbine

Krzysztof Rogowski[1,*], and *Martin O. L.* Hansen[2]

[1]Institute of Aeronautics and Applied Mechanics, Warsaw University of Technology, 00-665 Warsaw, Poland
[2]DTU Wind Energy, Technical University of Denmark, Lyngby, 2800 Kgs., Denmark

Abstract. This paper presents numerical results of a one-bladed vertical axis wind turbine using the k-ε RNG turbulence model. The rotor geometry was developed based on the research report available on the literature. The obtained results of aerodynamic blade loads and wake profiles downstream behind the rotor are in a good agreement with the experiment. This paper presents also the ability of calculating fluid flow parameters around the operating wind turbine by using the original approach – the actuator cell model (ACM).

1 Introduction

The depletion of global reserves of fossil fuels has resulted in an increase in ecological awareness of the society and an increase in interest in alternative energy sources. The use of wind energy, sometimes referred to as "indirect solar energy", is a promising technology capable of delivering a portion of energy demand in many regions of the world. In the future, wind turbines should be better adapted to the conditions of developing countries because wind energy is the cheapest source of clean energy offering carbon-free generation at the level of costs currently comparable to gas. Wind turbines very effectively provide centralized electricity because they are concentrated in large clusters (wind farms) thanks to which energy is supplied via local distribution grid to local energy companies. European countries such as Germany, Denmark or Spain are the main wind energy beneficiaries. In 2017 the percentage of electricity produced by wind power was already above 40% (44%) in DK and this number is increasing with new wind farms planned. The UK, which has the largest wind resources in Europe, aspires to bring down the price of wind energy. According to the global wind report published by the Global Wind Energy Council (GWEC), in 2017 about 54 GW of power was installed wind turbines. The report forecasts that this amount may increase to even 800 GW in 2021 [1].

The Darrieus vertical-axis wind turbine uses lift force to drive. Thanks to its high aerodynamic efficiency, this turbine is competitive compared to a classic propeller-type

* Corresponding author: krogowski@meil.pw.edu.pl

horizontal-axis wind turbine. It has been proven that the Darrieus wind turbine is one of the most effective wind energy conversion systems [2, 3].

Numerical models for wind turbines with a vertical axis of rotation have been developed for a long time. Strickland was one of the first who began to develop the vortex model for Darrieus-type rotors [2]. This class of aerodynamic models basing on vorticity equations much better predicted overall Darrieus wind turbine performance than commonly known one-dimension momentum based aerodynamic models [3].

Numerical methods of fluid dynamics have become an important tool in the study of aerodynamic properties of wind turbines [4–7]. However, these methods are limited due to the size of the rotor. They also have limitations due to overestimation of the critical angle of attack. CFD methods require a large computer infrastructure to carry out efficient research on a small three-dimensional rotor. It involves not only high computer costs but also software licences and electricity expenses. Therefore, most studies using CFD published until the end of 2018 only concern the modelling of the aerodynamic performance of rotors of VAWT in 2D. Another possibility would be to use system identification methods, however this approach is not possible due to high cost of the experiments [8, 9].

Currently there is a big focus on the offshore wind energy sector and his includes Darrieus wind turbines. One example is the DeepWind project, which involves the use of Darrieus floating rotors to generate electricity [10, 11]. This entails the need to create new analytical tools for the aerodynamic analysis of VAWTs.

The authors of this article intend to validate the classic CFD approach with available experimental results and then validate the original method using the numerical results and experimental results. This article aims to compare the velocity fields obtained using two numerical approaches: URANS with the ReNormalisation Group (*RNG*) *k-ε* model and the actuator cell model developed by Rogowski [12, 13].

2 Numerical model

The numerical model of a one-bladed rotor was developed based on the experiment of Strickland et al [2]. The experiment was performed in a water towing tank. The rotating single-bladed rotor was moved by means of a special chain mechanism.

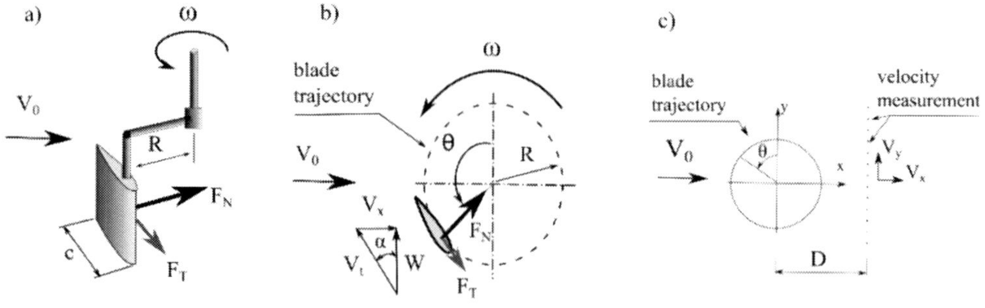

Fig. 1. One-bladed vertical-axis wind turbine (a); velocity vectors and aerodynamic blade loads (b); method of velocity measurement (c).

The rotor configuration consists of a blade constructed using the NACA 0012 profile with the chord length, *c* of 9.14 cm and the pitch angle of 0 deg. The intention of Strickland et al. [2] was to conduct a two-dimensional experiment, therefore the rotor blades were very long in comparison to the chord length. The rotor diameter $D = 2R$ was 1.22 m. During numerical simulations water was used as an operating fluid. The angular velocity of the rotor ω was 0.75 rad/s. The water velocity at the inlet was chosen to be 9.1 cm/s. This value corresponds to the tip speed ratio of 5.0. In the experiment the aerodynamic blade loads were measured using strain gage bridges. Moreover, during the experiment, wake velocity profiles were measured downstream behind the rotor. The rotor's silhouette is shown in Fig. 1a. Wake velocity profiles were also computed by CFD using checkpoints behind the wind turbine (Fig. 1c).

3 Method

In this work, the unsteady Reynolds-averaged Navier-Stokes (URANS) approach was used to calculate the flow parameters and aerodynamic loads on the blades. In addition, this paper presents the operation of the actuator cell model (ACM) developed by Rogowski [13] for determining flow parameters around the Darrieus-type rotor. Both the URANS and ACM approaches were implemented using the ANSYS Fluent package.

In the URANS approach, the k-ε RNG turbulence model was used for the calculations. It is two-equation turbulence model that adds two transport equations for the turbulence kinetic energy k and its dissipation ε. Figure 2a presents the concept of this approach. Inside the square computational domain a smaller domain with a circular shape rotates. Data between these domains is exchanged thanks to the interface.

The ACM model differs from the URANS approach in that the rotor blade is not modelled. An additional third very small computing domain is introduced. This additional third domain rotates during simulation around the rotor axis of rotation at the angular velocity equal to the rotor angular velocity. The aerodynamic blade loads are introduced into this third computational domain as momentum sources.

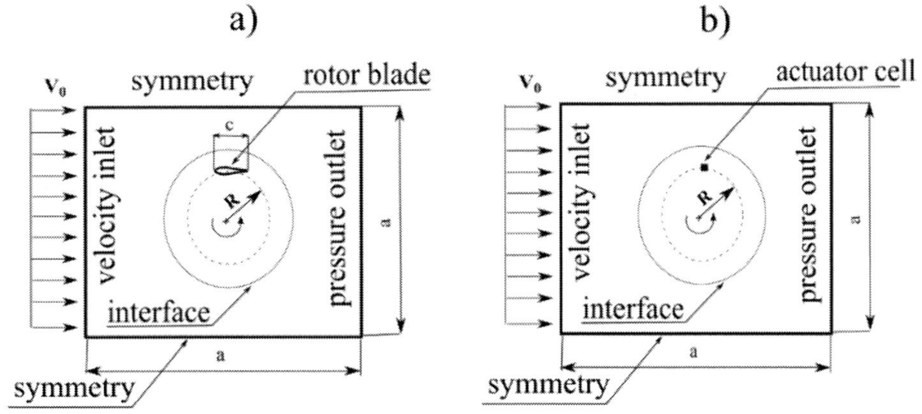

Fig. 2. RANS approach (a); Actuator cell model (b).

Both grids consist of unstructured quadrilateral elements in the whole domain. In the URANS approach the structured mech near the airfoil blade edge is generated. This structured grid provides a better determination of the flow parameters in the boundary layer. However, it gives more elements, which increases the cost of calculations. For all simulations presented here, the wall y+ was lower than 0.8. The total number of mesh cells for ACM approach is 43,707 elements whereas for URANS approach the number of elements is 131,958. Transient calculations require the determination of the time step size. In these calculations, the step size corresponds to the increment of azimuthal angle $\Delta\theta = 0.1$ deg. Figure 3a presents the mesh distribution for URANS model whereas the mesh for the ACM is given in Fig. 3b. Both grids were examined in terms of element density. More details on these grids can be found in [12, 13].

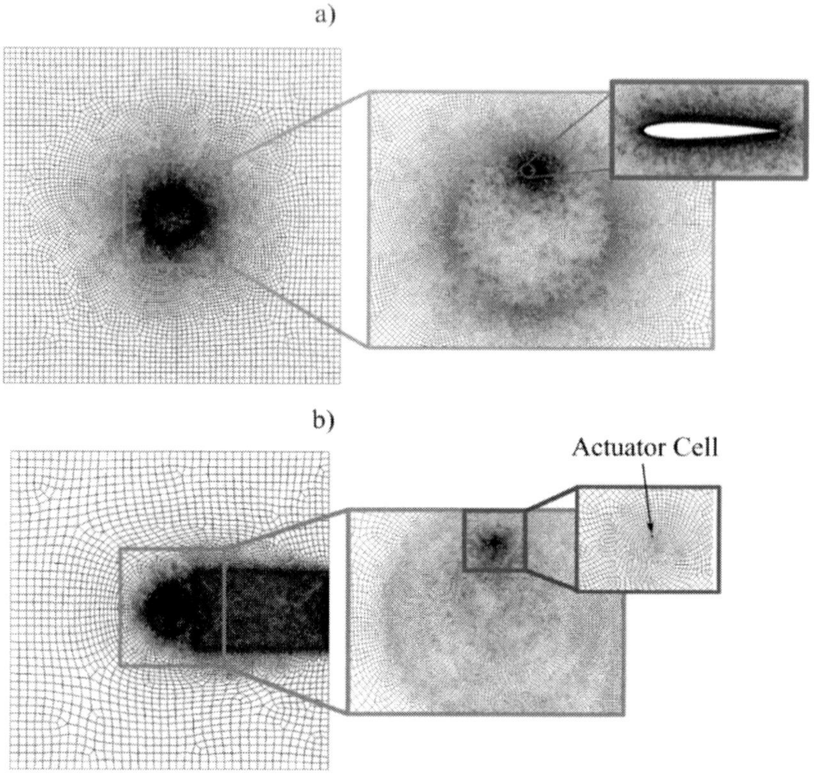

Fig. 3. Mesh for URANS simulations (a); Mech for ACM approach (b).

4 Results and discussion

Numerical results are given as aerodynamic blade loads and aerodynamic flow parameters. The non-dimensional tangential and normal force coefficients are expressed as:

$$CFN, CFT = \frac{F_{N,T}}{0.5\rho c V_0^2} \tag{1}$$

where: c is the chord length; V_0 is the velocity of the flow at the inlet; ρ is the water density and F_N and F_T are aerodynamic blade load components (Figs 1a and 1b). Figure 4 presents computed and measured normal and tangential blade load components. Almost in the entire azimuth range, the numerical results agree with the experimental results. The largest discrepancies are in the case of the tangential force in the azimuth range from 100 to 180 degrees. They result from the local crossing of critical angles of attack.

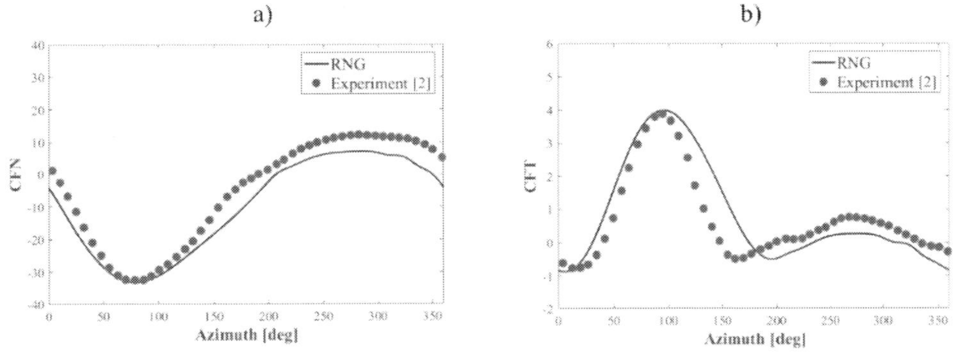

Fig. 4. Non-dimensional aerodynamic blade load coefficients as a function of azimuthal position: the normal component (a) and the tangential component (b).

The flow field at a distance of one rotor diameter behind the rotor was quantified by means of instantaneous velocity profiles for several azimuth positions. The numerical results were compared with the experimental results [2]. Velocity components are normalized by the velocity at the inlet V_0 ($u = V_x/V_0$). In the case of velocity profiles, two numerical approaches were used: the URANS model with the k-ε turbulence model and the ACM approach. In the ACM method, the momentum sources were calculated based on experimental loads. As can be seen in the Fig. 5, the differences in velocity profiles are very small for all three sets of data.

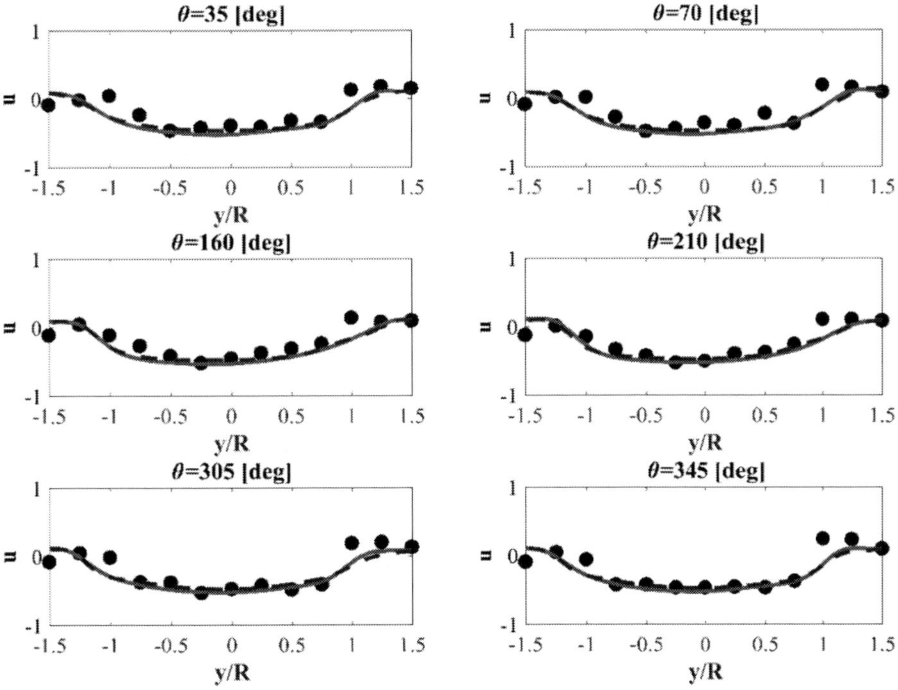

Fig. 5. Velocity profiles downstream behind the rotor: experiment [2] – black circles, k-ε RNG – solid red line and ACM approach – blue dotted line.

Both numerical approaches were also validated by comparing streacklines. Streaklines represent the trajectory of solid particles or particles of other fluid injected into the fluid. In the case of Strickland's experimental studies [2], a dye was injected into the water, which was then photographed. The injector was located at one point under the water surface at the trailing edge of the blade. In the case of numerical methods, massless particles were used. These particles come out of the trailing edge of the blade profile or in the case of the ACM method from the third the smallest domain (from the active mesh cell). The visualizations given in Figure 6 show that the trajectories of particles are strongly curved. Despite large dissipation of particles within one-two diameters downstream behind the rotor, the motion images of particles obtained by two numerical approaches are agree with each other and with the experiment.

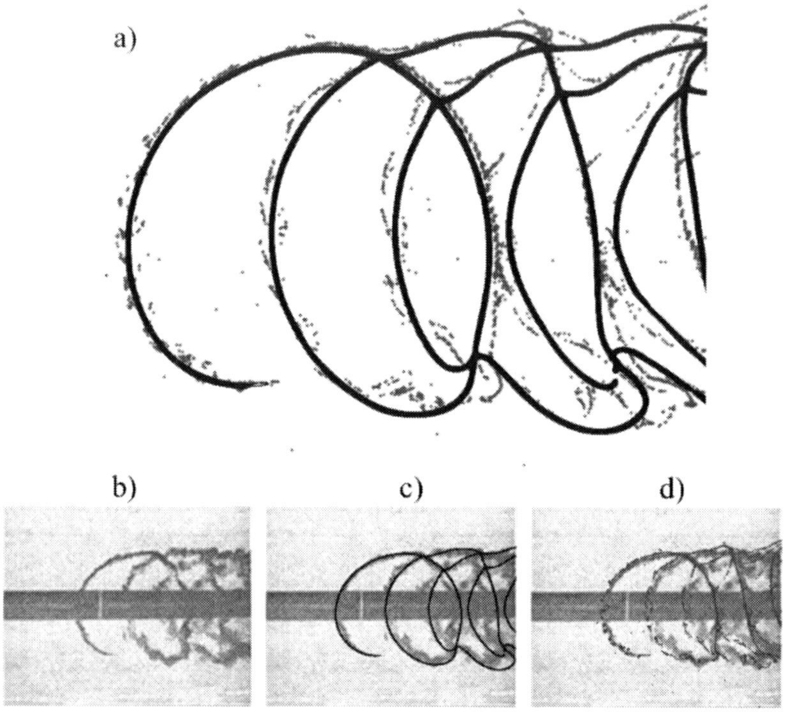

Fig. 6. Streaklines: the comparison between ACM (red colour) and URANS (blue colour) (a); experimental results [2] (b); the comparison between the experimental results [2] and URANS (blue colour) (c); the comparison between the experimental results [2] and ACM (red colour) (d).

Figure 7 shows contour maps of the component flow velocity V_x for two numerical approaches. The velocity component V_x is parallel to the undisturbed flow direction. The widths of the aerodynamic wakes on both numerical models are similar. The ACM model was coupled with the laminar model therefore some differences in velocity distributions are visible in Fig. 7.

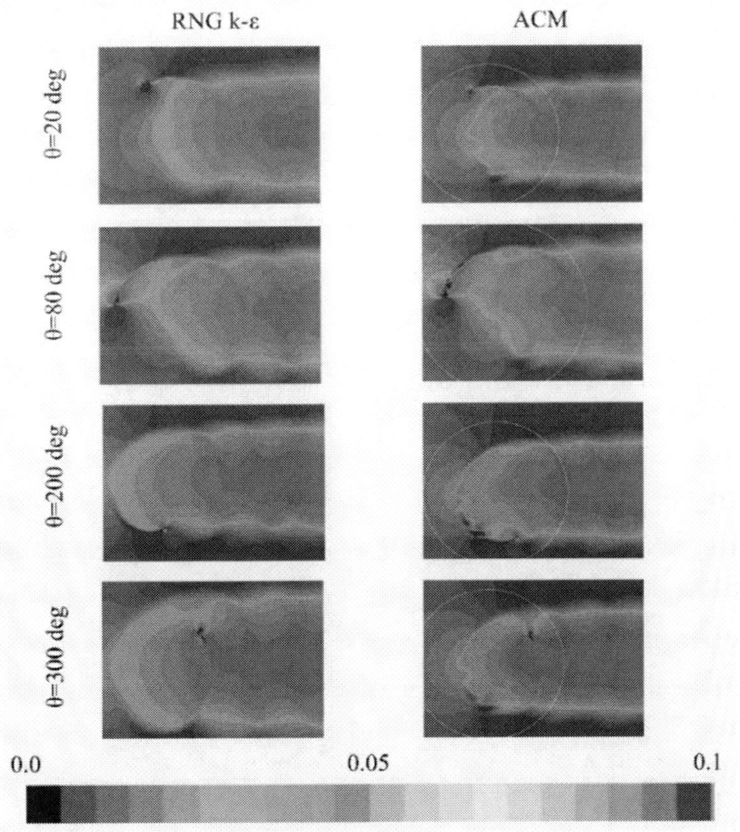

Fig. 7. Contour maps of velocity component V_x [m/s]. Comparison of results for *k-ε* RNG and ACM.

As assured in the Introduction, Darrieus's rotors are very effective devices. The efficiency of these devices can be measured using the so-called rotor power coefficient, C_P. It is commonly defined as the ratio of the energy produced by the turbine rotor divided by the total energy available in the wind at that velocity. Following the Paraschivoiu [3], the rotor power coefficient can be defined as the product of the averaged rotor torque coefficient, $\overline{C_m}$, and the tip speed ratio, *TSR*.

$$C_P = \overline{C_m} \cdot TSR \tag{2}$$

The tip speed ratio is defined as the ratio between the tangential velocity of the blade and the flow velocity. For a TSR value of 5, the tangential velocity of the blade is 5 times greater than the undisturbed flow rate. The rotor power coefficient computed in the present study using the RNG k-ε turbulence model is equal to 0.237. The result presented by Strickland et al. [2] by using their vortex model code is 0.248. The result obtained in these simulations is greater by 4.6% compared to the results of Strickland at al. [2].

5 Conclusions

The purpose of the paper was the analysis of the flow past the one-bladed vertical-axis wind turbine using two numerical approaches: URANS with the k-ε RNG turbulence model and the home-made actuator cell model.

The k-ε RNG model gives satisfactory results of aerodynamic forces for the single-bladed rotor of the Darrieus wind turbine. The velocity profiles obtained using the URANS and ACM approaches are quite good compared to the experimental results. The ACM method presented by the authors is a development method. In this work it was proved that the flow field calculated with it gives results in accordance with the classical URANS approach and with the experiment. The high value of the rotor power coefficient for the one-bladed rotor indicates the high energy potential of the Darrieus wind turbine. The high rotational velocity makes it possible to use this type of wind turbine to produce electricity and transmit it to the grid.

The presented numerical computations were performed in the Interdisciplinary Centre for Mathematical and Computational Modelling of the Warsaw University, Grant No. GB73-5.

The author, Krzysztof Rogowski would like to thank the National Science Center for the possibility of a scientific trip to Denmark where the method and results described in this article were consulted.

References

1. E. A. D. Kumara, N. K. Hettiarachchi, K. G. R. M. Jayathilake, Int. J. Sci. Res. Inno. Techn. **4**, 8, 56–67 (2017)
2. J. H. Strickland, B. T. Webster, T. Nguyen, *Vortex Model of the Darrieus Turbine: An Analytical and Experimental Study* (Tehnical raport SAN79-7058, 1979)
3. I. Paraschivoiu, *Wind Turbine Design: With Emphasis on Darrieus Concept* (Presses inter Polytechnique, 2002)
4. G. Bangga, G. Hutomo, R. Wiranegara, H. Sasongko, J. of Mech. Sc. and Techn. **31**, 1, 261–267 (2017)
5. K. Rogowski, M. O. L. Hansen, P. Lichota, J. of Appl. Fluid Mech. **11**, 4 (2018)
6. R. Rezaeiha, I. Kalkman, B. Blocken, Renewable Energy **107**, 373–385 (2017)
7. J. H. Wiśniewski, *Decreasing Wear of Large Vertical Axis Wind Turbines by Employing a Multi-Level Turbine* (Concept Renewable and Alternative Energy: Concepts, Methodologies, Tools, and Applications, 2016)
8. P. Lichota, D. A. Noreña, *A Priori Model Inclusion in the Multisine Maneuver Design* (17th International Carpathian Control Conference, IEEE, Tatrzańska Łomnica, Słowacja, 2016)
9. P. Lichota, K. Sibilski, P. Ohme, J. of Aircraft **54**, 2 (2017)
10. U. S. Paulsen, H. A. Madsen, K. A. Kragh, P. H. Nielsen, I. Baran, J. Hattel, E. Ritchie, K. Leban, H. G. Svendsen, P. A. Berthelsen, Energy Procedia, **53** (2014)
11. D. Verelst, H. A. Madsen, M. Borg, U. S. Paulsen, H. G. Svendsen, P. A. Berthelsen, Energy Procedia, **80** (2015)
12. K. Rogowski, M. O. L. Hansen, R. Maroński, P. Lichota, J. Phys. Conf. Ser. **753**, 022050 (2016)
13. K. Rogowski, J. Phys. Conf. Ser. **1101**, 012028 (2018)

Influence of the variability of the odour emission rate on its impact range: a case study of the selected industrial source

Elżbieta Romanik[1,*], *Yaroslav* Bezyk[1], *Marcin* Pawnuk[1], *Urszula* Miller[1], and *Agnieszka* Grzelka[1]

[1]Wroclaw University of Science and Technology, Faculty of Environmental Engineering, Wybrzeże Wyspiańskiego 27, 50-370 Wrocław, Poland

Abstract. Odour concentration measurements in a chosen industrial source were made in this study using the method of dynamic olfactometry. The two different scenarios considered the variation of the odour emission rate as input for the dispersion model were compared for the period 2017 (before installation of the equipment for gas treatment) and 2018 (after implementation of purifying technologies). In this paper the odour impact range was determined by applying model calculations conducted in the Polish reference dispersion model – OPERAT FB software for the grid size 2 x 2 km. The conducted research shows a significant improvement in the odour impact range of chosen industrial source in year 2018 compared to 2017.

1 Introduction

The odour nuisance is related to the emission of odours into the atmosphere. The three main groups of emission sources of odours can be distinguished. They include agriculture and animal husbandry, municipal waste management and industry [1]. Analysis of problems and citizens' complaints in Poland, related to the odour impact assessment, identified: 16% of the complaints lodged against industrial activities, 34% – about agriculture and breeding, 39% – as result of municipal management and 11% related to other sources [2]. Industrial sources of odour nuisance are characterized by the greatest variety. In this group, a wide range of industrial facilities such as agriculture and food processing industry or chemical industry can be distinguished. Due to the large diversity of products and raw materials, such as dry products as plastics, pharmaceuticals, paints, varnishes, rubber products, fertilizers, acids, products, and semi-finished products from crude oil, the odour nuisance associated with the activity of the chemical industry belongs to the most diverse [1].

In general, each of the industrial sources of odours differs both in terms of emitted odorant types and their amount. The amount of odours emission from production operations depends on a number of factors and remedies used to prevent odour nuisance. Industrial facilities are often located in the immediate vicinity of residential areas; therefore, it is necessary to use

* Corresponding author: elzbieta.romanik@pwr.edu.pl

appropriate techniques to prevent odour emissions into the atmosphere. Among the techniques allowing for the elimination of odorous pollution, the following techniques can be distinguished: combustion, adsorption, absorption, biological methods or deodorization [1]. The use of an appropriate deodorization system allows for effective reduction of the odour emissions of the facility, as well as to improve the quality of life of the local residents.

2 Methodology

2.1 Purpose and methodology of measurements

The odour emission measurements are carried out in accordance with the methodology described in VDI 3880:2011 [3] and PN-EN 13725:2007 [4] guidelines. Sampling campaigns are mainly performed during rainless weather that will guarantee the efficiency of measurements and repeatability of results. A vacuum sampler (marked CSD30 sampler) and PET-bags, made of chemically inert material, are generally used during samplings. All elements of collecting set are made of odourless materials that do not absorb odorants. According to the recommendations, the bags are previously conditioned. The measurements of weather data including temperature, humidity, and pressure in the sampling day are carried out using a TESTO 435-2 probe system. During the sampling time, the study facilities must be worked without any deflections.

After collection, without delay, the samples of odours are transported to the Olfactory Laboratory. Then, the odour samples are evaluated and quantified to determine odour concentration. Odour concentration measurements are made using the dynamic olfactometry method in accordance with the PN-EN 13725:2007 standards: 'Air quality - determination of odour concentration by dynamic olfactometry' [4]. The odour measuring equipment includes a dilution device – the four-station olfactometer TO8 with the necessary attachments. The obtained data is calculated on the basis of sensitivity tests determined from the geometric mean of all individual measurements. The unit of the odour concentration is reported as the European odour unit per cubic meter (ou_E/m^3).

2.2 Study site

The industrial facilities investigated in this study are located on the outskirts of a compact urban agglomeration in Poland. In the immediate vicinity, on the northern side of the Plant, an undeveloped area and further the railway line is located. On the eastern side, the Plant is adjacent to another industrial facility. Approximately 1 kilometer to the south of the Plant, there are residential area and services, industrial development and inbuilt areas. There are housing developments at a distance of about 30 m from the borders on the west side of the Plant. At the distance of the emission impact range up to 850 m (equal to fifty times of the emitter height) [5] there are no protected objects: national parks, health resorts, monuments/historical treasures or other areas being the subject of the protection in accordance to Polish Nature Conservation Low [6]. Waste gases from plant production activity are discharged to the main emitter that height is 17 m; the release cross-section of the chimney is particularly 2.5 x 1.2 m. After the system modification (after measurements from 2017), the exhaust air from the production part is centrally purified using a multi-stage adsorb vessel system.

2.3 Odour dispersion methods and models

The mechanisms of dispersion of odorants in the atmosphere are the same as the air based transport of other pollutants. Mathematical models of odour dispersion in the air are used to evaluate the environmental impact of odour-causing substances for various emission release scenarios. The accuracy of the odour dispersion model prediction depends on the multiple input parameters: odour emission data (exit velocity, plume rise, temperature, etc.), sources and surface characteristics (surface roughness, local topography, nearby buildings), meteorological conditions and pollution transformation in the atmosphere (wind speed, stability, mixing height, wind direction) [7]. Based on the dispersion model, it is possible to calculate the odour emission distribution from individual sources as well as area sources. The odour concentrations at receptor points in the study area can be evaluated by using available dispersion models.

Gaussian-type models are the most common dispersion models used in the atmospheric simulation of pollution transport [8]. In general, the most widely used models for simulation of odorous compounds are Gaussian plume (e.g. AERMOD) and Gaussian puffs (e.g. CALPUFF) distribution software. Each one has its own advantages and disadvantages associated with the steady-state approximations and the vertical particle movement due to gravity during the travel time [9]. The Operat FB model developed based on Gaussian plume, is mainly used in Poland for predicting odour concentration and calculating odour emission rates [5]. The model is based on knowledge and the assumption of the atmospheric transport of pollution, especially relating to the constant and uniform emission rates; constant parameters of the wind and speed direction; processes of vertical and crosswind diffusion; and the terrain domain.

3 Results and discussion

3.1 Emission calculations

The odour emission from the installation was calculated as mean odour concentration and exhaust gas flow rate on the main/cetral emitter (CE). The gas flow was specified for the standard conditions as required by PN 13725:2007 [4].

$$E = \frac{C \cdot V}{3600} \quad [ou_E/s] \tag{1}$$

where:

E – odour emission from emitter, $[ou_E/s]$;

C – odour concentration at the outlet, $[ou_E/m^3]$;

V – gas flow rate in the emitter, $[m^3/s]$.

On the basis of the olfactometric measurements, the average odour concentrations (expressed in ou_E/m^3 units) in collected gas samples were determined. The comparison of the obtained results for the years 2017–2018 is represented in Table 1.

The results of odour concentration measurements from 2018 show the high decreasing in odour level by 11896 ou_E/m^3 as compared with the samplings applied in 2017. The reduction of the odour concentration at the outlet of the emitter is caused by the introduction of modifications in the deodorization installation of waste gases from the Plant.

Table 1. Comparison of the odour concentrations measured in 2017 and 2018.

Emitter name	Date	Sample name	Odour concentration, [ou_E/m^3]	Average concentration, [ou_E/m^3]
Central emitter (CE)	2017	CE /6	12784	12127
		CE /8	11324	
		CE /11	12274	
	2018	CE/1	196	231
		CE/2	299	
		CE/3	196	

3.2 Input model data

Calculations of the odour dispersion in the Plant surrounding areas were carried out in the reference of the Regulation of the Minister of Environment dated 26 January 2010 on reference values for some substances in the air (Journal of Laws 2010, no. 16, item 87) [5]. Modelling criteria used in the study included the basic values of emission source parameters (emitter height and diameter, odour emission rate), as well as the current dispersion conditions (meteorological data and topographical features).

Modelling of odours dispersion was performed based on the following scenarios:

- the analysis performed in normal mode at the maximum system capacity before installing the gas deodorization system (on the year 2017).
- the analysis performed in normal mode at the maximum system capacity after installing the gas deodorization system (on the year 2018).

The comparison of the parameters of emitters, results of odour emission calculations and meteorological data obtained for 2017 and 2018 is represented in Table 2. The odours emissions from the central emitter (CE) were 68620.74 ou_E/s and 4899.56 ouE/s in 2017 and 2018, respectively. The average odour concentration in 2017 was 12127.33 ou_E/m^3, while in 2018 did not exceed 230.33 ou_E/m^3.

Table 2. Comparison of emitter parameters and results of odour emission calculations in 2017 and 2018.

Name of Emitter	Exhaust velocity [m/s]	Gas flow [m^3/s]	Air temperature [oC]	Relative humidity [%]	Atmospheric pressure [hPa]	Odor concentrations [ou_E/m^3]	Odour emission rate [ou_E/s]	Volume flow rate of odours under standard conditions ISO, Vr, 20 [ou_E/s]	Odour emission under standard conditions [ou_E/s]
CE (2017)	3.01	5.89	24.27	100.00	987.21	12127.33	71439.48	5.66	68620.74
CE (2018)	7.12	21.35	21.21	76.32	1013.31	230.33	4917.62	21.27	4899.56

Calculations of the odour impact range were carried out in grid format receptors with selected size and step of each grid cells and coefficient of aerodynamic roughness. The grid size was determined to be equal to 2000 m x 2100 m with a grid step at 50 m. The coefficient of aerodynamic roughness identified in the area within 50 times radius of the emitter height [5] was 0.353 m.

3.3 Results of the odour dispersion modelling

The OPERAT FB – a Polish reference dispersion model (according to Regulation of the Minister of the Environment of 26 January 2010 on reference values for certain substances in the air [5]) was applied in the analysis of the odour impact range. The environmental detection threshold for odours as a mixture for residential areas under the Polish conditions [10] is established within the range of values 1 ou_E/m^3. The maximum frequency of exceedances of the odour concentration at a level of 1 ou_e/m^3 is determined at 3% during the year [10]. The results of the obtained maximum odour concentration in the receptor grid are shown in Table 3. The maximum annual odour concentration in the testing area were 1.3863 and 0.0296 ou_E/m^3 for the year 2017 and 2018, respectively. The maximum frequency of the exceedance of the odour concentration of 1 ou_E/m^3 was 19.45% in 2017. In the year 2018 there were no exceedances of odour detection threshold in the calculation grid for industrial Plant.

Table 3. Comparison of the maximum odour concentration in the receptor network in 2017 and 2018.

Name of pollutant	Frequency of exceedances of 1 ou_E/m^3 per year [%]			Maximum annual concentrations in the receptor network [ou_E/m^3]			
	Calculated		Limit	Calculated		Limit	
	2017	2018		2017	2018	2017	2018
Odours	19.45	0.00	< 3	1.3863	0.0296	> 1	< 1

The results of the modelling of maximum odour concentration values outside the Plant border at additional points marked as residential buildings are compared in Table 4. The highest annual odour concentration occurred at the point B4; the provided value was up to 37.821 ou_E/m^3 for 2017 and 0.358 ou_E/m^3 for the year 2018.

Table 4. Comparison of odour concentrations calculated for additional points in the receptor network in 2017 and 2018.

No.	Name of point	2017 Max. conc. ou/m³	2017 Avg. conc. ou/m³	2017 Freq. of exceed., 1 ou/m³ [%]	2018 Max. conc. ou/m³	2018 Avg. conc. ou/m³	2018 Freq. of exceed., 1 ou/m³ [%]
1	B1	31.852	0.2371	5.61	0.303	0.0061	0.00
2	B2	27.050	0.1230	2.84	0.282	0.0033	0.00
3	B3	28.747	0.1198	2.74	0.287	0.0032	0.00
4	B4	37.821	0.2917	5.84	0.358	0.0068	0.00
5	B5	25.132	0.3229	6.34	0.282	0.0075	0.00
6	B6	21.257	0.1883	3.87	0.275	0.0048	0.00
7	B7	15.390	0.0997	1.83	0.233	0.0028	0.00
8	B8	21.035	0.1688	3.43	0.274	0.0048	0.00

The frequency of the exceeding of a given one-hour odour concentration threshold in the year 2017 and 2018 is shown in figure 1. Comparing the left and the right panel in fig. 1, the large differences in the impact distances of odour emission before and after modification purifying system can be seen. The exceedance level of 1 ou_E/m^3 and 3% is much more pronounced at calculation grid for 2017, in distance towards the north exceeds 400 m and approximately 500 m to the east of the Plant boundaries. In 2018, the frequency of exceedances of one-hour odour concentration has not occurred. Modelling of odour dispersion based on measurements of their emissions in 2018 shows the significant improvements in the frequency of exceeding compared to 2017 (when it was 19.45%).

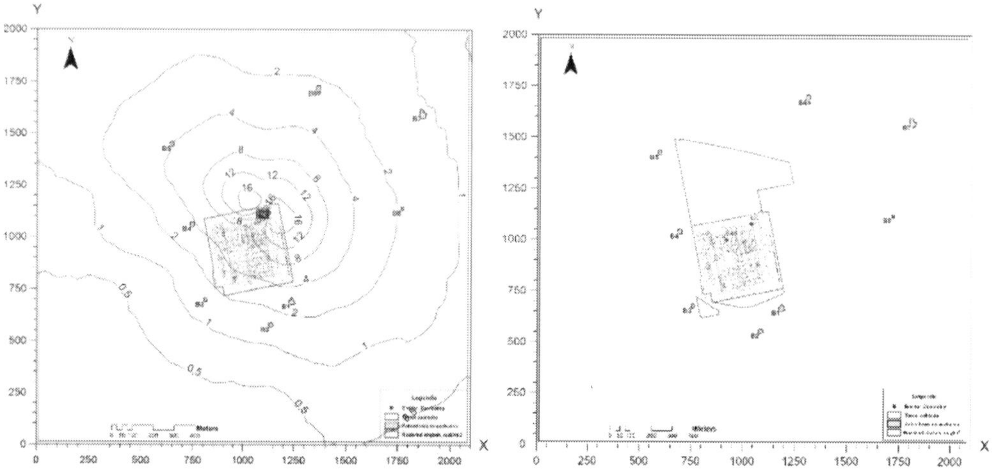

Fig. 1. The frequency of exceeding of the one-hour odour concentrations (exceedance threshold of 3%) in 2017 (left side) vs 2018 (right side).

The distribution of average odour concentrations in 2017 compared to 2018 is shown as isolines at the study grid in figure 2. The exceedance of the given odour threshold of 1 ou_E/m^3 is determined at the distance of approximately 50 m towards the east from the Plant boundaries and 80 m towards the north.

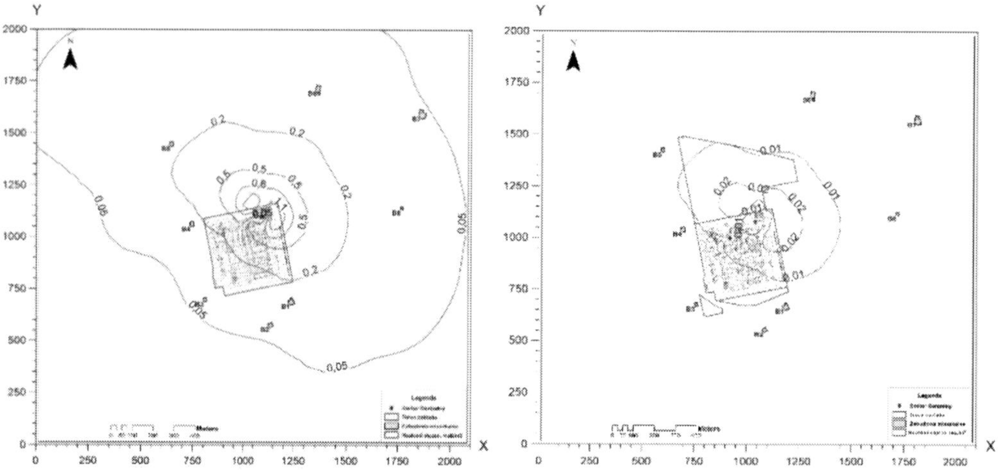

Fig. 2. The average odour concentrations, ou_E/m^3 (exceedance threshold of 1 ou_E/m^3) in 2017 (left side) vs 2018 (right side).

In figure 3, the distribution of maximum odour concentrations in 2017 is compared to 2018. Exceedances of 1 ou_E/m^3 limit have occurred in all directions of the calculation grid (over 1000 m of plant boundaries). The distribution of maximum odour concentration in 2018, has not exceeded the limit of 1 ou_E/m^3 at any point of the calculation grid. Maximum of annual average odour concentration calculated on the basis of measurements from 2018 amounted to 0.0296 ou_E/m^3 and emissions determined in 2017 was equal to 1.3863 ou_E/m^3. In addition, the distincly different odour impact range in a study period 2017–2018 is caused by a significant difference in the values of odour emmisions from the central emitter. The comparison of the odour impact range obtained for the year 2018 shows a significant reduction in the intensity of emission from the central emitter, in relation to the values in 2017.

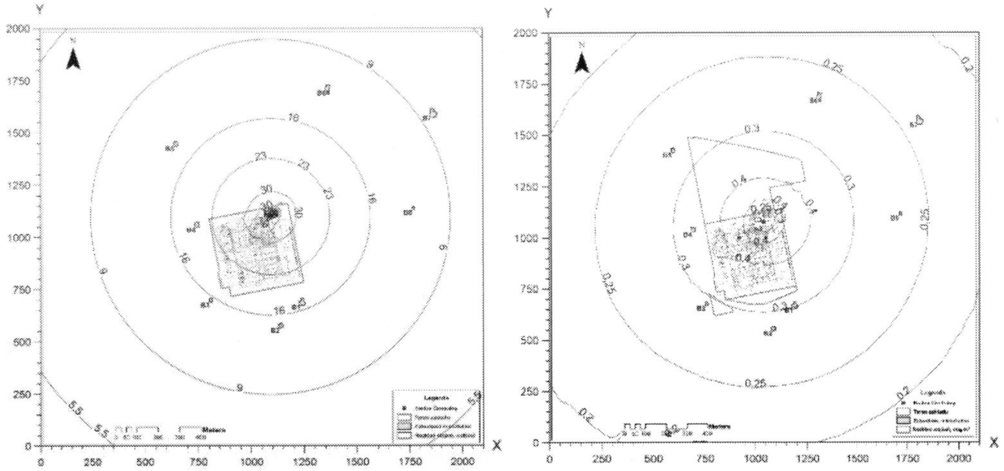

Fig. 3. The maximum odour concentrations, ou_E/m^3 (exceedance threshold of 1 ou_E/m^3) in 2017 (left side) vs 2018 (right side).

In case of the odour dispersion modelling the predicted concentrations in the environment were used for designating areas and populations affected by the odour nuisance. At the selected points within the study grid, where the residential buildings were located, the significant improvement in the air quality was observed in 2018 compared to 2017. The highest value of average annual maximum concentrations of odours was determined in the point B4, at the level of 37.821 ou_E/m^3 and 0.358 ou_E/m^3 in 2017 and 2018, respectively. According to the average concentrations of odours, the highest value was evaluated in the point B5. The average odour levels were found to be 0.3229 ou_E/m^3 in 2017, while in 2018 – 0.0075 ou_E/m^3. According to the emission ranges, results show that the development of solutions for the improvement of the existing production and deodorization conditions may prove helpful in limiting the odour dispersion.

4 Summary and conclusions

Most complaints related to chemical odours are associated with industrial facilities that should continue to improve its odour control performance. The emission of odours into the atmosphere from the industrial plants, especially in urban areas, can cause problems related to odour nuisance, and thus even negative health effects. In order to asses the odour impact before and after an industrial plant upgrade, the odour concentration measurements by dynamic olfactometry method was used in this study. The prediction of odour concentrations determined by the Polish reference dispersion model pointed to the direct impact of the modernization of production lines and the deodorization system in the analysed Plant. The model calculations of the odour impact range with respect to the emission values show that the distance of odour dispersion decreases with the emission decrease. The research results indicate that odour evaluation and dispersion modelling are valid tools in the evaluation of odour control improvements implemented by the operators of industrial facilities.

The study was co-financed within the framework of the project No. 0402/0095/18 with the specific subsidy granted for the Faculty of Environmental Engineering Wroclaw University of Science and Technology (W7) by the Minister of Science and Higher Education to conduct research and development work and related tasks contributing to the development of young scientists and doctoral students in 2018/2019.

References

1. M. Szklarczyk, J. Zwoździak, I. Sówka, *Industrial sources of odour emissions*, In: M. I. Szynkowska, J. Zwoździak (Eds.), *Współczesna problematyka odorów* (Wyd. Naukowo-Techniczne, Warszawa, 2010)
2. I. Sówka *Methods of identyfication of odour gases emitted from industrial pants* (Oficyna Wydawnicza Politechniki Wrocławskiej, Wrocław, 2011)
3. VDI 3883: *Olfactometry. Static sampling* (Verein Deutscher Ingenieure, 2011)
4. PN-EN 13725:2007: *Air quality. Determination of odour concentration, with the use of dynamic olfactometry method*
5. Regulation of the Minister of Environment dated 26 January 2010 on reference values for some substances in the air. (Journal of Laws 2010, no. 16, item 87).
6. The Act of Nature Convesration, of 16 April 2004 (Journal of Laws 2004 no. 92, item 880)
7. I. Sówka, Y. Bezyk, A. Grzelka, U. Miller, Ł. Pachurka, Wat. Sci. & Tech. **2017**, 2 (2017)
8. N. S. Holmes, L. Morawska, Atm. Env. **40**, 30 (2006)
9. A. S Rood, Atm. Env. **89** (2014)
10. Draft law on preventing odour nuisance from 27.02.2009 (Poland)

Analysis of flow resistance in bundles of power plant condensers

Artur Rusowicz[1], *Jakub* Kajurek[1,*], and *Kuat* Baubekov[2]

[1]Warsaw University of Technology, Faculty of Power and Aeronautical Engineering Institute of Heat Engineering, ul. Nowowiejska 21/25, 00-665 Warsaw, Poland
[2]S.Seifullin Kazakh AgroTechnical University, Faculty of Power Engineering, Department of Heat Power Engineering, Zhenis avenue 62, 010011 Astana, Republic of Kazakhstan,

Abstract. Shell-side pressure drop is a very important variable in the successful design of the condensers. The prediction of this pressure drop through the horizontal tube banks, with condensation, has long been a problem facing design engineers. Low pressure drop is a requirement in designing condensers for power plants. The paper presents a comparison of the various correlation to determine the pressure drop in the tube bundle. It is an important element for the verification of numerical simulations. Analysis of flow resistance for power plant condenser were made.

1 Introduction

A condenser is an important component that affects the efficiency and performance of power plants. Development of advanced numerical methods for shell-side flows in condensers is a critical step in improving current condenser design techniques. The advantage of numerical simulation is that they can provide a more detailed information on fluid flow and heat transfer in the tube bundle. This information may eliminate, in early stages of the design process, the problems related with flow induced vibration and flow distribution and improve overall heat transfer coefficients and increase the unit performance.

Numerical modelling of the power plant condensers play an important role in the design and the diagnosis of this type of heat exchangers [1–4]. The development of mathematical and physical models should be verified in the way of experiments [5, 6]. Due to the significant size of these heat exchangers and an important role in the energy system access to this data is limited.

The shell-side and water-side flows are treated as the steady state and the steam-side flow is assumed to behave as an ideal mixture made up of noncondensable gases and steam only. The steam is taken as saturated. The mixture of noncondensable gases and steam assumed to be perfect gas, although other equations of state could be considered. The two-dimensional steady-state porous medium conservation equations of mass, momentum, heat and mass transfer resistance are written in Cartesian coordinate system. Local volume porosity β is defined as ratio of the fluid volume and total volume of the corresponding element.
Mass conservation equation for the mixture:

* Corresponding author: jakub.kajurek@itc.pw.edu.pl

$$\frac{\partial(\beta\rho u)}{\partial x}+\frac{\partial(\beta\rho v)}{\partial y}=-\beta m \quad (1)$$

Momentum conservation equations for the mixture:

$$\frac{\partial}{\partial x}(\beta\rho uu)+\frac{\partial}{\partial y}(\beta\rho vu)=\frac{\partial}{\partial x}\left(\beta\mu_{eff}\frac{\partial u}{\partial x}\right)+\frac{\partial}{\partial y}\left(\beta\mu_{eff}\frac{\partial u}{\partial y}\right)-\beta\frac{\partial p}{\partial x}-\beta\dot{m}u-\beta F_u \quad (2)$$

$$\frac{\partial}{\partial x}(\beta\rho uv)+\frac{\partial}{\partial y}(\beta\rho vv)=\frac{\partial}{\partial x}\left(\beta\mu_{eff}\frac{\partial v}{\partial x}\right)+\frac{\partial}{\partial y}\left(\beta\mu_{eff}\frac{\partial v}{\partial y}\right)-\beta\frac{\partial p}{\partial y}-\beta\dot{m}u-\beta F_v \quad (3)$$

where x is the longitudinal coordinate to the flow direction, y is the perpendicular coordinate to flow direction, u is the velocity in the x direction, v is the velocity in the y direction, ρ is the density, p is the pressure, $\dot{m}$ is the mass flow rate, β is the local porosity, μ_{eff} is the effective viscosity, F is the local flow resistance.

The boundary conditions for the inlet, vent, walls and baffles and plane of symmetry are:

- *Inlet:* The velocity and pressure are specified at inlet boundary.
- *Walls*: The shell walls and baffles of condenser are assumed to be nonslip, impervious to flow and adiabatic. Thus, the normal velocity components are equal to zero.
- *Vent*: At the outlet of the condenser, the boundary condition is velocity of steam-air mixture, which is calculated on the basis of the characteristic of venting apparatus.
- *Plane of symmetry*: Along the centre line the derivatives with respect to the cross-steam direction of all field variables are set to zero.

The mathematical model of the steam flow 1–3 and the heat transfer in the power plant condenser is solved. As a result of this solution the values of mixture velocity and pressure are obtained. The whole procedure is repeated until the satisfactory accuracy is achieved. For solving system eqs. (1)–(3) the Stream line Upwind Petrov-Galerkin finite element method is applied. The calculations are performed in mesh using of 852 points and 1542 triangles in the main and cross flow direction. The geometric and operating parameters for steam condenser are given in tab. 1.

In the description of the heat exchange in the energy condensers there are three driving forces: the temperature difference, the difference of gas concentrations and a decrease in a vapor pressure associated with a hydraulic flow resistance. Article focuses on the description of the hydraulic flow resistance in a bundle of energy condensers. It compares the several available correlations that can be used to verify the numeric calculations. The flow resistances are one of the components affecting the total pressure drop in tube bundle. The second important element is the establishment of conducting compressible fluid flow. The value of the effective viscosity is assumed in the calculations μ_{eff}. The choice of this value also requires optimization as the literature gives different guidelines [7]. The article, however, focuses on the analysis of the flow resistance associated with the flow resistance forces.

Table 1. Geometric and operating parameters for a 50 MWe condenser.

Geometrical parameters	
Number of Tube Bundles	7642
Number of Tubes Per I Bundle	3812
Number of Tubes Per II Bundle	3830
Condenser Length [m]	6.58
Tube Outer Diameter [mm]	24
Tube Inner Diameter [mm]	22
Tube Pitch [mm]	32
Tube material	Brass
Operating parameters	
Inlet Temperature of Cooling Water [°C]	24
Inlet Velocity of Cooling Water [m/s]	1.57
Total Steam Condensation Rate [t/h]	140

2 Mathematical model

The literature uses different coefficients describing the flow resistance in a bundle of tubes. The most common are used depending on where the pressure drop is proportional to the kinetic energy of flowing fluid. The dependences on local flow resistance in a bundle of pipes are widely used [8]:

$$F_u = \xi \rho u U \tag{4}$$

$$F_v = \xi \rho v U \tag{5}$$

where U is the resultant speed of the steam. The pressure loss factor can be determined based on the relationship proposed by Rhodesa and Carlucci [9]:

$$\xi = 2\left(\frac{f}{P}\right)\left(\frac{P\beta}{P-d_o}\right)^2\left(\frac{1-\beta}{1-\beta_t}\right) \tag{6}$$

where P is the tube pitch, d_o is the tube outer diameter and f is the coefficient of resistance expressed as:

$$f = 0.619Re^{-0.198} \quad for\ Re < 8000 \tag{7}$$

$$f = 1.156Re^{-0.2647} \quad for\ 8000 \leq Re < 200000 \tag{8}$$

The local porosity β in Eq. (5) is defined as follows:

$$\beta = 1 - \frac{N}{N_t}(1-\beta_t) \tag{9}$$

where N is the number of analysed tubes in a row, N_t is the number of pipes in a row, β_t is the porosity within the tube bundle.

An alternative approach to describe the flow resistance in a bundle of tubes is to use the same as for porous deposits, Darcy's law:

$$F_u = \mu R_u u \tag{10}$$

$$F_v = \mu R_v v \tag{11}$$

where the flow resistance R_u and R_v are specified with empirical dependencies [10]:

$$R_u = d_o^{-2} G f \theta_u \tag{12}$$

$$R_v = d_o^{-2} G f \theta_v \tag{13}$$

G in above equations is the coefficient, which depends on the geometry of the pipe bundle:

$$G = -1.017 + \frac{0.3325}{\beta} + \frac{0.3574}{\beta^2} + \frac{0.01348}{\beta^3} \quad for\ Re > 20 \tag{14}$$

The resistance factor f is given as:

$$f = 350 Re^{0.0446} \quad for\ Re < 20 \tag{15}$$

$$f = 103 Re^{0.338} \quad for\ 20 < Re < 300 \tag{16}$$

$$f = 6.64 Re^{0.880} \quad for\ Re > 300 \tag{17}$$

The local porosity in Eq. (13) for hexagonal pitch in bundle is specified as:

$$\beta_h = 1 - \frac{\pi}{2\sqrt{3}} \left(\frac{d_o}{P}\right)^2 \tag{18}$$

while for square pitch tubes as:

$$\beta_k = 1 - \frac{\pi}{4} \left(\frac{d_o}{P}\right)^2 \tag{19}$$

In equations (11) and (12) θ_u and θ_v respectively are the correction factors that determine the condensation of steam i.e. effects of biphasic mixture on the flow resistance, which are the features of flow direction and the Reynolds number.

Based on experimental work in the Institute of Heat Engineering at Warsaw University of Technology flow resistance coefficients were proposed [11]:

• for a hexagonal bundle

$$f = 10.69 Re^{0.821} \quad for\ 300 < Re < 8000 \tag{20}$$

• for a square bundle

$$f = 4 Re^{0.951} \quad for\ 300 < Re < 8000 \tag{21}$$

• and from examination [10] for both types of bundle

$$f = 7.87 Re^{0.88} \quad for\ 300 < Re < 8000 \tag{22}$$

The unit pressure drop was determined as:

$$\frac{\Delta p}{\Delta x} = \frac{\mu}{d_o^2} f U \tag{23}$$

All above correlations are being used in two-dimensional calculations of energy condensers with isotropy assumption of oppositions of the flow in the bunch.

3 Analysis of the results

The calculations were carried out for the power plant condenser of the block 50 MW built from brass pipes about the outer diameter of 24 mm, wall thickness of 1 mm and for hexagonal scale of 32 mm. The following parameters were taken: steam pressure at the inlet to the condenser of 7778 Pa, temperature of the cooling water of 25°C, steam speed at the inlet to the condenser of 37.4 m/s and stream of steam of 140 t/h [6]. The cross section of the analysed condenser is described in Fig. 1a with emphasized A-A (hexagonal bunch of the pipes) and B-B (square bunch of the pipes) sections at which drops of the pressure were being analysed. The diameters were selected in a way allowing steam flow in vertical and horizontal direction in the bunch of pipes.

The pressure drops in the sections A-A and B-B were determined based on the numerical simulations with using correlation (9)–(12). Two different effective viscosities were used, which were equalled respectively to 100 and 1000 of the molecular viscosity. An average speed of the steam for section A-A was assumed to be 6 m/s, what corresponded to the individual drops of the pressure of 220 Pa/m for $\mu_{eff} = 100\mu_{mol}$ and of 333 Pa/m for $\mu_{eff} = 1000\mu_{mol}$ (Fig. 1b). For section B-B the individual drops of the pressure for both effective viscosities at the average speed of the steam were equalled to 290 Pa/m.

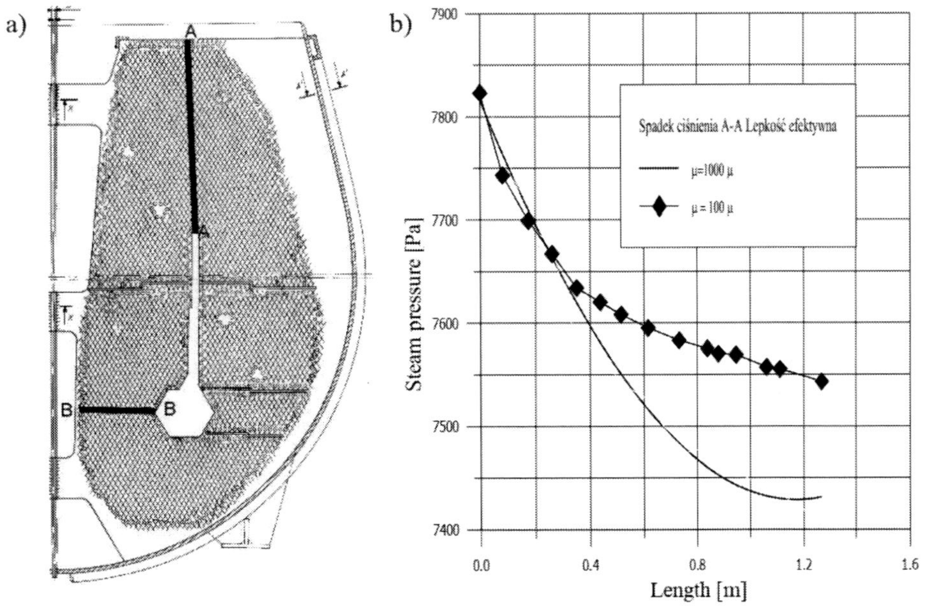

Fig. 1. a) cross-section of the condenser with sections A-A and B-B, b) pressure drops for section A-A.

The pressure drops calculated based on the correlations presented in section 2 are shown in Fig. 2. A good agreement between the numerical calculations with the empirical correlations can be observed. A pressure drops between 290–333 Pa/m for an average steam speed of 6 m/s was achieved nearly for all correlations. Only the result from Rhodesa and Carlucci equations shows quite a deviation. The pressure drops described by the correlations are strongly affected by the effective viscosity. The effective viscosity can be optimized in relation to steam balances, what was proposed for example in [10]. As a part of conducted analysis, it can be also stated that the pressure drops for the hexagonal bunches are higher than for the square. These results are also in accordance with the experimental results presented in [12].

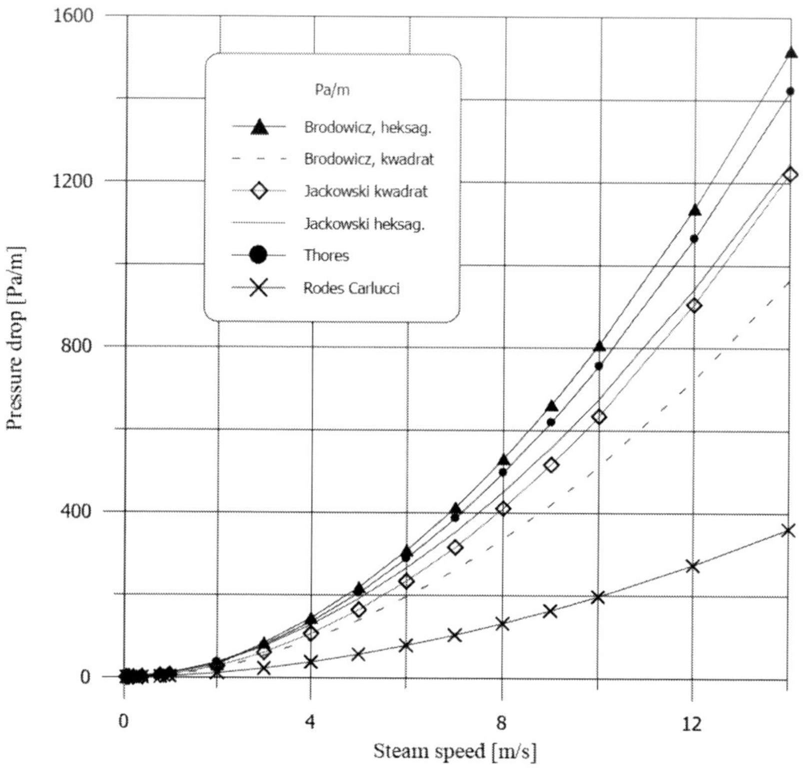

Fig. 4. The pressure drops obtained from different correlations.

4 Conclusion

In this paper the comparison of the various correlation for the pressure drop in the tube bundle is presented. The pressure drops were analysed for the hexagonal bunch of the pipes as well as for the square. The hexagonal arrangement was also analysed numerically. A good agreement between the results from numerical calculations and empirical correlations for the hexagonal bunch was achieved. It was also shown that the pressure drops for the hexagonal bunch are higher than for square.

References

1. R. Laskowski, A. Rusowicz, A. Grzebielec, Arch. Thermodyn. **36**, 3 (2015)
2. R. Laskowski, A. Rusowicz, A. Grzebielec, Przem. Chemiczny **94**, 10 (2015)

3. R. Laskowski, A. Smyk A. Rusowicz, A. Grzebielec, Entropy **18**, 12 (2016)
4. C. Zhang, A Bokil. Int. Journal of Heat and Mass Transfer **40**, 15 (1997)
5. A. Rusowicz, A. Grzebielec, R. Laskowski, Rynek Energii **134**, 1 (2018)
6. A. Rusowicz, R. Laskowski, A. Rusowicz, Thermal Science **21**, 1A (2017)
7. K. Brodowicz, K. M. Ostrowski, A. Rusowicz, D. Wierzbicki, Biuletyn ITC PW **80**, (1995)
8. T. Fujii, H. Uehara, K. Hirata, K. Oda, Int. Journal Heat Mass Transfer **15**, 2 (1972)
9. B. D. Rhodes, L. N. Carlucci, Int. Conf. on Numerical Method in Nuclear Engineering, Canadian Nuclear Society/American Nuclear Society, Montreal (1983)
10. K. Brodowicz, A. Czaplicki, Int. Heat Transfere Conf., San Francisco (1986)
11. L. Jackowski, Dissertation, Institute of Heat Engineering, Warsaw University of Technology (1987)
12. N. E. Todres, D. Ebeling-Koning, J. T. Robnison, Report No DOE/ER/12075-4FR (1985)

Energy saving in the formation of covered courtyards

Andrey Rymarov[1,*]

[1]National Research University Moscow State University of Civil Engineering, 129337, Yaroslavskoye Shosse, 26, Moscow, Russia

Abstract. In the city in the winter going outside the house you find yourself in a slippery and snowy weather with frosts and wind of different intensity. Forming covered with translucent roofs courtyards and streets can improve the quality of life of citizens. At the same time, in summer, at elevated outdoor temperatures, it is possible to create the required microclimate for people by blocking yards and streets and adding engineering systems that form the microclimate. Transport should be removed from such yards and streets, except for electric cars and bicycles, which will reduce the gas contamination and dust of these protected spaces. To form a year-round comfortable space for people in the covered yard, it is necessary to analyse the temperature regime. The most difficult is the thermal regime of the covered yard in the warm period of the year, when the heat flow from solar radiation in conjunction with the high temperatures of the outside air form the temperature of the indoor air in the covered yard. Thermal behaviour of the covered courtyard during the warmer period of the year considered in this article.

1 Introduction

In Russia, there are not enough places for a comfortable stay outside of houses and other buildings, as it is always necessary to protect yourself from changeable and often adverse weather throughout the year. The cold period of the year is known for its frosts with wind, snowfall and ice-covered roads [1], and the warm period of the city due to global warming in recent decades in Russia can be formed very hot temperature conditions in the daytime.

Blocked yard or street, or part of the street translucent ceiling, allows throughout the year and in the warm period of the year to get more comfortable conditions for people outside the premises and buildings. Such blocked yards or streets can connect neighbourhoods with parks and transportation systems and other places and buildings in the city. And, such yards can form the potential of energy saving for heating systems and heat consumption.

* Corresponding author: rymarov@list.ru

2 Methods of calculating the thermal conditions

Considered a residential building with a covered courtyard. When forming a covered yard, it is necessary to carry out calculations of its thermal regime to determine the dynamics of the change in the value of the internal air temperature in order to understand the level of comfort when a person is inside such a space.

In the warm period of the year, the flow of heat into the space of the covered yard occurs through the translucent overlap due to solar radiation and due to heat transfer, and when the air space of the covered yard is heated, the heat flows through the walls and Windows overlooking the courtyard will enter the premises of the building, warming them, and the heat flow will enter the ground of the blocked yard, increasing its temperature, which allows for calculations of the temperature regime in the closed yard. Passages to the courtyard through the arches should be closed with easily opened doors.

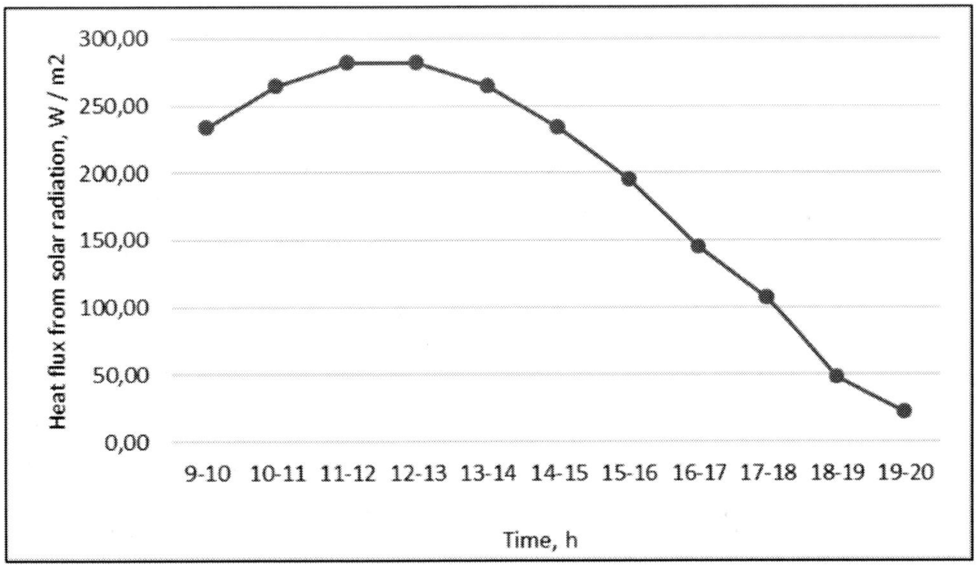

Fig. 1. Heat flux from solar radiation.

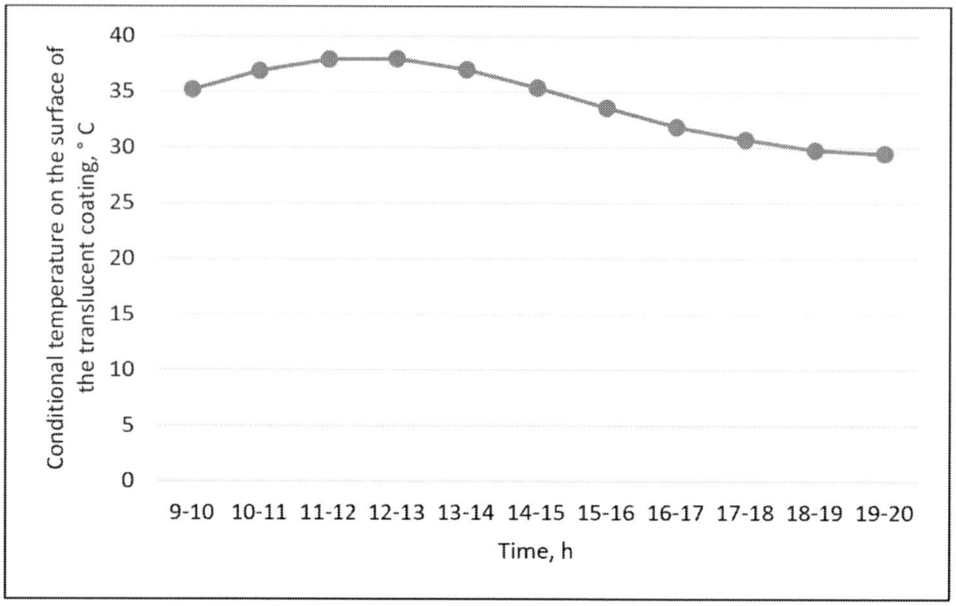

Fig. 2. Conditional temperature on the surface of the translucent coating for July.

Consider the thermal regime of the covered yard for a building located in the climatic conditions of the city of Moscow, which will allow us to understand how to manage its microclimate.

To analyse the thermal regime of the covered translucent roof of the courtyard in the warm period of the year, it is necessary to analyse the temperature change of the outside air in the warm period of the year, when the heat from solar radiation (figure 1) and the outside air temperature (figure 2 and 3) are maximum in the daytime.

Figure 1 shows how the heat flux from solar radiation entering the horizontal surface with a cloudless sky during the day from 9 to 20 hours, which is necessary to determine the heat flux coming through the translucent coating in the inner courtyard due to heat transfer and in the form of solar radiation on the walls, Windows and soil, warming them. The main source of heat input into the courtyard is heat from solar radiation, its value is determined based on existing climatic reference data, depending on the latitude and azimuth of the sun during the day. Based on data on heat fluxes from solar radiation, the calculations of the conditional temperature of the outside air at the surface of the translucent overlap during the daytime for July from 9 to 20 hours, and the results of the calculations are shown in figure 2.

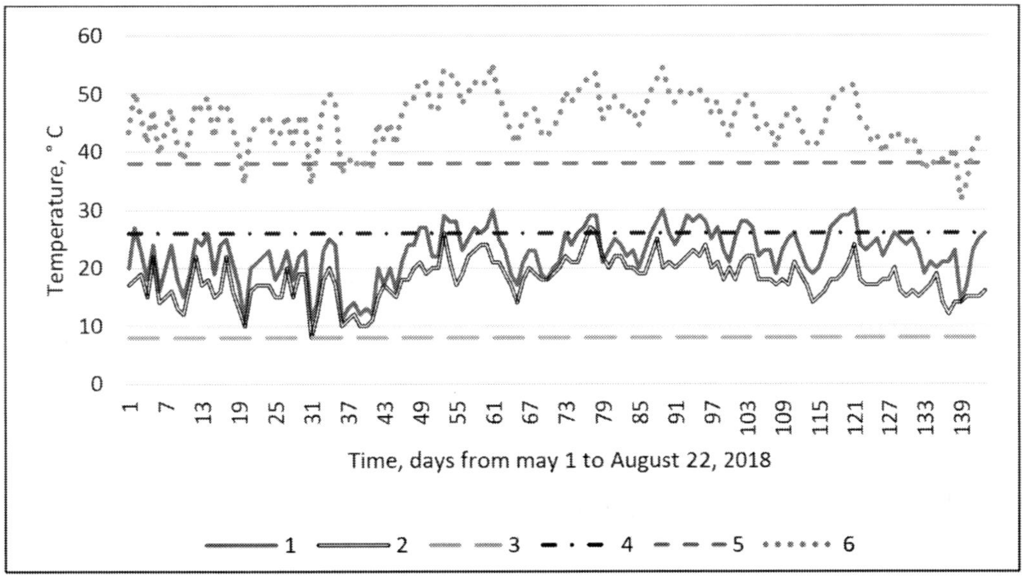

Fig. 3. Change in air temperature (o C) in time from May 1 to August 22, 2018: 1 – average daytime temperature, 2 – average night temperature, 3 – end temperature of the cold period of the year, 4 – temperature for calculations of air conditioning systems, 5 – maximum temperature for July, 6 – conditional temperature at the surface of the translucent structure for the horizontal surface.

The average temperature for day and night time is obtained according to the existing meteorological information. The temperature of +8°C in comparison with the average values of day and night outside air temperature shows that this period is warm and heating systems do not work. The temperature for the calculation of the air conditioning system shows that its value does not consider the flow of heat from solar radiation. The maximum outside temperature is closer to the nominal outside temperature, which considers the heat input from solar radiation, but still below it. From the graph analysis in Fig. 3 it can be seen that the conditional temperature of the outside air at the surface of the translucent structure for the horizontal surface has a maximum value on the temperature regime of the covered translucent roof of the courtyard in the warm period of the year, and in interaction with the penetrating through the translucent coating of solar radiation determine the temperature of the internal air in the courtyard.

Temperature values from 1 to 5 according to figure 3 are taken based on the normative document SP 131.13330.2012 "Construction climatology", operating in the territory of the Russian Federation. The calculation of the conditional temperature at the horizontally located surface of the translucent roof (figures 2 and 3) is carried out based on the technique described in the Handbook for the design of buildings [3].

3 Results of calculations

If the courtyard is blocked by a translucent roof with non-opening doors, the courtyard will turn into a kind of atrium, which can and should be used, for example, by filling the atrium space, for example, a winter garden with the necessary microclimate for the winter and summer periods of the year [4].

The overlap of the courtyard should have opening doors, which will allow in the warm period of the year to regulate the thermal regime in the airspace of the courtyard. If in the warm period of the year the indoor courtyard with translucent ceiling to keep permanently closed, in the daytime it will be very hot, and at night it will cool. If in the evening, night and morning hours of the day to open the sash cover, the courtyard will be ventilated and cool after sunset naturally.

Consider a building with a courtyard, schematic section of the building is shown in figure 4. The plan of the building is shown in figure 5. Heat flows in the courtyard (incoming and outgoing) are shown in figures 4 and 5.

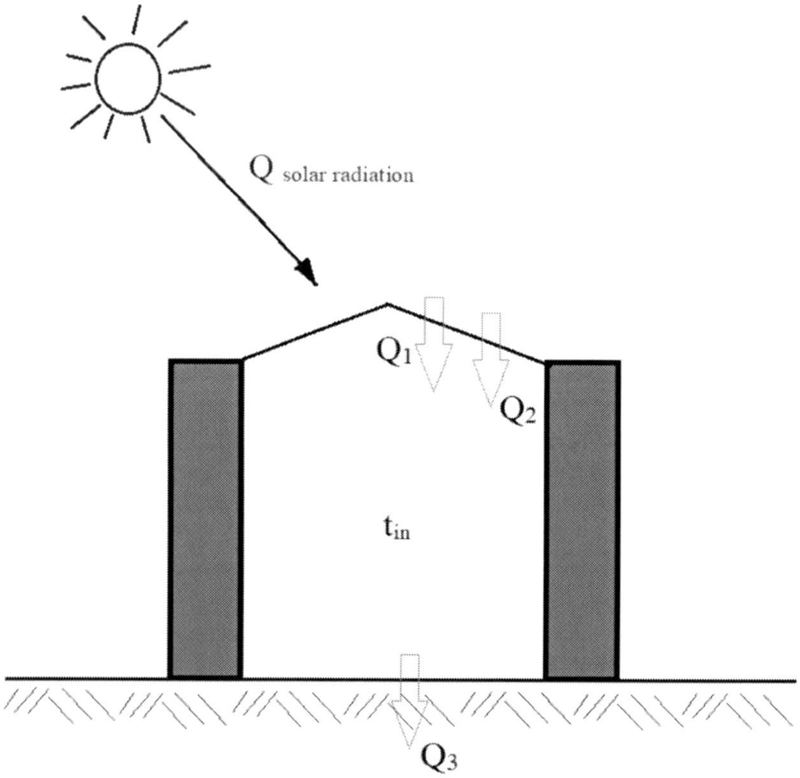

Fig. 4. The section of the building with the blocked yard with application of a translucent design.

To reduce the temperature of the air in the courtyard during the day, it is proposed to use an air-to-ground heat pump, which will cool the air in the courtyard space, directing the excess heat to the ground for accumulation, or an air-to-water heat pump, directing the excess heat to a heat-insulated water tank for accumulation. The accumulated heat can be used in the cold period of the year to heat the space of the courtyard, for the heating system, for hot water supply, etc. Figure 6 shows the heat pump application diagram.

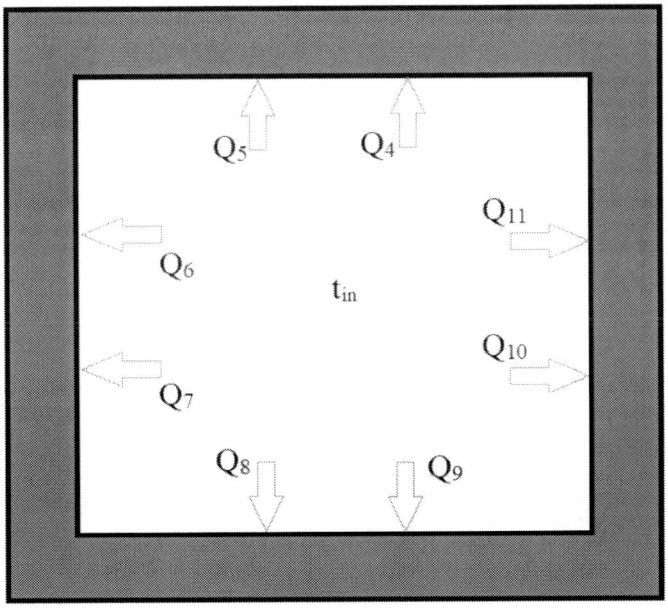

Fig. 5. The plan of the building with a blocked translucent design yard.

The covered yard is set with dimensions in the plan of 63×63 m along the perimeter of the covered yard there is a building with the length of the external facades of 87 m and the number of floors 5. The total area of Windows and walls facing the courtyard is 840 m^2 and 3393.6 m^2, respectively, and the area of the roof and ground in the covered courtyard is 3960 m^2. The volume of the covered yard is 213312 m^3 [1].

According to the existing Russian Federation standards set of rules 50.13330.2012 "Thermal protection of buildings" and set of rules 131.13330.2012 "Construction climatology" derived heat transfer coefficients of Windows of 2.04 W/ (m^2×°C), walls of 0.33 W/ (m^2×°C) and the translucent roof of 2.78 W/ (m^2×°C).

We will accept as the assumption that in all building in day and night time the temperature in the rooms turned to the courtyard equal +25°C at the expense of the working air conditioning system is maintained.

Heat flow in the daytime from the overlap enters the courtyard (Q_1 and Q_2), and then when the temperature of the internal air in the courtyard increases, the heat flow enters the premises of the building through the Windows and walls ($Q_4 - Q_{11}$) and into the ground (Q_3), as shown in figures 4 and 5.

At night, when there is no sun and the outside temperature drops, the courtyard begins to cool down and the heat flow begins to flow from the courtyard through the coating into the outside air. If necessary, you can open the sash cover and then the cooling process of the courtyard at night will increase.

When an air-to-ground or air-to-water heat pump is used, the excess heat from the courtyard air space (Q_{12}) moves throughout the day to the ground or water (Q_{13}) for accumulation and subsequent use (figure 6).

Balance equations of losses and heat inputs in the covered yard allow to determine the temperature of the internal air (t).

$$Q_1 + Q_2 = Q_3 + Q_4 + \cdots + Q_{11} \tag{1}$$

$$f(t) = f(Q_1 \dots . Q_{11}) \tag{2}$$

The balance equations of heat loss and heat input in the covered yard, considering the action of the heat pump, are supplemented by the absorption of excess heat (Q_{12}) by the evaporator of the heat pump.

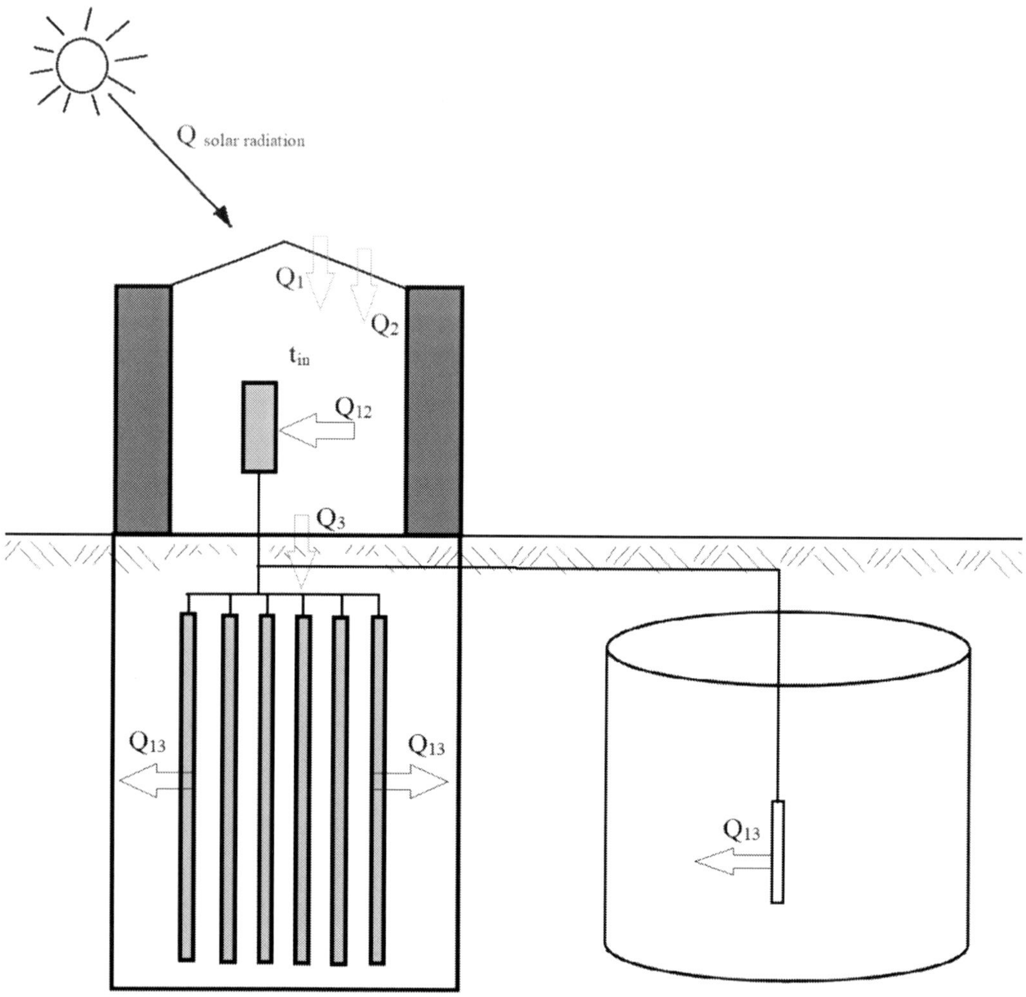

Fig. 6. The section of the building with the blocked translucent design the yard and with heat pumps of type air-ground and air-water.

The calculations showed that for the period from May 1 to August 22, 2018, the average value of the maximum amount of heat entering the courtyard during the day is 600 kW, and the average value of the maximum air temperature is 70°C. the Necessary temperature for a comfortable stay of people in the courtyard is taken to be 22–26°C, which is approximately 45°C (figure 7) below the temperature in the courtyard. To use the soil as a heat energy storage device, it is necessary to simulate the thermal regime of the soil [5], which will determine the accumulating capacity and volume of the soil mass, as well as the depth and frequency of wells for the placement of capacitors and determine the required number and power of heat pumps.

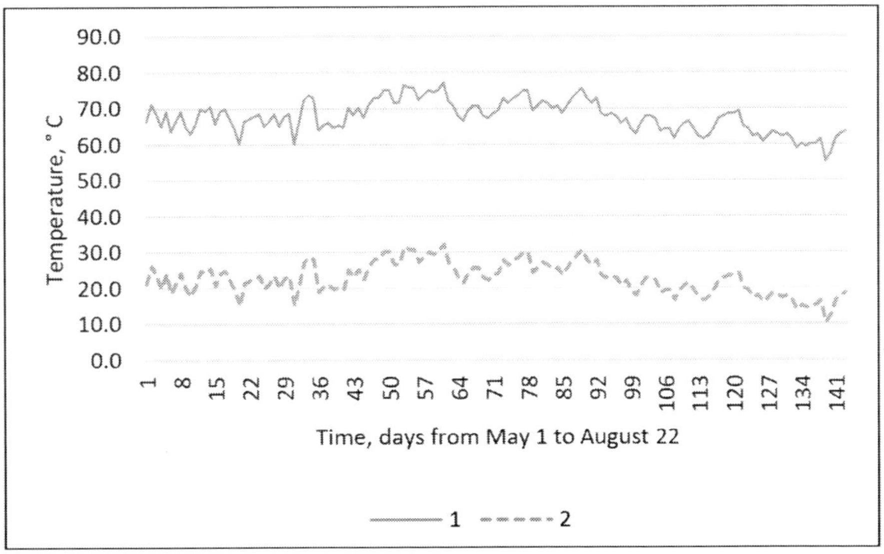

Fig. 7. Temperature change in the covered courtyard from May 1 to August 22, 2018 with the maximum value of heat coming from solar radiation (1) and the necessary temperature for a comfortable stay of a person in the courtyard (2).

The calculations are carried out for the temperature conditions of Moscow for the actual outdoor temperature values for the warm period of 2018.

4 Conclusion

The increased temperature in the indoor courtyard allows you to get a more comfortable space for people to relax in the winter, and save thermal energy in the warm season, accumulating it in the ground or water for later use. Translucent coating obscures the space of the yard, which may require additional lighting [6], it is necessary to clean the glass from snow in winter and dust in the warm season. Ventilation of a large volume of the courtyard is necessary from the condition of the number of people who may rest or spend time in the considered space of the courtyard.

The saved thermal energy can be used for any kind of heating of the courtyard in winter and other tasks of heat consumption by the building, which will lead to energy saving, which improves the environment and allows to solve many infrastructure problems.

References

1. A. Rymarov, M. Razakov, *The thermal mode of the covered yard* (IOP Conf. Series: Earth and Environmental Science **177**, 012025, 2018)
2. V. N. Bogoslovsky, *Directory of the designer. Ventilation and air conditioning* (Book 1, part 3, ed. N. Pavlov, Y. I. Schiller, Moscow, Stroyizdat, 319, 1992)
3. A. Rymarov, N. Parfenteva, K. Valančius, P. Sabina, M. Violeta, Journal of Environmental Engineering and Landscape Management, **26**, 3, 195–201 (2018)
4. O. Brukhanov, A. Rymarov, A. Malysheva, D Titkov, MATEC Web of Conferences **5**, Series "5th International Scientific Conference "Integration, Partnership and Innovation in Construction Science and Education", IPICSE 2016", 04028 (2016)
5. S. E. Gorgots, A. K. Solovyov, Lighting engineering, **2**, 51–52 (2006)

Assessment of the impact of the traffic situation in urban area on the amount of volatile organic compounds getting into the cabin of the car on the road

Maria Skrętowicz[1,*], and *Joanna* Świeściak[1]

[1]Wroclaw University of Science and Technology, Department of Automotive Engineering, ul. Braci Gierymskich 164, 52-640 Wrocław, Poland

Abstract. In the paper the levels of concentrations of volatile organic compounds getting into car cabin during the driving have been measured. Three series of the tests have been performed. Each series has been carried out in different road situation in terms of traffic intensity. For the tests exploited passenger car, Subaru Impreza was used. Before every series the background measurements was carried out and directly after that, the correct the correct tests were performed. Inside the cabin 14 different organic compounds have been identified. To analyse the change of the concentrations levels between the correct and background test have been calculated. The results indicate that the longer time of travel caused that the driver and passengers are more exposed to highest concentrations of VOCs.

1 Introduction

Traveling is the part of everyday human life. People spend several hours in a car every day driving to work and back home, shopping or attending to other matters. Spending this amount of time in the car makes microatmosphere of the vehicle specific environment with a significant influence on human health and well-being. Air pollution getting into the interior of the vehicle, both the necessary tightness of the vehicle affects negatively the quality of the atmosphere inside the passenger car. Depending on the length of the route, congestion and other components of the road traffic, the vehicle user is exposed temporarily or for a long time to such harmful pollution components as carbon dioxide, hydrocarbons from the VOCs group (e.g. benzene, toluene, xylenes) or particulate matters. The choice of proper ventilation or recirculation of the air plays a very important role. It largely influences the air quality in the car's microenvironment and the air humidity level

* Corresponding author: maria.skretowicz@pwr.edu.pl

[1]. In order to determine the relationship between the pollution of the microatmosphere of the vehicle and the road traffic, appropriate road traffic components that may affect the amount of pollution during the journey should be defined. In the paper traffic in urban conditions will be investigated.

Road components in urban conditions include physical and mathematical elements. The physical components include: road infrastructure, user or group of users (e.g. vehicle driver and passenger), while the mathematical: traffic description (algorithms describing the motion) [2].

Physical components can be relatively easily determined as follows:

- The road infrastructure is created by building structures (roads, streets, bridges, tunnels, etc.). They form intersections and interstitial sections. While intersections are a relatively problematic element of road infrastructure because of the decision-making by user at this facility, other tools are taken into account, such as road signs, speed limits, traffic light and traffic regulations. Road infrastructure is the reason for road traffic and it is the basis for traffic analysis.
- Road traffic is created by users, such as driver, passengers or group of users (students). Their goal is to move to a different location by choosing the appropriate mean of transport (car, bicycle, public transport or on foot, also counted as a means of transport).

Mathematical component of road transportation is defined by:

- Motion description – the parameters characterizing description of the movement are the speed and its changes, the length of the route and the time of movement. This is very important element of the road traffic characteristics – the data obtained from the traffic description enable the analysis and the traffic assessment. Different calculation methods and algorithms (e.g. regarding traffic lights) are used to describe the motion. The calculation methods include theoretical, simulative, empirical and forecasting methods.

In addition to the above components, the road traffic characteristics are strongly influenced by other factors that should be taken into account. Therefore, other important data to be included is the reason for traffic generation (e.g. business meeting), urban traffic model (e.g. morning rush hours) and the way to achieve the goal (e.g. minimum travel time). Additionally, the perception of a person influences the description of the movement. Decisiveness, errors, speed of human reaction or lack of information can lead to a significant change in the description of vehicle movement. In this case probability calculations should be defined.

The above factors indirectly affect pollution inside vehicles. The direct factors are mainly combustion products of fuels and engine oils (especially in traffic where street congestions often occur), leaks of operating fluids, dust on the surface and tribological processes occurring during vehicle operation, as well as upholstery elements made of plastics [3]. The above mentioned road traffic components also have a significant impact on the environment. Road infrastructure and physical road traffic components have a major impact on the level of ecosystem pollution. In addition to the exhaust gases in the air, toxic pollution also have an impact on the environment, which from roads and streets penetrate a larger area and triggered by weather conditions or surface disturbance (e.g. during pedestrian crossing on the outskirts of streets) [4]. In addition to this, the type of vehicle, the volume of its cabin, age and technical condition and the type of power unit affect the pollution inside the vehicle. Such a large number of factors affecting the amount of pollution entering the interior of the vehicle causes researches on this subject very broad and the relationships are constantly being discovered.

2 Methodology

2.1 Selection of the optimal measuring route

The right measuring route was the main component of this research. The measurement route was chosen to best reflect the conditions of everyday driving in the city under different conditions. The test scenario predicted various elements of city traffic, such as large, moderate and small traffic intensity with maximum speed allowed and during street congestions. The test also required continuous driving for one hour, due to the type of measurement of volatile pollutants.

Five types of measuring routes with different lengths and different traffic volume have been selected to extract an optimal route that will meet all requirements (Table 1). The korkowo.pl program was used to identify traffic congestion [5].

Table 1. Selected routes to carry out the tests.

Route number	The length of the route, km
Route No. 1	24.25
Route No. 2	33.69
Route No. 3	27.93
Route No. 4	28.80
Route No. 5	27.44

After consultation and careful analysis of the routes, route No. 2 was chosen. This route provided varied traffic conditions, such as maximum permitted speed in places with an increased number of traffic lanes, as well as the presence of traffic congestions from overlapping public transport routes with passenger cars. In addition, a long route No. 2 provided the required time needed to collect the appropriate samples. In order to verify the assumptions a four tests run was carried out. The first ride was made to confirm the time needed to perform the tests. Three more were made in three variants of urban traffic intensity (small, moderate and large traffic congestion). After performing the tests, the route No. 2 was modified and established to make the proper measurement. Route No. 2 after modifications is shown in Figure 1.

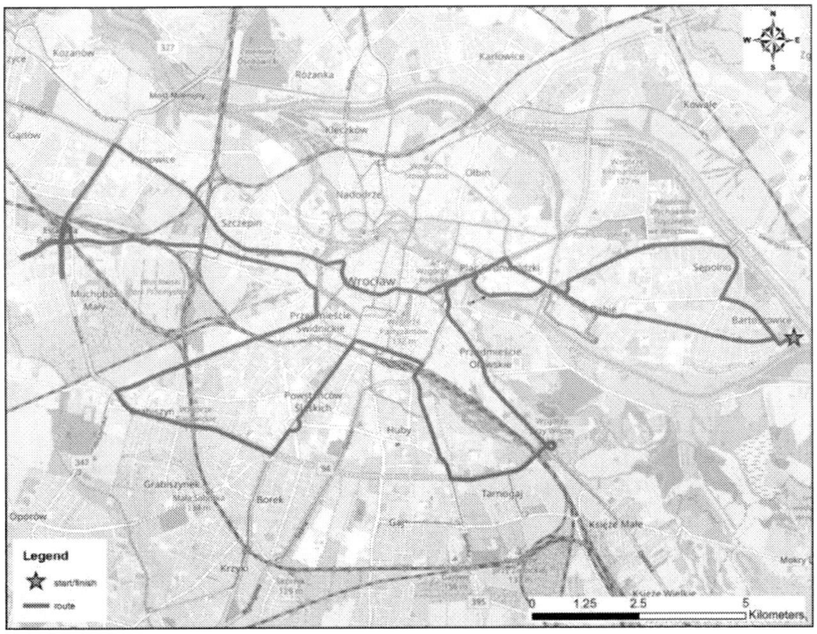

Fig. 1. Selected measuring route No. 2 after modifications (43 km) [6].

2.2 Vehicle characteristics and boundary conditions

The Subaru Impreza passenger car was used to test the air quality in the car cabin (Figure 2). This version is a type of station wagon and its cabin interior volume is 2.56 m^3 [7]. The vehicle was equipped with a ventilation system with air conditioning with a manually controlled 4-stage airflow. The ventilation system was equipped with a pollen filter, which at the time of testing was not replaced. The vehicle's age was 13 years. The interior of the vehicle is equipped with various upholstery elements, made of such materials as: plastics (dashboard, middle channel, door panels), sponges (seats, rear seat and parts of door panels) and textiles (upholstery, roof linings).

Fig. 2. Presentation of the vehicle used for the tests (source https://www.edmunds.com/subaru/impreza/2004/).

The boundary conditions were the appropriate settings of ventilation in the vehicle cabin. In order to maintain repeatable measurement conditions, fixed parameters of the ventilation system were established:

- closed windows,
- air conditioning switched off,
- fan speed set to level 2,
- the air temperature is set to cold,
- direction of blowing air directed to the head and legs.

2.3 Air quality measurement methodology

2.3.1 Measuring station

In order to collect air samples, an ASP II aspirator with two measurement channels was used. The tubes filled with active carbon was used as the carrier of the collected organic pollutants (VOCs). The test included sampling outside the vehicle and inside to obtain information on the amount of contaminants present in the cabin compared to the air pollution outside the vehicle. Measuring points were installed in the following places: the first one was mounted in the vehicle cabin at the height of the passenger's head and the second measuring point was located outside the vehicle near the air inlet to the cabin at the bottom of the windshield.

2.3.2 Plan of measurements

To perform the tests were provided three series of rides respectively at the low (series 1), moderate (series 2) and large (series 3) traffic intensity. In the first place before each test during the vehicle movement a stationary measurement was taken (background measurement). The background test was performed in order to obtain outside sample as a reference for later measurements. Background measurement was performed for three hours to collect 90 litres of air and adsorb them on active carbon in the aspiration tube. The air flow was set at 30 l/h. During the measurement, the car was closed to separate the ambient air in the cabin. After the completion of the background sample collection the aspiration tubes were replaced and the correct measurement started. The boundary conditions were retained as above for the period of actual measurement. The air flow in the aspirator has remained unchanged, but the time of the measurement was different, depending on the travel time. In Table 2 parameters for each of sampling series are presented.

Table 2. Sampling parameters.

Parameter	**Series 1**	**Series 2**	**Series 3**
Air flow, dm^3/h	30	30	30
Time of sampling, min	85	80	134
Air volume, dm^3	43,6	43.6	66.8
Temperature inside the cabin, °C	38.4	42.3	44.4
Outdoor temperature, °C	21	25	26.5

2.3.3 Chromatographic analysis

After sampling, the tubes with active carbon have been moved into the laboratory and chromatographic analysis were made. VOCs adsorbed on the active carbon have been extracted in carbon disulphide (CS_2), the solution has been cleaned and next given on the gas chromatograph Varian 450-GC with capillary column and autosampler. As the results of chromatographic analysis were concentrations of separated organic compounds such as benzene, toluene, ethylbenzene, etc. in ppm of the pollutant in 2 ml of CS_2. It was needed to calculate the concentrations of VOCs in the air in $\mu g/m^3$, taking into account the amount of air collected during the sampling.

3 Results and discussion

In Table 3 the obtained concentrations values (in $\mu g/m^3$) of identified in chromatographic analysis VOCs are presented. Two isomers of xylene (*para-* and *meta-*) are presented together, because their retention time overlap. The shorts used in the Table 3 mean: BI – background sampled inside the cabin, BO – background sampled outside the car, MI – correct measurement conducted inside the cabin, MO – correct measurement conducted outside the car. Due to destroyed in the laboratory the tube in series 3 BO, the concentrations in that sample have been not obtained.

Table 3. The results of the measurements.

Compound	**Series 1**				**Series 2**				**Series 3**			
	BI	BO	MI	MO	BI	BO	MI	MO	BI	BO	MI	MO
n-pentane	18.9	34.8	24.9	39.9	15.3	52.2	28.0	34.8	16.5	-	15.7	19.1
2-propanol	64.5	15.4	29.3	-	51.1	76.0	26.7	123	24.0	-	0.5	0.7
Benzene	2.6	-	4.0	-	1.0	13.1	3.6	5.9	1.3	-	5.6	0.2
2-butanol	36.6	1.0	0.9	-	35.9	-	-	-	17.3	-	-	-
Toluene	13.8	0.9	2.7	3.0	7.2	-	0.7	-	2.8	-	4.0	0.7
1-butanol	6.8	-	-	2.9	5.9	-	1.2	-	1.5	-	0.3	0.6
Ethylbenzene	1.7	-	6.0	-	-	-	-	-	-	-	-	-
p,m-xylene	2.1	-	-	2.8	1.9	-	4.3	-	0.6	-	2.6	-
Kumene	9.8	-	-	-	7.4	-	-	-	3.1	-	0.3	-
o-xylene	42.6	-	-	0.7	31.7	-	-	-	9.5	-	-	-
Propylbenzene	-	-	0.3	-	-	-	0.4	-	-	-	2.8	-
Mezytylene	73.8	-	0.9	-	58.1	-	1.1	-	33.6	-	1.8	-
p-cymene	7.7	-	0.5	-	7.7	-	-	-	2.8	-	0.4	-
Butylbenzene	6.5	-	-	1.3	17.3	-	0.6	-	0.9	-	0.9	0.9

In total, in all measurements 14 different organic compounds have been identified. The largest amount of VOCs have been detected in tests conducted inside the car cabin, both in background and correct measurements. As it can be seen, concentrations values of n-pentane (n-pentane represents the mixture of the lightest hydrocarbons) in each series are higher in samples taken outside than in samples taken inside the car cabin. generally, the highest concentrations inside the car cabin were received for n-pentane, 2-propanol, benzene, 2-butanol, toluene and propylbenzene.

To analyse how the concentration levels of the pollutants change during the route, it was needed to calculate the difference between concentrations measured in correct tests (MI) and background (BI). In Figure 3 the differences for each series have been shown.

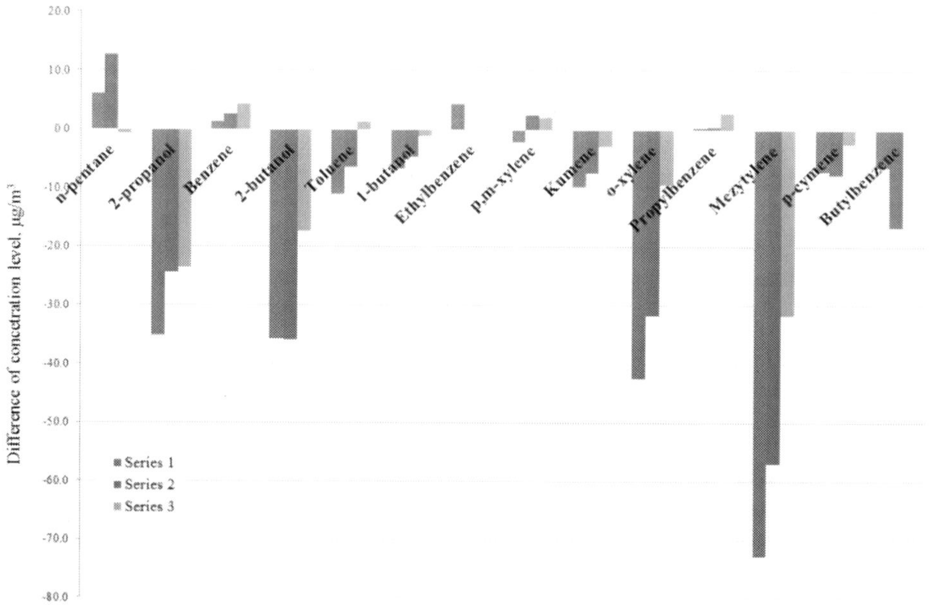

Fig. 3. Difference of concentration levels between correct and background measurements in each series.

For more pollutants levels of concentrations was lower in correct measurements than in background, therefore value of the difference of that pollutants was negative. It is caused by the methodology used during the background and correct measurements. In background tests ignition of the vehicle was switched off, so the fan wasn't working. In turn, during the correct measurement the fan was working with speed set to level 2. Therefore pollutants cumulated inside the cabin have been dispersed.

In the series 1 concentrations of four of identified organic compounds have grown during the route: n-pentane, benzene, ethylbenzene and propylbenzene. Worth attention is fact, that propylbenzene haven't been identified in background and appeared only in correct test, what means that it got into the cabin in traffic with ventilation air. In the series 2 pollutants, the concentration of which was increased were almost the same as in series 1 except ethylbenzene and p,m-xylene. In that series propylbenzene also haven't been identified in background test. In the series 3 the concentration of benzene, toluene, p,m-xylene and propylbenzene increased. What is important, difference in case of benzene between concentration level measured in correct and background test was positive in each series and the highest concentration was measured in series 3, when the traffic intensity was the largest. Generally, along with the increase of travel time (increase of traffic intensity), the negative difference decreases. For the toluene, the negative difference in series 2 is smaller than in series 1, in turn in series 3 the growth of concentration occurred. It indicate that despite dispersion of pollutants caused by turning on the fan on the beginning of correct measure, during the route organic compounds have got into the interior. In series 3 the temperature inside the cabin was the highest, therefore emission from the cabin equipment could take place, but tested car has been in operation for many years, so share of that type of emission was negligible.

4 Summary

In the paper measurements of volatile organic compounds inside the car cabin depending on the traffic intensity have been performed. Permanent route has been established and three measuring series have been carried out. Different road situations have been chosen, characterized by low, moderate and high traffic intensity. Due to the route haven't been changed in each series, with growth the traffic, the time of travel increased too. For the tests exploited passenger car, Subaru Impreza was used. Before every series the background measurements was carried out and directly after that, the correct the correct tests were performed.

Inside the cabin 14 different organic compounds have been identified. To analyse the change of the concentrations levels between the correct and background test have been calculated. The results indicate that the longer time of travel caused that the driver and passengers are more exposed to highest concentrations of VOCs. Although the fan switched on at the beginning of the correct tests caused the dispersion of cumulated pollutants, it is significantly visible the influence of the traffic intensity on the amount of the pollutants getting into the interior.

In the future, it is needed for conducting additional tests taking into account weather conditions, various settings of the vehicle's ventilation system or the state of the cabin air filter. That will allow to obtain comparative results, specify the reasons for increases and decreases in the concentration of harmful compounds and the development of methods to minimize their impact on the driver and passengers.

References

1. M. L. Grady, H. Jung, Y. Kim, J. K. Park, B. Ch. Lee, SAE Technical Paper,
2. M. Kruszyna, Oficyna Politechniki Wrocławskiej **56**, 24 (2013)
3. Z. Chłopek, *Ekologiczne aspekty motoryzacji i bezpieczeństwa ruchu drogowego* (Politechnika Warszawska, 2012)
4. A. W. Coffin, Journal of Transport Geography **15**, 5 (2007)
5. http://korkowo.pl/ [access: 07.02.2019]
6. http://www.navime.pl/ [access: 07.02.2019]
7. http://www.cars101.com/subaru/impreza/impreza2004.html [access:07.02.2019]

Reduction of chloride emission by thickening of metallurgical wastewater

Krzysztof Słowik[1,*], *Marcin* Stec[1], *Tomasz* Iluk[1], *Andrzej* Czaplicki[1], and *Mateusz* Kochel[1]

[1]Institute for Chemical Processing of Coal, ul. Zamkowa 1, 41-803 Zabrze, Poland

Abstract. Purification and thickening tests of metallurgical industrial wastewater from sintering dust dechlorination unit were carried out. Raw wastewater of pH 6.95 consisted primarily of Zn^{2+}, Pb^{2+}, Cd^{2+}, Tl^{+}, Cl^{-}, SO_4^{-2}, at concentrations of approx. 3,200; 60; 11,500; 100; 40,100 and 5,300 mg/dm^3 respectively. The heavy metals removal consisted in chemical precipitation and sedimentation, firstly by sodium hydroxide, and secondly by sodium sulphide solutions resulting in average removal to levels below 1mg/dm^3. Moreover, the experimental studies of evaporation process confirmed the possibility of chloride ions reducing, the distillate yield was approx. 80% of the initial effluent stream. The solid product obtained by two-stage evaporation was characterized by satisfactory purity. The results indicate maturity of the applied processes for wastewater treatment, allowing conditions close to zero-liquid-discharge to be achieved.

1 Introduction

Wastewater containing high concentrations of heavy metals, chloride and sulphate ions comes mainly from mining, the smelting industry, plating processes or pigments, as well as manufacture of alloys and batteries. The zinc and lead production using the Imperial Smelting Process is performed in Huta Cynku "Miasteczko Śląskie" S.A. The primary raw materials used for the production of zinc and lead ores are concentrates, while the secondary raw material is scrap from zinc alloys. Depending on the raw material used, production is carried out as a sulphide or oxide campaign. The main sources of wastewater are the wet gas treatment process and the slag granulation processes. Additional streams are rain water, cooling water and other process waters containing toxic heavy metals.

Regardless of the raw material used in the production process, all wastewater is treated in the central wastewater treatment plant to achieve an environmentally acceptable level of harmful substances, which are limited by environmental protection regulations [1, 2]. Toxic and carcinogenic properties of heavy metals like, Pb, Hg and Cd have been confirmed [3, 4] and when entering natural watercourses the heavy metals are quickly and irreversibly incorporated into the food chain. Therefore, in the case of insufficient purification, these wastewaters can be an important source of environmental pollution in the areas to which they are discharged. Removal of metals and sulphates from wastewater from the production of non-ferrous metals is successfully carried out by means of a combination of chemical

* Corresponding author: kslowik@ichpw.pl

precipitation, sedimentation, filtration, flotation, ultrafiltration, activated carbon filtration or reverse osmosis techniques [1].

Heavy metal-free wastewater still contains anions, which must also be reduced to acceptable values. The evaporation process is one of promising techniques allowing its application for wastewater treatment. It can be effective in solution thickening as well as for removing salts and other undesirable materials from wastewater in order to produce valuable solids and reusable high-quality water. There are various types of evaporators available on the market, e.g.: falling film, forced circulation, plate or thermal and mechanical vapour recompression system [5, 6]. As EPCON reports [7], the energy consumption is 8–40 kWh per tonne of evaporated water, and it is possible to treat volumes ranged from 100 kg/h to 100,000 kg/h for commercial applications with water recovery up to 99%.

As shown above, for the evaporation process, it is required to provide a significant amount of heat to evaporate the solvent, and to concentrate the solution. Therefore, the evaporation processes are particularly suitable for plants having waste heat streams.

Differences in waste heat parameters and wastewater composition force companies to develop systems which are tailored to their processes.

The investment and operating costs of such solutions are high, but where waste heat can be used, it is possible to reduce the high costs. Such solutions become more and more common in "zero liquid discharge applications".

Waste management, energy considerations and costs are some of the key factors that affect the search for new technological solutions that enable zero-waste production. This publication presents the results of a collaborative project entitled "Effective generation of electric power from metallurgical waste gases with simultaneous reduction of chloride ion emissions into the environment." The aim of the research was to remove heavy metals, chloride and sulphate anions from raw wastewater, as well as to recover water with the level of impurities enabling its reuse in the production process, in a combined process, utilizing waste heat from electricity production.

2 Materials and methods

The investigated wastewater was a raw effluent produced in a dechlorination unit in the plant operation with oxides used as the raw material. The raw filtrate was pumped from the leachate tank into a 1 m^3 IBC container and subsequently transported to the test site.

The pilot plant was located on site of HCM S.A. in Miasteczko Śląskie, and consisted of a batch reactor with a volume of 1,300 dm^3 equipped with a mechanical stirrer and a pH measurement system. It allows for cleaning of 900 dm^3 of raw wastewater in one test cycle. It was additionally equipped with three pumps: one for pumping raw wastewater from the container into the reactor, the second one for decanting the solution from the sludge after sedimentation, and the third one for filtrate circulation in the reactor and pumping the sludge into filtration bags. Solutions of reagents (sodium sulphide and sodium hydroxide) were stored in stirred tanks, from which they were supplied to the reactor by means of dosing pumps. The thickening and crystallization process was carried out in two steps. For preliminary concentration, a heat pump forced circulation evaporator, model PC F 0.7 FF, with distillate capacity of 700 dm^3 per day was used. In the second stage, the heat pump scraped evaporator, model R 150V3 FF1 V, with distillate capacity of 150 dm^3 per day was used. Both evaporators were delivered by Veolia Water Technologies Italia Spa. Additionally, the IBC containers were used for collecting and storing distillate water, concentrate and treated wastewater. The wastewater treatment plant is presented in figure 1.

Fig. 1. Wastewater treatment pilot plant view (left – chemical neutralization unit, right – evaporation and crystallization unit).

The physicochemical properties like pH, conductivity, temperature and redox potential of the samples obtained were determined directly during the experiments. The ELMETRON CPC-461 universal meter equipped with a glass combined pH electrode, a conductivity sensor with a platinum electrode and a type Pt-100 resistance temperature sensor was used.

The concentration of sulphide ions was determined by means of the EAg2S-305W type EUROSENSOR ion-selective electrode. As the reference electrode, the EAgClK-312 type electrode (EUROSENSOR) was used. The measurement of the potential difference between the electrodes was conducted using the ELMETRON CPC-461 meter.

A TCS Electronic Platform Scale was used to weigh chemical reagents and samples of distillate, concentrate and solid products. The sodium sulphide and hydroxide used in the treatment were of technical grade. The AXIS ATS 210 moisture analyser was used to determine the moisture content of materials as well as the salt content in concentrates at the temperature of 105°C. Additionally, the device was used for density determination by measuring the weight of a given volume of concentrate. In order to determine the concentration of heavy metals, chloride and sulphate ions, sampled probes were sent to the plant's laboratory.

The contamination level of salt mixture was determined using the iCAP 6500 DUO Thermo Scientific Inductively Coupled Plasma Optical Emission Spectrometer and the Dionex ICS-5000 ion chromatograph.

Figure 2 presents a diagram of the procedure applied for purification and concentration testing.

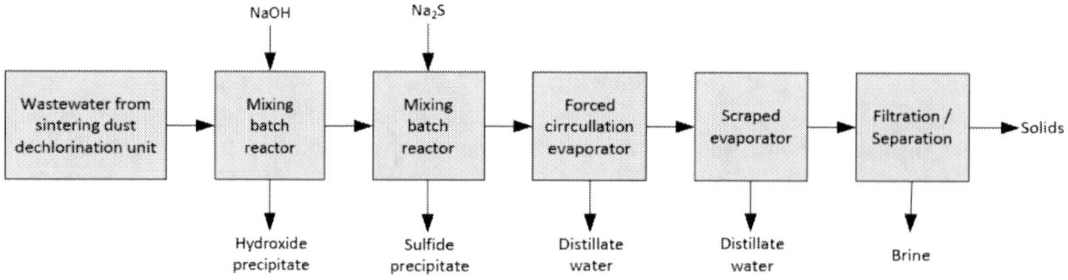

Fig. 2. Wastewater treatment process operations.

The chemical wastewater neutralization process was divided into two steps. In the first step, heavy metals (Pb, Zn, Cd) were precipitated as hydroxides using NaOH. The suspension thus obtained was allowed to sediment over the night (14 hours). The clarified solution was decanted to the intermediate container. The hydroxide mixture sludge was dewatered using fabric bag filters. The filtrate received from dewatering was subsequently added to the clarified solution. After the reactor cleaning, the pre-purified wastewater was pumped into the reactor for further treatment. In the second step of the heavy metals removal, a sodium sulphide solution was added to precipitate heavy metal sulphides (including thallium). Its dosing was controlled by measuring the potential difference of the cell with the ion-selective sulphide electrode – pS. Once the sodium sulphide dosing had been completed, the suspension was left for sedimentation. The clarified solution was decanted to the container, and the remaining precipitate was separated with fabric bag filters. The filtrate received from dewatering was subsequently added to the clarified solution. In the second step, the sedimentation process was supported by adding a flocculant. The wastewater cleaned of heavy metals was directed to the tank of the concentration and crystallization system.

The evaporation process was performed in two steps. The first one consisted in initial concentration of the feed wastewater in the first evaporator (model PC F 0.7 FF3) without crystallization. Thus produced, the streams of concentrate and distillate water were collected in dedicated containers. Subsequently, the concentrate was automatically sampled into the second evaporator (R 150V3 FF1 V) and evaporation process was continued with production of the solid. The device was operated periodically, allowing salt mixture and brine removal after experiment was finished. The solid product was dewatered using paper bag filters. The products thus obtained were weighted, and the samples were sent to the laboratory for qualitative and quantitative analysis. The stability of operating parameters for both evaporators were periodically checked.

3 Results and discussion

The wastewater treatment was carried out from 13.09.2018 to 26.09.2018. The results of the wastewater treatment are shown in Table 1. In the first treatment step, where the pH value exceeded 10.5, the process was stopped. It was found that chemical precipitation by NaOH including sedimentation and filtration significantly decreases the content of Zn, Pb and Cd, contrary to thallium. Thallium (I) hydroxide is well soluble [8], which is why it does not precipitate in the first treatment step. The remaining heavy metals, especially thallium, were precipitated by 1M sodium sulphide solution added in the second treatment step. The amount of the sodium sulphide added to the solution was controlled with the concentration of S^{2-} ions in accordance with a procedure proposed by Veeken et al. [9]. Its dosing was continued until the value of –496,8 mV. High reactivity of sulphide ions (S^{-2} and HS^{-}) with heavy metals and very low solubility of most heavy metal sulphides makes it possible to significantly improve the efficiency of wastewater treatment. This method requires precise control of sulphide ions concentration in the solution in order to prevent the possible odor of gaseous hydrogen sulphide. As for thallium and zinc after Na_2S addition, their concentration was satisfactorily reduced to the level fulfilling discharge standards.

Table 1. Results of treatment for Zn, Pb, Cd, Tl, Cl^-, SO_4^{2-} removal by 2 step chemical neutralization.

Treatment step	**Contaminant concentration, mg/l**					
	Zn^{2+}	Pb^{2+}	Cd^{2+}	Tl^{+}	Cl^{-}	SO_4^{2-}
Wastewater	3214.4	58.47	11,423.25	92.19	40,123	5,330
First neutralization	7.29	0.82	34.13	80.1	43,827	4,850
Second neutralization	0.74	0.91	1.59	0.73	42,592	4,371

The wastewater discharge standard of the Polish Ministry of Environment for Zn, Pb, Cd, and Tl is not more than 2; 0.5; 0.4; 1.0 mg/l, respectively, and the chloride and sulphate concentration should not excess 1,000 and 500 mg/l.

The received treatment results are in line with published data for processes utilizing similar treatment method namely: two-step heavy metals precipitation using caustic and sodium sulphide. Charemtanyarak [10] showed that precipitation using lime followed by Na_2S allowed reduction of zinc and cadmium to 1.0 and 1.5 mg/dm^3, respectively. The synthesized raw wastewater, used in the cited experiments, contained initially 450 mg/dm^3 of zinc and 150 mg/dm^3 of cadmium and the initial concentration was smaller by the order of magnitude, comparing to wastewater presented in this paper. Despite significantly smaller concentration in the raw wastewater, the achieved results of the treatment are similar.

Treatment by precipitation of wastewater having a high initial concentration of the cadmium was presented by Islamoglu et al. [11]. In their research, the original sample contained ~25000 mg/dm^3 of cadmium, and its concentration was reduced to ~8 mg/dm^3 by precipitation using Na_2S. The results are, again, in line with the data presented in this paper.

The acrylamide based flocculating agent solution was added for sedimentation of metal sulphides, which particularly favourably influenced the volume and grain size of the precipitated thallium sulphide.

The density values decreased slightly compared to raw wastewater, and they amounted to 1.068 g/cm^3 after the chemical neutralization steps. The conductivity value of the solutions did not change significantly, and it amounted to 175 mS/cm after the second treatment step. The pH value for the NaOH treated wastewater dropped from the initial value of 10.5 to 10.15 after its overnight conditioning, sedimentation and filtration. However, after the precipitation of the remaining heavy metals and thallium as sulphides in the second step, the pH value reached 10.32. The average content of solids in the wastewaters subject to examination was 10% by weight. The determined values of physicochemical parameters for raw and chemically treated wastewaters are presented in Table 2.

Table 2. Values of physicochemical parameters of treated wastewater.

Treatment step	Parameter			
	pH	Conductivity, mS/cm	Solids, %	Density, g/cm^3
Wastewater	6.19	164.36	9.1	1.074
First neutralization	10.15	163.26	10.9	1.069
Second neutralization	10.32	175.04	10.1	1.068

After chemical neutralization, anions with very high concentrations were still present in the treated wastewater. The initial wastewater concentration test in the first evaporator was conducted with the cycle time of 350 s (frequency of concentrate discharging from the evaporator). The operating parameters as well as Q^I – distillate stream, m_c^I – mass of concentrate (one discharge) and concentrate density – d_c^I were periodically controlled. The values are presented in Table 3 and confirm the process stability.

The determined average yield of the distillate was 27.58 kg/h (662 kg per day) at the average boiler pressure (vacuum) of 5.3 kPa, which confirms the declared nominal parameters for the evaporator used for the initial concentration of the treated wastewater. The produced distillate water was characterized by an average conductivity of 131 μS/cm. The average mass of concentrate per one discharge equalled 2.09 kg, which corresponded to 21.5 kg of concentrate per hour.

Table 3. Operating parameters in the first evaporation step.

PC F 0.7 FF3 operating parameters					
Duration time, min.	Q^I, kg/h	m_c^I, kg	d_c^I, g/cm^3	Boiler pressure, kPa	Distillate conductivity, µS/cm
37	27.96	2.12	1.139	5.3	140
170	27.48	1.94	1.131	5.2	129
240	27.48	2.28	1.126	5.3	130
300	27.48	2.16	1.123	5.4	126
356	27.48	1.96	1.124	5.3	128
Average	27.58	2.09	1.129	5.3	131

The second evaporator was operated for 7.25 hours, the boiler pressure (vacuum) and the distillate stream (Q^{II}) were controlled periodically and are presented in Table 4.

Table 4. Operating parameters in the second evaporation step.

R 150V3 FF1 V operating parameters		
Duration time, min.	Q^{II}, kg/h	Boiler pressure, kPa
75	6.00	4.5
200	5.76	4.5
260	5.60	4.0
369	5.36	4.0
419	5.52	n.a.
Average	5.51	4.3

n.a. – data not available

The determined average yield of the distillate was 5.51 kg/h (132.2 kg per day) at the average boiler pressure of 4.3 kPa. The values confirm the nominal parameters for the evaporator used for salt crystallization. The determined values of conductivity, content of solids and density of the concentrates obtained are provided in Table 5.

Table 5. Physicochemical parameters of concentrates.

Treatment step	Parameter		
	Conductivity, mS/cm	Solids, %	Density, g/cm^3
First evaporation	379.4	20.9	1.129
Second evaporation	472.0	29.7	1.262

The thickening of the solutions is indicated by the density increase from 1.068 to 1.129 g/cm^3 after the first step, subsequently reaching the value of 1.262 g/cm^3 for brine after the second evaporation. The content of solids in the solutions increased from 10.1% for chemically treated wastewater, to 20.9% after the first evaporation, and to 29.7% by weight for the brine thus produced. Also, the conductivity of the concentrated solutions increased from 175 to 379 after the first evaporation step, finally reaching 472 mS/cm for the brine.

The 22.26 kg of brine were produced during the crystallization test. The determined concentration of chlorides changed from the initial 42,592 to 98,710 after the first evaporation, and subsequently up to 200,645 mg/l in brine.

Results presented by M. Turek et al. [12] shows "Debiensko" Desalination Plant performance on concentration of coal-mine brine with a TDS of about 60 g/dm^3 in two stage evaporation process. From initial 32.8 g/dm^3 of Cl^- content in coal-brine the concentration increased to 175.9 g/dm^3 after first evaporator (falling film with vapour compression). After crystallization step the concentration of chlorides reached 205.6 g/dm^3 what is in well correspondence with our result.

The 5.32 kg of wet salt mixture was obtained with the moisture content of ~19%. The total concentration of anions and cations determined in the dried salts sample equalled 931 g/kg, and consisted primarily of: Cl^- – 519.9, SO_4^{-2} – 29.3, K^+ – 20.8, Na^+ – 360.5 g/kg. The concentration of other cations, i.e. Cd^{2+}, Pb^{2+}, Zn^{2+}, Tl^+, Ca^{2+} Mg^{2+}, was 10.3, 6.0, 1.0, 3.0, 92.5 and 58.8 mg/kg, respectively. Based on the values determined, the content of chloride ions accounted for 56%, which confirms the possibility of effective reduction of the environmental emissions of these ions. Based on the measured values, it was estimated that the yield of the distillate was ~ 81% of the initial wastewater stream.

4 Conclusions

Despite the very high content of heavy metal cations as well as chloride and sulphate anions in the treated wastewater, the proposed procedure allowed for a properly purified filtrate to be produced. The results indicated that the treatment technology for removal of heavy metals should use a combination of hydroxide and sulphide precipitation, particularly necessary for thallium removal. The effectiveness of using a vacuum evaporator system to recover fresh water as well as to remove chloride and sulphate ions from the tested wastewater was confirmed. No significant technical problems were encountered during the experimental work. The major benefits of the technology implementation are as follows:

- high quality water recovery,
- reduction of chloride and sulphide ions emission,
- overall zinc yield of the plant (recycle of precipitated hydroxide to the process).

It is also economically sound to apply the proposed technology under the technical and technological conditions of Huta Cynku Miasteczko Śląskie S.A. It is expected to trigger the following two effects:

- energy-saving effect, by utilising waste heat,
- environmental effect, by reducing the concentration and load of harmful substances discharged into the environment.

This research was conducted under agreement no. 1/BR/2016 concluded between HCM S.A. and the Institute for Chemical Processing of Coal for the performance of a project entitled "Effective generation of electric power from metallurgical waste gases with simultaneous reduction of chloride ion emissions into the environment" co-financed by the Polish National Centre for Research and Development under Sectoral Programme INNOSTAL no. POIR.01.02.00-00-0170/1, the 2014-2020 Smart Growth Operational Programme.

References

1. G. Cusano, M. R. Gonzalo, F. Farrell, R. Remus, S. Roudier, L. Delgado Sancho, Best Available Techniques (BAT) Reference Document for the Non-Ferrous Metals Industries, Industrial Emissions Directive 2010/75/EU Integrated Pollution Prevention and control, (2017)
2. Regulation of the Minister of the Environment of November 18, 2014 on the conditions to be met when introducing sewage into waters or into the ground, and on substances particularly harmful to the aquatic environment (in Polish).
3. J. Godt, F. Scheidig, C. Grosse-Siestrup, V. Esche, P. Brandenburg, A. Reich, D. A. Groneberg, J. Occup. Med Toxicol. **1**, 22 (2006)
4. P. B. Tchounwou, C. G. Yedjou, A. K. Patlolla, D. J. Sutton, Experientia Supplementum **101**, 133–164 (2012)

5. GEA engineering for a better world, available on: http://www.gea.com/en/applications/chemicals/treatment_of_emissions/treatment_emissions_evaporation.jsp.
6. EVALED® Evaporators - Evaporation technologies Veolia Group, available on: http://www.evaled.com/.
7. EPCON Evaporation Technology AS, available on: http://www.epcon.org/industrial-wastewater.html
8. R. H. Perry, D. W. Green, J. O. Maloney, *Perry's Chemical Engineers' Handbook* (1997)
9. H. Veeken, L. Akoto, L. W. H. Pol, J. Weijma, Water Res. **37**, 15, 3709–3717 (2003)
10. L. Charerntanyarak, Water Sci. Technol. **39**, 10–11, 135–138 (1999)
11. S. Islamoglu, L. Yilmaz, H. O. Ozbelge, Sep. Sci. Technol. **41**, 15, 3367–3385 (2006)
12. M. Turek, P. Dydo, R. Klimek, Desalination **184**, 439–446 (2005)

Analysis of exposure of inhabitants of Polish cities to air pollution with particulate matters with application of statistical and geostatistical tools

Izabela Sówka[1,*], *Alicja* Nych[1], *Dominik* Kobus[2], *Yarolsav* Bezyk[1], and *Maciej* Zathey[3]

[1]Wroclaw University of Science and Technology, Faculty of Environmental Engineering, Wyb. Wyspianskiego 27, 50-370 Wroclaw, Poland
[2]inFAIR, www.infair.eu, dominik.kobus@infair.eu
[3]Wroclaw University of Science and Technology, Faculty of Architecture, ul. Bolesława Prusa 53/55, 50-317 Wroclaw, Poland

Abstract. The aim of this research was the analysis of air quality in selected Polish cities in 2008-2018 combined with health assessment based on the methodology and software of World Health Organization (AirQ +) as well as an analysis of main directions of activities aimed at improving air quality in Polish cities. The results of calculations and spatial analysis of exposure indicate the year 2010 as the one with the highest concentrations of PM_{10} and $PM_{2.5}$, as well as Małopolskie and Śląskie Voivodships (agglomerations: Upper Silesia, Krakow and Rybnik-Jastrzębska) as the two most threatened by negative impact on human life and health. In a detailed analysis of the Wrocław agglomeration, for 2017, the estimated number of deaths of adults caused by short-term exposure to PM2.5 with a limiting concentration above 10 μg/m^3 (in accordance with WHO guidelines on air quality) was over 130 people, and the estimated number of patients hospitalized because of cardiovascular and respiratory diseases due to the above exposure was 160 and 80 cases, respectively. The analysis showed more than 80 cases of asthma attacks in children with asthma caused by short-term exposure and three death cases in infants due to long-term exposure to PM10.

1 Introduction

Air pollution is one of the most important determinants of health. Numerous epidemiological studies have demonstrated the relationship between air pollution and a wide range of adverse health effects in general population; effects ranged from subtle subclinical effects to premature death. Some groups – for example, elderly people, children, pregnant women and people with chronic illnesses (e.g. asthma) may be more at risk and may be more at risk from exposure to air pollution. In addition, some groups may be exposed to a higher level of outdoor air pollution, for example people living near busy traffic routes or in specific occupations or socio-economic groups [1].

* Corresponding author: izabela.sowka@pwr.edu.pl

In 2010, lung diseases were the cause of 6.9% of all hospitalizations in Poland, and death rates in 2010 due to respiratory illnesses amounted to 50.6 per 100,000 residents [2]. Because of general cardiovascular diseases, on the other hand, over one million people were treated [3]. Also, in the document "Policy paper for health care for 2014–2020" [4] published in 2014, respiratory illnesses were listed among the most important health problems of Polish citizens. They were also included in the group of health issues that generated the largest incapacity expenses in 2010 (7.5% of the total expenditure) [5]. According to the 2012 State Health Report [6], mortality due to cardio-pulmonary syndrome and other heart diseases (51.5 thousand deaths in 2010 – 30% of deaths due to the whole circulatory causes), is higher in Poland than throughout the European Union. Death rates in 2010 due to cardiorespiratory syndrome and other heart problems were in total 134.9 per 100,000. residents [2].

The aim of this study was to analyse air quality in selected Polish cities, combined with a health assessment based on the methodology and software of World Health Organization (AirQ +).

2 Research methodology

2.1 Air pollution monitoring data analysis

The analysis of variability of the concentrations of selected air pollutants (PM_{10} and $PM_{2.5}$) was carried out for urban agglomerations with populations of over 250,000, according to the classification adopted in the Regulation of the Minister of Environment of August 2, 2012 on zones in which air quality is assessed (Journal of Laws from 2012, item 914). Results from measurement stations included in the National Environmental Monitoring network, recorded in the period between 2008 and 2018 were taken into account and Air Quality Portal of the Chief Inspectorate for Environmental Protection was the source of data used for calculations. Data from 2018 should be treated as not finally approved yet by bodies responsible for monitoring.

The analysis took into account data from an urban and suburban background and industrial stations (located within agglomeration boundaries). Due to the significantly limited spatial representativeness, uneven distribution in agglomerations and lack of continuity in the period covered by the analysis, data from traffic stations was omitted. Prior to the processing of data, they were selected due to the quality level, as well as applied measurement methodologies. Annual series of measurement results with a completeness of at least 75%, obtained by the manual gravimetric method (being the reference method), as well as automatic methods in the absence of manual measurements in a given location were included. For the purposes of analysis, the 1-hour data series were aggregated to 24-hour data, with the condition of minimum 75% completeness of data in the day. Then, for each agglomeration, daily indicators for individual pollutants were calculated: daily average and maximum daily mean values for each day for all stations in the agglomeration. For each agglomeration an annual index was calculated, being an arithmetic mean of daily average indicators.

2.2 WHO model

The analysis of the impact (both short- and long-term) of selected air pollutants on health and lives of exposed people was carried out using AirQ + software developed by WHO. The effect of exposure to excessive occurrence in atmospheric air of particulate matter ($PM_{2.5}$ and PM_{10}) has been analysed. In particular, for 12 agglomerations throughout Poland, the effect of long-term $PM_{2.5}$ exposure on the total mortality was analysed. In addition, for the same agglomerations, the effect of exposure to $PM_{2.5}$ on cases of hospital admission for cardiovascular (HACD) and respiratory diseases (HARD) was evaluated. Results of concentrations of individual pollutants from the period between 2008 and 2018 were used for calculations. Additionally, for the Wrocław agglomeration, for 2017, analysis of long-term impact from PM_{10} dust on post neonatal infant mortality and short-term impact on the occurrence of asthma attacks in children with asthma was carried out. Moreover, effects of short-term $PM_{2.5}$ exposure on adults mortality, as well as HACD and HARD were analysed. Statistical data on population, deaths and baselines for hospitalization in the analysed areas was taken from the Central Statistical Office and Ministry of Health. The default values in the software were used for relative risk (which estimates the relationship between exposure and health problems and indicates the probability of disease development in the exposed group, compared to those who are not exposed), as well as the value of the limit concentration below which health impact is not analysed. Table 1 presents the relative risk for each health effect analysed in this paper.

Table 1. Relative risk (95% confidence interval) and used for investigating PM health effects.

Health effect	Relative risk (95% CI)
Short term effect $PM_{2.5}$ Mortality, all causes (adults age 30+ years)	**1.0123** (1.0045–1.0201)
Short term effect $PM_{2.5}$ Hospital admission for cardiovascular diseases	**1.0091** (1.0017–1.0166)
Short term effect $PM_{2.5}$ Hospital admission for respiratory diseases	**1.008** (1.0048–1.0112)
Short term effect PM_{10} Incidence of asthma symptoms in asthmatic children	**1.028** (1.006–1.051)
Long term effect PM_{10} Post neonatal infant Mortality, all cause	**1.04** (1.02–1.07)
Long term effect $PM_{2.5}$ Mortality, all causes (adults age 30+ years)	**1.062** (1.04–1.083)

2.3 Geostatistical analysis

A geostatistical analysis is a very useful approach that allows providing accessible information related to the ambient air quality monitoring network. Then, the geographical information system (GIS) as a set of powerful techniques of geoprocessing, mapping and spatial analysis is capable in terms of distribution and spatial patterns for each air pollutants [7].

The GIS estimated air pollution surfaces can be certainly useful for further analysis of the air pollution series in the urban area at different time scales and prediction purposes of air pollutants for exposure studies.GIS was used in the work to visualize the results of statistical analyses of estimated number of attributable cases of total mortality caused by long-term exposure on $PM_{2.5}$ in 2010 against the average of 2008–2018 and the estimated number of attributable cases of patients hospitalized for cardiovascular (HACD) and respiratory (HARD) diseases in Poland in 2010 caused by short-term exposure on $PM_{2.5}$.

3 Research results and discussion

3.1 Characteristics of dust pollution in selected Polish cities

Cities in Poland, including the largest agglomerations, show very high diversity in terms of air quality. This also applies to particulate matter, the concentration of which was analysed. The areas with lower concentrations of this pollution include the agglomerations: Trójmiejska and Szczecińska, while the highest concentrations occur in Rybnicko-Jastrzębska, Górnośląska and Krakowska. The annual variability course of PM_{10} and $PM_{2.5}$ concentrations indicates a declining trend, with the occurrence of peaks in 2010 in most cities (Fig. 1. – example for $PM_{2.5}$). This year was characterized, among other things, by such meteorological conditions in winter, that are conducive to an increased emission of particulate matter, mainly from the domestic heating sources – e.g. a large number of days with the minimum temperature below 10°C [8]. Such conditions, combined with low wind speed and the occurrence of the temperature inversion phenomenon, lead to an increase in emission and weakening dispersion conditions and as a consequence to higher pollution. The occurrence of smog episodes is quite common in autumn and winter in Poland, including in the analysed agglomerations [9–11].

In the last four years, a stable situation with respect to the average concentration of $PM_{2.5}$ can be observed in most cities, even with increases in some agglomerations.

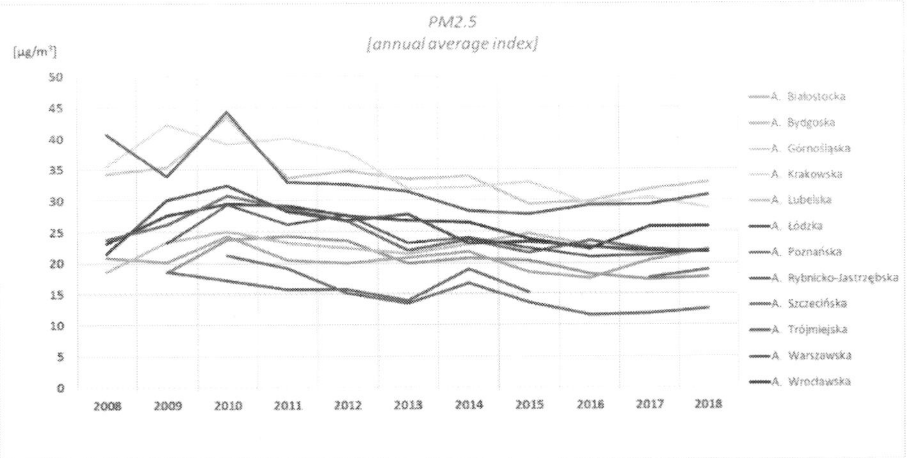

Fig. 1. Variability of annual $PM_{2.5}$ indices in agglomerations in 2008–2018 *(source: own elaboration based on data from the State Environmental Monitoring)*.

PM_{10} and $PM_{2.5}$ concentrations for all cities show large seasonal, weekly and daily variations. An example of data from 2017 for one of the stations located in the Wrocław Agglomeration is shown in Figure 2. January and February were characterized by very high concentrations, including deep episodes – average concentrations significantly differed in this period from the rest of the year. The concentrations on weekend days are noticeably lower, which can be attributed to smaller car traffic that impacts this station. Such an influence is also indicated by analysis of daily variation, where traffic peaks are clearly visible, but they are clearer (especially in the morning) for the larger fraction of measured dust (PM_{10}).

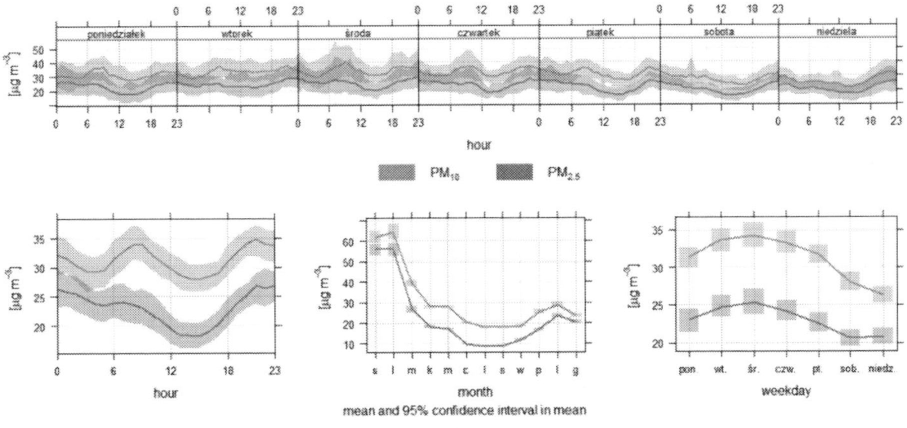

Fig. 2. Seasonal, weekly and daily PM_{10} concentration variability at the Wrocław-Korzeniowskiego station in 2017 *(source: own state-owned monitoring based on the State Environmental Monitoring)*.

3.2 Short and long-term exposure of inhabitants of Polish cities to dust concentrations $PM_{2.5}$ and PM_{10}

The results of calculations of the estimated number of deaths caused by long-term exposure to $PM_{2.5}$ above 10 $\mu g/m^3$ and the results of calculations of the estimated number of patients hospitalized for cardiovascular (HACD) and respiratory system diseases (HARD) in 2010 due to short-term exposure on $PM_{2.5}$ are showed on Figures 3 and 4.

Due to the highest average concentration in analysed period, the risk for 2010 was compared to the medium risk in 2008–2018. As predicted, in 2010 the estimated risk was higher. For the majority of agglomerations, this risk increased by more than half. The exception here were the Wrocław and Trójmiasto agglomerations, where the risk has increased on average two and three times, respectively. However, the highest estimated risk attributed to total mortality due to long-term exposure to PM2.5 in 2010 was calculated for the Górnośląska, Krakowska and Rybnicko-Jastrzębska agglomerations: 22.44%, 19.11% and 18.91%, respectively. The number of deaths that could be avoided was greatest in the agglomerations of Górny Śląsk, Warsaw and Krakow, and amounted to 3962, 2419 and 1422, respectively (Figure 3). Moreover, the estimated risk for both HACD and HARD is greatest in the Gornośląska and Krakowska agglomerations, and is respectively 3.84%, 3.23% for HACD and 3.28%, respectively, and 2.84% for HARD. The largest number of cases potentially excessively hospitalized due to the above health problems were calculated for the Upper Silesian and Warsaw agglomerations: 3222 and 1850 respectively for HACD and 1256 and 721 for HARD (Figure 4).

An analysis of calculations of the estimated number of cases assigned to specific contamination and health effects, which could potentially be avoided in short-term exposure to PM_{10} and $PM_{2.5}$, lowering their concentration below 10 $\mu g/m^3$ in 2017 in Wrocław agglomeration showed that reduction of dust concentration PM2.5 would give the most measurable results in the number of people hospitalized for cardiovascular diseases (over 160 cases), compared to the impact on hospitalizations due to the respiratory ones (about 80 hospitalizations) and total mortality in people over 30 years of age (about 130 deaths). In addition, an estimated more than 80 cases of asthma attacks in children with asthma would potentially avoid reducing PM_{10}. The long-term effect of PM10 dust on postneonatal infant mortality in this period (2017) is 3 excess deaths.

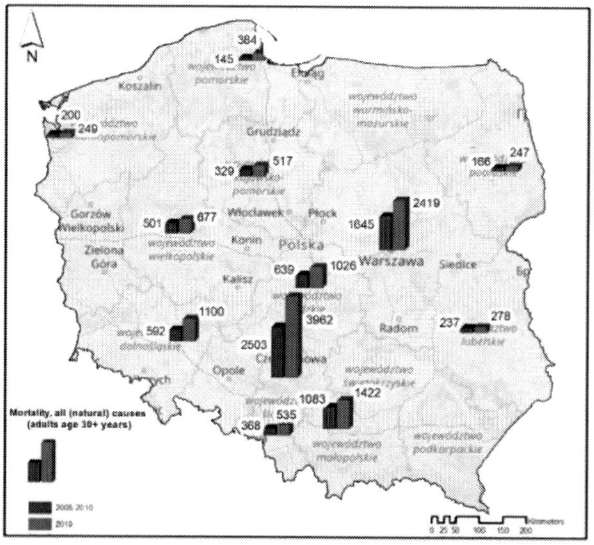

Fig. 3. Comparison of estimated number of attributable cases of total mortality caused by long-term exposure on $PM_{2.5}$ in 2010 against the medium of 2008–2018.

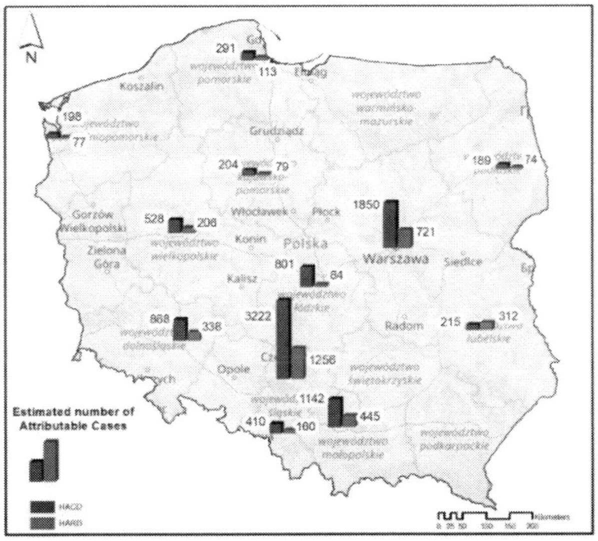

Fig. 4. Estimated number of attributable cases of patients hospitalized for cardiovascular (HACD) and respiratory (HARD) diseases in Poland in 2010 caused by short-term exposure on $PM_{2.5}$.

3.4 Analysis of main directions of activities aimed at improving air quality in Polish cities

Formal activities related to the improvement of air quality in Polish cities depend mainly on legal solutions and subsequent financing of their effects. In this context, cities use the law laid down by the Polish national parliament and regional councils. Local activities, being a commune or urban initiative, are complementary. Actions such as: a day without the car, tree planting or dissemination of information about threats resulting from exceeding limit values of air pollution are primarily aimed at educating and shaping social attitudes. However, they do not eliminate the causes of pollution.

In order to achieve the observance of environmental quality standards, in Poland, based on the Act – Environmental Protection Law (Journal of Laws 2001 No. 62 item 627 with later amendments), air protection programs adopted by way of a resolution by voivodship parliaments are created. They contain recommendations being an obligation addressed mainly to local governments and, therefore, also to cities. Unfortunately, this mechanism is ineffective as evidenced by the state of atmospheric pollution and the need to introduce additional restrictions. Therefore, on the basis of the amendment to the Act – Environmental Protection Law, which was passed on November 12, 2015, regional councils gained the possibility of introducing restrictions and prohibitions on the operation of installations in which fuels are burned - adoption of the so-called anti-smog resolutions (Table 3).

In parallel to the preparation of the anti-smog resolutions, social pressure has caused the preparation of system solutions at a central level in Poland. The ordinance of the Minister of Development and Finance on requirements for solid fuel boilers has been adopted, which formally eliminated from 01.07.2018 the sale of boilers that do not comply with EN 303-5: 2012. The second significant regulation, issued by the Minister of Energy, defines the quality requirements for solid fuels. It uses a two-year grace period for the use of coal fines below 3 mm in diameter. In this respect, the national provision is more liberal than regional solutions. This results in maintaining a trade in fuels whose combustion is prohibited by regulations adopted by the regional councils. This paradox sustains illegal combustion of high-emission solid fuels by 30 June 2020.

Table 3. A list of voivodships in which the regional councils have introduced restrictions and prohibitions regarding the operation of installations in which fuels are burned pursuant to art. 96 of the Act – Environmental Protection Law (drawn up on the basis of resolutions of regional councils included in the analysis).

	Voivodship	Separate regulations for cities	The date of the most stringent prohibition	Requirements for fuels and installations
1.	dolnośląskie	Resolution adopted on 30.11.2017		
		Wrocław	**01.07.2028**	**The target total elimination of solid fuels.**
		Spas	**01.07.2028**	**Target total elimination of solid fuels in seven locations, in four locations permitted installations in accordance with EC Regulation (EU) 2015/1185 with regard to ecodesign requirements.**
2.	lubuskie	Resolution adopted on 18.06.2018		
		Gorzów Wielkopolski	**01.01.2023**	**Class 5 installations according to PN-EN 303-5: 2012.**
		Zielona Góra	**01.01.2023**	**Class 5 installations according to PN-EN 303-5: 2012.**
3.	łódzkie	Resolution adopted on 24.10.2017		
4.	małopolskie	Resolution adopted on 23.01.2017		
		Kraków	**01.09.2019**	**The target total elimination of solid fuels.**
5.	mazowieckie	Resolution adopted on 24.10.2017		
6.	opolskie	Resolution adopted on 26.09.2017		
7.	podkarpackie	Resolution adopted on 23.04.2018		
8.	śląskie	Resolution adopted on 07.04.2017		
9.	wielkopolskie	Resolution adopted on 18.12.2017		
		Poznań	**01.01.2028**	**Installations in accordance with the EC Regulation (EU) 2015/1185 with regard to ecodesign requirements.**
		Kalisz	**01.01.2028**	**Class 5 installations according to PN-EN 303-5: 2012**
10.	zachodniopomorskie	Resolution adopted on 26.09.2018		

4 Conclusions

The analysis of research results and calculations carried out indicate higher concentrations of dust contamination, which may contribute to the deterioration of the quality of life of the inhabitants of Polish cities. Reduction of air pollution in Polish cities should therefore become a priority in activities in the field of environmental protection in Poland.

A real improvement of air quality in cities will be possible only after the transformation of heating systems and a widespread use of zero-emission or low-emission heat production technology with a simultaneous improvement of energy efficiency in housing, housing estates and transport systems. For this purpose, a radical change in the energy policy of the state is needed. In this context, the energy policy and the air quality policy pursued by regions and cities will be a secondary activity. Programs for co-financing changes in heat sources, which are mainly responsible for air pollution, should be considered as the initial stage of converting the heat protection system in Poland.

The investigations were co-financed within the framework of the order No. 0401/00581/18.

References

1. WHO Regional Office for Europe, *Health risk assessment of air pollution – general principles* (Copenhagen, 2016)
2. Ministerstwo Zdrowia, *Program Polityki Zdrowotnej* (2017)
3. Ministertwo Zdrowia, *Narodowy Program Wyrównywania Dostępności do Profilaktyki i Leczenia Chorób Układu Sercowo-Naczyniowego POLKARD na lata 2013–2016* (2014)
4. Ministerstwo Zdrowia, *Krajowe ramy strategiczne* (2015)
5. E. Karczewicz, A. Kania, *Wydatki na świadczenia z ubezpieczeń społecznych związane z niezdolnością do pracy w 2010 r.* (2012)
6. B. Wojtyniak, P. Goryński, B. Moskalewicz, *Sytuacja zdrowotna ludności Polski i jej uwarunkowania* (pod red. NIZP-PZH, Warszawa, 2011)
7. J. Gulliver, K. de Hoogh, D. Fecht, D. Vienneau, D. Briggs, Atmos. Environ, **45**, 7072–7080 (2011)
8. A. Chlebowska-Styś, I. Sówka, D. Kobus, Ł. Pachurka, E3S Web of Conferences **17**, 1–8 (2017)
9. D. Kobus, A. Nych, I. Sówka, E3S Web of Conferences **44**, 00070 (2018)
10. J. Iwanek, J. Strużewska, J. Kamiński, P. Durka, D. Kobus, J. Kostrzewa, T. Pecka, *Analiza wybranych epizodów wysokich stężeń pyłu PM10 z lat 2013–2016, Etap II, Epizody z lat 2015–2016* The work done for CIEP (Instytut Ochrony Środowiska – Państwowy Instytut Badawczy: IOŚ PIB, 2017)
11. K. Juda-Rezler, B. Toczko (ed.), *Pyły drobne w atmosferze. Kompendium wiedzy o zanieczyszczeniu powietrza pyłem zawieszonym w Polsce* (Inspekcja Ochrony Środowiska: IOŚ, Biblioteka Monitoringu Środowiska, Warszawa, 2016)
12. Rozporządzenie Ministra Rozwoju i Finansów z dnia 1 sierpnia 2017 r. w sprawie wymagań dla kotłów na paliwo stałe
13. Rozporządzenie Ministra Energii z dnia 27 września 2018 r. w sprawie wymagań jakościowych dla paliw stałych
14. Ustawa – Prawo ochrony środowiska Dz. U. 2001 Nr 62 poz. 627 z pózn. zm.

Energy balancing in ground heat exchanger for heat pump systems – a case study with simulations

Ewelina Stefanowicz[1,*], and *Małgorzata* Szulgowska-Zgrzywa[1]

[1]Wrocław University of Science and Technology, Wybrzeże Stanisława Wyspiańskiego 27, 50-370 Wrocław, Poland

Abstract. The negative impact of systems based on fossil fuel on the environment and the desire to promote sustainable development is increasingly encouraging building owners to invest in renewable energy sources. One of the possibilities of using a renewable energy source located in shallow soil layers is a brine-to-water heat pump. Ground energy can be used for both heating and cooling buildings. This article presents the advantages of balancing the supply and energy consumption from the ground for the large heat pump systems. Authors presented a case study and the simulation of the system operation with different cooling load share vs heating load. This was done on the basis of the analysis of data from several years of measurements from the actual installation as well as the simulation in Earth Energy Designer software. The results of the simulations shows that the most advantageous strategy for the operation of the analysed installation is to provide 50 to 75% of the cooling load in relation to the heating load.

1 Introduction

Poland is one of the European countries where air pollution caused by heat sources based on fossil fuel is the largest. The main goal for the Polish energy industry is the sustainable development of renewable energy sources. One of these technologies, using the renewable energy as a heat source is ground-source heat pumps (GSHPs). In Poland, there are still not many GSHPs installations due to high investment costs (in 2017, 5120 GHP installations [1]). The most common type of ground heat exchanger are single U-tube probes. The probes have a depth of 30 to 200 m, and the space between the pipe and the ground is filled with a material based on bentonite with high thermal conductivity. The most commonly used fluid is a 25% solution of propylene glycol. In practice, other factors are rarely used but there is potential to improve the GSHPs thermal parameters due to the use of nano-fluids or water as described in the previous article [2]. Due to the constantly growing interest in GSHPs, more and more detailed research is being conducted to improve the

* Corresponding author: ewelina.stefanowicz@pwr.edu.pl

efficiency of these installations. Incorrect operation of the lower source significantly affects the system efficiency, the coefficient of performance, and thus the system operating costs, extending the investment return time. The impact of an unbalanced GSHPs system on the environment is described in article [3]. It was emphasized that in the case of a building with unbalanced heating and cooling load, the temperature changes around the ground heat exchanger are more and more visible in subsequent years of operation. The use of these systems in buildings with unstable heating and cooling demand and with frequent peak load operation may even lead to freezing of the ground heat exchanger [4, 5]. Therefore, for large systems, the pilot drilling and the thermal response tests are in practice carried out to determine the effective thermal conductivity of the soil. An overview of existing heat exchange models for GSHPs was presented by Javed et al. [6]. Currently built GSHPs installations have a very diverse heating and cooling load. The research shows that long-term of large-scale, multiple-probes installations operation has less sensitivity to heat exchange disturbances in the lower source, e.g. groundwater flow, under the condition of a balanced heating load [5]. In other publications, researchers conduct an analysis of ground temperature changes based on a numerical model for different thermal loads, borehole configurations and indicate the necessity of analyses for changes in the distance between BHE's in further studies [7, 8]. The presented article shows a case study – analysis and simulations of the heat pump's lower source system. Based on the results of the measurements, the model of the lower energy source of the heat pump system in the Earth Energy Designer (EED) was calibrated. This model was used to analyse the operation of this installation with five different share of the cooling and the heating load.

2 Case study

2.1 Description of the installation

In this article, the authors analysed the operation of the heat pump's lower source installation operating for the heating and cooling system in a commercial and service building. The heat source in the installation is a brine-to-water heat pump cooperating with the lower source consisting of several independent drilling fields. Each field consists of 20 boreholes 130 m deep each. The pipe in the borehole have a diameter of 40 x 3.0 mm. The distance between the BHEs is 5.3 m. The heat carrier fluid in the installation is a 25% propylene glycol solution. The volumetric heat carrier fluid flow rate is 0.15 dm^3/s per borehole. Each borehole's field is connected to a wells. The measurement and analysis results, described below, were carried out for one of the three boreholes groups, i.e. for 40 probes. The thermal load of the soil is about 200 MWh per year for the each group. The cooling load is about 100 MWh per year.

2.1 Results of the measurements

In the analysed system the heating energy (Q_H) taken from the ground is balanced with the cooling energy (Q_C) delivered in the summer. On average, about 40% of the energy collected in the winter is given back to the ground in a regeneration process realized by the free-cooling (Fig. 1). The system of the heat pumps' low enthalpy energy sources works correctly. Despite the high heating load and a small spacing between boreholes, there is no significant decrease in the average glycol temperature in the following years, the value of this temperature is close to 11°C for all three years (Fig. 2). It should be emphasized, that in the first stage of investment implementation the vertical boreholes in this installation were intended to be used only as a source of thermal energy. However, due to the very small distances between the boreholes, there was a risk of freezing of the exchanger. In the case of operation of this installation only with a cooling load, the effect of freezing of the lower source would appear in the fifth year of operation, as shown in the Fig. 4. Unbalanced energy consumption from the ground may result in an irreversible environmental effect. The freezing ground around the boreholes will pose a serious threat to the plants and the direction of the natural groundwater flow. It is therefore necessary to properly balance the heating and cooling energy demand. That is why it was decided to change the strategy of exploiting the lower heat source. The analysis of the work of this system for various strategies of the energy balances is shown in the next part of this article.

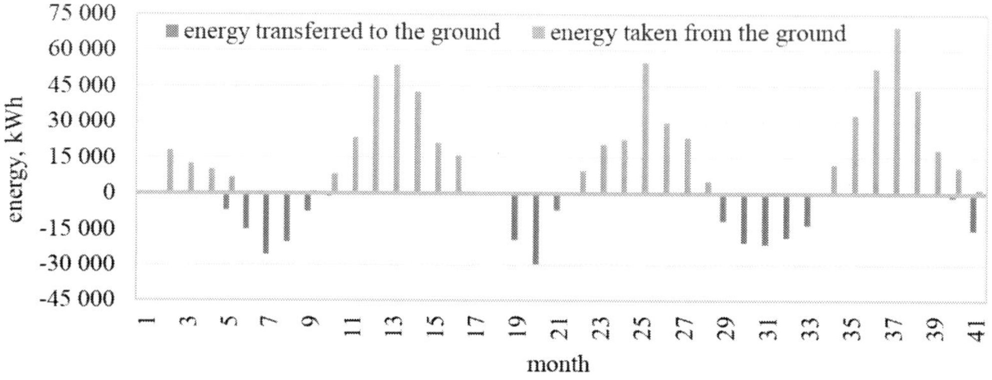

Fig. 1. The heating and cooling load of the building in analysed installation.

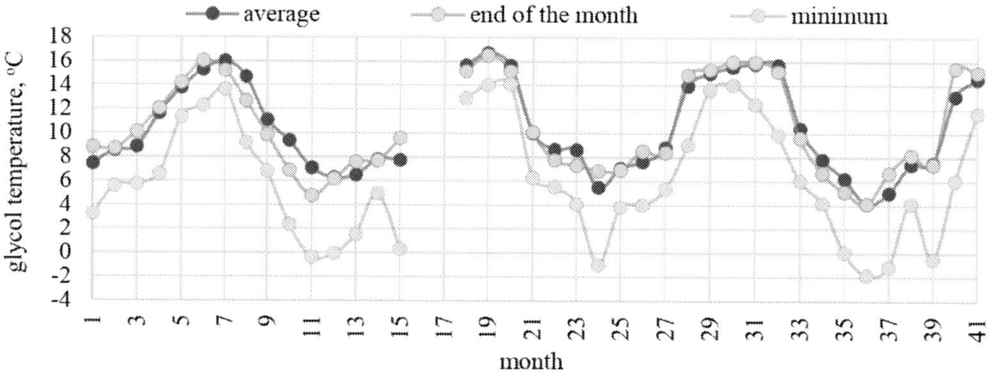

Fig. 2. The average glycol temperature (measurements).

2.2 Calibration of the lower source model

The construction of the ground heat exchanger, for the brine-to-water heat pump consisting of more than 10 boreholes, requires a detailed analysis of the ground parameters. In the case of the analysed installation, TRT test were carried out [9]. The results of the calculation, based on the linear infinite source model, provided the information about the effective thermal conductivity coefficient of the ground amounting 1.93 W/(mK). The undisturbed ground temperature was 10.85°C. The thermal capacity of the ground, estimated based on the literature, is 1.79 MJ/(m^2K). The model of this installation was prepared in Earth Energy Designer (EED). The model was calibrated based on the ground parameters and the measurements: the heating and cooling load and the glycol temperatures in the first year of system operation.

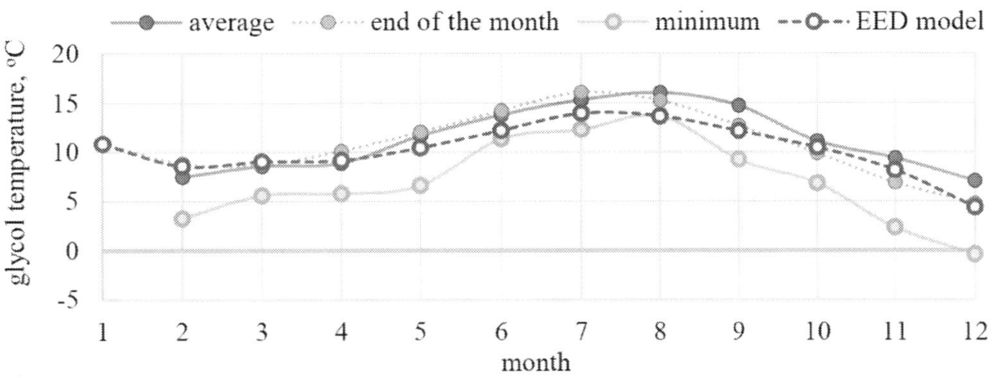

Fig. 3. The average glycol temperature, the simulation and the measurements in the borehole (2014).

The Fig. 3. shows the average heat carrier fluid temperature for the boreholes field (40 probes) during the first year of the system operation and the results of the simulation in the EED program. Earth Energy Designer is a flexible software for designing and analysing the ground heat exchangers installation. It takes into account the boreholes configuration, their technical parameters, ground and heat carrier fluid parameters as well as the heating and cooling load of the installation. The amount of the energy demand can be given in detail (for each month) or for whole year with a percentage breakdown for individual months. The program also allows to specify the time of operation and the first month of installation work. The model in EED reflects very well the actual work of the installation. The average relative uncertainties of the simulation of the average temperature of the heat-carrier fluid (T_{avg}) in individual months are 9%. The model shows the biggest mismatch in May and June (up to 14%) and in November (18%).

3 Analysis of the lower source configuration

3.1 Analysis variants

The purpose of the simulation studies described below is to determine the correct balance of the heating energy (Q_H) taken from the ground in winter and the cooling energy (Q_C) delivered to the ground during the summer. The following temperatures limits were adopted in these analysis: 0°C - is the minimum temperature of the heat carrier fluid in winter (for the protecting against freezing), and 20°C - is the maximum temperature of the heat carrier fluid in summer (for the protecting against overheating). Table 1. presents the parameters of subsequent simulations, i.e. the ratio of the heating energy (Q_H) to the regeneration energy (Q_C). Table 2 presents the monthly distribution of heating and cooling energy based on data from three years of actual work of the installation.

Table 1. Simulation parameters: the share of the cooling and the heating energy.

Simulation	1	2	3	4	5
Q_H	100%	100%	100%	100%	100%
Q_C	0%	25%	50%	75%	100%

Table 2. Simulation parameters: the monthly energy profile.

Month	I	II	III	IV	V	VI	VII	VIII	IX	X	XI	XII
Q_H	28%	15%	10%	5%	2%	0%	0%	0%	0%	5%	13%	21%
Q_C	0%	0%	0%	0%	11%	22%	27%	28%	11%	1%	0%	0%

Each simulation was carried out for two variants. In the first variant, the size of the installation corresponded to the size of the actual installation for which the model was prepared. If the analysis of the results indicated an unacceptable temperature of the working medium in subsequent years of operation (beyond the accepted limit values 0°C in winter and 20°C in summer) then the next stage of the analysis was to increase the distance between boreholes (increase the energy storage volume) so as to maintain the required temperature regime. It should be noted that the problem of limited space available for the installation of a ground heat exchanger is particularly important in the urban area, hence it was decided to discuss this issue. The results of the analyses and the discussion are presented below.

3.2 Simulation 1 – no cooling load

The first variant of the analysis presents the operation of the installation with a high heating load and no cooling load. As can be seen in the Fig. 4., the average heat carrier fluid temperatures drop below 0°C in the fifth year of system operation for a configuration in which space between the boreholes is 5.3 m. After the 10th year of system operation, the risk of the lower source freezing is significant. In order for the stable operation of the lower source throughout the entire operating period, the distance between the boreholes should be increased at the design and execution phase of the system up to 15 m.

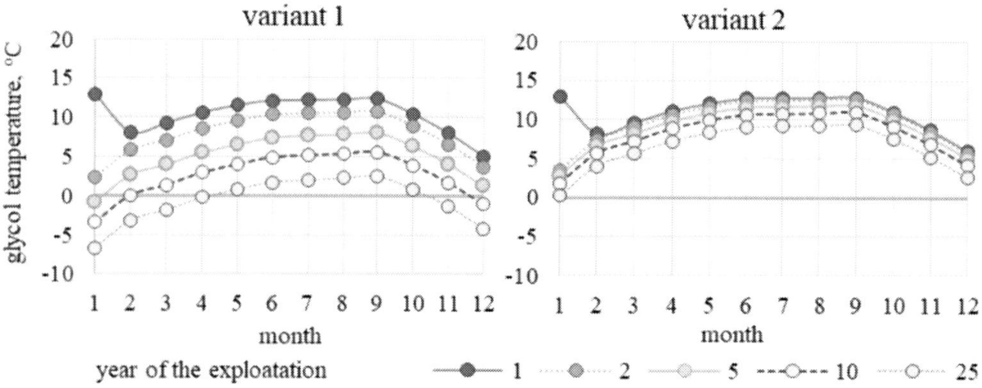

Fig. 4. Results for the simulation 1: variant 1 – distance between the borehole 5.3 m (area of the exchanger 759 m^2); variant 2 – distance between the borehole 15 m (area of the exchanger 6075 m^2).

The results of this simulation show the effect of lower source self-regeneration. If the distance between the boreholes is maintained, the ground source in the summer period is regenerated, thanks to which the temperature differences between the consecutive years of operation are significantly lower than in the case of the installation in which the distance between boreholes is 5.3 m. In order to enable self-regeneration of the ground source, the area increases from 758.5 m^2 to 6075 m^2.

3.3 Simulation 2 – 25% cooling load

The results of the analysis are shown in the Fig. 5. The first variant of the analysis shows the regeneration of the system with a 25% cooling load in relation to the heating load, with a spacing of 5.3 m between boreholes. Average fluid temperatures lower than 0°C appear in the 10th year of operation. This is a significant improvement over variant 1 in simulation 1. Increasing the distance between probes up to 11.5 m ensures stable operation of the system over the entire exploitation period, the average temperature of the glycol never drops below 0°C (variant 2). This results in an increase in the boreholes field's area from 758.5 m^2 to 3570.8 m^2.

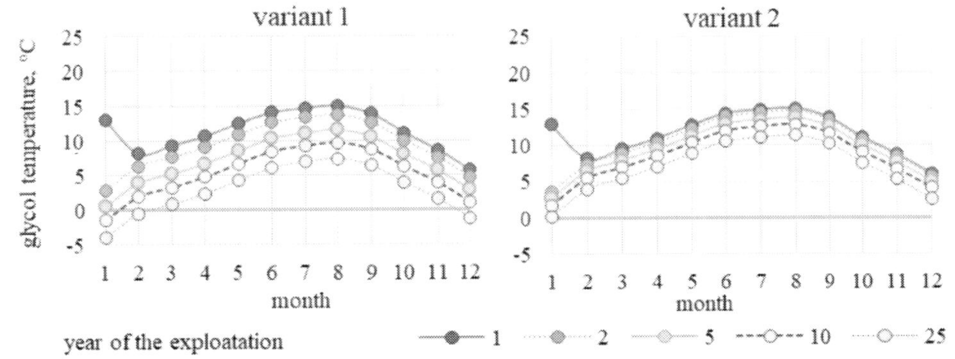

Fig. 5. Results for the simulation 2: variant 1 – distance between the borehole 5.3 m (area of the exchanger 759 m^2); variant 2 – distance between the borehole 11.5 m (area of the exchanger 3571 m^2).

3.4 Simulation 3 – 50% cooling load

The results of the analysis are shown in the Fig. 6. The work of the lower source with the regeneration corresponding to 50% of the heating load ensures stable operation of this installation. The working fluid temperature below 0°C appears only in the 25th year of operation, but the risk of the exchanger freezing at such values is negligible. To ensure the assumptions for the analysis are met (the glycol temperature throughout the analysis period above 0°C) it would be necessary to increase the distance between probes up to 7 m. The surface area of the ground heat exchanger would increase to 1323 m^2.

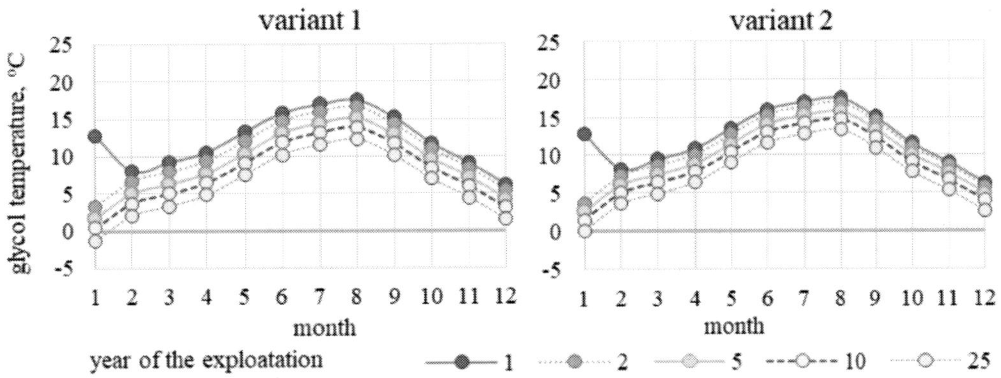

Fig. 6. Results for the simulation 3: variant 1 – distance between the borehole 5.3 m (area of the exchanger 759 m^2); variant 2 – distance between the borehole 7 m (area of the exchanger 1323 m^2).

3.5 Simulation 4 – 75% cooling load

The results of the analysis are shown in the Fig. 7. The work of the lower source with the regeneration corresponding to 75% of the heating load ensures stable operation of the installation. The effect of freezing the lower source does not appear. Theoretically, providing this amount of the regeneration energy to the system give the possibility to reduce the distance between the boreholes to 4 m. The surface area of the ground heat exchanger then decreases to 432 m². This is a very good alternative for the vertical boreholes installations with limited area.

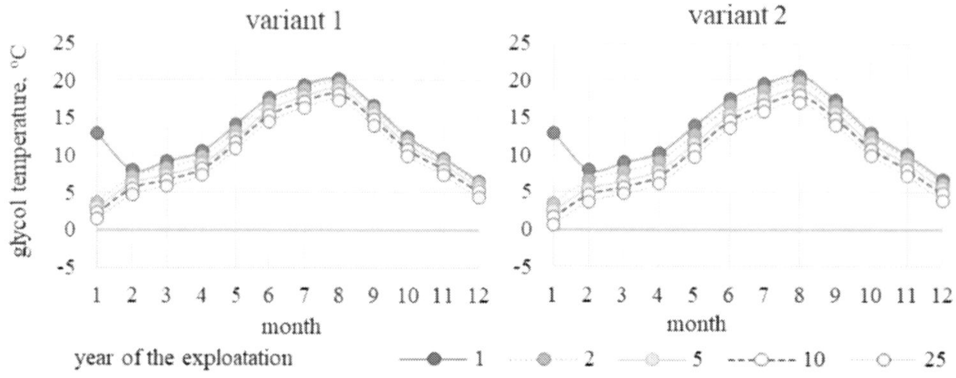

Fig. 7. Results for the simulation 4: variant 1 – distance between the borehole 5.3 m (area of the exchanger 759 m²); variant 2 – distance between the borehole 4 m (area of the exchanger 432 m²).

3.6 Simulation 5 – 100% cooling load

The results of the analysis are shown in the Fig. 8. Designing systems with the excess regeneration may cause the problem of overheating the ground around BHE in the summer. In the analysed case, the equalization of the energy balance between winter and summer periods causes that the average glycol temperature exceeds 20°C and even increasing the distance between boreholes up to 15 m does not allow to reduce this temperature.

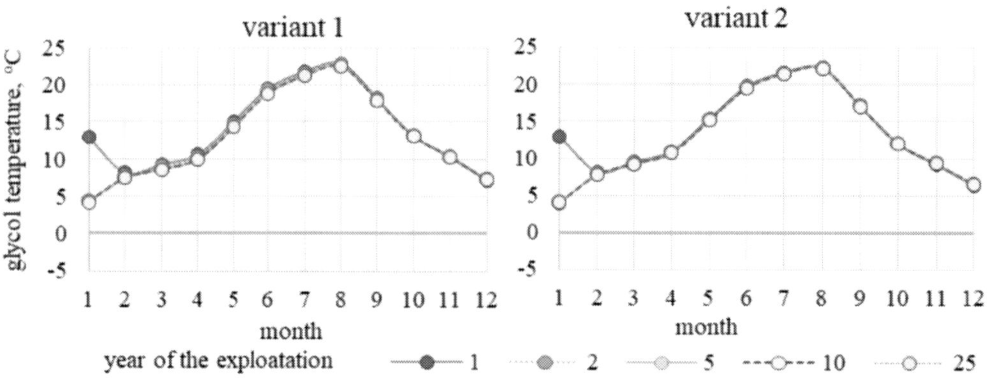

Fig. 8. Results for the simulation 5: variant 1 – distance between the borehole 5.3 m (area of the exchanger 759 m²); variant 2 – distance between the borehole 15 m (area of the exchanger 6075 m²).

It follows that in this case the intensive regeneration of the ground is not recommended. Ground humidity may be reduced and thus the heat exchange may be limited. An increase of the ground temperature around the BHE, especially in cities, where the area designated for ground heat exchanger is small, usually can lead to thermal intrusion and thus the conflict of interests of building owners in urban agglomerations [9].

4 Conclusions

Ground heat pumps are a very good alternative for heat sources based on fossil fuel. Especially for large installations, it is necessary to design them properly. Starting from the determination of soil parameters through a pilot drilling and the TRT test. An important element is the long-term operation analysis that allows to improve work parameters as shown in publications [2, 10]. It should be emphasized that in the case of large installations, if an appropriate proportion of heating and cooling energy loads and the size of the ground energy storage are not ensured, it may occur to irreversible environmental effects. Other studies also show the importance of balancing the heat load of the heat pump's lower source installation. Researchers analysed changes in ground temperature for buildings with different heating and cooling energy load (hospital, fast-food restaurant, residence, school) based on a mathematical model [11]. Analyses were also made in terms of the size of the energy store. The research results show the importance of accurate simulations, modelling and designing of lower sources, and indicate the importance of further analyses towards systems with extreme heating and cooling energy needs [12]. In this study, the authors made analyses for a system with a very high energy demand. The analysis criterion was to maintain the working medium temperature in all years of operation above 0°C in the winter and 20°C in the summer and to indicate the possibility of changing the distance between BHE's. Regeneration of the ground allows reducing the distance between BHE's without adversely affecting the ground temperature. Monitoring and analysis of the system's operation allow to notice disturbing signals and the analysis of the installation's operation allows to develop a long-term strategy for improving the operation of the lower energy source system. Importantly, even if the initial project of the installation did not provide for ground regeneration, it is possible to correct this situation. The most advantageous strategy for the operation of the analysed installation is to provide 50 to 75% of the cooling load in relation to the heating load. It will allow to improve the work parameters of the installation, i.e. it will ensure stable temperature of the heat carrier fluid both in the heating and cooling season. It will positively affect the heat pump's coefficient of performance, and thus the system operating costs. It should also be emphasized that the cooling energy obtained in the process of free-cooling is energy with very low production costs (related only to the operation of circulation pumps), which has a positive effect on the final building energy.

References

1. PORT PC, *The heat pump market in Poland in 2010-2017. Prospects for the development of the heat pump market by 2030* (on-line: portpc.pl/pdf/raporty/Raport_Rynkowy_PORT_PC_2018.pdf, access: 10 Jan 2019)
2. E. Stefanowicz, M. Szulgowska-Zgrzywa, N. Fidorów-Kaprawy, E3S Web of Conf. **44** (2018)
3. N. Daemi, M. M. Krol, Renewable Energy, **134** (2019)
4. D. Banks, *An Introduction to Thermogeology : Ground Source Heating and Cooling* (Blackwell, Oxford, 2008)
5. S. E. Dehkordi, R. A. Schincariol, B. Oloffson, Groundwater **53**, 558–571 (2014)
6. Javed, P. Pahlen, J. Claesson, *Vertical ground heat exchangers : a review of heat flow models* (Effstock2009, Stockholm, 2009)
7. X. Zheng, *Long-term Effects of Ground Source Heat Pumps on Underground Temperature* (Comsol Conf., Boston, 2013)

8. P. Bayer, M. De Paly, M. Beck, Appl. Energy, **136** (2014)
9. J. D. Spitler, S. E. A. Gehlin, Ren. and Sust. Energ. Rev. **50**, 1125 (2009)
10. E. Stefanowicz, N. Fidorów-Kaprawy, E3S Web of Conf. **17** (2017)
11. W. Choi, R. Ooka, Y. Nam, Sust. Cit. and Soc. **38**, 429 (2018)
12. Y. L. E. Law, S. B. Dworkin, Appl. Energy, **179** (2016)

Artificial neural network (ANN) approach for modeling of selected biogenic compounds in a mixture of treated municipal and dairy wastewater

Joanna Struk-Sokołowska[1], *Piotr* Ofman[1,*], and *Sevgi* Demirel[2]

[1]Bialystok University of Technology, Department of Civil and Enviromental Engineering Technology and Systems, 15-351 Bialystok, Poland
[2]Niğde Ömer Halisdemir University, Department of Environmental Engineering, 51240 Merkez/ Niğde, Turkey

Abstract. This paper presents artificial neural network (ANN) model of wastewater treatment plant, which was used for average monthly concentrations of $N\text{-}NH_4^+$, $N\text{-}NO_3^-$, $N\text{-}NO_2^-$, total Kiejdahl nitrogen (TKN), PO_4^{3-} and SO_4^{2-} approximation. ANN model was developed for wastewater treatment plant located in Bystre, Poland which treats municipal wastewater with a share of dairy wastewater. The object was chosen because of the unique location, in the Great Mazury Lakes area and the need for its special environmental protection. Input layer of developed ANN model consisted of BOD, COD, concentrations of total nitrogen and total phosphorus, total organic carbon, sulphates, wastewater temperature and pH., The developed model reflected extreme values observed during study period. Average error percentage with which output variables were approximated equalled to 35.35%; 8.99%; 21.23%; 5.08%; 10.99%; 3.02% respectively for $N\text{-}NH_4^+$, $N\text{-}NO_3^-$, $N\text{-}NO_2^-$, TKN, PO_4^{3-} and SO_4^{2-}.

1 Introduction

Chemical composition of surface waters is deteriorating due to discharges of treated wastewater from industry or major urban agglomerations among other reasons [1, 2]. The activated sludge method allows the reduction of compounds decomposable through biological and chemical ways with a satisfactory result but does not provide their complete removal from purified medium [3]. More often the design of wastewater treatment plants is based on computer simulations instead of expensive and time-consuming installations or pilot-scales experiments [4], especially those which are using activated sludge as their primary treatment method [5, 6]. The degree to which the wastewater will be treated in the classical mechanical-biological system is not constant and depends on factors difficult to estimate some of which are not related to activated sludge technological parameters [7, 8]. A high quality representative model can provide a favourable solution in the process control

* Corresponding author: p.ofman@pb.edu.pl

and helps to explain the real process performance and to develop a continuous control strategy for this type of technologies. Because of their reliable, robust and salient characteristics in capturing the non-linear relationships existing between variables (multi-input/output) in complex systems, it has become apparent that numerous applications of ANNs have been successfully conducted in various parts of environmental engineering field [9]. Available activated sludge process modelling software is extensive and allows for accurate description of unit processes that occur during the biological wastewater treatment process [10]. The current models of pollutant removal often concern the municipal and industrial wastewater treatment [11]. Unlike commercial software models, mathematical models created on the basis of empirical research conducted in facilities are usually the most accurate and allow for a fuller change reflection that occur during biological wastewater treatment [12] including biogenic compounds [13], suspensions and micro-pollutants. These play an increasingly important role in the aquatic environment [14] and their disinfection [15, 16]. Therefore, mathematical modelling of biological wastewater treatment plants can be one of optimized and novel method for biogenic compound outflow within treated wastewater. Dedicated model also allows to identify the main factors affecting treated wastewater quality, independent from technological parameters [17].

The aim of this study was the development of a model approximating average monthly concentrations of $N\text{-}NH_4^+$, $N\text{-}NO_3^-$, $N\text{-}NO_2^-$, total Kiejdahl nitrogen (TKN), PO_4^{3-} and SO_4^{2-} in treated wastewater from municipal WWTP with dairy wastewater inflow (23%). Prediction of these selected biogenic compounds in treated wastewater with artificial neural networks is rarely analysed by researchers and it is difficult to find information on this subject in literature. This paper is a continuation of research and results presented in the article by Skoczko et. al 2017.

2 Materials and methods

2.1 WWTP in Bystre

The wastewater treatment plant in Bystre, Poland was chosen because of the unique location, in the Great Mazury Lakes area (North Poland) and the need for its special environmental protection. Figure 1 presents WWTP in Bystre technological scheme. WWTP was commissioned in 1995. The facility was modernized in 2002-2003. The maximum daily capacity of the treatment plant is 14,000 $m^3 \cdot d^{-1}$, while the average daily volume of wastewater reaches 6,400 $m^3 \cdot d^{-1}$. The equivalent number of inhabitants (PE) for the facility is 98,615. The municipal wastewater treatment plant accepts dairy sewage from the District Dairy Cooperative in Giżycko. Wastewater delivered to the plant is deposited to a septic tank at the level of about 9,000 $m^3 \cdot year^{-1}$ to the catchment point located on site. Annual amount of municipal and dairy wastewater that inflows to WWTP are given in table 1.

Raw wastewater flows into WWTP by a gravity and pressure sewage system. Mechanical wastewater pre-treatment takes place in gratings building. The screenings accumulating on the grating are automatically directed to the auger conveyor. The next stage of mechanical treatment is the sand trap, in which, due to the release of wastewater flow, precipitation and sedimentation of the mineral sediment occurs. In addition, the grit chamber is aerated in order to obtain better float of floating parts. After mechanical treatment, wastewater flows into biological chamber system that performs basic biological treatment. In the first stage, wastewater flows into the defosfatation chamber (anaerobic), then into the denitrification chambers (anoxic), to which the nitrate stream is recirculated from the nitrification chambers (internal recirculation). Wastewater from the denitrification chamber flows to the nitrification (oxygen) chambers. In the denitrification and nitrification chambers, biological wastewater treatment takes place based on activated sludge biomass process. After treatment in biological reactors, the mixture of activated sludge and sewage flows through the overflows into separation chambers and further into two secondary settling tanks. In settling tanks, the activated sludge is separated from wastewater in slow flow conditions. The sludge accumulating on the bottom of the settler is picked up into the sludge funnel, moved to the recirculated and excessive sludge pumping station from where it is pumped to the biological treatment system (external recirculation) or as excessive sludge to the gravitational sludge thickener.

Table 1. Average flows and percentage in the mixture of municipal and dairy wastewater delivered to the WWTP in Bystre.

Year	Wastewater		Percentage
	Municipal	Dairy	
	$m^3 \cdot year^{-1}$	$m^3 \cdot year^{-1}$	%
2014	2 318 140	514 725	22.2
2015	2 386 413	567 748	23.8
Average	2 352 277	541 237	23.0

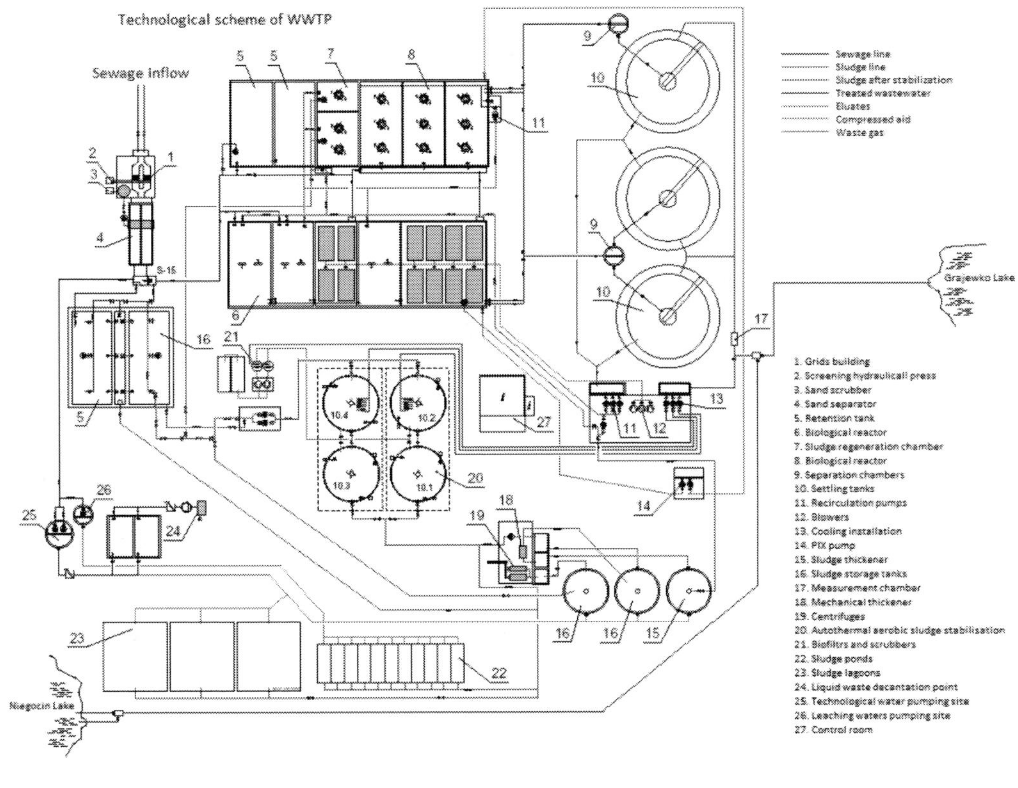

Fig. 1. Technological scheme of WWTP [17].

2.2 Wastewater samples

Analyses of BOD, COD, total nitrogen and total phosphorus, total organic carbon, sulfates, wastewater temperature, pH, $N\text{-}NH_4^+$, $N\text{-}NO_3^-$, $N\text{-}NO_2^-$, total Kiejdahl nitrogen (TKN),

PO_4^{3-} and SO_4^{2-} in raw and treated wastewater were carried out in accordance with APHA [18]. Analyzed samples of mixture of municipal and dairy wastewater were collected monthly in 2014 and 2015.

2.3 Mathematical model

The mathematical model describing the wastewater treatment process in Giżycko was developed based on the research results conducted in 2014 and 2015. The database was created from average concentration of individual parameters observed during study period. Model input neurons were represented by values of BOD, COD, concentrations of total nitrogen and total phosphorus, total organic carbon, sulfates, wastewater temperature and pH. Artificial neural network model was established based on the BFGS calculation algorithm. The learning process was automatically choosing the number of periods after which the cycle calculation stopped. In the case of the presented model the learning process lasted 138 epochs. The error function was the sum of squares and network itself has been prepared on the basis of multilayer perceptron. In total 336 single results were used for the construction of the model. Due to the size of database used for the calculations the training set contained all the laboratory analysis results. The algorithm was developed using licensed software Statistica 13.1 in Polish, working on Windows 10.

3 Results and discussion

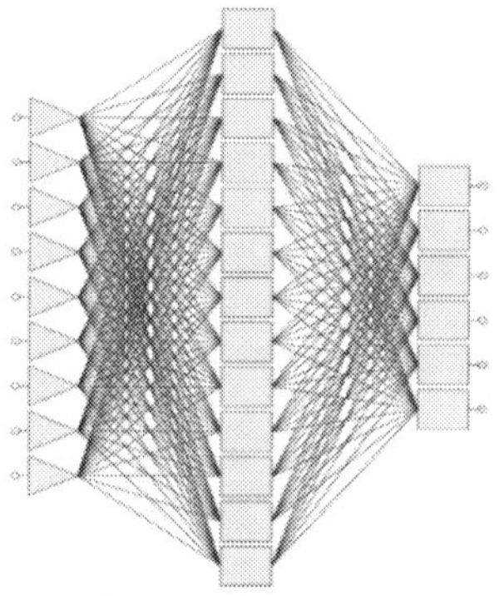

Fig. 2. Topology of artificial neural network

The best among the tested neural network topologies was 9-13-6 which means that none of the variables selected for calculation had been ruled out. Hidden layer of the model consisted of 13 neurons and the output layer of 6. The overall network learning quality was 0.97, and mean learning error equalled 2.55. Activation functions of the hidden layer and output were respectively the exponential function and sinus. Topology of developed mathematical model was shown in figure 1.

The obtained model had a good representation of predicted values with respect to the observed. Match of each variable was described by correlation coefficients in the range from 0.94 to 0.99. Determination coefficients reflecting the linear fit between actual and predicted values ranged from 0.88 to 0.97. The model in the greatest extent reflected changes of PO_4^{3-} in treated wastewater in terms of prediction accuracy the variables were concentration of $N\text{-}NH_4$, SO_4^{2-}, $N\text{-}NO_3$, $N\text{-}NO_2$ and TKN. Model prediction accuracy with graphical interpretation of R^2 coefficient of developed ANN model was shown in detail in figure 3. The variables adopted for analysis for the calculation method of the ANN model are shown in table 2.

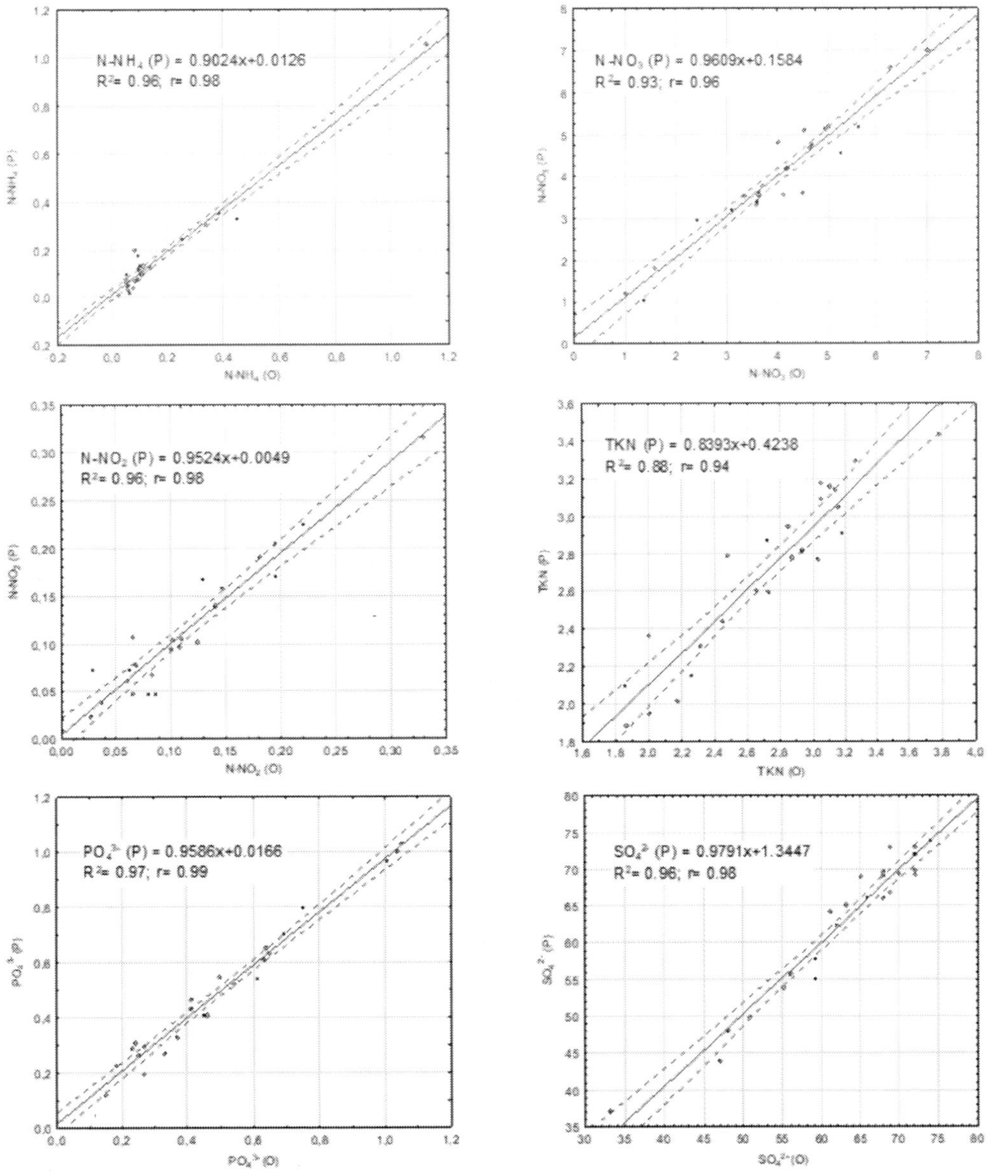

Fig. 3. R^2 graphical interpretation for approximated biogenic compounds in treated wastewater.

Table 2. Weights of input variables included in the SSN model.

Parameter	COD	TSS	P	TC	BOD	pH	N	SO_4^{2-}	°C
Rank	37.00	8.10	4.13	5.21	3.92	4.00	3.70	1.58	25.94

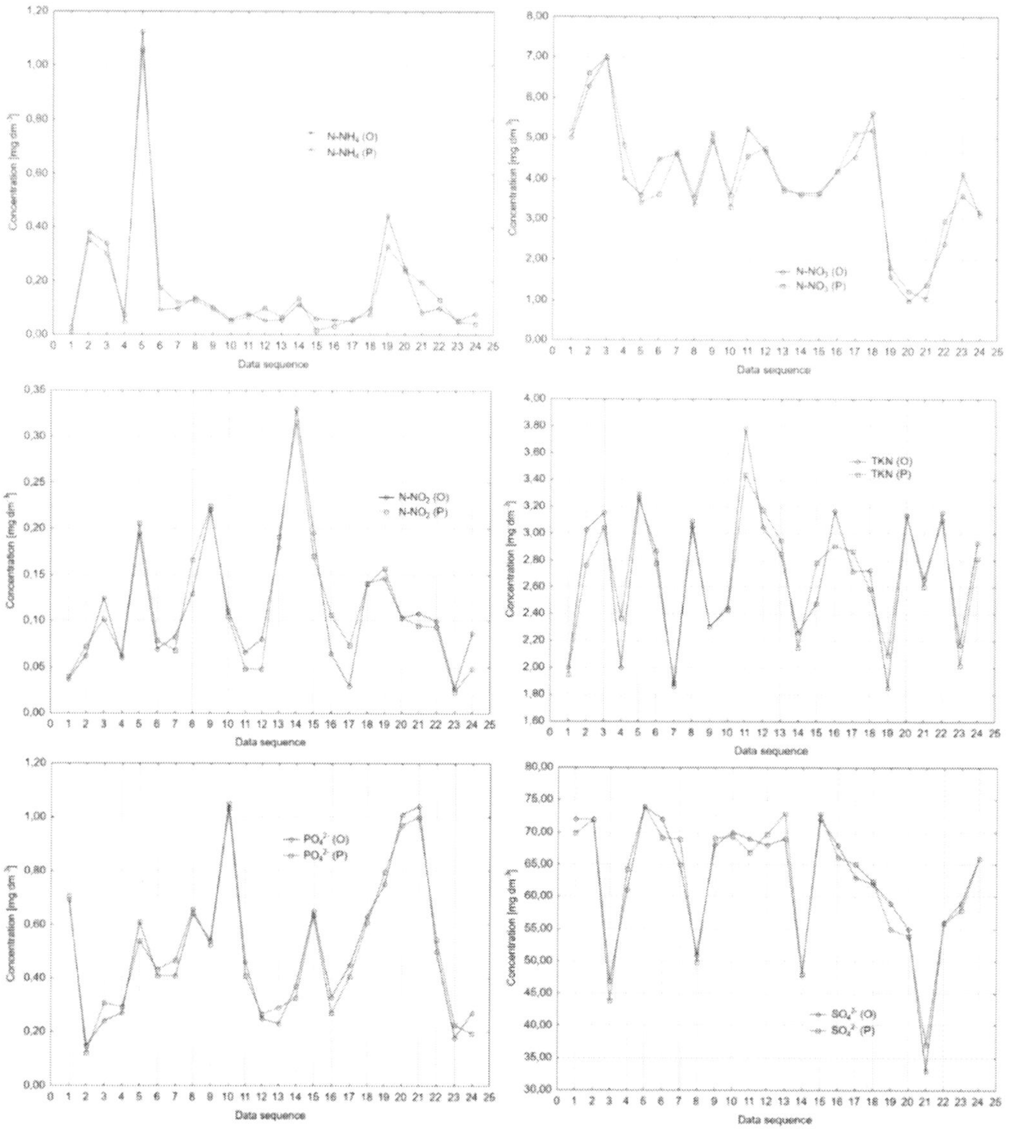

Fig. 4. Summary of observed and predicted values in treated wastewater.

Among the variables adopted for analysis for the calculation method (tab. 2) of the ANN model, the COD value (37.00) and wastewater temperature were the most significant (25.94). The high impact of these variables on the ANN model can be referred to the specificity of wastewater treatment plant for which the model was developed. In the case of dairy wastewater, the COD value flowing into the treatment plant determines mainly the effectiveness of further wastewater treatment [19, 20] whereas the temperature of wastewater affects both oxygen solubility, which is particularly important in oxygen chambers and determines the nutritional efficiency of microorganisms present in activated sludge. The rank of other variables were close to each other and varied from 1.58 to 8.10. The SO_4^{2-} concentration was the smallest factor affecting the calculation model of the model. The nature of the individual output variables predicted by artificial neural network model was presented in figure 4.

Baki and Aras [21] used the method of artificial neural networks to predict BOD values in treated wastewater. Input variables in the considered models were sewage flow, COD, TSS, sewage temperature, total phosphorus, total nitrogen and electrolytic conductivity. Depending on the ANN learning method, the authors observed different adjustment of the model to the observed values which varied from 0.65 to 0.84. Picos-Benitez et al [22] developed an artificial neural network model to approximate the COD values flowing from the model anaerobic reactor that purifies saline wastewater. In the created model, the authors used organic matter load, conductivity and wastewater temperature as input variables, while the COD value in treated wastewater was approximated with an absolute error of 9.226%. Good adjustment of the described models to the BOD and COD values in treated wastewater could be due to the fact that it was the only value approximated by the neural network. Szeląg and Studziński [23] used the artificial neural network algorithm to predict the concentration of $N\text{-}NH_4$ in wastewater purified from treatment plants with RLM 275,000 and an average daily flow of up to 72,000 $m^3 \cdot d^{-1}$. The absolute mean error obtained by the authors was 2.82% and it was smaller in comparison to the model presented in this work.

All treated wastewater quality parameters showed good fit to changes occurred in real conditions. Developed neural network model reflected well the extreme values observed during study period. The average error percentage with which output variables were approximated equalled 35.35%; 8.99%; 21.23%; 5.08%; 10.99%; 3.02% respectively for $N\text{-}NH_4^+$, $N\text{-}NO_3^-$, $N\text{-}NO_2^-$, TKN, PO_4^{3-} and SO_4^{2-}.

4 Conclusion

Artificial neural networks are used for finding the simplest relation between variables. Most often ANN models are used to predict 1 variable. This approach allows to describe the phenomena and dependencies between the output variable and the input variables on the basis of which the approximation is carried out. A more complex approach to modeling of wastewater treatment plants operation using artificial neural networks is the approximation of many variables at the same time. It allows the observation of general phenomena and allows to identify parameters affecting the whole of such a complex process as wastewater treatment. In the presented model finding such a relation enabled changes reflection in concentration of 6 variables simultaneously. The model of artificial neural network, whose task was the prediction of selected biogenic compounds, allowed for a good reflection of changes occurring during mechanical-biological wastewater treatment. The model did not use technological parameters of activated sludge, nor quantitative characteristics of raw wastewater inflowing to object. The basis for the algorithm development were only the relationships occurring between selected parameters of raw and treated wastewater. Among the input variables considered, the COD and wastewater temperature had the biggest influence on the accuracy of the model. These variables were characterized by significantly higher weights compared to the other parameters placed in input layer. Hence, it can be assumed that just the inclusion of COD and wastewater temperature in the ANN model allowed to obtain a match between observed and approximated values.

The research has been carried out in the framework of project No S/WBiIS/3/2014 of the Bialystok University of Technology, Poland and financed from the funds for science from the Ministry of Science.

References

1. M. Delkash, B. E. Bakhshayesh, H. Kazemian, Micropor. Mesopor. Mat. **214**, 224–241 (2015)
2. S. De Gisi, G. Lofrano, M. Grassi, M.Notarnicola, SM&T **9**, 10–40 (2016)

3. I. Evan-Ezra, M. Beliavski, S. Tarre, C. Dosoretz, M. Green, Desalination **272**, 85–89 (2011)
4. S. A. Dellana, D. West, Environ. Modell. and Softw. **24**, 96–106 (2009)
5. T. Deblonde, C. Cossu-Leguille, P. Hartemann, Int. J. Hyg. Environ. Health **214**, 442–448 (2011)
6. M. Pomiès, J. M. Choubert, C. Wisniewski, M. Coquery, Sci. Total Environ. **443**, 733–748 (2013)
7. K. Li, J. Wang, J. Liu, Y. Wei, M. Chen, J. Environ. Sci. **43**, 106–117 (2016)
8. P. Falås, A. Wick, S. Castronovo, J. Habermacher, T. A. Ternes, A. Joss, Water Res. **95**, 240–249 (2016)
9. K. Yetilmezsoy, S. Demirel, Journ. Hazardous Mat. **153**, 1288–1300 (2008)
10. R. Benintendi, Process Saf. Environ. **575**, 1–9 (2015)
11. W. Dąbrowski, B. Karolinczak, P. Malinowski, D. Boruszko, Water **11**, 180 (2019)
12. I. Skoczko, P. Ofman, E. Szatyłowicz, Annual Set The Environmental Protection **18**, 493–506 (2016)
13. Y. Lin, M. Guo, N. Shah, D. C. Stuckey, Bioresour. Technol. **215**, 227–238 (2016)
14. L. Rizzo, A. Fiorentino, M. Grassi, M. Guida, J. Environ. Chem. Eng. **3**, 122–128 (2015)
15. M. M. Amin, H. Hashemi, B. Bina, A. H. Movahhedian, H. Farrokhzadeh, M. Ghasemian, Desalination **260**, 70–78 (2010)
16. E. M. Seeger, M. Braeckevelt, N. Reiche, J. A. Müller, M. Kästner, Ecol. Eng. **95**, 635–644 (2016)
17. I. Skoczko, J. Struk-Sokołowska, P. Ofman, Annual Set The Environmental Protection **19**, 633–650 (2017)
18. E. W. Rice, R. B. Baird, A. D. Eaton, L. S. Clesceri, *Standard methods for the examination of water and wastewater* (22nd edition, American Public Health Association, American Water Works Association and Water Environment Federation, Washington, D.C., USA, 2012)
19. J. Struk-Sokołowska, A. Mielcarek, J. Wiater, J. Rodziewicz, Desalin. Water Treat. **105**, 1944–3994 (2018a)
20. J. Struk-Sokołowska, J. Rodziewicz, A. Mielcarek, Water Sci. Technol. **1**, 156–169 (2018b)
21. O. T. Baki, E. Aras, Membr. Water Treat. **9**, 455–462 (2018)
22. A. R. Picos-Benitez, J. D. Lopez-Hincapie, A. U. Chavez-Ramirez, A. Rodriguez-Garcia, Water Sci. Technol. **75**, 1351–1361 (2017)
23. B. Szeląg, J. Studziński, Pol. J. Environ. Stud. **26**, 2231–2238 (2017)

Univariate and multivariate DMS calibration for a single analyte

Andrzej Szczurek[1,*], *Monika* Maciejewska[2], and *Żaneta* Zajiczek[1]
[1]Faculty of Environmental Engineering, Wroclaw University of Science and Technology, Wybrzeże Wyspiańskiego 27, 50-370 Wrocław, Poland

Abstract. Differential mobility spectrometry (DMS) is a promising measurement technique. It is used in the detection of chemical warfare agents, explosives, drugs, and volatile organic compounds. The measurement principle is based on separation of gas-phase ions according to their differential mobility in alternating low and high electric fields. The DMS measurement result is a two dimensional spectrum of ion current displayed as a function of separation voltage and compensation voltage. The DMS spectral peaks, in terms of their height, location and width, are affected by gas sample composition, separation field and the gas flow rate. In this work, there is presented the calibration procedure which utilises the univariate and multivariate approach to differential ion mobility spectrum. We demonstrated the possibility of a successful retrieval of quantitative information using partial least squares regression as well as univariate linear regression. However, the multivariate approach outperformed the univariate one in terms of the quality of the model and the concentration prediction accuracy.

1 Introduction

Differential mobility spectrometry (DMS) is one of the most promising measurement techniques [1, 2]. A lot of experimental data concerning a rapidly growing number and variety of its applications had been published and discussed. DMS is a powerful tool widely used in the detection of chemical warfare agents, explosives, drugs, and volatile organic compounds [3–5]. The measurement principle of this technique is based on separation and characterization of gas-phase ions [1]. When ions move through a gas under the influence of an electric field, the combination of acceleration due to the field and deceleration due to collisions with gas molecules affects their movement. DMS separates ions based on their differential mobility in alternating low and high electric fields. The differential mobility of an ion depends on a number of properties, including the mass, shape, centre of mass, dipole moment, effects of clustering between ions and neutral gas molecules during separation.

In DMS, ions are carried by a gas between parallel or coaxial electrodes [1, 2]. The electric field alternates perpendicularly to the DMS channel. Ions oscillate between the two electrodes and may either exit the device or collide with an electrode depending on their differential mobility in high and low electric fields. Particular ions may be retained in

* Corresponding author: monika.maciejewska@pwr.edu.pl

equilibrium, inside the gap by applying a low strength constant compensation voltage (CV) to the electrodes. It corrects the migration of the ion caused by the differential mobility so that the ion can pass between the electrodes and be detected. Ions with differing drift velocities, due to their properties, can be passed through the gap at different compensation voltages. Scanning the CV produces a spectrum of the ion mixture.

The information about chemical species is included in a signal which is an ion current measured at the output of the filter gap. This current is a product of averaged ion density and carrier gas flow rate. Instrumental parameters and ion properties affect the ion density and hence the output current.

The DMS provides information about the analyzed gas in a spectral form. Differential mobility spectrum is defined as the current at the detection electrode with respect to the compensation voltage. The spectrum consists of a series of peaks, which are affected by instrumental parameters such as carrier gas flow rate, separating field amplitude and waveform, filter gap size and geometry, ion properties such as diffusion, mobility. In practice, the only parameter that is changed with time during a single analytical run is the compensation voltage. Other parameters are optimized and fixed beforehand. A spectral peak can be characterized by its position, height and width. These characteristics depend on many factors. The DMS peak height is affected by the analyte concentration, separation field and the gas flow rate. The DMS peak width (full width at half maximum) appears to be strongly dependent on the gas flow rate for the planar DMS and on the separation voltage amplitude for the coaxial DMS. The peak position depends on the concentration of an analyte. Hence, the calibration of a differential mobility spectrometer based on peak's characteristics is complex operation and it has to be done in a multidimensional space. In this work, there is presented the calibration procedure based on univariate and multivariate approach to differential mobility spectrum.

2 Experimental part

This work was based on laboratory experiments. They consisted in measurements of reference gas mixtures using differential ion mobility spectrometry. Reference gas mixtures were composed of clean air and the predefined amount of acetic acid. The examined concentration range was from 2.1 ppm to 41.7 ppm.

Measurements were realised with the dedicated setup, as shown in Fig. 1. The setup was composed of the module for preparation of clean air, module for preparation of reference gas mixtures and the measuring instrument, DMS spectrometer.

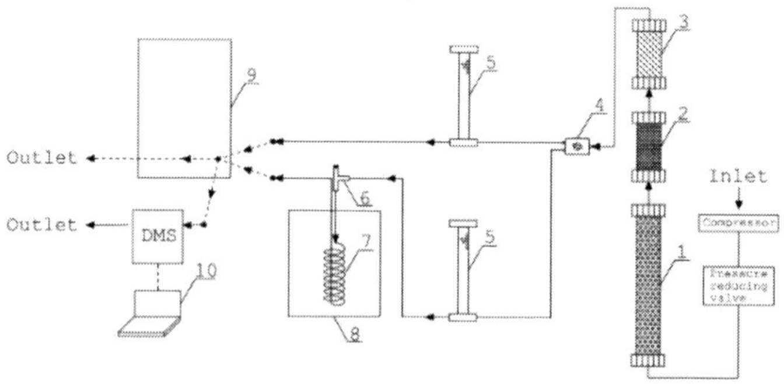

Fig. 1. Measurement setup: 1. filter (silica gel), 2. filter (soda lime and molecular sieves), 3. filter (activated charcoal), 4. switch, 5.rotameter; 6. analyte injection point, 7. coil, 8. thermostat, 9. Tedlar bag, 10. computer.

In the measurement setup, clean air was used as the carrier gas and for removing the analyte from the set-up, once the measurement was completed. The clean air was attained from indoor air passed through a set of filters. They were filled with silica gel, soda lime, molecular sieves and activated charcoal, see Fig. 1, which allowed for air drying and removal of volatile organic compounds. The air flow was enforced by the compressor. The pressure was adjusted with the pressure reducing valve.

Reference gas mixtures were prepared using an evaporation method. The module dedicated for preparation included: thermostat, steal coil with an inlet port for the liquid analyte, syringe, Tedlar bag, rotameters and gas tubing, see Fig. 1. Clean air was delivered to this module and the flow rate was controlled using rotameters. For preparing reference gas mixture, the determined amount of liquid analyte was injected, with a syringe, through the injection port to the air stream. The analyte was vaporised in the air inside the coil, which was maintained at the constant temperature. The air containing the analyte vapours was collected in the Tedlar bag. The concentration of reference gas mixture was determined based on the amount of the injected analyte, gas flow rate and the time of gas collection in the Tedlar bag. Before preparation of reference gas mixture and after the measurement, Tedlar bag was cleaned by filling it with clean air and emptying it, three times.

During measurement, the reference gas mixture from the Tedlar bag was directed to the DMS spectrometer and the ion mobility spectrum was recorded. Five measurements were done for each reference gas mixture. A single measurement lasted 30 s. During measurement, gas flow rate through the instrument was 0.5 dm^3/min.

The experiments consisted in measurements of reference gas mixtures composed of clean air and reference gas mixtures composed of clean air and acetic acid. The following concentrations of analyte were considered: 2.1, 4.2, 5.2, 10.4, 20.8 and 41.7 ppm. The measurement of each reference gas mixture which contained the analyte was preceded by the measurement of reference gas mixture which contained clean air. The humidity of the clean air was maintained at the level of 11% RH ± 2% RH (at 323K) throughout all experiment. All reference gas mixtures were prepared using the same Tedlar bag.

The DMS spectrometer applied in experiments was constructed in the Military Institute of Chemistry and Radiometry, Poland. [3]. Gas ionization in this device is achieved by the β-emitter, nickel electrode covered with radioactive nickel ^{63}Ni. The DMS chamber has planar geometry. The drift region is formed by two electrode-plates (5 x 25 mm) separated by a 0.5 mm gap. A high amplitude, high frequency, asymmetric voltage waveform, generated by HSV generator (2 MHz) is applied to electrodes, crosswise to the direction of the carrier gas flow. The separation voltage (SV) amplitude is from 100 to 1600 V (peak to peak), which corresponds to from 6,5 to 104 Td. The compensation voltage is scanned from -30 to 8 V. At the end of the drift zone there are located two detector electrodes (5 x 5 mm), for collecting positive and negative ions respectively. Two types of ions are neutralized in parallel. The internal gas flow through the detector is 2.0 L/min. The chamber entrance temperature is 318 K and chamber temperature is 323 K.

3 Methods

The result of measurement with DMS spectrometer is a two-dimensional spectrum of ion current. Actually, two spectra are attained during a measurement, one for negative ions and one for positive ions. The ion current is recorded as a function of the separation voltage and the compensation voltage. The two voltages were discretized as follows SV={100, 150, ..., 1600} and CV={-30, -29.926, ..., 7.888}, which resulted in a finite and countable set of combinations of (SV, CV). The ion current is recorded for each of these combinations.

Two approaches, univariate and multivariate were compared in respect of extracting the quantitative information about the analyte from DMS spectrum. The univariate approach consists in using one independent variable as the predictor of the response variable. In this work, it was represented by simple linear regression. Simple linear regression is used for determining the calibration function in traditional analytical solutions. The multivariate quantitative approach consists in using more than one independent variable as the predictor of the response variable(s). In this work, the multivariate approach was represented by partial least squares (PLS) regression. This technique is willingly used in spectrometric calibration [6], where multivariate measurement data is obtained.

The applied univariate regression model was [7]:

$$c_{AA} = a_0 + a_1 I_{(SV,CV)} + \varepsilon \tag{1}$$

where $I_{(SV,CV)}$ is the ion current at the collection electrode (associated with the particular combination of separation voltage and compensation voltage) recorded during the measurement of reference gas mixture, containing acetic acid at the concentration c_{AA}. The model coefficients are a_0 and a_1. The error of data fitting with the model is ε.

Partial least squares regression model consists of two equations [6]:

$$X = TP^T + E \tag{2}$$

$$Y = TQ^T + F \tag{3}$$

where X is the input data matrix (each column refers to one independent variable) and Y is the output data matrix (each column refers to one response variable). The score matrix T consists of latent variables. They are artificial variables, derived under the constraint of explaining as much as possible of the covariance between X and Y. P is the loading matrix which allows to transform loadings into independent variables. R is the loading matrix which allows to transform loadings into the dependent variables. E and F are error matrices.

We proposed that the independent variables in PLS model are ion currents associated with all compensation voltages CV and the defined separation voltage. While univariate model is based on a single element of two dimensional DMS spectrum, the multivariate model explores the one dimensional DMS spectrum.

The quality of the linear univariate models and multivariate models was evaluated using coefficient of determination (R^2). The accuracy of analyte concentration prediction with the models was determined using root mean square error (RMSE).

4 Results and discussion

The presented results are based on negative ions mobility spectra. Based on our analysis, the positive ions mobility spectra did not convey the useful information.

An important aspect of measurement is precision. The precision of DMS spectra was examined. The precision was determined as the spread of five spectra, obtained from the series of five measurements of one particular reference gas mixture. The precision was represented by the standard deviation and the coefficient of variation. As shown in Fig. 2, for all measurements the coefficient of variation was less than 3%, which indicates high precision of DMS spectra. Considering coefficient of variation, measurements of clean air were less precise than the measurements of reference gas mixtures containing analyte, while standard deviation indicated similar precision. This discrepancy could be an effect of grater ion currents during measurements of reference gas mixtures containing the analyte. Based on Fig. 2, DMS spectra for small concentrations were more precise compared with spectra for high concentrations. Clean air spectra were nearly equally precise. High precision of the readout is an important parameter of the measurements method which may be used for the quantitative assessment.

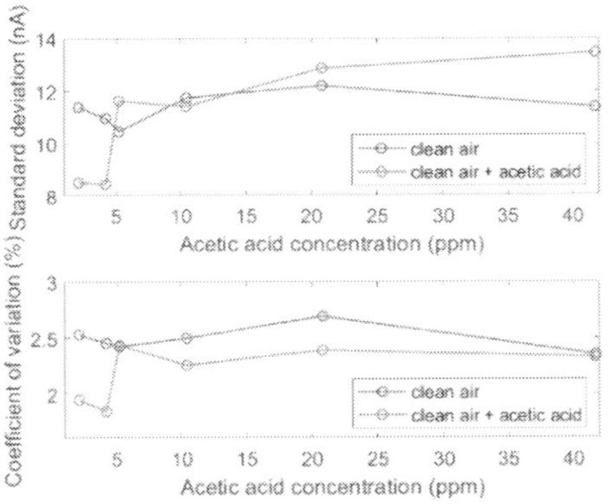

Fig. 2. Precision of DMS spectra in terms of standard deviation and coefficient of variation. Precision was determined for spectra of reference gas mixtures containing clean air and acetic acid in clean air. Measurements of the first mixtures preceded the measurements of the second mixtures, that is why the precisions are shown in pairs.

Based on theory of ion mobility [1, 2], the analyte ion peak position and height changes depending on gas sample composition. The change of peak height is by all means valuable, as a source of information about the analyte concentration. However, the change of peak location is a difficulty which needs to be overcome while developing a calibration function. The frequently encountered cause of peak shifts is sample humidity variation. The problem, has been referred on multiple occasions [4, 5, 8]. In the current study we indicate the problem of analyte peak location (not only height) change as a consequence of the variation of analyte concentration in reference gas mixture.

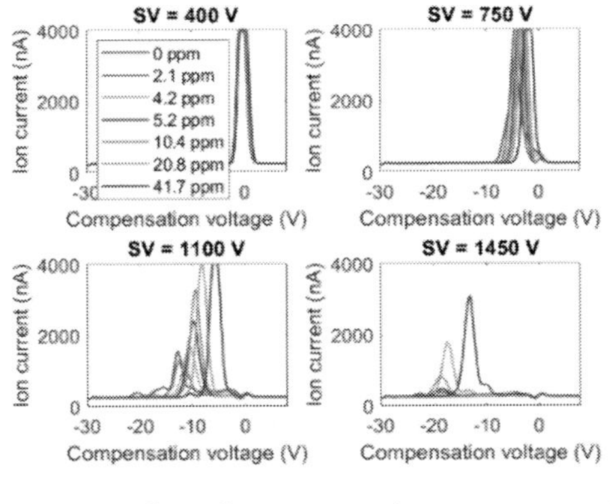

Fig. 3. Unidimensional DMS spectra for reference gas mixtures composed of clean air and acetic acid. Ion current recorded at fixed separation voltage, was shown as a function of the compensation voltage. The negative ions spectra are presented.

The overlap of peaks associated with different analyte concentrations is crucial when developing analyte quantification methods based on ion current associated with and individual combination of (SV,CV). For this approach, it would be ideal to have all concentrations peaks in the same place, and their amplitudes proportional to the analyte concentration. Although the method is still operational when all peaks partially overlap, its performance gradually decreases with the decreasing overlap. The magnitude of peaks separation is associated with the overall range of analyte concentration, which may be considered as the limiting factor for the univariate calibration method. In case of small concentration ranges, the overlap of extreme peaks is sufficient to achieve analyte quantification based on ion current associated with an individual combination of (SV,CV).

When broader concentration ranges are involved, the method becomes less effective. In such cases other multivariate spectrum analysis techniques have to be applied, which are capable to extract the quantitative information from the spectrum in spite of concentration dependent peak allocation [9].

The problem of analyte peak position change, related to analyte concentration, was illustrated in Fig. 3. We displayed one dimensional DMS spectra of reference gas mixtures containing various amounts of analyte. As shown, peaks associated with different concentrations of the same analyte have different positions in the spectrum. Additionally, the relative positions of ion current peaks depends on the magnitude of separation voltage. The degree of overlap between peaks associated with various concentrations of analyte is a function of both, compensation and separation voltages.

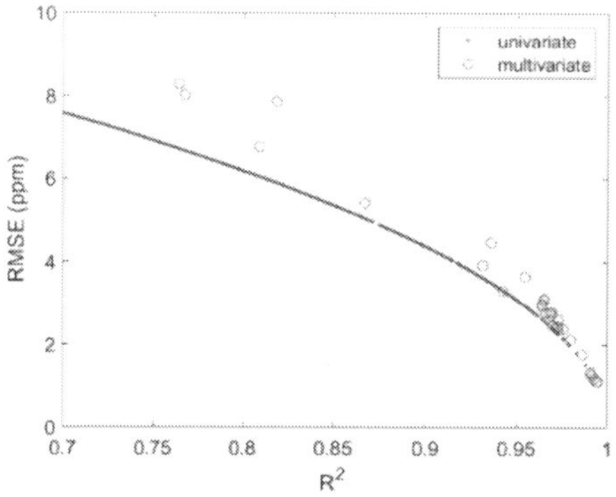

Fig. 4. Coefficient of determination (R^2) for univariate linear regression models and multivariate partial least squares regression models versus root mean squared errors (RMSE) of acetic acid concentration prediction with these models.

In Fig. 4, the model quality measure R^2 and concentration prediction error (RMSE) are shown. They characterise the developed univariate models (blue dots) and multivariate models (red circles). Based on the comparison, greatest values of coefficient of determination and smallest error values were attained in case of multivariate models.

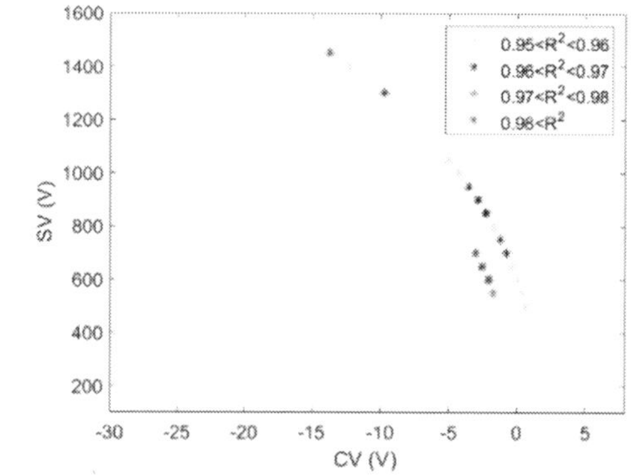

Fig. 5. Combinations of compensation voltage (CV) and separation voltage (SV). The ion current recorded at these combinations was a valuable source of information about acetic acid concentration in reference gas mixtures when using univariate linear regression.

The performance of univariate models was analysed in relation to spectral coordinates of the independent variable. The best performing univariate model was featured by $R^2 = 0.988$ and RMSE = 1.48 ppm. The model was associated with SV = 650 V and CV = -2.5391 V. In Fig. 5 we show (SV, CV) coordinates of independent variables of other, relatively well performing univariate models, based on the cut off criterion, $R^2 = 0.95$. They indicate parts of spectrum which are useful for quantitative determination of analyte, using univariate models. As shown in Fig. 5 the distribution of valuable combinations (SV, CV) in two dimensional spectrum is not random. However, it would be difficult if not impossible to anticipate their location based on the visual inspection of DMS spectra, without the appropriate data analysis.

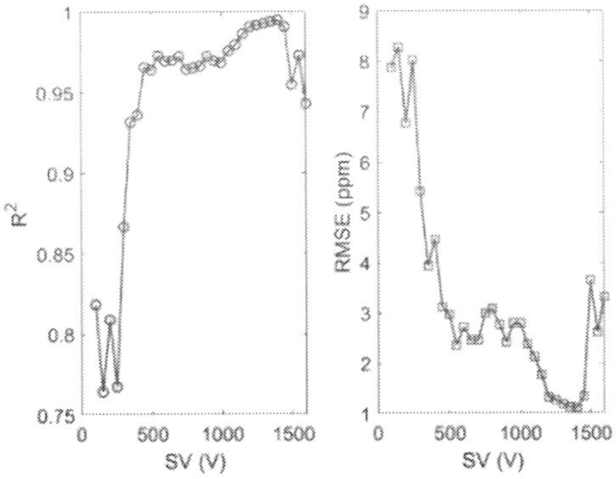

Fig. 6. Coefficient of determination (R2) for PLS regression models and root mean squared errors (RMSE) of acetic acid concentration prediction with these models. Parameters are displayed versus separation voltages which were the source of the independent variables for the model.

The performance of the multivariate models was examined in the domain of separation voltage. As shown in Fig. 6, the indicator of quantification model performance (R^2) and the indicator of concentration prediction performance (RMSE) were dependent on the separation voltage. For small separation voltages, the model performance was poor and concentration prediction errors were high. Both parameters improved with the increase of SV. Starting from SV=1200 V, coefficient of determination was greater than R^2=0.95 and for separation voltage in a range between SV = 1200 V and SV = 1450 V, coefficient of determination was greater than R^2 = 0.99. The best results (R^2 = 0.995 and RMSE = 1.1 ppm) were attained for SV = 1400. Interestingly, for separation voltages exceeding SV = 1450V model quality and prediction performance got worse.

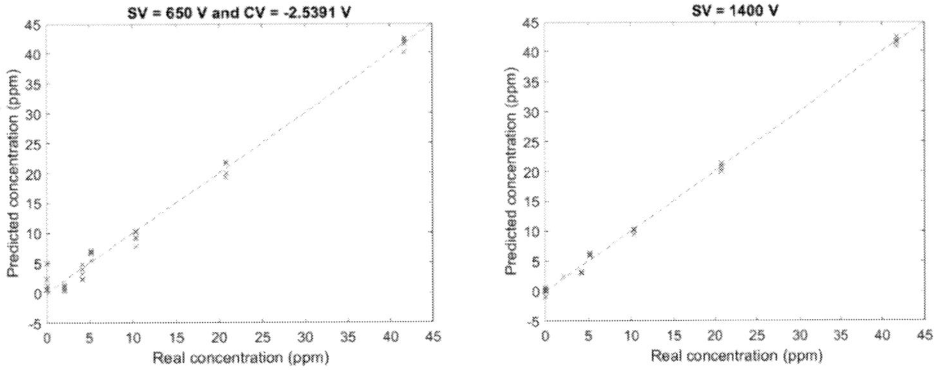

Fig. 7. Left: Diagnostic plot for the best univariate model. The independent variable was ion current recorded at the combination of compensation and separation voltage SV = 650 V and CV = -2.5391 V. Right: Diagnostic plot for the best multivariate model. The vector of independent variables consisted of ion currents recorded all compensation voltages while the separation voltage was SV = 1400 V.

The diagnostic plots for the best performing univariate model and the best performing multivariate model are compared in Fig. 7. The visual inspection reveals the supremacy of the multivariate model, which agrees with the evaluation based on the objective indicators R^2 and RMSE. It shall be admitted that difference between the level of indicators for best univariate and best multivariate model is not striking. However, the solution offered by multivariate approach is more reliable. There were several excellent PLS models and just one very good univariate model. It is also important that better results of multivariate model were particularly noticeable at small concentrations, which are most difficult to determine in general.

5 Conclusions

In this work we focussed on the DMS spectrum as a source of quantitative information about the analyte. There was compared the potential of the univariate and the multivariate approach to spectrum. The first utilised one explanatory variable – the ion current associated with a single combination of separation voltage and compensation voltage. The second extracted the information form multiple explanatory variables – ion currents associated with all compensation voltages and one separation voltage.

We demonstrated the possibility of a successful retrieval of quantitative information using PLS regression as well as univariate linear regression. However, the multivariate approach outperformed the univariate approach in terms of the quality of the model and the concentration prediction accuracy. The supremacy was visible at small concentrations, which are generally most difficult to determine.

The analysis allowed to identify regions of spectrum which are useful regarding quantitative determination of the analyte. The part of spectrum associated with separation voltage less than SV=500 V was useless for this purpose. Best performing univariate models utilised ion current recorded at smaller separation voltages than best performing multivariate models.

References

1. H. Borsdorf, G. A. Eiceman, Appl. Spectrosc. Rev. **41**, 323–375 (2006)
2. B. B. Schneider, E. G. Nazarov, F. Londry, P. Vouros, T. R. Covey, Mass Spectrom. Rev. **9999**, 1–51 (2015)
3. M. Maziejuk J. Puton M. Szyposzyńska, Z. Witkiewicz, Talanta **144**, 1201–1206 (2015)
4. T. N. Obee, M. Piech, J. V. Manteste, S. Dardona, Int. J. Ion Mobilit. Spec. **15**, 131–139 (2012)
5. N. Krylova, E. Krylov, G. A. Eiceman, J. A. Stone, J Phys Chem A. **107**, 19, 3648–54 (2003)
6. T. Naes, T. Isaksson, T. Fearn, T. Davies, A User-friendly Guide to Multivariate Calibration and Classification, NIR Publications, Chichester, UK (2004)
7. A. Szczurek, M. Maciejewska, Ż. Zajiczek, M. Maziejuk, ASEE17, E3S Web of Conferences **22**, 00173 (2017)
8. A. Szczurek, M. Maziejuk, M. Maciejewska, T. Pietrucha, T. Sikora, Sens Actuators B Chem. **240**, 1237–1244 (2017)
9. M. Maziejuk, A. Szczurek, M. Maciejewska, T. Pietrucha, M. Szyposzyńska, Talanta **152**, 137–146 (2016)

Electronic nose – an instrument for odour nuisances monitoring

Bartosz Szulczyński[1,*], and *Jacek* Gębicki[1]

[1]Department of Process Engineering and Chemical Technology, Faculty of Chemistry, Gdańsk University of Technology, ul. Gabriela Narutowicza 11/12, 80-233 Gdańsk, Poland

Abstract. An increasingly frequent problem of people living in urban agglomerations is the occurrence of odour nuisance. Although the source of these nuisances is different, their common feature is that they are a complex mixture of odour compounds with different odour thresholds. However, from a practical point of view, the most valuable would be a direct link between the odour intensity and the results of on-line analytical air monitoring. Such a possibility is created by the use of electronic noses (devices that are supposed to imitate the human sense of smell) to measure odours. The paper presents the use of an electronic nose combined with multiple liear regression model (MLR) to determine the odour intensity of the two-component mixture samples of commonly known odour compounds: trimethylamine (TMA) and triethylamine (TEA) in concentration range 50–200 ppm v/v. The obtained results were compared with the theoretical values determined using Zwaardemaker and euclidean additivity (EA) models. For high concentrations of substances in the mixtures (> 150 ppm v/v), the masking effect was observed.

1 Introduction

The ability to identify and distinguish odour substances for some time was possible only through the human sense of smell. With the development of science and technology, there is a growing interest in devices constructed and functioning in the image of human senses. Over the last 30 years, analytical systems have been developed that could in some ways replace human sense of smell [1–3]. Devices that have aroused interest due to the use of a wide range of chemical sensors in their construction are electronic noses. These devices, being analogues of the sense of human smell, allow their application in many fields of science and industry, such as: medical diagnostics, environmental protection, food and chemical industry or criminology [4–7]. Electronic noses enable a holistic analysis of the composition of the gas mixture, without the need to separate and identify its individual components [8, 9]. In comparison to other techniques used to analyze odours such as olfactometry or gas chromatography, electronic noses have additional advantages. In relation to olfactometric techniques, there is a lack of olfactory adaptation phenomenon and having trained personnel with a specific olfactory perception. In relation to

* Corresponding author: bartosz.szulczynski@pg.edu.pl

chromatographic techniques, electronic noses are characterized by a short analysis time and a lower price of the device. Electronic noses with their advantages and certain limitations complement the above-mentioned measuring techniques used for odour analysis.

The smell, which is a sensual impression, is relatively difficult to quantitative description. Four basic odour parameters are determined: odour concentration, odour threshold, odour intensity and hedonic tone. However, the values of these parameters determined for mixtures of odour substances are not additive. This is due to the occurrence of the odour interactions based on the mutual masking, strengthening or weakening of odours. Studies on the types of olfactory interactions have been conducted for a very long time, but so far have not led to the explanation of the mechanism of these processes. The objects of experimental research are usually air samples containing only two or three types of odorants [10, 11].

The paper presents the use of an electronic nose to determine the odour intensity of the two-component mixture samples of trimethylamine (TMA) and triethylamine (TEA) – two commonly known odour compounds. The results were compared with the theoretically determined values.

1.1 Odour intensity

The odour intensity is the intensity of the smell, i.e. the strength of the olfactory sensation, which is associated with the frequency of nerve impulses in a given olfactory receptor. The intensity depends on the amount of the odorants, which are inhaled with air. They come into contact with olfactory chemoreceptors, so the intensity depends on the concentration of the compound in the inhaled air. The most frequently used mathematical equations describe the dependence of the odour intensity on the concentration of odorants. They are the expression of psychophysical rights of a general nature and concern the perception of all sensory impressions. Two equations are commonly used to describe the odour intensity: Weber-Fechner law (1) and Stevens's power law (2).

$$I = k_{W-F} \cdot log \frac{c}{c_{OT}} \tag{1}$$

$$I = k_S \cdot c^n \tag{2}$$

where: I – odour intensity; k_{W-F}, k_S – equation coefficients; c – compound concentration; c_{OT} – odour threshold, n – empirical constant value.

Equations (1) and (2) are applicable to individual compound concentration. The combination of experimental values for pure chemical compounds does not allow to predict the intensity of their mixtures. So it is not an additive amount. This is due to odour interactions occurring in the odorant mixture, which contribute to strengthening, weakening or masking one fragrance with the other [12].

In the case of mixtures, mathematical models are used that allow to estimate the odour intensity of the mixture using intensity values for its individual components. One of these models is the Zwaardemaker model. It is often called the vector summation model. This takes on form (3) for a two-component mixture and form (4) for a three-component mixture. The coefficient α found in the equations (3) and (4) is approximately constant for one pair of mixture components.

$$I_{AB} = \sqrt{I_A^2 + I_B^2 + 2 \cdot I_A I_B \cdot \cos \alpha_{AB}} \tag{3}$$

$$I_{ABC} = \sqrt{I_A^2 + I_B^2 + I_C^2 + 2 \cdot I_A I_B \cdot \cos \alpha_{AB} + 2 \cdot I_B I_C \cdot \cos \alpha_{BC} + 2 \cdot I_A I_C \cdot \cos \alpha_{AC}} \tag{4}$$

Recently, studies on the determination of this coefficient for compounds from the group of aldehydes, esters or aromatic hydrocarbons were carried out by researchers who determined its value at the level of: 0.1, 0.2 and 0.3 respectively [13].

Another model used to estimate the odour intensity is the EA model (Euclidean Additivity). The intensity of the mixture is determined using equation (5). This is a special case of the Zwaardemaker model with the $\cos\alpha = 0$ value.

$$I_{AB} = \sqrt{I_A^2 + I_B^2} \quad (5)$$

1.2 Electronic nose

Active development of industry is inevitably connected with the emission of chemical compounds to the atmosphere. Taking into account standards and legal regulations regarding the emission of dangerous compounds into the environment, innovative technologies have been developed to identify and determine the concentrations of odorous volatile compounds [14]. In order to provide a more precise and objective assessment of the odorous compounds than in the case of sensory analysis, an electronic nose was constructed using chemical sensors. These sensors initially appeared individually, but with the development of technology began to create special gas sensor matrices, which are equipped with several, or even more, gas chemical sensors [15].

The electronic nose was first developed in 1982 by Dodd and Persuad to imitate the sense of human sense of smell. The device was designed mainly to detect volatile chemical compounds emitted from various sources. An electronic nose (e-nose) is structurally constructed from four blocks: sampling, detection, data processing and pattern recognition system (Figure 1).

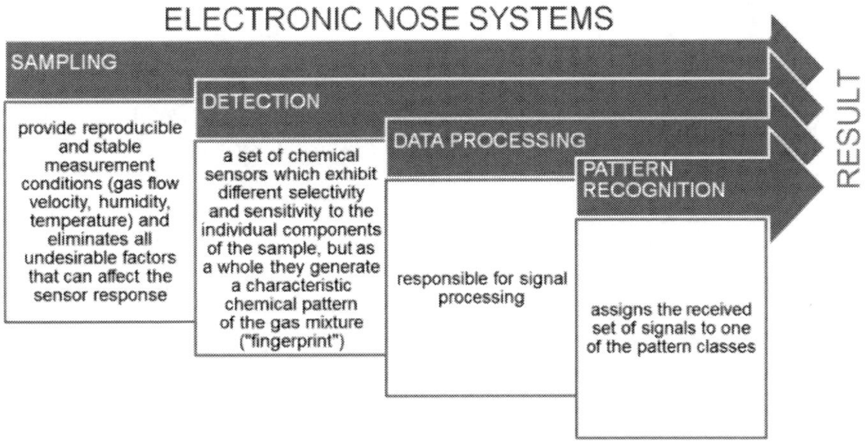

Fig. 1. Electronic nose systems.

The concept of an electronic nose is collinear with a human sense of smell. Its operation is based on the use of sensors matrix, which after exposure to odours generate a proper set of signals. The sampling system has the task of transferring volatile molecules from the atmosphere to the sensors array that are enclosed in the chamber under constant temperature and humidity conditions. In the detection zone, chemical gas sensors are located. They convert the chemical information into an analytically useful electric signal. The most common sensors used for building electronic noses are: electrochemical sensors (EC), semiconductor sensors (MOS), conductivity sensors with a layer of conductive polymers (CP), surface acoustic wave (SAW) sensors, piezoelectric sensors (QMB) and photoionization sensors (PID) [9].

The determination of odour intensity using an electronic nose requires the use of calibration models. To develop such models, a set of explanatory variables is used (signals from electronic nose sensors) and a set of dependent variables (values of the odour intensity expressed in the verbal scale). One of the most popular calibration techniques are: multiparameter linear regression (MLR), principle component regression (PCR), partial least squares regression (PLS) and artificial neural networks (ANN). These methods have been successfully applied to monitor changes in the aroma concentration in such processes as: biofiltration or wastewater treatment as well as to support measurements made using dynamic olfactometry or sensory analysis [16–23].

The presented research uses a self-designed and constructed prototype of an electronic nose equipped with eight chemical sensors: 1 PID sensor, 2 electrochemical sensors and 5 semiconductor sensors. The models of the sensors used are shown in Table 1.

Table 1. Types and models of sensors used in the e-nose prototype.

Sensor number	Model	Detected substances
S_1	MiniPID	volatile organic compounds
S_2	FECS44-100	ammonia, amines
S_3	FECS50-100	hydrogen sulfide
S_4	TGS2600	air pollutions
S_5	TGS2602	volatile organic compounds
S_6	TGS2603	air pollutions (triethylamine, mercaptanes)
S_7	TGS823	organic solvent vapours
S_8	TGS8100	hydrogen, ethanol

2 Experimental

A group of probants took part in the research, who were to evaluate the prepared samples in terms of odour intensity. The research team consisted of 6 people who were familiarized with the guidelines concerning sensory tests. They assessed each of he prepared samples using a seven-grade verbal scale: 0 – not perceptible, 1 – very weak, 2 – weak, 3 – distinct, 4 – strong, 5 – very strong, 6 – extremely strong.

Five samples of trimethylamine and trimethylamine were prepared in deionized water at the following concentration levels: 50, 100, 200, 400 and 800 ppm v/v. The samples were subjected to sensory analysis in terms of the odour intensity. A group of six probants performed the odour evaluation using a verbal-point scale, each measurement was repeated five times for each person and the concentration level. The average values of the odour intensity for aqueous solutions of the analyzed substances are presented in the Table 2.

Table 2. The average values of the odour intensity of trimethylamine (TMA) and triethylamine (TEA) aqueous solutions.

Concentration, ppm v/v	Average odour intensity (sensory analysis)	
	trimethylamine	triethylamine
50	0.8	0.2
100	1.4	0.9
200	2.4	1.6
400	2.8	2.5
800	3.6	3.2

Samples with the same concentrations were tested using an electronic nose prototype. Three analyzes were performed for each substance and concentration. To develop the calibration model, the maximal value of the sensor's signal after the exposition on a given sample was used. An example of the MOS sensor response is shown in the Figure 2.

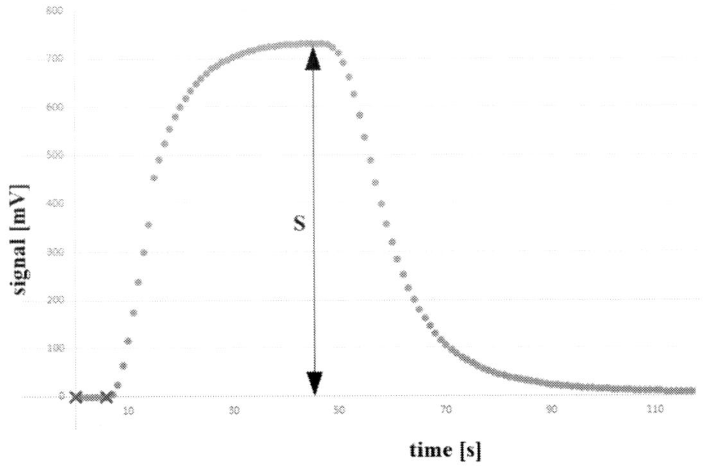

Fig. 2. MOS sensor signal (the interval between the red crosses indicates the baseline).

On the basis of the obtained signals using a multiparameter linear regression (MLR), an equation was determined that links the fragrance intensity of the sample to the signal from individual sensors [21]. After estimating the confidence interval of the model parameters statistically non-significant model parameters were removed from the equation. The equation takes the form of (6):

$$I = 12.7574 - 0.0143 \cdot S_2 - 0.0142 \cdot S_4 + 0.0118 \cdot S_6 + 0.0016 \cdot S_8 \quad (6)$$

3 Results and discussion

Twenty five binary mixtures of TMA and TEA were prepared. Next, the theoretical values of the fragrance intensity with the use of Zwaardemaker (I_{Zw}) and EA (I_{EA}) models were determined for them. For each sample an electronic nose analysis was also performed. The odour intensity was determined using the model (6). The obtained results are presented in the Table 3.

Analyzing the results presented in Table 4, it can be concluded that in the case of studied mixtures Zwaardemaker model and the EA model correlate to each other, giving similar results. However, it should be noted that in the case of high concentrations of the amines (trimethylamine and triethylamine) in the mixture, the difference between the results obtained using e-nose with MLR model and theoretical models are greater than 10% (relative error related to EA and Zwaardemaker model). This fact indicates the presence of odour interactions in the tested mixtures. In addition, all these values are negative. This indicates the occurrence of masking effect.

Table 3. Odour intensity values determined using Zwaardemaker model, EA model and using electronic nose.

Mixture composition		I_{Zw}	I_{EA}	$I_{e\text{-}nose}$	Error $\frac{I_{EA} - I_{e-nose}}{I_{EA}}$	Error $\frac{I_{Zw} - I_{e-nose}}{I_{Zw}}$
C_{TMA}, ppm v/v	C_{TEA}, ppm v/v					
50	50	0.84	0.82	0.80	-2%	-5%
50	75	1.07	1.00	1.05	5%	-2%
50	100	1.30	1.22	1.31	8%	1%
50	150	1.67	1.58	1.66	5%	-1%
50	200	1.95	1.85	1.91	3%	-2%
75	50	1.25	1.23	1.21	-1%	-3%
75	75	1.43	1.36	1.31	-3%	-8%
75	100	1.62	1.52	1.55	2%	-4%
75	150	1.94	1.82	1.87	3%	-4%
75	200	2.20	2.07	2.25	9%	2%
100	50	1.54	1.52	1.45	-4%	-6%
100	75	1.70	1.62	1.72	6%	1%
100	100	1.87	1.76	1.90	8%	2%
100	150	2.16	2.03	2.05	1%	-5%
100	200	2.40	2.25	1.76	-22%	-27%
150	50	1.95	1.93	2.00	4%	3%
150	75	2.09	2.01	2.06	2%	-1%
150	100	2.24	2.12	2.07	-3%	-8%
150	150	2.50	2.35	1.83	-22%	-27%
150	200	2.71	2.54	2.09	-18%	-23%
200	50	2.24	2.22	2.36	7%	6%
200	75	2.37	2.29	2.35	3%	-1%
200	100	2.51	2.39	2.57	7%	2%
200	150	2.75	2.59	2.21	-15%	-20%
200	200	2.95	2.77	2.38	-14%	-19%

4 Conclusions

The obtained results presents that electronic noses can be successfully used to estimate the fragrance intensity of air samples especially in areas where amine compounds are responsible for the occurrence of odour nuisances. Research conducted on binary odour compounds mixtures shows that simple theoretical models do not actually reflect all interactions occurring in the studied samples. Conformity of results was obtained only for low concentrations of substances in the mixture (< 150 ppm v/v). This is due to the limitation of models, especially the EA model, which does not take into account the interaction coefficient at all. Only two-component mixtures were used in the tests. In further studies, it is planned to compare the results with 3-5 component mixtures. This approach determines the use of more sophisticated pattern recognition algorithms, i.e. artificial neural networks or fuzzy logic systems. However, as shown in the paper, for simple systems it is possible to use a MLR model with relatively high accuracy ($R^2 = 0.917$).

In the conducted research, it was also shown that due to the convergence of results, the mechanisms of identifying fragrances implemented in the electronic nose are very similar to the mechanisms occurring in the human sense of smell. Additional advantages of using e-noses are the ability to omit the initial sample preparation stage, as well as the ability to work in the field and carry out measurements in the on-line mode. It is possible to optimize the number and type of sensors used to detect specific substances, which will allow to create devices dedicated to a specific application. Dynamic progress in the field of analytical devices from the group of electronic noses is expected to contribute to the development of odours analysis.

The investigations were financially supported by the Grant No. UMO-2015/19/B/ST4/02722 from the National Science Centre, Poland.

References

1. J. W. Gardner, P. N. Bartlett, Sens. Actuators B **18**, 211–220 (1994)
2. P. Boeker, Sens. Actuators B **204**, 2–17 (2014)
3. J. Gębicki, Trends Anal. Chem. **77**, 1–13 (2016)
4. R. M. Stuetz, R. A. Fenner, G. Engin, Water Res. **33**, 442–452 (1999)
5. S. Deshmukh, R. Bandyopadhyay, N. Bhattacharyya, R. A. Pandey, A. Jana, Talanta **144**, 329–340 (2015)
6. W. Cynkar, D. Cozzolino, B. Dambergs, L. Janik, M. Gishen, Sens. Actuators B **124**, 167–171 (2007)
7. T. Alizadeh, S. Zeynali, Sens. Actuators B **129**, 412–423 (2008)
8. L. Spinelle, M. Gerboles, G. Kok, S. Persijn, T. Sauerwald, Sensors **17**, 1520 (2017)
9. B. Szulczyński, J. Gębicki, Environments **4**, 21, (2017)
10. D. T. Hill, C. L. Barth, Trans. of ASAE **19**, 939–944 (1976)
11. T. C. Pearce, J. W. Gardner, Analyst **123**, 2057–2066 (1998)
12. J. Kośmider, Archiwum Ochrony Środowiska **29**, 17–30 (2003)
13. L. Yan, J. Liu, S. Jiang, Ch. Wu, K. Gao, Sensors **17**, 1624 (2017)
14. Z. Wei, J. Wang, W. Zhang, Food Chem. **177**, 89–96 (2015)
15. A. Loutfi, S. Coradeschi, G. K. Mani, P. Shankar, J. S. B. Rayappan, J. Food Eng. **144**, 103–111 (2015)
16. J. Gebicki, B. Szulczyński, M. Kaminski, Meas. Sci. Technol. **26**, 125103 (2015)
17. R. López, I. O. Cabeza, I. Giráldez, M. J. Díaz, Bioresour. Technol. **102**, 7984–7993 (2011)
18. B. Szulczyński, K. Armiński, J. Namieśnik, J. Gębicki, Sensors **18**, 519 (2018)
19. B. Szulczyński, J. Gębicki, J. Namieśnik, Chem. Pap. **72**, 527–532 (2018)
20. M. Delgado-Rodríguez, M. Ruiz-Montoya, I. Giraldez, R. López, E. Madejón, M. J. Díaz, Atmos. Environ. **51**, 278–285 (2012)
21. B. Szulczyński, P. Rybarczyk, J. Gębicki, Monatsh. Chem. **149**, 1693–1699 (2018)
22. J. H. Sohn, M. Dunlop, N. Hudson, T. I. Kim, Y. H. Yoo, Sens. Actuators B **135**, 455–464 (2009)
23. B. Szulczyński, J. Namieśnik, J. Gębicki, Sensors **17**, 2380 (2017)

Influence of mechanical ventilation and cooling systems on vibrations of high precision machines

Lukasz Scislo[1,*], and *Nina* Szczepanik-Scislo[2]

[1]Faculty of Electrical and Computer Engineering, Tadeusz Kosciuszko Cracow University of Technology, ul. Warszawska 24, 31-155 Krakow, Poland
[2]Faculty of Environmental Engineering, Tadeusz Kosciuszko Cracow University of Technology, ul. Warszawska 24, 31-155 Krakow, Poland

Abstract. The aim of the research was to describe the effects that air and water cooling systems can have on the dynamic behaviour of precise machines. Although much thought is paid to vibration isolation of the ground and isolation of individual effects of machines operating close to each other, it is often forgotten to model or to measure the effects that the ventilation or the machines cooling systems have on the machine itself. This can be especially important for high precision machines used for experimental research and medical equipment. The article shows the effects of ventilation and cooling system on the induction of additional resonant frequencies of the system for the high precision machine like a linear collider. This kind of machine requires special environmental conditions to assure proper beam stability. Due to the dynamic behaviour of typical machines, the presence of the new high amplitude frequencies in the 0-100 Hz range is very dangerous for its stability of work. In the case of high precision machines, it is not only a cause of not optimal working conditions but very often is a cause of serious problems.

1 Introduction

One of the visible effects of the dynamical behaviour of any kind of working machinery are the vibrations induced in the system. The increase of the vibration levels may be caused by misalignment of parts, not enough structure stiffness or wrong connections. The dynamic conditions may also change due to the changes in environmental conditions like air pressure, temperature, humidity, ventilation conditions or a combination of these parameters. This is why the dynamic behaviour monitoring of machines (measurements of the amplitude of vibration at certain frequencies) is especially beneficiary to provide definite information about the accuracy, stability of working conditions and possible critical problems with certain parts or the whole system. Vibration monitoring is also a way to examine the effect and possibility of resonance occurrence. This effect is often present when a machine part (e.g. casing, piping and supply lines, supporting structures) or a machine working in the

* Corresponding author: lscislo@pk.edu.pl

vicinity have similar dynamical behaviour at certain frequencies (resonant frequencies). Therefore, dynamic behaviour measurements and analysis are an effective, non-intrusive method to monitor machine condition during initialisations, planned or unplanned terminations and during normal operations [2], especially when using non-contact measurement methods [1]. This is also the way to perform determinations of the operating and mechanical condition of equipment, which further can be used for system evaluation [3]. In consequence, the complex vibration monitoring of past, current and estimation of future behaviour can be the ideal way to identify developing problems before they become too serious and cause unscheduled termination of machines operations [4]. The unexpected shutdown and additional repairs, services and calibrations can be avoided by scheduling regular monitoring of machine vibrations either on the basis of chosen intervals or in a continuous manner [5-6]. Consequently, increasing amplitudes of vibration levels can identify flawed maintenance practices including mistakes in the installation process, different material used for replacements, parts misalignments or wrong isolation from ground motion conditions [7-8]. It can also allow the detection of new environmental influence from nearby operating devices and systems or e.g. civil engineering operations [4, 9].

2 Methods

2.1 The test object

To evaluate vibration influence on the high precision machines a test object very sensitive to vibration was chosen. It was decided to perform the measurements in the test chamber of Compact Linear Collider (CLIC), which could be a device to open the way to new physics and additionally to proof or disproof current theories [10]. The collider is a two-beam acceleration design, which imposes restrictive requirements on its components to assure the beam stability. One of the most important technical challenges, this accelerator needs unprecedented ground motion suppression techniques. This is especially important having in mind that the nominal centre of mass energy of 3 TeV and generation a high accelerating gradient of 100 MV/m [10, 11]. One of the sources of dynamic misalignments are the ground motion conditions and their effect on supporting structure and the accelerator itself [12, 13]. The second source often comes from dynamical behaviour of the environment and additional equipment and system operating in the close distance.

2.1 Measurement equipment and setup

To perform vibration monitoring, especially to record ground motion, proper instrumentation was needed. This consisted of a dedicated sensor and an acquisition system with signal processor [14, 15]. When considering the ground motion the most common equipment is a geophone. A geophone consists of three independent sensing units, one in the vertical direction (Z), and the other two in horizontal directions (X and Y respectively) [11]. In case of dynamic behaviour examination of the specific equipment or structure, smaller sensors called accelerometers are often used. For this study two highly sensitive geophone sensors (G1, G2) shown in Fig.1, were used in order to perform ground vibration measurements in the test area. Additionally, three precise seismic accelerometers PCB 393B31 (S1, S2, S3) were used to measure the modules acceleration. To measure the dynamic response of the structure, two tri-axial accelerometers PBC 356B08 (A4, A5) were also placed on the structures girder and the magnet itself. General (ambient) vibration measurements were performed using 2 high precision geophones placed on the ground. They were located inside the laboratory chamber with the examined machine, while three of the seismic accelerometers were fixed on top of the module in vertical (2 seismic accelerometers, one on the structure and one on the girder) and horizontal direction (1 seismic accelerometer on the girder). Fig. 1 shows the placement of sensors within the chamber.

Fig. 1. Placement of the sensors on the ground, girder and structure (left) and on the magnet (right).

The measurement followed the scheme shown in Table 1. During a specific time of the day, different subsystems (air, water cooling systems) would be turned on or off, thus allowing to determine the difference in vibration levels, its contribution and excitation frequencies or each of the subsystems.

Table 1. Measurements scheme.

Mes. No.	Air cooling	Flow [% of max.]	Water cooling	Flow [% of max]
1	OFF	0	OFF	0
2	ON	50	OFF	0
3	OFF	0	ON	50
4	OFF	0	ON	100
5	ON	100	ON	100
6	ON	100	OFF	0

The principle of work of the air and water cooling system is shown in Fig 2.

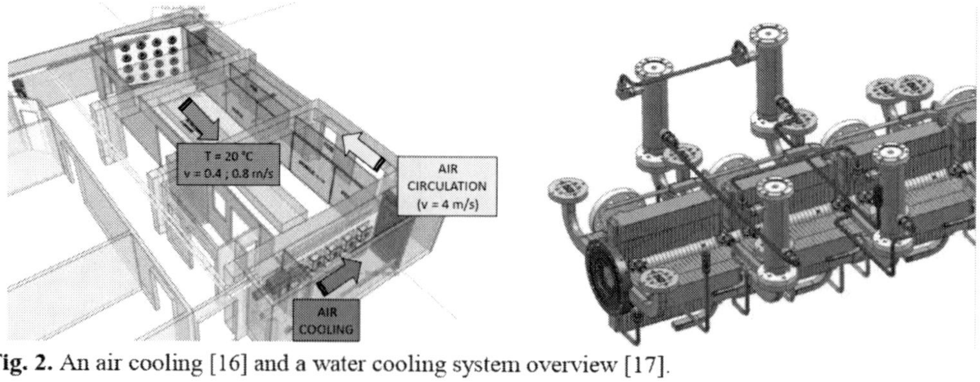

Fig. 2. An air cooling [16] and a water cooling system overview [17].

The maximal airflow was set to 0.8 m/s (100%). An additional measurement was taken with 50% of maximal airflow (0.4 m/s). In principle, the water cooling system should not cause problems in the measured frequency range (due to the high stiffness of the structure. However, additional measurements were taken to prove this assumption.

3 Discussion of results

When considering systems with dynamic conditions like in the study, it is beneficiary to use a sweep excitation in the frequency range. This is a useful indicator which allows a good overview of current dynamic conditions. It can also provide information about the effectiveness of vibration isolation [18]. This approach to the system diagnostics requires a detailed and accurate frequency analysis. It is especially dangerous for the given system if distortion of harmonic frequencies happens. This often results in matching this frequency value with one of the resonant ones and can result in magnification of amplitudes. If it occurs it may force the designer to change the responsible element or redesign the whole system. Moreover, ground vibration from sources such as working civil engineering machinery, landing planes or road traffic can be a big influence on the vibratory conditions measured and in some cases can be observed by the building occupants [4].

3.1 Ambient vibration measurement

Recording ambient vibrations of the examined area required an acquisition system, which records data in a time domain and later processed that date in the frequency domain. These dynamic conditions are the effect of different forces coming from working ventilation system, wind, traffic noise, some excitation coming from p and s-waves from earthquakes and other small changes in the environment. This why the ambient vibration measurement technique is beneficiary both to describe dynamical conditions of the environment but can also be used as a structural health monitoring [19].

These background vibration measurements were done in order to study the effects of the normal working conditions in the building and traffic effect from the nearby road. The measurements were done about 15 meters from public roads. Additionally, a small machining workshop was located in the neighbouring room, which also can be a source of additional vibrations.

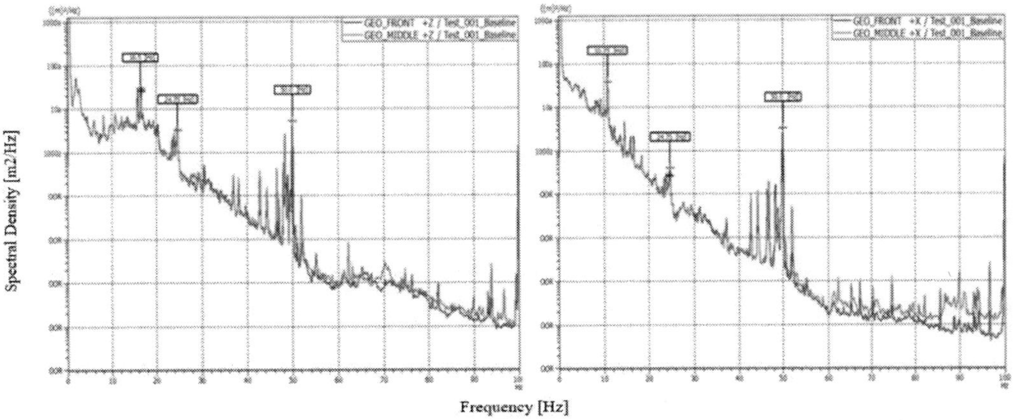

Fig. 3. Power Spectral Density for vertical (left) and horizontal (right) ground vibration in the examined room.

The measurement set up allowed the observation of dynamic conditions in the examined room. Furthermore, this was a start point to evaluate how ground vibration is transferred to objects, including the examined machine in the test area.

3.2 Integrated RMS vibration comparison between the baseline and situation of all system in operation

The integrated RMS (Root Mean Square) value was used to add up the total vibration in a spectrum, including ambient. The measurement data was taken in 20 minutes time blocks during the normal working day. During post-processing, the power spectral density was calculated. Integrated displacement from 1 Hz to 100 Hz in the vertical direction was calculated by the sensor (geophone) located on the ground (Fig. 4 left) and a sensor located on the examined structure (Fig. 5 right). Each time conditions were examined for the case without the presence of ventilation and water cooling system (ALL_OFF) and both systems working with maximal power (ALL_ON).

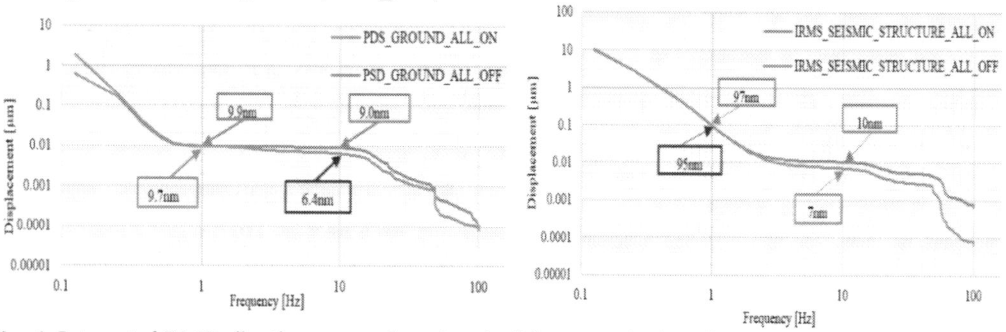

Fig. 4. Integrated RMS vibration comparison (vertical) between the baseline and situation of all system on (air cooling 100%, water cooling 100%). Geophone (G2) located on the ground (left) and a seismic sensor located on the structure (right).

Additionally, these values were compared with additional locations (Table 2) associated with low vibration conditions both on the surface and underground.

Table 2. IRMS results measured in different locations.

Measurement place	IRMS at 1 Hz [nm]	IRMS at 10 Hz [nm]
Examined room	9.7	6.4
Mechanical Measurement Laboratory (CERN)	17.4	16.7
Metrology Laboratory (CERN)	11.7	10.1
LHC tunnel (CERN)	1.9	0.5

As it is visible, the room had very good dynamic conditions in comparison with other surface placements. LHC tunnel, located 100 m below the surface is included here as a reference and has 10 times lower level of the typical building structures at the surface. This was one of the main reasons to place the experiments underground and assure the stability of the LHC work.

3.3 Ventilation and cooling system influence on the vibration levels (separate system in operation)

The following graphs show the results for vertical (Z) and lateral (X) direction. The measurements were made using uniaxial seismic accelerometers. Due to their high accuracy, they present reliable measurement results for changing conditions within the system. The baseline measurement was conducted without any system working with normal activity in the building and nearby road. The data was collected over a period of over 20 minutes.

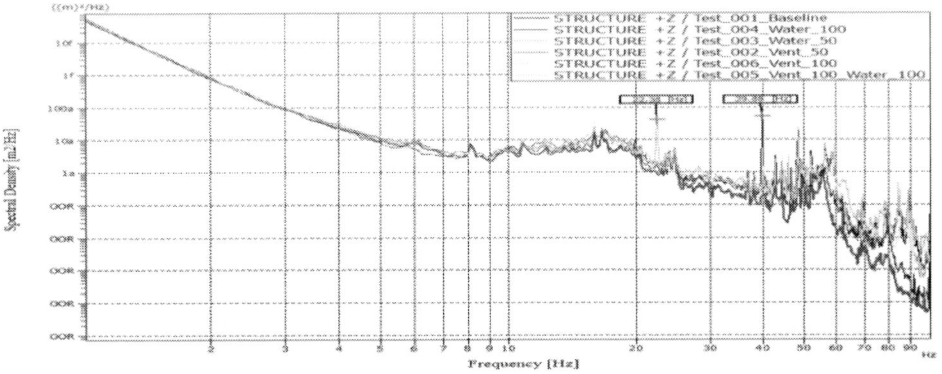

Fig. 5. Power Spectral Density for vertical (V) vibration of the module.

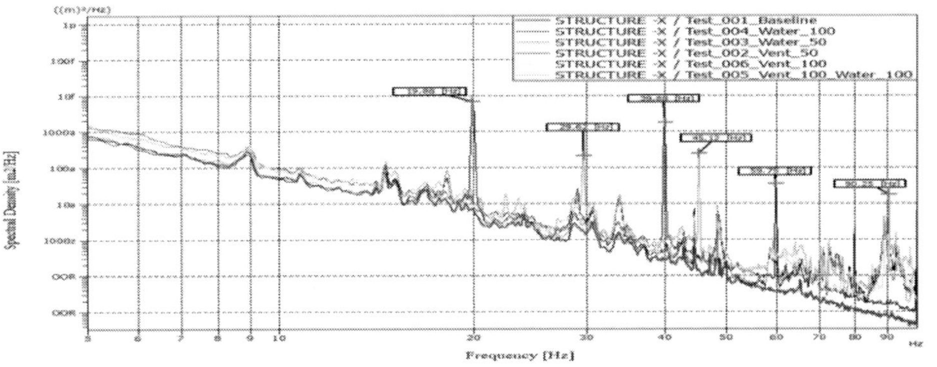

Fig. 6. Power Spectral Density for lateral (EW) vibration of the module.

As expected, new frequencies were observed due to the excitation from the ventilation and water cooling system (Fig. 5 and 6). The new picks were present with both settings of the airflow of the ventilation system (50%, 100%). Additional frequencies were observed while both systems were in operation (ventilation and water cooling). Specifically, the ventilation systems working at 50% of the power excites frequencies 19.88 Hz, 39.88 Hz, 59.75 Hz (Fig.6). Moreover, the ventilation system working on 100% of the power shows the excitation of additional frequencies at 29.62 Hz and 45.12 Hz (Fig.6).

Specific frequencies that were excited are clearly visible. Specifically, 22.38 Hz, 48.38 Hz, 56.25 Hz, 84 Hz (vertical direction) are the frequencies excited by the air and water cooling systems. Additional vibration may cause the misalignment of the beam and in consequences failure to produce the data for the experiments or even danger to the machine and operators.

 https://doi.org/10.1051/e3sconf/201910000080

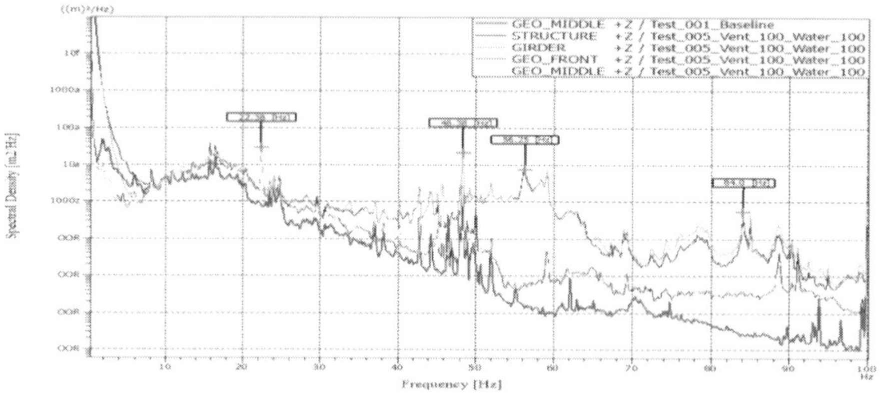

Fig. 7. Power Spectral Density for vertical (V) vibration of the module.

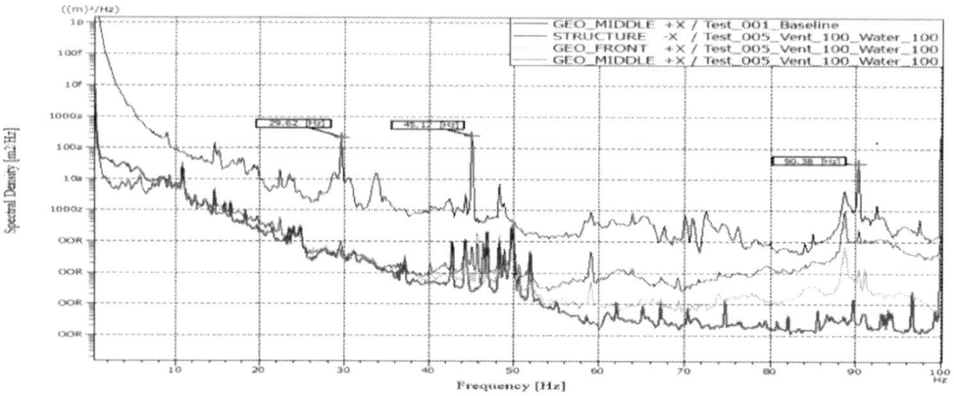

Fig. 8. Power Spectral Density for lateral (EW) vibration of the module.

4 Conclusion and observations

The goal of the measurements was to measure the levels of vibration caused to the module by its subsystems (air cooling, water cooling). An associated goal was to compare the level of vibration with all systems operating at full capacity. This allowed to observed the effect that those systems have on the experiments. The vibration of high precision equipment and research aperture is influenced by the operation of nearby machinery. This is visible in the PSD graphs corresponding to the ground vibration measured by the geophones located next to the examined module (Fig. 3). Additionally, there was an increase in the integrated level of vibration (compared at 1 Hz) between all systems disabled and all systems in operation (Fig. 4). This increase in vibration levels between the ground and the structure was expected but its level clearly proves the big influence of the cooling and ventilation systems on the modules dynamics. The comparison of PSD graphs for the situation with all subsystem turned off and all subsystems turned on (Fig. 7 and 8), show the expected increase of vibration levels in the latter case. Several peaks are visible in the graphs for the module corresponding to the natural frequencies of the structure and the girder. This situation not only had an influence on the results of the experiments, but also on the safety and stable operations of the whole structure. Prolonged exposure to this additional excitation will result also in fatigue problems. Therefore, the additional vibration will affect the alignment of the parts and cause the necessity to make corrections and repairs. Additionally, when evaluating both systems separately, the influence of the air cooling system on vibrations levels was greater than the influence of the water cooling system. However, the work of both systems simultaneously effects in additional excitations (Fig. 5). This allowed also to invalidate the assumption that due to its high stiffness, the water cooling system will have no influential effect on the structure. Finally, it was proven that the effects of cooling systems can have a large influence on the vibration levels and excitation of additional frequencies. Due to this fact, the problem of vibration must be taken into account during the design and integration of new machines. The observation of the additional frequencies is especially important in the 5–80 Hz spectrum where structures are prone to resonant effects due to the work of other machines in the vicinity. Not measuring the vibration influence on the high precision machines, like accelerators, will effect in the corruption of the experiments data and loss of accuracy. Moreover, for high precision research equipment, which requires the operation of advanced cooling and ventilation it is advisable to use constant monitoring of their influence on the whole system. In this article, it was proven that vibration monitoring is a valid chose for dynamic behaviour evaluation.

References

1. M. Guinchard, 9th International Particle Accelerator Conference, IPAC18, April 29–May 4, 2018, Vancouver, Canada. S. Koscielniak, Ed. (JACoW, [S.l.]), 2565–2567 (2018)
2. M. Tsypkin, IEEE International Electric Machines and Drives Conference, IEMDC 2011 (2011)
3. Z. K. Peng, F. L. Chu, Mech. Syst. Signal Process (2004)
4. M. Guinchard, 9th International Particle Accelerator Conference, IPAC18, April 29–May 4, 2018, Vancouver, Canada. S. Koscielniak, Ed. (JACoW, [S.l.]), 2568–2570 (2018)
5. C. Cempel, Vib. Phys. Syst. **25**, 97–102 (2012)
6. A. Panda, Š. Olejárová, J. Valíček, M. Harničárová, Int. J. Adv. Manuf. Technol. **97**, 401–411 (2018)
7. M. Kondo, S. Manabe, T. Takashige, Q. Rep. 105–111 (2016)
8. D. Goyal, B. S. Pabla, CIRP J. Manuf. Sci. Technol. **10**, 24–35 (2015)
9. R. B. Randall, *Vibration-Based Condition Monitoring: Industrial, Aerospace and Automotive Applications* (2010)
10. CERN, "CLIC collaboration, CLIC Conceptual Design Report," (2012)
11. G. Balik, B. Caron, D. Schulte, J. Snuverink, J. Pfingstner, Nucl. Instruments Methods Phys. Res. Sect. A Accel. Spectrometers, Detect. Assoc. Equip. **700**, 163–170 (2013)
12. A. Seryi, Particle Accelerator Conference, 2003. PAC (2003)
13. L. Malina, R. Corsini, D. Gamba, T. Persson, P. K. Skowroński, IPAC 2016 – Proceedings of the 7th International Particle Accelerator Conference (2016)
14. M. Jashank, D. C. Kala, Int. J. Comput. Appl. 15–18 (2013)
15. Z. Agioutantis, Rock Eng. Rock Mech. Struct. Rock Masses – Proc. EUROCK 2014, ISRM Eur. Reg. Symp. 965–970 (2014)
16. F. Duarte Ramos, F.-X. Nuiry, W. Klempt, *Experimental tests on the air cooling of the CLIC vertex detector* (2016)
17. A. Moilanen, M. Aicheler, A. Vamvakas, J. Väinölä, S. Doebert, *Finite Element Model for Thermal-Structural analysis of CLIC Lab Module type 0#2* (2017)
18. J. Burjánek, G. Gassner-Stamm, V. Poggi, J. R. Moore, D. Fäh, Geophys. J. Int. (2010)

The model of energy efficiency transportation of thermal and electric energy in the underground heat collector

Dmitriy Titkov[1,*]

[1]National Research University Moscow State University of Civil Engineering, 129337, Yaroslavskaya road, 26, Moscow, Russia

Abstract. The paper deals with modeling of the underground heat collector thermal conditions for utilities, calculating variations in temperature taking place in the soil mass around the collector as well as temperature variations in the air space of the latter. This leads to the reduction in the cost of heat and electricity transportation and a significant cost cutting of heat that is electricity supply of urban areas. The dependences of the thermal soil massiveness on the laying depth of the collector and changes in soil temperature in the course of time are presented.

1 The problem of heat and electricity transportation cost reduction by controlling thermal conditions of the underground heat collector

The problem of the underground heat collector energy efficiency providing is conditioned by the complexity of the engineering structure besides specific features of the surrounding soil and its thickness. This leads to many factors influencing conditions for the formation of the underground heat collector microclimate. These include the properties of the soil: thermal conductivity, thermal massiveness, heat absorption coefficient; outdoor air parameters and their change when passing through all layers (soil, enclosing structures of the collector) [1, 3, 4, 7]. At the same time, thermal conditions control in the air space of the collector based on the EE principles results in cost cutting of heat or energy supply to urban areas and cost reduction of the heat consumption by consumers.

The use of mathematical modeling methods in the study and prediction of the out-door air temperature effect on the thermal conditions of the underground heat collector allows taking into account thermal massiveness of the soil and enclosing structures, amplitude decay of variations in temperature and changes in the time of heat exchange processes taking place in the thickness of the soil, enclosing structures and space of the collector [2, 5, 6, 8, 9].The model must take into account the heat resistance of the enclosure and the fact that the heat exchange process between the collector and the outdoor air is not steady-state.

* Corresponding author: titkov1000@gmail.com

2 Background information for energy efficiency modeling

To build a mathematical model in the process of research, it is necessary to identify the effect of outdoor air temperature on the temperature conditions change in the soil mass and to describe the unsteady-state processes of temperature influence time lag in the course of time.

Let us consider the temperature change in the soil mass, thermal massiveness and time shift depending on the outdoor air temperature at a depth of up to 1.2 m in increments of 0.1 m (see Fig.1).

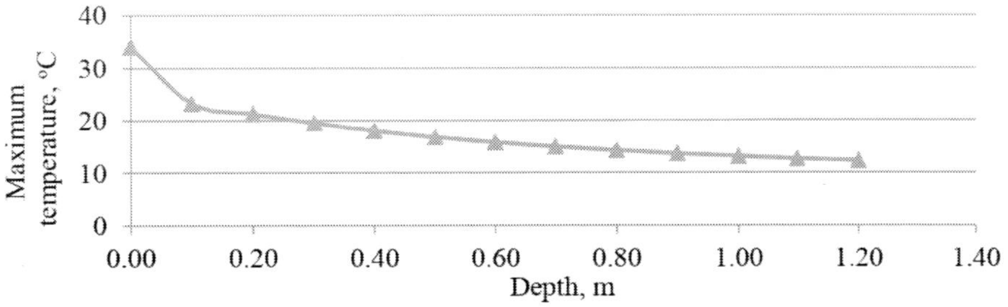

Fig. 1. Graph of the influence of the maximum outdoor temperature in the warm season on the soil at a depth of 1.2 m.

At a depth of 1.1–1.2 m, the temperature in the soil column stabilizes and approaches the steady-state process of temperature change.

Due to the large thermal massiveness and considerable thickness of the soil mass, the processes of heat and mass transfer are not steady-state. The time variation of the soil heat transfer processes with the outdoor air occurs with a shift in time (phase shift) [10].

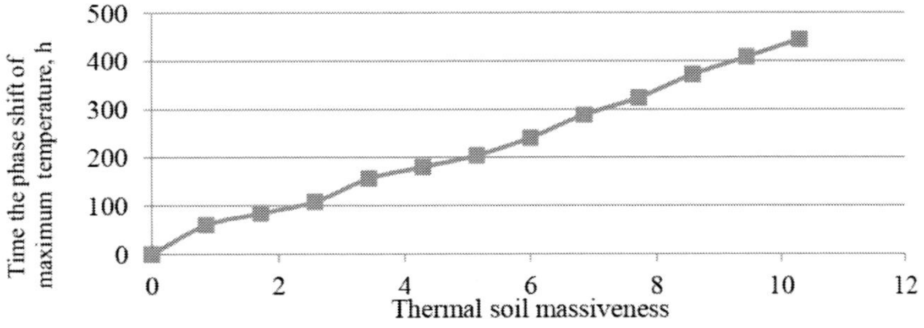

Fig. 2. Graph of time shift change (phase) according to the thermal massiveness of the soil.

The graph (see Fig. 2) also shows an increase in the time shift conditioned by the depth of the soil mass. It occurs with increasing depth, when the thermal massiveness increases as well and hence the time lag of the temperature conditions changes takes place [11].

Figure 3 is a graph presents the way of the time shift occurring according to the depth of the soil mass.

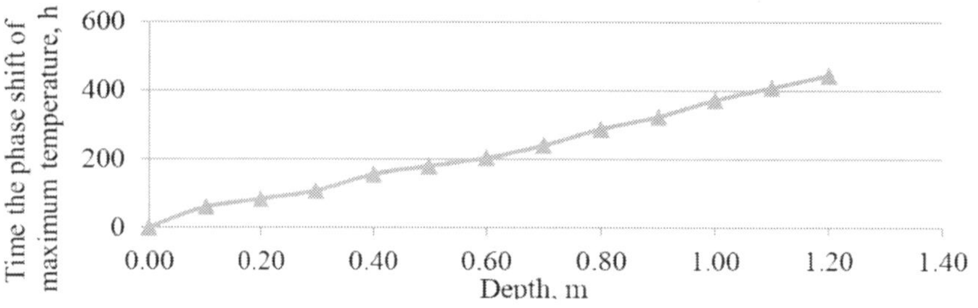

Fig. 3. Time shift graph depending on the depth of the soil mass.

With the change in depth, thermal inertia and thermal massiveness change as well and hence there is some shift in time of temperature change in the thickness of the soil [12].

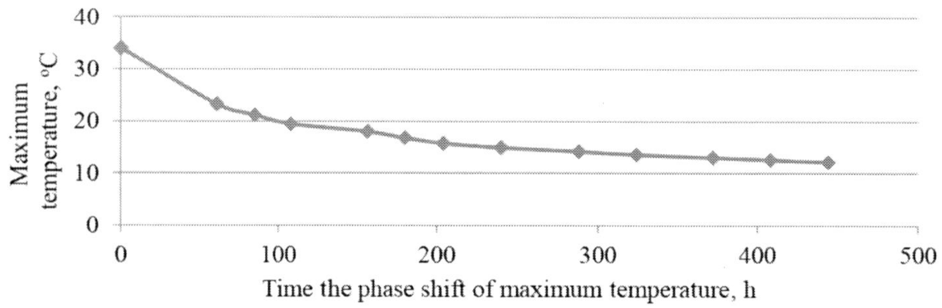

Fig. 4. Graph of the outdoor air temperature time shift alongside with the change of the temperature in the thickness of the soil column.

As can be seen from the graph in figure 4, the temperature change occurs with a time shift and the longer the phase shift the lower the temperature is, but within a certain period of time the temperature dependence on the time shift is equalized and approaches a relatively steady-state process [13]. Figure 5 and figure 6 have the graphs showing the influence of the outside air temperature alongside with the temperature in the soil column taking into account a time shift, being late in response of the temperature change depending on the depth with a step of 0.1 m [13].

Judging by the graphs in figure 5 and figure 6 it can be seen that the time shift and lag of temperature influence take place with each step of deepening into the soil. The graph in figure 5 shows the variations of the outdoor air temperature and allows seeing the air temperature change in the soil column comparing it with the change in the outdoor air temperature. Figure 6 presents the graphs of examining of the temperature conditions changes in the soil column without taking into account the graphs of the outdoor air temperature [14].

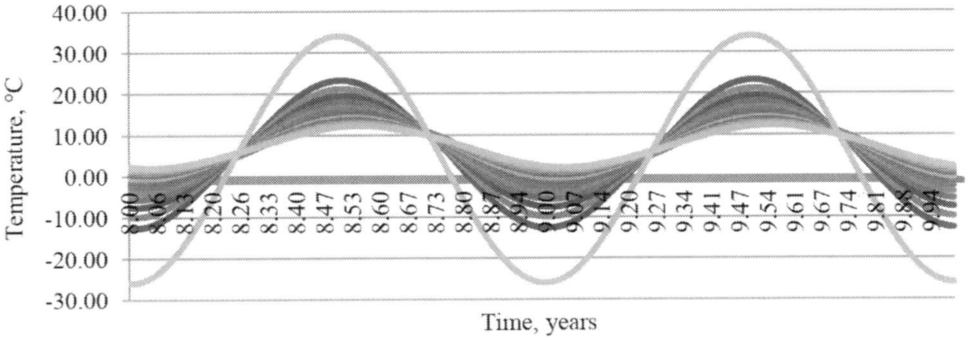

Fig. 5. Graph of influence of the outdoor air temperature alongside with the temperature in the soil column taking into account a time shift, being late in response of the temperature change depending on the depth with a step of 0.1 m. reflecting outdoor air temperature changes.

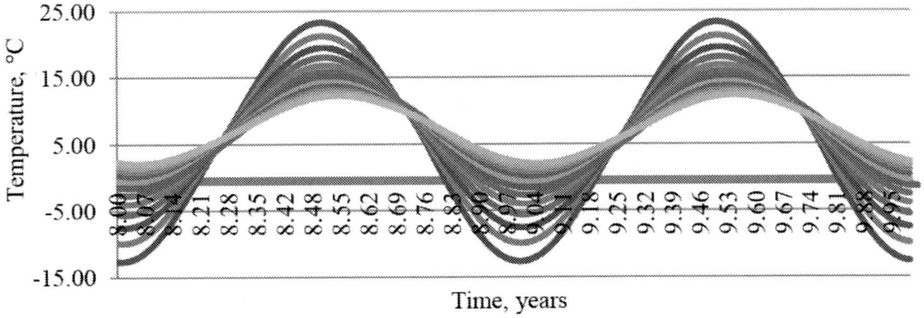

Fig. 6. Graph of influence of the outdoor air temperature alongside with the temperature in the soil column taking into account a time shift, being late in response of the temperature change.

Let us consider the temperature change in the soil column, thermal massiveness and time shift depending on the temperature of the outdoor air at a depth of 1.3 m to 2.4 m in increments of 0.1 m [11].

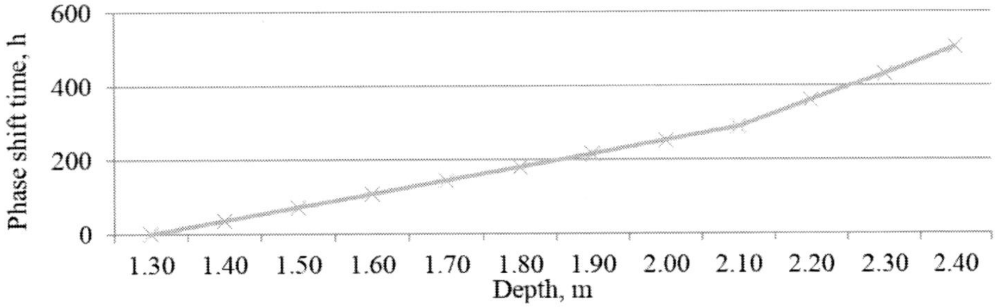

Fig. 7. Graph of the dependence of temperature variations changes in the course of time according to the laying depth of soil from 1.3 m to 2.4 m in increments of 0.1 m.

The graph shows that with each step of deepening there is an increasing deviation in the time lag of the temperature conditions influence taking place in the soil mass minding that the temperature change does not occur immediately, but with a certain period of time equal to the time of phase shift.

https://doi.org/10.1051/e3sconf/201910000081

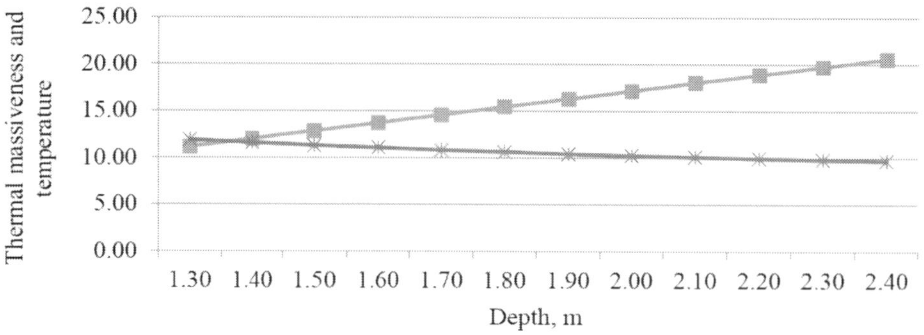

Fig. 8. Graphs of the thermal massiveness of soil and the temperature depending on the depth. (line 2 – thermal massiveness, line 5 – temperature).

The graphs show the thermal massiveness changes conditioned by further deepening while the decreasing of temperature in soil being minor. This is due to the thermal massiveness of the soil, being large enough and its increasing with each step of deepening [10].

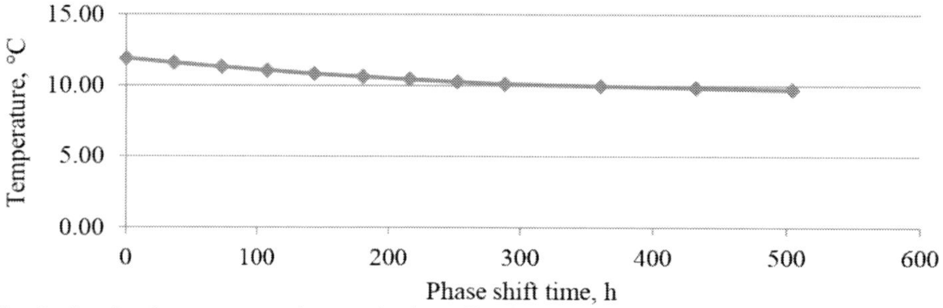

Fig. 9. Graph of temperature changes in the soil column depending on the time shift [10].

The graph in Fig. 9 shows that the longer the time lag, the lower the temperature is. However, with the time shift of about 300 h, the temperature drop is negligible and with the further time shift, the temperature change tends to a steady-state process. Let us consider the change in temperature in the thickness of the soil, thermal massiveness and time shift, depending on the temperature of the outdoor air at a depth of 2.5 m to 3.5 m in increments of 0.1 m [14]. It can be seen that temperature decay and variations together with time lag of outdoor air temperature effect on soil layer decrease due to increasing of thermal massiveness of soil.

Let us consider the temperature change in the soil column thermal massiveness and time shift depending on the temperature of the outdoor air at a depth of 3.5 m in increments of 0.1 m [14].

Fig. 10 present integrated graphs of the outdoor temperature effect and temperature variations decay, taking place in the soil column depending on time.

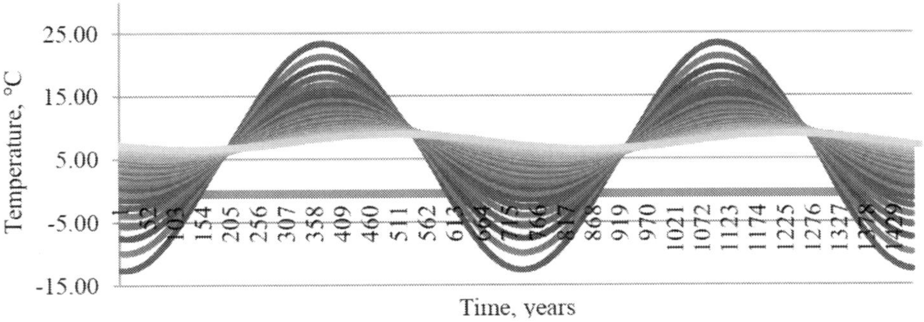

Fig. 10. Graphs of the outdoor temperature effect and temperature variations decay, taking place in the soil column depending on time.

These graphs give a complete picture of temperature variations and time delay temperature changes due to the increase in the thermal massiveness of the soil with each step of deepening. The graphs clearly show the shifts in the time of maximum temperatures and reflect a qualitative picture of the temperature variations amplitudes in the outdoor air as well as in the soil column within the period of 8–10 years. It can be concluded that in 3.5m or more soil mass deepening the outdoor air temperature effect can be neglected [11].

The type of soil examined in this paragraph was soil loam, being wet and having the following parameters: density (wet) $\rho_{gr.}$ = 2 710 kg/m^3; specific gravity $\gamma_{gr.}$ = 24525 N/m^3; coefficient of thermal conductivity (wet) $\lambda_{gr.}$ = 1,26 W/m×°C; the heat capacity $c_{gr.}$= 0.84 kJ/kg. (in accordance with state standard 25100-2011. Soils. Classification.) The climatic parameters of the outdoor air were taken for the city of Moscow in accordance with set of rules 131.13330.2012 construction climatology. The updated edition of building codes 23-01-99* (with Change N 2)).

3 Results of modelling and controlling the thermal conditions of the underground heat collector

Fortran programming language and Force 2.0 compiler were chosen for mathematical modeling. The initial data are the following: soil temperature (initial) tgr.0 °C; the temperature of internal air tair °C; coefficient of soil thermal conductivity λgr., W/m×°C; coefficient of thermal conductivity of enclosing structure layers λcon., W/ m×°C; coefficient of soil thermal conductivity cgr., kJ/kg; coefficient of thermal conductivity of enclosing structure layers ccon., kJ/kg; soil density ρgr., kg / m3; density of enclosing structure layers ρcon., kg / m3. A segment from the supply to the exhaust shaft, shown in Fig. 11, was taken as a design element of the underground collector. The segment length is 250 m and the volume of the design element for modelling is 0.1 m3. Having separated the under-ground collector into two symmetrical parts along the length, it is divided into segments. Along the length, axis k, the selected collector segment is divided into 2620 segments, along the width, axis i, - into 45 segments and in the depth from the ground surface, axis j, – into70 segments. In each of the segments the unsteady heat transfer and the effect of thermal massiveness on the delay of temperature fluctuations in time are calculated. When calculating this problem, we set the boundary conditions of the third type, set the temperature of the external and internal air, which varies over time and the processes of non-stationary heat transfer between the collector, soil and surfaces of enclosing structures and engineering communications, the air flow rate at the inlet to the supply shaft of the underground collector. The result of the calculation is the change of the collector thermal conditions over time and the rate change along the collector, estimation of the soil thermal massiveness affecting the temperature shift in time [12].

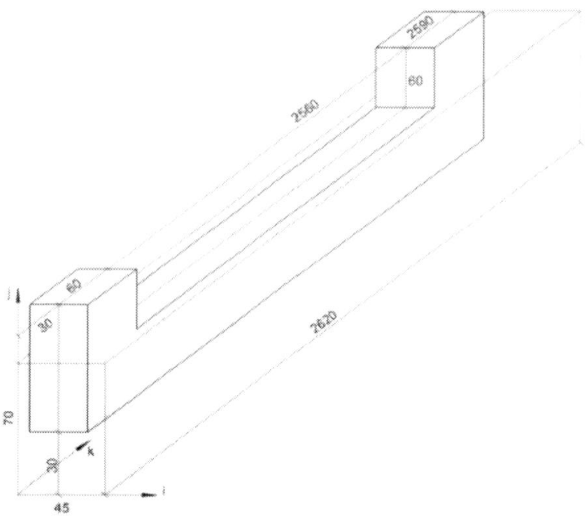

Fig. 11. Design part of the underground collector divided into segments of 0.1 m.

Fig.13 demonstrates the graphs of changing temperature in the collector depending on time and without regard to changing outer air temperature for better consideration of temperature changes within the collector.

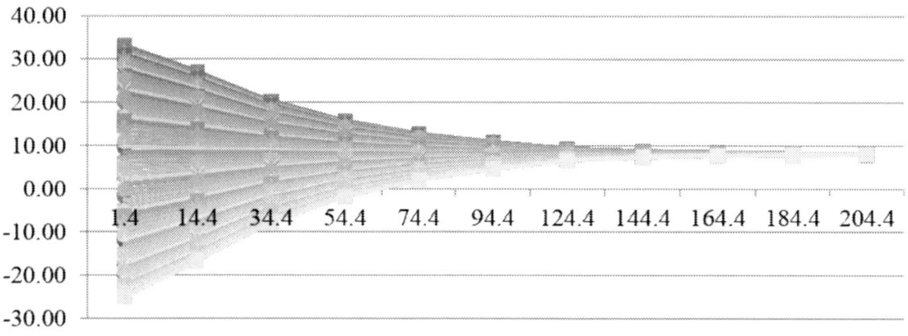

Fig. 12. Temperature field in the air along the underground collector in the range from -25°C to 32°C.

These graphs more clearly show the temperature changes within the underground collector in the range of small amplitude fluctuations, which allows seeing the shifts of the maximum temperature fluctuations over time. Graphs of Fig. 12 and 13 demonstrate how the temperature in the underground collector changes over time. The graphs show shifts of temperature maxima, they are not as pronounced as in the soil mass due to the low thermal massiveness of the air. However, it can be seen from the graphs that the temperature change is sinusoidal with some time delay and the influence of the outer inlet air has a noticeable change in thermal conditions only on a relatively small length of the collector in its initial sections. With the further passage of the air along the collector, the convective component of heat transfer and mass transfer decreases and the temperature stabilizes in the range of +8°C (± 0.5°C), which does not go beyond the requirements of the minimum temperature of internal air in the underground collector that does not have internal heat surpluses [13].

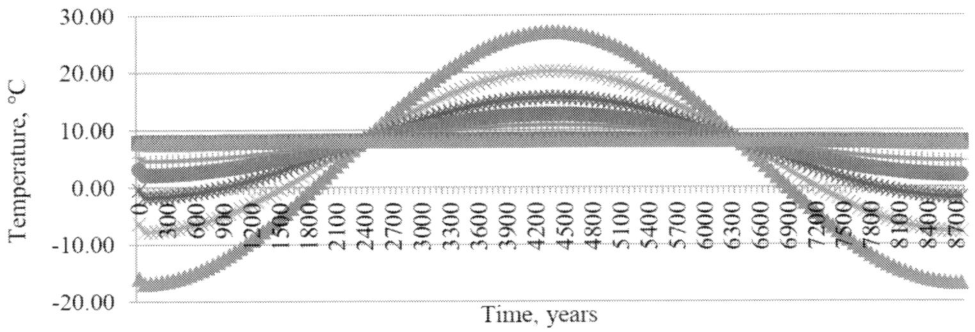

Fig. 13. Graphs of changing temperature in the collector depending on time and without regard to changing outer air temperature.

4 Summary

At present there are no software systems for modeling and calculating the thermal conditions of the underground collector, and it does not allow accurately calculate heat losses in the underground collector. The proposed model gives an opportunity to calculate the parameters of the underground collector thermal conditions, changing in time, which leads to lower transportation costs of thermal and electric energy and reduces energy consumption by consumers. With the help of the proposed model, it is possible to predict the processes of heat exchange in the soil mass and the effect of thermal massiveness depending on the type of soil, climatic region and effects of outer air parameters on the parameters of the collector thermal conditions in time. The problem of energy saving during transportation of heat and electricity is solved, which directly affects the reduction of energy consumption by consumers.

References

1. ASHRAE Guide And Data Book: Fundamentals and Air Conditioning Engineering, 912 (1963)
2. S. B. Carpenter, *Principal Plum Dispersion Models* (TVA Pow. Plants, 63 Annual Meeting, Air Pollution Assoc., June, 1970)
3. F. N. Frenkiel, R. E. Munn, *Turbulent Diffusion in Enviromental Pollution* (Advances in Geophysics Series, 18A and 18B, New-York, Academic Press, 1974)
4. A. N. Huber, *Incorporating Building. Terrain Wake Effects on StaskEnfluent* (Preprint Volume for the Joint Conf. on Applic. of Air Pollut. Metod., Amer. Meteorolog. Socity, Boston, MA, 1977)
5. O. Roberts, Proc. Roy. Soc. **104A**, 11–15 (1923)
6. E. A. Skaret, H.M. Mathisen, Enviroment International **8**, 473–481 (1982)
7. G. L. Tuve, Trans. ASHRAE **59**, 261–282 (1953)
8. J. E. Woods, R. R. Crawford, ASHRAE Transactions, 683–696 (1983)
9. J. E. Woods, Trans. ASHRAE, 559–571 (1975)
10. O. Brukhanov, A. Rymarov, A. Malysheva, D. Titkov, *Analysis of heat losses of underground tunnel for engineering utilities with available methods* (Matec Web of

Conferences 5. Cep. "5th Inter. Scien. Conf. "Integration, Partnership and Innovation in Construction Science and Education", 2016)

11. A. G. Rymarov, D. G. Titkov, *Peculiarities of calculation of thermal regime of underground communication collector internet bulletin wolghas* (Volgograd State University of Architecture and Civil Engineering, **4**, 40, 6, 2015)
12. A. G. Rymarov, D. G. Titkov, *Influence of the massity of the environment on the thermal regime of the underground collector for engineering communications, natural and technical sciences* (OOO "Izd." Sputnik + "Moscow **6**, 84, 563–564, 2015)
13. O. N. Brukhanov, A. G. Rymarov, D. G. Titkov, *Analysis of methods of calculation of thermal losses by underground mounting collector, natural and technical sciences* (OOO "Izd." Sputnik + "Moscow **8**, 86, 89–91, 2015)
14. A. G. Rymarov, D. G. Titkov, Science of Education **10-2**, 44–47 (2015)

Thermal efficiency of the building envelope with the air layer and reflective coatings

Adam Ujma[1,*], and *Nina* Umnyakova[2]

[1]Czestochowa University of Technology, Faculty of Civil Engineering, ul. Akademicka 3, 42-200 Czestochowa, Poland
[2]Research Institute of Building Physics of the Russian Academy of Architecture and Building Sciences (NIISF RAASN), 127238, Moscow, Lokomotivnyj pr. 21, Russia

Abstract. Currently, when calculating the heat-shielding qualities of external enclosing structures, the ability of fencing surfaces to reflect the heat flux by radiation is not taken into account. All this leads to an underestimation of heat-shielding characteristics in comparison with reality. Studies have shown. that a number of factors affect the thermal protection of air layers - the difference in temperature on the surfaces of the interlayer, its thickness. But the most important is the coefficient of elimination of the air gap surface. The use of reflective heat insulation with a low emissivity on one of the surfaces of the air gap makes it possible to increase its thermal protection 2.5–3 times. The article presents the calculations of thermal protection of air gaps with different emissivity coefficients of surfaces and conclusions about the effectiveness of the use of reflective insulation.

1 Introduction

Currently, the market all over world has a wide range of various effective insulation building materials based on mineral wool, polystyrene foam and others. Their thermal insulation properties depend on the thermal conductivity of the material. With an increasing in the thickness of the insulation, the heat flux passing through this heat-insulating material decreases. However, only the coefficient λ is taken into account in the calculations, and the ability of materials surfaces to reflect the flow of heat is nowhere taken into account. Moreover the industry in many countries produces innovative materials with reflective thermal properties.

The principle of operation of innovative heat-insulating materials with reflective film from aluminum foil is based on the ability of a shiny surface with a low emissivity to reflect the radiant component of the heat flux. For this reason, the effectiveness of reflective insulation does not depend on the thickness of the material, but only on the emission of the surface. In this case, it is necessary to have an air gap at the reflective surface, through which the radiant component of the heat flux passes. The combination of reflective insulation with heat-insulating material, for example, polyethylene foam, allows to obtain a qualitatively new product with high heat-shielding qualities.

*Corresponding author: aujma55@wp.pl

The analysis of thermal insulation capacity of such materials, applied in specific building structures, is undertaken, among others, by the papers [1–4]. These studies indicate that such materials show positive effects with regard to improvement of thermal insulation capacity of external envelopes wall and other constructions..

However, the widespread introduction of such materials into building practice presents great difficulties. Firstly, standard methods for assessing heat-shielding qualities are not suitable for them and it is necessary to develop new methods or modify existing ones, which requires both time and money. In addition, existing calculation methods are not always applicable to innovative materials. In particular, when using materials with reflective insulation, standard calculation methods do not take into account the reflection of the radiant component of the heat flux. As a result, at the stage of choosing materials, designers are not able to correctly assess the actual thermal protection of an innovative material or structure with these materials and correctly arrange it in the building envelope.

Conducted in NIISF RAACS (Russia) studies of reflective insulation made it possible to evaluate the effectiveness of its use. Experimental studies of closed air gaps with reflective thermal insulation and without it have shown that the presence of a shine reflective material on one of the surfaces of the air layer allows to increase its thermal resistance in 2.5–3 times. A method was developed for calculating the thermal resistance of closed air spaces with reflective thermal insulation, taking into account the multiple reflection of the radiant heat flux from its internal surfaces. Thermo-technical calculations of a multilayer wall with air gaps of various thickness in the presence of reflective insulation were carried out. The obtained theoretical and experimental studies have shown good convergence of the results.

2 The method of determining thermal parameters of the air layer

According to the standard [5], heat resistance of an unventilated air space R_g, with a length and width higher than the 10-fold thickness, is expressed by the following relation:

$$R_g = \frac{1}{h_a + h_r} \tag{1}$$

where:

h_a – is the convective/conduction heat transfer coefficient, W/(m² K)

h_r – is the radiation heat transfer coefficient, W/(m² K).

The value of the convective heat transfer coefficient h_a is assumed on the basis of parameters (Table 1).

Table 1. Value of the convective/conduction heat transfer coefficient h_a depending on the temperature difference on the air layer surfaces ΔT, based on the standard [5].

Direction of heat flow	**$\Delta T \leq 5$ K**	**$\Delta T > 5$ K**
	W/(m² K)	
horizontal	1.25	$0.73 \cdot \Delta T^{1/3}$
upwards	1.95	$1.14 \cdot \Delta T^{1/3}$
downwards	$0.12 \cdot d^{-0.44}$	$0.09 \cdot \Delta T^{0.187} \cdot d^{-0.44}$
d – thickness of the air layer, m		

According to the given data, the coefficient h_a only depends on the air layer thickness in case of a horizontal envelope, at a downward heat flow. Only for a temperature difference

ΔT higher than 5K, the coefficient h_a also depends on ΔT. This seems to be a major simplification since, according to the research data presented in Table 1, the intensity of conductive and convective heat exchange in an unventilated air layer depends on both the layer thickness d and the temperature difference on the surfaces thereof.

The data presented in the paper [6] show that the effect of the width of an air gap and the temperature difference at the limits of the air layer is significant. Even at a low temperature difference, the change of layer thickness from 2 cm to 5 cm causes the coefficient $\lambda_\lambda + \lambda_a$ to nearly double.

3 Heat transfer parameters in air spaces

The chart (Fig. 1) compares the heat transfer coefficient values, calculated on the basis of data from the studies [7–10], with values resulting from the standard parameters [5].

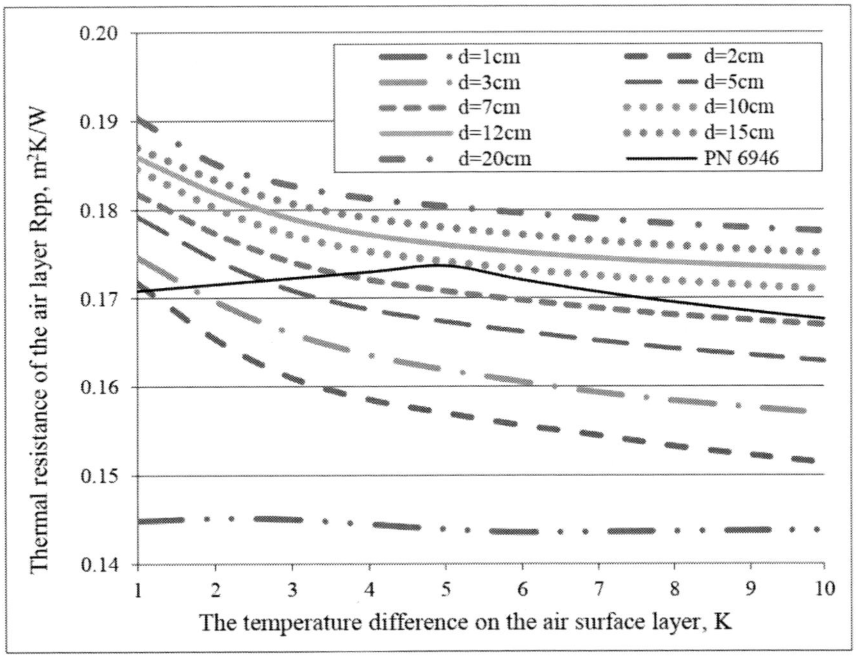

Fig.1. The nature of the change of value of the conductive and convective heat transfer coefficient depending on the temperature difference at the limits of the air layer and its thickness.

As the comparison of calculation results (Fig. 1) shows, the values given by the standard parameters overlap with the data of calculations for air layer thickness in the order of 7–10 cm, at a temperature difference above 5 K.

The standard parameters [5] give a constant value of the coefficient h_a with regard to the temperature difference $\Delta T < 5\ K$, disregarding the fact that the value of the coefficient h_a largely depends on the air layer thickness and the temperature difference ΔT.

In general, it can be stated that for air layer thickness lower than 7 cm, standard values are understated, and for thickness higher than 10 cm, they are overstated.

The value of the radiation heat transfer coefficient h_r, in accordance with the standard parameters [5], is determined on the basis of the relation:

$$h_r = E \cdot h_{r0} \tag{2}$$

where:
E – is the surface emittance, emissivity parallel surface, -

$$E = \varepsilon_{1-2} \tag{3}$$

h_{r0} – is the radiation coefficient for a black-body surface, W/(m^2 K)

$$h_0 = 4 \cdot \sigma \cdot T_m^3 \tag{4}$$

σ – is the Stefan-Boltzmann constant, 5.67·10^{-8} W/(m^{2}K^4)
T_m – is the average temperature of the emitting surface and its surroundings, K

The standard [5] lacks instructions concerning how the temperature at the surfaces limiting the air layer can be included in the calculations. The explanations mention T_m as the average temperature of the surface and its environment. Therefore, it can be assumed as the average temperature of both surfaces of the air layer.

The radiation heat transfer coefficient $h_{r(1-2)}$ between the surfaces limiting the air layer may be derived from the relation below:

$$h_{r(1-2)} = \frac{C_{1-2}[(\frac{\theta_1 + 273}{100})^4 - (\frac{\theta_2 + 273}{100})^4]}{\theta_1 - \theta_2} \tag{5}$$

where:
C_{1-2} – radiation heat constant surfaces air layer 1 and 2, W/(m^{2}K^4),
θ_1, θ_2 – temperature surfaces air layer 1 and 2, K

This relation was used to calculate how the value of the coefficient h_r (the designation $h_{r(1-2)} = h_r$ has been applied hereinafter) depends on the temperature of surfaces creating an unventilated air layer. Assuming the emissivity of both surfaces at the level of ε = 0.9, the value of the coefficient h_r was determined for three cases; temperature of the warmer surface: 19°C – which may correspond to the location of an air void in an external wall on the side of the interior; 9°C – example location of an air void in the middle of an envelope; 0°C – location of a void on the side of the exterior (Table 2).

The comparison of data in Table 2 shows that the value of the coefficient h_r decreases along with the increase of the temperature difference between the surfaces creating an air void. There is also a noticeable reduction in the value of the coefficient h_r along with the temperature drop on the surfaces of the air layer. This translates to an increase in the thermal resistance of the air layer, resulting from the phenomenon of thermal radiation between two surfaces creating an unventilated air layer.

The value of the coefficient h_r was also determined for a case when one of the surfaces limiting the air layer is coated with a low-emissivity material, such as aluminium foil. Three values of the radiation heat constant C were assumed: 0.2; 0.4; 0.6. Calculations were performed for a case of location of an air layer on the warmer side of an external wall (for θ_1 = 19°C), and in the second case, assuming location of the void on the cold side of the envelope (for θ_1 = 0°C). The results were compared in the chart (Fig. 2).

Table 2. Value of the radiation heat transfer coefficient h_r by depending on the temperature value θ_l and temperature difference on the air layer surfaces ΔT.

Temperature difference, ΔT	Value of the radiation heat transfer coefficient h_r by depending on the air temperature value θ_l		
K	0°C	9°C	19°C
	W/(m² K)		
1	3.76	4.15	4.60
2	3.74	4.12	4.58
3	3.72	4.10	4.55
4	3.70	4.08	4.53
5	3.68	4.03	4.51
6	3.66	4.04	4.49
7	3.64	4.01	4.46
8	3.62	3.99	4.44
9	3.60	3.97	4.42
10	3.58	3.95	4.39

The comparison shows that the reduction of radiation heat constant (emissivity) of one of the surfaces of the air void, as well as the drop of the surface temperature, result in significant reduction of the value of the coefficient h_r, evidencing an increase in a component of thermal resistance of such an envelope, connected with a drop in intensity of radiation heat exchange.

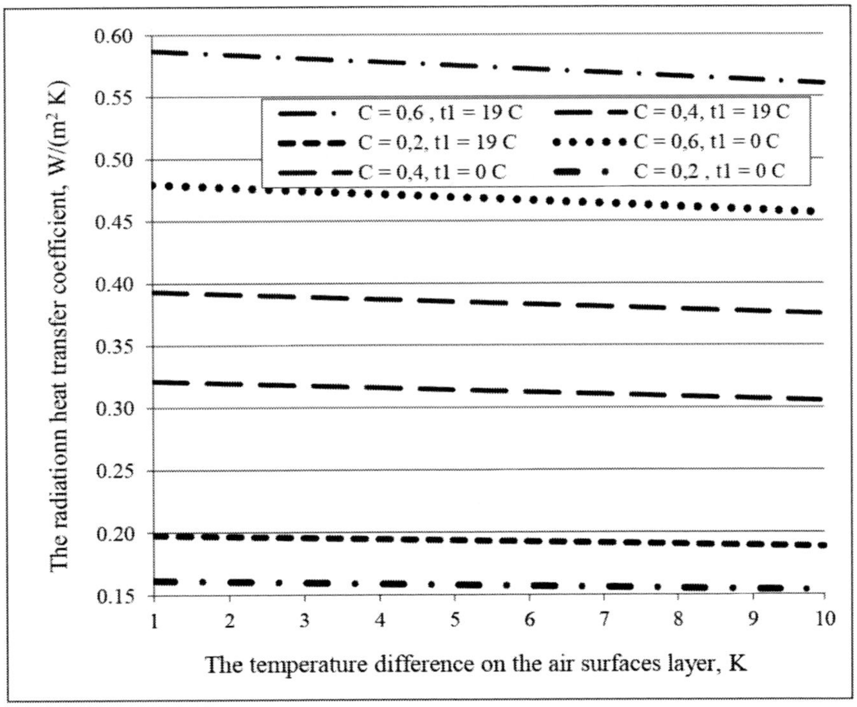

Fig. 2. The nature of the change of value of the radiation heat transfer coefficient h_r depending on the temperature difference at the limits of the air layer and radiation heat constant C for a parallel surfaces.

This also indicates the possibility of considerable reduction of the insulation properties of the envelope in case of loss of the assumed material radiation heat constant level, e.g. as a result of oxidation of the aluminium foil surface over time. This fact requires separate research and appropriate consideration in thermal insulation capacity assessment of external envelopes featuring structural solutions of such kind.

4 Conclusions

Investigations, fulfilled by the authors, show that the heat protection properties of closed air gaps depends mush on the thickness of the layer: with increasing thickness from 1 cm to 7 cm, the thermal resistance of the air gap increases by 0.25 m^2K/W from 0.145 m^2K/W up to 0.17 m^2K/W. Also, it should be mentioned that temperature difference has influence on heat protection of air gap: when the thickness of the air gap is 1 cm, the temperature difference between the sides of the air gap practically does not affect its thermal protection. With increasing thickness, the effect of the difference between temperatures on the surfaces of the air gap increases. The reason for this is an increase in convective heat exchange with the growth of the temperature difference between the surfaces of the air gap.

The greatest influence on the thermal protection of the air layers has emissivity surfaces. is exerted by the emissivity of the surfaces. When its value decreases from $C = 0.6\ W/(m^2K^4)$ to $C = 0.2\ W/(m^2K^4)$, the heat protection of the air gap increases by 4 times.

References

1. R. Wojcik, Insul. **11/12** (2014)
2. K. Patoka, Insul. **9** (2013)
3. D. W. Yarbrough, K. S. Teh, L. Ch. Haw, E. Salleh, S. Mat, M. Y. Sulaiman, J. of Pow. and Ener. Eng. **4** (2016)
4. S. Uvslokk, Ch. Schlemminger, S. Asphaug, Ener. Proc. **132** (2017)
5. PN-EN ISO 6946-2017-10 *Building components and building elements. Heat resistance and heat transfer coefficient. Calculation methods* (Polish Committee for Standardization, W. 2017)
6. N. Umnyakova, A. Ujma, Insul. **11/12** (2017)
7. GOST R 56734-2015 *Buildings and constructions. The calculation of thermal protection of walls with reflective insulation* (Standartinform, M. 2016)
8. K. F. Fokin, *Building heat engineering of elements of building envelope* (ABOK-PRESS, M, 2006)
9. N. P. Umnyakova, Stroitel'naya orbita, **2** (2013)
10. N. P. Umnyakova, Zhilishchnoye stroitel'stvo, **1/2** (2014)

Investigations of the flow field inside a square cyclone separator using DPIV and CFD

Marek Wasilewski[1,*], and *Lakhbir* Singh Brar[2]

[1]Opole University of Technology, Faculty of Production Engineering and Logistics, ul. Proszkowska 76, 45-758 Opole, Poland,
[2]Mechanical Engineering Department, Birla Institute of Technology, Mesra, Ranchi 835215, India

Abstract. This study used CFD and DPIV to investigate the flow field inside a square cyclone. For CFD, turbulent flow was modelled using 3D RANS equations (with an RSM complementary model). The results of the analysis indicate that the flow inside a square cyclone has two constituents: an outer, free flow and an inner, forced flow in the centre, as with cyclones with the traditional structure. Furthermore, the results indicate that the corners of a square cyclone generate additional vortexes that cause pressure drop to increase (but may, at the same time, potentially improve separation efficiency). The two methods provided highly consistent results. DPIV may be a very good method for validating results obtained with CFD, including for the analysis of flow fields.

1 Introduction

A review of subject literature and practical examples shows that there are many structural variants of cyclone separators. The most common ones are the solutions proposed by Stairmand, Lapple and Swift [1]. Many studies have been conducted concerning these geometries [e.g. 2–9]. However, there are relatively few studies that address a type of cyclone with an unusual geometry, i.e. the square cyclone. Square cyclones are a response to the Circulating fluidized bed (CFB) boiler technology. Along with the development of large CFB boilers, the oversized body of the traditional cyclone became a significant drawback due to the thickness of the fireproof wall, as it necessitates a long start-up time for the boiler. The first square cyclone to be introduced into CFB boilers was the Pyroflow COMPACT, produced by the Ahlstrom Corporation in the 1990s [10]. The square cyclone had a lower separation efficiency than its round counterpart, but it was smaller, had a shorter start-up time, and was simpler and less expensive to manufacture. Later on, a new variant of the square cyclone was developed and applied in CFB boilers that implemented water cooling and a curved inlet, which helped to accelerate the particles and, consequently, improve separation efficiency [11–13]. Fatahian et al. [14] proposed an unusual solution to improve the efficiency of the square cyclone. They introduced a laminarizer inside the inlet and the vortex finder and found that it had a positive effect on separation efficiency.

* Corresponding author: m.wasilewski@po.opole.pl

In the past, cyclone separators were selected for the purposes of particular installations usually through experimental research and empirical and semi-empirical models. Such research was expensive and time-consuming. With the development of computer technology and software, new structural designs of cyclone separators began to rely extensively on research involving the computational fluid dynamics (CFD), which today constitutes the most popular method in studies on the subject. Various numerical models are employed, including k-ε [e.g. 15–16], Reynolds Stress Model (RSM) [e.g. 3–6, 17–21] or Large-Eddy Simulation (LES) [e.g. 7, 8, 22, 23], depending on the assumed goals and scope of the study. However, this type of research needs to be validated against other methods in order to be considered reliable. Validation is most often performed based on traditional experimental research, which is limited by the fact that it only analyses separation efficiency and pressure drop. It is impossible to conduct an extensive analysis of the field distributions of the fluid phase flow or the trajectory of the solid phase.

This issue can be addressed by using Digital Particle Image Velocimetry (DPIV). Particle Image Velocimetry (PIV) was first applied outside the laboratory wind tunnel in the 1980s. At the time, obtaining a high-quality image took between two and three days. Today, thanks to modern high-speed CCD and CMOS digital cameras and high-capacity computer units, the image is ready within seconds [24, 25]. This is one of the reasons why the PIV (currently DPIV) is growing in popularity. The second reason is the possibility to obtain temporary velocity fields, which in turn allows for a quantitative detection of the spatial structures in unstable flows. The third reason is the need for experimental data used to verify CFD codes. The growing popularity of DPIV translates into an increasing number of studies (from many disciplines) that apply this method [e.g. 26–33]. Research on cyclone separators has not yet taken full advantage of DPIV: subject literature includes few studies on cyclones that use this method [34–36].

The aim of this study was to investigate of the flow field inside a square cyclone separator using DPIV and CFD. The study involved a cyclone with a nontraditional structure, i.e. a square cyclone. The DPIV method was verified (in relation to CFD research) in terms of vectors maps and mean tangential and axial velocities within selected planes. In addition, the values of pressure drop obtained through experimental research were compared to those obtained through CFD research.

2 Experimental and DPIV studies

For the purpose of the research, optical track was constructed (Fig. 1). The main component was the square cyclone separator (Fig. 2). Because the researched object needed to be made from a transparent material in order to conduct DPIV measurements, the model was made from poly(methyl methacrylate) (PMMA). The DPIV setup consisted of an Nd:YAG laser as the source of light, a CCD camera (placed at a distance allowing for full use of the available image resolution) that registered flow phenomena and an impulse generator to synchronize the other two devices. A seeding device was installed in front of the cyclone inlet to introduce seeding particles, i.e. Di-Ethyl-Hexyl-Sebacic-Acid-Ester (DEHS), into the measurement space. The setup was controlled through the Dantec Dynamic Studio 2015b software. The same setup was used to record and to process the images and analyze them digitally. Table 1 presents the technical parameters of the DPIV setup.

In addition, measurement equipment (flow meter, pressure sensors + differential pressure gauge) and a transportation system for the fluid phase (pipe, exhaust fan with frequency converter) were installed. A transportation system for the solid phase was also installed in order to model the real two-phase flow (in this study, the share of the solid phase amounted to 0.005 kg/1 m^3). Measurements were performed at four different inlet velocities: v_i = 7 m/s (*Re* = 233,808), v_i = 9 m/s (*Re* = 300,610), v_i = 10 m/s (*Re* = 334,011) and v_i = 12 m/s (*Re* = 334,011).

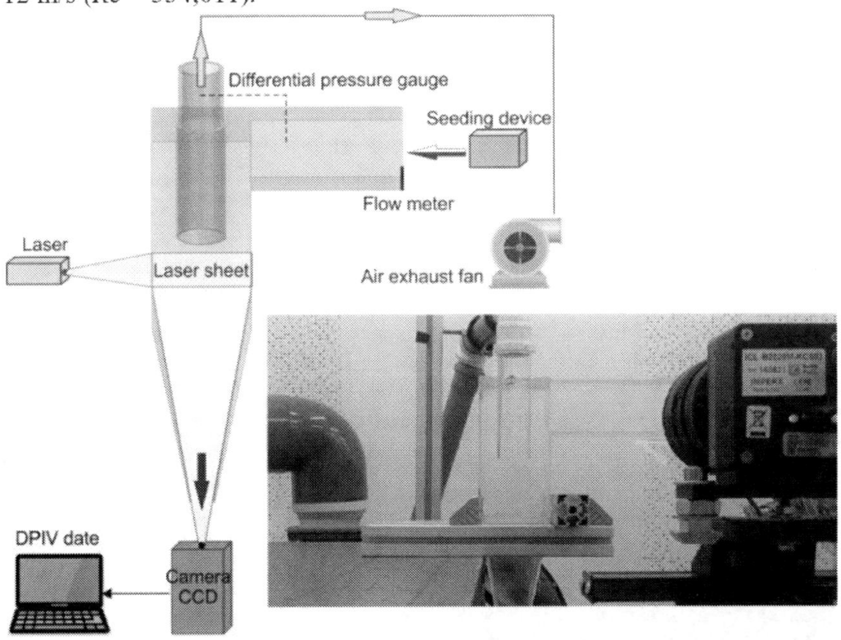

Fig. 1. Experimental set-up.

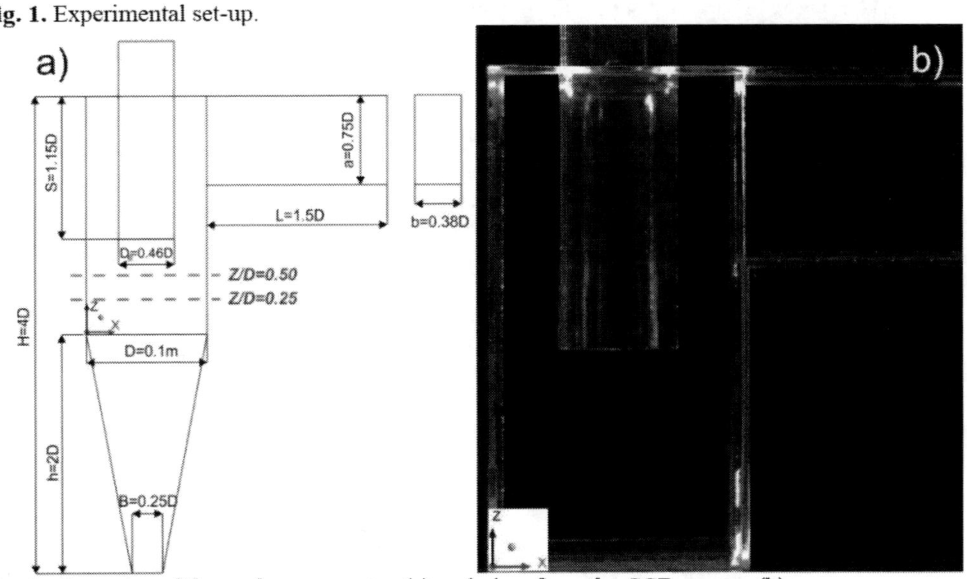

Fig. 2. Geometry of the cyclone separator (a) and view from the CCD camera (b).

Table 1. Technical parameters of DPIV system.

Parameter	**Description**
Camera CCD	Dantec Dynamics FlowSense EO 4M
Laser	Dantec Dynamics DualPower TR
Pulse generator	Berkeley Nucleonics Corp Model 575-8
Time between pulses, μs	25
Interrogation area size (horizontal x vertical), pix	32 x 32
DPIV algorithm	Adaptive Correlation with Local Neighbourhood
Detection method	Normalized median validation

3 CFD studies

The finite volume method based Fluent software (Version 18.2) was used to conduct the CFD research. The RANS method, coupled with the complementary RSM model, was used to model the turbulent flow. Detailed information about these models can be found in several studies as [3,18-21]. The solution was found by formulating equations for the pressure (Pressure Based Solution Method Segregated Solver). Differential equations were solved using the SIMPLE (Semi Implicit Method for Pressure Linked Equations) computational algorithm in order to correctly determine the coupling between the pressure and velocity fields and to maintain the momentum continuity equation. The second-order upwind interpolation method was used to determine the representative samples of the constituent values on the surface of the control volumes. The convergence condition was set at 10^{-6} for the continuity equation and 10^{-3} for the other equations. The flow conditions corresponded to the conditions applied in the experimental measurements. The 'velocity-inlet' boundary condition was used for the inlet and the 'outflow' boundary condition was used for the outlet.

A model was created based on the geometry of the cyclone separator that was used in the experimental measurements. A three-dimensional domain was applied. In the next stage, the computational domain was discretized. A sensitivity analysis was conducted in order to take into account the effect of the density of the computational mesh. Three mesh densities were generated (Mesh 1 – about 240,000 elements, Mesh 2 – about 325,000 elements, Mesh 1 – about 392,000 elements). In each case, the computational area was discretized using a hexagonal mesh with a varying density depending on the area of the device. The value of the pressure drop was validated with respect to the experimental measurements. For Mesh 1, the difference between the measurement methods was about 11%, regardless of the *Re*. For Meshes 2 and 3, the difference ranged from 8% to 9%. Mesh 2 was selected for a further analysis due to its minimal density that did not display large changes in the obtained values. Figure 3 compares the obtained values of Euler number (*Eu*) for Mesh 2 depending on the Reynolds number (*Re*).

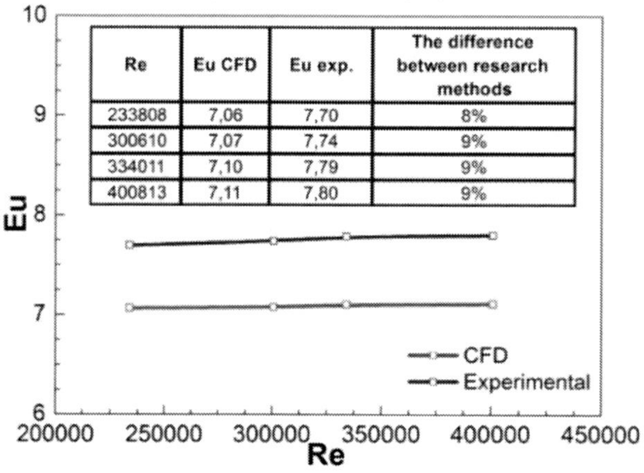

Re	Eu CFD	Eu exp.	The difference between research methods
233808	7,06	7,70	8%
300610	7,07	7,74	9%
334011	7,10	7,79	9%
400813	7,11	7,80	9%

Fig. 3. Comparative analysis of *Eu* values depending on the research method (Mesh 2).

4 Results and discussion

4.1 Vector of the flow field

The two-phase flow in cyclone separators is strongly turbulent. The cross-section of traditional, circular cyclone structures contains two vortexes: the outer (free) vortex where fluid swirls down till the bottom and the inner (forced) vortex where the fluid is directed upward toward the vortex finder. This reflects the Rankine vortex, which is very beneficial for the separation of solid particles. The vector field shown in Fig. 4a indicates that a similar phenomenon also occurs in a square cyclone. An outer downward vortex is clearly visible near the cyclone wall, whereas in the middle section, the fluid moves upwards towards the vortex finder. The same phenomenon also occurs for vectors obtained through DPIV and CFD. In the case of DPIV, a slightly higher velocity was observed near the axis of the vortex finder (close to the inlet section). The velocity in this area decreased towards the walls of the vortex finder (which may have been caused by strong turbulences and the fact that the measurement system did not take into account all seeding particles). With both methods, the vectors had nearly identical directions.

The vector field shown in Fig. 4b (*Z/D = 0.25)* indicates that the maximal values of velocity occurred near the cyclone walls, whereas lower velocities occurred in the cyclone axis. This is also confirmed by the presence of the Rankine vortex. The fact that the central vortex was not located in the centre of the cyclone may be due to an unsteady oscillating flow near a precessing vortex core (PVC) in the central region. Furthermore, small local vortexes appeared in the corners of the walls due to a sharp turn in the flow. The geometry of the square cyclone in this region enforces a sharp, abnormal turn of the two-phase mixture, causing intense collisions between the particles and between the walls and the particles. Corners are one of the primary regions that cause pressure drop. Corners were found to benefit particle separation, mainly due to the strongly fluctuating flow absorbing a large portion of the kinetic energy of both the particles and the fluid [37, 38].

Figure 5 complements the vector analysis by showing velocity vectors in the centre of the inlet section of the cyclone separator. In this case, after the mixture stream left the inlet section, it encountered a significant obstacle, i.e. the external wall of the vortex finder, which caused the velocity to increase considerably in this region. The results obtained with CFD and DPIV were also very similar in this case. The distribution of velocity vectors in the inlet section obtained with DPIV indicates variations in velocity magnitude along the entire width.

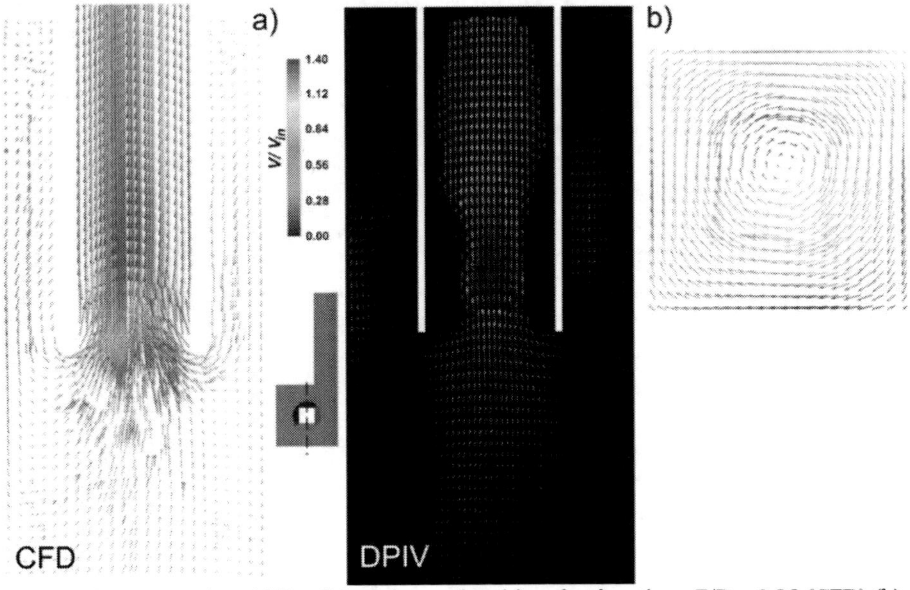

Fig. 4. Velocity vectors in middle of the inlet section (a) and at locations $Z/D = 0.25$ (CFD) (b).

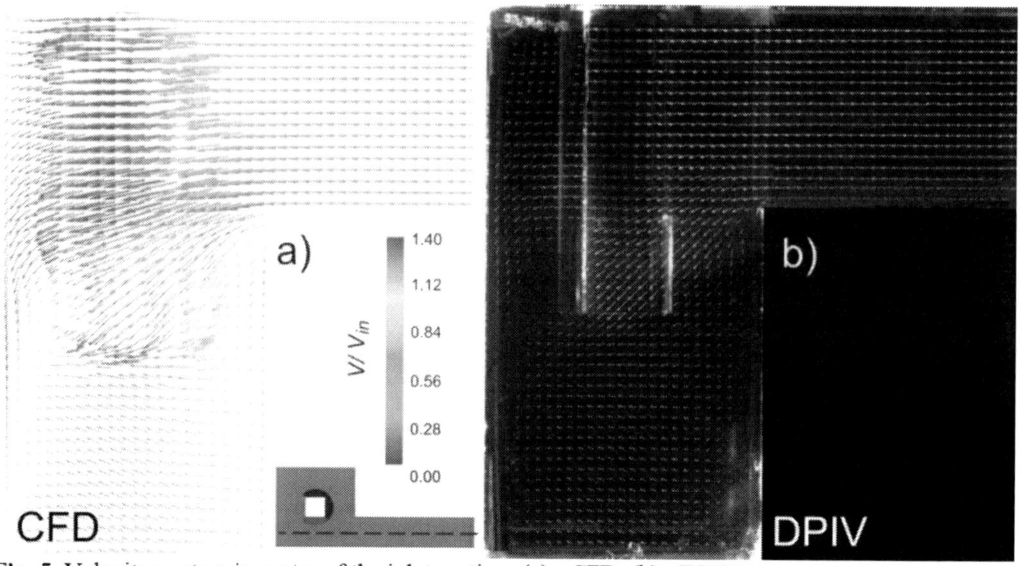

Fig. 5. Velocity vectors in center of the inlet section: (a) - CFD, (b) - DPIV.

4.2 Mean axial and tangential velocity

In traditional cyclone separators, axial velocity affects the movement of solid particles inside the cyclone: the fluid and the axial velocity of the external vortex cause them to move downwards and rotate near the walls. Figures 6 indicate that a similar phenomenon occurs in a square cyclone. Axial velocity is negative in the outer vortex region (a downward flow) and positive in the internal vortex region (an upward flow directed towards the vortex finder). The axial velocity contours presented here for both CFD and DPIV are very similar. However, note that DPIV yielded higher maximal and minimal values of the parameter.

A crucial constituent of the flow field in cyclone separators is tangential velocity, as it characterizes the efficiency of these devices by significantly affecting centrifugal force and, consequently, separation efficiency. The greater the centrifugal force in the radial direction, the more particles are directed towards the walls (and the fewer particles are directed towards the vortex finder). Figure 6b (bottom row) shows the mean tangential velocity profile in two planes ($Z/D = 0.5$ and $Z/D = 0.25$). In this case, the shape of the obtained curves also corresponded to cyclones with the traditional structure; this is consistent with the principles of rotational flow. The profiles in both sections are similar in shape (for both methods). They only differ in maximal and minimal values. Tangential velocity is much higher near the walls and approaches zero in the centre of the cyclone. The aim of the strongly eddying flow near the walls is to direct solid particles towards the walls, thus allowing for their separation. Positive values were observed on the left side, and negative values were observed on the right side. Tangential velocity in DPIV also reached higher maximal and minimal values.

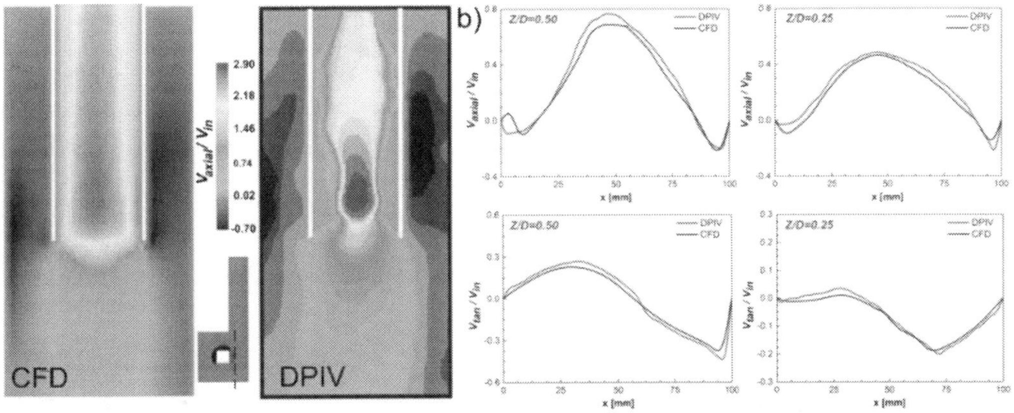

Fig. 6. Contour plots of mean axial velocity in the central section (a) and radial profiles (b) of mean axial velocity (top row) and mean tangential velocity (bottom row) at locations *Z/D = 0.5* and *Z/D = 0.25*.

5 Conclusion

Two methods were employed to research the flow field inside a square cyclone separator, i.e. CFD and DPIV. The former is commonly used to analyze flow inside cyclone separators. However, in order to be reliable, the obtained results must be validated through a comparison to experimental measurements, which is hindered by the lack of a possibility to compare the obtained flow field distributions. This study has led to the following conclusions:

- The obtained vector fields and profiles of axial and tangential velocities for a square cyclone separator indicate that the flow inside a square cyclone has two constituents: an external, free flow and an internal, forced flow in the centre, as with cyclones with the traditional structure;
- The corners of a square cyclone general additional vortexes that cause pressure drop to increase (but may, at the same time, potentially improve separation efficiency);
- Slightly higher maximal and minimal values of the analyzed parameters were obtained for DPIV;
- The obtained results are highly consistent between both methods (and, consequently, constitute good methods for researching flow in square cyclones and can be used to optimize it);
- DPIV may also be a very good method for validating results obtained with CFD, including for the analysis of flow fields and other parameters of flow.

As a future work, we plan to extend the work by employing both DPIV and CFD (preferably LES) to enhance the understanding of the flow physics and to optimize the geometry of square cyclone.

This research was financially supported by the scientific activity number 2018/02/X/ST8/01645 from the National Science Centre, Poland (Narodowe Centrum Nauki – NCN).

References

1. L. K. Wang, N. C. Pereira, Y. T. Hung, *Air Pollution Control Engineering* (Totowa, 2004)

2. J. Dirgo. D. Leith. Aerosol Sci. Tech. **4** (1985)
3. B. Wang, D. L. Xu, K. W. Chu, A. B. Yu, App. Math. Model. **30** (2006)
4. K. Elsayed, C. Lacor, Appl. Math. Model. **35** (2011).
5. L. S. Brar, R. P. Sharma, K. Elsayed, Powder Technol. **286** (2015)
6. C. W. Haig, A. Hursthouse, D. Sykes, S. Mcilwain, Appl. Math. Model. **40** (2016)
7. L. S. Brar, K. Elsayed, Sep. Purif. Technol. **207** (2018)
8. L. Brar, Today: Proceedings **5** (2018)
9. V. Kumar, K. Jha, J. Mech. Sci. Technol. **11** (2018)
10. R. Gamble, T. Hyppanen, T. Kauranen, Proceeding of the 12th International Conference on Fluidized Bed Combustion (1993)
11. G. X. Yue, Y. Li, X. X. Zhao, Proceeding of the 14th International Conference on Fluidized Bed Combustion (1997)
12. S. Wang, M. Fang, Z. Luo, X. Li, M. Ni K, Cen. Powder Technol. **102** (1999)
13. J. F. Lu, G. X. Yu, Q. Liu, Proceeding of the 15th International Conference on Fluidized Bed Combustion, 1999
14. H. Fatahian, E. Fatahian, M. E. Nimvari, Powder Technol. **339** (2018)
15. B. Dobrowolski, M. Skulska, J.Wydrych, Archiwum Energetyki **38** (2008)
16. M. Thulasiraman, K. Pitchandi, J. Mech. Sci. Technol. **29** (2015)
17. F. Mariani, F. Risi, C. Grimaldi, Sep. Purif. Technol. **179** (2017)
18. E. Kashani, A. Mohebbi, M. G. Heidari, Powder Technol. **327** (2018)
19. M. Wasilewski, L. S. Brar, Sep. Purif. Technol. **213** (2019)
20. D. Misiulia, K. Elsayed, A. G. Andersson, Sep. Purif. Technol. **185** (2017)
21. M. Wasilewski, G. Ligus, E3S Web of Conferences 00185 **44** (2018)
22. S. Pirker, C. Goniva, C. Kloss, S. Puttinger, J. Houben, S. Schneiderbauer, Powder Technol. **235** (2013)
23. L. S. Brar, K. Elsayed, Powder Technol. **311** (2017)
24. R. J. Adrian, J. Westerweel, Cambridge University Press (2011)
25. C. E. Willert, Apply. Sci. Res. **56** (1996)
26. K. H. Leea, C. H. Leea, C. S. Leeb, Fuel **83** (2004)
27. C. Iwaki, K. H. Cheong, H. Monji, G. Matsui, Exp. Fluids **37** (2004)
28. J. Westerweel. Exp. Fluids **29** (2000)
29. S. S. Paul, M. F. Tachie, S. J. Ormiston, Int. J. Heat Fluid Fl. **28** (2007)
30. Ch. Bin, G. Liejin, J. Therm. Sci. **9** (2012)
31. G. Ligus, Sz. Kołodziej, E3S Web of Conferences 01005 **19** (2017)
32. Sz. Kołodziej, G. Ligus, E3S Web of Conferences 01017 **19** (2017)
33. G. Ligus, M. Wasilewski, E3S Web of Conferences 00096 **44** (2018)
34. Z. Liu, Y. Zheng, L. Jia, J. Jiao, Q. Zhang, Chem. Eng. Sci. **61** (2006)
35. Z. Liu, J. Jiao, Y. Zheng, Q. Zhang, AIChE J. **52** (2006)
36. B. Zhang, S. Hui, International Conference on Power Engineering (2007)
37. Y. Su, Y. Mao, Chem. Eng. J. **121** (2006)
38. K. Qiu, J. Yan, X. Li, Journal of Engineering for Thermal Energy and Power **14** (1999)

Attempts to establish a regional probabilistic model of intense rainfall for the Upper and Middle Oder River basin

Marcin Wdowikowski[1,*], *Bartosz* Kaźmierczak[2], *Ondrej* Ledvinka[3], and *Matěj* Jedlička[3]

[1]Institute of Meteorology and Water Management - National Research Institute, Podlesna 61, 01-673 Warsaw, Poland
[2]Wrocław University of Science and Technology, Faculty of Environmental Engineering, Plac Grunwaldzki 13, 50-377 Wrocław, Poland
[3]Czech Hydrometeorological Institute, Hydrology Database and Water Budget Department, Na Sabatce 2050/17, 143 06 Prague 412, Czechia

Abstract. The main scope of this study was to present the possibility of developing a heavy rainfall regional model, based on a probabilistic approach. Following the methods and conclusions of the scientific and technical work of the research team in the Upper and Middle Oder basin, carried out on rainfall data between 1961 and 2010, it was found that the generalized exponential distribution enables a satisfactory probabilistic description of heavy rainfall, and the results of the models constructed on its basis are obtained in an accessible and not time-consuming way. The regional model presented in the article sufficiently describes the empirical heavy rainfall at eight analyzed meteorological stations, especially in the range of rainfall amounts occurring most frequently from a statistical point of view, i.e. with the p-value from 0.3 to 1.0.

1 Introduction

The safe design of urban areas drainage requires reliable information on rainfall conditions in a specific region, derived statistically from long-term meteorological measurements and observations [1–4]. The determination of probable rainfall is also necessary to identify areas exposed to the occurrence of floods, regardless of their nature (flash floods or urban floods) [5–9]. In Poland, there is still a deficiency of scientific works in the field of probabilistic modelling of intense rainfall; therefore, this article is an attempt to reduce this deficit. The paper is a brief description of a consistent continuation of the research activities undertaken by the authors in the works [8, 9]. Thus, for the description of the study area and methodological details we rather refer to the previously published works and indicate only the necessary differences or modifications here. At the beginning, before the preparation of this work, there were primarily the conclusions of previous studies [1, 2, 8–11]. This paper attends to the attempt if the development of the regional rainfall model is possible for data series with different precipitation characteristics at locations with different

* Corresponding author: marcin.wdowikowski@imgw.pl

elevation. In order to achieve the intended goal, the quantiles of the probability distribution best fitted to random variables representing eight individual stations will be compared with the quantiles of an averaged time series variable that theoretically represents the largest precipitation heights for the whole Upper and Middle Oder basin. The purpose of the proposed simplification is to answer the question whether the generalization of rainfall data, the quality of which is usually the subject of many scientific discussions, as the input to the rainfall model, can give satisfactory results from the viewpoint of engineering applications.

2 Study area and methods

Following the paper [9], the basic study material was composed of one-minute rainfall series from meteorological stations located in the Upper and Middle (also called Central) Oder basin covering the years from 1961 to 2010, provided by the Polish Institute of Meteorology and Water Management – National Research Institute (IMWM-NRI). Figure 1 (adopted from [9]) shows the eight stations located in the Oder basin with elevation depicted in meters a.s.l. The Upper and Middle Oder was limited by the profiles of water gauges in Chalupki, directly located on the Polish-Czech border (downstream of Bohumín), and Slubice, located near the Polish-German border. Due to these research assumptions, only the Polish part of the Upper Oder basin was taken into account. With respect to [9], some potential meteorological stations such as Cieszyn, Leszno, Grabik and Słubice (not shown in Fig. 1) had to be excluded from further analyses because of the violation of the data series homogeneity due to the analyzed long-term period.

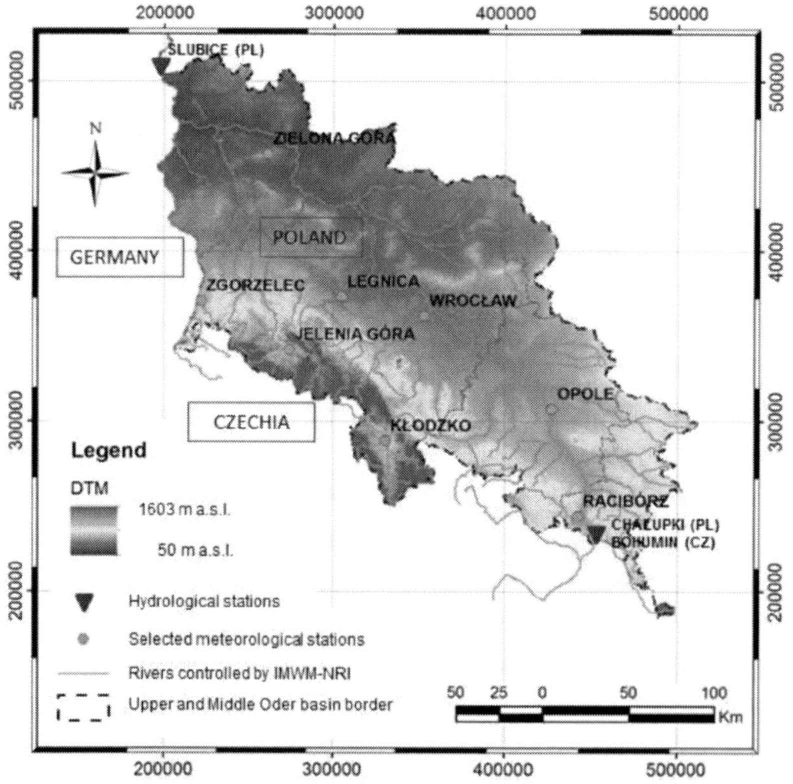

Fig. 1. Location of the eight analyzed meteorological stations of the IMWM-NRI within the Upper and Middle Oder basin (adopted from [9], based on the PUWG-92 coordinate system).

Probabilistic modelling of intense rainfall was made on the basis of the methodology developed and constantly improved by the research team of the Wrocław University of Science and Technology (Faculty of Environmental Engineering) and the IMWM-NRI, specifically for rainfall data from Wrocław (2009–2015) [1, 2, 10] and then from Legnica [11]. Simplified procedure for the development of a probabilistic model assumes the following tasks:

1. separation of maximum interval precipitation heights for rainfall duration from 5 minutes to 3 days based on the available precipitation events in the assumed 16 time periods (5, 10, 15, 30, 45, 60, 90, 120, 180, 360, 720, 1080, 1440, 2160, 2880 and 4320 minutes); this method is sometimes called the Partial Duration Series (PDS) approach,
2. selection of representative random samples of the largest precipitation values for the analyzed multi-year period, in this case 50 years of measurements,
   Important: For further analyses, the selection of the highest rainfall values can be made directly if one takes only the 50 largest values from the non-increasing sequence of all peaks found. In such a case, however, the independence of rainfall data above the accepted cut-off threshold (POT) must be maintained. Otherwise, the annual maximum precipitation (AMP) method should be used.
3. selection of the probability distribution that best fits the measurement data (through a numerical estimation of probability distribution parameters followed by a goodness-of-fit criterion assessment),
4. determination of the dependence of selected probability distributions parameters using the rainfall duration and the probability of exceedance.

Following the methods and results given in [1, 2, 8–11] and also the theoretical basis in [12], in this research, the four probability distributions were fitted to the data series from the eight stations, namely the Gumbel, Weibull, Log-Normal and Generalized Exponential Distribution (GED). The goodness of fit was provided by the BIC and rRMSE criteria. The results were presented in figures containing a comparison between the regional model and the eight station maximum empirical data series for chosen time periods (editorial frames of this paper did not allow to present all 16 time periods for eight stations; moreover, the comparison of all time periods for all analyzed stations would be unclear). All necessary numerical calculations were performed using the R-package statistical modules ('FAdistr', 'fExtremes', 'MASS' and 'bbmle') compiled in our own constructed scripts.

3 Results

With respect to the research methods described in Section 2, maximum rainfall data series were derived from the long-term period database representing the eight analyzed stations. As a result of this data separation, we got eight data frames combined with 50 maximum values of 16 rainfall durations (from 5 to 4320 minutes). In the next step, the averaging of particular rainfall durations from all eight stations was done in order to get regional data frame containing maximum values for the Oder River basin. The procedure of choosing the best-fitting probability distribution was applied to the maximum rainfall data prepared in described way. The results of the BIC and rRMSE criteria for each rainfall duration are summarized in Table 1. The smallest values (in bold) indicate the greatest correlation between the theoretical distribution and the regional average obtained using the eight measured data series. In general, for rainfall durations between 5 and 2160 minutes both criteria pointed out that GED is the best-fitting probability distribution. These results are in strict accordance with the conclusions of the work for Wrocław [1, 2, 10], Legnica [11] and several stations in the catchment area of the Lusatian Neisse River basin [8]. Despite major differences for the 2880- and 4320-minute rainfall duration, GED was adopted for further analysis as the best-fitting. In fact, for the purposes of designing and operating drainage systems of urbanized areas, the rainfall duration limited to one day is sufficient [1, 2]. The GED goodness of fit is shown in Fig. 2.

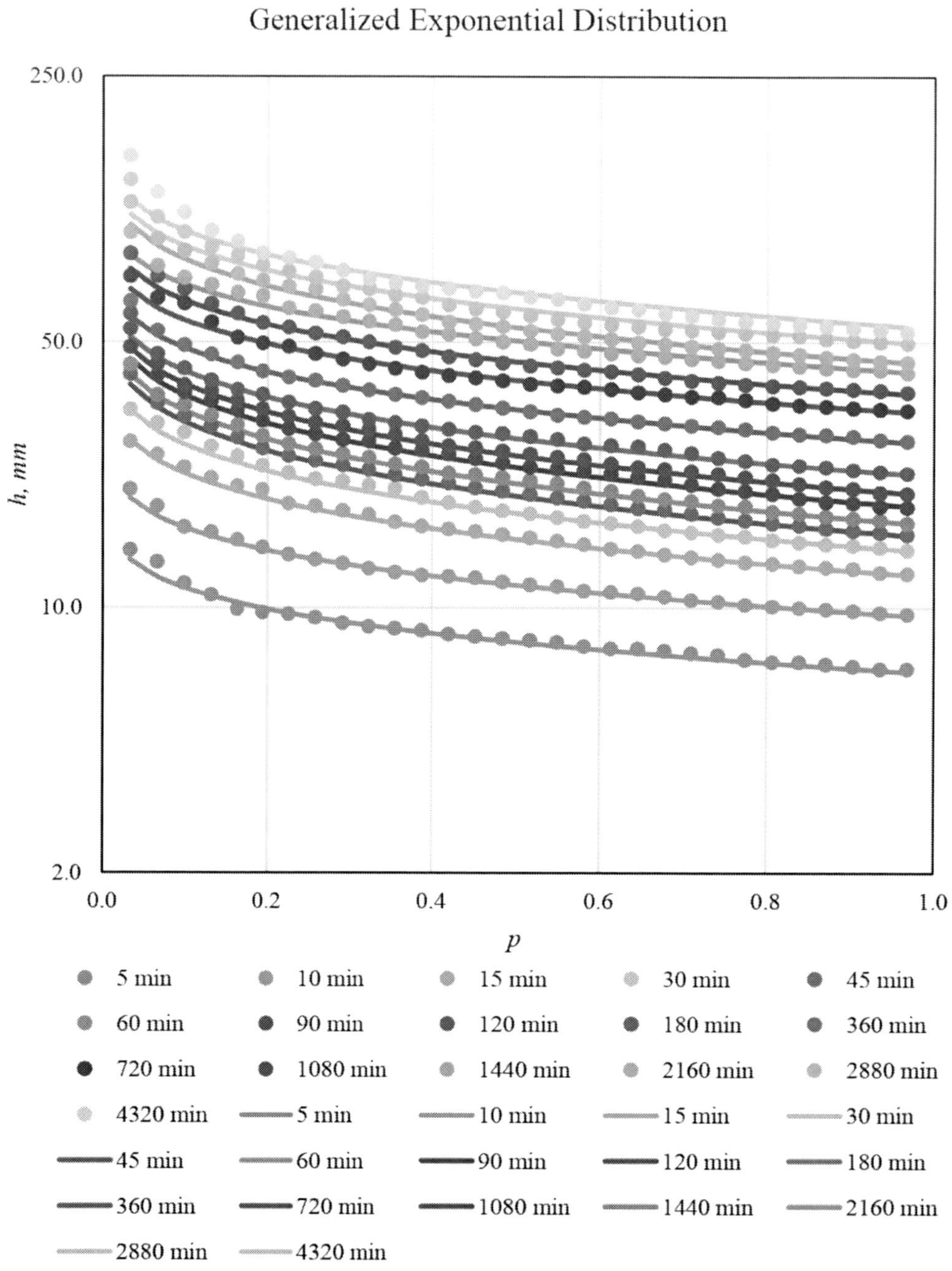

Fig. 2. Comparison of the GED theoretical maximum rainfall heights (lines) and their empirical regional-average counterparts (points) in the Upper and Middle Oder basin for all 16 rainfall durations. Vertical axis represents rainfall heights (h [mm]) while horizontal axis the probability of exceedance (p [-]).

It can be clearly seen in Fig. 2 that, for averaged measurements, GED was best fitted to rainfall durations t ranging from 5 to 1440 minutes. For $t \in [2160;4320]$, the differences ranged generally from 11% to 23%, and in the extreme case of the highest precipitation for the duration $t = 4320$ minutes almost 31%.

Table 1. BIC information criterion and rRMSE values for analyzed probability distributions and rainfall duration of averaged data set.

t	BIC				rRMSE			
[min]	GED	Gumbel	Log-Nor	Weibull	GED	Gumbel	Log-Nor	Weibull
5	182.90	190.60	185.60	**182.58**	**2.07**	4.62	2.19	2.24
10	215.64	226.50	218.84	**215.48**	**1.42**	5.13	1.77	1.66
15	**250.92**	266.50	255.92	251.20	**1.54**	6.17	3.08	1.60
30	**268.72**	288.24	273.52	269.06	**1.89**	7.15	4.02	1.97
45	**278.02**	303.52	286.72	278.90	1.68	7.52	4.90	**1.66**
60	**290.62**	310.86	296.96	291.18	**1.67**	7.20	3.68	**1.67**
90	**298.20**	319.44	304.42	298.78	1.80	7.32	3.59	**1.78**
120	**302.34**	325.28	308.52	302.90	2.07	7.43	3.72	**2.00**
180	**312.46**	329.94	316.90	312.76	2.19	6.69	2.54	**2.14**
360	325.64	337.20	327.96	**325.60**	1.66	5.78	**1.48**	2.04
720	**338.34**	352.12	340.72	338.48	2.27	6.29	**1.79**	2.32
1080	**351.10**	366.14	353.14	351.24	2.19	6.61	**1.68**	2.21
1440	**369.42**	379.48	370.20	369.50	2.60	6.04	**1.89**	2.65
2160	**372.14**	397.04	377.24	372.58	2.36	7.77	2.64	**2.04**
2880	394.48	408.50	**393.86**	394.44	3.80	7.73	**2.33**	3.26
4320	406.20	419.94	**393.86**	406.40	4.91	8.41	10.09	**4.12**

The final step of this research was to compare the GED quantile determined based on the regional-average data set with the quantiles corresponding to the empirical data for the eight analyzed stations. Figure 3 shows the correlation level between the GED theoretical regional-average rainfall and the empirical maximum rainfall for the eight meteorological stations.

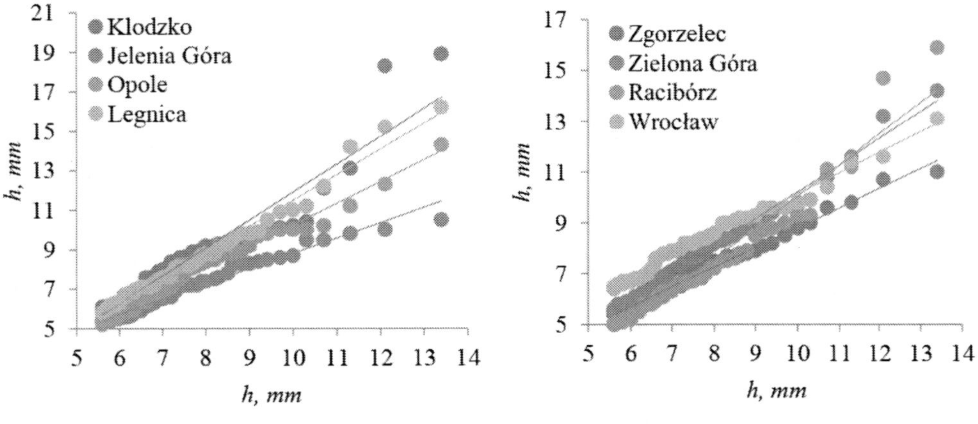

Fig. 3. Correlation between the GED regional-average rainfall height (h [mm]) and the empirical maximum rainfall height (h [mm]) (coloured dots and linear trend lines with R^2 described in text) for the eight meteorological stations of the IMWM-NRI for the rainfall duration $t = 5$ minutes.

Simple regression correlation factor indicated a high level of model conformity to measured data. The values of the R^2 coefficient, for the rainfall duration $t = 5$ minutes, were respectively (decreasing order): Zgorzelec 98.95%, Zielona Góra 98.91%, Legnica 98.42%,

Opole 98.35%, Wrocław 98.18%, Jelenia Góra 97.08%, Racibórz 94.70%, Kłodzko 90.64%. A strong correlation in the range of heavy rainfall for durations such as t = 5 minutes is very important for the drainage design purposes. Figure 4 shows that the differences between theoretical and measured data were primarily due to the least-frequent rainfall with the probability of exceedance up to p = 0.1. The largest differences in precipitation values were 5.5 mm and 6.2 mm for Kłodzko, 2.9 mm for Jelenia Góra and 2.4 mm for Zgorzelec. The remaining differences did not exceed 2.0 mm, and in the majority of cases 1.0 mm, which can be considered a very good result.

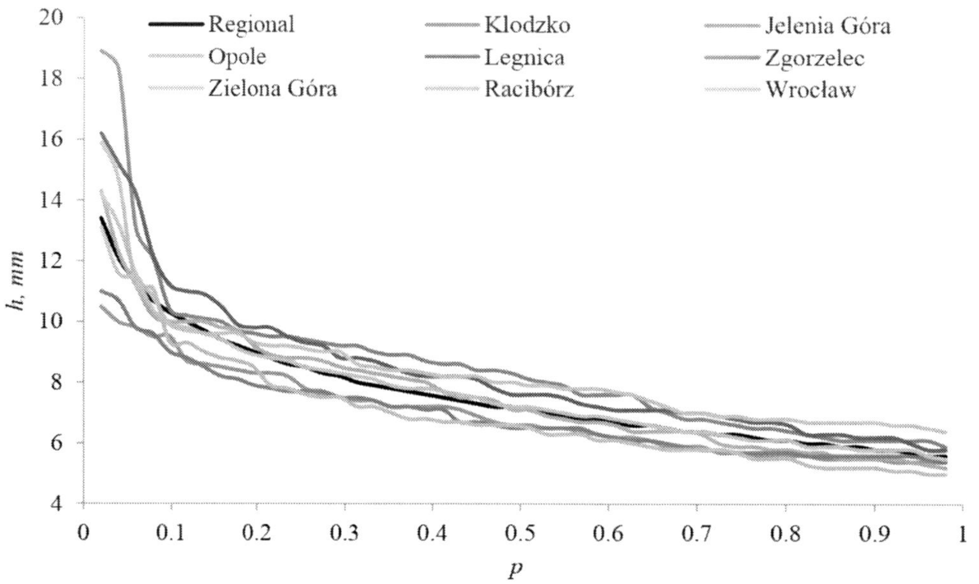

Fig. 4. The GED regional-average rainfall height (h [mm]) (black line) and the empirical maximum rainfall height (h [mm]) for the eight meteorological stations of the IMWM-NRI in the Upper and Middle Oder basin for the rainfall duration t = 5 minutes and their dependence on the probability of exceedance (p [-]).

The editing scope of the work does not allow to present, in a similar manner, all the results obtained for all analyzed rainfall durations. Therefore, Fig. 5 presents one of the negative examples for t = 1440 minutes. Significantly inferior-quality fit of the GED theoretical data to the measurement data for the duration of precipitation from a few to several hours was confirmed in the literature of the subject and has been confirmed in previous studies conducted for Wrocław [1, 2, 10], Legnica [11], several Lusatian Neisse basin stations [8] and 12 meteorological stations in the Upper and Middle Oder basin [9]. The example presented in Fig. 5 shows that the data from the Racibórz, Opole and Jelenia Góra stations are significantly different not only from the proposed regional model but also from other measurement stations, which, on the other hand, show satisfactory compatibility between themselves and the regional model. Distinct differences also apply to the least expected precipitation corresponding to the probability of exceedance p between 0.01 and 0.3. The differences here range from 22% to 39%, which gives a rainfall difference of up to 45.6 mm. The topographic location of these three stations is not without significance in the observed situations because it is a submontane area prone to the occurrence of intense rainfall that is hardly predictable. Also, in this area, floods harmful to the technical infrastructure, human property and the natural environment are very common [7].

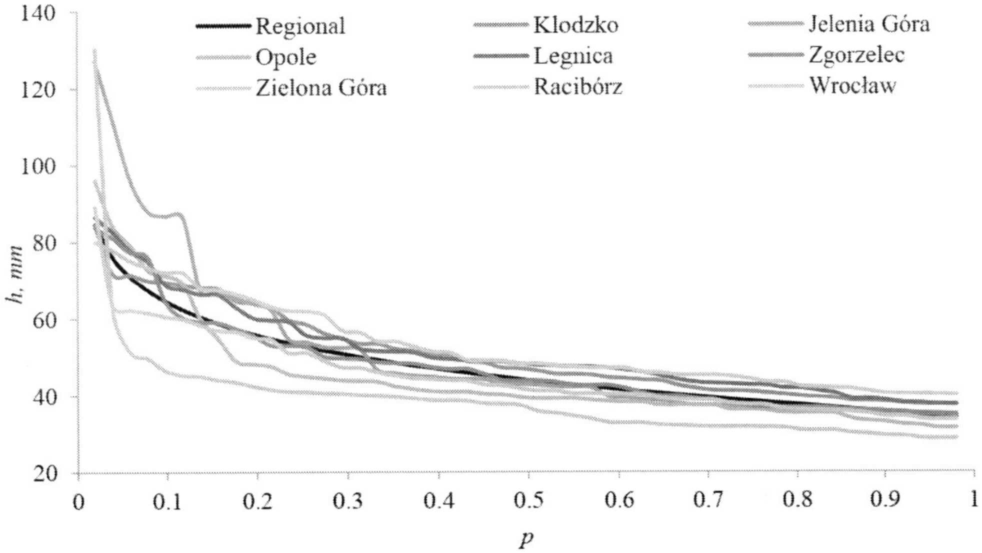

Fig. 5. The GED regional-average rainfall height (h [mm]) (black line) and the empirical maximum rainfall height (h [mm]) for the eight meteorological stations of the IMWM-NRI in the Upper and Middle Oder basin for the rainfall duration t = 1440 minutes and their dependence on the probability of exceedance (p [-]).

It is worth mentioning that the occurrence of extremely high rainfall values such as those from Racibórz (i.e. h = 130.3 mm for t = 1440 minutes or h = 227.4 mm for t = 4320 minutes) poses considerable difficulties when trying to describe the precipitation distribution. This is because most of the quality tests indicate such values as outliers. The statistical description of the precipitation data in mountain areas is difficult on the one hand, but necessary on the other in order to carry out the further research for various tasks, especially for the purpose of water supply and sewerage.

3 Conclusions

Eight meteorological stations and four probability distributions were used to describe the long-term rainfall data in the Oder River basin (1961–2010). Averaging of the maximum rainfall durations values allowed to create a regional model based on the GED distribution. An evaluation analysis led to the formulation of the following conclusions:

- general conclusion: we can provide well-fitting (to measured data) local probabilistic rainfall models, but it is much harder to develop a regional model for the terrain with varying elevation (in the analyzed Oder basin area from ~100 m to nearly 400 m a.s.l. – without higher parts of the Sudetes),
- the best-fitting probability distribution was GED,
- the differences between model and measured data were mainly higher in the extreme maximum values range (probability of exceedance p between 0.02 and 0.3).
- for Legnica, Kłodzko, Zielona Góra, Jelenia Góra and for most rainfall durations at Opole and Wrocław, regional model described measured data to a satisfactory degree,
- the largest differences between the data from the regional model and those from the eight stations were usually between 20–30%, and in extreme cases they did not exceed 40%,

- more and more detailed analyses (including the usage of other probability distributions and a spatial analysis) should be performed in the future for the Oder River basin for flash flood and urban flood protection purposes,
- if possible, similar analyses should also be done in the Czech parts of the Oder catchment and, especially, in its mountainous parts in both Czechia and Poland; maybe, in whole Poland (in order to find the regional schemes and also the probability distribution parameters similarities).

References

1. A. Kotowski, *The Principles of Safe Dimensioning of Sewer Systems* (Seidel-Przywecki, Warsaw, 2015)
2. A. Kotowski, B. Kaźmierczak, A. Dancewicz, *The Modeling of Precipitation for the Dimensioning of Sewer Systems* (Polish Academy of Sciences, Warsaw, 2010)
3. P. Licznar, K. Siekanowicz-Grochowina, M. Oktawiec, A. Kotowski, E. Burszta-Adamiak, Ochrona Środowiska **40**, 17 (2018)
4. P. Licznar, E. Burszta-Adamiak, A. Kotowski, K. Siekanowicz-Grochowina, M. Oktawiec, Ochrona Środowiska **40**, 21 (2018)
5. W. Szalińska, I. Otop, T. Tokarczyk, Meteorology Hydrology and Water Management **2**, 13 (2014)
6. L. Kuchar, W. Szalińska, S. Iwański, L. Jelonek, Meteorology Hydrology and Water Management **2**, 49 (2014)
7. G. Dumieński, M. Krzyżanowski, A. Tiukało E3S Web of Conferences **44**, 00037 (2018)
8. M. Wdowikowski, B. Kaźmierczak, O. Ledvinka, Meteorology Hydrology and Water Management **4**, 53 (2016)
9. M. Wdowikowski, A. Kotowski, P.B. Dąbek, B. Kaźmierczak, E3S Web of Conferences **17**, 00096 (2017)
10. B. Kaźmierczak, A. Kotowski, J Hydrol **525**, 345 (2015)
11. B. Kaźmierczak, M. Wdowikowski, Periodica Polytechnica Civil Engineering **60**, 305 (2016)
12. S. Coles, *An Introduction to Statistical Modeling of Extreme Values* (Springer Series in Statistics, Springer-Verlag, London, 2001)

Thermomodernization of buildings with the utilization of green walls

Małgorzata Wesołowska[1,*], and *Marta* Laska[1]

[1]Wroclaw University of Science and Technology, Faculty of Environmental Engineering, Wybrzeże Wyspiańskiego 27, 50-370 Wrocław, Poland

Abstract. The revitalization of historical buildings is a subject that is increasingly being discussed around the world. This is connected with the goal set by the European and thus Polish and local policy to reduce energy consumption in existing buildings. The publication presents an unconventional solution for thermal modernization of buildings with the utilization of green walls. The article presents the literature and legal acts review on green walls, their construction and impact on existing buildings. It describes the benefits and disadvantages resulting from the use of vegetation on the external walls. Additionally the possibility of using green walls at the historic facility located in Wroclaw (Poland) is presented.

1 Introduction

The construction sector is one of the largest energy consumers, not only in Europe, but also in the world [1]. Increasing emphasis is placed on reducing its consumption which leads to improvement of the environment quality, especially lowering the level of pollution in the atmosphere. For this purpose the European Parliament and the Council established the Directive on the energy performance and efficiency of buildings [2] (EPBD). This legal act relates to existing buildings, including historical buildings.

The article presents below the possibilities of utilization of vegetation on external walls as a non-conventional solution of existing buildings in urban area. The historic post-industrial facility, located in Wroclaw (Poland), being under the conservation protection is described as the example of such a solution. The issue associated with the use of vegetation on the walls, which may give a number of benefits, in particular for the urban tissue, are broadly described.

1.1 Thermomodernization

The majority of buildings in Europe are residential buildings, accounting for around 75% of the total floor space area of building stock [3]. Almost 40% of them are over 50 years old [4]. The literature [3] describes also the statistics of non-residential buildings where three quarters of the service floor area is covered by offices (30%), wholesale (27%) and education (16%). The Wroclaw's communal records of monuments itself has over 10 000 positions [5]. Many of these facilities require renovations, due to high energy consumption, a state

* Corresponding author: malgorzata.wesolowska@pwr.edu.pl

threatening the health or life of people staying in them and surrounded area, and to preserve cultural heritage.

Modernization of buildings is often connected with change of their functions and thus may lead to change in the requirements for indoor conditions. This results in the necessity of complex thermomodernization that may include the insulating the walls and roofs, improving the windows or glazing, modernization or changing the HVAC installation or the heat and cool source for high-efficiency systems, if it is technically, functionally and economically feasible [2], and also utilization of "solutions based on nature" such as green walls [2]. Possibilities of revitalization of such objects can be found in the literature [6, 7].

There are [8] three important aspects of facilities modernization, namely economic, environmental and social. The first one arises from the fact that historic buildings are constructed in old technologies and follow, much less restrictive regulations in terms of energy consumption, material used and thermal resistance of building envelope and air tightness. This results in high energy consumption and operating costs. The environmental aspect is closely related to the economic one, as higher energy performance of building leads to lower amount of energy needed for building operation and thus reduction of the pollution. The last - the social dimension of building modernization, is strictly connected with the improvement of building indoor conditions that influences the level of satisfaction from the thermal environment of building users their health and productivity [9]. Additionally green walls and roofs are recommended to strengthen the effect of well-being and relaxation of occupants [10, 11].

The fundamental activity leading to the thermal improvements is based on lowering heat losses through the building envelope and protection from overheating. When the buildings being under cultural heritage are taken under consideration and the historic preservation officer take care of the building, the traditional insulation techniques cannot be applied. Therefore the utilization of green walls and roofs seem to be the most desirable solution in renovation process. Solutions based on nature have also additional positive effect on the surrounded urban area in terms of lowering the local air pollution and noise level, increasing shading effect and counteracting the formation of urban heat islands.

2 Historic buildings under national and European law

In every European city with a rich past one can find historic buildings. It is usually assumed that a building with more than 50 years of existence, while maintaining integrity with the time, style and historical value, can be considered as a monument [1, 4].

The revitalization of the protected buildings in Poland and their surroundings have to follow the guidelines of the voivodship Conservator of Monuments, which are based on the polish law acts [12]. All decisions regarding interference in the facility should be made to preserve its historical elements. For this reason, thermomodernization measures may be considerably limited. Historic buildings are divided into four classes [13] as follows: class I includes buildings in which architectural details have been preserved, so their parts require a special protection; class II includes objects that do not require special care, where details have not been preserved; buildings with historical significance and architectural values as well as buildings located in architecturally protected areas belonging to class III; non-historic buildings constitute to class IV.

One of the main issues addressed in [2] is the financing of renovation projects. The cost of buildings renovations in 2006 amounted to 302 million EUR and was only 20 million EUR lower than the founds designated to new residential buildings [8]. Therefore, following [2] the Member States should put a special emphasis to support all activities, which can lead to lower energy consumption of existing buildings and support the research on new solutions to improve energy performance of historic buildings.

3 Green walls

Large cities in Europe improve in accordance with sustainable development. This idea applies to the use of vegetation on the facades of buildings, which nowadays is experiencing a renaissance. It is also conducive to leveling urban heat islands. The largest Polish cities, also strive to maintain as many green areas as possible. The efforts are visible in the acts of local law [14]. Krakow allocated the founds to support the development of small forms of greenery. The City Council of Wroclaw decided to exempt its residents from the real estate tax, when a biological cover with appropriate area was used on the partitions [15].

Restrictions related to conservation protection in many cases result in the lack of consent for the use of external thermal insulation composite systems. These enforce the need to use internal wall insulation (IWI) and thus to reduce the usable space, or other solutions like these indicated in the point 17 of the directive [2]. The publication draws attention to the use of natural solutions, namely green roofs and walls, as they have a positive impact on both, the building itself and the increase of thermal insulation resistance of external walls and thus the lowering the energy consumption of the building.

3.1 Construction of green walls

The green walls can be divided into two main categories due to their construction, namely green façades and living walls, Fig. 1.

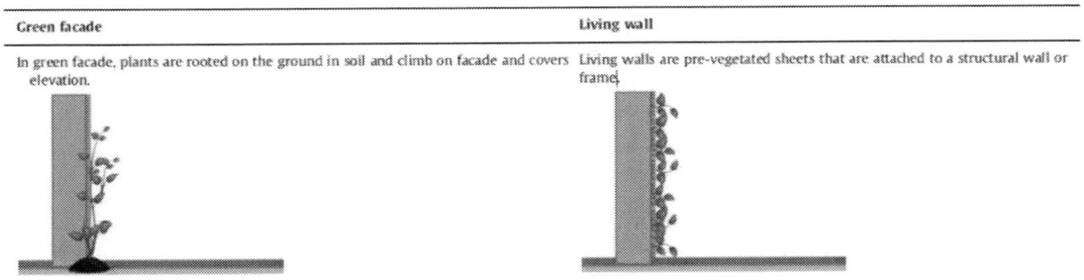

Fig. 1. Dichotomy of vertical greenery systems [16].

The main difference between these two solutions is the location of the plant substrate. The first assumes that climbing plants are rooted at the foot of the wall. The green façade vegetation climbs on the supporting structures directly or indirectly on the façade [17]. In the second solution the soil substrate in living walls is located in dedicated boxes on the entire surface of the façade. Therefore it requires supporting elements and the irrigation system, and thus is more expensive in comparison to green facades. Additionally the weight of living walls is within the range of 60–500 kg/m^2 [18].

3.2 Vegetation

The choice of plants for green walls depends on many factors. The main differences result from the construction of the wall and the local climate conditions. When green facades are under consideration it is necessary to plant vines, while for living walls heather vegetation should be planted. The list of plants suitable for Polish conditions is presented in Table 1.

Table 1. Plants used in green facades in a temperate climate and Polish conditions [19].

No.	Polish name	English name	Latin name	Requirements
1.	winobluszcz trójklapowy	creeper tricuspidata	*Parthenocissus tricuspidata*	No requirement of supports or frames. The plants have the ability to "stick" to the wall.
2.	winobluszcz pięciolistkowy	creeper quinquefolia	*Parthenocissus quinquefolia*	
3.	bluszcz pospolity	ivy	*Hedera helix*	
4.	hortensja pnąca	climbing hydrangea	*Hydrangea petiolaris*	
5.	milin	dropmore sp.	*Campsis* sp.	
6.	glicyna	glycine	*Wisteria sp.*	The plants circle around vertical elements, lines stretched on supports and appropriate frames.
7.	wiciokrzew	honeysuckle	*Lonicera sp.*	
8.	dławisz	celastros	*Celastrus sp.*	
9.	chmiel	hop	*Humulus sp.*	
10.	winobluszcz zaroślowy	thicket creeper	*Parthenocissus inserta*	The plants climb on rigid supports with steel or wooden rods. They attaches themselves to the wall with special mustaches and petioles.
11.	aktinidia	aktinidia	*Actintdia sp.*	
12.	kokornak	aristoloehia	*Aristolochia sp.*	
13.	rdest	fallopius	*Fallopa sp.*	
14.	męczennica	passifora	*Passifora sp.*	
15.	winnik	ampelopsis	*Ampelopsis sp.*	
16.	powojnik	clematis	*Clematis sp.*	

In addition, insolation is very important factor. A north wall is suitable for flowers that do not have excessive lighting requirements due to the prevailing partial shade and shade. The southern façade is exposed to direct solar radiation and thus high temperature. The façade facing east is suitable for most plants, mainly those that bloom and prefer the gentle sun exposure. The western one is ideally suited for blooming flowers due to the presence of sharp sun and high temperatures in the afternoon [20].

3.3 Advantages of green walls

The use of vegetation on the walls brings a number of benefits. This is especially important for heavily urbanized areas and with a dense network of urban tissue. In the case of historic buildings, it is often the only solution approved by the monument conservator.

3.3.1 Environmental benefits

The most important features of plants is the absorption of pollutants, called natural air filtration, and production of the oxygen. The photosynthesis leads to lowering the impact of the greenhouse effect in the urban area [16], the temperature, especially in the city centers and minimize the risk of heat island phenomenon occurrence. The Fig. 2. presents the temperature distribution in the area of impact walls with direct, indirect façade greenery and living wall system. Vegetation has a larger albedo (the light reflection coefficient) than most building materials, and thus can absorb more solar radiation heat instead of reflecting it [21].

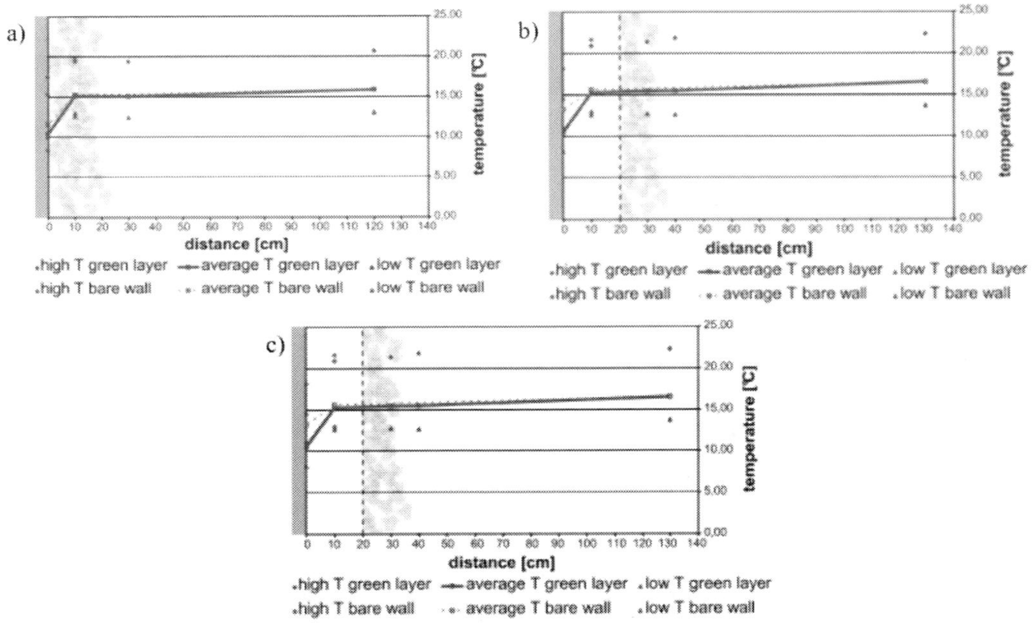

Fig. 2. The temperature distribution for: different types of green walls: a) direct façade greenery, b) indirect façade greenery, c) living wall system [22].

3.3.2 Energy savings

As aforementioned, the green wall acts as a thermal insulation for a building. The applied vegetation is a solar filter, absorption of sunlight during the reaction of photosynthesis and evapotranspiration is a process of heat reduction [18, 21]. In the literature, one can find the studies showing the temperature differences using a green façade with ivy and without, reaching even 30K [23]. This results in reducing the energy demand for cooling in the summer.

Table 2. The influence of a biological cover on the heat transfer coefficient for an external wall [23].

Coefficient of heat transfer through the external wall, U [W/m²·K]		Indicator for improving wall insulation [%]
Without plant cover	With climbing plants	
0.32	0.31	3.12
0.54	0.51	5.56
1.24	1.16	6.45
2.06	1.74	15.53
2.53	2.06	18.58

The use of evergreen plant species also gives benefits in the winter, which is presented in Table 2. The table lists the heat transfer coefficients for partitions, which does not have a vegetation cover and where the 35 cm of ivy is applied. A significant improvement in insulation performance is particularly visible for partitions having an initial high heat transmittance coefficient (U). The vegetation, used to build green walls, also has a positive effect on the moisture level of the wall and thus its physical condition and indoor humidity.

It also prevents against scattering rain and protects the wall against dampness that has direct impact on the heat transfer coefficient, and hence the energy needed to heat the building [23].

3.3.3 Social benefits

In the past, green walls were used to improve the attractiveness of the buildings. It has been known for centuries that closeness to nature has a positive effect on the humans psyche and improves their well-being. Greenery also improve the microclimate, air quality and create the sound barrier inside buildings and in their surrounded area [16, 23].

3.4 Disadvantages of green walls

Apart number of advantages the green solutions are not free from imperfections. Uncontrolled development of the biologically active cover may cause overgrowing of gutters and window openings. Thus such a façade need special and regular care. One should also take into consideration the patchy foliage of plants, which may affect thermal losses and the increased risk of moisture penetration. In addition, some creepers have a slow increase, which makes the effectiveness of green facades grow with their age. Also the designing of green walls is particularly important and need to consider proper plant choice and its correct arrangement [19].

4 The vegetation on the existing buildings

Wroclaw is a city with a rich history. In its area one can find the number of buildings being under the conservation protection and thus has a great potential in utilization of the solutions based on greenery. The following part of the paper describes the possibility of using green walls on the post-industrial building from the beginning of the 20th century being under protection of the conserver officer [24].

4.1 Thermomodernization of external walls

The exemplary building, due to the historic importance cannot be insulated with the traditional methods. The use of IWI system is questionable as there would be necessity to interfere in the valuable internal surfaces and would be connected with the increased risk of water condensation.

Fig. 3. Facade of the considered building (photo: M. Wesołowska) [21].

Therefore the implementation of the green facade is recognized as an appropriate solution and without negative influence on the valuable history wall. The solution is safe for construction and in accordance with the local spatial development plan imposing the condition of maintaining a biologically active area.

The inventory of the building has shown that two of the walls cannot be used as the green walls, due to the presence of a gates and waste storage. The designated area that could be covered with vegetation is over 400 m^2 on the southern wall and about 230 m^2 on the east. This gives the total potential of 650 m^2 of green insulation.

4.2 Thermal analysis

As a result of the calculations, the insulation of vertical walls in the building was determined. The value of the heat transfer coefficient varies from 1.2 $W/m^2{\cdot}K$ to 2.0 $W/m^2{\cdot}K$. The largest surface of the external walls is a partition with U-value of 1.2 $W/m^2{\cdot}K$. Following the literature (see Table 2), due to lack of external insulation, application of green climbing plants can bring the improvement in this heat transfer coefficient even more than 6% and lower its value to about 1.16 $W/m^2{\cdot}K$. The improvement of the thermal insulation maybe is not spectacular and will bring the heating cost reduction about 1000 PLN per annum, however the investment costs of the solution are negligible and additional advantages during the summer season and the environmental impact are unquestioned.

The vegetation will cause a reduction in the surface temperature of the wall. Depending on the adopted green façade solution, it is possible to lower it to even 1.7°C [18]. This, in turn, can have a positive effect on indoor comfort temperature, lowering the cool demand and, if AC system is applied, lowering energy costs.

5 Summary

Thermomodernization of buildings with the utilization of green facades is particularly important issue for buildings under monument protection. Applying both direct and indirect systems, there is no need to interfere in the wall with traditional thermal insulation methods. These activities will not cause the loss of the historical value of the building. It is safe option for the historic façades.

The use of vegetation on the walls gives a number of benefits. The biggest one is the reduction of heat and cool demand and thus the energy consumption of the building by increasing the insulation of the external walls. This solution improves moisture condition of the wall and thus has a positive effect on the building construction itself and quality of indoor air. In addition, the reduction of carbon dioxide and improvement of external air quality is particularly important for urban areas. Vegetation helps to eliminate the effects of the urban heat island, which is an increasing problem in dense urban tissue.

The work was partially financed by The Faculty of Environmental Energy, Wroclaw University of Science and Technology, Poland. Ref no. 0401/0055/18. The authors woud like to thank to Tratwa Association.

References

1. L. Mazzarella, Energy Build. **95**, 23–31 (2015)
2. Dyrective of the European Parliament ond of the Council, **844** (2018)
3. https://ec.europa.eu/energy/en/eu-buildings-factsheets. (Accessed: 29th January 2019)
4. A. L. Webb, Renew. Sustain. Energy Rev. **77**, 748–759 (2017)
5. *BIP Urzędu Miejskiego Wrocławia, Gminna Ewidencja Zabytków Miasta Wrocławia. Available* at: http://bip.um.wroc.pl/artykuly/204/zabytki-wroclawia. (Accessed: 20th February 2019)

6. M. Baborska-Narożny, Czas. Tech. Archit. **12**, 275–279 (2012)
7. S. Kaczmarek, Social aspects of reconstruction of old industrial regions in Europe, Prace naukowe Uniwersytetu Śląskiego w Katowicach, **1722**, 120–126 (1998)
8. D. Anastaselos, E. Giama, A. M. Papadopoulos, Energy Build. **41**, 1165–1171 (2009)
9. M. Laska, E3S Web of Conf., **17**, 00050 (2017)
10. M. Mansor, I. Said, I. Mohamad, Asian J. Environ.-Beh. Stud. **2** (2017)
11. M. Demuzere, J. Environ. Manage. **146**, 107–115 (2014)
12. Ustawa z dnia 23 lipca 2003 r. o ochronie zabytków i opiece nad zabytkami, Dz.U. 2003 Nr **162** poz. 1568
13. M. Drozdowicz, M. Laska, Rynek Instal. **10**, 1–6 (2016)
14. *Plan adaptacji miasta Krakowa do zmian klimatu do roku 2030*
15. *UCHWAŁA RADY MIEJSKIEJ WROCŁAWIA w sprawie zwolnień od podatku od nieruchomości powierzchni użytkowych lokali mieszkalnych w ramach projektu intensyfikacji powstawania terenów zieleni w obrębie Miasta Wrocławia*, **XV/268/15**
16. T. Safikhani, A. M. Abdullah, D. R. Ossen, M. Baharvand, Renew. Sustain. Energy Rev. **40**, 450–462 (2014)
17. A. B. Besir, E. Cuce, Renew. Sustain. Energy Rev. **82**, 915–939 (2018)
18. E. Cuce, Appl. Energy, **194**, 247–254 (2017)
19. E. Trzaskowska, Teka Kom. Arch. Urb. Stud. Kraj. – OL PAN VI, 110–121 (2010)
20. A. Kania, Zasady projektowania i wykonywania zielonych dachów i żyjących ścian. Poradnik dla gmin. SWISS CONTRIBUTION, **91** (2013)
21. G. Pérez, J. Coma, I. Martorell, L. F. Cabeza, Renew. Sustain. Energy Rev. **39**, 139–165 (2014)
22. K. Perini, M. Ottelé, A. L. A. Fraaij, E. M. Haas, R. Raiteri, Build. Environ. **46**, 2287–2294 (2011)
23. D. Fabianowski, Zieleń miejska, **8** (2011)
24. M. Wesołowska, *Building energy audit and analysis of possibility of using green facades improving condition and humidity of the object*, praca magisterska (Wroclaw University of Science and Technology, 2018)

Role of flow-through lakes in reducing pollutants outgoing from rural areas

Marlena Wiktorska[1,*], and *Sławomir* Szymczyk[1]

[1]University of Warmia and Mazury, Department of Water Resources, Climatology and Environmental Management, Plac Łódzki 2, 10-719 Olsztyn, Poland

Abstract: This study assessed the water quality in Lake Wydmińskie (Warmińsko-Mazurskie Voivodeship) based on the content of both organic and inorganic pollutants flowing into the lake with water from drainage systems. The study demonstrated that the water flowing into the lake through drainage ditches that drain the improved part of the catchment area of the lake, was a potential source of diffuse pollutants and excess chemical components were deposited in the lake which could lead to its degradation. The most polluted water was found in the main tributary and the lake bay fed by its water. For this reason, it was concluded that the greatest problem was the area-wide sources of pollutants outgoing from agricultural areas. It was also found that water flowing out of the lake was of a considerably better quality than water flowing into the lake through the ditches. This clearly indicates the accumulative role of the lake on the route of pollutant migration from agricultural sources.

1 Introduction

Environmental conditions of each lacustrine system are determined by the type of the lake and its exposure to various environmental factors. Therefore, the quality of surface waters is not only determined by natural processes (meteorological conditions, erosion and weathering, etc.) but also by anthropogenic (urban, industrial, and agricultural) impact [1]. The lacustrine ecosystem is exceptionally sensitive to external factors, and it is very easy to disturb its state of equilibrium. Population growth, agricultural practices and wastewater flowing from urban areas have led to an increased nutrient outflow to water bodies which, in many cases, resulted in accelerated eutrophication [2]. Lakes and water bodies, particularly urban ones, without adequate protection become pollutant absorbers, which can lead to their degradation [3]. An excess of nutrients and organic carbon in the soil and water due to industrial, agricultural and other human activities results in widespread environmental problems [4]. Organic matter *(OM)* serves numerous geochemical and ecological functions in inland waters and may affect water quality. Various parameters of aqueous *OM* are measured using various methods, as no single analytical tool can provide final structural or functional information about it [5]. Waters outgoing from agricultural

*Corresponding author: marlena.wiktorska@uwm.edu.pl

https://doi.org/10.1051/e3sconf/201910000086

areas supply the lake with huge amounts of inorganic anions, such as phosphates, nitrates and ammonia, organic nitrogen compounds and heavy metals [6]. The chemical composition of water in small bodies and lakes provides important information on the sources of dissolved ions and hydrogeochemical processes and on anthropogenic activities affecting the lacustrine environment [7]. Many natural factors (physical, biological, geographical, hydrological and meteorological) impact the chemistry of water bodies [8]. However, the greatest hazard is posed by the intensification of agriculture [9].

2 Materials and methods

The analysis of the quality of water in Lake Wydmińskie as well as in its tributaries and the outflow, was carried out from November 2015 to October 2018. The measurements and sampling were performed at monthly frequency. The lake is located in north-eastern Poland (N: 54°00' E: 22°00') in the Warmińsko-Mazurskie Voivodeship, in the commune of Wydminy. The maximum depth of the lake is 9.8 m, while the average depth is 2.98 m. The area of the lake is relatively large, as the lake surface occupies an area of 336.6 ha (actual value), and the geodetic area is 377.46 ha (determined by average value of the isobath 0 m). The shape of the lake is very elongated (Fig. 1).

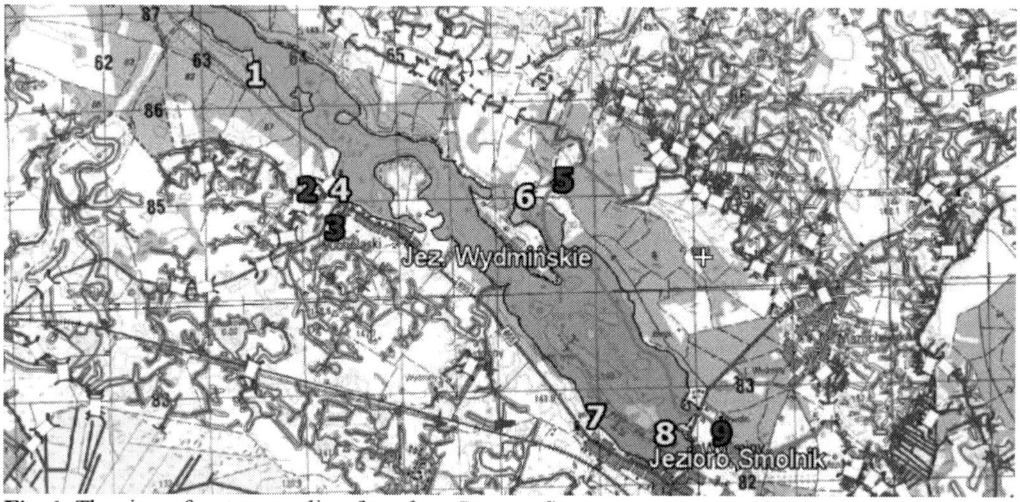

Fig. 1. The sites of water sampling (based on Geoportal).

The total catchment area of Lake Wydmińskie is approx. 34 km^2. Within the total catchment area, agricultural land is predominant, particularly in the catchment basin of the main tributary in the vicinity of the village of Mazuchówka. The immediate catchment area of the lake occupies a relatively small area and is mainly overgrown by forests. The catchment area is drained through a system of watercourses and drainage ditches. The monitoring point in the part of the catchment area located in the forest (1) is a water intake point located in an area overgrown with forest vegetation. Tributary I (2) which flows into the lake on the edge of Sucholaski is a drainage ditch that discharges water from a small part (with an area of approx. 2.4 km^2) in the western part of its drainage catchment. This is a typically agricultural area. Over the section near the outlet, the ditch channel is shallow (0.5–0.7 m of depth) and the bed is covered with stones and gravel. Tributary II (3) is located directly in Sucholaski and drains a small part of the catchment area of approx. 1.5 km^2. It is a typically agricultural catchment area with a large proportion of meadows and pastures as well as forest plantings on post-agricultural areas. Water from agricultural land is discharged through the drainage system that flows into a small pond from which water flows into the lake through a ditch. Over the section near the outlet, the drain channel is shallow (0.3–0.5 m) and stony. The monitoring point in the locality of Sucholaski (4) was located near a publicly accessible village beach. This is an area potentially susceptible to high anthropogenic pressure due to its proximity to rural buildings and its recreational nature. The largest tributary of the lake is the main drainage ditch (5) which drains the eastern part of the catchment area of approx. 18.5 km^2. This is an extensive area with a dense system of feeder ditches. The catchment area is typically agricultural, with a large proportion of meadows and pastures. Near the outlet to the lake, the main ditch bed has a width of approx. 3.5 m and a depth of over 0.5 m. The ditch banks are overgrown with common reed (*Phragmites australis)* and broadleaf cattail (*Typha latifolia*) and the water surface during the entire growing period was completely covered with the common duckweed (*Lemna minor*). The monitoring point in the bay (6) was located behind the outlet of the main tributary. The area of the bay is heavily overgrown with common reed rushes. The monitoring point on the bridge (7) linking the shore of the lake with its largest island is situated near the housing development in Wydminy. Water samples at this point were collected from the pelagic zone. The monitoring point on the island (8) was located at the village beach, in the immediate vicinity of the housing development in Wydminy. Water samples at this point were collected from the littoral zone. The outflow from the lake (9) is located in Wydminy. It is a regulated watercourse with a length of approx. 2 km. Its channel at the place of measurement has a width of approx. 2.5–3 m, the depth is over 0.5 m, and the bed is covered with stones and gravel. Directly in the field, at the designated monitoring points in water, the following were determined: total dissolved substances (*TDS*), dissolved oxygen (*DO*) content, water temperature and the pH values (using a YSI 6600 multiparameter sonde). Water samples were collected to polyethylene 3 dm^3 bottles; after the stabilisation, the samples were delivered to the laboratory where the following were determined: 5-day Biological Oxygen Demand (BOD_5) using the OxiToP®Control method, the Chemical Oxygen Demand (COD_{Cr}) using the dichromate method, dry residue content after evaporation and drying at a temperature of 105°C, and the residue on ignition (ash components) after combustion in a muffle furnace at 550°C [10]. The obtained results were subjected to a statistical analysis using Statistica software ver. 13 PL; the normality of data distribution was determined using the Shapiro-Wilk test at $p \leq 0.05$; diagrams of average values along with non-outlying maximum and minimum values were developed (for the analysis of pH values, the median was assumed to be the basis); homogeneous groups were determined based on the (post-hoc) Duncan's test. Homogeneous groups are groups with average values that do not differ statistically. This test determined these groups and defined between which monitoring points there are statistically significant differences in the level of analyzed pollutants.

2 Results

The variability of the values of analysed indices of pollutants of water in Lake Wydmińskie and its tributaries and the outflow was determined by the type of water (drainage ditches and the lake) and primarily by the impact of the method of managing the lake catchment area (Fig. 2–5). Total Dissolved Solids (*TDS*) is a water quality parameter that determines the concentration of dissolved organic and inorganic chemical substances in water. High values of this index can indicate poor living conditions of organisms in the analysed aquatic environment as well as a poor water quality for crop irrigation and unsuitability of the water for consumption [11]. In the lake (except the bay) and in the outflow, they were at a level ranging from approx. 0.25 $g \cdot dm^{-3}$ to approx. 0.33 $g \cdot dm^{-3}$. In the tributaries and in the bay, they were significantly higher (from approx. 0.38 $g \cdot dm^{-3}$ to 0.54 $g \cdot dm^{-3}$) (Fig. 2). None of these values exceeds the range of limit values specified by the Food and Agriculture

Organizations of the United Nations (FAO) [12] (2 g·dm^{-3}); in addition, it was observed, only for tributary II, that the standards specified by the World Health Organization (WHO) [13] (0.5 g·dm^{-3}) were slightly exceeded. This is a watercourse with a rocky bed and an intense water flow; for this reason, the chemical pollutants contained in the water do not accumulate in sediments but are transported to the lake. However, a high imbalance between the *TDS* value in the water of the lake and its outflow and in the water of tributaries and the bay should be noted. Water in the lake and in the outflow is characterised by significantly lower values of this index. No significant differences in the *TDS* values were observed between particular measuring points located in most parts of the lake (except the bay), which indicates no variation in the intensity of the pressure of particular types of catchment area on the lake. Water collected from the points located in the vicinity of Sucholaski and Wydminy was characterised by relatively low *TDS* values, which may indicate slight human pressure exerted by these locations, even though an increased impact of human activities could be expected from them. The water least loaded with dissolved substances was found in the part of the lake with the lowest human pressure, with a forest area located in the immediate catchment area of the lake. The water outgoing from the lake also contained relatively low concentrations of dissolved substances, even though that part of the lake is located under a busy regional road which is a potential source of linear pollutants. The seasonal variability of water pH values is determined by the weather conditions which, in turn, determines the availability of the main ions in water. The pH affects both the solubility of many substances and the activity of most biological systems. The low pH value facilitates the transition of many substances, e.g. heavy metals, sulphur compounds, and nutrients into active forms [14, 15]. The pH values in the analysed water were similar at all monitoring points (Fig. 3); however, it is necessary to note their very high variability whose extreme values exceeded the limits specified by the WHO (6.5–8.5) and the FAO (6.0–8.5). This indicates a periodic excess of chemical pollutants, both acidic and alkaline. Particularly unfavourable is a decrease in the pH value below 6.5, which in combination with oxygen deficits results in the predominance of anaerobic reduction processes in the aquatic ecosystem, leading to the formation of substances adversely affecting aquatic organisms.

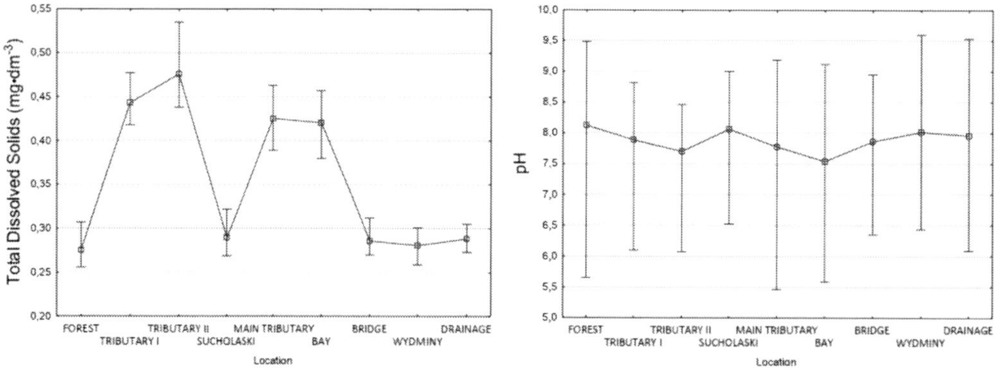

Fig. 2. The effect of the catchment area management method on the *TDS* values and the pH in the water of tributaries, outflow and in particular parts of the lake.

Dry residue is an indicator of the contents of solids and substance dissolved and suspended in water, while the residue on ignition determines the mineral substance content. The difference between these indices allows the organic compound content of water to be approximately estimated. The trend in the dry residue and residue on ignition contents was similar to that for the *TDS* index, with significantly higher values noted in the lake tributaries and in the bay (Fig. 3). The water in the monitored (analysed) parts of the lake and in the outflow were characterised by a lower content and did not differ from each other significantly (dry residue ranged from approx. 250 mg·dm^{-3} to approx. 350 g·dm^{-3}, and the mineral residue from approx. 150 mg·dm^{-3} to approx. 250 mg·dm^{-3}). Based on the above indices, it can be assumed that irrespective of the type (flowing waters and the lake) of the analysed waters, organic compounds accounted for an average of approx. 30% of all substances. Significantly higher values of the analysed indices in water of all drainage ditches draining the improved agricultural catchment area and of the bay in relation to other parts of the lake indicate the very great contribution of agriculture to the loading of lake water with mineral and organic substances (dry residue ranged from approx. 375 mg·dm^{-3} to approx. 600 mg·dm^{-3}, and the mineral residue from approx. 260 mg·dm^{-3} to approx. 450 mg·dm^{-3}). Skonieczek et al. [16] also demonstrated in their study a clear effect of the method of managing the catchment area on the contents of dry residue and of mineral substance residue on ignition. The highest content was found in the water of tributaries from the drainage system that represent the agricultural catchment area (dry residue of 635–656 mg·dm^{-3}; mineral substances: 512–572 mg·dm^{-3}), while at other points, the values ranged from 233 to 470 mg·dm^{-3} and from 161 to 386 mg·dm^{-3}, respectively. The results obtained by these authors are in line with the trends revealed in the results of own research. On the other hand, Sobczyńska-Wójcik [17] found that the lacustrine ecosystem of Nowe Włóki exhibited a stable nature in terms of dry residue and mineral residue contents (the average dry residue content ranged from 210 to 235 mg·dm^{-3}, and the ash component content ranged from 130 to 157 mg·dm^{-3}). Similar conclusions can be drawn based on the results of own research.

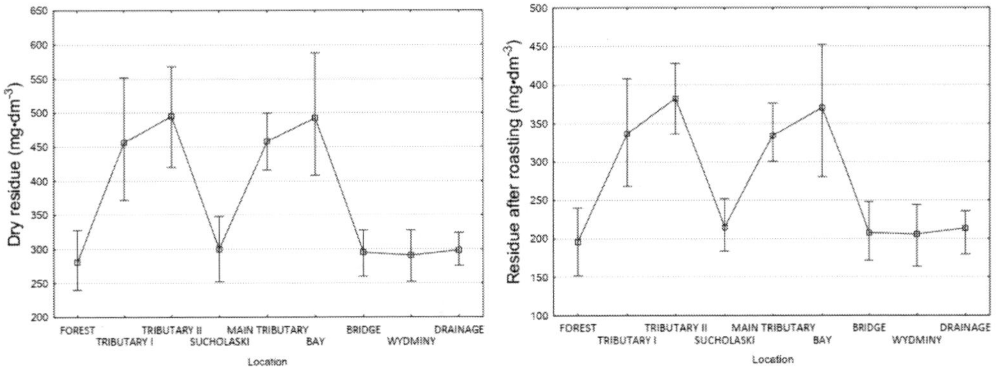

Fig. 3. The dry residue and residue on ignition contents in the water of tributaries, outflow and in particular parts of the lake.

Chemical Oxygen Demand (COD) and Biological Oxygen Demand (BOD) are traditional, commonly used indicators for the assessment of the organic matter content of water. They are used to determine the organic enrichment by oxygen-absorbing substances [5]. The difference between these indicators is that the COD comprises both the biodegradable and non-degradable fractions, while the BOD is limited to the determination of the amounts of readily biodegradable substances [18]. Having analysed the distribution of COD values during the study period in the water of Lake Wydmińskie, significantly higher values of this indicator were found in the lake tributaries and in the bay fed by the main tributary. It follows from the above that the tributaries may pose a potential hazard to the analysed water body through the deposition of pollutants. The COD values in the water of tributary I range from approx. 45 mg·dm^{-3} to 80 mg·dm^{-3}, tributary II (from approx. 31 mg·dm^{-3} to 65 mg·dm^{-3}), the main tributary from approx. 50 mg·dm^{-3} to approx. 98 mg·dm^{-3}, and in the water of the bay from approx. 55 mg·dm^{-3} to 89 mg·dm^{-3} (Fig. 4). In other monitoring points located in various parts of the lake and in the outflow, the index values ranged from approx. 30 mg·dm^{-3} to 40 mg·dm^{-3}. Diffuse pollutants are transferred from agricultural land to drainage ditches and large water reservoirs, which results in significant deterioration in the quality of water in rivers and lakes [19]. This type of pollutants causes serious problems due to their large scale and diffuse nature. Moreover, their migration in the environment is more difficult to control than that of point pollutants [20]. A similar trend was also noted for the BOD; however, the values of this indicator in water of two smaller tributaries, as opposed to the COD, did not differ significantly from these values in water in other points of the lake; moreover, tributary II was characterised by the lowest BOD value of all monitoring points. This indicates that the water of this watercourse contained a much lower readily biodegradable fraction represented by the BOD indicator than at other monitoring points.

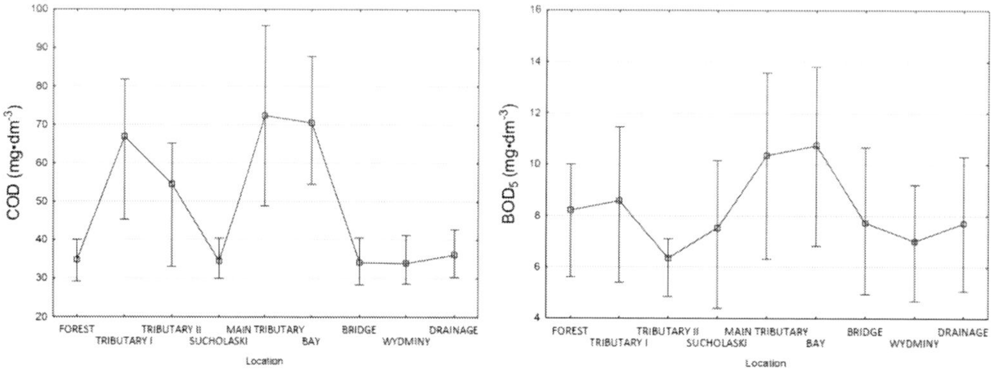

Fig. 4. COD and BOD_5 values in water of the tributaries, outflow and in particular parts of the lake.

Water temperature is the most important indicator of the ecological environment of the lake and is used to determine changes between various factors of water quality (it affects *inter alia* the dissolved oxygen content of water, the pH and the development of phytoplankton), which provide a basis for inspections and the improvement of ecological environments of lakes [21]. No statistically significant differences were noted in the average water temperature between the analysed monitoring points, although it was observed that in the water of tributaries and the bay, the temperature was lower than at other points. Tributary III was characterised by the lowest temperature; however, it should be noted that in the summer periods, it was dried up, which prevented measurements and disturbed the comparative assessment. The low levels of oxygen dissolved in water (0–8 mg·dm^{-3}) are due to severe water pollution, mainly with organic matter and are reflected in high values of Biological or Chemical Oxygen Demand in the water. The high levels of dissolved oxygen (12–15 mg·dm^{-3}) may result from an excessive growth of algae and macrophytes, which is associated with intensive photosynthesis processes. In turn, the average range of the values for oxygen dissolved in water (8–12 mg·dm^{-3}) is usually an indicator of a properly functioning ecosystem [11]. In almost all water analyses the dissolved oxygen content was similar to values which indicate a properly functioning ecosystem. Only the waters in the main tributary and in the bay were characterised by a significantly lower content of dissolved oxygen (which did not exceed 10 mg·dm^{-3}, while average values amounted to approx. 7 mg·dm^{-3}). This indicates a large amount of substances absorbing oxygen in biochemical processes in the water, which is confirmed by the COD and BOD_5 values.

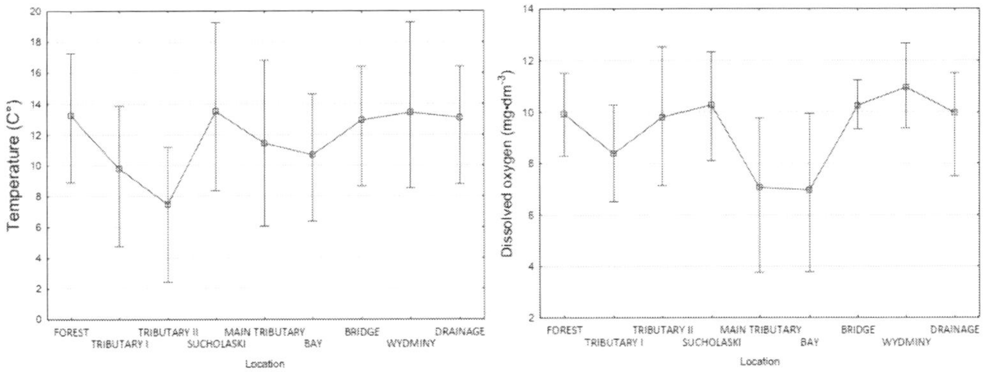

Fig. 5. Water temperature and the dissolved water oxygen content of the tributaries, outflow and particular parts of the lake.

Homogeneous groups contained in table 1 confirm the statistically significant trends that were identified in results in the graphs (Fig. 2–5). In most cases, there was obtained a homogeneous groups consisting of tributaries lakes and bay, which was characterized by significantly higher impurities in comparison to the second homogeneous group consisting of the other monitoring points.

Table 1. Homogeneous groups of the analysed indices based on a (post-hoc) Duncan test.

No.	**Location**	***TDS***	**DR**	**RoI**	***COD***	***BOD_5***	**Temp.**	***DO***
1	Forest	0.281a	287a	203a	33.84a	7.7ab	12.92a	9.95ab
2	Tributary I	0.446bc	460bc	333b	67.62bc	7.8ab	9.74a	8.37bc
3	TributaryII	0.474c	513c	396c	58.18b	6.1a	7.46a	9.80ab
4	Sucholaski	0.290a	299a	220a	33.73a	7.1ab	12.95a	10.33ab
5	M.tributary	0.423b	447b	324b	77.45c	9.7bc	11.05a	7.01c
6	Bay	0.416b	492bc	362bc	74.77c	11.0c	10.55a	6.90c
7	Bridge	0.283a	298a	214a	33.55a	6.7ab	12.25a	10.28ab
8	Wydminy	0.289a	302a	220a	34.44a	7.1ab	12.93a	11.01a
9	Drainage	0.290a	301a	221a	35.93a	7.4ab	12.87a	9.98ab

Legend: a, b, c – homogeneous groups

3 Conclusions

The study demonstrated that the management method of the immediate catchment area of the lake (the forest and the high-density residential housing development in Sucholaski and Wydminy) did not significantly differentiate the analysed indices of water quality (pH, *TDS*, *COD*, *BOD_5*, dry residue and residue on ignition). This was probably due to the properly functioning wastewater management in these localities and to the order maintained on the beaches. It was found that water of the poorest quality was flowing into the lake through drainage ditches. Moreover, the following relationship was demonstrated: the greater the area being drained (drainage catchments of particular ditches), the worse parameters of the incoming water. The significantly lower values of the analysed indices found in the water outgoing from the lake, compared to its tributaries, clearly indicate the accumulative role in relation to the mineral and organic substances outgoing from agricultural areas. Consequently, this situation results in the deterioration of water quality, an increase in sediment accumulation and slow overgrowth of the water body, which was particularly noticeable in the lake bay fed by water of the main tributary.

References

1. G. Papatheodorou, G. Demopoulou, N. Lambrakis, Ecol. Model **193** (2006)
2. P. Choudhary, J. Routh, G. J. Chakrapani, Environ. Earth Sci. **60** (2010)
3. M. V. Prasanna, S. Chidambaram, T. V. Gireesh, T. V. Jabir Ali, Environ. Earth. Sci. **63**, 1 (2010)
4. A. S. Commault, G. Lear, S. Bouvier, L. Feilerd, J. Karacs, R. J. Weld, Biochem. Eng. J. **109** (2016)
5. M. Sepp, T. Kõiv, P. Nõges, T. Nõges, Sci. Total Environ. **610–611** (2018)
6. F. A. Khan, A.A. Ansari, The Botanical Review **71**, 4 (2005)
7. V. B. Singh, A. L. Ramanathan, A. Mandal, Arabian J. Geosci. **9**, 308 (2016)
8. M. Hayashi, W. L. Quinton, J. Pietroniro, J. J. Gibson. J. Hydrol. **296**, 1–4 (2004)
9. O. F. Schoumans, W. J. Chardon, M. E. Bechmann, C. Gascuel-Odoux, G. Hofman, B. Kronvang, G. H. Rubæk, B. Ulén, J. M, Dorioz, Sci. Total Environ. **468–469** (2014)
10. APHA. *Standard methods for the examination of water and wastewater* (21st ed. Washington, DC, New York: American Public Health Association, 2005)
11. Food and Agriculture Organization of the United Nations, *Water Quality for Agriculture, Irrigation and Drainage* (by Ayers RS, Westcot DW, 1994)
12. World Health Organization, *Guidelines for Drinking-water Quality* (2004)
13. E. Sienkiewicz, M. Gąsiorowski, Sci. Total Environ. **605–606** (2017)
14. V. Gopal, H. Achyuthan, M. Jayaprakash, Acta Ecol. Sin. **38** (2018)
15. P. Skonieczek, J. Koc, M. Duda, PECO **7**, 1 (2013)
16. K. Sobczyńska-Wójcik, PECO **3**, 2 (2009)
17. L. Przywara, A. Adamiec, M. Kuglarz, K. Grübel, PECO **11**, 1 (2017)
18. A. G. B. Blankenberg, K. Haarstad, A. K. Sovik, Desalination **226**, 1–3 (2008)
19. P. J. A. Withers, C. Neal, H. P. Jarvie, D. G. Doody, Sustainability **6**, 9 (2014)
20. K. Yang, Z. Yu, Y. Luo, Y. Yang, L. Zhao, X. Zhou, Sci. Total Environ. **624** (2018)

Seasonal changes in hydraulic flow conditions in overgrown lowland river

Krzysztof Wolski[1,*]

[1]Wroclaw University of Environmental and Life Sciences, The Faculty of Environmental Engineering and Geodesy, Institute of Environmental Engineering, ul. Norwida 25, 50-375 Wroclaw, Poland

Abstract. The paper presents the results of numerical research on the impact of seasonal changes in vegetation on the hydraulic conditions prevailing in the bed of overgrowing lowland river on the example of Ślęza River (Poland). Hydrological characterization of the analyzed section of the river was made on the basis of data from the years 2006–2017 from the Ślęza water gauge. Based on them, the growing season and characteristic flows for modelling were determined. A two-dimensional hydrodynamic model (2HD) was built in two scenarios corresponding to the beginning and the end of the vegetation period for an about 50 m long section of the river. The bathymetric data of the bed and the plant distribution were measured in the field for both scenarios. The models were calibrated using a coefficient of roughness based on the velocities and surface elevations measured directly in the field. Various characteristic flows were then tested on calibrated and verified models for both scenarios. Changes in hydraulic conditions in both scenarios were compared by means of velocities and shear stress occurring in the riverbed. The obtained results indicate a significant impact of plants on the hydraulic conditions in the riverbed. The existence of plants causes the accumulation of backwater and change of local velocity distributions while maintaining medium and maximum velocities in the riverbed.

1 Introduction

Studies on the impact of aquatic vegetation on the flow conditions in watercourses are extremely complex. It is influenced by the amount of possibility of interactions between plants, hydrological and fluvial conditions as well as morphology of the riverbed [1, 2]. One of the key variables is the spatial and temporal complexity of the existing vegetation. The occurring systems of plants are characterized by high species diversity, heterogeneity of occurrence and density, and - what is important for this work – in seasonal variability [3]. Seasonal variability of vegetation does not occur equally in the whole world due to the climatic conditions. It is most visible in temperate zones with distinct seasons of hot and cold year, where the temperature drops below the border to favourable vegetation conditions (depending on plants, 5–15°C). Together with the distinct vegetative and non-vegetative season, the resistance induced by vegetation is significantly different due to the morphological and biomechanical characteristics of plants [4, 5].

* Corresponding author: krzysztof.wolski@upwr.edu.pl

Differences in coefficient of roughness occurring in particular seasons of the year were indicated in many works [6–8]. In addition to identifying the roughness itself, many researchers have been comprehensively approaching the changes taking place in the riverbed, characterizing seasonal changes in hydraulic conditions occurring in the channel [9–11].

Marcinkowski et al. [12] also prove a significant effect of vegetation phases on the conditions prevailing in the anastomosing river system on the example of Narew River. They indicate that the roughness coefficients differ between the seasons (higher values in the growing season) and that there are significant differences in the distribution of water flow in the multichannel anastomosing rivers.

The vegetative phases of plants and their influence on flow conditions were also compared for the Polish conditions Walczak et al. [13]. They showed a large impact of plant vegetation phases on flow resistance and water velocity in the riverbed. They noticed that the greater convergence of the Manning's roughness coefficients n (MRCn) accepted from the tables [14] can be observed in the growing season than in the non-vegetative period.

Although in many of the above mentioned works the idea of a parameter correcting the coarseness coefficient appeared depending on the vegetative phase of plants (or the seasons), in engineering practice, where the determination of MRCn is an important tool in hydraulic modelling, such solutions are not used.

Understanding the impact of plants on hydraulic and morphological conditions in the riverbed is necessary in the case of ecological river regulation (restoration of rivers) and NBS (nature-based solutions), where specifically selected plants in terms of species and density can perform specific functions such as flow deflectors [15, 16].

In this study, the MRCn was spatially determined on the basis of field measurements on two dates (at the beginning and end of the growing season). A two-dimensional hydrodynamic model was built, taking into account the natural conditions of the riverbed (shaping the bed and specific MRCn). Based on calibrated the MRCn of the model, the hydraulic conditions prevailing in the riverbed for different scenarios at the beginning and end of the vegetation period were compared for different flows.

2 Study area

The research was conducted for the Ślęza River around its 12.3 kilometre. The river basin for the analysed sections has an area of about 900 km^2. The modelled section of the Ślęza River is located in the geographical location 16°58'36"East longitude and 51°04'53"North latitude. The Ślęza River is a left-bank tributary of the Odra River. Its length is about 85.1 km, and the catchment area equals ca. 975 km^2. The location of the surveyed segment of the river is presented in Fig. 1.

3 Hydrological data

Based on average daily data from the years 2006–2017 for the Ślęza station made available by the Institute of Meteorology and Water Management – National Research Institute, the hydrological characteristics of the section under investigation were performed. The Ślęza gauge station is located approx. 5 km above the analysed section of the watercourse (Fig. 1). The station has been operating since 2006, flows are measured constantly, while the water temperature was measured in years 2006–2015. Figure 2 shows measured mean daily water temperatures and average daily flows in the indicated multi-year period.

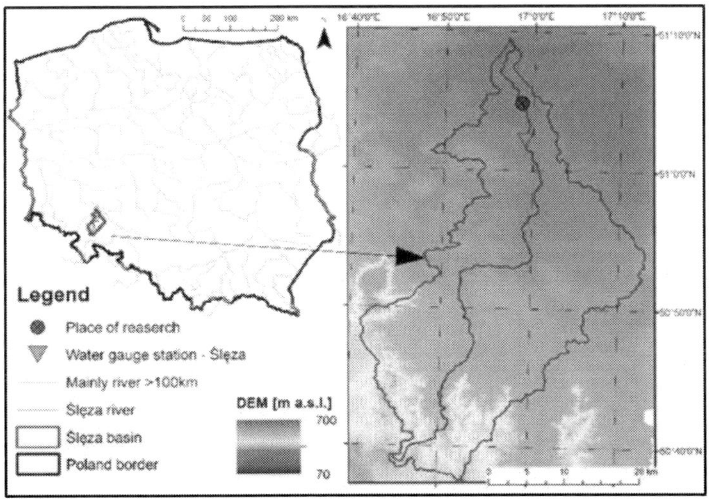

Fig. 1. Area of study – place of research on the Ślęza Basin against DEM (right) and in reference to Poland (left).

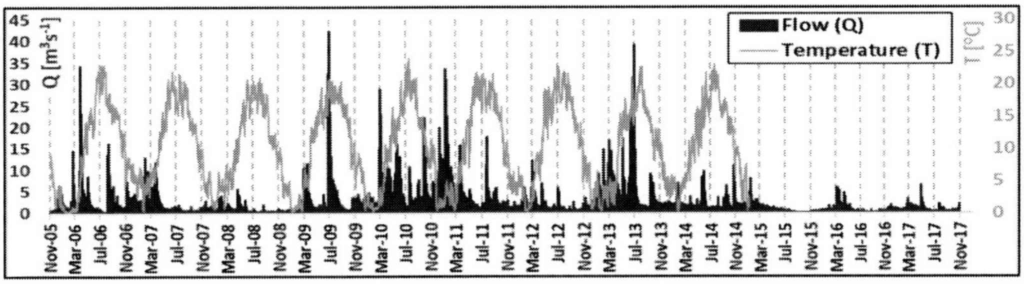

Fig. 2. The course of the daily water temperature and daily flows in the years 2006–2017 on the Ślęza water gauge (Ślęza River).

Water temperatures indicate a characteristic sinusoidal course closely related to the seasons of the year. The characteristics of water temperatures in individual months are shown in table 1. On its basis, the duration of the vegetation season for plants has been determined. It was assumed that the vegetation season is valid in months when the average water temperature is greater than 10 degrees Celsius, from April to October. The calculated monthly averages and the median of water temperatures in each month are similar, which indicates the normal distribution of the data sets. Characteristics of flows occurring in particular months were made together with the division into the vegetation season and the non-vegetation season. The results are presented in table 1.

There is a noticeable difference in the size of flows in the vegetation and non-vegetation season (mean flow in vegetative period is more than 30% smaller than mean flow in whole year). The calculated values of average flow in particular months are in each case greater than the median of flows, which indicates the deviation of results from the normal distribution and that single occurrences of larger flows occur relatively rarely, but affect the determined average flow.

For the verification purposes, histograms of the frequency of occurrence of flows for the vegetation season and comparative for the whole year were made. It was shown that low flows (below 1.5 m^3s^{-1}) occur in the vegetation season in 60% of the time, while 49% in the entire year. This confirms that the longest flows lasting in the growing season are low flows. Therefore six flow values were adopted for comparative analyses in the models: $Q_1 = 0.5$ m^3s^{-1}, $Q_2 = 1.0$ m^3s^{-1}, $Q_3 = 1.5$ m^3s^{-1}, $Q_4 = 2.0$ m^3s^{-1}, $Q_5 = 3.0$ m^3s^{-1} and $Q_6 = 5.0$ m^3s^{-1}.

Table 1. The minimum, maximum, average and median values of monthly and seasonal flow and water temperatures on the Ślęza water gauge (Ślęza River) in the years 2006–2017.

Month	Temperature *T*, °C				Flow *Q*, m^3s^{-1}			
	min	max	mean	median	min	max	mean	median
January	0.1	7.2	2.4	1.8	0.5	33.6	3.5	2.3
February	0.2	7.6	2.4	2.3	0.7	26.1	3.5	2.3
March	0.4	11.2	5.0	5.0	0.9	34.3	4.9	3.7
April	3.0	17.2	10.4	10.4	0.8	18.0	3.0	2.2
May	7.0	20.8	14.9	14.6	0.6	17.3	2.6	1.4
June	11.8	22.2	17.6	17.8	0.2	42.4	2.7	1.0
July	15.0	24.0	19.3	19.2	0.2	37.5	2.2	0.9
August	14.4	22.7	18.3	18.6	0.1	16.2	1.5	0.9
September	10.4	19	14.9	15.0	0.1	22.2	1.7	0.9
October	4.8	16.6	10.7	10.8	0.1	20.4	1.7	1.0
November	1.0	11.6	6.8	7.2	0.2	7.5	1.9	1.6
December	0.4	7.8	3.4	3.4	0.4	19.9	2.4	1.8
Years 2006–2015 (CP)	**0.1**	**24.0**	**10.4**	**10.4**	**0.1**	**42.4**	**2.6**	**1.6**
Vegetative period (VP)					**0.1**	**42.4**	**2.2**	**1.2**
Non-vegetative period (NP)					**0.2**	**34.3**	**3.2**	**2.1**

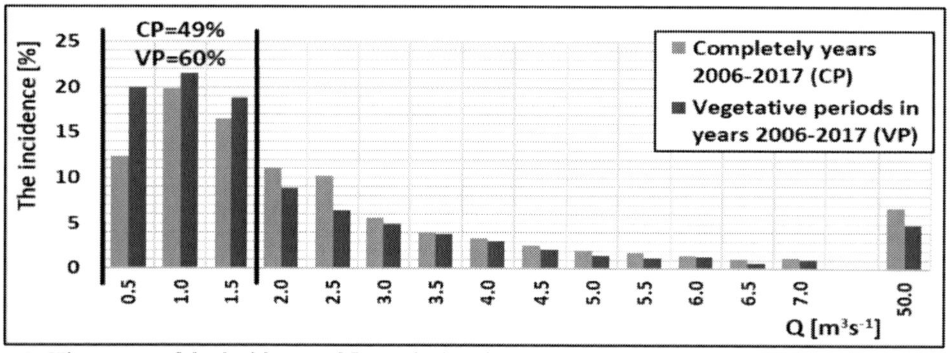

Fig. 3. Histogram of the incidence of flows during the vegetation season and throughout the year on the Ślęza water gauge (Ślęza River) in the years 2006–2018.

4 Methods

4.1 Field measurements

Field measurements were carried out throughout 2018. Figure 4 shows what the research stand looked like in four phases of vegetation in 2018. It can be noticed a very large variation in the number of plants in the riverbed in individual months. The main measurement work was carried out in two series on April 14–19 in the initial vegetation season and on October 11–14 in the final vegetation season. In 2018, for almost the entire summer during the maximum vegetation there were very low flows, which in combination with the developed vegetation did not allow for stable and reliable measurements.

As part of the measurements, the bathymetry of the bed was mapped in the 1 m x 1 m grid (Fig. 5) as well as the development and density of vegetation (on a scale from 1–3). At the same time, the water velocities in the hydrometric divisions were measured using the Mini Current Meter (SEBA M1) and in the cross-sections profiles using an acoustic Doppler current profiler (StreamPro ADCP). The measured water flow rates were then used to calibrate the numerical model.

Fig. 4. Measuring station in 2018 (dates in the drawing).

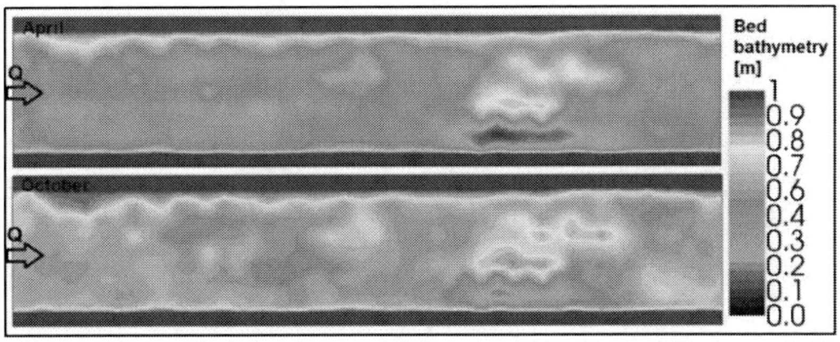

Fig. 5. Distribution of the bed bathymetry in April (scenario 1) and in October (scenario 2).

During measurements in the riverbed there were characteristic groups of species among which typical macrophytes can be distinguished: helophytes (rush – *Phragmites australis*, *Glyceria maxima*, *Scirpus lacustris* and *Sparganium erectum*), elodeids (submerged – *Elodea canadensis* and *Myriophyllum L.*) and aquatic perennials (growing on shallowing – *Alisma plantago-aquatica L.* and *Rumex hydrolapathum Huds.*). The spring season was dominated by helophytes - withered clusters of last year shoots and young sprouts growing in them, and to a lesser extent, elodeids. In the autumn, the number and density of helophytes have increased significantly, some of which stems are already in a dry form. The intense growth of helophytes has stopped the elodeids, which occur only in few trims, but occupy less space than in the spring. The emergence of aquatic perennials in shallow stands with a soft swamp substrate is characteristic. On the basis of the inventories made, a spatial matrix of existing plants was made which, at the stage of hydrodynamic model calibration, constituted the basis for spatial MRCn in the riverbed.

4.2 Hydrodynamic model

For hydrodynamic modelling, the Nays2DH model implemented in the IRIC environment was used (https://i-ric.org). Nays2DH is a computational model for simulating unsteady horizontal two-dimensional (2D) flow, sediment transport, and morphological changes of bed and banks in rivers which was developed by Shimizu et al. [17, 18].

The two-dimensional model is based on the GRID mesh characterizing the dimensions of the analysed section of the channel (11 m wide and 47 m in length) and spatial resolution for the calculation of 0.2 m x 0.2 m. The GRID mesh prepared in this way has been applied to the bed bathymetry on the basis of field values additionally interpolated to GRID mesh in the ArcGIS environment using the Natural Neighbour method and a pre-prepared MRCn distribution interpolated using the Kriging method also in the ArcGIS environment. The calculations were made for the steady flow conditions for the simulation time $T = 660$ s with the time step $\Delta t = 0.02$ s. As the upper boundary condition, a constant inflow was established with the intensity specified in a given scenario, while free outflow was established as the bottom boundary condition. In the model for the purposes of this study, the calculation of the bed deformation was excluded, which allows to assess additional morphological changes.

The calibration was carried out, according to the diagram in Fig. 6, for both scenarios separately due to different bathymetry and different flows measured during field tests. The elevation of surface and the velocities measured in the field were checked. The calibration consisted in such selection of the factor of roughness (N) that after multiplication by Plants matrix ($P_{i,j}$), the MRCn matrix ($n_{i,j}$) was obtained, which after entering the model will give the results of the velocity and elevation of surface water similar, as much as possible, to field measurements.

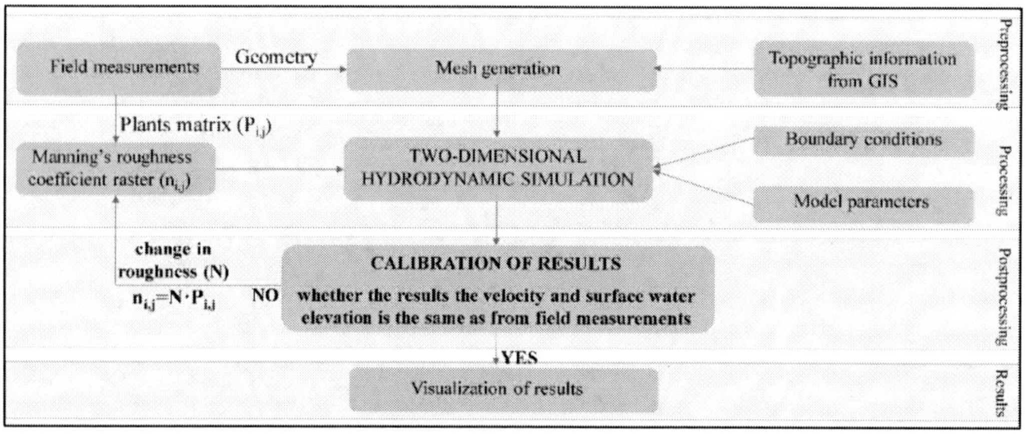

Fig. 6. Calibration diagram used in the test.

5 Results

As part of the performed calibrations, the MRCn matrix was obtained (n) for both scenarios, which are presented in Fig. 7. The much smaller n values for Scenario 1 (April) than for Scenario 2 (October) are noticeable. The value of the coefficient calculated for the whole section for the scenario 1 are: n_{mean} = 0.039 ms$^{-1/3}$, n_{max} = 0.069 ms$^{-1/3}$, and for the scenario 2: n_{mean} = 0.055 ms$^{-1/3}$, n_{max} = 0.088 ms$^{-1/3}$.

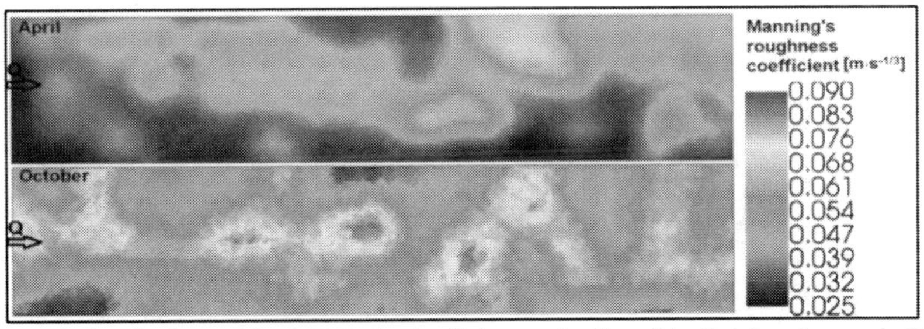

Fig. 7. Calibrated distribution of the MRCn in April (scenario 1) and in October (scenario 2).

Selected results of the simulations carried out for 6 different flows are presented in table 2. The obtained results point to the significant impact of plants primarily on the elevation of surface in autumn conditions compared to spring ones. Differences are H = 0.10–0.13 m and decrease with increasing flow rates (Fig. 8A). Analysis of the whole riverbed in terms of velocity did not show significant differences in the maximum velocity in the bed nor the mean velocity. The mean velocity differences are u_{mean} = -0.002–0.009 ms^{-1} and increase very slightly with the flow rate (Fig. 8B), while the maximum velocity u_{max} = -0.0094–-0.013 ms^{-1} and decrease with the flow rate (Fig. 8C). The table also show analysis results for the maximum total shear stress, showing that the difference between scenario 1 and scenario 2 one is S = -1.6–3.4 Nm^{-2} and increases with the flow rate (Fig. 8D).

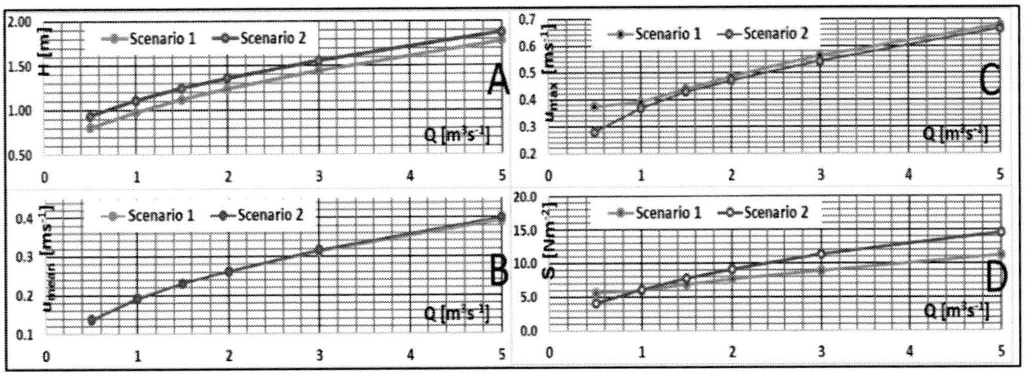

Fig. 8. Graph of the flow rate and: A. average surface of elevation (H), B. mean flow velocity (u_{mean}), C. Maximum flow velocity (u_{max}), D. shear stress (S).

Table 2. Selected results of simulations carried out for 6 flows (Q – flow [m^3s^{-1}], u_{mean} – mean velocity [ms^{-1}], u_{max} – maximal velocity [ms^{-1}], S – shear stress [Nm^{-2}], Δ – differences between scenarios 1 and 2).

	Q		*H*			u_{mean}			u_{max}			*S*		
	1	2	1	2	Δ	1	2	Δ	1	2	Δ	1	2	Δ
Qk	1.12	0.65	1.01	1.00	-	0.204	0.156	-	0.396	0.314	-	6.3	4.7	-
Q1	0.5	0.5	0.80	0.94	0.13	0.139	0.137	-0.002	0.374	0.28	-0.094	5.7	4.1	-1.6
Q2	1	1	0.98	1.11	0.13	0.194	0.191	-0.003	0.392	0.368	-0.024	6.0	6.1	0.1
Q3	1.5	1.5	1.12	1.24	0.13	0.231	0.231	0	0.438	0.429	-0.009	6.9	7.8	0.8
Q4	2	2	1.24	1.36	0.12	0.261	0.262	0.001	0.486	0.471	-0.015	7.7	9.1	1.5
Q5	3	3	1.44	1.56	0.11	0.312	0.317	0.005	0.562	0.542	-0.02	8.9	11.3	2.4
Q6	5	5	1.78	1.88	0.10	0.391	0.4	0.009	0.677	0.664	-0.013	11.2	14.6	3.4

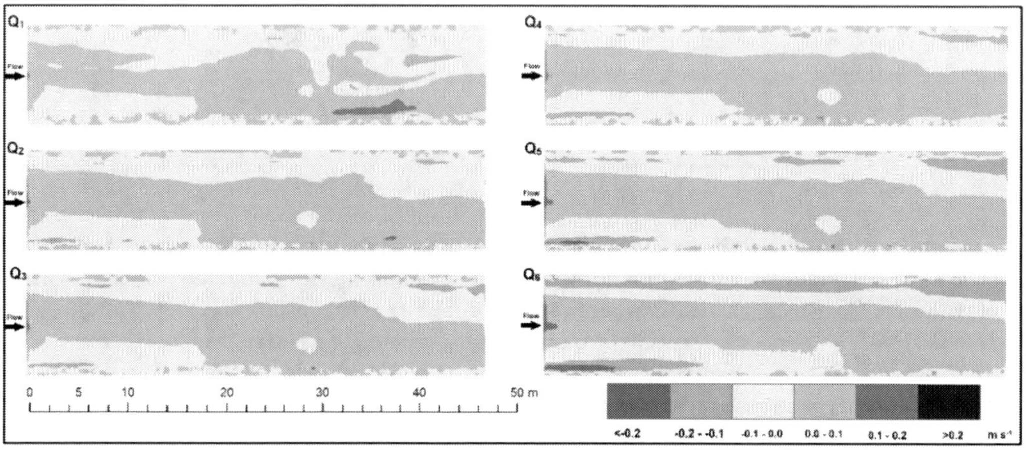

Fig. 9. Distribution of velocity variation between scenario 1 and scenario 2 for different flow rates.

The shown changes between scenario 1 and 2 in medium and maximum velocities in terms of the whole riverbed constituted very small values, therefore spatial velocity changes were also examined. It was done by calculating the differences in each computational cell of the hydrodynamic model. The differences between individual flows are from -0.35 ms^{-1} to 0.27 ms^{-1} and increase with the flow rate (Fig. 9).Positive velocity differences (spring flows higher than in autumn) occur in the central part of the riverbed, while the opposite tendency occurs at the banks.

6 Conclusions

The analyses carried out have unequivocally confirmed the impact of plants on seasonal changes in hydraulic conditions in the overgrown lowland river bed. It has been shown that the development of plants affects the increase of the MRCn in the channel and the change in spatial velocity distribution, despite the persistence of similar values of medium and maximum velocity. It has also been proved that the presence of plants affects the accumulation of water, which decreases with increasing flow rate. In addition, the differences in shear stress between spring and autumn conditions have been shown to increase with the flow rate. The examined plant impact is important for hydraulic changes at low and medium flows, which in the studied conditions of the Ślęza River occur for the longest time during the year. The established hydraulic changes are followed by the morphological changes in the riverbed, which may be the subject of further work.

The study was performed with Nays2DH and IRIC software both on free license. The data were provided by the Institute of Meteorology and Water Management – National Research Institute.

References

1. E. J. Hickin, Can.Geogr. **28** (1984)
2. R. C. Grabowski, A. M. Gurnell, River Rec. Applic. **32** (2016)
3. A. M. Gurnell, Earth Surf. Process. Landforms, **40** (2015)
4. A. M. Łoboda, R. J. Bialik, M. Karpiński, Ł. Przyborowski, Aquat. Bot. **147** (2018)
5. A. M. Łoboda, R. J. Bialik, M. Karpiński, Ł. Przyborowski, Polish J. Environ. Stud. **28** (2019)
6. F. M. Bakry, T. K. Gates, A. F. Khattab, J. Irrig. Drain. Eng. **118** (1992)
7. T. Helmiö, J. Järvelä, Boreal Environ. Res. **9** (2004)
8. L. De Doncker, P. Troch, R. Verhoeven, K. Bal, P. Meire, J. Quintelier, Environ. Fluid Mech. **9** (2009)
9. P. D. Champion, C. C. Tanner, Hydrobiologia, **441** (2000)
10. M. T. O'Hare, C. McGahey, N. Bissett, C. Cailes, P. Henville, P. Scarlett,. J. Hydrol. **385** (2010)
11. S. Song, B. Schmalz, Y. P. Xu, N. Fohrer, Water Resour. Manag. **31** (2017)
12. P. Marcinkowski, A. Kiczko, T. Okruszko, Water, **10** (2018)
13. N. Walczak, M. Hammerling, M. Spychała, J. Nieć, J. Ecol. Eng. **16**, 5 (2015)
14. V.T. Chow, *Open-Channel Hydraulics* (1959)
15. T. Kałuża, A. Radecki-Pawlik, K. Szoszkiewicz, K. Plesiński, B. Radecki-Pawlik, I. Laks, Sci. Total Environ. **627** (2018)
16. P. M. Rowiński, K. Västilä, J. Aberle, J. Järvelä, M. B. Kalinowska, J. Ecohydrology and Hydrobiology, **18** (2018)
17. C. L. Jang, Y. Schimizu, J. Hydraul. Eng. **131**, 7 (2005)
18. Y. Shimizu, H. Takebayashi, T. Innoue, M. Hamaki, T. Iwasaki, M. Nabi, *IRIC Software. Changing River Science. Nays2DH Solver Manual* (2011)

The impact of chemical contaminants on biocenosis (ecotoxycological studies)

Magdalena Wróbel[1,*], *Agnieszka* Stojanowska[1], *Martyna* Nosarzewska[1], *Radosław* Rutkowski, and *Justyna* Rybak[1]

[1]Wrocław University of Science and Technology, Faculty of Environmental Engineering, Wybrzeże Wyspiańskiego 27, 50-370 Wrocław, Poland

Abstract. The analysis of the impact of PCC "Rokita" chemical industry on the natural environment has been studied. Samples of water, soil and sediment taken nearby industry were analyzed with ecotoxicological tests.. We used *Avena sativa* for study the impact of industry on producers and to evaluate soil quality. Additionally the commercial test, Phytotoxkit, test was conducted. Water quality was assessed using *Tubifex tubifex* and *Daphnia magna*/ Sediment's quality was assessed with the application of Ostracodtoxkit F test (*Heterocypris incongruens*). Conducted ecotoxicological tests proved that the studied area surrounding PCC "Rokita" is still contaminated with organic toxic compounds we observed mortality in water and soil of organisms tested. The fighest impact of toxicants was recorded for sediments suggesting accumulation of toxicants over many years.

1 Introduction

Nowadays, the chemical industries are developing extremely fast. Along with the expanding range of new products, pollution produced by factories is also growing which may have a negative impact on human and the entire biocenosis [1]. This impact comes from the interaction between contamination and the organism, population or biocenosis, which leads to the mortality of organisms, inhibition of growth or physiological functions. What is more, accumulated pollution can be additionally transformed into toxic products during metabolism process [2].

PCC "Rokita" was founded in 1946, in place of a former German factory, in order to produce various chemical products for war purposes. Due to its rapid development in terms of production and employment "Rokita" became one of the largest chemical plants in Poland in sixties of XX century. In 1976, an explosion occurred due to the connection of a tank with an oxide of ethyl to tanks with ammonia at the border station between Poland and the Federal Republic of Germany. The age XXI brought major changes due to the modernization concerning the process of chlorine formation. Environmental-friendly membrane technology was used, instead of the mercury installation which was harmful to the environment [3].

To determine the environmental contamination by toxic substances, chemical analyzes are used. Ecotoxicological tests are also frequently used, which allow to detect the presence

* Corresponding author: magdalena.wrobel@pwr.edu.pl

and determine the influence of chemical compounds on living organisms. Thanks to them it is possible to obtain the reliable data and results regarding the quality of the environment [4].

The main aim of the presented work was to analyze the impact of PCC "Rokita" on the surrounding environment and therefore on human health. In order to evaluate the toxicity of the studied area, ecotoxicological tests of water, sediment and soil were conducted.

2 Materials and methods

All study sites were located near PCC "Rokita"in Brzeg Dolny, Lower Silesia (Poland). The profile of industry has changed over time depending on market demands [5]. This industry has been chosen to study because of its great impact on environment in the past, what is more some serious accidents took place (i.e. explosion, chlorine leakage in 2017). The contaminants that entered the environment were a serious threat to the ecosystem. Thus, after the improvement of working conditions, with the beginning of XXI century we could also expect the improvement of the environment surrounding this industry as PCC "Rokita" which primarily deals with the production of chlorine, chlorinated products – organic compounds.

2.1 Materials

Description of sites presented in Table 1.

Table 1. Description of sites.

Sample	Site	Coordinates
Odra river	the site was located near the PCC Rokita municipal plant, at a distance of about 800 m from the plant	51°15'36.9"N 16°44'01.3"E
Soil	the site was located at the gate of the plant from the west	51°16'25.8"N 16°43'52.0"E

2.2 Methods

2.2.1. Tubifex tubifex

Tubifex tubifex (Oligochaeta, Tubificidae), called sludge worm, is a cosmopolitan freshwater oligochaete. Its presence suggest healthy conditions of the aquatic environment [6]. Tests using *T. tubifex* were carried out to check the toxicity of water collected from the Odra river (according to standards ASTM E1706-04, ASTM E1706-95b) [7].

The solutions of water were prepared at concentrations of 100%, 50%, 25%, 12.5%, 6.25%, and control samples. Ten sludge worms were exposed to each concentration (tests were done in three replicates for each concentration). After 24 hours, 48 h the mortality of organisms was analyzed. Worms were considered dead when they were completely immobilized.

2.2.2. Avena sativa

In ecotoxicological studies of the soil, mainly plants are used [2]. In this paper we have chosen *Avena sativa*according to the polish standard PN-ISO 11269-2:2001) [8]. The samples were prepared in the following concentrations: 100%, 50%, 25%, 12.5%, 6.25% and reference soil. The samples of the soil were placed in the small containers and then ten oat seeds were planted in each pot. The test was carried out in triplicate for each concentration. After seven days we assessed the quality of seed germinated and measured the shoot and root length.

2.2.3. Ostracodtoxkit F

Ostracod exposure tests were conducted in the form of 6 replicates per site sample and 1 control per sample by using reference sediment according to the standard operating procedures of the test kit. Ostracod neonates of the benthic ostracod crustacean *Heterocypris incongruens* were exposed to test sediments (6 days in darkness at 25°C ± 2°C). At the end of the test period, all surviving ostracods were counted and measured with micrometre cover slip. Growth inhibition of surviving ostracods exposed 6 days to studied sediments was determined in relation to (non-toxic) reference sediment. The percentage mortality was also determined [9].

2.2.4. Seed germination and early growth microbiotest with higher plants

Phytotoxkit test measures the decrease (or the absence) of seed germination and of the growth of the young roots after a few days of exposure of seeds of selected higher plants to toxicants or to contaminated soils.

The test was conducted in accordance with the procedure recommended by the manufacturer MicroBioTests Inc. 90 ml of reference soil and tested soils were placed in triplicates on the lower compartment of the test plates (21×15.5×0.8 cm) and were hydrated with assessed volume of distilled water needed for complete hydration. The bottom part of the test plates were covered by a filter paper and then test seeds; monocotyl *Sorghum saccharatum* (Sorgho) were placed in a row at 1 cm below the middle ridge of the plates. The plates prepared in this way were incubated in a horizontal position at 25°C in darkness for 72 h. Then, photos of the test plates were taken with a digital camera. Root lengths were subsequently measured with the aid of the *Image J* program [10].

2.2.5. Daphnia magna

Daphnia magna is commonly used to assess acute toxicity in a 24–48h assay based on immobility or mortality of the test organisms. Daphnids were obtained from the pet shop and the species was positively identified before testing. Test organisms were at similar age and size Using a wide bore pipette, thirty actively swimming daphnids were carefully placed into petri dishes. Then, containers were filled with undiluted samples of tested waters (breeding water was used as a control). After 24 h incubation at 20°C in darkness, the mortality and survival of daphnids were recorded.

3 Results and discussion

3.1 Results of toxicity testing: water

In the table 2. the mortality of the sludge worms *T. tubifex* in water is shown.

The biggest mortality was observed at the concentration of 100% (after 72 hours more than 50 % of all worms were dead), which indicates the toxicity of the water from Odra river. At the concentration of 50% more than 60% of worms did not show any movement. At the concentrations of 25%, 12.5% and 6.25% the mortality was not strong and did not exceed 50%. In control, *T. tubifex* showed typical movement and remained active until the end of test period. Additionally, we assessed the toxicity of water with *D. magna*.

Table 2. The number of alive worms after 24, 48 and 72 hours.

Concentration, %	24 h	48 h	72 h
100	20	15	14
50	25	20	18
25	30	25	22
12.5	30	27	25
6.25	30	27	26
control	30	30	-

Table 3. Percentage of mortality during 24 h of the experiment.

Sample	Mortality, %
Conrol	-
Tested water	30

This results shows that Odra river is slightly toxic (percentage mortality is 30%), but the calculation of LC_{50} was impossible.

3.2 Results of toxicity testing: soil

The results of ecotoxicological test with *A. sativa* on soil are shown in the table 4.

Table 4. Results of shoot and root length in *A. sativa*.

Concentration (%)	replicate 1 (cm)	replicate 2 (cm)	Replicate 3 (cm)	Average shoot length (cm)	Average root length (cm)	Numbers of germinated seeds (max. 30)
100	2.0	3.0	4.6	3.2	1.4	13
50	11.5	14.7	12.1	12.8	5.5	21
25	15.2	17.0	12.8	15.0	6.1	19
12.5	14.6	11.6	14.3	13.5	5.9	20
6.25	16.0	17.0	15.7	16.2	6.3	16
control/reference sample	17.5	17.5	16.0	17.0	6.7	30

At the concentration of 100% more than 50% of the seeds did not germinate. Percentage of germinated seeds varied. The average shoot length was only about 3 cm and average root length was 1.4 cm in the contaminated soil (100% concentration). ID_{50} was 100 % The shoot length in the reference soil was 17 cm on average and 6.7 for roots, respectively.

We also assessed the soil toxicity with the application of Phytotoxkit. The results of this seed germination and early growth microbiotest with *Sorghum saccharatum* are shown in Table 5.

Table 5. The percentage of growth inhibition of roots for *Sorghum saccharatum*.

Sample	Percentage growth inhibition of roots, %
Control	39.31
Tested soil	31.38

Phytotoxkit shows that the length of sorghum roots was slightly smaller in the tested sample, although the calculation of IC_{50} was not possible.

3.3 Results of toxicity testing: sediments

In order to assess the chronic toxicity of Odra sediments Ostracodtoxcit F was performed. We have measured the length of *H. incongruens* on the 1st and 6th day of the experiment and the percentage of mortality and percentage of growth inhibition was determined. The results are shown in Table 6.

Table 6. The results of Ostracodtoxkit F.

Site	The average length of the body day 0 [mm]	The average length of the body day 6 [mm]	Mortality [%]	Growth inhibition [%]
Control	232	833	-	-
Odra river	232	492	35	57

This test shows that the sediment from the river is toxic as we recorded smaller body length of organisms tested (492mm on average comparing to control 833 mm) and we recorded inhibition growth (57 %) and quite high crustacean mortality (35%).

3.4 Discusion

To sum up the obtained results, the acute toxicity was observed for water samples with *T. tubifex* in raw water, although other concentrations did not show such high impact of pollutants on water quality. Additional test with *D. magna* did not allow to calculate LC_{50} but the results were lower comparing to control samples. Tests conducted on soil also revealed the impact of contaminants on plant growth, especially for *A. sativa*, plants which was selected for studies. Similarly, the calculation of ID_{50} was possible only for 100% of soil concentration for shoot and root length of *A. sativa*. Additional test, done with commercial Phytotoxkit showed only slight inhibition of root growth in *S. saccharatum*. The promising results were obtained when assessing the sediments toxicity (derived from Odra river). The growth inhibition (57%) and mortality (35%) of *H. incongruens* when applying Ostracodtoxcit F suggesting the existing impact of accumulated previously used chemicals in sediments on water biocenosis, as PCC "Rokita" mainly dealt and still deals with the production of chlorine and chlorinated products, organic compounds. Although the chemical industry is constantly evolving, and takes its impact on the natural environment into consideration [3, 11]. On the other hand, chemical industries, such as PCC "Rokita" annually produce thousands of substances with an effect unknown to the environment and humans. The performed tests in this study indicate that toxicity around the area of this industry still exists, but it is not as high as used to be in the past. According to the report on the state of the environment for Lower Silesian Voivodship [12, 13], PCC "Rokita" maintains a tendency to decrease concentrations of various pollutants and it is constantly improving technologically, which might be the reason why toxicity in this area is not as severe as used to be before [14, 15].

4 Conclusions

1. We observed water toxicity in studied area as the mortality of the *T. tubifex* was observed. The LC_{50} was calculated for 100 % concentration (pure water without dilutions) only. The mortality of other studied species *D. magna* did not exceed 50% of the population
2. For soil, we recorded inhibition of shoot and root growth for *A. sativa*. The IC_{50} was calculated for pure soil (no dilution) In commercial test, Phytotoxkit, only slight root inhibition growth was observed comparing to control for *S. saccharatum*.
3. Sediments studies based on commercial test Ostracodtoxkit F revealed a negative impact on growth and survival of tested animals, which suggest that organic compounds accumulated in decades could have still impact on water environment.
4. The conducted ecotoxicological tests showed that the studied area near PCC "Rokita" S.A. still needs to be monitored and cleaned.

The investigations were co-financed within the framework of order No. 0401/0056/18 with a specific subsidy granted to the Faculty of Environmental Engineering, Wroclaw University of Science and Technology (W-7), by the Minister of Science and Higher Education.

References

1. A. Sobczak, Environmental Medicine **15**, 7–17 (2012)
2. T. Traczewska, *Biologiczne metody oceny skażenia środowiska* (15–34, 2011)
3. D. Łaba, Chemia Przemysłowa **3**, 76–77 (2012)
4. A. Valavanidis, T. Vlachogianni, Science Advances on Environmental Chemistry, Toxicology and Ecotoxicology, **1–29** (2015)
5. J. Rogala, *Historia Zakładów Chemicznych Rokita S. A.* (1–6, 1996)
6. K. A. Beauchamp, R. D. Kathman, T. S. McDowell, R. P. Hedrick, Molecular Phylogenetics and Evolution **19**, 216–224 (2001)
7. C. Amiard-Triquet, J. C. Amiard, C. Mouneyrac, *Aquatic ecotoxicology Advancing Tools for Dealing with Emerging Risks* (2015)
8. PN-ISO 11269-2:2001 Jakość gleby – Oznaczanie wpływu zanieczyszczeń na florę glebową – Część 2: Wpływ zanieczyszczonej gleby na wschody i wczesny wzrost roślin wyższych

9. OSTRACODTOXKIT F Chronic "direct contact" toxicity test for freshwater sediments. Standard Operational Procedure (2017)
10. PHYTOTOXKIT F test in assessment of toxicity of two types of sewage sludge standard operational procedure (2017)
11. Y. C. Cloete, B. F. Shaddock, A. Nel, Water SA. **43**, 409 (2017)
12. L. A. Maranho, L. T. Maranho, R. G. Botelho, V. L. Tornisielo, Arch. Environ. Contam. Toxicol. **65**, 319–320 (2014)
13. A. Antosz, E. Banach, L. Barański, R. M. Ciechanowicz-Kusztal, Raport o stanie środowiska w województwie dolnośląskim w 2001 roku, **2**, 5–8 (2001)
14. M. Bełcik, A. Trusz-Zdybek, E. Galas, K. Piekarska, Atmospheric Environment **95**, 620–628 (2014)
15. A. Trusz, M. Wolf-Baca, A. Siedlecka, Inżynieria Ekologiczna **19**, 4, 51–57 (2018)

Research and evaluation of water quality in outdoor swimming pools

Joanna Wyczarska-Kokot[1], *Anna* Lempart[1,*], and *Marta* Marciniak[1]

[1]Institiute of Water and Wastewater Technology, Silesian University of Technology, ul. Konarskiego 18, 44-100 Gliwice, Poland

Abstract. The article presents the assessment of the quality of water in outdoor pools, based on the results of bacteriological and physicochemical tests and their comparison with the admissible values in this respect. The analyzes were carried out during two bathing seasons, in 2017 and 2018. The quality of water from sports, recreational and children's pools was analyzed, as well as water introduced into the pools from the circulation systems. The decisive role in the assessment of pool water quality is played by the results of microbiological tests and the assessment of the sanitary condition of the facility. In pool water and circulatory water samples, the number of indicator microorganisms (CFU) did not exceed the limit values. The problem was the very high levels of chloroform (> 0.03 mg/L) and THM (> 0.1 mg/L), lower than the required redox values (750 mV) and nitrate content, increasing over the bathing season. Considering that the analyzed pools are seasonal and recreational outdoor pools, the observed excessive concentrations of DBP, with a short exposure, did not pose a direct threat to the health of bathers. However, they indicate the need to supervise the processes of water treatment and disinfection or their modernization.

1 Introduction

Public awareness of a healthy and active lifestyle translates, inter alia, into increased interest in water sports. Swimming is one of the most popular aquatic activities. Just like natural water, public pool water may contain microbiological and chemical contaminants. Possible sources of these contaminants in swimming pools include contamination from swimmers (saliva, sweat, cosmetics and urine), from the water supplied to the pool, from circulation, water treatment and disinfection system [1–5]. For outdoor pools, additional pollutants are creams, suntan oils and pollutants coming from leaves and dust in the surrounding environment or from rain water [6, 7].

Therefore, the quality of the water in pools, and thus the health safety of swimmers, will be influenced by (apart from the effectiveness of treatment): the way of swimming pool operation, cooperation of the hydraulic system with the treatment and disinfection system, the number of swimmers and their intensity of swimming (pool load) [3, 8–10].

* Corresponding author: anna.lempart@polsl.pl

Both swimmers and staff of the swimming pool facilities may be exposed to contamination by accidental ingestion of water, inhalation of water aerosols and absorption through skin [11–15].

Lack of proper quality control of the pool water can result in health risks for swimmers and staff and the spread of infections. For the above reasons, ensuring the safety of pool users is a priority. For this purpose, documents regulating the admissible values of physicochemical and microbiological parameters of swimming pool water were issued. The most important document in most countries is the regulation in force. In Poland, it is the Ministry of Health's regulation of 2016, on the requirements of water in swimming pools [16]. A more detailed one is a German norm, a European-like standard which also specifies methods for treating swimming pool water [17]. The general guidelines on the basis of which legal regulations were created in individual countries are specified in the document of the World Health Organization [18]. It should be emphasized that the quality of water in both outdoor and indoor swimming pools is subject to legal regulations.

The aim of this work is to assess the quality of water in outdoor pools, during the entire bathing season, based on the results of bacteriological and physicochemical tests and their comparison with the admissible values.

2 Characteristics of research objects

The tested facility includes: recreational swimming pool (RP), children's pool (CP), sports swimming pool (SP) and a water playground. The facility has two closed circulations of water treatment: circuit I for the basin of a sports swimming pool (SP) and a water playground, and circuit II for a recreational pool (RP) and a pool for children (CP). The circuits are equipped with retention tanks with a capacity of 30 m^3, and the filtration systems. The water was filtered with the help of the open vacuum washed with diatomaceous earth filters. The filters were rinsed every 3 days. In order to increase the filtration efficiency, a coagulant (0.5% aluminum hydroxychloride solution) was added before the filters. The water disinfection was carried out with 13% sodium hypochlorite solution. If the pH of the water needed adjustment, a 30% solution of sulfuric acid was used. In order to prevent algal blooms, an algicide was periodically added. The basic technical parameters for the tested pools are summarized in Table 1.

Table 1. Technical parameters of the tested swimming pools.

Parameter	Unit	Circuit I		Circuit II	
		Sport pool (SP)	Water playground	Recreational pool (RP)	Children pool (CP)
Dimensions of the pool basin	m×m	24.95×12.37	24.9×20.8	24.35×12.2	24.9×20.8
Depth of the pool basin	m	1.2	0.0	0.8–1.2	0.1–0.6
Usable area	m^2	285	90	268	420
Volume of the pool basin	m^3	342	0.0	268	147
Max. attendance	person/h	68	30	72	80
The time of one water change in the pool basin	h	2.70	0.43	1.54	0.50
Filtration flow	m^3/h	190		470	

3 Materials and research methodology

In order to assess the water quality in the tested outdoor swimming pools, the following parameters were analyzed: nitrate content (spectrophotometric method; spectrophotometer DR 3900 with RFID technology, Hach®), free chlorine and combined chlorine (colorimetric method in situ, Pocket ColorimeterTM II, Hach®), chloroform and total trihalomethane - THM (gas chromatography method, Agilent Technologies GC7890B chromatograph with MSD5977A mass detector), turbidity (nephelometric method, TN 100 turbidity meter, Eutech®), the pH of water, redox and temperature (potentiometric method, sensION meter + MM150 DL, Hach®), oxidisability – COD (titrimetric method in an acid environment in accordance with PN-EN ISO 8467:2001), number of colony forming units (CFU) Pseudomonas aeruginosa (method according to PN-EN ISO 16266:2009), the Escherichia coli CFU number (method according to PN-EN ISO 9308-1:2014-12/A1:2017-04) and the Legionella sp. CFU number (method according to PN-EN ISO 11731:2017).

Sampling took place in accordance with the guidelines of the Polish standard PN-ISO 5667-5:2003. Water samples were taken from swimming pools (SP, CP, RP) and test cocks were installed on pipelines of circuits I and II (SC1, SC2), supplying treated water to swimming pools. Samples were taken from a depth of about 30 cm below the water surface. During the bathing season in 2017, samples were taken 7 times (7 measuring days), and in 2018 six times (6 measurement days), taking into account the beginning and the end of the bathing season. The authors, guided by the experiments from previous studies, took samples at several characteristic basin points and used a mixed sample for analysis [8].

The results of the analysis were compared with the guidelines included in the documents specifying the requirements for water in swimming pools [16–18].

4 Results and discussion

The pool water pH values should be in the range of 6.5 to 7.6, which is very important from the point of view of the efficiency and speed of disinfection, coagulation and the creation of disinfection by-products [19,20]. In the analyzed samples, the pH of the water did not differ from the recommended values and ranged from 6.90 to 7.60. The average pH values of water in the following sampling points were: 7.22 in SC1, 7.32 in SC2, 7.26 in CP, 7.33 in RP and 7.30 in SP. There were no significant differences between the water pH values depending on the type of swimming pool (for children, recreational or sports) or water treatment system (circuit I, circuit II).

The redox potential values of water in pool basins at pH values between 6.5 and 7.3 should not be less than 750 mV, and at pH values between 7.3 and 7.6 not less than 770 mV [16, 17]. In the analyzed water samples, both in the 2017 and 2018 bathing season, the redox values met the above requirement in only 3 out of 66 samples. The average redox values in the 2017 season were from 669 mV (SC2) to 715 mV (SC1), and in the season 2018 from 671 mV (CP) to 701 mV (SP). Similarly as in the case of measuring the pH of water, no significant differences were found between the redox values depending on the type of pool or water treatment system. From the experience of authors and literature data [9, 13, 20] it appears that maintaining high redox values, especially in open basins, requires supporting NaOCl disinfection with, for example, chlorine dioxide and/or ozone.

The turbidity of water entering the basin trough the circulation system should not be higher than 0.3 NTU, and the water in the basin should not be higher than 0.5 NTU [16, 17]. In the 2017 bathing season, this requirement was met in all samples. In the season 2018 in samples of water collected in the peak season (2–23.08.2018), during hot weather and maximum attendance, the turbidity of water entering the SP pool was 0.48 NTU, and the one in the basin 0.52 NTU. The turbidity of water introduced into the CP and RP basin was

0.63 NTU, and in the basins 1.30 and 0.98 NTU, respectively. To reduce such high turbidity values, the filter rinsing process was intensified in the following days. The filtration cycle was shortened from 3 to 2 days, thus obtaining water turbidity at the level of 0.15 NTU to 0.21 NTU.

Recommended concentrations of free chlorine in open water and with hydromassage devices should be from 0.7 to 1.0 mg Cl_2/L. There is no recommendation for water entering the basin trough the circulation system [16–18]. Both in the 2017 and 2018 bathing season, free chlorine concentrations were in accordance with the guidelines. In the 2017 season, free chlorine concentrations in CP pool were 0.70–0.96 mg Cl_2/L, in RP pool 0.40–0.89 mg Cl_2/L, in SP pool 0.60–1.99 mg Cl_2/L. In the 2018 season, free chlorine concentrations in CP pool were 0.72–0.95 mg Cl_2/L, in RP pool 0.50–0.95 mgCl_2/L, in SP pool 0.58–0.99 mg Cl_2/L. Constant dosing of stabilized sodium hypochlorite and measurement, and control of free chlorine concentration ensured its constant level in particular basins.

The concentrations of combined chlorine in the tested samples were very small and, with the exception of water samples collected on 3.08.2017, did not exceed the recommended values (Fig. 1). This was not the case with chloroform content (Fig. 2). In 25 samples tested in 2017, in as many as 22 the chloroform content significantly exceeded 0.03 mg/L (from 0.036 to 0.150 mg/L). In 2018, in 19 out of 30 samples the chloroform content was above the limit value (from 0.032 to 0.110 mg/L). The presence of too high chloroform content was also reflected in the excess THM content (Fig. 3). High levels of DBP may indicate irregularities in the water treatment process, especially in the process of its disinfection. It is necessary to check the water disinfection phases, change or improve the applied technologies, e.g. apply an additional oxidizing agent.

In the analyzed pool waters, special attention was also paid to the content of nitrates. In the tests carried out at the beginning of the 2017 and 2018 season, the nitrate content ranged from 0.89 to 2.40 mg NO_3^-/L. It systematically increased and at the end of the seasons ranged from 10.00 to 19.00 mg NO_3^-/L, with a permissible content of 20.00 mg NO_3^-/L (Fig. 4). High nitrate content in the pool water indicates insufficient exchange of circulating water for fresh make-up water (usually from the water supply network) and the potential for creating DBP.

a)

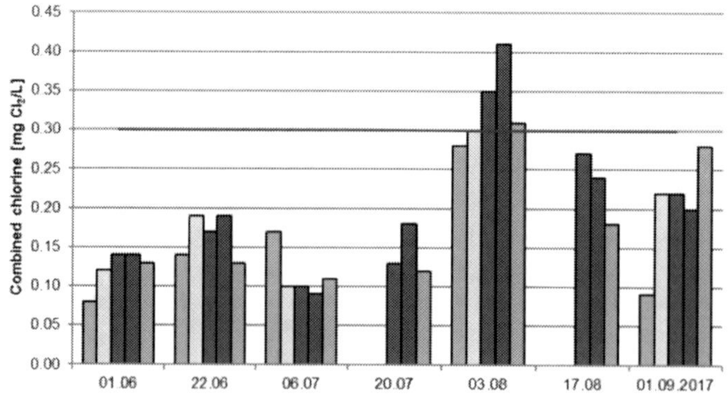

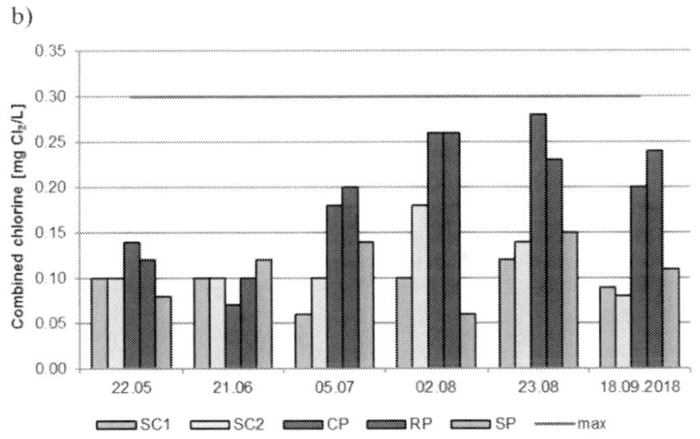

Fig. 1. Combined chlorine in tested swimming pool waters a) in 2017, b) in 2018.

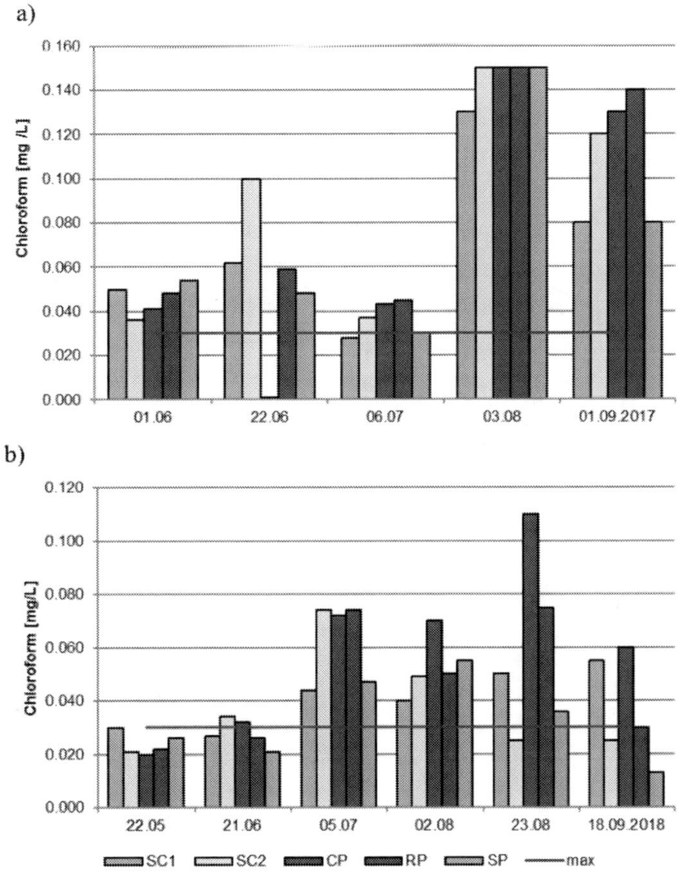

Fig. 2. Chloroform in tested swimming pool waters a) in 2017, b) in 2018.

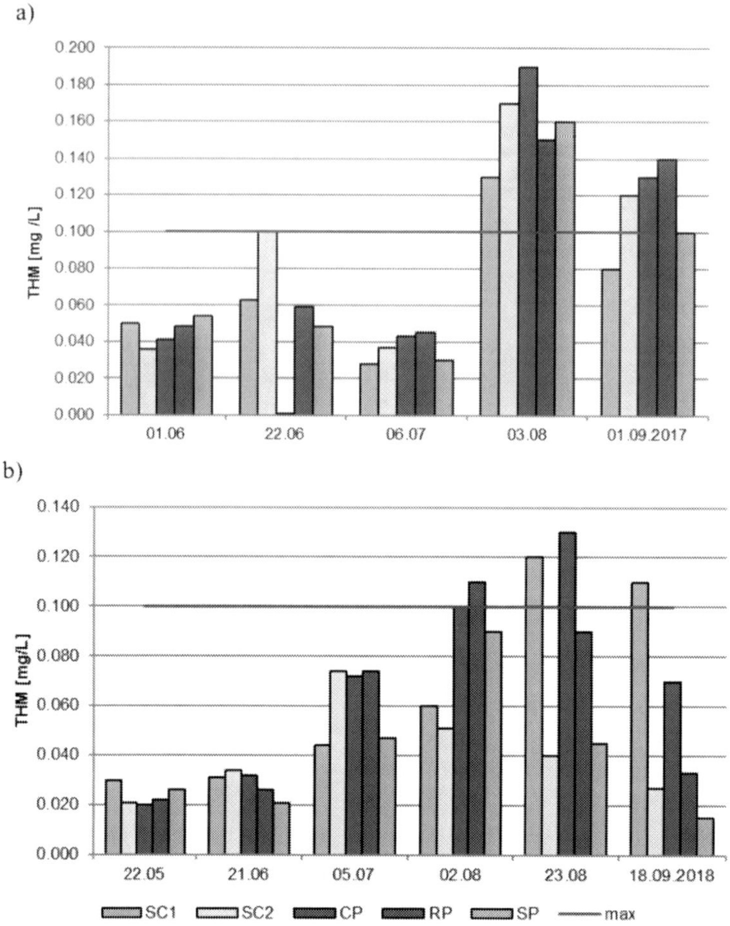

Fig. 3. THM in tested swimming pool waters a) in 2017, b) in 2018.

a)

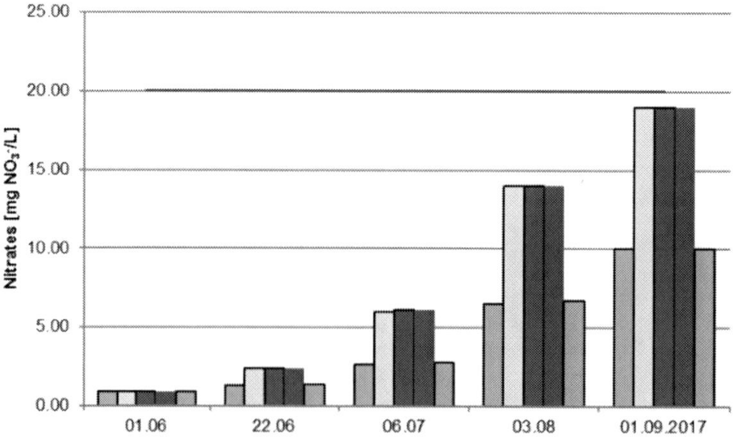

b)

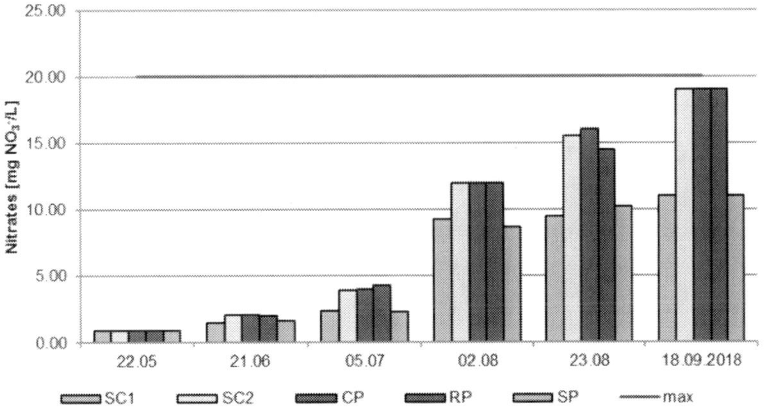

Fig. 4. Nitrates in tested swimming pool waters a) in 2017, b) in 2018.

5 Conclusions

The quality of pool water depends on various factors, such as the concentration of water pollutions and their volatility, the number of bathers and their behaviour, the ventilation system, the water and air temperature and the water circulation system.

The results of physicochemical tests of pool water quality parameters, carried out by sanitary inspection bodies, are treated as an auxiliary in assessing the degree of its contamination.

The decisive role is played by the results of microbiological tests and the assessment of the sanitary condition of the facility.

In the test cycle in 2017 and 2018, in the pool water samples (CP, RP, SP) and in the water samples from the circulation system (SC1, SC2), no CFU of *Escherichia coli*, *Pseudomonas aeruginosa* and *Legionella sp.* were found, and the total number of microorganisms in 36°C after 48 hours did not exceed 100 CFU/1mL in pool water samples and 20 CFU/1mL in the water samples from the circulation system.

Considering that the analyzed pools are seasonal and recreational outdoor swimming pools, the observed excessive values of disinfection by-products (DBP), i.e. chloroform and THM at short exposure, did not pose a direct threat to the health of bathers. However, they indicate the need to supervise the processes of water treatment and disinfection or their modernization. In such cases, the most common recommendation is to use at least two-stage disinfection of circulating water (irradiation with UV/ozonizing/dosing chlorine dioxide+dosing with NaOCl) or using additional filtration through a bed of active carbon.

It needs to be remembered that testing swimming pool water quality for the content of DBP should be connected not only with the control of free chlorine, but also with redox potential and nitrogen content.

This work was supported by the Ministry of Science and Higher Education of the Republic of Poland within statutory funds and within funds for young scientists BKM-508/RIE-4/2018 awarded to the Institute of Water and Wastewater Engineering Silesian University of Technology in 2018.

References

1. T. L. L. Teo, H. M. Coleman, S. J. Khan, Environ. Int. **76**, 16–31 (2015)

2. M. M. A. El-Salam, Environ. Monit. Assess. **184**, 7395–7406 (2012)
3. R. Tardif, M. Rodriguez, C. Catto, G. Charest-Tardif, S. Simard, J. Environ. Sci. **58**, 163–172 (2017)
4. J. De Laat, W. Feng, D. A. Freyfer, F. Dossier-Berne, Water Res. **45**, 1139–1146 (2011)
5. Q. Zheng, L. K. Jmaiff Blackstock, W. Deng, H. Wang, X. C. Le, X. F Li, J. Environ. Sci. **53**, 322–325 (2017)
6. F. Yang, Z. Yang, H. Li, F. Jia, Y. Yang, Environ. Sci. Water Res. Technol. **4**, 218–225 (2018)
7. W. L. Bradford, Int. J. Aquat. Res. Educ. **8**, 168–181 (2014)
8. J. Wyczarska-Kokot, A. Lempart, M. Dudziak, Ecol. Chem. Eng. A. **24**, 217–226 (2017)
9. H. Ilyas, I. Masih, J. van der Hoek, Water, **10**, 797 (2018)
10. J. H. Jacobs, S. Spaan, G. B. G. J. van Rooy, C. Meliefste, V. A. C. Zaat, J. M. Rooyackers, D. Heederik, Eur. Respir. J. **29**, 690–698 (2007)
11. T. Manasfi, B. Coulomb, J. L. Boudenne, Int. J. Hyg. Environ. Health. **220**, 591–603 (2017)
12. C. M. Villanueva, S. Cordier, L. Font-Ribera, L. A. Salas, P. Levallois, Curr. Environ. Health Rpt. **2**, 107–115 (2015)
13. M. Z. Afifi, E. R. Blatchley, Water Res. **68**, 771–783 (2015)
14. A. Gomà, R. de Lluis, J. Roca-Ferrer, J. Lafuente, C. Picado, Environ. Res. **152**, 315–321 (2017)
15. S. Simard, R. Tardif, M. J. Rodriguez, Water Res. **47**, 1763–1772 (2013)
16. Polish Ordinance of the Minister of Health, Official Journal 2015, item. 2016 (2015)
17. DIN 19643, *Water Treatment for Swimming and Bathing Pools* (2012)
18. Guidelines for Safe Recreational Water Environments. Vol. **2**: *Swimming Pools and Similar Environments*, WHO (2006)
19. K. M. S. Hansen, S. Willach, M. G. Antoniou, H. Mosbæk, H. J. Albrechtsen, H. R. Andersen, Water Res. **46**, 6399–6409 (2012)
20. J. Wyczarska-Kokot, Desalin. Water Treat. **134**, 7–14 (2018)

The use of slotted – displacement ceiling diffuser in rooms with stationary workplaces with computer equipment

Agnieszka Zając[1,*]

[1]Wroclaw University of Science and Technology, Faculty of Environmental Engineering, Department of Air Conditioning, Heating, Gas Engineering and Air Protection, ul. Norwida 4/6, 50-373 Wroclaw, Poland

Abstract. This paper presents a specification of premises with a stationary workstations. An analysis of thermal loads occurring in a public utility rooms equipped with a computer, electronic and multimedia equipment was carried out. Attention was drawn to an annual occurrence of a positive heat balances in an occupied workstations and heat losses in winter time in unoccupied premises. For an air distribution a slotted displacement ceiling diffuser was proposed, used for mixing ventilation (MV) in up-up type of air exchange in room. The results of measurements in the form of air flows in the area of its operation are provided. The graphs show the graphical distribution of air velocities and temperatures in the vertical plane passing through the transverse axis of the air diffuser. The study focused on one of the representative airflow of supply air and the behaviour of the air stream during heating and cooling was presented.

1 Introduction

Nowadays, specialised workstations fitted with a computer equipment are often designed. Such workplaces are located in buildings for educational purpose (computer rooms at universities, in schools, in educational centres), in offices (training rooms in banks, telecommunications), in entertainment (gaming rooms, internet cafes) and even production (rooms for regulation and control of technological processes or monitoring and supervision rooms).

Such computer rooms require efficiently operating ventilation or air conditioning installations [1]. The presence of a large number of people and computer equipments causes a significant deterioration of air quality, and in the absence of efficient air-conditioning contributes to the reduction of user concentration and reduced performance [2]. While favourable microclimate conditions positively affect the well-being, improve the reception of information and, as a result, the desired increase in work efficiency [3, 4]. Creation of conditions in which a person would feel a state of thermal comfort is the main goal of installing air conditioning devices [2].

* Corresponding author: agnieszka.zajac@pwr.edu.pl

The most important factors of the room's microclimate are the thermodynamic air parameters and velocity in people's occupancy area [5]. In workplaces, both the temperature and a relative humidity of the air should be shaped. This means that the air should be heating and humidification in a winter time, and a cooling and dehumidification in the summer, in HVAC systems.

The condition of people's well-being in public premises with the use of mechanical ventilation, or air conditioning, is determined by the correct calculation of the ventilating air stream and the proper solution for distributing this air in the room [6]. With an improperly designed flow, a significant part of the supplied air may not take part in the assimilation of unnecessary heat gains. The result of this is the creation in the room of the so called air dead zones or areas with uneven temperature distribution. The current state of air conditioning technology allows to prepare air of any parameters, solve the most complicated installation systems and link them with automatic control devices and control and measurement apparatus. However, it is difficult to obtain a proper and effective distribution of air in the room [1, 7].

In computer rooms, the most reasonable is to bring fresh air directly into the people's area [8–10]. This solution is related to a down-up type of air exchange in rooms (displacement ventilation, DV or stratum ventilation, SV) [11]. In the case of computer rooms, this solution can be difficult to implement. This applies in particular to architectural and construction limitations related to distribution of air installations and the installation of a supply elements. Often, there are several computers installed in a relatively small area, where the height of the rooms does not exceed 3.0–3.5 m. In such situation, the use of a part of the height for installing an additional floor to make the space for running air installations (air ducts) may be unacceptable. In addition, maintaining the cleanliness of floors, so that the air does not entrain dust particles, is often difficult to implement, especially when the air-conditioned room is a room in an academic building [12, 13]. Students/pupils change every 45 or 90 minutes and often enter the room almost straight from the street.

The air flow from the top to the bottom is unfavourable when the supplied air mixes with the conventionally carried out streams of used air (air which assimilated heat and moisture from people and other sources) [14]. However, it is possible to resolve the supply elements and such arrangement of diffusers and exhausters in the room so that convection currents from occupants and electronic equipment are used and the thermal sensations of the users are correct. The description of the ceiling diffuser with a limited and properly directed air stream intended for the organisation of up-up air exchange is contained in this paper [15].

2 Analysis of rooms with stationary workplaces with computer equipment

Thermal loads of sample computer rooms located on the last floor of one of the Wroclaw University of Science and Technology's buildings were analyzed [16]. In each workstation there are 12 computers installed in the area of 30 m^2 (Fig. 1). Thus, there is 2.5 m^2 per one computer station. The distance between the workstations is 1.25 m. Due to thermal loads resulting from the location of external walls, the rooms were divided into two zones: with windows on the south-east (SE) and north-west (NW) side. The share of window surfaces in the total habitable and usable area of the room is 15%. Carrying out a detailed analysis of all annual thermal loads, depending on the internal conditions, is an important factor influencing the selection of the air conditioning system [17]. The heat balance values in these rooms change during their use and depend on both internal conditions (number of people and equipment and lighting included) as well as external (seasons and day, exterior wall positioning, window type and area, etc.). In the case of premises with an external walls one should expect both – heat gains from solar radiation in a summer and heat losses in a winter time. The architectural and construction solutions of the entire building as well as individual rooms have a significant impact on the heat gains [18]. Of particular importance are the thermal accumulation of the building and the types of transparent divisions used along with their exposure [19]. These factors cause fluctuations in heat gains in individual rooms.

Fig. 1. Localization of computer workstations, supply diffusers and exhaust grills in one of many rooms in Wrocław University of Science and Technology building a) projection of the room, b) vertical section.

The graphics below show the course of heat gains, during the use of the room, from individual sources in a summer time (Fig. 2) and the course of heat gains and losses in a winter time (Fig. 3). When using all computers the heat gains definitely exceed losses and must be assimilated by air conditioning. However, in the period of breaks in the use of the rooms and the lack of heat gains, the occurring heat losses must be compensated. The analysis of thermal loads shows that with the maximum heat gain in the conditions of fixed operation, the basic sources of heat are:
- computers that represent 60% of the total heat gain,
- users, about 20% of the total heat gain,
- solar radiation through transparent partitions about 10%,
- other, for example, electric lighting and non-transparent partitions about 10%.

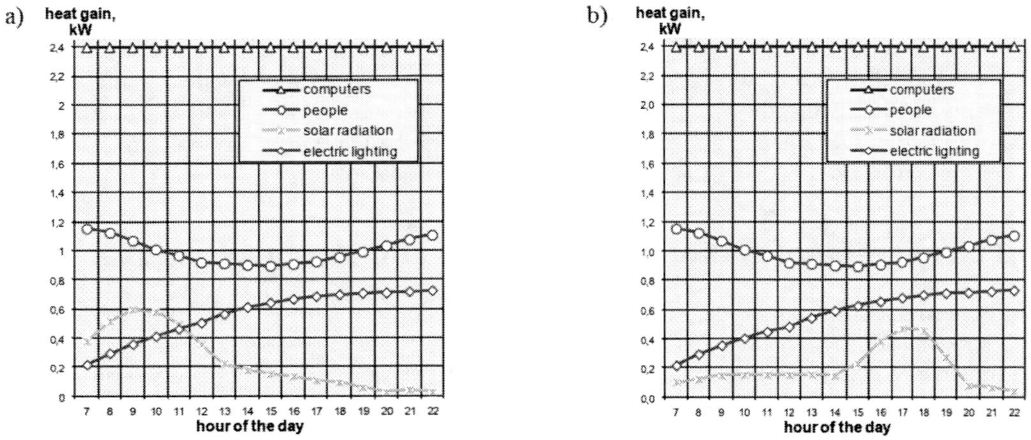

Fig. 2. The course of heat gains from individual sources in a summer time (June) in rooms with windows on the: a) south-east (SE) side and b) north-west (NW) side.

The quoted heat shares are similar for computer rooms with a similar area per one workplace. It should be noted that the gains of heat from electronic devices and people are changing to a small extent. However, heat gains from other sources are variable. The maximum heat gains in each room are 4.5 kW. Their occurrence time is similar in premises of a given zone, while for premises of a different zones the time of their occurrence is different. In winter time, in unused rooms, at calculated outdoor air temperatures of the winter period (at -18°C), heat balances may be negative and amount to -0.75 kW. Then, the heat losses occurring must be covered by air systems or central heating. When using the computer room (at the same time and under the same conditions) heat gains definitely exceed the losses, and the heat balance comes to 50–70% of the value from the summer period. It should be noted that the heat gain from solar radiation will only be included on sunny days. On cloudy days, these gains will not affect the heat balance.

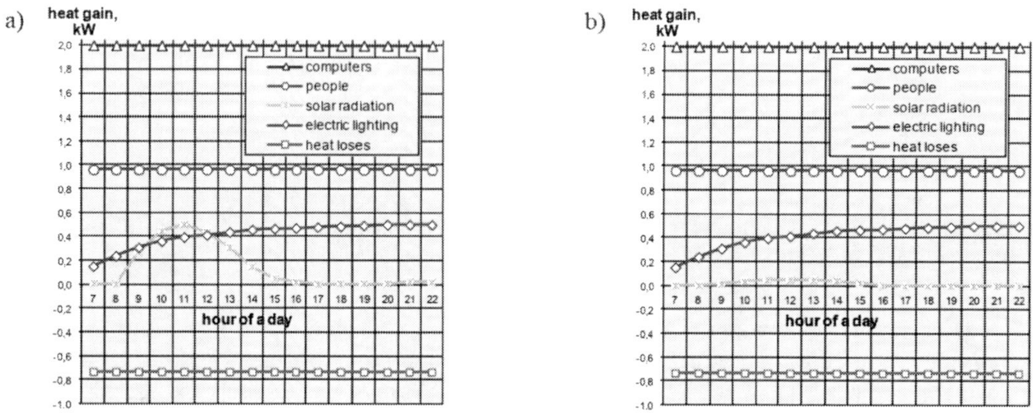

Fig. 3. The course of heat gains from individual sources and heat losses in winter (December) in rooms with windows on the: a) south-east (SE) side and b) north-west (NW) side.

For the analysed computer room, the volume of the air supplied from the ceiling side should be between 1650 m^3/h to 2200 m^3/h and from the floor side not less than 4500 m^3/h. This is due to the accepted temperature increases (difference between exhaust and supply air temperature) in the occupied zone (people's area), in order to meet the thermal comfort conditions and do not feel drafts. In each case it provides huge air change rates. The size of flow rates directly affects the dimensions of the distribution ducts and the ventilation or air handling unit (AHU). If the additional temperature gradient were taken into account in the calculations, in the case of high rooms, which would be created above the heat sources, the needed air flow rate could be reduced to 2700 m^3/h for each room by the ceiling side air supply. And using the creation of convection currents from people and electronic equipment in the organization of up-up type of air exchange with a slotted – displacement ceiling diffuser, the air flow rate could be reduced to 1350 m^3/h.

3 Measurement station with slotted – displacement diffuser

The design of the prototype slotted – displacement ceiling diffuser was discussed, among others in [16]. Fig. 4 includes a construction graphic drawing and a view of the diffuser, from which air flows through the centre of perforated surface and two slots, pre-limiting and directing supply air. The air flow rate supplying from the central part of the diffuser is characterized by high inductance and marked turbulence. On the outside, it is limited by air slots that prevent from spreading to the users' heads. By reducing the intense mixing, the air stream is brought above the floor, without spreading from side to side, and thus only a small air suction takes place. It begins to stay on the floor, and when meeting heat sources, assimilates their excess and due to convective streams, rises up, straight to the exhaust installation. Such effect of this diffuser has been validated by research in real conditions on a full-scale measurement room in which the real processes were simulated. The advantages of displacement ventilation and down – up type of air exchange by supplying air to the room zone with the smallest unit heat gains and its exhaust from heat sources are used.

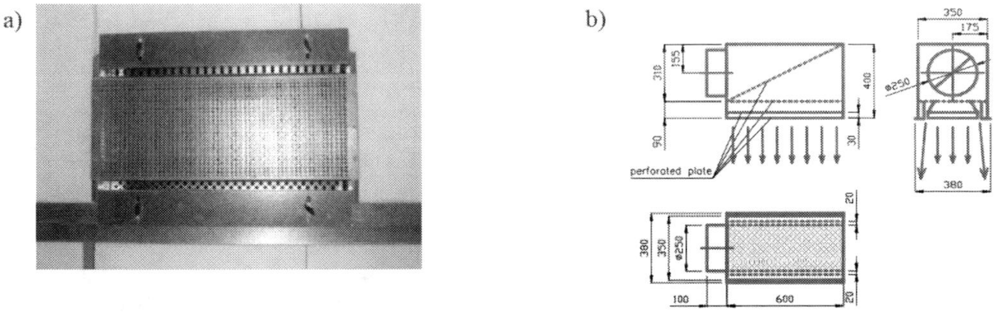

Fig. 4. a) A view, b) construction graphic drawing of the slotted - displacement ceiling diffuser with 20 mm wide slot.

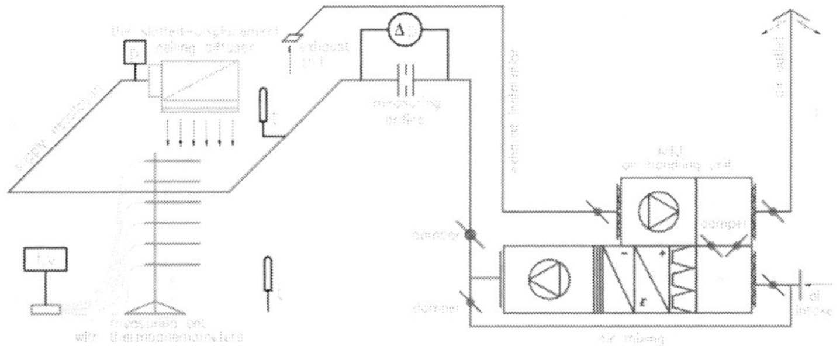

Fig. 5. Graphic diagram of the measurement station with the slotted - displacement ceiling diffuser.

The schematic graphic diagram of the measurement station is shown above in Fig. 5. The ventilation air installation of the researched workstation consists of a supply and exhaust part. The external (outdoor) air or the mixture of external and recycling air (from the room), is prepared in the AHU and transported to the computer room and supplied by a single slotted – displacement diffuser. Depending on the needs, the air will be heated in the electric heater, cooled in a glycol cooler or without heat treatment, introduced into the supply air duct. The air from the room is removed through the ceiling, which is contacted by grids located in the false ceiling.

4 Air velocity and temperature measurement results

The study used experimental measurements in a full-scale test room with dimensions of 4 m width x 3 m length and 2.85 m height and a HVAC control system. The air velocity and temperature measurements were carried out at different distances from air supply diffuser on measurement station. Sixteen measurement levels were determined at sixteen heights from the floor, as presented in Fig. 9 a) (one measurement session consisted of eight heights – left, while measurements in the second session were carried out after moving the tripod core, with fixed probes, by 0.15 m down – right). The measurements were carried out mainly in the longitudinal and transversal axes of the diffuser containing 29 vertical measuring lines. Measuring grid points have been spaced at intervals of 0.05 or 0.1 m. During the tests, the air flow rates (from 325 to 625 m^3/h every 75 m^3/h), the width of the slot limiting the stream (from 0 to 45 mm every 5 mm), and the degrees of perforation of the outer plate and the internal plates levelling the flow (from 20 to 50% every 10%) were changed. With larger airflow than 625 m^3/h in the area close to the floor around the people (occupant zone), the velocity exceeded the value of 0.5 m/s. The share of air flowing through the slots in relation to the total volume of air flow varied from 30% to 60%. The difference in the result of the subtraction between supply air and room temperature did not exceed 1 K. Such conditions are in fact rare. At very low outside air temperatures, the ventilation/air conditioning unit can also fulfils fully or partially the duty of space heating. Thus, the supply air must have a higher temperature [20]. However, for most of the year it is necessary to assimilate the heat generated by the electronic equipment and users. The supply air temperature must therefore be lower than in the room [21].

The obtained measurement results in the form of a graphic air velocity and temperature distribution (in vertical planes passing through the transverse axis of the diffuser) for air flow rate of 550 m^3/h are depicted on the diagrams (Figs. 7 and 8). The velocity distribution for the transverse axis was chosen because it represents the air flow towards the heads, neck and back of the users/occupants. Extending the stream in the longitudinal direction is not so important. It can only cause an increase in velocity in the transition zone when the diffusers are placed too close to each other. The optimal distance of diffusers, resulting from the width of two computer stations is 2.5 m which guarantees maintaining the velocity below 0.3 m/s in people's area.

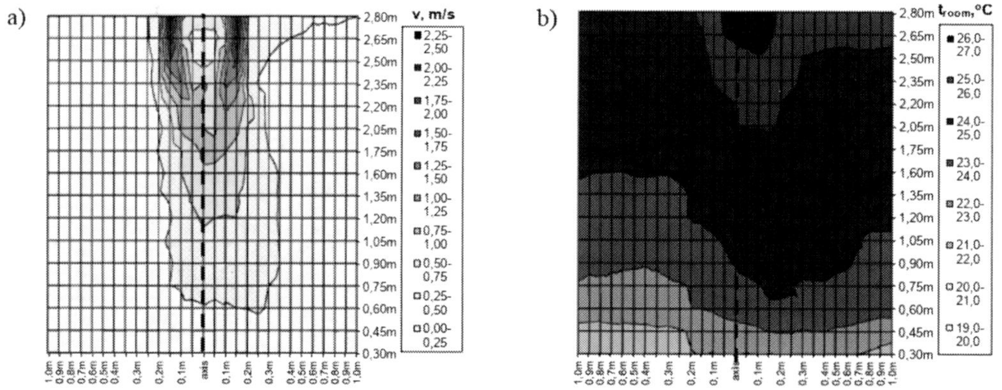

Fig. 7. Distribution of air a) velocity b) temperature in the area of operation of the diffuser during heating of the room (supply air flow rate 550 m^3/h, supply air temperature 27°C).

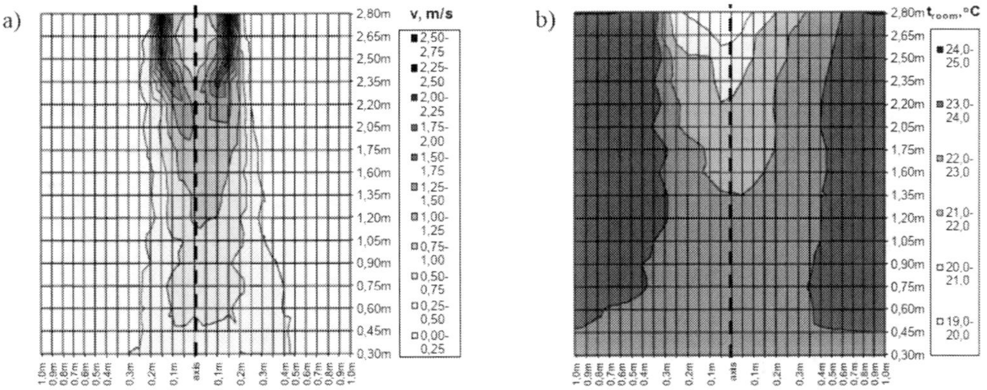

Fig. 8. Distribution of air a) velocity b) temperature in the area of operation of the diffuser during cooling of the room (supply air flow rate 550 m³/h, supply air temperature 19°C).

The measurements were conducted with attached simulation systems for heat gains from people and computers. Heat sources were placed near the outer walls, away from the diffuser, as it seen in Fig. 9b. The diagrams do not show a cross-section through the entire room. The focus was only on the stream of air supplied from the diffuser. The air velocity distribution and temperature over the heat sources and local convective flows arising there are not shown because it did not come within the scope of research. The range of measurements performed is schematically shown in Fig. 9b.

The average velocity of supply air from the centre perforated part was in range of 0.45 to 0.85 m/s, while from the limiting gaps it was from 1.65 to 3.25 m/s, in the range of tested air flow rates 325–625 m³/h and with a constant gap width of 20 mm. A velocity of 0.4–0.8 m/s was noted under the diffuser in the people's area. In addition to the area of operation of the diffuser, i.e. where people are continuously present, the velocity did not exceed 0.3 m/s. The range of increased velocities can be observed in the transition between computer stations, where a person is temporarily staying and moving. A shorter, higher air flow should be perceived by users as refreshing [22]. High velocity zone, so-called zone of influence of the diffuser, was from 0.3 to 0.6 m.

The slotted – displacement diffuser will be useful in rooms with clearly marked passageways, where there will not be a stationary workstation directly under the diffuser.

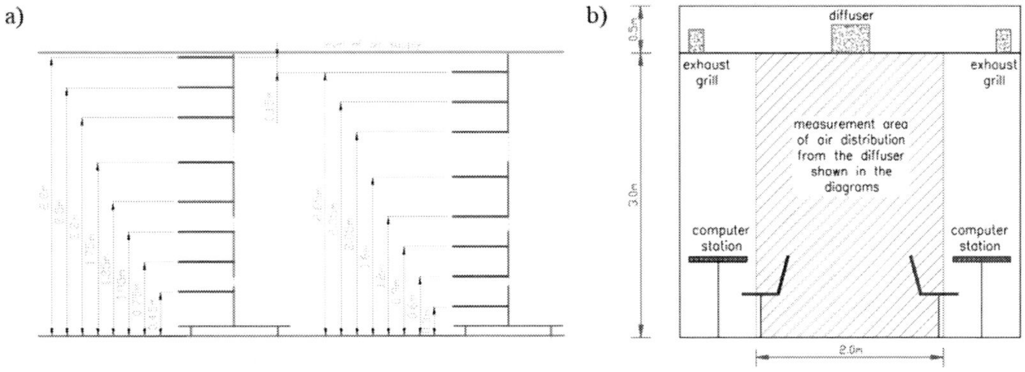

Fig. 9. a) Location of the measurement probes on the stand, b) Measuring range of the stream flowing out of the diffuser into the room (the area of operation of the diffuser).

5 Conclusion

In currently constructed buildings, one may see the premises saturated with an electronic, computer and multimedia equipment, which means that heat gains occur throughout the year. In addition, energy efficient technology for construction of new buildings and good insulation of the external partitions (walls, roof) significantly limit the heat loss or heat gain by heat transfer. Heat deficiencies can occur only in a winter time calculation conditions, and in rooms where internal heat sources are turned off (computers, electric lighting) and where there are no users. Based on the above in 24/7 use of computer rooms, especially without external partitions, there is only a small change to the overall heat balance throughout a year.

In the case of premises with an external walls, insufficiently insulated, both – heat gains from solar radiation in summer and heat losses in winter vary substantially. It should be noted that the heat gains from electronic devices and people are changing in a small extent, they are almost stable. However, heat gains from other sources are variable and depend on external conditions.

It is proposed to use the air exchange system with the slotted – displacement ceiling diffusers of the original construction as an alternative solution to the most recommended systems with supply from a floor level. This paper presents selected results of measurements of the diffuser in non-isothermal conditions. The temperature difference between supplied and room air in both (heating and cooling) cases was about 5 K. The graphical distribution of velocity and air temperature shows the steady and limited air distribution. Although the air flows from the central part of the diffuser are characterised by high inductance and marked turbulence, they are limited by supply air slots preventing from spreading to the users' heads. The advantages of displacement ventilation and the organization of down-up type air exchange can be used due to supplying of air into the room zone with the smallest unit heat gains and its exhausting over concentrated heat sources. Such air movement gives good results in shaping the microclimate of premises with variable and high thermal loads. The measuring diffuser allows for the assimilation of heat similar to that for overhead mixing ventilation (MV) while meeting all the advantages of a positive displacement ventilation (DV) in the area of workplaces and continuous residence of people.

The work was supported by The Faculty of Environmental Engineering, Wroclaw University of Science and Technology, Poland. No. 0401/0055/18.

References

1. G. Cao, H. Awbi, R. Yao, Y. Fan, K. Sirén, R. Kosonen, J. Zhang, Build Environ. **73**, 16 (2014)
1. X. Shan, J. Zhou, W. C. Chang, E. H. Yang, Build Environ. **102**, 10 (2016)
2. M. Puteh, M. H. Ibrahim, M. Adnan, C. N. Cheahmad, N. M. Noh, Procedia Soc. Behav. Sci. **46**, 5 (2012)
3. Y. Wang, F. Y. Zhao, J. Kuckelkorn, D. Liu, L. Q. Liu, X. C. Pan, Energy **64**, 11 (2014)
4. ASHRAE Handbook, *HVAC Application* (2011)
5. M. H. Fathollahzadeh, G. Heidarinejad, H. Pasdarshahri, Build Environ. **90**, 9 (2015)
6. H. B. Awbi, Front Mech. Eng. **1**, 4 (2015)

7. F. Bauman, A. Daly, ASHRAE (2003)
8. R. Becker, M. Paciuk, Build Environ. **44**, 13 (2009)
9. M. A. Bos, J. A. Love, Build Environ. **69**, 8 (2013)
10. R. Li, S. C. Sekhar, A. K. Melikov, Build Environ. **45**, 8 (2010)
11. F. Causone, F. Baldin, B. W. Olesen, S. P. Corgnati, Energy Build. **42**, (2010)
12. A. Lipczynska, J. Kaczmarczyk, A. K. Melikov, Build Environ. **92**, 12 (2015)
13. H. Ho, L. Rosario, M. M. Rahman, Build Environ. **46**, 13 (2011)
14. A. Zając, E3S Web of Conferences **44**, 8 (2018)
15. A. Zając, *Doctoral thesis*, PRE **6**, 144 (2008)
16. W. P. Jones, *Air conditioning engineering* (2001)
17. H. H. Sait, Energ. Convers. Manage. **66**, 10 (2013)
18. H. W. Li Danny, L. Yang, J. C. Lam, Energy **54**, 10 (2013)
19. H. Amai, A. Novoselac, Build Environ. **109**, 11 (2016)
20. B. W. Olesen, A. Simone, M. Krajcík, F. Causone, M. DeCarli, Int. J. Vent. **9**, 13 (2011)
21. N. M. Mateus, G. Carrilho da Graca, Build Environ. **114**, 13 (2017)
22. R. Tomasi, M. Krajcik, A. Simone, B. W. Olesen, Energy Build. **60**, 10 (2013)

Concept of reducing harmful emissions by road transport vehicles in the tourist route Karpacz–Jelenia Góra

Pawel Zajac[1,*], *Szymon* Haladyn[2], and *Stanislaw* Kwasniowski[1]

[1]Wrocław University of Technology, Faculty of Mechanical Engineering, Wybrzeże Wyspiańskiego 27, 50-370 Wroclaw, Poland
[2]Wrocław University of Technology, "Logistics" Students' Scientific Circle functioning at the Faculty of Mechanical Engineering, Wybrzeże Wyspiańskiego 27, 50-370 Wrocław, Poland

Abstract. The research carried out by the "Logistics" Students' Scientific Circle functioning at the Faculty of Mechanical Engineering of the Wrocław University of Technology shows that every day, between the communes which cross railway lines 308 and 340, 1911 persons move in circular migrations solely due to occupational activity. Some of them could travel by rail. Eliminating 500 cars a day (i.e. 250 outwards and return travels on the route Kowary/Karpacz–Jelenia Góra) would reduce annual carbon dioxide emissions by approximately 460 tonnes. Further reductions in CO_2 emissions in transport can be achieved thanks to the use of railways by tourists, of which - in the area of the communes of Karpacz, Kowary and Mysłakowice, i.e. an area of outstanding recreational valors - there was over 300,000 in 2016 [3]. Rational train running interval was determined and a timetable proposed, taking into account routes leading beyond Jelenia Góra. The paper justifies the need for the introduction of complementary communication provided by electrobuses. The concept of reactivating traffic in this area should be complemented by the construction of a *Park&Ride* car park. The possibility of further development of the transport system in the area in question was also proposed.

1 Introduction

Hydrocarbons are oxidised in the engine of the vehicle. The main products of the reaction are: carbon dioxide (approx. 86%) and water (approx. 13%). The quantity of carbon dioxide (molar mass $M_{CO_2} = 44.01 \left[\frac{g}{mol}\right]$, molar mass of carbon: $M_C = 12 \left[\frac{g}{mol}\right]$) produced per kilogram of conventional fuel is:

$$m = \frac{44{,}01 \cdot 0{,}86}{12} = 3{,}15\ [kg] \tag{1}$$

Assuming an average consumption of 6 litres of conventional fuel per 100 km driven by a passenger car (approx. $4.3\ kg$), the aforementioned vehicles will emit 4.2 million kg of

* Corresponding author: pawel.zajac@pwr.edu.pl

carbon dioxide on the route in question. This indicator clearly shows the advisability of taking pro-ecological actions and supporting them by the European Union and its members in the sustainable development policy, and is also desirable in the era of electromobility. The idea of electromobility of passenger transport in this region is also supported by the situation in which Zakopane has found itself, namely: the excessive smog that contradicts the idea of tourism development and proper recreation. The Jelenia Góra Valley should do everything that is possible to avoid a similar situation.

In the area of the analyzed lines no. 308 and 340 tourist traffic develops intensively, hotel and leisure facilities are being built. Karpacz – like Zakopane – is facing inefficiency of the road system, and the peripheral location (on the border of almost three countries: Poland–Germany–Czech Republic) makes it a location excluded in terms of transport. Restoring rail passenger traffic would make it possible to change this negative state and would attract Polish and German tourists to the Karpacz region, both for one-day or weekend stays and longer trips.

Karpacz's transport system is overloaded during holiday periods; cars parked on narrow streets make it difficult to move around (snow removal in winter), preventing tourists from enjoying the charm of the resort. Further development of Karpacz as a holiday resort is hindered by the shortage of parking spaces. Constructing a P+R car park in the immediate vicinity of the railway stop on the initial section of the Jelenia Góra - Karpacz line will allow to reduce the number of cars entering the town, to reach the Jelenia Góra Valley by private means of transport, and then leave it and continue the travel by train.

Moreover, it is worthwhile to launch the new Mysłakowice Dolne stop at kilometre 7+360. This is desirable in view of the urban development of the areas along the railway line and, consequently, the need to cover the new urbanized areas with railway transport.

Ultimately, a possibility exists of restoring rail traffic to Kowary and, in order to improve the transport coverage of the region in question, it is proposed to consider the construction of a new four-kilometre railway line from Kowary to Karpacz.

In the era of electromobility and implementation of the sustainable development policy – especially in areas of high recreational and touristic value – it would be advisable to base passenger traffic on environmentally friendly means of transport. For this reason, the project of revitalisation combined with electrification of the line and construction of the *Park&Ride* car park seems to be justified [1].

2 Theoretical travel

The basis for the development of communication solutions, especially the timetable, are traction calculations based on the train motion equation.

These calculations, so-called theoretical or minimal time travel, allow to determine the trend of changes in vehicle speed versus its traction properties and external conditions. On its basis it is possible to determine the technical travel times of individual sections, which are necessary to propose the timetable; also, energy consumption may be determined and the possibilities of its recovery in the case of electrodynamic braking estimated. This calculation is based on the train **Błąd! Nie można odnaleźć źródła odwołania.**

$$m \cdot k \cdot \frac{dv}{dt} = p(v, S) \cdot Q$$ **(Błąd! W dokumencie nie ma tekstu o podanym stylu.)**

where: k – rotational mass coefficient for a laden train being $k = 1{,}06\ [-]$, m – train mass $[kg]$ $(m = 172000\ kg)$, Q – train weight $[kN]$, p – unit acceleration force $[N/kN]$, comprising of the combination of the following unit forces: tractive effort $f(v)$, braking effort $b(v)$, train resistance effort $w(v)$ and resistance resulting from railway reductions. $i_m(S)$.

2.1 Tractive effort

Tractive effort F t is the tractive force of the traction unit 31WE. In the traction calculations the unit tractive effort **Błąd! Nie można odnaleźć źródła odwołania.** is used, expressing the tractive effort versus unit weight, expressed as $[kN]$ [3].

$$f = \frac{F}{Q} \left[\frac{N}{kN}\right] \tag{3}$$

The calculations assume that braking effort is defined by formula (4).

$$B = \mu_i \cdot m \cdot N_i \; [kN] \tag{4}$$

where: μ_i – friction coefficient between the brake lining and the brake disc, m – number of brake linings in the vehicle $[-]$, dla 31WE $m = 40\;[-]$, N_i – pressure of shoe lining on the brake discs. The friction coefficient relationship μ_i versus speed for a disc brake [4] was adopted after Sawczuk and Jüngst. The shoe lining pressure on the brake discs is determined from the formula **Błąd! Nie można odnaleźć źródła odwołania.**.

$$N_i = N_{nom} \cdot \xi_i \; [kN] \tag{5}$$

where: N_{nom} – nominal pressure on the lining, $N_{nom} = 17{,}16\;[kN]$, ξ_i – pressure drop coefficient at a given braking degree.

The pressure drop coefficient at a given braking degree is calculated from formula **Błąd! Nie można odnaleźć źródła odwołania.**.

$$\xi_i = \frac{\Delta p_i}{\Delta p_{maks}} \tag{6}$$

where: Δp_{maks} – pressure drop in the service line during emergency braking, $\Delta p_{maks} = 0.5\;[MPa]$, Δp_i – pressure drop in the service line for individual braking degrees; in the case of the theoretical travel performed, full (2nd degree) braking was applied – $\Delta p_2 = 0.22\;[MPa]$.

In order to be able to take into account the braking weight, the concept of unit braking effort defined by formula **Błąd! Nie można odnaleźć źródła odwołania.** is introduced.

$$b = \frac{B \cdot \delta}{Q} \; [N/kN] \tag{7}$$

where: δ – percentage of braking weight $[-]$, Q – train weight $[kN]$. The braking weight percentage value is determined using the formula **Błąd! Nie można odnaleźć źródła odwołania.**.

$$\delta = \frac{\varepsilon}{Q} \; [-] \tag{8}$$

where: ε – braking weight $[kN]$ expressed as: $\varepsilon = m_h \cdot Q$, the braking mass of 31WE multiple unit is $m_h = 281\;[t]$.

2.2 Basic resistance effort

Calculations of the multiple unit resistance effort were based on the guidelines developed at the Railway Institute in Warsaw (formerly CNTK) given in [5].

$$W = 0{,}67 \cdot v^2 + 25{,}8 \cdot v + 2618\ [kN] \quad (9)$$

2.3 Additional resistance effort (from profile)

In order to carry out the calculations, the profile of the analyzed railway line was reduced by reducing the elevation and gradients, as well as curves of tracks. The so-called profile resistance [6] is then determined. For reduction of resistance effort on curves, the relation **Błąd! Nie można odnaleźć źródła odwołania.** was assumed [3].

$$i_ł = \frac{\sum l_ł \cdot \frac{690}{r_ł}}{\Delta l}\ [‰]\ lub\ \left[\frac{N}{kN}\right] \quad (10)$$

where: $l_ł$ – curve length at a given reduction step $[m]$, $r_ł$ – curve radius $[m]$, Δl – reduction step value$[m]$. The reduction of a railway line consists in summing the values of all local resistance components occurring at a given reduction step **Błąd! Nie można odnaleźć źródła odwołania.**. Fig. 4 shows the profile characteristics of the analyzed line.

$$i_m = i_w + i_ł\ [‰] \quad (11)$$

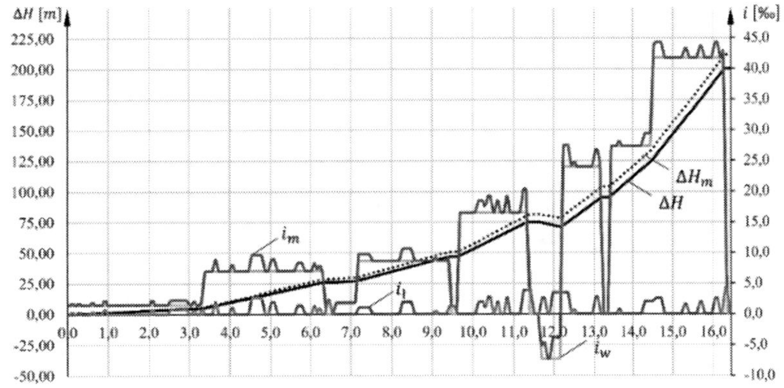

Fig. 1. Reduction of railway line Jelenia Góra–Karpacz.

2.4 Calculation of theoretical travel

The calculations assume that the train moves at maximum speed $v = 60\ [km/h]$. A fragment of the calculation results is shown in Fig. 5. Calculations were carried out for the whole route Jelenia Góra–Karpacz, outwards and return. The calculations showed that the time needed for the multiple unit to complete the route to Karpacz (without stoppages, but taking into account stops and acceleration) is 18.77 $[min]$, return travel: 18.80 $[min]$.

 https://doi.org/10.1051/e3sconf/201910000091

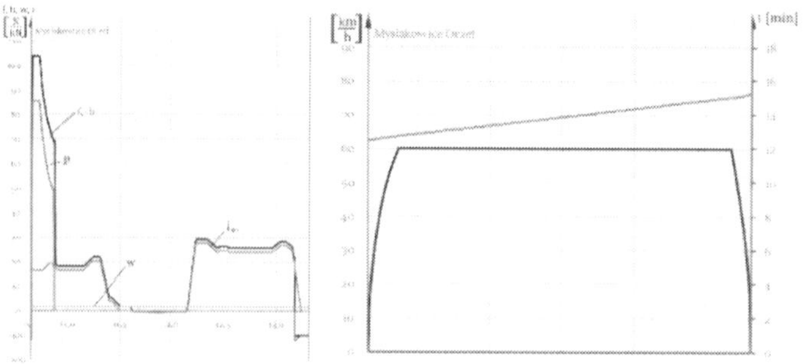

Fig. 2. Portion of theoretical ride 31WE emu from Jelenia Góra to Łomnica Średnia.

3 Proposed transport offer

3.1 Scheduled travel time

The realization of theoretical travel leads to the determination of the technical travel time of the train, which determines the timetable to the greatest extent. It is - according to the Instruction [7] – a component A of the train's travel time. The operating reserve is component B (for a passenger train it is 5 $[min]$ for each 100 $[km]$ of travel). Component D is the sum of stopping times at stops and intermediate stations, but the minimum commercial standstill time for an electric multiple unit is 0^3 $[min]$. The total scheduled train travel time between Jelenia Góra - Karpacz and Karpacz - Jelenia Góra is presented in Table 1.

Table 1. Transfer time for route Jelenia Góra–Karpacz.

Portion of line		Transfer to Karpacza				Transfer to J. Góry			
		A $[min]$	B $[min]$	A+B $[min]$	D $[min]$	A $[min]$	B $[min]$	A+B $[min]$	D $[min]$
Jelenia Góra	Łomnica Średnia	3^7	0^1	3^8	0^3	3^7	0^2	3^9	-
Łomnica Średnia	Łomnica Dolna	1^7	0^1	1^8	0^3	1^8	0^1	1^9	0^3
Łomnica Dolna	Łomnica	2^0	0^1	2^1	0^3	2^0	0^1	2^1	0^3
Łomnica	Mysłakowice Dolne	1^6	0^1	1^7	0^3	1^6	0^1	1^7	0^3
Mysłakowice Dolne	Mysłakowice	2^3	0^1	2^4	0^5	2^3	0^1	2^4	0^3
Mysłakowice	Mysłakowice Orzeł	1^4	0^1	1^5	0^3	1^4	0^1	1^5	0^5
Mysłakowice Orzeł	Miłków	3^0	0^1	3^1	0^3	3^0	0^1	3^1	0^3
Miłków	Karpacz	3^5	0^2	3^7	-	3^4	0^1	3^5	0^3
	$\sum$	19^2	0^9	20^1	2^3	19^2	0^9	20^1	2^3
	$\sum$	-	-	22^4		-	-	22^4	

Source:own elaboration

The analysis showed that the scheduled, minimum time of train travel between Jelenia Góra and Karpacz is $t_T = 22.4$ $[min]$ and is consistent with the time needed to complete the return ride.

3.2 Route capacity

The minimum time required to complete the outward and return journey is equal to the value given by the formula **Błąd! Nie można odnaleźć źródła odwołania.**.

$$t_m = \sum t_{j_i} + \sum t_{j_j} + \sum_{k=1}^{n} t_{p_{1k}} + \sum_{k=1}^{n} t_{p_{2k}} + \sum_{l=1}^{n} t_{z_l} + t_r \; [min] \qquad (12)$$

where: n – number of intermediate commercial travel time between stopovers on the way outwards $[min]$, t_{j_j} – travel time between stopovers on the return way $[min]$, $t_{p_{1k}}$ – stopover time between stops on the way outwards $[min]$, $t_{p_{2k}}$ – stopover time between stops on the return way $[min]$, t_z – locomotive changeover time $[min]$, t_r – time reserve confirming to the guidelines contained in [7] $[min]$.

In the considered case of transport route from Jelenia Góra to Karpacz, the minimum time needed for the outwards and return travel, after filling the determined data into formula **Błąd! Nie można odnaleźć źródła odwołania.** and taking into account the six-minute time needed to complete the locomotive changeover is $t_m = 56.8 \; [min]$. This means that it is possible to run trains cyclically in sixty minutes.

In the event of traffic revival on the analyzed route, a half-hour cycle could be introduced – this solution would require recreation of the passing siding on the route in Mysłakowice, which would enable the passing of trains on the single-track line.

3.3 Timetable assumptions

The basic route is assumed to be a Jelenia Góra–Karpacz–Jelenia Góra shuttle. Moreover, during periods of increased tourist traffic (winter/summer holidays, public holidays, long weekends) it is planned to launch trains on a route running beyond Karpacz – namely to Wrocław. At a later stage, consideration should be given to launching connections also to Węgliniec.

It is reasonable for trains to run from around 5.00 am to around 10.30 pm. These times – being turning points in a way - are determined by the working hours of business. An important factor is also the need to ensure connections with Jelenia Góra for young students.

From the point of view of tourist traffic, it is particularly desirable to provide a possibility of reaching the place of rest in the late morning (about 10:00 a.m.) and returning from it at about 6:00 p.m. Establishing the above time frames would allow the residents of Wrocław situated about 140 $[km]$ apart for comfortable arranging of one-day and weekend trips.

3.4 Proposed timetable

It is proposed to start a total of 18 pairs of shuttle trains on the route of Jelenia Góra–Karpacz –Jelenia Góra per working day, which translates into train operation of 592.56 $[wzkm]$. It is planned to implement travels at hourly rate; the first connection would be made from Jelenia Góra station before 5.00 a.m., while the last arrival at the destination station is expected after 11.00 p.m.

For public holidays it is assumed that 12 pairs of trains on the route Jelenia Góra– Karpacz –Jelenia Góra and one pair of connections on the route Wrocław Główny–Karpacz–Wrocław Główny will be launched. Train operation will be reduced compared to working day to a level of 427.96 $[wzkm]$.

The projected timetable is presented in the form of Table 246 of the Network Train Schedule (Table 2).

4 Buffer parking arrangement

The road system of the eastern resorts of the Giant Mountains suffers – especially during the tourist season - from significant overload. Difficulties are caused particularly by parked cars. As a result, it is justified to construct a *Park&Ride* car park located in the immediate vicinity of one of the initial access points to the Carpathian line. It is proposed to provide parking spaces for 500 vehicles in the first stage. According to the standard proposed by Wł. Czarnecki in [8], such a car park would occupy an area of about 1.2 ha depending on the layout of parking spaces and internal roads used. The analysis of the spatial development status showed that an adequate area is located in the immediate vicinity of the Łomnica Średnia passenger stop.

Historical and terrain conditions have determined the course of railway lines 308 and 340, making it impossible for the railway to reach all areas which are significant population centres, so that sometimes a passenger stop or station is located at a disadvantageous distance from traffic generators, and the access route - from the point of view of potential railway customers – is excessively long.

In order to increase the accessibility of the railway connection between Jelenia Góra and Karpacz, in the two cases described above it is proposed to launch a complementary bus service (connected with the trains) on the routes: Karpacz (railway station)–Bierutowice– the outwards and return travel time is 32 minutes, and Mysłakowice (railway station)–Kostrzyca–Kowary – the outwards and return travel time is 26 minutes; hence it is possible to extend some of the routes towards the power tunnel.

Table 2. Projected timetable.

246 JELENIA GÓRA - MYSŁAKOWICE - KARPACZ

Wrocław Główny o.							§ X												
Jelenia Góra p.							§ X												
termin kursowania	⊛	⊛		⊛		⊛	Ⓒ										⊛		⊛
przewoźnik - kategoria	KD - Os	KD - Os	KD - Os	KD - Os	KD - Os	KD - Os	KD - Os	KD - Os	KD - Os	KD - Os	KD - Os	KD - Os	KD - Os	KD - Os	KD - Os	KD - Os	KD - Os	KD - Os	KD - Os
informacje dodatkowe																			
numer pociągu	60500/1	60502/3	60504/5	60506/7	60508/9	60510/1	60512/3	60514/5	60516/7	60518/9	60520/1	60522/3	60524/5	60526/7	60528/9	60530/1	60532/3	60534/5	60536/7
JELENIA GÓRA o.	§ 04:42	§ 05:42	06:42	§ 07:42	08:42	§ 09:42	§ 09:42	10:42	11:42	12:42	13:42	14:42	15:42	16:42	17:42	18:42	§ 19:42	20:42	§ 22:17
Łomnica Średnia |	§ 04:46	§ 05:46	06:46	§ 07:46	08:46	§ 09:46	§ 09:46	10:46	11:46	12:46	13:46	14:46	15:46	16:46	17:46	18:46	§ 19:46	20:46	§ 22:21
Łomnica Dolna |	§ 04:48	§ 05:48	06:48	§ 07:48	08:48	§ 09:48	§ 09:48	10:48	11:48	12:48	13:48	14:48	15:48	16:48	17:48	18:48	§ 19:48	20:48	§ 22:23
Łomnica |	§ 04:50	§ 05:50	06:50	§ 07:50	08:50	§ 09:50	§ 09:50	10:50	11:50	12:50	13:50	14:50	15:50	16:50	17:50	18:50	§ 19:50	20:50	§ 22:25
Mysłakowice Dolne ↓	§ 04:52	§ 05:52	06:52	§ 07:52	08:52	§ 09:52	§ 09:52	10:52	11:52	12:52	13:52	14:52	15:52	16:52	17:52	18:52	§ 19:52	20:52	§ 22:27
Mysłakowice p.	§ 04:55	§ 05:55	06:55	§ 07:55	08:55	§ 09:55	§ 09:55	10:55	11:55	12:55	13:55	14:55	15:55	16:55	17:55	18:55	§ 19:55	20:55	§ 22:30
Mysłakowice o.	§ 04:55	§ 05:55	06:55	§ 07:55	08:55	§ 09:55	§ 09:55	10:55	11:55	12:55	13:55	14:55	15:55	16:55	17:55	18:55	§ 19:55	20:55	§ 22:30
Mysłakowice Orzeł |	§ 04:57	§ 05:57	06:57	§ 07:57	08:57	§ 09:57	§ 09:57	10:57	11:57	12:57	13:57	14:57	15:57	16:57	17:57	18:57	§ 19:57	20:57	§ 22:32
Miłków ↓	§ 05:00	§ 06:00	07:00	§ 08:00	09:00	§ 10:00	§ 10:00	11:00	12:00	13:00	14:00	15:00	16:00	17:00	18:00	19:00	§ 20:00	21:00	§ 22:35
Karpacz p.	§ 05:04	§ 06:04	07:04	§ 08:04	09:04	§ 10:04	§ 10:04	11:04	12:04	13:04	14:04	15:04	16:04	17:04	18:04	19:04	§ 20:04	21:04	§ 22:39

termin kursowania	⊛	⊛		⊛										⊛	Ⓒ		⊛		⊛
przewoźnik - kategoria	KD - Os	KD - Os	KD - Os	KD - Os	KD - Os	KD - Os	KD - Os	KD - Os	KD - Os	KD - Os	KD - Os	KD - Os	KD - Os	KD - Os	KD - Os	KD - Os	KD - Os	KD - Os	KD - Os
informacje dodatkowe																			
numer pociągu	60600/1	60602/3	60604/5	60606/7	60608/9	60610/1	60612/3	60614/5	60616/7	60618/9	60620/1	60622/3	60624/5	60626/7	60628/9	60630/1	60632/3	60634/5	60636/7
Karpacz o.	§ 05:11	§ 06:11	07:11	§ 08:11	09:11	10:11	11:11	12:11	13:11	14:11	15:11	16:11	17:11	§ 18:11	§ 18:11	19:11	§ 20:11	21:11	§ 22:46
Miłków |	§ 05:14	§ 06:14	07:14	§ 08:14	09:14	10:14	11:14	12:14	13:14	14:14	15:14	16:14	17:14	§ 18:14	§ 18:14	19:14	§ 20:14	21:14	§ 22:49
Mysłakowice Orzeł ↓	§ 05:18	§ 06:18	07:18	§ 08:18	09:18	10:18	11:18	12:18	13:18	14:18	15:18	16:18	17:18	§ 18:18	§ 18:18	19:18	§ 20:18	21:18	§ 22:53
Mysłakowice p.	§ 05:20	§ 06:20	07:20	§ 08:20	09:20	10:20	11:20	12:20	13:20	14:20	15:20	16:20	17:20	§ 18:20	§ 18:20	19:20	§ 20:20	21:20	§ 22:55
Mysłakowice o.	§ 05:20	§ 06:20	07:20	§ 08:20	09:20	10:20	11:20	12:20	13:20	14:20	15:20	16:20	17:20	§ 18:20	§ 18:20	19:20	§ 20:20	21:20	§ 22:55
Mysłakowice Dolne |	§ 05:23	§ 06:23	07:23	§ 08:23	09:23	10:23	11:23	12:23	13:23	14:23	15:23	16:23	17:23	§ 18:23	§ 18:23	19:23	§ 20:23	21:23	§ 22:58
Łomnica |	§ 05:25	§ 06:25	07:25	§ 08:25	09:25	10:25	11:25	12:25	13:25	14:25	15:25	16:25	17:25	§ 18:25	§ 18:25	19:25	§ 20:25	21:25	§ 23:00
Łomnica Dolna |	§ 05:27	§ 06:27	07:27	§ 08:27	09:27	10:27	11:27	12:27	13:27	14:27	15:27	16:27	17:27	§ 18:27	§ 18:27	19:27	§ 20:27	21:27	§ 23:02
Łomnica Średnia ↓	§ 05:29	§ 06:29	07:29	§ 08:29	09:29	10:29	11:29	12:29	13:29	14:29	15:29	16:29	17:29	§ 18:29	§ 18:29	19:29	§ 20:29	21:29	§ 23:04
JELENIA GÓRA p.	§ 05:33	§ 06:33	07:33	§ 08:33	09:33	10:33	11:33	12:33	13:33	14:33	15:33	16:33	17:33	§ 18:33	§ 18:33	19:33	§ 20:33	21:33	§ 23:08
Jelenia Góra o.															§ X				
Wrocław Główny p.															§ X				

5 Resumption of traffic on the Mysłakowice–Kowary route and construction of the Kowary–Karpacz line

The launch of a similar number of train pairs running between Jelenia Góra and Kowary, parallel to the Jelenia Góra–Karpacz connections, could generate an oversupply of connections between Jelenia Góra and Mysłakowice. On the other hand, changing trains at the station in Mysłakowice does not allow for rational use of the rolling stock and would be unattractive for travellers. Consideration should therefore be given to the possibility – much like the existing bus network - of running a train route to Karpacz so that it also passes through Kowary. This would require the construction of less than four-kilometre section of a single track, connecting the Kowary arch with the railway line no. 340 between Miłków and Karpacz.

Variants 1, 2, 3A and 3B implement the connection between Karpacz and Kowary in the conventional way, i.e. assuming the maintenance of the terminus in Karpacz (which is not a problem, as the rolling stock designed to operate this connection is of two-way type). They differ in the way of passing through the town of Ściegny and in the place of inclusion of the proposed line into railway line no. 340, the differences being determined by landscape and occurrence of population centres.

A different approach was taken for the preparation of variant 4. A loop arrangement covering Mysłakowice–Kowary–Karpacz–Mysłakowice was conceived. However, such a solution requires a departure from the concept of renewing traffic to the existing railway station.

6 Summary

The time of travel on the route being the subject of this paper, both on the route to Karpacz as well as to Jelenia Góra, will be 22^4 $[min]$. This makes it possible to run trains with a 60 minute cycle.

The proposed timetable assumes starting 18 pairs of shuttle trains per working day, 12 pairs of courses between Jelenia Góra–Karpacz–Jelenia Góra and one pair of connections between Wrocław–Karpacz–Wrocław, arranged in such a way that it is possible for the inhabitants of Wrocław to make even a one-day trip to Karpacz.

The complementary transport line planned to be launched, based on low-floor, modern electric buses, increases the availability of railway services, generating additional passenger streams, especially tourist streams (in relation to Bierutowice) and those related to daily commutes (circular migrations) – mainly on the route to Kowary. This communication – due to the use of electric and non-emission rolling stock – will contribute to the improvement of the natural environment in the Giant Mountains.

Restoration of railway traffic on the route Jelenia Góra–Mysłakowice–Karpacz will contribute to increasing the tourist and sightseeing attractiveness of this part of the Giant Mountains, it will have a positive impact on the activation and increase in the mobility of the local community. Revitalization combined with electrification of the railway line will increase the number of tourists visiting Karpacz.

References

1. Railway line nr 308 Jelenia Góra – Kamienna Góra - dolnyslask.org.pl [on-line] [access on April 1 2018] https://dolny-slask.org.pl/554495,Linia_kolejowa_nr_308_Jelenia_Gora__Kamienna_Gora.html
2. J. Podolski, J. Kacprzak, J. Myslek, *Principles of electric traction* (Communication and Communication Publishing, Warszawa, 1980)
3. A. Chwieduk, T. Dyr, *Train traffic design* (Printing Works of the Radom Polytechnic, Radom, 1997)
4. W. Sawczuk, M. Jüngst, Brake discs for rail vehicles (In: TTS. Rail Transport Technique **12**, 2016)
5. J. Madej, *The theory of rail vehicle traffic* (Warsaw University of Technology Publishing House, Warszawa, 2012)

6. J. Kwasnikowski, Elements of motion theory and rationalization of running a train (Scientific Publisher of the Institute for Sustainable Technologies – PIB, Radom, 2013)
7. PKP Polskie Linie Kolejowe S.A., Instruction on train timetables (Ir-11), Warszawa 2015
8. W. Czarnecki, Planning of cities and settlements. Volume II. (Places of work and residence, State Scientific Publisher, Warszawa, 1966)
9. Head Office of Geodesy and Cartography, Geoportal 2, access on April 3 2018, http://mapy.geoportal.gov.pl/imap
10. Central Statistical Office, Local Data Bank, access 26.09.2018, https://bdl.stat.gov.pl/BDL/start

Risk analysis associated with secondary water pollution in selected agglomerations of the Silesian Voivodship

Izabela Zimoch, and *Ewelina* Bartkiewicz[1,*]

[1]Silesian University of Technology, Faculty of Energy and Environmental Engineering
Institute of Water and Wastewater Engineering, ul. Konarskiego 18, 44-100 Gliwice, Poland

Abstract. Water quality is the most important parameter related to water intended for human consumption. The main task of water supply companies is to keep water quality at the possible highest level. Water, which consumers receive in their homes, is the final product of a multi-stage processing, which aims to obtain water with the required physico-chemical and biological composition. Nevertheless, during transport through the Water Supply System (WSS), water reacts with various factors, sometimes causing secondary pollution. An important threat associated with water quality reduction is disinfection by-products (DBP), formed during the oxidation of organic compounds and some inorganic compounds found in water. DBPs has carcinogenic properties, that's why they are threats for human health.The article presents the results of water quality analysis and risk assessments. In research the real WSS, which supplies water to selected Silesian agglomeration, was analyzed. The work uses laboratory tests of selected water parameters. Statistical methods were used to determine the causes of water quality deterioration.

1 Introduction

Ensuring high water quality intended for consumption is one of the main tasks for waterworks. Low water quality causes a number of threats to human health [1–4]. To ensure the water stability in the distribution system, a disinfection process is used. The aim of disinfection is to destroy living and spore forms of pathogenic organisms. The most commonly used disinfectants are various forms of chlorine, because of its high oxidation potential properties and low costs. Chlorination also limits microbial contamination in the water supply system. However, the organic compounds present in the water and some of the inorganic compounds, react with the free chlorine forming the so-called disinfection by-products (DBPs). Currently, more than a thousand forms of DBPs [5] are known, the most frequent DBPs include THMs and haloacetic acids (HAA). The first of these are regulated by the Regulation of the Minister of Health of 7 December 2017 on the quality of water intended for human consumption, in which the permissible

* Corresponding author: ewelina.bartkiewicz@polsl.pl

concentration of THMs (sum of trichloromethane, bromodichloromethane, dibromochloromethane and tribromethane) in water is specified. Disinfection by-products have become a serious health threat due to their potential teratogenicity, carcinogenicity and mutagenicity [6–11]. In table No. 1 are listed examples of diseases caused by DBPs exposure to the human body. DBPs also affect the reproductive system and the course of pregnancy [5]. THMs, and other DBPs, can get into the body by water consumption, as well by inhalation and contact with the skin [12–14]. These compounds can be accumulate in the body, and the effects of DBPs impact can be noticeable even after several years.

Two mechanisms are responsible for creating the DBPs in the water distribution system: (1) reduction, oxidation or disproportionation of the disinfectant and (2) oxidation reaction of compounds present in the water by the disinfectant [15, 16]. Disinfection by-products are formed when their precursors react with the disinfectant. The main precursors to DBP present in water are natural organic substances (NOMs), which are generally associated with the quality of surface waters [17]. Seasonal variation in the formation of DBP is observed, many studies suggest that these compounds are usually higher in the summer because the organic matter content in the water source increases [18]. An important element in the creation of DBPs is the type and dose of disinfectant. Less DBPs are formed using ozone than with chlorine dioxide [19]. An issue worth considering in the creation of DBPs is the spatial variability of these compound in the distribution system. In the literature, only a few works can be found devoted to this task. Problems with determining the spatial variability of DBPs in the water distribution system are related to the structure, flow time and hydraulic conditions [20]. Water supply system operating conditions (including water flow and flow time) are parameters that depend on many factors and are difficult to determine, therefore tests based on hydraulic parameters are subject to a certain error.

The paper presents the spatial and seasonal variability of THM concentration in the distribution system. The carcinogenic risk for THMs exposure via ingestion pathway was determined based on the possessed data.

Table 1. Individual-based studies related to chlorination byproduct exposure [10].

Author (ref); Year	**Cancer sites; years of diagnosis**	**No. Of cases**	**Study location**	**Exposure timing**
[21]	Bladder, liver, kidney; 1963–75	81, 45, 31	Western Maryland (USA)	Years in 1963 domicile
[22]	Bladder;1978	2 805	10 locations in USA	Lifetime
[23]	Bladder; 1975–92	294	Western Maryland (USA)	Years in 1975 domicile
[24]	Bladder; 1988–89	327	Colorado (USA)	Age 20 to interview

[25]	Colon; 1978–80	200	North Carolina (USA)	1953–78
[26]	Colon	347	Wisconsin (USA)	Lifetime
[27]	Pancreas; 1975–89	101	Western Maryland (USA)	Water source in 1975

2 Research object

The study subject is the selected subsystem of the biggest collective Water Distribution System (WDS) in Poland, which is located in the southern-west of Poland in Silesian region. Analyzed WDS is composed of Water Treatment Plant A (WTP A), pumping station B (PS B) and storage tanks C (ST C). Figure 1 shows scheme of analyzed WDS.

WTP A is the second largest (in terms of water production) station of the main water supply system. The average daily water production of this station is 50 000 cubic meter and it provides water to about 334000 people. WTP A is based on surface water captured from the Soła cascade basin. The water treatment system is based on basic processes such as coagulation, filtration and disinfection. The process of coagulation is guided on the filters with aluminum sulphate (contact coagulation). The filters have been designed as a pressure with a sand layer. Water disinfection is carried out using chlorine gas. Pumping station B is located 32 km from WTP A. The water that flows into pumping station is subjected to another disinfection, this time with sodium hypochlorite. Pumping station B, depending on water demand, transports water in the amount of 45000 cubic meter per day. Storage tanks C are located 12 km from the pumping station B. In storage tanks C the water disinfection is also carried out with sodium hypochlorite. The C tanks transport water in three directions. For the purpose of research, tanks C are the final object. The analyzed main network consists of reinforced concrete pipes with a diameter of 1500 mm (section between WTP A and PS B) and steel pipes with a diameter of 1400 mm (section between PS B and ST C).The water flow rate is in the range of 0.3–0.32 m/s for the section between WTP and PS and 0.28–0.34 m/s for the section between PS and ST. The average water demand for this area is about 17,000 meters per day, thus supplying around 115 000 people.

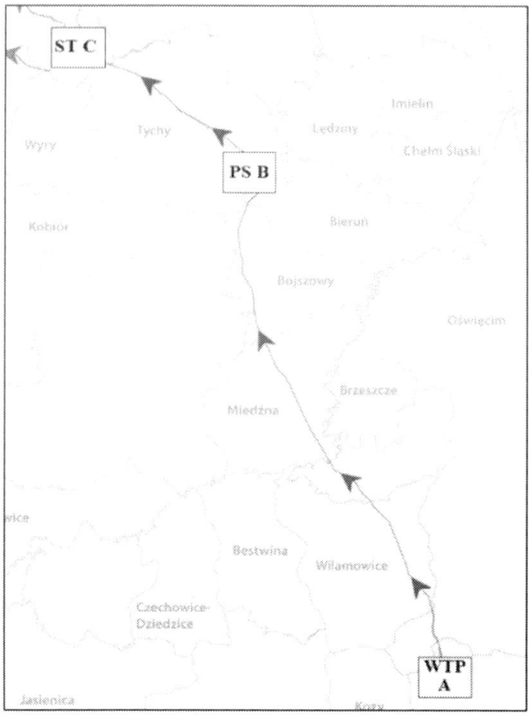

Fig. 1. Scheme of Water Supply System.

3 Research methodology

3.1 Water quality

Drinking water samples for analysis were collected between January 2016 and December 2018. Samples were collected at WTP A, PS B and ST C, four times a month (average number of samples per month) as part of water quality control. Water tests included physicochemical and biological parameters of water. Figure 2 shows the variability of temperature, pH and THM from the period of 3 years. At Table 2 we gathered basic statistics of chloroform, bromodichloromethane, dibromochloromethane, bromoform and sum of THMs.

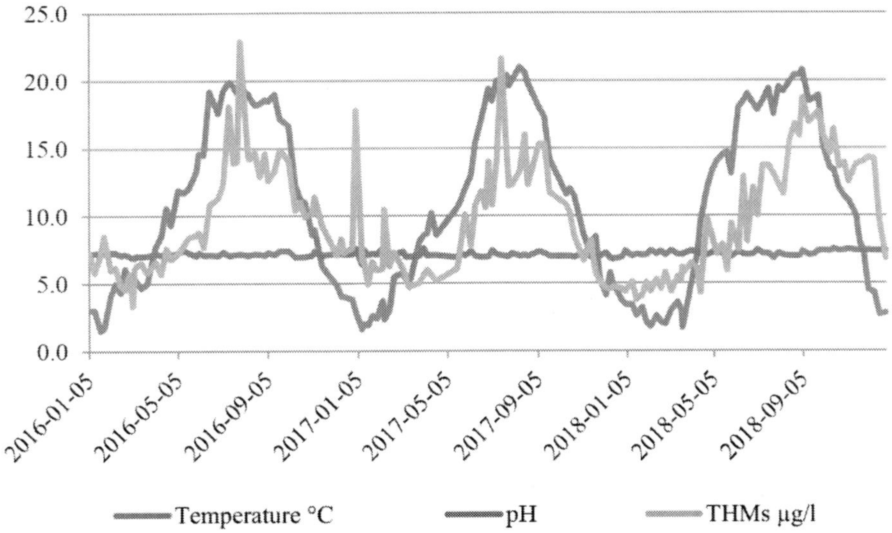

Fig. 2. Variability of selected water parameters in the period 2016–2018 at WTP A.

Table 2. Distributions of THMs concentrations.

Parameters		**$CHCl_3$**	**$CHCl_2Br$**	**$CHClBr_2$**	**$CHBr_3$**	**THMs**
WTP A	Min	0.00	0.00	0 .00	0.00	3.30
	Max	19.8	7.30	1.40	1.00	22.90
	Mean	8.68	1.47	0.35	0.07	9.93
	SD	4.29	1.23	0.25	0.14	4.41
PS B	Min	7.50	1.40	0.00	0.00	10.10
	Max	30.70	5.70	1.90	0.30	34.70
	Mean	16.83	2.70	0.47	0.06	19.55
	SD	4.37	0.78	0.27	0.08	5.06
ST C	Min	3.30	1.10	0.00	0.00	4.40
	Max	33.9	5.40	1.50	0.30	38.10
	Mean	19.15	2.99	0.51	0.06	22.16
	SD	5.82	1.00	0.23	0.08	6.67

Figure 3 shows the seasonal variation of THM concentration in water. The highest THM concentrations occurred in summer and autumn.

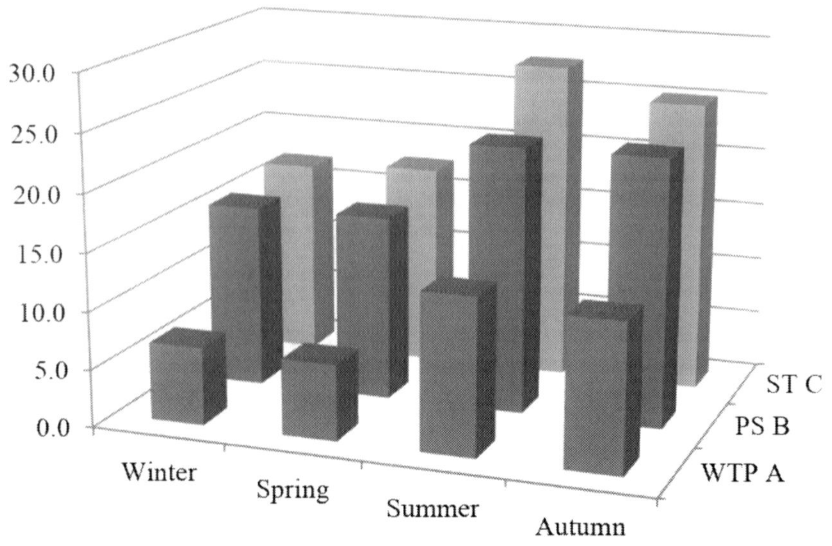

Fig. 3. Average THMs concentrations at points WTP A, PS B and ST C for each season.

3.2 Risk assessment

Based on THMs data, a human health risk assessment was conducted to determine the carcinogenic risk associated with exposure to THMs via ingestion pathway.

For calculation of lifetime daily exposure for ingestion pathway the following equation was used [28]:

$$CD\ I_{ing} = \frac{C_w \cdot IR \cdot EF \cdot CF}{BW \cdot AT} \tag{1}$$

where: $CD\ I_{ing}$ is chronic daily intake, mg/kg/d, C_W is the chemicals concentration in water, µg/L, *IR* the ingestion rate, L/day, *EF* the exposure frequency, days/year or events/year, *ED* the exposure duration, year, CF is the mass conversion factor from µg to mg, *BW* the body weight, kg, and *AT* the average time, days. The risk assessment associated with THMs exposure via ingestion pathway was calculated using following equation [29]:

$$\text{Risk} = \text{CD I}_{\text{ing}} \cdot \text{CSF}_{\text{oral}} \tag{2}$$

where: CSF_{oral} is cancer slope factor, mg/kg/d. The minimum, maximum and mean THMs concentrationsat PS B and ST C points were used for the calculations, which will represent THMs concentrations respectively at section WTP A – PS B and PS B – ST C. At Table 3 we gathered all exposure parameters and population characteristics. Ingestion rate was chosen as the minimum recommended amount of water that a person should consume. Exposure duration and body weight was recommended by USEPA [28] as 70 years exposure and average weight for male 70 kg and for female 60 kg. Exposure frequency is the number of days, where THMs has appeared. CSF_{oral} values were taken from [30].

Table 3. Exposure parameters and population characteristics.

Parameters	Symbol	Unit	Distribution	Value
$CHCl_3$, $CHCl_2Br$, $CHClBr_2$ and $CHBr_3$concentration	C_w	µm/L	Mean: PS B	Table 1
			Mean: ST C	Table 1
Ingestion rate	IR	L/day	Min	2.0
Exposure frequency	EF	day/year	Max	365
Exposure duration	ED	year	Mean	70
Mass conversion factor	CF	-		0.001
Body weight	BD	kg	Mean: Male	70
			Mean: Female	60
Average time	AT	days	Mean	70x365
Cancer slope factor for$CHCl_3$	CSF_{oral}	mg/kg·d		0.0310
Cancer slope factor for $CHCl_2Br$	CSF_{oral}	mg/kg·d		0.0620
Cancer slope factor for $CHClBr_2$	CSF_{oral}	mg/kg·d		0.0840
Cancer slope factor for $CHBr_3$	CSF_{oral}	mg/kg·d		0.0079

4 Result and discussion

The maximum THM concentration was in ST C and is equal 38.10.The standard deviation value indicates a high variability of the THM content in water. During the study period, there were no exceeded THM concentrations specified in the Regulation of the Minister of Health of 7 December 2017.The highest THM concentrations occurred in summer and autumn caused by the variability of water quality in these periods (algae blooms, leaf fall). Chloroform was the major contributor to the total concentration of THMs. The highest concentration of chloroform varied between 3.30 and 33.90 µg/L and occurs in ST C. Chloroform is the most common factor affecting THMs concentration in water. Based on the research, it was found that chloroform occurs in every sample, as well as bromodichloromethane. Whereas, dibromochloromethane and bromoformare present in about 90% of samples.

The chronic daily intake of THMs via ingestion pathway is shown in Table 4 and in Table nr 5 is shown results for cancer risk. The highest lifetime risk of $2.74 \cdot 10^{-5}$ was obtain for female in section PS B – ST C. The lowest lifetime cancer risk was obtained for bromoform, values in order of 10^{-8}. The average lifetime cancer risk for total THMs in both section was $2.41 \cdot 10^{-5}$, which is higher than EPA recommended cancer risk value about 24 times. The percentage contribution of average cancer risks: 71.5% for $CHCl_3$, 22.9% for $CHCl_2Br$, 5.50% for $CHClBr_2$ and 0.10% for $CHBr_3$.

Table 4. Chronic daily intake of THMs via ingestion pathway.

Section	Gender	$CHCl_3$	$CHCl_2Br$	$CHClBr_2$	$CHBr_3$	Sum
WTP A – PS B	Male	$48.09 \cdot 10^{-5}$	$7.71 \cdot 10^{-5}$	$1.35 \cdot 10^{-5}$	$0.17 \cdot 10^{-5}$	$57.33 \cdot 10^{-5}$
	Female	$56.11 \cdot 10^{-5}$	$9.00 \cdot 10^{-5}$	$1.58 \cdot 10^{-5}$	$0.20 \cdot 10^{-5}$	$66.88 \cdot 10^{-5}$
PS B – ST C	Male	$54.72 \cdot 10^{-5}$	$8.56 \cdot 10^{-5}$	$1.46 \cdot 10^{-5}$	$0.17 \cdot 10^{-5}$	$64.91 \cdot 10^{-5}$
	Female	$63.83 \cdot 10^{-5}$	$9.98 \cdot 10^{-5}$	$1.71 \cdot 10^{-5}$	$0.20 \cdot 10^{-5}$	$75.73 \cdot 10^{-5}$

Table 5. Cancer risk associated with THMs exposure via ingestion pathway.

Section	Gender	$CHCl_3$	$CHCl_2Br$	$CHClBr_2$	$CHBr_3$	Sum
WTP A – PS B	Male	$1.49 \cdot 10^{-5}$	$0.48 \cdot 10^{-5}$	$0.11 \cdot 10^{-5}$	$1.34 \cdot 10^{-8}$	$2.08 \cdot 10^{-5}$
	Female	$1.74 \cdot 10^{-5}$	$0.56 \cdot 10^{-5}$	$0.13 \cdot 10^{-5}$	$1.56 \cdot 10^{-8}$	$2.43 \cdot 10^{-5}$
PS B – ST C	Male	$1.70 \cdot 10^{-5}$	$0.53 \cdot 10^{-5}$	$0.12 \cdot 10^{-5}$	$1.36 \cdot 10^{-8}$	$2.35 \cdot 10^{-5}$
	Female	$1.98 \cdot 10^{-5}$	$0.62 \cdot 10^{-5}$	$0.14 \cdot 10^{-5}$	$1.59 \cdot 10^{-8}$	$2.74 \cdot 10^{-5}$

5 Conclusions

The work involved analysis of water quality in terms of the occurrence of disinfection by-products. The spatial and seasonal distribution of pollutants was taken into account. The analysis shows that the highest THMs concentrations occurred in summer and autumn. It was also found that the greater the distance from the WTP, the higher the THMs concentrations. It was also found that the most common factor affecting the occurrence of THM is chloroform. At research the lifetime cancer risk via ingestion pathway was evaluated. The result show that women are more vulnerable than men. The highest cancer risk was obtained for section, which is the farthest from the WTP. The highest degree of risk is associated with chloroform, which is $1.73 \cdot 10^{-5}$ (mean value).

The quality of the analyzed water is good, during the period of research there were no exceedances of the THM concentration recorded in the Regulation of the Minister of Health. The average risk associated with exposure to THMs was $2.41 \cdot 10^{-5}$, which is a low result, however, it exceeds the value recommended by the EPA.

This work was supported by Ministry of Science and Higher Education Republic of Poland within statutory funds as well as BKM-508/RIE-4/2018 research.

References

1. D. Baruaand, W. B. Greenough, *Cholera* (Springer Science Business Media, LLC, New York, 1992)
2. S. Ishii, M. J. Sadowsky, Microbes Environ. **23** (2008)
3. S. Khan, M. Shahnaz, N. Jehan, S. Rehman, M. T. Shah, I. Din, J. Cleaner Prod. **60** (2013)
4. I. Zimoch, E. Lobos, Environ. Prot. Eng. **38** (2012)
5. Y. Wang, G. Zhu, B. Engel, Ecotoxicol. Environ. Saf. **170** (2018)
6. J. J. Rook, Water Treat. Exam. **23** (1974)
7. S. E. Hrudey, L. C. Backer, A. R. Humpage, S. W. Krasner, D. S. Michaud, L. E. Moore, J. Toxicol. Environ. Health B Crit. Rev. **18** (2015)
8. S. E. Hrudey, J. Fawell, Water Sci. Technol.-Water Supply **15** (2015)
9. M. Kumari, Ecotoxicol. Environ. Saf. **113** (2015)
10. K. P. Cantor, Cancer Causes and Control (1997)
11. I. Zimoch, E. Lobos, Desalin. Water Treat. **57** (2016)
12. D. A. Savitz, P. C. Singer, A. H. Herring, Am. J. Epidemiol. **164** (2006)
13. Y. M. Tan, K. H. Lao, H. J. Clewell, J. Expo Sci. Environ. Epidemiol. **17** (2007)
14. S. Chowdhury, Sci. Total Environ. **463–464** (2013)

15. K. N. Shammas, K. L. Wang, *Water Engineering – Hydraulics, Distribution and Treatment* (1st Ed., John Wiley & Sons, 2016)
16. I. Zimoch, Ochr. Środ. **33** (2011)
17. H. Y. Zhai, X. R. Zhang, Environ. Sci. Technol. **45** (2011)
18. M. J. Rodriguez, J. B. Serodes, P. Levallois, Water Res. **38** (2004)
19. G. Hua, D. A. Reckhow, Water Res. **41** (2007)
20. C. M. Shanks, Jean-B. Sérodes, M. J. Rodriguez, Water Res. **47** (2013)
21. Wilkins Jr, Am. J. Epidemiol. **114** (1981)
22. K. P Cantor, R. Hoover, P. Hartge, JNCI, **79** (1987)
23. D. M Freedman, K. P. Cantor, N. L. Lee, Cancer Causes Control, **8** (1997)
24. M. A. McGeehin, J. S. Reif, J. Becher, E. J Mangione, Am. J. Epidemiol. **138** (1993)
25. R. L Jolley, R. J. Bull, W. P. Davis, S. Katz, M. H. Jr Roberts, V. A. Jacobs, *Water Chlorination: Chemistry, Environmental Impact and Health Effects* (Vol. 5, Lewis Publishers, Inc., Chelsea, MI (USA), 1985)
26. T. B. Young, D. A. Wolf, M. S. Kanarek, Int. J. Epidemiol. **16** (1987)
27. C. B. IJsselmuiden, C. Gaydos, B. Feighner, Am. J. Epidemiol. **136** (1992)
28. USEPA. Guidelines for exposure assessment. Risk assessment Forum, US Environmental Protection Agency; 1992. Washington DC, EPA/600/Z-92/001
29. H. Amjad, I. Hashmi, M. S. Ur Rehman, M. Ali Awan, S. Ghaffar, Z. Khan, Ecotoxicol. Environ. Saf. **91** (2013)
30. Integrated Risk Information System (IRIS) (2009). USEPA (Electronic data base). Web link: http://www.epa.gov/iris/ (2009)

24-hour nutrient mass balance of small storage reservoir included in municipal rainwater drainage system

Maciej Ziułkiewicz[1], *Michał* Górecki[2,*], *Anna* Fortuniak[1], *Aneta* Walas[3], and *Rafał* Grulke[3]

[1]Laboratory of Geology, University of Łódź, 90-139 Łódź, ul. Narutowicza 88, Poland
[2]Fourth year of doctoral studies, the Faculty of Geographical Sciences, University of Łódź, 90-139 Łódź, ul. Narutowicza 88, Poland
[3]Third year of full-time studies in the field of Geomonitoring, the Faculty of Geographical Sciences, University of Łódź, 90-139 Łódź, ul. Narutowicza 88, Poland

Abstract. The paper presents the results of research, which aimed at tracing the quantitative and qualitative effect of the "Tomaszowska" urban retention reservoir (located in the Łódź city) on the Olechówka River flowing through it over a period of one day. For this purpose, two measurement-control points were established located directly above the inflow of river to the reservoir, and below the weir. Hydrological and hydrochemical measurements were performed every two hours. Results obtained in this research present the reservoir as an object that in terms of the inflowing river, among others, in a daily balance of pollutants retains mineral and biogenic compounds, and releases dissolved and suspended organic compounds, and also changes the structure of pollution (in the case of total dissolved solids and total nitrogen, increasing the contribution of their organic forms).

1 Introduction

The Olechówka River in the administrative boundaries of Łódź fulfils the function of a recipient of stormwaters. Several retention reservoirs exist along its course, fulfilling different economic functions. From the point of view of the state of water quality in the river, the "Tomaszowska" reservoir is the most interesting. It was established for the regulation of discharges of the river below the outlet of a large stormwater collector from the Widzew borough, and for the protection of water quality in the "Młynek" reservoir located below, retaining water for recreational purposes.

Research on the Olechówka River showed that the "Tomaszowska" reservoir is an important element affecting river water quality in conditions of water level increases caused by precipitation [1], as well as during the flow of the same part of river water from the „sources" to the mouth – with consideration of the time of water retention in the reservoir [2].

* Corresponding author: michal.gorecki@geo.uni.lodz.pl

The research on the "Tomaszowska" reservoir was based on results of hydrochemical research conducted in 2017 involving the determination of the daily variability of pollution indicators in two profiles established on the course of the Olechówka River, representing the weakest and strongest transformed sections of its channel [3]. Research implemented in 2018 aimed at tracing the quantitative and qualitative effect of the urban retention reservoir on the river flowing through it over a period of one day – with no consideration of the retention time. Like in previous years, the works were implemented with the participation of students of major Geomonitoring at the Faculty of Geographic Sciences of the University of Łódź.

The consideration of the role of the retention reservoir in the determination of the daily variability of hydrochemical parameters of river waters in the current research aims at capturing this among other factors determining water quality in the urban catchment.

2 Study object

The "Tomaszowska" reservoir was established in 1972 through damming of the Olechówka River a concrete weir on km 9+350 of the course of the river. The reservoir occupies an area of 1.20 ha, and at a constant damming ordinate on the weir (199.5 m a.s.l.), it has a volume of approximately 17.5 thousand m^3 [4]. The catchment of the reservoir with an area of 15.5 km^2 is occupied by: a metropolitan residential district Olechów, industrial and warehouse areas of Dąbrowa, and a cargo railway station PKP Łódź-Olechów (Fig. 1). The catchment sealing coefficient varies from 0.2 for residually preserved agricultural areas to 0.6 for the residential district. The weighted average for the catchment of the "Tomaszowska" reservoir is 0.45. The Olechówka River channel on the inflow to the reservoir features heavy bank enforcements (UTC–1) in the form of concrete slabs, and biological enforcements were applied below the reservoir (UB) [4]. On the section from the beginning of the river to the "Tomaszowska" reservoir is located smaller reservoir named « Upper Olechówka » with automatically regulated outflow. Moreover are total of 12 outlets of drainage ditches (2) and sewage collectors (10) function along the course of the Olechówka River [2].

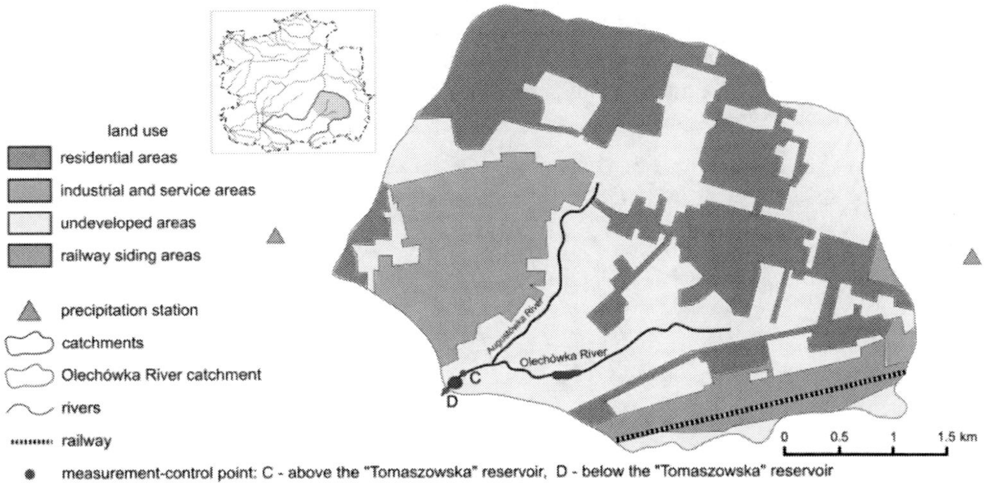

Fig. 1. Land development in the catchment of the "Tomaszowska" reservoir and location of measurement-control points.

3 Study methodology

The research in 2018 was conducted over one day, from 8:00 on 23 May to 8:00 of the following day at a measurement-control point (mcp) located directly above the inflow of the Olechówka River to the reservoir (C), and at a mcp located below the weir (D) (Fig. 1). Water samples were collected at those sites every two hours for determinations of ions NH_4^+, NO_2^-, NO_3^-, and PO_4^{3-}, total nitrogen (TN), and total phosphorus (TP), as well as total suspension (Susp.), oxidability (COD-Mn), and total dissolved solids (TDS). During sampling, the following physical-chemical water parameters were recorded: temperature (Temp.), reaction (pH), standardized electrolytic conductivity (SEC), and dissolved oxygen (DO). Hydrochemical observations were supplemented by the measurement of discharge rate (Q) in the Olechówka River by means of a current meter. The research was first implemented at a mcp C, and then after approximately 0.5 h at a mcp D. A two-hour time step was applied in measurements and analyses, because it was assumed that during low water stages caused by a longer period with no precipitation in the catchment, fluctuations of values of hydrochemical parameters will not be high [3], and that the conditions of low flow can be maintained in a river for a long time, hence the quality of its waters is an important factor affecting the health condition of water organisms [5]. Chemical analyses of the collected river water samples were performed at the laboratory of Laboratory of Geology and at the laboratory of the Collective Sewage Treatment Plant in Łódź. The statistical calculations of the obtained results employed Statistica 13.1. software.

4 Results

Discharges in the Olechówka River were within the following ranges: 9.4–27.4 $L \cdot s^{-1}$ at a mcp C, and 16.0–23.2 $L \cdot s^{-1}$ at a mcp D (Fig. 2). Considering the volume of the reservoir, this provides for an average time of water retention of 10 days and 19 hours. Between 8:00 on 23 May and 8:00 on 24 May, a total of 1356 m^3 of water was supplied to the reservoir, and 1622 m^3 was discharged, i.e. 266 m^3 more. Notice the evident increase in the amount of inflowing water that occurred between 18:00 and 22:00. According to data provided by the Department of Meteorology and Climatology of the University of Łódź, no atmospheric precipitation was recorded at any of the precipitation stations located in immediate vicinity the catchment of the "Tomaszowska" reservoir over 4 days preceding the hydrochemical research.

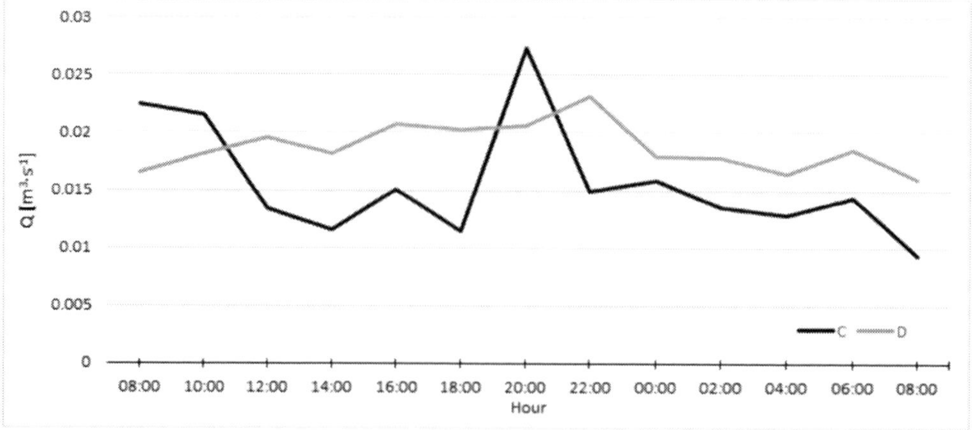

Fig. 2. Changes in discharge rate (Q) in the Olechówka River above (C) and below (D) the "Tomaszowska" reservoir during the study day.

Waters flowing out of the reservoir were warmer, better oxygenated, and contained more suspension (Table 1). The remaining parameters were higher in inflowing waters. The reservoir evidently stabilised the physical-chemical daily variability of the Olechówka River. An exception was the slightly higher variability of the pH and variability of Susp. and COD-Mn on the outflow.

Table 1. Physical-chemical characteristics of the waters of the Olechówka River above (C) and below (D) the "Tomaszowska" reservoir.

	mcp C			mcp D		
parameter	**Mean**	**SD**	**CV[%]**	**Mean**	**SD**	**CV[%]**
Temp. [°C]	17.14	2.60	**15.1**	20.45	1.53	7.5
SEC [$\mu S \cdot cm^{-1}$]	606.60	58.10	**9.6**	508.80	23.14	4.5
pH [-]	7.56	0.18	2.4	7.53	0.31	**4.1**
DO [$mg \cdot dm^{-3}$]	5.29	0.63	**11.9**	6.68	0.30	4.5
Susp. [$mg \cdot dm^{-3}$]	5.38	1.50	27.9	9.00	3.76	**41.8**
COD-Mn [$mgO_2 \cdot dm^{-3}$]	9.09	0.47	5.1	8.48	1.27	**15.0**
NH_4^+ [$mg \cdot dm^{-3}$]	0.44	0.61	**137.5**	0.23	0.04	15.6
NO_2^- [$mg \cdot dm^{-3}$]	0.16	0.03	**21.0**	0.17	0.03	16.3
NO_3^- [$mg \cdot dm^{-3}$]	3.57	1.37	**38.4**	1.67	0.61	36.4
TN [$mg \cdot dm^{-3}$]	3.83	2.36	**61.7**	2.45	0.17	7.1
PO_4^{3+} [$mg \cdot dm^{-3}$]	0.89	0.14	**15.6**	0.69	0.09	13.5
TP [$mg \cdot dm^{-3}$]	0.26	0.12	**47.2**	0.17	0.02	10.4

Notice: Highest values are in highlighted

The hydrochemical interpretation of the variability of Q in the Olechówka River employed similarity analysis by means of the Ward agglomeration method. The degree was identified to which the hydrochemical state of waters from particular sampling hours shows mutual similarity in terms of: pH, SEC, DO, Susp., COD-Mn, TP, NH_4^+, and NO_3^- concentrations (Fig. 3). They are elements which according to earlier studies [1–3] reflected the specificity of pollution of the Olechówka River in its urban catchment.

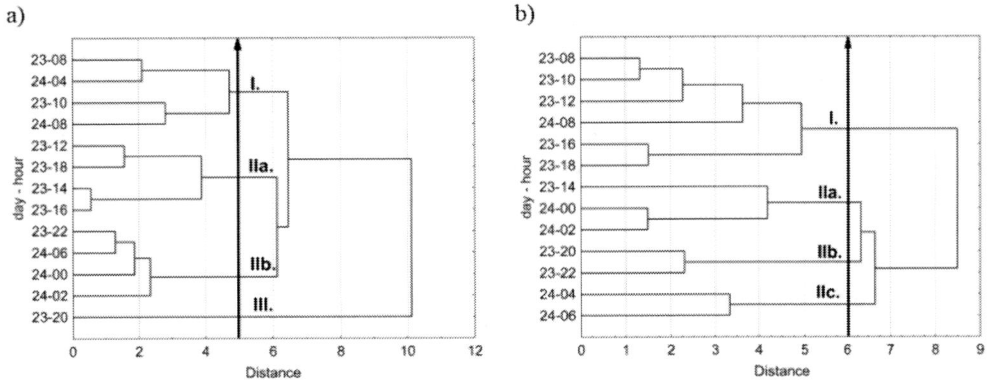

Fig. 3. Dendrograms of similarity of moments of water sampling in the Olechówka River on the inflow to the "Tomaszowska" reservoir (mcp C – **a**), and on the outflow (mcp D – **b**). Vectors point to the designation of groups of cases based on the analysis of the graph of the agglomeration course [6], Roman digits with a letter index mark selected groups of cases.

Groups of cases designated in the dendrograms were considered as hydrochemical states of the waters of the Olechówka River characteristic on a given measurement day for the site above and below the reservoir. Their representation on the graph of river discharge fluctuations at both mcp is presented in Figure 4, and mean values of hydrochemical parameters for each designated state are provided in Table 2.

a)

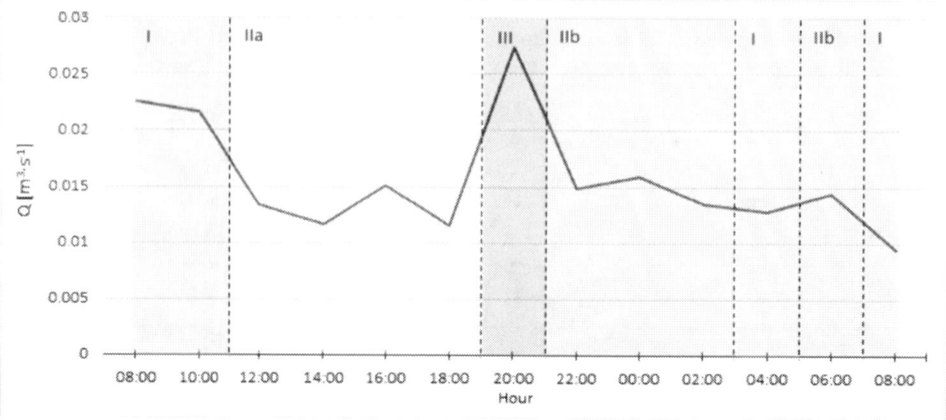

b)

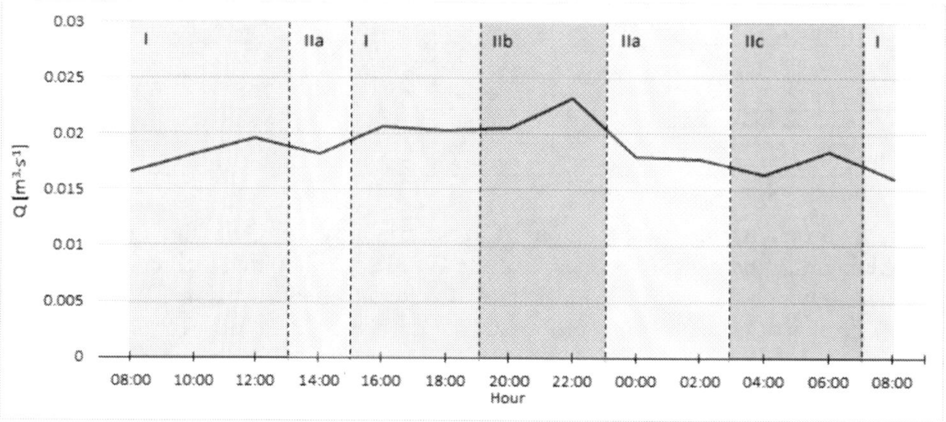

Fig. 4. Hydrochemical states of the waters of the Olechówka River above (a) and below (b) the "Tomaszowska" reservoir during the study day 23–24.05.2018.

In the waters of the Olechówka River inflowing to the reservoir, the daily hydrochemical axis is the water level increase at 20:00. At that time, the water reached the worst quality parameters (maximum values of almost all parameters, except for oxygen the concentration of which reached the minimum). Before the water level increase, the water showed the best quality parameters - reaching minimum values of indices, except for DO and Temp. that reached daily maximums. Directly after the water level increase, the state evidently deteriorated. This dominant course of quality fluctuations also overlaps with the effect of the earlier water level increase (at 8:00–10:00) with a lower degree of water contamination than in the case of the water level increase from 20:00.

The warmest waters flew out of the reservoir between 20:00 and 22:00. With the highest amount of water flowing through the weir, the maximum oxygenation and highest load of Susp. also occurred (bioseston?). They were waters with the minimum concentration of NH_4^+ and the highest concentration of NO_3^- (evidence of effective nitrification during the day). The coolest waters flew out of the reservoir between 4:00 and 6:00. The lowest DO concentration occurred at the time (after the night), as well as the highest SEC, lowest pH, and highest daily concentrations of TN, TP, NH_4^+, and COD-Mn. In the morning

(8:00–12:00) and afternoon (16:00–18:00), the waters show the lowest SEC, and the lowest concentrations of TN, TP, and NO_3^-.

Table 2. Mean values of water quality parameters in the Olechówka River calculated for the designated hydrochemical states.

state	**Q** $[m^3 \cdot s^{-1}]$	**T** [°C]	**SEC** $[\mu S \cdot cm^{-1}]$	**pH** [-]	**DO** $[mg \cdot dm^{-3}]$	**Susp.** $[mg \cdot dm^{-3}]$	**COD-Mn** $[mgO_2 \cdot dm^{-3}]$	**TN** $[mg \cdot dm^{-3}]$	**TP** $[mg \cdot dm^{-3}]$	NH_4^+ $[mg \cdot dm^{-3}]$	NO_3^- $[mg \cdot dm^{-3}]$
Inflow to the reservoir (mcp C)											
I	0.016	17.1	602	7.6	5.2	5	9.1	3.83	0.26	0.443	3.569
IIa	0.013	20.6	582	7.6	5.8	4	8.9	2.84	0.20	0.176	2.420
IIb	0.015	15.1	600	7.7	4.9	5	9.6	3.24	0.24	0.377	3.143
III	0.027	17.0	728	7.8	4.7	9	9.2	11.50	0.65	2.438	6.021
Outflow from the reservoir (mcp D)											
I	0.019	20.7	498	7.6	6.7	8	8.8	2.36	0.15	0.228	1.257
IIa	0.018	20.4	517	7.7	6.7	7	7.0	2.53	0.18	0.239	2.073
IIb	0.022	21.2	529	7.5	6.9	16	8.8	2.44	0.17	0.176	2.614
IIc	0.017	18.8	545	7.3	6.4	9	9.4	2.62	0.19	0.283	1.357

From the reservoir the water level decreased two hours later in reference to inflow. The excess waters leaving the reservoir (mcp C – *IIb*) were more than 4°C warmer than excess waters flowing into the reservoir (mcp D – *III*), they had SEC lower by 200 $\mu S \cdot cm^{-1}$, DO concentration higher by more than 2 $mg \cdot dm^{-3}$, higher Susp. concentration, somewhat lower COD-Mn, and considerably lower concentrations of biogenic indices.

The measurement of Q in the river with the determination of concentrations of quality indicators permits the identification of pollution loads supplied and discharged from the reservoir. Their daily balance (Fig. 5) suggests that the "Tomaszowska" reservoir retains mineral and biogenic compounds, and releases organic compounds measured by TDS, COD-Mn, and Susp. Due to the structure of the weir allowing discharge of water from the surface layer of the reservoir, the last parameter seems to represent bioseston.

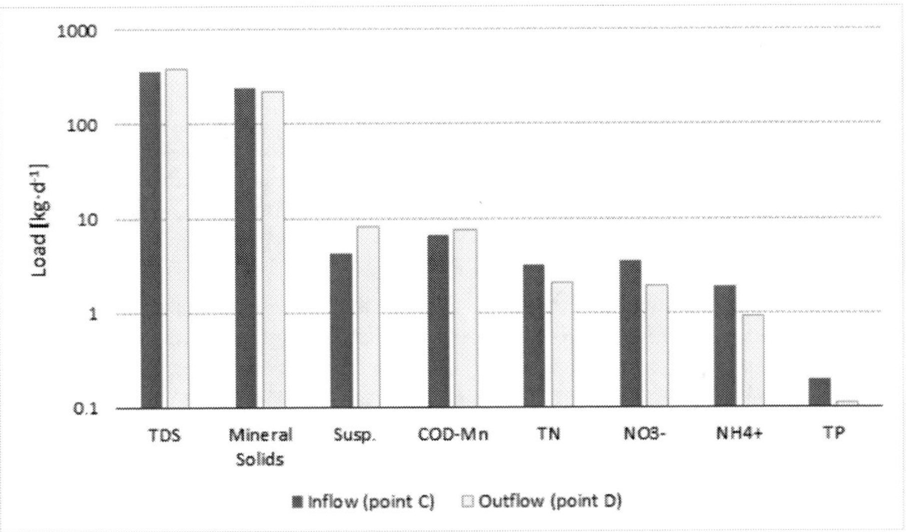

Fig. 5. Daily load of pollutants supplied and discharged from the "Tomaszowska" reservoir through the Olechówka River.

The mean daily composition of TDS, with separation of mineral and organic substances, and composition of TN, with designation of mineral nitrogen and dissolved nitrogen, is presented in Table 3.

Table 3. Mean daily composition of TDS and TN with designation of mineral and organic components in the waters of the Olechówka River above (C) and below (D) the "Tomaszowska" reservoir.

study site	mineral solids	organic components	mineral nitrogen	organic nitrogen
inflow (C)	66.7%	33.3%	72.5%	27.5%
outflow (D)	59.2%	40.8%	55.7%	44.3%

The reservoir slightly changes the structure of runoff of TDS in the Olechówka River, inconsiderably increasing the contribution of dissolved organic substances. In the case of TN the change is more evident in favour of organic nitrogen which almost balances the mineral forms.

5 Conclusions and discussion

Daily comparative research on water flowing in and out of a reservoir is assumed to provide a possibility to capture the state of waters flowing through the reservoir over a considerably shorter time than in the case of comparison of its volume and mean outflow rate. The research conducted in 2016 [2], taking the time into account, showed a significant increase in the load of the Olechówka River with pollutants discharged from the reservoir. For example, among others a two-fold increase in TN, TP, and NH_4^+ concentrations occurred, as well as a four-fold increase in COD-Mn, and approximately 40-fold increase in Susp. Results obtained in 2018 present the reservoir as an object that in terms of the inflowing of the Olechówka River:

- reduces daily fluctuations of values of hydrochemical parameters, with the exception of Susp., COD-Mn, and pH;
- modifies the hydrochemical image of water level increases, reducing the load of the waters of the Olechówka River with dissolved, organic, and biogenic substances, with the exception of Susp.;
- in a daily balance of pollutants retains mineral and biogenic compounds, and releases dissolved and suspended organic compounds;
- changes the structure of pollution, in the case of TDS and TN, increasing the contribution of their organic forms.

The captured changes in the quality of the waters of the Olechówka River above and below the "Tomaszowska" reservoir are in accordance with the observations by [7] and [8] regarding fluctuations of concentrations and loads of biogenic substances in small retention reservoirs in urban and suburban areas. The "Tomaszowska" reservoir proved to be an effective "biogenerator" assimilating mineral forms of nutrients, and simultaneously releasing products of their biochemical transformations by aquatic vegetation to the Olechówka River in the form of dissolved and suspended organic matter.

The water level increase recorded at 20:00 on 23 May deserves attention. It was not caused by precipitation, because no precipitation was recorded in the nearby meteorological stations. It was probably result of discharge of contaminants of a different origin from one of a dozen sewage collectors having their outlets in the Olechówka River channel above the "Tomaszowska" reservoir. The load of pollutants discharged in the Olechówka River at the time constituted a considerable part of the daily load: 35% TN and 30% TP, 21% Susp., 14% TDS, and 13.5% substances represented by COD-Mn. This observation confirms the research experience from 2017 [3], when the most dynamic hydrochemical fluctuations were determined to occur in the Olechówka River in the evening and by night.

In terms of the recorded dynamics of fluctuations of hydrochemical parameters, the study sites on the Olechówka River located above and below the "Tomaszowska" reservoir corresponded with the dynamics of changes documented in 2017 in the profile at the Zygmunta Street - representing a more natural section of the Olechówka River [3]. With the exception of pH and COD-Mn which below the "Tomaszowska" reservoir were characterised by the highest daily dynamics of fluctuations in all four profiles of the Olechówka River (studied jointly in 2017 and 2018), the most hydrochemically variable profile in a daily scale was that at the Podmokła Street, located somewhat above the inflow of the Olechówka River to the Jasień River.

References

1. M. Ziułkiewicz, A. Fortuniak, A. Waack-Zając, M. Górecki, S. Grzędzińska, B. Małecka, Monografie Komitetu Gospodarki Wodnej Polskiej Akademii Nauk **39**, 279–291 (2016)
2. T. Bagrowicz, A. Fortuniak, M. Górecki, M. Lewandowska, M. Ziułkiewicz, E3S Web of Conferences **17**, 00002 (2017)
3. A. Fortuniak, K. Hałas, M. Górecki, K. Szewczyk, A. Wojtania, M. Ziułkiewicz, E3S Web of Conferences **44**, 00042 (2018)
4. M. Zalewski, J. Bocian, M. Zawilski, G. Sakson, I. Kujawa, *Projekt generalny rzeki Olechówki. Projekt generalny rzek dla miasta Łodzi – zlewnie rzeki Ner, rzeki Miazgi i rzeki Bzury –R-2* (Biuro Inżynierii Wodnej, Środowiska i Melioracji AQUAPROJEKT, Łódź, 2000)
5. R.B. Grayson, C.J. Gippel, B.L. Finlayson, B.T. Hart, J. Hydrol. **199**, 121–134 (1997)
6. A. Stanisz, *Przystępny kurs statystyki z zastosowaniem STATISTICA PL na przykładach z medycyny, t. 3, Analizy wielowymiarowe* (StatSoft, Kraków, 2007)
7. R. Gołdyn, *Zmiany biologicznych i fizyczno-chemicznych cech jakości wody rzecznej pod wpływem jej piętrzenia we wstępnych, nizinnych zbiornikach zaporowych* (Wydawnictwo Naukowe UAM, Poznań, 2000)
8. A.R. Ignatius, T.C. Rasmussen, J. Hydrol. Reg. Stud. **8**, 145–161 (2016)

Seasonality analysis of water losses in a selected collective water supply system

Jakub Żywiec[1,*], and *Barbara* Tchórzewska-Cieślak[1]

[1]Rzeszow University of Technology, Faculty of Civil and Environmental Engineering and Architecture, Department of Water Supply and Sewage Systems, al. Powstańców Warszawy 12, 35-959 Rzeszów, Poland

Abstract. The aim of the study is to analyse the seasonality of water losses from a collective water-supply system (CWSS) on the example of a provincial city located in south eastern Poland. The analysis was based on exploitation data obtained from the water-supply company for the period 2012–2016. For the analysis of seasonality a method based on average homogeneous sub-periods was used, with a relative seasonality index and the absolute levels of seasonal fluctuations being determined. Water losses from the analysed CWSS are also characterised using unit water-volume indicators within the water supply network. Over the research period, the volume of water produced and sold was found to decline, even as the number of inhabitants using the network increased. On the other hand, there is a growing demand for water used for own purposes by company, as related directly to their development. Water losses and values determined for the water-loss volume indicator both show a downward trend for the analysed CWSS. It is further observed that, in January, March and May, the loss of water is above average, while the volume of water lost in the other months is below average.

1 Introduction

Ongoing climate change requires the management of natural resources to be both rational and economical. Particular attention should be paid to resources of fresh water, because of their declining availability as a consequence of urbanisation, industrial and agricultural development and growth in the human population. According to a UN Resolution of 28 June 2010, access to drinking water is nothing less than a fundamental human right. Poland has some of the most limited water resources of any European country [1], making the saving of water at the levels of both the consumer and the water-supply company a matter of considerable importance. Analyses of losses from collective water-supply systems (CWSS) have featured in many scientific publications [2–9], but the matter remains a current one, worthy of further research, and the limitation of water losses need to be a priority for operators of water-supply networks.

Loss of water is a problem for every water-supply system, reflecting the difference between the volumes of water produced and sold and the amount of water used for

* Corresponding author: j.zywiec@prz.edu.pl

companies' own purposes. Such losses of water can be classed as [5]: apparent losses (resulting from inaccuracies of measurement devices and measurement inconsistencies, but also the theft of water) or real losses (as the sum of water lost in reservoirs, water connections and via leaks, due to failures of mains and distribution-network pipelines). 80–100% of total losses are reflected by real losses [4], in particular those relating to pipeline failures. The exploitation of a water-supply network in terms of water losses is aimed to limit them, to minimise their potential negative effects on unit water costs, company profitability and the excessive use of water resources. Water supply companies have therefore to apply a proper renovation and repair strategy, ensuring that the costs of removing sources of leakage do not exceed the monetary value of the water wasted. Every system has its level of unavoidable water loss, the elimination of which would be uneconomical. These relate directly to the length of the network, the number and length of water connections, and the pressure in the analysed system. The current trend is to determine an optimal level of losses, a process that entails complicated economic analysis including costs: of producing and distributing water, of active monitoring of the network, and of removing leaks, as well as the resource and environmental costs of water being wasted [5, 6, 10–12].

Like many other phenomena, the loss of water is characterised by a certain unevenness and variability over time. Knowledge of these relationships is a valuable hint for operators, allowing amounts of water lost to be linked with other operating parameters of the water-supply network during the year, such as pressure, the failure rate of pipelines, or external factors like air temperature [13]. This information can be used in planning the schedule for renovation and repairs, in a manner that ensures priority renewal for the pipelines failing most often, with a view to preventing breakdown and limiting amounts of water lost from the supply network.

The main purpose of this paper is to analyse seasonal fluctuations in losses of water and their characteristics, by water-volume indicators within the water supply network in the selected CWSS.

2 Research methodology

The occurrence of seasonal fluctuations was examined using a method based on average homogeneous sub-periods [14, 15]. This assumes seasonality indicators S_i (relative index) being determined in line with formula (1):

$$S_i = \frac{\bar{y}_i \cdot d}{\sum_{i=1}^{d} \bar{y}_i} \cdot 100\ \% \qquad (1)$$

$$\bar{y}_i = \frac{\sum_{i=1}^{d} k_i}{d} \qquad (2)$$

where:
S_i – the value of the seasonal fluctuations index for the i-th sub-period,
$\bar{y}_i$ – the arithmetic mean for the size of the phenomenon studied in homogeneous sub-periods (months),
d – the number of the sub-period (12 months),
k_i – the size of the phenomenon studied in the i-th sub-period.

The correctness of calculations is determined by reference to the sum of S_i indicators, which should equal 1200, for the monthly seasonality analysis.

$$\sum_{i=1}^{12} S_i = 1200 \tag{3}$$

Another measure describing seasonal fluctuations is the absolute seasonality index (absolute level of seasonal fluctuations). This is expressed in the same units as the studied phenomenon. This index is determined for each analysed sub-period in line with formula (4):

$$\pm g_i = S_i \cdot \bar{y} - \bar{y} \tag{4}$$

$$\bar{y} = \frac{\sum_{i=1}^{12} \bar{y_i}}{12} \tag{5}$$

where:
g_i – the absolute level of seasonal fluctuations,
$\bar{y}$ – the arithmetic mean for the magnitude of the phenomenon studied.

The sum of the absolute seasonal fluctuation levels g_i should be equal to 0 where analysis is performed properly.

$$\sum_{i=1}^{12} g_i = 0 \tag{6}$$

In the study, values of unit indicators of water volumes in the water-supply network were presented, as basic parameters allowing the sizes of water losses and the hydraulic load of the network to be determined. We can distinguish among these [4, 6, 11, 15]:
- the water-produced volume indicator:

$$q_{wtl} = \frac{V_{wtl} \cdot 1000}{LM \cdot 365}, \frac{dm^3}{M \cdot d} \tag{7}$$

where:
V_{wtl} – the volume of water produced, m^3,
LM – the number of inhabitants using the network.

- the water-sold volume indicator:

$$q_{sp} = \frac{V_{sp} \cdot 1000}{LM \cdot 365}, \frac{dm^3}{M \cdot d} \tag{8}$$

where:
V_{sp} – the volume of water sold, m^3.

- the water used for own purposes volume indicator:

$$q_{cw} = \frac{V_{cw} \cdot 1000}{LM \cdot 365}, \frac{dm^3}{M \cdot d} \tag{9}$$

where:
V_{cw} – the volume of water used by the company for its own purposes, m^3.

- the water-loss volume indicator:

$$q_{str} = \frac{V_{str} \cdot 1000}{LM \cdot 365}, \frac{dm^3}{M \cdot d} \tag{10}$$

where:
V_{str} – the volume of water losses, m^3.

$$V_{str} = V_{wtl} - V_{sp} - V_{cw}\ , m^3 \tag{11}$$

- the non-profit water volume indicator:

$$q_{nd} = \frac{(V_{wtl} - V_{sp}) \cdot 1000}{LM \cdot 365}, \frac{dm^3}{M \cdot d} \tag{12}$$

- the hydraulic load indicator for the water-supply network:

$$q_o = \frac{V_{wtl}}{L_{M+R} \cdot 365}, \frac{m^3}{km \cdot d} \tag{13}$$

where:
L_{M+R} - the summed length of the mains and distribution network, km.

- the indicator of water-loss volume in relation to the length of the network:

$$q_{strL} = \frac{V_{str}}{L \cdot 365}, \frac{m^3}{km \cdot d} \tag{14}$$

where:
L - the total length of the network, km.

3 Research object

The analysis was performed for the collective water supply system (CWSS) of a city located in south-eastern Poland, with an area of 120.4 km^2. About 200,000 inhabitants use the water-supply network, and the average demand for water is of 36,000 m^3/d. The total length of the network is 1006.6 km (in 2016), of which 54.9 km is the mains network, 585.5 km the distribution network, and 365.2 km of water connections. The research was based on exploitation data provided by the water-supply company in the period 2012-2016, including: volumes of water produced, water sold and water used by the company for its own purposes present in annual and monthly statements, numbers of inhabitants using the water-supply network and length of the network.

Table 1. Operating data in the period 2012–2016

Year	V_{wtl}, m^3	V_{sp}, m^3	V_{cw}, m^3	LM	Length of network: main, km	Length of network: distribution, km	Lenght of water connection, km
2012	13123532	9719587	1589406	167155	49.8	520.5	323.8
2013	12995329	9750269	1639039	171827	49.8	524.8	324.0
2014	13079211	9861071	1734834	180839	49.8	529.0	324.0
2015	13819560	10084749	1867238	181709	49.8	540.4	324.3
2016	13160958	10113934	1895640	183310	54.9	586.5	365.2

4 Results

For the seasonal analysis of water losses, the month was selected as a homogeneous sub-period in each analysed year (2012–2016). The sizes of water losses in the different months are as shown in Fig. 1.

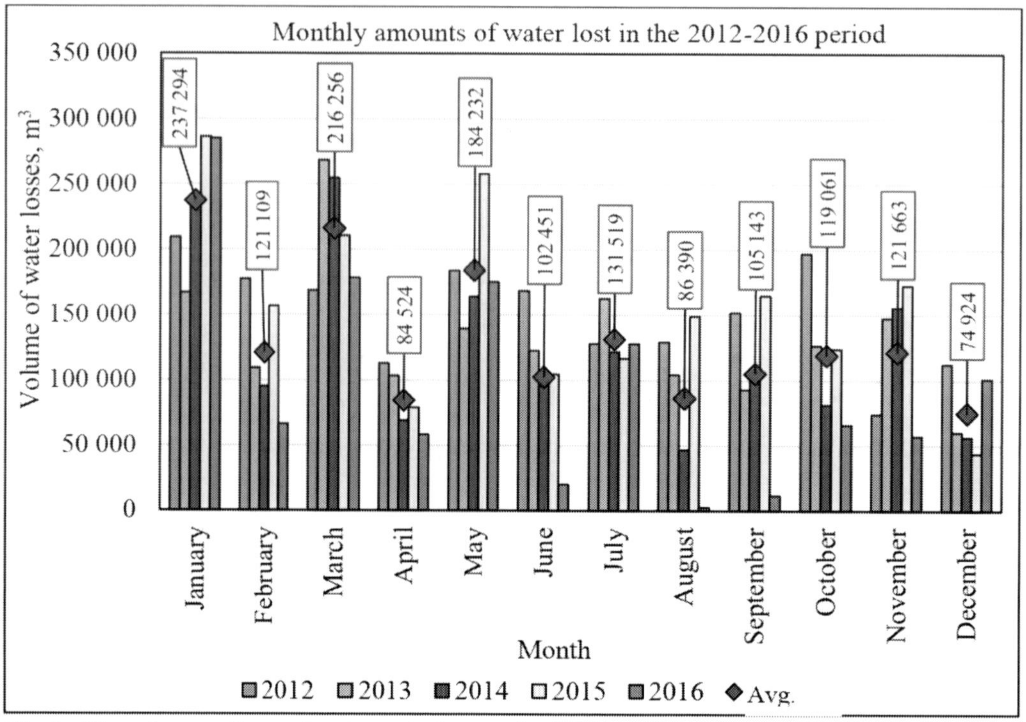

Fig. 1. The volumes of water lost in particular months (sub-periods).

Table 2 shows the calculation of values for the relative seasonal fluctuation index S_i and absolute seasonal fluctuation levels g_i. For monthly sub-periods of seasonal fluctuations, the sum of values of the relative seasonality index S_i should be $\Sigma S_i = 1200$, while the sum of absolute seasonal fluctuations levels $\Sigma g_i = 0$. This proves the correctness of calculations.

Table 2. Relative and absolute seasonal fluctuations in water losses over the 2012–2016 period.

Month	Year					Sum	$\bar{y}_i$	S_i, %	g_i
	2012	2013	2014	2015	2016				
I	209332	167177	238323	286351	285287	1186470	237294	179.7	105247
II	177554	109349	95228	156853	66561	605545	121109	91.7	-10938
III	168981	268560	254619	210668	178452	1081280	216256	163.8	84209
IV	112707	103442	69132	78991	58346	422618	84524	64.0	-47523
V	183959	139456	164182	258118	175444	921159	184232	139.5	52185
VI	168703	122662	95844	104854	20191	512254	102451	77.6	-29596
VII	128227	162768	121792	116760	128049	657596	131519	99.6	-528
VIII	129299	104517	46554	149018	2560	431948	86390	65.4	-45657
IX	151811	93019	103931	165237	11716	525714	105143	79.6	-26904
X	197239	126645	81290	124031	66100	595305	119061	90.2	-12986
XI	74173	147878	155771	172958	57534	608314	121663	92.1	-10384
XII	112554	60548	56640	43734	101144	374620	74924	56.7	-57123
							Sum:	1200	0

Fluctuations in values for the seasonality index S_i show differences in volumes of water lost in each sub-period as compared with the average monthly loss in the analysed period $\bar{y} = 132{,}047$ m^3 ($S_i = 100\%$). Fig. 2. shows the direct influence of seasonal fluctuations ($S_i - 100\%$) and absolute fluctuation levels (g_i) on the sizes of water losses characterising each month.

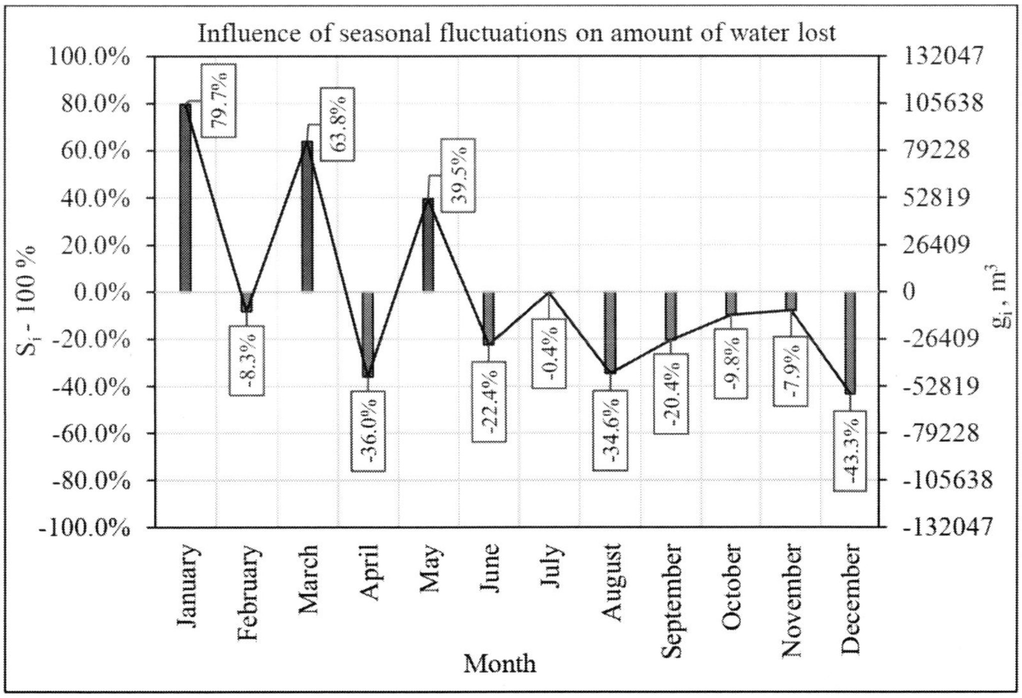

Fig. 2. The influence of relative seasonal fluctuations and absolute fluctuation levels on amounts of water lost.

The absolute levels of seasonal fluctuations g_i quantify differences in amounts of water lost in comparison with the average $\bar{y}$ ($g_i = 0$). The index g_i adopts a unit for the analysed phenomena, its sign revealing whether in a given sub-period the amount was above or below the average value in the examined period.

The highest seasonality impact on the amount of water lost is observed in January, for which the relative seasonality index $S_i = 179.7\%$, as corresponding with an absolute level of seasonal fluctuations g_i equal to 105,247 m^3. This means that the size of the water loss in this month is of 79.9 %, and is 105 247 m^3 greater than the average value for the analysed period, $\bar{y} = 132{,}047$ m^3 (100%). Increased losses of water in relation to the average value were also observed in March and May. In other months: February, April, June, July, August, September, October, November and December, the amounts of water lost are lower than the average value for the analysed period. The highest value of reduction in size of the loss of water in comparison with the average value, was observed in December (at 43.3%). This corresponds to a value of the absolute level of seasonal fluctuations $g_i = - 57{,}123$ m^3.

In the first half of the year, an increase in values for the seasonality index is to be observed (meaning that the amount of water lost is above average). One of the main causes of water loss is leakage associated with failure of the water-supply network. Losses increase in size as a result of greater numbers of failures, during the winter-spring period especially. This is followed by a periodic decrease in the sizes of water losses due to ongoing repair work being carried out on failing pipelines. The second half of the year is then characterised by reduced (below-average) amounts of water lost (with the seasonality index $S_i < 100\%$ and absolute levels of seasonal fluctuations being negative). Repair and replacement work along failing pipelines improves the network's technical condition, reducing both numbers of failures and amounts of water lost in summer and autumn. Knowledge regarding the seasonality of this phenomenon may be helpful as renovation and repair plans will be prepared. A proper renewal strategy will help reduce sizes of water losses, preventing failures of pipes and uncontrolled leakage.

Table 3 presents calculations of amounts of water lost, water volume indicators in the water-supply network and hydraulic load indicators. Fig. 3 presents changes in values for the selected unit indicators over the analysed period.

Table 3. Unit water-volume indicators and the hydraulic load indicator on the network.

Year	V_{str}, m^3	q_{wtl}, $\frac{dm^3}{M \cdot d}$	q_{sp}, $\frac{dm^3}{M \cdot d}$	q_{cw}, $\frac{dm^3}{M \cdot d}$	q_{str}, $\frac{dm^3}{M \cdot d}$	q_{nd}, $\frac{dm^3}{M \cdot d}$	q_o, $\frac{m^3}{km \cdot d}$	q_{strL}, $\frac{m^3}{km \cdot d}$
2012	1 814 539	215.1	159.3	26.1	29.7	55.8	63.0	5.6
2013	1 606 021	207.2	155.5	26.1	25.6	51.7	62.0	4.9
2014	1 483 306	198.2	149.4	26.3	22.5	48.8	61.9	4.5
2015	1 867 573	208.4	152.1	28.2	28.2	56.3	64.2	5.6
2016	1 151 384	196.7	151.2	28.3	17.2	45.5	56.2	3.1

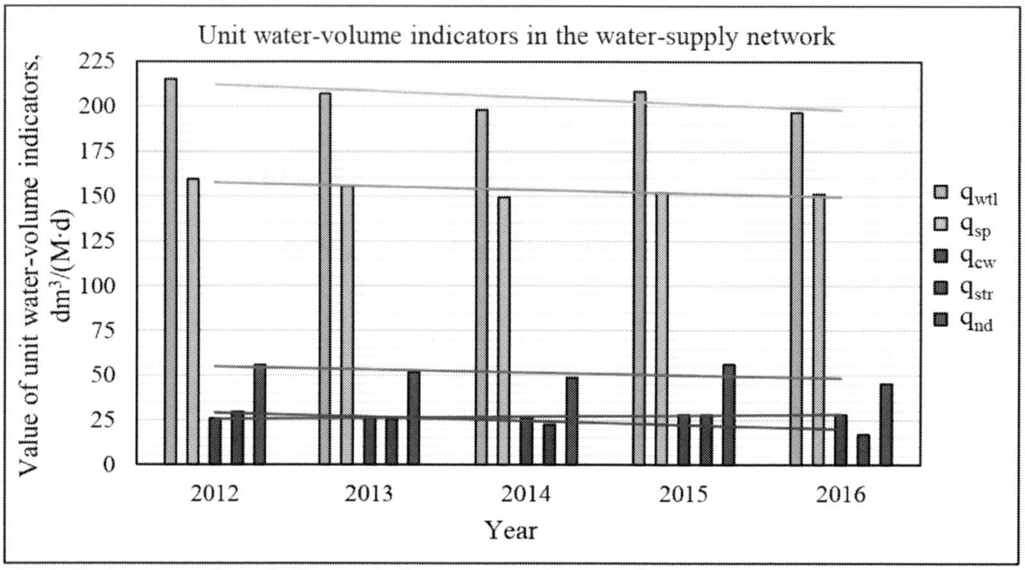

Fig. 3. Unit water-volume indicators in the water-supply network.

The analysis performed showed a decrease (defined by the trend line) for the water-produced volume indicator q_{wtl} and the water-sold volume indicator q_{sp} with observable growing number of inhabitants using the water supply. This phenomenon inscribe in global trend of environment-friendly lifestyle, which include water saving among the other things. Values noted for both the water-loss volume indicator q_{str} and the non-profit water-volume indicator q_{nd} were in decline over the examined period, due to achieved reductions in amounts lost that reflect an improved technical condition of the network thanks to the repair or replacement of failing pipelines. The indicator concerning water used for companies' own purposes – q_{cw} – is characterised by a slight increase, which relates directly to the development of the water-supply network in the period under analysis. With the reduction in the hydraulic load on the network, as described by the q_o indicator, values for the indicator of volume of water lost in relation to the length of the network have also displayed a downward trend.

5 Conclusion

The growing population, progressive urbanization, development of industry and agriculture, and climate change all threaten the size of water resources, making it important for the latter to be used properly and rationally. Water should be saved at the consumption, production and distribution stages. However, as water losses from a water-supply network should be minimised to a level that is optimal, the phenomenon requires deep, multifaceted analysis.

Based on the seasonality studies, it was observed that the year may be divided into two periods. The first of these includes winter and spring months and is characterised by above-average losses of water ($\bar{y}$ = 132,047 m^3). In contrast, in the summer and autumn months water losses are of below-average size. The highest value for the seasonality index was the S_i = 179.7% recorded in January, while the lowest was the S_i = 56.7% noted for December. The g_i index quantifies the difference in the loss of water as compared with the average figure. The lowest value for g_i (of -57,123 m^3) was obtained for December, while the highest was g_i = + 105 247 m^3, noted for January. The information obtained may be used by the operator of the analysed water-supply network for planning the schedule of pipeline repairs. Also noteworthy is the simplicity and universality of the methodology used, which allows for use it in characterising other phenomena in a CWSS or other municipal systems.

Analysis of water-volume indicators in the water-supply network reveals a decrease in production volume (q_{wtl}), water consumption (q_{sp}) and water losses (q_{str}). As the water-supply network developed, there was an increase in the value of the water use for own purposes volume indicator (q_{cw}). Declining trend in turn describes the change in values for the network hydraulic load indicator (q_o), and the value of water-loss volume in relation to the length of the network indicator (q_{strL}).

References

1. M. Gutry-Korycka, A. Sadurski, Z. W. Kundzewicz, J. Pociask-Karteczka, L. Skrzypczyk, Nauka **1**, 77–98 (2014)
2. T. AL-Washali, S. Sharma, M. Kennedy, Water Resour. Manag. **30**, 14, 4985–5001 (2016)
3. B. B. Azevedo, T. A. Saurin, Water Resour. Manag. **32**, 9, 2919–2936 (2018)
4. H. Hotloś, *Ilościowa ocena wpływu wybranych czynników na parametry i koszty eksploatacji sieci wodociągowych* (Monografia **49**, Oficyna Wydawnicza Politechniki Wrocławskiej, Wrocław, 2007)
5. E. Ociepa, M. Mrowiec, I. Deska, Proceedings of ECOpole **10**, 1, 247–255 (2016)
6. K. Pietrucha-Urbanik, Advances in Intelligent Systems and Computing **470**, 355–362 (2016)
7. K. Pietrucha-Urbanik, A. Studziński, Gaz, Woda i Technika Sanitarna **10**, 452–454 (2012)
8. A. Studziński, K. Pietrucha-Urbanik, A. Mędrala, JCEEA **4**, 193–201 (2014)
9. E. Renaud, J. Pillot, C. Aubrun, E. Bréjoux, B. Augeard, Water Sci. Tech-W Sup. **18**, 1, 251–258 (2018)
10. M. Kwietniewski, J. Rak, *Niezawodność infrastruktury wodociągowej i kanalizacyjnej w Polsce* (Komitet Inżynierii Lądowej i Wodnej PAN, Warszawa, 2010)
11. D. Szpak, B. Tchórzewska-Cieślak, Advances in Intelligent Systems and Computing **470**, 479–489 (2016)
12. J. S. Rogers, Y. Kleiner, B. J. Adams, Water Resour. Res. **34**, 8, 2039–2051 (1998)
13. H. Hotloś, Ochr. Sr. **2**, 57–62 (2013)
14. M. Sobczyk, *Statystyka* (Wydawnictwo Naukowe PWN, Warszawa, 2002)
15. I. Piegdoń, B. Tchórzewska-Cieślak, E3S Web Conferences **44**, 00140 (2018)

AUTHOR INDEX

Alioshko Sergii 113, 430
Andrianov Alexey 1, 167
Anielak Anna M. 470
Anisimov Sergey....72, 486
Badowska Ewa 8
Bagrowicz Tomasz 16
Bandzierz Dawid 8
Bartkiewicz Ewelina 727
Bartyna-Zielinska Malgorzata....384
Báthory Csongor 25
Baubekov Kuat....559
Bednarski Mateusz 135
Bezyk Yaroslav....551, 590
Bieniek Artur 33
Boguszewski Karol....337
Brillowska-Dabrowska Anna 159
Broszkiewicz-Suwaj Ewa....320
Cepinski Wojciech 42
Chmielewska Agnieszka 50
Chutrtong Jaruwan 58
Ciapala Bartlomiej....65
Cichon Aleksandra 72, 486
Cichowicz Robert....80
Cwilichowska Natalia 96
Cybulski Zefiryn 203
Czajczynska Dina....87
Czajka Krzysztof M. 87
Czaplicki Andrzej 582
Czardybon Agata 534
Czatzkowska Malgorzata 518
Demirel Sevgi....607
Dobó Zsolt....25
Domowicz Andrzej....369
Dudkiewicz Edyta 105
Dudziak Mariusz....345
Fialko Nataliia....113, 430
Fieducik Jolanta....120
Fortuniak Anna....736
Gallas Dawid 414
Gaska Krzysztof....304
Gebicki Jacek 159, 623
Generowicz Agnieszka....195
Gil Pawel....129
Girard Martin 399, 407
Gis Maciej....135
Gmurkowska Renata....143
Gnedash Georgii 113, 430
Godlewski Robert 399
Golba Radoslaw 501
Golebiowska Justyna 151
Górecki Michal....736
Gospodarek Milena....159
Govorova Zhanna 1, 167
Grelich Paulina 173
Grulke Rafal....736
Grzelka Agnieszka.... 180, 551
Gurgenidze David.... 251
Gwozdziej-Mazur Joanna.... 187
Hajduga Gabriela.... 195, 304
Haladyn Szymon.... 718
Hansen Martin O. L..... 543
Harnisz Monika.... 454, 477, 518
Holc Dorota.... 203
Ignasiak Karina.... 534
Iluk Tomasz 582
Isajenko Krzysztof.... 399, 407
Iwanski Slawomir.... 320
Jachimowicz Piotr 477
Jadwiszczak Piotr.... 437, 446
Janda Anna 211
Janiak Kamil.... 235
Janowski Miroslaw.... 65
Jasinski Remigiusz 218
Jedlicka Matej 661
Jelonek Leszek.... 320
Jouhara Hussam.... 87
Jurasz Magdalena 227
Jurga Anna 96, 235
Kaczmarczyk Michal 243
Kajurek Jakub.... 559
Kalabegishvili Mirian 251
Kazmierczak Bartosz 462, 661
Kiss Márton L..... 25
Klimiashvili Levan 251
Knapik Maciej 259
Kobus Dominik.... 590
Kochel Mateusz 582
Kokurewicz Lukasz.... 268
Kolodynski Wojciech 279
Komorowska-Kaufman M..... 203
Koniuszewska Izabela 518
Kopczynski Artur.... 286, 294
Korzeniewska Ewa 454, 477
Koval Viktor 304
Krawczak Ewelina.... 312
Krylów Malgorzata 195
Krzyzynska Renata 87
Kuchar Leszek.... 320
Kuchmacz Jan 33
Kudlek Edyta.... 345
Kularbphettong Kunyanuth.... 58
Kuzma Joanna.... 96, 235
Kwarciak-Kozlowska Anna.... 328
Kwasniowski Stanislaw.... 718
Laska Marta 105, 669
Lasocki Jakub.... 337
Ledvinka Ondrej 661
Lempart Anna 345, 701
Leoniak Maciej.... 268
Licznar Pawel.... 462

AUTHOR INDEX

Ligaj Kinga 353
Ligus Grzegorz 361
Lijewski Piotr 369
Lis Anna 376
Lój-Pilch Magdalena 392
Lopuszynska Agata 384
Lukaszek-Chmielewska Aneta 399, 407
Maciejewska Monika 615
Madrecka Beata 203
Marciniak Marta 701
Marcinkowski Tadeusz 211
Matczuk Agnieszka 422
Mazur Janusz 16
Merkisz Jerzy 414
Mika Lukasz 33
Mikhno Inesa 304
Mikulik Jerzy 227
Miller Urszula 180, 551
Miodonski Stanislaw 268
Moszczynska Marta 173
Musz-Pomorska Anna 353, 422
Muszynski-Huhajlo Mateusz 268
Navrodska Raisa 113, 430
Nguyen Thien 96
Niederwieser Tobias 235
Niemierka Elzbieta 437, 446
Niestepski Sebastian 454, 477
Nosarzewska Martyna 694
Nych Alicja 590
Ofman Piotr 607
Oktawiec Michal 462
Orlinski Piotr 135
Orlinski Tomasz 470
Osinska Adriana 454, 477
Pacak Anna 72, 486
Palotás Árpád Bence 25
Panchal Jatinkumar 361
Pandelidis Demis 72, 486
Pawnuk Marcin 493, 551
Piasecki Adam 501
Piechurski Krzysztof 509
Piekarska Katarzyna 279
Pilarska Agnieszka 501
Piotrowska Barbara 399, 407
Piszko Pawel 96
Potorski Jacek 518
Presich Georgii 430
Pruss Alina 203
Puchlik Monika 526
Rejdak Michal 534
Robak Jolanta 534
Rogowski Krzysztof 543
Romanik Elzbieta 180, 551
Roszczyk Pawel 294
Rusowicz Artur 559
Rutkowski Radoslaw 694
Rybak Justyna 694
Rybarczyk Piotr 159
Rymaniak Lukasz 369
Rymarov Andrey 566
Sadura Konrad 422
Scislo Lukasz 630
Shevcuk Svitlana 113, 430
Siedlecki Maciej 414
Singh Brar Lakhbir 653
Skretowicz Maria 574
Slowik Krzysztof 582
Sokolnicka Barbara 369, 414
Sówka Izabela 493, 590
Spodyniuk Nadiia 376
Stanczyk Grzegorz 173
Stawarz Olga 399, 407
Stec Marcin 582
Stefanowicz Ewelina 598
Stelegowski Artur 80
Stojanowska Agnieszka 694
Strub Daniel 279
Struk-Sokolowska Joanna 607
Swiesciak Joanna 574
Swietochowski Kamil 187
Szalanski Pawel 42
Szczepanik-Scislo Nina 630
Szczerbak Krystian 173
Szczurek Andrzej 615
Szulczynski Bartosz 623
Szulgowska-Zgrzywa Malgorzata 509, 598
Szymczyk Slawomir 677
Szymlet Natalia 369, 414
Tchórzewska-Cieslak Barbara 744
Titkov Dmitriy 638
Trohák Attila 25
Ujma Adam 251, 647
Ulewicz Malgorzata 113
Umnyakova Nina 647
Walas Aneta 736
Wasilewski Marek 653
Wdowikowski Marcin 462, 661
Wesolowska Malgorzata 669
Widomski Marcin K. 353, 422
Wiktorska Marlena 677
Wlodarczyk Renata 328
Wojtkowski Karol 399, 407
Wolski Krzysztof 685
Wróbel Magdalena 694
Wyczarska-Kokot Joanna 701
Wystalska Katarzyna 328
Zajac Agnieszka 709
Zajac Daniel 361
Zajac Pawel 718
Zajiczek Zaneta 615

AUTHOR INDEX

Zakrzewska Anita 392
Zaleski Janusz 462
Zathey Maciej...590
Zdyb Agata 312
Zelazna Agnieszka 151
Zielewicz Ewa...392
Zimoch Izabela..727
Ziulkiewicz Maciej 16, 736
Zywiec Jakub 744